AAPS Advances in the Pharmaceutical Sciences Series

Series Editor
Yvonne Perrie, Strathclyde Institute of Pharmacy and Biomedical Sciences
University of Strathclyde, Glasgow, UK

The AAPS Advances in the Pharmaceutical Sciences Series, published in partnership with the American Association of Pharmaceutical Scientists, is designed to deliver volumes authored by opinion leaders and authorities from around the globe, addressing innovations in drug research and development, and best practice for scientists and industry professionals in the pharma and biotech industries. Indexed in Reaxys
SCOPUS Chemical Abstracts Service (CAS) SCImago
EMBASE

More information about this series at https://link.springer.com/bookseries/8825

Elizabeth C. M. de Lange
Margareta Hammarlund-Udenaes
Robert G. Thorne

Editors

Drug Delivery to the Brain

Physiological Concepts, Methodologies and Approaches

Second Edition

Editors
Elizabeth C. M. de Lange
Research Division of Systems
Pharmacology and Pharmacy
Leiden University
Leiden, The Netherlands

Margareta Hammarlund-Udenaes
Department of Pharmacy
Uppsala University
Uppsala, Sweden

Robert G. Thorne
Biology Discovery
Denali Therapeutics
South San Francisco, CA, USA

ISSN 2210-7371 ISSN 2210-738X (electronic)
AAPS Advances in the Pharmaceutical Sciences Series
ISBN 978-3-030-88772-8 ISBN 978-3-030-88773-5 (eBook)
https://doi.org/10.1007/978-3-030-88773-5

This Springer imprint is published by the registered company Springer Nature Switzerland AG
The registered company address is: Gewerbestrasse 11, 6330 Cham, Switzerland

Preface

Seven years have passed since the publication of the first edition of this book. So, it is very reassuring to see the significant progress that has been made in the field as we now release a second edition. This research area has come a long way since a group of researchers at the NIH first used electron microscopy to unequivocally establish that tight junctions between brain endothelial cells form a critical aspect of the blood-brain barrier (Brightman and Reese 1969; Reese and Karnovsky 1967). This finding ushered in an era of research and science investigating precisely what factors determine transport across the blood-brain barrier (BBB) as well as the blood-cerebrospinal fluid barriers (BCSFB). For example, many studies over the subsequent decades have been devoted to understanding how these barriers to the central nervous system (CNS) are formed; how cerebral endothelial cells interact with astrocytes, pericytes, immune cells, and neurons (collectively defined as the neurovascular unit) to dynamically regulate BBB function; and how diseases and acute insults affect the barriers and their regulation.

Along with all of this work, there has been a steady rise in the level of interest and research surrounding drug delivery to the brain. Indeed, several key advances have occurred since the publication of the last edition, also with respect to the development and approval of biologics (peptides, proteins, oligonucleotides, and gene therapy vectors) for neuroscience indications.

Back in 2014, the only approved biologic directed to a CNS target that was also unquestionably delivered into the CNS was ziconotide (Prialt), a ~3 kDa peptide toxin originally isolated from the cone snail that is administered intrathecally to treat severe, chronic pain (with approval in the USA in 2004 and in Europe in 2005). Toward the end of 2016, FDA approval was granted for nusinersen (Spinraza), a ~7 kDA $2'-O$-methoxyethyl-modified antisense oligonucleotide administered as an intrathecal bolus to treat spinal muscular atrophy. Approval was next granted in 2017 for cerliponase alfa (Brineura), a ~59 kDa recombinant enzyme delivered intraventricularly for the treatment of neuronal ceroid lipofuscinosis type 2, a pediatric neurodegenerative lysosomal storage disorder also known as Batten disease. The year 2019 witnessed two firsts for onasemnogene abeparvovec-xioi (Zolgensma) as it simultaneously represented both the first CNS-directed gene therapy to be approved and the first ever approved biologic administered systemically with the ability to cross the BBB to reach the brain. The future now looks brighter than ever for delivering large molecules to the brain and spinal cord. Large molecule delivery to the CNS is of course an important focus for this book.

However, it must be recognized that nearly all currently approved CNS drugs besides the examples above are small molecular weight pharmaceuticals. Although lipophilicity has earlier been emphasized in predicting brain entry, there is increasing awareness that other factors are critical for small molecule drugs to reach their required target site concentrations within the brain. The combination of general BBB diffusion, influx/efflux transport, and carrier-mediated transport, together with plasma and intra-brain distribution, all play an important role in the success or failure of CNS drugs. It is also important to consider the effect that disease conditions may have on these factors.

Where do we stand today? As we go to print, the field has now witnessed the first approval of a systemic monoclonal antibody therapy for Alzheimer's disease (aducanumab). This development alone may be signaling a major change in the way biopharma, the academic world, regulatory bodies, and other key stakeholders approach the testing and clinical development of new therapies for the brain and spinal cord. Aducanumab represents the first biomarker-based drug approval for any neurodegenerative disease. This achievement has the potential to give rise to shorter development paths for future drugs, generate new natural history data for Alzheimer's patients, and facilitate a range of new opportunities in the form of combination trials going forward. New CNS drugs have historically suffered from considerably lower success rates during development than those for non-CNS indications, partly due to transporter protection of the brain but also due to poor understanding of CNS mechanisms of disease and biodistribution. Why do CNS drugs suffer from these low success rates during development? Some of the reasons undoubtedly include: (i) our still incomplete understanding of the brain and its many functions, (ii) a propensity for CNS drugs to suffer from off-target side effects, (iii) a poor track record for many CNS drugs when it comes to pre-clinical predictions of clinical challenges, (iv) a much larger influence of transporters than in other organs/tissues, (v) a shortage of validated biomarkers for assessing therapeutic efficacy in treating neurological and psychiatric disorders, and (vi) a lack of studies integrating more than one aspect of the problem. Drug delivery issues obviously present a key challenge, so it is encouraging that clinical trials have increasingly focused on delivery aspects. As was apparent during production of the first edition of this book, it is quite clear that better ideas, technology, and mathematical modeling have substantially progressed our understanding with respect to CNS delivery and are needed to translate into better clinical trials and improved clinical success in the coming years.

The ability to achieve consistent, targeted delivery to the CNS target site has remained a major, largely unmet challenge, but this book attests to the potential we have to address this hurdle in the years ahead. The field has seen a critical mass of dedicated, multidisciplinary scientists from all over the world come together in recent years with shared purpose and commitment to making significant progress in this vitally important research area, as evidenced by joint scholarly output, passionately communicated science at conferences, and rapidly growing national and international societies. This provides perhaps the greatest cause for optimism, because our future success in developing new ideas, technology, and understanding related to CNS barriers/drug delivery will likely require just such cooperation and collegiality, as well as strong collaborative efforts between academic centers, federal research bodies, and biopharma.

Lastly, an important reason for producing a book such as this is to also hopefully provide an introduction to the field to promising young scientists who have not yet decided how to direct their careers. We hope the second edition of this book supports their curiosity and investigation and provides some assistance in identifying CNS barriers and drug delivery science as a field with interesting questions and exceptionally worthy goals.

Leiden, The Netherlands   Elizabeth C. M. de Lange
Uppsala, Sweden   Margareta Hammarlund-Udenaes
South San Francisco, CA, USA   Robert G. Thorne

Book Structure

Over the past few decades, great strides have been made in each of the five parts this book has been divided into. The basic physiology of the BBB and BCSFB has been defined and the manner in which the brain handles drugs is much better appreciated (Part I). Increasingly elegant *in vitro*, *in vivo*, and pharmacokinetic concepts have been applied to the study of drug transport across the BBB and intra-brain distribution, and mathematical models to even predict CNS pharmacokinetics in multiple physiological compartments (Part II). Industry experience in developing CNS drugs has deepened and a better appreciation of the critical factors that lead to development success or failure has been attained (Part III). Many strategies for improved CNS delivery, often focused upon delivering biologics into the brain, have been proposed, developed, and tested with varying degrees of success and optimism for near-term clinical application (Part IV). There have also been major developments in our understanding of barrier changes in disease conditions and how these changes affect CNS drug delivery (Part V).

Each of the chapters contained in this book have been written by experts in the field, carefully chosen so that the book brings diverse, cutting-edge viewpoints and state-of-the-art summaries from scientists representing both academic and industry perspectives. In addition to providing detailed coverage of the different topic areas, chapters also include a description of future challenges and unresolved questions combined with a concluding section in some chapters entitled "Points for Discussion." The "Points for Discussion" contain further questions and observations intended to stimulate discussion among a group of people in either a classroom or small group setting; they may also prove useful as assignments for a graduate-level survey course. In addition to wide ranging coverage of physiological concepts relevant to CNS drug delivery, the book also contains a detailed review of brain structure, function, blood supply, and fluids in the Appendix, written as a concise, detailed "crash course" covering relevant background for the book's content.

We have designed this book to be useful for a wide audience, from graduate or professional students being exposed to this research area for the first time to established academic and industry scientists looking to learn about the state-of-the-art, to experts already performing CNS drug delivery research or working in related areas.

We hope that it succeeds in introducing some of the major questions faced by the field as well as in stimulating new thoughts on how to answer them!

References

Brightman MW, Reese TS (1969) Junctions between intimately apposed cell membranes in the vertebrate brain. J Cell Biol 40 (3):648–677

Reese TS, Karnovsky MJ (1967) Fine structural localization of a blood-brain barrier to exogenous peroxidase. Journal of Cell Biology 34:207–217

Acknowledgments

There are many people and experiences to acknowledge in putting such a book together. We are indebted first and foremost to the dynamic and supportive international community of CNS barriers and drug delivery scientists whom we have had the pleasure of interacting with over many years now. Many of our colleagues and collaborators from this community kindly agreed and worked hard to contribute chapters for this book. We also gratefully acknowledge the professionalism, support, and collaborative spirit of our colleagues at Springer throughout the entire process, from project conception to completion.

We therefore dedicate this book to all colleagues and friends in the field, as well as to future scientists who are to further address the challenging and difficult but rewarding tasks to finding better treatment for CNS diseases, and most importantly to all patients suffering from diseases within the CNS.

Contents

Contributors

N. Joan Abbott Institute of Pharmaceutical Science, Blood-Brain Barrier Group, King's College London, London, UK

Anuska V. Andjelkovic Departments of Neurosurgery, and Pathology, University of Michigan, Ann Arbor, MI, USA

Bjoern Bauer Department of Pharmaceutical Sciences, College of Pharmacy, University of Kentucky, Lexington, KY, USA

Ulrich Bickel Department of Pharmaceutical Sciences, School of Pharmacy, Texas Tech University Health Sciences Center, Amarillo, TX, USA

Luke H. Bradley Department of Neuroscience, University of Kentucky College of Medicine, Lexington, KY, USA

Yujia Alina Chan Stanley Center for Psychiatric Research, Broad Institute of MIT and Harvard, Cambridge, MA, USA

Hsueh Yuan Chang Department of Pharmaceutical Sciences, School of Pharmacy and Pharmaceutical Sciences, State University of New York at Buffalo, Buffalo, NY, USA

Xavier Declèves Faculty of Pharmacy, University of Paris, Therapeutic Optimisation in Neurophychopharmacology Team « The Blood-Brain Barrier in Brain Patho-physiology and Therapy », Inserm UMRS-1144, Paris, France

Elizabeth C. M. de Lange Research Division of Systems Pharmacology and Pharmacy, Leiden University, Leiden, The Netherlands

Benjamin E. Deverman Stanley Center for Psychiatric Research, Broad Institute of MIT and Harvard, Cambridge, MA, USA

Diana E. M. Dolman Institute of Pharmaceutical Science, King's College London, London, UK

William F. Elmquist Brain Barriers Research Center, Department of Pharmaceutics, University of Minnesota, Minneapolis, MI, USA

Koji L. Foreman Department of Chemical and Biological Engineering, University of Wisconsin, Madison, WI, USA

David Fortin Department of Neurosurgery and Neuro-oncology, Université de Sherbrooke, Sherbrooke, QS, Canada

Markus Fridén Inhalation Product Development, Pharmaceutical Technology & Development, Operations, AstraZeneca, Gothenburg, Sweden

Pieter J. Gaillard 2-BBB Medicines BV, Leiden, the Netherlands

Jean-François Ghersi-Egea FLUID Team and BIP Facility, Lyon Neuroscience Research Center, INSERM U1028, CNRS UMR5292, Lyon 1 University, Lyon, France

Ryohei Goto Faculty of Pharmaceutical Sciences, Tohoku University, Sendai, Japan

Johann Mar Gudbergsson Neurobiology Research and Drug Delivery, Department of Health Science and Technology, Aalborg University, Aalborg, Denmark

Roger N. Gunn Department of Brain Sciences, Imperial College London, London, UK

Invicro, London, UK

Margareta Hammarlund-Udenaes Translational PKPD Research Group, Department of Pharmacy, Uppsala University, Uppsala, Sweden

Arsalan S. Haqqani Human Health Therapeutics Research Centre, National Research Council Canada, Ottawa, Canada

Anika M. S. Hartz Sanders-Brown Center on Aging, University of Kentucky, Lexington, KY, USA

Yang Hu Discovery ADME, Drug Discovery Science, Boehringer Ingelheim RCV GmbH & Co KG, Vienna, Austria

Kullervo Hynynen Physical Sciences Platform, Sunnybrook Research Institute, Department of Medical Biophysics, University of Toronto, Toronto, ON, Canada

Institute of Biomaterials and Biomedical Engineering, University of Toronto, Toronto, ON, Canada

Shingo Ito Department of Pharmaceutical Microbiology, Faculty of Life Sciences, Kumamoto University, Kumamoto, Japan

Kasper Bendix Johnsen Department of Health Technology, Technical University of Denmark, Lyngby, Denmark

Richard F. Keep Departments of Neurosurgery, Molecular and Integrative Physiology, University of Michigan, Ann Arbor, MI, USA

Elisa E. Konofagou Department of Biomedical Engineering and Department of Radiology, Columbia University, New York, NY, USA

Niyanta N. Kumar Pharmacokinetics, Pharmacodynamics & Drug Metabolism, Merck Co Inc, West Point, PA, USA

Dominique Lesuisse Tissue Barriers, Rare and Neurologic Diseases, Sanofi, France

Jeffrey J. Lochhead Department of Pharmacology, University of Arizona Colelge of Medicine, Tucson, AZ, USA

Irena Loryan Translational PKPD Research Group, Department of Pharmacy, Uppsala University, Uppsala, Sweden

Takeshi Masuda Department of Pharmaceutical Microbiology, Faculty of Life Sciences, Kumamoto University, Kumamoto, Japan

Dallan McMahon Physical Sciences Platform, Sunnybrook Research Institute, Toronto, ON, Canada

Afroz S. Mohammad Brain Barriers Research Center, Department of Pharmaceutics, University of Minnesota, Minneapolis, MI, USA

Torben Moos Neurobiology Research and Drug Delivery, Department of Health Science and Technology, Aalborg University, Aalborg, Denmark

Geetika Nehra Sanders-Brown Center on Aging, College of Medicine, University of Kentucky, Lexington, KY, USA

Behnam Noorani Department of Pharmaceutical Sciences, School of Pharmacy, Texas Tech University Health Sciences Center, Amarillo, TX, USA

Sumio Ohtsuki Department of Pharmaceutical Microbiology, Faculty of Life Sciences, Kumamoto University, Kumamoto, Japan

Sean P. Palecek Department of Chemical and Biological Engineering, University of Wisconsin, Madison, WI, USA

Jane E. Preston Institute of Pharmaceutical Science, King's College London, London, UK

Andreas Reichel Bayer Pharma AG, Research Pharmacokinetics, Berlin, Germany

Jaap Rip 20Med Therapeutics BV, Enschede, The Netherlands

Ramakrishna Samala Department of Pharmaceutical Sciences, School of Pharmacy, Texas Tech University Health Sciences Center, Amarillo, TX, USA

Jann N. Sarkaria Department of Radiation Oncology, Mayo Clinic, Rochester, MI, USA

Julia A. Schulz Department of Pharmaceutical Sciences, College of Pharmacy, University of Kentucky, Lexington, KY, USA

Mallory J. Senslik Translational Imaging Biomarkers, MRL, Merck & Co Inc, West Point, PA, USA

Dhaval Shah Department of Pharmaceutical Sciences, School of Pharmacy and Pharmaceutical Sciences, State University of New York at Buffalo, Buffalo, NY, USA

Eric V. Shusta Department of Chemical and Biological Engineering, Department of Neurological Surgery, University of Wisconsin, Madison, WI, USA

Quentin Smith Department of Pharmaceutical Sciences, School of Pharmacy, Texas Tech University Health Sciences Center, Amarillo, TX, USA

Danica B. Stanimirovic Human Health Therapeutics Research Centre, National Research Council Canada, Ottawa, Canada

Nathalie Strazielle Brain-i, Lyon, France

Stina Syvänen Department of Public Health and Caring Sciences, Uppsala University, Uppsala, Sweden

Masanori Tachikawa Graduate School of Pharmaceutical Sciences, Tohoku University, Sendai, Japan

Graduate School of Biomedical Sciences, Tokushima University, Tokushima, Japan

Surabhi Talele Brain Barriers Research Center, Department of Pharmaceutics, University of Minnesota, Minneapolis, MI, USA

Tetsuya Terasaki Graduate School of Pharmaceutical Sciences, Tohoku University, Sendai, Japan

School of Pharmacy, Faculty of Health Sciences, University of Eastern Finland, Joensuu, Finland

Robert G. Thorne Denali Therapeutics, South San Francisco, CA, USA

Department of Pharmaceutics, College of Pharmacy, University of Minnesota, Minneapolis, MN, USA

Helen Thorsheim Department of Pharmaceutical Sciences, School of Pharmacy, Texas Tech University Health Sciences Center, Amarillo, TX, USA

Yasuo Uchida Graduate School of Pharmaceutical Sciences, Tohoku University, Sendai, Japan

Takuya Usui Graduate School of Pharmaceutical Sciences, Tohoku University, Sendai, Japan

Jianming Xiang Department of Neurosurgery, University of Michigan, Ann Arbor, MI, USA

Siti R. Yusof HICoE Centre for Drug Research, Universiti Sains Malaysia, Minden, Penang, Malaysia

Lei Zhang Medicine Design, Pfizer Inc, Cambridge, MA, USA

Ningna Zhou Department of Pharmacology, Yunnan University of Tranditional Chinese Medicine, Kunming, China

Part I
Physiology and Basic Principles for Drug Handling by the CNS

Chapter 1
Anatomy and Physiology of the Blood-Brain Barriers*

N. Joan Abbott

Abstract This chapter covers the three main barrier layers separating blood and the CNS: the endothelium of the brain vasculature, the epithelium of the choroid plexus secreting cerebrospinal fluid (CSF) into the ventricles and the arachnoid epithelium forming the middle layer of the meninges on the brain surface. There are three key barrier features at each site that control the composition of brain fluids and regulate CNS drug permeation: (i) physical barriers result from features of the cell membranes and of the tight junctions restricting the paracellular pathway through intercellular clefts; (ii) transport barriers result from membrane transporters mediating solute uptake and efflux, together with vesicular mechanisms mediating transcytosis of larger molecules such as peptides and proteins; and (iii) enzymatic barriers result from cell surface and intracellular enzymes that can modify molecules in transit. Brain fluids (CSF and brain interstitial fluid) are secreted, flow through particular routes and then drain back into the venous system; this fluid turnover aids central homeostasis and also affects CNS drug concentration. Several CNS pathologies involve changes in the barrier layers and the fluid systems. Many of these aspects of physiology and pathology have implications for drug delivery.

Keywords Blood brain barrier · Blood-CSF barrier · Permeability · Tight junctions · Choroid plexus · Drug delivery · Neurovascular unit · Glia · Pericyte · Central nervous system pathology

1.1 Neural Signalling and the Importance of CNS Barrier Layers

The brain and spinal cord (central nervous system, CNS) are the control centres of the body, generating central programmes, coordinating sensory input and motor output and integrating many of the activities of peripheral organs and tissues. CNS

N. J. Abbott (✉)
Institute of Pharmaceutical Science, Blood-Brain Barrier Group, King's College London, London, UK
e-mail: joan.abbott@kcl.ac.uk

© American Association of Pharmaceutical Scientists 2022
E. C. M. de Lange et al. (eds.), *Drug Delivery to the Brain*, AAPS Advances in the Pharmaceutical Sciences Series 33, https://doi.org/10.1007/978-3-030-88773-5_1

neurons use chemical and electrical signals for communication, requiring precise ionic movements across their membranes. This is particularly critical at central synapses generating graded synaptic potentials and somewhat less so along axons signalling via all-or-none action potentials. Hence precise control (homeostasis) of the CNS microenvironment is crucial for reliable neural signalling and integration. It has been argued that this was one of the strongest evolutionary pressures driving the development of cellular barriers at the interfaces between the blood and the CNS, since animals with better CNS regulation would have more reliable, efficient and rapid neural signalling, giving selective advantage in finding and remembering food sources, catching prey and avoiding predators (Abbott 1992). These cellular barriers at the interfaces act as key regulatory sites, controlling ion and molecular flux into and out of the CNS, while the resident cells of the CNS including neurons and their associated glial cells, the macroglia (astrocytes, oligodendrocytes) and microglia contribute to local regulation of the composition of the interstitial (or extracellular) fluid (ISF, ECF) (for reviews see Abbott et al. 2010; Nicholson and Hrabětová 2017). The molecular flux control at CNS barriers includes delivering essential nutrients, removing waste products and severely restricting the entry of potentially toxic or neuroactive agents and pathogens. The barrier layers also act as the interface between the central and peripheral immune systems, exerting strong and selective control over access of leucocytes from the circulation (Engelhardt and Coisne 2011; Greenwood et al. 2011; Ransohoff and Engelhardt 2012; Engelhardt et al. 2017).

Three main barrier sites can be identified (Fig. 1.1): the endothelium of the brain microvessels (forming the blood-brain barrier, BBB) (Reese and Karnovsky 1967), the epithelium of the choroid plexus (specialised ependyma) secreting cerebrospinal fluid (CSF) into the cerebral ventricles (Becker et al. 1967) and the epithelium of the arachnoid mater covering the outer brain surface above the layer of subarachnoid CSF (Nabeshima et al. 1975); the choroid plexus and arachnoid form the blood-CSF barrier (BCSFB) (Abbott et al. 2010). The endothelium forms the largest interface (based on surface area) between blood and CNS and hence represents the major site for molecular exchange and the focus for drug delivery; the choroid plexus also plays a critical role, while the properties of the arachnoid membrane suggest it plays a relatively minor role in exchange. (Note that recent microanatomical studies clarify the role of arachnoid granulations in CSF drainage (Sokołowski et al. 2018; Kutomi and Takeda 2020).) At each of these sites (endothelium, choroid plexus and arachnoid), intercellular tight junctions (*zonulae occludentes*) restrict diffusion of polar solutes through the cleft between cells (paracellular pathway), forming the 'physical barrier'. In brain endothelium, blood-arachnoid barrier (Uchida et al. 2020) and choroid plexus, solute carriers on the apical and basal membranes together with ecto- and endo-enzymes regulate small solute entry and efflux. In brain endothelium, mechanisms of adsorptive and receptor-mediated transcytosis allow restricted and regulated entry of certain large molecules (peptides, proteins) with growth factor and signalling roles within the CNS. Finally, the endothelial and choroid plexus barriers help regulate the innate immune response and the recruitment of leucocytes, contributing to the surveillance and the reactive functions of the central immune cell population. Thus, these interface layers work together as physical, transport, enzymatic (metabolic) and immunological barriers (for reviews

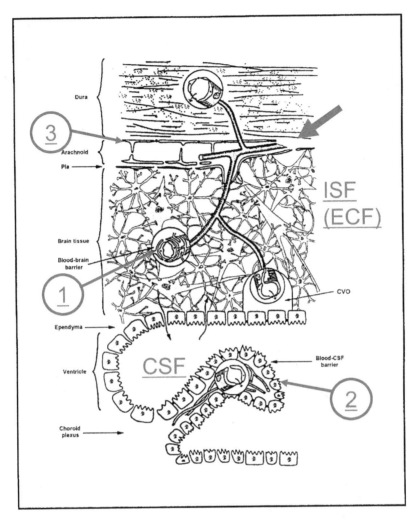

Fig. 1.1 Location of barrier sites in the CNS. Blood enters the brain via surface arteries (red arrow, top). Barriers between blood and neural tissue are present at three main sites: (1) the brain endothelium forming the blood-brain barrier (BBB), (2) the choroid plexus epithelium which secretes cerebrospinal fluid (CSF) and (3) the arachnoid epithelium forming the middle layer of the meninges. At each site, the physical barrier results from tight junctions that reduce the permeability of the paracellular pathway (intercellular cleft). In circumventricular organs (CVO), containing neurons specialised for neurosecretion and/or chemosensitivity, the endothelium is leaky. This allows tissue-blood exchange, but as these sites are separated from the rest of the brain by an external glial barrier and from CSF by a barrier at the ependyma, CVOs do not form a leak across the BBB. ISF (ECF): interstitial or extracellular fluid. Figure based on Segal MB and Zlokovic BV 1990 Fig. 1.1, p2 in 'The Blood-Brain Barrier, Amino Acids and Peptides' (Kluwer), modified by A Reichel. Reproduced from Abbott et al. 2003 Lupus 12:908, and with permission of Springer

see Abbott and Friedman 2012; Abbott 2013). The barrier functions are not fixed but dynamic, able to respond to a variety of regulatory signals from the blood and the brain side, and can be significantly disturbed in many CNS and systemic pathologies. This chapter will focus on the physical, transport and enzymatic barrier functions of the blood-brain barrier and the choroid plexus, as most relevant to CNS drug delivery. As this chapter is meant primarily to provide an introduction and overview, references to key reviews are interspersed with those to original findings; more detailed background may be obtained by consulting sources within the reviews cited.

1.2   The Brain Endothelium and the Neurovascular Unit

The brain capillaries supply blood in close proximity to neurons (maximum diffusion distances typically 8–25 μm); hence the activities of the BBB are key to brain homeostasis. The brain endothelium of the BBB acts within a cellular complex, the neurovascular unit (NVU) (Fig. 1.2) (Abbott et al. 2010; Muoio et al. 2014; Iadecola 2017), composed at a local level of grey matter of the segment of capillary, its

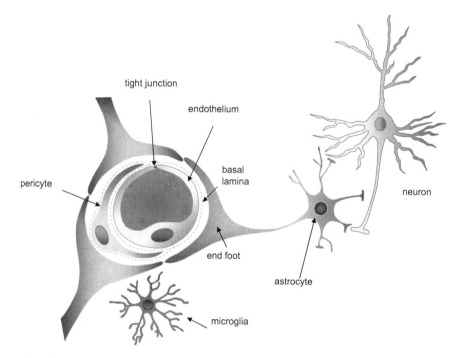

Fig. 1.2 The neurovascular unit (NVU). The NVU is composed of several cell types in close association, working together to maintain an optimal neuronal microenvironment. Cerebral endothelial cells forming the BBB make tight junctions which restrict the paracellular pathway. Pericytes partially envelope the endothelial cells and share a common basal lamina with them. Astrocytes ensheath the microvessel wall. Pericytes and astrocytes are important in barrier induction and maintenance, and astrocytes provide links to neurons. Microglia are CNS-resident immune cells with highly motile cellular processes, some of which can contact the astrocyte basal lamina. By S Yusof and NJ Abbott, from Abbott (2013) with permission

associated pericytes, perivascular astrocytes, basement membranes and microglial cells, the resident immune cells of the CNS (Ransohoff and Perry 2009; Mäe et al. 2011; Bohlen et al. 2019). Together this cellular complex supports a small number of neurons within that NVU module (Iadecola and Nedergaard 2007; Abbott et al. 2010). Recent advances in understanding show that the NVU concept should be expanded to a 'multidimensional' network of signalling between local capillary networks 'in which mediators released from multiple cells engage distinct signalling pathways and effector systems across the entire cerebrovascular network in a highly orchestrated manner' (Iadecola 2017). There are also implications for development and pathology (Dalkara and Alarcon-Martinez 2015).

Several functions of the BBB can be identified and their roles in CNS homeostasis highlighted (Abbott et al. 2010; Abbott 2013). By regulating ionic and molecular traffic and keeping out toxins, the barrier contributes to neuronal longevity and the health and integrity of neural network connectivity (Iadecola 2017). Ionic homeostasis is essential for normal neural signalling. Restricting protein entry limits the innate immune response of the brain and the proliferative potential of the CNS microenvironment. Separating the neurotransmitter pools of the peripheral nervous system (PNS) and CNS minimises interference between signalling networks using the same transmitters while allowing 'non-synaptic' signalling by agents able to move within the protected interstitial fluid (ISF) compartment. Regulating entry of leucocytes allows immune surveillance with minimal inflammation and cellular damage. Finally, the system is well organised for endogenous protection and 'running repairs' (Liu et al. 2010; Tian et al. 2011; Ransohoff and Brown 2012; Daneman 2012; Posada-Duque and Cardona-Gómez 2020). The other cells of the NVU, especially the astrocytes, pericytes and microglia, together with components of the extracellular matrix (ECM), contribute to these activities (Errede et al. 2021).

Given the key role of circulating leucocytes in patrolling, surveillance and repair of the CNS, it has been proposed that these cells, plus the glycocalyx at the endothelial surface (Haqqani et al. 2011; Okada et al. 2020), should be included in an 'extended NVU' (Neuwelt et al. 2011) (Fig. 1.3). Current research on the cell/cell interactions involved is revealing further details of the complexity of the NVU and its critical role in maintaining a healthy BBB. Damage to the endothelial glycocalyx in lungs and other organs including the brain during the Covid-19 (SARS-CoV-2) pandemic (2020) may have contributed to severity of this pandemic disease (Okada et al. 2020).

1.3 Nature and Organisation of the Membranes of the Barrier Layers

Many powerful techniques are being applied to increase molecular understanding of barrier function (Redzic 2011; Pottiez et al. 2011; Daneman 2012; Saunders et al. 2013), including biophysical investigation of the lipid membranes, quantitative proteomics, imaging at close to the level of individual molecules and use of genetic mutants and siRNA to test the roles of individual components.

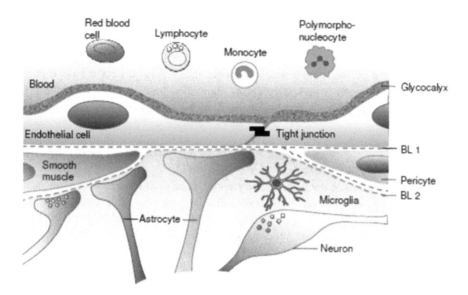

Fig. 1.3 The 'extended' NVU at the level of the microvessel wall, incorporating the glycocalyx and circulating cells. Recent work has highlighted the importance of the glycocalyx on the luminal endothelial surface for endothelial function and the role of circulating leucocytes in monitoring and interacting with this surface. By S Yusof and NJ Abbott, modified from Abbott et al. (2010) with permission

The outer cell membranes (plasmalemma) of the barrier layers, like other mammalian cell membranes, consist of a lipid bilayer with embedded protein, the 'fluid mosaic' model of the membrane. The membrane lipids include glycerophospholipids, sterols and sphingolipids. The hydrophilic polar heads of phospholipids form a continuous layer at the outer and inner leaflets of the membrane, with hydrophobic chains extending into the core of the membrane; the outer leaflet contains mainly zwitterionic phosphatidylcholine (PC) and phosphatidylethanolamine (PE), while the inner leaflet contains mainly negatively charged phosphatidylserine. PC and PE are the main phospholipids in brain endothelium at 20% and 30%, respectively, with cholesterol at ~20% (Krämer et al. 2002). Under physiological conditions, the lipid bilayer is in a liquid crystalline state. The high percentage of PE and cholesterol in brain endothelium helps to increase its packing density (Gatlik-Landwojtowicz et al. 2006; Seelig 2007) which affects the way molecules partition into and diffuse through the membrane. At the molecular level, there is continual motion of the phospholipid tails within the membrane, creating transient gaps that permit flux of small gaseous molecules (oxygen, CO_2) and small amounts of water (Abbott 2004; Dolman et al. 2005; MacAulay and Zeuthen 2010). Many lipophilic agents including drugs permeate well through the lipid bilayer (Bodor and Buchwald 2003) (Fig. 1.4). However, the tight lipid packing restricts permeation of certain hydrophobic molecules including many drugs and regulates access to particular membrane transport proteins such as the ABC (ATP-binding cassette) efflux transporters,

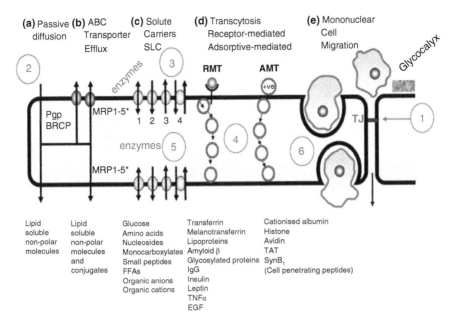

Fig. 1.4 Routes across the brain endothelium. Diagram of brain endothelium showing (numbered red circles) the tight junctions (1) and cell membranes (2) forming the 'physical barrier', transporters (3) and vesicular mechanisms (4) (forming the 'transport barrier'), enzymes forming the 'enzymatic barrier' (5) and regulated leucocyte traffic (6) the 'immunologic barrier'. (**a**) Solutes may passively diffuse through the cell membrane and cross the endothelium; a higher lipid solubility and several other physicochemical factors favour this process. (**b**) Active efflux carriers (ABC transporters) may intercept some of these passively penetrating solutes and pump them out. Pgp and BCRP are strategically placed in the luminal membrane of the BBB endothelium. MRPs 1–5 are inserted into either luminal or abluminal membranes, with some species differences in the polarity and the MRP isoforms expressed. (**c**) Carrier-mediated influx via solute carriers (SLCs) may be passive or primarily or secondarily active and can transport many essential polar molecules such as glucose, amino acids and nucleosides into the CNS. The solute carriers (black numbers) may be bidirectional, the direction of net transport being determined by the substrate concentration gradient (1), unidirectional either into or out of the cell (2/3), or involve an exchange of one substrate for another or be driven by an ion gradient (4). In this last case, the direction of transport is also reversible depending on electrochemical gradients. (**d**) RMT requires receptor binding of ligand and can transport a variety of macromolecules such as peptides and proteins across the cerebral endothelium (transcytosis). AMT appears to be induced in a non-specific manner by positively charged macromolecules and can also transport across the endothelium. Both RMT and AMT appear to be vesicular-based systems which carry their macromolecule content across the endothelial cells. (**e**) Leucocyte entry is strictly regulated; under some conditions leucocytes may cross the endothelium by diapedesis either through the endothelial cells or via modified tight junctions. Tight junction modulation can result from signals from cells associated with the NVU or be induced pharmacologically. Modified from Abbott et al. (2010), with permission

P-glycoprotein (Pgp) (Aänismaa et al. 2008) and breast cancer resistance protein (BCRP) (Fig. 1.4).

In certain regions of cell membranes, zones enriched in cholesterol and sphingolipids form dynamic microdomains termed 'lipid rafts'; these 10–200 nm heterogeneous structures are associated with a variety of proteins and play roles in cell polarisation, endocytosis, signal transduction, adhesion, migration and links to the cytoskeleton, among others. In brain endothelium, such rafts (Cayrol et al. 2011) have documented functions in leucocyte adhesion and trafficking, junctional molecular architecture and localisation and function of transporters (Dodelier-Devilliers et al. 2009). A subset of rafts form caveolae, with high expression of caveolin-1, and can be further classified by function in *scaffolding* for junctional proteins and adhesion to basal lamina, immune cell *adhesion* and recruitment and transendothelial *transport*. Certain environmental pollutants such as polychlorinated biphenyl (PCB) induce disruption of BBB tight junction occludin and endothelial barrier function; activation of matrix metalloprotease MMP-2 in lipid rafts is involved in the reduction of occludin levels by BCB (Eum et al. 2015).

1.4 Tight Junctions in Brain Endothelium and Barrier Epithelia: Structure and Restrictive Properties

The tight junctions of the CNS barrier layers forming the 'physical' barrier (Fig. 1.4) involve a complex 3-D organisation of transmembrane proteins (claudins, occludin), spanning the cleft to create the diffusional restriction and coupling on the cytoplasmic side to an array of adaptor and regulatory proteins linking to the cytoskeleton (Cording et al. 2013). Adherens junctions, while not themselves restricting paracellular permeability, are important in formation and stabilisation of tight junctions (Paolinelli et al. 2011; Daneman 2012).

The brain endothelial tight junctions are capable of restricting paracellular ionic flux to give high transendothelial electrical resistance (TEER) in vivo of >1000 ohm. cm^2, while choroid plexus tight junctions are leakier, although the complex frond-like morphology of the in vivo mammalian plexus makes TEER harder to measure. TEER of ~150 ohm.cm^2 has been recorded across the simpler bullfrog choroid plexus. The brain endothelium shows high expression of the 'barrier-forming' claudin 5, together with claudin 3 and 12, while in choroid plexus the 'pore-forming' claudin 1 dominates, with detectable claudin 2, 3 and 11 (Strazielle and Ghersi-Egea 2013). Metastatic cells may migrate through the cerebral endothelium either through the tight junctions (paracellular) or across the cells (transcellular) initiated by filopodia extended from the cells (Herman et al. 2019).

For the arachnoid epithelium the situation is less clear; the arachnoid barrier layer is closely apposed to the dura and difficult to isolate intact. It has recently proved possible to culture arachnoid cells in vitro, which express claudin 1 and generate a TEER of ~160 ohm.cm^2 with restriction of larger solute permeation (Lam et al. 2011, 2012; Janson et al. 2011). The perineurium forming part of the outer sheath of peripheral nerves is a continuation of the arachnoid layer of the spinal meninges and easier to study than the arachnoid; a TEER of ~480 ohm.cm^2 (Weerasuriya et al. 1984) and expression of claudin 1 have been observed (Hackel et al. 2012). The pattern of barrier properties is consistent with the brain endothelium exerting the most stringent effect on paracellular permeability, while the choroid plexus with a major role in secreting CSF is leakier; the arachnoid epithelium appears to create a barrier of intermediate tightness.

Several junctional proteins, especially occludin and ZO-1, show considerable dynamic activity (half times 100–200 s) (Shen et al. 2008) while maintaining overall junctional integrity and selectivity. Many modulators from both the blood and the brain side can cause junctional opening, some via identified receptor-mediated processes (Abbott et al. 2006; Fraser 2011), possibly aiding repair and removal of debris, but in healthy conditions this is local and transient and does not significantly disturb the homeostatic function of the barrier. Indeed, the presence of endogenous 'protective' molecules and mechanisms able to tighten the barriers is increasingly recognised as important in protection and maintenance at the barrier sites (Bazan et al. 2012; Cristante et al. 2013). Recent studies have highlighted the possible role of microRNAs in barrier protection (Reijerkerk et al. 2013), and astrocyte-derived fatty-acid-binding protein 7 protects BBB integrity through a caveolin-1/MMP signalling pathway following traumatic brain injury (Rui et al. 2019).

1.5 Small Solute Transport at the Barrier Layers

Many BBB solute carriers (SLCs) with relatively tight substrate specificities have been described (Abbott et al. 2010; Redzic 2011; Neuwelt et al. 2011; Parkinson et al. 2011; Zaragoza 2020), mediating entry of major nutrients such as glucose, amino acids, nucleosides, monocarboxylates and organic anions and cations and efflux from the brain of some metabolites (Fig. 1.4). Among the group of ABC (efflux) transporters, Pgp (ABCB1) and BCRP (ABCG2) are the dominant players on the apical (blood-facing) membrane, especially Pgp in rodents and BCRP in primates, but the expression levels, localisation and roles of the multidrug-resistant associated proteins (MRPs, ABCC group) are less clear (Shawahna et al. 2011) (Fig. 1.4). ABC transporters have broader substrate specificity than the SLCs, making analysis of their structure-activity relationship (SAR) difficult (Demel et al. 2009). Synergistic activity between Pgp and BCRP has been observed (Kodaira et al. 2010), and ABC transporters and cytochrome P450 (CYP) enzymes together generate an active metabolic barrier within the NVU (Declèves et al. 2011). Differences between species and between in vitro models may make it difficult to draw firm conclusions in comparative studies (Shawahna et al. 2013; Breuss et al. 2020).

There are many differences between the transporters and enzymes expressed in the different barrier layers, suggesting they play different but complementary roles in regulation of molecular flux (Strazielle and Ghersi-Egea 2013; Saunders et al. 2013; Yasuda et al. 2013; Zaragoza 2020). The transporters present include considerable overlap in function/apparent redundancy at each site, reflecting their evolutionary history (Dean and Annilo 2005) and ensuring maintained function in case of loss or defect of a single transporter.

1.6   Vesicular Transport and Transcytosis

Classification of types of vesicular transport by cells is complex, but it is clear that certain features of endocytosis and transcytosis in the highly polarised brain endothelium are different from those of less polarised endothelia such as that of skeletal muscle. Non-specific fluid-phase endocytosis and transcytosis are downregulated in the brain compared with non-brain endothelium. However, for certain endogenous peptides and proteins, two main types of vesicle-mediated transfer have been documented in the BBB: receptor-mediated transcytosis (RMT) and adsorptive mediated transcytosis (AMT) (Abbott et al. 2010) (Fig. 1.4). There appears to be some overlap in function between caveolar and clathrin-mediated vesicular routes and likely involvement of other types of molecular entrapment, engulfment and transendothelial movement that are less well characterised (Mayor and Pagano 2007; Strazielle and Ghersi-Egea 2013). Recent studies using manufactured non-ionic surfactant vesicles (NISVs) show that decoration with glucosamine can enhance delivery across the BBB in vivo and in vitro (Woods et al. 2020); these NISVs hold promise for drug delivery.

Electron microscopy of the choroid plexus shows a variety of vesicular and tubular profiles, but the epithelium appears to be specialised for secretion rather than transcytosis (Strazielle and Ghersi-Egea 2013).

1.7   Routes for Permeation Across Barrier Layers
and Influence on Drug Delivery

Many of these routes for permeation across the brain endothelium (Fig. 1.4) can be used for drug delivery; several classical CNS drugs are sufficiently lipid-soluble to diffuse through the endothelial cell membranes to reach the brain ISF (Bodor and Buchwald 2003). However, for less lipophilic agents with slower permeation and hence longer dwell time in the lipid bilayer, activity of ABC efflux transporters can significantly reduce CNS access (Seelig 2007; Turunen et al. 2008; Aänismaa et al. 2008). As barrier tightness, transporter expression/activity and vesicular mechanisms can be altered in pathology, it is difficult to predict CNS distribution and pharmacokinetics of drugs in individual patients, particularly where barrier dysfunction may change both regionally and in time during the course of pathologies such as epilepsy, stroke and cancer (Stanimirovic and Friedman 2012).

1.8 Development, Induction, Maintenance and Heterogeneity of the BBB

Study of BBB evolution, development and maintenance gives valuable insights into both normal physiology and the changes that can occur in pathology. Studies in invertebrates and lower vertebrates especially archaic fish provide strong evidence that the first barrier layers protecting the CNS were formed by specialised glial cells at the vascular-neural interface and that as the intracerebral vasculature became more complete and complex, the barrier was increasingly supported by pericytes and endothelium. Later there was a shift to the dominant modern vertebrate pattern, where the endothelium forms the principal barrier layer (Bundgaard and Abbott 2008). Interestingly, the pericytes and astrocytes still remain closely associated with the brain endothelium, reflecting their evolutionary history and contributing to the NVU.

In the development of the mammalian brain, the endothelium of the ingrowing vessel sprouts develops basic restrictive barrier properties under the influence of neural progenitor cells (NPCs) (Liebner et al. 2008; Daneman et al. 2009), with pericytes subsequently refining the phenotype by downregulating features characteristic of non-brain endothelium; later, astrocytes help upregulate the full differentiated BBB phenotype (Daneman et al. 2010; Armulik et al. 2010; Stebbins et al. 2019).

Some of the signalling mechanisms involved in this induction are known, including the Wnt/β-catenin (Liebner et al. 2008) and sonic hedgehog pathways (Alvarez et al. 2011), and some of them may be involved in maintaining barrier integrity in the adult. It is clear that endothelial cells and pericytes are in turn involved in signalling to astrocytes, to regulate the expression of ion and water channels, receptors, transporters and enzymes on the astrocyte endfeet, so that mutual induction and maintenance is involved in sustaining the critical features of barrier and NVU function (Abbott et al. 2006). This regulation extends to the microanatomy and microenvironment of the perivascular space created by the extracellular matrix/basal lamina components of the endothelial-pericyte-astrocyte complex (Liebner et al. 2011; Stebbins et al. 2019). Microglial cell processes are found among the astrocyte endfeet (Mathiisen et al. 2010), suggesting roles in monitoring and influencing the local cellular organisation and function; indeed, microglial cells have been shown to regulate leucocyte traffic (reviewed in Daneman 2012). Specific perivascular nerve fibres associated with cerebral microvessels are involved in regulation of vascular tone (Hamel 2006). However, less is known about microglial and neuronal induction of barrier properties, and the signalling pathways involved in barrier maintenance on a minute-by-minute basis are relatively unexplored.

The NVU contains several mechanisms for protection of the BBB against minor damage such as local oxidative stress, e.g. by tightening the barrier (Abbott et al. 2006) and presence of detoxifying transporters and enzymes (Strazielle and Ghersi-Egea 2013), but this field is expanding with recognition that some of the 'protectins', protective agents identified in peripheral tissues, are also active in the brain

(Bazan et al. 2012). Recently the protein annexin-A1/lipocortin has been shown to be involved in the anti-inflammatory and neuroprotective effects of microglia (McArthur et al. 2010) and to act as an endogenous BBB tightening agent (Cristante et al. 2013; Wang et al. 2017). Improved understanding of the mechanisms for 'self-repair' within the NVU to correct minor local damage is likely to prove critical in future development of therapies that treat CNS disorders at much earlier phases of the pathology than currently possible, with expected major gains in efficacy.

There are several phenotypic and functional differences between the endothelial cells of different segments of the cerebral microvasculature (reviewed in Ge et al. 2005; Patabendige et al. 2013). Compared with arteriolar or venular endothelium, cerebral capillary endothelium has a more complex pattern of tight junction strands in freeze-fracture images consistent with tighter tight junctions and higher expression of solute transporters including efflux transporters and of certain receptors involved in transcytosis. Arteriolar endothelium shows higher expression of certain enzymes and absence of P-glycoprotein and in a few regions shows bidirectional transcytosis of tracers such as horseradish peroxidase, creating a local protein 'leak'. The post-capillary venule segment is specialised for regulation of leucocyte traffic and control of local inflammation. Some differences between the vascular beds of different brain regions have been observed at both micro- and macro-levels, but in general their significance is unclear.

1.9 Beyond the Barrier: The Fluid Compartments of the ISF and CSF

The cells of the brain, chiefly neurons and macroglia (astrocytes and oligodendrocytes) but also microglia, the resident immune cells of the brain, are bathed by an ionic medium similar to plasma, but containing very low protein and slightly more Mg^{2+}, less K^+ and Ca^{2+} (Somjen 2004). This extracellular or interstitial fluid (ECF, ISF) occupies around 20% of the brain volume (Sykova and Nicholson 2008). The ventricles and subarachnoid space contain cerebrospinal fluid (CSF), secreted by the choroid plexuses of the lateral, third and fourth ventricles, and with a daily turnover in humans of two to four times per day (Silverberg et al. 2003). The outflow pathways include arachnoid granulations and outpouchings of the arachnoid membrane into veins in the dura, but some CSF also drains along cranial nerves (especially olfactory) and blood vessel sheaths to the lymph nodes of the neck. Species differences have been reported in the relative importance of these drainage routes (Johanson et al. 2008).

The origin and dynamics of the ISF are less well understood. The brain microvessels have the ionic transport mechanisms and channels and low but sufficient water permeability to generate ISF as a secretion (Fig. 1.5), and calculations show that a proportion of ISF water may come from glucose metabolism of the brain, aided by aquaporin 4 (AQP4) water channels in the perivascular endfeet membranes of

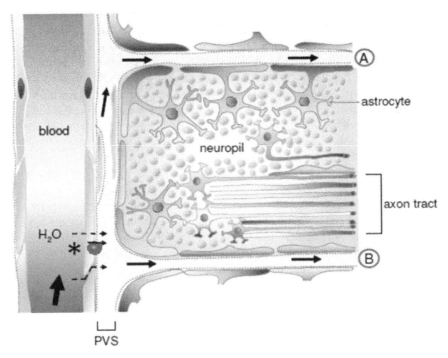

Fig. 1.5 Proposed sites of generation of ISF and routes for ISF flow. A large fraction of ISF is proposed to be formed by brain capillary endothelium, driven by the ionic gradient set up by the abluminal Na, K, ATPase (* circle + arrow). Water follows passively either through the endothelial cell membranes or via the tight junctions (dashed arrows). Driven by this hydrostatic pressure gradient and with the addition of some CSF from the subarachnoid space, ISF moves by bulk flow through low resistance pathways formed by perivascular spaces (PVS, predominantly around larger vessels including arterioles and arteries, venules and veins), connecting with (**a**) glial-lined boundary zones between blocks of neuropil and (**b**) regions adjacent to axon tracts. The narrow spaces between cells within the neuropil appear to be too narrow to permit significant bulk flow. Not to scale. Modified by S Yusof from Abbott (2004), with permission

astrocytes (Abbott 2004; Dolman et al. 2005). Within the neuropil, the small blocks of tissue demarcated by the lattice of fine microvessels, in which neural communication occurs, the distances from the vessel to the furthest neuron are small, typically <30 μm, so that diffusion within the neuropil is an effective means of ionic and molecular movement. Indeed, many studies in isolated brain slices and in situ confirm the local diffusive behaviour of test molecules injected into the brain (Thorne and Nicholson 2006; Wolak and Thorne 2013). However, superimposed on this local diffusion is the possibility for flow of ISF over longer distances, especially along perivascular spaces (Abbott et al. 2018). Convective flow of ISF through the delicate synaptic networks of the brain would be potentially damaging and has not been convincingly demonstrated in spite of claims of the 'glymphatic' hypothesis (Taoka and Naganawa 2020).

There is considerable historical evidence for flowing ISF, capable of clearing waste products including large molecules such as β-amyloid from the interstitium moving via routes offering the least resistance to flow, along axon tracts and blood vessels (Abbott 2004; Weller et al. 2008, 2009). Careful studies of clearance of tracer molecules injected into the parenchyma give a figure for clearance half-time of 2–3 h (Groothuis et al. 2007), around ten times faster than reported earlier (Cserr et al. 1981). Most of this flow can be accounted for by fluid secretion across cerebral capillary endothelium (Abbott 2004) (Fig. 1.5), but recent studies add to a body of earlier evidence showing that a proportion of CSF from the subarachnoid space can flow into the brain along periarterial (Virchow-Robin) spaces, contributing to ISF, with return out along nerve fibre tracts and blood vessels (Abbott 2004; Iliff et al. 2012; Yang et al. 2013). However, there is some controversy over whether arteries (Weller et al. 2008, 2009) or veins (Iliff et al. 2012) are chiefly responsible for the ISF outflow route from the brain parenchyma. In any event, with the flow largely confined to major extracellular 'highways' in the tissue, the rate of turnover will be similar to that of CSF. Thus ISF and CSF can be regarded as parallel fluids maintaining a continuous flow through the low resistance pathways of the brain (ISF) and through the ventricles and subarachnoid space (CSF), capable of some mixing hence with some shared roles, but also many distinct and complementary functions. Between them, the CSF and ISF contribute to maintaining tissue buoyancy, waste removal, circulation of secretory products such as vitamins and hormones from choroid plexuses, non-synaptic or 'distance' signalling ('volume transmission') and providing routes for immune surveillance without disturbing neuronal networks (Strazielle and Ghersi-Egea 2013; Dalakas et al. 2020).

1.10   Changes in BBB and BCSFB in Pathology

The BBB is altered in many CNS pathologies, including stroke, vascular dementia, Alzheimer's disease, Parkinson's disease, multiple sclerosis, amyotrophic lateral sclerosis, hypoxia, ischaemia, diabetes mellitus and epilepsy (reviewed in Abbott et al. 2006, 2010; Friedman 2011; Abbott and Friedman 2012; Daneman 2012; Stanimirovic and Friedman 2012; Potschka 2012; Michalicova et al. 2020). Even mild traumatic brain injury (mTBI) may result in changes in BBB integrity and function (Rawlings et al. 2020). Changes can include upregulation of luminal adhesion molecules, increased adhesion and transmigration of leucocytes, increased leakiness of tight junctions, extravasation of plasma proteins via paracellular or transcellular routes and altered expression of drug transporters. Given the importance of the BBB in CNS homeostasis, it is clear that gross barrier dysfunction is likely to be associated with disturbance of neural signalling, in both the short and the long term (Abbott and Friedman 2012). In many pathologies, a combination or sequence of events may make the barrier vulnerable, including hypoxia, infection, activation of the clotting system and inflammation, components of the diet and environmental toxins, and genetic factors may also contribute (Shlosberg et al. 2010).

Inflammation and free radicals are now recognised to play major roles in many or even most of the pathologies with BBB disturbance, but the aetiology and sequence of changes are generally unclear, and in many cases, it is not known whether changes occur simultaneously or as part of an inflammatory cascade (Friedman 2011; Kim et al. 2013). Certain brain regions are more often affected, including the hippocampus and cerebral cortex grey matter, but again the reasons are uncertain.

The Coronavirus (SARS-CoV-2, Covid-19), first reported in humans in Wuhan, China, in Dec 2019, subsequently developed into a worldwide pandemic. A number of clinical and laboratory studies have followed on its sites and modes of action; Covid-19 has serious effects on the vasculature in multiple organ systems including the cerebral vasculature. In vitro evidence suggests that the viral spike proteins S1 and S2 trigger a pro-inflammatory response in brain endothelial cells that may contribute to altered BBB function (Buzhdygan et al. 2020).

For minor damage, the cells of the NVU aided by recruitment of leucocytes may effect a repair, and short- and long-term changes in protective mechanisms including upregulation of efflux transporters and enzymes may be involved. Certainly several types of altered cell/cell interaction can be detected in pathology, particularly between endothelium and astrocytes, but also with powerful roles played by microglia, the tissue-resident macrophage, changing from a relatively quiescent and static process-bearing morphology to a more amoeboid and migratory form, secreting a different repertoire of cytokines and chemokines (Saijo and Glass 2011; Smith et al. 2012; Daneman 2012). In certain pathological neuroinflammatory and neurodegenerative conditions, there is unexpected phenotypic convergence between CNS microglia (the tissue-resident macrophages of the brain) and peripheral macrophages, suggesting that the two cell types act synergistically, boosting their mutual activities and therapeutic potential (Grassivaro et al. 2021).

Agents released from most of the cells of the NVU in pathology can modulate brain endothelial tight junctions, with several inflammatory mediators increasing barrier permeability and a few agents able to counter or reverse this (Abbott et al. 2006). Potentiating effects of several cytokines including IL-1β and TNFα on the 'first line' of inflammatory mediators (e.g. bradykinin) have been documented (Fraser 2011). At the molecular level, a great many signalling pathways can be identified, regulating both the expression and activity of barrier features, particularly well documented for the effects of xenobiotics, neurotransmitters and inflammation on Pgp (Miller 2010). Recent identification of a number of microRNAs (miRNAs) shown to influence angiogenesis (Caporali and Emanueli 2011) vascular functions (Hartmann and Thum 2011) and BBB physiology/pathology (Reijerkerk et al. 2013; Mishra and Singh 2013) adds a further level of complexity. Furthermore, new information on a whole family of secreted and information-carrying extracellular vesicles including exosomes (György et al. 2011; Haqqani et al. 2013) adds to the repertoire of ways in which a cell or group of cells can influence other cells nearby or further away. Indeed, the flow pathways allowing circulation of the brain ISF have suitable properties for this kind of non-neural communication (Abbott 2004; Abbott et al. 2018) and could also play an important part in the dissemination of CNS pathologies (multiple sclerosis, cancers) that start at a relatively restricted locus.

The choroid plexus and CSF/ISF flow system are also affected by ageing and by many pathologies, including tumours, infection, trauma, ischaemia, neurodegenerative disease and hydrocephalus (Johanson et al. 2008; Serot et al. 2012). Many of these affect the anatomy, connectivity and outflow routes of the fluid flow systems, but changes in the physiology of the choroid plexus and the resulting disturbance in generation and composition of CSF are also critical. Ageing is associated with a reduction in CSF production and in secretion of many choroid plexus-derived proteins, particularly important for the zones of neurogenesis close to the ventricular wall.

1.11   Implications for Drug Delivery

The anatomy and physiology of the CNS barriers and fluid systems described here have many implications for drug delivery, whether for agents designed to act in the CNS or for those with peripheral targets where the aim is to minimise CNS side effects. Clearly barrier changes in ageing and pathology will affect distribution and delivery of both CNS and peripheral drugs. Improved experimental methods and models, molecular and pharmacokinetic modelling and new developments in understanding barrier function help in measuring and predicting the concentration of drugs at the active site. The expanding field of 'biologic' therapeutics, large molecules with specific actions in the CNS, poses new challenges but is also giving novel insights into mechanisms and ways to improve CNS drug delivery of complex molecules. Many of these issues will be discussed in further chapters in this volume.

1.12   Points for Discussion

- Why is it important to understand the different properties of the three main barrier layers (Sect. 1.1)? What kinds of technique can be used to establish the relative importance of each in determining CNS distribution of a particular drug?
- Several 'key functions' of the BBB are listed (1.2). Is it possible to put these in order of importance for brain function?
- Much of the BBB and choroid plexus literature is devoted to documenting changes in pathology. Why has maintenance of healthy function received less attention?
- What models and techniques would you propose for a new study on cell/cell interaction within the NVU?
- What is the glycocalyx of the brain endothelium, and what properties of the cells is it most likely to influence?
- Why is it important to know about the organisation of the BBB lipid membrane in modelling drug permeation? Why are potential substrates for ABC transporters particularly affected by the membrane composition?

- Why is it difficult to establish how water moves across the BBB and choroid plexus?
- How does knowledge of BBB development help in understanding barrier function?
- What is the significance of heterogeneity in function, e.g. between the capillary and post-capillary venule segments of the cerebral microvasculature?

Acknowledgements I am grateful to Dr. Siti R Yusof for help with artwork and many colleagues for discussions.

References

Aänismaa P, Gatlik-Landwojtowicz E, Seelig A (2008) P-glycoprotein senses its substrates and the lateral membrane packing density: consequences for the catalytic cycle. Biochemistry 47:10197–10207

Abbott NJ (1992) Comparative physiology of the blood-brain barrier. In: MWB B (ed) Physiology and pharmacology of the blood-brain barrier, Handbook of experimental pharmacology, vol 103. Springer, Heidelberg, pp 371–396

Abbott NJ (2004) Evidence for bulk flow of brain interstitial fluid: significance for physiology and pathology. Neurochem Int 45:545–552

Abbott NJ (2013) Blood-brain barrier structure and function and the challenges for CNS drug delivery. J Inherit Metab Dis 36:437–449

Abbott NJ, Friedman A (2012) Overview and introduction: the blood-brain barrier in health and disease. Epilepsia 53(Suppl 6):1–6

Abbott NJ, Mendonca LL, Dolman DE (2003) The blood-brain barrier in systemic lupus erythematosus. Lupus 12:908–915

Abbott NJ, Rönnbäck L, Hansson E (2006) Astrocyte-endothelial interactions at the blood-brain barrier. Nat Rev Neurosci 7:41–53

Abbott NJ, Patabendige AA, Dolman DE, Yusof SR, Begley DJ (2010) Structure and function of the blood-brain barrier. Neurobiol Dis 37:13–25

Abbott NJ, Pizzo ME, Preston JE, Janigro D, Thorne RG (2018 Mar) The role of brain barriers in fluid movement in the CNS: is there a 'glymphatic' system? Acta Neuropathol 135(3):387–407

Alvarez JI, Dodelet-Devillers A, Kebir H, Ifergan I, Fabre PJ, Terouz S, Sabbagh M, Wosik K, Bourbonnière L, Bernard M, van Horssen J, de Vries HE, Charron F, Prat A (2011) The Hedgehog pathway promotes blood-brain barrier integrity and CNS immune quiescence. Science 334:1727–1731

Armulik A, Genové G, Mäe M, Nisancioglu MH, Wallgard E, Niaudet C, He L, Norlin J, Lindblom P, Strittmatter K, Johansson BR, Betsholtz C (2010) Pericytes regulate the blood-brain barrier. Nature 468:557–561

Bazan NG, Eady TN, Khoutorova L, Atkins KD, Hong S, Lu Y, Zhang C, Jun B, Obenaus A, Fredman G, Zhu M, Winkler JW, Petasis NA, Serhan CN, Belayev L (2012) Novel aspirin-triggered neuroprotectin D1 attenuates cerebral ischemic injury after experimental stroke. Exp Neurol 236:122–130

Becker NH, Novikoff AB, Zimmerman HM (1967) Fine structure observations of the uptake of intravenously injected peroxidase by the rat choroid plexus. J Histochem Cytochem 15:160–165

Bodor N, Buchwald P (2003) Brain targeted drug delivery; experiences to date. Am J Drug Deliv 1:13–26

Bohlen CJ, Friedman BA, Dejanovic B, Sheng M (2019 Dec 3) Microglia in brain development, homeostasis, and neurodegeneration. Annu Rev Genet 53:263–288

Breuss MW, Mamerto A, Renner T, Waters ER (2020) The evolution of the mammalian ABCA6-like genes: analysis of phylogenetic, expression and population genetic data reveals complex evolutionary histories. Genome Biol Evol evaa179. https://doi.org/10.1093/gbe/evaa179. Epub ahead of print

Bundgaard M, Abbott NJ (2008) All vertebrates started out with a glial blood-brain barrier 4-500 million years ago. Glia 56:699–708

Buzhdygan TP, DeOre BJ, Baldwin-Leclair A, Bullock TA, McGary HM, Khan JA, Razmpour R, Hale JF, Galie PA, Potula R, Andrews AM, Ramirez SH (2020 Oct 11) The SARS-CoV-2 spike protein alters barrier function in 2D static and 3D microfluidic in-vitro models of the human blood-brain barrier. Neurobiol Dis 146:105131

Caporali A, Emanueli C (2011) MicroRNA regulation in angiogenesis. Vasc Pharmacol 55:79–86

Cayrol R, Haqqani AS, Ifergan I, Dodelet-Devillers A, Prat A (2011) Isolation of human brain endothelial cells and characterization of lipid raft-associated proteins by mass spectroscopy. Methods Mol Biol 686:275–295

Cording J, Berg J, Käding N, Bellmann C, Tscheik C, Westphal JK, Milatz S, Günzel D, Wolburg H, Piontek J, Huber O, Blasig IE (2013) In tight junctions, claudins regulate the interactions between occludin, tricellulin and marvelD3, which, inversely, modulate claudin oligomerization. J Cell Sci 26:554–564

Cristante E, McArthur S, Mauro C, Maggioli E, Romero IA, Wylezinska-Arridge M, Couraud PO, Lopez-Tremoleda J, Christian HC, Weksler BB, Malaspina A, Solito E (2013) Identification of an essential endogenous regulator of blood-brain barrier integrity, and its pathological and therapeutic implications. Proc Natl Acad Sci USA 110:832–841

Cserr HF, Cooper DN, Suri PK, Patlak CS (1981) Efflux of radiolabeled polyethylene glycols and albumin from rat brain. Am J Phys 240:F319–F328

Dalakas MC, Alexopoulos H, Spaeth PJ (2020 Nov) Complement in neurological disorders and emerging complement-targeted therapeutics. Nat Rev Neurol 16(11):601–617

Dalkara T, Alarcon-Martinez L (2015 Oct 14) Cerebral microvascular pericytes and neurogliovascular signaling in health and disease. Brain Res 1623:3–17

Daneman R (2012) The blood-brain barrier in health and disease. Ann Neurol 72:648–672

Daneman R, Agalliu D, Zhou L, Kuhnert F, Kuo CJ, Barres BA (2009) Wnt/beta-catenin signaling is required for CNS, but not non-CNS, angiogenesis. Proc Natl Acad Sci USA 106:641–646

Daneman R, Zhou L, Kebede AA, Barres BA (2010) Pericytes are required for blood-brain barrier integrity during embryogenesis. Nature 468:562–566

Dean M, Annilo T (2005) Evolution of the ATP-binding cassette (ABC) transporter superfamily in vertebrates. Annu Rev Genomics Hum Genet 6:123–142

Declèves X, Jacob A, Yousif S, Shawahna R, Potin S, Scherrmann JM (2011) Interplay of drug metabolizing CYP450 enzymes and ABC transporters in the blood-brain barrier. Curr Drug Metab 12:732–741

Demel MA, Krämer O, Ettmayer P, Haaksma EE, Ecker GF (2009) Predicting ligand interactions with ABC transporters in ADME. Chem Biodivers 6:1960–1969

Dodelet-Devillers A, Cayrol R, van Horssen J, Haqqani AS, de Vries HE, Engelhardt B, Greenwood J, Prat A (2009) Functions of lipid raft membrane microdomains at the blood-brain barrier. J Mol Med (Berl) 87:765–774

Dolman D, Drndarski S, Abbott NJ, Rattray M (2005) Induction of aquaporin 1 but not aquaporin 4 messenger RNA in rat primary brain microvessel endothelial cells in culture. J Neurochem 93:825–833

Engelhardt B, Coisne C (2011) Fluids and barriers of the CNS establish immune privilege by confining immune surveillance to a two-walled castle moat surrounding the CNS castle. Fluids Barriers CNS 8:4. https://doi.org/10.1186/2045-8118-8-4

Engelhardt B, Vajkoczy P, Weller RO (2017) The movers and shapers in immune privilege of the CNS. Nat Immunol 18(2):123–131. https://doi.org/10.1038/ni.3666

Errede M, Girolamo F, Virgintino D (2021) High-resolution confocal imaging of pericytes in human fetal brain microvessels. Methods Mol Biol 2206:143–150

Eum SY, Jaraki D, András IE, Toborek M (2015 Sep 15) Lipid rafts regulate PCB153-induced disruption of occludin and brain endothelial barrier function through protein phosphatase 2A and matrix metalloproteinase-2. Toxicol Appl Pharmacol 287(3):258–266

Fraser PA (2011) The role of free radical generation in increasing cerebrovascular permeability. Free Radic Biol Med 51:967–977

Friedman A (2011) Blood-brain barrier dysfunction, status epilepticus, seizures, and epilepsy: a puzzle of a chicken and egg? Epilepsia 52(Suppl 8):19–20

Gatlik-Landwojtowicz E, Aänismaa P, Seelig A (2006) Quantification and characterization of P-glycoprotein-substrate interactions. Biochemistry 45:3020–3032

Ge S, Song L, Pachter JS (2005) Where is the blood-brain barrier ... really? J Neurosci Res 79:421–427

Grassivaro F, Martino G, Farina C (2021 Apr) The phenotypic convergence between microglia and peripheral macrophages during development and neuroinflammation paves the way for new therapeutic perspectives. Neural Regen Res 16(4):635–637

Greenwood J, Heasman SJ, Alvarez JI, Prat A, Lyck R, Engelhardt B (2011) Review: Leucocyte-endothelial cell crosstalk at the blood-brain barrier: a prerequisite for successful immune cell entry to the brain. Neuropathol Appl Neurobiol 37:24–39

Groothuis DR, Vavra MW, Schlageter KE, Kang EW, Itskovich AC, Hertzler S, Allen CV, Lipton HL (2007) Efflux of drugs and solutes from brain: the interactive roles of diffusional transcapillary transport, bulk flow and capillary transporters. J Cereb Blood Flow Metab 27:43–56

György B, Szabó TG, Pásztói M, Pál Z, Misják P, Aradi B, László V, Pállinger E, Pap E, Kittel A, Nagy G, Falus A, Buzás EI (2011) Membrane vesicles, current state-of-the-art: emerging role of extracellular vesicles. Cell Mol Life Sci 68:2667–2688

Hackel D, Krug SM, Sauer RS, Mousa SA, Böcker A, Pflücke D, Wrede EJ, Kistner K, Hoffmann T, Niedermirtl B, Sommer C, Bloch L, Huber O, Blasig IE, Amasheh S, Reeh PW, Fromm M, Brack A, Rittner HL (2012) Transient opening of the perineurial barrier for analgesic drug delivery. Proc Natl Acad Sci USA 109:E2018–E2027

Hamel E (2006) Perivascular nerves and the regulation of cerebrovascular tone. J Appl Physiol 100:1059–1064

Haqqani AS, Hill JJ, Mullen J, Stanimirovic DB (2011) Methods to study glycoproteins at the blood-brain barrier using mass spectrometry. Methods Mol Biol 686:337–353

Haqqani AS, Delaney CE, Tremblay TL, Sodja C, Sandhu JK, Stanimirovic DB (2013) Method for isolation and molecular characterization of extracellular microvesicles released from brain endothelial cells. Fluids Barriers CNS 10:4. https://doi.org/10.1186/2045-8118-10-4

Hartmann D, Thum T (2011) MicroRNAs and vascular (dys)function. Vasc Pharmacol 55:92–105

Herman H, Fazakas C, Haskó J, Molnár K, Mészáros Á, Nyúl-Tóth Á, Szabó G, Erdélyi F, Ardelean A, Hermenean A, Krizbai IA, Wilhelm I (2019 Apr) Paracellular and transcellular migration of metastatic cells through the cerebral endothelium. J Cell Mol Med 23(4):2619–2631

Iadecola C (2017 Sep 27) The neurovascular unit coming of age: a journey through neurovascular coupling in health and disease. Neuron 96(1):17–42

Iadecola C, Nedergaard M (2007) Glial regulation of the cerebral microvasculature. Nat Neurosci 10:1369–1376

Iliff JJ, Wang M, Liao Y, Plogg BA, Peng W, Gundersen GA, Benveniste H, Vates GE, Deane R, Goldman SA, Nagelhus EA, Nedergaard M (2012) A paravascular pathway facilitates CSF flow through the brain parenchyma and the clearance of interstitial solutes, including amyloid β. Sci Transl Med 4:147ra111. https://doi.org/10.1126/scitranslmed.3003748

Janson C, Romanova L, Hansen E, Hubel A, Lam C (2011) Immortalization and functional characterization of rat arachnoid cell lines. Neuroscience 177:23–34

Johanson CE, Duncan JA 3rd, Klinge PM, Brinker T, Stopa EG, Silverberg GD (2008) Multiplicity of cerebrospinal fluid functions: new challenges in health and disease. Cerebrospinal Fluid Res 5:10. https://doi.org/10.1186/1743-8454-5-10

Kim SY, Buckwalter M, Soreq H, Vezzani A, Kaufer D (2013) Blood-brain barrier dysfunction-induced inflammatory signaling in brain pathology and epileptogenesis. Epilepsia 53(Suppl 6):37–44

Kodaira H, Kusuhara H, Ushiki J, Fuse E, Sugiyama Y (2010) Kinetic analysis of the cooperation of P-glycoprotein (P-gp/Abcb1) and breast cancer resistance protein (Bcrp/Abcg2) in limiting the brain and testis penetration of erlotinib, flavopiridol, and mitoxantrone. J Pharmacol Exp Ther 333:788–796

Krämer SD, Schütz YB, Wunderli-Allenspach H, Abbott NJ, Begley DJ (2002) Lipids in blood-brain barrier models in vitro II: influence of glial cells on lipid classes and lipid fatty acids. In Vitro Cell Dev Biol 38:566–571

Kutomi O, Takeda S (2020) Identification of lymphatic endothelium in cranial arachnoid granulation-like dural gap. Microscopy (Oxf) dfaa038

Lam CH, Hansen EA, Hubel A (2011) Arachnoid cells on culture plates and collagen scaffolds: phenotype and transport properties. Tissue Eng Part A 17:1759–1766

Lam CH, Hansen EA, Janson C, Bryan A, Hubel A (2012) The characterization of arachnoid cell transport II: paracellular transport and blood-cerebrospinal fluid barrier formation. Neuroscience 222:228–238

Liebner S, Corada M, Bangsow T, Babbage J, Taddei A, Czupalla CJ, Reis M, Felici A, Wolburg H, Fruttiger M, Taketo MM, von Melchner H, Plate KH, Gerhardt H, Dejana E (2008) Wnt/beta-catenin signaling controls development of the blood-brain barrier. J Cell Biol 183:409–417

Liebner S, Czupalla CJ, Wolburg H (2011) Current concepts of blood-brain barrier development. Int J Dev Biol 55:467–476

Liu DZ, Ander BP, Xu H, Shen Y, Kaur P, Deng W, Sharp FR (2010) Blood-brain barrier breakdown and repair by Src after thrombin-induced injury. Ann Neurol 67:526–533

MacAulay N, Zeuthen T (2010) Water transport between CNS compartments: contributions of aquaporins and cotransporters. Neuroscience 168:941–956

Mäe M, Armulik A, Betsholtz C (2011) Getting to know the cast – cellular interactions and signaling at the neurovascular unit. Curr Pharm Des 17:2750–2754

Mathiisen TM, Lehre KP, Danbolt NC, Ottersen OP (2010) The perivascular astroglial sheath provides a complete covering of the brain microvessels: an electron microscopic 3D reconstruction. Glia 58:1094–1103

Mayor S, Pagano RE (2007) Pathways of clathrin-independent endocytosis. Nat Rev Mol Cell Biol 8:603–612

McArthur S, Cristante E, Paterno M, Christian H, Roncaroli F, Gillies GE, Solito E (2010) Annexin A1: a central player in the anti-inflammatory and neuroprotective role of microglia. J Immunol 185:317–328

Michalicova A, Majerova P, Kovac A (2020 Sep 30) Tau protein and its role in blood-brain barrier dysfunction. Front Mol Neurosci 13:570045

Miller DS (2010) Regulation of P-glycoprotein and other ABC drug transporters at the blood brain barrier. Trends Pharmacol Sci 31:246–254

Mishra R, Singh SK (2013) HIV-1 Tat C modulates expression of miRNA-101 to suppress VE-cadherin in human brain microvascular endothelial cells. J Neurosci 33:5992–6000

Muoio V, Persson PB, Sendeski MM (2014 Apr) The neurovascular unit – concept review. Acta Physiol (Oxf) 210(4):790–798

Nabeshima S, Reese TS, Landis DMD, Brightman MW (1975) Junctions in the meninges and marginal glia. J Comp Neurol 164:127–169

Neuwelt EA, Bauer B, Fahlke C, Fricker G, Iadecola C, Janigro D, Leybaert L, Molnár Z, O'Donnell ME, Povlishock JT, Saunders NR, Sharp F, Stanimirovic D, Watts RJ, Drewes LR (2011) Engaging neuroscience to advance translational research in brain barrier biology. Nat Rev Neurosci 12:169–182

Nicholson C, Hrabětová S (2017) Brain extracellular space: the final frontier of neuroscience. Biophys J 113(10):2133–2142. https://doi.org/10.1016/j.bpj.2017.06.052

Okada H, Yoshida S, Hara A, Ogura S, Tomita H (2020 Aug) Vascular endothelial injury exacerbates coronavirus disease 2019: the role of endothelial glycocalyx protection. Microcirculation 13:e12654

Paolinelli R, Corada M, Orsenigo F, Dejana E (2011) The molecular basis of the blood brain barrier differentiation and maintenance. Is it still a mystery? Pharmacol Res 63:165–171

Parkinson FE, Damaraju VL, Graham K, Yao SY, Baldwin SA, Cass CE, Young JD (2011) Molecular biology of nucleoside transporters and their distributions and functions in the brain. Curr Top Med Chem 11:948–972

Patabendige A, Skinner RA, Morgan L, Abbott NJ (2013) A detailed method for preparation of a functional and flexible blood-brain barrier model using porcine brain endothelial cells. Brain Res pii: S0006–8993(13)00519–2. https://doi.org/10.1016/j.brainres.2013.04.006

Posada-Duque RA, Cardona-Gómez GP (2020 Sep) CDK5 Targeting as a therapy for recovering neurovascular unit integrity in alzheimer's disease. J Alzheimers Dis 28

Potschka H (2012) Role of CNS efflux drug transporters in antiepileptic drug delivery: overcoming CNS efflux drug transport. Adv Drug Deliv Rev 64:943–952

Pottiez G, Duban-Deweer S, Deracinois B, Gosselet F, Camoin L, Hachani J, Couraud PO, Cecchelli R, Dehouck MP, Fenart L, Karamanos Y, Flahaut C (2011) A differential proteomic approach identifies structural and functional components that contribute to the differentiation of brain capillary endothelial cells. J Proteome 75:628–641

Ransohoff RM, Brown MA (2012) Innate immunity in the central nervous system. J Clin Invest 122:1164–1171

Ransohoff RM, Engelhardt B (2012) The anatomical and cellular basis of immune surveillance in the central nervous system. Nat Rev Immunol 12:623–635

Ransohoff RM, Perry VH (2009) Microglial physiology: unique stimuli, specialized responses. Annu Rev Immunol 27:119–145

Rawlings S, Takechi R, Lavender AP (2020 Oct 1) Effects of sub-concussion on neuropsychological performance and its potential mechanisms: a narrative review. Brain Res Bull 165:56–62

Redzic Z (2011) Molecular biology of the blood-brain and the blood-cerebrospinal fluid barriers: similarities and differences. Fluids Barriers CNS 8:3. https://doi.org/10.1186/2045-8118-8-3

Reese TS, Karnovsky MJ (1967) Fine structural localization of a blood-brain barrier to exogenous peroxidase. J Cell Biol 34:207–217

Reijerkerk A, Lopez-Ramirez MA, van Het Hof B, Drexhage JA, Kamphuis WW, Kooij G, Vos JB, van der Pouw Kraan TC, van Zonneveld AJ, Horrevoets AJ, Prat A, Romero IA, de Vries HE (2013) MicroRNAs regulate human brain endothelial cell-barrier function in inflammation: implications for multiple sclerosis. J Neurosci 33:6857–6863

Rui Q, Ni H, Lin X, Zhu X, Li D, Liu H, Chen G (2019) Astrocyte-derived fatty acid-binding protein 7 protects blood-brain barrier integrity through a caveolin-1/MMP signaling pathway following traumatic brain injury. Exp Neurol 322:113044. https://doi.org/10.1016/j.expneurol.2019.113044

Saijo K, Glass CK (2011) Microglial cell origin and phenotypes in health and disease. Nat Rev Immunol 11:775–787

Saunders NR, Daneman R, Dziegielewska KM, Liddelow SA (2013) Transporters of the blood-brain and blood-CSF interfaces in development and in the adult. Mol Asp Med 34:742–752

Seelig A (2007) The role of size and charge for blood-brain barrier permeation of drugs and fatty acids. J Mol Neurosci 33:32–41

Serot JM, Zmudka J, Jouanny P (2012) A possible role for CSF turnover and choroid plexus in the pathogenesis of late onset Alzheimer's disease. J Alzheimers Dis 30:17–26

Shawahna R, Uchida Y, Declèves X, Ohtsuki S, Yousif S, Dauchy S, Jacob A, Chassoux F, Daumas-Duport C, Couraud PO, Terasaki T, Scherrmann JM (2011) Transcriptomic and quantitative proteomic analysis of transporters and drug metabolizing enzymes in freshly isolated human brain microvessels. Mol Pharm 8:1332–1341

Shawahna R, Decleves X, Scherrmann JM (2013 Jan) Hurdles with using in vitro models to predict human blood-brain barrier drug permeability: a special focus on transporters and metabolizing enzymes. Curr Drug Metab 14(1):120–136

Shen L, Weber CR, Turner JR (2008) The tight junction protein complex undergoes rapid and continuous molecular remodeling at steady state. J Cell Biol 181:683–695

Shlosberg D, Benifla M, Kaufer D, Friedman A (2010) Blood-brain barrier breakdown as a therapeutic target in traumatic brain injury. Nat Rev Neurol 6:393–403

Silverberg GD, Mayo M, Saul T, Rubenstein E, McGuire D (2003) Alzheimer's disease, normal-pressure hydrocephalus, and senescent changes in CSF circulatory physiology: a hypothesis. Lancet Neurol 2:506–511

Smith JA, Das A, Ray SK, Banik NL (2012) Role of pro-inflammatory cytokines released from microglia in neurodegenerative diseases. Brain Res Bull 87:10–20

Sokołowski W, Barszcz K, Kupczyńska M, Czubaj N, Skibniewski M, Purzyc H (2018) Lymphatic drainage of cerebrospinal fluid in mammals - are arachnoid granulations the main route of cerebrospinal fluid outflow? Biologia (Bratisl) 73(6):563–568

Somjen GG (2004) Ions in the brain: normal function, seizures and stroke. Oxford University Press, Oxford

Stanimirovic DB, Friedman A (2012) Pathophysiology of the neurovascular unit: disease cause or consequence? J Cereb Blood Flow Metab 32:1207–1221

Stebbins MJ, Gastfriend BD, Canfield SG, Lee MS, Richards D, Faubion MG, Li WJ, Daneman R, Palecek SP, Shusta EV (2019) Human pluripotent stem cell-derived brain pericyte-like cells induce blood-brain barrier properties. Sci Adv 5(3):eaau7375

Strazielle N, Ghersi-Egea JF (2013) Physiology of blood-brain interfaces in relation to brain disposition of small compounds and macromolecules. Mol Pharm 10:1473–1491

Sykova E, Nicholson C (2008) Diffusion in brain extracellular space. Physiol Rev 88:1277–1340

Taoka T, Naganawa S (2020 Nov) Neurofluid dynamics and the glymphatic system: a neuroimaging perspective. Korean J Radiol 21(11):1199–1209

Thorne RG, Nicholson C (2006) In vivo diffusion analysis with quantum dots and dextrans predicts the width of brain extracellular space. Proc Natl Acad Sci USA 103:5567–5572

Tian W, Sawyer A, Kocaoglu FB, Kyriakides TR (2011) Astrocyte-derived thrombospondin-2 is critical for the repair of the blood-brain barrier. Am J Pathol 179:860–868

Turunen BJ, Ge H, Oyetunji J, Desino KE, Vasandani V, Güthe S, Himes RH, Audus KL, Seelig A, Georg GI (2008 Nov 15) Paclitaxel succinate analogs: anionic and amide introduction as a strategy to impart blood-brain barrier permeability. Bioorg Med Chem Lett 18(22):5971–5974

Uchida Y, Goto R, Takeuchi H, Łuczak M, Usui T, Tachikawa M, Terasaki T (2020 Feb) Abundant expression of OCT2, MATE1, OAT1, OAT3, PEPT2, BCRP, MDR1, and xCT transporters in blood-arachnoid barrier of pig and polarized localizations at CSF- and blood-facing plasma membranes. Drug Metab Dispos 48(2):135–145

Wang Z, Chen Z, Yang J, Yang Z, Yin J, Zuo G, Duan X, Shen H, Li H, Chen G (2017 Jul) Identification of two phosphorylation sites essential for annexin A1 in blood-brain barrier protection after experimental intracerebral hemorrhage in rats. J Cereb Blood Flow Metab 37(7):2509–2525

Weerasuriya A, Spangler RA, Rapoport SI, Taylor RE (1984) AC impedance of the perineurium of the frog sciatic nerve. Biophys J 46:167–174

Weller RO, Subash M, Preston SD, Mazanti I, Carare RO (2008) Perivascular drainage of amyloid-beta peptides from the brain and its failure in cerebral amyloid angiopathy and Alzheimer's disease. Brain Pathol 18:253–266

Weller RO, Djuanda E, Yow HY, Carare RO (2009) Lymphatic drainage of the brain and the pathophysiology of neurological disease. Acta Neuropathol 117:1–14

Wolak DJ, Thorne RG (2013) Diffusion of macromolecules in the brain: implications for drug delivery. Mol Pharm 10:1492–1504

Woods S, O'Brien LM, Butcher W, Preston JE, Georgian AR, Williamson ED, Salguero FJ, Modino F, Abbott NJ, Roberts CW, D'Elia RV (2020 Aug 10) Glucosamine-NISV delivers antibody across the blood-brain barrier: optimization for treatment of encephalitic viruses. J Control Release 324:644–656

Yang L, Kress BT, Weber HJ, Thiyagarajan M, Wang B, Deane R, Benveniste H, Iliff JJ, Nedergaard M (2013) Evaluating glymphatic pathway function utilizing clinically relevant intrathecal infusion of CSF tracer. J Transl Med 11:107 [Epub ahead of print] PubMed PMID: 23635358

Yasuda K, Cline C, Vogel P, Onciu M, Fatima S, Sorrentino BP, Thirumaran RK, Ekins S, Urade Y, Fujimori K, Schuetz EG (2013) Drug transporters on arachnoid barrier cells contribute to the blood-cerebrospinal fluid barrier. Drug Metab Dispos 41:923–931

Zaragozá R (2020) Transport of amino acids across the blood-brain barrier. Front Physiol 11:973. https://doi.org/10.3389/fphys.2020.00973. PMID: 33071801; PMCID: PMC7538855

Chapter 2
Increasing Brain Exposure of Antibodies

Dominique Lesuisse

Abstract The blood-brain barrier (BBB) with its network of highly tight and non-fenestrated endothelial cells, along with efflux transporters, remains a huge obstacle for biomolecules such as antibodies. This explains why some huge medical needs remain to be addressed for difficult targets for which biologics are the main modality in therapeutic area such as neurosciences or oncology (i.e., CNS lymphoma or glioblastoma). Several strategies are currently studied to enhance brain exposure of antibodies, including receptor-mediated transcytosis, nanotechnologies and charge, focused ultrasound, and intranasal delivery. This chapter will review most work in this area.

Keywords Antibody · Receptor-mediated transcytosis · Transferrin · Transferrin receptor · Insulin · Insulin receptor · Adsorptive-mediated transcytosis · Intranasal delivery · Focused ultrasounds

Abbreviations

AD	Alzheimer's disease
AMT	Adsorptive-mediated transcytosis
ARIA-E	Amyloid-related imaging abnormalities-edema
AUC	Area under the concentration curve
BBB	Blood-brain barrier
BDNF	Brain-derived neurotrophic factor
BLI	Bilayer interferometry
CDR	Complementarity-determining regions
CHO	Chinese hamster ovary
CNS	Central nervous system

D. Lesuisse (✉)
Tissue Barriers, Rare and Neurologic Diseases, Sanofi, Paris, France
e-mail: Dominique.lesuisse@sanofi.com

© American Association of Pharmaceutical Scientists 2022
E. C. M. de Lange et al. (eds.), *Drug Delivery to the Brain*, AAPS Advances
in the Pharmaceutical Sciences Series 33, https://doi.org/10.1007/978-3-030-88773-5_2

CNTF	Cytokine ciliary neurotrophic factor
cTfRMAb	Chimeric MAb against the mouse TfR
ECD	Extracellular domain
ELISA	Enzyme-linked immunosorbent assay
EPO	Erythropoietin
Fab	Fragment antigen binding
FcRn	Neonatal Fc receptor
FIR	First infusion reaction
GDNF	Glial-derived neurotrophic factor
GFR	Growth factor receptor
Her2	Human epidermal growth factor receptor 2
HIR	Human insulin receptor
HIRMAb	Engineered MAb against the HIR
HRP	Horse radish peroxidase
ID	Injected dose
IHC	Immunohistochemistry
IL1, IL-6	Interleukins 1 and 6
iv	Intravenous
KDa	Kilodalton
MAb	Monoclonal antibody
MPS	Mucopolysaccharidosis
MSD	Mesoscale discovery
MTH	Molecular Trojan horse
Nab	Neutralizing antibody
NGF	Nerve-derived growth factor
NMDAR	N-methyl-D-aspartate receptor
PBS	Phosphate-buffered saline
PBSB	Phosphate-buffered saline containing 1% bovine serum albumin
PD	Pharmacodynamic
PK	Pharmacokinetic
PS	Permeability surface area
RGMa	Repulsive guidance molecule A
RMT	Receptor-mediated transcytosis
ScFv	Single-chain variable fragment
sdAb	Single-domain antibody
SLC	Solute carrier
SPR	Surface plasmon resonance
TfR	Transferrin receptor
TNF	Tumor necrosis factor
V_HH	Camelid single-domain antibody
VNAR	Variable domain of new antigen receptors
WT	Wild type
%ID/g	Percent of injected dose per gram

2.1 Introduction

Passive immunotherapy, i.e., treatment with therapeutic antibodies, is increasingly perceived as a potential therapeutic solution of choice for difficult or intractable targets (i.e., protein-protein interactions or aggregated proteins) due to their potential for high-affinity protein binding and exquisite selectivity. However, the use for CNS disorders such as Alzheimer's (AD), Parkinson's, Huntington's diseases, or brain cancers (Kumar et al. 2018a) has been very limited so far owing to the presence of the blood-brain barrier (BBB).

Brain is indeed a highly protected tissue. The endothelial cells lining the blood vessels that are at the interface of blood and brain are, as opposed to the ones at the periphery, extremely tight, non-fenestrated, and equipped with many efflux systems. This BBB is only permeable to very small lipophilic compounds but is actively preventing most molecules to enter and especially large or polar molecules such as biotherapeutics and antibodies (Banks 2016; Obermeier et al. 2013) (Fig. 2.1).

This explains why some large medical needs remain to be addressed specially for difficult targets for which biologics are the main modalities in therapeutic areas such as neurosciences, oncology (such as CNS lymphoma or glioblastoma), or rare diseases. This is also explaining why so few biologics are in development in CNS. The biologics that are on the market, like the various interferons or antibodies such as Tysabri® (natalizumab), Lemtrada® (alemtuzumab), Ocrevus® (ocrelizumab) or the more recent Kesimpta® (ofatumumab), Emgality® (galcanezumab), Ajovy® (fremanezumab), Vyepti® (eptinezumab), Aimovig® (erenumab), Enspryng® (satralizumab), and Uplizna® (inebilizumab), are most certainly acting peripherally. Peptides or proteins such as Prialt® (ziconotide) or Brineura® (cerliponase alfa) and oligonucleotides or siRNAs such as Spinraza® (nusinersen),

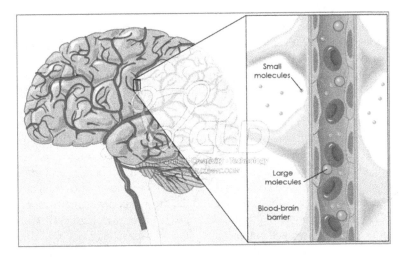

Fig. 2.1 The Blood Brain Barrier (BBB). https://ib.bioninja.com.au/options/option-a-neurobiology-and/a2-the-human-brain/blood-brain-barrier.html

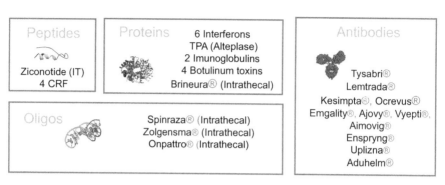

Fig. 2.2 Biotherapeutics approved in CNS indications. *CRF* corticotropin releasing factors, *t-PA* tissue plasminogen activator

Zolgensma® (onasemnogene abeparvovec), or Onpattro® (patisiran) are given intrathecally. Aside from the recent highly debated FDA approval of Aduhelm® (aducanumab), no biologic has been launched in Alzheimer's or Parkinson's diseases (Fig. 2.2).

Antibodies are very large (average 150 kDa) polar molecules that will have very slow diffusion rates within tissues, even more so in the brain protected by the BBB. The tissue to blood ratio of antibodies is generally in the range of 10–50% (Ryman and Meibohm 2017), while for highly protected brain tissue, this ratio is reported on an average of 0.1% (Pepinsky et al. 2011). However, most of these reports arise from rodent studies, which BBB is not the same as human (Deo et al. 2013; Friden et al. 2009). Regarding data in humans, several human IgGs have been reported with a similar range of brain penetration, with endogenous IgG subclasses in healthy volunteers found at 800–1000-fold lower levels in CSF than in serum (Kaschka et al. 1979) and rituximab demonstrating 1000-fold lower CSF exposure than plasma (Rubenstein et al. 2003); however, CSF exposures are poor predictors of brain levels. A recent study by Yadav (Yadav et al. 2017) showed that while iv-administered anti-BACE1 resulted in CSF antibody concentrations of ~0.1% of serum concentrations, terminal brain antibody concentrations only reached ~0.01–0.02% of serum concentrations. Such concentration will be in most cases insufficient to lead to a significant target engagement in the brain and a therapeutic effect. In fact, it is probably the reason why most anti-amyloid antibodies (crenezumab, bapineuzumab, solanezumab) have been discontinued. Bapineuzumab doses have been severely limited in the clinic by the occurrence of ARIA-E (amyloid-related imaging abnormalities-edema), an invalidating cerebrovascular side effect of these therapies (van Dyck 2018). For other antibodies such as crenezumab or solanezumab, which also bind to the most abundant soluble monomeric Aβ in CSF and brain interstitial fluid, the brain free IgG concentration is extremely low, further limiting engagement on aggregated Aβ in amyloid plaques in the brain. Gantenerumab and aducanumab have enhanced specificity for the aggregated vs. monomeric Ab, therefore increasing the free IgG bioavailability in the brain even if high doses (10 mg/kg and over) are used in clinical studies. Hopes were raised recently with aducanumab, which was initially discontinued in March 2019 after a futility analysis, but recent analysis of all clinical data available since showed a reduction in cognitive decline at the highest dose in one of the two phase III studies, in line with CSF biomarker activity. Aducanumab has been reported to have 13-fold higher brain to plasma AUC ratio than the 0.1% frequently reported for systemically administered antibodies (Sevigny et al. 2016). This value should be used cautiously, because it was obtained in a transgenic amyloid mouse model where abundant amyloid plaques (target epitope) in the brain should drastically enhance the brain accumulation compared to a WT mouse model devoid of target. This has been shown, for instance, with bapineuzumab demonstrating significantly higher brain levels in mouse models of amyloid than in age-matched nontransgenic mice (Bard et al. 2012), while in clinical studies a more standard value of CSF/plasma ratio of 0.1% was reported. The recent clinical results and FDA approval of aducanumab could open a breach in the long history of failures in the field; however, they were obtained at the highest dose and were associated to ARIA-E in a significant number of patients. Strategies to increase brain exposure of antibodies and biotherapeutics in general will be key to success in these areas.

2.2 Overview of Strategies to Increase Brain Exposure of Antibodies

Leaving aside invasive modes of brain delivery which likely won't be able to deliver to most of the brain tissue (Jones and Shusta 2009) and will not be discussed here, one can divide the main strategies to increase brain exposure of biotherapeutics into four. Ferrying the biotherapeutics by a ligand or antibody against a receptor, which performs *transcytosis* (sometimes referred to as the "Trojan horse" approach), is certainly the most used strategy when coming to antibodies. Various *formulations including charge and nanotechnologies* have been used to facilitate brain crossing. Certain chemicals or *focused ultrasounds* have been reported to temporarily open the BBB to let biotherapeutics through. *Intranasal administration* bypasses the BBB and enables biotherapeutics to follow the olfactory axon bundles directly into the brain via the cribriform plate. These last three strategies have been less applied to antibodies.

Because several of these strategies are directly inspired from the brain's own mechanisms of importing various endo- or exogenous nutrients, xenobiotics, or toxins, these will first be shortly reviewed.

This chapter will review the main advances of each of these strategies on increasing brain exposure of *antibodies*. As a result of high activity in the anti-amyloid antibodies field, several examples will be focused on this target.

2.3 Brain Transport Mechanisms

The physiological way that the brain imports most of its vitamins, nutrients, or proteins or the pathological way that the brain's viral or bacterial infection can occur has been most inspirational for the scientists trying to increase the brain exposure of biotherapeutics. These different mechanisms (Goulatis and Shusta 2017) are depicted in Fig. 2.3. Very small and hydrophilic compounds could use the paracellular way (Fig. 2.3-A), but this pathway is minimal due to very tight intercellular junctions in the brain endothelial cells (Di et al. 2013). Most lipophilic small molecule drugs such as benzodiazepines or cholinesterase inhibitors use the transcellular passive diffusion through the lipid membranes (Huttunen et al. 2019) (Fig. 2.3-B). Endogenous solute carriers (SLCs) are membrane-bound transport proteins that promote the influx into the brain of essential substances, such as amino acids, sugars, vitamins, electrolytes, nucleosides, bile acids, and even macromolecules like proteins (Fig. 2.3-C) (Nalecz 2017). It is increasingly recognized that some of these transporters (e.g., LAT1 (Rankovic 2015), Glut1, MCT1) are involved in the brain uptake of several small molecule drugs (Dobson and Kell 2008) such as gamma aminobutyric acid analogues or L-dopa and that brain uptake of some poorly brain penetrant small molecules has been improved by coupling them to a prodrug recognizing one of these transporters. Receptor-mediated transcytosis (RMT, Fig. 2.3-E) is a specific endogenous process allowing brain transport of proteins such as insulin or insulin-like growth factor, transferrin, low-density lipoprotein receptor-related proteins 1 and 2, or heparin-binding epidermal growth factor-like growth factor (Jones and Shusta 2007). Polycationic substance can bind to the plasma membrane and lead to a process of endocytosis called adsorptive-mediated transcytosis (AMT, Fig. 2.3-F). Histones and wheat germ agglutinin use this mechanism (Herve et al. 2008). Diapedesis (Fig. 2.3-G) and fluid-phase endocytosis (Fig. 2.3-H) are used by cells, virus, or particles (Smith and Gumbleton 2006). Soluble plasma proteins such as albumin or IgGs are using fluid-phase or bulk-phase transcytosis, but while this process represents the main pathway of transfer through endothelial cells to peripheral tissues, it is very limited in a healthy BBB (Herve et al. 2008). Some cytokines such as IL-6 (Banks et al. 1994), IL-1 (Banks et al. 1991), TNF (Banks et al. 1995), NGF, CNTF, BDNF (Poduslo and Curran

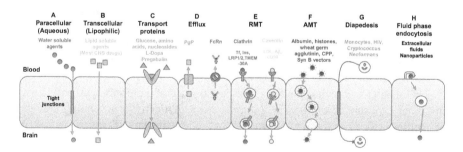

Fig. 2.3 Brain transport mechanisms

1996), or erythropoietin (Banks et al. 2004) have been reported to be transported to the BBB by either saturable or non-saturable transport including extracellular pathways. However, the %ID/g of the brain remains modest and the transporters have not been identified.

Brain endothelial cells are equipped with many efflux pumps (MDR1, BCRP, MRP4, others; Fig. 2.3-D) (Kusuhara and Sugiyama 2001a, b) that can exfiltrate several xenobiotics. For antibodies, much faster efflux from the brain to blood was observed compared to albumin or dextrans after intracerebral injection in rats. The neonatal Fc receptor (FcRn, Fig. 2.3-D) was hypothesized as the likely culprit as the process was competitively inhibited by Fc but not (Fab)2 fragments (Zhang and Pardridge 2001). On the other hand, experiments using labeled IgGs in FcRn-deficient mice showed no major difference in brain to plasma AUC ratios compared to WT (wild type) mice suggesting that FcRn does not contribute significantly to the BBB transport in mice (Garg and Balthasar 2009).

It can be seen from Fig. 2.3 that the only mechanisms enabling transport through the BBB of large biological or molecular entities are the last four mechanisms (Fig. 2.3-E, F, G, H). To make use of these mechanisms to ferry drugs and biotherapeutics into the brain, several groups have looked for ligands (antibodies, peptides or small molecules) specific of some of these paths, with the idea of either directly fusing them to the biotherapeutic or to a nanoformulation encapsulating it.

2.4 Trojan Horse Approach

The technology which is making use of an endogenous brain transporter receptor to ferry an antibody across the BBB has been referred to as Trojan horse (Pardridge 2002) as it is a ligand of these receptors which cargos the biotherapeutic into the brain. Several receptors have been used. The main ones are *receptors mediating transcytosis* such as insulin, transferrin, lipoprotein-related proteins, or IgF1 receptors. At present, the most convincing data on brain enhancement of biotherapeutics have been making use of this mechanism. Other types of transporters such as *SLCs* have been reported too, and some recent examples have uncovered the potential of transporters such as Glut1 or LAT1 to ferry antibodies across the BBB.

Two main strategies have been used to ferry an antibody across the BBB according to the ligands of these receptors used: *antibodies* or *peptides* (Fig. 2.4). To date, no *small molecule* has been reported to be able to carry an antibody into the brain. The first strategy is making use of antibodies against these receptors engineering them into bispecific constructs recognizing, on one hand, the transcytosis receptor and, on the other hand, the therapeutic antibody (Fig. 2.4-A). The second strategy is fusing peptides to the therapeutic antibody (Fig. 2.4-B).

We will review both strategies. The bulk of literature is making use of antibodies against these receptors as carriers, but a few examples are reported with peptide or protein ligands of these receptors.

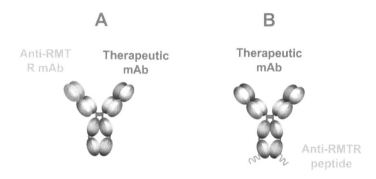

Fig. 2.4 Two strategies to ferry an antibody to the brain using the Trojan Horse approach

Fig. 2.5 Formats where an anti-TfR antibody (blue) is fused to therapeutic antibodies fragments (green, Anti-Aβ); In orange: proteins such as IL1 RA, TNFα decoy receptor, iduronate sulfatase, GDNF

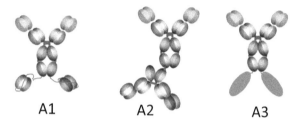

Full antiTfR mAb's fused to Ab or Ab fragments

2.4.1 Using Antibodies Ligands of Transporter Receptors to Carry Antibodies to the Brain

2.4.1.1 Transferrin Receptor

Transferrin receptor (TfR), an endogenous mechanism of iron transport, whose expression is enriched on brain endothelial cells, has now been widely recognized as a mechanism enabling brain transport of various cargos (Johnsen et al. 2019). To ferry antibodies into the brain using the transferrin receptor, different bispecific formats have been reported leading to various levels of brain exposure enhancement. These formats can be roughly categorized into five bins (Figs. 2.5, 2.6, 2.7, 2.8 and 2.9). The first one is a series of anti-TfR antibodies to which are fused the therapeutic proteins, as depicted into Fig. 2.5. This format has been reported with anti-amyloid β ScFvs (Boado et al. 2010) (Fig. 2.5-A1) or (Fab)2 fragments (Sehlin et al. 2017) (Fig. 2.5-A2). Therapeutic proteins not belonging to antibodies have also been reported in this format such as IL1 receptor antagonist (Webster et al. 2017), a TNF alpha decoy receptor protein (Chang et al. 2017; Sumbria et al. 2013), GDNF (Zhou et al. 2010), lysosomal enzymes such as iduronate sulfatase (Sonoda et al. 2018), and EPO (Chang et al. 2018) all fused to the C-terminal portion of the heavy chains of an anti-TfR antibody (Fig. 2.5-A3), but they will not be discussed in this chapter focusing on antibodies.

Fig. 2.6 Formats where the therapeutic antibody (green, anti-Aβ) is fused to anti-TfR (blue) fragments

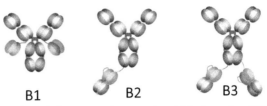

B1   B2   B3

Full anti therapeutic target mAb's fused to TfR Ab fragments

Fig. 2.7 Formats bivalent for both TfR (blue) and therapeutic target (green, anti-Aβ)

C1   C2

Bispecific Ab's bivalent for the therapeutic target and for TfR

Alternatively, the antibody against the therapeutic target can be fused to anti-TfR ScFvs (Fig. 2.6-B1) or Fab fragments (Fig. 2.6-B2 and B3) (Niewoehner et al. 2014). This format has been mainly reported with anti-Aβ amyloid antibodies.

Bispecific antibodies such as DVDs (dual variable domains) (Wu et al. 2007) or TBTIs (tetravalent bispecific tandem IgG's) (Do et al. 2020; Rao and Li 2007) (Fig. 2.7-C1 and C2) have also been reported. They have been engineered with a second Ig variable domain fused to the first one and have their carrying paratopes against TfR and a therapeutic target. DVDs have been prepared with a few therapeutic targets such as anti-β amyloid but also anti-Her2 (human epidermal growth factor receptor 2), RGMa (repulsive guidance molecule A), and TNF (tumor necrosis factor) (Karaoglu Hanzatian et al. 2018). Of note, these three first categories of formats are bivalent for the therapeutic target and have the potential to keep their maximal affinity/avidity for their target which is undoubtedly an advantage specially in the field of aggregated proteins such as amyloid (Fuller et al. 2015).

Bispecific antibodies monovalent for the therapeutic target and for TfR such as *D1* have also been reported (Yu et al. 2011) (Fig. 2.8). The studies using this format have been key to establish the scope of the TfR technology (vide infra).

New formats are still arising (Fig. 2.9), such as the ATV platform (Fig. 2.9-E) reported by Kariolis (Kariolis et al. 2020) where the affinity for the TfR has been engineered in the Fc domain of a therapeutic antibody against either BACE1, Tau, α-synuclein, or TREM-2. The tribody (Syvanen et al. 2017) (Fig. 2.9-F) is another format linking two anti-Aβ ScFvs to an anti-TfR Fab. Finally, formats where a therapeutic antibody (anti-TrkB) (Clarke et al. 2020) or an ScFv fragment of the therapeutic antibody (rituximab) (Stocki et al. 2019) are fused to an anti-TfR single-chain antibody from shark (VNAR) have been reported (Fig. 2.9-G1–4). These

Fig. 2.8 Formats
monovalent for both TfR
(blue) and therapeutic
target (green, anti-Aβ)

D1

Bispecific Ab's monovalent for the therapeutic
target and for TfR

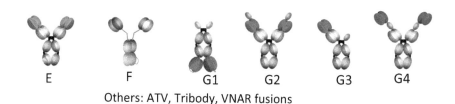

E   F   G1   G2   G3   G4

Others: ATV, Tribody, VNAR fusions

Transferrin receptor; Anti beta amyloïde; BACE1; **Rituximab**; TrkB

Fig. 2.9 Other formats. Anti-TfR (blue), anti-Aβ (green), BACE1, Tau, αSyn, TREM2 (orange), Rituximab (red)

single-chain antibodies have also been reported to increase brain exposure of proteins when engineered into Fc constructs.

All these formats have been shown to enhance brain exposure of their therapeutic antibodies in mice to various extents. Direct comparisons are difficult as no head-to-head format comparisons have been reported and the PK parameters shown are not necessarily the same (Cmax or concentration at specific time points, brain to plasma AUCs or at specific time points, %ID/g brain or even PS product); neither are the studies performed at the same doses which can also impact the result as the process of receptor-mediated transcytosis is saturable or using the same animal models or brain handling procedures. In addition, some of the studies are also performed using radiolabeled constructs which could in certain cases yield overestimated results as the resolution between vessels and brain parenchyma can most of the time not be assessed and/or the stability of the labeling is often not known. Altogether the reported brain enhancements seen in rodent PKs could be in the range of 10–12-fold for the best constructs.

Affinity/Valency for TfR and Brain Exposure

As binding properties of all these constructs have been evaluated using several technologies and reporting either Kds (using SPR or BLI) or EC_{50}s (ELISA MSD or HRP), comparison between them across the literature is difficult. In addition, even when Kds are reported, the results could differ, and read affinity or avidity depends on what has been immobilized on the chip from the TfR or the bispecific. Furthermore, owing to the difficulties to have the extracellular part of the TfR in its native state and its aggregation/dimerization propensity, binding of the constructs on cells (over)expressing the TfR has been reported to be more reliable (Karaoglu Hanzatian et al. 2018).

Several views on TfR-binding affinities for best brain exposures can be found across the literature. Pardridge, one of the seminal authors describing this technology, reports that constructs with high affinity for transferrin receptors can generate high brain uptake (>1–2% ID/g) in mouse (Pardridge 2015). Several authors have shown a degree of correlation between transferrin binding and brain exposure. Yu et al. (Yu et al. 2011) produced several anti-TfR antibodies with a range of binding affinities for TfR from 1.7 to 111 nM by engineering alanine mutations into the CDR (complementarity-determining regions). They were able to show that the highest brain exposures were obtained with the high-affinity anti-TfR antibodies after non-saturating (trace) dosing and the low-affinity anti-TfR antibodies after therapeutic dosing. They explained this apparent discrepancy with a model where at non-saturating dose a higher-affinity antibody will bind more receptors at the luminal side of the BBB resulting in more association to the endothelium, while at therapeutic dose it is the easier dissociation of the low-affinity antibody from the endothelium which will cause higher brain exposure (Fig. 2.10). This was also

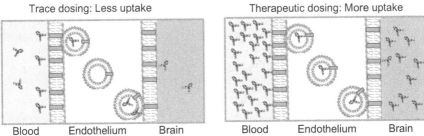

Fig. 2.10 Model explaining why high affinity for TfR antibodies display more uptake at tracing dose and less uptake at therapeutic dose and vice versa. (Figure adapted from Yu et al. 2011)

shown with several affinity variants of 8D3 (Webster et al. 2017), an anti-mouse TfR and OX26 (Thom et al. 2018) an anti-rat TfR antibody.

The same observation held for the bispecific format (Fig. 2.8-D1) with the high-affinity anti-TfR^A/BACE1 antibody showing lower brain exposure than the low-affinity anti-TfR^D/BACE1 antibody at therapeutic dosing (Bien-Ly et al. 2014).

In the case of the DVD constructs, Hanzatian et al. (2018) engineered heavy and light chains of an anti-mouse TfR mAb (AB221) while keeping the CDR's constant to produce two variants with lower affinities vs TfR (3.1 and 13.6 nM vs 0.12 nM) which demonstrated around fourfold higher brain exposure at 24 h than the initial anti-TfR mAb. When the initial AB221 was inserted into a DVD format carrying the variable domain of bapineuzumab in the outer position, the affinity was lowered of about tenfold (1.2 nM) and led to a ~ threefold brain enhancement after 20 mg/kg iv injection compared to a control DVD. The correlation between binding affinity to TfR and brain exposure was not clearly marked though and varied also as a function of the therapeutic targets exemplified; however, one-point kinetics might be mis-leading to compare brain exposures.

For constructs such as *B2* and *B3* (Fig. 2.6), the double TfR Fab anti-Aβ con-struct of higher affinity for TfR demonstrated lower brain exposure than its single Fab analog (Niewoehner et al. 2014). The higher brain exposure observed with *B2* was also correlated to its monovalency for transferrin receptor.

On the other hand, in the case of VNAR fusion constructs with rituximab, it was shown that high (sub-nM)-affinity TfR binding did not impede brain transport and that bivalent and monovalent for TfR constructs such as *G2* and *G3* (Fig. 2.9) exposed the brain similarly (Stocki et al. 2019). Interestingly some constructs show-ing strong nM binding for mTfR were unable to penetrate the brain after in vivo administration to mice.

Altogether these reports highlight the complexity and multiparametricity of the field with several other factors at stake in determining brain exposure of these con-structs including their format and Fc nature (Sun et al. 2019), the epitope of the TfR that is recognized, and the dose and the duration of plasma circulation, to name a few.

Cell Trafficking of Anti-TfR Bispecific Antibodies

These observed brain exposures were linked to the intracellular trafficking and sort-ing of the constructs. In fact, three potential endocytic sorting routes can be taken by these anti-TfR antibodies upon binding with endothelial cells at the BBB: recy-cling to the luminal side (Fig. 2.11-1), transcytosis to the parenchyma (Fig. 2.11-2), or sorting in the lysosome and degradation (Fig. 2.11-3) (Bien-Ly et al. 2014).

Bien-Ly et al. (2014) et al. demonstrated by very elegant experiments comparing high- and low-affinity bispecific constructs with two different labelings (^{111}In and ^{125}I) that greater degradation occurred with the high-affinity construct leading to lower transcytosis and brain exposure.

On the other hand, in addition to affinity, avidity for the TfR might well be at stake, and Niewoehner et al. (Niewoehner et al. 2014) after analysis of the

Fig. 2.11 Three potential endocytic sorting routes of an anti-transferrin receptor antibody. (Taken from Bien-Ly et al. 2014. Copyright from Creative common)

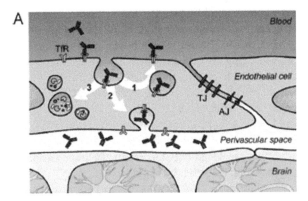

Fig. 2.12 Proposed pathway for differential intracellular sorting of monovalent and bivalent anti-TfR Fab fusions. Whereas monovalent anti-TfR sFab fusions undergo transcytosis across the BBB, bivalent anti-TfR dFab fusions lead to TfR dimerization and lysosomal degradation. (Taken from Bell et al. 2014. Copyright from Elsevier)

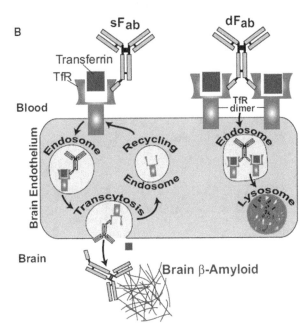

intracellular sorting and trafficking of both constructs concluded that whereas the monovalent fusion underwent transcytosis across the BBB, the bivalent fusion resulted in receptor dimerization and sorting/degradation into the lysosome (Fig. 2.12).

This reported lysosomal degradation is also accounting for the observed higher downregulation of the TfR after treatment with the higher-affinity/avidity anti-TfR antibodies. This was reported for bispecific *D1* (Fig. 2.7) where Yu et al. (2011) showed that high-affinity constructs led to a dose-dependent reduction of brain TfR levels in vivo yielding lower exposure of the construct after repeated administration. The high-affinity constructs were shown to reduce cell surface membrane TfR levels on treated bEnd3 cells. The same effect was observed with the bivalent construct *B3*, whereas "brain shuttle" *B2* did not alter the TfR content.

The cell trafficking of antibodies against transcytotic receptors has also been studied using in vitro models of transcytosis and helped understand additional parameters at stake for successful transcytosis. These models traditionally used in transwell systems, where an endothelial cell monolayer (either cell lines, or primary endothelial cells, or even iPSC-derived endothelial cells) is lined at the top in the presence of other cells from the neurovascular unit (astrocytes and pericytes), allow to determine the extent and kinetics of the drug to cross the monolayer and move from the apical side (top, blood lumen) to the basolateral side (bottom, brain parenchyma) (Fig. 2.13) (Helms et al. 2016).

A study using HCMEC/D3 cell line compared a few antibodies against different transcytotic receptors (Sade et al. 2014): An antibody against IgF1R was shown to be exclusively recycled to the apical side of the transwell. On the other hand, anti-TfR antibodies showed various levels of transcytosis according to their relative affinities at extracellular and endosomal pH (7.4 and 5.5, respectively). An antibody with reduced affinity at pH 5.5 showed significant transcytosis, while pH-independent

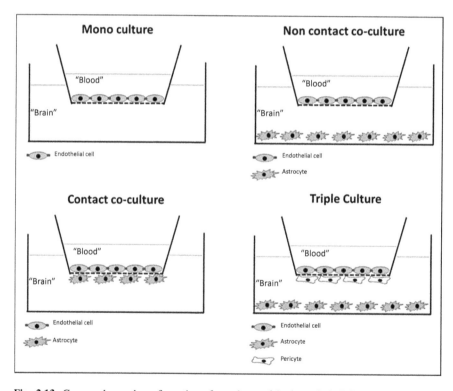

Fig. 2.13 Commonly used configurations for culture of brain endothelial cells. (Taken from Helms et al. 2016. Copyright from Sage Publication)

antibodies of comparable affinities at pH 7.4 remained associated with intracellular vesicular compartments and were finally targeted for degradation.

Target Engagement

Demonstration of a functional effect on the target has been nicely achieved in the case of bispecific BACE1/TfR antibodies. The target BACE1 enzyme, cleaving APP into Aβ amyloid peptides, is present in the WT mouse allowing to have a direct assessment of PK/PD relationship of the constructs. At 25 mg/kg, a bispecific construct BACE1/TfR (Fig. 2.8-D1) was able to produce a significant reduction in the brain Aβ1–40, while the monospecific BACE1 antibody did not display any effect indicating that the concentration of a BACE1 antibody in the brain strongly contributes to the potency (Yu et al. 2011).

Plaque binding in the brain of amyloid transgenic mouse models was also demonstrated with several bispecific Aβ/TfR constructs. The 8D3-F(ab')$_2$ (Fig. 2.5-A2) was located around insoluble amyloid deposits in tg-ArcSwe mice by nuclear track emulsion and Congo staining (Sehlin et al. 2016). The "brain shuttle" (Fig. 2.6-B2) displayed a 55-fold higher plaque decoration than the parent anti-Aβ mAb31 based on fluorescence after injection in PS2APP transgenic mice (Niewoehner et al. 2014). In life fluorescence analysis of a bispecific anti-Aβ/TfR TBTI (Fig. 2.7-C1 (Do T-MA 2020)) in WT mice (left) compared to APP mice (right) showed that fluorescence of the antibody is trapped in the brain of the transgenic mice at a time when it has disappeared from the blood and brain of the WT mice. Nicely, the fluorescence was seen where the plaques are known to be (Fig. 2.14).

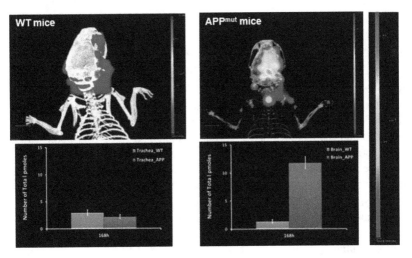

Fig. 2.14 Brain vs tracheal vascular areas in repeated FLIT session (168 h). AF750-TBTI3 after injection of 57 nmol/kg, iv in WT mice (left) vs APPmut mice(right) (Do et al. 2020)

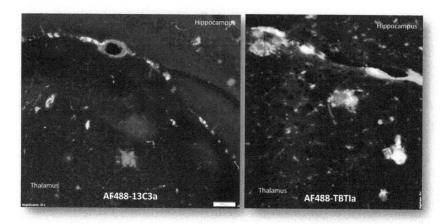

Fig. 2.15 Ex-vivo fluorescent imaging within thalamic sub-area of APPmut mice injected iv (retro-orbital) with 2 × 20 mg/kg of AF488-TBTI3a or AF488-13C3a. Mice were also injected with Angiospark 680 vessel marker. Cryostat sections were stained with Congo red. Both injected Abs were detectable in green, vessels in pink and mature amyloid plaques in red. Scale bar equals 100μm (Do et al. 2020)

The same TBTI was also analyzed by histology in the brain after injection in APP[mut] mice in comparison with the monospecific anti-Aβ antibody 13C3. Ex vivo fluorescent imaging on brain sections demonstrated by analysis of colocalization with the vessel marker Angiospark (Vasquez et al. 2011) and amyloid plaque Congo red staining that AF488-13C3a was restricted into the vessels while AF488-TBTI3a was also found around thalamic plaques (Fig. 2.15).

Efficacy

As several bispecific antibodies have been illustrated with anti-amyloid antibodies, most of the functional effects of the bispecific constructs have been shown in amyloid mouse models. The first results were obtained with the cTfRMab-ScFv fusion protein *A2* demonstrating a 57–61% decrease in the Aβ amyloid plaque burden in the cortex and hippocampus of 12–15-month-old PSAPP AD transgenic mice after daily subcutaneous administration of a 5 mg/kg dose (Sumbria et al. 2013). This therapeutic effect was comparable in size to the ones observed with 20 mg/kg doses of conventional anti-amyloid mAbs. However, the fusion protein had to be administered daily, while the effects reported with the monospecific anti-Aβ antibodies were obtained after weekly treatment. As the comparators mAbs Ab9 and gantenerumab, the fusion protein did not lower the brain concentration of urea- or formic acid-soluble Aβ1–40 and Aβ1–42 amyloid peptide. This was attributed to the very high concentrations of these species in the brain (2 μM and 8 μM, respectively) compared to the concentration of the fusion protein (~7 nM).

A nice demonstration of efficacy in an amyloid mouse model was also shown with the monovalent "brain shuttle" (Fig. 2.6-B2) where a significant reduction in plaque number both in the cortex and hippocampus was observed with a low weekly iv dose (2,67 mg/kg) comparable to a 20 mg/kg dose of the monospecific anti-Aβ antibody (mAb31) (Niewoehner et al. 2014). In this report, no effect was shown on brain pool of Aβ.

Finally, an effect on both plaque (number and surface) and cortical Aβ could be demonstrated with a fivefold lower dose of the TBTI construct Aβ/TfR (Fig. 2.7-C1) compared to a 10 mg/kg ip weekly dose of the monospecific anti-Aβ antibody 13C3 (Do T-MA 2020).

There might be some other subtle parameters governing efficacy of these bispecifics as it was reported that an unintended increase in brain amyloid had been observed upon treatment with a bispecific anti-TfR/anti-β amyloid DVD antibody (Fig. 2.7, C1 or C2) (Webster and Stanimirovic 2015), although it is quite possible that these are linked to the nature of the anti-Aβ antibody used, 3D6, which recognizes both monomeric and aggregated Aβ (Hanzatian et al. 2015).

Translation to Humans

One of the challenges of TfR-based technology is the fact that no anti-TfR extracellular domain (ECD) IgG has been identified with cross-reactivity between rodent and NHP/human species. Bridging studies have therefore to be established using human anti-TfR antibodies and mouse genetically modified to carry the human extracellular portion of their TfR and/or direct NHP PK evaluation. This bridging is not always straightforward as shown by Yu et al. (2014) where the lowest affinity of two anti-TfR/BACE1 antibodies cross-reactive for human and cynomolgus TfR displayed the higher brain exposure in mice genetically modified to express the extracellular domain of the human TfR and the lowest brain exposure in the NHP. This could complicate pharmacology and toxicology of the development candidates and could require double transgenic mice models.

Recently some lower molecular weight TfR-binding proteins have been shown to be rodent/human cross-reactive. This is the case for the VNARs (Stocki et al. 2019) where binding affinities for human, mouse, rat, and cynomolgus monkey TfR1 ECDs are in the same range. A potential reason for this might be the ability of VNARs to access buried epitopes. The future will tell if comparable brain enhancement of these constructs can be demonstrated in NHP as the ones observed in rodents.

Safety

When talking about the safety of these constructs, two aspects need to be considered: safety related, on one hand, to the therapeutic target and, on the other hand, to the transcytotic receptor.

Anti-Aβ amyloid antibodies have been plagued by target-related occurrence of cerebral microhemorrhages, namely, amyloid-related imaging abnormalities (ARIA) which are vasogenic edema/effusions and/or microhemorrhage and hemosiderosis detected by MRI signal alterations (van Dyck 2018). The 3-month administration of daily sc doses of the ScFv construct A2 (Fig. 2.5) caused no cerebral microhemorrhage contrarily to a 20 mg/kg weekly treatment of a conventional Aβ antibody, which was attributed to the fact that the ARIA are linked to the plasma Aβ peptide elevation causing BBB disruption. This plasma peptide elevation was not observed with the fusion constructs, which were rapidly cleared from plasma (Sumbria et al. 2013).

TfR is expressed in erythroid blood cell lines and internalizes a complex of transferrin with iron. To avoid liabilities linked to interference with iron transport and to prevent competition of the constructs with the high circulating levels of transferrin (2.6 mg/ml), all the constructs reported above have been engineered to be noncompetitive with the endogenous transferrin or in some cases with the hemochromatosis-associated protein (HFE) (Yu et al. 2014) binding sites. This was shown by direct competition studies, but also in certain studies via epitope mapping demonstrating that the TfR paratope recognition site was located on the apical part of the TfR extracellular domain distant from the transferrin binding site (Niewoehner et al. 2014). Nevertheless, significant effects on reticulocyte counts and red blood cells could be observed with several anti-TfR bispecific constructs and more specifically when they have high affinity for TfR (Sun et al. 2019). A comprehensive study on safety related to TfR has shown that in mice acute clinical symptoms are not observed with lower-affinity TfR constructs (Couch et al. 2013). An effect on reticulocyte count could be seen with high-affinity TfR antibodies, but this effect was only observed at higher doses with lower-affinity TfR antibodies (no effect at 1 and 5 mg/kg). This effect was shown to be effector function and complement-dependent and could be mitigated by Fc engineering to lower the effector functions. Altogether, there was no evidence of sustained decrease of mature red cell mass or change in serum iron parameters in any tissue at 7 days post fourth dose (25 mg/kg). Zhou has reported no adverse findings after chronic dosing of anti-TfR/GDNF fusion in mice (2 mg/kg twice a week, 12 weeks) (Zhou et al. 2011). This could be linked to lower effector functions of the constructs where a large protein has been fused to the Fc portion of the TfR antibody (Fig. 2.5-A3), potentially hindering its interaction with its partners. In fact, Weber recently investigated the role of Fc effector function in vitro and in an Fcγ receptor (FcγR)-humanized mouse model (Weber et al. 2018). They confirmed that Fc-effector dead anti-TfR antibodies eliminated the strong first infusion reactions (FIR) observed with a conventional IgG1 anti-TfR antibody. Interestingly, they observed no FIR linked to TfR binding in the periphery with their "brain shuttle" construct (Fig. 2.6-B2) while binding the target Aβ in brain did not abrogate the Fc-FcγR binding on microglia cells to induce plaque clearance. They explained this result by an inverted binding mode of the TfR-Fab on the construct preventing optimal interaction between Fc and FcγR.

Chronic dosing of a high-affinity anti-TfR twice weekly for 4 weeks in primates showed at the higher dose (30 mg/kg) decreased blood reticulocytes and anemia (Pardridge et al. 2018a). Immunohistochemistry of the brain tissue at the end of the experiment revealed several events suggesting brain inflammation, and moderate axonal/myelin degeneration was observed in the sciatic nerve. Further studies will be needed to determine if this neuropathology is induced by the antibody effector function or by its high affinity for the TfR. In another study in non-human primate, high- and low-affinity bispecific anti-TfR/BACE1 antibodies gave no signs of reticulocytes lowering. This difference to the mouse results was explained by the much higher number of TfR-positive circulating cells in mice (1.1%) than monkey (0.2%) and human (0%), suggesting in humans the site of maturation is largely retained in bone marrow (Yu et al. 2014).

Clinics

TfR-based technology has clearly enabled increased brain uptake of several biologics, including antibodies. The potential is high for CNS indications including brain tumors and metastases. Even if the safety linked to modulating TfR has not been totally cleared, several specialties are already in development using this technology. The most advanced development is a fusion between an anti-TfR antibody and a lysosomal enzyme iduronate sulfatase (JR-141) for the treatment of mucopolysaccharidosis (MPS) (Sonoda et al. 2018). A phase III trial (NCT03568175) is ongoing in 2019 to evaluate the efficacy of JR-141 on changes in the systemic and CNS symptoms over a 12-month period for the treatment of MPS II. Another composition based on nanocapsules containing an anti-tumoral plasmid and targeted to the brain with transferrin receptor-binding ScFvs is undergoing phase II trials to assess safety and efficacy in combination with temozolomide, gemcitabine, or paclitaxel in patients with metastatic pancreatic cancer or confirmed glioblastoma (NCT02340156 and NCT02340117). An anti-TfR probody (Polu and Lowman 2014), proteolytically activated antibody engineered to remain inert until activated locally in diseased tissue, CX-2029, was reported to be in phase I/II non-randomized, open-label trial (NCT03543813) to evaluate safety, tolerability, PK, PD, and anti-tumor activity of patients with metastatic or locally advanced unresectable solid tumors or diffuse large B-cell lymphoma. All the above developments are not with therapeutic antibodies. Very recently, Roche has announced that they are engaging their "brain shuttle" applied to gantenerumab in humans (Reuters Health News N 2019). This will be the first antibody in development using this technology.

2.4.1.2   Insulin Receptor

As for transferrin receptor, insulin receptor antibodies have been used to ferry biotherapeutics across the brain, and humanized anti-IR antibodies (HIRMab) have shown good brain exposure after administration in the non-human primate (Boado et al. 2012). Most of the applications so far have been with fusions with therapeutic proteins such as a human neurotrophin BDNF (drain-derived neurotrophic factor) (Boado et al. 2007a), human GDNF (glial-derived neurotrophic factor) (Zhou et al. 2011), or human EPO (erythropoietin) (Chang et al. 2018) and decoy receptors, such as the human type II tumor necrosis factor receptor (TNFR) extracellular domain (Chang et al. 2017), with lysosomal enzymes such as human iduronidase (IDUA) (Boado et al. 2012), iduronate 2-sulfatase (IDS) (Lu et al. 2010), arylsulfatase A (ASA) (Boado et al. 2013), N-sulfoglucosamine sulfohydrolase (SGSH) (Boado et al. 2014), and N-acetyl-alpha-glucosaminidase (NAGLU) (Pardridge 2017).

Only one application can be found with an antibody, in this case a fusion protein such as A1 (Fig. 2.5) where an ScFv anti-amyloid fragment was fused to the C-terminal of a HIRMab. These antibodies do not cross-react with the rodent IR therefore precluding full in vivo evaluation in chronic mouse amyloid models. Nevertheless, a functional effect could be demonstrated when the fusion protein was injected in the frontal cortex and hippocampus of mice and showed lower amyloid plaque compared to the ipsilateral site of injection (Boado et al. 2007b). However, several reports point to the complexity of insulin transport. Even though transcriptomic studies had shown mRNA enrichment in BECs versus the liver and lung, a recent proteomic study showed low levels of protein in mouse BECs. In fact, a mouse anti-IR antibody was unable to demonstrate brain exposure enhancement in mouse compared to an anti-TfR antibody (Zuchero et al. 2016). In vitro, despite expression of InsR and binding of insulin to bovine and murine brain endothelial cells, no transport of insulin across the BBB was observed (Hersom et al. 2018). And even though a saturable transport of insulin could be established in vivo, this transport still occurred in mice lacking expression of InsR on brain endothelial cells suggesting insulin could be transported into the brain independently of InsR (Rhea et al. 2018). Understanding how insulin is transported across the BBB will certainly be key in developing therapeutics to further increase CNS concentrations.

Even if some adverse effects such as pancreatic lesions presumably linked to this insulin receptor were observed after administration of a GDNF fusion in non-human primates (Ohshima-Hosoyama et al. 2012) and high chronic dosing (30 mg/kg) of the fusion proteins can lead to weak insulin agonist properties and hypoglycemia (Boado et al. 2012), some of these anti-insulin receptor fusion proteins are already in the clinics such as valanafusp alpha, a human insulin receptor antibody-iduronidase fusion in patients with mucopolysaccharidosis type I, and displayed no effect on plasma glucose for up to 24 h after 0.3–6 mg/kg drug infusion of anti-insulin (Pardridge et al. 2018b).

2.4.1.3   CD98

Recently Zuchero et al. (Zuchero et al. 2016) revisited transcriptomic and proteomic analyses of mouse brain endothelial cells. It was found that some previously reported targets such as InsR and LRP, or new targets identified through microarray profiling of BECs such as ldlrad3 or CD320, were in fact poorly represented at the mouse BBB as far as protein were concerned. In fact, antibodies against these receptors failed after mouse injection to localize to brain vasculature and showed enhanced brain exposure compared to a control IgG. On the other hand, the proteomic study highlighted some proteins such as CD98, basigin, and Glut1 as highly expressed. These targets had not been previously studied for RMT (even though Glut1 had been previously hypothesized to be involved in the brain uptake of glucose-decorated liposomes (Du et al. 2014). They were investigated first through generation of monospecific antibodies of high and low affinities (Fig. 2.16-A) and evaluation in mouse PK. While Glut1 and basigin antibodies demonstrated comparable brain levels after injection in mice as the ones obtained with TfR antibodies (1.5–4-fold vs control IgG), CD98hc demonstrated the most robust data with 9- and 11-fold enhancement observed with the high- and low-affinity anti-CD98[A] (1.5 nM) and anti-CD98[B] (4.6 nM), respectively.

CD98hc (SLC3A2) is the heavy chain of the L-type amino acid transporter (LAT1), a transmembrane heterodimeric protein involved in the transport of large, neutral, aromatic, or branched amino acids from the blood to the brain (Singh and Ecker 2018). LAT1 has also been reported to be involved in brain transport of several clinically used amino acid mimetic drugs and prodrugs, such as L-dopa, gabapentin (Dickens et al. 2013), and melphalan (Uchino et al. 2002). To further assess the potential of CD98hc in a pharmacodynamic study, Zuchero et al. (2016) engineered bispecific antibodies with one arm recognizing CD98hc with high (A, 4 nM) and low (B, 164 nM) affinities and the other arm BACE1 (analogously to the initial work performed with TfR – vide supra) (Fig. 2.16-B). PK study showed that the lower-affinity anti-CD98[B]/BACE1 produced overall better peripheral and brain exposures. Quantitation of brain Aβ after treatment with the bispecific antibodies is a direct way to assess the acute pharmacodynamic effect of the constructs in parallel to their pharmacokinetics. The bispecific anti-CD98/anti-BACE1 construct demonstrated 30–45% less brain amyloid-β after 50 mg/kg injection than the IgG-treated control mice confirming brain parenchymal enhancement. Additional studies demonstrated that the anti-CD98 antibodies induced no receptor downregulation and did not interfere with its transport functions confirming the potential of this mechanism to enhance brain exposure of antibodies.

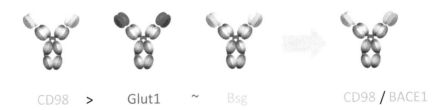

CD98 > Glut1 ~ Bsg   CD98 / BACE1

Fig. 2.16 (**a**) Antibodies against CD98, Glut1 and basigin. Best brain exposures are observed with CD98. (**b**) Bispecific anti-CD98/BACE1 construct (Zuchero et al. 2016)

2.4.1.4 TMEM30

Aside from rational selection of transporter receptors based on their known presence in the BBB and ability to import endogenous proteins, a few approaches have been reported to identify new mechanisms (Stanimirovic et al. 2014). The main goal is the identification of a brain-specific transporter as the ones evoked above are mostly ubiquitous and lead to all body distribution and potential safety issues.

One such approach has been screening combinatorial antibody libraries or fragments for their capacity of binding, uptake, transcytosis, or even direct in vivo selection of the clones exposed in the brain (Farrington et al. 2014; Muruganandam et al. 2002; Webster et al. 2016). A non-immune llama sdAb (single-domain antibody) phage display library derived from the $V_H H$ of the heavy-chain IgGs was subtractively panned through successive binding to human lung and brain microvascular endothelial cells. After four rounds, the resulting clones were further selected for their ability to cross a monolayer of human brain endothelial cells and finally for their brain exposure after iv injection in mice. FC5 and FC44 were shown to transmigrate across the human in vitro BBB model and demonstrated higher brain exposure after injection in mice than a control (Muruganandam et al. 2002). The sdAb FC5 was reengineered with an Fc in various constructs mono- or bivalent for FC5 or C- or N-terminal to the Fc (Fig. 2.17) (Farrington et al. 2014). Bi-FC5-hFc gave a 30-fold higher CSF exposure than a control with the same format after injection in rats.

TMEM30A, the β-subunit of the membrane P4-ATPase flippase ATP8b1, was later identified as the target of FC5 (Haqqani et al. 2013). FC5 was engineered into a bispecific construct with an anti-mGluR1 antibody after deriving it into an ScFv (FC5-ScFv). The ScFv was incorporated at the N-terminal portion of an mGluR1 antibody (BBB-mGluR1), while two control fusion antibodies were generated, a control ScFv fusion with the mGluR1 antibody (Con-mGluR1) and a control antibody fused to the FC5 ScFv (BBB-Nip) (Fig. 2.18). Co-injection of these three ScFv fusions to rat showed respective 3.5- and 10.5-fold increase of brain exposure of the BBB-Nip (no antigen in the rat) and BBB-mGluR1. BBB-mGluR1 was able to dose-dependently suppress thermal hyperalgesia in rats after 10, 30, and 60 mg/kg iv injections (Webster et al. 2016).

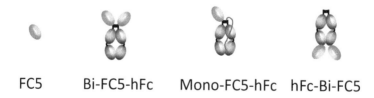

FC5   Bi-FC5-hFc   Mono-FC5-hFc   hFc-Bi-FC5

Fig. 2.17 Engineering of the VHH FC5 into various hFc constructs

FC5-ScFv   mGluR1   BBB- mGluR1   Con-mGluR1   BBB-Nip

Fig. 2.18 Constructs prepared from the FC5 V_HH: *Blue* anti-TfR; *yellow*: Anti-mGluR1; *grey* controls

2.4.1.5   IGF1R

Recently a series of camelid single-domain antibodies (V_HHs) were raised against different epitopes of the extracellular domain of human IGF1R away from the IGF1 binding site, humanized, and evaluated for BBB crossing in models in vitro and in vivo. Selected V_HHs demonstrated efficient saturable, energy-dependent transport across the human BBB model in vitro (Ribecco-Lutkiewicz et al. 2018) and highly enhanced brain and CSF exposure in rats (Stanimirovic et al. 2014, 2018). Future will tell if this mechanism is translatable to human as the IGF receptor at the human BBB differs from the IGFR at the animal BBB (Pardridge 2007).

2.4.2   Using Peptide Ligands of Transporter Receptors to Carry Antibodies to the Brain

An abundant literature reports of peptides able to increase brain exposure of various cargos such as peptides, nanoparticles, or oligonucleotides. A non-comprehensive overview of such peptide ligands and their sometimes-putative brain transporters or receptors is presented in Box 2.1 (the sketches refer to the mechanisms displayed in Fig. 2.3). Several peptides able to recognize receptors that are capable of *receptor-mediated transcytosis* across the brain have been identified, such as transferrin and several derived peptides, Angiopep2, melanotransferrin, or other peptides targeting LRP1 or LDL receptors. Several charged peptides such as TAT, SynB1, and penetratin target the process of *adsorptive-mediated transcytosis*. Peptides have been reported to use *carrier-mediated transport* to cross the BBB (Poduslo et al. 1994), but very few have been used to increase the transport of a drug. One of them could be glutathione (Gaillard et al. 2014); however the carrier has not been clearly identified and it was recently reported to use the N-methyl-D-aspartate receptor (NMDAR) (Fatima et al. 2019). Finally, several peptides often derived from bacteria or toxins and using other or non-identified mechanisms have been reported. This is the case for RVG peptide from rabies virus glycoprotein binding to the nicotinic acetylcholine receptor which showed great promise for carrying proteins and oligonucleotides into the brain (Huey et al. 2017), but there are many others (Box 2.1).

Box 2.1: Peptides with Reported Brain-Enhancing Properties Classified According to Their Mechanisms: Receptor-Mediated Transcytosis, Adsorptive-Mediated Transcytosis, Carrier-Mediated Transport, and Other Mechanisms (see Fig. 2.3 for Mechanisms Description)

RECEPTOR-MEDIATED TRANSCYTOSIS

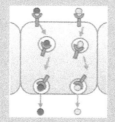

Transferrin receptor:
Transferrin (Ulbrich et al. 2009), **THR (THRPPMWSPVWP** (Lee et al. 2001) **and retroinverso analogs** (Tang et al. 2019), **T7 (HAIYPRH)** (Du et al. 2013; Liu et al. 2014; Wu et al. 2012), **cyclic peptide CRTIGPSVC** (Staquicini et al. 2011), **peptide N (CLPFFD) and Tf2 (CGGGHKYLR)** (Santi et al. 2017), **lactoferrin** (Hu et al. 2009)

LRP1 or LRP2 receptor:
Angiopep2 (Demeule et al. 2008; Regina et al. 2008; Shen et al. 2011), **PS80** (Ramge et al. 2000), **ApoE3** (Mulik et al. 2010), **ApoE141–50** (Shabanpoor et al. 2017), **ApoEII** (Bockenhoff et al. 2014), **ApoA** (Kreuter et al. 2007), **ApoB** (Spencer and Verma 2007), **K16ApoE** (Sarkar et al. 2014), **COG133** **(ApoE 133–149 LRVRLASHLRKLRKRLL)** (van Rooy et al. 2011), **melanotransferrin** (Gabathuler 2010; Gabathuler et al. 2005), **DSSHAFTLDELR** (Vitalis and Gabathuler 2014), **Raptor** (Prince et al. 2004)

LDL receptor:
Peptide 22 (Ac[cMPRLRGC]c-NH2 (Malcor et al. 2012)), **VH434 and VH4127** (Molino et al. 2017), **LRP2** (Pan et al. 2004)

IGF receptor: IgFII (Stefano et al. 2009)

Leptin receptor: leptin30 (leptin61–90) (Barrett et al. 2009; Liu et al. 2010), **g21 (leptin 12–32)** (Tosi et al. 2012)

ADSORPTIVE-MEDIATED TRANSCYTOSIS

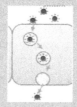

Cell-penetrating peptides (CPP's): TAT (Schwarze et al. 1999), **SynB1** (Rousselle et al. 2003), **Syn-B3** (Rousselle et al. 2001), **penetratin** (Xia et al. 2012), **dNP2 (KIKKVKKKGRK)** (Lim et al. 2015)

CARRIER-MEDIATED TRANSPORT

Active Na + —dep uptake transporter: glutathione (Lindqvist et al. 2013) **transport**

OTHER MECHANISMS

Nicotinic acetylcholine receptor: RVG (29 aa, from rabies virus) (Huey et al. 2017)**, RDP (39 aa)** (Fu et al. 2013)**, CDX (16 aa)** (Zhan et al. 2011)**, D-peptide** (Wei et al. 2015)

Unknown/other mechanisms: g7 (Tosi et al. 2007)**, TGN** (Li et al. 2011) **(TGNYKALHPHNG)** (Li et al. 2013)**, F3 peptide (CKDEPQRRSARL SAKPAPPKPEPKPKKAPAKK)** (Hu et al. 2013)**, odorranalectin** (Wu et al. 2012)**, LNP** (Yao et al. 2015)

From toxins, bacteria, virus:

- **g23 (HLNILSTLWKYR) (GM1, cholera toxin B)** (Georgieva et al. 2012; Stojanov et al. 2012)
- **CMR197 (HB-EGF) (diphteria toxin receptor)** (Gaillard et al. 2005)
- **NB03B** (Bode et al. 2017) **(Semliki Forest virus peptide)**
- **Tet1** (Kwon et al. 2010) **(Tetanus toxin)**
- **EPRNEEK** (Liu et al. 2014) **(*Streptococcus pneumoniae*, laminin receptor)**

From venoms:

- **Apamin** (Oller-Salvia et al. 2013)**, MiniAp4** (Prades et al. 2015) **(bee), MiniCTX3** (Diaz-Perlas et al. 2018) **(scorpion)**

These peptides have mostly been reported in the targeting of small molecules or other biotherapeutics (peptides, oligonucleotides) with most of the examples applied to nanoparticle formulations. Only a few of these peptides have been applied to ferrying an antibody across the BBB. Most examples can be found in the field of cancer metastases. *Angiopep2*, a ligand of LRP1 (low-density lipoprotein receptor-related protein 1) receptor, has been shown when conjugated to trastuzumab to reduce tumor growth in a Her2-positive orthotopic tumor model (Regina et al. 2015). Trastuzumab has also been conjugated to *melanotransferrin*, a protein ligand of the same receptor and shown to be able to reduce the number of preclinical human HER2+ breast cancer metastases in the brain compared to control groups. Tumors which remained after treatment were smaller than the control groups (Nounou et al. 2016). These two examples demonstrate efficacy in these models; however, the brain exposures of the conjugates have not been fully quantified even though Nounou et al. (Nounou et al. 2016) show using radioactive proteins and conjugates that the brain area/blood ratios are 10–225-fold higher for the conjugates than for trastuzumab alone. On the other hand, the status of the BBB in these tumors might be at least partially altered or compromised, which is clearly shown also by Nounou et al (Nounou et al. 2016) with the ratios in tumors being significantly higher than in normal brain for both trastuzumab and its melanotransferrin conjugate.

To evaluate the potential of some of these peptides to carry an antibody into a healthy brain, an anti-amyloid antibody was fused or conjugated to a selection of these peptides (*Angiopep2, melanotransferrin-derived peptide, RVG,* and *g7* peptides) (Lesuisse 2019) (Fig. 2.20). The two first peptides were selected based on the previous reports that they can increase the exposure of antibodies (vide supra). The RVG peptide is a 29-aa peptide derived from rabies virus glycoprotein (virus showing high degree of neurotropism in vivo) (Huey et al. 2017). In addition to showing the brain enhancing with various cargos (siRNA, SM, proteins) (Huey et al. 2017), several reports highlighted brain specificity, a precious property for such technology (Kumar et al. 2007). The peptide itself demonstrated transcytosis in human and murine models of BBB, in an active and competitive fashion vs bungarotoxin, an acetyl nicotinic ligand, suggesting translatability of the mechanism across species (Smith n.d.). The g7 peptide is derived from an opioid sequence (Tosi et al. 2010). g7-targeted nanoparticles encapsulating loperamide, an opioid agonist excluded from the brain, elicited a high analgesic effect, corresponding to 13–15% of the total loperamide dose injected by the g7-nanoparticles (Tosi et al. 2007). Solid-lipid nanoparticles capped with g7 peptide and encapsulating a fluorophore showed clear parenchymal uptake after iv injection in mice using dynamic fluorescence microscopy. In comparison, after injection of the non-g7-capped nanoparticles, the fluorescence remained confined in blood capillaries (Puech et al. n.d.) (Fig. 2.19). RVG and g7 peptides have not been reported in enhancing brain exposure of antibodies.

Various constructs starting from a murine Aβ anti-amyloid antibody were produced using either conjugation chemistry on the Fc part of the antibody or by production of a fusion on the N-terminal heavy chains of the antibody (Fig. 2.20). When engaged into mouse PK, the constructs failed to show appreciable brain enhancement compared to the naked anti-amyloid antibody.

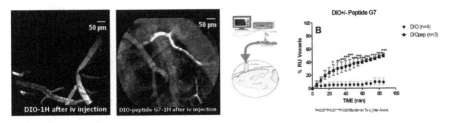

Fig. 2.19 Live imaging: IV tail vein injection of 100 μl/i.v. in anaesthetized Swiss mouse. Fluorescence reading using a Cell Vizio instrument (Lesuisse 2019)

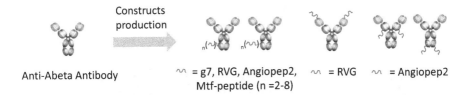

Fig. 2.20 Constructs between a murine anti-Abeta amyloid antibody and RVG, g7, melanotransferrin-derived peptide (DSSHAFTLDELR) and Angiopep2

The above results could be linked to the specific conditions, the chemistry of conjugation, or the linkers used in fusions and conjugations, but they point to the complexity of cargoing molecules as large as antibodies across the BBB using small-size peptides. Peptides also display reduced affinity for their receptors compared to antibodies which could also account for the above lack of brain exposure enhancement. This probably also explains why no small molecule ligand of transporter receptors has been reported to be able to enhance brain exposure of an antibody conjugated to it. In addition, the effects of brain enhancement of these peptides are often demonstrated in the context of nanoparticles with hundreds of peptides exposed at their surfaces potentially leading to an avidity or cooperativity in the binding to their antigens.

2.5 Charge: Adsorptive-Mediated Transcytosis

It has been reported since more than two decades that adding positive charges to proteins enables their brain penetration (Herve et al. 2008). IgGs or fragments have been cationized by covalent coupling of hexamethylenediamine or putrescine raising their isoelectric point largely above 7 and resulting in higher brain exposure vs the native protein. The extent of brain exposure increase is modest, such as a two to threefold for cationized [125I]-bovine IgG vs native [125I]-bovine IgG (Triguero et al. 1990). Aβ amyloid has also been used to illustrate the potential of this strategy for imaging applications. Polyamine-modified anti-Aβ (Fab)2 have been shown to label

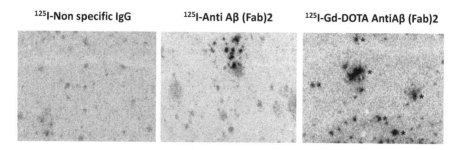

Fig. 2.21 Photomicrographs of histological sections of hippocampus processed for anti-Aβ immunohistochemistry with a rabbit polyclonal antibody and emulsion autoradiography with 4 weeks of exposure. (Taken from Muthu Ramakrishnan, Pharm Sci (Li et al. 2016). Copyright from Springer Nature)

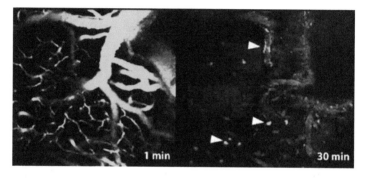

Fig. 2.22 Maximum intensity projection of fluorescence in an 8-month-old Tg4510 mouse at different timepoints after iv injection at 10 mg/kg. (Taken from Li et al. Journal of Controlled Release 243 (2016) 1–10 (Cramer et al. 2012). Copyright from Elsevier)

amyloid deposits in cortex and hippocampus of AD transgenic mouse brain following iv injection more efficiently than their non-charged counterparts (Fig. 2.21) (Ramakrishnan et al. 2008).

Charged anti-Aβ V_HH have been used to demonstrate in vivo imaging of amyloid deposits using two-photon microscopy (Li et al. 2016) (Fig. 2.22).

There are caveats to these adsorptive-mediated transcytosis approaches (Fig. 2.3-F). Cationization could alter the mAb biological activity if the functionalization has been linked to residues such as Asp or Glu that are essential for binding their antigens or could change their conformation. Another serious limitation is that the mode of penetration by interaction with negatively charged membranes will lead to random organ and tissue distribution and potentially yield toxicity and immunogenicity.

2.6 Nanotechnologies

Nanotechnologies, when adequately targeted to the brain, have been widely reported to increase brain exposure of small molecules or siRNA (Cramer et al. 2012) (even though none are presently developed for brain enhancement). However, the interest for large molecules such as antibodies is much less obvious. Here again some examples are found in the amyloid area. Enhanced glutathione PEGylated liposomal brain delivery of an Aβ single-domain antibody fragment (Fig. 2.23-A) displayed increased half-lives and higher brain exposures in an amyloid mouse model of AD (Rotman et al. 2015) (Table 2.1). In this case the nanoparticle was addressed to brain via glutathione.

Another report describes a liposomal formulation exposing an Aβ antibody on its surface (Fig. 2.23-B). The results of the paper show that in contrast to "older" animals (16 months) where an effect could be observed, there was a complete lack of AD improvement in "aged" (10 months) mice. This is in good agreement with lack of brain penetration of this non-brain-addressed formulation in adult animals where the BBB is not yet compromised (Ordonez-Gutierrez et al. 2017).

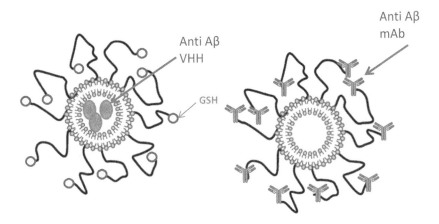

Fig. 2.23 (**a**) Liposomes encapsulating an anti-Amyloid $V_H H$ decorated with glutathione for brain targeting; (**b**) Liposomes exposing anti Aβ antibody at their surface

Table 2.1 Half-lives and percent of injected dose in the brain of APP mice after injection of free Aβ $V_H H$-DTPA-^{111}In and brain-targeted liposomes GSH-PEG Aβ $V_H H$-DTPA-^{111}In

	Free Aβ VHH-DTPA-^{111}In		GSH-PEG Aβ VHH-DTPA-^{111}In	
	WT	APP/PS1	WT	APP/PS1
$t_{1/2}$ (h)	3.83	4.36	13,8	15,2
Cerebellum Suv (%ID/g tissue per g mouse)	0.001 ± 0.000	0.001 ± 0.000	0.015 ± 0.006	0.094 ± 0.031

DTPA Diethylenetriaminepentaacetic acid (Rotman et al. 2015)

Recently Wen and colleagues showed that rituximab, an antibody against CD20, when encapsulated within a biodegradable, phosphorylcholine-based cross-linked zwitterionic polymer, displayed time-release behavior leading to higher brain exposure and improved therapeutic efficacy against CNS metastases in a mouse lymphoma model. The efficacy was further increased when these nanocapsules were conjugated at their surface with CXCL13, a chemokine ligand of CXCR5 highly expressed on tumor metastases (Wen et al. 2019). In another recent report, nimotuzumab and trastuzumab, antibodies against the epidermal growth factor (EGFR) and the human epidermal growth factor2 receptor (HER2), respectively, were encapsulated through cross-linking peptides containing acetylcholine and choline residues and displayed enhanced brain exposure presumably mediated by interaction with their receptor and transporter (Han et al. 2019).

The main challenges with these strategies are the very complex development and quality control of such formulations. The chemistry of conjugation of mAbs can be cumbersome, and addressing proteins on nanoparticles and their characterization further complicate these preparations. In addition to reaching the brain, as shown above, these formulations need to be targeted with a brain-specific agent which will further complicate the preparation and characterization.

2.7 BBB Disruption

2.7.1 Chemical Modulation of the BBB Permeability

Temporary BBB disruption can be induced by several agents such as mannitol or bradykinin and allow delivery of large molecules such as dextrans into the brain. However, the extent of enhancement is modest and no application to antibodies has been reported (Jones and Shusta 2009). Recently, it was shown that mice lacking CD73, which are unable to produce extracellular adenosine, are protected from EAE and that blockade of the A2A adenosine receptor (AR) inhibits T cell entry into the CNS (Carman et al. 2011). Following treatment with a broad-spectrum AR agonist, an intravenously administered anti-β-amyloid antibody 6E10 was observed to enter the CNS and bind β-amyloid plaques in a transgenic mouse model of AD while in the same conditions 6E10 alone did not show any labeling of the plaques. Another recent report describes cyclic peptides designed to disrupt the cadherin-cadherin interactions at the adherens junctions to increase the porosity of the BBB paracellular pathways (Ulapane et al. 2019). These new cyclic peptides when co-injected with a fluorescently labeled IgG mAb led to a ~ twofold increase in brain exposure in mice.

2.7.2   Focused Ultrasounds

Low-energy ultrasounds when applied under MRI guidance in parallel to injected microbubbles have the potential to temporarily open the BBB (Meng et al. 2017). Figure 2.24 (Abdul Razzak et al. 2019) outlines the main principle of BBB disruption using MRI-guided microbubble-assisted focused ultrasound technique (MB-FUS).

Very few examples of MRI-guided ultrasound can be found with antibodies. Rodent applications to herceptin (Kinoshita et al. 2006) and to an Aβ amyloid antibody BAM-10 (Jordao et al. 2010) have been reported. For this latter case though FUS alone is probably active on its own on plaque as it has been shown to reduce plaque presumably by inducing microglial activation. Focused ultrasounds seem particularly well adapted for brain tumors, glioblastoma as it has the potential to deliver the biotherapeutics in very specific area of the brain. Application to diseases such as AD might be more cumbersome as the pathology is spread over a much larger area. In addition, the duration of opening decreases rapidly, and mAbs are large molecules and would need to be injected rapidly after disruption.

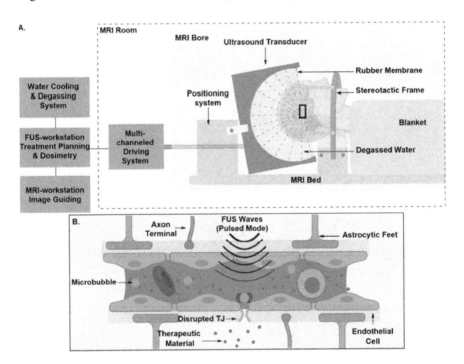

Fig. 2.24 (**a**) The patient's head is rested in a semi-spherical ultrasound transducer integrated into an MRI scanner. The transducer is attached to a mechanical positioning system. The focused ultrasound and the magnetic resonance parameters are remotely controlled by electronic interfaces. The patient's head is immobilized by a stereotactic frame. Overheating of the scalp, skull and brain tissue is minimized by the use of a water interface, which also acts as an acoustic coupler. (**b**) Pretreatment of the patient with microbubbles harnesses the acoustic power and concentrates it to the blood vessel, which attenuates acoustic power levels. Microbubbles move in the direction of the FUS wave propagation and under the influence of the FUS waves they oscillate, microstreaming the medium surrounding them, inducing mechanical stress that disrupts the TJs between ECs. (Taken from Kinoshita et al. (2006). Copyright from Creative Common)

The technology is disruptive and careful monitoring should be exerted. In fact, BBB disruption has been linked with several side effects such as seizures and others more severe like stroke. In particular, amyloid angiopathy may affect BBB opening and closure and increase the risk of intracranial hemorrhage. Altogether, this technology is still early even though it has reached the clinic with small molecules in GBM, and low-frequency FUS to open the BBB around gliomas (Mainprize et al. 2019), in the frontal lobe of patients with AD (Lipsman et al. 2018), and in the primary motor cortex of subjects with ALS (Abrahao et al. 2019) has now been shown safe and transient in pilot clinical trials.

2.8   Intranasal Administration

Intranasal delivery of therapeutics involves spraying therapeutics into the upper part of the nasal cavity to enable them to follow the olfactory axon bundles directly into the brain through the cribriform plate (Fig. 2.25) (Katare et al. 2017).

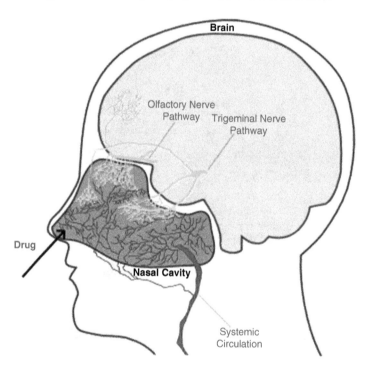

Fig. 2.25 Pathways for transport of drugs from the nasal cavity to the brain and systemic circulation. Nerves and nerve endings have been depicted in green, while blood vessels are depicted in red. (Taken from Kumar et al. (2018b). Copyright from Elsevier)

Application of intranasal administration to antibodies has been seldom. A recent article (Kumar et al. 2018b) reports of brain concentrations of an IgG after such administration. Brain exposures 30 min after 2.5 mg/rat intranasal administration of a radiolabeled [^{125}I]-IgG1 have been recorded (Table 2.2).

Table 2.2 Distribution of an antibody after intranasal administration (2.5 mg/rat) (Kumar et al. 2018b)

Organ	Concentrations (pM)
Blood	2663.77 ± 357.96
Olfactory bulbs	4418.54 ± 1862.50
Frontal cortex	547.44 ± 100.56
Caudoputamen	277.31 ± 63.95
Motor cortex	275.13 ± 48.4
Posterior hippocampus	276.28 ± 53.39
Cerebellum	249.73 ± 52.55

The remaining challenges of this mode of administration are on one hand the translatability between species with the very different proportions of the olfactory bulbs between rodent and human (Fig. 2.26). In addition, given the anatomical differences between species and in order to ensure proper delivery to the upper part of the nasal cavity, specific devices would need to be used for each animal model.

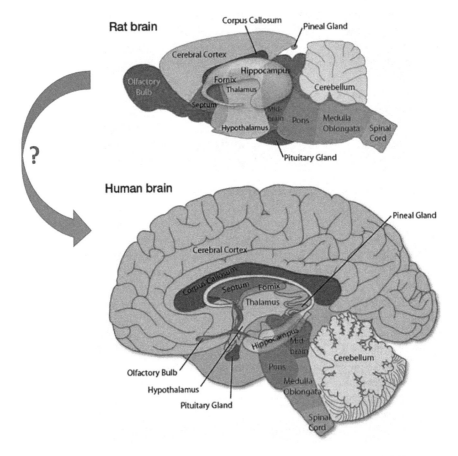

Fig. 2.26 Rat and human anatomy of the brain: the olfactory bulbs are shown in dark blue

2.9 Conclusions and Perspectives

Of all technologies presented above, receptor-mediated enhancement is the most advanced with a few constructs using either insulin or transferrin receptors in ongoing clinical development. Several challenges however remain in the field. Aside from the specific safety linked to the transcytotic receptor used and the therapeutic target (discussed in the above paragraphs), it is not totally clear at this stage how safe it will be to dramatically increase the brain exposure of antibodies or other biotherapeutics.

Using rodent models for translation to disease states is still challenging, and full understanding of BBB in health and disease is yet in its infancy (Deo et al. 2013). One of the prerequisites also of this approach is that the transcytotic receptor protein level and internalizing function are preserved in the targeted disease. This has, for instance, been demonstrated in the case of the TfR for Aβ and Tau neuropathologies using both mice transgenic amyloid models and human brain postmortem samples (Bourassa et al. 2019).

These bispecific constructs may present several challenges of scaling up and cost of goods. A potential solution could be the use of fusions to smaller proteins binding transcytotic receptors. Single-chain antibodies have already shown promises (Stocki et al. 2019), but even smaller moieties such as DARPins (Pluckthun 2015), Anticalins® (Rothe and Skerra 2018), or Nanofitins (Goux et al. 2017) could be explored.

Until now research has mainly focused on enhancing the influx of antibodies into the brain. Some recent reports suggest that lowering the efflux of antibodies from the brain could be a useful strategy, through, for instance, modulation of their glycosylation states (Finke et al. 2017). Another area of interest is the modulation of the Fc part of the antibodies to increase their circulation time in the periphery and increase their opportunities to use the extracellular pathways (Banks 2016). Brain retention through specific binding to a brain antigen could also be an emerging strategy with the recent example of anti-MOG antibodies fused to a lysosomal enzyme ASM and showing >tenfold brain exposures compared to the control (Nakano et al. 2019). The safety of this strategy also remains to be established.

One of the challenges in passive immunotherapy approaches such as anti-amyloid is the need to maintain levels above therapeutic dose through long-term treatment requiring patient engagement and compliance, as well as a significant cost of goods. Vector-mediated transfer of genes encoding antibodies with the potential to provide sustained concentrations has recently been explored with several recent examples in the field of anti-amyloid antibodies. However, the BBB remains a major challenge, and most of the reports to date use disruptive modes of administration such as intracranial, intraparenchymal, or ICV (Elmer et al. 2019). The rise of new brain-penetrant AAV might bring some potential future solutions (Ellsworth et al. 2019).

So far, the mechanisms identified for brain enhancement have been mostly ubiquitously expressed leading to exposure in other tissues than the brain. Identification of brain-specific mechanisms remains the ultimate unreached goal and should be the focus of future efforts in the field.

Acknowledgments The author wishes to acknowledge Patricia Senneville for help in putting this document together.

References

Abdul Razzak R, Florence GJ, Gunn-Moore FJ (2019) Approaches to CNS drug delivery with a focus on transporter-mediated transcytosis. Int J Mol Sci 20

Abrahao A et al (2019) First-in-human trial of blood-brain barrier opening in amyotrophic lateral sclerosis using MR-guided focused ultrasound. Nat Commun 10:4373

Banks WA (2016) From blood-brain barrier to blood-brain interface: new opportunities for CNS drug delivery. Nat Rev Drug Discov 15:275–292

Banks WA, Ortiz L, Plotkin SR, Kastin AJ (1991) Human interleukin (IL) 1 alpha, murine IL-1 alpha and murine IL-1 beta are transported from blood to brain in the mouse by a shared saturable mechanism. J Pharmacol Exp Ther 259:988–996

Banks WA, Kastin AJ, Gutierrez EG (1994) Penetration of interleukin-6 across the murine blood-brain barrier. Neurosci Lett 179:53–56

Banks WA, Kastin AJ, Broadwell RD (1995) Passage of cytokines across the blood-brain barrier. Neuroimmunomodulation 2:241–248

Banks WA, Jumbe NL, Farrell CL, Niehoff ML, Heatherington AC (2004) Passage of erythropoietic agents across the blood-brain barrier: a comparison of human and murine erythropoietin and the analog darbepoetin alfa. Eur J Pharmacol 505:93–101

Bard F et al (2012) Sustained levels of antibodies against Abeta in amyloid-rich regions of the CNS following intravenous dosing in human APP transgenic mice. Exp Neurol 238:38–43

Barrett GL, Trieu J, Naim T (2009) The identification of leptin-derived peptides that are taken up by the brain. Regul Pept 155:55–61

Bell RD, Ehlers MD (2014) Breaching the blood-brain barrier for drug delivery. Neuron 81:1–3

Bien-Ly N et al (2014) Transferrin receptor (TfR) trafficking determines brain uptake of TfR antibody affinity variants. J Exp Med 211:233–244

Boado RJ, Zhang Y, Zhang Y, Pardridge WM (2007a) Genetic engineering, expression, and activity of a fusion protein of a human neurotrophin and a molecular Trojan horse for delivery across the human blood-brain barrier. Biotechnol Bioeng 97:1376–1386

Boado RJ, Zhang Y, Zhang Y, Xia CF, Pardridge WM (2007b) Fusion antibody for Alzheimer's disease with bidirectional transport across the blood-brain barrier and abeta fibril disaggregation. Bioconjug Chem 18:447–455

Boado RJ, Zhou QH, Lu JZ, Hui EK, Pardridge WM (2010) Pharmacokinetics and brain uptake of a genetically engineered bifunctional fusion antibody targeting the mouse transferrin receptor. Mol Pharm 7:237–244

Boado RJ, Hui EK, Lu JZ, Pardridge WM (2012) Glycemic control and chronic dosing of rhesus monkeys with a fusion protein of iduronidase and a monoclonal antibody against the human insulin receptor. Drug Metab Dispos 40:2021–2025

Boado RJ, Lu JZ, Hui EK, Sumbria RK, Pardridge WM (2013) Pharmacokinetics and brain uptake in the rhesus monkey of a fusion protein of arylsulfatase a and a monoclonal antibody against the human insulin receptor. Biotechnol Bioeng 110:1456–1465

Boado RJ, Ka-Wai Hui E, Zhiqiang L, J. & Pardridge, W.M. (2014) Insulin receptor antibody-iduronate 2-sulfatase fusion protein: pharmacokinetics, anti-drug antibody, and safety pharmacology in rhesus monkeys. Biotechnol Bioeng 111:2317–2325

Bockenhoff A et al (2014) Comparison of five peptide vectors for improved brain delivery of the lysosomal enzyme arylsulfatase A. J Neurosci 34:3122–3129

Bode GH et al (2017) An in vitro and in vivo study of peptide-functionalized nanoparticles for brain targeting: the importance of selective blood-brain barrier uptake. Nanomedicine 13:1289–1300

Bourassa P, Alata W, Tremblay C, Paris-Robidas S, Calon F (2019) Transferrin receptor-mediated uptake at the blood-brain barrier is not impaired by Alzheimer's disease neuropathology. Mol Pharm 16:583–594

Carman AJ, Mills JH, Krenz A, Kim DG, Bynoe MS (2011) Adenosine receptor signaling modulates permeability of the blood-brain barrier. J Neurosci 31:13272–13280

Chang R et al (2017) Blood-brain barrier penetrating biologic TNF-alpha inhibitor for Alzheimer's disease. Mol Pharm 14:2340–2349

Chang R et al (2018) Brain penetrating bifunctional erythropoietin-transferrin receptor antibody fusion protein for Alzheimer's disease. Mol Pharm 15:4963–4973

Clarke E (et al) (2020) TrkB agonist antibody delivery to the brain using a TfR1 specific BBB shuttle provides neuroprotection in a mouse model of Parkinson's disease. BioRxiv preprint first posted online March 12, 2020

Couch JA et al (2013) Addressing safety liabilities of TfR bispecific antibodies that cross the blood-brain barrier. Sci Transl Med 5(183ra157):181–112

Cramer S, Rempe R, Galla HJ (2012) Exploiting the properties of biomolecules for brain targeting of nanoparticulate systems. Curr Med Chem 19:3163–3187

Demeule M et al (2008) Identification and design of peptides as a new drug delivery system for the brain. J Pharmacol Exp Ther 324:1064–1072

Deo AK, Theil FP, Nicolas JM (2013) Confounding parameters in preclinical assessment of blood-brain barrier permeation: an overview with emphasis on species differences and effect of disease states. Mol Pharm 10:1581–1595

Di L, Rong H, Feng B (2013) Demystifying brain penetration in central nervous system drug discovery. Miniperspective. J Med Chem 56:2–12

Diaz-Perlas C et al (2018) From venoms to BBB-shuttles. MiniCTX3: a molecular vector derived from scorpion venom. Chem Commun (Camb) 54:12738–12741

Dickens D et al (2013) Transport of gabapentin by LAT1 (SLC7A5). Biochem Pharmacol 85:1672–1683

Do T-MA (2020) Mabs

Do TM et al (2020) Tetravalent bispecific tandem antibodies improve brain exposure and efficacy in an amyloid transgenic mouse model. Mol Ther Method Clin Dev 19:58–77

Dobson PD, Kell DB (2008) Carrier-mediated cellular uptake of pharmaceutical drugs: an exception or the rule? Nat Rev Drug Discov 7:205–220

Du W et al (2013) Transferrin receptor specific nanocarriers conjugated with functional 7peptide for oral drug delivery. Biomaterials 34:794–806

Du D et al (2014) The role of glucose transporters in the distribution of p-aminophenyl-alpha-d-mannopyranoside modified liposomes within mice brain. J Control Release 182:99–110

Ellsworth JL et al (2019) Clade F AAVHSCs cross the blood brain barrier and transduce the central nervous system in addition to peripheral tissues following intravenous administration in nonhuman primates. PLoS One 14:e0225582

Elmer BM et al (2019) Gene delivery of a modified antibody to Abeta reduces progression of murine Alzheimer's disease. PLoS One 14:e0226245

Farrington GK et al (2014) A novel platform for engineering blood-brain barrier-crossing bispecific biologics. FASEB J 28:4764–4778

Fatima S et al (2019) Identification and evaluation of glutathione conjugate gamma-l-glutamyl-l-cysteine for improved drug delivery to the brain. J Biomol Struct Dyn 1(11)

Finke JM et al (2017) Antibody blood-brain barrier efflux is modulated by glycan modification. Biochim Biophys Acta Gen Subj 1861:2228–2239

Friden M et al (2009) Structure-brain exposure relationships in rat and human using a novel data set of unbound drug concentrations in brain interstitial and cerebrospinal fluids. J Med Chem 52:6233–6243

Fu A, Zhao Z, Gao F, Zhang M (2013) Cellular uptake mechanism and therapeutic utility of a novel peptide in targeted-delivery of proteins into neuronal cells. Pharm Res 30:2108–2117

Fuller JP et al (2015) Comparing the efficacy and neuroinflammatory potential of three anti-abeta antibodies. Acta Neuropathol 130:699–711

Gabathuler R (2010) Development of new peptide vectors for the transport of therapeutic across the blood-brain barrier. Ther Deliv 1:571–586

Gabathuler R, Arthur G, Kennard ML et al (2005) Development of a potential protein vector (NeuroTrans) to deliver drugs across the blood-brain barrier. In: de Boer AG (ed) Drug transporters and the diseased brain, int congress series, Esteve Foundation symposium XI, Sagaro (Girona), Spain. Elsevier 1277:171–184

Gaillard PJ, Visser CC, de Boer AG (2005) Targeted delivery across the blood-brain barrier. Expert Opin Drug Deliv 2:299–309

Gaillard PJ et al (2014) Pharmacokinetics, brain delivery, and efficacy in brain tumor-bearing mice of glutathione pegylated liposomal doxorubicin (2B3-101). PLoS One 9:e82331

Garg A, Balthasar JP (2009) Investigation of the influence of FcRn on the distribution of IgG to the brain. AAPS J 11:553–557

Georgieva JV et al (2012) Peptide-mediated blood-brain barrier transport of polymersomes. Angew Chem Int Ed Engl 51:8339–8342

Goulatis LI, Shusta EV (2017) Protein engineering approaches for regulating blood-brain barrier transcytosis. Curr Opin Struct Biol 45:109–115

Goux M et al (2017) Nanofitin as a new molecular-imaging agent for the diagnosis of epidermal growth factor receptor over-expressing Tumors. Bioconjug Chem 28:2361–2371

Han L et al (2019) Systemic delivery of monoclonal antibodies to the central nervous system for brain tumor therapy. Adv Mater 31:e1805697

Hanzatian DKG, Mueller AD, Mueller BK, Mueller R, Klein C (2015) Blood-brain barrier (BBB) penetrating dual specific binding proteins for treating brain and neurological diseases. WO 2015/191934 A2

Haqqani AS et al (2013) Method for isolation and molecular characterization of extracellular microvesicles released from brain endothelial cells. Fluids Barriers CNS 10:4

Helms HC et al (2016) In vitro models of the blood-brain barrier: an overview of commonly used brain endothelial cell culture models and guidelines for their use. J Cereb Blood Flow Metab 36:862–890

Hersom M et al (2018) The insulin receptor is expressed and functional in cultured blood-brain barrier endothelial cells but does not mediate insulin entry from blood to brain. Am J Physiol Endocrinol Metab 315:E531–E542

Herve F, Ghinea N, Scherrmann JM (2008) CNS delivery via adsorptive transcytosis. AAPS J 10:455–472

Hu K et al (2009) Lactoferrin-conjugated PEG-PLA nanoparticles with improved brain delivery: in vitro and in vivo evaluations. J Control Release 134:55–61

Hu Q et al (2013) F3 peptide-functionalized PEG-PLA nanoparticles co-administrated with tLyp-1 peptide for anti-glioma drug delivery. Biomaterials 34:1135–1145

Huey R, Hawthorne S, McCarron P (2017) The potential use of rabies virus glycoprotein-derived peptides to facilitate drug delivery into the central nervous system: a mini review. J Drug Target 25:379–385

Huttunen J, Gynther M, Vellonen KS, Huttunen KM (2019) L-Type amino acid transporter 1 (LAT1)-utilizing prodrugs are carrier-selective despite having low affinity for organic anion transporting polypeptides (OATPs). Int J Pharm 571:118714

Johnsen KB, Burkhart A, Thomsen LB, Andresen TL, Moos T (2019) Targeting the transferrin receptor for brain drug delivery. Prog Neurobiol 181:101665

Jones AR, Shusta EV (2007) Blood-brain barrier transport of therapeutics via receptor-mediation. Pharm Res 24:1759–1771

Jones AR, Shusta EV (2009) Antibodies and the blood-brain barrier. In: An Z (ed) Chapter 21: Therapeutic monoclonal antibodies: from bench to clinic. Wiley

Jordao JF et al (2010) Antibodies targeted to the brain with image-guided focused ultrasound reduces amyloid-beta plaque load in the TgCRND8 mouse model of Alzheimer's disease. PLoS One 5:e10549

Karaoglu Hanzatian D et al (2018) Brain uptake of multivalent and multi-specific DVD-Ig proteins after systemic administration. MAbs 10:765–777

Kariolis MS et al (2020) Brain delivery of therapeutic proteins using an Fc fragment blood-brain barrier transport vehicle in mice and monkeys. Sci Transl Med 12

Kaschka WP, Theilkaes L, Eickhoff K, Skvaril F (1979) Disproportionate elevation of the immunoglobulin G1 concentration in cerebrospinal fluids of patients with multiple sclerosis. Infect Immun 26:933–941

Katare YK et al (2017) Intranasal delivery of antipsychotic drugs. Schizophr Res 184:2–13

Kinoshita M, McDannold N, Jolesz FA, Hynynen K (2006) Noninvasive localized delivery of Herceptin to the mouse brain by MRI-guided focused ultrasound-induced blood-brain barrier disruption. Proc Natl Acad Sci U S A 103:11719–11723

Kreuter J et al (2007) Covalent attachment of apolipoprotein A-I and apolipoprotein B-100 to albumin nanoparticles enables drug transport into the brain. J Control Release 118:54–58

Kumar P et al (2007) Transvascular delivery of small interfering RNA to the central nervous system. Nature 448:39–43

Kumar NN et al (2018a) Passive immunotherapies for central nervous system disorders: current delivery challenges and new approaches. Bioconjug Chem 29:3937–3966

Kumar NN et al (2018b) Delivery of immunoglobulin G antibodies to the rat nervous system following intranasal administration: distribution, dose-response, and mechanisms of delivery. J Control Release 286:467–484

Kusuhara H, Sugiyama Y (2001a) Efflux transport systems for drugs at the blood-brain barrier and blood-cerebrospinal fluid barrier (Part 2). Drug Discov Today 6:206–212

Kusuhara H, Sugiyama Y (2001b) Efflux transport systems for drugs at the blood-brain barrier and blood-cerebrospinal fluid barrier (Part 1). Drug Discov Today 6:150–156

Kwon EJ et al (2010) Targeted nonviral delivery vehicles to neural progenitor cells in the mouse subventricular zone. Biomaterials 31:2417–2424

Lee JH, Engler JA, Collawn JF, Moore BA (2001) Receptor mediated uptake of peptides that bind the human transferrin receptor. Eur J Biochem 268:2004–2012

Lesuisse D (2019) Brain exposure enhancement of antibodies. Paper presented at the AAPS workshop, Washington, April 27–29, 2019

Li J et al (2011) Targeting the brain with PEG-PLGA nanoparticles modified with phage-displayed peptides. Biomaterials 32:4943–4950

Li J et al (2013) Brain delivery of NAP with PEG-PLGA nanoparticles modified with phage display peptides. Pharm Res 30:1813–1823

Li T et al (2016) Camelid single-domain antibodies: a versatile tool for in vivo imaging of extracellular and intracellular brain targets. J Control Release 243:1–10

Lim S et al (2015) dNP2 is a blood-brain barrier-permeable peptide enabling ctCTLA-4 protein delivery to ameliorate experimental autoimmune encephalomyelitis. Nat Commun 6:8244

Lindqvist A, Rip J, Gaillard PJ, Bjorkman S, Hammarlund-Udenaes M (2013) Enhanced brain delivery of the opioid peptide DAMGO in glutathione pegylated liposomes: a microdialysis study. Mol Pharm 10:1533–1541

Lipsman N et al (2018) Blood-brain barrier opening in Alzheimer's disease using MR-guided focused ultrasound. Nat Commun 9:2336

Liu Y et al (2010) A leptin derived 30-amino-acid peptide modified pegylated poly-L-lysine dendrigraft for brain targeted gene delivery. Biomaterials 31:5246–5257

Liu Y et al (2014) A bacteria deriving peptide modified dendrigraft poly-l-lysines (DGL) self-assembling nanoplatform for targeted gene delivery. Mol Pharm 11:3330–3341

Lu JZ, Hui EK, Boado RJ, Pardridge WM (2010) Genetic engineering of a bifunctional IgG fusion protein with iduronate-2-sulfatase. Bioconjug Chem 21:151–156

Mainprize T et al (2019) Blood-brain barrier opening in primary brain tumors with non-invasive MR-guided focused ultrasound: a clinical safety and feasibility study. Sci Rep 9:321

Malcor JD et al (2012) Chemical optimization of new ligands of the low-density lipoprotein receptor as potential vectors for central nervous system targeting. J Med Chem 55:2227–2241

Meng Y et al (2017) Focused ultrasound as a novel strategy for Alzheimer disease therapeutics. Ann Neurol 81:611–617

Molino Y et al (2017) Use of LDL receptor-targeting peptide vectors for in vitro and in vivo cargo transport across the blood-brain barrier. FASEB J 31:1807–1827

Mulik RS, Monkkonen J, Juvonen RO, Mahadik KR, Paradkar AR (2010) ApoE3 mediated poly(butyl) cyanoacrylate nanoparticles containing curcumin: study of enhanced activity of curcumin against beta amyloid induced cytotoxicity using in vitro cell culture model. Mol Pharm 7:815–825

Muruganandam A, Tanha J, Narang S, Stanimirovic D (2002) Selection of phage-displayed llama single-domain antibodies that transmigrate across human blood-brain barrier endothelium. FASEB J 16:240–242

Nakano R et al (2019) A new technology for increasing therapeutic protein levels in the brain over extended periods. PLoS One 14:e0214404

Nalecz KA (2017) Solute carriers in the blood-brain barier: safety in abundance. Neurochem Res 42:795–809

Niewoehner J et al (2014) Increased brain penetration and potency of a therapeutic antibody using a monovalent molecular shuttle. Neuron 81:49–60

Nounou MI et al (2016) Anti-cancer antibody trastuzumab-melanotransferrin conjugate (BT2111) for the treatment of metastatic HER2+ breast cancer tumors in the brain: an in-vivo study. Pharm Res 33:2930–2942

Obermeier B, Daneman R, Ransohoff RM (2013) Development, maintenance and disruption of the blood-brain barrier. Nat Med 19:1584–1596

Ohshima-Hosoyama S et al (2012) A monoclonal antibody-GDNF fusion protein is not neuroprotective and is associated with proliferative pancreatic lesions in parkinsonian monkeys. PLoS One 7:e39036

Oller-Salvia B, Teixido M, Giralt E (2013) From venoms to BBB shuttles: synthesis and blood-brain barrier transport assessment of apamin and a nontoxic analog. Biopolymers 100:675–686

Ordonez-Gutierrez L et al (2017) ImmunoPEGliposome-mediated reduction of blood and brain amyloid levels in a mouse model of Alzheimer's disease is restricted to aged animals. Biomaterials 112:141–152

Pan W et al (2004) Efficient transfer of receptor-associated protein (RAP) across the blood-brain barrier. J Cell Sci 117:5071–5078

Pardridge WM (2002) Drug and gene targeting to the brain with molecular Trojan horses. Nat Rev Drug Discov 1:131–139

Pardridge WM (2007) Blood-brain barrier delivery. Drug Discov Today 12:54–61

Pardridge WM (2015) Blood-brain barrier drug delivery of IgG fusion proteins with a transferrin receptor monoclonal antibody. Expert Opin Drug Deliv 12:207–222

Pardridge WM (2017) Delivery of biologics across the blood-brain barrier with molecular Trojan horse technology. BioDrugs 31:503–519

Pardridge WM, Boado RJ, Patrick DJ, Ka-Wai Hui E, Lu JZ (2018a) Blood-brain barrier transport, plasma pharmacokinetics, and neuropathology following chronic treatment of the rhesus monkey with a brain penetrating humanized monoclonal antibody against the human transferrin receptor. Mol Pharm 15:5207–5216

Pardridge WM, Boado RJ, Giugliani R, Schmidt M (2018b) Plasma pharmacokinetics of Valanafusp alpha, a human insulin receptor antibody-Iduronidase fusion protein, in patients with mucopolysaccharidosis type I. BioDrugs 32:169–176

Pepinsky RB et al (2011) Exposure levels of anti-LINGO-1 Li81 antibody in the central nervous system and dose-efficacy relationships in rat spinal cord remyelination models after systemic administration. J Pharmacol Exp Ther 339:519–529

Pluckthun A (2015) Designed ankyrin repeat proteins (DARPins): binding proteins for research, diagnostics, and therapy. Annu Rev Pharmacol Toxicol 55:489–511

Poduslo JF, Curran GL (1996) Permeability at the blood-brain and blood-nerve barriers of the neurotrophic factors: NGF, CNTF, NT-3, BDNF. Brain Res Mol Brain Res 36:280–286

Poduslo JF, Curran GL, Berg CT (1994) Macromolecular permeability across the blood-nerve and blood-brain barriers. Proc Natl Acad Sci U S A 91:5705–5709

Polu KR, Lowman HB (2014) Probody therapeutics for targeting antibodies to diseased tissue. Expert Opin Biol Ther 14:1049–1053

Prades R et al (2015) Applying the retro-enantio approach to obtain a peptide capable of overcoming the blood-brain barrier. Angew Chem Int Ed Engl 54:3967–3972

Prince WS et al (2004) Lipoprotein receptor binding, cellular uptake, and lysosomal delivery of fusions between the receptor-associated protein (RAP) and alpha-L-iduronidase or acid alpha-glucosidase. J Biol Chem 279:35037–35046

Puech F, BO, Aubin N et al Sanofi, IMI COMPACT, unpublished results

Ramakrishnan M et al (2008) Selective contrast enhancement of individual Alzheimer's disease amyloid plaques using a polyamine and Gd-DOTA conjugated antibody fragment against fibrillar Abeta42 for magnetic resonance molecular imaging. Pharm Res 25:1861–1872

Ramge P et al (2000) Polysorbate-80 coating enhances uptake of polybutylcyanoacrylate (PBCA)-nanoparticles by human and bovine primary brain capillary endothelial cells. Eur J Neurosci 12:1931–1940

Rankovic Z (2015) CNS drug design: balancing physicochemical properties for optimal brain exposure. J Med Chem 58:2584–2608

Rao EMV, Li D (2007) Antibodies that bind IL-4 and/or IL-13 and their uses. WO2009052081

Regina A et al (2008) Antitumour activity of ANG1005, a conjugate between paclitaxel and the new brain delivery vector Angiopep-2. Br J Pharmacol 155:185–197

Regina A et al (2015) ANG4043, a novel brain-penetrant peptide-mAb conjugate, is efficacious against HER2-positive intracranial tumors in mice. Mol Cancer Ther 14:129–140

Reuters Health News N, (2019)

Rhea EM, Rask-Madsen C, Banks WA (2018) Insulin transport across the blood-brain barrier can occur independently of the insulin receptor. J Physiol 596:4753–4765

Ribecco-Lutkiewicz M et al (2018) A novel human induced pluripotent stem cell blood-brain barrier model: applicability to study antibody-triggered receptor-mediated transcytosis. Sci Rep 8:1873

Rothe C, Skerra A (2018) Anticalin((R)) proteins as therapeutic agents in human diseases. BioDrugs 32:233–243

Rotman M et al (2015) Enhanced glutathione PEGylated liposomal brain delivery of an anti-amyloid single domain antibody fragment in a mouse model for Alzheimer's disease. J Control Release 203:40–50

Rousselle C et al (2001) Enhanced delivery of doxorubicin into the brain via a peptide-vector-mediated strategy: saturation kinetics and specificity. J Pharmacol Exp Ther 296:124–131

Rousselle C et al (2003) Improved brain uptake and pharmacological activity of dalargin using a peptide-vector-mediated strategy. J Pharmacol Exp Ther 306:371–376

Rubenstein JL et al (2003) Rituximab therapy for CNS lymphomas: targeting the leptomeningeal compartment. Blood 101:466–468

Ryman JT, Meibohm B (2017) Pharmacokinetics of monoclonal antibodies. CPT Pharmacometrics Syst Pharmacol 6:576–588

Sade H et al (2014) A human blood-brain barrier transcytosis assay reveals antibody transcytosis influenced by pH-dependent receptor binding. PLoS One 9:e96340

Santi M et al (2017) Rational design of a transferrin-binding peptide sequence tailored to targeted nanoparticle internalization. Bioconjug Chem 28:471–480

Sarkar G, Curran GL, Sarkaria JN, Lowe VJ, Jenkins RB (2014) Peptide carrier-mediated non-covalent delivery of unmodified cisplatin, methotrexate and other agents via intravenous route to the brain. PLoS One 9:e97655

Schwarze SR, Ho A, Vocero-Akbani A, Dowdy SF (1999) In vivo protein transduction: delivery of a biologically active protein into the mouse. Science 285:1569–1572

Sehlin D et al (2016) Antibody-based PET imaging of amyloid beta in mouse models of Alzheimer's disease. Nat Commun 7:10759

Sehlin D, Fang XT, Meier SR, Jansson M, Syvanen S (2017) Pharmacokinetics, biodistribution and brain retention of a bispecific antibody-based PET radioligand for imaging of amyloid-beta. Sci Rep 7:17254

Sevigny J et al (2016) The antibody aducanumab reduces Abeta plaques in Alzheimer's disease. Nature 537:50–56

Shabanpoor F et al (2017) Identification of a peptide for systemic brain delivery of a morpholino oligonucleotide in mouse models of spinal muscular atrophy. Nucleic Acid Ther 27:130–143

Shen J et al (2011) Poly(ethylene glycol)-block-poly(D,L-lactide acid) micelles anchored with angiopep-2 for brain-targeting delivery. J Drug Target 19:197–203

Singh N, Ecker G (2018) Insights into the structure, function, and ligand discovery of the large neutral amino acid transporter 1, LAT1. Int J Mol Sci 19

Smith J, GM Cardiff University, IMI COMPACT, unpublished results

Smith MW, Gumbleton M (2006) Endocytosis at the blood-brain barrier: from basic understanding to drug delivery strategies. J Drug Target 14:191–214

Sonoda H et al (2018) A blood-brain-barrier-penetrating anti-human transferrin receptor antibody fusion protein for neuronopathic Mucopolysaccharidosis II. Mol Ther 26:1366–1374

Spencer BJ, Verma IM (2007) Targeted delivery of proteins across the blood-brain barrier. Proc Natl Acad Sci U S A 104:7594–7599

Stanimirovic D, Kemmerich K, Haqqani AS, Farrington GK (2014) Engineering and pharmacology of blood-brain barrier-permeable bispecific antibodies. Adv Pharmacol 71:301–335

Stanimirovic DB, Sandhu JK, Costain WJ (2018) Emerging technologies for delivery of biotherapeutics and gene therapy across the blood-brain barrier. BioDrugs 32:547–559

Staquicini FI et al (2011) Systemic combinatorial peptide selection yields a non-canonical iron-mimicry mechanism for targeting tumors in a mouse model of human glioblastoma. J Clin Invest 121:161–173

Stefano JE et al (2009) In vitro and in vivo evaluation of a non-carbohydrate targeting platform for lysosomal proteins. J Control Release 135:113–118

Stocki PWK, Jacobsen CLM, et al (2019) High efficiency blood-brain barrier transport using a VNAR targeting the transferrin receptor. BioRxiv preprint first posted online October 28, 2019

Stojanov K et al (2012) In vivo biodistribution of prion- and GM1-targeted polymersomes following intravenous administration in mice. Mol Pharm 9:1620–1627

Sumbria RK, Hui EK, Lu JZ, Boado RJ, Pardridge WM (2013) Disaggregation of amyloid plaque in brain of Alzheimer's disease transgenic mice with daily subcutaneous administration of a tetravalent bispecific antibody that targets the transferrin receptor and the Abeta amyloid peptide. Mol Pharm 10:3507–3513

Sun J, Boado RJ, Pardridge WM, Sumbria RK (2019) Plasma pharmacokinetics of high-affinity transferrin receptor antibody-erythropoietin fusion protein is a function of effector attenuation in mice. Mol Pharm 16:3534–3543

Syvanen S et al (2017) A bispecific Tribody PET radioligand for visualization of amyloid-beta protofibrils - a new concept for neuroimaging. NeuroImage 148:55–63

Tang J et al (2019) A stabilized retro-inverso peptide ligand of transferrin receptor for enhanced liposome-based hepatocellular carcinoma-targeted drug delivery. Acta Biomater 83:379–389

Thom G et al (2018) Enhanced delivery of Galanin conjugates to the brain through bioengineering of the anti-transferrin receptor antibody OX26. Mol Pharm 15:1420–1431

Tosi G et al (2007) Targeting the central nervous system: in vivo experiments with peptide-derivatized nanoparticles loaded with Loperamide and Rhodamine-123. J Control Release 122:1–9

Tosi G et al (2010) Sialic acid and glycopeptides conjugated PLGA nanoparticles for central nervous system targeting: in vivo pharmacological evidence and biodistribution. J Control Release 145:49–57

Tosi G et al (2012) Can leptin-derived sequence-modified nanoparticles be suitable tools for brain delivery? Nanomedicine (Lond) 7:365–382

Triguero D, Buciak J, Pardridge WM (1990) Capillary depletion method for quantification of blood-brain barrier transport of circulating peptides and plasma proteins. J Neurochem 54:1882–1888

Uchino H et al (2002) Transport of amino acid-related compounds mediated by L-type amino acid transporter 1 (LAT1): insights into the mechanisms of substrate recognition. Mol Pharmacol 61:729–737

Ulapane KR, Kopec BM, Siahaan TJ (2019) Improving in vivo brain delivery of monoclonal antibody using novel cyclic peptides. Pharmaceutics 11

Ulbrich K, Hekmatara T, Herbert E, Kreuter J (2009) Transferrin- and transferrin-receptor-antibody-modified nanoparticles enable drug delivery across the blood-brain barrier (BBB). Eur J Pharm Biopharm 71:251–256

van Dyck CH (2018) Anti-amyloid-beta monoclonal antibodies for Alzheimer's disease: pitfalls and promise. Biol Psychiatry 83:311–319

van Rooy I, Mastrobattista E, Storm G, Hennink WE, Schiffelers RM (2011) Comparison of five different targeting ligands to enhance accumulation of liposomes into the brain. J Control Release 150:30–36

Vasquez KO, Casavant C, Peterson JD (2011) Quantitative whole body biodistribution of fluorescent-labeled agents by non-invasive tomographic imaging. PLoS One 6:e20594

Vitalis TZ, Gabathuler R (2014, October 30) Fragments of p97 and uses thereof. US 2014/0322132 A1

Weber F et al (2018) Brain shuttle antibody for Alzheimer's disease with attenuated peripheral effector function due to an inverted binding mode. Cell Rep 22:149–162

Webster CI, Stanimirovic DB (2015) A gateway to the brain: shuttles for brain delivery of macromolecules. Ther Deliv 6:1321–1324

Webster CI et al (2016) Brain penetration, target engagement, and disposition of the blood-brain barrier-crossing bispecific antibody antagonist of metabotropic glutamate receptor type 1. FASEB J 30:1927–1940

Webster CI et al (2017) Enhanced delivery of IL-1 receptor antagonist to the central nervous system as a novel anti-transferrin receptor-IL-1RA fusion reverses neuropathic mechanical hypersensitivity. Pain 158:660–668

Wei X et al (2015) A D-peptide ligand of nicotine acetylcholine receptors for brain-targeted drug delivery. Angew Chem Int Ed Engl 54:3023–3027

Wen J et al (2019) Sustained delivery and molecular targeting of a therapeutic monoclonal antibody to metastases in the central nervous system of mice. Nat Biomed Eng 3:706–716

Wu C et al (2007) Simultaneous targeting of multiple disease mediators by a dual-variable-domain immunoglobulin. Nat Biotechnol 25:1290–1297

Wu H et al (2012) A novel small Odorranalectin-bearing cubosomes: preparation, brain delivery and pharmacodynamic study on amyloid-beta(2)(5)(−)(3)(5)-treated rats following intranasal administration. Eur J Pharm Biopharm 80:368–378

Xia H et al (2012) Penetratin-functionalized PEG-PLA nanoparticles for brain drug delivery. Int J Pharm 436:840–850

Yadav DB et al (2017) Widespread brain distribution and activity following i.c.v. infusion of anti-beta-secretase (BACE1) in nonhuman primates. Br J Pharmacol 174:4173–4185

Yao H et al (2015) Enhanced blood-brain barrier penetration and glioma therapy mediated by a new peptide modified gene delivery system. Biomaterials 37:345–352

Yu YJ et al (2011) Boosting brain uptake of a therapeutic antibody by reducing its affinity for a transcytosis target. Sci Transl Med 3:84ra44

Yu YJ et al (2014) Therapeutic bispecific antibodies cross the blood-brain barrier in nonhuman primates. Sci Transl Med 6:261ra154

Zhan C et al (2011) Micelle-based brain-targeted drug delivery enabled by a nicotine acetylcholine receptor ligand. Angew Chem Int Ed Engl 50:5482–5485

Zhang Y, Pardridge WM (2001) Mediated efflux of IgG molecules from brain to blood across the blood-brain barrier. J Neuroimmunol 114:168–172

Zhou QH, Boado RJ, Lu JZ, Hui EK, Pardridge WM (2010) Monoclonal antibody-glial-derived neurotrophic factor fusion protein penetrates the blood-brain barrier in the mouse. Drug Metab Dispos 38:566–572

Zhou QH, Boado RJ, Hui EK, Lu JZ, Pardridge WM (2011) Chronic dosing of mice with a transferrin receptor monoclonal antibody-glial-derived neurotrophic factor fusion protein. Drug Metab Dispos 39:1149–1154

Zuchero YJ et al (2016) Discovery of novel blood-brain barrier targets to enhance brain uptake of therapeutic antibodies. Neuron 89:70–82

Chapter 3
Brain Delivery of Therapeutics via Transcytosis: Types and Mechanisms of Vesicle-Mediated Transport Across the BBB

Arsalan S. Haqqani and Danica B. Stanimirovic

Abstract Brain delivery of therapeutic antibodies and biologics is restricted due to the presence of the blood-brain barrier (BBB). However, their delivery can be improved with the use of "carrier" antibodies that target receptors on the luminal surface of the BBB which initiate a process termed receptor-mediated transcytosis (RMT). This review describes key steps and transcellular pathways various BBB-crossing antibodies undertake to deliver therapeutic cargos into the brain via RMT. The pathway is initiated with the receptor-mediated endocytosis through clathrin- and/or caveolin-dependent or independent pathways. Once internalized the antibodies are routed to various endosomal compartments where decisions are made regarding their fate during endosomal protein sorting process. During this process antibodies with specific attributes will be either discarded and degraded in lysosomes or rerouted into compartments destined for release on the abluminal surface of the brain endothelial cells. Different RMT receptors may engage different shuttling pathways between the luminal and abluminal sides of the BBB. Based on this knowledge, antibodies can be engineered to add attributes that facilitate preferential routing through pathways that result in enhanced BBB crossing.

Keywords Early endosomes · Multivesicular bodies · Exosomes · Clathrin · Caveolin · Receptor mediated · Endocytosis · Exocytosis · Transcytosis · Late endosomes · Lysosomes

A. S. Haqqani · D. B. Stanimirovic (✉)
Human Health Therapeutics Research Centre, National Research Council Canada, Ottawa, ON, Canada
e-mail: Arsalan.Haqqani@nrc-cnrc.gc.ca; danica.stanimirovic@nrc-cnrc.gc.ca

© American Association of Pharmaceutical Scientists 2022
E. C. M. de Lange et al. (eds.), *Drug Delivery to the Brain*, AAPS Advances in the Pharmaceutical Sciences Series 33, https://doi.org/10.1007/978-3-030-88773-5_3

Abbreviations

BBB	blood-brain barrier
BEC	brain endothelial cells
CavE	caveolin-mediated endocytosis
CCP	clathrin-coated pit
CCV	clathrin-coated vesicle
CIE	caveolin- and clathrin-independent endocytosis
CLIC	clathrin-independent carriers
CME	clathrin-mediated endocytosis
CNS	central nervous system
EE	early endosome
EV	extracellular vesicle
FcRn	neonatal Fc receptor
GEEC	GPI-anchored protein-enriched endocytic compartments
GPI	glycosylphosphatidylinositol
IGF1R	insulin-like growth factor receptor 1
ILV	intraluminal vesicle
Lat1	large neutral amino-acid transporter CD98
LE	late endosome
MVB	multivesicular bodies
RME	receptor-mediated endocytosis
RMT	receptor-mediated transcytosis
RV	recycling vesicle
TfR	transferrin receptor
TGN	trans-Golgi network

3.1 Introduction

Therapeutic antibodies have emerged as a novel class of targeted and efficacious biopharmaceuticals, supported by the advancements made in production and downstream processing technologies (Schiel et al. 2014; Ecker et al. 2015). However, the development of antibody therapeutics for diseases of the central nervous system (CNS) remains challenging, because access of therapeutic antibodies to the brain tissue is highly restricted by a tightly sealed layer of endothelial cells in brain microvessels that form the blood-brain barrier (BBB). Improved delivery into the brain can be achieved by using BBB carrier antibodies that bind to receptors expressed on the luminal surface of brain endothelial cells (BEC), shuttle to, and release at the abluminal side in a process termed receptor-mediated transcytosis (RMT). These BBB-crossing antibodies can be engineered into various formats of bi- or multi-specific antibodies where the BBB carrier "arm" enables delivery of the therapeutic antibody "arm" to its target within the brain (Stanimirovic et al. 2014).

Whereas enhanced brain delivery and pharmacological actions on brain targets have been shown for several BBB carriers in experimental animal models, the knowledge of key transcellular pathways they engage while translocating from the luminal to the abluminal side of BECs is still sparse. Further understanding of intracellular compartments and molecular networks BBB-crossing antibodies mobilize during transcytosis is necessary to inform antibody engineering that favor more efficient release pathways.

In this chapter, we describe details of some of the known and emerging pathways involved in the RMT of BBB-crossing antibodies against different BBB receptors.

3.2 Receptor-Mediated Transcytosis

RMT is a multistep process that involves receptor-mediated endocytosis (RME) of macromolecules at one surface of a polarized cell, followed by their endosomal sorting, and eventual exocytosis at another surface (usually the opposite side) of the cell. Naturally occurring macromolecules utilize the RMT process to bypass various physiological barriers in the body. The informative examples include transferrin and insulin proteins that engage their respective brain endothelial cell receptors, transferrin receptor (TfR), and insulin receptor (IR), to gain access to brain parenchyma via a transcellular transport. As a result, RMT receptors are attractive targets to develop molecular Trojan horses for delivery of macromolecule therapeutics across the BBB. Antibodies and peptides to several RMT receptors (Table 3.1) have been developed including various antibody formats against TfR (Pardridge et al. 1991; Yu et al. 2011, 2014; Niewoehner et al. 2014), humanized IgG against IR (Coloma

Table 3.1 Mechanisms of endocytosis, endosomal trafficking, and exocytosis of BBB crossing of antibodies targeting RMT receptors

RMT receptor	Endocytosis	Endosomal trafficking	Exocytosis
TfR	Predominantly CME. CavE for receptor recycling	EE and MVB (low affinity or monovalent ligand); LE, lysosomes (high affinity or bivalent ligand)	Sorting tubules, recycling vesicles, MVB/exosomes, others
IR	Both CME and CavE	EE and MVB	MVB/exosomes, others
LRP1	Both CME and CavE	EE and MVB	MVB/exosomes, others
TMEM30A complex	Predominantly CME	EE and MVB (enhanced for fc-containing ligands)	Recycling vesicles, MVB/exosomes, others
CD98	Likely CLIC/GEEC	Unknown	Unknown
IGF1R	CME	EE and MVB (monovalent single-domain antibodies)	MVB/exosomes, others
GLUT1	Unknown	Unknown	Unknown

et al. 2000; Boado et al. 2010), antibodies against the heavy subunit of the large neutral amino-acid transporter CD98 (Lat1) (Zuchero et al. 2016), LRP1-targeting Angiopep2 polypeptide (Xin et al. 2011), species cross-reactive camelid single-domain antibody FC5 that binds a glycosylated epitope of TMEM30A complex (Abulrob et al. 2005; Stanimirovic et al. 2014; Farrington et al. 2014; Webster et al. 2016), and humanized camelid antibodies against IGF1R (Stanimirovic et al. 2017; Ribecco-Lutkiewicz et al. 2018). To better understand how these carriers cross the BBB, we need to dissect various steps involved in the RMT pathway, namely, endocytosis, endosomal sorting, and exocytosis.

3.2.1 Endocytosis

Endocytosis is the uptake of proteins, lipids, extracellular ligands, and soluble molecules, such as nutrients, from the cell surface into the cell interior by endocytic vesicles. While small molecules are absorbed into cells through passive diffusion or transporter-mediated pathways, most macromolecules enter the BBB through endocytosis. The main types of endocytosis include macropinocytosis and micropinocytosis; the latter further distinguishes clathrin-mediated endocytosis (CME), caveolin-mediated endocytosis (CavE), and caveolin- and clathrin-independent endocytosis (CIE). A majority of anti-RMT receptor antibodies have been shown to engage the CME pathway (also traditionally referred to as the RME pathway) to enter the BBB, although other pathways could be engaged through various antibody/ligand displays. The graphical depiction of various endocytosis pathways described in more detail in the subsequent sections is shown in Fig. 3.1.

Macropinocytosis Macropinocytosis is a regulated form of endocytosis that permits non-selective internalization of solute molecules, nutrients, and antigens from extracellular fluids. It is an actin-dependent process initiated from surface membrane ruffles that give rise to large endocytic vesicles of 200–5000 nm in size, known as macropinosomes (Recouvreux and Commisso 2017). The macropinocytosis route is thought to be an effective mechanism for delivery of natural or synthetic particles such as exosomes and nanoparticles, typically ranging in size between 50 and 300 nm and containing plasmid DNA, siRNA, or proteins as payloads (Itakura et al. 2015; Ha et al. 2016; Chen et al. 2016a; Desai et al. 2019). Although smaller macromolecules such as antibodies present in the extracellular fluids may randomly internalize into cells during the macropinocytosis of larger particles, there is a lack of evidence for selective (receptor-mediated) uptake of antibodies via this pathway at the BBB (Itakura et al. 2015; Kähäri et al. 2019). On the other hand, since exosomes may utilize macropinocytosis as one way of entering the BBB (Chen et al. 2016a) and display/contain several RMT receptors (Haqqani et al. 2013), anti-RMT receptor antibodies bound to exosomes may also enter the BBB via the macropinocytosis pathway.

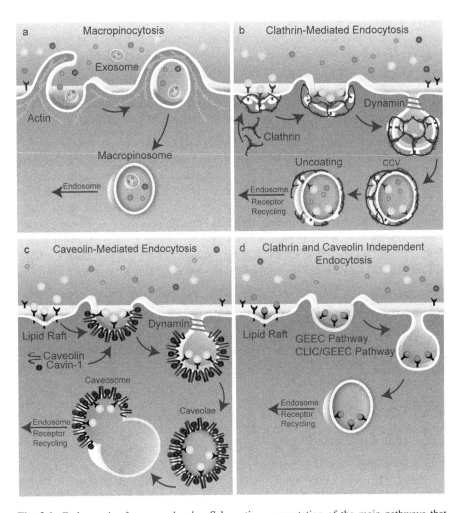

Fig. 3.1 *Endocytosis of macromolecules.* Schematic representation of the main pathways that macromolecules (such as antibodies) can undertake to enter cells. (**a**) Macromolecules present in the extracellular fluids may internalize randomly during the macropinocytosis of larger particles or as bound to exosomes and give rise to endocytic vesicles called macropinosomes. (**b**) Clathrin-mediated endocytosis involves binding of macromolecules to their receptors followed by formation of clathrin-coated pits that bud into endocytic vesicles called clathrin-coated vesicles (CCV), taking in both the receptor and the bound macromolecule. (**c**) Caveolin-mediated endocytosis process involves formation of cave-like surface invagination following macromolecule-receptor interaction that internalizes into endocytic vesicles called caveolae. (**d**) Endocytosis that neither involves clathrin nor caveolin mechanisms usually occurs via the formation of flotillin-regulated lipid rafts resulting in endocytic vesicles called clathrin-independent carriers (CLIC) or GPI-anchored protein-enriched endocytic compartments (GEEC). Once internalized via these endocytic pathways, these vesicles are routed to various endosomes for further sorting

Clathrin-Mediated Endocytosis CME is the most extensively studied and best understood type of endocytosis. It is also the main pathway for RME because the process is activated when a ligand binds to its receptor on the cell surface. CME itself is a multistep process that starts, following receptor activation, with the formation of clathrin-coated pits (CCPs) on the inner surface of the plasma membrane and involves recruitment of a large endocytic protein machinery, consisting of clathrin and over 50 additional cytosolic proteins. The pit then buds into endocytic vesicle of 85–150 nanometer in diameter called clathrin-coated vesicle, taking in both the receptor and the bound ligand. The vesicle then undergoes un-coating and fuses with early endosomes to release its contents (Conner and Schmid 2003).

Known RMT receptors and BBB-crossing antibodies against these receptors have been shown to internalize primarily through the CME pathway. TfR, the most studied RMT receptor, has been shown to co-localize with clathrin pits/protein by a variety of methods, including immunochemistry, live imaging, subcellular fractionation, and proteomics (Liu et al. 2010; Mayle et al. 2012; Villaseñor et al. 2017; Haqqani et al. 2018a, b). However, the BBB crossing efficiency of TfR antibodies varies depending on their design and affinity; for example, high-affinity bivalent TfR antibodies show poor exocytosis and abluminal release, whereas medium-affinity and monovalent TfR antibodies demonstrate efficient transcytosis and improved brain exposure (Niewoehner et al. 2014; Bien-Ly et al. 2014; Webster et al. 2017; Thom et al. 2018b; Haqqani et al. 2018b). Interestingly, immunofluorescence and live imaging demonstrated that both a weak and a strong BBB-crossing anti-TfR antibodies (bivalent dFab and monovalent sFab, respectively) co-localized with clathrin protein (Sade et al. 2014; Villaseñor et al. 2017). Similarly, bivalent anti-TfR OX26 antibodies of varying affinities and BBB-crossing efficiencies were all shown to co-localize with clathrin fractions using targeted quantitative mass spectrometry after subcellular fractionation of the rat brain endothelial cells (Haqqani et al. 2018b). These studies collectively suggest that the initial step of internalization through CME is common for all TfR antibodies regardless of their transcytosing efficiency, which is likely determined by the subsequent differential sorting through different intracellular routes.

IR has also been shown to co-localize with CME pathway by electron microscopic autoradiography in combination with inhibitors of CCP formation (Fan et al. 1982; Paccaud et al. 1992). However, IR may also internalize via non-CME pathways (McClain and Olefsky 1988; Gustavsson et al. 1999; Fagerholm et al. 2009). Similarly, Angiopep2, a polypeptide shown to cross BBB likely by engaging LRP1, was shown to use both CME and non-CME pathways. An uptake of the fluorescently labeled Angiopep2 into BECs was only moderately reduced in the presence of inhibitors of CCP formation (Xin et al. 2011). FC5, a BBB-crossing single-domain antibody engaging RMT receptor complex containing TMEM30A, was shown to internalize via clathrin-coated vesicles, blocked by inhibitors of CME pathway (Abulrob et al. 2005); in addition, both the receptor and the antibody co-localized with clathrin fractions (Abulrob et al. 2005; Haqqani et al. 2018a) by immunostaining and quantitative mass spectrometry.

Collectively these studies suggest that the CME pathway is the most common route that RMT receptors and their antibodies take to enter cells via endocytosis.

Caveolin-Mediated Endocytosis Caveolae are usually defined as small cave-like surface invaginations of 50–100 nm in diameter and have been shown to mediate vesicular transport and cell signaling (Sprenger et al. 2006). Caveolae are not present in all cell types but are found abundantly in ECs and aid in regulating numerous endothelial functions such as transcytosis, vascular permeability, and angiogenesis and can serve as docking sites for glycolipids and GPI-linked proteins, as well as for various receptors and signaling molecules (Sprenger et al. 2006). Caveolin-1, the main protein component of these structures, functions as a scaffolding protein and as a potential cholesterol sensor, regulating raft polymerization and lipid trafficking (Pohl et al. 2004; Song et al. 2007). The CavE pathway has been implicated in BBB transcytosis of IR and LRP1 ligands. In a series of experiments using cell fractionation, western blotting, and immunoprecipitation (Fagerholm et al. 2009), IR internalization was shown to occur via CavE pathway and to be insensitive to inhibitors of CCP formation. Similarly, cellular uptake of the fluorescently labeled anti-LRP1 polypeptide Angiopep2 was reduced by >70% in the presence of inhibitors of caveolae (Xin et al. 2011).

Interestingly, while anti-TfR antibodies have been shown to use the CME pathway for internalization/initialization of the RMT process, several studies have demonstrated the role of caveolin in recycling of various receptors, including TfR, on the apical side of polarized epithelial cells (Pol et al. 1999; Gagescu et al. 2000; Hansen et al. 2003; Lapierre et al. 2007; Leyt et al. 2007). The receptor recycling to the apical side is an essential step in maintaining their levels at the luminal membranes in order to allow continuous entry and shuttling of ligands through polarized cells.

CavE pathway has also been implicated in the transcytosis of other macromolecules such as lipids, likely regulated by a protein called major facilitator super family domain containing 2a (Mfsd2a). Ben-Zvi and co-workers identified Mfsd2a in a BBB-specific gene screen and demonstrated that Mfsd2a(−/−) mice have a leaky BBB with a dramatic increase in CNS-endothelial-cell vesicular "bulk" transcytosis from embryonic stages through to adulthood (Ben-Zvi et al. 2014). Furthermore, through unbiased lipidomic analysis in Mfsd2a transgenic mice, they demonstrated that Mfsd2a may act by suppressing lipid transcytosis likely via downregulation of caveola formation in CNS endothelial cells (Andreone et al. 2017).

Caveolin and Clathrin-Independent Endocytosis (CIE) Macromolecule endocytosis has also been shown to occur through membranes that do not contain either clathrin or caveolin protein. The molecular understanding of the steps involved in the CIE pathway is still in its infancy relative to the vast information known for CME and CavE pathways. The main CIE mechanism that has emerged is the clathrin-independent carrier (CLIC) pathway, also known as the glycosylphosphatidylinositol (GPI)-anchored protein-enriched endocytic compartments (GEEC). The CLIC/GEEC pathway internalizes GPI-anchored proteins, CD44, and some integrins as well as large volumes of fluid and extracellular material that do not have surface receptors (Ferreira and Boucrot 2018). The endocytosis process likely involves a formation of lipid rafts that are regulated by scaffolding protein flotillins, which are believed to stabilize lipid-raft microdomains in phagocytic, caveolin, and non-caveolin-containing membranes (Dermine et al. 2001; Vercauteren et al. 2011). However, there is limited evidence of RMT receptors (or their antibodies) internalizing via the CLIC/GEEC pathway. While both IR and LRP1 have been shown to co-localize with flotillins (Roura et al. 2014; Boothe et al. 2016), the endocytosis is more likely occurring via the CavE pathway as discussed above. A glycoprotein CD98 (SLC3A2) which hetero-dimerizes with SLC7A5 to form large neutral amino acid transporter LAT1 highly enriched in the BBB has been shown recently to shuttle anti-CD98 antibodies into the brain **in vivo** (Zuchero et al. 2016). This receptor likely utilizes the CLIC/GEEC internalization pathway since it is a GPI-anchored protein. In fact, CD98 has been shown to internalize via CIE pathway with novel downstream sorting mechanisms that may be independent of the widely known sorting at the EE (Eyster et al. 2009).

3.2.2 Sorting Through the Endosomes

Once receptors and their associated macromolecules are internalized via one of the endocytosis pathways, they are routed to various endosomes where decisions are made regarding their fate during processes known as endosomal protein sorting, graphically shown in Figs. 3.2 and 3.3. The main sorting stations in the cells include the early and late endosomes (Scott et al. 2014).

Early Endosome All internalized vesicles are first fused to a common early endosome (EE), which functions as the first key sorting station in the cell. Here the cell makes a major decision: Are the cargo and membrane components of the vesicles worth keeping or should they be sent to late endosome (LE)/lysosome for degradation? If the cargo is to be degraded, it goes through the process of early-to-late endosome maturation. This involves the cargo being concentrated in specific regions of the EE membranes that are pinched off to form endosomes that mature into multivesicular bodies (MVBs) and eventually fuse with LEs (Scott et al. 2014; van Weering and Cullen 2014). However, if the cargo does not need to be degraded, it is concentrated in a network of *tubular* EE subdomains leading to the formation of sorting tubules (Maxfield and McGraw 2004), which are recycled back to the plasma membranes or to the biosynthetic pathway at the level of the trans-Golgi network (TGN). The events in the sorting processes in EEs have been studied in detail at the molecular level and shown to involve an array of protein complexes that direct trafficking events to the appropriate destination (see reviews Scott et al. 2014; van Weering and Cullen 2014; Naslavsky and Caplan 2018). There is strong evidence that RMT receptors, including TfR (Sade et al. 2014; Niewoehner et al. 2014; Bien-Ly et al. 2014; Haqqani et al. 2018b), IR (Hunker et al. 2006), LRP1 (Tian et al. 2015; Haqqani et al. 2018a), and TMEM30A (Haqqani et al. 2018a), predominantly co-localize with EEs, especially when incubated with their respective antibodies that are strong BBB crossers. It is still not well understood how the cell decides whether a specific RMT receptor or its bound ligand should be sent for

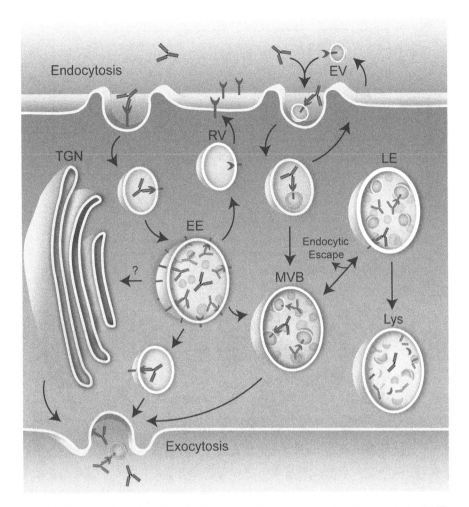

Fig. 3.2 *Endosomal sorting of antibodies during the receptor-mediated transcytosis (RMT).* Shown is a schematic depiction of intracellular trafficking pathways triggered by anti-RMT receptor antibodies at the BBB. Once the antibody binds to its receptor, expressed on the luminal membranes, it triggers internalization of the antibody-RMT receptor complex into endocytic vesicles via one of the endocytosis pathways, including extracellular vesicle (EV)- based endocytosis. While most endocytic vesicles fuse to early endosomes (EE), others (such as those containing EVs) may fuse to multivesicular bodies (MVB). In EE, it is decided whether the antibody will recycle back to the luminal side, be degraded, or undergo exocytosis at the abluminal side. Typically recycling vesicles (RV) will recycle the RMT receptor (with or without antibody) back to the luminal side, whereas MVB will receive cargo from EE for degradation or exocytosis. For degradation, the cargo is sent to late endosomes (LE) and lysosomes. Exocytosis may occur through multiple routes from EE: directly from vesicles (e.g., sorting tubules), via trans-Golgi network (TGN), or via a direct fusion of MVBs with abluminal membrane

degradation or rerouted for exocytosis, although some factors that may favor BBB cells to exocytose rather than degrade antibodies have been identified and will be discussed in the next section.

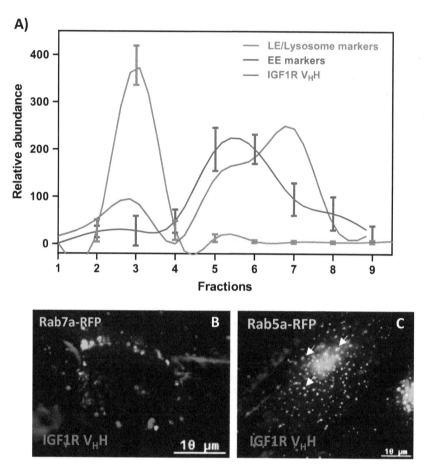

Fig. 3.3 IGF1R V_HH co-localization with endocytic vesicles in BEC. (**a**) Co-localization of the BBB-crossing IGF1R V_HH antibody with markers of early endosomes (EE) and late endosomes (LE)/lysosomes in subcellular fractions of SV-ARBEC cells as determined by mass spectrometry. Graph shows relative levels of the antibody, EE markers (e.g., Rab5a, Eea1), and LE/lysosome markers of late endosomes (e.g, Rab7, Lamp1, Lamp2) in each cellular fraction. (**b**) Co-immunofluorescence detection of IGF1R V_HH antibody and Rab5a and Rab7a markers

Multivesicular Bodies MVBs are spherical endosomal organelles containing a number of intraluminal vesicles (ILVs) formed by inward budding of the limiting membrane into the endosomal lumen (Zhang et al. 2019). MVBs have traditionally been considered intermediate endosomes between EE and LE, as they are formed from maturation of EE-released ILV-containing vacuoles that may eventually fuse with LE to deliver the content for degradation (van Weering and Cullen 2014; Naslavsky and Caplan 2018). MVBs are now known to have multiple subpopulations (van Niel et al. 2001; White et al. 2006; Tauro et al. 2013; Chen et al. 2016b; Haqqani et al. 2018a) and to be involved in numerous additional endocytic and trafficking functions including biogenesis and routing of ILVs to and from the plasma membrane to membranes of other organelles (Von Bartheld and Altick 2011; Colombo et al. 2014). The ILVs when released extracellularly are referred to as exosomes, which have recently emerged as natural therapeutic-delivery vehicles; a number of studies have shown that exosomes can cross the BBB and deliver therapeutics into the brain (Zhuang et al. 2011; Alvarez-Erviti et al. 2011; Chen et al. 2016a; Matsumoto et al. 2017). Through proteomic analysis of exosomes derived from BEC, we found that they are enriched with known RMT receptors including TfR, IR, TMEM30A, and others (Haqqani et al. 2013). We have proposed that a subpopulation of MVBs may play a key role as "transcytosing endosomes" trafficking between the apical and basolateral membranes and helping transport exosome-bound ligands from the luminal to the abluminal side of BEC (Haqqani et al. 2013, 2018a).

Late Endosomes and Lysosomes LE functions as a second trafficking hub in the endosomal system and as a last sorting station in the membrane trafficking cycle to and from lysosomes (Huotari and Helenius 2011; Raposo and Stoorvogel 2013; Bissig and Gruenberg 2014). In fact, live-cell imaging has shown that LEs and lysosomes frequently interact by "kiss-and-run" events and by direct fusion, resulting in the formation of hybrid organelles, in which the degradation of endocytosed macromolecules occurs and from which lysosomes are re-formed. Although LEs and lysosomes can be distinguished by their physical properties and ultrastructure (Scott et al. 2014), two organelles are difficult to differentiate molecularly – both contain highly sialylated membrane proteins LAMP1 and LAMP2 that form a protective glycocalyx lumen against degradative enzymes. Receptors, ligands, and other proteins that need to be downregulated are sorted out of the EE and fused to LE via intraluminal vesicles (Scott et al. 2014). Several studies have shown that higher co-localization of RMT receptors or their ligands with LE markers is associated with their lysosomal degradation at the BBB (Sade et al. 2014; Niewoehner et al. 2014; Haqqani et al. 2018a, b), a mechanism that is considered key for regulating surface expression of RMT receptors. However, not every cargo from LE is sent to lysosomes for degradation, because the LE are empowered to make the last decision; for example, in response to incoming signals via other pathways, LE can divert the cargo to other destinations, including the TGN, MVBs, plasma membrane, or even to cytoplasm via endosomal escape (Huotari and Helenius 2011; Raposo and Stoorvogel 2013; Bissig and Gruenberg 2014; Scott et al. 2014; Tashima 2018). Similar to EE, it is not well understood what regulates LE fusion with lysosomes to enact the final degradation of a specific RMT receptor or its bound ligand.

Antibody Trafficking Through the Endosomes There is a compelling body of evidence showing that poor BBB-crossing anti-RMT receptor antibodies are targeted for degradation through the LEs and lysosomes, while efficient BBB-crossing antibodies predominantly traffic through the EEs. Comparing intracellular localization of a poor BBB-crossing (high-affinity) anti-TfR[A] antibody and an efficient BBB-crossing (low-affinity) anti-TfR[D] antibody using immunofluorescence studies, Watts and co-workers showed that while both antibodies co-localize with EE marker

EEA1, poorly crossing anti-TfR^A showed more co-localization with lysosomal marker LAMP1 compared with efficient crosser anti-TfR^D antibody (Bien-Ly et al. 2014). Similarly, using high-resolution imaging, Freskgård and co-workers demonstrated that a non-BBB-crossing (bivalent) anti-TfR dFab antibody is preferentially co-localized with LAMP1, compared to an efficient BBB-crossing (monovalent) anti-TfR sFab antibody (Niewoehner et al. 2014). We recently evaluated localization of a number of bivalent anti-TfR OX26 affinity variants showing varying BBB-crossing efficiency in subcellular fractions of the rat brain endothelial cells using both targeted quantitative mass spectrometry and immunofluorescence (Haqqani et al. 2018b). While the parental high-affinity OX26$_5$ along with TfR co-localized with multiple LE and lysosomal markers, the medium-affinity OX26$_{76}$ and OX26$_{108}$ antibodies, along with TfR, routed predominantly into the early/recycling endosomes and demonstrated efficient BBB crossing (Haqqani et al. 2018b).

Additional evidence supporting the relevance of trafficking through the EE for BBB crossing comes from the extensive characterization of species cross-reactive camelid single-domain antibody, FC5 (Tanha et al. 2002; Abulrob et al. 2005; Farrington et al. 2014; Haqqani et al. 2018a). FC5 has been shown to deliver various therapeutic payloads, including peptides and antibodies, to their CNS targets (Farrington et al. 2014; Webster et al. 2016). By examining subcellular distribution of FC5 in rat brain endothelial cells using both targeted quantitative mass spectrometry and immunofluorescence, FC5 enrichment was observed in EEs and a subpopulation of molecularly distinct MVBs with a small proportion being routed to LEs and lysosomes (Haqqani et al. 2018a). Interestingly, FC5 fusion to Fc further enhanced the EE/MVB enrichment, reduced LE/lysosome levels, and increased BBB crossing (Haqqani et al. 2018a). In contrast, a low level of internalized non-BBB-crossing single-domain antibodies, with or without Fc, showed enrichment in LEs/lysosomes and depletion from EEs (Haqqani et al. 2018a).

Similar studies with the BBB-crossing camelid V_HH against insulin-like growth factor receptor 1 (IGF1R) (Stanimirovic et al. 2017; Ribecco-Lutkiewicz et al. 2018) revealed a slightly different routing path. IGF1R V_HH, after internalizing rat BEC via a CEM pathway, co-localized with the high-density, EE marker-containing subcellular fractions, with further enrichment in the higher density fractions, previously identified as a subset of MVBs (Fig. 3.3a; Haqqani et al. 2018a). No co-localization of internalized IGF1R V_HH with the LE marker Rab7a (Fig. 3.3b), and a significant co-localization with Rab5a-containing vesicles (Fig. 3.3c), indicative of EE, was also observed by immunofluorescence detection.

These results collectively strengthen the hypothesis that the lysosomal degradation is a key downstream mechanism by which BECs restrict antibody access to the brain and that BBB-crossing antibodies bypass this pathway and instead follow the EE/MVB route toward exocytosis.

3.2.3 Exocytosis to the Abluminal Side

The last step of the RMT process involves BBB-crossing antibodies exiting the endosomal pathway and being released on the basolateral side of the barrier. This process is probably the least understood among different RMT steps. Molecules in the EEs that do not need to be degraded are concentrated in a network of tubular EE subdomains leading to the formation of sorting tubules, which are destined for the plasma membrane, MVBs, or TGN, thereby avoiding lysosomal degradation (Maxfield and McGraw 2004; Grant and Donaldson 2009). In fact, using live-cell imaging, it was recently shown that an efficient BBB-crosser anti-TfR sFab localized to sorting tubules, whereas non-BBB-crosser anti-TfR dFab had been size-excluded from these tubules due to receptor cross-linking facilitated by a bivalent receptor binding (Villaseñor et al. 2017). Based on these and our own observations, we postulate that these sorting tubules are either (i) recycled back to the plasma membranes, (ii) evolve to ILV-containing vacuoles and fuse to the MVBs, or (iii) fuse to the TGN. Although recycling of vesicles to plasma membrane is usually believed to be back to apical membranes, similar mechanism may unfold for their movement to the basolateral side for transcytosis. Consistent with this assumption, we have shown that the BBB-crossing FC5 antibody co-localizes with recycling and exocytosing MVBs (Haqqani et al. 2018a), which is different to anti-TfR sFab that was found to undergo transcytosis by avoiding receptor cross-linking and lysosomal degradation (Villaseñor et al. 2017). Mechanisms of exocytosis from both MVBs and TGN have been previously described (Jaiswal et al. 2009; Von Bartheld and Altick 2011; Colombo et al. 2014). MVBs may directly fuse with the basolateral membranes and release the RMT receptor-bound antibodies to the abluminal side of the barrier. On the other hand, the TGN is well known to secrete newly synthesized molecules via exocytotic and secretory vesicles which fuse to the plasma membranes and release their content (Jaiswal et al. 2009). Similar mechanisms may also be involved in exocytosis of antibodies via TGN. A summary of reported pathways for endocytosis, trafficking, and exocytosis for antibodies targeting BBB RMT receptors is shown in Table 3.1.

3.3 Antibody Attributes That Favor Transcytosis: Designing more Efficient BBB Carriers

Increasing transcytosis efficiency of carrier antibodies developed against BBB RMT receptors could be accomplished by antibody engineering strategies that direct the antibody into endocytic pathways favoring transcytosis instead of lysosomal degradation. Through TfR and FC5 antibody engineering efforts, several antibody attributes that increase the efficiency of BBB crossing have been identified. Many of these are based on specific structure-function relationships that guide antibody docking and binding to its receptor, whereas some others are based on the intracellular milieu that antibody faces while traveling through endocytic pathways. Some of these factors include ligand-receptor affinity, pH sensitivity of ligand-receptor interactions, antibody valency, Fc format, and antibody position in the construct (Niewoehner et al. 2014; Bien-Ly et al. 2014; Villaseñor et al. 2017; Haqqani et al. 2018a, b). Here we describe evidence that these factors have resulted in increased BBB permeability, although it should be noted that the factors might be receptor specific since different receptors undertake different RMT pathways for transporting ligands across the BBB (Table 3.1).

Ligand-Receptor Affinity A number of studies have demonstrated that manipulating the binding affinity between the carrier antibody and its RMT receptor results in enhanced BBB permeability of the carrier. The strongest evidence exists for anti-TfR antibodies, where several studies have shown that the high-affinity binding to TfR results in receptor cross-linking and lysosomal degradation, whereas a moderate-affinity binding to TfR results in enhanced antibody transcytosis. Watts and co-workers compared BBB crossing of two bispecific antibodies with different binding affinities to TfR, where each antibody had an anti-TfR arm and an anti-BACE1 arm; the low-affinity anti-TfR^D antibody showed a significantly enhanced BBB crossing compared to the high-affinity anti-TfR^A antibody as demonstrated by labeling experiments both in *in vitro* and *in vivo* (Bien-Ly et al. 2014). In addition, live imaging and co-localization experiments demonstrated that high-affinity antibody facilitated degradation of TfR by directing it to lysosomes, resulting in downregulation of TfR in the BBB and reduced brain exposure to a second dose of the BBB-crossing, low-affinity TfR antibody (Bien-Ly et al. 2014). Similarly, in studies with affinity variants of the rat-specific anti-TfR antibody OX26 using a label-free mass spectrometry method that allows simultaneous quantification of antibodies, their receptors, and endosomal markers (Haqqani et al. 2018b), lowering the affinity of OX26 antibody resulted in rerouting of both the TfR and the antibody away from LE and lysosomes and toward the EE/recycling vesicles. OX26 antibodies with affinity range of 70–100 nM displayed a significantly higher BBB transcytosis in a BBB model *in vitro* (Haqqani et al. 2018b), as well as higher brain penetration in animal studies (Thom et al. 2018a), compared to a parental OX26 having affinity of 5 nM. Other studies have been able to similarly improve the BBB penetration of anti-TfR antibodies in different formats by lowering their affinities (Webster et al. 2017; Johnsen et al. 2018; Karaoglu Hanzatian et al. 2018). It is important to note that the optimal affinity range for maximal transcytosis is different for each TfR antibody, likely because each antibody engages different receptor epitopes. Medium-affinity TfR antibodies also show improved serum pharmacokinetics, resulting in longer brain exposure (Yu et al. 2011; Thom et al. 2018b). However, lowering affinities below the optimal range results in poor receptor engagement and low brain exposure (Yu et al. 2011; Thom et al. 2018b). These studies demonstrate that the optimization of binding affinities between the carrier antibody and its RMT receptor may result in improved efficiency of transcytosis and enhanced brain delivery.

Antibody Valency Many membrane receptors exist as dimers either at resting state or they dimerize in response to mono- or bivalent ligand binding (De Meyts et al. 1995; Terrillon and Bouvier 2004; Eckenroth et al. 2011). The latter may result in activation of the receptor, leading to signaling cascades and subsequent physiological effects mediated by the receptor. The latter is not a desirable action for BBB carrier antibodies, which aim not to disturb physiological activation/function of the receptor. Among RMT receptors, TfR, IR, and IGF1R are known to dimerize either at resting state or in response to ligand exposure (De Meyts et al. 1995; Eckenroth et al. 2011). To avoid receptor cross-linking and activation by bivalent antibodies, both monovalent and bivalent antibodies have been developed and tested for TfR, FC5, and IGF1R. Freskgård and co-workers engineered a high-affinity anti-TfR antibody at the C-terminus of an anti-amyloid beta antibody in either a bivalent (dFab) or monovalent (sFab) format (Niewoehner et al. 2014). While the bivalent dFab antibody failed to cross the BBB and led to lysosomal degradation, the monovalent sFab antibody exhibited facilitated BBB crossing, localization in sorting tubules, and reduction of amyloid deposits in a mouse model of Alzheimer's disease (Niewoehner et al. 2014; Villaseñor et al. 2017). Similarly, a monovalent fusion of IGF1R V_HH to Fc resulted in improved BBB transcytosis in vitro, compared to the bivalent IGF1R V_HH-Fc (unpublished observation).

Influence of antibody valency on BBB transcytosis has also been tested for FC5 (Farrington et al. 2014; Haqqani et al. 2018a). TMEM30A, a putative FC5 receptor, is not known to dimerize but is presented as a heteromeric flippase complex of multiple proteins (Wang et al. 2018). When monomeric FC5 V_HH was compared with monovalent FC5Fc or bivalent FC5Fc, the bivalent format showed enhanced BBB permeability in vitro and improved brain exposure and pharmacodynamic effects in vivo (Farrington et al. 2014). Bivalent FC5Fc also displayed stronger partitioning in EE and MVBs in BEC compared to monovalent FC5Fc (Haqqani et al. 2018a). Thus, engineering antibody valency is an important strategy to consider when designing BBB-crossing antibodies, as it could trigger either desired facilitation of receptor traffic or undesired receptor cross-linking, activation, and degradation. These studies also underscore that the nature of receptor-antibody interaction is unique for each antibody-receptor pair and that emerging learnings about factors that facilitate transcytosis cannot be broadly applied to all BBB carriers.

Ligand-Receptor Interaction in Acidic pH It has been observed that soon after internalization, many receptors that need to be recycled are uncoupled from their ligands at acidic pH in different endosomal compartments (such as EEs and MVBs) during the sorting processes (Goldstein et al. 1985; Scott et al. 2014), while the ligand may continue to sort to other destinations. To test whether such phenomenon may also facilitate antibody transcytosis, an anti-TfR antibody with reduced affinity at pH 5.5 was developed; this antibody demonstrated significant transcytosis, while pH-independent antibodies of comparable affinities at pH 7.4 remained associated with intracellular vesicular compartments (Sade et al. 2014). Therefore, another strategy to improve BBB crossing is to develop antibody variants that have different affinity interactions with the RMT receptor at different pHs.

Fc Format While an Fc domain of IgG is known to prolong circulatory half-life of antibodies through binding, internalization, and recycling in endothelial cells mediated by the neonatal Fc receptor (FcRn) (Giragossian et al. 2013), the presence of Fc domain has also been shown to enhance BBB permeability of BBB-crossing FC5 V_HH. When expressed in fusion with the human Fc in either monovalent or bivalent format, FC5 demonstrated improved BBB transcytosis *in vitro*, enhanced CSF levels, and improved pharmacodynamic potency *in vivo* compared to FC5 V_HH without the Fc (Farrington et al. 2014; Haqqani et al. 2018a). While the *in vivo* enhancements were largely due to prolonging of circulatory half-life, the increased BBB transcytosis *in vitro* might be due partially to FcRn-based rescue from intracellular lysosomal degradation (Lencer and Blumberg 2005). Thus, the addition of Fc to single-domain or single-chain antibodies (or non-antibody ligands) against RMT receptors may not only help extend systemic pharmacokinetics but also improve the efficiency of BBB transcytosis.

Antibody Position in the Construct A position of the anti-RMT receptor antibody in the bispecific construct may affect the efficiency of its transcytosis. For example, a placement of the FC5 on the C-terminus of the Fc or an antibody cargo resulted in low BBB transcytosis; however, FC5 fused to the N-terminal of Fc (Farrington et al., 2014) or heavy (or light) chain of an antibody (Webster et al., 2016) retained its ability to shuttle cargo across the BBB, suggesting that the N-terminus of FC5 is important for conformational antigen binding that triggers transcytosis.

3.4 Conclusions

In conclusion, we have described some of the key steps involved in the RMT process and different sorting pathways undertaken by various BBB-crossing antibodies as they "travel" through the BBB. It is apparent that the RMT process is a complex set of cross-communicating pathways comprising of various endocytosing, sorting, and exocytosing sub-pathways. We have assigned individual route(s) to some of the known RMT receptors, which they utilize for transporting ligands across the BBB (Table 3.1). Through better understanding of the RMT of antibodies, several key antibody attributes that facilitate abluminal release have been discovered and engineered to improve their BBB-crossing ability. With discovery of new RMT receptors and development of new carrier antibodies, we believe that these factors may serve as an initial guide for improving brain penetration of bispecific antibody therapeutics.

Acknowledgments We thank Caroline Sodja for designing and drawing the schematics in Figs. 3.1 and 3.2 and Christie E. Delaney and Ewa Baumann for their contributions to Fig. 3.3.

References

Abulrob A, Sprong H, Van Bergen en Henegouwen P, Stanimirovic D (2005) The blood-brain barrier transmigrating single domain antibody: mechanisms of transport and antigenic epitopes in human brain endothelial cells. J Neurochem 95:1201–1214. https://doi.org/10.1111/j.1471-4159.2005.03463.x

Alvarez-Erviti L, Seow Y, Yin H et al (2011) Delivery of siRNA to the mouse brain by systemic injection of targeted exosomes. Nat Biotechnol 29:341–345. https://doi.org/10.1038/nbt.1807

Andreone BJ, Chow BW, Tata A et al (2017) Blood-brain barrier permeability is regulated by lipid transport-dependent suppression of caveolae-mediated transcytosis. Neuron 94:581–594.e5. https://doi.org/10.1016/j.neuron.2017.03.043

Ben-Zvi A, Lacoste B, Kur E et al (2014) Mfsd2a is critical for the formation and function of the blood–brain barrier. Nature 509:507–511. https://doi.org/10.1038/nature13324

Bien-Ly N, Yu YJ, Bumbaca D et al (2014) Transferrin receptor (TfR) trafficking determines brain uptake of TfR antibody affinity variants. J Exp Med 211:233–244. https://doi.org/10.1084/jem.20131660

Bissig C, Gruenberg J (2014) ALIX and the multivesicular endosome: ALIX in wonderland. Trends Cell Biol 24:19–25. https://doi.org/10.1016/j.tcb.2013.10.009

Boado RJ, Hui EK-W, Lu JZ et al (2010) Selective targeting of a TNFR decoy receptor pharmaceutical to the primate brain as a receptor-specific IgG fusion protein. J Biotechnol 146:84–91. https://doi.org/10.1016/j.jbiotec.2010.01.011

Boothe T, Lim GE, Cen H et al (2016) Inter-domain tagging implicates caveolin-1 in insulin receptor trafficking and Erk signaling bias in pancreatic beta-cells. Mol Metab 5:366–378. https://doi.org/10.1016/j.molmet.2016.01.009

Chen CC, Liu L, Ma F et al (2016a) Elucidation of exosome migration across the blood-brain barrier model in vitro. Cell Mol Bioeng 9:509–529. https://doi.org/10.1007/s12195-016-0458-3

Chen Q, Takada R, Noda C et al (2016b) Different populations of Wnt-containing vesicles are individually released from polarized epithelial cells. Sci Rep 6:35562. https://doi.org/10.1038/srep35562

Coloma MJ, Lee HJ, Kurihara A et al (2000) Transport across the primate blood-brain barrier of a genetically engineered chimeric monoclonal antibody to the human insulin receptor. Pharm Res 17:266–274

Colombo M, Raposo G, Théry C (2014) Biogenesis, secretion, and intercellular interactions of exosomes and other extracellular vesicles. Annu Rev Cell Dev Biol 30:255–289. https://doi.org/10.1146/annurev-cellbio-101512-122326

Conner SD, Schmid SL (2003) Regulated portals of entry into the cell. Nature 422:37–44. https://doi.org/10.1038/nature01451

De Meyts P, Ursø B, Christoffersen CT, Shymko RM (1995) Mechanism of insulin and IGF-I receptor activation and signal transduction specificity. Receptor dimer cross-linking, bell-shaped curves, and sustained versus transient signaling. Ann N Y Acad Sci 766:388–401. https://doi.org/10.1111/j.1749-6632.1995.tb26688.x

Dermine JF, Duclos S, Garin J et al (2001) Flotillin-1-enriched lipid raft domains accumulate on maturing phagosomes. J Biol Chem 276:18507–18512. https://doi.org/10.1074/jbc.M101113200

Desai AS, Hunter MR, Kapustin AN (2019) Using macropinocytosis for intracellular delivery of therapeutic nucleic acids to tumour cells. Philos Trans R Soc Lond Ser B Biol Sci 374:20180156. https://doi.org/10.1098/rstb.2018.0156

Eckenroth BE, Steere AN, Chasteen ND et al (2011) How the binding of human transferrin primes the transferrin receptor potentiating iron release at endosomal pH. Proc Natl Acad Sci 108:13089–13094. https://doi.org/10.1073/pnas.1105786108

Ecker DM, Jones SD, Levine HL (2015) The therapeutic monoclonal antibody market MAbs 7:9–14. https://doi.org/10.4161/19420862.2015.989042

Eyster CA, Higginson JD, Huebner R et al (2009) Discovery of new cargo proteins that enter cells through clathrin-independent endocytosis. Traffic 10:590–599. https://doi.org/10.1111/j.1600-0854.2009.00894.x

Fagerholm S, Ortegren U, Karlsson M et al (2009) Rapid insulin-dependent endocytosis of the insulin receptor by caveolae in primary adipocytes. PLoS One 4:e5985. https://doi.org/10.1371/journal.pone.0005985

Fan JY, Carpentier JL, Gorden P et al (1982) Receptor-mediated endocytosis of insulin: role of microvilli, coated pits, and coated vesicles. Proc Natl Acad Sci U S A 79:7788–7791. https://doi.org/10.1073/pnas.79.24.7788

Farrington GK, Caram-Salas N, Haqqani AS et al (2014) A novel platform for engineering blood-brain barrier-crossing bispecific biologics. FASEB J 28:4764–4778. https://doi.org/10.1096/fj.14-253369

Ferreira APA, Boucrot E (2018) Mechanisms of carrier formation during clathrin-independent endocytosis. Trends Cell Biol 28:188–200. https://doi.org/10.1016/j.tcb.2017.11.004

Gagescu R, Demaurex N, Parton RG et al (2000) The recycling endosome of Madin-Darby canine kidney cells is a mildly acidic compartment rich in raft components. Mol Biol Cell 11:2775–2791. https://doi.org/10.1091/mbc.11.8.2775

Giragossian C, Clark T, Piché-Nicholas N, Bowman CJ (2013) Neonatal Fc receptor and its role in the absorption, distribution, metabolism and excretion of immunoglobulin G-based biotherapeutics. Curr Drug Metab 14:764–790

Goldstein JL, Brown MS, Anderson RG et al (1985) Receptor-mediated endocytosis: concepts emerging from the LDL receptor system. Annu Rev Cell Biol 1:1–39. https://doi.org/10.1146/annurev.cb.01.110185.000245

Grant BD, Donaldson JG (2009) Pathways and mechanisms of endocytic recycling. Nat Rev Mol Cell Biol 10:597–608. https://doi.org/10.1038/nrm2755

Gustavsson J, Parpal S, Karlsson M et al (1999) Localization of the insulin receptor in caveolae of adipocyte plasma membrane. FASEB J 13:1961–1971

Ha KD, Bidlingmaier SM, Liu B (2016) Macropinocytosis exploitation by cancers and cancer therapeutics. Front Physiol 7:381. https://doi.org/10.3389/fphys.2016.00381

Hansen GH, Pedersen J, Niels-Christiansen L-L et al (2003) Deep-apical tubules: dynamic lipid-raft microdomains in the brush-border region of enterocytes. Biochem J 373:125–132. https://doi.org/10.1042/BJ20030235

Haqqani AS, Delaney CE, Tremblay T-L, et al (2013) Method for isolation and molecular characterization of extracellular microvesicles released from brain endothelial cells. Fluids Barriers CNS 10:4. https://doi.org/https://doi.org/10.1186/2045-8118-10-4

Haqqani AS, Delaney CE, Brunette E et al (2018a) Endosomal trafficking regulates receptor-mediated transcytosis of antibodies across the blood brain barrier. J Cereb Blood Flow Metab 38:727–740. https://doi.org/10.1177/0271678X17740031

Haqqani AS, Thom G, Burrell M et al (2018b) Intracellular sorting and transcytosis of the rat transferrin receptor antibody OX26 across the blood-brain barrier *in vitro* is dependent on its binding affinity. J Neurochem. https://doi.org/10.1111/jnc.14482

Hunker CM, Kruk I, Hall J et al (2006) Role of Rab5 in insulin receptor-mediated endocytosis and signaling. Arch Biochem Biophys 449:130–142. https://doi.org/10.1016/J.ABB.2006.01.020

Huotari J, Helenius A (2011) Endosome maturation. EMBO J 30:3481–3500. https://doi.org/10.1038/emboj.2011.286

Itakura S, Hama S, Ikeda H et al (2015) Effective capture of proteins inside living cells by antibodies indirectly linked to a novel cell-penetrating polymer-modified protein a derivative. FEBS J 282:142–152. https://doi.org/10.1111/febs.13111

Jaiswal JK, Rivera VM, Simon SM (2009) Exocytosis of post-Golgi vesicles is regulated by components of the endocytic machinery. Cell 137:1308–1319. https://doi.org/10.1016/j.cell.2009.04.064

Johnsen KB, Bak M, Kempen PJ et al (2018) Antibody affinity and valency impact brain uptake of transferrin receptor-targeted gold nanoparticles. Theranostics 8:3416–3436. https://doi.org/10.7150/thno.25228

Kähäri L, Fair-Mäkelä R, Auvinen K et al (2019) Transcytosis route mediates rapid delivery of intact antibodies to draining lymph nodes. J Clin Invest 129:3086–3102. https://doi.org/10.1172/JCI125740

Karaoglu Hanzatian D, Schwartz A, Gizatullin F et al (2018) Brain uptake of multivalent and multi-specific DVD-Ig proteins after systemic administration. MAbs 10:765–777. https://doi.org/10.1080/19420862.2018.1465159

Lapierre LA, Avant KM, Caldwell CM et al (2007) Characterization of immunoisolated human gastric parietal cells tubulovesicles: identification of regulators of apical recycling. Am J Physiol Gastrointest Liver Physiol 292:G1249–G1262. https://doi.org/10.1152/ajpgi.00505.2006

Lencer WI, Blumberg RS (2005) A passionate kiss, then run: exocytosis and recycling of IgG by FcRn. Trends Cell Biol 15:5–9. https://doi.org/10.1016/j.tcb.2004.11.004

Leyt J, Melamed-Book N, Vaerman J-P et al (2007) Cholesterol-sensitive modulation of transcytosis. Mol Biol Cell 18:2057–2071. https://doi.org/10.1091/mbc.e06-08-0735

Liu AP, Aguet F, Danuser G, Schmid SL (2010) Local clustering of transferrin receptors promotes clathrin-coated pit initiation. J Cell Biol 191:1381–1393. https://doi.org/10.1083/jcb.201008117

Matsumoto J, Stewart T, Sheng L et al (2017) Transmission of α-synuclein-containing erythrocyte-derived extracellular vesicles across the blood-brain barrier via adsorptive mediated transcytosis: another mechanism for initiation and progression of Parkinson's disease? Acta Neuropathol Commun 5:71. https://doi.org/10.1186/s40478-017-0470-4

Maxfield FR, McGraw TE (2004) Endocytic recycling. Nat Rev Mol Cell Biol 5:121–132. https://doi.org/10.1038/nrm1315

Mayle KM, Le AM, Kamei DT (2012) The intracellular trafficking pathway of transferrin. Biochim Biophys Acta 1820:264–281. https://doi.org/10.1016/j.bbagen.2011.09.009

McClain DA, Olefsky JM (1988) Evidence for two independent pathways of insulin-receptor internalization in hepatocytes and hepatoma cells. Diabetes 37:806–815. https://doi.org/10.2337/diab.37.6.806

Naslavsky N, Caplan S (2018) The enigmatic endosome – sorting the ins and outs of endocytic trafficking. J Cell Sci 131. https://doi.org/10.1242/jcs.216499

Niewoehner J, Bohrmann B, Collin L et al (2014) Increased brain penetration and potency of a therapeutic antibody using a monovalent molecular shuttle. Neuron 81:49–60. https://doi.org/10.1016/j.neuron.2013.10.061

Paccaud JP, Siddle K, Carpentier JL (1992) Internalization of the human insulin receptor. The insulin-independent pathway. J Biol Chem 267:13101–13106

Pardridge WM, Buciak JL, Friden PM (1991) Selective transport of an anti-transferrin receptor antibody through the blood-brain barrier in vivo. J Pharmacol Exp Ther 259:66–70

Pohl J, Ring A, Ehehalt R et al (2004) Long-chain fatty acid uptake into adipocytes depends on lipid raft function. Biochemistry 43:4179–4187. https://doi.org/10.1021/bi035743m

Pol A, Calvo M, Lu A, Enrich C (1999) The "early-sorting"; endocytic compartment of rat hepatocytes is involved in the intracellular pathway of caveolin-1 (VIP-21). Hepatology 29:1848–1857. https://doi.org/10.1002/hep.510290602

Raposo G, Stoorvogel W (2013) Extracellular vesicles: exosomes, microvesicles, and friends. J Cell Biol 200:373–383. https://doi.org/10.1083/jcb.201211138

Recouvreux MV, Commisso C (2017) Macropinocytosis: a metabolic adaptation to nutrient stress in cancer. Front Endocrinol (Lausanne) 8:261. https://doi.org/10.3389/fendo.2017.00261

Ribecco-Lutkiewicz M, Sodja C, Haukenfrers J et al (2018) A novel human induced pluripotent stem cell blood-brain barrier model: applicability to study antibody-triggered receptor-mediated transcytosis. Sci Rep 8. https://doi.org/10.1038/s41598-018-19522-8

Roura S, Cal R, Gálvez-Montón C et al (2014) Inverse relationship between raft LRP1 localization and non-raft ERK1,2/MMP9 activation in idiopathic dilated cardiomyopathy: potential impact in ventricular remodeling. Int J Cardiol 176:805–814. https://doi.org/10.1016/j.ijcard.2014.07.270

Sade H, Baumgartner C, Hugenmatter A et al (2014) A human blood-brain barrier transcytosis assay reveals antibody transcytosis influenced by pH-dependent receptor binding. PLoS One 9:e96340. https://doi.org/10.1371/journal.pone.0096340

Schiel JE, Mire-Sluis A, Davis D (2014) Monoclonal antibody therapeutics: the need for biopharmaceutical reference materials. In: Schiel JE, Davis DL, Borisov OV (eds) State-of-the-art and emerging technologies for therapeutic monoclonal antibody characterization, Monoclonal antibody therapeutics: structure, function, and regulatory space, vol 1. American Chemical Society, Washington, DC, pp 1–34

Scott CC, Vacca F, Gruenberg J (2014) Endosome maturation, transport and functions. Semin Cell Dev Biol 31:2–10. https://doi.org/10.1016/j.semcdb.2014.03.034

Song L, Ge S, Pachter JS (2007) Caveolin-1 regulates expression of junction-associated proteins in brain microvascular endothelial cells. Blood 109:1515–1523. https://doi.org/10.1182/blood-2006-07-034009

Sprenger RR, Fontijn RD, van Marle J et al (2006) Spatial segregation of transport and signalling functions between human endothelial caveolae and lipid raft proteomes. Biochem J 400:401–410. https://doi.org/10.1042/BJ20060355

Stanimirovic D, Kemmerich K, Haqqani AS, Farrington GK (2014) Engineering and pharmacology of blood-brain barrier-permeable bispecific antibodies. Adv Pharmacol 71:301–335. https://doi.org/10.1016/bs.apha.2014.06.005

Stanimirovic D, Kemmerich K, Haqqani AS, et al (2017) Insulin-like growth factor 1 receptor -specific antibodies and uses thereof. Patents US2017015748, US2017015749, US2017022277

Tanha J, Dubuc G, Hirama T et al (2002) Selection by phage display of llama conventional V(H) fragments with heavy chain antibody V(H)H properties. J Immunol Methods 263:97–109

Tashima T (2018) Effective cancer therapy based on selective drug delivery into cells across their membrane using receptor-mediated endocytosis. Bioorg Med Chem Lett 28:3015–3024. https://doi.org/10.1016/j.bmcl.2018.07.012

Tauro BJ, Greening DW, Mathias RA et al (2013) Two distinct populations of exosomes are released from LIM1863 colon carcinoma cell-derived organoids. Mol Cell Proteomics 12:587–598. https://doi.org/10.1074/mcp.M112.021303

Terrillon S, Bouvier M (2004) Roles of G-protein-coupled receptor dimerization. EMBO Rep 5:30–34. https://doi.org/10.1038/sj.embor.7400052

Thom G, Burrell M, Haqqani AS et al (2018a) Enhanced delivery of galanin conjugates to the brain through bioengineering of the anti-transferrin receptor antibody OX26. Mol Pharm 15. https://doi.org/10.1021/acs.molpharmaceut.7b00937

Thom G, Burrell M, Haqqani AS, et al (2018b) Affinity-dependence of the blood-brain barrier crossing and brain disposition of the anti- transferrin receptor antibody OX26. Mol Pharm (in press)

Tian X, Nyberg S, Sharp PS et al (2015) LRP-1-mediated intracellular antibody delivery to the central nervous system. Sci Rep 5:11990. https://doi.org/10.1038/srep11990

van Niel G, Raposo G, Candalh C et al (2001) Intestinal epithelial cells secrete exosome-like vesicles. Gastroenterology 121:337–349. https://doi.org/10.1053/gast.2001.26263

van Weering JRT, Cullen PJ (2014) Membrane-associated cargo recycling by tubule-based endosomal sorting. Semin Cell Dev Biol 31:40–47. https://doi.org/10.1016/j.semcdb.2014.03.015

Vercauteren D, Piest M, van der Aa LJ et al (2011) Flotillin-dependent endocytosis and a phagocytosis-like mechanism for cellular internalization of disulfide-based poly(amido amine)/DNA polyplexes. Biomaterials 32:3072–3084. https://doi.org/10.1016/j.biomaterials.2010.12.045

Villaseñor R, Schilling M, Sundaresan J et al (2017) Sorting tubules regulate blood-brain barrier transcytosis. Cell Rep 21:3256–3270. https://doi.org/10.1016/j.celrep.2017.11.055

Von Bartheld CS, Altick AL (2011) Multivesicular bodies in neurons: distribution, protein content, and trafficking functions. Prog Neurobiol 93:313–340. https://doi.org/10.1016/j. pneurobio.2011.01.003

Wang J, Molday LL, Hii T et al (2018) Proteomic analysis and functional characterization of P4-ATPase phospholipid flippases from murine tissues. Sci Rep 8:10795. https://doi. org/10.1038/s41598-018-29108-z

Webster CI, Caram-Salas N, Haqqani AS et al (2016) Brain penetration, target engagement, and disposition of the blood-brain barrier-crossing bispecific antibody antagonist of metabotropic glutamate receptor type 1. FASEB J 30:1927–1940. https://doi.org/10.1096/fj.201500078

Webster CI, Hatcher J, Burrell M et al (2017) Enhanced delivery of IL-1 receptor antagonist to the central nervous system as a novel anti–transferrin receptor-IL-1RA fusion reverses neuropathic mechanical hypersensitivity. Pain 158:660–668. https://doi.org/10.1097/j. pain.0000000000000810

White IJ, Bailey LM, Aghakhani MR et al (2006) EGF stimulates annexin 1-dependent inward vesiculation in a multivesicular endosome subpopulation. EMBO J 25:1–12. https://doi. org/10.1038/sj.emboj.7600759

Xin H, Jiang X, Gu J et al (2011) Angiopep-conjugated poly(ethylene glycol)-co-poly(ε-caprolactone) nanoparticles as dual-targeting drug delivery system for brain glioma. Biomaterials 32:4293–4305. https://doi.org/10.1016/j.biomaterials.2011.02.044

Yu YJ, Zhang Y, Kenrick M et al (2011) Boosting brain uptake of a therapeutic antibody by reducing its affinity for a transcytosis target. Sci Transl Med 3:84ra44. https://doi.org/10.1126/scitranslmed.3002230

Yu YJ, Atwal JK, Zhang Y et al (2014) Therapeutic bispecific antibodies cross the blood-brain barrier in nonhuman primates. Sci Transl Med 6:261ra154. https://doi.org/10.1126/scitranslmed.3009835

Zhang Y, Liu Y, Liu H, Tang WH (2019) Exosomes: biogenesis, biologic function and clinical potential. Cell Biosci 9:19. https://doi.org/10.1186/s13578-019-0282-2

Zhuang X, Xiang X, Grizzle W et al (2011) Treatment of brain inflammatory diseases by delivering exosome encapsulated anti-inflammatory drugs from the nasal region to the brain. Mol Ther 19:1769–1779. https://doi.org/10.1038/mt.2011.164

Zuchero YJY, Chen X, Bien-Ly N et al (2016) Discovery of novel blood-brain barrier targets to enhance brain uptake of therapeutic antibodies. Neuron 89:70–82. https://doi.org/10.1016/j. neuron.2015.11.024

Chapter 4
Blood-Arachnoid Barrier as a Dynamic Physiological and Pharmacological Interface Between Cerebrospinal Fluid and Blood

Yasuo Uchida, Ryohei Goto, Takuya Usui, Masanori Tachikawa, and Tetsuya Terasaki

Abstract The blood-arachnoid barrier (BAB) consists of arachnoid epithelial cells linked by tight junctions, and forms one of the interfaces between blood and cerebrospinal fluid (CSF). The BAB was long believed to be impermeable to water-soluble substances and to play a largely passive role until our in vivo studies demonstrated that it is an active interface. Our quantitative proteomic analyses revealed that multiple transporters (OAT1, OAT3, P-gp, BCRP, MATE1, OCT2, PEPT2, etc.) are expressed more abundantly at the BAB than at the blood-cerebrospinal fluid barrier, their membrane localizations are polarized in the BAB, and there are regional differences between the cerebral and spinal cord BAB. These findings would provide a better understanding about the central nervous system kinetics of drugs and endogenous compounds, which cannot be explained by blood-brain and blood-cerebrospinal fluid barriers. Here, we introduce the BAB transport systems and discuss the physiologically and pharmacologically crucial roles of the BAB.

Keywords Blood-arachnoid barrier · Transporter · Protein expression level · Regional difference · Transporter localization · Species difference · In vivo contribution · Cerebrospinal fluid · Blood-cerebrospinal fluid barrier · quantitative targeted absolute proteomics

Y. Uchida (✉) · T. Usui · M. Tachikawa · T. Terasaki
Division of Membrane Transport and Drug Targeting, Graduate School of Pharmaceutical Sciences, Tohoku University, Sendai, Japan

Faculty of Pharmaceutical Sciences, Tohoku University, Sendai, Japan
e-mail: yasuo.uchida.c8@tohoku.ac.jp

R. Goto
Faculty of Pharmaceutical Sciences, Tohoku University, Sendai, Japan

© American Association of Pharmaceutical Scientists 2022
E. C. M. de Lange et al. (eds.), *Drug Delivery to the Brain*, AAPS Advances in the Pharmaceutical Sciences Series 33, https://doi.org/10.1007/978-3-030-88773-5_4

Abbreviations

BAB	Blood-arachnoid barrier
BBB	Blood-brain barrier
BCSFB	Blood-cerebrospinal fluid barrier
CNS	Central nervous system
CSF	Cerebrospinal fluid
LC-MS/MS	Liquid chromatography-tandem mass spectrometry
qTAP	Quantitative targeted absolute proteomics

4.1 Introduction

It has long been believed that the transport of various substances, including drugs, between CSF and blood is regulated by choroid plexus epithelial cells, which form the blood-cerebrospinal fluid barrier (BCSFB) and express a variety of transporters. However, the CSF volume in total ventricles accounts for only about 15% and 5% of the total CSF volume in human and rat, respectively, and about 80% of the CSF volume exists in subarachnoid space, which is far away from the ventricles (Thorne 2014). Because the choroid plexuses exist in ventricles, it is likely that other regulatory mechanisms contribute to the barrier function between CSF and blood, especially in the subarachnoid space.

The blood-arachnoid barrier (BAB) is another CSF-to-blood interface, consisting of arachnoid epithelial cells linked by tight junctions, and the arachnoid epithelial cells face the subarachnoid CSF in the brain and spinal cord (Nabeshima et al. 1975). Therefore, the BAB could potentially have a marked influence on the concentrations of a variety of substances including drugs in the CSF, especially in the subarachnoid CSF. But, about 30 years ago, it was claimed that the "arachnoid membrane is impermeable to water-soluble substances and its role in forming blood−CSF barrier is largely passive" (Spector and Johanson 1989), and until recently, this was believed to be the case.

In 2013, Yasuda et al. dramatically undermined this theory, showing that several transporters are expressed at human arachnoid mater and mouse leptomeninges at the mRNA level; they also immunohistochemically detected P-gp and bcrp in mouse arachnoid mater cells, but not other meningeal tissue (Yasuda et al. 2013). These findings suggested that the BAB might be an active functional barrier regulating drug pharmacokinetics in the subarachnoid CSF. However, in order to establish definitively whether or not the BAB is an active functional barrier, it remained necessary to determine the in vivo functional contributions of the BAB transporters to pharmacokinetics in the CSF and to measure the expression levels of individual transporters in the BAB at the protein level.

To address these questions, we have conducted a program of quantitative proteomic studies and in vivo functional analyses of the BAB transporters. We used quantitative targeted absolute proteomics (qTAP) methodology to show that multiple important transporters are expressed in porcine (Uchida et al. 2020) and rat (Zhang et al. 2018) leptomeninges more abundantly than in choroid plexus, and furthermore, we confirmed the contributions of oat1, oat3 (Zhang et al. 2018), and oatp1a4 (Yaguchi et al. 2019) to the clearance of organic anions from the CSF at the BAB by means of in vivo intracisternal administration method. We also identified the localizations of these transporters at the CSF-facing or blood (dura)-facing plasma membrane of arachnoid epithelial cells (Uchida et al. 2020) and demonstrated marked differences in the protein abundances of the transporters among different regions of the BAB in the brain and spinal cord. Overall, our work has demonstrated that the BAB is an important active functional interface in the central nervous system (CNS). In this chapter, we summarize these findings and discuss the physiological and pharmacological significance of the BAB in regulating the concentrations of drugs and endogenous compounds in the CSF.

4.2 Quantitative Protein Expression Profile of Transporters at the BCSFB: Interspecies Difference Between Human and Experimental Animals (Rat, Dog, and Pig) and Regional Difference Among Four Ventricular Choroid Plexus

To understand the physiological and pharmacological roles of the BAB, we need to comprehensively compare the subtypes and protein expression levels of transporters expressed at the BAB with those at the BCSFB. Before introducing the result of such comparative studies in the next section, we shall first consider the quantitative protein expression profile of transporters at the BCSFB.

Figure 4.1 shows how the protein expression levels of transporters in human choroid plexus differ from those in experimental animals such as rat, dog, and pig. In rat choroid plexus, transporters involved in the BCSFB transport of organic anions, such as oatp1a5, pept2, oatp1c1, oat3, and mrp4, are more abundantly expressed than in human choroid plexus (Fig. 4.1a) (Uchida et al. 2015). Interestingly, although PEPT2 has been considered to be an important transporter contributing to the clearance of neuropeptides and beta-lactam antibiotics from the CSF on the basis of rodent studies, its level was under the limit of quantification in humans. MATE1 is a transporter involved in the efflux of a variety of organic cations, including drugs, in the kidney. It was not detected in rat choroid plexus, but was abundantly expressed in human choroid plexus. The CSF concentrations of various cationic neurotoxins produced in the CNS, such as creatinine and N-methylnicotinamide, are maintained at low (nontoxic) level under normal conditions (Williams and Ramsden 2005; Tachikawa et al. 2008). Because MATE1 can transport these organic cations (Terada et al. 2006; Tanihara et al. 2007), it may contribute to the elimination of these cationic neurotoxins from the CSF across the human BCSFB.

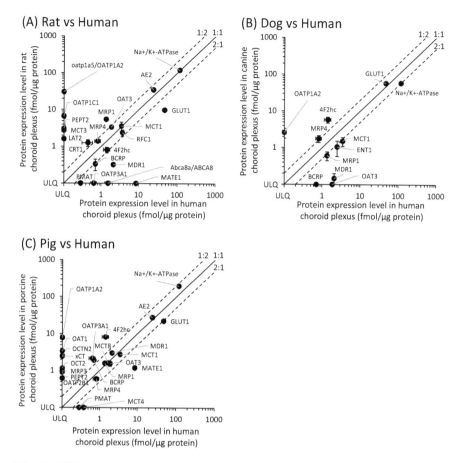

Fig. 4.1 Differences in the protein expression levels of transporters in choroid plexus between human and experimental animals (rat, dog, and pig)

Protein expression levels of transporters in choroid plexus from rat (**a**), dog (**b**), and pig (**c**) were compared with those of humans. Rat data were obtained for the plasma membrane fraction of choroid plexus pooled from lateral, third, and fourth ventricles. Dog data are average protein expression levels in the whole tissue lysate of choroid plexus separately isolated from lateral (minor contamination with third ventricular choroid plexus) and fourth ventricles. Pig data are average protein expression levels in the plasma membrane fractions of choroid plexus separately isolated from right lateral, left lateral, third, and fourth ventricles. Human data were obtained for the plasma membrane fraction of choroid plexus isolated from the fourth ventricle of a single donor at 5 h 6 min postmortem (donor information: 92 years old, Caucasian, male, diagnosed with depression and dementia, and cardiovascular abnormality as the cause of death). Values are mean ± SD except for pig data (mean ± SEM). *ULQ* under the limit of quantification. Human and rat data are from a previous report (Uchida et al. 2015). Canine and porcine data are from two previous reports (Braun et al. 2017; Uchida et al. 2020)

In drug development, the dog is a widely used experimental animal for toxicity study of candidate compounds before proceeding to human clinical trials (Bailey et al. 2013). It is considered that the quantification of CSF concentrations in dog is useful to predict the efficacy and adverse effects of candidate compounds in human CNS. However, the quantitative protein expression profile of canine choroid plexus is very different from that of human choroid plexus (Fig. 4.1b) (Uchida et al. 2015; Braun et al. 2017). In particular, the protein expression level of BCRP, which is a major efflux transporter of multiple lipophilic drugs, was under the limit of quantification, and the anionic drug transporter OATP1A2 was abundantly expressed in canine choroid plexus. Relative to the dog, the porcine choroid plexus had more transporters whose protein expression levels are within a twofold range of those in human choroid plexus (Fig. 4.1c) (Uchida et al. 2015, 2020). However, porcine choroid plexus also expresses several drug transporters that are not detected in human choroid plexus, such as OATP1A2, OAT1, OCTN2, MRP3, PEPT2, and OATP2B1. Nevertheless, it should be noted that the protein expression levels of transporters in human choroid plexus (Fig. 4.1) were obtained from a single 92-year-old donor with depression and dementia and may not be typical. Further analysis of samples from multiple donors with a normal brain is urgently needed.

The regional differences in the protein expression levels (in units of fmol/μg protein) of 16 transporters in choroid plexus of the four different ventricles (right lateral, left lateral, third, and fourth) were within a twofold range for most of the transporters in pig, except for OAT1, OAT3, and MRP4 (2.44-, 2.01-, and 2.06-fold differences, respectively) (Uchida et al. 2020).

4.3 Absolute Protein Expression Amounts and CSF/Blood-Side Localizations of Transporters at the BAB: Comparison with BCSFB

To understand the difference in the physiological and pharmacological roles between the BAB and BCSFB, we need to compare the quantitative protein expression profile in the BAB with that of the BCSFB. Porcine biology, including genomics, anatomy, physiology, and disease progression, reflects human biology more closely than is the case for many other experimental animals (Walters et al. 2011; Patabendige et al. 2013). The protein expression levels of transporters at the BCSFB and BBB in pig reflect those in humans to some extent, although not completely (Fig. 4.1c) (Uchida et al. 2011b; Kubo et al. 2015). Therefore, we selected the pig as a model animal to determine the absolute protein expression amounts of transporters at the BAB and to compare the obtained values with those at the BCSFB. To understand the difference in the transport capacities of individual transporters between the BAB and BCSFB, the total protein expression amount of each transporter at the whole BAB or BCSFB (total of four ventricular BCSFBs) per one porcine cerebrum was calculated and compared (Fig. 4.2). These results showed that various important transporters are abundantly expressed at the BAB; the protein amounts of OAT1, OAT3, PEPT2, OCT2, and MATE1 were 8.9-, 7.8-, 5.8-, 90-, and 33-fold greater than those at the BCSFB, respectively (Fig. 4.2) (Uchida et al. 2020).

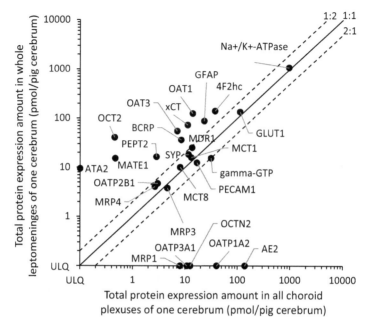

Fig. 4.2 Differences in the total protein expression amounts of transporters between choroid plexus and leptomeninges in one porcine cerebrum
Protein expression levels (fmol/μg protein) were determined by quantitative targeted absolute proteomics (qTAP) in the plasma membrane fractions of porcine leptomeninges isolated from four regions of the cerebrum and the choroid plexus isolated from four ventricles, and the data were used to calculate the total protein expression amounts in whole leptomeninges or all choroid plexuses of one porcine cerebrum (pmol/pig cerebrum). The data are from a previous report (Uchida et al. 2020). The calculation method is described in the text. Values are mean ± SEM. *ULQ* under the limit of quantification

The protein expression levels of OAT3 in choroid plexus of pig and rodent are similar (rat, 3.37 fmol/μg protein; pig, 1.54 fmol/μg protein) (Uchida et al. 2015, 2020). The uptake of fluorescein, an oat3 substrate (Wolman et al. 2013), by choroid plexus isolated from oat3-knockout mice is significantly smaller than that in the case of wild-type mice (Sweet et al. 2002). This suggests that the protein expression level of oat3 at the BCSFB in mice (assuming it is the same as in rat, 3.37 fmol/μg protein) is high enough to make a significant contribution to the in vivo transport of its substrates. Because the OAT3 level at the BAB is 7.76-fold greater than that at the BCSFB (Fig. 4.2), OAT3 is considered to contribute to the clearance of substrates from the CSF at the BAB. As is the case of oat3, it has also been shown that pept2 contributes to BCSFB transport by means of uptake experiments using choroid plexus isolated from pept2-knockout and wild-type mice (Ocheltree et al. 2004). Because the PEPT2 level at the BAB is 5.79-fold greater than that at the BCSFB (Fig. 4.2), PEPT2 is also considered to contribute to BAB transport. Importantly, most of the transporters that have > twofold greater protein amounts at the BAB than at the BCSFB are more abundantly expressed than PEPT2 (Fig. 4.2). This suggests that many transporters contribute significantly to substrate transport at the BAB.

In order to discuss the roles of individual transporters at the barrier tissues, it is critical to know the localizations of the transporters. We have previously established proteomics methodology to quantitatively determine the localizations of many transporters simultaneously without the need to use antibodies; this involves plasma membrane separation by means of different sucrose density gradient ultracentrifugations, followed by the simultaneous quantifications of many target transporters in the separated membrane fractions using qTAP (Kubo et al. 2015). This method was applied to the cerebral leptomeninges and enabled us to simultaneously determine the CSF-facing or blood (dura)-facing localizations of 14 transporters at the BAB (Fig. 4.3) (Uchida et al. 2020). Basically, the localizations of most transporters were the same as those at the BCSFB, but notably, some transporters showed opposite localization. Among transporters of organic anions, two abundant transporters, OAT1 and OAT3 (Fig. 4.2), were localized at the CSF-facing plasma membrane (Fig. 4.3), and so it is considered that the BAB plays a role in transporting organic anions in the CSF-to-blood direction, as is the case in the BCSFB. On the other hand, MRP3, which was present in a similar protein amount to MRP4, was localized at the CSF-facing plasma membrane at the BAB. MRP3 can pump out drugs such as the CNS-acting agent morphine-3-glucuronide from the intracellular to extracellular space (van de Wetering et al. 2007), and therefore, it may play an important role in drug delivery to the CNS (Fig. 4.3).

It is noteworthy that MATE1 and OCT2 were abundantly expressed in the blood (dura)- and CSF-facing plasma membranes at the BAB, respectively (Figs. 4.2 and 4.3). They can transport various organic cations, including drugs, bidirectionally (Konig et al. 2011). Metformin is a substrate of these two transporters, and the CSF concentration of metformin is about 15-fold higher than the plasma concentration (Labuzek et al. 2010). It should be also noted that, although the brain parenchymal concentration is also higher than plasma concentration, the CSF concentration is the highest (Labuzek et al. 2010). Metformin is not a substrate of any of the transporters so far identified at the BBB. Therefore, it is likely that metformin may actively pass from the circulating blood to the CSF via MATE1 and OCT2 at the BAB. This may imply that the BAB is a potential route of drug delivery to the CNS.

4.4 Drug Efflux Transporters P-gp/MDR1 and BCRP/ABCG2

It is important to know how P-gp and BCRP influence drug distribution in the CSF and CNS, because these two transporters are more abundantly expressed at the BAB than the BCSFB (Fig. 4.2) and their localizations at the BAB are opposite to those at the BCSFB (Fig. 4.3).

Contrary to our data, Yasuda et al. (2013) reported that BCRP is expressed at both the CSF- and blood-facing plasma membranes of the arachnoid epithelial cells (Yasuda et al. 2013). BCRP mediates drug efflux from the brain at the luminal

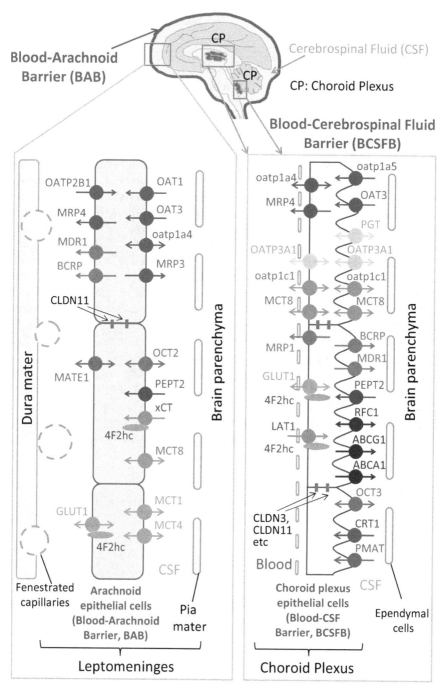

Fig. 4.3 The localizations of transporters at the BAB and BCSFB

The localizations of transporters at the BAB and BCSFB are taken from three previous reports (Tachikawa et al. 2014; Yaguchi et al. 2019; Uchida et al. 2020). The expression levels of claudins and tight junction proteins are from two previous reports (Uchida et al. 2015, 2019)

membrane of the BBB and drug influx into CSF in brain ventricles at the CSF-facing plasma membrane of choroid plexus epithelial cells. Knockout of the BCRP gene in mice increases the penetration of substrates into the brain while decreasing penetration into ventricular CSF (Shen et al. 2009). These results suggest that the brain concentrations of substrates are regulated by BCRP at the BBB, while ventricular CSF concentrations are regulated by BCRP at the BCSFB. We have already shown that the corresponding transporters expressed at the BAB regulate substrate concentrations in cisternal CSF (Zhang et al. 2018; Yaguchi et al. 2019). The CSF-to-plasma concentration ratios of the BCRP-selective substrates daidzein and genistein in cisterna magna at steady state are 3.96- and 2.54-fold larger in bcrp-knockout mice than in wild-type mice, respectively (Kodaira et al. 2011), and this supports the idea that BCRP limits the penetration of substrates into CSF at the blood-facing plasma membrane of the BAB. It should be noted that Yasuda et al. determined the BCRP localization by using immortalized arachnoid epithelial cells (Yasuda et al. 2013), and immortalization might have disrupted the polarized localization of BCRP, since it would potentially impair cell-to-cell tight junctions, which maintain the polarity of membrane transporters.

In dog, the protein expression level of BCRP at the BBB is 9.13-fold greater than that in rat (Hoshi et al. 2013; Braun et al. 2017), suggesting that the efflux activity of BCRP at the BBB could be greater than that in rat. However, the unbound cisternal CSF-to-plasma concentration ratios of the BCRP-selective substrates dantrolene and daidzein are 6.03- and 4.19-fold higher than those in rat, respectively (Kodaira et al. 2011; Braun et al. 2017). This supports the idea that the efflux system at the BBB will not play a determinant role of the drug distribution in the CSF of subarachnoid space.

The protein abundances of BCRP and P-gp (7.85 and 5.42 fmol/µg protein, respectively) in the plasma membrane fraction of cerebral leptomeninges were 7.5-fold and 3.8-fold smaller than those in porcine brain capillaries, respectively (Zhang et al. 2017; Uchida et al. 2020). In the steady state, the bcrp-knockout/wild-type mouse CSF-to-plasma concentration ratios for daidzein and genistein are smaller than the brain-to-plasma concentration ratios (Kodaira et al. 2011). The corresponding ratios of mdr1a/1b-knockout/wild-type mouse for the P-gp-selective substrates quinidine and verapamil are also smaller than the brain-to-plasma concentration ratios. Therefore, in the steady state, the differences in the contributions of BCRP and P-gp to limiting substrate distributions between brain parenchyma and cisternal CSF can be explained by the differences in their protein expression levels between the BBB and BAB. This suggests that substrate concentrations in brain parenchyma are regulated by P-gp and BCRP at the BBB, while those in the subarachnoid space are regulated by P-gp and BCRP at the BAB.

4.5   Species Difference in the Protein Expression Levels of Transporters at the BAB Between Rat and Pig

To understand how the BAB function in rodents correlates with that in large animals, the protein expression levels of transporters at the BAB were compared between rat and pig (Fig. 4.4). Most transporters showed > twofold differences in their protein expression levels, suggesting the existence of significant interspecies differences. The elimination of organic anionic neurotoxins from the CNS is essential to maintain the homeostasis of brain function. We have previously shown that oat3 at the BBB eliminates anionic neurotoxins such as the major catecholamine metabolite homovanillic acid and uremic toxins in rodents (Mori et al. 2003; Deguchi et al. 2006). However, in contrast to rodents, protein expression of OAT3 at the BBB of human beings and large animals, including monkey and dog, has not been detected (Ito et al. 2011; Uchida et al. 2011b; Braun et al. 2017). The protein expression level of OAT3 in choroid plexus is also smaller in human than in rat

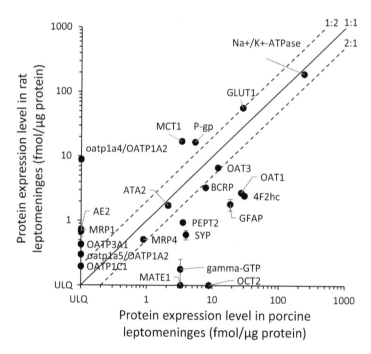

Fig. 4.4 Differences in the protein expression amount of transporters in the plasma membrane fraction of brain leptomeninges between rat and pig

Rat data were obtained for the plasma membrane fraction of leptomeninges pooled from whole brain, and values are mean ± SD. Porcine data are average protein expression levels quantified in the plasma membrane fractions of leptomeninges separately isolated from four different cerebral lobes (frontal, parietal, occipital, and temporal), and values are mean ± SEM. *ULQ* under the limit of quantification. Rat and porcine data are from two previous reports (Zhang et al. 2018; Uchida et al. 2020)

(Uchida et al. 2015). Therefore, it remains unclear how organic anionic neurotoxins are eliminated from the CNS in human beings and large animals. Figure 4.4 shows that OAT1 and OAT3 are abundantly expressed in the leptomeninges of porcine cerebrum, and their levels (27.2 and 12.1 fmol/µg protein) are 9.97- and 1.82-fold greater than those in rat leptomeninges, respectively (Zhang et al. 2018; Uchida et al. 2020). Furthermore, OAT1 and OAT3 were 8.94- and 7.76-fold more abundant in the whole BAB than in the total choroid plexuses per porcine cerebrum, respectively (Fig. 4.2). These results suggest that the contribution of BAB to the elimination of organic anions from CNS is large in pig, compared to rodent. Homovanillic acid is preferentially transported by human OAT1 over OAT3 in in-vitro-transfected cell lines (Shen et al. 2018), and OAT1 is about twofold more abundant than OAT3 at the porcine BAB (Figs. 4.2 and 4.4). Therefore, OAT1 at the BAB may play an important role in the elimination of homovanillic acid produced in the CNS in large animals, including human beings.

Among the oatp family, oatp1a4 is abundantly expressed in the rat BAB (Fig. 4.4). In the BBB, oatp1a4 plays an important role in the elimination of neurosteroids from the brain and the delivery of anionic drugs into the brain (Asaba et al. 2000; Ohtsuki et al. 2007; Ose et al. 2010). Although oatp1a4 at the BAB may have a similar role, the protein expression level of OATP1A2, which is a functional homolog of rodent oatp1a4, was under the limit of quantification in porcine BAB (Fig. 4.4). OATP1A2 expression is also very different in choroid plexus of human beings and large animals (not detected in human choroid plexus, but abundantly expressed in porcine and canine choroid plexus; Fig. 4.1).

Further study will be needed to clarify the protein expression levels of transporters, including OAT1, OAT3, and OATP1A2, in the human BAB in order to understand the similarities and differences in the BAB transport systems of human beings and various experimental animals.

4.6   In Vivo Contributions of Oat1, Oat3, and Oatp1a4 at the BAB to the Clearance of Organic Anions from CSF

To confirm that the BAB is an active functional barrier, we examined whether oat1, oat3, and oatp1a4 at the BAB contribute to the in vivo clearance of organic anions from the CSF, because oat1 and oat3 are much more abundant at the BAB than at the BCSFB in both rat (Zhang et al. 2018) and pig (Fig. 4.2), and oatp1a4 is more abundant than oat1 and oat3 at the BAB, even though its functional homolog OATP1A2 was not detected at the porcine BAB (Fig. 4.4). First, to exclude the influence of BCSFB transport systems, we used the intracisternal administration method to inject substances directly into the cisterna magna; we confirmed that substances injected into the cisterna magna do not reach the CSF region close to the choroid plexus by using fluorescent rhodamine 123 (Zhang et al. 2018). Then, employing the validated intracisternal administration method, we demonstrated that the organic anion transporter substrate para-aminohippuric acid (PAH) is significantly more rapidly eliminated from the cisternal CSF than the impermeable marker inulin (Zhang et al. 2018). Cephalothin shows the greatest difference of IC_{50} between rat oat1 (0.57 mM) and rat oat3 (0.08 mM). In the presence of 3 mM cephalothin (inhibiting both oat1 and oat3), the residual concentration of PAH in the cisternal CSF at 15 min after intracisternal administration showed no significant difference from that of inulin (Fig. 4.5), suggesting that the elimination of PAH from cisternal CSF can be fully accounted for by uptake via oat1 and oat3 at the BAB. Only 17% of PAH elimination was inhibited by 0.2 mM cephalothin (selectively inhibiting oat3), suggesting that PAH elimination from cisternal CSF is predominantly mediated by oat1 (Fig. 4.5). As shown in Fig. 4.4, the protein expression levels of OAT1 and OAT3 in the porcine BAB are 9.97- and 1.82-fold greater than those in rat, respectively. This suggests that the contributions of OAT1 and OAT3 to the clearance of substrates from CSF at the porcine BAB are greater than those in rat. The protein levels in human BAB may be similar to those in the porcine BAB, but

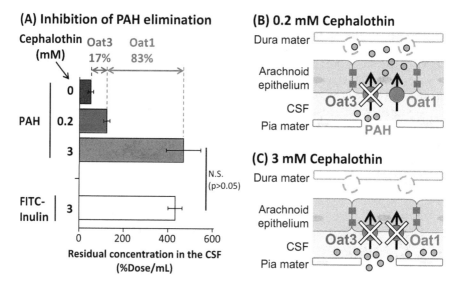

Fig. 4.5 Inhibitory effects of cephalothin on para-aminohippuric acid (PAH) elimination from the CSF at 15 min after intercisternal injection in rats

(**a**) Preadministration of different concentrations of cephalothin (0.2 or 3 mM) inhibited the uptake of PAH (a substrate of Oats) compared with the buffer-administered control. Each value is the mean ± SEM (n = 5–6). No significant difference between the %dose/mL values of PAH and FITC-inulin (a membrane-impermeable substance) was seen in the presence of 3 mM cephalothin. Cephalothin inhibits oat1 and oat3 with IC_{50} values of 0.57 mM and 0.08 mM, respectively. Based on the assumption that 0.2 mM cephalothin can selectively inhibit oat3 and 3 mM cephalothin can inhibit both oat1 and oat3, the contributions of oat1 and oat3 to the PAH elimination were calculated to be 83% and 17%, respectively. Data are from a previous report (Zhang et al. 2018). (**b**) PAH elimination when only oat3 is inhibited by 0.2 mM cephalothin. (**c**) PAH elimination when both oat1 and oat3 are inhibited by 3 mM cephalothin

further work is needed to establish the contributions of OAT1 and OAT3 to substrate clearance.

As with oat, the oatp substrate SR101 was also rapidly eliminated from the cisternal CSF after intracisternal administration, and the elimination was inhibited by not only taurocholate (a broad-spectrum inhibitor of oatps) but also digoxin (a strong substrate/inhibitor of oatp1a4, but not oatp1a1, oat1, or oat3, among the potential transporters of organic anion SR101) (Yaguchi et al. 2019). This suggests that oatp1a4 at the BAB contributes to the clearance of substrates from the cisternal CSF. The cisterna magna is close to the brain and spinal cord parenchymal tissues, and oatp1a4 is also expressed in brain and spinal cord capillaries. As shown in Fig. 4.6, SR101 administered into cisternal CSF did not diffuse to the spinal cord parenchymal tissue, but was accumulated in the arachnoid membrane (Figs. 4.6a and 6C), although it diffused to the parenchyma in the presence of taurocholate

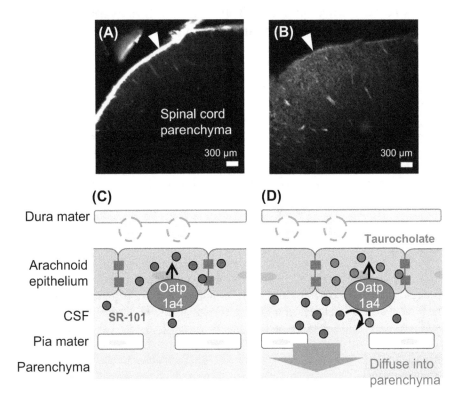

Fig. 4.6 Effect of taurocholate (oatp substrate/inhibitor) on the distribution of SR-101 (oatp1a4 substrate) in the cervical spinal cord proximal to the SR-101 injection site 20 min after intracisternal injection in rats. SR-101 was administered via cisterna magna puncture without (**a, c**) or with (**b, d**) preadministration of taurocholate. In the control, the fluorescence signals of SR-101 were predominantly detected in the leptomeninges at the surface of the spinal cord (**a, c**). In contrast, the fluorescence intensity of SR-101 was diminished in the leptomeninges, but increased in the parenchyma of the spinal cord in the presence of taurocholate (**b, d**). Scale bars: 300 μm. Arrowhead indicates the arachnoid membrane. Data are from a previous report (Yaguchi et al. 2019)

(Figs. 4.6b and d). These results support the idea that the elimination of SR101 is mediated by oatp1a4 at the BAB, but not at the parenchymal capillaries. It is also suggested that oatp1a4 at the BAB plays an important role in suppressing the distribution of substrates to spinal cord parenchyma by its rapid clearance activity from the CSF.

These in vivo studies establish that the BAB is not a passive barrier, but an active functional barrier. Furthermore, the apparent uptake clearances of PAH (26.5 µL/min) (Zhang et al. 2018) and SR101 (13.4 µL/min) (Yaguchi et al. 2019) after intracisternal administration were much greater than the CSF bulk flow (2.9 µL/min) (Suzuki et al. 1985), suggesting the existence of more rapid transport system at the BAB than the CSF flow.

4.7 Protein Expression Levels of Transporters at the BAB in the Spinal Cord Region: Comparison with Brain BAB and BCSFB

Drug concentrations in CSF are widely used as a surrogate marker for drug concentrations in brain interstitial fluid (ISF). Thus, it is important to clarify whether or not drug concentrations in CSF taken from the lumbar region are identical to those in brain ISF and, if there is a difference, to establish the molecular mechanisms involved. Actually, it has been simulated that the relationship of drug concentrations in brain extracellular fluid and subarachnoid CSF in humans varies in time-dependent manner (Yamamoto et al. 2017). For this purpose, it is necessary to quantify the protein abundances of the transporters not only at the brain BAB and BCSFB but also at the spinal cord BAB. Here we summarize the results of studies on the protein expression levels of BAB transporters in the cervical, thoracic, and lumbar spinal cords in pig.

Table 4.1 shows the protein concentrations in the plasma membrane fractions of leptomeninges isolated from cervical, thoracic an,d lumbar spinal cords in units of fmol/µg protein. However, the use of these units means that regional differences in the protein expression levels of transporters (among different spinal cord regions or between spinal cord and brain, etc.) cannot be correctly assessed, because of differences in the purity of arachnoid epithelial cells contained in the isolated leptomeninges and the purity of plasma membrane in the plasma membrane fraction among different regions. To overcome this problem, we used the equations and parameters shown in Fig. 4.7 and Table 4.2, respectively, to convert the values of protein concentration (fmol/µg protein, Table 4.1) to protein expression level per 1 cm^2 of leptomeninges (pmol/cm^2, Table 4.3). This unit conversion was also conducted for the cerebral leptomeninges, and the protein expression levels were compared among seven different regions of leptomeninges in the spinal cord and cerebrum (Table 4.3). Most transporters showed smaller protein expression levels in the three spinal cord leptomeninges, as compared with the four cerebral leptomeninges. The level of

Table 4.1 Protein concentrations of transporters in 1 μg protein of plasma membrane fractions of leptomeninges isolated from the three spinal cord regions in pig (units: fmol/μg protein)

Molecule	Protein expression level (fmol/μg protein)		
	Cervical (upper)	Thoracic (middle)	Lumbar (lower)
Organic anion transporters			
OAT1	2.82 ± 0.28	4.33 ± 0.38	2.47 ± 0.21
OAT2	ULQ(<1.52)	ULQ(<1.31)	ULQ(<1.33)
OAT3	1.81 ± 0.21	2.31 ± 0.53	1.47 ± 0.22
OAT6	ULQ(<0.777)	ULQ(<0.757)	ULQ(<0.748)
MRP2	ULQ(<0.494)	ULQ(<0.584)	ULQ(<0.492)
MRP3	0.458 ± 0.031	0.379 ± 0.162	0.434 ± 0.162
MRP4	ULQ(<0.255)	0.431 ± 0.043	ULQ(<0.407)
OATP1A2	ULQ(<1.91)	ULQ(<2.22)	ULQ(<2.00)
OATP2B1	ULQ(<0.691)	0.562 ± 0.089	ULQ(<0.530)
OATP1B3	ULQ(<0.974)	ULQ(<0.803)	ULQ(<1.22)
OATP3A1	ULQ(<0.327)	ULQ(<0.379)	ULQ(<0.217)
Lipophilic drug transporters			
MDR1/P-gp	0.841 ± 0.247	1.73 ± 0.33	0.800 ± 0.286
BCRP/ABCG2	1.34 ± 0.40	2.83 ± 0.38	1.22 ± 0.56
MRP1	ULQ(<1.29)	ULQ(<1.72)	ULQ(<1.43)
Thyroid hormone transporters			
MCT8	1.48 ± 0.09	1.67 ± 0.09	1.46 ± 0.18
OATP1C1	ULQ(<1.11)	ULQ(<1.18)	ULQ(<1.80)
Peptide transporters			
PEPT1	ULQ(<0.248)	ULQ(<0.251)	ULQ(<0.303)
PEPT2	0.736 ± 0.180	1.66 ± 0.43	0.684 ± 0.274
Glucose transporter			
GLUT1	4.57 ± 0.12	7.49 ± 0.40	4.38 ± 0.31
Amino acid transporters			
xCT	3.56 ± 0.75	5.87 ± 0.40	3.71 ± 0.97
4F2hc	5.54 ± 0.54	8.91 ± 0.75	5.20 ± 0.40
ATA2	ULQ(<1.24)	ULQ(<1.17)	ULQ(<1.06)
Organic cation transporters			
PMAT	ULQ(<2.43)	ULQ(<2.65)	ULQ(<2.16)
MATE1	0.582 ± 0.164	1.41 ± 0.23	0.662 ± 0.188
MATE2K	ULQ(<2.87)	ULQ(<2.53)	ULQ(<2.17)
OCT1	ULQ(<0.593)	ULQ(<0.602)	ULQ(<0.623)
OCT2	1.14 ± 0.28	2.80 ± 0.32	0.552 ± 0.113
OCT3	ULQ(<1.27)	ULQ(<0.878)	ULQ(<0.560)
OCTN1	ULQ(<1.97)	ULQ(<1.84)	ULQ(<1.68)
OCTN2	ULQ(<0.686)	ULQ(<0.765)	ULQ(<0.603)
Monocarboxylate transporters			
MCT1	2.32 ± 0.52	2.45 ± 0.21	2.08 ± 0.28
MCT2	ULQ(<1.00)	ULQ(<1.27)	ULQ(<0.897)

(continued)

Table 4.1 (continued)

Molecule	Protein expression level (fmol/μg protein)		
	Cervical (upper)	Thoracic (middle)	Lumbar (lower)
MCT3	ULQ(<0.577)	ULQ(<0.745)	ULQ(<0.563)
MCT4	ULQ(<1.30)	ULQ(<1.31)	ULQ(<1.24)
Markers			
AE2	ULQ(<1.67)	ULQ(<2.59)	ULQ(<2.04)
Na$^+$/K$^+$-ATPase	44.0 ± 4.9	68.2 ± 9.5	40.1 ± 5.8
GFAP	25.5 ± 2.7	26.7 ± 1.9	34.7 ± 4.3
SYP	ULQ(<0.457)	ULQ(<0.551)	ULQ(<0.450)
Gamma-GTP	1.16 ± 0.16	1.91 ± 0.11	0.705 ± 0.131
PECAM1	ULQ(<0.329)	0.661 ± 0.088	ULQ(<0.531)

Plasma membrane fractions were prepared from freshly isolated leptomeninges of cervical, thoracic, and lumbar spinal cords of pig. The plasma membrane fractions were digested with Lys-C and trypsin. Using the digested peptide samples spiked with internal standard peptides, all of the target molecules were quantified by LC-MS/MS. Four sets of transitions were used for each peptide pair (target peptide and the corresponding internal standard peptide). Values are mean ± SEM (3–4 transitions × 3 measurements). *ULQ* under the limit of quantification. The values of the LQ are shown in parenthesis

P-gp in the lumbar region was 1.8- to 3.4-fold smaller than in the cerebral leptomeninges (Table 4.3).

Uric acid, an end product of purine metabolism, is associated with hypertension and metabolic syndrome and acts as an antioxidant in the CNS (Bowman et al. 2010). The CSF concentration is about tenfold lower than the plasma concentration (Bowman et al. 2010). Uric acid is a substrate of OAT1, OAT3, MRP4, and BCRP, and these four transporters are thought to transport substrates in the CSF-to-blood direction at the BAB, based on their membrane localizations and transport properties. In humans, the uric acid concentration in the lumbar CSF is about twofold higher than that in the cisternal CSF (Degrell and Nagy 1990), where the concentrations of substances would potentially be influenced by the transport systems of both the brain BAB and cervical BAB. The protein expression levels of OAT1, OAT3, MRP4, and BCRP at the lumbar BAB are almost identical to those at the cervical BAB, but much smaller than those at the cerebral BAB (Table 4.3). Therefore, the weaker activities of these transporters resulting from their smaller expression levels at the lumbar BAB than the BAB close to the cisternal CSF (cerebral and cervical) may account for the higher concentration of uric acid in the lumbar CSF than in the cisternal CSF.

The molecular mechanisms responsible for the CSF concentration gradient of substances between ventricles and the other subarachnoid space are also of interest. Using the equations and parameters shown in Fig. 4.7 and Table 4.2, respectively, we calculated the total protein expression amounts of individual transporters at the cervical, thoracic, or lumbar BAB (pmol/lumbar BAB, etc.), and these results are listed in Table 4.4, together with the total protein expression amounts in each ventricular choroid plexus (pmol/right lateral BCSFB, etc.) and whole cerebral BAB

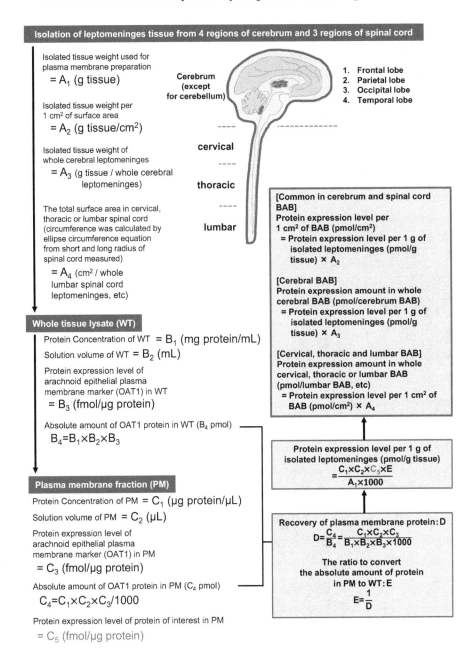

Fig. 4.7 Conversion of the unit of protein expression level of transporters from "fmol/μg protein" (red) to "pmol/cm²" and "pmol/whole BAB" in each region

Individual parameters are shown in Table 4.2. For the PM of spinal cord leptomeninges, the protein expression levels in units of fmol/μg protein are shown in Table 4.1. For the PM of cerebral leptomeninges, the protein expression levels in units of fmol/μg protein are from a previous report (Uchida et al. 2020). OAT1 is used as an arachnoid epithelial plasma membrane marker

Table 4.2 Experimental parameters used to convert the units of protein expression from fmol/μg protein to pmol/cm^2 and pmol/whole BAB in each region in pig

	Spinal cord leptomeninges			Cerebral leptomeninges			
	Cervical	Thoracic	Lumbar	Frontal	Parietal	Occipital	Temporal
[C$_1$] Protein concentration of PM (μg protein/μl)	4.97	4.01	6.08	2.81	2.47	2.63	2.30
[C$_2$] Solution volume of PM (μl)	100	100	100	100	100	100	100
[C$_3$] Protein expression level of arachnoid epithelial plasma membrane marker (OAT1) in PM (fmol/μg protein)	2.82 ± 0.28	4.33 ± 0.38	2.47 ± 0.21	30.5 ± 1.4	25.4 ± 0.7	25.5 ± 0.8	27.5 ± 1.2
[C$_4$] Absolute amount of OAT1 protein in PM (pmol)	1.40	1.74	1.50	8.56	6.27	6.70	6.32
[B$_1$] Protein concentration of WT (mg protein/ml)	0.520	0.750	0.670	1.40	1.28	1.50	1.07
[B$_2$] Solution volume of WT (ml)	31.0	32.5	30.0	29.0	27.5	26.5	30.0
[B$_3$] Protein expression level of arachnoid epithelial plasma membrane marker (OAT1) in WT (fmol/μg protein)	2.10 ± 0.25	2.20 ± 0.18	1.89 ± 0.22	3.70 ± 0.20	4.06 ± 0.06	3.49 ± 0.06	3.74 ± 0.01
[B$_4$] Absolute amount of OAT1 protein in WT (pmol)	33.8	53.6	38.0	150	143	139	120
[D] Recovery ratio of plasma membrane protein	0.0414	0.0324	0.0395	0.0570	0.0439	0.0483	0.0527
[E] The ratio to convert the absolute amount of protein in PM to WT	24.1	30.9	25.3	17.5	22.8	20.7	19.0
[A$_1$] Isolated tissue weight used for plasma membrane preparation (g tissue)	0.70	0.70	0.70	0.70	0.62	0.65	0.68
[A$_2$] Isolated tissue weight per 1 cm^2 of surface area (g tissue/cm^2)	0.0212	0.0199	0.0185	0.0205	0.0211	0.0204	0.0177
[A$_4$] The total surface area in cervical, thoracic, or lumbar spinal cord (cm^2/whole lumbar spinal cord leptomeninges, etc); calculated by [F$_1$] × [F$_4$]	45.9	104.9	34.8	–	–	–	–
[A$_3$] Isolated tissue weight of whole cerebral leptomeninges (g tissue / whole cerebral leptomeninges) (calculated value for cervical, thoracic, and lumbar leptomeninges; [A$_2$]×[A$_4$])	0.972	2.09	0.644	0.598 (Whole leptomeninges of one cerebrum)			
Correction factor to convert the units from "fmol/μg protein" in plasma membrane fraction to "pmol/g tissue" (mg protein/g tissue)	17.13	17.69	21.97	7.04	9.07	8.38	6.42

Correction factor to convert the units from "fmol/μg protein" in plasma membrane fraction to "pmol/cm²" (mg protein/cm²)	0.363	0.352	0.406	0.144	0.191	0.171	0.114
$[F_1]$ Length of spinal cord (cm)	15.5	38.9	14.9	–	–	–	–
$[F_2]$ Short radius of spinal cord (cm)	0.328	0.311	0.250	–	–	–	–
$[F_3]$ Long radius of spinal cord (cm)	0.594	0.533	0.478	–	–	–	–
$[F_4]$ Circumference of spinal cord (cm); calculated by ellipse circumference equation from $[F_2]$ and $[F_3]$	2.96	2.70	2.34	–	–	–	–

These parameters are measured values or values calculated from the measured values. Capital letters in parenthesis for individual parameters are those used in Fig. 4.7. PM, plasma membrane fraction. WT, whole tissue lysate. The data for cerebral leptomeninges are from a previous report (Uchida et al. 2020)

Table 4.3 Protein expression levels of transporters in 1 cm² surface area of leptomeninges in the three spinal cord regions and four cerebral lobes in pig (units: pmol/cm²)

| Molecule | Protein expression level (pmol/cm²) | | | | | | |
| | Spinal cord leptomeninges | | | Cerebral leptomeninges | | | |
	Cervical (upper)	Thoracic (middle)	Lumbar (lower)	Frontal	Parietal	Occipital	Temporal
Organic anion transporters							
OAT1	1.03 ± 0.10	1.52 ± 0.13	1.00 ± 0.09	4.40 ± 0.20	4.86 ± 0.13	4.35 ± 0.13	3.12 ± 0.13
OAT3	0.656 ± 0.075	0.813 ± 0.186	0.596 ± 0.088	2.26 ± 0.16	1.92 ± 0.07	1.71 ± 0.07	1.44 ± 0.08
MRP3	0.166 ± 0.011	0.134 ± 0.057	0.176 ± 0.066	0.121 ± 0.008	0.143 ± 0.010	0.133 ± 0.010	0.110 ± 0.004
MRP4	ULQ(<0.0925)	0.152 ± 0.015	ULQ(<0.165)	0.134 ± 0.006	0.107 ± 0.013	0.195 ± 0.009	0.107 ± 0.007
OATP2B1	ULQ(<0.251)	0.198 ± 0.031	ULQ(<0.216)	0.132 ± 0.006	0.184 ± 0.011	0.174 ± 0.008	0.139 ± 0.010
Lipophilic drug transporters							
MDR1/P-gp	0.305 ± 0.090	0.610 ± 0.118	0.325 ± 0.116	0.804 ± 0.049	1.11 ± 0.04	0.873 ± 0.031	0.591 ± 0.037
BCRP/ABCG2	0.486 ± 0.145	0.995 ± 0.133	0.495 ± 0.226	1.11 ± 0.04	1.46 ± 0.05	1.32 ± 0.04	0.948 ± 0.032
Thyroid hormone transporters							
MCT8	0.537 ± 0.032	0.588 ± 0.032	0.593 ± 0.073	0.282 ± 0.016	0.460 ± 0.017	0.377 ± 0.013	0.233 ± 0.003
Peptide transporter							
PEPT2	0.267 ± 0.066	0.585 ± 0.151	0.278 ± 0.111	0.475 ± 0.029	0.637 ± 0.024	0.643 ± 0.025	0.440 ± 0.025
Glucose transporter							
GLUT1	1.66 ± 0.04	2.64 ± 0.14	1.78 ± 0.13	3.89 ± 0.04	6.15 ± 0.16	5.19 ± 0.34	2.87 ± 0.10
Amino acid transporters							
xCT	1.29 ± 0.27	2.07 ± 0.14	1.51 ± 0.40	1.92 ± 0.07	3.17 ± 0.08	2.55 ± 0.13	2.09 ± 0.06
4F2hc	2.01 ± 0.20	3.14 ± 0.26	2.11 ± 0.16	3.96 ± 0.05	6.01 ± 0.09	4.79 ± 0.16	3.99 ± 0.03
ATA2	ULQ(<0.450)	ULQ(<0.411)	ULQ(<0.430)	0.303 ± 0.018	0.400 ± 0.065	ULQ(<0.245)	0.243 ± 0.018
Organic cation transporters							
MATE1	0.211 ± 0.060	0.496 ± 0.082	0.269 ± 0.076	0.761 ± 0.032	0.531 ± 0.027	0.574 ± 0.018	0.205 ± 0.009
OCT2	0.414 ± 0.102	0.984 ± 0.112	0.224 ± 0.046	1.65 ± 0.07	1.47 ± 0.06	1.41 ± 0.05	0.917 ± 0.043
Monocarboxylate transporter							

MCT1	0.842 ± 0.190	0.863 ± 0.075	0.846 ± 0.115	0.430 ± 0.030	0.700 ± 0.021	0.614 ± 0.058	0.383 ± 0.021
Markers							
Na$^+$/K$^+$-ATPase	16.0 ± 1.8	24.0 ± 3.3	16.3 ± 2.4	35.8 ± 1.8	43.8 ± 3.5	39.8 ± 2.3	26.9 ± 1.5
GFAP	9.28 ± 0.97	9.38 ± 0.66	14.1 ± 1.8	2.45 ± 0.19	3.83 ± 0.13	3.98 ± 0.07	1.62 ± 0.18
SYP	ULQ(<0.166)	ULQ(<0.194)	ULQ(<0.183)	0.649 ± 0.046	0.678 ± 0.051	0.680 ± 0.060	0.438 ± 0.029
Gamma-GTP	0.421 ± 0.057	0.674 ± 0.040	0.286 ± 0.053	0.446 ± 0.018	0.721 ± 0.023	0.537 ± 0.032	0.352 ± 0.009
PECAM1	ULQ(<0.120)	0.233 ± 0.032	ULQ(<0.216)	0.381 ± 0.028	0.607 ± 0.043	0.367 ± 0.021	0.315 ± 0.016

The protein expression levels (pmol/cm^2) were calculated using the data (fmol/µg protein) in Table 4.1 and reported data (Uchida et al. 2020), as described in Fig. 4.7. Values are mean ± SEM. ULQ, under the limit of quantification. The values of the LQ are shown in parenthesis

Table 4.4 Total protein expression amounts of transporters in the cerebral, cervical, thoracic, and lumbar BAB and the right lateral, left lateral, third, and fourth ventricular BCSFB per one pig (units: pmol per pig)

Molecule	Leptomeninges Cervical (upper)	Thoracic (middle)	Lumbar (lower)	Cerebrum	Choroid plexus Right lateral	Left lateral	Third	Fourth
Unit	pmol/cervical BAB	pmol/thoracic BAB	pmol/lumbar BAB	pmol/cerebral BAB	pmol/right lateral BCSFB	pmol/left lateral BCSFB	pmol/third BCSFB	pmol/fourth BCSFB
Organic anion transporters								
OAT1	47.2 ± 4.6	159 ± 14	34.8 ± 3.1	125 ± 3	3.37 ± 0.17	4.01 ± 0.20	2.13 ± 0.10	4.46 ± 0.17
OAT3	30.1 ± 3.4	85.3 ± 19.5	20.8 ± 3.1	54.8 ± 1.8	2.47 ± 0.45	2.34 ± 0.37	0.811 ± 0.059	1.44 ± 0.40
MRP3	7.61 ± 0.50	14.1 ± 6.0	6.13 ± 2.30	3.80 ± 0.12	1.42 ± 0.13	1.48 ± 0.13	0.448 ± 0.018	1.21 ± 0.12
MRP4	ULQ(<4.24)	15.9 ± 1.6	ULQ(<5.75)	4.07 ± 0.21	0.945 ± 0.074	0.669 ± 0.113	0.327 ± 0.023	0.710 ± 0.091
OATP1A2	ULQ(<31.8)	ULQ(<81.8)	ULQ(<28.3)	ULQ(<8.43)	13.9 ± 0.5	14.3 ± 0.8	2.49 ± 0.14	9.34 ± 0.58
OATP2B1	ULQ(<11.5)	20.8 ± 3.3	ULQ(<7.52)	4.72 ± 0.16	1.00 ± 0.09	1.00 ± 0.07	0.264 ± 0.021	0.722 ± 0.073
OATP3A1	ULQ(<5.46)	ULQ(<14.0)	ULQ(<3.06)	ULQ(<1.63)	4.02 ± 0.18	3.72 ± 0.17	0.699 ± 0.024	2.51 ± 0.12
Lipophilic drug transporters								
MDR1/P-gp	14.0 ± 4.1	64.0 ± 12.4	11.3 ± 4.0	25.1 ± 0.9	3.81 ± 0.17	4.11 ± 0.18	1.54 ± 0.05	4.17 ± 0.17
BCRP/ABCG2	22.3 ± 6.7	104 ± 14	17.2 ± 7.9	36.1 ± 0.9	2.50 ± 0.20	2.30 ± 0.25	1.06 ± 0.04	2.55 ± 0.25
MRP1	ULQ(<21.5)	ULQ(<63.7)	ULQ(<20.2)	ULQ(<5.62)	2.55 ± 0.20	2.43 ± 0.22	0.506 ± 0.038	2.56 ± 0.18
Thyroid hormone transporter								
MCT8	24.6 ± 1.5	61.7 ± 3.4	20.6 ± 2.5	10.0 ± 0.6	2.61 ± 0.14	2.76 ± 0.09	0.590 ± 0.015	2.05 ± 0.09
Peptide transporter								
PEPT2	12.2 ± 3.0	61.4 ± 15.8	9.68 ± 3.87	16.4 ± 0.5	1.10 ± 0.11	1.26 ± 0.10	0.473 ± 0.027	ULQ(<0.512)
Glucose transporter								
GLUT1	76.1 ± 1.8	277 ± 15	62.0 ± 4.5	134 ± 8	36.6 ± 1.0	35.4 ± 1.7	7.94 ± 0.13	31.8 ± 0.4
Amino acid transporters								
xCT	59.1 ± 12.4	217 ± 15	52.6 ± 13.9	72.8 ± 3.3	3.10 ± 0.23	3.16 ± 0.02	1.19 ± 0.07	3.74 ± 0.56

4F2hc	92.2 ± 9.2	329 ± 27	73.5 ± 5.6	140 ± 5	9.52 ± 0.54	10.2 ± 0.3	4.11 ± 0.15	13.4 ± 0.4
ATA2	ULQ(<20.6)	ULQ(<43.1)	ULQ(<15.0)	9.40 ± 0.69	ULQ(<1.76)	ULQ(<1.84)	ULQ(<0.494)	ULQ(<1.40)
Organic cation transporters								
MATE1	9.67 ± 2.75	52.0 ± 8.6	9.37 ± 2.65	15.2 ± 0.9	ULQ(<0.530)	ULQ(<0.548)	0.464 ± 0.042	ULQ(<0.204)
OCT2	19.0 ± 4.7	103 ± 12	7.80 ± 1.60	40.6 ± 1.2	ULQ(<0.440)	ULQ(<0.546)	0.452 ± 0.021	ULQ(<0.295)
OCTN2	ULQ(<11.4)	ULQ(<28.2)	ULQ(<8.53)	ULQ(<2.65)	4.26 ± 0.22	4.67 ± 0.18	0.879 ± 0.045	2.78 ± 0.19
Monocarboxylate transporter								
MCT1	38.6 ± 8.7	90.5 ± 7.9	29.5 ± 4.0	15.8 ± 0.9	4.11 ± 0.61	4.35 ± 0.18	1.23 ± 0.06	3.42 ± 0.16
Markers								
AE2	ULQ(<27.7)	ULQ(<95.6)	ULQ(<28.9)	ULQ(<12.2)	42.9 ± 0.7	42.0 ± 3.4	9.92 ± 0.59	44.0 ± 1.0
Na^+/K^+-ATPase	733 ± 82	2518 ± 349	567 ± 83	1074 ± 37	305 ± 11	314 ± 14	81.0 ± 3.0	258 ± 14
GFAP	426 ± 45	984 ± 69	491 ± 61	87.8 ± 8.0	5.13 ± 0.29	6.18 ± 0.16	2.41 ± 0.06	9.46 ± 0.47
SYP	ULQ(<7.61)	ULQ(<20.3)	ULQ(<6.37)	18.2 ± 0.8	4.03 ± 0.33	3.55 ± 0.42	0.785 ± 0.082	3.19 ± 0.32
Gamma-GTP	19.3 ± 2.6	70.7 ± 4.2	10.0 ± 1.9	15.3 ± 0.9	11.1 ± 0.5	10.9 ± 0.1	2.33 ± 0.07	6.71 ± 0.50
PECAM1	ULQ(<5.50)	24.4 ± 3.4	ULQ(<7.52)	12.4 ± 0.8	5.00 ± 0.06	5.52 ± 0.30	1.58 ± 0.07	4.69 ± 0.27

The total protein expression amounts (units, pmol/cervical BAB, etc.) in cervical, thoracic, or lumbar BAB in one pig were calculated using the data (fmol/μg protein) in Table 4.1, as described in Fig. 4.7. The total protein expression amounts (units, pmol/cerebral BAB, etc) in cerebral BAB and each ventricular BCSFB in one pig are reported values (Uchida et al. 2020). Values are mean ± SEM. *ULQ* under the limit of quantification. LQ values are shown in parenthesis (pmol/cerebral BAB). The BAB in the three spinal cord regions expressed much greater amounts of many transporters, such as OAT1, OAT3, MRP3, P-gp, BCRP, PEPT2, MATE1, and OCT2, than did choroid plexus (Table 4.4). This is similar to the quantitative relationship between cerebral BAB and choroid plexus. As described above, in large animals, homovanillic acid is considered to be eliminated from the CNS via OAT1 at the BAB, because the BBB does not express OAT3, which is the main route of homovanillic acid clearance from the brain in rodents. In dog, the cisternal and lumbar CSF concentrations of homovanillic acid are about 20-fold smaller than the ventricular concentration (Moir et al. 1970). The total protein expression amounts of OAT1 at the cerebral, cervical, and lumbar BAB were much greater than those at the BCSFB (Table 4.4). The concentration of homovanillic acid in cisterna magna is increased eightfold by oral administration of probenecid, an inhibitor of OAT1 (Guldberg et al. 1966). In contrast, probenecid did not increase the ventricular concentration very much (1.5-fold) (Guldberg et al. 1966). These results support the idea that the transport capacities of OAT1 are much larger at the BAB in cerebrum and spinal cord than those at the BCSFB.

4.8 Can the CSF Concentrations of P-gp and BCRP Substrates Reflect the Brain ISF Concentration?

Verapamil and quinidine are moderate and good substrates of human P-gp, respectively, and the differences between brain ISF-to-unbound plasma concentration ratio ($K_{p,uu,brain-ISF}$) and cisternal CSF-to-unbound plasma concentration ratio ($K_{p,uu,cisternal-CSF}$) for these compounds at steady state are within a threefold range in cynomolgus monkey (2.3- and 2.6-fold higher $K_{p,uu,cisternal-CSF}$ than $K_{p,uu,brain-ISF}$ for verapamil and quinidine, respectively; $K_{p,uu,cisternal-CSF}$, 0.183 for verapamil and 0.169 for quinidine) (Nagaya et al. 2014). Thus, cisternal CSF concentrations in cynomolgus monkey might be similar to the brain ISF concentrations within a threefold range in the case of P-gp substrates, although the cisternal CSF concentrations could be slightly higher because of the higher efflux activity of P-gp at the BBB than that at the cisternal region of the BAB. However, in the lumbar CSF, the lumbar CSF-to-unbound plasma concentration ratio ($K_{p,uu,lumbar-CSF}$) of verapamil at steady state is 1.13 in humans (Friden et al. 2009), and the inhibition of P-gp activity increases the distribution of nelfinavir (a P-gp substrate) into the brain parenchyma, but not into the lumbar CSF (Kaddoumi et al. 2007). Therefore, it seems that P-gp does not contribute to the efflux transport of substrates at the lumbar BAB. This may be partially explained by the 1.8- to 3.4-fold smaller expression level of P-gp at the lumbar BAB than at the cerebral BAB (Table 4.3). The cisternal CSF concentration might be regulated by P-gp at the cerebral BAB (more abundant) as well as the cervical BAB, and this could explain why the concentrations of P-gp substrates in the cisternal CSF are smaller than those in the lumbar CSF.

The $K_{p,uu,cisternal-CSF}$ of the BCRP substrates dantrolene and daidzein in dogs was 8.3- and 7.3-fold higher than the unbound brain-to-plasma concentration ratio ($K_{p,uu,brain}$) at steady state, respectively (Braun et al. 2017). Thus, compared to P-gp substrates, the cisternal CSF concentrations of BCRP substrates may be considerably larger than the unbound brain concentrations. Because the lumbar BAB has a 1.9- to 3.0-fold smaller expression level of BCRP than the cerebral BAB in pig (Table 4.3), it is possible that the $K_{p,uu,lumbar-CSF}$ of the BCRP substrates is larger than the $K_{p,uu,cisternal-CSF}$.

To understand the relevance of these findings to humans, it will be important to clarify the protein expression levels of transporters, including P-gp and BCRP, at the human BAB, as well as the porcine BAB. The key questions are whether the protein expression levels at the human BAB can help explain the drug concentrations in the lumbar CSF and whether the brain ISF concentrations can be extrapolated from the lumbar CSF concentrations on the basis of the differences in the protein expression levels at the BBB and lumbar BAB.

4.9   Perspectives

The development of candidate drugs that can penetrate into the human CNS remains a very challenging area, in part because BBB permeability screening using in vitro models is of limited predictive value (Bagchi et al. 2019). Furthermore, animal models are unlikely to fully reflect disease-associated changes of drug penetration into the CNS in patients. The CNS contains four barriers, the BBB, BCSFB, BAB, and the blood-spinal cord barrier (BSCB, consisting of spinal cord capillary endothelial cells), and we believe a breakthrough in CNS drug delivery research would require a detailed understanding of the transport mechanisms at these barriers in humans. To achieve this, it is necessary to comprehensively identify transporters and receptors not only at the gene level but also at the functional protein level and to quantitatively elucidate the expression levels and functions of the identified molecules. For example, the tight junction molecule claudin-11, which was not identified at the gene level or in a rodent study, was identified by protein-level analysis using human material for the first time and shown to contribute to tight junction formation at the human BBB to same extent as the conventional tight junction molecule claudin-5; furthermore, claudin-11 plays a causal role in the disruption of CNS barriers in multiple sclerosis (Uchida et al. 2019).

In 2012, Aebersold's group at ETH Zurich, in Switzerland, developed SWATH-MS (sequential window acquisition of all theoretical fragment ion spectra mass spectrometry), which revolutionized comprehensive quantitative proteomics (Gillet et al. 2012). It is greatly superior to the conventional shotgun method in terms of reproducibility, comprehensiveness, quantitative accuracy, and sensitivity (it can cover even low-abundance membrane transporters). The use of SWATH-MS in combination with the in silico peptide selection criteria that we previously established (Kamiie et al. 2008) further improves the data reliability. Application of advanced SWATH-MS to the comprehensive quantitative analysis of plasma membrane proteins at the four CNS barriers in humans makes possible the comprehensive identification of transporters and receptors, including functionally unknown molecules, and thereafter, the potential importance of the identified molecules can be assessed based on absolute protein quantification. Furthermore, we have already developed methodology to reconstruct in vivo transport function from in vitro model data by taking account of the differences in the protein expression levels of transporters between in vivo and in vitro systems (Uchida et al. 2011a). This reconstruction technique has been validated in monkey and an animal disease model (Uchida et al. 2014a, b). Therefore, a similar approach should be applicable to reconstruct the in vivo transport functions of transporters, including newly identified transporters, at the CNS barrier in various pathological conditions in humans. For this purpose, it should be possible to utilize formalin-fixed paraffin-embedded pathological specimens, which are available in many medical institutions, including hospitals, and cover an enormous range of diseases. Such studies should open up many new possibilities for CNS drug development and delivery.

Conflict of Interest
The authors declare no conflict of interest.

Acknowledgments The studies mentioned in this chapter were supported in part by Grants-in-Aids from the Japanese Society for the Promotion of Science (JSPS) for Young Scientists (A) [KAKENHI: 16H06218], Scientific Research (B) [KAKENHI: 17H04004], Bilateral Open Partnership Joint Research Program (between Finland and Japan), Fostering Joint International Research (A) [KAKENHI: 18KK0446], and Early-Career Scientists [KAKENHI: 19 K16438]. This study was also supported in part by Grants-in-Aids from the Ministry of Education, Culture, Sports, Science and Technology (MEXT) for Scientific Research on Innovative Areas [KAKENHI: 18H04534] and from Mochida Memorial Foundation for Medical and Pharmaceutical Research.

References

Asaba H, Hosoya K, Takanaga H, Ohtsuki S, Tamura E, Takizawa T, Terasaki T (2000) Blood-brain barrier is involved in the efflux transport of a neuroactive steroid, dehydroepiandrosterone sulfate, via organic anion transporting polypeptide 2. J Neurochem 75:1907–1916

Bagchi S, Chhibber T, Lahooti B, Verma A, Borse V, Jayant RD (2019) In-vitro blood-brain barrier models for drug screening and permeation studies: an overview. Drug Des Devel Ther 13:3591–3605

Bailey J, Thew M, Balls M (2013) An analysis of the use of dogs in predicting human toxicology and drug safety. Altern Lab Anim 41:335–350

Bowman GL, Shannon J, Frei B, Kaye JA, Quinn JF (2010) Uric acid as a CNS antioxidant. J Alzheimers Dis 19:1331–1336

Braun C, Sakamoto A, Fuchs H, Ishiguro N, Suzuki S, Cui Y, Klinder K, Watanabe M, Terasaki T, Sauer A (2017) Quantification of transporter and receptor proteins in dog brain capillaries and choroid plexus: relevance for the distribution in brain and CSF of selected BCRP and P-gp substrates. Mol Pharm 14:3436–3447

Degrell I, Nagy E (1990) Concentration gradients for HVA, 5-HIAA, ascorbic acid, and uric acid in cerebrospinal fluid. Biol Psychiatry 27:891–896

Deguchi T, Isozaki K, Yousuke K, Terasaki T, Otagiri M (2006) Involvement of organic anion transporters in the efflux of uremic toxins across the blood-brain barrier. J Neurochem 96:1051–1059

Friden M, Winiwarter S, Jerndal G, Bengtsson O, Wan H, Bredberg U, Hammarlund-Udenaes M, Antonsson M (2009) Structure-brain exposure relationships in rat and human using a novel data set of unbound drug concentrations in brain interstitial and cerebrospinal fluids. J Med Chem 52:6233–6243

Gillet LC, Navarro P, Tate S, Rost H, Selevsek N, Reiter L, Bonner R, Aebersold R (2012) Targeted data extraction of the MS/MS spectra generated by data-independent acquisition: a new concept for consistent and accurate proteome analysis. Mol Cell Proteomics 11(O111):016717

Guldberg HC, Ashcroft GW, Crawford TB (1966) Concentrations of 5-hydroxyindolylacetic acid and homovanillic acid in the cerebrospinal fluid of the dog before and during treatment with probenecid. Life Sci 5:1571–1575

Hoshi Y, Uchida Y, Tachikawa M, Inoue T, Ohtsuki S, Terasaki T (2013) Quantitative atlas of blood-brain barrier transporters, receptors, and tight junction proteins in rats and common marmoset. J Pharm Sci 102:3343–3355

Ito K, Uchida Y, Ohtsuki S, Aizawa S, Kawakami H, Katsukura Y, Kamiie J, Terasaki T (2011) Quantitative membrane protein expression at the blood-brain barrier of adult and younger cynomolgus monkeys. J Pharm Sci 100:3939–3950

Kaddoumi A, Choi SU, Kinman L, Whittington D, Tsai CC, Ho RJ, Anderson BD, Unadkat JD (2007) Inhibition of P-glycoprotein activity at the primate blood-brain barrier increases the distribution of nelfinavir into the brain but not into the cerebrospinal fluid. Drug Metab Dispos 35:1459–1462

Kamiie J, Ohtsuki S, Iwase R, Ohmine K, Katsukura Y, Yanai K, Sekine Y, Uchida Y, Ito S, Terasaki T (2008) Quantitative atlas of membrane transporter proteins: development and application of a highly sensitive simultaneous LC/MS/MS method combined with novel in-silico peptide selection criteria. Pharm Res 25:1469–1483

Kodaira H, Kusuhara H, Fujita T, Ushiki J, Fuse E, Sugiyama Y (2011) Quantitative evaluation of the impact of active efflux by p-glycoprotein and breast cancer resistance protein at the blood-brain barrier on the predictability of the unbound concentrations of drugs in the brain using cerebrospinal fluid concentration as a surrogate. J Pharmacol Exp Ther 339:935–944

Konig J, Zolk O, Singer K, Hoffmann C, Fromm MF (2011) Double-transfected MDCK cells expressing human OCT1/MATE1 or OCT2/MATE1: determinants of uptake and transcellular translocation of organic cations. Br J Pharmacol 163:546–555

Kubo Y, Ohtsuki S, Uchida Y, Terasaki T (2015) Quantitative determination of luminal and abluminal membrane distributions of transporters in porcine brain capillaries by plasma membrane fractionation and quantitative targeted proteomics. J Pharm Sci 104:3060–3068

Labuzek K, Suchy D, Gabryel B, Bielecka A, Liber S, Okopien B (2010) Quantification of metformin by the HPLC method in brain regions, cerebrospinal fluid and plasma of rats treated with lipopolysaccharide. Pharmacol Rep 62:956–965

Moir AT, Ashcroft GW, Crawford TB, Eccleston D, Guldberg HC (1970) Cerebral metabolites in cerebrospinal fluid as a biochemical approach to the brain. Brain 93:357–368

Mori S, Takanaga H, Ohtsuki S, Deguchi T, Kang YS, Hosoya K, Terasaki T (2003) Rat organic anion transporter 3 (rOAT3) is responsible for brain-to-blood efflux of homovanillic acid at the abluminal membrane of brain capillary endothelial cells. J Cereb Blood Flow Metab 23:432–440

Nabeshima S, Reese TS, Landis DM, Brightman MW (1975) Junctions in the meninges and marginal glia. J Comp Neurol 164:127–169

Nagaya Y, Nozaki Y, Kobayashi K, Takenaka O, Nakatani Y, Kusano K, Yoshimura T, Kusuhara H (2014) Utility of cerebrospinal fluid drug concentration as a surrogate for unbound brain concentration in nonhuman primates. Drug Metab Pharmacokinet 29:419–426

Ocheltree SM, Shen H, Hu Y, Xiang J, Keep RF, Smith DE (2004) Role of PEPT2 in the choroid plexus uptake of glycylsarcosine and 5-aminolevulinic acid: studies in wild-type and null mice. Pharm Res 21:1680–1685

Ohtsuki S, Ito S, Matsuda A, Hori S, Abe T, Terasaki T (2007) Brain-to-blood elimination of 24S-hydroxycholesterol from rat brain is mediated by organic anion transporting polypeptide 2 (oatp2) at the blood-brain barrier. J Neurochem 103:1430–1438

Ose A, Kusuhara H, Endo C, Tohyama K, Miyajima M, Kitamura S, Sugiyama Y (2010) Functional characterization of mouse organic anion transporting peptide 1a4 in the uptake and efflux of drugs across the blood-brain barrier. Drug Metab Dispos 38:168–176

Patabendige A, Skinner RA, Morgan L, Abbott NJ (2013) A detailed method for preparation of a functional and flexible blood-brain barrier model using porcine brain endothelial cells. Brain Res 1521:16–30

Shen J, Carcaboso AM, Hubbard KE, Tagen M, Wynn HG, Panetta JC, Waters CM, Elmeliegy MA, Stewart CF (2009) Compartment-specific roles of ATP-binding cassette transporters define differential topotecan distribution in brain parenchyma and cerebrospinal fluid. Cancer Res 69:5885–5892

Shen H, Nelson DM, Oliveira RV, Zhang Y, McNaney CA, Gu X, Chen W, Su C, Reily MD, Shipkova PA, Gan J, Lai Y, Marathe P, Humphreys WG (2018) Discovery and validation of pyridoxic acid and homovanillic acid as novel endogenous plasma biomarkers of Organic Anion Transporter (OAT) 1 and OAT3 in cynomolgus monkeys. Drug Metab Dispos 46:178–188

Spector R, Johanson CE (1989) The mammalian choroid plexus. Sci Am 261:68–74

Suzuki H, Sawada Y, Sugiyama Y, Iga T, Hanano M (1985) Saturable transport of cimetidine from cerebrospinal fluid to blood in rats. J Pharmacobiodyn 8:73–76

Sweet DH, Miller DS, Pritchard JB, Fujiwara Y, Beier DR, Nigam SK (2002) Impaired organic anion transport in kidney and choroid plexus of organic anion transporter 3 (Oat3 (Slc22a8)) knockout mice. J Biol Chem 277:26934–26943

Tachikawa M, Kasai Y, Takahashi M, Fujinawa J, Kitaichi K, Terasaki T, Hosoya K (2008) The blood-cerebrospinal fluid barrier is a major pathway of cerebral creatinine clearance: involvement of transporter-mediated process. J Neurochem 107:432–442

Tachikawa M, Uchida Y, Ohtsuki S, Terasaki T (2014) Recent progress in blood–brain barrier and blood–CSF barrier transport research: pharmaceutical relevance for drug delivery to the brain. In: Hammarlund-Udenaes M, de Lange EC, Thorne RG (eds) Drug delivery to the brain – physiological concepts, methodologies and approaches. Springer, New York, pp 23–62

Tanihara Y, Masuda S, Sato T, Katsura T, Ogawa O, Inui K (2007) Substrate specificity of MATE1 and MATE2-K, human multidrug and toxin extrusions/H(+)-organic cation antiporters. Biochem Pharmacol 74:359–371

Terada T, Masuda S, Asaka J, Tsuda M, Katsura T, Inui K (2006) Molecular cloning, functional characterization and tissue distribution of rat H+/organic cation antiporter MATE1. Pharm Res 23:1696–1701

Thorne RG (2014) Appendix: primer on central nervous system structure/function and the vasculature, ventricular system, and fluids of the brain. In: Hammarlund-Udenaes M, de Lange EC, Thorne RG (eds) Drug delivery to the brain – physiological concepts, methodologies and approaches. Springer, New York, pp 685–707

Uchida Y, Ohtsuki S, Kamiie J, Terasaki T (2011a) Blood-brain barrier (BBB) pharmacoproteomics: reconstruction of in vivo brain distribution of 11 P-glycoprotein substrates based on the BBB transporter protein concentration, in vitro intrinsic transport activity, and unbound fraction in plasma and brain in mice. J Pharmacol Exp Ther 339:579–588

Uchida Y, Ohtsuki S, Katsukura Y, Ikeda C, Suzuki T, Kamiie J, Terasaki T (2011b) Quantitative targeted absolute proteomics of human blood-brain barrier transporters and receptors. J Neurochem 117:333–345

Uchida Y, Ohtsuki S, Terasaki T (2014a) Pharmacoproteomics-based reconstruction of in vivo P-glycoprotein function at blood-brain barrier and brain distribution of substrate verapamil in pentylenetetrazole-kindled epilepsy, spontaneous epilepsy, and phenytoin treatment models. Drug Metab Dispos 42:1719–1726

Uchida Y, Wakayama K, Ohtsuki S, Chiba M, Ohe T, Ishii Y, Terasaki T (2014b) Blood-brain barrier pharmacoproteomics-based reconstruction of the in vivo brain distribution of P-glycoprotein substrates in cynomolgus monkeys. J Pharmacol Exp Ther 350:578–588

Uchida Y, Zhang Z, Tachikawa M, Terasaki T (2015) Quantitative targeted absolute proteomics of rat blood-cerebrospinal fluid barrier transporters: comparison with a human specimen. J Neurochem 134:1104–1115

Uchida Y, Sumiya T, Tachikawa M, Yamakawa T, Murata S, Yagi Y, Sato K, Stephan A, Ito K, Ohtsuki S, Couraud PO, Suzuki T, Terasaki T (2019) Involvement of Claudin-11 in disruption of blood-brain, -spinal cord, and -arachnoid barriers in multiple sclerosis. Mol Neurobiol 56:2039–2056

Uchida Y, Goto R, Takeuchi H, Łuczak M, Usui T, Tachikawa M, Terasaki T (2020) Abundant expression of OCT2, MATE1, OAT1, OAT3, PEPT2, BCRP, MDR1 and xCT transporters in blood-arachnoid barrier of pig, and polarized localizations at CSF- and blood-facing plasma membranes. Drug Metab Dispos. Epub ahead of print

van de Wetering K, Zelcer N, Kuil A, Feddema W, Hillebrand M, Vlaming ML, Schinkel AH, Beijnen JH, Borst P (2007) Multidrug resistance proteins 2 and 3 provide alternative routes for hepatic excretion of morphine-glucuronides. Mol Pharmacol 72:387–394

Walters EM, Agca Y, Ganjam V, Evans T (2011) Animal models got you puzzled?: think pig. Ann N Y Acad Sci 1245:63–64

Williams AC, Ramsden DB (2005) Autotoxicity, methylation and a road to the prevention of Parkinson's disease. J Clin Neurosci 12:6–11

Wolman AT, Gionfriddo MR, Heindel GA, Mukhija P, Witkowski S, Bommareddy A, Vanwert AL (2013) Organic anion transporter 3 interacts selectively with lipophilic beta-lactam antibiotics. Drug Metab Dispos 41:791–800

Yaguchi Y, Tachikawa M, Zhang Z, Terasaki T (2019) Organic anion-transporting polypeptide 1a4 (Oatp1a4/Slco1a4) at the blood-arachnoid barrier is the major pathway of sulforhodamine-101 clearance from cerebrospinal fluid of rats. Mol Pharm 16:2021–2027

Yamamoto Y, Danhof M, de Lange ECM (2017) Microdialysis: the key to physiologically based model prediction of human CNS target site concentrations. AAPS J 19:891–909

Yasuda K, Cline C, Vogel P, Onciu M, Fatima S, Sorrentino BP, Thirumaran RK, Ekins S, Urade Y, Fujimori K, Schuetz EG (2013) Drug transporters on arachnoid barrier cells contribute to the blood-cerebrospinal fluid barrier. Drug Metab Dispos 41:923–931

Zhang Z, Uchida Y, Hirano S, Ando D, Kubo Y, Auriola S, Akanuma SI, Hosoya KI, Urtti A, Terasaki T, Tachikawa M (2017) Inner blood-retinal barrier dominantly expresses breast cancer resistance protein: comparative quantitative targeted absolute proteomics study of CNS barriers in pig. Mol Pharm 14:3729–3738

Zhang Z, Tachikawa M, Uchida Y, Terasaki T (2018) Drug clearance from cerebrospinal fluid mediated by organic anion transporters 1 (Slc22a6) and 3 (Slc22a8) at arachnoid membrane of rats. Mol Pharm 15:911–922

Chapter 5
Quantitative and Targeted Proteomics of the Blood-Brain Barrier: Species and Cell Line Differences

Shingo Ito, Takeshi Masuda, and Sumio Ohtsuki

Abstract Proteomics is a powerful tool for comprehensive comparison of protein expression using quantitative proteomics as well as for determining the absolute expression levels of target proteins by quantitative targeted absolute proteomics (QTAP). Such proteomic techniques have been used in blood-brain barrier (BBB) research, and the output of these approaches has yielded substantial information. This chapter introduces two proteomic applications for understanding BBB models. One of them is QTAP, which is used for assessing species differences in a variety of different BBB proteins, including both transporters (e.g., ABC and SLC family members) and receptors. Analysis of protein expression levels of transporters such as MDR1/Mdr1a in isolated brain microvessels has demonstrated significant species-level differences. Quantitative plasma membrane proteomics is another technique used for comparing BBB model cell lines, which also introduces methodologies for plasma membrane preparations. The expression profile of membrane proteins in cultured cells provides helpful information and new insights for assessing such cells, particularly for in vitro BBB model systems.

Keywords Quantitative proteomics · Targeted proteomics · Protein expression · ABC transporter · SLC transporter · Receptor · Plasma membrane · Species differences · Cell line differences

5.1 Introduction

Proteomics is becoming an important and essential approach for the comprehensive identification of proteins in the blood-brain barrier (BBB) research as well as in other areas of life sciences research. Currently, proteomics can be used to obtain

S. Ito · T. Masuda · S. Ohtsuki (✉)
Department of Pharmaceutical Microbiology, Faculty of Life Sciences,
Kumamoto University, Kumamoto, Japan
e-mail: sohtsuki@kumamoto-u.ac.jp

quantitative information on protein expression levels via mass spectrometry and can be classified into two main types: quantitative and targeted proteomics (Veenstra 2007). Quantitative proteomics is an unbiased approach that is used to comprehensively compare the relative protein levels among multiple sample groups. Targeted proteomics is a biased approach that is used to measure only targeted proteins. While quantitative proteomics can provide significantly higher proteomic information than targeted proteomics, the sensitivity of targeted proteomics is better. Furthermore, targeted proteomics can be used to determine the absolute amounts of targeted proteins by introducing internal standard peptides along with the samples, and this method is known as quantitative targeted absolute proteomics (QTAP) (Ohtsuki et al. 2013). There are various applications of these techniques in BBB research. From the drug development point of view, quantitative proteomics has been utilized for predicting drug distribution in the human brain, which is rather challenging (Ohtsuki et al. 2011). In this chapter, we focus on two proteomic applications for understanding BBB models: QTAP for assessing differences in BBB proteins at the species level and quantitative plasma membrane proteomics for comparing protein expression in BBB model cell lines.

5.2 Profile of Transporter and Receptor Proteins in the Microvessels of the Human Brain

It is important to elucidate the molecular basis of transport function at the human BBB to understand the distribution of drugs and endogenous compounds in the human brain and to identify species-level differences. Uchida et al. investigated the expression of ABC and SLC transporter proteins from isolated human brain microvessels using QTAP (Uchida et al. 2011a). Brain microvessels were isolated from six frozen human cerebral cortices obtained from Caucasian patients and one frozen human cerebral cortex from a Japanese patient. The absolute protein expression levels of 34 ABC transporters, 66 SLC transporters, and eight receptors were measured in the microvessels of the isolated human brain using QTAP. Among the ABC transporters, ABCG2 (8.14 fmol/μg protein) was the most abundant ABC transporter, and MDR1/ABCB1 (6.06 fmol/μg protein) was the second most highly expressed ABC transporter. The mRNA expression profile of each brain cell type is available in the Brain RNAseq database (Zhang et al. 2016). The mRNA expression of ABCG2 (FPKM 48.2) in human brain endothelial cells was greater than that of MDR1 (FPKM 18.5), which supports that expression of ABCG2 in human brain microvascular endothelial cells was greater than that in MDR1.

Among the measured SLC transporters, GLUT1/SLC2A1 (glucose transporter 1, 139 fmol/μg protein) was the most abundant SLC transporter in isolated human brain microvessels. High levels of EAAT1/SLC1A3 (excitatory amino acid transporter 1, 24.5 fmol/μg protein) were detected. GLUT3/14/SLC2A3/14 (4.40 fmol/μg protein), 4F2hc/SLC3A2 (3.47 fmol/μg protein), BGT1/SLC6A12 (betaine-GABA transporter, 3.16 fmol/μg protein), CAT1/SLC7A1 (cationic amino acid transporter 1, 1.13 fmol/μg protein), and MCT1/SLC16A1 (monocarboxylate transporter 1, 2.27 fmol/μg protein) were also detected at levels greater than 1 fmol/μg protein in the isolated human brain microvessels. From the mRNA expression profile obtained from the Brain RNAseq database, the mRNA of GLUT1 (FPKM 79.5), CAT1 (FPKM 15.8), and MCT1 (FPKM 17.1) was found to be predominantly expressed in the brain endothelial cells, suggesting that the amount of these proteins reflected their expression levels in microvascular endothelial cells. In contrast, EAAT1 mRNA was predominantly detected in mature astrocytes (FPKM 972), and its expression in endothelial cells was low (FPKM 17). Since it is possible that the astrocytes in the isolated brain microvessels are contaminated, the EAAT1 levels measured using QTAP are likely to be inclusive of their levels in astrocytes.

The amounts of LAT1/SLC7A5 (large neutral amino acids transporter 1, 0.431 fmol/μg protein), RFC/SLC19A1 (reduced folate carrier, 0.763 fmol/μg protein), and ENT1/SLC29A1 (equilibrative nucleoside transporter 1, 0.568 fmol/μg protein) were below 1 fmol/μg protein. The mRNA of LAT1 (FPKM 22.7) was predominantly expressed in human endothelial cells, but the mRNA of RFC (FPKM 1.81) and ENT1 (FPKM 1.5) was widely expressed in the brain cells. Therefore, it is suggested that although LAT1 mRNA is predominantly expressed in the microvascular endothelial cells of humans, its protein expression is low. Expression of drug transporter proteins such as peptide transporters (PEPTs), organic anion transporters (OATs), organic anion-transporting polypeptides (OATPs), organic cation transporters (OCTs), organic cation/carnitine transporters (OCTNs), and multidrug and toxic compound extrusions (MATEs) could not be detected.

Among the transcytosis receptors, transferrin receptor (TFRC, 2.34 fmol/μg protein) insulin receptor (INSR, 1.09 fmol/μg protein), and low-density lipoprotein receptor-related protein 1 (LRP1, 1.51 fmol/μg protein) were detected in the isolated human brain microvessels. The mRNA expression level of TFRC (FPKM 21.9) was found to be the highest in human brain endothelial cells compared to INSR (FPKM 2.72) and LRP1 (FPKM 1.13), suggesting that among these receptors, TFRC1 is predominantly expressed in the human BBB.

5.2.1   Species-Level Differences in Isolated Brain Microvessels: ABC Transporters

In a previous study, species-level differences in mRNA expression of ABC transporters were identified in the brain microvessels of various species, including mice, rats, pigs, cows, and humans (Warren et al. 2009). To date, information on the protein expression of drug transporters at the BBB by QTAP for species such as mice (Kamiie et al. 2008; Uchida et al. 2013), rats (Hoshi et al. 2013), dogs (Braun et al. 2017), marmosets (Hoshi et al. 2013), cynomolgus monkeys (Ito et al. 2011), and humans (Uchida et al. 2011a) have been published. The protein expression levels for each species are summarized in Table 5.1.

Among the ABC transporters, there is cooperative action of MDR1/ABCB1 and ABCG2 at the brain barrier to prevent the entry of drugs into the brain by pumping them out from the endothelial cells into the circulating blood. The prediction of the drug distribution pattern in the human brain greatly depends on species-level differences in these ABC transporters at the BBB. Furthermore, information on their absolute protein expression levels is essential for choosing appropriate animal models and for interpreting the results obtained using these models in preclinical studies. QTAP revealed species-level differences in the expression of MDR1 and ABCG2 in the isolated brain microvessels. In mouse brain microvessels, the expression of Abcg2 was approximately 70% lower than that of Mdr1a/Abcb1a. In contrast, in humans, the expression of ABCG2 was 1.3-fold greater than that of ABCB1, which was 36% and 32% of Mdr1a in mice and rats, respectively. The expression level of ABCB1 in monkeys is similar to that in humans and is 31% of Mdr1a expression in mice.

It is possible that the lower protein expression of MDR1 in human and monkey brain microvessels leads to the prediction of higher brain distribution of MDR1 substrates in primates than in rodents. In fact, a previous PET analysis study has reported that the brain distribution of [^{11}C]GR205171 and [^{18}F]altanserin, which are MDR1 substrates, was 8.6- and 4.5-fold greater in humans than in rodents, respectively (Syvanen et al. 2009). The extent of penetration of [^{11}C]verapamil and [^{11}C]GR205171 into the brains of monkeys was also 4.1- and 2.8-fold greater than in rodents, respectively (Syvanen et al. 2009). As shown in Table 5.1, the brains of marmosets show similar expression levels of MDR1 and ABCG2 as humans. This suggests that marmosets are an appropriate model to predict drug distribution patterns in the human brain rather than in rodents. In dogs, ABCG2 expression was greater than MDR1 expression as in humans. However, it should be noted that ABCG2 expression in dogs is 5.6-fold greater than in humans.

The compensation of Mdr1a/b and Abcg2 expression was investigated in Mdr1a/b(–/–) double knockout, Abcg2(–/–) knockout, and Mdr1a/b(–/–) Abcg2(–/–) triple knockout mice (Agarwal et al. 2012). There was no significant difference in the expression of Abcg2 between wild-type and Mdr1a/b(–/–) double knockout mice. Similarly, there was no difference in the expression of Mdr1a between wild-type and Abcg2(–/–) knockout mice. Furthermore, a similar trend was observed in the expression of the other transporter and receptor proteins that were measured in the isolated brain microvessels in the Mdr1a/b(–/–) double knockout, Abcg2(–/–) knockout, and Mdr1a/b(–/–) Abcg2(–/–) triple knockout and wild-type mice. Thus, it was concluded that there are no compensatory changes in the protein expression of transporters and receptors in these knockout mice.

Table 5.1 Protein expression levels of transporters, receptors, and tight junction proteins in the isolated brain capillaries of different species

Gene symbols/alias	Protein expression (fmol/μg protein)							
	Mouse (ddy)	Mouse (C57BL/6 J)	Rat (SD)	Rat (Wistar)	Dog	Marmoset	Monkey (Chinese adult cynomolgus)	Human
	Cerebrum	Cerebrum	Cerebrum	Cerebrum	Cerebrum	Cerebrum	Cerebrum	Cerebrum
	Mean ± SEM	Mean ± SEM	Mean ± SEM	Mean ± SEM	Mean ± SEM	Mean ± SEM	Mean ± SEM	Mean ± SEM
ABC transporters								
ABCB1/MDR1 (Mdr1a: mouse and rat)	15.5 ± 0.8	17.8 ± 1.2	19.0 ± 2.0	19.2 ± 1.1	6.71 ± 1.82	6.48 ± 1.31	5.12 ± 0.91	6.06 ± 1.69
ABCC1/MRP1	Not measured	<LOQ	<LOQ	<LOQ	<LOQ	<LOQ	<LOQ	<LOQ
ABCC4/MRP4	1.59 ± 0.07	1.51 ± 0.27	1.60 ± 0.29	1.46 ± 0.08	<LOQ	0.320 ± 0.057	0.303 ± 0.008	0.195 ± 0.069
ABCG2/BCRP	4.02 ± 0.29	5.48 ± 0.37	4.15 ± 0.29	5.74 ± 0.50	45.2 ± 10.8	16.5 ± 1.4	14.2 ± 1.4	8.14 ± 2.26
SLC transporters								
SLC2A1/GLUT1	90.0 ± 2.9	101 ± 4	84.0 ± 4.1	98.2 ± 7.0	209 ± 64	145 ± 20	131 ± 22	139 ± 46
SLC3A2/4F2hc	16.4 ± 0.3	Not measured	<LOQ	<LOQ	24.7 ± 5.0	3.69 ± 0.30	Not measured	3.47 ± 0.83
SLC7A5/LAT1	2.19 ± 0.09	1.17 ± 0.36	3.41 ± 0.74	2.58 ± 0.84	<LOQ	not measured	<LOQ	0.431 ± 0.091
SLC15A2/PEPT2	Not measured	Not measured	Not measured	Not measured	1.73 ± 0.50	Not measured	Not measured	<LOQ
SLC16A1/MCT1	23.7 ± 0.9	13.7 ± 0.5	11.6 ± 0.6	13.5 ± 0.8	<LOQ	3.04 ± 0.35	0.755 ± 0.373	2.27 ± 0.85
SLC22A2/OCT2	Not measured	Not measured	Not measured	Not measured	±	Not measured	<LOQ	<LOQ
SLC22A8/OAT3	1.97 ± 0.07	2.29 ± 0.40	2.13 ± 0.49	1.37 ± 0.18	<LOQ	not measured	<LOQ	<LOQ
SLCO1A2/OATP1A2 (Slco1a4: mouse)	2.11 ± 0.12	Not measured	Not measured	Not measured	<LOQ	Not measured	0.725 ± 0.046	<LOQ
SLCO2B1/OATP2B1	Not measured	Not measured	Not measured	Not measured	Not measured	<LOQ	<LOQ	<LOQ

(continued)

Table 5.1 (continued)

Gene symbols/alias	Protein expression (fmol/μg protein)							
	Mouse (ddy)	Mouse (C57BL/6 J)	Rat (SD)	Rat (Wistar)	Dog	Marmoset	Monkey (Chinese adult cynomolgus)	Human
	Cerebrum	Cerebrum	Cerebrum	Cerebrum	Cerebrum	Cerebrum	Cerebrum	Cerebrum
	Mean ± SEM	Mean ± SEM	Mean ± SEM	Mean ± SEM	Mean ± SEM	Mean ± SEM	Mean ± SEM	Mean ± SEM
SLC29A1/ENT1	Not measured	Not measured	Not measured	Not measured	0.581 ± 0.342	Not measured	0.541 ± 0.072	0.568 ± 0.134
Receptors								
TFRC	Not measured	5.22 ± 0.47	6.74 ± 0.39	8.93 ± 1.16	18.0 ± 0.8	Not measured	Not measured	2.34 ± 0.76
INSR	Not measured	1.13 ± 0.18	0.785 ± 0.111	1.15 ± 0.34	<LOQ	0.656 ± 0.157	1.46 ± 0.22	1.09 ± 0.21
LRP1	Not measured	1.37 ± 0.33	1.09 ± 0.14	1.16 ± 0.21	1.49 ± 0.76	Not measured	1.29 ± 0.05	1.51 ± 0.26
Tight junction proteins								
Claudin-5	Not measured	8.07 ± 1.47	7.91 ± 0.9	7.00 ± 0.80	Not measured	8.03 ± 0.98	7.17 ± 0.77	Not measured
Membrane markers								
Na$^+$/K$^+$ ATPase	39.4 ± 1.01	39.0 ± 0.9	68.6 ± 4.5	36.2 ± 5.5	70.5 ± 10.8	31.5 ± 3.5	36.1 ± 8.7	35.1 ± 12.6
Reference	Kamiie et al. 2008	Uchida et al. 2013	Hoshi et al. 2013	Hoshi et al. 2013	Braun et al. 2017	Hoshi et al. 2013	Ito et al. 2011	Uchida et al. 2011

<LOQ: Below the lower limit of quantification

5.2.2 Species-Level Differences in Isolated Brain Microvessels: SLC Transporters

Various SLC transporters are expressed in the BBB, and one of their important roles is to supply nutrients, such as glucose and amino acids, to the brain. The glucose transporters GLUT1 (SLC2A1) and GLUT3/14 (SLC2A3/14) were detected in isolated brain capillary endothelial cells (Uchida et al. 2011a). The protein expression level of GLUT1 was 32-fold greater than that of GLUT3/14 (139 vs. 4.4 fmol/µg protein), suggesting that GLUT1 is mainly involved in glucose transport in the human BBB. The maximal velocity of glucose transport across the human BBB was reported to be 0.4–2.0 µmol/min/g of the brain and that across the mouse BBB was 1.42 µmol/min/g of the brain (Pardridge 1983; Gruetter et al. 1996). The similarity in the transport rates in humans and mice is consistent with the GLUT1 protein levels in the BBB (80.4–216 and 82.1–101 fmol/µg protein, respectively) (Uchida et al. 2011a, 2013).

Among the glucose transporters, GLUT3/14 showed a remarkable interspecies difference. Although GLUT3/14 has been quantified in isolated human and monkey brain capillaries, it has not been performed in mouse brain capillaries. Brain RNAseq showed that the mRNA expression of GLUT3 in human endothelial cells (FPKM 19.3) was the same as that in neurons (FPKM 11.6). Furthermore, mRNA expression of GLUT3 was detected in immortalized human brain microvascular endothelial cells (hCMEC/D3 cells) (Meireles et al. 2013). In contrast, mRNA expression of GLUT3 in mouse brain endothelial cells (FPKM 0.983) was much lower than that in mouse neurons (FPKM 32.9). Therefore, GLUT3 is expressed in the primate BBB and might be involved in glucose transport from the blood to the brain.

MCT1 (SLC16A1) is a proton-coupled monocarboxylate transporter that mediates the transport of lactate and ketone bodies across the BBB. Lactate and ketone bodies can also provide energy to the brain like glucose. Although GLUT1 showed similar protein expression levels among species, the protein expression of MCT1 in mouse and rat was more than 3.8-fold greater than that in marmosets, monkeys, and humans (Table 5.1). While the rate of glucose consumption in the mouse brain was estimated to be higher than that of the human brain (van Gelder 1989), the glucose transport rate and GLUT1 expression at the BBB were similar in both humans and mice as mentioned above. Thus, the higher expression of MCT1 in rodents is likely to be responsible for the supply of lactate and ketone bodies as energy sources to the brain to support higher energy demand in the brain.

LAT1 (SLC7A5) and 4F2hc (SLC3A2) can form a heterodimer and function as a transporter for large neutral amino acids, such as leucine, tryptophan, tyrosine, and phenylalanine. Protein expression levels of LAT1 and 4F2hc in the human brain microvessels were 20% and 21% of those in mice, respectively (Table 5.1). PET analysis has indicated that the rate of cerebral protein synthesis in the human brain (0.345–0.614 nmol/min/g) is lower than that in the rodent brain (3.38 nmol/min/g) (Hawkins et al. 1989). The concentration of serotonin, the precursor for which is tryptophan, is lower in the human brain than in the mouse brain (20 ng/g vs. 679 ng/g brain) (Irifune et al. 1997; Young et al. 1994). Based on these results, it was suggested that the supply of large neutral amino acids across the BBB is slower in humans than in mice due to lower expression of LAT1 and 4f2hc in human brain microvessels.

OAT3 (SLC22A8) is an organic anion transporter involved in the brain-to-blood efflux transport of anionic drugs and neurotransmitter metabolites across the rodent BBB (Mori et al. 2003, 2004). Although OAT3 was detected in mouse and rat brain microvessels, it was found to be under detection limits in human, monkey, and dog brain microvessels (Table 5.1). This species-level difference was supported by the RNAseq data. While the mRNA expression of OAT3 was selectively higher in the microvascular endothelial cells (FPKM 188) in the mouse brain, its expression in human microvascular endothelial cells was lower (FPKM 0.225). The differences in the protein and mRNA expressions suggest that OAT3 plays a weaker role in the brain-to-blood efflux of anionic compounds across the BBB in humans than in mice. Similarly, ASCT2 (SLC1A5), TAUT (SLC6A6), and Oatp2 (Slco1a4) were detected in mouse brain microvessels but not in human brain microvessels.

5.2.3   Species-Level Differences in Isolated Brain Microvessels: Receptors

TFRC, INSR, and LRP1 are expressed at the BBB; therefore, their antibodies and ligand peptides can be considered as BBB-permeable carriers for drug delivery to the brain. The protein expression levels of TFRC in humans are 45%, 30%, and 13% of that in mouse, rat, and dog, respectively (Table 5.1). Although TFRC is a promising target for the delivery of macromolecules across the BBB, differences in its protein expression suggest the possibility that prediction of delivery efficiency using anti-TFRC antibodies to the human brain of animal models based on the delivery efficiency in animal models might be overestimated. The protein expression levels of INSR and LRP1 were similar among the species (Table 5.1).

5.3   Necessities of Plasma Membrane Proteome Analysis for BBB Research

BBB transport functions are mainly regulated by plasma membrane proteins, such as transporters, receptors, and tight junction proteins. To maintain CNS health, selective permeability of endogenous metabolites by the BBB is required, and these various plasma membrane proteins cooperate with each other for this purpose. In case of the tight junction as the physical barrier of the BBB, a study has shown that knockout of the claudin-5 gene in mice showed only a minor effect on the BBB physical barrier, despite the fact that claudin-5 is considered to be an essential protein at the tight junctions (Nitta et al. 2003). This indicates that the tight junctions might be formed of other proteins in addition to claudin-5. Moreover, for the transport functions of the BBB, the concentration of small compounds in the CNS is regulated by multiple transporter proteins that are responsible for efflux or influx actions.

The subcellular localization of plasma membrane proteins is dynamically changed, which affects their functions. It has been reported that the protein level of ABCG2 on the plasma membrane is highly correlated with its transporter activity (Liu et al. 2017). The expression level of plasma membrane proteins is relatively low within the proteome. The mRNA levels do not correlate completely with protein expression levels (Ohtsuki et al. 2012). Therefore, to understand the role of the expression of plasma membrane proteins in the BBB functions, it is important to enrich the plasma membrane fraction from the cells and comprehensively analyze the membrane proteins by quantitative and targeted proteomic approaches.

5.4 Methodologies of Plasma Membrane Preparation

To perform plasma membrane proteomics, it is necessary to collect the plasma membrane fraction using a compatible sample preparation method for proteomic analysis. To date, several plasma membrane enrichment protocols have been reported. Sucrose density gradient centrifugation was reported by Boone et al. (Boone et al. 1969) and has been applied to analyze various cell types and tissues. This protocol has been established since long, and it is possible to obtain other cellular organelles at the same time. However, it requires a gradient maker as well as an ultracentrifuge. Affinity chromatography-based purification protocols use antibodies or lectins to target cell surface proteins (Lee et al. 2008; Pahlman et al. 1979; Lawson et al. 2006). The enrichment efficiency of plasma membranes by these protocols was greater than that of sucrose density centrifugation. However, the plasma membrane fractions include several contaminants of antibodies or lectin proteins, which affect the accurate quantification of proteins in the subsequent steps. In addition, these traditional separation protocols require a large number of cells (at least 10^9 cells as the starting material). On the other hand, large-scale proteome profiling methods require lesser amounts of starting materials owing to the continual advancements in LC-MS techniques.

Recently, we have demonstrated that the Plasma Membrane Extraction Kit (BioVison, USA) with a modified protocol is effective for the enrichment of the plasma membrane fraction from as low as 5×10^6 cells using HEK293 cells and human BBB model cell lines, and it is a proteomics compatible approach (Fig. 5.1) (Masuda et al. 2019). In this method, Na$^+$/K$^+$ ATPase, which was used as a plasma membrane marker protein, was highly enriched in the plasma membrane fraction, whereas GM130 and COX4 proteins, which are organelle markers, were effectively eliminated from the plasma membrane fraction. One of the tips for using this kit is to use a fully thawed cell pellet. It is important to uniformly break the cells to obtain a highly pure plasma membrane fraction. A partially frozen cell pellet results in a lack of uniformity in cell disruption. Moreover, this kit does not require ultracentrifugation and yields a few micrograms of plasma membrane proteins, which is sufficient for recent large-scale quantitative proteomics (Fig. 5.1). In addition, the plasma membrane fraction can be collected as a pellet, which is directly applied to sample preparation for LC-MS/MS-based proteomics.

5.5 Comparison of Plasma Membrane Proteome Between Two BBB Cell Lines

In a previous study by Kubo et al., luminal-rich and abluminal-rich fractions of the plasma membrane of brain microvessels were prepared from 100 g of porcine brain using Ficoll density gradient centrifugation (Kubo et al. 2015). From the values quantified for each fraction by QTAP, they successfully determined the luminal- and abluminal-distribution ratios of membrane transporter proteins in brain microvessels comprehensively. Although plasma membrane proteomics of intact brain microvessels can provide important information, large amounts of brain are necessary for sample preparation. Due to this limitation in analyzing the human BBB,

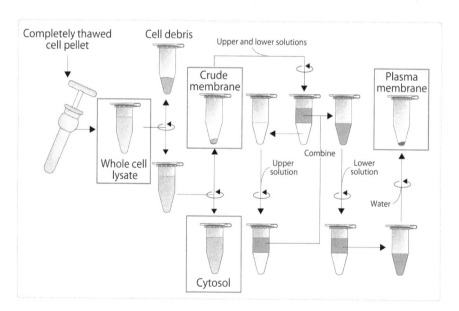

Fig. 5.1 Schematic representation of plasma membrane preparation procedure using the Plasma Membrane Extraction Kit

The subcellular fractions, including the cytosol, crude membranes, and plasma membrane fractions, were prepared from cultured cells. Detailed information regarding the preparation procedure has been described in our previous report (Masuda et al. 2019)

in vitro human BBB cell models have been established and used not only for the development of drugs targeting the CNS but also for studying BBB biology. Several groups have reported different approaches to study the BBB model cell line, such as immortalizing the human brain microvascular endothelial cells (HBMEC) (Kamiichi et al. 2012; Stins et al. 2001; Weksler et al. 2005) and deriving them from human embryonic, pluripotent, or cord blood hematopoietic stem cells (Boyer-Di Ponio et al. 2014; Cecchelli et al. 2014; Lippmann et al. 2012). The expression of major signature proteins of BBB, such as GLUT1, LAT1, MCT1, MDR1, ZO-1, and occludin, was confirmed by mRNA and/or protein expression analysis. In addition, it was validated that all these models have selective permeability and physical barriers, which are observed in the in vivo BBB. The uptake and efflux activities were measured using marker compounds, such as acetylated low-density lipoproteins, rhodamine, and verapamil. Interestingly, stem cell-derived BBB models show relatively high transendothelial electrical resistance values (TEER) compared to immortalized BBB model cell lines (Helms et al. 2016). However, MDR1 expression was insufficient in stem cell-derived models (Kurosawa et al. 2018; Ohshima et al. 2019). To assist researchers in selecting the most appropriate cell line for specific purposes, information on the large-scale plasma membrane protein expression profile of these BBB model cell lines would be beneficial.

hCMEC/D3 and HBMEC/ciβ cells are brain microvascular endothelial cell lines and were established by immortalization through transduction of a human telomerase reverse transcriptase (hTERT) subunit and a simian virus 40 large T antigen (SV40T) (Weksler et al. 2005) or temperature-sensitive SV40T (tsSV40T) (Kamiichi et al. 2012). Both cell lines showed similar TEER values (5–20 Ω cm^2) and sodium fluorescein permeability values (1–3×10^{-5} cm/s) (Eigenmann et al. 2013; Furihata et al. 2015). Recently, to assist researchers in selecting the most appropriate cell line, large-scale quantitative proteomic data of the plasma membrane fractions was reported (Masuda et al. 2019). The plasma membrane fractions enriched by the Plasma Membrane Extraction Kit were subjected to comparative proteomics, and 2350 proteins were quantified. This dataset contains 345 plasma membrane proteins, and the expression levels of 100 and 35 out of the 345 proteins were significantly increased or decreased in hCMEC/D3 to HBMEC/ciβ, respectively.

As shown in Fig. 5.2, hCMEC/D3 cells expressed higher levels of amino acid transporters (SNAT1, SNAT2, SNAT5, ASCT1, CAT1, and LAT1), ABC transporters (BCRP, MDR1, and MRP4), and GLUT1 than HBMEC/ciβ. The expression level of TFRC was also 4.56-fold greater in hCMEC/D3 cells. In another report, the absolute amounts of plasma membrane proteins were compared between hCMEC/D3 cells and human brain microvessels (Ohtsuki et al. 2013). MDR1 levels in hCMEC/D3 cells were similar to those in human brain microvessels, indicating that hCMEC/D3 cells are more suitable for efflux assays than HBMEC/ciβ cells. In addition to P-gp, the protein expression profile suggested that the sensitivity and dynamic range of the ABC transporter-mediated efflux assay and TFRC-mediated uptake assay is higher in hCMEC/D3 cells than in HBMEC/ciβ cells. The internalization of plasma membrane proteins was comprehensively identified in hCMEC/D3 cells by the combination of surface biotinylation and quantitative proteomics

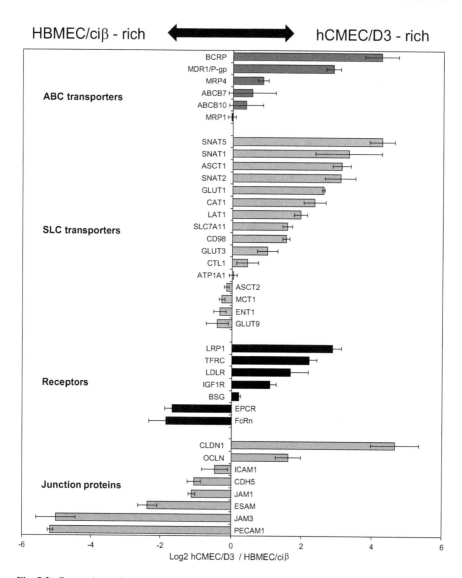

Fig. 5.2 Comparison of protein expression levels between two types of cells used as BBB models The ratios of the proteins expressed in hCMEC/D3 cells were compared to that in HBMEC/ciβ cells. The original protein expression data have been reported in our previous paper (Masuda et al. 2019)

(Ito et al. 2020). Among the identified internalized proteins, TFRC showed the most abundant levels in the internalization fraction.

In contrast, junction proteins, such as PECAM1, JAM1, JAM3, and ESAM as well as neonatal Fc receptor (FcRn), were highly expressed in HBMEC/ciβ cells compared to hCMEC/D3 cells. FcRn mediates the clearance of IgG from the CNS to the blood across the BBB (Zhang and Pardridge 2001). For efficient pharmacological treatment with antibodies, it is important to understand the FcRn-mediated clearance of IgG from the brain, which might indicate that HBMEC/ciβ cells are a more suitable model cell line for antibody retention assays. In addition, HBMEC/ciβ cells might form better tight junctions than hCMEC/D3 cells as tight junction protein levels in HBMEC/ciβ cells were higher (2.19-fold on an average of eight proteins) than in hCMEC/D3 cells. From these results, it was concluded that HBMEC/ciβ cells might be a more suitable BBB model cell line than hCMEC/D3 for IgG-clearance assay and for integrity assays of tight junctions. Similarly, plasma membrane proteome analysis is also an effective technique for assessing the effect of culture conditions on cellular functions as a BBB model.

5.6   Conclusion

This chapter summarizes the application of quantitative and targeted proteomics for assessing animal and cultured cell models for human BBB. However, this is only one aspect of the applications, and proteomics can also be used for identifying novel targets for drug development and brain delivery at the BBB. The unique characteristics of the protein are due to their posttranslational modifications, such as phosphorylation. Phospho-proteomics has also been conducted to understand the posttranslational regulation of MDR1 activity at the BBB (Hoshi et al. 2019). Currently, highly sensitive proteomic methods are available, and the expression levels of a number of proteins can be compared from samples whose amounts are as low as 1 μg. For the progress of BBB proteomics, one of the important challenges is the purification of brain microvessels. As mentioned above, contamination of other brain cells cannot be excluded in the proteomic data from isolated brain microvessels. Furthermore, at least five mouse brains are necessary to isolate brain microvessels. To overcome this, protein expression can be confirmed using immunohistochemical analysis, but it is difficult to obtain antibodies for specific modifications in the target proteins. Recently, we developed a new method to isolate brain microvessels from single frozen mouse brains with higher purity than the standard isolation method (Ogata et al. 2021). This method can produce multiple sample preparations in parallel using a bead homogenizer. Despite higher purity, contaminants from other brain cells were still observed. Therefore, purification methods need to be improved for brain microvessels to obtain better omics data, including proteomics.

The absolute amounts of transporters obtained using QTAP are also important information for predicting drug distribution in the brain. Such prediction studies have been reported in terms of in vitro to in vivo extrapolation from the liver and intestine using the relative expression factor method. Uchida et al. reported that the transporter activity of MDR1/Mdr1a in the mouse BBB and the drug distribution in the mouse and monkey brain can be reconstructed using the absolute amounts of MDR1/Mdr1a in vitro and in vivo (Uchida et al. 2011b, 2014a, b). Since proteins are functional molecules in our body, proteomics will help us to promote BBB research.

Points of Discussion

- How do the function and molecular expression differ in the BBB of different brain regions?
- How does the BBB proteome change as a function of age and sex?
- How is the BBB proteome altered in disease conditions, and what are the species differences in the changes?
- What kind of transporters are involved in organic anion and cation transport across the human BBB?
- Which subtypes of ABCC/MRP are expressed in the human BBB, and which one has the highest contribution to BBB transport?
- What types of membrane proteins are involved in macromolecular transport across the BBB?
- How can the purity of isolated brain microvessels for omics analysis be improved?
- How can the polarized proteome in brain microvascular endothelial cells be addressed?

References

Agarwal S, Uchida Y, Mittapalli RK, Sane R, Terasaki T, Elmquist WF (2012) Quantitative proteomics of transporter expression in brain capillary endothelial cells isolated from P-glycoprotein (P-gp), breast cancer resistance protein (Bcrp), and P-gp/Bcrp knockout mice. Drug Metab Dispos 40(6):1164–1169. https://doi.org/10.1124/dmd.112.044719

Boone CW, Ford LE, Bond HE, Stuart DC, Lorenz D (1969) Isolation of plasma membrane fragments from HeLa cells. J Cell Biol 41(2):378–392. https://doi.org/10.1083/jcb.41.2.378

Boyer-Di Ponio J, El-Ayoubi F, Glacial F, Ganeshamoorthy K, Driancourt C, Godet M, Perriere N, Guillevic O, Couraud PO, Uzan G (2014) Instruction of circulating endothelial progenitors in vitro towards specialized blood-brain barrier and arterial phenotypes. PLoS One 9(1):e84179. https://doi.org/10.1371/journal.pone.0084179

Braun C, Sakamoto A, Fuchs H, Ishiguro N, Suzuki S, Cui Y, Klinder K, Watanabe M, Terasaki T, Sauer A (2017) Quantification of transporter and receptor proteins in dog brain capillaries and choroid plexus: relevance for the distribution in brain and CSF of selected BCRP and P-gp substrates. Mol Pharm 14(10):3436–3447. https://doi.org/10.1021/acs.molpharmaceut.7b00449

Cecchelli R, Aday S, Sevin E, Almeida C, Culot M, Dehouck L, Coisne C, Engelhardt B, Dehouck MP, Ferreira L (2014) A stable and reproducible human blood-brain barrier model derived from hematopoietic stem cells. PLoS One 9(6):e99733. https://doi.org/10.1371/journal.pone.0099733

Eigenmann DE, Xue G, Kim KS, Moses AV, Hamburger M, Oufir M (2013) Comparative study of four immortalized human brain capillary endothelial cell lines, hCMEC/D3, hBMEC, TY10, and BB19, and optimization of culture conditions, for an in vitro blood-brain barrier model for drug permeability studies. Fluids Barriers CNS 10(1):33. https://doi.org/10.1186/2045-8118-10-33

Furihata T, Kawamatsu S, Ito R, Saito K, Suzuki S, Kishida S, Saito Y, Kamiichi A, Chiba K (2015) Hydrocortisone enhances the barrier properties of HBMEC/cibeta, a brain microvascular endothelial cell line, through mesenchymal-to-endothelial transition-like effects. Fluids Barriers CNS 12:7. https://doi.org/10.1186/s12987-015-0003-0

Gruetter R, Novotny EJ, Boulware SD, Rothman DL, Shulman RG (1996) 1H NMR studies of glucose transport in the human brain. J Cereb Blood Flow Metab 16(3):427–438. https://doi.org/10.1097/00004647-199605000-00009

Hawkins RA, Huang SC, Barrio JR, Keen RE, Feng D, Mazziotta JC, Phelps ME (1989) Estimation of local cerebral protein synthesis rates with L-[1-11C]leucine and PET: methods, model, and results in animals and humans. J Cereb Blood Flow Metab 9(4):446–460. https://doi.org/10.1038/jcbfm.1989.68

Helms HC, Abbott NJ, Burek M, Cecchelli R, Couraud PO, Deli MA, Forster C, Galla HJ, Romero IA, Shusta EV, Stebbins MJ, Vandenhaute E, Weksler B, Brodin B (2016) In vitro models of the blood-brain barrier: an overview of commonly used brain endothelial cell culture models and guidelines for their use. J Cereb Blood Flow Metab 36(5):862–890. https://doi.org/10.1177/0271678X16630991

Hoshi Y, Uchida Y, Tachikawa M, Inoue T, Ohtsuki S, Terasaki T (2013) Quantitative atlas of blood-brain barrier transporters, receptors, and tight junction proteins in rats and common marmoset. J Pharm Sci 102(9):3343–3355. https://doi.org/10.1002/jps.23575

Hoshi Y, Uchida Y, Tachikawa M, Ohtsuki S, Couraud PO, Suzuki T, Terasaki T (2019) Oxidative stress-induced activation of Abl and Src kinases rapidly induces P-glycoprotein internalization via phosphorylation of caveolin-1 on tyrosine-14, decreasing cortisol efflux at the blood-brain barrier. J Cereb Blood Flow Metab 271678X18822801. https://doi.org/10.1177/0271678X18822801

Irifune M, Fukuda T, Nomoto M, Sato T, Kamata Y, Nishikawa T, Mietani W, Yokoyama K, Sugiyama K, Kawahara M (1997) Effects of ketamine on dopamine metabolism during anesthesia in discrete brain regions in mice: comparison with the effects during the recovery and sub-anesthetic phases. Brain Res 763(2):281–284. https://doi.org/10.1016/s0006-8993(97)00510-6

Ito K, Uchida Y, Ohtsuki S, Aizawa S, Kawakami H, Katsukura Y, Kamiie J, Terasaki T (2011) Quantitative membrane protein expression at the blood-brain barrier of adult and younger cynomolgus monkeys. J Pharm Sci 100(9):3939–3950. https://doi.org/10.1002/jps.22487

Ito S, Oishi M, Ogata S, Uemura T, Couraud PO, Masuda T, Ohtsuki S (2020) Identification of cell-surface proteins endocytosed by human brain microvascular endothelial cells in vitro. Pharmaceutics 12(6). https://doi.org/10.3390/pharmaceutics12060579

Kamiichi A, Furihata T, Kishida S, Ohta Y, Saito K, Kawamatsu S, Chiba K (2012) Establishment of a new conditionally immortalized cell line from human brain microvascular endothelial cells: a promising tool for human blood-brain barrier studies. Brain Res 1488:113–122. https://doi.org/10.1016/j.brainres.2012.09.042

Kamiie J, Ohtsuki S, Iwase R, Ohmine K, Katsukura Y, Yanai K, Sekine Y, Uchida Y, Ito S, Terasaki T (2008) Quantitative atlas of membrane transporter proteins: development and application of a highly sensitive simultaneous LC/MS/MS method combined with novel in-silico peptide selection criteria. Pharm Res 25(6):1469–1483. https://doi.org/10.1007/s11095-008-9532-4

Kubo Y, Ohtsuki S, Uchida Y, Terasaki T (2015) Quantitative determination of luminal and abluminal membrane distributions of transporters in porcine brain capillaries by plasma membrane fractionation and quantitative targeted proteomics. J Pharm Sci 104(9):3060–3068. https://doi.org/10.1002/jps.24398

Kurosawa T, Tega Y, Higuchi K, Yamaguchi T, Nakakura T, Mochizuki T, Kusuhara H, Kawabata K, Deguchi Y (2018) Expression and functional characterization of drug transporters in brain microvascular endothelial cells derived from human induced pluripotent stem cells. Mol Pharm 15(12):5546–5555. https://doi.org/10.1021/acs.molpharmaceut.8b00697

Lawson EL, Clifton JG, Huang F, Li X, Hixson DC, Josic D (2006) Use of magnetic beads with immobilized monoclonal antibodies for isolation of highly pure plasma membranes. Electrophoresis 27(13):2747–2758. https://doi.org/10.1002/elps.200600059

Lee YC, Block G, Chen H, Folch-Puy E, Foronjy R, Jalili R, Jendresen CB, Kimura M, Kraft E, Lindemose S, Lu J, McLain T, Nutt L, Ramon-Garcia S, Smith J, Spivak A, Wang ML, Zanic M, Lin SH (2008) One-step isolation of plasma membrane proteins using magnetic beads with immobilized concanavalin A. Protein Expr Purif 62(2):223–229. https://doi.org/10.1016/j.pep.2008.08.003

Lippmann ES, Azarin SM, Kay JE, Nessler RA, Wilson HK, Al-Ahmad A, Palecek SP, Shusta EV (2012) Derivation of blood-brain barrier endothelial cells from human pluripotent stem cells. Nat Biotechnol 30(8):783–791. https://doi.org/10.1038/nbt.2247

Liu H, Huang L, Li Y, Fu T, Sun X, Zhang YY, Gao R, Chen Q, Zhang W, Sahi J, Summerfield S, Dong K (2017) Correlation between membrane protein expression levels and transcellular transport activity for breast cancer resistance protein. Drug Metab Dispos 45(5):449–456. https://doi.org/10.1124/dmd.116.074245

Masuda T, Hoshiyama T, Uemura T, Hirayama-Kurogi M, Ogata S, Furukawa A, Couraud PO, Furihata T, Ito S, Ohtsuki S (2019) Large-scale quantitative comparison of plasma transmembrane proteins between two human blood-brain barrier model cell lines, hCMEC/D3 and HBMEC/cibeta. Mol Pharm 16(5):2162–2171. https://doi.org/10.1021/acs.molpharmaceut.9b00114

Meireles M, Martel F, Araujo J, Santos-Buelga C, Gonzalez-Manzano S, Duenas M, de Freitas V, Mateus N, Calhau C, Faria A (2013) Characterization and modulation of glucose uptake in a human blood-brain barrier model. J Membr Biol 246(9):669–677. https://doi.org/10.1007/s00232-013-9583-2

Mori S, Takanaga H, Ohtsuki S, Deguchi T, Kang YS, Hosoya K, Terasaki T (2003) Rat organic anion transporter 3 (rOAT3) is responsible for brain-to-blood efflux of homovanillic acid at the abluminal membrane of brain capillary endothelial cells. J Cereb Blood Flow Metab 23(4):432–440. https://doi.org/10.1097/01.WCB.0000050062.57184.75

Mori S, Ohtsuki S, Takanaga H, Kikkawa T, Kang YS, Terasaki T (2004) Organic anion transporter 3 is involved in the brain-to-blood efflux transport of thiopurine nucleobase analogs. J Neurochem 90(4):931–941. https://doi.org/10.1111/j.1471-4159.2004.02552.x

Nitta T, Hata M, Gotoh S, Seo Y, Sasaki H, Hashimoto N, Furuse M, Tsukita S (2003) Size-selective loosening of the blood-brain barrier in claudin-5-deficient mice. J Cell Biol 161(3):653–660. https://doi.org/10.1083/jcb.200302070

Ogata S, Ito S, Masuda T, Ohtsuki S (2021) Efficient isolation of brain capillary from a single frozen mouse brain for protein expression analysis. J Cereb Blood Flow Metab 41(5):1026–1038. https://doi.org/10.1177/0271678X20941449

Ohshima M, Kamei S, Fushimi H, Mima S, Yamada T, Yamamoto T (2019) Prediction of drug permeability using in vitro blood-brain barrier models with human induced pluripotent stem cell-derived brain microvascular endothelial cells. Biores Open Access 8(1):200–209. https://doi.org/10.1089/biores.2019.0026

Ohtsuki S, Uchida Y, Kubo Y, Terasaki T (2011) Quantitative targeted absolute proteomics-based ADME research as a new path to drug discovery and development: methodology, advantages, strategy, and prospects. J Pharm Sci 100(9):3547–3559. https://doi.org/10.1002/jps.22612

Ohtsuki S, Schaefer O, Kawakami H, Inoue T, Liehner S, Saito A, Ishiguro N, Kishimoto W, Ludwig-Schwellinger E, Ebner T, Terasaki T (2012) Simultaneous absolute protein quantification of transporters, cytochromes P450, and UDP-glucuronosyltransferases as a novel approach for the characterization of individual human liver: comparison with mRNA levels and activities. Drug Metab Dispos 40(1):83–92. https://doi.org/10.1124/dmd.111.042259

Ohtsuki S, Ikeda C, Uchida Y, Sakamoto Y, Miller F, Glacial F, Decleves X, Scherrmann JM, Couraud PO, Kubo Y, Tachikawa M, Terasaki T (2013) Quantitative targeted absolute proteomic analysis of transporters, receptors and junction proteins for validation of human cerebral microvascular endothelial cell line hCMEC/D3 as a human blood-brain barrier model. Mol Pharm 10(1):289–296. https://doi.org/10.1021/mp3004308

Pahlman S, Ljungstedt-Poahlman I, Sanderson A, Ward PJ, Grant A, Hermon-Taylor J (1979) Isolation of plasma-membrane components from cultured human pancreatic cancer cells by immuno-affinity chromatography of anti-beta 2M sepharose 6MB. Br J Cancer 40(5):701–709. https://doi.org/10.1038/bjc.1979.250

Pardridge WM (1983) Brain metabolism: a perspective from the blood-brain barrier. Physiol Rev 63(4):1481–1535. https://doi.org/10.1152/physrev.1983.63.4.1481

Stins MF, Badger J, Sik Kim K (2001) Bacterial invasion and transcytosis in transfected human brain microvascular endothelial cells. Microb Pathog 30(1):19–28. https://doi.org/10.1006/mpat.2000.0406

Syvanen S, Lindhe O, Palner M, Kornum BR, Rahman O, Langstrom B, Knudsen GM, Hammarlund-Udenaes M (2009) Species differences in blood-brain barrier transport of three positron emission tomography radioligands with emphasis on P-glycoprotein transport. Drug Metab Dispos 37(3):635–643. https://doi.org/10.1124/dmd.108.024745

Uchida Y, Ohtsuki S, Katsukura Y, Ikeda C, Suzuki T, Kamiie J, Terasaki T (2011a) Quantitative targeted absolute proteomics of human blood-brain barrier transporters and receptors. J Neurochem 117(2):333–345. https://doi.org/10.1111/j.1471-4159.2011.07208.x

Uchida Y, Ohtsuki S, Kamiie J, Terasaki T (2011b) Blood-brain barrier (BBB) pharmacoproteomics: reconstruction of in vivo brain distribution of 11 P-glycoprotein substrates based on the BBB transporter protein concentration, in vitro intrinsic transport activity, and unbound fraction in plasma and brain in mice. J Pharmacol Exp Ther 339(2):579–588. https://doi.org/10.1124/jpet.111.184200

Uchida Y, Tachikawa M, Obuchi W, Hoshi Y, Tomioka Y, Ohtsuki S, Terasaki T (2013) A study protocol for quantitative targeted absolute proteomics (QTAP) by LC-MS/MS: application for inter-strain differences in protein expression levels of transporters, receptors, claudin-5, and marker proteins at the blood-brain barrier in ddY, FVB, and C57BL/6J mice. Fluids Barriers CNS 10(1):21. https://doi.org/10.1186/2045-8118-10-21

Uchida Y, Ohtsuki S, Terasaki T (2014a) Pharmacoproteomics-based reconstruction of in vivo P-glycoprotein function at blood-brain barrier and brain distribution of substrate verapamil in pentylenetetrazole-kindled epilepsy, spontaneous epilepsy, and phenytoin treatment models. Drug Metab Dispos 42(10):1719–1726. https://doi.org/10.1124/dmd.114.059055

Uchida Y, Wakayama K, Ohtsuki S, Chiba M, Ohe T, Ishii Y, Terasaki T (2014b) Blood-brain barrier pharmacoproteomics-based reconstruction of the in vivo brain distribution of P-glycoprotein substrates in cynomolgus monkeys. J Pharmacol Exp Ther 350(3):578–588. https://doi.org/10.1124/jpet.114.214536

van Gelder NM (1989) Brain taurine content as a function of cerebral metabolic rate: osmotic regulation of glucose derived water production. Neurochem Res 14(6):495–497. https://doi.org/10.1007/bf00964908

Veenstra TD (2007) Global and targeted quantitative proteomics for biomarker discovery. J Chromatogr B Analyt Technol Biomed Life Sci 847(1):3–11. https://doi.org/10.1016/j.jchromb.2006.09.004

Warren MS, Zerangue N, Woodford K, Roberts LM, Tate EH, Feng B, Li C, Feuerstein TJ, Gibbs J, Smith B, de Morais SM, Dower WJ, Koller KJ (2009) Comparative gene expression profiles of ABC transporters in brain microvessel endothelial cells and brain in five species including human. Pharmacol Res 59(6):404–413. https://doi.org/10.1016/j.phrs.2009.02.007

Weksler BB, Subileau EA, Perriere N, Charneau P, Holloway K, Leveque M, Tricoire-Leignel H, Nicotra A, Bourdoulous S, Turowski P, Male DK, Roux F, Greenwood J, Romero IA, Couraud PO (2005) Blood-brain barrier-specific properties of a human adult brain endothelial cell line. FASEB J 19(13):1872–1874. https://doi.org/10.1096/fj.04-3458fje

Young LT, Warsh JJ, Kish SJ, Shannak K, Hornykeiwicz O (1994) Reduced brain 5-HT and elevated NE turnover and metabolites in bipolar affective disorder. Biol Psychiatry 35(2):121–127. https://doi.org/10.1016/0006-3223(94)91201-7

Zhang Y, Pardridge WM (2001) Mediated efflux of IgG molecules from brain to blood across the blood-brain barrier. J Neuroimmunol 114(1–2):168–172. https://doi.org/10.1016/s0165-5728(01)00242-9

Zhang Y, Sloan SA, Clarke LE, Caneda C, Plaza CA, Blumenthal PD, Vogel H, Steinberg GK, Edwards MS, Li G, Duncan JA 3rd, Cheshier SH, Shuer LM, Chang EF, Grant GA, Gephart MG, Barres BA (2016) Purification and characterization of progenitor and mature human astrocytes reveals transcriptional and functional differences with mouse. Neuron 89(1):37–53. https://doi.org/10.1016/j.neuron.2015.11.013

Chapter 6
Drug Metabolism at the Blood-Brain and Blood-CSF Barriers

Jean-François Ghersi-Egea, Nathalie Strazielle, and Xavier Declèves

Abstract Drug metabolism is in most cases a detoxication process allowing the organism to inactivate and eliminate foreign substances to which it is exposed. While the liver is the main site of drug metabolism, drug metabolizing enzymes that catalyze functionalization and conjugation reactions have been detected in the brain, and several of these enzymes are notably enriched at blood-brain interfaces. This chapter summarizes the principles of drug metabolism, reviews the molecular and functional evidence for drug metabolizing enzyme location at both the blood-brain and blood-CSF barriers, and discusses their functional significance for modulating cerebral drug delivery and brain exposure to small molecular weight drugs or toxins.

Keywords Drug metabolizing enzymes · Blood-brain barrier · Blood-CSF barrier · Neuropharmacology · Cytochrome P450 · Conjugation enzymes · Drug transporters

6.1 Introduction and History of Cerebral Drug Metabolism

Drug metabolism is a process whereby xenobiotics (either exogenous non-nutrient organic compounds including pharmacological molecules or environmental toxics) are enzymatically transformed in the body to form one or several metabolites. Usually, biotransformation reactions largely take place in the liver and strongly influence the transport and partitioning of a compound within the body, its toxicity, and its rate and route of elimination. About 50 multispecific enzymes catalyze the

J.-F. Ghersi-Egea
Inserm U1028, CNRS UMR 5292, Lyon Neuroscience Research Center, Lyon-1 University, Lyon, France

N. Strazielle
Brain-i, Lyon, France

X. Declèves (✉)
Inserm U1144, Université de Paris, Therapeutic optimisation in Neuropsychopharmacology, Team The Blood-Brain Barrier in Brain Pathophysiology and Therapy, Paris, France
e-mail: xavier.decleves@parisdescartes.fr

© American Association of Pharmaceutical Scientists 2022   141
E. C. M. de Lange et al. (eds.), *Drug Delivery to the Brain*, AAPS Advances in the Pharmaceutical Sciences Series 33, https://doi.org/10.1007/978-3-030-88773-5_6

biotransformation of xenobiotics in human. Additional enzymes usually involved in endogenous metabolism also participate to the biotransformation of some selected drugs. Drug metabolism is a multiphase process (Fig. 6.1). Phase 0 corresponds to the penetration of the drug in the metabolizing cell and may request influx transporters (Doring and Petzinger 2014). Phase I is a functionalizing phase and most of the time is oxidative, while it can also be reductive in some instances (Cashman 2000; Ghersi-Egea et al. 1998; Nebert and Russell 2002). Functionalization enzymes generate metabolites which are usually more polar than the parent compounds and thus more readily eliminated. Phase II of drug metabolism corresponds to conjugation processes whereby a hydrophilic moiety, such as a glucuronic acid, a sulfate, or a cysteine-bearing molecule (e.g., glutathione), is bound to the parent drug or to the phase I metabolite (Duffel et al. 2001; Eaton and Bammler 1999; King et al. 2000). Fig. 6.2 summarizes the main pathways of, and enzymes involved in, drug metabolism. Phase III of metabolism involves transport processes mediating the efflux of phases I and II metabolites out of the producing cells and their further excretion from the body. The best known phase III transport proteins belong to the ATP-binding cassette (ABC) subfamily C of transporters and are often referred to as multidrug-related resistance proteins (Chaves et al. 2014; Slot et al. 2011; Strazielle and Ghersi-Egea 2015). They accept a large range of drug conjugates as substrates. Overall these different metabolic and transport steps allow the biotransformation of drugs to polar metabolites that are readily excreted out of the body. The produced

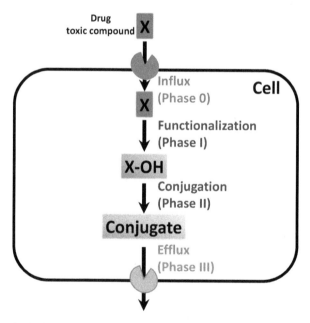

Fig. 6.1 General principle of drug metabolism. A small molecular weight xenobiotic (X) is taken up by diffusion or by phase 0 transmembrane influx transporters belonging to the solute carrier superfamily (SLC). Once in the cell, the compound is functionalized by phase I drug metabolizing enzymes to produce a hydroxylated metabolite (X-OH) that is further conjugated by phase II conjugation drug metabolizing enzymes. The conjugated metabolite is efflux by phase III transmembrane efflux transporters belonging to the ATP-binding cassette (ABC) transporter superfamily

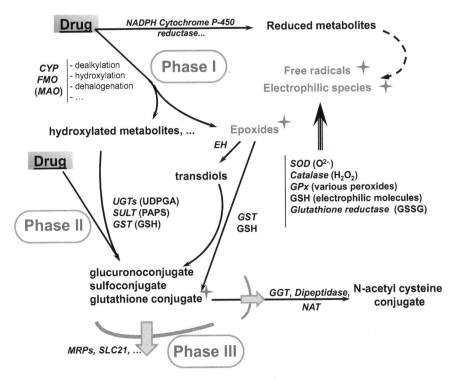

Fig. 6.2 Main pathways of drug metabolism. Functionalization enzymes (phase I) generate metabolites which are more polar than the parent compounds and more readily excretable. They include the numerous isoforms of cytochrome P450 (CYP), also called mixed function oxidase, the flavin monoamine oxidases (FMO), and also more specific oxidation enzymes such as monoamine oxidases (MAO). They also include enzymes such as NADPH-cytochrome P450 reductase responsible for xenobiotic (e.g., nitro-compounds) reduction. Oxidative and reductive processes can lead to the formation of reactive metabolites such as epoxides or can generate oxygen-derived free radicals. These are in turn inactivated by epoxide hydrolases (EH), conjugation to glutathione, or reactive oxygen species-inactivating enzymes. Phase II of drug metabolism corresponds to a conjugation process whereby a hydrophilic moiety, such as a glucuronic acid, a sulfate, or a cysteine-bearing molecule (e.g., glutathione), is bound to the drug or to the phase I metabolites. This biotransformation is catalyzed by uridine diphosphoglucuronosyl transferases (UGTs), glutathione S-transferases (GSTs), or sulfotransferases (SULTs). Glutathione conjugates can be further metabolized along the mercapturic acid pathway involving two ectoenzymes, gamma-glutamyltranspeptidase (GGT) and dipeptidase and N-acetyltransferase (NAT). Usually inactive or less active than the parent compounds, the produced metabolites in some instances can be pharmacologically more active or toxic. Red stars refer to classes of metabolites that can be potentially harmful. Names or abbreviations of enzymes appear in italic. Phase III of drug metabolism refers to the efflux of the conjugates out of the cells, which primarily involves ABC transporters of the multidrug resistance-associated protein family (MRPs) and of the SLC21 family of organic anion transport (OAT) proteins. Other abbreviations: SOD superoxide dismutase, GPx glutathione peroxidase, GSH and GSSG reduced and oxidized glutathione, respectively. Modified from (Strazielle et al. 2004)

metabolites are usually inactive or less active than the parent compounds. Yet, in some instances, they can be pharmacologically more active, such as for morphine-6-glucuronide which has stronger analgesic properties than morphine (Christrup 1997). They can also be more toxic as exemplified by the high carcinogenicity of hydroperoxide metabolites of benzo[a]pyrene (Gelhaus et al. 2011) (Fig. 6.2). A specific feature of drug metabolism is that at least in the liver the expression of many phase I and phase II isoenzymes and some phase III transporters can be transcriptionally induced upon exposure to drugs or other exogenous compounds. This occurs through different mechanisms. For instance, binding of polycyclic aryl hydrocarbons to the cytosolic aryl hydrocarbon receptor (AhR) induces the translocation of this receptor into the nucleus, which subsequently activates an enhancer DNA element called xenobiotic responsive element (XRE) present in the promoter of a number of drug metabolizing enzyme genes. Other xenobiotics such as phenobarbital, dexamethasone, and fibrates interact with nuclear receptors such as the constitutive androstane receptor (CAR), the pregnane X receptor (PXR), or else the peroxisome proliferator activated receptor (PPAR), respectively (Aleksunes and Klaassen 2012; Tolson and Wang 2010; Xu et al. 2005). Finally, electrophilic compounds can induce the nuclear translocation of the nuclear factor (erythroid-derived 2)-like 2 (Nrf2). The activation of the nrf2 pathway enhances the transcription of genes bearing the antioxidant response element (ARE), which include genes of glutathione-S-transferases (GSTs) and enzymes involved in the antioxidant cellular machinery (Calkins et al. 2009; Higgins and Hayes 2011). The overall benefit of these induction mechanisms is an increase in the protective activities toward drugs or xenobiotics to which cells are exposed.

Following the pioneer discovery of hepatic drug metabolism as a major process for xenobiotic detoxication in mammals, an era of research on extrahepatic sites for drug metabolism opened. Brain, as other organs, was scanned for drug metabolizing enzyme (DME) activities. These enzymatic activities often measured in tissue homogenates were found to be low to very low in the whole brain compared with the liver and have been considered insignificant, until the complexity and specificity of the morphological and cellular organization of this organ was taken into account to refine the findings.

The brain is constituted of numerous anatomically differentiated structures, whose parenchymal tissue is composed of intermingled cells of different types, namely, neurons, glial cells including astrocytes, myelin-producing oligodendrocytes, as well as microglial cells bearing immune functions. Besides, the brain has an internal circulatory system of its own. The cerebrospinal fluid (CSF) circulates through the ventricular cavities lined by the ependyma into various membrane-filled cisterns and subarachnoid spaces of the brain before being resorbed in the venous circulation. Exchanges between the brain and the periphery are controlled by specific cellular interfaces between the blood and the different brain structures. Within the neuropil, the blood-brain barrier (BBB) is located at the endothelial wall of the brain microvessels, while the epithelium of the choroid plexuses and the arachnoid cells form a barrier between the blood and the CSF in ventricles (BCSFB) and subarachnoid spaces (arachnoid barrier, AB), respectively (Fig. 6.3). Cells at both the BBB and BCSFB/AB are sealed by tight junctions, so that only those drugs which are lipophilic enough to cross lipid membranes have access to the brain if they are not efficiently effluxed into the bloodstream by ABC transporters located in these barriers. Finally, the ependyma bordering the ventricles and the pia-glia limitans

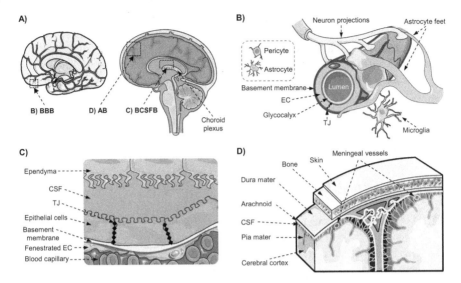

Fig. 6.3 Human blood-brain interfaces. (**a**) There are three main interfaces regulating the exchanges between blood and brain (left), either directly to the parenchyma or through the cerebrospinal fluid (CSF) (right). (**b**) The blood-brain barrier (BBB) is formed mainly by the brain microvascular endothelial cells (EC), attached by tight junctions (TJ), but their specialized phenotype and function are regulated and maintained by the neurovascular unit (NVU) formed by the basement membrane and neighboring cells including brain vascular pericytes, astrocytes, neurons, and microglia. (**c**) The blood-cerebrospinal fluid barrier (BCSFB) is formed by the tightly jointed epithelial cells of the choroid plexus (CP), which cover the fenestrated EC of the CP capillaries. (**d**) The meninges are composed of three layers: the outermost fibrous sheet of dura mater, the arachnoid mater, and the pia, the latter two enclosing CSF in the subarachnoid space; the arachnoid cells present tight junctions and form the blood-arachnoid barrier (AB). Adapted from (Gomez-Zepeda et al. 2019)

forming the interface between the subarachnoid CSF and the neuropil regulate to some extent the exchanges between the CSF and the neuropil (Fig. 6.3).

Given this extensive heterogeneity, differences in the expression levels of DME were therefore searched among cerebral regions and cell subpopulations. Both phase I and phase II enzymes were found to be heterogeneously distributed among regions and between neurons and glial cells (Bhamre et al. 1993; Miksys and Tyndale 2002; Minn et al. 1991; Monks et al. 1999; Teissier et al. 1998). In addition, a clear enrichment in cytochrome P450 (CYP)-dependent monooxygenases, monoamine oxidases (MAOs), epoxide hydrolases, several phase II enzymes, and antioxidant enzymes was demonstrated in the cells forming the blood-brain interfaces, both in rodent and in human. Some of these enzymes were proved to be sensitive to exogenous inducers (Ghersi-Egea et al. 1994; Ghersi-Egea et al. 1988; Ghersi-Egea et al. 1993; Hansson et al. 1990; Johnson et al. 1993; Riachi et al. 1988; Tayarani et al. 1989; Volk et al. 1991). This suggested a putative new function of these enzymes as a metabolic barrier between the blood and the brain working together with efflux transporters to limit brain entry of xenobiotics (reviewed in Ghersi-Egea et al. 1995).

Since these pioneer works, various studies explored this potential new barrier aspect of blood-brain interfaces. They initiated the identification of phase I and II enzyme isoforms and aimed at establishing the functional relevance of drug metabolism at the barriers. This paper describes our current understanding of drug metabolism at the BBB, BCSFB, and AB and explores the mechanisms regulating the expression of DME in these interfaces.

6.2   Current Status

6.2.1   The Blood-Brain Barrier

6.2.1.1   Anatomical and Functional Features of the Blood-Brain Barrier

The BBB is formed by the endothelial cells lining the brain capillaries and microvessels (Ballabh et al. 2004; Cardoso et al. 2010) (Zhang and Harder 2002) (Fig. 6.3). These cells are the main determinants of the BBB phenotype in humans and other animals (Khan 2005) and are referred to as brain microvessel endothelial cells (BMVECs) in this review. Only brain microvessels possess the properties of a fully efficient BBB, since the degree of leakiness across the endothelium varies inversely to the vessel diameter (Hawkins and Davis 2005). Although BMVECs are responsible for the BBB phenotype in vivo, these cells are in dynamic contact with other cells such as astrocytes, pericytes, and neurons that form the neurovascular unit. The mature BBB phenotype is believed to result from the particular interaction between the BMVECs and these other cells in the surroundings (Ballabh et al. 2004; Calabria and Shusta 2008; Cardoso et al. 2010; Lee et al. 2006). The walls of the brain microvessels are mainly lined with endothelial cells, and BMVECs are fundamentally different from the endothelial cells lining the vessels in peripheral tissues (Choi and Kim 2008). The BMVECs have narrow junctional complexes (tight and *adherens* junctions), reducing gaps or spaces between cells and restricting free passive diffusion of blood-borne substances by paracellular route into the brain interstitial fluid (Zlokovic 2008). BMVEC tightness is known to be 50- to 100-fold higher than that in peripheral microvessels (Abbott 2002). It also provides this endothelium with a particularly high transendothelial electrical resistance (TEER) of 1500 to 2000 Ω cm^2 (Hawkins and Egleton 2006). BMVECs also differ from peripheral endothelial cells by several factors: a) the uniform thickness of their cytoplasm, b) absence or restricted fenestrae, c) poor endocytotic activities, d) continuous basement membrane, e) negatively charged surface, and f) a large number of mitochondria (Ballabh et al. 2004; Cardoso et al. 2010; de Boer and Gaillard 2006; Persidsky et al. 2006). In the following paragraph, the review focuses only on phase I and phase II drug metabolizing enzymes at the BBB. Transporters potentially involved in phase III of drug metabolism are described in other chapters of this book.

6.2.1.2   Molecular Characterization, Relative Expression, and Functional Significance of Drug Metabolizing Enzymes at the Blood-Brain Barrier

Phase I of Drug Metabolism   One of the primary functions of peripheral metabolism is mostly to render substrates more polar, thus more water soluble, facilitating their removal from the body via excretion into the urine or bile. In the BBB, the presence of phase I DME in BMVECs raises the question of their role in physiology and pharmacology. Rendering substrates more polar is probably not the primary function of phase I DME in the BBB. We may hypothesize that metabolism at the BBB may be more considered as a mechanism of brain protection by inactivating pharmacologically active compounds or toxic substances, thus preventing their access to the neuropil. While this is the general case, the opposite can happen in the case of prodrugs where an inactive parent compound may be transformed into a pharmacologically active metabolite as exemplified in the introduction. These metabolites can be beneficial to the brain if they are derived from a prodrug and harmful if they are toxic metabolites.

MAOs are phase I DME evidenced in the 1960s that metabolize neuroactive monoamines like adrenaline, norepinephrine, dopamine, serotonin, and their precursors and are thus important for controlling neurochemical signaling in the brain (Van Gelder 1968). MAOs are present in the mitochondria of BMVECs which contain up to five times more mitochondria than vascular endothelial cells in the periphery (Betz et al. 1980). The MAOs at the BBB may be considered as a second line of defense together with luminal drug transporters for the brain against chemical assault (Minn et al. 1991). They may also protect the brain from exogenous pyridine derivatives (Riachi and Harik 1988). The expression pattern of the genes encoding the two MAO subunits (*A* and *B*) was established in freshly isolated human brain microvessels (Shawahna et al. 2011). The metabolic hyperactivity of the BMVECs may explain the high concentration of *MAOA* transcripts quantified in the isolated human microvessels which were almost six times more abundant than *MAOB* transcripts. MAOA and MAOB activity was also demonstrated in bovine BMVEC (Baranczyk-Kuzma et al. 1986).

The CYP superfamily contains a substantial number of enzymes that mainly catalyze phase I oxidative reactions. These CYPs are responsible for the transformation of at least 60% of the FDA-approved small molecule drugs. Although they are present mainly in the liver, some extrahepatic isoforms are expressed in the gut, kidneys, and brain and may be important for inactivating drugs and toxicants. CYP activity and expression have first been evidenced in isolated rat and human brain microvessels (Ghersi-Egea et al. 1994). Later, expression of CYP46A1, CYP2J2, CYP2U1, CYP1B1, CYP2E1, and CYP2D6 was demonstrated in the whole human brain (Dutheil et al. 2008; Dutheil et al. 2009), while the expression profile of the genes encoding the main CYP isoforms was established in freshly isolated human brain microvessels (Dauchy et al. 2008). The main CYPs responsible for metabolizing most of the drugs in the liver (CYP1A2, CYP2B6, CYP2C9, CYP2C19,

CYP2D6, and CYP3A4) were absent from the BBB (Decleves et al. 2011). The gene expression profile of CYPs showed that *CYP1B1* and *CYP2U1* were the main isoforms significantly expressed in isolated brain microvessels (over 90% of all CYP mRNAs quantified). *CYP1B1* levels were 15 times more abundant in the brain microvessels than in the cerebral cortex. These transcriptomic data were confirmed at protein levels using a targeted absolute quantitative proteomic approach. Using isolated human brain microvessels from brain biopsies sampled as far as possible of the disease focus in patients suffering from epilepsia or glioma, CYP1B1 and CYP2U1 were detected at similar levels (0.45 fmol/µg total proteins) among the 13 CYP proteins studied (Shawahna et al. 2011). Despite the small amounts of their mRNAs, some CYP isoforms, like CYP1A1 and CYP3A4, are of special interest. Ghosh and collaborators colocalized by immunohistochemistry CYP3A4 with von Willebrand factor (vWF) in endothelial cells isolated from epileptic patients (Ghosh et al. 2010) and showed that CYP3A4 expression was correlated with frequency of seizures and antiepileptic therapy (Williams et al. 2019). This suggests that some CYPs such as CYP3A4 can be induced at the BBB in this disease state and may explain failure in the treatment of epilepsia. The high expression of *CYP2U1* observed by gene expression profile analysis (Dauchy et al. 2008) confirms the expression of this CYP observed previously at the genomic and proteomic level at the BBB (Karlgren et al. 2004). Since no drugs have been identified as metabolized by CYP2U1, its role in the detoxification of drugs is still poorly understood. CY2U1 may be implicated in the metabolism of endogenous compounds like arachidonic acid into hydroxyeicosatetraenoic acid and thus may help to regulate cerebral blood flow. CYP1B1 is implicated in the metabolism of some xenobiotics but is well known in the metabolism of endogenous compounds like estradiol, melatonin, and arachidonic acid derivatives (Vasiliou and Gonzalez 2008). It is also readily induced via the regulatory pathway mediated by AhR (see Sect. 6.2.1.3). This raises the question of the influence of CYP1B1-mediated metabolic pathways on tampering the integrity of the BBB. Substances activating AhR can penetrate the BMVECs because they are highly lipophilic and not substrates of ABC efflux transporters. CYP1B1 expressed in cells derived from human ovaries and intestine can be induced by cigarette smoke (Josserand et al. 2006; Vidal et al. 2006), a process that may occur also at the BBB. While epoxide hydrolase, well expressed at the BBB, is usually a detoxifying enzyme inactivating carcinogenic epoxides, the sequential action of CYP1B1 and epoxide hydrolase may substantially increase the number of reactive metabolites like diol epoxides (Jacob et al. 2011), potentially deleterious for BBB integrity. Finally, Wnt/β-catenin signaling is crucial for blood-brain barrier (BBB) development and maintenance. Interestingly, β-catenin was shown to influence endothelial metabolism by transcriptionally regulating the murine Cyp1b1. As Cyp1b1 generated retinoic acid as well as 20-hydroxyeicosatetraenoic acid that regulated the P-glycoprotein and BBB junction proteins, it was suggested that Wnt/β-catenin signaling could modulate BBB properties through Cyp1b1 transcription (Ziegler et al. 2016).

Phase II of Drug Metabolism Phase I reactions often render substrates sufficiently polar to undergo excretion. However, many other substrates need additional phase II metabolism in which they are conjugated to a polar molecule to make them sufficiently water-soluble to undergo excretion. While most substrates undergo phase I followed by phase II metabolism, some are directly conjugated and eliminated without any phase I reaction. Phase II reactions leading to more polar phase II metabolites are carried out by enzymes belonging to the following main families (Fig. 6.2): UDP-glucuronosyltransferases (UGTs), glutathione S-transferases (GSTs), and sulfotransferases (SULTs). Two other conjugation enzyme families are N-acetyltransferases (NATs) and methyltransferases (MTs).

UGT transcripts or proteins were not detected in freshly isolated human brain microvessels (Shawahna et al. 2011), and UGT activity toward planar compounds such as 1-naphthol was not detected in human brain capillaries (Ghersi-Egea et al. 1993), suggesting the absence of glucuronidation at the human BBB. On the contrary, homogenates of rat brain microvessels have been found to be enriched in this UGT activity as compared to whole brain homogenate (Ghersi-Egea et al. 1988). Similarly, results reported by Benzi and collaborators, based on in situ brain perfusion in the monkey, indicated that this organ contained efficient glucuronidation (Benzi et al. 1967). This set of data suggests therefore interspecies differences. UGTs seem to be important for the conjugation of drugs in hepatic and intestinal tissues. While the presence of some isoforms in the brain, and particularly in neurons, could modulate the concentrations of neurotherapeutics like morphine within the brain, UGTs do not seem to interfere with the entry of drugs at the BBB. The UGT1A6 and UGT2B7 in human neurons seem to account for the glucuronidation of the neurotransmitter serotonin and endogenous morphine. Interestingly, UGT1A4 expression was observed at the BBB and in cultures of brain endothelial cells of patients with resistant epilepsia, suggesting that UGT1A4 expression can be upregulated in the epileptic region (Ghosh et al. 2013). Expression of UGT in the brain and in particular at the BBB is nicely reviewed (Ouzzine et al. 2014).

GSTs are dimeric proteins that also form a multigenic family of membrane-bound and cytosolic enzymes. α (GSTA), μ (GSTM), and π (GSTP) classes of cytosolic GSTs are considered to be mainly involved in drug metabolism and detoxication pathways (Hayes et al. 2005). Expression of GST in the human brain was first evidenced by Carder et al. by immunochemistry (Carder et al. 1990). This pioneer work already detected high levels of GST α and π in brain cortex as well as in the BBB. Measurable quantities of GST mRNAs and proteins were confirmed at the human BBB using isolated brain microvessels, GSTP1 being the most abundant GST enzyme followed by GSTM2–5 and GSTT1 (Shawahna et al. 2011). These findings are consistent with those of previous studies showing considerable expression of GSTs from the α, π, and μ isoforms in postmortem human brain tissues (Listowsky et al. 1998). Some GST isoforms, like GSTA4, are more abundant in fetal and adult human brains than in the liver. In the rat BBB, GSTpi colocalizes with Abcc2/Mrp2, the regulation of both genes being coordinated by the pregnane X receptor (PXR) (Bauer et al. 2008). The contribution of Abcc2 to the rodent BBB efflux processes remains however a matter of debate, and we detected neither ABCC2/MRP2 transcripts nor proteins in human microvessels (Shawahna et al. 2011). The high concentration of GSTs at the human BBB may be due to the need to neutralize oxidative compounds. GSTP1 has also been detected in the cerebral capillary endothelium of a sample obtained from epileptic patients (Shang et al. 2008). Although glutathione can interact directly with electrophiles, GST-mediated conjugation is quite often found in several tissues, including the CNS. The concentration of glutathione may differ from one brain region to another depending on the developmental stage of the neurons, with concentrations being higher in newly developed neurons, suggesting that it is involved in neuroprotection (Sun et al. 2006). As glutathione is negatively charged at physiological pH, it cannot penetrate the cell membrane. Its presence in the cytoplasm of BMVECs is due to the ability of selected cells to synthesize glutathione. Glutathione and glutathione conjugates are often transported from the cytoplasm to the mitochondria by SLC transporters (OATPs) and often extruded by phase III ABC efflux pumps, ABCCs (MRPs) and ABCG2/BCRP. We have found considerable amounts of the human gamma-glutamyl transpeptidase (GGT) protein (Shawahna et al. 2011). This is the only enzyme that can cleave the γ-glutamyl bond of glutathione pointing to an active γ-glutamyl cycle at the BBB (Meister 1974). Thus, GSTs at the BBB may neutralize reactive oxygen species (ROS) involved in oxidative stress. They could also be involved in drug-resistant epilepsy, preventing the accumulation of antiepileptic drugs by conjugating them with glutathione in the cerebral cortex where the epileptic foci are located.

SULTs are well-characterized phase II metabolizing enzymes that were discovered in the 1960s. They catalyze the sulfation of numerous endogenous and exogenous substrates. There are two forms of SULT; the membrane-associated SULTs are generally implicated in protein sulfation in the Golgi apparatus, and the cytosolic SULTs catalyze the sulfation of a wide range of soluble substrates including xenobiotics. SULT1A1 and SULT1A3 are expressed in the human brain, and a phenol SULT activity was first observed at the end of the 1980s in bovine BMVEC (Baranczyk-Kuzma et al. 1989). Low levels of *SULT1A1* transcripts were detected in isolated human brain microvessels (Shawahna et al. 2011). Since SULTs are involved in the conjugation of numerous substrates including hormones and steroids, they play a key role in the metabolism of aromatic monoamines including catecholamine neurotransmitters, neurosteroids, and catecholamine metabolites in the CNS (Rivett et al. 1982). Within the brain SULT isoforms are believed to be mainly localized within the neurons. In addition to controlling the activity of thyroid hormones and neurosteroids, they are also implicated in the synthesis of chondroitin sulfate, keratan sulfate, and the proteoglycans that are involved in cell-cell interactions and differentiation. Lastly, SULTs may be implicated in the metabolism of drugs like acetaminophen and methyldopa (Gamage et al. 2006).

Some MTs like catechol-O-methyltransferase (COMT) and thiopurine methyltransferase (TPMT) are ubiquitous enzymes, being distributed throughout the body including the CNS. While COMT and TPMT metabolize exogenous substrates, their main function is to catalyze the *O*- and *S*-methylation of endogenous substrates like catecholamines and purines (Gottwald et al. 1997; McLeod et al. 2000). COMT was initially found in glia, but immunoreactivity investigations have also detected it in neurons (Karhunen et al. 1995). COMT activity was detected in primary cultures of rat BMVEC but was low in bovine BMVEC (Baranczyk-Kuzma et al. 1986) as compared to that in the whole bovine gray matter. TPMT and hista- mine *N*-methyltransferase (HNMT) are soluble enzymes usually found in the cyto- sol of brain endothelial cells and neurons (Nishibori et al. 2000; Stanulla et al. 2009). The gene expression and protein level of these three MTs (COMT, HNMT, and TPMT) was easily quantified in freshly isolated human brain microvessels (Shawahna et al. 2011), but no quantitative data are available for their activity in human BMVEC and their expression profile in animal species.

6.2.1.3   Regulation of Drug Metabolizing Enzymes at the Blood-Brain Barrier

Certain ABC efflux transporters and DMEs have shown common transcriptional regulatory pathways as it was first described for ABCB1/MDR1/P-glycoprotein and CYP3A4 (Synold et al. 2001). The transcription factors able to upregulate CYP as well as phase II DMEs (UGT, GST, SULT, etc.) and phase III transporters are PXR, CAR, AhR, PPAR, and Nrf2. This is particularly true for tissues involved in the pharmacokinetics of drugs such as the liver and the gut, but limited investigations have been carried out to understand regulatory mechanisms and key effectors of these pathways at the BBB. PXR has been shown to be present and functional at the BBB of transgenic mice expressing human (hPXR), but few data are available on the presence of PXR at the human BBB. Zastre and collaborators showed that the ABCB1 gene in a human cerebral microvessel endothelial cell line (hCMEC/D3) was upregulated by PXR agonists (Zastre et al. 2009). Only low amounts of *CAR* and *PXR* genes were detected in freshly isolated human brain microvessels (Dauchy et al. 2008). The low levels of PXR transcripts are not in agreement with reports showing that hPXR induced P-gp activity at the BBB of transgenic mice (Bauer et al. 2006) and that pig PXR, which is very similar to hPXR, induced P-gp in pig- cultured brain endothelial cells (Ott et al. 2009). However, hPXR expression is increased in human BMVEC in brain samples of resistant epileptic patients, demon- strating that the disease state may increase PXR and subsequently CYP3A4 expres- sion (Ghosh et al. 2017). Similarly, it was recently shown that knocking down PXR in the bEnd.3 murine BMVEC cells decreased the expression of both P-gp and CYP3A. CAR activation has been shown to induce the expression of ABC trans- porters in isolated rat brain microvessels and in the hCMEC/D3 human brain endo- thelial cell line (Chan et al. 2010; Wang et al. 2010b). Unfortunately, induction of DMEs via PXR or CAR activation has not yet been studied in vivo at the human BBB. In contrast to PXR and CAR, high levels of *AhR* transcripts were found in rat (Jacob et al. 2011) and human BMVECs (Dauchy et al. 2009). AhR is another tran- scriptional factor implied in the regulation of certain genes involved in drug metab- olism. AhR does not belong to the nuclear receptor superfamily, unlike PXR and

CAR. However, AhR belongs to a family known as basis helix-loop-helix/Per-ARNT-Sim (bHLH/PAS). This family includes also ARNT (AhR nuclear transloca-tor), which heterodimerizes with AhR to form an active transcription-initiating complex. Similar to the nuclear receptors, AhR in the nucleus regulates the tran-scription of the target genes (Barouki et al. 2007). AhR ligands are hydrophobic in nature and can be endogenous or exogenous. Xenobiotics able to activate AhR are mainly polycyclic aromatic hydrocarbons such as dioxins (environmental pollut-ants), benzo[a]pyrene (tobacco), and β-naphthoflavone but also some medications such as omeprazole (Denison and Nagy 2003). The list of genes regulated by AhR differs from that of PXR and CAR. However, some similarities are observed as AhR appears to be involved in regulating the expression of ABC transporters like ABCC3/MRP3 and ABCG2/BCRP. The CYP1A1/CYP1A2 and CYP1B1 are the best known and most studied AhR target genes, and these isoforms are able to metabolize many procarcinogens into reactive metabolites. Therefore, any prolonged exposure to AhR ligands, including many environmental pollutants, may lead to an increased formation of reactive metabolites to cause toxicity. Activation of AhR by dioxin, one of the most potent AhR ligands, strongly induced Cyp1a1 and Cyp1b1 in iso-lated rat brain microvessels (Jacob et al. 2011; Wang et al. 2010a). Interestingly, CYP1B1 has been shown as one of the main CYPs expressed at the human BBB, but its function at this location remains poorly understood (see Sect. 6.2.1.2). More recently, it was shown in hCMEC/D3 cells that activation of AhR by dioxin (TCDD) increased CYP1B1 expression in hCMEC/D3 cells without altering those of ABCB1 and ABCG2 (Jacob et al. 2015). We hypothesize that ligands of AhR, including polycyclic aromatic hydrocarbons (PAHs) like coplanar polychlorinated biphenyls (PCBs) and benzo[a]pyrene, may induce some AhR target genes, including *CYP1A1* and *CYP1B1*, at the human BBB. The role of CYP1B1 as a metabolic activator of toxic pollutants to form potentially neurotoxic metabolites remains to be deter-mined. Several studies have demonstrated the important role of Nrf2 as a factor protecting the BBB and CNS. Nrf2 indeed upregulates the expression of TJ, pro-motes redox metabolic functions, and produces ATP with mitochondrial biogenesis (Sivandzade et al. 2019). However, there is still no data on the effect of Nrf2 activa-tion or inactivation on the expression of DME at the BBB.

6.2.2   The Blood-Cerebrospinal Fluid Barrier and the Ependyma

6.2.2.1   Anatomical and Functional Features of the Choroidal Blood-CSF Barrier

The bulk of CSF is secreted by the choroid plexuses. CSF represents 50% of the extracellular fluid of the brain in human. It flows through the ventricular system, then into the midbrain and hindbrain cisterns, velae, and subarachnoid spaces before being absorbed into the venous blood via the arachnoid villi or drained into the lymphatic system. Exchanges between the CSF and fluid-filled extracellular spaces of the brain parenchyma are not restricted as cells forming most of the ependymal ventricular wall or the external *glia limitans* lack tight junctions (Fig. 6.3). Drug metabolism at these places may however impact on the distribution of xenobiotics in the brain (see infra). The BCSFB lies at the choroid plexus epithelium and, downstream of CSF flow, at the arachnoid membrane. The former site is therefore mainly involved in CSF drug delivery. The choroid plexus-CSF system adds a degree of complexity to the mechanisms that set the cerebral bioavailability of both endogenous and exogenous bioactive compounds. The CSF circulatory pathway and the interplay between BBB, BCSFB, and CSF have been described elsewhere (Ghersi-Egea et al. 2009a; Strazielle and Ghersi-Egea 2013, 2016).

The brain contains four choroid plexuses, located in the two lateral, the third, and the fourth ventricles. The different choroid plexuses display a somewhat different gross anatomy but are all organized as an ensemble of villi formed by a monolayer of epithelial cells surrounding a highly vascularized conjunctive core (Fig. 6.3). The choroid plexuses display the highest local cerebral blood flow among brain structures. The fenestrated vessels present in the choroidal stroma are highly permeable even to polar solutes, and the actual barrier between blood and CSF is located at the epithelium whose cells are sealed by tight junctions (Strazielle and Ghersi-Egea 2000). Besides the production of CSF from plasma by a tightly regulated secretory process (Praetorius and Damkier 2017), the choroid plexuses also fulfill neuroendocrine functions by secreting various biologically active polypeptides and hormone carrier proteins and participate to the neuroimmune surveillance of the brain (reviewed in (Chodobski and Szmydynger-Chodobska 2001, Ghersi-Egea et al. 2018). Choroid plexus functions also include the selective blood-to-CSF entry of required molecules such as inorganic anions, nutrients, and hormones (Damkier et al. 2010; Redzic et al. 2005; Schmitt et al. 2011), as well as the CSF-to-blood export of toxic compounds and metabolites (Kusuhara and Sugiyama 2004; Strazielle et al. 2004). These transport processes are facilitated by several factors. The BCSFB is located between two circulating fluids. The surface area of exchange is enhanced by the organization of the choroid plexus into numerous villi and by the anatomical peculiarities of the choroidal epithelium which develops an extended apical brush border and basolateral interdigitations (Keep and Jones 1990). Nonetheless, the molecular exchanges between the blood and the CSF across the choroidal epithelium are tightly regulated. Like at the BBB, the presence of tight junctions that link the epithelial cells together strongly reduces the nonspecific paracellular leakage (Kratzer et al. 2012). Different types of influx and efflux transport systems account for the selectivity and directionality of solute transport. Relevant to efflux transport proteins, the choroid plexuses express high levels of basolaterally located transporters of the multidrug resistance-related ABCC protein family, which participate to the low brain penetration of various drugs (Gazzin et al. 2008; Leggas et al. 2004; Wijnholds et al. 2000).

The detoxication reactions that take place at the choroid plexuses represent another neuroprotective facet of CP functions toward toxic compounds and may also decrease the delivery of some drugs into the brain.

6.2.2.2 Molecular Characterization, Relative Expression, and Function of Drug Metabolizing Enzymes in Choroid Plexuses and Ependyma

Choroid plexuses appear to be a major site of drug metabolism in the brain. In rat, the choroidal specific enzymatic activities of enzymes such as epoxide hydrolases (EHs), UGTs, or GSTs do reach hepatic levels (Ghersi-Egea et al. 1994; Strazielle and Ghersi-Egea 1999). As in liver, some DME activities in choroid plexus are inducible by foreign compounds (Leininger-Muller et al. 1994). This metabolic detoxication capacity is another function shared by the choroid plexuses and the liver, besides their ability to synthesize and secrete the thyroid hormone carrier transthyretin (Schreiber and Richardson 1997). They are, however, differences in drug metabolism between the two organs, particularly in phase I proteins.

Phase I of Drug Metabolism Some CYP-dependent monooxygenase activities have been measured in isolated rat choroid plexuses (Ghersi-Egea et al. 1994), albeit at lower levels than in the liver. The molecular identification and localization of CYPs in the choroid plexus is only partial. Immunohistochemical and in situ hybridization studies identified CYP1A1, but not 1A2 in rat and mouse choroid plexuses following induction by β-naphthoflavone or the carcinogenic 3-methylcholanthrene. The enzyme was located at the choroidal vessel walls rather than at the BCSFB proper. It was shown to metabolize heterocyclic amines into reactive intermediates, a metabolic activity that is deleterious in this instance (Brittebo 1994; Dey et al. 1999; Morse et al. 1998). CYP1A1 was not detected prior to inductive treatment. Immunohistochemical evidence for the localization of a CYP2B1/2-like protein in the rat and mouse choroid plexus has been reported (Miksys et al. 2000a; Volk et al. 1991). An antibody raised against isoforms of the CYP2D subfamily generated a strong signal in the rat choroid plexus, possibly associated with the endothelium (Miksys et al. 2000b). CYP2B and 2D proteins metabolize a large range of xenobiotics including centrally acting drugs, but no relevant metabolic activities have yet been measured in the choroidal tissue. A thorough transcriptomic evaluation of the choroidal expression of CYP isoforms in rat revealed that of the several dozen CYP identified in this species, only four isoforms were present in the choroidal tissue, suggesting that CYP plays only a marginal role in detoxification at the choroid plexus (Kratzer et al. 2013). No data on CYP expression in human choroid plexus have been reported so far. Of note, the activity of NADPH-cytochrome P450 reductase, the enzyme that provides the electrons necessary to the activity of microsomal CYPs, is sizably measured in rat choroid plexus homogenate and in choroid plexus epithelial cells (Strazielle and Ghersi-Egea 1999). Besides its role in electron transfer to CYPs, this enzyme can generate free radicals by CYP-independent reductive metabolism of drugs, a mechanism that participates in the toxicity of compounds able to undergo a single electron reduction (Ghersi-Egea et al. 1998). Besides CYP-dependent monooxygenases, flavin-containing monooxygenases (FMO) also play an important role in phase I metabolism of foreign chemicals, including psychoactive drugs (Cashman 2000). FMO1 mRNA has been localized in the mouse choroidal epithelium by in situ hybridization (Janmohamed et al. 2004), and both FMO1 and FMO3 mRNAs have been identified in rat choroid plexus (Kratzer et al. 2013). This calls for more information concerning the choroidal function of these different FMO isoforms.

Among all brain structures in rat, the choroid plexus displays the highest level of EH activity toward carcinogenic epoxides. The brain vessels have the second highest level of activity (Ghersi-Egea et al. 1994, Strazielle and Ghersi-Egea 1999). Inactivation of carcinogenic epoxides is mainly attributed to the membrane-bound form of EH, Ephx1, whose expression is high in the choroid plexus (Kratzer et al. 2013). The enzymatic and transcriptomic data match immunohistochemical data showing in mice that the highest signal for this isoform is associated with the choroidal epithelium (Marowsky et al. 2009). One study investigated the localization of the soluble form of EH (sEH) in human brain. sEH is involved in the metabolism of lipid-derived biologically active endogenous epoxides rather than that of carcinogenic xenobiotics. A high immunohistochemical signal toward sEH was also associated with the human choroid plexus (Sura et al. 2008), while Marowsky et al. (2009) did not specifically report such localization in the mouse brain.

Finally, other choroidal enzymes with narrower substrate specificity, such as MAOs or alcohol dehydrogenases, can also participate in the phase I of drug metabolism (reviewed in (Strazielle et al. 2004).

Phase II of Drug Metabolism In the rat, the choroidal activity of the UGT isoenzyme(s) responsible for the conjugation of planar compounds is high, reaching the hepatic level. This enzymatic activity is located in the epithelium and is inducible by exogenous polycyclic aromatic hydrocarbons as it is in the liver (Ghersi-Egea et al. 1994; Leininger-Muller et al. 1994; Strazielle and Ghersi-Egea 1999). It is likely to be catalyzed by one or several UGT1A isoenzymes. High levels of mRNAs were detected in the rat choroid plexus with probes common to all genes of the UGT1A subfamily (Kratzer et al. 2013). There may however be species differences in the conjugation capacity to glucuronic acid in the choroid plexus, especially between rodent, primate, and human as discussed for the BBB (see Sect. 6.2.1.2).

Detoxication by sulfoconjugation appears to be active in both human and rodent choroid plexuses. High levels of SULT1A1 activity toward phenolic compounds and of SULT1A1 protein were reported in fetal human choroid plexus by comparison to other brain structures (Richard et al. 2001). Adult material was not tested in this study. *SULT1A1* mRNA level is high in choroidal material of both developing and adult rat (Kratzer et al. 2013). Additional functional studies are needed to precisely evaluate the impact of sulfoconjugation in the detoxication properties of the BCSFB.

Immunohistochemical evidence for the presence of the three main, alpha, mu, and pi classes of GSTs involved in drug metabolism and detoxication in the rodent choroid plexuses has been reported a long time ago (Cammer et al. 1989; Johnson et al. 1993; Philbert et al. 1995). Immunoreactivity of GST pi has also been demonstrated in human choroid plexus (Carder et al. 1990), and the use of a transgenic reporter mouse for human GSTP1 confirmed its high expression in the choroid plexus (Henderson et al. 2014). A high GSTalpha 4 (GST 8–8) mRNA enrichment has been reported in the rat choroid plexus (Liang et al. 2004). The conjugation to GSH of 1-chloro-2,4-dinitrobenzene (CDNB), which is a substrate for several cytosolic forms and the microsomal form of GSTs, is higher by one order of magnitude in choroid plexus than in brain parenchyma in the newborn rat. This activity is found associated with the epithelial cells in rat choroid plexus and is also high in human choroidal tissue (Ghersi-Egea et al. 2006; Strazielle and Ghersi-Egea 1999). The multiplicity of the GST isoenzymes, displaying differential substrate specificities, was confirmed by transcriptomic analysis, and this explains that glutathione-dependent enzymatic metabolism in the choroidal epithelium inactivates a broad spectrum of noxious compounds (Kratzer et al. 2018). The choroidal epithelium reconstituted in vitro prevents GST substrates presented at the blood-facing membrane to reach the apical, CSF-facing medium (Ghersi-Egea et al. 2006). The definitive proof that choroidal GSTs act as a metabolic barrier that prevents GST substrates from entering into the CSF was obtained in postnatal rats using a functional knockdown model for choroidal glutathione conjugation (Kratzer et al. 2018). Finally choroid plexus epithelial cells have the ability to efficiently take up glutathione precursors and to synthesize and recycle GSH (Burdo et al. 2006; Lee et al. 2012; Monks et al. 1999; Tate et al. 1973). These data point out an important role of GST-dependent detoxication pathways at the BCSFB.

GST mRNA, protein, and activity have also been detected or shown to be enriched in the ependyma lining the ventricle in rat (Abramovitz et al. 1988; Cammer et al. 1989; Kratzer et al. 2018; Liang et al. 2004; Philbert et al. 1995), mouse (Beiswanger et al. 1995), and human (Carder et al. 1990). This suggests that GST-dependent detoxification forms a second line of defense at the interface between the CSF and the neuropil.

Phase III of Drug Metabolism Different transporters of the ABCC family, including MRP1 and MRP4, are ideally located at the basolateral, blood-facing membrane in both rodent and humans to export conjugated metabolites into the systemic circulation (Gazzin et al. 2008; Ginguene et al. 2010; Leggas et al. 2004). Other basolateral transporters such as Oatp1a4 (Oatp2) may also transport drug conjugates at the basolateral membrane. The expression and functional significance of drug transporters at the BCFSB have been reviewed elsewhere (e.g., Ghersi-Egea et al. 2009b, Leslie et al. 2005, Strazielle and Ghersi-Egea 2005; other chapters in this book). The fate of metabolites produced within the ependyma is presently unknown.

Antioxidant Systems In addition to being a cosubstrate for GSTs, reduced glutathione is also active as a main intracellular antioxidant molecule. The reduced/oxidized glutathione redox cycle is active in choroid plexus which in rat displays enriched levels of glutathione reductase activity compared with the neuropil. Glutathione is also substrate for the glutathione peroxidases, and peroxidase activities are ten to 30 times higher in choroid plexuses than in brain parenchyma in adult and developing rats (Saudrais et al. 2018; Tayarani et al. 1989). Glutathione peroxidase 1 and 4 and their companion enzyme glutathione reductase are well expressed in choroid plexus epithelial cells. Experiments using live choroid plexuses isolated from developing rats showed that these enzymes are highly efficient to control the levels of hydrogen peroxide and possibly other peroxides in the cerebrospinal fluid. This is important as hydroperoxide released at low concentration is involved in several cell signaling pathways, especially during brain development, and at high concentration generates deleterious oxidative stress susceptible to compromise brain development and function (reviewed in (Saudrais et al. 2018). Although catalase is also well expressed in the choroidal tissue (Tayarani et al. 1989), choroidal peroxidase activities are more potent than catalase to detoxify extracellular hydroperoxide and act also to prevent blood-borne hydroperoxide to reach the cerebrospinal fluid (Saudrais et al. 2018).

Finally, superoxide dismutase also displays significantly higher activities in the choroid plexus than in brain tissue in rat (Tayarani et al. 1989). Altogether the choroid plexuses appear to possess a powerful machinery to fight reactive chemical species including reactive oxygen species.

6.2.2.3 Expression of Drug Metabolizing Enzymes and Phase III Transporters in the Arachnoid Blood-CSF Barrier

Little data are available regarding drug metabolism and transport at the arachnoid barrier. Analyses are complicated by the membranous nature of the meningeal layers, intermingled with large and small meningeal vessels, potentially generating non-specific immunohistochemical signals, and preventing the identification of the cellular source for mRNA signals obtained from meningeal extracts. A pioneer work identified high levels of epoxide hydrolase activities in pia-arachnoid membrane homogenates in rat, which clearly differed from the lower activities measured in the dura matter or cerebral vessels (Ghersi-Egea et al. 1994). Less spectacular drug metabolizing enzyme activities also recorded in the pia-arachnoid fractions included GST, UGT, and CYP activities. In this work, arachnoid was not distinguished from pia membranes and included some meningeal vessels. Since then mRNAs for CYP 1A1, 1A2, and 1B1 were identified in meninges of pigs, together with a CYP-dependent oxygenase activity (Nannelli et al. 2009). Low levels of GST and antioxidant enzyme activities were also reported in these preparations. The relative proportion of arachnoid versus pia and dura membranes, or vessels, was not specified in the study. A more recent study using microarrays reported CYP1B1 and, relative to phase III of metabolism, ABCC1, ABCC4, and BCRP mRNA expression in leptomeninges (i.e., pia-arachnoid membranes) of newborn mouse (Yasuda et al. 2013). No phase II enzymes were reported. Immunohistochemistry confirmed the presence of ABCG2 in arachnoid, but not pia matter in mouse. The same microarray study identified CYP 1B1, 3A5, 2D6, 2B6, and BCRP gene expression in human arachnoid. Finally, two other studies focused on meningiomas, also identified CYP1A1 mRNA in autopsic "meninges" used as control tissue (Talari et al. 2018) and GST-pi protein in autopsic arachnoid used as control tissue (Cui et al. 2014). Further investigation is clearly required to understand the relevance of the arachnoid membrane in cerebral drug bioavailability and metabolism.

6.2.2.4 Pharmacotoxicological Significance and Regulation of Drug Metabolism at the Blood-CSF Barrier

As for the BBB, most of the data available about drug metabolism in the BCSFB are related to the molecular identity, level of expression, cellular localization of the enzymes, and specific activities measured in homogenates or subcellular fractions. In vivo data demonstrating drug metabolizing enzyme activities in the choroid plexus are scarce. One work demonstrated carcinogen metabolic activation in the choroidal endothelium through CYP1A1-mediated metabolism following induction with β-naphthoflavone (Brittebo 1994; Granberg et al. 2003). The metabolite irreversibly bound to the site of production and could thus be detected. A second work allowed to visualize GSH conjugation in situ in the rat brain following intracerebroventricular injection of a GST substrate whose conjugate is fluorescent (Kratzer et al. 2018). In vivo evidence that choroidal drug metabolism significantly changes the CSF bioavailability of drugs is difficult to obtain. Blood-to-CSF concentration ratios for substrates and metabolites need be measured over short periods of drug exposure to prevent the potential interference of transport and metabolism at the BBB, neuropil, and extracerebral sites and to avoid complications due to CSF circulation and CSF-brain extracellular fluid exchanges. In addition, the volume of CSF samples available for analysis is small in rodents, especially when investigating postnatal stages of development. One successful example has been the in vivo demonstration that choroidal GSTs influence the CSF bioavailability of GST substrates (Kratzer et al. 2018; see above). To overcome these limitations, cellular models of the BCSFB have been developed to address transport and metabolism across this barrier independently of other brain and peripheral parameters. Such a model has been developed in rat and has been validated for transport and metabolic studies (Strazielle et al. 2003; Strazielle and Ghersi-Egea 1999). It was used to show that the choroidal epithelium acts as a blood-to-CSF metabolic barrier toward selected xenobiotics through conjugation via either a UGT-dependent pathway (Strazielle and Ghersi-Egea 1999) or a GST-dependent pathway (Ghersi-Egea et al. 2006; Kratzer et al. 2018). In the latter case, combining in vivo and in vitro approaches enabled to show that the barrier was efficient even toward high concentrations of the substrate, as long as the intracellular glutathione pool was not limiting. Following glutathione depletion, the efficacy of the barrier became dependent on the rate of glutathione neosynthesis by the choroidal epithelial cells, a synthetic pathway that could be enhanced by exposing the cells to drugs such as N-acetylcysteine. Both glutathione and glucuronoconjugates were mainly effluxed at the basolateral membrane, by mechanisms likely to involve MRP/ABCC transporters. These data showed that at least some of the choroidal enzymatic equipment is pharmacotoxicologically efficient. Additional in situ imaging and in vivo metabolic studies are eagerly needed to precisely delineate the role of choroidal metabolism in reducing the entry of xenobiotics into the CSF and brain or increasing their elimination rate from the brain and in participating to the overall neuroprotective function of blood-brain interfaces.

As in the liver, choroidal DMEs may be induced by a wide range of xenobiotics including drugs. Examples of choroidal induction have been published only for AhR ligands such as β-naphthoflavone and carcinogenic compounds like 3-methylcholantrene (e.g., Leininger-Muller et al. 1994; Morse et al. 1998). UGT and GST activity, respectively measured following treatment of rats with phenobarbital and diallyl sulfide, two inducers of the CAR pathway, were not increased in choroid plexuses, while they were induced in the liver (Koehn et al. 2019; Leininger-Muller et al. 1994). Other inducing mechanisms activated by drugs or oxidative stress are likely functional at the blood-CSF barrier, because transcription factors such as PXR or Nrf2 are expressed in the choroid plexus (D'Angelo et al. 2013; Kratzer et al. 2013). In this regard, the Nrf2 pathway involved in both drug metabolism and antioxidant mechanisms was shown to be inducible in choroid plexus epithelial cells in vitro (Xiang et al. 2012). Such induction mechanisms may increase the neuroprotective functions associated with the BCSFB.

6.2.2.5   Drug Metabolism Associated with the Blood-CSF Barrier during Development

The efficacy of blood-brain interfaces in protecting neural cells during the critical period of brain development has been a subject of debate throughout the last decades (Ek et al. 2012; Johansson et al. 2008). More recently, evidence both in rodent and in human for an early and efficient establishment of the tight junctions that seal the cells forming the blood-brain and blood-CSF barriers has been gathered. This prevents non-specific paracellular leakage between blood and brain during fetal development (Ek et al. 2006; Kratzer et al. 2012). The BCSFB in particular appears to follow a specific pattern of early maturation during brain development. The choroid plexus appears early during the embryonic life and seems to acquire an "adult" morphological and functional phenotype earlier than most brain structures. This highlights the special role of the choroidal tissue in regulating blood-brain exchanges during development (Dziegielewska et al. 2001). With respect to metabolic capacities toward drugs and toxic compounds, the choroid plexuses already possess high detoxification capacities in the newborn rat (Kratzer et al. 2018; Strazielle and Ghersi-Egea 1997). Overall GST activities are higher in newborn than in adult rat choroid plexuses and are also very high in choroidal tissue from fetal human brain (Ghersi-Egea et al. 2006, Kratzer et al. 2018). Yet, the developmental profile of enzyme expression differs from one GST class to another (Beiswanger et al. 1995; Carder et al. 1990; Kratzer et al. 2018). The choroid plexus-to-brain tissue ratio of GPX activity is highest in developing animals (Saudrais et al. 2018). High levels of SULT1A1 are also clearly associated with the choroidal tissue in developing human brain (Richard et al. 2001). The protein level of Mrp1/Abcc1 that can export drug conjugates is already high in the choroid plexus of developing animals, by contrast to the protein level of the prototypic BBB efflux transporter P-glycoprotein which increases during postnatal development in microvessels (Gazzin et al. 2008). Taken together, these data point to active glutathione-dependent detoxification functions of the BCSFB in developing individuals. They also suggest that other neuroprotective processes are especially active at the choroid plexuses during brain development, but additional work is needed to explore this hypothesis.

6.3 Future Challenge

While the presence of specific drug metabolizing enzymes has been clearly established at both the blood-brain and blood-CSF barriers, definite proofs that their activity can influence either the cerebral bioavailability or the neurotoxicity of drugs and xenobiotics remain scarce. Designing in vivo experiments and pharmacokinetic models oriented toward the study of cerebral drug metabolism is therefore mandatory to assess the significance of such metabolic pathways. This should be done in both adult and developing animals, owing to the substantial metabolic activity of the BCSFB during brain development.

No information is available concerning the level of drug metabolizing enzyme expression in brain microvessels from developing animals or from fetal/neonate human. The establishment of developmental expression profiles for relevant enzymes at both barriers will allow appreciating the degree of maturity of these protective interfaces in the developing brain.

An in-depth molecular characterization of choroidal DME isoforms is still needed to build a comprehensive view of drug metabolism at blood-brain interfaces, and species differences need to be assessed to appreciate the predictive value of experimental pharmacokinetic models used to determine the influence of metabolism on the cerebral bioavailability of drugs in adult and pediatric patients.

Finally, as detoxication processes appear to contribute to the neuroprotective functions of the blood-brain interfaces, their importance in protecting the brain in pathological situations, e.g., following exposure to environmental toxins or following oxidative insults, needs to be explored more thoroughly. The pharmacological enhancement of these metabolic functions could be a strategy to improve neuroprotection in a pathophysiological context. This could be explored by assessing whether the induction pathways known to be efficient in the liver can be activated and function in the blood-brain interfaces.

6.4 Conclusions

Evidence for the presence and activity of several phase I and phase II drug metabolizing enzymes at blood-brain interfaces has been gathered over the past decades. The functional significance of drug metabolizing enzyme activities at the BBB and the fate of the produced metabolites remain to be explored. This is particularly true for drugs whose metabolites produced within blood-brain interfaces can reach the brain parenchyma and be active. Therefore, the cooperativity of phase I and II DME and SLC/ABC transporters should be studied to better understand brain pharmacokinetics-pharmacodynamics (PKPD) relationships for drugs and their active metabolites. The high choroidal specific activities of selected drug metabolizing enzymes, concurrent with the efficient efflux of metabolites by multispecific transporters at the BCSFB, confer a function of metabolic barrier and detoxication to the choroid plexus. The inducibility of these enzymatic systems in the BBB and BCSFB opens the interesting possibility to pharmacologically enhance neuroprotection at blood-brain interfaces.

Acknowledgments Supported by ANR-10-IBHU-0003 Cesame grant and Fondation pour la Recherche Médicale FRM to JFGE. The authors thank Dr. David Gomez-Zepeda for his help in the manuscript and figures.

References

Abbott NJ (2002) Astrocyte-endothelial interactions and blood-brain barrier permeability. J Anat 200:629–638

Abramovitz M, Homma H, Ishigaki S, Tansey F, Cammer W, Listowsky I (1988) Characterization and localization of glutathione-S-transferases in rat brain and binding of hormones, neurotransmitters, and drugs. J Neurochem 50:50–57

Aleksunes LM, Klaassen CD (2012) Coordinated regulation of hepatic phase I and II drug-metabolizing genes and transporters using AhR-, CAR-, PXR-, PPARalpha-, and Nrf2-null mice. Drug Metab Dispos 40:1366–1379

Ballabh P, Braun A, Nedergaard M (2004) The blood-brain barrier: an overview: structure, regulation, and clinical implications. Neurobiol Dis 16:1–13

Baranczyk-Kuzma A, Audus KL, Borchardt RT (1986) Catecholamine-metabolizing enzymes of bovine brain microvessel endothelial cell monolayers. J Neurochem 46:1956–1960

Baranczyk-Kuzma A, Audus KL, Borchardt RT (1989) Substrate specificity of phenol sulfotransferase from primary cultures of bovine brain microvessel endothelium. Neurochem Res 14:689–691

Barouki R, Coumoul X, Fernandez-Salguero PM (2007) The aryl hydrocarbon receptor, more than a xenobiotic-interacting protein. FEBS Lett 581:3608–3615

Bauer B, Yang X, Hartz AM, Olson ER, Zhao R, Kalvass JC, Pollack GM, Miller DS (2006) In vivo activation of human pregnane X receptor tightens the blood-brain barrier to methadone through P-glycoprotein up-regulation. Mol Pharmacol 70:1212–1219

Bauer B, Hartz AM, Lucking JR, Yang X, Pollack GM, Miller DS (2008) Coordinated nuclear receptor regulation of the efflux transporter, Mrp2, and the phase-II metabolizing enzyme, GSTpi, at the blood-brain barrier. J Cereb Blood Flow Metab 28:1222–1236

Beiswanger CM, Diegmann MH, Novak RF, Philbert MA, Graessle TL, Reuhl KR, Lowndes HE (1995) Developmental changes in the cellular distribution of glutathione and glutathione S-transferases in the murine nervous system. Neurotoxicology 16:425–440

Benzi G, Berte F, Crema A, Frigo GM (1967) Cerebral drug metabolism investigated by isolated perfused brain in situ. J Pharm Sci 56:1349–1351

Betz AL, Firth JA, Goldstein GW (1980) Polarity of the blood-brain barrier: distribution of enzymes between the luminal and antiluminal membranes of brain capillary endothelial cells. Brain Res 192:17–28

Bhamre S, Bhagwat SV, Shankar SK, Williams DE, Ravindranath V (1993) Cerebral flavin-containing monooxygenase-mediated metabolism of antidepressants in brain: immunochemical properties and immunocytochemical localization. J Pharmacol Exp Ther 267:555–559

Brittebo EB (1994) Metabolism-dependent binding of the heterocyclic amine Trp-P-1 in endothelial cells of choroid plexus and in large cerebral veins of cytochrome P450-induced mice. Brain Res 659:91–98

Burdo J, Dargusch R, Schubert D (2006) Distribution of the cystine/glutamate antiporter system xc- in the brain, kidney, and duodenum. J Histochem Cytochem 54:549–557

Calabria AR, Shusta EV (2008) A genomic comparison of in vivo and in vitro brain microvascular endothelial cells. J Cereb Blood Flow Metab 28:135–148

Calkins MJ, Johnson DA, Townsend JA, Vargas MR, Dowell JA, Williamson TP, Kraft AD, Lee JM, Li J, Johnson JA (2009) The Nrf2/ARE pathway as a potential therapeutic target in neurodegenerative disease. Antioxid Redox Signal 11:497–508

Cammer W, Tansey F, Abramovitz M, Ishigaki S, Listowsky I (1989) Differential localization of glutathione-S-transferase Yp and Yb subunits in oligodendrocytes and astrocytes of rat brain. J Neurochem 52:876–883

Carder PJ, Hume R, Fryer AA, Strange RC, Lauder J, Bell JE (1990) Glutathione S-transferase in human brain. Neuropathol Appl Neurobiol 16:293–303

Cardoso FL, Brites D, Brito MA (2010) Looking at the blood-brain barrier: molecular anatomy and possible investigation approaches. Brain Res Rev 64:328–363

Cashman JR (2000) Human flavin-containing monooxygenase: substrate specificity and role in drug metabolism. Curr Drug Metab 1:181–191

Chan GN, Hoque MT, Cummins CL, Bendayan R (2010) Regulation of P-glycoprotein by orphan nuclear receptors in human brain microvessel endothelial cells. J Neurochem 118:163–175

Chaves C, Shawahna R, Jacob A, Scherrmann JM, Decleves X (2014) Human ABC transporters at blood-CNS interfaces as determinants of CNS drug penetration. Curr Pharm Des 20:1450–1462

Chodobski A, Szmydynger-Chodobska J (2001) Choroid plexus: target for polypeptides and site of their synthesis. Microsc Res Tech 52:65–82

Choi YK, Kim KW (2008) Blood-neural barrier: its diversity and coordinated cell-to-cell communication. BMB Rep 41:345–352

Christrup LL (1997) Morphine metabolites. Acta Anaesthesiol Scand 41:116–122

Cui GQ, Jiao AH, Xiu CM, Wang YB, Sun P, Zhang LM, Li XG (2014) Proteomic analysis of meningiomas. Acta Neurol Belg 114:187–196

Damkier HH, Brown PD, Praetorius J (2010) Epithelial pathways in choroid plexus electrolyte transport. Physiology (Bethesda) 25:239–249

D'Angelo B, Ek CJ, Sandberg M, Mallard C (2013) Expression of the Nrf2-system at the blood-CSF barrier is modulated by neonatal inflammation and hypoxia-ischemia. J Inherit Metab Dis 36:479–490

Dauchy S, Dutheil F, Weaver RJ, Chassoux F, Daumas-Duport C, Couraud PO, Scherrmann JM, De Waziers I, Decleves X (2008) ABC transporters, cytochromes P450 and their main transcription factors: expression at the human blood-brain barrier. J Neurochem 107:1518–1528

Dauchy S, Miller F, Couraud PO, Weaver RJ, Weksler B, Romero IA, Scherrmann JM, De Waziers I, Decleves X (2009) Expression and transcriptional regulation of ABC transporters and cytochromes P450 in hCMEC/D3 human cerebral microvascular endothelial cells. Biochem Pharmacol 77:897–909 de Boer AG, Gaillard PJ (2006) Blood-brain barrier dysfunction and recovery. J Neural Transm 113:455–462

Decleves X, Jacob A, Yousif S, Shawahna R, Potin S, Scherrmann JM (2011) Interplay of drug metabolizing CYP450 enzymes and ABC transporters in the blood-brain barrier. Curr Drug Metab 12:732–741

Denison MS, Nagy SR (2003) Activation of the aryl hydrocarbon receptor by structurally diverse exogenous and endogenous chemicals. Annu Rev Pharmacol Toxicol 43:309–336

Dey A, Jones JE, Nebert DW (1999) Tissue- and cell type-specific expression of cytochrome P450 1A1 and cytochrome P450 1A2 mRNA in the mouse localized in situ hybridization. Biochem Pharmacol 58:525–537

Doring B, Petzinger E (2014) Phase 0 and phase III transport in various organs: combined concept of phases in xenobiotic transport and metabolism. Drug Metab Rev 46:261–282

Duffel MW, Marshal AD, McPhie P, Sharma V, Jakoby WB (2001) Enzymatic aspects of the phenol (aryl) sulfotransferases. Drug Metab Rev 33:369–395

Dutheil F, Beaune P, Loriot MA (2008) Xenobiotic metabolizing enzymes in the central nervous system: contribution of cytochrome P450 enzymes in normal and pathological human brain. Biochimie 90:426–436

Dutheil F, Dauchy S, Diry M, Sazdovitch V, Cloarec O, Mellottee L, Bieche I, Ingelman-Sundberg M, Flinois JP, de Waziers I, Beaune P, Decleves X, Duyckaerts C, Loriot MA (2009) Xenobiotic-metabolizing enzymes and transporters in the normal human brain: regional and cellular mapping as a basis for putative roles in cerebral function. Drug Metab Dispos 37:1528–1538

Dziegielewska KM, Ek J, Habgood MD, Saunders NR (2001) Development of the choroid plexus. Microsc Res Tech 52:5–20

Eaton DL, Bammler TK (1999) Concise review of the glutathione S-transferases and their significance to toxicology. Toxicol Sci 49:156–166

Ek CJ, Dziegielewska KM, Stolp H, Saunders NR (2006) Functional effectiveness of the blood-brain barrier to small water-soluble molecules in developing and adult opossum (Monodelphis domestica). J Comp Neurol 496:13–26

Ek CJ, Dziegielewska KM, Habgood MD, Saunders NR (2012) Barriers in the developing brain and Neurotoxicology. Neurotoxicology 33:586–606

Gamage N, Barnett A, Hempel N, Duggleby RG, Windmill KF, Martin JL, McManus ME (2006) Human sulfotransferases and their role in chemical metabolism. Toxicol Sci 90:5–22

Gazzin S, Strazielle N, Schmitt C, Fevre-Montange M, Ostrow JD, Tiribelli C, Ghersi-Egea JF (2008) Differential expression of the multidrug resistance-related proteins ABCb1 and ABCc1 between blood-brain interfaces. J Comp Neurol 510:497–507

Gelhaus SL, Harvey RG, Penning TM, Blair IA (2011) Regulation of benzo[a]pyrene-mediated DNA- and glutathione-adduct formation by 2,3,7,8-tetrachlorodibenzo-p-dioxin in human lung cells. Chem Res Toxicol 24:89–98

Ghersi-Egea JF, Minn A, Siest G (1988) A new aspect of the protective functions of the blood-brain barrier: activities of four drug-metabolizing enzymes in isolated rat brain microvessels. Life Sci 42:2515–2523

Ghersi-Egea JF, Perrin R, Leininger-Muller B, Grassiot MC, Jeandel C, Floquet J, Cuny G, Siest G, Minn A (1993) Subcellular localization of cytochrome P450, and activities of several enzymes responsible for drug metabolism in the human brain. Biochem Pharmacol 45:647–658

Ghersi-Egea JF, Leninger-Muller B, Suleman G, Siest G, Minn A (1994) Localization of drug-metabolizing enzyme activities to blood-brain interfaces and circumventricular organs. J Neurochem 62:1089–1096

Ghersi-Egea JF, Leininger-Muller B, Cecchelli R, Fenstermacher JD (1995) Blood-brain interfaces: relevance to cerebral drug metabolism. Toxicol Lett 82-83:645–653

Ghersi-Egea JF, Maupoil V, Ray D, Rochette L (1998) Electronic spin resonance detection of superoxide and hydroxyl radicals during the reductive metabolism of drugs by rat brain preparations and isolated cerebral microvessels. Free Radic Biol Med 24:1074–1081

Ghersi-Egea JF, Strazielle N, Murat A, Jouvet A, Buenerd A, Belin MF (2006) Brain protection at the blood-cerebrospinal fluid interface involves a glutathione-dependent metabolic barrier mechanism. J Cereb Blood Flow Metab 4:6

Ghersi-Egea JF, Gazzin S, Strazielle N (2009a) Blood-brain interfaces and bilirubin-induced neurological diseases. Curr Pharm Des 15:2893–2907

Ghersi-Egea JF, Monkkonen KS, Schmitt C, Honnorat J, Fevre-Montange M, Strazielle N (2009b) Blood-brain interfaces and cerebral drug bioavailability. Rev Neurol (Paris) 165:1029–1038

Ghersi-Egea JF, Strazielle N, Catala M, Silva-Vargas V, Doetsch F, Engelhardt B (2018) Molecular anatomy and functions of the choroidal blood-cerebrospinal fluid barrier in health and disease. Acta Neuropathol 135:337–361

Ghosh C, Gonzalez-Martinez J, Hossain M, Cucullo L, Fazio V, Janigro D, Marchi N (2010) Pattern of P450 expression at the human blood-brain barrier: roles of epileptic condition and laminar flow. Epilepsia 51:1408–1417

Ghosh C, Hossain M, Puvenna V, Martinez-Gonzalez J, Alexopolous A, Janigro D, Marchi N (2013) Expression and functional relevance of UGT1A4 in a cohort of human drug-resistant epileptic brains. Epilepsia 54:1562–1570

Ghosh C, Hossain M, Solanki J, Najm IM, Marchi N, Janigro D (2017) Overexpression of pregnane X and glucocorticoid receptors and the regulation of cytochrome P450 in human epileptic brain endothelial cells. Epilepsia 58:576–585

Ginguene C, Champier J, Maallem S, Strazielle N, Jouvet A, Fevre-Montange M, Ghersi-Egea JF (2010) P-glycoprotein (ABCB1) and breast cancer resistance protein (ABCG2) localize in the microvessels forming the blood-tumor barrier in ependymomas. Brain Pathol 20:926–935

Gomez-Zepeda D, Taghi M, Scherrmann JM, Decleves X, Menet MC (2019) ABC transporters at the blood-brain interfaces, their study models, and drug delivery implications in gliomas. Pharmaceutics 12

Gottwald MD, Bainbridge JL, Dowling GA, Aminoff MJ, Alldredge BK (1997) New pharmacotherapy for Parkinson's disease. Ann Pharmacother 31:1205–1217

Granberg L, Ostergren A, Brandt I, Brittebo EB (2003) CYP1A1 and CYP1B1 in blood-brain interfaces: CYP1A1-dependent bioactivation of 7,12-dimethylbenz(a)anthracene in endothelial cells. Drug Metab Dispos 31:259–265

Hansson T, Tindberg N, Ingelman-Sundberg M, Kohler C (1990) Regional distribution of ethanol-inducible cytochrome P450 IIE1 in the rat central nervous system. Neuroscience 34:451–463

Hawkins BT, Davis TP (2005) The blood-brain barrier/neurovascular unit in health and disease. Pharmacol Rev 57:173–185

Hawkins BT, Egleton RD (2006) Fluorescence imaging of blood-brain barrier disruption. J Neurosci Methods 151:262–267

Hayes JD, Flanagan JU, Jowsey IR (2005) Glutathione transferases. Annu Rev Pharmacol Toxicol 45:51–88

Henderson CJ, McLaren AW, Wolf CR (2014) In vivo regulation of human glutathione transferase GSTP by chemopreventive agents. Cancer Res 74:4378–4387

Higgins LG, Hayes JD (2011) The cap'n'collar transcription factor Nrf2 mediates both intrinsic resistance to environmental stressors and an adaptive response elicited by chemopreventive agents that determines susceptibility to electrophilic xenobiotics. Chem Biol Interact 192:37–45

Jacob A, Hartz AM, Potin S, Coumoul X, Yousif S, Scherrmann JM, Bauer B, Decleves X (2011) Aryl hydrocarbon receptor-dependent upregulation of Cyp1b1 by TCDD and diesel exhaust particles in rat brain microvessels. Fluids Barriers CNS 8:23

Jacob A, Potin S, Chapy H, Crete D, Glacial F, Ganeshamoorthy K, Couraud PO, Scherrmann JM, Decleves X (2015) Aryl hydrocarbon receptor regulates CYP1B1 but not ABCB1 and ABCG2 in hCMEC/D3 human cerebral microvascular endothelial cells after TCDD exposure. Brain Res 1613:27–36

Janmohamed A, Hernandez D, Phillips IR, Shephard EA (2004) Cell-, tissue-, sex- and developmental stage-specific expression of mouse flavin-containing monooxygenases (Fmos). Biochem Pharmacol 68:73–83

Johansson PA, Dziegielewska KM, Liddelow SA, Saunders NR (2008) The blood-CSF barrier explained: when development is not immaturity. BioEssays 30:237–248

Johnson JA, el Barbary A, Kornguth SE, Brugge JF, Siegel FL (1993) Glutathione S-transferase isoenzymes in rat brain neurons and glia. J Neurosci 13:2013–2023

Josserand V, Pelerin H, de Bruin B, Jego B, Kuhnast B, Hinnen F, Duconge F, Boisgard R, Beuvon F, Chassoux F, Daumas-Duport C, Ezan E, Dolle F, Mabondzo A, Tavitian B (2006) Evaluation of drug penetration into the brain: a double study by in vivo imaging with positron emission tomography and using an in vitro model of the human blood-brain barrier. J Pharmacol Exp Ther 316:79–86

Karhunen T, Tilgmann C, Ulmanen I, Panula P (1995) Neuronal and non-neuronal catechol-O-methyltransferase in primary cultures of rat brain cells. Int J Dev Neurosci 13:825–836

Karlgren M, Backlund M, Johansson I, Oscarson M, Ingelman-Sundberg M (2004) Characterization and tissue distribution of a novel human cytochrome P450-CYP2U1. Biochem Biophys Res Commun 315:679–685

Keep RF, Jones HC (1990) A morphometric study on the development of the lateral ventricle choroid plexus, choroid plexus capillaries and ventricular ependyma in the rat. Brain Res Dev Brain Res 56:47–53

Khan E (2005) An examination of the blood-brain barrier in health and disease. Br J Nurs 14:509–513

King CD, Rios GR, Green MD, Tephly TR (2000) UDP-glucuronosyltransferases. Curr Drug Metab 1:143–161

Koehn LM, Dziegielewska KM, Mollgard K, Saudrais E, Strazielle N, Ghersi-Egea JF, Saunders NR, Habgood MD (2019) Developmental differences in the expression of ABC transporters at rat brain barrier interfaces following chronic exposure to diallyl sulfide. Sci Rep 9:5998

Kratzer I, Vasiljevic A, Rey C, Fevre-Montange M, Saunders N, Strazielle N, Ghersi-Egea JF (2012) Complexity and developmental changes in the expression pattern of claudins at the blood-CSF barrier. Histochem Cell Biol. In press

Kratzer I, Liddelow SA, Saunders NR, Dziegielewska KM, Strazielle N, Ghersi-Egea JF (2013) Developmental changes in the transcriptome of the rat choroid plexus in relation to neuroprotection. Fluids Barriers CNS 10:25

Kratzer I, Strazielle N, Saudrais E, Monkkonen K, Malleval C, Blondel S, Ghersi-Egea JF (2018) Glutathione conjugation at the blood-CSF barrier efficiently prevents exposure of the developing brain fluid environment to blood-borne reactive electrophilic substances. J Neurosci 38:3466–3479

Kusuhara H, Sugiyama Y (2004) Efflux transport systems for organic anions and cations at the blood-CSF barrier. Adv Drug Deliv Rev 56:1741–1763

Lee SW, Kim WJ, Park JA, Choi YK, Kwon YW, Kim KW (2006) Blood-brain barrier interfaces and brain tumors. Arch Pharm Res 29:265–275

Lee A, Anderson AR, Rayfield AJ, Stevens MG, Poronnik P, Meabon JS, Cook DG, Pow DV (2012) Localisation of novel forms of glutamate transporters and the cystine-glutamate antiporter in the choroid plexus: implications for CSF glutamate homeostasis. J Chem Neuroanat 43:64–75

Leggas M, Adachi M, Scheffer GL, Sun D, Wielinga P, Du G, Mercer KE, Zhuang Y, Panetta JC, Johnston B, Scheper RJ, Stewart CF, Schuetz JD (2004) Mrp4 confers resistance to topotecan and protects the brain from chemotherapy. Mol Cell Biol 24:7612–7621

Leininger-Muller B, Ghersi-Egea JF, Siest G, Minn A (1994) Induction and immunological characterization of the uridine diphosphate-glucuronosyltransferase conjugating 1-naphthol in the rat choroid plexus. Neurosci Lett 175:37–40

Leslie EM, Deeley RG, Cole SP (2005) Multidrug resistance proteins: role of P-glycoprotein, MRP1, MRP2, and BCRP (ABCG2) in tissue defense. Toxicol Appl Pharmacol 204:216–237

Liang T, Habegger K, Spence JP, Foroud T, Ellison JA, Lumeng L, Li TK, Carr LG (2004) Glutathione S-transferase 8-8 expression is lower in alcohol-preferring than in alcohol-nonpreferring rats. Alcohol Clin Exp Res 28:1622–1628

Listowsky I, Rowe JD, Patskovsky YV, Tchaikovskaya T, Shintani N, Novikova E, Nieves E (1998) Human testicular glutathione S-transferases: insights into tissue-specific expression of the diverse subunit classes. Chem Biol Interact 111-112:103–112

Marowsky A, Burgener J, Falck JR, Fritschy JM, Arand M (2009) Distribution of soluble and microsomal epoxide hydrolase in the mouse brain and its contribution to cerebral epoxyeicosatrienoic acid metabolism. Neuroscience 163:646–661

McLeod HL, Krynetski EY, Relling MV, Evans WE (2000) Genetic polymorphism of thiopurine methyltransferase and its clinical relevance for childhood acute lymphoblastic leukemia. Leukemia 14:567–572

Meister A (1974) An enzymatic basis for a blood-brain barrier? The gamma-glutamyl cycle-background and considerations relating to amino acid transport in the brain. Res Publ Assoc Res Nerv Ment Dis 53:273–291

Miksys SL, Tyndale RF (2002) Drug-metabolizing cytochrome P450s in the brain. J Psychiatry Neurosci 27:406–415

Miksys S, Hoffmann E, Tyndale RF (2000a) Regional and cellular induction of nicotine-metabolizing CYP2B1 in rat brain by chronic nicotine treatment. Biochem Pharmacol 59:1501–1511

Miksys S, Rao Y, Sellers EM, Kwan M, Mendis D, Tyndale RF (2000b) Regional and cellular distribution of CYP2D subfamily members in rat brain. Xenobiotica 30:547–566

Minn A, Ghersi-Egea JF, Perrin R, Leininger B, Siest G (1991) Drug metabolizing enzymes in the brain and cerebral microvessels. Brain Res Brain Res Rev 16:65–82

Monks TJ, Ghersi-Egea JF, Philbert M, Cooper AJ, Lock EA (1999) Symposium overview: the role of glutathione in neuroprotection and neurotoxicity. Toxicol Sci 51:161–177

Morse DC, Stein AP, Thomas PE, Lowndes HE (1998) Distribution and induction of cytochrome P450 1A1 and 1A2 in rat brain. Toxicol Appl Pharmacol 152:232–239

Nannelli A, Rossignolo F, Tolando R, Rossato P, Longo V, Gervasi PG (2009) Effect of beta-naphthoflavone on AhR-regulated genes (CYP1A1, 1A2, 1B1, 2S1, Nrf2, and GST) and anti-oxidant enzymes in various brain regions of pig. Toxicology 265:69–79

Nebert DW, Russell DW (2002) Clinical importance of the cytochromes P450. Lancet 360:1155–1162

Nishibori M, Tahara A, Sawada K, Sakiyama J, Nakaya N, Saeki K (2000) Neuronal and vascular localization of histamine N-methyltransferase in the bovine central nervous system. Eur J Neurosci 12:415–426

Ott M, Fricker G, Bauer B (2009) Pregnane X receptor (PXR) regulates P-glycoprotein at the blood-brain barrier: functional similarities between pig and human PXR. J Pharmacol Exp Ther 329:141–149

Ouzzine M, Gulberti S, Ramalanjaona N, Magdalou J, Fournel-Gigleux S (2014) The UDP-glucuronosyltransferases of the blood-brain barrier: their role in drug metabolism and detoxication. Front Cell Neurosci 8:349

Persidsky Y, Ramirez SH, Haorah J, Kanmogne GD (2006) Blood-brain barrier: structural components and function under physiologic and pathologic conditions. J Neuroimmune Pharmacol 1:223–236

Philbert MA, Beiswanger CM, Manson MM, Green JA, Novak RF, Primiano T, Reuhl KR, Lowndes HE (1995) Glutathione S-transferases and gamma-glutamyl transpeptidase in the rat nervous systems: a basis for differential susceptibility to neurotoxicants. Neurotoxicology 16:349–362

Praetorius J, Damkier HH (2017) Transport across the choroid plexus epithelium. Am J Physiol Cell Physiol 312:C673–C686

Redzic ZB, Preston JE, Duncan JA, Chodobski A, Szmydynger-Chodobska J (2005) The choroid plexus-cerebrospinal fluid system: from development to aging. Curr Top Dev Biol 71:1–52

Riachi NJ, Harik SI (1988) Strain differences in systemic 1-methyl-4-phenyl-1,2,3,6-tetrahydropyridine neurotoxicity in mice correlate best with monoamine oxidase activity at the blood-brain barrier. Life Sci 42:2359–2363

Riachi NJ, Harik SI, Kalaria RN, Sayre LM (1988) On the mechanisms underlying 1-methyl-4-phenyl-1,2,3,6-tetrahydropyridine neurotoxicity. II. Susceptibility among mammalian species correlates with the toxin's metabolic patterns in brain microvessels and liver. J Pharmacol Exp Ther 244:443–448

Richard K, Hume R, Kaptein E, Stanley EL, Visser TJ, Coughtrie MW (2001) Sulfation of thyroid hormone and dopamine during human development: ontogeny of phenol sulfotransferases and arylsulfatase in liver, lung, and brain. J Clin Endocrinol Metab 86:2734–2742

Rivett AJ, Eddy BJ, Roth JA (1982) Contribution of sulfate conjugation, deamination, and O-methylation to metabolism of dopamine and norepinephrine in human brain. J Neurochem 39:1009–1016

Saudrais E, Strazielle N, Ghersi-Egea JF (2018) Choroid plexus glutathione peroxidases are instrumental in protecting the brain fluid environment from hydroperoxides during postnatal development. Am J Physiol Cell Physiol 315:C445–C456

Schmitt C, Strazielle N, Richaud P, Bouron A, Ghersi-Egea JF (2011) Active transport at the blood-CSF barrier contributes to manganese influx into the brain. J Neurochem 117:747–756

Schreiber G, Richardson SJ (1997) The evolution of gene expression, structure and function of transthyretin. Comp Biochem Physiol B Biochem Mol Biol 116:137–160

Shang W, Liu WH, Zhao XH, Sun QJ, Bi JZ, Chi ZF (2008) Expressions of glutathione S-transferase alpha, mu, and pi in brains of medically intractable epileptic patients. BMC Neurosci 9:67

Shawahna R, Uchida Y, Decleves X, Ohtsuki S, Yousif S, Dauchy S, Jacob A, Chassoux F, Daumas-Duport C, Couraud PO, Terasaki T, Scherrmann JM (2011) Transcriptomic and quantitative proteomic analysis of transporters and drug metabolizing enzymes in freshly isolated human brain microvessels. Mol Pharm

Sivandzade F, Bhalerao A, Cucullo L (2019) Cerebrovascular and neurological disorders: protective role of NRF2. Int J Mol Sci 20

Slot AJ, Molinski SV, Cole SP (2011) Mammalian multidrug-resistance proteins (MRPs). Essays Biochem 50:179–207

Stanulla M, Schaeffeler E, Moricke A, Coulthard SA, Cario G, Schrauder A, Kaatsch P, Dordelmann M, Welte K, Zimmermann M, Reiter A, Eichelbaum M, Riehm H, Schrappe M, Schwab M (2009) Thiopurine methyltransferase genetics is not a major risk factor for secondary malignant neoplasms after treatment of childhood acute lymphoblastic leukemia on Berlin-Frankfurt-Munster protocols. Blood 114:1314–1318

Strazielle N, Ghersi-Egea JF (1997) Drug metabolism in newborn rat choroid plexus from lateral, third and fourth ventricle. Dev Anim Vet Sci 27:895–901

Strazielle N, Ghersi-Egea JF (1999) Demonstration of a coupled metabolism-efflux process at the choroid plexus as a mechanism of brain protection toward xenobiotics. J Neurosci 19:6275–6289

Strazielle N, Ghersi-Egea JF (2000) Choroid plexus in the central nervous system: biology and physiopathology. J Neuropathol Exp Neurol 59:561–576

Strazielle N, Ghersi-Egea JF (2005) Factors affecting delivery of antiviral drugs to the brain. Rev Med Virol 15:105–133

Strazielle N, Ghersi-Egea JF (2013) Physiology of blood-brain interfaces in relation to brain disposition of small compounds and macromolecules. Mol Pharm 10:1473–1491

Strazielle N, Ghersi-Egea JF (2015) Efflux transporters in blood-brain interfaces of the developing brain. Front Neurosci 9:21

Strazielle N, Ghersi-Egea JF (2016) Potential pathways for CNS drug delivery across the blood-cerebrospinal fluid barrier. Curr Pharm Des 22:5463–5476

Strazielle N, Belin MF, Ghersi-Egea JF (2003) Choroid plexus controls brain availability of anti-HIV nucleoside analogs via pharmacologically inhibitable organic anion transporters. AIDS 17:1473–1485

Strazielle N, Khuth ST, Ghersi-Egea JF (2004) Detoxification systems, passive and specific transport for drugs at the blood-CSF barrier in normal and pathological situations. Adv Drug Deliv Rev 56:1717–1740

Sun X, Shih AY, Johannssen HC, Erb H, Li P, Murphy TH (2006) Two-photon imaging of glutathione levels in intact brain indicates enhanced redox buffering in developing neurons and cells at the cerebrospinal fluid and blood-brain interface. J Biol Chem 281:17420–17431

Sura P, Sura R, Enayetallah AE, Grant DF (2008) Distribution and expression of soluble epoxide hydrolase in human brain. J Histochem Cytochem 56:551–559

Synold TW, Dussault I, Forman BM (2001) The orphan nuclear receptor SXR coordinately regulates drug metabolism and efflux. Nat Med 7:584–590

Talari NK, Panigrahi MK, Madigubba S, Phanithi PB (2018) Overexpression of aryl hydrocarbon receptor (AHR) signalling pathway in human meningioma. J Neuro-Oncol 137:241–248

Tate SS, Ross LL, Meister A (1973) The -glutamyl cycle in the choroid plexus: its possible function in amino acid transport. Proc Natl Acad Sci U S A 70:1447–1449

Tayarani I, Cloez I, Clement M, Bourre JM (1989) Antioxidant enzymes and related trace elements in aging brain capillaries and choroid plexus. J Neurochem 53:817–826

Teissier E, Fennrich S, Strazielle N, Daval JL, Ray D, Schlosshauer B, Ghersi-Egea JF (1998) Drug metabolism in in vitro organotypic and cellular models of mammalian central nervous system: activities of membrane-bound epoxide hydrolase and NADPH-cytochrome P-450 (c) reductase. Neurotoxicology 19:347–355

Tolson AH, Wang H (2010) Regulation of drug-metabolizing enzymes by xenobiotic receptors: PXR and CAR. Adv Drug Deliv Rev 62:1238–1249

Van Gelder NM (1968) A possible enzyme barrier for gamma-aminobutyric acid in the central nervous system. Prog Brain Res 29:259–271

Vasiliou V, Gonzalez FJ (2008) Role of CYP1B1 in glaucoma. Annu Rev Pharmacol Toxicol 48:333–58. https://doi.org/10.1146/annurev.pharmtox.48.061807.154729

Vidal JD, VandeVoort CA, Marcus CB, Lazarewicz NR, Conley AJ (2006) In vitro exposure to environmental tobacco smoke induces CYP1B1 expression in human luteinized granulosa cells. Reprod Toxicol 22:731–737

Volk B, Hettmannsperger U, Papp T, Amelizad Z, Oesch F, Knoth R (1991) Mapping of phenytoin-inducible cytochrome P450 immunoreactivity in the mouse central nervous system. Neuroscience 42:215–235

Wang X, Hawkins BT, Miller DS (2010a) Aryl hydrocarbon receptor-mediated up-regulation of ATP-driven xenobiotic efflux transporters at the blood-brain barrier. FASEB J

Wang X, Sykes DB, Miller DS (2010b) Constitutive androstane receptor-mediated up-regulation of ATP-driven xenobiotic efflux transporters at the blood-brain barrier. Mol Pharmacol 78:376–383

Wijnholds J, deLange EC, Scheffer GL, van den Berg DJ, Mol CA, van der Valk M, Schinkel AH, Scheper RJ, Breimer DD, Borst P (2000) Multidrug resistance protein 1 protects the choroid plexus epithelium and contributes to the blood-cerebrospinal fluid barrier. J Clin Invest 105:279–285

Williams S, Hossain M, Ferguson L, Busch RM, Marchi N, Gonzalez-Martinez J, Perucca E, Najm IM, Ghosh C (2019) Neurovascular drug biotransformation machinery in focal human epilepsies: brain CYP3A4 correlates with seizure frequency and antiepileptic drug therapy. Mol Neurobiol 56:8392–8407

Xiang J, Alesi GN, Zhou N, Keep RF (2012) Protective effects of isothiocyanates on blood-CSF barrier disruption induced by oxidative stress. Am J Physiol Regul Integr Comp Physiol 303:R1–R7

Xu C, Li CY, Kong AN (2005) Induction of phase I, II and III drug metabolism/transport by xenobiotics. Arch Pharm Res 28:249–268

Yasuda K, Cline C, Vogel P, Onciu M, Fatima S, Sorrentino BP, Thirumaran RK, Ekins S, Urade Y, Fujimori K, Schuetz EG (2013) Drug transporters on arachnoid barrier cells contribute to the blood-cerebrospinal fluid barrier. Drug Metab Dispos 41:923–931

Zastre JA, Chan GN, Ronaldson PT, Ramaswamy M, Couraud PO, Romero IA, Weksler B, Bendayan M, Bendayan R (2009) Up-regulation of P-glycoprotein by HIV protease inhibitors in a human brain microvessel endothelial cell line. J Neurosci Res 87:1023–1036

Zhang C, Harder DR (2002) Cerebral capillary endothelial cell mitogenesis and morphogenesis induced by astrocytic epoxyeicosatrienoic acid. Stroke 33:2957–2966

Ziegler N, Awwad K, Fisslthaler B, Reis M, Devraj K, Corada M, Minardi SP, Dejana E, Plate KH, Fleming I, Liebner S (2016) Beta-catenin is required for endothelial Cyp1b1 regulation influencing metabolic barrier function. J Neurosci 36:8921–8935

Zlokovic BV (2008) The blood-brain barrier in health and chronic neurodegenerative disorders. Neuron 57:178–201

Part II
PK Concepts and Methods for Studying CNS Drug Delivery

Chapter 7
Pharmacokinetic Concepts in Brain Drug Delivery

Margareta Hammarlund-Udenaes

Abstract This chapter presents the pharmacokinetic principles of blood-brain barrier (BBB) transport and the intra-brain distribution of small molecular drugs, in order to provide a basis for understanding drug delivery to the brain from a clinically relevant perspective. The most important concentrations to measure when determining drug distribution are those of the unbound drug, because it is the unbound drug that causes the pharmacological effect by interacting with the target. Therefore, this chapter also discusses the pharmacokinetic basis, the kind of information provided, and the in vivo relevance of the methods used to obtain reliable, therapeutically useful estimates of brain drug delivery. The main factors governing drug distribution to the brain are the permeability of the BBB to the drug (influx clearance), the extent of nonspecific binding to brain tissue, and the efflux clearance of the drug. The ratio of the influx and efflux clearances provides an estimation of the extent of drug equilibration across the BBB, described by the partition coefficient of unbound drug, $K_{p,uu,brain}$. This parameter is important, as active uptake and/or efflux transporters influence the brain concentrations of unbound drug in relation to those in plasma. The advantage of using $K_{p,uu,brain}$ during the drug discovery process lies in its ability to predict the potential success of drugs intended for action within the brain or, conversely, of those with few or no side effects in the brain.

Keywords Rate · Extent · Kp,uu · Interstitial · Intracellular · Species

Abbreviations

$[plasma]_u/[brain]_u$	Ratio of plasma to brain unbound drug concentrations
A_{brain}	Amount of drug per g brain tissue excluding blood
A_{slice}	Amount of drug per g of brain slice
$A_{tot.brain\_inc\_blood}$	Amount of drug per g brain tissue including blood
$AUC_{tot,brain}$	Area under the total brain concentration-time curve

M. Hammarlund-Udenaes (✉)
Translational PKPD Group, Department of Pharmacy, Uppsala University, Uppsala, Sweden
e-mail: mhu@farmaci.uu.se

© American Association of Pharmaceutical Scientists 2022   173
E. C. M. de Lange et al. (eds.), *Drug Delivery to the Brain*, AAPS Advances
in the Pharmaceutical Sciences Series 33, https://doi.org/10.1007/978-3-030-88773-5_7

$AUC_{tot,plasma}$	Area under the total plasma concentration-time curve
$AUC_{u,brainISF}$	Area under the unbound brain ISF concentration-time curve
$AUC_{u,plasma}$	Area under the unbound plasma concentration-time curve
BBB	Blood-brain barrier
BBMEC cells	Bovine brain microvessel endothelial cells
BCSFB	Blood-cerebrospinal fluid barrier
Caco-2	Human epithelial colorectal adenocarcinoma cells
C_{buffer}	Concentration of drug in the buffer (brain slice method)
C_i	Apparent concentration of drug in a peripheral brain compartment i
CL_{act_efflux}	Active efflux clearance from brain to blood at the BBB ($\mu l/min/g\_brain$)
CL_{act_uptake}	Active uptake clearance from blood to brain at the BBB ($\mu l/min/g\_brain$)
CL_{bulk_flow}	Clearance by bulk flow from brain ISF to CSF ($\mu l/min/g\_brain$)
CL_i	Intercompartmental clearance between brain ISF and the peripheral brain compartment i
CL_{in}	Net influx clearance of drug to the brain ($\mu l/min/g\_brain$), also called permeability clearance
$CL_{metabolism}$	Metabolic clearance of drug in the brain or at the BBB ($\mu l/min/g\_brain$)
CL_{out}	Net efflux clearance of drug from the brain ($\mu l/min/g\_brain$)
$CL_{passive}$	Passive diffusional clearance of drug at the BBB
CNS	Central nervous system
CSF	Cerebrospinal fluid
$C_{tot,blood}$	Total concentration of drug in blood
$C_{tot,plasma}$	Total concentration of drug in plasma
$C_{u,brainISF}$	Concentration of drug in the brain ISF (by definition unbound)
$C_{u,cell}$	Average concentration of unbound drug in brain cells
$C_{u,plasma}$	Unbound concentration in plasma
$C_{u,ss,brainISF}$	Unbound steady-state concentration in brain ISF
$C_{u,ss,plasma}$	Unbound steady-state concentration in plasma
ECF	Extracellular fluid in the brain (also called ISF, interstitial fluid)
$f_{u,brain}$	Fraction of unbound drug in brain homogenate
$f_{u,brain,corrected}$	Fraction of unbound drug in brain homogenate after correction for pH partitioning based on the pKa(s) of the drug
$f_{u,D}$	Fraction of unbound drug in diluted brain homogenate
$f_{u,plasma}$	Fraction of unbound drug in plasma
GI	Gastrointestinal
ICF	Intracellular fluid in the brain
ISF	Interstitial fluid in the brain (also called ECF, extracellular fluid)
K_i	Inhibition constant

K_{in}	In situ brain perfusion unidirectional transfer constant (a clearance estimate equal to PS or CL_{in}) ($\mu l/min/g\_brain$)
$K_{p,brain}$	Partition coefficient (ratio) of total brain to total plasma drug concentrations
$K_{p,u,brain}$	Ratio of total brain drug concentration to plasma unbound drug concentration
$K_{p,uu,brain}$	Ratio of brain ISF to plasma unbound drug concentrations
$K_{p,uu,cell}$	Ratio of brain ICF to ISF unbound drug concentrations
$K_{p,uu,CSF}$	Ratio of CSF to plasma unbound drug concentrations
logBB	Logarithm of the ratio of total brain to total plasma drug concentrations (equal to K_p)
MDCK cells	Madin-Darby canine kidney cells
Mdr1	Gene encoding for P-glycoprotein
P_{app}	Unidirectional apparent permeability coefficient measured in the apical-to-basolateral direction (cm/s)
PBS	Phosphate-buffered saline
PET	Positron emission tomography
P-gp	P-glycoprotein
PS	Permeability surface area product (in this context equal to net influx clearance to the brain) ($\mu l/min/g\_brain$)
V_{blood}	Volume of blood in brain tissue
V_f	Volume of buffer film remaining around the sampled brain slice
V_i	Apparent volume of distribution of a peripheral brain compartment i
V_{ISF}	Physiological (and apparent) volume of ISF
$V_{u,brain}$	Volume of distribution of unbound drug in brain (ml/g_brain)

7.1 Introduction

The delivery of drugs from blood to brain takes place across the brain capillary endothelial cells comprising the blood-brain barrier (BBB). This is depicted in Fig. 7.1 in a classical electron micrograph of a capillary, the extremely thin endothelial cell layer and the brain parenchymal cells. Despite its thinness, the BBB is a very important organ that controls the brain environment in relation to blood, picking up nutrients, discarding waste products, and hindering influx of potentially harmful substances, including many drugs. The large surface area of the BBB and the high rate of blood flow to the brain ensure fast delivery of drugs to the brain (see Chap. 1 and Appendix for anatomical and physiological details of the BBB), but do not always ensure adequate drug concentrations within the brain.

This fact, together with earlier often inadequate methods used for measuring brain drug delivery, has caused problems in central nervous system (CNS) drug

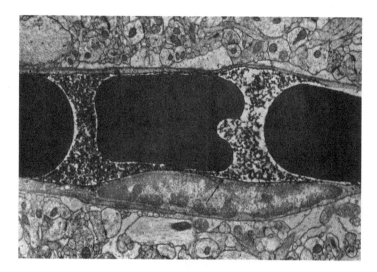

Fig. 7.1 An electron micrograph of a brain capillary with three erythrocytes, endothelial cell walls comprising the BBB, and brain parenchymal cells. The black color indicates intravenously administered peroxidase that does not pass the endothelial cells. The micrograph shows the two membranes of the BBB, the luminal membrane facing the blood and the abluminal membrane facing the brain parenchyma (x20 000). From Reese and Karnovsky with permission from the publisher (Reese and Karnovsky 1967)

discovery and development, due to measuring mainly total drug concentrations in the brain. The methods used in the industry are developing rapidly; these methods are discussed further in other chapters. This chapter focuses on the pharmacokinetic principles of small molecular drug delivery to the brain, on the rate and extent of drug transport as two separate factors governing drug delivery to the brain and on the pharmacokinetic parameters needed to describe this.

Figure 7.2 provides a more schematic drawing of how drugs are distributed across the BBB and into the brain. As depicted, it is only the unbound drug molecules, i.e., those that are not bound to plasma proteins that are able to transverse membranes, in this case the BBB. The rate at which the drug enters the brain interstitial fluid (ISF, also called extracellular fluid (ECF)) depends on the permeability of the BBB to the particular molecule. Together with the passive and active uptake and efflux processes at the BBB, this will determine how much drug enters the brain ISF. The drug molecules will then be further distributed to and equilibrated within the brain cells, specific and nonspecific binding sites, and organelles, depending on the physicochemical interactions between the drug and the tissue.

Drug transport between blood and cerebrospinal fluid (CSF) takes place at the blood-CSF barrier (BCSFB). There is also some exchange between CSF and brain ISF. Transport from CSF to ISF involves passive diffusion, while transport from ISF to CSF involves both passive diffusion and bulk flow of ISF, including possible influence of the "glymphatic" flow (Cserr et al. 1977; Nicholson and Sykova 1998; Abbott et al. 2018; Iliff et al. 2012). See also Chap. 1. The pH of blood is 7.4, while

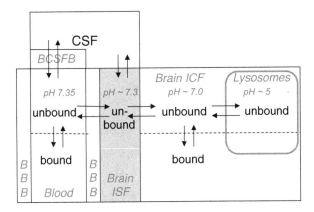

Fig. 7.2 Schematic illustration of drug distribution and equilibration across the BBB and other membranes within the brain parenchyma and unbound drug and drug bound to tissue components. The physiological volumes of the intra-brain compartments are brain interstitial fluid (ISF) 0.2 ml/g_brain and brain intracellular fluid (ICF) 0.8 ml/g_brain, of which the lysosomal compartment is 0.01 ml/g_brain. The figure is adapted from Hammarlund-Udenaes et al. (Hammarlund-Udenaes et al. 2008) with permission from the publisher

that of brain ISF is around 7.3, of the cell cytosol is 7, and in lysosomes is around 5.2. These pH differences influence drug equilibration, with basic drugs accumulating more in low-pH organelles, especially in the lysosomes. By definition, the concentrations in brain ISF are unbound, as are the concentrations in the intracellular fluid (ICF). The extent of nonspecific binding is generally quantitatively much greater than that of specific binding to receptors or other target sites.

It is only the unbound drug that is in contact with receptor or other target sites, and experimental data show that these concentrations are best correlated with clinical effects or side effects in the brain (Hammarlund-Udenaes 2010; Watson et al. 2009; Kalvass et al. 2007b; Large et al. 2009). The site of action of the particular drug will determine whether brain ISF or brain ICF concentration is the more important in relation to the pharmacodynamic measurement. It has been clearly shown for dopamine agonists and other drugs that the unbound drug brain concentrations are much more closely related to receptor occupancy than the total brain concentrations or the concentrations of unbound drug in the blood (Watson et al. 2009; Stevens et al. 2012). This is clearly shown in Fig. 7.3, which depicts the receptor occupancy of several dopamine antagonists in relation to their plasma, total brain, and unbound drug brain concentrations.

The amount of drug to be delivered to the brain to achieve the desired effect is of course always an issue when deciding on the dose to be administered. However, a trade-off between side effects and the desired effects also needs to be taken into consideration. For drugs that are very efficiently effluxed at the BBB, there will be much lower unbound concentrations in brain ISF than in plasma. This is advantageous if peripheral effects and avoidance of CNS side effects are desired, but is less suitable if CNS effects are desired and peripheral side effects are to be avoided.

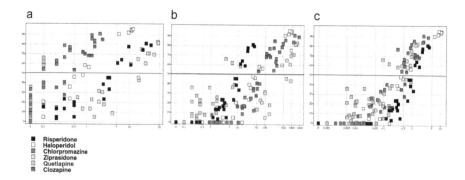

Fig. 7.3 Relationships between receptor occupancy and concentrations of neuroleptics normalized for their in vitro affinity for rat striatal D_2 receptors. a) Total plasma concentrations, b) total brain concentrations, and c) unbound brain concentrations, illustrating the clear advantage of unbound brain concentrations when comparing drugs. Reprinted from Watson et al. 2009 () with permission from the publisher

For measurements based on pharmacokinetic principles, drug delivery can be described by three distinctly different parameters. Two of these are important components of the transport of the drug across the BBB, and the third describes the intra-brain distribution of the drug. The first parameter describes the **rate of drug delivery to the brain** based on the permeability surface area product (PS), which in pharmacokinetic literature is often called the net influx clearance (CL_{in}, μl/min/g_ brain). This describes the unidirectional net drug transport from blood to brain. The second parameter is the **extent of delivery**, which can be described either by the total drug concentrations in the brain and plasma or by unbound drug concentrations at steady state. The total drug concentration ratio between the brain and plasma is termed $K_{p,brain}$. Another way of describing the same parameter is logBB, which is also used for computational approaches (Abraham et al. 1995; Norinder and Haeberlein 2002; Young et al. 1988; Norinder et al. 1998; Mensch et al. 2010a; Sun 2004; Shityakov et al. 2013; Muehlbacher et al. 2011; Fan et al. 2010). The unbound drug concentration ratio between brain ISF and plasma is termed $K_{p,uu,brain}$ (Gupta et al. 2006). The relationship between the unbound and total drug concentrations in plasma is described by the fraction of drug that is not bound to plasma proteins, $f_{u,plasma}$. There are two alternative measurements in brain parenchymal tissue that can be used to describe **intra-brain distribution**, the third parameter. This parameter correlates unbound to total drug concentrations in the brain. $f_{u,brain}$ is the fraction of unbound drug in the brain based on brain homogenate measurements (Kalvass and Maurer 2002), and $V_{u,brain}$ is the unbound volume of distribution within the brain in ml/g_brain tissue based on brain slice measurements (; Friden et al. 2007, 2009a; Kakee et al. 1996; Loryan et al. 2013). It should be noted that this volume term is not the same as those determined from in situ brain perfusion or PET studies. In the coming sections, these three parameters will be described in more detail. In vitro and in vivo methods used for determining brain drug delivery are further described in other chapters.

7.2 Historical Aspects on Studying Brain Drug Delivery

Several expressions have been used to describe drug delivery to the brain in the literature: permeation (Tamai and Tsuji 2000; Abbott et al. 2008), brain penetration (Schinkel et al. 1996), extent of brain penetration (Liu et al. 2008), CNS penetration (Summerfield et al. 2006), BBB penetration (Gunn et al. 2012), brain delivery (Pardridge et al. 1992), and CNS distribution (Dai et al. 2005; Kalvass et al. 2007a). The expressions used for the total brain to total plasma concentration ratio also vary: [brain]/[plasma] (Kalvass and Maurer 2002), K_p (a classical expression in pharmacokinetics for the partition coefficient between tissue and plasma (Rowland and Tozer 2011)), $K_{p,brain}$ (Gupta et al. 2006), and B/P (Maurer et al. 2005). Expressions for the brain to blood (or vice versa) unbound drug concentration ratios have been described as $K_{p,uu}$ (Gupta et al. 2006), $K_{p,free}$ (Liu et al. 2005), and $[plasma]_{,u}/[brain]_{,u}$ (Kalvass et al. 2007a).

Kalvass and Maurer made a seminal contribution in 2002 by initiating investigation into how to find out whether drugs are actively effluxed at the BBB (Kalvass and Maurer 2002), after P-gp had been found in the BBB (Tsuji et al. 1992; Cordon-Cardo et al. 1989; Thiebaut et al. 1989) and after the development of the P-gp knockout mouse model (Schinkel et al. 1996). They introduced the in vitro brain homogenate binding method in this context and simplified the estimation of extent of drug binding from diluted brain homogenate samples. The ratio of the fraction of unbound drug in plasma to that in the brain ($f_{u,plasma}/f_{u,brain}$) was compared with the ratio of total brain to plasma concentrations ($K_{p,brain}$). Kalvass and Maurer concluded that, if the two ratios are the same, the drug will be transported across the BBB mainly by passive means. Efflux was indicated by differences between the ratios, i.e., this was an indirect way of describing BBB transport properties. We know today that the ratio of $f_{u,plasma}/f_{u,brain}$ itself as an indication of partitioning between brain and blood is misrepresentative, as the main cause of deviations in $K_{p,brain}$ from this ratio is active transport at the BBB. The authors also compared CSF concentrations to brain and plasma concentrations and found that CSF concentrations overpredicted brain exposure for P-gp substrates.

Maurer et al. continued the work with a comparison of plasma and brain concentrations for 33 compounds (Maurer et al. 2005). Differences in $f_{u,plasma}/f_{u,brain}$ within a threefold range were allowed, to cope with experimental errors and differences considered of little consequence for pharmacology or pharmacokinetics. The authors stated that "Because the brain to plasma ratio (K_p) is determined largely by nonspecific binding, efforts to optimize this parameter may actually lead to an unproductive or counterproductive design of drugs that are unnecessarily basic, lipophilic, and simply have a greater degree of nonspecific partitioning into brain tissue" (Maurer et al. 2005). This has proven to be a very relevant statement, which partly explains the poor success rate in developing new drugs for CNS diseases (Kaitin 2008; Kola and Landis 2004). They also surmised that the underprediction of tissue distribution of bases, but not of neutral compounds and acids, based on

$f_{u,brain}$ values could be the result of disruption of the subcellular acidic organelles such as lysosomes during homogenization.

Data from the literature were used by Kalvass and coauthors to compare more drugs, using the correlations developed earlier by Kalvass and Maurer (Kalvass et al. 2007a; Kalvass and Maurer 2002). They commented that $K_{p,brain}$ was still (in 2007) used to optimize brain delivery (values of ≥ 1 were arbitrarily given an interpretation of good brain delivery and values $\ll 1$ of poor brain delivery) and issued another warning that this classification could be misleading, as $K_{p,brain}$ is also influenced by the relative extent of binding to plasma proteins and brain tissue (Kalvass et al. 2007a). A ratio based on plasma to brain concentrations of unbound drug was proposed ([plasma]$_u$/[brain]$_u$), and a log-log graph which plotted the in vivo P-gp efflux ratio vs [plasma]$_u$/[brain]$_u$ was developed. Their conclusions on the BBB transport of the studied drugs were based on the quadrant into which the drug fell. This way of estimating BBB transport is further discussed by Avdeef in his book (Avdeef 2012). Kalvass et al. found indications of active uptake at the BBB and also found that efflux transport mediated by transporters other than P-gp was not able to be accurately predicted by the P-gp efflux ratios in Mdr1a(+/+) and Mdr1a(−/−) mice. For ten of the 34 drugs studied, the extent of efflux in vivo was greater than could be explained by P-gp, and active uptake into the brain was indicated for three drugs. Thus, the in vivo P-gp efflux ratio for knockout and wild-type mice was not sufficient to predict brain delivery, and the [plasma]$_u$/[brain]$_u$ ratio was better predictive than the P-gp efflux ratio alone (Kalvass et al. 2007a). Despite this, most drug companies continue to trust P-gp efflux ratios in vivo or in vitro as the parameter of choice.

Concepts around the BBB transport of drugs were developed further by our group, with the proposal of the term $K_{p,uu}$ by Gupta et al. to succinctly describe the brain ISF to blood concentration ratio for unbound drug (Gupta et al. 2006). Before the publication of this expression in 2006, the efficiency of net active efflux or uptake for individual drugs had been described as the "ratio of unbound brain to unbound blood concentrations" (Bouw et al. 2000, 2001; Xie et al. 2000; ; Tunblad et al. 2003; 2004a, b, 2005 Bostrom et al. 2005). The approach thus separated BBB transport properties from protein binding in plasma and binding to brain constituents, treating the three parameters as independent, individual properties of the drugs. It was indicated that the permeability of the brain to the drug (PS, CL_{in}) and the extent of equilibration across the BBB ($K_{p,uu,brain}$) were not correlated (Hammarlund-Udenaes 2000; Hammarlund-Udenaes et al. 2008; Hammarlund-Udenaes et al. 1997). The brain slice technique was also developed for studies of nonspecific binding to brain tissue in a high-throughput model and was compared with the brain homogenate method (Friden et al. 2007, 2009a, 2011 Loryan et al. 2013).

Doran et al. concluded that most CNS drugs have some degree of P-gp-mediated transport and that this does not hamper their clinical use (Doran et al. 2005). They studied the total brain to plasma, CSF to plasma, and CSF to brain concentration ratios in Mdr1a(+/+) and Mdr1a(−/−) mice without taking into account differences between the drugs in nonspecific binding in the brain. They found that despite being a good P-gp substrate, risperidone has sufficient clinical effect in the CNS because of its high potency; the question of the correct dose in relation to peripheral side effects is also pertinent here.

At around the same time, Liu and coworkers published on properties that govern the equilibration of drug concentrations between brain and blood (Liu et al. 2005). They concluded that rapid permeation alone does not guarantee rapid equilibration. What is required for rapid equilibration is a combination of rapid permeation and low brain tissue binding. The authors used permeability as a surrogate for efflux clearance, although they are not strictly interchangeable. Nonetheless, the combination of efflux clearance from the brain and the extent of brain binding determines the equilibration time across the BBB (Hammarlund-Udenaes et al. 1997; Liu et al. 2005; Syvanen et al. 2006).

Liu et al. proposed a direct extrapolation of $f_{u,plasma}$ to describe $f_{u,brain}$ as they (Liu et al. 2005) and others (Kalvass and Maurer 2002; Maurer et al. 2005) found a good correlation between the two ($r^2 = 0.69$ (Liu et al. 2005)). Although the use of $f_{u,plasma}$ for $f_{u,brain}$ has not been evaluated any further, its use can be questioned today if a good estimation of $K_{p,uu,brain}$ is the goal. Even a twofold difference between the two will result in a twofold difference in the value of $K_{p,uu,brain}$ and could skew information on the parameter needed for selection of the best drug candidates (see further Sect. 7.3.2.2).

Liu and Chen also discussed the extent and rate of brain penetration by looking at ways to increase the $K_{p,uu,brain}$ by reducing the efflux clearance or increasing the influx clearance (Liu and Chen 2005). In this paper, $K_{p,brain}$ was considered unsuitable for evaluation of the potential success of a candidate as a CNS drug. Liu et al. later proposed strategies for studying transporters at the BBB, including: "1) Drug discovery screens should be used to eliminate good P-gp substrates for CNS targets. Special consideration could be given to moderate P-gp substrates as potential CNS drugs based on a high unmet medical need and the presence of a large safety margin. 2) Selection of P-gp substrates as drug candidates for non-CNS targets can reduce their CNS-mediated side effects" (Liu et al. 2008).

Several articles in the area have also been published by Summerfield and coworkers. In one study, they used Mdr1a/b(+(+) and Mdr1a/b(−/−) mice to investigate total brain to blood ratios ($K_{p,brain}$) in vitro, covering a wide range of physicochemical properties (Summerfield et al. 2006). They also compared $f_{u,brain}$ and $f_{u,blood}$. They concluded that the in vitro estimation of $f_{u,brain}/f_{u,blood}$ overpredicted the K_p observed in vivo because the in vitro ratio assumes that the concentrations in brain and blood are equal, while in reality they are not, because of active transport in the BBB. In their next study, they investigated 50 marketed drugs and compared in situ brain perfusion permeability with in vitro permeability and then correlated these parameters with physicochemical information (Summerfield et al. 2007). In their 2008 publication they studied species differences in plasma and brain binding and found a good correlation in brain binding between rat, pig, and humans, thereby improving the prediction of drug distribution to the brain in humans; they also published a table defining PET and pharmacokinetic expressions (Summerfield et al. 2008). The use of PET and in vitro equilibrium dialysis to assess BBB transport of candidate drugs in CNS drug development was advocated in a later publication (Gunn et al.

2012). An integrated approach involving permeability, active efflux, and brain distribution, and focusing on unbound drug, was proposed by Jeffrey and Summerfield (Jeffrey and Summerfield 2010). In a later paper, they state that "Assessing the equilibration of the unbound drug concentrations across the blood-brain barrier ($K_{p,uu}$) has progressively replaced the partition coefficient based on the ratio of the total concentration in brain tissue to blood (K_p)" (Summerfield et al. 2016).

Hakkarainen et al. compared the in vitro apparent permeability coefficient (P_{app}) from three cell culture systems with in vivo microdialysis measuring $K_{p,uu,brain}$ for nine drugs (Hakkarainen et al. 2010). Unfortunately, the use of an in vitro microdialysis probe recovery method in this otherwise thorough paper potentially affected the accurate measurement of the ISF concentrations and thus the $K_{p,uu,brain}$ values. When the results for two P-gp substrates were omitted, the authors found an extremely good correlation between the permeability of BBMEC cells and the microdialysis results (r = 0.99) and noted that the lower the permeability, the lower the $K_{p,uu,brain}$. When the drugs known to be P-gp substrates were included, the relationship became nonsignificant, as would be expected since lower $K_{p,uu,brain}$ values indicate more active efflux and are not correlated with permeability per se, as discussed above and below.

7.3 Parameters Describing Drug Delivery to the Brain

7.3.1 Rate of Brain Drug Delivery

7.3.1.1 What and Why

Permeability as a measurement of drug delivery to the brain has historically been the most common way of optimizing drug delivery to this area. Permeability measurements give an estimate of the unidirectional rate of transport of a drug across the BBB in situ or in a cell model in vitro. Rather than telling us how much drug has equilibrated across the BBB at steady state, these measurements tell us how fast the drug is transported across the BBB into the brain.

Permeability measurements are based on the tradition of studying gastrointestinal (GI) absorption. Physiological differences between the GI tract and the BBB, however, make this concept less translatable. Many articles have compared permeability values from in silico predictions, in vitro cell models, in situ methods, and in vivo methods (Summerfield et al. 2007; Abbott et al. 2008; Bickel 2005; Friden et al. 2009b; Hammarlund-Udenaes et al. 2009; Chen et al. 2011; Avdeef and Sun 2011; Avdeef 2011; Di et al. 2009; 2012; Broccatelli et al. 2012; Fan et al. 2010; Liu et al. 2004; Lanevskij et al. 2013; Mensch et al. 2010a, b; Levin 1980; Garberg et al. 2005; Abbott 2004b). Quite commonly, methods measuring the rate of permeation are compared with those measuring the extent of permeation (Pardridge 2004; Hakkarainen et al. 2010).

7.3.1.2 Methods and Relationships

Permeability is described by the rate of permeation in cm/s, obtained by dividing the PS value estimated from in situ brain perfusion (called K_{in}) by the luminal surface area of the vascular space, estimated to be 150 cm^2/g_brain in vivo in rats (Fenstermacher et al. 1988), or by dividing by the surface area of the cell culture in vitro. The in vitro measurement is called P_{app}, the apparent permeability coefficient. In vitro methods include BBB-specific cell models from different origins, as well as Caco-2 or MDCK cells (please see other chapters in this book).

The in situ brain perfusion method is a very elegant way of rapidly determining permeability in an animal model (Takasato et al. 1984; Smith and Allen 2003; Banks et al. 1997). It can also be performed in genetically modified mice to study the influence of active transporters (Dagenais et al. 2000). Examples of CL_{in} (K_{in}) values from in situ brain perfusion and microdialysis studies are given in Table 7.1. It can clearly be seen, when Mdr1a(+/+) and Mdra1a(−/−) mice are compared, that CL_{in} is decreased in the presence of P-glycoprotein (P-gp). CL_{in} therefore describes the net influx clearance across the BBB. In general, the permeability of the BBB to a drug appears to be less critical to drug delivery than the influence of active efflux transporters. More about the pharmacokinetic aspects and relationships of the transport processes at the BBB can be found in Sect. 7.3.5.

7.3.2 Extent of Brain Drug Delivery

7.3.2.1 What and Why

The extent of drug delivery to the brain is based on steady-state measurements of the ratios of total concentrations in brain and plasma (the partition coefficient $K_{p,brain}$ or logBB), total concentrations in brain and unbound concentrations in plasma ($K_{p,u,brain}$), or unbound concentrations in brain ISF and plasma ($K_{p,uu,brain}$). In comparison to absorption from the GI tract, the amount of drug delivered to the brain can be compared with the bioavailability of drug in the brain, although the determining forces are somewhat different.

The most important advantage of using $K_{p,uu,brain}$ instead of $K_{p,brain}$ lies in its ability to, during the drug discovery process, predict the success of drugs intended for action within the brain or, conversely, for the avoidance of side effects in the brain. $K_{p,uu,brain}$ is the parameter that most closely relates to the drug's pharmacodynamic profile, if the receptors are situated facing the brain ISF. If the relevant receptors are intracellular, further investigations are required (see Sect. 7.3.4 and, in more detail, Chap. 13). The $K_{p,uu,brain}$ value is not influenced by plasma protein binding and brain parenchymal binding that would otherwise confound its interpretation. It gives a concrete value to the net result of passive and active transport across the BBB.

Table 7.1 Examples of in situ/in vivo CL_{in} values obtained by in situ brain perfusion or microdialysis

Drug	CL_{in} (µl/ min/g _brain)	CL_{in} in Mdr1a (−/−) mice (µl/min/ g_brain)	Species	References
Alfentanil	1940	2290	Mouse	Zhao et al. (2009)
Antipyrine	492	–	Rat	Avdeef and Sun (2011)
Atenolol	1.8	–	Rat	Avdeef and Sun (2011)
Cimetidine	7	11	Mouse	Zhao et al. (2009)
Colchicine	9	19	Mouse	Zhao et al. (2009)
Diazepam	2500	2500	Mouse	Zhao et al. (2009)
DPDPE	0.547	6.36	Mouse	Dagenais et al. (2004)
Fentanyl	1840	2280	Mouse	Dagenais et al. (2004)
Fexofenadine	3	13	Mouse	Zhao et al. (2009)
Imipramine	1860	–	Rat	Avdeef and Sun (2011)
Loperamide	100	1030	Mouse	Dagenais et al. (2004)
Methadone	420	1090	Mouse	Dagenais et al. (2004)
Morphine	10.4	12.9	Mouse	Dagenais et al. (2004)
Morphine	11.4	–	Rat	Bouw et al. (2000) and Tunblad et al. (2004b)
Morphine-3-glucuronide	0.11	–	Rat	Xie et al. (2000)
Morphine-6-glucuronide	1.66	–	Rat	Bouw et al. (2001) and Tunblad et al. (2005)
Oxycodone	1910	–	Rat	Bostrom et al. (2006)
Phenytoin	334	347	Mouse	Zhao et al. (2009)
Quinidine	34	541	Mouse	Zhao et al. (2009)
Ritonavir	23	80	Mouse	Zhao et al. (2009)
Sufentanil	340	295	Mouse	Zhao et al. (2009)
Terfenadine	1740	2020	Mouse	Zhao et al. (2009)
Valproate	243	181	Mouse	Zhao et al. (2009)
Verapamil	315	1370	Mouse	Zhao et al. (2009)

When $K_{p,uu,brain}$ is combined with the target binding properties of the drug, it is possible to estimate the required plasma concentrations, and thus the doses, for pharmacological success. There is no clear cutoff point below which a drug is not suitable for action within the brain, but the lower the $K_{p,uu,brain}$ value, the higher is the dose required to obtain pharmacologically relevant concentrations in the brain given similar potency. The trade-off is more between a dose that can be administered in relation to clinical effect vs side effects and a dose that is economically defendable.

7.3.2.2 Methods and Relationships

The $K_{p,brain}$ ratio can be determined by measuring steady-state drug concentrations or the area under the concentration-time curves in brain tissue, excluding capillary blood concentrations ($AUC_{tot,brain}$), and plasma ($AUC_{tot,plasma}$) after a single dose:

$$K_{p,brain} = \frac{AUC_{tot,brain}}{AUC_{tot,plasma}} \tag{7.1}$$

Measuring the AUC after a single dose is comparable to taking samples of brain and blood at one time point during steady state. The AUCs can then be substituted by the steady-state drug concentrations.

$K_{p,uu,brain}$ can be determined directly from microdialysis samples from brain and plasma sites or by measuring total brain and plasma concentrations at steady state combined with plasma protein binding (giving the fraction of unbound drug in plasma, $f_{u,plasma}$) and brain slice or brain homogenate measurements of nonspecific binding to brain parenchyma (Friden et al. 2007, 2009a, 2010; Loryan et al. 2016):

$$K_{p,uu,brain} = \frac{AUC_{u,brainISF}}{AUC_{u,plasma}} = \frac{AUC_{tot,brain}}{AUC_{tot,plasma} {}^* V_{u,brain} {}^* f_{u,plasma}} \tag{7.2a}$$

Here, $AUC_{u,brainISF}$ describes the concentrations of unbound drug in brain ISF, and $AUC_{u,plasma}$ describes the concentrations of unbound drug in plasma. $V_{u,brain}$ measured with the brain slice method may be replaced by $1/f_{u,brain}$ after correction for pH partitioning if a brain homogenate is used to determine the nonspecific brain binding, as described in Eq. 7.2b.

$$K_{p,uu,brain} = \frac{AUC_{tot,brain} {}^* f_{u,brain,corrected}}{AUC_{tot,plasma} {}^* f_{u,plasma}} \tag{7.2b}$$

Thus, $V_{u,brain}$ is similar but not equal to $1/f_{u,brain}$, which can result in different results if pH partitioning is not compensated for (Friden et al. 2011). More about the similarities and differences between these parameters is given in Sect. 7.3.3 and in Chap. 13. As the combined method involves measuring three individual parameters, the experimental error in each of them will affect the $K_{p,uu,brain}$ estimate (Kalvass et al. 2007a). Here, the uncertainty propagation method can be used (Loryan et al. 2017; Yusof et al. 2019).

The concentration of drug in brain ISF is determined by diffusion, transport, metabolism, and binding processes, as described in Fig. 7.1. The differential equations describing the equilibration across the BBB between unbound drug in plasma and the brain ISF compartment are:

$$\frac{V_{ISF} {}^* dC_{u,brainISF}}{dt} = CL_{in} {}^* C_{u,plasma} - \left(CL_{out} + CL_i\right) * C_{u,brainISF} + CL_i {}^* C_i \tag{7.3}$$

$$\frac{V_i^* dC_i}{dt} = CL_i^* \left(C_{u,brainISF} - C_i \right) \qquad (7.4)$$

V_{ISF} describes both the physiological volume of the ISF and the apparent volume of distribution in the ISF, as it is assumed that there is no binding in this compartment. CL_{in} and CL_{out} describe the net influx and efflux clearance across the BBB. CL_{in} is equivalent to PS. V_i and C_i are the apparent volume of and drug concentration in a possible deeper brain compartment i, and CL_i is the intercompartmental clearance between this compartment and the ISF. The plasma unbound drug concentration ($C_{u,plasma}$) is the driving force for the brain concentrations. Further equations necessary to describe the plasma concentration-time profile are beyond the scope of this chapter.

At steady state, there is no change in concentration in brain ISF, $dC_{u,brainISF}/dt = 0$, and the drug concentrations in plasma ($C_{u,ss,plasma}$) and brain ($C_{u,ss,brainISF}$) are in equilibrium. If $C_{u,brainISF} = C_i$, which can be assumed since C_i describes a hypothetical compartment, the relationship in Eq. 7.3 becomes:

$$CL_{in}^* C_{u,ss,plasma} = CL_{out}^* C_{u,ss,brainISF} \qquad (7.5)$$

As $K_{p,uu,brain}$ is a steady-state parameter, it is not influenced by the further partitioning of the drug into brain cells:

$$K_{p,uu,brain} = \frac{C_{u,ss,brainISF}}{C_{u,ss,plasma}} = \frac{CL_{in}}{CL_{out}} \qquad (7.6)$$

It can be seen in Eq. 7.6 that $K_{p,uu,brain}$ is determined by the relative size of the net influx and efflux clearances. This means that influx and efflux clearances can both be small and large and still result in the same $K_{p,uu,brain}$. This explains why the permeability per se is not the most important parameter for estimating the extent of drug delivery to the brain. While rapid delivery to and elimination from the brain is clinically important for, for example, anesthetic drugs, the steady-state concentration in the brain is more important than the rate of delivery to the brain when a drug is to be administered repeatedly over time. The range of CL_{in} values within which brain delivery is still sufficient can, therefore, be quite wide. This is exemplified in Table 7.1 by the good clinical effects of morphine despite its low permeability clearance vs the lack of clinical effect of loperamide despite its higher permeability clearance. This phenomenon is also illustrated in Fig. 7.4.

Equation 7.6 can be further developed to include the different processes governing the uptake and elimination of drug from brain ISF:

$$K_{p,uu,brain} = \frac{CL_{in}}{CL_{out}} = \frac{CL_{passive} + CL_{act\_uptake} - CL_{act\_efflux}}{CL_{passive} - CL_{act\_uptake} + CL_{act\_efflux} + CL_{bulk\_flow} + CL_{metabolism}} \qquad (7.7)$$

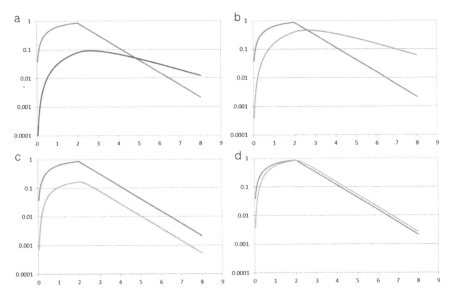

Fig. 7.4 Illustration of the absolute values of CL_{in} and CL_{out} and their relationships with the resulting brain concentration-time profile of unbound drug with time on the x-axis and concentration on the y-axis. The blue line, similar in all parts of the figure, describes the unbound drug concentration in blood after a short intravenous infusion of a fictive drug. The other lines describe the brain unbound drug concentrations. The relative values of CL_{in} and CL_{out} are in a) $CL_{in} = 1$, $CL_{out} = 5$ giving a $K_{p,uu,brain}$ of 0.2; b) $CL_{in} = 5$ and $CL_{out} = 5$, giving a $K_{p,uu,brain}$ of 1.0; c) $CL_{in} = 10$ and $CL_{out} = 50$ giving a $K_{p,uu,brain}$ of 0.2; and d) $CL_{in} = 50$ and $CL_{out} = 50$, giving a $K_{p,uu,brain}$ of 1.0. In a) and b), CL_{out} values together with the size of $V_{u,brain}$ (the same in all simulations) result in a longer half-life for the drug in the brain than in blood. In c) and d), the half-life in the brain follows that in blood because of the more rapid processes in the brain than in blood. A comparison of (**a**) and (**c**), (**b**) and (**d**), respectively, shows that the $K_{p,uu,brain}$ is the same, independent of a tenfold difference in CL_{in} and independent of differences in half-lives in the brain

$CL_{passive}$ is the passive diffusional clearance across the BBB, which is assumed to be equal in both directions. CL_{act_efflux} describes the active efflux transport back across the BBB to the plasma (Syvanen et al. 2006). CL_{act_uptake} describes the active uptake transport across the BBB into the brain. Both active transport parameters can include one or several transporter functions and can, if of interest, be further divided into the individual processes. CL_{bulk_flow} is the bulk flow of ISF from brain to CSF, reported to be 0.1–0.3 µl/min/g_brain (Cserr et al. 1977; Rosenberg et al. 1980; Abbott 2004a). $CL_{metabolism}$ describes the elimination of a drug through metabolism within the brain.

Equation 7.7 assumes that $CL_{passive}$ is the same, independent of direction of transport across the BBB. In reality, this may not be correct for the two membranes of the BBB (luminal vs abluminal), as a result of different fluid flow rates and diffusion properties. The equation suggests that active efflux of a drug will reduce CL_{in} and that active uptake will reduce CL_{out}. An experimental illustration of this is provided by the distinct effect of P-gp on CL_{in} that was found by Dagenais et al. (Dagenais et al. 2004). They used in situ brain perfusion methodology in Mdr1a(+/+) and Mdr1a (−/−) mice. The PS of loperamide increased tenfold from 100 µl/min/g_ brain in Mdr1a(+/+) mice to 1030 µl/min/g_brain in Mdr1a (−/−) mice (Table 7.1). It should also be borne in mind that CL_{in} and CL_{out} are the net clearances across both the luminal and abluminal membranes of the brain endothelial cells when, in reality, transporters are usually situated in either the apical or basolateral membrane and are rarely situated in both membranes.

If the only method of transport is passive, or if the influx and elimination processes are of the same magnitude, the unbound concentrations in the brain will equal those in plasma when equilibrium is reached between the two sites. $K_{p,uu,brain}$ will be smaller than unity if efflux dominates the transport process (Gupta et al. 2006; Hammarlund-Udenaes et al. 2008) and greater than unity if active uptake dominates (Bostrom et al. 2006; Hammarlund-Udenaes et al. 2008; Sadiq et al. 2011; Kurosawa et al. 2017). The relationships and their interpretation are further described in Table 7.2.

Most drugs seem to be effluxed at the BBB. This can be seen in Fig. 7.5, which provides the $K_{p,uu,brain}$ values for a selection of drugs that are acids, bases, neutrals, and zwitterions (Friden et al. 2009b).

7.3.3 Intra-Brain Distribution

7.3.3.1 What and Why

Estimation of the extent of nonspecific binding of a drug to brain tissue is necessary in order to relate the total brain concentrations, which are easily measured, to the unbound drug concentrations, which are more difficult to measure but more valuable for optimizing drug treatment. This is an intra-brain measurement and is not related to BBB function.

7.3.3.2 Methods and Relationships

The three methods by which intra-brain distribution can be estimated include microdialysis in the brain in conjunction with a brain sample to provide total brain concentrations at steady state (Wang and Welty 1996; Hammarlund-Udenaes 2013), the brain homogenate method (Kalvass and Maurer 2002; Mano et al. 2002), and the brain slice method (Kakee et al. 1996; Friden et al. 2009a, 2010; Loryan et al. 2013). The microdialysis and brain slice methods result in an estimate of $V_{u,brain}$ in ml/g_ brain tissue, while the brain homogenate method results in an estimate of $f_{u,brain}$.

Table 7.2 Relationship between the rate and the extent of equilibration across the BBB. More than one transporter may be acting on the drug and transport can be in either direction. Further examples from a combination of iv infusion, brain slice, and plasma protein binding measurements can be found in Fridén et al. (Friden et al. 2009b)

Parameter value	Relationship	Interpretation	In vivo examples
$K_{p,uu} \approx 1$	$CL_{in} \approx CL_{out}$	Net influx and efflux clearances are similar either because the drug is only passively transported across the BBB or because the active influx and efflux rates are similar. Note that the absolute sizes of the clearances are not important, only the relationship between the two.	Codeine (Xie and Hammarlund-Udenaes 1998) Diazepam (Dubey et al. 1989) Olanzapine, haloperidol (Loryan et al. 2016)
$K_{p,uu} < 1$	$CL_{in} < CL_{out}$	Elimination processes from the brain are more efficient than influx processes. This may be because of more active efflux transport at the BBB, metabolism within the brain parenchyma, or bulk flow (the latter requires clearances to be quite low, as bulk flow is 0.1–0.3 µl/min/g_brain).	Morphine (Bouw et al. 2000; Tunblad et al. 2003; Bostrom et al. 2008) Risperidone and paliperidone (Doran et al. 2012; Liu et al. 2009; Loryan et al. 2016) Ofloxacin, perfloxacin (Ooie et al. 1997) 6-Mercaptopurine, probenecid (Deguchi et al. 2000) Atenolol, methotrexate, paclitaxel (Friden et al. 2009b; Hu et al. 2017; Chen et al. 2017; Westerhout et al. 2014) Quinidine, indinavir, dexamethasone (Uchida et al. 2011a) Quinidine (Westerhout et al. 2013)
$K_{p,uu} > 1$	$CL_{in} > CL_{out}$	Influx processes across the BBB are quantitatively more efficient than efflux/metabolism/bulk flow processes. This can only be accomplished if the drug is actively transported from blood to brain.	Oxycodone (Bostrom et al. 2006) Diphenhydramine (Sadiq et al. 2011) Nicotine (Tega et al. 2013) Varenicline (Kurosawa et al. 2017)

Microdialysis

Microdialysis can be used to determine both $K_{p,uu,brain}$ and $V_{u,brain}$. In order to calculate $V_{u,brain}$, it is necessary to measure total brain concentrations at steady state at the same time as obtaining the concentration of unbound drug in brain ISF by microdialysis.

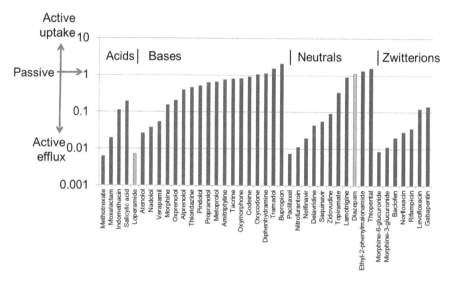

Fig. 7.5 $K_{p,uu,brain}$ values from a combined study of $K_{p,brain}$, $f_{u,plasma}$, and $V_{u,brain}$ in rats. $K_{p,uu,brain}$, to the extent that it can be extrapolated to humans, indicates the clinical usefulness of the drug for action in the brain. The brain ISF concentrations are similar ($K_{p,uu,brain} \approx 1$), lower ($K_{p,uu,brain} < 1$) or higher ($K_{p,uu,brain} > 1$) than the unbound concentrations in plasma. Data from Fridén et al. (Friden et al. 2009b)

The expression $V_{u,brain}$ was introduced by Wang and Welty in their microdialysis study of gabapentin influx and efflux across the BBB (Wang and Welty 1996). The paper was seminal for improving understanding of how the BBB transport of drugs can be evaluated (Hammarlund-Udenaes et al. 2008). $V_{u,brain}$ can be described by Eq. 7.8:

$$V_{u,brain} = \frac{A_{tot,brain\_incl\_blood} - V_{blood} {}^{*}C_{tot,blood}}{C_{u,brainISF}} \tag{7.8}$$

where $A_{tot,brain\_incl\_blood}$ is the amount of drug present per g brain, obtained from chemical analysis of the brain tissue sample. It is then necessary to subtract the amount of drug in the brain capillaries in order to obtain the amount present in the brain tissue itself. V_{blood} is the physiological volume of blood present in the brain tissue sample, and $C_{tot,blood}$ is the total concentration of the drug in the blood. The volume used here is critical for correct estimation of $V_{u,brain}$ (Friden et al. 2010).

Brain Homogenate

The brain homogenate method results in an estimate of $f_{u,brain}$. In short, this method uses fresh or frozen brain homogenate that is diluted with phosphate-buffered saline (PBS) and equilibrated across a dialysis membrane. The method is described in detail in Chap. 13. Samples of buffer and homogenate are analyzed, and the fraction of unbound drug in the original sample is calculated using Eq. 7.9 to compensate for the dilution:

$$Undiluted\_f_{u,brain} = \frac{\frac{1}{D}}{\left(\left(\frac{1}{f_{u,D}}\right)-1\right)+\frac{1}{D}} \tag{7.9}$$

D is the dilution factor for the brain tissue sample, and $f_{u,D}$ is the fraction of unbound drug in the diluted brain homogenate sample.

There are several advantages associated with the brain homogenate method: It is easy to carry out, using the same equipment as that used for plasma protein binding, high-throughput methodology can be used, and the process can be based on frozen tissue. However, it should be borne in mind that homogenizing the sample can expose sites that normally do not bind the drug in vivo (Liu and Chen 2005). Furthermore, membrane structures are destroyed by homogenization. This excludes the measurement of the influence of possible transport processes and pH differences between the brain parenchymal cells and organelles.

The brain homogenate method was used by Di et al. to compare $f_{u,brain}$ values between species, with subsequent important potential for using animal brain homogenates to estimate the nonspecific binding of drugs in human brain (Di et al. 2011). Summerfield had earlier studied species differences between rat, pig, and humans regarding binding to brain tissue (Summerfield et al. 2008). Human brain regional differences in binding including the influence of disease, in comparison to binding in the rat, were studied by Gustafsson et al. highlighting the need of case-by-case evaluation of regional brain binding in translational CNS research (Gustafsson et al. 2019).

Brain Slice

The brain slice method results in an estimate of $V_{u,brain}$ in ml/g_brain tissue. This method, which provides information that is relevant for issues such as nonspecific binding of drug to tissues, lysosomal trapping, and active uptake of drug into cells, is described in detail in Chap. 13. The brain slice method has been optimized for high-throughput of drugs, using cassettes of five to ten drugs that can be studied simultaneously, although it is important that the total combined concentration of drugs in buffer in the cassette does not exceed 1 μM (Friden et al. 2007, 2009a Loryan et al. 2013).

$V_{u,brain}$ is obtained by dividing the total brain concentration found in the slices by the buffer concentration, which is assumed to describe the ISF unbound concentration. Equation 7.10 is adapted from Eq. 7.8 to the in vitro situation:

$$V_{u,brain} = \frac{A_{slice} - V_f{}^* C_{buffer}}{C_{buffer}{}^* \left(1 - V_f\right)}$$

(7.10)

A_{slice} is the amount of drug per gram of slice and C_{buffer} is the concentration of drug in the buffer. V_f is the volume of buffer film that remains around the sampled slice due to incomplete absorption of buffer by the filter paper. Fridén et al. confirmed the value of V_f as 0.094 ml/g_slice (Friden et al. 2009a), in agreement with the original observation by Kakee et al. (Kakee et al. 1996).

7.3.3.3 Interpretations and Caveats

Relevant physiological volumes in brain tissue include the volume of brain ISF at 0.2 ml/g_brain (Nicholson and Phillips 1981; Nicholson and Sykova 1998) and the volume of total brain water at 0.8 ml/g_brain (Reinoso et al. 1997). Thus, drugs with values of $V_{u,brain}$ lower than 0.8 ml/g_brain are predominantly distributed outside the brain cells, with minimal binding to proteins or membranes (e.g., moxalactam, which has a $V_{u,brain}$ of 0.46 ml/g_brain (Friden et al. 2010)). As the values for $V_{u,brain}$ increase further above 0.8 ml/g_brain, intracellular distribution and binding to proteins or membranes also increase (e.g., loperamide, which has a $V_{u,brain}$ of 370 ml/g_brain (Friden et al. 2010)). $V_{u,brain}$ varies between 0.2 and 3000 ml/g_brain for the drugs studied to date. Table 7.3 provides examples of known $V_{u,brain}$ values and the interpretations that can be made based on this information; currently, the highest value is for thioridazine (Friden et al. 2009b).

When using $V_{u,brain}$ to determine $K_{p,uu,brain}$ (Eq. 7.2a), Fridén et al. indicated that the value of V_{blood} from the literature (Eq. 7.8) may be too high (Friden et al. 2010). This appeared especially true for drugs with low $K_{p,brain}$ values. A low $K_{p,brain}$ can be the result of either very efficient efflux at the BBB or a level of plasma protein binding that greatly exceeds the nonspecific binding of the drug in the brain. The latter situation causes a problem when the value for V_{blood} used in Eq. 7.8 is too high. An improved method was developed for this estimation (Friden et al. 2010). It should be noted that the remaining brain vascular space can vary with the method used to sacrifice the animal.

7.3.4 Intracellular Drug Distribution

The intracellular concentrations of drugs cannot be measured directly. However, information on the intracellular distribution of the drug can be obtained by combining brain slice and homogenate data (Friden et al. 2007, 2009a, 2011; Loryan et al. 2014). $K_{p,uu,cell}$ describes the steady-state ratio of intracellular to brain ISF concentrations of unbound drug, assuming an average concentration ratio for all cell types within the brain. In the drug discovery process, this will extend the available

Table 7.3 Interpretation of $V_{u,brain}$ information. For practical purposes, the value of 0.8 ml/g_brain can be approximated to 1 ml/g_brain. The values were obtained using the brain slice method in rats; for further descriptions, see Friden et al. (Friden et al. 2009b) and Loryan et al. (Loryan et al. 2013)

Parameter value	Interpretation	Examples (ml/g_brain)
$V_{u,brain}$ < 0.8 ml/g_brain	Restricted distribution of the drug to the interstitial fluid. Probably very low entrance into cells and very little binding to proteins or membranes.	Morphine-3-glucuronide (0.7) Moxalactam (0.6)
$V_{u,brain}$ ≈ 0.8 ml/g_brain	Free distribution of the drug in ISF and intracellular fluid and/or slight binding to proteins or membranes.	Salicylic acid (1.0) Zidovudine (1.1)
$V_{u,brain}$ > 0.8 ml/g_brain	Binding to proteins or membranes or distribution to subcellular organelles such as lysosomes. The higher the value, the more drug is bound or distributed.	Amitriptyline (310) Atenolol (2.5) Diazepam (20, 17.8) Digoxin (33.1) Gabapentin (4.6) Indomethacin (14) Levofloxacin (1.7) Loperamide (370) Nelfinavir (860) Oxycodone (4.2) Paclitaxel (769) Paroxetine (714) Thioridazine (3333, 2650) Verapamil (54, 47)

information about the distribution of new chemical entities and will help in selecting optimal drug candidates. It is important to measure $K_{p,uu,cell}$ and subsequently estimate the average concentration of unbound drug in brain cells ($C_{u,cell}$), in relation to the pharmacodynamic measurements when the drug has an intracellular site of action or when information about possible active transport processes at the ISF-cellular interface is required. This is also relevant, going even one step further into lysosomal distribution, when predicting and understanding possible side effects due to lysosomal accumulation (Loryan et al. 2017).

$K_{p,uu,cell}$ is calculated as:

$$K_{p,uu,cell} = \frac{C_{u,cell}}{C_{u,brainISF}} = V_{u,brain} * f_{u,brain} \tag{7.11}$$

$V_{u,brain}$ is determined from brain slice experiments and $f_{u,brain}$ is determined from equilibrium dialysis of brain homogenates. The details of how to estimate $K_{p,uu,cell}$ and the further division of this parameter into cytosolic and lysosomal components are further described in Chap. 13. Maurer et al. have mentioned lysosomal accumulation as a possible reason for differences in the distribution of acidic, neutral, and basic drugs between homogenates and in vivo measurements in tissues other than the brain (Maurer et al. 2005). This appears also to be important in brain tissue when comparing brain slice data with data from brain homogenates (Friden et al. 2011; Loryan et al. 2014).

7.3.5 Combining Rate, Extent, and Intra-Brain Drug Distribution in Brain Pharmacokinetics

It will be obvious by now that the three main properties of brain drug delivery, CL_{in}, $K_{p,uu,brain}$, and $V_{u,brain}$, describe three individual properties of a drug. Figure 7.6 provides the $V_{u,brain}$ and $K_{p,uu,brain}$ values for 41 drugs (Friden et al. 2009b).

It can be seen from the figure that these two properties are not correlated. Two examples in the figure highlight this: loperamide and diazepam. The very low $K_{p,uu,brain}$ of loperamide (0.007) indicates that only 0.7% of the concentration of unbound loperamide in plasma will be present in brain ISF and thus that the efflux of loperamide at the BBB is very efficient. At the same time, loperamide has a high affinity to brain tissue, with a $V_{u,brain}$ of 370 ml/g_brain. The transport of diazepam at the BBB, on the other hand, is mainly passive, with a $K_{p,uu,brain}$ close to 1 and a lower affinity, with a $V_{u,brain}$ of 12 ml/g_brain. Similarly, the permeability clearance has little in common with the size of $K_{p,uu,brain}$. As discussed earlier (Eq. 7.6), the

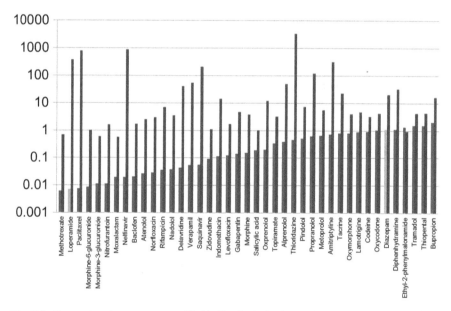

Fig. 7.6 Connections between nonspecific binding in the brain, as shown by $V_{u,brain}$ values (ml/g_brain) and $K_{p,uu,brain}$ ratios for 41 drugs. The scale on the logarithmic y-axis shows the experimentally obtained values for $K_{p,uu,brain}$ and $V_{u,brain}$. The drugs are sorted according to their $K_{p,uu,brain}$ value from smallest to largest. The individual $V_{u,brain}$ values are plotted alongside the $K_{p,uu,brain}$ values and show that there is very little correlation between the two parameters. Data are from Fridén et al. (Friden et al. 2009b)

influx and efflux clearances can both be small and large but can still result in the same $K_{p,uu,brain}$.

The time for drug concentrations to reach equilibrium between brain and blood, on the other hand, is determined by the efflux clearance and the extent of intra-brain binding ($V_{u,brain}$), giving rise to an intrinsic half-life in the brain, which can be shorter or longer than that in plasma. If the plasma half-life is longer than the intrinsic half-life, it will also determine the half-life in brain, which will be equal to that in plasma, and the intrinsic half-life will not be possible to observe. Thus, the unbound drug concentration in plasma is the driving force for the half-life in the brain, and the pharmacokinetic profile in plasma is therefore an important determinant of the concentration-time profile in the brain. Only when elimination of the drug is slower from the brain than from plasma will the intrinsic half-life in the brain be observable. Thus, the plasma concentration-time profile is important for the resulting pharmacodynamics in the brain, be it effects or side effects.

The determinants of the concentration-time profile of a drug in the brain are comparable to the parameters determining the pharmacokinetics in plasma: The plasma concentration-time profile is similarly determined by the absorption and elimination rates and the extent of binding to tissues. The relative unbound concentrations in brain and plasma are determined by the transport process that dominates the movement of the drug at the BBB. This may either be active efflux, active influx, or passive transport as discussed earlier. CL_{in} therefore only influences the brain concentrations (cf bioavailability) in relation to the efflux clearance, but will not influence the concentration-time profile, including the time to reach equilibrium, a fact that may be hard to grasp.

Active efflux of a drug will not only decrease CL_{in} but will also increase CL_{out}, as described in Eq. 7.7, thus increasing the rate of the equilibration processes across the BBB, although this depends on how the efflux transporter functions. If it only hinders influx (the so-called vacuum cleaner model), the efflux from the brain parenchyma will not be influenced, and the active process will not influence the brain elimination half-life (Syvanen et al. 2006). It is, however, more likely that the transporter will both hinder influx and increase efflux (e.g., P-gp). In this case, the part that increases efflux will subsequently affect the elimination process and therefore the time to reach equilibrium across the BBB, while the part that hinders influx will not affect the elimination process and therefore neither the time to equilibrium.

Equilibrium across the BBB is thus reached more quickly for strong P-gp substrates than for drugs that are weaker substrates or that are only passively transported, but otherwise have similar properties. Active efflux also has an important influence on the time aspects of equilibration across the BBB in the studies comparing drug uptake into the brains of Mdr1a/b(−/−) and Mdr1a/b(+/+) mice. Equilibration is expected to take longer in Mdr1a/b (−/−) mice. When sampling at a specific time after a single dose, this can influence the difference between the two groups of mice. Possible differences in equilibration time therefore need to be taken into consideration.

Padowski and Pollack have discussed the theoretical effects of P-gp on the time to equilibrium across the BBB (Padowski and Pollack 2011), and the theoretical consequences of active uptake and efflux have also been discussed by several authors (Golden and Pollack 1998; Hammarlund-Udenaes et al. 1997; Syvanen et al. 2006). Liu and Chen have suggested that the parameters determining the half-life of equilibration are the permeability of the BBB to the respective drug and the extent of binding in the brain (Liu and Chen 2005). As explained in this chapter, they are more clearly described as the efflux permeability and the extent of binding in the brain. The slower of the two half-lives in the plasma and brain will determine the observed half-life in the brain.

Cooperation between P-gp and breast cancer resistance protein (BCRP) in increasing the efficiency of the efflux process at the BBB has been clearly described by Kusuhara and Sugiyama (Kusuhara and Sugiyama 2009). The presence and con-tributions of other, including as yet unknown, transporters should also be included in speculations about the fate of drugs at the BBB (Hammarlund-Udenaes et al. 2008; Kalvass et al. 2007a; Agarwal et al. 2012).

As stated earlier, measurement of unbound drug concentrations in plasma is not enough to determine the unbound concentrations in the brain. Binding to brain parenchymal tissue is too different from binding to plasma proteins to allow predic-tion of one from the other. The presence of active transport at the BBB does not allow the ratio of the fraction of unbound drug in plasma to that in brain ($f_{u,plasma}/f_{u,brain}$) to be used to predict brain penetration, as discussed in Sect. 7.2.

7.4 CSF Pharmacokinetics vs Brain ISF Pharmacokinetics

The CSF is an accessible sampling site for measuring human brain concentrations of unbound drug, given that CSF concentrations follow brain concentrations. However, the role of the CSF as an alternative site for measuring unbound brain concentrations is still under discussion and has not been well established. De Lange and Danhof proposed that the CSF may be of limited value in the prediction of unbound brain concentrations (de Lange and Danhof 2002). There are both similari-ties and differences in drug concentrations between brain ISF and CSF. The BCSFB, situated between the epithelial cells of the choroid plexus, is different from the BBB as a transport site for drugs, and the cells have different origins (epithelial vs endo-thelial), which could influence transporter expression (Fig. 7.2). The relevant ques-tion for drug discovery is whether the transporter functions in the BBB are similar enough to those in the BCSFB to allow the extrapolation of CSF data to obtain data on the exposure of the brain to unbound drug.

While CSF sampling could be useful in the selection of drug candidates for entry into development programs, Lin cautions that CSF concentrations could differ from brain unbound drug concentrations (Lin 2008). Fridén et al. demonstrated the cor-relations between rat CSF and rat brain ISF concentrations for 41 compounds (Friden et al. 2009b). In this study, 33 of the $K_{p,uu,brain}$ values were within a ± three-fold range of the $K_{p,uu,CSF}$ values, which is considered quite good ($r^2 = 0.80$). However, Fig. 7.7a shows that the regression line deviates from the line of identity for these

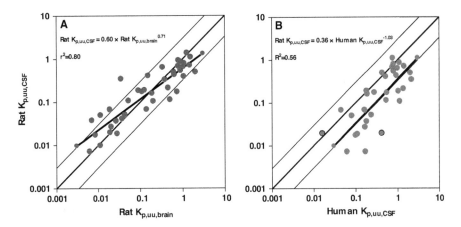

Fig. 7.7 (**a**) Correlations between rat $K_{p,uu,brain}$ and $K_{p,uu,CSF}$ for 41 drugs. The middle diagonal line is the line of identity. The two parallel lines show a threefold difference in range from the line of identity. (**b**) Correlations between $K_{p,uu,CSF}$ in humans (x-axis, data from Shen et al. (Shen et al. 2004)) and $K_{p,uu,CSF}$ in rats (y-axis). Although there is a good correlation between the species, there is a threefold deviation from the line of identity. Reprinted with permission from Fridén et al. (Friden et al. 2009b). Copyright 2009 American Chemical Society

compounds. CSF concentrations were lower than the unbound brain concentrations at high $K_{p,uu,brain}$ values and higher at low $K_{p,uu,brain}$ values. This confirms earlier work by Kalvass and Maurer, who found that unbound brain concentrations were over-predicted by CSF concentrations for drugs with low $K_{p,uu,brain}$ values (Kalvass and Maurer 2002). While the results from Fridén et al. support the use of $K_{p,uu,CSF}$ for comparisons of brain exposure between drugs (Friden et al. 2009b), it should be borne in mind that other drugs could behave differently and that individual drug concentrations could deviate from the predicted value quite extensively.

Differences in the location and expression of P-gp between the BBB and the BCSFB could explain the concentration differences at low $K_{p,uu,brain}$ values. P-gp and BCRP are located in the luminal membranes of the endothelial cells in the BBB. According to an early report, P-gp was thought to be located in the apical membrane of the epithelial cells of the choroid plexus, which would result in substrates being transported toward the CSF (Rao et al. 1999). This has, however, been questioned (Sun et al. 2003). It seems unlikely that P-gp would transport substrates into the CSF in the epithelial cells of the BCSFB and in the opposite direction, into the blood, at the BBB. If this was the case, the CSF would be an even less suitable site of measurement for estimating brain ISF concentrations. Although studies have shown less efficient P-gp functioning at the BCSFB than at the BBB, the findings do not actually support the transport of drugs toward the CSF . The reason for the differences in P-gp function may have been found by Gazzin et al. who measured the relative content of P-gp and Mrp1 protein in rat and human brain capillaries and choroid plexus (Gazzin et al. 2008). They showed that the P-gp content in rat choroid plexus homogenates was only 0.5% of that in brain endothelial cells, while the opposite trend was seen with Mrp1 – the microvessel content was only 4% of that in the choroid plexus. Human data showed a similar picture. Thus, although it is present at the BCSFB, P-gp seems to have a significantly smaller role than at the BBB because of its lower expression.

The correlation between human and rat $K_{p,uu,CSF}$ is unexpectedly good; however, the threefold deviation from the line of identity, with higher CSF to plasma concentration ratios in humans than in rats (Fig. 7.7b), is an issue not yet explained (Friden et al. 2009b; Shen et al. 2004).

Issues on differences in time between dosage and sampling, and the sites of sampling, in humans vs rodents should also be taken into consideration when studying the use of CSF sampling to estimate drug distribution to the brain. The timing aspects of CSF concentration-time profiles vs brain ISF profiles have been studied by Westerhout et al. using a multiple microdialysis probe approach in rats (Westerhout et al. 2012). It takes only slightly longer to reach similar concentrations of acetaminophen in rat CSF from the cisterna magna and 3rd/fourth ventricles than in brain ISF, although the difference is extended for CSF from the subarachnoid space furthest away from the brain ISF, which is of relevance when sampling CSF in humans. Westerhout et al. developed a physiological pharmacokinetic model for multiple brain compartments, based on these rat data. After translation of the model by changing the physiological parameters to those in humans, they were able to successfully predict lumbar CSF data on acetaminophen comparable to those available from humans. The model also predicted human ISF concentration-time profiles (Westerhout et al. 2012), further developed into a generic model for nine drugs (Yamamoto et al. 2017).

In summary, it appears that CSF is an adequate sampling site for obtaining a preliminary understanding of unbound brain concentrations, provided to be at steady state, with the caveat of taking into account deviations at low and high $K_{p,uu,brain}$ values. The results support the use of $K_{p,uu,CSF}$ for reasonable comparisons of brain exposure to drugs. However, it should be borne in mind that individual drugs could deviate quite extensively from the general correlation.

7.5   Drug Interactions at the BBB

Because transporters play such an important role at the BBB in controlling the traffic of drug molecules into and out of the brain, they may also be targets of clinically significant drug interactions, however rather unlikely (Kalvass et al. 2013). Unfortunately, interaction studies at the BBB in humans are few (Bauer et al. 2012, 2015; Matsuda et al. 2017). Cyclosporin is the most potent P-gp inhibitor on the market, doubling the brain concentrations of verapamil and loperamide (Sasongko et al. 2005; Hsiao and Unadkat 2012). Quinidine also inhibits P-gp in humans, causing a 20% reduction in the response to CO_2 (opiate-induced respiratory depression) when administered with loperamide (Sadeque et al. 2000).

The $K_{p,uu,brain}$ value of a drug can give information on its interaction potential at the BBB (Hammarlund-Udenaes et al. 2008). For a $K_{p,uu,brain}$ close to unity, the interaction potential is likely to be very low, given that the drug is mainly passively transported. The lower the $K_{p,uu,brain}$, the higher the theoretical possibility of an interaction with other drugs, depending on whether the low $K_{p,uu,brain}$ was caused by efflux via a single transporter or if there are several transporters acting on one drug. Inhibition of the main efflux transporter would thus result in increased brain concentrations, while an interaction at an uptake transporter would decrease brain concentrations. In practice, it appears that interactions at the BBB are very rare, irrespective of the direction of active transport (Sadiq et al. 2011; Sasongko et al. 2005; Liu et al. 2008). This low incidence of interaction is possibly the result of relatively low concentrations of both victim drug and perpetrator in plasma. For example, the inhibition constant K_i for an interaction between diphenhydramine and oxycodone at the uptake transporter in cell cultures was much higher than the maximum possible clinical concentration (Sadiq et al. 2011). This is quite different from the situation in the gastrointestinal tract and the liver after oral administration, where much higher concentrations are present and the likelihood of an interaction is subsequently much greater.

7.6 Species Comparisons

Species differences in the extent of drug transport at the BBB are the result of differences in transporter expression and the capacity/specificity of substrates. It is well known that the expression of P-gp and BCRP proteins in humans is different from that in other species; for example, BCRP content is higher than P-gp content in humans, and P-gp content is higher than BCRP content in rats/mice (Ito et al. 2011; Uchida et al. 2011b, 2020). This could explain the differences in the results obtained when studying three PET tracers that are P-gp substrates in several species (Syvanen and Hammarlund-Udenaes 2010).

While the behavior of morphine at the BBB is very similar in rats, pigs, and humans ($K_{p,uu,brain}$ values are about 0.3–0.6), the $K_{p,uu,brain}$ in sheep deviates from this somewhat (1.2–1.9, depending on age) (Bengtsson et al. 2009; Ederoth et al. 2004; Tunblad et al. 2003, 2004a; Bouw et al. 2000;). This could be because of differences in transporter expression between sheep and the other species, i.e., possibly a lack of an efflux transporter.

There is a clear need for further translational studies between experimental animals and humans to learn more about species differences in transporter function at the BBB.

7.7   Current Status and Future Challenges

The understanding of the pharmacokinetics of drug delivery to the brain has developed rapidly, although there is still some confusion on rate vs extent measurements and methods and what they describe. There are today ways of measuring unbound concentrations in the brain using high-throughput methodology. In vivo studies have shown that there are still transport proteins acting as efflux or uptake transporters at the BBB that have not yet been identified. The presence and actions of transporters other than P-gp therefore need to be included in the thinking on brain penetration.

The scientific community and the drug industry are continuously striving to find correlations that will simplify measurements and enable prediction of successful new CNS drugs. There is, however, a difference between finding a correlation coefficient that is good enough versus predicting the fate of an individual compound based on this correlation or based on measuring a substitute parameter. The use of log-log comparisons and correlation coefficients could actually hide important information. Considering what we now know about individual BBB transport properties, it is actually easier to select new compounds that have high and low $K_{p,uu}$ values and assign them to potential clinical use depending on whether the desired effect is therapeutic efficacy or the avoidance of side effects in the CNS. Other aspects, such as peripheral side effects and affinity to target, are also included in the decision-making process. It is recommended to put as much effort into the decision on the kind of measurements to be made, as to put the efforts into finding correlations between measures that may or may not be clinically relevant. The area of BBB transport of drugs illustrates the time lag between new scientific findings and adoption of these findings in the drug industry. Shortening this time lag would significantly improve the success rate in drug discovery/development.

More research is needed before we can extrapolate information from animal studies to prediction of clinically relevant brain drug delivery. Some progress has recently been made in demonstrating the expression of transporters at the BBB for different species, but in vivo examples are needed to confirm these findings and more experimental studies are required. When we have identified most of the transporters, there is a real chance that predictive science will be able to help in the selection of good compounds for use in the CNS. There is also a need for better predictive disease models, understanding of disease mechanisms, and understanding of how disease states can influence drug transport into the brain, although these are beyond the scope of this chapter.

In an era of increased use of peptides and proteins, there is hope that some of these compounds will be available to the brain. The task before us, of understanding and improving their uptake into the brain from a quantitative and mechanistic perspective, is vast. A greater understanding and quantitative investigation of the role of nanocarrier delivery of drugs to the brain is also required. The achievement of successful delivery by these means in humans will require biocompatible carriers, and these should be a particular focus.

7.8 Conclusions

The rate and extent of drug delivery to the brain are two individual properties that are not numerically related. Data on intra-brain distribution are required to obtain the full brain delivery picture in relation to total (unbound plus bound) drug concentrations. The pharmacokinetic relationship between the permeability of the BBB (influx clearance) and the extent of drug delivery to the brain explains why the permeability per se is of lesser importance for brain drug delivery. Recent findings have confirmed the great value of focusing measurements on the extent of delivery of unbound drug to the brain. This is governed by the net flux of drug across the BBB and ultimately determines the clinical success rate when receptor occupancy is taken into account.

7.9 Points for Discussion

- What are the reasons for extent of delivery being more clinically relevant than rate of delivery for estimating the delivery of drugs into the brain?
- What are the essential processes governing the net influx and efflux clearances at the BBB, CL_{in}, and CL_{out}?
- For which purposes can $V_{u,brain}$ measurements be used?
- In what way could estimation of CNS exposure of drugs by the use of ratio of total brain to total plasma drug concentrations be flawed?
- How does the exchange of drugs between blood and CSF differ from the exchange between blood and brain ISF?
- How is the CSF concentration of the drug related to the brain interstitial fluid concentration? Discuss the rationale of using a surrogate approach for approximation of brain interstitial fluid concentration in preclinical and clinical studies (i.e., using other measurements than the direct ones).
- How may the understanding of intracellular distribution of drug contribute to establishment of a link between PK and PD?
- What are the clinically relevant sites of drug-drug interaction regarding brain drug delivery?
- What are the key components of interspecies differences in brain drug delivery?

References

Abbott NJ (2004a) Evidence for bulk flow of brain interstitial fluid: significance for physiology and pathology. Neurochem Int 45:545–552

Abbott NJ (2004b) Prediction of blood–brain barrier permeation in drug discovery from in vivo, in vitro and in silico models. Drug Discov Today Technol 1:407–416

Abbott NJ, Dolman DE, Patabendige AK (2008) Assays to predict drug permeation across the blood-brain barrier, and distribution to brain. Curr Drug Metab 9:901–910

Abbott NJ, Pizzo ME, Preston JE, Janigro D, Thorne RG (2018) The role of brain barriers in fluid movement in the CNS: is there a 'glymphatic' system? Acta Neuropathol 135:387–407

Abraham MH, Chadha HS, Mitchell RC (1995) Hydrogen-bonding. Part 36. Determination of blood brain distribution using octanol-water partition coefficients. Drug Des Discov 13:123–131

Agarwal S, Uchida Y, Mittapalli RK, Sane R, Terasaki T, Elmquist WF (2012) Quantitative proteomics of transporter expression in brain capillary endothelial cells isolated from P-gp, BCRP, and P-gp/BCRP knockout mice. Drug Metab Dispos

Avdeef A (2011) How well can in vitro brain microcapillary endothelial cell models predict rodent in vivo blood-brain barrier permeability? Eur J Pharm Sci 43:109–124

Avdeef A (2012) Absorption and drug development. Solubility, permeability and charge state. Wiley

Avdeef A, Sun N (2011) A new in situ brain perfusion flow correction method for lipophilic drugs based on the pH-dependent crone-Renkin equation. Pharm Res 28:517–530

Banks WA, Jaspan JB, Kastin AJ (1997) Effect of diabetes mellitus on the permeability of the blood-brain barrier to insulin. Peptides 18:1577–1584

Bauer M, Zeitlinger M, Karch R, Matzneller P, Stanek J, Jager W, Bohmdorfer M, Wadsak W, Mitterhauser M, Bankstahl JP, Loscher W, Koepp M, Kuntner C, Muller M, Langer O (2012) Pgp-mediated interaction between (R)-[11C]verapamil and tariquidar at the human blood-brain barrier: a comparison with rat data. Clin Pharmacol Ther 91:227–233

Bauer M, Karch R, Zeitlinger M, Philippe C, Romermann K, Stanek J, Maier-Salamon A, Wadsak W, Jager W, Hacker M, Muller M, Langer O (2015) Approaching complete inhibition of P-glycoprotein at the human blood-brain barrier: an (R)-[11C]verapamil PET study. J Cereb Blood Flow Metab 35:743–746

Bengtsson J, Ederoth P, Ley D, Hansson S, Amer-Wahlin I, Hellstrom-Westas L, Marsal K, Nordstrom CH, Hammarlund-Udenaes M (2009) The influence of age on the distribution of morphine and morphine-3-glucuronide across the blood-brain barrier in sheep. Br J Pharmacol 157:1085–1096

Bickel U (2005) How to measure drug transport across the blood-brain barrier. NeuroRx 2:15–26

Bostrom E, Simonsson US, Hammarlund-Udenaes M (2005) Oxycodone pharmacokinetics and pharmacodynamics in the rat in the presence of the P-glycoprotein inhibitor PSC833. J Pharm Sci 94:1060–1066

Bostrom E, Simonsson US, Hammarlund-Udenaes M (2006) In vivo blood-brain barrier transport of oxycodone in the rat: indications for active influx and implications for pharmacokinetics/pharmacodynamics. Drug Metab Dispos 34:1624–1631

Bostrom E, Hammarlund-Udenaes M, Simonsson US (2008) Blood-brain barrier transport helps to explain discrepancies in in vivo potency between oxycodone and morphine. Anesthesiology 108:495–505

Bouw MR, Gardmark M, Hammarlund-Udenaes M (2000) Pharmacokinetic-pharmacodynamic modelling of morphine transport across the blood-brain barrier as a cause of the antinociceptive effect delay in rats--a microdialysis study. Pharm Res 17:1220–1227

Bouw MR, Xie R, Tunblad K, Hammarlund-Udenaes M (2001) Blood-brain barrier transport and brain distribution of morphine-6-glucuronide in relation to the antinociceptive effect in rats--pharmacokinetic/pharmacodynamic modelling. Br J Pharmacol 134:1796–1804

Broccatelli F, Larregieu CA, Cruciani G, Oprea TI, Benet LZ (2012) Improving the prediction of the brain disposition for orally administered drugs using BDDCS. Adv Drug Deliv Rev 64:95–109

Chen H, Winiwarter S, Friden M, Antonsson M, Engkvist O (2011) In silico prediction of unbound brain-to-plasma concentration ratio using machine learning algorithms. J Mol Graph Model 29:985–995

Chen X, Slattengren T,. de Lange ECM, Smith DE, Hammarlund-Udenaes M (2017) Revisiting atenolol as a low passive permeability marker. Fluids Barriers CNS 14:30

Cordon-cardo C, O'Brien JP, Casals D, Rittman-Grauer L, Biedler JL, Melamed MR, Bertino JR (1989) Multidrug-resistance gene (P-glycoprotein) is expressed by endothelial cells at blood-brain barrier sites. Proc Natl Acad Sci U S A 86:695–698

Cserr HF, Cooper DN, Milhorat TH (1977) Flow of cerebral interstitial fluid as indicated by the removal of extracellular markers from rat caudate nucleus. Exp Eye Res 25(Suppl):461–473

Dagenais C, Rousselle C, Pollack GM, Scherrmann JM (2000) Development of an in situ mouse brain perfusion model and its application to mdr1a P-glycoprotein-deficient mice. J Cereb Blood Flow Metab 20:381–386

Dagenais C, Graff CL, Pollack GM (2004) Variable modulation of opioid brain uptake by P-glycoprotein in mice. Biochem Pharmacol 67:269–276

Dai H, Chen Y, Elmquist WF, Yang H, Wang Q, Elmquist WF (2005) Distribution of the novel antifolate pemetrexed to the brain. J Pharmacol Exp Ther 315:222–229

de Lange EC, Danhof M (2002) Considerations in the use of cerebrospinal fluid pharmacokinetics to predict brain target concentrations in the clinical setting: implications of the barriers between blood and brain. Clin Pharmacokinet 41:691–703

Deguchi Y, Yokoyama Y, Sakamoto T, Hayashi H, Naito T, Yamada S, Kimura R (2000) Brain distribution of 6-mercaptopurine is regulated by the efflux transport system in the blood-brain barrier. Life Sci 66:649–662

Di L, Kerns EH, Bezar IF, Petusky SL, Huang Y (2009) Comparison of blood-brain barrier permeability assays: in situ brain perfusion, MDR1-MDCKII and PAMPA-BBB. J Pharm Sci 98:1980–1991

Di L, Umland JP, Chang G, Huang Y, Lin Z, Scott DO, Troutman MD, Liston TE (2011) Species independence in brain tissue binding using brain homogenates. Drug Metab Dispos 39:1270–1277

Di L, Artursson P, Avdeef A, Ecker GF, Faller B, Fischer H, Houston JB, Kansy M, Kerns EH, Kramer SD, Lennernas H, Sugano K (2012) Evidence-based approach to assess passive diffusion and carrier-mediated drug transport. Drug Discov Today 17:905–912

Doran A, Obach RS, Smith BJ, Hosea NA, Becker S, Callegari E, Chen C, Chen X, Choo E, Cianfrogna J, Cox LM, Gibbs JP, Gibbs MA, Hatch H, Hop CE, Kasman IN, Laperle J, Liu J, Liu X, Logman M, Maclin D, Nedza FM, Nelson F, Olson E, Rahematpura S, Raunig D, Rogers S, Schmidt K, Spracklin DK, Szewc M, Troutman M, Tseng E, Tu M, van Deusen JW, Venkatakrishnan K, Walens G, Wang EQ, Wong D, Yasgar AS, Zhang C (2005) The impact of P-glycoprotein on the disposition of drugs targeted for indications of the central nervous system: evaluation using the MDR1A/1B knockout mouse model. Drug Metab Dispos 33:165–174

Doran AC, Osgood SM, Mancuso JY, Shaffer CL (2012) An evaluation of using rat-derived single-dose neuropharmacokinetic parameters to project accurately large animal unbound brain drug concentrations. Drug Metab Dispos 40:2162–2173

Dubey RK, Mcallister CB, Inoue M, Wilkinson GR (1989) Plasma binding and transport of diazepam across the blood-brain barrier. No evidence for in vivo enhanced dissociation. J Clin Invest 84:1155–1159

Ederoth P, Tunblad K, Bouw R, Lundberg CJ, Ungerstedt U, Nordstrom CH, Hammarlund-Udenaes M (2004) Blood-brain barrier transport of morphine in patients with severe brain trauma. Br J Clin Pharmacol 57:427–435

Fan Y, Unwalla R, Denny RA, Di L, Kerns EH, Diller DJ, Humblet C (2010) Insights for predicting blood-brain barrier penetration of CNS targeted molecules using QSPR approaches. J Chem Inf Model 50:1123–1133

Fenstermacher J, Gross P, Sposito N, Acuff V, Pettersen S, Gruber K (1988) Structural and functional variations in capillary systems within the brain. Ann N Y Acad Sci 529:21–30

Friden M, Gupta A, Antonsson M, Bredberg U, Hammarlund-Udenaes M (2007) In vitro methods for estimating unbound drug concentrations in the brain interstitial and intracellular fluids. Drug Metab Dispos 35:1711–1719

Friden M, Ducrozet F, Middleton B, Antonsson M, Bredberg U, Hammarlund-Udenaes M (2009a) Development of a high-throughput brain slice method for studying drug distribution in the central nervous system. Drug Metab Dispos 37:1226–1233

Friden M, Winiwarter S, Jerndal G, Bengtsson O, Wan H, Bredberg U, Hammarlund-Udenaes M, Antonsson M (2009b) Structure-brain exposure relationships in rat and human using a novel data set of unbound drug concentrations in brain interstitial and cerebrospinal fluids. J Med Chem 52:6233–6243

Friden M, Ljungqvist H, Middleton B, Bredberg U, Hammarlund-Udenaes M (2010) Improved measurement of drug exposure in the brain using drug-specific correction for residual blood. J Cereb Blood Flow Metab 30:150–161

Friden M, Bergstrom F, Wan H, Rehngren M, Ahlin G, Hammarlund-Udenaes M, Bredberg U (2011) Measurement of unbound drug exposure in brain: modeling of pH partitioning explains diverging results between the brain slice and brain homogenate methods. Drug Metab Dispos 39:353–362

Garberg P, Ball M, Borg N, Cecchelli R, Fenart L, Hurst RD, Lindmark T, Mabondzo A, Nilsson JE, Raub TJ, Stanimirovic D, Terasaki T, Oberg JO, Osterberg T (2005) In vitro models for the blood-brain barrier. Toxicol In Vitro 19:299–334

Gazzin S, Strazielle N, Schmitt C, Fevre-Montange M, Ostrow JD, Tiribelli C, Ghersi-Egea JF (2008) Differential expression of the multidrug resistance-related proteins ABCb1 and ABCc1 between blood-brain interfaces. J Comp Neurol 510:497–507

Golden PL, Pollack GM (1998) Rationale for influx enhancement versus efflux blockade to increase drug exposure to the brain. Biopharm Drug Dispos 19:263–272

Gunn RN, Summerfield SG, Salinas CA, Read KD, Guo Q, Searle GE, Parker CA, Jeffrey P, Laruelle M (2012) Combining PET biodistribution and equilibrium dialysis assays to assess the free brain concentration and BBB transport of CNS drugs. J Cereb Blood Flow Metab 32:874–883

Gupta A, Chatelain P, Massingham R, Jonsson EN, Hammarlund-Udenaes M (2006) Brain distribution of cetirizine enantiomers: comparison of three different tissue-to-plasma partition coefficients: K(p), K(p,u), and K(p,uu). Drug Metab Dispos 34:318–323

Gustafsson S, Sehlin D, Lampa E, Hammarlund-Udenaes M, Loryan I (2019) Heterogeneous drug tissue binding in brain regions of rats, Alzheimer's patients and controls: impact on translational drug development. Sci Rep 9:5308

Hakkarainen JJ, Jalkanen AJ, Kaariainen TM, Keski-Rahkonen P, Venalainen T, Hokkanen J, Monkkonen J, Suhonen M, Forsberg MM (2010) Comparison of in vitro cell models in predicting in vivo brain entry of drugs. Int J Pharm 402:27–36

Hammarlund-Udenaes M (2000) The use of microdialysis in CNS drug delivery studies. Pharmacokinetic perspectives and results with analgesics and antiepileptics. Adv Drug Deliv Rev 45:283–294

Hammarlund-Udenaes M (2010) Active-site concentrations of chemicals - are they a better predictor of effect than plasma/organ/tissue concentrations? Basic Clin Pharmacol Toxicol 106:215–220

Hammarlund-Udenaes M (2013) Microdialysis in CNS PKPD research: unraveling unbound concentrations. In: Müller M (ed) Microdialysis in drug development. Springer, New York

Hammarlund-Udenaes M, Paalzow LK, de Lange EC (1997) Drug equilibration across the blood-brain barrier--pharmacokinetic considerations based on the microdialysis method. Pharm Res 14:128–134

Hammarlund-Udenaes M, Friden M, Syvanen S, Gupta A (2008) On the rate and extent of drug delivery to the brain. Pharm Res 25:1737–1750

Hammarlund-Udenaes M, Bredberg U, FRIDEN, M. (2009) Methodologies to assess brain drug delivery in lead optimization. Curr Top Med Chem 9:148–162

Hsiao P, Unadkat JD (2012) P-glycoprotein-based loperamide-cyclosporine drug interaction at the rat blood-brain barrier: prediction from in vitro studies and extrapolation to humans. Mol Pharm 9:629–633

Hu Y, Rip J, Gaillard PJ, de Lange ECM, Hammarlund-Udenaes M (2017) The impact of liposomal formulations on the release and brain delivery of methotrexate: an in vivo microdialysis study. J Pharm Sci 106:2606–2613

Iliff JJ, Wang M, Liao Y, Plogg BA, Peng W, Gundersen GA, Benveniste H, Vates GE, Deane R, Goldman SA, Nagelhus EA, Nedergaard M (2012) A paravascular pathway facilitates CSF flow through the brain parenchyma and the clearance of interstitial solutes, including amyloid beta. Sci Transl Med 4:147ra111

Ito K, Uchida Y, Ohtsuki S, Aizawa S, Kawakami H, Katsukura Y, Kamiie J, Terasaki T (2011) Quantitative membrane protein expression at the blood-brain barrier of adult and younger cynomolgus monkeys. J Pharm Sci 100:3939–3950

Jeffrey P, Summerfield S (2010) Assessment of the blood-brain barrier in CNS drug discovery. Neurobiol Dis 37:33–37

Kaitin KI (2008) Obstacles and opportunities in new drug development. Clin Pharmacol Ther 83:210–212

Kakee A, Terasaki T, Sugiyama Y (1996) Brain efflux index as a novel method of analyzing efflux transport at the blood-brain barrier. J Pharmacol Exp Therapeutics 277:1550–1559

Kalvass JC, Maurer TS (2002) Influence of nonspecific brain and plasma binding on CNS exposure: implications for rational drug discovery. Biopharm Drug Dispos 23:327–338

Kalvass JC, Maurer TS, Pollack GM (2007a) Use of plasma and brain unbound fractions to assess the extent of brain distribution of 34 drugs: comparison of unbound concentration ratios to in vivo p-glycoprotein efflux ratios. Drug Metab Dispos 35:660–666

Kalvass JC, Olson ER, Cassidy MP, Selley DE, Pollack GM (2007b) Pharmacokinetics and pharmacodynamics of seven opioids in P-glycoprotein-competent mice: assessment of unbound brain EC50,u and correlation of in vitro, preclinical, and clinical data. J Pharmacol Exp Ther 323:346–355

Kalvass JC, Polli JW, Bourdet DL, Feng B, Huang SM, Liu X, Smith QR, Zhang LK, Zamek-Gliszczynski MJ, International Transporter, C (2013) Why clinical modulation of efflux transport at the human blood-brain barrier is unlikely: the ITC evidence-based position. Clin Pharmacol Ther 94:80–94

Kola I, Landis J (2004) Can the pharmaceutical industry reduce attrition rates? Nat Rev Drug Discov 3:711–715

Kurosawa T, Higuchi K, Okura T, Kobayashi K, Kusuhara H, Deguchi Y (2017) Involvement of proton-coupled organic cation antiporter in Varenicline transport at blood-brain barrier of rats and in human brain capillary endothelial cells. J Pharm Sci 106:2576–2582

Kusuhara H, Sugiyama Y (2009) In vitro-in vivo extrapolation of transporter-mediated clearance in the liver and kidney. Drug Metab Pharmacokinet 24:37–52

Lanevskij K, Japertas P, Didziapetris R (2013) Improving the prediction of drug disposition in the brain. Expert Opin Drug Metab Toxicol

Large CH, Kalinichev M, Lucas A, Carignani C, Bradford A, Garbati N, Sartori I, Austin NE, Ruffo A, Jones DN, Alvaro G, Read KD (2009) The relationship between sodium channel inhibition and anticonvulsant activity in a model of generalised seizure in the rat. Epilepsy Res 85:96–106

Levin VA (1980) Relationship of octanol/water partition coefficient and molecular weight to rat brain capillary permeability. J Med Chem 23:682–684

Lin JH (2008) CSF as a surrogate for assessing CNS exposure: an industrial perspective. Curr Drug Metab 9:46–59

Liu X, Chen C (2005) Strategies to optimize brain penetration in drug discovery. Curr Opin Drug Discov Devel 8:505–512

Liu X, Tu M, Kelly RS, Chen C, Smith BJ (2004) Development of a computational approach to predict blood-brain barrier permeability. Drug Metab Dispos 32:132–139

Liu X, Smith BJ, Chen C, Callegari E, Becker SL, Chen X, Cianfrogna J, Doran AC, Doran SD, Gibbs JP, Hosea N, Liu J, Nelson FR, Szewc MA, van Deusen J (2005) Use of a physiologically based pharmacokinetic model to study the time to reach brain equilibrium: an experimental analysis of the role of blood-brain barrier permeability, plasma protein binding, and brain tissue binding. J Pharmacol Exp Ther 313:1254–1262

Liu X, Chen C, Smith BJ (2008) Progress in brain penetration evaluation in drug discovery and development. J Pharmacol Exp Ther 325:349–356

Liu X, van Natta K, Yeo H, Vilenski O, Weller PE, Worboys PD, Monshouwer M (2009) Unbound drug concentration in brain homogenate and cerebral spinal fluid at steady state as a surrogate for unbound concentration in brain interstitial fluid. Drug Metab Dispos 37:787–793

Loryan I, Friden M, Hammarlund-Udenaes M (2013) The brain slice method for studying drug distribution in the CNS. Fluids Barriers CNS 10:6

Loryan I, Sinha V, Mackie C, van Peer A, Drinkenburg W, Vermeulen A, Morrison D, Monshouwer M, Heald D, Hammarlund-Udenaes M (2014) Mechanistic understanding of brain drug disposition to optimize the selection of potential neurotherapeutics in drug discovery. Pharm Res 31:2203–2219

Loryan I, Melander E, Svensson M, Payan M, Konig F, Jansson B, Hammarlund-Udenaes M (2016) In-depth neuropharmacokinetic analysis of antipsychotics based on a novel approach to estimate unbound target-site concentration in CNS regions: link to spatial receptor occupancy. Mol Psychiatry

Loryan I, Hoppe E, Hansen K, Held F, Kless A, Linz K, Marossek V, Nolte B, Ratcliffe P, Saunders D, Terlinden R, Wegert A, Welbers A, Will O, Hammarlund-Udenaes M (2017) Quantitative assessment of drug delivery to tissues and association with Phospholipidosis: A case study with two structurally related diamines in development. Mol Pharm 14:4362–4373

Mano Y, Higuchi S, Kamimura H (2002) Investigation of the high partition of YM992, a novel antidepressant, in rat brain - in vitro and in vivo evidence for the high binding in brain and the high permeability at the BBB. Biopharm Drug Dispos 23:351–360

Matsuda A, Karch R, Bauer M, Traxl A, Zeitlinger M, Langer O (2017) A prediction method for P-glycoprotein-mediated drug-drug interactions at the human blood-brain barrier from blood concentration-time profiles, validated with PET data. J Pharm Sci 106:2780–2786

Maurer TS, Debartolo DB, Tess DA, Scott DO (2005) Relationship between exposure and non-specific binding of thirty-three central nervous system drugs in mice. Drug Metab Dispos 33:175–181

Mensch J, Jaroskova L, Sanderson W, Melis A, Mackie C, Verreck G, Brewster ME, Augustijns P (2010a) Application of PAMPA-models to predict BBB permeability including efflux ratio, plasma protein binding and physicochemical parameters. Int J Pharm 395:182–197

Mensch J, Melis A, Mackie C, Verreck G, Brewster ME, Augustijns P (2010b) Evaluation of various PAMPA models to identify the most discriminating method for the prediction of BBB permeability. Eur J Pharm Biopharm 74:495–502

Muehlbacher M, Spitzer GM, Liedl KR, Kornhuber J (2011) Qualitative prediction of blood-brain barrier permeability on a large and refined dataset. J Comput Aided Mol Des 25:1095–1106

Nicholson C, Phillips JM (1981) Ion diffusion modified by tortuosity and volume fraction in the extracellular microenvironment of the rat cerebellum. J Physiol 321:225–257

Nicholson C, Sykova E (1998) Extracellular space structure revealed by diffusion analysis. Trends Neurosci 21:207–215

Norinder U, Haeberlein M (2002) Computational approaches to the prediction of the blood-brain distribution. Adv Drug Deliv Rev 54:291–313

Norinder U, Sjoberg P, Osterberg T (1998) Theoretical calculation and prediction of brain-blood partitioning of organic solutes using MolSurf parametrization and PLS statistics. J Pharm Sci 87:952–959

Ooie T, Terasaki T, Suzuki H, Sugiyama Y (1997) Kinetic evidence for active efflux transport across the blood-brain barrier of quinolone antibiotics. J Pharmacol Exp Therapeutics 283:293–304

Padowski JM, Pollack GM (2011) Influence of time to achieve substrate distribution equilibrium between brain tissue and blood on quantitation of the blood-brain barrier P-glycoprotein effect. Brain Res 1426:1–17

Pardridge WM (2004) Log(BB), PS products and in silico models of drug brain penetration.[comment]. Drug Discov Today 9:392–393

Pardridge WM, Boado RJ, Black KL, Cancilla PA (1992) Blood-brain barrier and new approaches to brain drug delivery. West J Med 156:281–286

Rao VV, Dahlheimer JL, Bardgett ME, Snyder AZ, Finch RA, Sartorelli AC, Piwnica-Worms D (1999) Choroid plexus epithelial expression of MDR1 P glycoprotein and multidrug resistance-associated protein contribute to the blood-cerebrospinal-fluid drug-permeability barrier. Proc Natl Acad Sci U S A 96:3900–3905

Reese TS, Karnovsky MJ (1967) Fine structural localization of a blood-brain barrier to exogenous peroxidase. J Cell Biol 34:207–217

Reinoso RF, Telfer BA, Rowland M (1997) Tissue water content in rats measured by desiccation. J Pharmacol Toxicol Methods 38:87–92

Rosenberg GA, Kyner WT, Estrada E (1980) Bulk flow of brain interstitial fluid under normal and hyperosmolar conditions. Am J Physiol 238:F42–F49

Rowland M, Tozer T (2011) Clinical pharmacokinetics and pharmacodynamics. Concepts and applications., Baltimore and Philadephia, Lippincott, Williams & Wilkins

Sadeque AJ, Wandel C, He H, Shah S, wood, A. J. (2000) Increased drug delivery to the brain by P-glycoprotein inhibition. Clin Pharmacol Ther 68:231–237

Sadiq MW, Borgs A, Okura T, Shimomura K, Kato S, Deguchi Y, Jansson B, Bjorkman S, Terasaki T, Hammarlund-Udenaes M (2011) Diphenhydramine active uptake at the blood-brain barrier and its interaction with oxycodone in vitro and in vivo. J Pharm Sci 100:3912–3923

Sasongko L, Link JM, Muzi M, Mankoff DA, Yang X, Collier AC, Shoner SC, Unadkat JD (2005) Imaging P-glycoprotein transport activity at the human blood-brain barrier with positron emission tomography. Clin Pharmacol Ther 77:503–514

Schinkel AH, Wagenaar E, Mol CA, Van Deemter L (1996) P-glycoprotein in the blood-brain barrier of mice influences the brain penetration and pharmacological activity of many drugs. J Clin Invest 97:2517–2524

Shen DD, Artru AA, Adkison KK (2004) Principles and applicability of CSF sampling for the assessment of CNS drug delivery and pharmacodynamics. Adv Drug Deliv Rev 56:1825–1857

Shityakov S, Neuhaus W, Dandekar T, Forster C (2013) Analysing molecular polar surface descriptors to predict blood-brain barrier permeation. Int J Comput Biol Drug Des 6:146–156

Smith QR, Allen DD (2003) In situ brain perfusion technique. Methods Mol Med 89:209–218

Stevens J, Ploeger BA, Hammarlund-Udenaes M, Osswald G, van der Graaf PH, Danhof M, de Lange EC (2012) Mechanism-based PK-PD model for the prolactin biological system response following an acute dopamine inhibition challenge: quantitative extrapolation to humans. J Pharmacokinet Pharmacodyn

Summerfield SG, Stevens AJ, Cutler L, del Carmen Osuna M, Hammond B, Tang SP, Hersey A, Spalding DJ, Jeffrey P (2006) Improving the in vitro prediction of in vivo central nervous system penetration: integrating permeability, P-glycoprotein efflux, and free fractions in blood and brain. J Pharmacol Exp Ther 316:1282–1290

Summerfield SG, Read K, Begley DJ, Obradovic T, Hidalgo IJ, Coggon S, Lewis AV, Porter RA, Jeffrey P (2007) Central nervous system drug disposition: the relationship between in situ brain permeability and brain free fraction. J Pharmacol Exp Ther 322:205–213

Summerfield SG, Lucas AJ, Porter RA, Jeffrey P, Gunn RN, Read KR, Stevens AJ, Metcalf AC, Osuna MC, Kilford PJ, Passchier J, Ruffo AD (2008) Toward an improved prediction of human in vivo brain penetration. Xenobiotica 38:1518–1535

Summerfield SG, Zhang Y, Liu H (2016) Examining the uptake of central nervous system drugs and candidates across the blood-brain barrier. J Pharmacol Exp Ther 358:294–305

Sun H (2004) A universal molecular descriptor system for prediction of logP, logS, logBB, and absorption. J Chem Inf Comput Sci 44:748–757

Sun H, Dai H, Shaik N, Elmquist WF, BUNGAY, P. M. (2003) Drug efflux transporters in the CNS. Adv Drug Deliv Rev 55:83–105

Syvanen S, Hammarlund-Udenaes M (2010) Using PET studies of P-gp function to elucidate mechanisms underlying the disposition of drugs. Curr Top Med Chem

Syvanen S, Xie R, Sahin S, Hammarlund-Udenaes M (2006) Pharmacokinetic consequences of active drug efflux at the blood-brain barrier. Pharm Res 23:705–717

Takasato Y, Rapoport SI, Smith QR (1984) An in situ brain perfusion technique to study cerebrovascular transport in the rat. Am J Physiol 247:H484–H493

Tamai I, Tsuji A (2000) Transporter-mediated permeation of drugs across the blood-brain barrier. J Pharm Sci 89:1371–1388

Tega Y, Akanuma S, Kubo Y, Terasaki T, Hosoya K (2013) Blood-to-brain influx transport of nicotine at the rat blood-brain barrier: involvement of a pyrilamine-sensitive organic cation transport process. Neurochem Int 62:173–181

Thiebaut F, Tsuruo T, Hamada H, Gottesman MM, Pastan I, Willingham MC (1989) Immunohistochemical localization in normal tissues of different epitopes in the multidrug transport protein P170: evidence for localization in brain capillaries and crossreactivity of one antibody with a muscle protein. J Histochem Cytochem 37:159–164

Tsuji A, Terasaki T, Takabatake Y, Tenda Y, Tamai I, Yamashima T, Moritani S, Tsuruo T, Yamashita J (1992) P-glycoprotein as the drug efflux pump in primary cultured bovine brain capillary endothelial cells. Life Sci 51:1427–1437

Tunblad K, Jonsson EN, Hammarlund-Udenaes M (2003) Morphine blood-brain barrier transport is influenced by probenecid co-administration. Pharm Res 20:618–623

Tunblad K, Ederoth P, Gardenfors A, Hammarlund-Udenaes M, Nordstrom CH (2004a) Altered brain exposure of morphine in experimental meningitis studied with microdialysis. Acta Anaesthesiol Scand 48:294–301

Tunblad K, Hammarlund-Udenaes M, Jonsson EN (2004b) An integrated model for the analysis of pharmacokinetic data from microdialysis experiments. Pharm Res 21:1698–1707

Tunblad K, Hammarlund-Udenaes M, Jonsson EN (2005) Influence of probenecid on the delivery of morphine-6-glucuronide to the brain. Eur J Pharm Sci 24:49–57

Uchida Y, Ohtsuki S, Kamiie J, Terasaki T (2011a) Blood-brain barrier (BBB) pharmacoproteomics: reconstruction of in vivo brain distribution of 11 P-glycoprotein substrates based on the BBB transporter protein concentration, in vitro intrinsic transport activity, and unbound fraction in plasma and brain in mice. J Pharmacol Exp Ther 339:579–588

Uchida Y, Ohtsuki S, Katsukura Y, Ikeda C, Suzuki T, Kamiie J, Terasaki T (2011b) Quantitative targeted absolute proteomics of human blood-brain barrier transporters and receptors. J Neurochem 117:333–345

Uchida Y, Yagi Y, Takao M, Tano M, Umetsu M, Hirano S, Usui T, Tachikawa M, Terasaki T (2020) Comparison of absolute protein abundances of transporters and receptors among blood-brain barriers at different cerebral regions and the blood-spinal cord barrier in humans and rats. Mol Pharm 17:2006–2020

Wang Y, Welty DF (1996) The simultaneous estimation of the influx and efflux blood-brain barrier permeabilities of gabapentin using a microdialysis-pharmacokinetic approach. Pharm Res 13:398–403

Watson J, Wright S, Lucas A, Clarke KL, Viggers J, Cheetham S, Jeffrey P, Porter R, Read KD (2009) Receptor occupancy and brain free fraction. Drug Metab Dispos 37:753–760

Westerhout J, Ploeger B, Smeets J, Danhof M, de Lange EC (2012) Physiologically based pharmacokinetic modeling to investigate regional brain distribution kinetics in rats. AAPS J 14:543–553

Westerhout J, Smeets J, Danhof M, Lange DE, E. C. (2013) The impact of P-gp functionality on non-steady state relationships between CSF and brain extracellular fluid. J Pharmacokinet Pharmacodyn 40:327–342

Westerhout J, van Den Berg DJ, Hartman R, Danhof M, de Lange EC (2014) Prediction of methotrexate CNS distribution in different species - influence of disease conditions. Eur J Pharm Sci 57:11–24

Xie R, Hammarlund-Udenaes M (1998) Blood-brain barrier equilibration of codeine in rats studied with microdialysis. Pharm Res 15:570–575

Xie R, Bouw MR, Hammarlund-Udenaes M (2000) Modelling of the blood-brain barrier transport of morphine-3-glucuronide studied using microdialysis in the rat: involvement of probenecid-sensitive transport. Br J Pharmacol 131:1784–1792

Yamamoto Y, Valitalo PA, van den Berg DJ, Hartman R, van den Brink W, Wong YC, Huntjens DR, Proost JH, Vermeulen A, Krauwinkel W, Bakshi S, Aranzana-Climent V, Marchand S, Dahyot-Fizelier C, Couet W, Danhof M, van Hasselt JG, de Lange EC (2017) A generic multi-compartmental CNS distribution model structure for 9 drugs allows prediction of human brain target site concentrations. Pharm Res 34:333–351

Young RC, Mitchell RC, Brown TH, Ganellin CR, Griffiths R, Jones M, Rana KK, Saunders D, Smith IR, Sore NE et al (1988) Development of a new physicochemical model for brain penetration and its application to the design of centrally acting H2 receptor histamine antagonists. J Med Chem 31:656–671

Yusof SR, Mohd Uzid M, Teh EH, Hanapi NA, Mohideen M, Mohamad Arshad AS, Mordi MN, Loryan I, Hammarlund-Udenaes M (2019) Rate and extent of mitragynine and 7-hydroxymitragynine blood-brain barrier transport and their intra-brain distribution: the missing link in pharmacodynamic studies. Addict Biol 24:935–945

Zhao R, Kalvass JC, Pollack GM (2009) Assessment of blood-brain barrier permeability using the in situ mouse brain perfusion technique. Pharm Res 26:1657–1664

Chapter 8
In Vitro Models of CNS Barriers

N. Joan Abbott, Siti R. Yusof, Andreas Reichel, Diana E. M. Dolman, and Jane E. Preston

Abstract In vitro models of the blood-brain barrier provide valuable mechanistic information and useful assay systems for drug discovery and delivery. However, it is important to take into account issues including species differences and to what extent features of the in vivo BBB are retained in cell culture. The history and applications of a primary cells, immortalized cell lines, and stem cell-derived BBB models are reviewed, with evaluation of their strengths and weaknesses, in selecting and optimizing a suitable model for particular applications. Understanding of the unstirred water layers gives insights into the "intrinsic permeability" of the membrane, and proteomic and transcriptomic studies have expanded the characterization of the barrier function. Technologies to derive brain endothelium from human stem cells create 3D models of the neurovascular unit, and miniaturize "organ-on-a-chip" flow systems give great promise for the future. All these technologies are crucial to translate BBB research to viable treatment options for patients.

Keywords Blood-brain barrier · In vitro models · Endothelia · Astrocyte · Pericyte · TEER · Primary cells · bEND.3 · hCMEC/D3 · iPSC · IVIVC

N. J. Abbott · D. E. M. Dolman · J. E. Preston (✉)
Institute of Pharmaceutical Science, King's College London, London, UK
e-mail: joan.abbott@kcl.ac.uk; jane.preston@kcl.ac.uk

S. R. Yusof
HICoE Centre for Drug Research, Universiti Sains Malaysia, Penang, Malaysia
e-mail: sryusof@usm.my

A. Reichel
Bayer AG, Pharma R&D, Research Pharmacokinetics, Berlin, Germany
e-mail: andreas.reichel@bayer.com

© American Association of Pharmaceutical Scientists 2022
E. C. M. de Lange et al. (eds.), *Drug Delivery to the Brain*, AAPS Advances in the Pharmaceutical Sciences Series 33, https://doi.org/10.1007/978-3-030-88773-5_8

8.1   Introduction

From the earliest demonstration of restricted exchange between the blood and the brain (Ehrlich 1885) leading to the modern understanding of the blood-CNS barriers, animal experiments and clinical observations have provided valuable information about the physiology and pathology of the barrier layers. However, obtaining mechanistic information from such studies at the cellular and molecular level is complex and time-consuming, and it is often difficult to obtain sufficient spatial and temporal resolution. The situation was dramatically improved by the introduction of in vitro methods (reviewed in Joó 1992).

8.1.1   Background and Early History

The first successful isolation of cerebral microvessels (Siakotos and Rouser 1969; Joó and Karnushina 1973) prepared the way for development of in vitro models of the blood-brain barrier (BBB), which have contributed to current understanding of its physiology, pharmacology, and pathophysiology (reviewed in Joó 1992). Methods have also been developed for in vitro models of the choroid plexus and of the arachnoid epithelium (blood-CSF barrier, BCSFB). However, this proliferation of in vitro models and techniques causes problems for attempts at comparison between models and transferability of results obtained with different models and makes it hard for scientists entering the field to select an optimal model for their particular interests. This chapter gives an overview of the current status of the most widely used in vitro CNS barrier models, with an update on an earlier review (Abbott et al. 2014; Reichel et al. 2003), and offers guidance in model selection for specific applications, including permeability assay for drugs and "new chemical entities" (NCEs).

Isolated brain microvessels were the first model system for studying the BBB in vitro, offering new opportunities to investigate physiological and pathological processes at the cellular, subcellular, and molecular level (Pardridge 1998). A new generation of in vitro models emerged with the first successful isolation of viable brain endothelial cells (BECs), which could be maintained in cell culture (Brendel et al. 1974; Panula et al. 1978; Bowman et al. 1981; see Joo 1992). There followed a number of advances which allowed improved isolation of endothelial cells from brain capillaries with minimal contamination from cells of arterioles and venules, both improving the "barrier phenotype" of the endothelial monolayer and minimizing the contamination by smooth muscle cells, pericytes, and glia (Krämer et al. 2001). The first successful growth of endothelial cells on filters (Fig. 8.1a) allowed measurement of transendothelial permeability, and adopting technology developed for epithelia (Grasset et al. 1984) allowed monitoring of transendothelial electrical resistance (TEER) as a measure of tightness to small ions (Rutten et al. 1987; Hart et al. 1987). Many of the techniques for understanding ways to improve the yield,

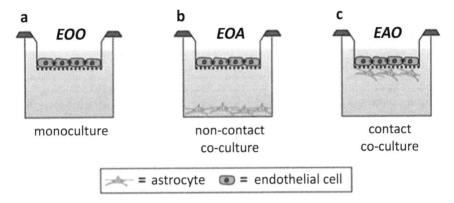

Fig. 8.1 Configurations for brain endothelial cell-astrocyte co-culture models. The three-letter label indicates cell location, in the following order: on the top of filter, on the underside of filter, and in the base of well. Thus panel (**a**) shows a typical monolayer culture with endothelial cells E on top of the filter and no other cell types present; hence EOO, (**b**) shows noncontact co-culture with astrocytes A or mixed glia in the base of the well (EOA) and (**c**) shows "contact" (note that depending on the size of the filter pores and time in co-culture, the glia may or may not actually send fine processes through the filter to contact the endothelial cells) co-culture with astrocytes growing on the underside of the filter, with no cells in the base of the well (EAO). (Redrawn by R Thorne, based on Nakagawa et al. 2009, with permission)

viability, and expression of differentiated phenotype benefited from parallel developments in growing epithelial cells especially Caco-2 (Wilson 1990).

Protocols for isolating and maintaining brain endothelial cells have been described for a large number of species including mouse, rat, cow, sheep, pig, monkey, and human, typically producing confluent cell monolayers after about 9 days in culture (Garberg 1998; Deli et al. 2005). However, with passage, cultured BECs tend to show diminished characteristics of the in vivo BBB, e.g., tight junctional complexity, specific transporters, enzymes, and vesicular transport, reverting toward the "default" non-brain endothelial phenotype characteristic of early BBB development (Daneman et al. 2010b). DeBault and Cancilla (1980) first reported that many of these BBB features can be at least partly reestablished by co-culturing the BECs with astrocytes in arrangements allowing either direct contact or noncontact humoral exchange. Co-cultures with astrocytes followed with improved BBB phenotype (Fig. 8.1b, c) (Dehouck et al. 1990; Rubin et al. 1991; Kasa et al. 1991; see Cecchelli et al. 1999). It should be noted that a complication of contact co-culture (Fig. 8.1c) during transport studies is the continuing presence of the astrocytes. Lipophilic compounds in particular may become trapped in the astrocytes, and many drugs are metabolized by enzymes highly expressed in the astrocyte layer (Dutheil et al. 2010). However, it is argued that the close association of endothelium and astrocytes mimics that in vivo, hence providing a good model for studying flux across the "combined barrier."

During the next stage of development, some of the more sophisticated primary cultured models became so complex to prepare and maintain that they were not practical for routine assays; this was at least partly the motivation for the generation of much simpler models employing immortalized cell lines. The ready availability of molecular biological techniques led to creation of immortalized and transfected CNS barrier cell line models (Reichel et al. 2003 Deli et al. 2005). However, unlike the well-accepted Caco-2 cell line employed for studies of intestinal absorption, or Madin-Darby canine kidney (MDCK) cells used as reliable epithelial models, there were no uniformly satisfactory cell line models for studying the BBB and other CNS barriers in vitro, mainly because of the poor development of tight junctions and hence generation of models on filters that were too leaky for study of transendo-thelial or transepithelial permeation. Most recently, BBB in vitro models have been derived from human stem cells (Lippmann et al. 2012; Cecchelli et al. 2014). These successfully generate very tight monolayers with endothelial-like phenotype (Le Roux et al. 2019), although they also express epithelial-like adhesion molecules and transporters which may complicate interpretation (Lu et al. 2021). Attempts to rein-troduce lost BBB features, or silence non-BBB features in immortalized or stem cell models by means of transfection/transduction, are a promising prospect with mixed success so far (Gericke et al. 2020; Lu et al. 2021) but with great future potential. Rather, molecular techniques allowing more subtle manipulation of cells for experi-mental purposes (e.g., to introduce imaging tracers, Huber et al. 2012) are proving practical and popular.

In vitro systems generally do not express fully the in vivo properties of the BBB, so specific modifications continued to be introduced to study particular aspects of BBB function. As the in vitro systems developed differed with respect to isolation procedures, cell culture conditions and configuration (mono-/co-culture), and the cell type (origin and species), attempts were made in a European Union Concerted Action Programme (1993–1997) to standardize the most popular models to facili-tate comparison of the data collated from different laboratories (Garberg 1998; de Boer and Sutanto 1997). ECVAM, European Centre for Validation of Alternative Methods, also sponsored comparison between different in vitro BBB and epithelial models as CNS drug permeability assay systems (Garberg et al. 2005; see also Avdeef 2011). However, since no consensus emerged as to the "best model," most groups have continued to improve, optimize, and extend the range of applications of the models they selected or developed for historical and practical reasons. Indeed, over the last 15 years, significant progress has been made to the point that scientists new to the field have range of good and practical options (see Table 8.2) and can make informed choices. Some key landmarks in development of in vitro CNS bar-rier models are shown in Table 8.1.

8.1.2   Criteria for Useful In Vitro CNS Barrier Models

The ideal in vitro CNS barrier model would preserve in a reproducible way all the features of the in vivo equivalent and be straightforward and inexpensive to prepare. The features to reproduce would include all aspects of the "physical, transport, and

Table 8.1 Landmarks in development of in vitro BBB models

Landmark advance	References
Isolation of brain microvessels	Siakotos and Rouser (1969), Joó and Karnushina (1973)
Growth of brain endothelial cells in culture	Panula et al. (1978), Bowman et al. (1981)
Growth of brain endothelial cells on filters, TEER measurement (bovine, human)	Rutten et al. (1987), Hart et al. (1987)
Development of immortalized cell line models mouse, rat, bovine, porcine, human	1988 onwards; see text and Table 8.2
Clonal bovine brain endothelial cell culture to avoid contaminating pericytes, co-culture with astrocytes (base of well) TEER >600 $\Omega.cm^2$	Dehouck et al. (1990)
Addition of differentiating factors to medium to improve BBB phenotype (bovine, porcine)	Rubin et al. (1991) (CPT-cAMP), Hoheisel et al. (1998) (hydrocortisone)
"Dynamic" BBB model with intraluminal flow (DIV-BBB)	Stanness et al. (1996, 1997)
Tight porcine brain endothelial cell layer without astrocytes, TEER 700 (up to 1,500) $\Omega.cm^2$	Franke et al. (1999, 2000)
Further option for co-culture—astrocytes on the underside of filter, tighter layer (bovine)	Gaillard and de Boer (2000)
Confocal microscopy method for transport studies in isolated brain microvessels	Miller et al. (2000)
Conditionally immortalized rat, mouse cell lines from the brain and retina endothelium, choroid plexus	Terasaki and Hosoya (2001)
First BBB genomics screen, isolated rat brain microvessels	Li et al. (2001, 2002)
Addition of puromycin to kill contaminating pericytes (rat)	Perrière et al. (2005)
Introduction of hCMEC/D3 human immortalized brain endothelial cell line	Weksler et al. (2005)
Quantitative proteomics of brain endothelium	Kamiie et al. (2008)
Tri-culture models—endothelium, pericytes, astrocytes	Nakagawa et al. (2009)
Transcriptome analysis of purified brain endothelium	Daneman et al. (2010a)
Method to measure and correct for unstirred water layers, paracellular permeability for cells on filters, allowing improved in vitro-in vivo correlation (IVIVC)	Avdeef (2011)
Human stem cell-derived BBB models introduced	Lippmann et al. (2012), Cecchelli et al. (2014)
Microfluidic BBB model prototypes	Booth and Kim (2012), Prabhakarpandian et al. (2013), Griep et al. (2013)
3D spheroid development. Self-assembly of endothelial sphere surrounding a monolayer of pericytes and astrocyte core	Urich et al. (2013)
"Organ-on-a-chip" technology. Miniaturized, 3D microfluidic flow system for co-culture with neurons and astrocytes	Brown et al. (2015) and Adriani et al. (2017)

Table 8.2 The most widely used immortalized cell lines and primary cell models of the BBB

Immortalized cell line	Species, transfection	1st publication	Recent references	Number of citations to March 2021	Number of citations 2019–2020
bEND.3†	Mouse (3)	Williams et al. (1989); Montesano et al. (1990)	Zhang et al. (2021); Wainwright et al. (2020); *García-Salvador et al. (2020)	595	145
hCMEC/D3†	Human (5)	Weksler et al. (2005)	Laksitorini et al. (2020); Fatima et al. (2020); *Veszelka et al. (2018)	434	136
iPSC, BLEC	Human stem cell derived	Lippmann et al. (iPSC) *2014; Cecchelli et al. (BLEC) 2014	Li et al. (2021); Nishihara et al. (2020); *Raut et al. (2021)	111	55
RBE4	Rat (2)	Roux et al. (1994)	Baumann et al. (2021); Sadeghzadeh et al. (2020);	186	11
bEND5†	Mouse (3)	Wagner and Risau (1994)	Devraj et al. (2020)	27	5
TR-iBRB2	Rat retina (4)	Hosoya et al. (2001)	Akanuma et al. (2018)	50	2
cEND†	Mouse (3)	Förster et al. (2005)	Ittner et al. (2020)	18	1
GP8.3	Rat (1)	Greenwood et al. (1996)	Veszelka et al. (2018)	27	1
GPNT†	Rat (1)	Régina et al. (1999)	Regan et al. (2021)	20	1
TR-BBB13	Rat (4)	Hosoya et al. (2000)	Tachikawa et al. (2020)	14	1
MBEC4	Mouse (1)	Shirai et al. (1994)	Mizutani et al. (2016)	44	0
HBMEC/ciβ	Human (4)	Kamiichi et al. (2012)	Masuda et al. (2019)	3	1
Primary cells	**Species**	**1st publication**	**Recent references**		**Number of citations 2019–2020**
	Mouse†	DeBault et al. (1979); Hansson et al. (1980)	Liu et al. (2020); Puscas et al. (2019); *Wuest et al. (2013)		36

(continued)

Table 8.2 (continued)

	Rat†	Bowman et al. (1981)	Luo et al. (2020); Ohshima et al. (2019); *Watson et al. (2013)		31
	Human†	Dorovini-Zis et al. (1991)	Nascimento Conde et al. (2020); Devraj et al. (2020); *Li et al. (2015)		21
	Porcine	Mischeck et al. (1989)	Woods et al. (2020); Di Marco et al. (2019); *Gericke et al (2019)		14
	Bovine	Dorovini-Zis et al. (1984)	Goldeman et al. (2020); Kristensen et al. (2020)		12

Transfection vectors/method: (1) SV40 large T antigen; (2) adenovirus El A gene; (3) Polyon, virus middle T antigen; (4) temperature-sensitive SV40 large T antigen; (5) sequential lentiviral transduction of hTERT and SV40 large T antigen. †Commercially available. *Source data for Fig. 8.3 showing TEER vs. permeability

enzymatic barrier" functions outlined in Chap. 1 and, where relevant, also their immunological features. Replicating the in vivo environment can retain or upregulate BBB features, e.g., by providing luminal medium flow to mimic blood flow shear stress; co-culturing with multiple cells of the neurovascular unit including neurons, pericytes, and astrocytes; and culturing in 3D capillary-like tubes (Booth and Kim 2012; Adriani et al. 2017). However, in the context of this volume, the models should also provide easy to use, readily available and reproducible assay tools for the reliable prediction of the penetration of compounds including drugs into the CNS in relation to both the route and rate of brain entry.

Thus far, no single BBB or BCSFB model fulfills these stringent requirements. However, satisfactory results may be obtained with models expressing the most critical features of the BBB or BCSFB in vivo that are relevant for the particular interest of the study. This means that it is important that users undertake basic model characterization to include the specific BBB feature(s) for which the model is then applied.

8.1.3 The Physical Barrier and Tight Junctions: Monitoring CNS Barrier Tightness In Vitro

The expression of functional tight junctions between the BECs is one of the most critical features due to their consequences for the function of the BBB. In the in vivo BBB, complex and extensive tight junctions contribute significantly to the control over CNS ion and molecular penetration. This is achieved by (1) very severe restriction of the paracellular pathway, (2) limiting flux of permeant molecules to transendothelial pathways, (3) associated expression of specific carrier systems for hydrophilic solutes essential for the brain (e.g., nutrients), and (4) permitting polarized expression of receptors, transporters, and enzymes at either the luminal or abluminal cell surface allowing the BBB to act as a truly dynamic interface between the body periphery (blood) and the central compartment (brain), capable of vectorial transport of certain solutes.

As discussed in Chap. 1, the tight junctions of the choroid plexus epithelium and arachnoid express different claudins than those of brain endothelium and are leakier than those of the BBB; however, their presence in the epithelial barrier layers has a similar effect on the properties of these epithelia, e.g., in polarization of function and regulation of transepithelial transport.

8.1.3.1 Methods to Measure Barrier Permeability and TEER

In vitro models to be used for transendothelial/transepithelial drug permeation studies need to have sufficiently restrictive tight junctions to impede paracellular permeation, mimicking the in vivo situation. Paracellular permeability can be assessed using inert extracellular tracers (Avdeef 2011, 2012). For tighter layers, small tracer molecules can be used, such as radiolabelled sucrose (MW 342, hydrodynamic radius r: 4.6 Å or 0.46 nm) or mannitol (MW 182, r 3.6 Å), or fluorescent markers such as Lucifer yellow (LY; MW 443, r 4.2 Å) or sodium fluorescein (MW 376, r 4.5 Å). For leakier layers, larger tracers used such as inulin, dextrans, and serum albumin are used to characterize paracellular pathways. However, the use of these tracers is labor-intensive and time-consuming, inevitably involving additional assays and analytical delays, and has poor time resolution, and fluorescent tracers may interfere with analysis or permeation of, for example, fluorescent substrates of membrane transporters.

For less invasive monitoring, measurement of transendothelial/epithelial electrical resistance (TEER) is simpler, gives a real-time readout, and has a variety of applications: (1) to monitor the status of the barrier layer, especially for cells grown on opaque filters where visual inspection of confluence is not possible; (2) to determine the culture day on which optimum tightness is reached for experiments; (3) in quality control of cells grown on filters, establishing the baseline permeability of cell monolayers on individual filters to allow exclusion of poor monolayers that fall below a satisfactory threshold tightness; and (4) to follow changes in resistance over time, e.g., to follow the effects of particular growth conditions or a drug or pharmacological agent on barrier integrity and tight junction function.

Before measuring TEER, it is important to know that choice of filter membrane will impact TEER measurement in several ways, regardless of the tightness of cell monolayer. First, smaller filters will give lower TEER due to the edge "effect." Cells cannot make a tight junction at the circumference of the filter where they meet the polystyrene at the edge, and this contributes to paracellular leak, particularly in filters with a small surface area relative to the circumference (e.g., in 24-well formats; Stone et al. 2019). Second, clear Transwell filters have fewer membrane pores per cm^2 compared to translucent filters, regardless of the pore size (e.g., Falcon and Costar "Snapwell" Transwell insert pore densities; clear ~ $1 \times 10^6/cm^2$; translucent ~ $1 \times 10^8/cm^2$), which results in increased TEER not entirely mitigated by subtraction of TEER across a "blank" filter. Third, increasing pore size above 1 μm allows cells to migrate through the pores from the apical side to form a second layer of cells on the underside of the filter, resulting in increased TEER (Wuest et al. 2013).

Two main types of TEER system are used (Fig. 8.2; Benson et al. 2013). The first and simplest is the voltohmmeter (VO) (Fig. 8.2a), where a pair of current and voltage electrodes in "chopstick" array is used. In the second, more recently developed instruments use the method of impedance spectroscopy (IS) (Fig. 8.2b). Permit monitoring of both TEER across cell layers and IS allows continuous analysis over hours to days and also gives information about the electrical capacitance which can reveal additional features of the barrier properties such as cell shape and the degree of cell-substrate adhesion. The earliest IS devices involved growing cells on solid microstructured electrodes, so these systems were not suitable for use in association with drug permeability screening. More recently developed systems permit use of cells grown on porous filters and simultaneous monitoring of multiple filters, e.g., in a 12- or 24-well format.

8.1.3.2 TEER Measurement Based on Ohm's Law: V = IR (Voltage = Current × Resistance)

In the most widely used VO applications (Fig. 8.2a), such as the WPI (World Precision Instruments) "EVOM" system (and Millipore/Millicell equivalent), an AC (alternating current) square wave, here at 12.5 Hz, is passed between voltage electrodes in either side of the cell layer, the resulting current is measured, and the ohmic resistance R is derived. When multiplied by the surface area of the filter membrane, this gives TEER in $\Omega.cm^2$. A few papers in the literature give the units of TEER as "Ω/cm^2" which is incorrect, and this suggests that the authors do not fully understand the theory or methodology. An AC voltage source is preferred over DC as the latter can have polarizing effects on the electrodes or damage the cells. Earlier designs of chopstick electrode pairs (e.g., WPI STX2) were flexible, making it difficult to place the electrodes at a constant distance apart. Recent improvements in design give fixed electrode spacing (e.g., STX100C) and hence better reproducibility. The "Endohm" chamber system with large plate electrodes to fit in the filter cup (above) and the well (below) the cells on the filter, can sample a larger area of membrane including the more uniform central area and can give more reproducible readings (Cohen-Kashi Malina et al. 2009; Helms et al. 2010, 2012; Patabendige et al. 2013a, b); however, the "plunger" action of inserting the upper electrode can disturb the cells, particularly brain endothelial cells, which are much thinner and more fragile than the CNS barrier epithelial cells.

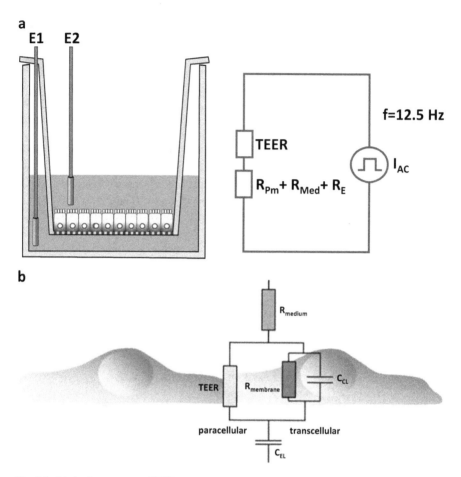

Fig. 8.2 Methods to measure TEER. (**a**) Resistance measurement in voltohmmeter (VO) system using "chopstick" electrodes. The electrodes (E1, E2) in either side of the cell monolayer on the porous filter are used to determine the electrical resistance. The ohmic resistance across the cell layer (TEER), the cell culture medium in the upper and lower compartments (R_{Med}), the membrane of the filter inserts (R_{Pm}), and electrode-medium interface (R_E) all contribute to the total electrical resistance. I_{AC}, alternating square wave current. (**b**) Measurement of TEER and capacitance in impedance spectroscopy (IS) system. Equivalent circuit diagram showing the contribution of the transcellular and paracellular pathways to the total impedance, Z, of the cellular system. *TEER* transendothelial electrical resistance, C_{EL} capacitance of the electrodes, C_{CL} capacitance of the cell layer, R_{medium} ohmic resistance of the medium, and $R_{membrane}$ ohmic resistance of the cell membranes. For tight endothelia and epithelia, TEER is dominated by the transcellular pathway. TEER is determined from the circuit analysis using Z measured at different frequencies of alternating current. (From Benson et al. 2013, with permission)

8.1.3.3 Impedance Spectroscopy Systems

An IS device (Fig. 8.2b) that has proven reliable in the context of BBB and choroid plexus epithelial (CPE) models is the "cellZscope" system (nanoAnalytics), available in different formats capable of accommodating 6, 12, or 24 filter inserts and giving continuous readout of TEER (Benson et al. 2013). The system is computer-controlled, and TEER and capacitance are derived from an electric equivalent circuit model within the software. There is an optimum frequency range appropriate for deriving TEER and capacitance. One drawback of this system is the indirect method for calculating TEER, which relies on the use of the equivalent circuit and certain assumptions about the way current will flow through the system at different frequencies. A nanoAnalytics technical note comparing TEER measured with the cellZscope system and with chopstick electrodes shows good correspondence when the system parameters are set correctly, in particular when impedance at low frequencies is used (f<1kHz; Cacopardo et al. 2019). However, there are some discrepancies in the impedance literature measuring TEER across cultured choroid plexus epithelial (CPE) cells. Wegener et al. (1996, 2000) grew porcine CPE cells on gold film electrodes and recorded TEER 100–150 $\Omega.cm^2$, rising to 210 $\Omega.cm^2$ in the presence of the differentiating agent 250 uM CPT-cAMP, while other studies reported TEER >1500 $\Omega.cm^2$ in serum-free medium (reviewed in Angelow et al. 2004). Using a VO device, Strazielle and Ghersi-Egea (1999) recorded 187 $\Omega.cm^2$ in primary rat CPE, while Baehr et al. (2006) reported 100–150 $\Omega.cm^2$ in pig choroid plexus and commented this would be equivalent to ~600 $\Omega.cm^2$ in an impedance system. Using a VO system with a stable continuous subcultivatable porcine CPE cell line, Schroten et al. (2012) reported TEER >600 $\Omega.cm^2$. In general, the values up to ~600 $\Omega.cm^2$ fit better with evidence for leakier tight junctions in CPE than BBB (Bouldin and Krigman 1975), but it is clear that more "side-by-side" comparisons of VO and IS systems using a particular in vitro model would be helpful to clarify the situation.

8.1.3.4   Relation Between Permeability and TEER

Since 1990 steady progress has been made in the standard (flat filter) in vitro system, to the point where some of the best are able to reach a level of tightness approaching the in vivo BBB (>1000–2000 $\Omega.cm^2$) which is essential for the ionic homeostasis of the brain interstitial fluid required for neuronal function. For assessing solute and drug transport across the BBB, the tighter the monolayer, the better the resolution (dynamic range) for distinguishing between transendothelial permeability and paracellular "leak." Dynamic range can be established experimentally from the permeability ratio between a high and low permeant compounds, e.g., propranolol vs. sucrose. High dynamic range gives better discrimination and rank-ordering of compounds with similar physical chemical properties within a series. However, even models with medium-range tightness are capable of providing adequate resolution for certain applications, particularly if the models show reproducible tightness reflected in consistent values for solute permeability. Recent improvements in understanding, separating, and correcting for the components of in vitro systems that affect cell permeation (unstirred water layer/aqueous boundary layer and porosity of paracellular pathway) also provide ways to determine the true transcellular endothelial permeability, P_c (Avdeef 2011, 2012).

TEER effectively measures the resistance to ion flow ("charge" transfer) across the cell layer, carried by the chief charge carriers in body fluids and physiological saline solutions, Na^+ and Cl^-. The *conductance* "*g*" is the reciprocal of resistance ($g = 1/R$) and is a combined measure of both the ionic permeability of the cell layer and the total number (concentration) of available ions. *Permeability* ($cm.s^{-1}$) is the ability of a solute (including ions) to move through a membrane channel or pore, i.e., is a measure of "mass" transfer and is a property of the membrane or cell layer. Hence conductance is related to permeability.

Traditionally BBB groups have measured either the apparent permeability of the monolayer (P_{app}) or the endothelial permeability P_e, corrected for permeability of the filter. Since TEER is inversely related to permeability, a plot of permeability vs. TEER will give a falling exponential curve. Measuring TEER and permeability of a paracellular marker (e.g., sucrose, mannitol, some small fluorescent tracers) on the same filter with an attached monolayer are useful ways of monitoring the status and reproducibility of the preparation, both for quality control and for experimental studies (Gaillard and de Boer 2000; Lohmann et al. 2002).

Where the monolayer properties including P_{app} are reproducible, a TEER above ~150 $\Omega.cm^2$ may be sufficient to ensure P_{app} for small- to medium-sized molecules is relatively independent of TEER, i.e., giving accurate values for P_{app} (Gaillard and de Boer 2000), or even lower TEER may be suitable to determine P_{app} of macromolecules (Wainwright et al. 2020). Indeed, many groups have adopted a quality threshold of 200–250 $\Omega.cm^2$ for permeability assays of small drug molecules, which is not easy to achieve in some primary and immortalized cell line BBB models (Fig. 8.3). Lohmann et al. (2002) using monocultured porcine brain endothelial cells and measuring TEER with an impedance system found TEER in the range 300–1500 $\Omega.cm^2$; P_e was quite variable at low TEER so they set a threshold of 600 $\Omega.cm^2$ for cells to be used for experiments. It is clear that the appropriate threshold should be selected for the particular cell model, TEER measuring system used, and type of study.

8.1.4 Barrier Features Related to Transporters, Enzymes, Transcytosis, and Immune Responses

As with TEER, reasonable compromises may also be made with other aspects of the BBB. Indeed, it is generally accepted that for a particular application, the model needs only to be characterized for those features which are both relevant and critical for the point of interest. For example, for an in vitro BBB system useful to screen small drug compounds for their CNS penetrability, the model needs to be sufficiently tight and should possess relevant polarized carrier and efflux systems in order to produce useful information. Similarly, for examination of transendothelial or transepithelial permeation of large molecules and nanocarrier systems where vesicular routes may be involved, it is important that the cell system chosen reflects the specialized features of such transport in the polarized in vivo barrier system.

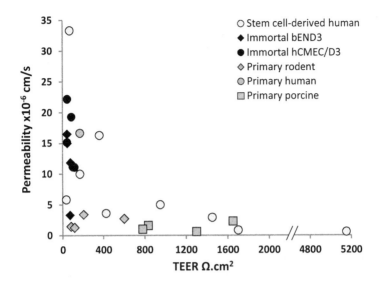

Fig. 8.3 Relationship between small molecule permeability and TEER in BBB in vitro models from multiple laboratories. Permeability is to one of sucrose (MW 342), sodium fluorescein (MW 376), mannitol (MW 182), or LY (MW 457), calculated as $P_{app} \times 10^{-6}$ cm/s or as $P_e \times 10^{-6}$ cm/s (P_e is essentially equivalent to P_{app} when the filter is freely permeable to the compound of interest). Data collated from recent models providing concurrent TEER and permeability data for their model. Cecchelli et al. (2014); García-Salvador (2020); Gericke et al. (2020); Helms et al. (2016); Le Roux et al. (2019); Li et al. (2015); Lippmann et al. (2014); Martins et al. (2016); Matsumoto et al. (2020); Patabendige et al. (2013b); Rand et al. (2021); Raut et al. (2021); Santa-Maria et al. (2021); Seok (2013); Smith et al. (2007); Veszelka et al. (2018); Watson et al. (2013); Wuest et al. (2013); Yamashita (2020); Zolotoff et al. (2020). Selected data sources are also referenced in Table 8.2

However, for many drug permeability projects, the model may not need to show the full complement of immunological responses which will only be necessary in those systems used to study the immune response of the CNS barriers. The existing in vitro model systems have very different levels of characterization and have generally been chosen for utility in a particular area of research interest.

8.2 Current Status: Overview of Current In Vitro BBB Models

Isolated brain capillaries can be used in suspension or fixed onto glass slides. By contrast, all cell-based systems require specific growth surface coatings and cell culture media for growing BECs. Although the cell preparations and culture conditions are all based on the same principle, in order to obtain functional in vitro BBB models, several small but significant differences between the systems, as well as preferences between laboratories, have been introduced (Garberg 1998; de Boer and Sutanto 1997), an ongoing process as shown by recent papers (Thomsen et al. 2017; Veszelka et al. 2018; Stone et al. 2019). In the following sections, current in vitro models of the BBB are briefly surveyed; for greater detail on specific systems, the reader is referred to the corresponding key publications.

8.2.1   Isolated Brain Capillaries

Brain capillaries can be isolated from animal as well as human autopsy brains using mechanical and/or enzymatic procedures (Pardridge 1998; Miller et al. 2000). Typically, the capillary fragments consist of endothelial cells ensheathed by a basement membrane containing pericytes to which remnants of astrocytic foot processes and nerve endings may cling. Often preparations contain small venules and precapillary arterioles and hence smooth muscle cells. Isolated brain capillaries are metabolically active, although a significant loss of ATP and hence activity during the isolation procedure has been reported (Pardridge 1998). As the luminal surface of isolated brain microvessels cannot easily be accessed in vitro, most studies investigate the abluminal properties and function of the BBB. The technique has been used with porcine, rat, and mouse microvessels and has given detailed insights into the cellular and molecular mechanisms regulating transport at the BBB and blood-spinal cord barrier, especially for P-glycoprotein (Pgp) (Miller 2010; Campos et al. 2012).

After isolation, brain microvessels can be stored frozen at -70°C, thereby providing a versatile tool for several applications and a viable source for the cultivation of brain microvessel endothelial cells (Audus et al. 1998). In earlier studies, isolated brain capillaries were used to examine receptor- and adsorptive-mediated endocytosis and solute transporter systems (Pardridge 1998; Fricker 2002). Confocal and live imaging microscopy has expanded possible studies, for example, transendothelial transport of fluorescent substrates for drug transporters (Miller et al. 2000) (Fig. 8.4) and regulation of MRP, BCRP, and PgP function in human, porcine, and rodent capillaries by glutamate (Bauer et al. 2008; Salvamoser et al. 2015; Luna-Munguia et al. 2015).

Isolated capillaries have also proven a valuable resource to characterize BBB mRNA and key transport protein expression, comparing different species and luminal vs. abluminal polarization (Shawahna et al. 2011, Ito et al. 2011a, Uchida et al. 2011b; Hoshi et al. 2013; Kubo et al. 2015) (see Sect. 8.2.6.2). Isolated brain capillaries from both animals and human with a neurological disorders or genetic alteration are contributing to elucidation of the role of the BBB in CNS pathophysiology (Wang et al. 2012; Hartz et al. 2012).

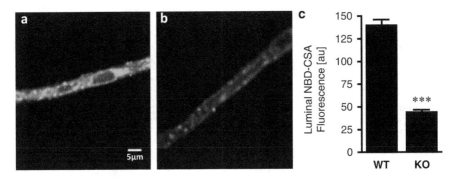

Fig. 8.4 Isolated mouse brain capillaries to study P-glycoprotein function. P-glycoprotein transport function measured as luminal accumulation of fluorescent Pgp-specific substrate NBD-CS. NBD-cyclosporin A in isolated brain capillaries from (**a**) wild-type and (**b**) CF-1 Pgp-deficient mice. (**c**) Image analysis. Methods: Brain capillaries were isolated from wild-type (CF1TM) and CF-1 P-glycoprotein-deficient mice (KO; CF1-Abcblamds). P-glycoprotein transport activity was determined by exposing capillaries to 2 fM NBD-CSA for 1 h and measuring luminal fluorescence using confocal microscopy and image analysis. (Data are mean ± SEM for 7 capillaries for each preparation of 20 mice; shown are arbitrary units (0–255). Statistics: *** P<0.001 (Student t-test). Hartz AMS and Bauer B, unpublished data, with permission)

8.2.2   Primary and Low Passage Brain Endothelial Cells

Apart from isolated brain microvessels, the system next closest to in vivo is primary BECs which are isolated from or grow out of brain capillary fragment: Primary as well as low passage BECs retain many of the endothelial and BBB-specific characteristics of the BBB in vivo; however, these features are often downregulated or even lost with increasing passage if not re-induced. The most successful way to retain BBB features is through co-culture with inducing cells such as astrocytes, pericytes or neurons either in noncontact formation (Fig. 8.1b) or in contact formation (Fig. 8.1c). In addition, most protocols modify the culture medium once cells reach confluence, withdrawing serum to reduce proliferation and encourage cell-cell contact, including cAMP to encourage basement membrane formation and glucocorticoids to improve tight junction protein expression (Hoheisel et al. 1998; Thomsen et al. 2017).

Rat and Mouse Models Due to the much higher yield of BECs from bovine and porcine brains compared to rat brains (up to 200 million cells per porcine brain, compared to 1–2 million cells per rat brain), the former species currently represent the most popular source for in vitro BBB models both in academia and industry. However, primary cultured rat and mouse systems continue to be useful for investigation of pharmacology and transport, in studies where specific antibodies for larger species are lacking, and for comparison with standard in vivo rodent (rat, mouse) models used for PKPD analysis. The increasing availability of high-quality BECs from commercial sources has also added to the consistency and continued use of these models.

The use of primary rodent models for transendothelial permeability measurements was until recently limited by the relatively leaky monolayers generated (TEER 150–200 Ω.cm^2 due to the small flaws caused by contaminating pericytes, which are less of a problem in the bovine and porcine systems) (Patabendige et al. 2013a, b). However, Watson et al. (2013) showed that improvements in methods through generation of purer rat primary cultures, co-culture with mixed glia from the same species ("syngenic" culture), and short trypsinization times can give higher TEER of up to 600 Ω.cm^2. Inclusion of puromycin to eliminate contaminating pericytes from the monolayer is a relatively simpler procedure to generate consistent monolayers with suitable TEER ~200 Ω.cm^2 and low paracellular permeability $P_e \sim 3 \times 10^{-6}$ cm/s (Stone et al. 2019).

Recent studies with primary rat and mouse BECs have focused on in vitro pathological models to mirror in vivo rodent studies, for example, the effects of stroke (Venkat et al. 2021; Kong et al. 2021), inflammation and T cell migration (Hamminger et al. 2021), and demyelination syndrome (Scalisi et al. 2021).

Bovine Models Bovine BEC cultures are widely used, but differences between the procedures have developed historically in different BBB groups. Pioneered by Bowman et al. (1983) and later modified by Audus and Borchardt (1986) in the USA, bovine BECs are typically isolated by a combination of mechanical and enzymatic protocols and originally grown in monoculture (Miller et al. 1992) with early studies showing TEER in the range 160–200 Ω.cm^2 and sucrose permeability $10 - 20 \times 10^{-6}$ cm/s (Raub et al. 1992; Shah et al. 2012).

In Europe, several modifications to the protocol have greatly enhanced the model's BBB properties. The group of Cecchelli and coworkers (Dehouck et al. 1990; Cecchelli et al. 1999) pioneered the omission of enzymatic steps in the bovine BEC isolation, using instead micro-trypsinization and subculturing of endothelial cell islands (clones) that grow out of brain capillaries selectively attached to a defined extracellular matrix. The most recent protocols use BECs after a single passage, supplemented with dexamethasone and cAMP plus phosphodiesterase inhibitor (Eigenmann et al. 2016; Kristensen et al. 2020; Goldeman et al. 2020). BECs can reach TEERs of 600 Ω.cm^2 in monoculture, increasing to 1000–2000 Ω.cm^2 in contact co-culture with rat astrocytes (Fig. 8.1c). Co-culture also aids in reducing paracellular permeability (mannitol $P_{app} < 1 \times 10^{-6}$ cm/s; Tornabene et al. 2019) and in halting or countering the loss of specific BBB markers (Goldeman et al. 2020).

The model has been successfully used to study BBB transport (e.g., Wallace et al. 2011) and rank-order compounds according to their BBB permeability (Lundquist et al. 2002; Eigenmann et al. 2016); higher throughput variants of the model have been introduced for drug screening and toxicity testing (Culot et al. 2008; Vandenhaute et al. 2012), and it is one of the few models which have proven suitable for the study of receptor-mediated transcytosis (Candela et al. 2010).

Porcine Models Galla and coworkers (Hoheisel et al. 1998; Franke et al. 1999, 2000) developed a model based on porcine BECs (PBEC model) cultured without serum or astrocytic factors but in the presence of the tight junction protein differentiating agent hydrocortisone. In their hands, this model gives among the highest TEER values measured in vitro thus far (400–1500 Ω.cm^2 with VO monitoring, or higher in IS systems, with sucrose permeability down to $1^{-4} \times 10^{-6}$ cm/s). The model has been used as a screening tool for CNS penetration of small drugs (Lohmann et al. 2002) and nanocarriers (Qiao et al. 2012) and for a number of mechanistic studies of BBB transporters and cell-cell interaction in the neurovascular unit (NVU). Using this model, Cohen-Kashi Malina et al. (2009, 2012) showed an increased TEER of the PBECs, from 415 Ω.cm^2 in monoculture to 1112 Ω.cm^2 in contact co-culture (Fig. 8.1c). The model was sufficiently tight and polarized to examine the role of endothelial and glial cells in glutamate transport from the brain to blood (Cohen-Kashi Malina et al. 2012).

A different PBEC method originally developed by Louise Morgan and the group of Rubin (Eisai Laboratories, London), based on a method for bovine BECs (Rubin et al. 1991), was reintroduced by Skinner et al. (2009) using serum-free medium and supplements hydrocortisone and cAMP plus phosphodiesterase inhibitor. Further optimization including a growth phase with plasma-derived serum rather than fetal serum and noncontact co-culture with rat astrocytes (Patabendige et al. 2013a, b; Nielsen et al. 2017) gave maximum TEER of 2400 Ω.cm^2; permeability to LY was <1 $\times$ 10^{-6} cm sec^{-1} at average TEER 1249 Ω.cm^2 (Nielsen et al. 2017). The P_{app} is uniformly low in BECs with TEER >500 Ω.cm^2, so for this model a threshold is set to 500 Ω.cm^2 to be used for experiments. Interestingly, co-culture with porcine pericytes reduced TEER compared to culture with porcine astrocytes (Thomsen et al. 2015), which underlines the complexity of cell-cell interactions.

The model shows good functional and polarized expression of transport proteins (Patabendige et al. 2013a; Kubo et al. 2015), tight junctions, enzymes, and receptors (see Nielsen et al. 2017). The model has been used to study receptor-mediated transcytosis (RMT) for interleukin-1 (Skinner et al. 2009) and LRP-1 and RAGE substrates (Wainwright et al. 2020) and more recently for studies of nanoparticle delivery of monoclonal antibodies to the brain (Woods et al. 2020) and effect of inhibition of Pgp, MRP5, and BCRP on amyloid clearance from brain to blood (Shubbar and Penny 2020).

Human Models The limited availability of human brain tissue makes primary human BECs a precious tool for the study of the human BBB at the cellular and molecular level (Dorovini-Zis et al. 1991). The source material usually derives either from autopsies or biopsies (e.g., temporal lobectomy of epilepsy patients), and the most popular applications are studies related to the BBB in CNS diseases. Commercial human brain endothelial cells are increasingly available, although batch-batch variation may pose problems. Human BEC monolayers are fragile in culture, contributing to low TEER values of 120–180 Ω.cm^2 (Mukhtar and Pomerantz 2000; Giri et al. 2002). Co-culture with combinations of NVU cells including pericytes and neurons does not necessarily increase TEER but does speed up response to dexamethasone supplement and increases sensitivity to oxygen-glucose deprivation (Stone et al. 2019).

A great advantage of primary human models is the ability to generate cultures from tissue originating from patient pathology samples (Giri et al. 2002) and to mimic pathology and interrogate cell signaling in a human model including SARS-CoV-2 infection (Larochelle et al. 2012; Liu and Dorovini-Zis 2012; Sugimoto et al. 2020; Nascimento Conde et al. 2020). In addition, these models have also been used to study drug transport (Riganti et al. 2013) and nanoparticle permeation (Gil et al. 2012).

8.2.3   Immortalized Brain Endothelial Cell Lines

Primary cultured BECs have been successfully used as in vitro model of the BBB; however, their widespread and routine use has been restricted mainly by the time-consuming and often difficult preparation of the system which limits the continuous and homogeneous supply of biological assay material. Therefore, attempts have been made by several laboratories to immortalize primary BECs, thereby avoiding the lengthy process of cell isolation.

The first generation of immortalized CNS barrier cell lines (first publication 1988–2000) involved introducing genes such as polyomavirus T antigen (bEND.3 cells), adenovirus ETA gene (RBE4), or SV40 large T antigen (many) (Table 8.2). Subsequently, conditionally immortalized cell lines have been established by using transgenic mice and rats harboring the temperature-sensitive SV40 large T antigen gene (tsA58 T antigen gene) (Terasaki and Hosoya 2001; Terasaki et al. 2003). The advantage is that only small amounts of tissue are needed to establish a cell line, and the cell lines generated show better maintenance of in vivo functions proliferate well and reach confluence in 3–5 days. The gene is stably expressed in all tissues, and cell cultures can easily be immortalized by activating the gene at 33 °C (Ribeiro et al. 2010). The technique has been used to generate both brain endothelial and choroid plexus cell lines.

Of immortalized brain endothelial cell lines introduced in 1988–2000, several have proven reliable and popular and are still in use (Table 8.2). The models have been characterized to varying degrees, but all shared a common weakness, i.e., insufficient tightness when grown as a cell monolayer on a porous membrane. Innovations to improve tightness have focused on the same interventions used for primary cells: co-culture with inducing cells and addition of glucocorticoids such as hydrocortisone or dexamethasone (see Sect. 8.2.2). The situation more recently has significantly improved, as detailed further below, and the most recent addition to BBB models, human stem cell-derived endothelial-like cells, has enormous promise to combine human cells with a tight monolayer and stability through multiple passages.

Bovine and Porcine Cell Lines As good primary cultured bovine and porcine BECs are now routinely produced in several groups, the use of immortalized bovine and porcine models showing more restricted features (Reichel et al. 2003) is less widespread.

Rat and Mouse Cell Lines One of the first, and still most widely used, immortalized in vitro models is the mouse bEND.3 cell line derived originally from BALB/c mouse brain endothelia infected with the polyomavirus middle T oncogene (Williams et al. 1989; Montesano et al. 1990). The ease of availability and use, consistent generation of monolayers, and ability to compare with mouse WT and KO in vivo studies make this a popular choice. The bEND.3 cell line expresses the relevant tight junctions and transport proteins but does not generate high TEER, possibly because of inherent proteolytic activity (Montesano et al. 1990). TEER is typically 40–50 $\Omega.cm^2$ in monoculture, increasing to 70–80 $\Omega.cm^2$ in co-culture with astrocytes, and permeability to LY or sodium fluorescein ranges from 3 to 15×10^{-6} cm/s (Seok et al. 2013; Martins et al. 2016; García-Salvador et al. 2020). Attempts to improve culture systems using puromycin, for example, have not yielded success (Puscas et al. 2019).

Most recently, bEND.3 cells have been used for the study of brain delivery of large molecules or nanoparticles (Zhang et al. 2021; Wainwright et al. 2020), drug screening in comparison with in vivo mouse data (Puscas et al. 2019), and stroke models (Baumann et al. 2021).

In an interesting breakthrough, Förster et al. (2005) returned to the earlier cell transduction technology used for bEND3 and bEND5 to generate mouse cEND cell: which uniquely among immortalized brain endothelial cell lines can produce tight monolayers, with reported TEER up to >800 $\Omega.cm^2$. The details of the immortalization method have been published, and the cells have been used for studies on the involvement of glucocorticoids on tight junction regulation and on hypoxia and multiple sclerosis (Burek et al. 2012).

For rat, the RBE4 and GP8/GPNT cell lines are still in use, although less frequent in the last 2 years (Table 8.2), and have proven useful for a broad array of topics ranging from mechanistic transport studies to receptor-mediated modulation and inflammatory responses. Many of the currently available immortalized rat and mouse cell lines, especially conditionally immortalized lines, have been generated in Japan and are widely used, often in parallel in vivo/in vitro studies, especially for identification and examination of carrier-mediated transport (Ito et al. 2011b, c; Lee et al. 2012; Tega et al. 2013).

Human Cell Lines Immortalization of human BECs has proven much more difficult than for BECs of other species, but several human cell line models are reported (Reichel et al. 2003; Deli et al. 2005) suitable for examination of the physiology, pharmacology, and pathology of the human BBB in vitro and as a screening tool for CNS penetration.

The most widely used is the hCMEC/D3 cell line (Table 8.2) introduced by Weksler et al. (2005, 2013), building on the author's prior experience developing rat RBE4, GP8.3, and GPNT cell lines. hCMEC/D3 cells are contact-inhibited, can reach confluence in as quickly as 48 h, and retain features for up to ~30 passages (Weksler et al. 2005; Schrade et al. 2012) making them a robust laboratory tool. Like most cell line models, TEER is typically low, around 35–50 $\Omega.cm^2$, however, modifications including astrocyte co-culture, addition of simvastatin, hydrocortisone, or lithium activating the Wnt system, can elevate TEER to 90–200 $\Omega.cm^2$ with permeability to LY or sodium fluorescein between 10 and 20 × 10^{-6} cm/s (Förster et al. 2008; Schrade et al. 2012; Veszelka et al. 2018; Gericke et al. 2020; García-Salvador et al. 2020).

hCMEC/D3 has rapidly been adopted as an immortalized model of choice for studies where TEER is not a major issue, e.g., macromolecule and nanoparticle uptake and transport (Markoutsa et al. 2011; Yamaguchi et al. 2020), pathology, and cell signaling (Ito et al. 2017; Alam et al. 2020). A review by Weksler et al. (2013) summarizes many of the useful applications of the model and gives a balanced view of its strengths and weaknesses.

Human Stem Cell Derived Developing a stable, human BBB model is essential to fully investigate CNS drug delivery and pathophysiological targets that are translatable to patients. Since primary and immortalized human BECs have limitations as discussed above, efforts have been made to develop a suitable BBB model using human stem cells. These have the advantage of a human genotype and so the added potential for generating cells from patients to study diseases (Raut et al. 2021) or personalized drug interactions. A disadvantage is the cells do not originate from brain endothelium, but rather they are pluripotent or hematopoietic stem cells in origin, which must be differentiated and induced to express BBB features and suppress non-BBB features.

The induced pluripotent stem cell model (iPSC) was developed by the Shusta group involving initial differentiation of stem cells into endothelial cells and co-culture with neural cells providing Wnt/β-catenin signaling and then purification and further maturation of the endothelial cells to develop a full BBB-like phenotype (Lippmann et al. 2012, 2013, 2014). Multiple laboratories are applying these methods, and the cells reliably generate TEERs of 1000–2000 $\Omega.cm^2$ with the highest reported over 5000 $\Omega.cm^2$ and low paracellular permeability of 0.5–2 × 10^{-6} cm/s (Lippmann et al. 2014; Le Roux et al. 2019; Raut et al. 2021).

Brain-like endothelial cells (BLEC) derive from CD34+ hematopoietic stem cells isolated from cord blood which makes them relatively easily harvested and available (Cecchelli, et al. 2014). Cells are differentiated and then co-cultured with pericytes to induce BBB features. The TEERs are superior to immortalized or primary human BBB models but are variable between laboratories ranging from 40 to 360 $\Omega.cm^2$ with permeability of 5–15 × 10^{-6} cm/s (to LY or sodium fluorescein; Rand et al. 2021; Santa-Maria et al. 2021). Applying shear stress using a flow system (see Sect. 8.2.4) improves TEER to >400 $\Omega.cm^2$ (Santa-Maria et al. 2021), making these cells suitable for the more complex methods described below.

There has been rapid progress on signaling and transcription factors to differentiate stem cells into a mature BBB phenotype (Lu et al. 2021; Roudnicky et al. 2020a), and as a consequence, this may benefit other BBB models. For example, factors identified that increase expression of claudin-5 in hPSCs were applied to primary human cells to improve BBB phenotype (Roudnicky et al. 2020b); changes in gene expression following pericyte co-culture with BLECs have identified molecular processes in BBB formation (Heymans et al. 2020).

An area for future work is to characterize these models, looking for "non-BBB" features and ensuring they are downregulated so that erroneous interpretations about BBB function are not made. For example, the iPSC model expresses some epithelial adhesion proteins and transporters, but these can be downregulated with endothelial transcription factors (Lu et al. 2021). The reproducibility and transferability of these models will also be critical features in the future, but these models show enormous promise in taking the field forward.

Non-BBB Cell Lines It is generally difficult to make BEC cell lines switch from the exponential growth phase after cell seeding to a more static phase of cell differentiation after the cells have reached confluence. Therefore, most immortalized cell lines are less applicable for studies requiring a tight and stable in vitro barrier, but they have proven useful for mechanistic and biochemical studies requiring large amounts of biological material as described above. However, the insufficient tightness of immortalized BEC lines renders them unsuitable for use in simple BBB permeability screens. Therefore, some groups have turned to other cell lines which, although of non-brain origin, either express sufficient brain endothelial features for functional and permeation studies such as ECV304/C6 (Hurst and Fritz, 1996; Neuhaus et al. 2009; Wang et al. 2011) or prove on validation to be useful predictors of passive and Pgp-mediated CNS penetrability of compounds, such as MDCK cells engineered to overexpress human Pgp (MDCK-MDR1) and Caco-2 cells (Summerfield et al. 2007; Hellinger et al. 2012).

8.2.4 Complex BBB Models: 3D Models, Dynamic Flow, and Microfluidics

It would be expected for in vitro models retaining complex features of the in vivo NVU that they would be more successful in showing a functional BBB phenotype. In cell culture models, the inclusion of pericytes can be beneficial, depending on the differentiation state of the pericytes (Thanabalasundaram et al. 2011). Not all in vitro models are reported to respond positively to pericytes (co-culture does not improve TEERs in primary human or porcine models; Stone et al. 2019; Thomsen et al. 2015), but many examples of barrier-inducing and stabilizing effects of pericytes on BBB function have been demonstrated (Fig. 8.5) (e.g., Nakagawa et al. 2009; Vandenhaute et al. 2011), and a practical commercial rat tri-culture model is available. A more complex model development is the "spheroid" or brain organoid, which is a sphere of endothelial cells surrounding a monolayer of pericytes and astrocyte core. These 3D cell systems spontaneously self-organize in a hanging droplet culture plate or in a well with ultralow attachment (Urich et al. 2013;

Kumarasamy and Sosnik 2021) and have been used to study nanoparticle uptake, for example. While these recapitulate the cell-cell interactions, it is difficult to determine detailed BBB function.

Another example of more closely mimicking the in vivo environment is growing BECs in porous tubes with luminal flow and external astrocytes to aid barrier induction (Stanness et al. 1996, 1997; Janigro et al. 1999). This "dynamic in vitro" (DIV) BBB model (Fig. 8.6) proved an important innovation and convincingly demonstrates not only improved junctional tightness but also other BBB features reflecting the differentiating effects of flow. There is growing interest in combining 3D, tri-culture, and flow in a single miniaturized "microfluidic" platform capable of mimicking more closely the in vivo conditions, but with less cell volume and need for reagents. Pioneering studies established the feasibility of the method and scope for miniaturization (Booth and Kim 2012; Griep et al. 2013; Prabhakarpandian et al. 2013), with BBB cell line models RBE4, bEND3, and hCMEC/D3. The positive effects of flow in DIV and microfluidic systems can be demonstrated in primary cells, immortalized cells, and more recently stem cell-derived models, for example, TEER is improved up to 500 $\Omega.cm^2$ in primary human cells and BLEC (Cucullo et al. 2011; Santa-Maria et al. 2021), 1000 $\Omega.cm^2$ in hCMEC/D3 (Partyka et al. 2017), and 4000 $\Omega.cm^2$ in iPSCs (Grifno et al. 2019). However, the complexity of the geometry (multiple hollow fibers) in this model and the assumptions made in calculating TEER from the current measured make it difficult to compare TEER values with those from flat filter configurations.

Despite the undoubted improvement in BBB characteristics with these systems, these models are more difficult to set up and maintain than standard mono- or co-cultured models (Fig. 8.1) and have not yet been fully assessed for the whole range of BBB features including vesicular transport (Naik and Cucullo 2012; Abbott 2013). There is also wide variation between groups in the BEC cells used and the species and types of co-cultured cells; a recent review by Bhalerao et al. (2020) gives an excellent overview of the challenges in comparing between groups. Many questions could be addressed in such systems, including the contribution of differential flow rates/shear stress to the observed heterogeneity of endothelial cytoarchitecture and function in different segments of the vasculature (Ge et al. 2005;

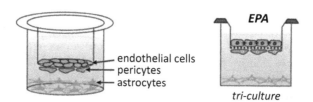

Fig. 8.5 Configuration for tricellular BBB co-culture model, reflecting the organization of the neurovascular unit (NVU). As for Fig. 8.1b, but here with addition of pericytes. Endothelial cells E on the top of the filter, pericytes P on the underside of the filter, and astrocytes A in the base of the well (EPA arrangement). (Redrawn by R Thorne, based on Nakagawa et al. 2009, with permission)

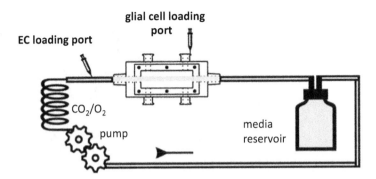

Fig. 8.6 Dynamic in vitro BBB model, DIV-BBB. Diagram showing cartridge containing replaceable bundle of hollow porous polypropylene fibers (capillary tubes) (yellow) suspended in the chamber and in continuity with a medium source through a flow path consisting of gas-permeable silicon tubing. A servo-controlled variable-speed pulsatile pump generates flow from the medium source through the capillary tube bundle and back. The circulatory pathway feeds both endothelial cells (EC) growing on the luminal surface of the capillary tubes and glia growing abluminally on their outer walls. The model has been used to assess the effects of flow on endothelial physiology, pathophysiology, and leukocyte trafficking. (From Cucullo et al. 2002, with permission)

Macdonald et al. 2010; Saubaméa et al. 2012; Paul et al. 2013; *cf* Ballermann et al. 1998). Given the complexity of the microfluidics chambers, these are not likely to be suitable for high-throughput permeability assays at least in the short term, but meanwhile the generation of detailed mechanistic information is likely to be the most valuable output. An important advantage will be the ability to test barrier cells from different species and with different pathologies, under equivalent conditions.

8.2.5   Application of In Vitro Models for BBB Drug Permeability Assay

A realistic in vitro assay system for screening and optimizing NCEs should combine as many features as possible of the in vitro BBB yet be suitable for medium to-high-throughput screening. Most pharmaceutical/biotech companies already have screens for intestinal permeability (generally Caco-2) and, for "Pgp-liability," often MDCK-MDR1 cells (Summerfield et al. 2007), so a convenient and pragmatic system is to expect early-stage screening on such models and later refinement in a more "brain-like" system. A possible "screening cascade" involving early in silico modeling, then non-brain epithelial models, and finally CNS barrier models may be practical (Abbott 2004). However, given the very different morphologies of endothelial cells and the epithelial cells Caco-2 and MDCK, especially in cell thickness, luminal membrane microstructure, glycocalyx composition, junctional structure, and organelle content (Fig. 8.7) together with physiological differences in transcytosis mechanisms, transporter, and enzyme function, caution still needs to be applied in such a sequential screen (see also Lohmann et al. 2002).

Rat BBB   **VB-Caco-2**   **MDCK-MDR1**

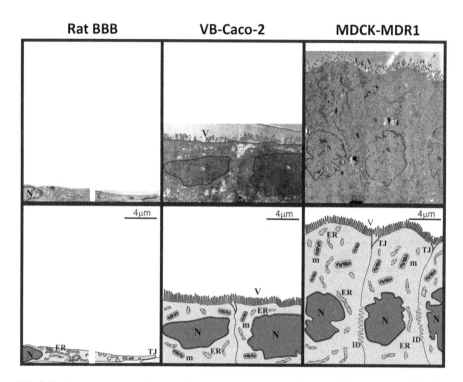

Fig. 8.7 Electron micrographs of cell cultured rat brain endothelium, VB-Caco-2 and MDCE-MDR1 cell cytoarchitecture, with drawings below. VB-Caco-2 cells were created by growing Caco-2 cells in 10 nM vinblastine (VB, Pgp substrate) for at least six passages to elevate P-g expression. (*ER* endoplasmic reticulum, *ID* interdigitations, *m* mitochondrion, *N* nucleus, *TJ* intercellular tight junctions, *V* microvilli. From Hellinger et al. 2012, with permission)

Most studies for CNS-specific permeability screening have focused on the BBB as the largest surface area blood-CNS interface, closest to neurons, but there is growing awareness of the need for assay systems of the choroid plexus reflecting especially the transport and enzymatic importance of this barrier (Strazielle and Ghersi-Egea 2013). A medium- to high-throughput BBB system using bovine endothelial cells exposed to glial-conditioned medium is available (Culot et al. 2008), and primary cultured porcine cells are also suitable either as monocultures or cocultures with astrocytes (Patabendige et al. 2013a). Hellinger et al. (2012) compared a rat tri-culture model (TEER ~200 $\Omega.cm^2$) with Caco-2 and MDCK-MDR1 cells in screening ten compounds (selected for predominantly passive permeation, efflux transport, or both) and concluded that for passive permeability and Pgp-liability, the epithelial layers gave better resolution, while the BBB model would have advantage in reflecting other in vivo BBB transporters. However, with a more limited drug set, Mabondzo et al. (2010) using human primary BECs concluded they were better than Caco-2 cells at correlating with in vivo human PET ligand uptake (detailed below), which may reflect important differences in species, drug set, or the culture protocols of the in vitro systems used.

8.2.6 In Vitro-In Vivo Correlations (IVIVC)

Since the earliest in vitro BBB permeability assays (e.g., Dehouck et al. 1990; Cecchelli et al. 1999), there has been interest in comparing the performance of the in vitro models against permeability data generated in vivo, typically by constructing an in vitro vs. rodent in vivo permeability plots and determining the correlation (in vitro-in vivo correlation, IVIVC). Rodent in vivo data used have been either measurements of Brain Uptake Index (BUI) or permeability data derived from in situ brain perfusion, the K_{in} (unidirectional influx coefficient), or the derived P_c (transcellular permeability). However, the relatively leaky tight junctions in vitro (high paracellular permeability) and the presence of unstirred water layers (or aqueous boundary layers, ABL; Youdim et al. 2003) weaken the correlation (Avdeef 2011).

Despite these limitations, reasonable correlations can be generated, especially for primary cells. For example, IVIVC using primary mouse BECs vs. in vivo mouse brain-to-blood ratio gave better correlation than bEnd.3 cell line vs. in vivo ($r^2 = 0.765$ primary cells; $r^2 = 0.019$ bEND.3; Puscas et al. 2019). The increasing availability of agents suitable for human positron emission tomography (PET) imaging now allows comparison of in vitro human BBB models with human brain uptake. For example, Mabondzo et al. (2010) compared transport of seven drugs across primary human BECs co-cultured with syngenic astrocytes to human brain PET-MRI data and showed excellent correlation ($r^2 = 0.90$) and being better than Caco-2 vs. human brain PET ($r^2 = 0.17$). Le Roux et al. (2019) similarly showed good human IVIVC correlation for eight PET ligands, using iPSC-derived BECs ($r^2 = 0.83$). It will clearly be important to extend these studies to a wider drug library and compare other in vitro BBB models to human PET data.

8.2.6.1 Unstirred Water Layer, Paracellular Permeability, and Intrinsic Permeability Calculation

Building on quantitative biophysical models validated in epithelia and applying his software pCEL-X, Avdeef (2011) used literature values (to 2008) of permeability from several different in vitro BBB and epithelial models and deconvoluted the apparent permeability P_e of the endothelial barrier into its three components: P_{ABL}, P_C, and P_{para}, (ABL, transcellular and paracellular permeabilities, respectively). Finally, P_0, the intrinsic (charge-corrected) permeability, was calculated from P_C by incorporating the pK_a value(s) of the molecule. Figure 8.8 shows the log-log IVIVC of P_0 data from monocultured porcine brain endothelium vs. P_0 data from rodent in situ brain perfusion studies. The correlation coefficient r^2 for the IVIVC (0.58) was greater than that for the uncorrected in vitro data, P_e vs. P_C in situ (0.33). The porcine BBB model also performed better than bovine, rodent, and human models in this study. By applying the method to permeability data from the tightest current

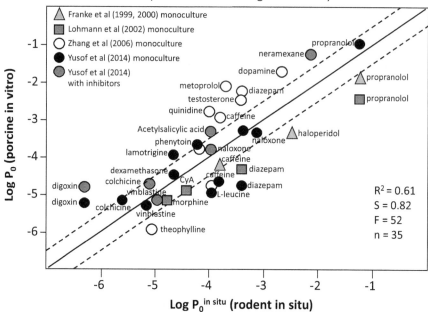

Fig. 8.8 In vitro-in vivo correlation analysis (IVIVC). Intrinsic transcellular permeability (P_0) data were compared with in situ brain perfusion data from rodent. P_{app} data were corrected for aqueous boundary layer (ABL) permeability, paracellular permeability, filter restriction, and possible uptake of the charged species. In situ brain perfusion data from rodent were collected from the literature and analyzed using the pKa FLUX method to derive P_0 (Dagenais et al. 2009; Suzuki et al. 2010; Avdeef 2011, 2012). The predictions for in situ BBB permeation of acetylsalicylic acid and neramexane (calculated from pCEL-X) and dexamethasone and metoprolol (Caco-2 values) were used in the analysis (underlined). The solid line is the linear regression with r^2 value of 0.61. The dashed line is the reference "line of identity." (Modified from Yusof et al. 2014 with permission)

in vitro BBB models, the correlations are expected to improve. The method helps to identify the most reliable in vitro models for predicting in vivo permeability and to correct the data obtained from leakier models.

8.2.6.2 Transcriptomics, Proteomics, and PKPD Modeling

Transcriptome examination and quantitative proteomics of freshly isolated brain capillaries and purified brain endothelial cells have helped determine the degree to which in vitro models reflect the in vivo condition and how closely models from other species resemble the phenotype of the human CNS barriers (Kamiie et al. 2008; Daneman et al. 2010a; Ohtsuki et al. 2011; Hoshi et al. 2013; Al Feteisi et al. 2018; Chaves et al. 2020). These techniques are also revealing changes in BEC protein and mRNA expression related to disease models such as stroke and seizures (Tornabene et al. 2019; Munji et al. 2019; Gerhartl et al. 2020) and shedding light on the role of miRNAs (Kalari et al. 2016). In the future, it should be possible to combine information from in vivo and in vitro studies (Ito et al. 2011b, c) with quantitative proteomics (Uchida et al. 2011a, b, 2013; Kubo et al. 2015) to generate data for PKPD and "physiologically based pharmacokinetic" (PBPK) modeling and for prediction of human CNS free drug concentrations (Shawahna et al. 2013), based on data including information generated in in vitro models from different species (Ball et al. 2012). The ultimate aim will be to permit reliable in vitro-in vivo extrapolation (IVIVE) to human brain (Ball et al. 2013). BBB-specific transcriptome databases, such as the BBBomics hub http://bioinformaticstools.mayo.edu/bbbomics/ (Kalari et al. 2016) and the European Brain Barriers Training Network BBBhub http://bbbhub.unibe.ch/ (Heymans et al. 2020), will be valuable resources in this endeavor.

8.2.7   How to Select an Appropriate In Vitro BBB Model

It is clear that a wide range of models are available for studies of the BBB relevant to normal physiology and pathological situations and to test and optimize CNS delivery of appropriate therapies. Careful selection with a variety of controls in place can give valuable information about the role of the BBB in pathology and the rate and extent of entry of therapeutics into the CNS. These models are helping to refine a variety of formulations and constructs to improve their value in a range of diseases.

For scientists starting a new BBB project without prior experience, collaboration with an established group or groups is recommended, including adopting their well-characterized cell or cell line models if these are suitable for the application planned (Table 8.3).

8.2.8   Epithelial CNS Barriers

8.2.8.1   Choroid Plexus Epithelial (CPE) Cells

The choroid plexus is relatively straightforward to isolate with cell viability maintained for several hours, permitting studies of uptake and efflux, but without defined polarity (Gibbs and Thomas 2002). When polarity of transport is important, perfusion and isolation of sheep choroid plexus permits studies of vectorial transport across the epithelium (Preston et al. 1989). Primary culture models of rodent, porcine, and human CPE have been developed (see Baehr et al. 2006), but the most readily available human material is from fetal material or CP papilloma, which may not accurately reflect normal function (Redzic 2013). Resistances of 100–600 $\Omega.cm^2$ have been

Table 8.3 How to select an appropriate in vitro BBB model (see text)

Property of interest	Recommended cell model(s)	Check
Transendothelial permeability of small compounds (<500 MW), detecting both passive and transporter-mediated flux	Primary cultured cells: —Without astrocytes: porcine —With astrocytes: bovine, porcine —With astrocytes and pericytes: bovine, rat, porcine	Check TEER; aim for high TEER and high dynamic range, giving better discrimination and rank-ordering within a drug series
ABC efflux transporters	Primary cultured system showing in vivo pattern, polarity/localization (bovine, porcine)	Check relative expression compared to human BBB, may permit prediction of PK in human
Transporters mediating brain entry or exit of small compounds via SLCs, ABC transporters	Many models including cell lines show sufficient expression, suitable for uptake and efflux studies	Check expression of transport system of interest; compare with in vivo or primary culture
Metabolic enzymes affecting drug permeation	Many models show sufficient expression	Check model has been characterized for enzymes
Receptor-mediated endocytosis and transcytosis	Primary cultured cells with astrocytes, found critical for full expression and function	Check receptor expression and polarity show features of BBB-type transcytosis rather than "default" non-BBB phenotype
Non-BBB characteristics		Check for absence of epithelial features, e.g., cadherins, transporters, and extensive caveolae

observed (see also Sect. 8.1.3.4), some models are tight enough for demonstration of CSF secretion, and the models have been used for a variety of studies of transport, metabolism, and leukocyte traffic (Redzic 2013; Strazielle and Ghersi-Egea 2013; Monnot and Zheng 2013). A stable continuous subcultivatable porcine cell line PCP-R, (Schroten et al. 2012) and some immortalized cell lines (human Z310, Monnot and Zheng 2013; rat TR-CSFB3, Terasaki and Hosoya 2001) have been introduced. The models have generally not been used for drug permeability screening.

8.2.8.2 Arachnoid Epithelial Cells

It has recently been proven possible to culture arachnoid cells in vitro, which express claudin 1 and generate a TEER of ~160 $\Omega.cm^2$ with restriction of larger solute permeation (Lam et al. 2011, 2012; Janson et al. 2011). Characterization of the expression patterns of drug transporters and enzymes in arachnoid tissue and arachnoid barrier (AB) cells shows expression of both Pgp and breast cancer resistance protein (BCRP); an immortalized cell line of AB cells showed Pgp expression on the apical (dura-facing) membrane and BCRP on both apical and basal (CSF-facing) membranes (Yasuda et al. 2013). Microarray analysis of mouse and human arachnoid tissue showed expression of many drug transporters and some drug metabolizing enzymes. The consistency across in vitro models and isolated tissue makes it likely that these proteins contribute to the blood-CSF barrier function and confirms that useful in vitro models can be generated and applied to examine these functions in detail.

8.3   Future Directions and Challenges

It is clear that in vitro models will continue to play important roles in generating mechanistic information about cellular and intercellular events in CNS barrier layers, capable of informing a range of applications in health and disease, drug discovery, and drug delivery. Some emerging technologies and their combination offer clear future directions—the challenge will be to make them effective and advance understanding.

We need:

1. Generation of reliable and tight in vitro models of the human BBB, choroid plexus, and arachnoid barriers, reproducing the in vivo condition.
2. Development of an accepted "industry standard" in vitro BBB model, robust, reliable, predictive of human drug PK, and capable of operation in medium- to high-throughput screening of NCEs.
3. Better understanding of TEER measurement in different systems, with accepted calibration protocols, reference thresholds, and intersystem correlations.
4. Better integration of in silico, in vitro, and in vivo models to provide complementary information and more complete characterization of permeability routes and transport systems; we need more projects designed with parallel in vitro and in vivo assessment.
5. More computational modeling with software optimized for CNS barrier models, before, during, and after experiments to better understand and correct for artifacts in permeability-measuring systems.
6. Microfluidics platforms integrating flow, TEER, and other sensors and permitting advanced live cell imaging, suitable for studies of a single barrier cell type or co-cultures reflecting the in vivo condition as within the NVU.

8.4   Conclusions

In the ~40 year history of in vitro CNS barrier models, there have been a number of major advances and of course also many false starts, with natural evolution of the field by which useful, reliable, and informative models become more widely used, building up the critical mass of basic information from which new developments can take off. Groups developing and adopting in vitro models can learn from the history and current status of the field to ensure that further progress is soundly based and effective and results reliable and applicable between laboratories and across the field. New investigators have available a range of good models and excellent tools and increasingly will work by collaboration to apply them. Exciting times!

8.5 Points for Discussion

1. Imagine a new project in your lab that requires an in vitro model; (a) define the requirements of the model, (b) decide on the most suitable model(s) to use, and justify this choice.
2. Why are leakier BBB models (TEER <200 $\Omega.cm^2$) less suitable for transendothelial permeability screening?
3. For transendothelial permeability measurement, why is it useful to (a) measure the TEER of each filter with cells and (b) make parallel measurements of TEER and permeability of a paracellular marker (P_{app} or P_e), ideally in each experimental run?
4. What are appropriate paracellular markers for the model(s) you selected in (1)?
5. Why has it proven difficult to develop good primary cultured human BBB models?
6. What is an unstirred water layer (aqueous boundary layer, ABL), and why is it a problem for in vitro but not in vivo BBB studies? For transendothelial permeation, which types of compound are most affected by the ABL? If the ABL is not considered, minimized, and corrected for, how would transendothelial permeability measurements be affected?
7. How can in vitro models from different species contribute to prediction of drug PK in human brain interstitial fluid using a process of PBPK?
8. As an exercise, design a microfluidic chamber suitable for studies of transendothelial and transepithelial permeability using CNS barrier cells. What additional features would it provide not generally available for "flat" ("transwell") filter systems? In what ways could these features be important?
9. What are the main differences in generating an in vitro BBB model from human stem cells and from freshly isolated human brain microvessels? How would you select the most "BBB-like" clones from a variety of clones generated from stem cells using different growth conditions and media?

Acknowledgments We are grateful for discussions and comments from many colleagues especially Prof. Maria Deli, Dr. Alex Avdeef, and Prof. Margareta Hammarlund-Udenaes and for redrawing of Fig. 8.1 by Dr. Robert Thorne and for Fig. 8.4 from Dr. Anika Hartz and Dr. Bjoern Bauer.

References

Abbott NJ (2004) Prediction of blood-brain barrier permeation in drug discovery from in vivo, in vitro and in silico models. Drug Discov Today Technol 1:407–416

Abbott NJ (2013) Blood-brain barrier structure and function and the challenges for CNS drug delivery. J Inherit Metab Dis 36:437–449

Abbott NJ, Dolman DEM, Yusof SR, Reichel A (2014) In vitro models of CNS barriers. In: Hammarlund-Udenaes M, de Lange E, Thorne R (eds) Drug delivery to the brain, AAPS advances in the pharmaceutical sciences series, vol 10. Springer, New York

Adriani G, Ma D, Pavesi A, Kamm RD, Goh EL (2017) A 3D neurovascular microfluidic model consisting of neurons, astrocytes and cerebral endothelial cells as a blood–brain barrier. Lab Chip 17(3):448–459. https://doi.org/10.1039/C6LC00638H

Akanuma SI, Yamakoshi A, Sugouchi T, Kubo Y, Hartz AMS, Bauer B, Hosoya KI (2018) Role of l-type amino acid transporter 1 at the inner blood-retinal barrier in the blood-to-retina transport of gabapentin. Mol Pharm 15(6):2327–2337. https://doi.org/10.1021/acs.molpharmaceut.8b00179

Al Feteisi H, Al-Majdoub ZM, Achour B, Couto N, Rostami-Hodjegan A, Barber J (2018) Identification and quantification of blood-brain barrier transporters in isolated rat brain microvessels. J Neurochem 146(6):670–685. https://doi.org/10.1111/jnc.14446

Alam C, Hoque MT, Sangha V, Bendayan R (2020) Nuclear respiratory factor 1 (NRF-1) upregulates the expression and function of reduced folate carrier (RFC) at the blood-brain barrier. FASEB J 34(8):10516–10530. https://doi.org/10.1096/fj.202000239RR

Angelow S, Zeni P, Galla HJ (2004) Usefulness and limitation of primary cultured porcine choroid plexus epithelial cells as an in vitro model to study drug transport at the blood-CSF barrier. Adv Drug Deliv Rev 56:1859–1873

Audus KL, Borchardt RT (1986) Characterisation of an in vitro blood-brain barrier model system for studying drug transport and metabolism. Pharm Res 3:81–87

Audus KL, Rose JM, Wang W, Borchardt RT (1998) Brain microvessel endothelial cell culture systems. In: Pardridge WM (ed) Introduction to the blood-brain barrier: methodology, biology and pathology. Cambridge University Press, Cambridge

Avdeef A (2011) How well can in vitro brain microcapillary endothelial cell models predict rodent in vivo blood-brain barrier permeability? Eur J Pharm Sci 43:109–124

Avdeef A (2012) Absorption and drug development: solubility, permeability and charge state, 2nd edn. Wiley, Hoboken, NJ

Baehr C, Reichel V, Fricker G (2006) Choroid plexus epithelial monolayers - a cell culture model from porcine brain. Cerebrospinal Fluid Res 3:13

Ball K, Bouzom F, Scherrmann JM, Walther B, Declèves X (2012) Development of a physiologically based pharmacokinetic model for the rat central nervous system and determination of an in vitro-in vivo scaling methodology for the blood-brain barrier permeability of two transporter substrates, morphine and oxycodone. J Pharm Sci 101:4277–4292

Ball K, Bouzom F, Scherrmann JM, Walther B, Declèves X (2013) Physiologically based pharmacokinetic modelling of drug penetration across the blood-brain barrier – towards a mechanistic IVIVE-based approach. AAPS J 15:913–932

Ballermann BJ, Dardik A, Eng E, Liu A (1998) Shear stress and the endothelium. Kidney Int Suppl 67:S100–S108

Bauer B, Hartz AM, Pekcec A, Toellner K, Miller DS, Potschka H (2008) Seizure-induced upregulation of P-glycoprotein at the blood-brain barrier through glutamate and cyclooxygenase-2 signaling. Mol Pharmacol 73(5):1444–1453. https://doi.org/10.1124/mol.107.041210

Baumann J, Tsao CC, Huang SF, Gassmann M, Ogunshola OO (2021) Astrocyte-specific hypoxia-inducible factor 1 (HIF-1) does not disrupt the endothelial barrier during hypoxia in vitro. Fluids Barriers CNS 18(1):13. https://doi.org/10.1186/s12987-021-00247-2

Benson K, Cramer S, Galla HJ (2013) Impedance-based cell monitoring: barrier properties and beyond. Fluids Barriers CNS 10(1):5. https://doi.org/10.1186/2045-8118-10-5

Bhalerao A, Sivandzade F, Archie SR, Chowdhury EA, Noorani B, Cucullo L (2020) In vitro modeling of the neurovascular unit: advances in the field. Fluids Barriers CNS 17(1):22. https://doi.org/10.1186/s12987-020-00183-7

Booth R, Kim H (2012) Characterization of a microfluidic in vitro model of the blood-brain barrier (μBBB). Lab Chip 12:1784–1792

Bouldin TW, Krigman MR (1975) Differential permeability of cerebral capillary and choroid plexus to lanthanum ion. Brain Res 99:444–448

Bowman PD, Betz AL, Ar D, Wolinsky JS, Penney JB, Shivers RR, Goldstein GW (1981) Primary culture of capillary endothelium from rat brain. In Vitro 17(4):353–362

Bowman PD, Ennis SR, Rarey KE, Betz AL, Goldstein GW (1983) Brain microvessel endothelial cells in culture: a model for study of blood-brain barrier permeability. Ann Neurol 14:396–402

Brendel K, Meezan E, Carlson EC (1974) Isolated brain microvessels: a purified, metabolically active preparation from bovine cerebral cortex. Science 185:953–955

Brown JA, Pensabene V, Markov DA, Allwardt V, Neely MD, Shi M, Britt CM, Hoilett OS, Yang Q, Brewer BM, Samson PC, McCawley LJ, May JM, Webb DJ, Li D, Bowman AB, Reiserer RS, Wikswo JP (2015) Recreating blood-brain barrier physiology and structure on chip: A novel neurovascular microfluidic bioreactor. Biomicrofluidics 9(5):054124. https://doi.org/10.1063/1.4934713

Burek M, Salvador E, Förster CY (2012) Generation of an immortalized murine brain microvascular endothelial cell line as an in vitro blood brain barrier model. J Vis Exp 29(66):e4022. https://doi.org/10.3791/4022

Cacopardo L, Costa J, Giusti S, Buoncompagni L, Meucci S, Corti A, Mattei G, Ahluwalia A (2019) Real-time cellular impedance monitoring and imaging of biological barriers in a dual-flow membrane bioreactor. Biosens Bioelectron 140:111340. https://doi.org/10.1016/j.bios.2019.111340

Campos CR, Schröter C, Wang X, Miller DS (2012) ABC transporter function and regulation at the blood-spinal cord barrier. J Cereb Blood Flow Metab 32:1559–1566

Candela P, Gosselet F, Saint-Pol J, Sevin E, Boucau MC, Boulanger E, Cecchelli R, Fenart L (2010) Apical-to-basolateral transport of amyloid-β peptides through blood-brain barrier cells is mediated by the receptor for advanced glycation end-products and is restricted by P-glycoprotein. J Alzheimers Dis 22:849–859

Cecchelli R, Dehouck B, Descamps L, Fenart L, Buée-Scherrer V, Duhem C, Lundquist S, Rentfel M, Torpier G, Dehouck MP (1999) In vitro model for evaluating drug transport across the blood-brain barrier. Adv Drug Deliv Rev 36:165–178

Cecchelli R, Aday S, Sevin E, Almeida C, Culot M, Dehouck L, Coisne C, Engelhardt B, Dehouck MP, Ferreira L (2014) A stable and reproducible human blood-brain barrier model derived from hematopoietic stem cells. PLoS One 9(6):e99733. https://doi.org/10.1371/journal.pone.0099733

Chaves C, Do TM, Cegarra C, Roudières V, Tolou S, Thill G, Rocher C, Didier M, Lesuisse D (2020) Non-Human Primate Blood-Brain Barrier and In Vitro Brain Endothelium: From Transcriptome to the Establishment of a New Model. Pharmaceutics 12(10):967. https://doi.org/10.3390/pharmaceutics12100967

Cohen-Kashi Malina K, Cooper I, Teichberg VI (2009) Closing the gap between the in-vivo and in-vitro blood-brain barrier tightness. Brain Res 1284:12–21

Cohen-Kashi-Malina K, Cooper I, Teichberg VI (2012) Mechanisms of glutamate efflux at the blood-brain barrier: involvement of glial cells. J Cereb Blood Flow Metab 32:177–189

Cucullo L, McAllister MS, Kight K, Krizanac-Bengez L, Marroni M, Mayberg MR, Stanness KA, Janigro D (2002) A new dynamic in vitro model for the multidimensional study of astrocyte-endothelial cell interactions at the blood-brain barrier. Brain Res 951:243–254

Cucullo L, Hossain M, Puvenna V, Marchi N, Janigro D (2011) The role of shear stress in blood-brain barrier endothelial physiology. BMC Neurosci 12:40

Culot M, Lundquist S, Vanuxeem D, Nion S, Landry C, Delplace Y, Dehouck MP, Berezowski V, Fenart L, Cecchelli R (2008) An in vitro blood-brain barrier model for high throughput (HTS) toxicological screening. Toxicol In Vitro 22:799–811

Dagenais C, Avdeef A, Tsinman O, Dudley A, Beliveau R (2009) P-glycoprotein deficient mouse in situ blood–brain barrier permeability and its prediction using an in combo PAMPA model. Eur J Pharm Sci 38(2):121–137. https://doi.org/10.1016/j.ejps.2009.06.009

Daneman R, Zhou L, Agalliu D, Cahoy JD, Kaushal A, Barres BA (2010a) The mouse blood-brain barrier transcriptome: a new resource for understanding the development and function of brain endothelial cells. PLoS One 5(10):e13741. https://doi.org/10.1371/journal.pone.0013741

Daneman R, Zhou L, Kebede AA, Barres BA (2010b) Pericytes are required for blood-brain barrier integrity during embryogenesis. Nature 468:562–566

De Boer AG, Sutanto W (eds) (1997) Drug transport across the blood-brain barrier. Harwood, Amsterdam

DeBault LE, Cancilla PA (1980) Gamma-glutamyl transpeptidase in isolated brain endothelial cells: induction by glial cells in vitro. Science 207:653–655

DeBault LE, Kahn LE, Frommes SP, Cancilla PA (1979) Cerebral microvessels and derived cells in tissue culture: isolation and preliminary characterization. In Vitro 15(7):473–487

Dehouck MP, Méresse S, Delorme P, Fruchart JC, Cecchelli R (1990) An easier, reproducible, and mass-production method to study the blood-brain barrier in vitro. J Neurochem 54:1798–1801

Deli MA, Abrahám CS, Kataoka Y, Niwa M (2005) Permeability studies on in vitro blood-brain barrier models: physiology, pathology, and pharmacology. Cell Mol Neurobiol 25:59–127

Devraj G, Guérit S, Seele J, Spitzer D, Macas J, Khel MI, Heidemann R, Braczynski AK, Ballhorn W, Günther S, Ogunshola OO, Mittelbronn M, Ködel U, Monoranu CM, Plate KH, Hammerschmidt S, Nau R, Devraj K, Kempf VAJ (2020) HIF-1α is involved in blood-brain barrier dysfunction and paracellular migration of bacteria in pneumococcal meningitis. Acta Neuropathol 140(2):183–208. https://doi.org/10.1007/s00401-020-02174-2

Di Marco A, Gonzalez Paz O, Fini I, Vignone D, Cellucci A, Battista MR, Auciello G, Orsatti L, Zini M, Monteagudo E, Khetarpal V, Rose M, Dominguez C, Herbst T, Toledo-Sherman L, Summa V, Muñoz-Sanjuán I (2019) Application of an in Vitro Blood-Brain Barrier Model in the Selection of Experimental Drug Candidates for the Treatment of Huntington's Disease. Mol Pharm 16(5):2069–2082. https://doi.org/10.1021/acs.molpharmaceut.9b00042

Dorovini-Zis K, Bowman PD, Betz AL, Goldstein G (1984) Hyperosmotic arabinose solutions open the tight junctions between brain capillary endothelial cells in tissue culture. Brain Res 302(2):383–386. https://doi.org/10.1016/0006-8993(84)90254-3

Dorovini-Zis K, Prameya R, Bowman PD (1991) Culture and characterization of microvascular endothelial cells derived from human brain. Lab Invest 64(3):425–436

Dutheil F, Jacob A, Dauchy S, Beaune P, Scherrmann JM, Declèves X, Loriot MA (2010) ABC transporters and cytochromes P450 in the human central nervous system: influence on brain pharmacokinetics and contribution to neurodegenerative disorders. Expert Opin Drug Metab Toxicol 6:1161–1174

Ehrlich P (1885) Das Sauerstoffbeduerfnis des Organismus. Eine Farbenanalytische Studie. Hirschwald, Berlin, In

Eigenmann DE, Dürig C, Jähne EA, Smieško M, Culot M, Gosselet F, Cecchelli R, Helms HCC, Brodin B, Wimmer L, Mihovilovic MD, Hamburger M, Oufir M (2016) In vitro blood-brain barrier permeability predictions for GABAA receptor modulating piperine analogs. Eur J Pharm Biopharm 103:118–126. https://doi.org/10.1016/j.ejpb.2016.03.029

Fatima N, Gromnicova R, Loughlin J, Sharrack B, Male D (2020) Gold nanocarriers for transport of oligonucleotides across brain endothelial cells. PLoS One 15(9):e0236611. https://doi.org/10.1371/journal.pone.0236611

Förster C, Silwedel C, Golenhofen N, Burek M, Kietz S, Mankertz J, Drenckhahn D (2005) Occludin as direct target for glucocorticoid-induced improvement of blood-brain barrier properties in a murine in vitro system. J Physiol 565:475–486

Förster C, Burek M, Romero IA, Weksler B, Couraud PO, Drenckhahn D (2008) Differential effects of hydrocortisone and TNFalpha on tight junction proteins in an in vitro model of the human blood-brain barrier. J Physiol 586(7):1937–1949. https://doi.org/10.1113/jphysiol.2007.146852

Franke H, Galla HJ, Beuckmann CT (1999) An improved low-permeability in vitro-model of the blood-brain barrier: transport studies on retinoids, sucrose, haloperidol, caffeine and mannitol. Brain Res 818:65–71

Franke H, Galla HJ, Beuckmann CT (2000) Primary cultures of brain microvessel endothelial cells: a valid and flexible model to study drug transport through the blood-brain barrier in vitro. Brain Res Brain Res Protoc 5:248–256

Fricker G (2002) Drug transport across the blood-brain barrier. In: Pelkonen O, Baumann A, Reichel A (eds) Pharmacokinetic challenges in drug discovery. Springer, Berlin

Gaillard PJ, de Boer AG (2000) Relationship between permeability status of the blood-brain barrier and in vitro permeability coefficient of a drug. Eur J Pharm Sci 12:95–102

Garberg P (1998) In vitro models of the blood-brain barrier. Altern Lab Anim 26(6):821–847

Garberg P, Ball M, Borg N, Cecchelli R, Fenart L, Hurst RD, Lindmark T, Mabondzo A, Nilsson JE, Raub TJ, Stanimirovic D, Terasaki T, Oberg JO, Osterberg T (2005) In vitro models for the blood-brain barrier. Toxicol In Vitro 19:299–334

García-Salvador A, Domínguez-Monedero A, Gómez-Fernández P, García-Bilbao A, Carregal-Romero S, Castilla J, Goñi-de-Cerio F (2020) Evaluation of the Influence of Astrocytes on In Vitro Blood-Brain Barrier Models. Altern Lab Anim 48(4):184–200. https://doi.org/10.1177/0261192920966954

Ge S, Song L, Pachter JS (2005) Where is the blood-brain barrier … really? J Neurosci Res 79:421–427

Gerhartl A, Pracser N, Vladetic A, Hendrikx S, Friedl HP, Neuhaus W (2020) The pivotal role of micro-environmental cells in a human blood-brain barrier in vitro model of cerebral ischemia: functional and transcriptomic analysis. Fluids Barriers CNS 17(1):19. https://doi.org/10.1186/s12987-020-00179-3

Gericke B, Römermann K, Noack A, Noack S, Kronenberg J, Blasig IE, Löscher W (2020) A face-to-face comparison of claudin-5 transduced human brain endothelial (hCMEC/D3) cells with porcine brain endothelial cells as blood-brain barrier models for drug transport studies. Fluids Barriers CNS 17(1):53. https://doi.org/10.1186/s12987-020-00212-5

Gibbs JE, Thomas SA (2002) The distribution of the anti-HIV drug, 2'3'-dideoxycytidine (ddC), across the blood-brain and blood-cerebrospinal fluid barriers and the influence of organic anion transport inhibitors. J Neurochem 80:392–404

Gil ES, Wu L, Xu L, Lowe TL (2012) β-Cyclodextrin-poly(β-amino ester) nanoparticles for sustained drug delivery across the blood-brain barrier. Biomacromolecules 13:3533–3541

Giri R, Selvaraj S, Miller CA, Hofman F, Yan SD, Stern D, Zlokovic BV, Kalra VK (2002) Effect of endothelial cell polarity on beta-amyloid-induced migration of monocytes across normal and AD endothelium. Am J Physiol Cell Physiol 283:C895–C904

Goldeman C, Ozgür B, Brodin B (2020) Culture-induced changes in mRNA expression levels of efflux and SLC-transporters in brain endothelial cells. Fluids Barriers CNS 17(1):32. https://doi.org/10.1186/s12987-020-00193-5

Grasset E, Pinto M, Dussaulx E, Zweibaum A, Desjeux JF (1984) Epithelial properties of human colonic carcinoma cell line Caco-2: electrical parameters. Am J Physiol 247:C260–C267

Greenwood J, Pryce G, Devine L, Male DK, dos Santos WL, Calder VL, Adamson P (1996) SV40 large T immortalised cell lines of the rat blood-brain and blood-retinal barriers retain their phenotypic and immunological characteristics. J Neuroimmunol 71:51–63

Griep LM, Wolbers F, de Wagenaar B, ter Braak PM, Weksler BB, Romero IA, Couraud PO, Vermes I, van der Meer AD, van den Berg A (2013) BBB ON CHIP: microfluidic platform to mechanically and biochemically modulate blood-brain barrier function. Biomed Microdevices 15:145–150

Grifno GN, Farrell AM, Linville RM, Arevalo D, Kim JH, Gu L, Searson PC (2019) Tissue-engineered blood-brain barrier models via directed differentiation of human induced pluripotent stem cells. Sci Rep 9(1):13957. https://doi.org/10.1038/s41598-019-50193-1

Hamminger P, Marchetti L, Preglej T, Platzer R, Zhu C, Kamnev A, Rica R, Stolz V, Sandner L, Alteneder M, Kaba E, Waltenberger D, Huppa JB, Trauner M, Bock C, Lyck R, Bauer J, Dupré L, Seiser C, Boucheron N, Engelhardt B, Ellmeier W (2021) Histone deacetylase 1 controls CD4+ T cell trafficking in autoinflammatory diseases. J Autoimmun 119:102610. https://doi.org/10.1016/j.jaut.2021.102610

Hansson E, Sellström A, Persson LI, Rönnbäck L (1980) Brain primary culture - a characterization. Brain Res 188(1):233–246

Hart MN, VanDyk LF, Moore SA, Shasby DM, Cancilla PA (1987) Differential opening of the brain endothelial barrier following neutralization of the endothelial luminal anionic charge in vitro. J Neuropathol Exp Neurol 46:141–153

Hartz AM, Bauer B, Soldner EL, Wolf A, Boy S, Backhaus R, Mihaljevic I, Bogdahn U, Klünemann HH, Schuierer G, Schlachetzki F (2012) Amyloid-β contributes to blood-brain barrier leakage in transgenic human amyloid precursor protein mice and in humans with cerebral amyloid angiopathy. Stroke 43:514–523

Hellinger E, Veszelka S, Tóth AE, Walter F, Kittel A, Bakk ML, Tihanyi K, Háda V, Nakagawa S, Duy TD, Niwa M, Deli MA, Vastag M (2012) Comparison of brain capillary endothelial cell-based and epithelial (MDCK-MDR1, Caco-2, and VB-Caco-2) cell-based surrogate blood-brain barrier penetration models. Eur J Pharm Biopharm 82:340–351

Helms HC, Waagepetersen HS, Nielsen CU, Brodin B (2010) Paracellular tightness and claudin-5 expression is increased in the BCEC/astrocyte blood-brain barrier model by increasing media buffer capacity during growth. AAPS J 12:759–770

Helms HC, Madelung R, Waagepetersen HS, Nielsen CU, Brodin B (2012) In vitro evidence for the brain glutamate efflux hypothesis: brain endothelial cells cocultured with astrocytes display a polarized brain-to-blood transport of glutamate. Glia 60:882–893

Helms HC, Abbott NJ, Burek M, Cecchelli R, Couraud PO, Deli MA, Förster C, Galla HJ, Romero IA, Shusta EV, Stebbins MJ, Vandenhaute E, Weksler B, Brodin B (2016) In vitro models of the blood-brain barrier: An overview of commonly used brain endothelial cell culture models and guidelines for their use. J Cereb Blood Flow Metab 36(5):862–890. https://doi.org/10.1177/0271678X16630991

Heymans M, Figueiredo R, Dehouck L, Francisco D, Sano Y, Shimizu F, Kanda T, Bruggmann R, Engelhardt B, Winter P, Gosselet F, Culot M (2020) Contribution of brain pericytes in blood-brain barrier formation and maintenance: a transcriptomic study of cocultured human endothelial cells derived from hematopoietic stem cells. Fluids Barriers CNS 17(1):48. https://doi.org/10.1186/s12987-020-00208-1

Hoheisel D, Nitz T, Franke H, Wegener J, Hakvoort A, Tilling T, Galla HJ (1998) Hydrocortisone reinforces the blood-brain barrier properties in a serum free cell culture system. Biochem Biophys Res Commun 247:312–315

Hoshi Y, Uchida Y, Tachikawa M, Inoue T, Ohtsuki S, Terasaki T (2013) Quantitative atlas of blood-brain barrier transporters, receptors, and tight junction proteins in rats and common marmoset. J Pharm Sci 102(9):3343–3355. https://doi.org/10.1002/jps.23575

Hosoya KI, Takashima T, Tetsuka K, Nagura T, Ohtsuki S, Takanaga H, Ueda M, Yanai N, Obinata M, Terasaki T (2000) mRNA expression and transport characterization of conditionally immortalized rat brain capillary endothelial cell lines; a new in vitro BBB model for drug targeting. J Drug Target 8:357–370

Hosoya K, Tomi M, Ohtsuki S, Takanaga H, Ueda M, Yanai N, Obinata M, Terasaki T (2001) Conditionally immortalized retinal capillary endothelial cell lines (TR-iBRB) expressing differentiated endothelial cell functions derived from a transgenic rat. Exp Eye Res 72:163–172

Huber O, Brunner A, Maier P, Kaufmann R, Couraud PO, Cremer C, Fricker G (2012) Localization microscopy (SPDM) reveals clustered formations of P-glycoprotein in a human blood-brain barrier model. PLoS One 7(9):e44776. https://doi.org/10.1371/journal.pone.0044776

Hurst RD, Fritz IB (1996) Properties of an immortalised vascular endothelial/glioma cell co-culture model of the blood-brain barrier. J Cell Physiol 167:81–88

Ito K, Uchida Y, Ohtsuki S, Aizawa S, Kawakami H, Katsukura Y, Kamiie J, Terasaki T (2011a) Quantitative membrane protein expression at the blood-brain barrier of adult and younger cynomolgus monkeys. J Pharm Sci 100:3939–3950

Ito S, Ohtsuki S, Katsukura Y, Funaki M, Koitabashi Y, Sugino A, Murata S, Terasaki T (2011b) Atrial natriuretic peptide is eliminated from the brain by natriuretic peptide receptor-C-mediated brain-to-blood efflux transport at the blood-brain barrier. J Cereb Blood Flow Metab 31:457–466

Ito S, Ohtsuki S, Nezu Y, Koitabashi Y, Murata S, Terasaki T (2011c) 1α,25-Dihydroxyvitamin D3 enhances cerebral clearance of human amyloid-β peptide(1-40) from mouse brain across the blood-brain barrier. Fluids Barriers CNS 8:20. https://doi.org/10.1186/2045-8118-8-20

Ito S, Yanai M, Yamaguchi S, Couraud PO, Ohtsuki S (2017) Regulation of Tight-Junction Integrity by Insulin in an In Vitro Model of Human Blood-Brain Barrier. J Pharm Sci 106(9):2599–2605. https://doi.org/10.1016/j.xphs.2017.04.036

Ittner C, Burek M, Störk S, Nagai M, Förster CY (2020) Increased Catecholamine Levels and Inflammatory Mediators Alter Barrier Properties of Brain Microvascular Endothelial Cells in vitro. Front Cardiovasc Med 7:73. https://doi.org/10.3389/fcvm.2020.00073

Janigro D, Leaman SM, Stanness KA (1999) Dynamic modeling of the blood-brain barrier: a novel tool for studies of drug delivery to the brain. Pharm Sci Technolo Today 2:7–12

Janson C, Romanova L, Hansen E, Hubel A, Lam C (2011) Immortalization and functional characterization of rat arachnoid cell lines. Neuroscience 177:23–34

Joó F (1992) The cerebral microvessels in culture, an update. J Neurochem 58:1–17

Joó F, Karnushina I (1973) A procedure for the isolation of capillaries from rat brain. Cytobios 8:41–48

Kalari KR, Thompson KJ, Nair AA, Tang X, Bockol MA, Jhawar N, Swaminathan SK, Lowe VJ, Kandimalla KK (2016) BBBomics-human blood brain barrier transcriptomics hub. Front Neurosci 10:71. https://doi.org/10.3389/fnins.2016.00071

Kamiichi A, Furihata T, Kishida S, Ohta Y, Saito K, Kawamatsu S, Chiba K (2012) Establishment of a new conditionally immortalized cell line from human brain microvascular endothelial cells: a promising tool for human blood-brain barrier studies. Brain Res 1488:113–122. https://doi.org/10.1016/j.brainres.2012.09.042

Kamiie J, Ohtsuki S, Iwase R, Ohmine K, Katsukura Y, Yanai K, Sekine Y, Uchida Y, Ito S, Terasaki T (2008) Quantitative atlas of membrane transporter proteins: development and application of a highly sensitive simultaneous LC/MS/MS method combined with novel in-silico peptide selection criteria. Pharm Res 25:1469–1483

Kasa P, Pakaski M, Joó F, Lajtha A (1991) Endothelial cells from human fetal brain microvessels may be cholinoceptive, but do not synthesize acetylcholine. J Neurochem 56:2143–2146

Kong LY, Li Y, Rao DY, Wu B, Sang CP, Lai P, Ye JS, Zhang ZX, Du ZM, Yu JJ, Gu L, Xie FC, Liu ZY, Tang ZX (2021) miR-666-3p mediates the protective effects of mesenchymal stem cell-derived exosomes against oxygen-glucose deprivation and reoxygenation-induced cell injury in brain microvascular endothelial cells via mitogen-activated protein kinase pathway. Curr Neurovasc Res. https://doi.org/10.2174/1567202618666210319152534

Krämer SD, Abbott NJ, Begley DJ (2001) Biological models to study blood-brain barrier permeation. In: Testa B, van de Waterbeemd H, Folkers G, Guy R (eds) Pharmacokinetic optimization in drug research: biological, physicochemical and computational strategies. Wiley-VCH, Weinheim

Kristensen M, Kucharz K, Felipe Alves Fernandes E, Strømgaard K, Schallburg Nielsen M, Cederberg Helms HC, Bach A, Ulrikkaholm Tofte-Hansen M, Irene Aldana Garcia B, Lauritzen M, Brodin B (2020) Conjugation of Therapeutic PSD-95 Inhibitors to the Cell-Penetrating Peptide Tat Affects Blood-Brain Barrier Adherence, Uptake, and Permeation. Pharmaceutics 12(7):661. https://doi.org/10.3390/pharmaceutics12070661

Kubo Y, Ohtsuki S, Uchida Y, Terasaki T (2015) Quantitative Determination of Luminal and Abluminal Membrane Distributions of Transporters in Porcine Brain Capillaries by Plasma

Membrane Fractionation and Quantitative Targeted Proteomics. J Pharm Sci 104(9):3060–3068. https://doi.org/10.1002/jps.24398

Kumarasamy M, Sosnik A (2021) Heterocellular spheroids of the neurovascular blood-brain barrier as a platform for personalized nanoneuromedicine. iScience 24(3):102183. https://doi.org/10.1016/j.isci.2021.102183

Laksitorini MD, Yathindranath V, Xiong W, Parkinson FE, Thliveris JA, Miller DW (2020) Impact of Wnt/β-catenin signaling on ethanol-induced changes in brain endothelial cell permeability. J Neurochem. https://doi.org/10.1111/jnc.15203

Lam CH, Hansen EA, Hubel A (2011) Arachnoid cells on culture plates and collagen scaffolds: phenotype and transport properties. Tissue Eng Part A 17:1759–1766

Lam CH, Hansen EA, Janson C, Bryan A, Hubel A (2012) The characterization of arachnoid cell transport II: paracellular transport and blood-cerebrospinal fluid barrier formation. Neuroscience 222:228–238

Larochelle C, Cayrol R, Kebir H, Alvarez JI, Lécuyer MA, Ifergan I, Viel É, Bourbonnière L, Beauseigle D, Terouz S, Hachehouche L, Gendron S, Poirier J, Jobin C, Duquette P, Flanagan K, Yednock T, Arbour N, Prat A (2012) Melanoma cell adhesion molecule identifies encephalitogenic T lymphocytes and promotes their recruitment to the central nervous system. Brain 135:2906–2924

Le Roux GL, Jarray R, Guyot AC, Pavoni S, Costa N, Théodoro F, Nassor F, Pruvost A, Tournier N, Kiyan Y, Langer O, Yates F, Deslys JP, Mabondzo A (2019) Proof-of-concept study of drug brain permeability between in vivo human brain and an in vitro iPSCs-human blood-brain barrier model. Sci Rep 9:16310. https://doi.org/10.1038/s41598-019-52213-6

Lee NY, Choi HO, Kang YS (2012) The acetylcholinesterase inhibitors competitively inhibited an acetyl L-carnitine transport through the blood-brain barrier. Neurochem Res 37:1499–1507

Li JY, Boado RJ, Pardridge WM (2001) Blood-brain barrier genomics. J Cereb Blood Flow Metab 21:61–68

Li JY, Boado RJ, Pardridge WM (2002) Rat blood-brain barrier genomics. II. J Cereb Blood Flow Metab 22:1319–1326

Li Y, Zhou S, Li J, Sun Y, Hasimu H, Liu R, Zhang T (2015) Quercetin protects human brain microvascular endothelial cells from fibrillar β-amyloid1-40-induced toxicity. Acta Pharm Sin B 5(1):47–54. https://doi.org/10.1016/j.apsb.2014.12.003

Li Y, Terstappen GC, Zhang W (2021) Differentiation of Human Induced Pluripotent Stem Cells (hiPSC) into endothelial-type cells and establishment of an in vitro blood-brain barrier model. Methods Mol Biol. https://doi.org/10.1007/7651_2021_363

Lippmann ES, Azarin SM, Kay JE, Nessler RA, Wilson HK, Al-Ahmad A, Palecek SP, Shusta EV (2012) Derivation of blood-brain barrier endothelial cells from human pluripotent stem cells. Nat Biotechnol 30:783–791

Lippmann ES, Al-Ahmad A, Palecek SP, Shusta EV (2013) Modeling the blood-brain barrier using stem cell sources. Fluids Barriers CNS 10(1):2. https://doi.org/10.1186/2045-8118-10-2

Lippmann ES, Al-Ahmad A, Azarin SM, Palecek SP, Shusta EV (2014) A retinoic acid-enhanced, multicellular human blood-brain barrier model derived from stem cell sources. Sci Rep 4(1):1–10. https://doi.org/10.1038/srep04160

Liu KK, Dorovini-Zis K (2012) Differential regulation of CD4+ T cell adhesion to cerebral microvascular endothelium by the chemokines CCL2 and CCL3. Int J Mol Sci 13:16119–16140

Liu Y, Huber CC, Wang H (2020) Disrupted blood-brain barrier in 5×FAD mouse model of Alzheimer's disease can be mimicked and repaired in vitro with neural stem cell-derived exosomes. Biochem Biophys Res Commun 18:S0006-291X(20)30342-9. https://doi.org/10.1016/j.bbrc.2020.02.074

Lohmann C, Hüwel S, Galla HJ (2002) Predicting blood-brain barrier permeability of drugs: evaluation of different in vitro assays. J Drug Target 10:263–276

Lu TM, Houghton S, Magdeldin T, Durán JGB, Minotti AP, Snead A, Sproul A, Nguyen DT, Xiang J, Fine HA, Rosenwaks Z, Studer L, Rafii S, Agalliu D, Redmond D, Lis R (2021) Pluripotent stem cell-derived epithelium misidentified as brain microvascular endothelium requires ETS factors to acquire vascular fate. Proc Natl Acad Sci USA 118(8):e2016950118. https://doi.org/10.1073/pnas.2016950118

Luna-Munguia H, Salvamoser JD, Pascher B, Pieper T, Getzinger T, Kudernatsch M, Kluger G, Potschka H (2015) Glutamate-mediated upregulation of the multidrug resistance protein 2 in porcine and human brain capillaries. J Pharmacol Exp Ther 352(2):368–378. https://doi.org/10.1124/jpet.114.218180

Lundquist S, Renftel M, Brillault J, Fenart L, Cecchelli R, Dehouck MP (2002) Prediction of drug transport through the blood-brain barrier in vivo: a comparison between two in vitro cell models. Pharm Res 19:976–981

Luo H, Saubamea B, Chasseigneaux S, Cochois V, Smirnova M, Glacial F, Perrière N, Chaves C, Cisternino S, Declèves X (2020) Molecular and Functional Study of Transient Receptor Potential Vanilloid 1-4 at the Rat and Human Blood-Brain Barrier Reveals Interspecies Differences. Front Cell Dev Biol 8:578514. https://doi.org/10.3389/fcell.2020.578514

Mabondzo A, Bottlaender M, Guyot AC, Tsaouin K, Deverre JR, Balimane PV (2010) Validation of in vitro cell-based human blood-brain barrier model using clinical positron emission tomography radioligands to predict in vivo human brain penetration. Mol Pharm 7(5):1805–1815. https://doi.org/10.1021/mp1002366

Macdonald JA, Murugesan N, Pachter JS (2010) Endothelial cell heterogeneity of blood-brain barrier gene expression along the cerebral microvasculature. J Neurosci Res 88:1457–1474

Markoutsa E, Pampalakis G, Niarakis A, Romero IA, Weksler B, Couraud PO, Antimisiaris SG (2011) Uptake and permeability studies of BBB-targeting immunoliposomes using the hCMEC/D3 cell line. Eur J Pharm Biopharm 77(2):265–274. https://doi.org/10.1016/j.ejpb.2010.11.015

Martins JP, Alves CJ, Neto E, Lamghari M (2016) Communication from the periphery to the hypothalamus through the blood-brain barrier: An in vitro platform. Int J Pharm 499(1-2):119–130. https://doi.org/10.1016/j.ijpharm.2015.12.058

Masuda T, Hoshiyama T, Uemura T, Hirayama-Kurogi M, Ogata S, Furukawa A, Couraud PO, Furihata T, Ito S, Ohtsuki S (2019) Large-Scale Quantitative Comparison of Plasma Transmembrane Proteins between Two Human Blood-Brain Barrier Model Cell Lines, hCMEC/D3 and HBMEC/ciβ. Mol Pharm 16(5):2162–2171. https://doi.org/10.1021/acs.molpharmaceut.9b00114

Matsumoto J, Dohgu S, Takata F, Iwao T, Kimura I, Tomohiro M, Aono K, Kataoka Y, Yamauchi A (2020) Serum amyloid A-induced blood-brain barrier dysfunction associated with decreased claudin-5 expression in rat brain endothelial cells and its inhibition by high-density lipoprotein in vitro. Neurosci Lett 738:135352. https://doi.org/10.1016/j.neulet.2020.135352

Miller DS (2010) Regulation of P-glycoprotein and other ABC drug transporters at the blood-brain barrier. Trends Pharmacol Sci 31:246–254

Miller DW, Audus KL, Borchardt RT (1992) Application of cultured endothelial cells of the brain microvasculature in the study of the blood-brain barrier. J Tiss Cult Meth 14:217–224

Miller DS, Nobmann SN, Gutmann H, Toeroek M, Drewe J, Fricker G (2000) Xenobiotic transport across isolated brain microvessels studied by confocal microscopy. Mol Pharm 58:1357–1367

Mischeck U, Meyer J, Galla HJ (1989) Characterization of gamma-glutamyl transpeptidase activity of cultured endothelial cells from porcine brain capillaries. Cell Tissue Res 256(1):221–226

Mizutani T, Ishizaka A, Nihei C (2016) Transferrin Receptor 1 Facilitates Poliovirus Permeation of Mouse Brain Capillary Endothelial Cells. J Biol Chem 291(6):2829–2836. https://doi.org/10.1074/jbc.M115.690941

Monnot AD, Zheng W (2013) Culture of choroid plexus epithelial cells and in vitro model of blood-CSF barrier. Methods Mol Biol 945:13–29

Montesano R, Pepper MS, Möhle-Steinlein U, Risau W, Wagner EF, Orci L (1990) Increased proteolytic activity is responsible for the aberrant morphogenetic behavior of endothelial cells expressing the middle T oncogene. Cell 62(3):435–445

Mukhtar M, Pomerantz RJ (2000) Development of an in vitro blood-brain barrier model to study molecular neuropathogenesis and neurovirologic disorders induced by human immunodeficiency virus type 1 infection. J Hum Virol 3:324–334

Munji RN, Soung AL, Weiner GA, Sohet F, Semple BD, Trivedi A, Gimlin K, Kotoda M, Korai M, Aydin S, Batugal A, Cabangcala AC, Schupp PG, Oldham MC, Hashimoto T, Noble-

Haeusslein LJ, Daneman R (2019) Profiling the mouse brain endothelial transcriptome in health and disease models reveals a core blood-brain barrier dysfunction module. Nat Neurosci 22(11):1892–1902. https://doi.org/10.1038/s41593-019-0497-x

Naik P, Cucullo L (2012) In vitro blood-brain barrier models: current and perspective technologies. J Pharm Sci 101:1337–1354

Nakagawa S, Deli MA, Kawaguchi H, Shimizudani T, Shimono T, Kittel A, Tanaka K, Niwa M (2009) A new blood-brain barrier model using primary rat brain endothelial cells, pericytes and astrocytes. Neurochem Int 54:253–263

Nascimento Conde J, Schutt WR, Gorbunova EE, Mackow ER (2020) Recombinant ACE2 expression is required for SARS-CoV-2 to infect primary human endothelial cells and induce inflammatory and procoagulative responses. mBio 11(6):e03185-20. https://doi.org/10.1128/mBio.03185-20

Neuhaus W, Germann B, Plattner VE, Gabor F, Wirth M, Noe CR (2009) Alteration of the glycocalyx of two blood-brain barrier mimicking cell lines is inducible by glioma conditioned media. Brain Res 1279:82–89

Nielsen SSE, Siupka P, Georgian A, Preston JE, Tóth AE, Yusof SR, Abbott NJ, Nielsen MS (2017) Improved method for the establishment of an in vitro blood-brain barrier model based on porcine brain endothelial cells. J Vis Exp 127:56277. https://doi.org/10.3791/56277

Nishihara H, Gastfriend BD, Soldati S, Perriot S, Mathias A, Sano Y, Shimizu F, Gosselet F, Kanda T, Palecek SP, Du Pasquier R, Shusta EV, Engelhardt B (2020) Advancing human induced pluripotent stem cell-derived blood-brain barrier models for studying immune cell interactions. FASEB J 34(12):16693–16715. https://doi.org/10.1096/fj.202001507RR

Ohshima M, Kamei S, Fushimi H, Mima S, Yamada T, Yamamoto T (2019) Prediction of drug permeability using in vitro blood-brain barrier models with human induced pluripotent stem cell-derived brain microvascular endothelial cells. Biores Open Access 8(1):200–209. https://doi.org/10.1089/biores.2019.0026

Ohtsuki S, Uchida Y, Kubo Y, Terasaki T (2011) Quantitative targeted absolute proteomics-based ADME research as a new path to drug discovery and development: methodology, advantages, strategy, and prospects. J Pharm Sci 100:3547–3559

Panula P, Joó F, Rechardt L (1978) Evidence for the presence of viable endothelial cells in cultures derived from dissociated rat brain. Experientia 34:95–97

Pardridge WM (1998) Isolated brain capillaries: an in vitro model of blood-brain barrier research. In: Pardridge WM (ed) Introduction to the blood-brain barrier: methodology, biology and pathology. Cambridge University Press, Cambridge UK

Partyka PP, Godsey GA, Galie JR, Kosciuk MC, Acharya NK, Nagele RG, Galie PA (2017) Mechanical stress regulates transport in a compliant 3D model of the blood-brain barrier. Biomaterials 115:30–39. https://doi.org/10.1016/j.biomaterials.2016.11.012

Patabendige A, Skinner RA, Abbott NJ (2013a) Establishment of a simplified in vitro porcine blood-brain barrier model with high transendothelial electrical resistance. Brain Res 1521:1–15

Patabendige A, Skinner RA, Morgan L, Abbott NJ (2013b) A detailed method for preparation of a functional and flexible blood-brain barrier model using porcine brain endothelial cells. Brain Res 1521:16–30

Paul D, Cowan AE, Ge S, Pachter JS (2013) Novel 3D analysis of Claudin-5 reveals significant endothelial heterogeneity among CNS microvessels. Microvasc Res 86:1–10

Perrière N, Demeuse P, Garcia E, Regina A, Debray M, Andreux JP, Couvreur P, Scherrmann JM, Temsamani J, Couraud PO, Deli MA, Roux F (2005) Puromycin-based purification of rat brain capillary endothelial cell cultures. Effect on the expression of blood-brain barrier-specific properties. J Neurochem 93:279–289

Prabhakarpandian B, Shen MC, Nichols JB, Mills IR, Sidoryk-Wegrzynowicz M, Aschner M, Pant K (2013) SyM-BBB: a microfluidic blood brain barrier model. Lab Chip 13:1093–1101

Preston JE, Segal MB, Walley GJ, Zlokovic BV (1989) Neutral amino acid uptake by the isolated perfused sheep choroid plexus. J Physiol 408:31–43

Puscas I, Bernard-Patrzynski F, Jutras M, Lécuyer MA, Bourbonnière L, Prat A, Leclair G, Roullin VG (2019) IVIVC assessment of two mouse brain endothelial cell models for drug screening. Pharmaceutics 11(11):587. https://doi.org/10.3390/pharmaceutics11110587

Qiao R, Jia Q, Hüwel S, Xia R, Liu T, Gao F, Galla HJ, Gao M (2012) Receptor-mediated delivery of magnetic nanoparticles across the blood-brain barrier. ACS Nano 6:3304–3310

Rand D, Ravid O, Atrakchi D, Israelov H, Bresler Y, Shemesh C, Omesi L, Liraz-Zaltsman S, Gosselet F, Maskrey TS, Beeri MS, Wipf P, Cooper I (2021) Endothelial Iron Homeostasis Regulates Blood-Brain Barrier Integrity via the HIF2α-Ve-Cadherin Pathway. Pharmaceutics 13(3):311. https://doi.org/10.3390/pharmaceutics13030311

Raub TJ, Kuentzel SL, Sawada GA (1992) Permeability of bovine brain microvessel endothelial cells in vitro: barrier tightening by a factor released from astroglioma cells. Exp Cell Res 199:330–340

Raut S, Patel R, Al-Ahmad AJ (2021) Presence of a mutation in PSEN1 or PSEN2 gene is associated with an impaired brain endothelial cell phenotype in vitro. Fluids Barriers CNS 18(1):3. https://doi.org/10.1186/s12987-020-00235-y

Redzic ZB (2013) Studies on the human choroid plexus in vitro. Fluids Barriers CNS 10(1):10. https://doi.org/10.1186/2045-8118-10-10

Regan JT, Mirczuk SM, Scudder CJ, Stacey E, Khan S, Worwood M, Powles T, Dennis-Beron JS, Ginley-Hidinger M, McGonnell IM, Volk HA, Strickland R, Tivers MS, Lawson C, Lipscomb VJ, Fowkes RC (2021) Sensitivity of the natriuretic peptide/cGMP system to hyperammonaemia in rat C6 glioma cells and GPNT brain endothelial cells. Cells 10(2):398. https://doi.org/10.3390/cells10020398

Régina A, Romero IA, Greenwood J, Adamson P, Bourre JM, Couraud PO, Roux F (1999) Dexamethasone regulation of P-glycoprotein activity in an immortalized rat brain endothelial cell line, GPNT. J Neurochem 73:1954–1963

Reichel A, Begley DJ, Abbott NJ (2003) An overview of in vitro techniques for blood-brain barrier studies. Methods Mol Med 89:307–324

Ribeiro MM, Castanho MA, Serrano I (2010) In vitro blood-brain barrier models–latest advances and therapeutic applications in a chronological perspective. Mini Rev Med Chem 10:262–270

Riganti C, Salaroglio IC, Pinzòn-Daza ML, Caldera V, Campia I, Kopecka J, Mellai M, Annovazzi L, Couraud PO, Bosia A, Ghigo D, Schiffer D (2013) Temozolomide down-regulates P-glycoprotein in human blood-brain barrier cells by disrupting Wnt3 signaling. Cell Mol Life Sci

Roudnicky F, Kim BK, Lan Y, Schmucki R, Küppers V, Christensen K, Graf M, Patsch C, Burcin M, Meyer CA, Westenskow PD, Cowan CA (2020a) Identification of a combination of transcription factors that synergistically increases endothelial cell barrier resistance. Sci Rep 10(1):3886. https://doi.org/10.1038/s41598-020-60688-x

Roudnicky F, Zhang JD, Kim BK, Pandya NJ, Lan Y, Sach-Peltason L, Ragelle H, Strassburger P, Gruener S, Lazendic M, Uhles S, Revelant F, Eidam O, Sturm G, Kueppers V, Christensen K, Goldstein LD, Tzouros M, Banfai B, Modrusan Z, Graf M, Patsch C, Burcin M, Meyer CA, Westenskow PD, Cowan CA (2020b) Inducers of the endothelial cell barrier identified through chemogenomic screening in genome-edited hPSC-endothelial cells. Proc Natl Acad Sci USA 117(33):19854–19865. https://doi.org/10.1073/pnas.1911532117

Roux F, Durieu-Trautmann O, Chaverot N, Claire M, Mailly P, Bourre JM, Strosberg AD, Couraud PO (1994) Regulation of gamma-glutamyl transpeptidase and alkaline phosphatase activities in immortalized rat brain microvessel endothelial cells. J Cell Physiol 159:101–113

Rubin LL, Hall DE, Porter S, Barbu K, Cannon C, Horner HC, Janatpour M, Liaw CW, Manning K, Morales J, Tanner LI, Tomaselli KJ, Bard F (1991) A cell culture model of the blood-brain barrier. J Cell Biol 115:1725–1735

Rutten MJ, Hoover RL, Karnovsky MJ (1987) Electrical resistance and macromolecular permeability of brain endothelial monolayer cultures. Brain Res 425:301–310

Sadeghzadeh M, Wenzel B, Gündel D, Deuther-Conrad W, Toussaint M, Moldovan RP, Fischer S, Ludwig FA, Teodoro R, Jonnalagadda S, Jonnalagadda SK, Schüürmann G, Mereddy VR, Drewes LR, Brust P (2020) Development of Novel Analogs of the Monocarboxylate Transporter Ligand FACH and Biological Validation of One Potential Radiotracer for Positron

Emission Tomography (PET) Imaging. Molecules 25(10):2309. https://doi.org/10.3390/molecules25102309

Salvamoser JD, Avemary J, Luna-Munguia H, Pascher B, Getzinger T, Pieper T, Kudernatsch M, Kluger G, Potschka H (2015) Glutamate-Mediated Down-Regulation of the Multidrug-Resistance Protein BCRP/ABCG2 in Porcine and Human Brain Capillaries. Mol Pharm 12(6):2049–2060. https://doi.org/10.1021/mp500841w

Santa-Maria AR, Walter FR, Figueiredo R, Kincses A, Vigh JP, Heymans M, Culot M, Winter P, Gosselet F, Dér A, Deli MA (2021) Flow induces barrier and glycocalyx-related genes and negative surface charge in a lab-on-a-chip human blood-brain barrier model. J Cereb Blood Flow Metab. https://doi.org/10.1177/0271678X21992638

Saubaméa B, Cochois-Guégan V, Cisternino S, Scherrmann JM (2012) Heterogeneity in the rat brain vasculature revealed by quantitative confocal analysis of endothelial barrier antigen and P-glycoprotein expression. J Cereb Blood Flow Metab 32:81–92

Scalisi J, Balau B, Deneyer L, Bouchat J, Gilloteaux J, Nicaise C (2021) Blood-brain barrier permeability towards small and large tracers in a mouse model of osmotic demyelination syndrome. Neurosci Lett 746:135665. https://doi.org/10.1016/j.neulet.2021.135665

Schrade A, Sade H, Couraud PO, Romero IA, Weksler BB, Niewoehner J (2012) Expression and localization of claudins-3 and -12 in transformed human brain endothelium. Fluids Barriers CNS 9:6. https://doi.org/10.1186/2045-8118-9-6

Schroten M, Hanisch FG, Quednau N, Stump C, Riebe R, Lenk M, Wolburg H, Tenenbaum T, Schwerk C (2012) A novel porcine in vitro model of the blood-cerebrospinal fluid barrier with strong barrier function. PLoS One 7(6):e39835. https://doi.org/10.1371/journal.pone.0039835

Seok SM, Kim JM, Park TY, Baik EJ, Lee SH (2013) Fructose-1,6-bisphosphate ameliorates lipopolysaccharide-induced dysfunction of blood-brain barrier. Arch Pharm Res 36(9):1149–1159. https://doi.org/10.1007/s12272-013-0129-z

Shah KK, Yang L, Abbruscato TJ (2012) In vitro models of the blood-brain barrier. Methods Mol Biol 814:431–449

Shawahna R, Uchida Y, Declèves X, Ohtsuki S, Yousif S, Dauchy S, Jacob A, Chassoux F, Daumas-Duport C, Couraud PO, Terasaki T, Scherrmann JM (2011) Transcriptomic and quantitative proteomic analysis of transporters and drug metabolizing enzymes in freshly isolated human brain microvessels. Mol Pharm 8:1332–1341

Shawahna R, Decleves X, Scherrmann JM (2013) Hurdles with using in vitro models to predict human blood-brain barrier drug permeability: a special focus on transporters and metabolizing enzymes. Curr Drug Metab 14:120–136

Shirai A, Naito M, Tatsuta T, Dong J, Hanaoka K, Mikami K, Oh-hara T, Tsuruo T (1994) Transport of cyclosporin A across the brain capillary endothelial cell monolayer by P-glycoprotein. Biochim Biophys Acta 1222:400–404

Shubbar MH, Penny JI (2020) Therapeutic drugs modulate ATP-Binding cassette transporter-mediated transport of amyloid beta(1-42) in brain microvascular endothelial cells. Eur J Pharmacol 874:173009. https://doi.org/10.1016/j.ejphar.2020.173009

Siakotos AN, Rouser G (1969) Isolation of highly purified human and bovine brain endothelial cells and nuclei and their phospholipid composition. Lipids 4:234–239

Skinner RA, Gibson RM, Rothwell NJ, Pinteaux E, Penny JI (2009) Transport of interleukin-1 across cerebromicrovascular endothelial cells. Br J Pharmacol 156:1115–1123

Smith M, Omidi Y, Gumbleton M (2007) Primary porcine brain microvascular endothelial cells: biochemical and functional characterisation as a model for drug transport and targeting. J Drug Target 15(4):253–268. https://doi.org/10.1080/10611860701288539

Stanness KA, Guatteo E, Janigro D (1996) A dynamic model of the blood-brain barrier "in vitro". Neurotoxicology 17:481–496

Stanness KA, Westrum LE, Fornaciari E, Mascagni P, Nelson JA, Stenglein SG, Myers T, Janigro D (1997) Morphological and functional characterization of an in vitro blood-brain barrier model. Brain Res 771:329–342

Stone NL, England TJ, O'Sullivan SE (2019) A Novel Transwell Blood Brain Barrier Model Using Primary Human Cells. Front Cell Neurosci 13:230. https://doi.org/10.3389/fncel.2019.00230

Strazielle N, Ghersi-Egea JF (1999) Demonstration of a coupled metabolism-efflux process at the choroid plexus as a mechanism of brain protection toward xenobiotics. J Neurosci 19:6275–6289

Strazielle N, Ghersi-Egea JF (2013) Physiology of blood-brain interfaces in relation to brain disposition of small compounds and macromolecules. Mol Pharm 10:1473–1491

Sugimoto K, Ichikawa-Tomikawa N, Nishiura K, Kunii Y, Sano Y, Shimizu F, Kakita A, Kanda T, Imura T, Chiba H (2020) Serotonin/5-HT1A Signaling in the Neurovascular Unit Regulates Endothelial CLDN5 Expression. Int J Mol Sci 22(1):254. https://doi.org/10.3390/ijms22010254

Summerfield SG, Read K, Begley DJ, Obradovic T, Hidalgo IJ, Coggon S, Lewis AV, Porter RA, Jeffrey P (2007) Central nervous system drug disposition: the relationship between in situ brain permeability and brain free fraction. J Pharmacol Exp Ther 322:205–313

Suzuki T, Ohmuro A, Miyata M, Furuishi T, Hidaka S, Kugawa F, Fukami T, Tomono K (2010) Involvement of an influx transporter in the blood–brain barrier transport of naloxone. Biopharm Drug Dispo 31(4):243–252. https://doi.org/10.1002/bdd.707

Tachikawa M, Murakami K, Akaogi R, Akanuma SI, Terasaki T, Hosoya KI (2020) Polarized hemichannel opening of pannexin 1/connexin 43 contributes to dysregulation of transport function in blood-brain barrier endothelial cells. Neurochem Int 132:104600. https://doi.org/10.1016/j.neuint.2019.104600

Tega Y, Akanuma S, Kubo Y, Terasaki T, Hosoya K (2013) Blood-to-brain influx transport of nicotine at the rat blood-brain barrier: involvement of a pyrilamine-sensitive organic cation transport process. Neurochem Int 62:173–181

Terasaki T, Hosoya K (2001) Conditionally immortalized cell lines as a new in vitro model for the study of barrier functions. Biol Pharm Bull 24:111–118

Terasaki T, Ohtsuki S, Hori S, Takanaga H, Nakashima E, Hosoya K (2003) New approaches to in vitro models of blood-brain barrier drug transport. Drug Discov Today 8:944–954

Thanabalasundaram G, Schneidewind J, Pieper C, Galla HJ (2011) The impact of pericytes on the blood-brain barrier integrity depends critically on the pericyte differentiation stage. Int J Biochem Cell Biol 43:1284–1293

Thomsen LB, Burkhart A, Moos T (2015) A Triple Culture Model of the Blood-Brain Barrier Using Porcine Brain Endothelial cells, Astrocytes and Pericytes. PLoS One 10(8):e0134765. https://doi.org/10.1371/journal.pone.0134765

Thomsen MS, Birkelund S, Burkhart A, Stensballe A, Moos T (2017) Synthesis and deposition of basement membrane proteins by primary brain capillary endothelial cells in a murine model of the blood-brain barrier. J Neurochem 140(5):741–754. https://doi.org/10.1111/jnc.13747

Tornabene E, Helms HCC, Pedersen SF, Brodin B (2019) Effects of oxygen-glucose deprivation (OGD) on barrier properties and mRNA transcript levels of selected marker proteins in brain endothelial cells/astrocyte co-cultures. PLoS One 14(8):e0221103. https://doi.org/10.1371/journal.pone.0221103

Uchida Y, Ohtsuki S, Kamiie J, Terasaki T (2011a) Blood-brain barrier (BBB) pharmacoproteomics: reconstruction of in vivo brain distribution of 11 P-glycoprotein substrates based on the BBB transporter protein concentration, in vitro intrinsic transport activity, and unbound fraction in plasma and brain in mice. J Pharmacol Exp Ther 339:579–588

Uchida Y, Ohtsuki S, Katsukura Y, Ikeda C, Suzuki T, Kamiie J, Terasaki T (2011b) Quantitative targeted absolute proteomics of human blood-brain barrier transporters and receptors. J Neurochem 117:333–345

Uchida Y, Tachikawa M, Obuchi W, Hoshi Y, Tomioka Y, Ohtsuki S, Terasaki T (2013) A study protocol for quantitative targeted absolute proteomics (QTAP) by LC-MS/MS: application for inter-strain differences in protein expression levels of transporters, receptors, claudin-5, and marker proteins at the blood-brain barrier in ddY, FVB, and C57BL/6J mice. Fluids Barriers CNS 10(1):21. https://doi.org/10.1186/2045-8118-10-21

Urich E, Patsch C, Aigner S, Graf M, Iacone R, Freskgård PO (2013) Multicellular self-assembled spheroidal model of the blood brain barrier. Sci Rep 3:1500. https://doi.org/10.1038/srep01500

Vandenhaute E, Dehouck L, Boucau MC, Sevin E, Uzbekov R, Tardivel M, Gosselet F, Fenart L, Cecchelli R, Dehouck MP (2011) Modelling the neurovascular unit and the blood-brain barrier with the unique function of pericytes. Curr Neurovasc Res 8:258–269

Vandenhaute E, Sevin E, Hallier-Vanuxeem D, Dehouck MP, Cecchelli R (2012) Case study: adapting in vitro blood-brain barrier models for use in early-stage drug discovery. Drug Discov Today 17:285–290

Venkat P, Ning R, Zacharek A, Culmone L, Liang L, Landschoot-Ward J, Chopp M (2021) Treatment with an Angiopoietin-1 mimetic peptide promotes neurological recovery after stroke in diabetic rats. CNS Neurosci Ther 27(1):48–59. https://doi.org/10.1111/cns.13541

Veszelka S, Tóth A, Walter FR, Tóth AE, Gróf I, Mészáros M, Bocsik A, Hellinger É, Vastag M, Rákhely G, Deli MA (2018) Comparison of a Rat Primary Cell-Based Blood-Brain Barrier Model With Epithelial and Brain Endothelial Cell Lines: Gene Expression and Drug Transport. Front Mol Neurosci 11:166. https://doi.org/10.3389/fnmol.2018.00166

Wagner EF, Risau W (1994) Oncogenes in the study of endothelial cell growth and differentiation. Semin Cancer Biol 5:137–145

Wainwright L, Hargreaves IP, Georgian AR, Turner C, Dalton RN, Abbott NJ, Heales SJR, Preston JE (2020) CoQ10 Deficient Endothelial Cell Culture Model for the Investigation of CoQ10 Blood-Brain Barrier Transport. J Clin Med 9(10):3236. https://doi.org/10.3390/jcm9103236

Wallace BK, Foroutan S, O'Donnell ME (2011) Ischemia-induced stimulation of Na-K-Cl cotransport in cerebral microvascular endothelial cells involves AMP kinase. Am J Physiol Cell Physiol 301:C316–C326

Wang Q, Luo W, Zhang W, Liu M, Song H, Chen J (2011) Involvement of DMT1+IRE in the transport of lead in an in vitro BBB model. Toxicol In Vitro 25:991–998

Wang S, Qaisar U, Yin X, Grammas P (2012) Gene expression profiling in Alzheimer's disease brain microvessels. J Alzheimers Dis 31:193–205

Watson PM, Paterson JC, Thom G, Ginman U, Lundquist S, Webster CI (2013) Modelling the endothelial blood-CNS barriers: a method for the production of robust in vitro models of the rat blood-brain barrier and blood-spinal cord barrier. BMC Neurosci 14(1):59

Wegener J, Sieber M, Galla HJ (1996) Impedance analysis of epithelial and endothelial cell monolayers cultured on gold surfaces. J Biochem Biophys Methods 32:151–170

Wegener J, Hakvoort A, Galla HJ (2000) Barrier function of porcine choroid plexus epithelial cells is modulated by cAMP-dependent pathways in vitro. Brain Res 853:115–124

Weksler BB, Subileau EA, Perrière N, Charneau P, Holloway K, Leveque M, Tricoire-Leignel H, Nicotra A, Bourdoulous S, Turowski P, Male DK, Roux F, Greenwood J, Romero IA, Couraud PO (2005) Blood-brain barrier-specific properties of a human adult brain endothelial cell line. FASEB J 19:1872–1874

Weksler B, Romero IA, Couraud PO (2013) The hCMEC/D3 cell line as a model of the human blood brain barrier. Fluids Barriers CNS 10(1):16. https://doi.org/10.1186/2045-8118-10-16

Williams RL, Risau W, Zerwes HG, Drexler H, Aguzzi A, Wagner EF (1989) Endothelioma cells expressing the polyoma middle T oncogene induce hemangiomas by host cell recruitment. Cell 57:1053–1063

Wilson G (1990) Cell culture techniques for the study of drug transport. Eur J Drug Metab Pharmacokinet 15:159–163

Woods S, O'Brien LM, Butcher W, Preston JE, Georgian AR, Williamson ED, Salguero FJ, Modino F, Abbott NJ, Roberts CW, D'Elia RV (2020) Glucosamine-NISV delivers antibody across the blood-brain barrier: Optimization for treatment of encephalitic viruses. J Control Release 324:644–656. https://doi.org/10.1016/j.jconrel.2020.05.048

Wuest DM, Wing AM, Lee KH (2013) Membrane configuration optimization for a murine in vitro blood-brain barrier model. J Neurosci Methods 212(2):211–221. https://doi.org/10.1016/j.jneumeth.2012.10.016

Yamaguchi S, Ito S, Masuda T, Couraud PO, Ohtsuki S (2020) Novel cyclic peptides facilitating transcellular blood-brain barrier transport of macromolecules in vitro and in vivo. J Control Release 321:744–755. https://doi.org/10.1016/j.jconrel.2020.03.001

Yamashita M, Aoki H, Hashita T, Iwao T, Matsunaga T (2020) Inhibition of transforming growth factor beta signaling pathway promotes differentiation of human induced pluripotent stem cell-derived brain microvascular endothelial-like cells. Fluids Barriers CNS 17(1):36. https://doi.org/10.1186/s12987-020-00197-1

Yasuda K, Cline C, Vogel P, Onciu M, Fatima S, Sorrentino BP, Thirumaran RK, Ekins S, Urade Y, Fujimori K, Schuetz EG (2013) Drug transporters on arachnoid barrier cells contribute to the blood-cerebrospinal fluid barrier. Drug Metab Dispos 41:923–931

Youdim KA, Avdeef A, Abbott NJ (2003) In vitro trans-monolayer permeability calculations: often forgotten assumptions. Drug Discov Today 8:997–1003

Yusof SR, Avdeef A, Abbott NJ (2014) In vitro porcine blood-brain barrier model for permeability studies: pCEL-X software pKa(FLUX) method for aqueous boundary layer correction and detailed data analysis. Eur J Pharm Sci 65:98–111. https://doi.org/10.1016/j.ejps.2014.09.009

Zhang Y, Li CS, Ye Y, Johnson K, Poe J, Johnson S, Bobrowski W, Garrido R, Madhu C (2006) Porcine brain microvessel endothelial cells as an in vitro model to predict in vivo blood-brain barrier permeability. Drug Metab Dispos 34:1935–1943

Zhang Y, He J, Shen L, Wang T, Yang J, Li Y, Wang Y, Quan D (2021) Brain-targeted delivery of obidoxime, using aptamer-modified liposomes, for detoxification of organophosphorus compounds. J Control Release 329:1117–1128. https://doi.org/10.1016/j.jconrel.2020.10.039

Zolotoff C, Voirin AC, Puech C, Roche F, Perek N (2020) Intermittent hypoxia and its impact on Nrf2/HIF-1α expression and ABC transporters: an in vitro human blood-brain barrier model study. Cell Physiol Biochem 54(6):1231–1248. https://doi.org/10.33594/000000311

Chapter 9
Human In Vitro Blood-Brain Barrier Models Derived from Stem Cells

Koji L. Foreman, Sean P. Palecek, and Eric V. Shusta

Abstract In vitro blood-brain barrier (BBB) models have significant utility in understanding the BBB in health and disease. While human BBB models using primary and immortalized brain endothelial cells have been described, issues regarding the retention of BBB properties and scalability have hampered their use for certain applications. Differentiation of stem cells to endothelial cells exhibiting BBB properties offers advantages including high barrier, human relevance, and ease of scaling. In this chapter, we will introduce stem cells and how they are being used to model the blood-brain barrier. We will focus predominantly on the characteristics of BBB-like endothelial cells, which express BBB markers and exhibit barrier and transporter activities, differentiated from human pluripotent stem cells (hPSCs). We will also discuss the incorporation of co-culture with other stem cell-derived or primary cells of the neurovascular unit (NVU) into stem cell-derived BBB models, as well as microfluidic and suspension culture models. These stem cell-derived BBB models have applications in modeling human BBB development, studying the roles of disease on BBB function, drug discovery, and development of strategies for neurotherapeutic delivery.

Keywords Human pluripotent stem cells · Blood-brain barrier · Models

K. L. Foreman · S. P. Palecek
Department of Chemical and Biological Engineering, University of Wisconsin, Madison, WI, USA
e-mail: klforeman@wisc.edu; sppalecek@wisc.edu

E. V. Shusta (✉)
Department of Chemical and Biological Engineering, University of Wisconsin, Madison, WI, USA

Department of Neurological Surgery, University of Wisconsin, Madison, WI, USA
e-mail: eshusta@wisc.edu

© American Association of Pharmaceutical Scientists 2022
E. C. M. de Lange et al. (eds.), *Drug Delivery to the Brain*, AAPS Advances in the Pharmaceutical Sciences Series 33, https://doi.org/10.1007/978-3-030-88773-5_9

9.1   Introduction

The blood-brain barrier (BBB) is a major roadblock in the development of new therapeutics targeting central nervous system diseases such as brain tumors and neurodegenerative diseases (Pardridge 2005). Human in vitro BBB models offer platforms for systematic study of BBB development and its role in disease, as well as for drug discovery. However, primary and immortalized human brain microvascular endothelial cells (BMECs), which typically are the basis for these models, often possess significantly diminished BBB characteristics. For example, transendothelial electrical resistance (TEER), a commonly used metric of paracellular permeability, is frequently abnormally low. TEER of these models typically falls below $100\ \Omega\text{-cm}^2$ (Helms et al. 2016; Weksler et al. 2005); however, it is estimated that the true physiologic TEER can be above $5000\ \Omega\text{-cm}^2$ in vivo (Butt et al. 1990; Srinivasan et al. 2015). Human BBB models with low TEER values ($\sim100\text{--}200\ \Omega\text{-cm}^2$) allow aberrant paracellular passage of larger proteins and increased penetration of small molecules (Mantle et al. 2016), making these models insufficient for predictive in vitro drug permeability studies. Primary human BMECs are a finite resource that cannot be easily scaled for scientific or drug development studies. The use of primary animal BMECs can alleviate scaling and availability pressures; however, animal models possess numerous key differences compared to human BBB performance, particularly with respect to efflux transporter expression, a key component of BBB resistance (Aday et al. 2016; H. W. Song et al. 2020). To address issues such as human relevance, scalability, and model fidelity, researchers have recently begun to build BBB models using human stem cell sources. These models are comprised of stem cell-derived BMEC-like cells possessing BBB properties which can be combined with supporting cells of the NVU such as stem cell-derived astrocytes, pericytes, neurons, and microglia. Below we will outline the stem cell sources, cellular differentiation strategies, and multicellular constructs that can be used to investigate the BBB in health and disease.

9.2   Stem Cell Sources for BBB Modeling

With the development of stem cell isolation, culture, and differentiation technologies, scalable human BBB models with physiological properties have become possible. A stem cell is defined as a cell that is capable of self-renewing and differentiating into more specialized cell types (Chagastelles and Nardi 2011). This self-renewal capacity means that the stem cells can be expanded in the undifferentiated state and serve as a source for large quantities of the various cell types, enabling large-scale studies and screens. Having a single self-renewing precursor can also help limit the batch-to-batch variability intrinsic to human primary cells. Moreover, stem cells can be isolated from individual patients, allowing the study of disease conditions with a genetic basis. Finally, the transition from an undifferentiated stem cell to one possessing BBB properties could offer insights regarding the mechanisms of human BBB induction and maintenance.

There are many different types of stem and progenitor cells found throughout development, only some of which reside in the adult. Here, we will largely focus on human pluripotent stem cells (hPSCs), which include human embryonic stem cells (hESCs) and induced pluripotent stem cells (iPSCs). hPSCs possess the theoretical capacity for infinite self-renewal (although in practice, they are limited by eventual acquisition of genetic mutations) and can differentiate into any cell type in the adult (Nagy et al. 1990). hESCs are derived from the inner cell mass of human blastocysts, as first demonstrated in 1998 by the Thomson lab (Thomson 1998). iPSCs are generated by reprogramming somatic cell types, such as fibroblasts or lymphocytes, by inducing expression of core transcription factors that regulate pluripotency to produce an hESC-like cell (Takahashi and Yamanaka 2006; Yu et al. 2007). As will be discussed below, iPSCs enable a whole new spectrum of patient-specific BBB modeling, with particular application in studying how human genetic diseases influence the BBB.

Umbilical cord blood-derived stem cells have also been used to generate human BMEC-like cells. While these cord blood-derived adult stem cells are substantially more fate restricted than hPSCs, they have the capacity to differentiate to endothelial cells (Bailey et al. 2004), and some attempts have been made to induce BBB properties in these endothelial cells (Boyer-Di Ponio et al. 2014; Cecchelli et al. 2014).

It is important to note that while stem cell-derived BBB models recapitulate some aspects of BBB protein expression and phenotypes, they are not perfect facsimiles of the in vivo BBB. They differentiate under conditions inspired by human BBB development, but do not receive all cues found in the developing brain. Moreover, the culture conditions in vitro differ from the in vivo microenvironment. Thus, in matching a model to a specific application, it is important to consider particular attributes the model lacks or possesses relative to the in vivo BBB. Below, we will discuss the various routes that researchers have employed for the differentiation of BMEC-like cells and the use of these cells in modeling the BBB.

9.3 Differentiation of Stem Cells to BMEC-Like Cells

The incorporation of stem cell-derived BMEC-like cells into BBB models has been enabled by the development of several BMEC differentiation protocols, as discussed in detail below. From here forward, we will refer to all the different forms of stem cell-derived BMEC-like cells simply as BMECs. This is not meant to imply, however, that they are perfect facsimiles of in vivo or in vitro human BMECs.

9.3.1 hPSC-Derived BMECs

9.3.1.1 Co-differentiation with Neural Progenitors

As the endothelial cells of the perineural vascular plexus invade the developing neural tube during development, they acquire a BBB phenotype in part by interactions with developing neural cells (Bauer and Bauer 1997). A stem cell co-differentiation technique was developed to mimic this interaction to generate hPSC-BMECs (Lippmann et al. 2012). Initially, hPSCs are seeded and allowed to expand for 3 days prior to inducing differentiation. Next, the hPSCs are exposed to basal medium supplemented with amino acids, a serum replacement cocktail, and β-mercaptoethanol that drives differentiation to a mixed population containing neural and endothelial cells, among other cell types (Fig. 9.1). This co-differentiation results in colonies of PECAM-1 positive endothelial cells surrounded by βIII tubulin and nestin expressing neural tracts that somewhat resemble the in vivo cellular environment (Lippmann et al. 2012, 2013). During this phase, endothelial cells gain GLUT-1 expression in concert with nuclear β-catenin localization, likely a consequence of Wnt signaling from the co-differentiating neural cells (Lippmann et al. 2012). These data suggested the importance of Wnt signaling in induction of BBB properties during the differentiation process, in agreement with BBB development in mice (Cho et al. 2017a; R. Daneman et al. 2009; Liebner et al. 2008; Stenman et al. 2008).The co-differentiating cell mixture is then switched to endothelial medium to facilitate selective outgrowth of the endothelial cells (Lippmann et al. 2012).

At the end of this growth phase, the endothelial cells exhibited expression of BBB markers including tight junction associated proteins ZO-1, occludin, and claudin-5; endothelial markers PECAM-1, von Willebrand factor, and VE-cadherin; and efflux transporters MRP1, BCRP, and P-gp (Table 9.1). Passaging and subculture of the cell mixtures onto an endothelial-selective matrix of collagen IV and fibronectin yields a nearly pure monolayer of hPSC-BMECs (Lippmann et al. 2012). The monocultured hPSC-BMECs also demonstrate strong tight junction formation with TEER values up to 500 Ω-cm^2 (Stebbins et al. 2016), which is below in vivo levels (Butt et al. 1990; Srinivasan et al. 2015) but equivalent to or greater than most other models (Helms et al. 2016). The hPSC-BMECs also possess polarized P-gp, MRP1, and BCRP efflux transporter activities (Katt et al. 2016; Lippmann et al. 2012; Stebbins et al. 2016; Wilson et al. 2015). The barrier quality can vary greatly depending on initial stem cell seeding density and stem cell line, so careful optimization is required to generate cells that express BMEC markers and exhibit BBB phenotypes (Katt et al. 2016; Qian et al. 2017; Wilson et al. 2016). hPSC-BMECs at both the subculture stage and post-differentiation can be cryopreserved, though addition of a ROCK inhibitor post-thaw is necessary to restore barrier function and increase cell attachment and survival (Wilson et al. 2016).

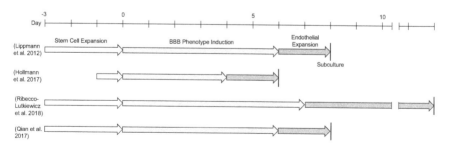

Fig. 9.1 Progression of differentiation stages for some common methods of hPSC-BMEC induction. Note that the media compositions are different in each protocol

Table 9.1 Summary of human stem cell-derived BMEC properties

References	Endothelial proteins	Tight junction-associated proteins	Efflux transporters	Other BBB proteins	TEER reported	Efflux activity measured	Additional assays
Lippmann et al. (2012); Stebbins et al. (2016)	vWF, VE-cad, PECAM-1	ZO-1, Ocln, Cldn-5	P-gp, BCRP, MRP1	GLUT-1	100–2000 Ω-cm² (RA dependent)	P-gp, BCRP, MRP1	Tube formation, small/large molecule, and drug permeability
Hollmann et al. (2017); Neal et al. (2019)	PECAM-1, VE-cad	Ocln, Cldn-5		GLUT-1	2000–8000 Ω-cm²	P-gp, MRP	Small molecule permeability, TEER duration
Ribecco-Lutkiewicz et al. (2018)	PECAM-1, vWF	ZO-1, Ocln, Cldn-5	P-gp	GLUT-1	400–1500 Ω-cm²		Drug permeability, sucrose permeability
Qian et al. (2017)	PECAM-1, VE-cad, Flk-1	ZO-1, Ocln, Cldn-5	P-gp, BCRP, MRP1	GLUT-1	2000–3000 Ω-cm²	P-gp, BCRP, MRP1	Tube formation, small molecule permeability
Delsing et al. (2018)	vWF, VE-cad, PECAM-1	ZO-1, Ocln, Cldn-5		GLUT-1	~100 Ω-cm²	P-gp, BCRP	Drug permeability, small molecule permeability
Praça et al. (2019)	vWF, VE-cad, Tie2	ZO-1, Ocln, Cldn-5	P-gp	GLUT-1	~60 Ω-cm²	P-gp	
Cecchelli et al. (2014)	JAM-A	ZO-1, Ocln, Cldn-5	P-gp, BCRP, MRP1, MRP2, MRP4, MRP5	RAGE, LRP1	~60 Ω-cm²		Small and large molecule and drug permeability
Boyer-Di Ponio et al. (2014)	VE-cad	ZO-1, Ocln, Cldn-5, Cldn-3	P-gp	GLUT-1		P-gp	Small and large molecule permeability, tube formation

Cldn-5 = Claudin-5, Cldn-3 = Claudin-3, Ocln = occludin, VE-cad = VE-cadherin, vWF = von Willebrand factor, RA = retinoic acid

Addition of retinoic acid (RA) during the endothelial expansion and subculture phases of BMEC differentiation improved expression of endothelial markers and increased barrier tightness (Lippmann et al. 2014). RA is critical for the patterning of the central nervous system along the anterior-posterior axis (Schubert et al. 2006) and has been implicated in BBB specification in vivo, possibly through regulation of Wnt signaling (Bonney et al. 2016, 2018; Mizee et al. 2013; Obermeier et al. 2013). RA addition during the endothelial expansion phase of BMEC differentiation was shown to improve multiple BBB phenotypes (Lippmann et al. 2014). Supplementation significantly increased TEER, peaking between 1000 and 3000 Ω-cm^2 for monocultured BMECs, corresponding with increased continuity of tight junction proteins occludin and claudin-5. The cultures also exhibited earlier induction and increased expression of VE-cadherin. Efflux transporter transcripts were also upregulated, but this did not translate to improved efflux transport activity as assessed by substrate accumulation assays (Lippmann et al. 2014). In addition, inhibition of TGFβ signaling can result in increased TEER as well as improved expression of some endothelial markers (Motallebnejad et al. 2019; Yamashita et al. 2020).

9.3.1.2 Accelerated Co-differentiation

The accelerated co-differentiation protocol is a variant of the previously described method that reduces differentiation duration from 13 to 7 days (Fig. 9.1). The stem cell expansion period is reduced from 3 days to 1. The co-differentiation occurs in the same basal medium, but rather than using a serum replacement mixture, a combination of L-ascorbic acid 2-phosphate sesquimagnesium salt, sodium selenite, sodium bicarbonate, human insulin, and human holotransferrin is used instead (Hollmann et al. 2017). This modified medium is applied for 4 days in comparison with 6 days in the UM protocol. The following endothelial expansion and subculture phases, including media compositions, remained the same. The resulting cells require 4 fewer days to generate but display similar barrier properties as assessed by TEER, efflux substrate accumulation assays, and immunocytochemistry (Table 9.1).

Further modifications of this accelerated protocol demonstrated that B-27 and N-2 media supplements, originally designed for neural differentiations (Bottenstein and Sato 1979; Brewer and Cotman 1989), can be applied in place of the platelet poor plasma-derived serum in the endothelial expansion phase (Neal et al. 2019). The B-27 supplement performed as well or better than supplementation with serum with respect to efflux transporter activity and TEER. In addition, reducing the supplement to only insulin, transferrin, and selenium was sufficient to differentiate cells having similar barrier properties. Taken together, these approaches provide a rapid and chemically defined serum-free method to differentiate hPSCs to BMEC-like cells.

9.3.1.3 Differentiation in Low Osmolarity Medium

Control of osmolarity during the differentiation has also been suspected to affect the quality of the differentiation. Therefore, during the induction phase, the basal medium was replaced with a low osmolarity alternative (Ribecco-Lutkiewicz et al. 2018). The differentiation continues in endothelial medium supplemented with platelet poor plasma-derived serum and bFGF (Fig. 9.1). The authors claim that unlike the aforementioned protocols, the extended low osmolarity protocol generated BMECs without co-culture, though neural cell presence was not explored. There is no selective matrix used for purification, and the method can take up to 21 days before an optimal barrier forms. The resulting hPSC-BMECs demonstrate TEER up to 1500 Ω-cm^2 and could be used to discriminate BBB passage of antibodies for targeted drug delivery. The cells express the vascular markers von Willebrand factor and CD31; tight junction associated proteins occludin, claudin-5, and ZO-1; as well as the transporters P-gp and GLUT1 (Table 9.1). An RNA microarray for transport and BBB-related transcripts showed good transcriptional alignment with primary human BMECs. As of this writing, this method has only been applied to a single iPSC line.

9.3.1.4 Directed Differentiation Models

To reduce the variability in the differentiation methods, researchers have worked to eliminate the need for both serum products and co-differentiation. Sequential activation of biologically relevant pathways was used to drive the entire cell population through endothelial differentiation and BBB phenotype induction. Induction of canonical Wnt signaling is used to drive cells toward mesoderm, and RA signaling is used to induce BBB phenotypes. Following an hPSC expansion period, Wnt signaling is activated to drive cells toward a mesoderm fate using 1 day of treatment with CHIR99021, a small molecule GSK3 inhibitor (Kempf et al. 2016; Lian et al. 2014; Naujok et al. 2014). After treatment, most of the cell population expresses brachyury, an indicator of a primitive streak-like state, suggesting that these cells, like BMECs in vivo, are mesoderm-derived (Lian et al. 2014; Qian et al. 2017). Next, the population is treated with basal medium supplemented with B-27 (Fig. 9.1). During this stage, cells progress through mesoderm to endothelial progenitors that express VEGFR2 and CD34 (Qian et al. 2017). At day 6, the cells undergo an expansion and BMEC specification phase in endothelial medium supplemented with RA and B-27 supplement. Serum products are inherently variable and undefined, so in addition to generating a mostly pure population of cells that appear to differentiate in register, the method is also chemically defined which helps limit batch-to-batch variability from reagents (Hollmann et al. 2017). The resulting cells demonstrate TEER around 4000 Ω-cm^2; possess polarized P-gp, MRP1, and BCRP activity as assessed by substrate transport assays; and have low permeability to fluorescent small molecules (Grifno et al. 2019; Qian et al. 2017) (Table 9.1). Similar to the protocol described in Sect. 9.3.1.1, hPSC-BMECs can be frozen on day 8 during the subculture phase. ROCK inhibitor is also required to ensure adequate survival post-thaw (Grifno et al. 2019). Since this protocol shares some similarities to in vivo development, it may be useful investigating specific factors involved in driving BMEC fate.

9.3.1.5 Induction of BMEC Properties in Endothelial Progenitors

Two approaches for BMEC differentiation have been reported which first generate endothelial progenitors from hPSCs using general endothelial cell differentiation protocols and subsequently induce BMEC phenotypes by factor addition and co-culture with NVU cells (Delsing et al. 2018; Praça et al. 2019). The first demonstration of this approach treated hPSCs with a proprietary mesoderm induction medium with additional VEGF, Activin A, BMP4, and CHIR99021 for 3 days (Delsing et al. 2018). Cells were then driven to a vascular fate using proprietary vascular specification medium supplemented with VEGF and SB431542, a TGFβ signaling inhibitor, for 8 days. The resulting endothelial progenitors were then sorted via magnetic activated cell sorting for CD31 expression and added to Transwells containing hPSC-derived pericytes and primary human astrocytes and neurons. Compared to monoculture, co-culture improved TEER to a peak near 100 Ω-cm^2, increased expression of GLUT1 and occludin, and decreased caveolin 1 expression, demonstrating some acquisition of BBB character (Table 9.1). There was no observable increase in P-gp activity or resistance to small molecule transfer, and TEER remained substantially below other hPSC-BMEC models, indicating less robust BBB functions.

In the second approach, immature endothelial progenitors were differentiated from hPSCs using medium containing BMP4, FGF-β, and VEGF to generate a mixed population of endothelial progenitors and non-endothelial cells. The population was sorted for PECAM-1 positive cells via magnetic activated cell sorting. The purified endothelial progenitor population was then treated with endothelial growth medium that contained VEGF for one passage, followed by treatment with VEGF, Wnt3a, and RA over four passages (Praça et al. 2019). This combination is similar to previously described protocols which combined Wnt and retinoic acid signaling. In vivo, VEGF and RA are released by or present in the neural tube during the invasion of nascent endothelial cells (Obermeier et al. 2013). Wnt7a, but not Wnt3a, is suspected in BMEC specification in vivo (R. Daneman et al. 2009; Stenman et al. 2008), although any Wnt activation may be sufficient to elicit the transition in vitro. The resultant cells demonstrate expression of tight junction and endothelial cell markers including claudin-5, occludin, VE-cadherin, CD31, von Willebrand factor, and GLUT-1. However, TEER does not exceed 60 Ω-cm^2, similar to non-tissue-specific endothelial cells (Tschugguel et al. 1995). In addition, the cells demonstrated P-gp activity through substrate accumulation assays, but this required the presence of pericyte co-culture. Taken together, the findings from these endothelial progenitor-based protocols indicate the difficulty in inducing BBB barrier properties onto immature endothelial cells as well as the value of pericytes in barrier induction. However, the similarity to in vivo developmental processes and high endothelial nature could make this a useful model for studying BBB development and identifying cues that can induce more robust BBB properties.

9.3.2 Genomic Comparison of hPSC-BMECs Using the Various Protocols

Above, and in Table 9.1, we have compared some of the phenotypes of the hPSC-derived BMECs arising from various differentiation protocols. In addition, cells differentiated via several of the methods have been compared on a transcriptome level to in vitro cultured primary BMECs and acutely isolated human BMECs. Despite the differences between the co-differentiation protocol described in Sect. 9.3.1.1 and the defined protocol described in Sect. 9.3.1.5, BMECs generated using these two methods were similar on a transcriptome-wide scale, mirroring their similarities in BBB phenotype (Qian et al. 2017). When compared to in vitro primary human BMECs, the transcriptomes of hPSC-BMECs generated by co-differentiation correlated more strongly with endothelial cells than neural epithelial cells as assessed by principal component analysis and hierarchical clustering of the 500 most variable genes (Vatine et al. 2019). The directed differentiation method described in Sect. 9.3.1.5 was found to have a strong positive correlation with co-differentiated hPSC-BMECs and clustered closer to primary human BMECs than mesoderm, endoderm, or ectodermal cells via hierarchical clustering (Qian et al. 2017). hPSC-derived endothelial progenitors co-cultured with astrocytes, neurons, and hPSC-pericytes saw induction of barrier-related protein transcripts, but at a lower level than those observed in co-differentiated hPSC-BMECs in co-culture (Delsing et al. 2018). However, the BMECs derived from endothelial progenitors had much higher expression of endothelial transcripts, such as CD31 and VE-cadherin, than those generated from the co-differentiation route. Instead, the co-differentiated hPSC-BMECs had a much higher expression of epithelial-associated transcripts (Delsing et al. 2018). Similarly, a recent study also noted the expression of epithelial associated transcripts, in co-differentiated BMECs (Vatine et al. 2019). Given the mixed epithelial and endothelial signatures of BMECs derived through co-differentiation or directed differentiation routes, BMECs derived from induction of endothelial progenitors may prove more effective at modeling endothelial attributes such as inflammatory response or studying cues involved in endothelial development, despite their subpar barrier and transport properties (Kim et al. 2017; Martins Gomes et al. 2019). Identifying routes to hPSC-BMECs that can preserve both strong BBB phenotypes and strong endothelial phenotypes remains an active area of research in the field.

9.3.3 Cord Blood Progenitor Cell Models

In addition to numerous hPSC-BMEC models described above, others have used umbilical cord-derived progenitor cells to produce BMEC-like cells. A population of CD34+ mononuclear cells was isolated from cord blood, and when exposed to an endothelial cell growth medium, these expanding colonies progressed into endothelial cell progenitors that expressed CD31, VE-cadherin, and von Willebrand factor (Cecchelli et al. 2014). To induce BBB properties in these endothelial progenitors, they were exposed to co-culture with bovine brain pericytes (Cecchelli et al. 2014). The resulting endothelial cells displayed membrane polarization and P-gp activity; tight junction associated proteins occludin, claudin-5, and ZO-1 were expressed; and TEER was ~100–200 Ω-cm^2 (Table 9.1). Application of selective agonists or ligands once again implicated Wnt signaling in development of BBB characteristics due to pericyte co-culture, which aligns with in vitro (Lippmann et al. 2012) and in vivo (R. Daneman et al. 2009; Stenman et al. 2008) studies.

A similar protocol employed outgrowth of cord blood mononuclear cells in endothelial growth medium to generate cord blood endothelial progenitors (Boyer-Di Ponio et al. 2014). In an effort to induce BBB properties, these endothelial progenitors were co-cultured with primary rat astrocytes. Compared to monoculture endothelial progenitors, resultant BMEC-like cells possessed increased occludin, GLUT-1, and P-gp expression and improved resistance to small molecule transport (Boyer-Di Ponio et al. 2014) (Table 9.1). Since these models are derived from donor umbilical cord blood, they are less amenable to the modeling of genetic disease than BMECs differentiated from patient-sourced iPSCs. However, this model has been used to measure the rates of transmigration of different CD4+ T helper cell subpopulations across the BBB, and comparison with transmigration across a model of the choroid plexus allowed the authors to conclude that different T helper cell populations may enter the CNS through different routes (Nishihara et al. 2020).

9.4 Co-culture Models

Thus far, we have described sourcing for stem cell-derived BMECs. Monocultures of BMECs are the simplest models and are oftentimes well suited for experimental goals. However, monoculture BMEC models lack several cellular components of the NVU. Therefore, there are situations where BMECs are combined with other NVU cell types in various co-culture architectures, particularly if the addition of NVU cells is required to induce a specialized property in the BMECs or if one wishes to examine intercellular communication between cell types of the NVU. In the CNS, various cells of the NVU induce and maintain BBB phenotypes in BMECs (Engelhardt and Liebner 2014; Gastfriend et al. 2018; Hawkins and Davis 2005; Obermeier et al. 2013) (Fig. 9.2).

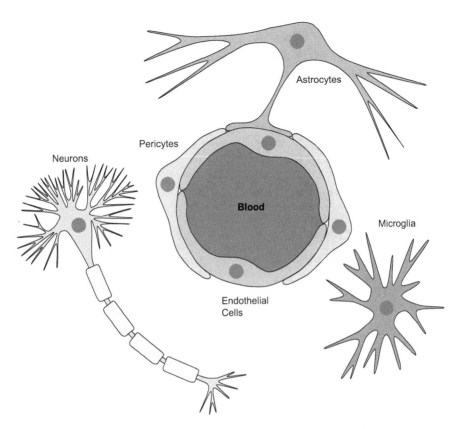

Fig. 9.2 The NVU is composed of BMECs that possess the properties most often associated with BBB function as well as surrounding cells such as pericytes which share a basement membrane with BMECs, astrocytes whose endfeet are intimately associated with microvessels, neurons, and microglia which communicate without contact with BMECs

Co-culture is often performed in a Transwell setting with some combination of neurons, pericytes, and astrocytes in the basolateral chamber and BMECs on the porous filter (Fig. 9.3). These models are simple and scalable and allow paracrine signaling via release of soluble factors, but the cells are not able to form cell-cell contacts even when co-cultured cells are plated on the opposite side of a BMEC-containing filter. There are many variations of models using the Transwell configuration with stem cell-derived BMECs and different combinations of NVU cell types and sources. We will examine each co-cultured cell type in more detail below.

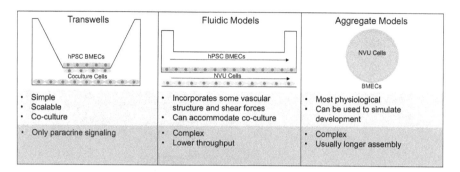

Fig. 9.3 Examples of the different architectures of BBB models. Transwells can be utilized in monoculture or co-culture formats, with hPSC-BMECs on the top of the membrane and co-cultured cells on the bottom of the membrane and/or on the bottom of the well. The fluidic model chip shows a co-culture enabled chip, though single channel chips are also widely employed. The aggregate model gives an example of an aggregated BBB spheroid with a shell of BMECs organized around a core of NVU cells. However, not all aggregates exhibit this localization

9.4.1 Neurons

Neurons are frequently added to the lower chamber of transwell-based BBB models (Table 9.2). Most approaches to date have employed hPSC-derived neurons that are generated from the EZ-sphere neuron differentiation method (Ebert et al. 2013). Using this approach, hPSCs are placed in a neural progenitor growth medium with additional bFGF, epidermal growth factor, and heparin. To generate the EZ-spheres that are reminiscent of primary human neurospheres, the cells are maintained in low attachment conditions, resulting in spheres of nestin expressing neural progenitors. These cells can be disassociated and treated with basal medium supplemented with B-27 without vitamin A to induce differentiation to neurons (Canfield et al. 2017). Co-culture of neurons with hPSC-BMECs has been shown to improve TEER above monoculture, and the impacts were additive with astrocyte co-culture (Canfield et al. 2017) (Table 9.2).

9.4.2 Astrocytes

Astrocytes are also often co-cultured with stem cell-derived BMECs because of their important roles in BBB induction and maintenance (Obermeier et al. 2013). Since astrocyte end feet contact the vasculature in vivo, there have been some attempts to keep astrocytes in close proximity and encourage both paracrine and juxtacrine signaling (Vatine et al. 2019). Astrocytes, like neurons, have been gener-ated in multiple ways; however, most co-cultures have employed EZ-spheres to generate hPSC-astrocytes. The EZ-spheres can be treated with medium with RA and N-2 supplement (Sareen et al. 2014), resulting in a cell population that, when

Table 9.2 Summary of hPSC-BMEC and NVU cell co-cultures

References	NVU cells	Architecture	Changes observed
Lippmann et al. (2012)	r A	Transwell	Increased TEER
Lippmann et al. (2014)	h NPC, h P	Transwell	Increased TEER
Hollmann et al. (2017)	h A/P	Transwell	Increased TEER with all cell types, further improvement with both
Neal et al. (2019)	hPSC A	Transwell	Increased TEER
Cecchelli et al. (2014)	b P	Transwell	Used to drive cord blood-derived cells to BBB phenotypes like decreased permeability and increased efflux transporter activity
Boyer-Di Ponio et al. (2014)	r A	Transwell	Used to drive cord blood-derived cells to BBB phenotypes like decreased permeability
Delsing et al. (2018)	hPSC mesoderm P, h A, h N	Transwell	Induction of BBB properties in endothelial progenitors, improvement of TEER, improved resistance to small molecules
Praça et al. (2019)	h P	Transwell	Necessary to see efflux transport activity in patterned hPSC-EPC-derived BMECs
Canfield et al. (2017)	hPSC N/A, r A	Transwell	TEER improvements, continuous tight junctions, no improvement of P-gp activity or increase in tight junction protein levels
Canfield et al. (2019)	hPSC N/A/P	Transwell	As previous, with pericyte-induced reduction of non-receptor-mediated transcytosis
Stebbins et al. (2019)	hPSC neural crest P	Transwell	Improvement in TEER, resistance to small molecule permeability, lowered non-receptor-mediated transcytosis
Faal et al. (2019)	hPSC mesoderm and neural crest P, h neural crest P	Transwell	Improvements in TEER with mesoderm and neural crest pericytes, tube formation assay
Bradley et al. (2019)	h A	Transwell	Investigation of the effects of region-specific astrocytes on TEER, small molecule permeability, and tight junction continuity
Nishihara et al. (2020)	h T cells	Transwell	Selective T cell transmigration was observed
Appelt-Menzel et al. (2017)	hPSC NSC, h A, h P, h NSC	Transwell	Increased TEER and assessed transcript and protein level changes
Jamieson et al. (2019)	hPSC mesoderm P	Direct + fluidic	No TEER change, rescue of TEER under stress
Vatine et al. (2019)	hPSC A/N	Fluidic	Resistance to blood-to-brain transport in chip

(continued)

Table 9.2 (continued)

References	NVU cells	Architecture	Changes observed
Park et al. (2019)	h P/A	Fluidic	Improved BBB relevant transcripts when applied with hypoxia
Motallebnejad et al. (2019)	hPSC A	Fluidic	No changes in TEER observed on co-culture with astrocytes suspended in 3D hydrogel
Blanchard et al. (2020)	hPSC mesoderm P, hPSC A,	Direct	Improved TEER compared to monocultured controls

r = rat, h = human, b = bovine, hPSC = human pluripotent stem cell derived, A = astrocyte, P = pericyte, N = neuron, NSC = neural stem cell, NPC = neural progenitor cell

disassociated and maintained in basal medium with N-2, forms astrocytes (Canfield et al. 2017). Though astrocytes are not present during the initial stages of development when invasion and BBB patterning occurs (Obermeier et al. 2013), primary rat astrocytes have been used to induce BMEC phenotypes in cord blood-derived stem cells (Boyer-Di Ponio et al. 2014). Co-culture of BMECs with hPSC-derived or primary astrocytes, often in addition to other NVU cell types, has been shown to result in increased TEER, resistance to small molecule permeability, and improved tight junction continuity (Canfield et al. 2017; Hollmann et al. 2017; Lippmann et al. 2012) (Table 9.2). A recent study has also shown that brain region-specific hPSC-astrocytes demonstrate different barrier inductive properties as assessed by TEER and small molecule permeability (Bradley et al. 2019). However, co-culture of hPSC-derived BMECs with a mixture of astrocytes and neurons does not appear to affect efflux transporter expression and activity (Canfield et al. 2017).

9.4.3 Pericytes

Pericytes are found throughout the body in close association with capillaries. In the brain, pericytes have been shown to be indispensable to BBB function. Mouse experiments have indicated a marked increase in BBB permeability that correlates with a loss in pericyte number (Armulik et al. 2010; Daneman et al. 2010). In addition, as described above, bovine brain pericytes have been used to induce BBB phenotypes in endothelial progenitors in vitro (Cecchelli et al. 2014). Moreover, co-culture of pericytes with hPSC-derived endothelial progenitors prior to endothelial maturation was necessary for acquisition of efflux transport activity (Praça et al. 2019) (Table 9.2).

When mesoderm-derived pericytes were directly plated on hPSC-BMECs to simulate the close proximity of pericytes to the vasculature, there were not significant changes in the barrier properties. However, the pericytes were found to self-organize to the basolateral side of the hPSC-BMECs (Jamieson et al. 2019). While most pericytes in the body are derived from mesodermal lineages, forebrain pericytes are derived from the neural crest, a multipotent ectodermal cell population

(Adams and Bronner-Fraser 2009). Until recently, there were no established differentiation protocols for brain-specific pericyte-like cells. Two techniques have recently been reported to generate neural crest-derived hPSC-pericyte-like cells. Both approaches begin by generating neural crest stem cells. One neural crest protocol accomplishes this through application of CHIR99021, B-27 supplement, and bovine serum albumin over 5 days (Leung et al. 2016). The second neural crest protocol employs 15 days of basal medium supplemented with L-ascorbic acid 2-phosphate magnesium, bFGF, heparin, sodium selenium, insulin, transferrin, CHIR99021, sodium bicarbonate, and TGFβ and BMP pathway inhibitors to generate neural crest stem cells (Stebbins et al. 2019). The resulting neural crest stem cells can then be further differentiated to brain pericyte-like cells using a proprietary pericyte medium (Faal et al. 2019) or basal medium supplemented L-ascorbic acid 2-phosphate magnesium, sodium selenium, insulin, sodium bicarbonate, transferrin, heparin, and fetal bovine serum (Stebbins et al. 2019). Neural crest-derived pericyte-like cells express many of the "pericyte-specific" transcripts identified by single cell RNA sequencing of in vivo brain pericytes (Stebbins et al. 2019; Vanlandewijck et al. 2018); however, there are differences between in vivo pericytes and neural crest-derived pericytes, as expected (Faal et al. 2019; Stebbins et al. 2019). Neural crest-derived brain pericyte-like cells increased resistance to paracellular transport, as assessed by TEER and small molecule permeability, and decreased nonspecific transcytosis in co-cultured hPSC-BMECs (Stebbins et al. 2019). Also, despite the transcriptional differences between neural crest- and mesoderm-derived pericytes, both improved TEER compared to monoculture in a side by side comparison (Faal et al. 2019). While TEER increases and reduction of permeability are seen with co-culture with other NVU cell types such as neurons and astrocytes, reduction in nonspecific transcytosis was observed only with pericyte co-culture (Canfield et al. 2019; Stebbins et al. 2019), indicating the importance of including pericytes in modeling applications where the mechanisms of transcellular transport are important. This reduction in transcytosis persisted for 24 hours after removal of pericytes from co-culture (Canfield et al. 2019) (Table 9.2).

9.4.4 Multiple Cell Co-culture

To better mimic the in vivo NVU and account for cross talk between the multiple cell types of the NVU, researchers have performed multicellular co-cultures with hPSC-BMECs. These models often yield additive improvements in tight junction continuity and TEER (Canfield et al. 2017, 2019; Hollmann et al. 2017), as well as retention of the unique nonspecific transcytosis phenotype induced by hPSC-brain pericytes (Canfield et al. 2019) (Table 9.2). Other studies have reported more minimal impact of co-cultured NVU cells (Appelt-Menzel et al. 2017). The application of pericytes, astrocytes, and neurons has also been used to induce modest BBB phenotypes in hPSC-derived endothelial progenitor cells (Delsing et al.

2018). In choosing the cell types to include in a BBB model, the added complexity of obtaining and organizing these cells should be balanced by improvements in model performance resulting from each cell type.

9.5 Application of Physiologically Relevant Structures and Forces to Stem Cell-Derived BBB Models

It is challenging to generate in vitro BBB models that capture all of the crucial elements of the in vivo niche. Although co-culture approaches described above allow paracrine signaling among NVU cells, there are other components missing, such as direct cell contact and shear flow forces. As microfluidics and organ-on-a-chip technologies have developed, more realistic models can now be developed. Microfluidic devices have been seeded with the stem cell-derived NVU cells to model human brain microvessels in a more physiologic environment. Moreover, stacked flow channels allow precise spatial and temporal regulation of the fluidic environment, maintaining optimal conditions on the vascular and parenchymal sides of the chips (Fig. 9.3). These models offer improved fidelity at the expense of increased complexity and cost.

BBB organ-on-a-chip models typically consist of two overlaid channels, separated by a non-cell permeable membrane which allows soluble factor exchange. Different NVU cell types are seeded in each channel to simulate the vasculature and parenchyma, much like the transwell co-cultures described above. Brain parenchyma is simulated by a combination of astrocytes, neurons, and pericytes (Appelt-Menzel et al. 2017; Vatine et al. 2019; Wang et al. 2017) (Table 9.2). Some models have successfully demonstrated extension of astrocyte end feet through the membrane, more closely representing astrocyte-BMEC interactions in vivo (Vatine et al. 2019). These models have also been shown to successfully localize blood proteins to the apical side and restrict passage of small molecules (Grifno et al. 2019; Vatine et al. 2019; Wang et al. 2017).

In efforts to improve performance of BBB fluidic chips, design elements have been added to simulate other in vivo conditions. For instance, hPSC-pericytes have been embedded inside the gel to mimic the position of pericytes in vivo (Jamieson et al. 2019). In addition, to more accurately mimic the environment in the developing brain, hypoxia was applied to differentiating hPSC-BMECs followed by shear stress after differentiation. These conditions increased TEER and expression of endothelial markers and transporters in the resulting BMECs (Park et al. 2019). Hydrogel scaffolds have been constructed to create small capillaries lined with hPSC-BMECs and NVU cells embedded in the surrounding hydrogel. Such constructs can capture the size of microvessels, flow shear stress, and physical positioning of BMECs relative to NVU cells (Blanchard et al. 2020; Faley et al. 2019; Grifno et al. 2019; Jamieson et al. 2019). Other work has focused on replacing the commonly used transwell membrane with more biologically relevant collagen I hydrogels (Ruano-Salguero and Lee 2018). In addition, iPSC-BMECs, iPSC-mural cells, and iPSC-astrocytes have been combined in a Matrigel scaffold, and the combination allowed to mature for up to a month in an effort to better mimic the three-dimensional structure of the NVU (Blanchard et al. 2020). These models are useful when trying to mimic the behavior of cells within the in vivo BBB, understanding the mechanisms by which environmental cues regulate BBB behavior, as well as potentially providing a more accurate model for drug penetration. Of course, complexity of these models will impact cost, throughput, and robustness. Thus, the experimental questions must warrant the added complexity of creating these advanced models.

9.6 Stem Cell-Derived Aggregate BBB Models

Several recent reports have described efforts toward the development of 3D cell aggregate models of the brain and NVU. In these models, cell types of interest are formed into small free-floating aggregates, reminiscent of an organ or tissue (Fig. 9.3). A spheroid is assembled from somatic cells post-differentiation, while an organoid is a co-differentiated 3D structure that spontaneously self-organizes (M. A. Lancaster and Knoblich 2014). These models provide direct cell-cell contact and 3D structures similar to those observed in vivo which fluidic and transwell models struggle to generate. In addition, organoid models most closely recapitulate BBB developmental stages, allowing us to study human CNS development in vitro and elucidate critical pathways regulating BBB induction.

Some neurovascular spheroids have been created entirely from primary cells (Bergmann et al. 2018). Frequently these spheroids are constructed from primary astrocytes, pericytes, and endothelial cells (Bergmann et al. 2018; Cho et al. 2017b; Urich et al. 2013) though more complicated combinations utilizing neurons and microglia have also been developed (Nzou et al. 2018). These spheroids spontaneously organize with a layer of endothelial cells surrounding the brain parenchymal cells. With endothelial cells on the exterior, the spheroids can be exposed to a compound of interest, and accumulation within the brain "parenchyma" can be measured (Bergmann et al. 2018; Cho et al. 2017b). Recently, neurovascular spheroids have been assembled from stem cell-derived cells (Song et al. 2019), including human bone marrow-derived mesenchymal stromal cells (MSCs), hPSC-derived neural progenitor cells, and hPSC-derived endothelial cells. The neurovascular spheroids were formed by fusion of neural, endothelial, and MSC aggregates. The neural progenitors differentiate into astrocytes and neurons during spheroid maturation, demonstrated by the presence of mature neuron and astrocyte markers such as βIII-tubulin, S100β, GFAP, and vimentin. Imaging post-aggregation showed CD31 expressing endothelial cells were dispersed throughout the spheroids but did not form a perfusable vasculature. A similar approach to form neurovascular spheroids first seeded human neural progenitors in a polyethylene glycol-based scaffold (Schwartz et al. 2015). Following spontaneous neuronal and glial formation and organization, hPSC-derived endothelial cells were added, followed 3 days later by hPSC-derived microglial precursors. The constructs contained a mixed population of astrocytes, neurons, as well as a capillary network identified by CD31 expressing cells. It is not clear whether these capillaries exhibited BBB barrier phenotypes. Another study integrated hPSC-derived endothelial cells into hPSC-derived neural organoids before implantation into mouse brains to perfuse and enlarge the organoid (Pham et al. 2018). Despite the presence of mouse vasculature, the organoids contained capillary-like networks composed of cells expressing human-specific CD31. It is not clear whether these vessels exhibit BBB phenotypes.

Organoids are derived from a single source that proliferates and undergoes morphogenesis to generate tissue-like structures. While unvascularized brain organoids derived from hPSCs were among the first organoid systems developed (Lancaster et al. 2013), assembly of a brain organoid with an integrated NVU may be especially difficult because the cells of vascularized brain tissue originate from multiple germ layers (McCauley and Wells 2017). Thus, it may be necessary to assemble the organoids from multiple progenitors to mimic the invasion of the vascular plexus into the developing neural tube. A recent report attempted to generate an NVU organoid (Cakir et al. 2019). hPSCs were gene edited to express ETV2, a transcription factor that regulates vascular development, under control of a doxycycline-inducible promoter. These cells were aggregated with unmodified hPSCs and then treated with neural induction medium. Next, doxycycline was applied in situ to reprogram the ETV2-inducible cells to an endothelial fate. The BBB phenotype of the integrated endothelial cells was examined to a limited extent, but further work is needed to better understand their properties. Generation of vascularized organoids is still in its infancy, but these models offer the ability to understand mechanisms of BBB morphogenesis during brain development and to model the roles of disease on NVU structure.

9.7 Applications of Stem Cell-Derived BBB Models

9.7.1 Drug Permeability

Many of the stem cell-derived BBB models described in this chapter possess well-developed tight junctions and exhibit substantial barrier properties. TEER is the standard metric for measuring tight junction quality in vitro, and through systematic study, it was found that ~1000 Ω-cm^2 is sufficiently high to prevent aberrant paracellular leakage of both small and large molecules (Mantle et al. 2016). Thus, these models have been used to examine correlation between a drug permeability measured in vitro to the drug uptake in vivo, thereby demonstrating their potential utility in drug permeability estimation and screening (Cecchelli et al. 2007; Lippmann et al. 2012; Mantle et al. 2016; Wang et al. 2017). For example, candidate drugs have been screened in an hPSC-BMEC model to quickly eliminate those that may be unable to cross the BBB at therapeutically relevant rates (Vatine et al. 2017). Also, fluorescently conjugated cancer targeting molecules, which could be potentially used for intraoperative brain cancer imaging, were evaluated for their permeabilities across hPSC-BMECs. The data indicated that the fluorophore conjugated molecules, unlike the parent molecule, had lower BBB permeability because of selective MRP and BCRP efflux (Clark et al. 2016). Such information can help guide further drug development.

9.7.2 Studying Human BBB Development

One of the major benefits of stem cell-derived BBB models is the ability to elucidate the molecular and cellular mechanisms of human BBB development and maintenance of human BBB properties. For example, the co-culture of hPSC-BMECs with hPSC-derived or primary brain pericytes resulted in a reduction of nonspecific transcytosis (Canfield et al. 2019; Stebbins et al. 2019), while astrocyte or neuron co-culture did not have the same effect (Canfield et al. 2017, 2019). Also, hPSC-BMEC differentiations can be used to explore human BBB development in a tractable in vitro setting. As discussed in Sect. 9.3.1.1, induction of GLUT1 expression in differentiating BMECs correlated with nuclear localization of β-catenin (Lippmann et al. 2012). Conversely, Wnt inhibition during endothelial progenitor induction increased permeability compared to vehicle controls (Cecchelli et al. 2014). These data suggest that, like in mouse, Wnt signaling is important to the development of the human BBB. In addition, RA has been shown to improve the quality of the BMEC barrier (Lippmann et al. 2014; Praça et al. 2019), consistent with findings in mouse models (Mizee et al. 2013). Subsequent analysis of specific retinoic acid receptor agonists and antagonists revealed that RARα, RARγ, and RXRα activation mimicked the effects of RA application (Stebbins et al. 2018).

Stem cell-derived in vitro BBB models also permit investigation of factors that induce and regulate BMEC phenotype. In many types of endothelial cells, shear stress from blood flow is well known to cause a change in morphology and orientation with direction to the flow (Davies 1995; Iba and Sumpio 1991; Malek and Izumo 1996; Ye et al. 2015). However, it has been observed that hPSC-BMECs do not undergo any substantial alignment or morphological change with respect to shear stress (DeStefano et al. 2017). This is consistent with immortalized BMECs (Reinitz et al. 2015) and seems to indicate that BMECs respond uniquely to shear stress, although the physiologic significance of this difference is not yet understood.

9.7.3   Modeling BBB Disease

Another major benefit of stem cell-derived models over other in vitro BBB models is the ability to model genetically linked human diseases by employing patient-sourced or gene-edited iPSCs. While many CNS diseases are thought to be associated with BBB dysfunction, it has been difficult to mechanistically investigate this dysfunction in vitro. Since iPSC lines can be generated from patients with disease-related genotypes, subsequent differentiation into iPSC-BMECs and other NVU cells can allow in vitro analysis of human cells carrying disease-causing mutations, which often result in cells that exhibit disease-associated phenotypes (Bosworth et al. 2018). Genome editing tools can also be used to either introduce or correct mutations to an iPSC line, allowing direct exploration of the impacts of a single mutation on the BBB without the confounding factors of patient-to-patient variability or line-to-line iPSC variability.

As an example, psychomotor retardation, a disease caused by inactivating mutations of thyroid hormone transporter MCT8 (Anık et al. 2014), results in a thyroid hormone deficit in the brain, leading to serious developmental deficits. A clear genetic link to molecular transport, which can readily be quantified in vitro, made this disease an ideal test case for patient-sourced iPSC-derived BBB models. While MCT8-deficient iPSC-derived neural cells demonstrated thyroid hormone responsiveness, transport across the iPSC-BMECs was impaired (Vatine et al. 2017), suggesting that access to thyroid hormone in the developing brain was restricted due to a dysfunctional BBB. In another study, analysis of different neurodegenerative disease state-derived iPSC lines (Alzheimer's disease, Huntington's disease, Parkinson's disease, and amyotrophic lateral sclerosis) consistently showed that disease iPSC-BMECs exhibited impaired efflux transport activity when compared to healthy controls. Reductions in tight junction quality were also observed in the disease models (Katt et al. 2019), and these data suggest that a weakened BBB could contribute to disease progression (Katt et al. 2019; Lim et al. 2017). To explore childhood cerebral adrenoleukodystrophy, characterized by BBB breakdown through a well-known mutation in *ABCD1*, hPSC-BMECs from patient and healthy iPSC lines were used and demonstrated a reduction in tight junction quality as measured by increased frayed tight junctions and a reduction in TEER (Lee et al. 2018). Patient-sourced iPSCs were also used to investigate BBB function in Huntington's disease (HD), a debilitating neurodegenerative disease caused by a mutation in the HTT gene (Myers 2004). HD BMECs exhibited discontinuous tight junctions as well as dysregulation in angiogenesis and activation of TGFβ1, Wnt3a, GLI2, and ANGPT2 signaling cascades suggesting significant BBB dysfunction (Lim et al. 2017). As mentioned previously, model complexity should be justified by the experimental question. For example, BBB dysfunction in Alzheimer's disease (AD) is thought to be a result of interactions between several cells of the NVU, warranting a multicellular three-dimensional BMEC model. The resulting model using iPSCs carrying different *APOE* alleles was employed to assess the deposition of amyloid β, a protein known to accumulate in AD progression, and the authors concluded that the *APOE4* genotype led to increased amyloid accumulation as a result of pericyte effects (Blanchard et al. 2020).

In addition to modeling genetic diseases, stem cell-derived models of the BBB have been used to analyze the mechanism of brain invasion in bacterial meningitis. The application of Group B *Streptococcus*, a known pathogen responsible for meningitis, to a monolayer of hPSC-BMECs resulted in a reduction of protein and transcript levels of occludin, claudin-5, and ZO-1, three critical tight junction associated proteins (Kim et al. 2017), as well as in P-gp, a major efflux transporter (Kim et al. 2019). Another study applied *Neisseria meningitidis*, the pathogen responsible for a subtype of meningococcal meningitis, to hPSC-BMECs (Martins Gomes et al. 2019). *Neisseria* infection resulted in almost total loss of TEER within 48 h, as well as increased permeability to small molecules. This impact was directly observed via immunocytochemistry and corresponded to a reduction in TJP1and CLDN5 transcripts. Infection also increased secretion of the cytokines RANTES, IFN-γ, and IL-8 after 24 h, indicating an inflammatory response of the hPSC-BMECs.

9.8   Conclusion

Stem cell models of the human BBB provide advantages compared to other in vitro models or in vivo animal studies. In particular, human stem cells provide a renewable source of human BMEC-like cells and other cell types in the NVU, enabling a consistent product and scalable models. hPSCs in particular facilitate modeling genetic diseases that influence the BBB.

Human stem cell-derived BBB models have been implemented in various configurations. Transwell models offer ease of use and high expandability which are useful for screening studies. While these models can capture interactions between BMECs and other NVU cell types mediated by soluble factors, direct contact between cells and flow are difficult to implement in transwells. Fluidic models capture dynamic flow of the in vivo microenvironment and enable precise spatial and temporal control over application of soluble factors; however, they are not as easily scaled. Brain spheroids and organoids offer the potential to study BBB morphogenesis; however, assessing the barrier properties of the BMECs is more complicated than other configurations. To date, proof-of-concept studies of generating vascular networks in neural aggregates indicate promise for these models, but additional work is needed to recapitulate development of bona fide neurovascular organoids containing mesoderm- and ectoderm-derived cell types.

hPSC-derived BBB models can advance efforts for accurate, human relevant in vitro screening of novel neurotherapeutics. The high passive resistance of these models and polarized expression of BBB transporters make them especially adept at eliminating prospective candidates that cannot penetrate the human BBB and developing trans-BBB delivery strategies. hPSC-derived BBB models also provide a useful tool for studying molecular and cellular mechanisms of BBB induction and regulation, complementing animal models. Disease models using patient-derived or gene-edited iPSCs have thus far focused on genetic diseases impacting BMECs and microbial disease. There is significant potential in expanding these disease models to investigate the roles of different NVU cell types in more complex neurodegenerative diseases.

9.9 Points for Discussion

- Given the artificial nature of in vitro conversion of hPSC to BMEC-like cells, what is the best way to validate resultant cells?
- When are monocultured BMECs sufficient? What applications may require co-culture?
- What are the benefits and drawbacks of differentiating all cell types in a multicellular model from the same hiPSC line? What are the benefits and drawbacks of using primary cells for some or all of your cell types instead?
- In what situations would you consider using a microfluidic device? What factors would be more easily studied? What impacts could it have on cell behavior?
- In a 3D aggregate model of the BBB, how would you assess barrier properties of endothelial cells inside of the aggregates?
- In the following situations, which stem cell model could be used and why? What experiment would you design? What would be appropriate controls?

  - Assess BBB permeability of a novel therapeutic.
  - Assess the impact of a growth factor on BBB permeability.
  - Identify what factors, or combination of factors, are involved in BBB property acquisition.
  - Determine if a blood-borne compound causes disruption of BMEC-NVU signaling.
  - Investigate the link between a heritable astrocyte disease and brain endothelial permeability.

Acknowledgments The authors would like to acknowledge Benjamin D. Gastfriend, Moriah E. Katt, and Martha E. Floy for their contributions and suggestions. This work was funded in part by the National Institutes of Health (NS103844 to EVS and SPP and NS107461 to SPP and EVS) and the National Science Foundation (1703219 to EVS and SPP).

References

Adams MS, Bronner-Fraser M (2009) Review: the role of neural crest cells in the endocrine system. Endocr Pathol 20(2):92–100. https://doi.org/10.1007/s12022-009-9070-6
Aday S, Cecchelli R, Hallier-Vanuxeem D, Dehouck MP, Ferreira L (2016) Stem cell-based human blood–brain barrier models for drug discovery and delivery. Trends Biotechnol 34(5):382–393. https://doi.org/10.1016/j.tibtech.2016.01.001

Anık A, Kersseboom S, Demir K, Çatlı G, Yiş U, Böber E et al (2014) Psychomotor retardation caused by a defective thyroid hormone transporter: report of two families with different *MCT8* mutations. Horm Res Paediatr 82(4):261–271. https://doi.org/10.1159/000365191

Appelt-Menzel A, Cubukova A, Günther K, Edenhofer F, Piontek J, Krause G et al (2017) Establishment of a human blood-brain barrier co-culture model mimicking the neurovascular unit using induced Pluri- and multipotent stem cells. Stem Cell Rep 8(4):894–906. https://doi.org/10.1016/j.stemcr.2017.02.021

Armulik A, Genové G, Mäe M, Nisancioglu MH, Wallgard E, Niaudet C et al (2010) Pericytes regulate the blood–brain barrier. Nature 468(7323):557–561. https://doi.org/10.1038/nature09522

Bailey AS, Jiang S, Afentoulis M, Baumann CI, Schroider DA, Olson SB et al (2004) Transplanted adult hematopoietic stems cells differentiate into functional endothelial cells. Blood 103(1)

Bauer H-C, Bauer H (1997) Neural induction of the blood–brain barrier: still an enigma. Cell Mol Neurobiol 20(1):13–28

Bergmann S, Lawler SE, Qu Y, Fadzen CM, Wolfe JM, Regan MS et al (2018) Blood–brain-barrier organoids for investigating the permeability of CNS therapeutics. Nat Protoc 13(12):2827–2843. https://doi.org/10.1038/s41596-018-0066-x

Blanchard JW, Bula M, Davila-Velderrain J, Akay LA, Zhu L, Frank A et al (2020) Reconstruction of the human blood–brain barrier in vitro reveals a pathogenic mechanism of APOE4 in pericytes. Nat Med 26(6):952–963. https://doi.org/10.1038/s41591-020-0886-4

Bonney S, Harrison-Uy S, Mishra S, MacPherson AM, Choe Y, Li D et al (2016) Diverse functions of retinoic acid in brain vascular development. J Neurosci 36(29):7786–7801. https://doi.org/10.1523/JNEUROSCI.3952-15.2016

Bonney S, Dennison BJC, Wendlandt M, Siegenthaler JA (2018) Retinoic acid regulates endothelial β-catenin expression and Pericyte numbers in the developing brain vasculature. Front Cell Neurosci 12:476. https://doi.org/10.3389/fncel.2018.00476

Bosworth AM, Faley SL, Bellan LM, Lippmann ES (2018) Modeling neurovascular disorders and therapeutic outcomes with human-induced pluripotent stem cells. Front Bioeng Biotechnol 5:87. https://doi.org/10.3389/fbioe.2017.00087

Bottenstein JE, Sato GH (1979) Growth of a rat neuroblastoma cell line in serum-free supplemented medium. Proc Natl Acad Sci 76(1):514–517. https://doi.org/10.1073/pnas.76.1.514

Boyer-Di Ponio J, El-Ayoubi F, Glacial F, Ganeshamoorthy K, Driancourt C, Godet M et al (2014) Instruction of circulating endothelial progenitors in vitro towards specialized blood-brain barrier and arterial phenotypes. PLoS One 9(1):e84179. https://doi.org/10.1371/journal.pone.0084179

Bradley RA, Shireman J, McFalls C, Choi J, Canfield SG, Dong Y et al (2019) Regionally specified human pluripotent stem cell-derived astrocytes exhibit different molecular signatures and functional properties. Development 146(13):dev170910. https://doi.org/10.1242/dev.170910

Brewer GJ, Cotman CW (1989) Survival and growth of hippocampal neurons in defined medium at low density: advantages of a sandwich culture technique or low oxygen. Brain Res 494(1):65–74. https://doi.org/10.1016/0006-8993(89)90144-3

Butt AM, Jones HC, Abbott NJ (1990) Electrical resistance across the blood-brain barrier in anaesthetized rats: a developmental study. J Physiol 429:47–62

Cakir B, Xiang Y, Tanaka Y, Kural MH, Parent M, Kang Y-J et al (2019) Engineering of human brain organoids with a functional vascular-like system. Nat Methods 16(11):1169–1175. https://doi.org/10.1038/s41592-019-0586-5

Canfield SG, Stebbins MJ, Morales BS, Asai SW, Vatine GD, Svendsen CN et al (2017) An isogenic blood-brain barrier model comprising brain endothelial cells, astrocytes, and neurons derived from human induced pluripotent stem cells. J Neurochem 140(6):874–888. https://doi.org/10.1111/jnc.13923

Canfield SG, Stebbins MJ, Faubion MG, Gastfriend BD, Palecek SP, Shusta EV (2019) An isogenic neurovascular unit model comprised of human induced pluripotent stem cell-derived brain microvascular endothelial cells, pericytes, astrocytes, and neurons. Fluids Barriers CNS 16(1):25. https://doi.org/10.1186/s12987-019-0145-6

Cecchelli R, Berezowski V, Lundquist S, Culot M, Renftel M, Dehouck M-P, Fenart L (2007) Modelling of the blood–brain barrier in drug discovery and development. Nat Rev Drug Discov 6(8):650–661. https://doi.org/10.1038/nrd2368

Cecchelli R, Aday S, Sevin E, Almeida C, Culot M, Dehouck L et al (2014) A stable and reproducible human blood-brain barrier model derived from hematopoietic stem cells. PLoS One 9(6):e99733. https://doi.org/10.1371/journal.pone.0099733

Chagastelles PC, Nardi NB (2011) Biology of stem cells: an overview. Kidney Int Suppl 1(3):63–67. https://doi.org/10.1038/kisup.2011.15

Cho C, Smallwood PM, Nathans J (2017a) Reck and Gpr124 are essential receptor cofactors for Wnt7a/Wnt7b-specific signaling in mammalian CNS angiogenesis and blood-brain barrier regulation. Neuron 95(5):1056-1073.e5. https://doi.org/10.1016/j.neuron.2017.07.031

Cho C-F, Wolfe JM, Fadzen CM, Calligaris D, Hornburg K, Chiocca EA et al (2017b) Blood-brain-barrier spheroids as an in vitro screening platform for brain-penetrating agents. Nat Commun 8(1):15623. https://doi.org/10.1038/ncomms15623

Clark PA, Al-Ahmad AJ, Qian T, Zhang RR, Wilson HK, Weichert JP et al (2016) Analysis of cancer-targeting Alkylphosphocholine analogue permeability characteristics using a human induced pluripotent stem cell blood–brain barrier model. Mol Pharm 13(9):3341–3349. https://doi.org/10.1021/acs.molpharmaceut.6b00441

Daneman R, Agalliu D, Zhou L, Kuhnert F, Kuo CJ, Barres BA (2009) Wnt/ -catenin signaling is required for CNS, but not non-CNS, angiogenesis. Proc Natl Acad Sci 106(2):641–646. https://doi.org/10.1073/pnas.0805165106

Daneman R, Zhou L, Kebede AA, Barres BA (2010) Pericytes are required for blood–brain barrier integrity during embryogenesis. Nature 468(7323):562–566. https://doi.org/10.1038/nature09513

Davies PF (1995) Flow-mediated endothelial mechanotransduction. Physiol Rev 75(3):519–560. https://doi.org/10.1152/physrev.1995.75.3.519

Delsing L, Dönnes P, Sánchez J, Clausen M, Voulgaris D, Falk A et al (2018) Barrier properties and transcriptome expression in human iPSC-derived models of the blood-brain barrier: barrier properties and transcriptome expression in human iPSC-derived models. Stem Cells 36(12):1816–1827. https://doi.org/10.1002/stem.2908

DeStefano JG, Xu ZS, Williams AJ, Yimam N, Searson PC (2017) Effect of shear stress on iPSC-derived human brain microvascular endothelial cells (dhBMECs). Fluids Barriers CNS 14(1):20. https://doi.org/10.1186/s12987-017-0068-z

Ebert AD, Shelley BC, Hurley AM, Onorati M, Castiglioni V, Patitucci TN et al (2013) EZ spheres: a stable and expandable culture system for the generation of pre-rosette multipotent stem cells from human ESCs and iPSCs. Stem Cell Res 10(3):417–427. https://doi.org/10.1016/j.scr.2013.01.009

Engelhardt B, Liebner S (2014) Novel insights into the development and maintenance of the blood–brain barrier. Cell Tissue Res 355(3):687–699. https://doi.org/10.1007/s00441-014-1811-2

Faal T, Phan DTT, Davtyan H, Scarfone VM, Varady E, Blurton-Jones M et al (2019) Induction of mesoderm and neural crest-derived Pericytes from human pluripotent stem cells to study blood-brain barrier interactions. Stem Cell Rep 12(3):451–460. https://doi.org/10.1016/j.stemcr.2019.01.005

Faley SL, Neal EH, Wang JX, Bosworth AM, Weber CM, Balotin KM et al (2019) iPSC-derived brain endothelium exhibits stable, long-term barrier function in perfused hydrogel scaffolds. Stem Cell Rep 12(3):474–487. https://doi.org/10.1016/j.stemcr.2019.01.009

Gastfriend BD, Palecek SP, Shusta EV (2018) Modeling the blood–brain barrier: beyond the endothelial cells. Curr Opin Biomed Engin 5:6–12. https://doi.org/10.1016/j.cobme.2017.11.002

Grifno GN, Farrell AM, Linville RM, Arevalo D, Kim JH, Gu L, Searson PC (2019) Tissue-engineered blood-brain barrier models via directed differentiation of human induced pluripotent stem cells. Sci Rep 9(1):13957. https://doi.org/10.1038/s41598-019-50193-1

Hawkins BT, Davis TP (2005) The blood-brain barrier/neurovascular unit in health and disease. Pharmacol Rev 57(2):173–185. https://doi.org/10.1124/pr.57.2.4

Helms HC, Abbott NJ, Burek M, Cecchelli R, Couraud P-O, Deli MA et al (2016) In vitro models of the blood–brain barrier: an overview of commonly used brain endothelial cell culture models and guidelines for their use. J Cereb Blood Flow Metab 36(5):862–890. https://doi.org/10.117 7/0271678X16630991

Hollmann EK, Bailey AK, Potharazu AV, Neely MD, Bowman AB, Lippmann ES (2017) Accelerated differentiation of human induced pluripotent stem cells to blood–brain barrier endothelial cells. Fluids Barriers CNS 14(1):9. https://doi.org/10.1186/s12987-017-0059-0

Iba T, Sumpio BE (1991) Morphological response of human endothelial cells subjected to cyclic strain in vitro. Microvasc Res 42(3):245–254. https://doi.org/10.1016/0026-2862(91)90059-K

Jamieson JJ, Linville RM, Ding YY, Gerecht S, Searson PC (2019) Role of iPSC-derived pericytes on barrier function of iPSC-derived brain microvascular endothelial cells in 2D and 3D. Fluids Barriers CNS 16(1):15. https://doi.org/10.1186/s12987-019-0136-7

Katt ME, Xu ZS, Gerecht S, Searson PC (2016) Human brain microvascular endothelial cells derived from the BC1 iPS cell line exhibit a blood-brain barrier phenotype. PLoS One 11(4):e0152105. https://doi.org/10.1371/journal.pone.0152105

Katt ME, Mayo LN, Ellis SE, Mahairaki V, Rothstein JD, Cheng L, Searson PC (2019) The role of mutations associated with familial neurodegenerative disorders on blood–brain barrier function in an iPSC model. Fluids Barriers CNS 16(1):20. https://doi.org/10.1186/s12987-019-0139-4

Kempf H, Olmer R, Haase A, Franke A, Bolesani E, Schwanke K et al (2016) Bulk cell density and Wnt/TGFbeta signalling regulate mesendodermal patterning of human pluripotent stem cells. Nat Commun 7(1):13602. https://doi.org/10.1038/ncomms13602

Kim BJ, Bee OB, McDonagh MA, Stebbins MJ, Palecek SP, Doran KS, Shusta EV (2017) Modeling group B *streptococcus* and blood-brain barrier interaction by using induced pluripotent stem cell-derived brain endothelial cells. mSphere 2(6):e00398-17. https://doi.org/10.1128/mSphere.00398-17

Kim BJ, McDonagh MA, Deng L, Gastfriend BD, Schubert-Unkmeir A, Doran KS, Shusta EV (2019) Streptococcus agalactiae disrupts P-glycoprotein function in brain endothelial cells. Fluids Barriers CNS 16(1):26. https://doi.org/10.1186/s12987-019-0146-5

Lancaster MA, Knoblich JA (2014) Organogenesis in a dish: modeling development and disease using organoid technologies. Science 345(6194):1247125–1247125. https://doi.org/10.1126/science.1247125

Lancaster MA, Renner M, Martin C-A, Wenzel D, Bicknell LS, Hurles ME et al (2013) Cerebral organoids model human brain development and microcephaly. Nature 501(7467):373–379. https://doi.org/10.1038/nature12517

Lee CAA, Seo HS, Armien AG, Bates FS, Tolar J, Azarin SM (2018) Modeling and rescue of defective blood–brain barrier function of induced brain microvascular endothelial cells from childhood cerebral adrenoleukodystrophy patients. Fluids Barriers CNS 15(1):9. https://doi.org/10.1186/s12987-018-0094-5

Leung AW, Murdoch B, Salem AF, Prasad MS, Gomez GA, García-Castro MI (2016) WNT/β-catenin signaling mediates human neural crest induction via a pre-neural border intermediate. Development 143(3):398–410. https://doi.org/10.1242/dev.130849

Lian X, Bao X, Al-Ahmad A, Liu J, Wu Y, Dong W et al (2014) Efficient differentiation of human pluripotent stem cells to endothelial progenitors via small-molecule activation of WNT signaling. Stem Cell Rep 3(5):804–816. https://doi.org/10.1016/j.stemcr.2014.09.005

Liebner S, Corada M, Bangsow T, Babbage J, Taddei A, Czupalla CJ et al (2008) Wnt/β-catenin signaling controls development of the blood–brain barrier. J Cell Biol 183(3):409–417. https://doi.org/10.1083/jcb.200806024

Lim RG, Quan C, Reyes-Ortiz AM, Lutz SE, Kedaigle AJ, Gipson TA et al (2017) Huntington's disease iPSC-derived brain microvascular endothelial cells reveal WNT-mediated Angiogenic and blood-brain barrier deficits. Cell Rep 19(7):1365–1377. https://doi.org/10.1016/j.celrep.2017.04.021

Lippmann ES, Azarin SM, Kay JE, Nessler RA, Wilson HK, Al-Ahmad A et al (2012) Derivation of blood-brain barrier endothelial cells from human pluripotent stem cells. Nat Biotechnol 30(8):783–791. https://doi.org/10.1038/nbt.2247

Lippmann ES, Al-Ahmad A, Palecek SP, Shusta EV (2013) Modeling the blood–brain barrier using stem cell sources. Fluids Barriers CNS 10(1):2. https://doi.org/10.1186/2045-8118-10-2

Lippmann ES, Al-Ahmad A, Azarin SM, Palecek SP, Shusta EV (2014) A retinoic acid-enhanced, multicellular human blood-brain barrier model derived from stem cell sources. Sci Rep 4(1):4160. https://doi.org/10.1038/srep04160

Malek AM, Izumo S (1996) Mechanism of endothelial cell shape change and cytoskeletal remodeling in response to fluid shear stress. J Cell Sci 109:713–726

Mantle JL, Min L, Lee KH (2016) Minimum Transendothelial electrical resistance thresholds for the study of small and large molecule drug transport in a human *in vitro* blood–brain barrier model. Mol Pharm 13(12):4191–4198. https://doi.org/10.1021/acs.molpharmaceut.6b00818

Martins Gomes SF, Westermann AJ, Sauerwein T, Hertlein T, Förstner KU, Ohlsen K et al (2019) Induced pluripotent stem cell-derived brain endothelial cells as a cellular model to study Neisseria meningitidis infection. Front Microbiol 10:1181. https://doi.org/10.3389/fmicb.2019.01181

McCauley HA, Wells JM (2017) Pluripotent stem cell-derived organoids: using principles of developmental biology to grow human tissues in a dish. Development 144(6):958–962. https://doi.org/10.1242/dev.140731

Mizee MR, Wooldrik D, Lakeman KAM, van het Hof B, Drexhage JAR, Geerts D et al (2013) Retinoic acid induces blood-brain barrier development. J Neurosci 33(4):1660–1671. https://doi.org/10.1523/JNEUROSCI.1338-12.2013

Motallebnejad P, Thomas A, Swisher SL, Azarin SM (2019) An isogenic hiPSC-derived BBB-on-a-chip. Biomicrofluidics 13(6):064119. https://doi.org/10.1063/1.5123476

Myers RH (2004) Huntington's disease genetics. NeuroRx 1(2):8

Nagy A, Gocza E, Diaz EM, Prideaux VR, Ivanyi E, Markkl'La M, Rossant J (1990) Embryonic stem cells alone are able to support fetal development in the mouse. Development 110:815–821

Naujok O, Lentes J, Diekmann U, Davenport C, Lenzen S (2014) Cytotoxicity and activation of the Wnt/beta-catenin pathway in mouse embryonic stem cells treated with four GSK3 inhibitors. BMC Res Notes 7(1):273. https://doi.org/10.1186/1756-0500-7-273

Neal EH, Marinelli NA, Shi Y, McClatchey PM, Balotin KM, Gullett DR et al (2019) A simplified, fully defined differentiation scheme for producing blood-brain barrier endothelial cells from human iPSCs. Stem Cell Rep 12(6):1380–1388. https://doi.org/10.1016/j.stemcr.2019.05.008

Nishihara H, Soldati S, Mossu A, Rosito M, Rudolph H, Muller WA et al (2020) Human CD4+ T cell subsets differ in their abilities to cross endothelial and epithelial brain barriers in vitro. Fluids Barriers CNS 17(1):3. https://doi.org/10.1186/s12987-019-0165-2

Nzou G, Wicks RT, Wicks EE, Seale SA, Sane CH, Chen A et al (2018) Human cortex spheroid with a functional blood brain barrier for high-throughput neurotoxicity screening and disease modeling. Sci Rep 8(1):7413. https://doi.org/10.1038/s41598-018-25603-5

Obermeier B, Daneman R, Ransohoff RM (2013) Development, maintenance and disruption of the blood-brain barrier. Nat Med 19(12):1584–1596. https://doi.org/10.1038/nm.3407

Pardridge WM (2005) The blood-brain barrier: bottleneck in brain drug development. NeuroRx 2(1):12

Park T-E, Mustafaoglu N, Herland A, Hasselkus R, Mannix R, FitzGerald EA et al (2019) Hypoxia-enhanced blood-brain barrier Chip recapitulates human barrier function and shuttling of drugs and antibodies. Nat Commun 10(1):2621. https://doi.org/10.1038/s41467-019-10588-0

Pham MT, Pollock KM, Rose MD, Cary WA, Stewart HR, Zhou P et al (2018) Generation of human vascularized brain organoids. Neuroreport 29(7):588–593. https://doi.org/10.1097/WNR.0000000000001014

Praça C, Rosa SC, Sevin E, Cecchelli R, Dehouck M-P, Ferreira LS (2019) Derivation of brain capillary-like endothelial cells from human pluripotent stem cell-derived endothelial progenitor cells. Stem Cell Rep 13(4):599–611. https://doi.org/10.1016/j.stemcr.2019.08.002

Qian T, Maguire SE, Canfield SG, Bao X, Olson WR, Shusta EV, Palecek SP (2017) Directed differentiation of human pluripotent stem cells to blood-brain barrier endothelial cells. Sci Adv 3(11):e1701679. https://doi.org/10.1126/sciadv.1701679

Reinitz A, DeStefano J, Ye M, Wong AD, Searson PC (2015) Human brain microvascular endothelial cells resist elongation due to shear stress. Microvasc Res 99:8–18. https://doi.org/10.1016/j.mvr.2015.02.008

Ribecco-Lutkiewicz M, Sodja C, Haukenfrers J, Haqqani AS, Ly D, Zachar P et al (2018) A novel human induced pluripotent stem cell blood-brain barrier model: applicability to study antibody-triggered receptor-mediated transcytosis. Sci Rep 8(1):1873. https://doi.org/10.1038/s41598-018-19522-8

Ruano-Salguero JS, Lee KH (2018) Efflux pump substrates shuttled to cytosolic or vesicular compartments exhibit different permeability in a quantitative human blood–brain barrier model. Mol Pharm 15(11):5081–5088. https://doi.org/10.1021/acs.molpharmaceut.8b00662

Sareen D, Gowing G, Sahabian A, Staggenborg K, Paradis R, Avalos P et al (2014) Human induced pluripotent stem cells are a novel source of neural progenitor cells (iNPCs) that migrate and integrate in the rodent spinal cord: human neural progenitor cells. J Comp Neurol 522(12):2707–2728. https://doi.org/10.1002/cne.23578

Schubert M, Holland ND, Laudet V, Holland LZ (2006) A retinoic acid-Hox hierarchy controls both anterior/posterior patterning and neuronal specification in the developing central nervous system of the cephalochordate amphioxus. Dev Biol 296(1):190–202. https://doi.org/10.1016/j.ydbio.2006.04.457

Schwartz MP, Hou Z, Propson NE, Zhang J, Engstrom CJ, Costa VS et al (2015) Human pluripotent stem cell-derived neural constructs for predicting neural toxicity. Proc Natl Acad Sci 112(40):12516–12521. https://doi.org/10.1073/pnas.1516645112

Song L, Yuan X, Jones Z, Griffin K, Zhou Y, Ma T, Li Y (2019) Assembly of human stem cell-derived cortical spheroids and vascular spheroids to model 3-D brain-like tissues. Sci Rep 9(1):5977. https://doi.org/10.1038/s41598-019-42439-9

Song HW, Foreman KL, Gastfriend BD, Kuo JS, Palecek SP, Shusta EV (2020) Transcriptomic comparison of human and mouse brain microvessels. Sci Rep 10(1):12358. https://doi.org/10.1038/s41598-020-69096-7

Srinivasan B, Kolli AR, Esch MB, Abaci HE, Shuler ML, Hickman JJ (2015) TEER measurement techniques for in vitro barrier model systems. J Lab Autom 20(2):107–126. https://doi.org/10.1177/2211068214561025

Stebbins MJ, Wilson HK, Canfield SG, Qian T, Palecek SP, Shusta EV (2016) Differentiation and characterization of human pluripotent stem cell-derived brain microvascular endothelial cells. Methods 101:93–102. https://doi.org/10.1016/j.ymeth.2015.10.016

Stebbins MJ, Lippmann ES, Faubion MG, Daneman R, Palecek SP, Shusta EV (2018) Activation of RARα, RARγ, or RXRα increases barrier tightness in human induced pluripotent stem cell-derived brain endothelial cells. Biotechnol J 13(2):1700093. https://doi.org/10.1002/biot.201700093

Stebbins MJ, Gastfriend BD, Canfield SG, Lee M-S, Richards D, Faubion MG et al (2019) Human pluripotent stem cell-derived brain pericyte-like cells induce blood-brain barrier properties. Sci Adv 5

Stenman JM, Rajagopal J, Carroll TJ, Ishibashi M, McMahon J, McMahon AP (2008) Canonical Wnt signaling regulates organ-specific assembly and differentiation of CNS vasculature. Science 322(5905):1247–1250. https://doi.org/10.1126/science.1164594

Takahashi K, Yamanaka S (2006) Induction of pluripotent stem cells from mouse embryonic and adult fibroblast cultures by defined factors. Cell 126(4):663–676. https://doi.org/10.1016/j.cell.2006.07.024

Thomson JA (1998) Embryonic stem cell lines derived from human blastocysts. Science 282(5391):1145–1147. https://doi.org/10.1126/science.282.5391.1145

Tschugguel W, Zhegu Z, Gajdzik L, Maier M, Binder BR, Graf J (1995) High precision measurement of electrical resistance across endothelial cell monolayers. Pflugers Arch Eur J Physiol 430(1):145–147. https://doi.org/10.1007/BF00373850

Urich E, Patsch C, Aigner S, Graf M, Iacone R, Freskgård P-O (2013) Multicellular self-assembled spheroidal model of the blood brain barrier. Sci Rep 3(1):1500. https://doi.org/10.1038/srep01500

Vanlandewijck M, He L, Mäe MA, Andrae J, Ando K, Del Gaudio F et al (2018) A molecular atlas of cell types and zonation in the brain vasculature. Nature 554(7693):475–480. https://doi.org/10.1038/nature25739

Vatine GD, Al-Ahmad A, Barriga BK, Svendsen S, Salim A, Garcia L et al (2017) Modeling psychomotor retardation using iPSCs from MCT8-deficient patients indicates a prominent role for the blood-brain barrier. Cell Stem Cell 20(6):831-843.e5. https://doi.org/10.1016/j.stem.2017.04.002

Vatine GD, Barrile R, Workman MJ, Sances S, Barriga BK, Rahnama M et al (2019) Human iPSC-derived blood-brain barrier chips enable disease modeling and personalized medicine applications. Cell Stem Cell 24(6):995-1005.e6. https://doi.org/10.1016/j.stem.2019.05.011

Wang YI, Abaci HE, Shuler ML (2017) Microfluidic blood-brain barrier model provides in vivo-like barrier properties for drug permeability screening: microfluidic BBB model mimics in vivo properties. Biotechnol Bioeng 114(1):184–194. https://doi.org/10.1002/bit.26045

Weksler BB, Subileau EA, Perrière N, Charneau P, Holloway K, Leveque M et al (2005) Blood-brain barrier-specific properties of a human adult brain endothelial cell line. FASEB J 19(13):1872–1874. https://doi.org/10.1096/fj.04-3458fje

Wilson HK, Canfield SG, Hjortness MK, Palecek SP, Shusta EV (2015) Exploring the effects of cell seeding density on the differentiation of human pluripotent stem cells to brain microvascular endothelial cells. Fluids Barriers CNS 12(1):13. https://doi.org/10.1186/s12987-015-0007-9

Wilson HK, Faubion MG, Hjortness MK, Palecek SP, Shusta EV (2016) Cryopreservation of brain endothelial cells derived from human induced pluripotent stem cells is enhanced by rho-associated coiled coil-containing kinase inhibition. Tissue Eng Part C Methods 22(12):1085–1094. https://doi.org/10.1089/ten.tec.2016.0345

Yamashita M, Aoki H, Hashita T, Iwao T, Matsunaga T (2020) Inhibition of transforming growth factor beta signaling pathway promotes differentiation of human induced pluripotent stem cell-derived brain microvascular endothelial-like cells. Fluids Barriers CNS 17(1):36. https://doi.org/10.1186/s12987-020-00197-1

Ye M, Sanchez HM, Hultz M, Yang Z, Bogorad M, Wong AD, Searson PC (2015) Brain microvascular endothelial cells resist elongation due to curvature and shear stress. Sci Rep 4(1):4681. https://doi.org/10.1038/srep04681

Yu J, Vodyanik MA, Smuga-Otto K, Antosiewicz-Bourget J, Frane JL, Tian S et al (2007) Induced pluripotent stem cell lines derived from human somatic cells. Science 318:5

Chapter 10
Drug Delivery to the Brain: Physiological Concepts, Methodologies, and Approaches

Ramakrishna Samala, Behnam Noorani, Helen Thorsheim, Ulrich Bickel, and Quentin Smith

Abstract An important property in any central nervous system (CNS) drug is the ability to cross into the brain and reach therapeutic concentrations at safe and acceptable systemic doses. Multiple parameters influence drug availability to the brain. One of the most important of these is the blood-brain barrier (BBB). The vasculature of the brain differs from that of other organs of the body in that it greatly restricts the exchange of most solutes into the brain from the systemic circulation. Equilibration, which only requires seconds to minutes for low molecular weight drugs in the interstitial fluid of most tissues of the body, can require days to weeks for many agents in the brain. The restricted neurovascular exchange is based upon the unique properties of the endothelial cell membranes lining the brain blood vessels which limit the passive diffusion of many polar solutes into the brain and avidly pump out a broad array of polar and nonpolar agents through a series of active efflux transporters.

This chapter presents a conceptual overview of the primary methods to assess brain drug distribution in vivo, providing an insider's guide to many of the critical steps to use the methods appropriately. Then, two case examples are provided in detail illustrating application and interpretation of specific methods. The entire chapter is written with a perspective of providing an "insider's view" of the level of drug necessary to reach therapeutic action in the brain. Several parameters are broadly used to explain CNS drug passage and equilibration. One of these is the cerebrovascular permeability-surface area product (PS), which reflects how rapidly a solute can cross in or out of the brain. Another is the brain distribution volume or partition coefficient ($K_{p,brain}$), which characterizes the extent (either high or low) that a drug equilibrates in the brain. Because most drugs bind or associate reversibly to proteins, lipids, and other biologic macromolecules, a third parameter is the fraction to which a solute travels freely in the tissue or blood (f_u, the free or unbound frac-

R. Samala · B. Noorani · H. Thorsheim · U. Bickel · Q. Smith (✉)
Department of Pharmaceutical Sciences, School of Pharmacy, Texas Tech University Health Sciences Center, Amarillo, TX, USA
e-mail: Ramakrishna.samala@ttuhsc.edu; Behnam.Noorani@ttuhsc.edu; Helen.Thorsheim@ttuhsc.edu; Ulrich.Bickel@ttuhsc.edu; Quentin.Smith@ttuhsc.edu

© American Association of Pharmaceutical Scientists 2022
E. C. M. de Lange et al. (eds.), *Drug Delivery to the Brain*, AAPS Advances in the Pharmaceutical Sciences Series 33, https://doi.org/10.1007/978-3-030-88773-5_10

tion). This parameter can be used to calculate the free and bound drug concentrations from the total concentration that is measured by many analytical methods. Together with the time course of drug in the circulation, the above parameters can be used to predict drug total, free, and bound concentrations in brain tissue at all time points after administration. This information can then be used to calculate biologic activity if the binding constant (K_D) of the receptor or the inhibitory constant (K_i) of the signaling process is known. Specific methods, such as in situ brain perfusion, brain efflux index, and in situ brain microdialysis, are valuable to dissect the specific mechanisms operational at the barrier that mediate or regulate drug transport across the brain endothelial cell membranes. In the end, the investigator has a broad array of approaches to assess drug availability to the brain and to make recommendations that would improve outcomes. In some cases, such as for drugs that act in other tissues, the desire may be to limit brain exposure to avoid adverse drug reactions. A specific focus of the chapter is to promote accurate measurements and avoid nonspecific approaches that are error bound and have led to a lot of confusion in the field.

Keywords Blood-brain barrier · Brain partition coefficient · Permeability-surface area product · Unbound drug concentration · In situ brain perfusion · Brain efflux index · Brain microdialysis

10.1 Introduction

The blood-brain barrier (BBB) system represents a series of dynamic cellular interfaces at the border between the blood stream and the brain interstitial fluid (ISF) as well as cerebrospinal fluid (CSF), which together control drug access to the central nervous system (CNS) (Chap. 1). The BBB system is often called the "gatekeeper to the CNS," determining the ability of drugs and other compounds to cross into and reach therapeutic concentrations in brain interstitial fluid. As brain interstitial fluid is in direct contact with the extracellular faces of many neuronal and glial plasma membrane receptors, it is the key concentration that correlates best with much of in vivo CNS neuronal signaling (Kalvass et al. 2007; Liu et al. 2009) using receptors on the extracellular membranes. Some drugs also work intracellularly, in which case the intracellular compartment is needed. The barrier is thought to be critical to insulate the brain from circulating toxins, metals, and neuroactive substances and to provide a stable microenvironment for higher synaptic communication.

The BBB is located in mammals at the vascular endothelium that lines the blood vessels of the brain. The BBB phenotype is comprised of a series of properties which are not static but can change with diet, development, gender, disease, drug or toxin exposure, and age (Cardoso et al. 2010). An example of this is provided later in the chapter illustrating the remarkable changes in barrier integrity in primary and metastatic brain tumors that has critical impact to CNS cancer therapy. The hallmark of the brain vascular endothelial cells, like their counterparts in other organs, is their cellular plasma membranes, which are highly permeable to many lipophilic solutes, such as the metabolic gases, oxygen and carbon dioxide, and the neutral lipophilic drugs, antipyrine, diazepam, ethanol, and caffeine (Smith 2003). However, unlike the vasculature of many organs, the endothelium of the brain lacks open paracellular channels and fenestra which are the primary conduits for vascular exchange for many small polar solutes (MW < 5000). As a consequence, in the brain, plasma-to-tissue interstitial fluid equilibration times ($t_{1/2}$) for small polar solutes, such as sodium, sucrose, and inulin, can fall in the range of hours to days, instead of being in the range of seconds to minutes, as in other organs (Crone 1963). Such low BBB permeability impedes exchange and dampens fluctuations in brain interstitial fluid concentration achieved from vascular exposure to solutes. Further, the BBB expresses a number of other critical properties, such as active efflux transport, enzymatic catabolism, and cerebrospinal fluid (CSF) sink effect, which, when coupled with the low passive permeability of the barrier, markedly reduce brain equilibrium exposure to a wide range of compounds, including many therapeutic drugs. The restriction is not absolute; all drugs cross the BBB to some extent, which is determined by their properties and the properties of the barrier. Thus, the critical question is to what extent they cross the barrier and equilibrate into the brain. In fact, most drugs cross the BBB poorly and show restricted exposure to the CNS, which is a major problem in CNS drug development. A major advance in the last 20 years is the discovery of the amazing important role of active efflux transport in the overall function of the BBB.

Because of the complexity of the BBB and the realization that many BBB properties are not constant but vary with disease, development, and drug exposure, it has been difficult to develop a small cassette of in vitro or in silico models that adequately predict drug transport and availability to the CNS. As a consequence, in vivo testing of brain drug uptake and equilibration is still considered the "gold standard" of any CNS drug discovery program.

As illustrated in Fig. 10.1, many steps are required to convert measured drug levels in tissue to the true active (unbound) drug concentration in the key area of interest ($[Drug]_u$ or C_u). Similarly, traveling from an in vitro measurement of drug effect, such as the IC_{50} in extracellular fluid, to the true inhibitory constant (K_i) within the cell in vivo necessitates considering factors such as cell line differences, drug protein binding, cell-cell interactions, and the local environment within the cell (e.g., pH differences in tumor cells compared to healthy cells). The free (unbound) concentration ($[Drug]_u$ or C_u) in the brain provides critical insight on whether sufficient drug is present to produce biologic effect, as predicted by the binding K_D or inhibitory K_i constants of the receptor or enzyme system. The data is often expressed as a $C_u/K_{D\,or\,i}$ ratio where K is the concentration that provides 50% binding or activation/inhibition of the receptor pathway. For good activity, a brain concentration is needed exceeding $0.5 \times K_i$ and rising to $3–10 \times K_i$ for strong effect. Many drugs act at the nanomolar level. However, total drug concentration in the brain or plasma usually is deceptive because a large portion of drug may be tied up in nonselective binding to tissue or blood constituents, thus, the axiom of pharmacology that the

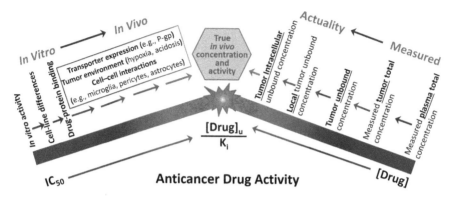

Fig. 10.1 Schematic diagram of the multiple steps necessary to move from measurements of drug concentration in plasma and extracellular *in vitro* IC$_{50}$ toward knowledge of active drug concentration ([Drug]$_u$) in a tumor and the drug's inhibitory constant (K$_i$) at the site of activity

free drug concentration, at the site of action, is the best measure of drug activity. These concepts are more fully discussed in Sect. 10.4.

For some systemically acting drugs, brain penetration is desired to be minimal, to avoid CNS adverse effects, e.g., CNS sedation caused by antihistamine drugs used to treat allergies. Thus, with medicinal chemistry, the properties of the drug are modulated to give the appropriate delivery for the desired outcome.

In this chapter, the primary techniques used to measure in vivo drug availability to the brain and CSF are reviewed, assessing strengths and weaknesses of different methods, with the goal of guiding biomedical researchers in the design of their CNS drug screening program. Case studies are provided at the end of the chapter to illustrate application of the methods and important issues to consider.

Many newcomers to the field ask the question, "Does a compound cross the BBB?" As noted briefly before, in truth, all compounds studied to date cross the BBB to some extent, albeit for some to a very limited degree. The real questions are, "To what extent can a compound cross the BBB and reach active concentrations in brain at safe and tolerable doses?" and "If a compound shows poor brain distribution, how can this be improved?" as well as "How do permeability and distribution of a compound change with disease?" This chapter aims to guide investigators through these critical questions, highlighting the latest methods used to assess brain uptake and exposure, to probe limiting factors that can be modified to improve brain distribution and effect, and to highlight that in a number of disease processes, distribution is highly heterogeneous, which may greatly impact therapeutic outcome.

10.2   Current Status

10.2.1   Two Parameters Commonly Used in In Vivo Brain Drug Distribution Experiments

In many cases, in vivo CNS drug exposure experiments are designed along classic pharmacokinetic lines to measure two primary parameters—the rate at which a compound crosses into the brain (i.e., BBB permeability-surface area product—PS) and the extent to which the compound distributes within the CNS (i.e., brain distribution volume or partition coefficient—$K_{p,brain}$) (Hammarlund-Udenaes et al. 2008; Liu et al. 2008). These two parameters have parallels to the clearance and volume of distribution of systemic blood pharmacokinetics. As the systemic drug clearance and volume of distribution can be used to calculate the concentration of drug in serum or blood at a time point after i.v. administration, the BBB PS and $K_{p,brain}$ can be used to calculate the total drug in the brain at any point after systemic administration given a defined serum exposure.

The BBB PS for influx (PS_{in}) is generally determined from the initial rate of drug uptake into the brain or from pharmacokinetic analysis of the time course of brain total drug concentration relative to that in plasma (Ohno et al. 1978; Smith and Rapoport 1986; Duncan et al. 1991; Liu et al. 2005). Measured BBB PS_{in} ranges for compounds >5 orders of magnitude from $\leq 1 \times 10^{-6}$ ml/s/g (inulin) to $>5 \times 10^{-1}$ ml/s/g for diazepam and is strongly influenced by solute lipophilicity and size, as well as the presence of BBB influx or efflux transport (Rapoport et al. 1979; Smith 2003). It is defined in terms of free drug in the circulation and thus differs somewhat from the BBB K_{in}, which is defined in terms of serum total drug or solute concentration, as noted below. Just as there is a PS_{in} for influx for a given drug, there is also a matching PS_{out} for free drug efflux and often they are not equal. PS_{in} can differ from PS_{out} due to facilitated or active transport, enzymatic conversion, or bulk flow. In many cases, $PS_{out} > PS_{in}$. The matching BBB efflux parameter for total drug is k_{out}.

Equilibrium drug distribution within the brain ($K_{p,brain}$), on the other hand, is more complex and is influenced by multiple parameters, including BBB active transport, enzymatic conversion, tissue binding, cellular transport, drug ionization, and lysosomal pH trapping (Friden et al. 2009a, b). $K_{p,brain}$ is commonly determined from the steady-state ratio of total drug concentration in the brain divided by that in serum or plasma:

$$K_{p,brain} = C_{tot,brain} / C_{tot,plasma} \qquad (10.1)$$

where $C_{tot,brain}$ = total drug concentration in brain and $C_{tot,plasma}$ = total drug concentration in plasma. $K_{p,brain}$ can also be determined from the ratio of area-under-the-curve drug concentration in the brain divided by that in serum or plasma. Both approaches are commonly used in the literature (Kemper et al. 2003):

$$K_{p,brain} \approx \int C_{tot,brain} \, dt \, / \int C_{tot,plasma} dt \qquad (10.2)$$

At steady state, looking only at the average brain concentration or with constant rate i.v. infusion, just a single parameter is necessary. Once $K_{p,brain}$ is measured, the steady-state $C_{tot,brain}$ can be calculated as $K_{p,brain} \times C_{tot,plasma}$. It should be noted that $K_{p,brain}$ is a comparison between the brain and plasma and cannot provide full information on the role of the BBB, because of the fundamental differences between the brain tissue and plasma. A better comparison can be achieved by evaluating unbound drug levels (as described in Sects. 10.2.3 and 10.3.2) or to compare brain drug concentration to that in the brain without barrier (i.e., non-barrier brain regions) or to a systemic tissue with similar drug or solute of interest binding properties. Drug binding can be measured in vitro, but drugs can differ greatly in their binding in plasma to plasma proteins versus binding to tissue proteins and lipids in organs of interest. An example of this is the comparison of the K_p of ibuprofen or naproxen (NSAIDs) to that of one of the vinca alkaloids (like vincristine or vinorelbine) or to taxanes, such as paclitaxel or docetaxel. Ibuprofen and naproxen bind very highly to serum albumin but show little or very limited binding to brain tissue (Mandula et al. 2006). As a result, due to the high level of bound NSAID in serum, the brain-to-plasma ratio for ibuprofen and naproxen at steady state is quite low ($K_p = 0.001–0.02$) even though the drugs cross the barrier readily and have a fairly high free drug BBB PS. In contrast, the vinca alkaloids and taxanes bind more highly to tissue elements, leading to far greater tissue K_p values than 1.0. Values in systemic tissues exceed 50–100, reflecting the high presence of bound drug at those sites at steady state. The brain is an exception because of active efflux at the BBB. Reported K_p values in the brain or neural tissues for taxanes and vinca alkaloids more commonly fall in the range of 0.5–1.0. Administration of drug efflux transport inhibitors or use of animals with genetic knockout of specific BBB efflux pumps leads to K_p values in the brain that are in line with that predicted by the tissue binding elements. For paclitaxel, the rise in K_p can be 10–50-fold. Thus, the careful investigator should use caution in the interpretation of any brain-to-plasma or -blood ratio measurement. Often, such values have been taken to show poor BBB penetration. The examples given above show the fallacy of such a view. The brain-to-plasma ratio gives some insight into the tissue level of drug at a given time point relative to plasma after a defined administration method. That drug level can be processed for bound and free agent, and estimates are provided for the level of active species in the brain. In the end, the comparison of the free drug level achieved relative to that needed for drug activity (such as provided by the K_D or K_i) is a more relevant assessment of the question whether adequate drug reaches the brain.

The above analysis assumed a constant level of drug in the circulation feeding the brain so that the brain or reference tissue could come to equilibration or steady-state levels with respect to the drug. Such experiments are often difficult to carry out and frequently do not mimic well the fluctuation seen with normal human delivery and dosage, where plasma and tissue concentrations rise and fall in concert with time and systemic elimination. Ultimately, to describe the full time course of drug concentration in the brain under nonsteady-state conditions requires two parameters after the most common drug administration approaches used in in vivo models (e.g., i.v. bolus or infusion, i.p. or oral dose administration). Most drugs in humans are given orally in discrete doses, and thus there will be maximum drug concentration in the brain followed by a declining curve as the drug is eliminated. Whether given in repeat oral doses to steady state, brain concentration will fluctuate between a maximum and a minimum during the dosing interval, depending upon the brain K_p, BBB PS, dose, and dosing interval as well as the plasma concentration. BBB PS is valuable as it gives insight on the time required to obtain steady-state distribution in the brain and thus an accurate estimate of $K_{p,brain}$ (Liu et al. 2005).

The halftime for brain equilibration can be calculated as:

$$t_{1/2} = \ln 2 / \left(PS / K_{pbrain} \right) \tag{10.3}$$

Brain equilibration requires $4\text{–}5 \times t_{1/2}$. Brain exposure requires time for equilibration as $K_{p\ brain}$ varies from 1.0 to >50 due to tight binding to tissue proteins. Passage across the BBB may be similar, but more time is required to reach steady state for high $K_{p\ brain}$ drugs as a consequence of the high levels accumulated in the brain due to binding.

10.2.2 Systemic Administration Method

The "gold standard" approach for in vivo characterization of the pharmacokinetics of drug penetration and distribution in the brain is the intravenous administration method (Fig. 10.2). Drug or test compound is delivered directly into the circulation by bolus injection or constant rate infusion. The time course of drug concentration is determined in the brain and serum by LC-MS/MS or other quantitative analytical technique. The data are then fit to a kinetic model to provide BBB PS and $K_{p,\ brain}$ (Ohno et al. 1978; Duncan et al. 1991; Liu et al. 2005) (Fig. 10.3). Exposure time can be manipulated to allow analysis of initial brain uptake, equilibrium distribution, or brain efflux, depending upon the mode of drug delivery into the circulation and penetration characteristics of the compound under study. Autoradiography or scanning mass spectrometry can be utilized to map drug distribution within brain lesions in diseases, such as stroke, infection, inflammation, neurodegenerative disease, or brain tumors (Fig. 10.4) (Lockman et al. 2010; Taskar et al. 2012). With disease-induced damage to the BBB, drug concentrations within brain parenchyma can become highly heterogeneous with values ranging >100 fold. Under such a condition, the average concentration within the lesion reflects only part of the variation observed under the disease condition. The systemic administration technique is the reference standard because it is based upon the in vivo BBB and CNS

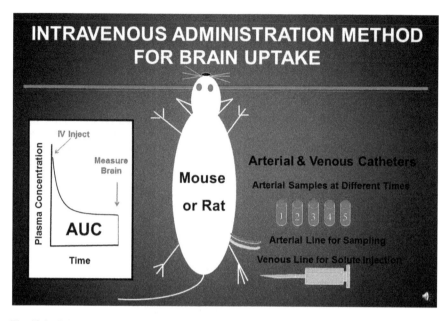

Fig. 10.2 Schematic diagram illustrating the intravenous (*i.v.*) administration method where a test drug is delivered as an *i.v.* bolus injection into the venous circulation and then arterial blood samples are collected at different times until brain drug concentration measurement. The plasma drug area-under-the-curve concentration integral (AUC) is calculated using the trapezoidal rule. Brain concentration, after vascular correction, is related to the time course of plasma concentration to calculate the brain tissue $K_p = AUC_{brain}/AUC_{plasma}$

parenchymal tissue, with all their complexity and unique differences. Most preclinical studies utilize rodents, such as rats or mice. Eventually, extension to primates or humans can be made with in vivo imaging, such as positron emission tomography. Care must be employed to distinguish metabolites from intact drug and drug within the brain from that residing within the blood circulation in the brain vasculature (Lockman et al. 2010). Further, in some cases, it is important to distinguish drug that has crossed the BBB from that sitting bound to the vascular endothelium or concentrated within the BBB endothelial space (Thomas et al. 2009). However, the small volume of the vascular endothelium within the brain tissue (<1 μL/g) limits the impact of BBB partitioning on overall brain tissue pharmacokinetics. Depending upon the drug and where it acts within the CNS, results can be expressed as BBB PS and $K_{p, brain}$ for whole brain or for specific regions within the CNS. In normal healthy brain, BBB PS varies only about three- to fivefold across the CNS and for many solutes correlates with regional differences in capillary surface area.

Fig. 10.3 Example of time course of drug concentration in brain, brain tumor and plasma following vascular injection of paclitaxel (Taxol) to immune compromised NuNu mice. Points equal mean ± SD for n = 3–6 animals. Kp is calculated from total drug concentration measurements as $AUC_{tissue}/AUC_{plasma}$

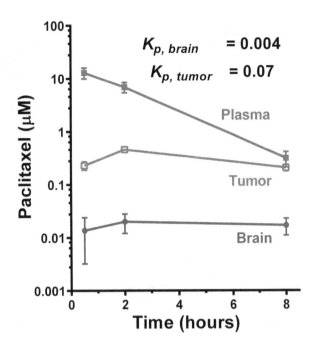

10.2.3 Free Vs. Total Drug in BBB Kinetic Analyses and Brain Microdialysis

In many cases, it is valuable to distinguish the concentration of drug that is free or unbound from the total drug concentration in the brain and serum, because the free drug concentration at the receptor site correlates best with pharmacodynamic models of activity. Many drugs bind significantly to proteins and lipids based upon lipophilicity and other factors. In such cases, total concentration can differ highly from the free drug concentration, which is usually the driving force for drug diffusion and equilibration. The fraction of drug that is unbound (f_u) can be determined from total concentration by ex vivo equilibrium dialysis or ultrafiltration (Hammarlund-Udenaes et al. 2008). Once f_u is measured, the free concentration (C_u) can be calculated as:

$$C_u = f_u \times C_{tot}. \tag{10.4}$$

C_u can also be measured directly in the brain by microdialysis (de Lange et al. 1997; Hammarlund-Udenaes et al. 2008).

The unbound drug concentration in the brain interstitial fluid is useful for comparison with pharmacodynamic studies (Kalvass et al. 2007; Liu et al. 2009) as well as for evaluation of the role of the BBB in hindering brain drug equilibration. The steady-state ratio of unbound drug concentration in the brain to that in the circulation is termed $K_{p,uu}$ and is calculated as:

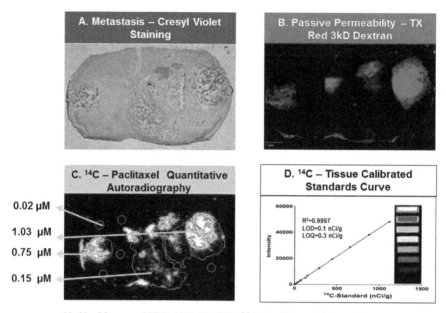

NuNu Mouse; MDA-MB-231BR, ^{14}C-Paclitaxel 10 mg/kg, i.v.; 2 hr

Fig. 10.4 Regional difference in brain and brain tumor uptake of paclitaxel at 2 h after i.v. injection of Taxol to immune-compromised NuNu mice, as determined using quantitative autoradiography. Coronal sections (20 μm) were prepared from frozen brain tissue using a cryostat. Matching sections were analyzed for brain distribution of ^{14}C-paclitaxel using autoradiography or stained for tumor using Cresyl violet. Texas red dextran was administered 10–15 min prior to death to measure BBB permeability. The vasculature of the brain was washed out after death using transcardial perfusion (1 min)

$$K_{p,uu} = C_{u,brain} / C_{u,plasma} = K_{p\,brain} \times f_{u,brain} / f_{u,plasma} \qquad (10.5)$$

A $K_{p,uu}$ value significantly less than 1.0 indicates restriction in brain drug availability as a consequence of the BBB, most commonly a result of active efflux transport, enzymatic breakdown, or CSF sink effect. In some cases, $K_{p,uu}$ exceeds 1.0, which is usually attributed to active influx transport at the BBB. A $K_{p,uu}$ value of ~1.0 indicates good brain drug bioavailability, especially when accompanied by a reasonably rapid BBB PS for rapid equilibration.

10.2.4   Brain Vascular Correction

Most analytical methods measure total brain drug concentration, and thus a vascular correction is usually required in order to obtain the amount of drug that has crossed the BBB. This correction can be made either by (a) washing out residual blood in brain vasculature at the end of the experiment or (b) subtracting the vascular content, calculated as product of the brain blood volume and the blood drug concentration at the end of the experiment. Brain blood volume is usually measured using radiotracers, such as [^{125}I]albumin, [^{14}C]dextran, or [^{3}H]inulin, which, under normal conditions, minimally cross the BBB. The vascular correction is most important at early uptake times, where blood concentration is frequently large (such as after bolus i.v. injection) and parenchymal brain concentrations are small. Brain vascular volume varies from 0.005 to 0.025 mL/g, and thus $K_{p,brain}$ values >0.25 are minimally influenced by vascular contribution. Section 10.3.1 presents additional considerations when evaluating BBB permeability and the contribution of the vasculature.

10.2.5   Influence of Flow on Initial Brain Uptake and BBB PS

In most studies, the initial rate of drug uptake into the brain does not directly measure BBB PS but the transfer coefficient (K_{in}) for drug uptake into the brain. K_{in} is related to PS as it gives an excellent index of the ease with which a solute can move from plasma into the brain, but it is not permeability (P). This is because, if the solute is sufficiently permeant at the BBB, the flow rate by which the solute is presented to the brain can influence its initial rate of brain uptake. The net result is that brain uptake also depends upon cerebral blood flow (F) which varies between brain regions and with neuronal activity. Brain uptake is also influenced by the capillary surface area (S). Renkin (1959) and Crone (1963) modeled this flow dependence using the Krogh single capillary model and derived the following relationship between flow (F), capillary permeability (P), the free fraction of drug in plasma ($f_{u,plasma}$), and capillary surface area (S):

$$K_{in} = F\left[1 - e^{-(fu,\ plasma \times PS/F)}\right] \tag{10.6}$$

(Mandula et al. 2006; Parepally et al. 2006). Thus, in most experiments, K_{in} is directly measured and BBB PS is calculated. Two limiting conditions in Eq. 10.6 are worth noting: (a) when F >> PS, $K_{in} \to$ PS, and (b) when PS >> F, $K_{in} \to$ F. For practical purposes, BBB $K_{in} \approx$ PS with less than 10% error when F > 5 × PS, and $K_{in} \approx$ F with less than 10% error when PS > 2.3 × F. Thus, K_{in} is an acceptable estimate of BBB PS when F/PS > 5. When F/PS < 5, K_{in} is influenced by both PS and F.

10.2.6   In Situ Brain Perfusion and Brain Efflux Index

In some situations, additional information is required regarding the mechanisms involved that restrict drug uptake into the brain at the BBB. With the normal in vivo approach, limits are placed on the degree to which an investigator can control or change brain blood flow, free drug concentration, or block transport or metabolic mechanisms. Knockout animals are available for several key BBB transporters, such as p-glycoprotein, breast cancer resistance protein, multidrug resistance protein-4, and organic acid transporting polypeptide. However, with current transporter knockouts, the alteration is not just at the BBB but at all sites within the CNS that usually express the transporter. This can lead to complexity in evaluating the separate role of the BBB in overall brain distribution of the compound.

As alternatives to direct in vivo analysis, in situ perfusion and brain efflux index methods are available for more specific studies of BBB transport. These approaches complement the standard i.v. administration method but allow greater flexibility in studying factors that may alter transport. The in situ perfusion method utilizes the in vivo structure of the BBB and cerebral tissues and simply superimposes its own vascular perfusion fluid as replacement to the animal's circulating blood (Fig. 10.5). The particular key advantage of this method is the facile control of perfusate solute concentration which can be altered over a much greater range than generally tolerated in vivo. The concentration dependence of transport is readily measured, as are the effects of ion concentrations and inhibitors. The perfusion approach was originally developed for rats (Takasato et al. 1984) but has been expanded to mice and used in multiple studies of BBB drug transport (e.g., Andre et al. 2013).

Mechanisms of brain-to-plasma efflux can also be investigated with the brain efflux index technique of Kakee et al. (1996a, b). This method involves direct microinjection of test solute and impermeant reference tracers into the brain. At various times thereafter, the ratio of test tracer to impermeant reference marker (R) is determined in the brain and expressed as a "brain efflux index" (BEI) value, defined as:

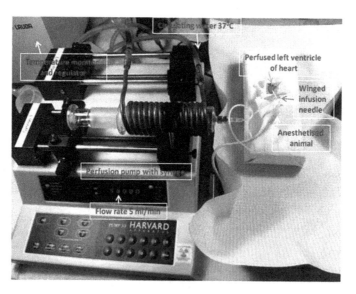

Fig. 10.5 Typical set up for *in situ* brain perfusion with syringe, infusion pump, temperature control and circulating water bath

$$BEI = 100 \times \left(1 - \left[\left(R\right)_{brain} / \left[\left(R\right)_{injectate}\right]\right.\right) \tag{10.7}$$

From this data a rate coefficient for efflux (k_{out}) from the brain is calculated and is converted to a transfer coefficient (i.e., clearance) for efflux, as $K_{out} = k_{out} \times K_{pu}$ where K_{pu} is determined from the steady-state distribution of test solute versus free drug concentration in brain slices in vitro. Caution must be exercised with the technique as BBB damage from needle tract injections may alter BBB transport or blood flow. Similarly, solute dilution in brain parenchyma is significant (>50 fold), forcing the necessity to use high levels (e.g., >100 mM) of competitor or transport inhibitor in the injection solution in some experiments. Finally, because of the transient nature of the experimental design, it is hard to precisely know the free drug concentration in interstitial fluid at the BBB as a function of time. K_{pu} assumes equilibration, but under normal circumstances, the efflux index is determined under conditions of changing drug concentration. Regardless, the brain efflux index method provides valuable insight on the role and characterization of a number of efflux transport systems at the BBB.

10.2.7 Cerebrospinal Fluid

The CSF is in direct contact with brain interstitial fluid and can be used to gain some insight as to drug distribution in the CNS. CSF is particularly useful when measured at steady state, where it provides a reasonable (± two- to threefold) estimate of the drug concentration in brain interstitial fluid. As CSF is low in protein content, it provides an estimate of free drug concentration in brain interstitial fluid. Under nonsteady-state conditions, the CSF provides a poor index of brain concentration due to multiple factors influencing its formation and circulation within the brain. This is particularly true when the BBB is disrupted in one location, whereas the CSF is sampled from distant locations, such as from the lumbar space.

10.3 Case Study: BBB Permeability Measured with Hydrophilic Low Molecular Weight Markers

10.3.1 Introduction

BBB dysfunction and breakdown can manifest both as contributing factors and as consequences of neurological disorders, resulting in the transfer of harmful substances from the blood into the brain and disturbed transport mechanisms. These aspects are covered extensively in other chapters of this book (*see Part V*). As a baseline for evaluating pathological conditions, it is necessary to measure the functional integrity of the BBB, and this is typically accomplished by quantifying the brain uptake of markers, which show low permeability at the intact barrier.

There are many technical and conceptual pitfalls associated with the experimental application of supposedly paracellular markers and the subsequent interpretation of data. One important aspect is that these markers can serve two distinct purposes. The first purpose is that, due to their characteristically low BBB permeability, these substances are often used as so-called vascular markers (see Sect. 10.2.4). A common example would be the simultaneous measurement of more permeable agents (e.g., drugs). When used as a vascular marker, it is assumed that the extent of brain uptake of that compound during a short experimental time period (1 min or less) can be neglected without compromising the evaluation. Under that premise, any concentration measured in whole-brain tissue represents brain intravascular space (with reference to concentration in whole blood) or brain plasma volume (with reference to plasma concentration). Such intravascular space values can then be used to correct brain concentrations of other substances, before calculating their BBB permeability. The second purpose is to determine the true permeability values of the BBB markers themselves. As pointed out above (see Sects. 10.1 and 10.2.1), all compounds cross the BBB to a certain degree, which is depending on their physico-chemical characteristics. In the case of drugs, the issue is whether that is sufficient to elicit effects. In the case of markers, which typically are devoid of pharmacological activity, the crucial question is whether the measured concentration, or the K_{in} value, is real or maybe an artifact. In the following we will discuss the frequently used BBB markers fluorescein, sucrose, and mannitol, point out common mistakes in their application, and present suitable experimental approaches.

10.3.2   Sodium Fluorescein and the Importance of Protein Binding

Sodium fluorescein (FL) is a low molecular weight (376 Da), highly fluorescent xanthene dye. FL exists at physiological pH in its monoanionic and dianionic form (Lavis et al. 2007), has a hydrophilic log D value of −1.3 at pH 7.4 (Zanetti-Domingues et al. 2013), and has low BBB permeability, which supposedly depends on the paracellular route (van Bree et al. 1988). Because of these physicochemical characteristics and the excellent spectral properties as a fluorophore, FL has been used extensively as an in vivo marker to determine the changes in BBB permeability in different diseases (Kaya and Ahishali 2011). In most experiments, investigators use the absolute brain concentrations of FL as a measure of permeability (Nishioku et al. 2010; Oppenheim et al. 2013; Tress et al. 2014; Cao et al. 2015). However, this approach may result in inaccuracies if the systemic exposure of the marker, such as area under the plasma concentration-time curve (AUC), is also altered by the disease or the intervention. FL is subject to metabolism in the form of glucuronidation in the liver. Furthermore, a significant fraction of FL circulating in the blood is bound to plasma proteins (Li and Rockey 1982). Because only the free drug is expected to cross the membranes, including the BBB, a change in the free fraction of FL in plasma may also affect the degree of accumulation of the marker in the brain (Mandula et al. 2006). The impact of these factors on brain uptake has been investigated in rats using a disease model of hepatic ischemia/reperfusion injury (IR), known as Pringle maneuver, alone or in combination with partial hepatectomy (HxIR) (Miah et al. 2015). The hypothesis was that these surgical interventions would increase the BBB permeability to FL. Pharmacokinetics after IV administration of FL were analyzed by HPLC with fluorescence detection, and brain uptake was evaluated by the single time point approach. As depicted in Fig. 10.6, IR and HxIR increased the plasma AUC compared to sham operated control animals after 15 min or 8 h of liver reperfusion. Additionally, the free fractions of FL in plasma, f_u, as determined by ultrafiltration, were significantly higher in the HxIR groups ($p < 0.01$) with 8 h reperfusion than in their respective sham groups (Table 10.1). With respect to brain concentrations, these were significantly increased in the IR and HxIR groups, as shown in Fig. 10.7a and b. Accordingly, brain uptake clearance K_{in} was increased when it was calculated using the AUC based on total plasma concentrations of FL (Fig. 10.7c and d). However, applying AUC values based on free plasma concentrations, there was no longer a difference in K_{in} values between the ischemia/reperfusion groups and controls (Fig. 10.7e and f). Therefore, without rigorous pharmacokinetic analysis, one would have come to the erroneous conclusion that the surgical intervention acutely compromised the BBB.

10.3.3 Sucrose and Mannitol: The Importance of Highly Specific Analytical Assays

Due to its favorable properties, the disaccharide sucrose has long been the widely accepted standard for the precise measurement of paracellular BBB permeability. The characteristics include a molecular weight (342 Da) similar to many small-molecule drugs, hydrophilicity (log P −3.6), neutral charge, absence of protein binding, lack of affinity to any known influx or efflux transporters at the BBB, and high metabolic stability in the circulation. Until recently, only radiolabeled versions, in particular [^{14}C]sucrose, have been used as permeability probes, and quantification was performed by simple liquid scintillation counting of tissue homogenates and plasma samples, without chromatographic separation. Nonspecific measurement of total radioactivity may be one of the factors contributing to vast differences in K_{in} values reported in the literature, covering at least a 45-fold range (Hladky and Barrand 2018). A few earlier publications had pointed out that radiolabeled sucrose tracer can contain low amounts (≈2%) of more lipophilic compounds (Preston and Haas 1986; Preston et al. 1998), which may distort BBB permeability values. This was recently confirmed by Miah et al. (2017), who developed a highly specific and sensitive LC-MS/MS method for stable isotope labeled [^{13}C$_{12}$]sucrose,

Fig. 10.6 Plasma
concentration - time
courses of sodium
fluorescein in the Sham,
IR, and HxIR animals
15 min (**a**) or 8 h (**b**) after
reperfusion (n = 6–7/
group). Symbols and bars
represent mean and SD
values, respectively. From
Miah et al. (2015), with
permission

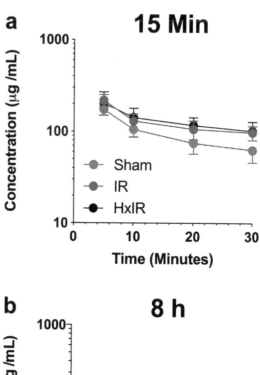

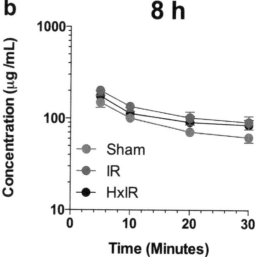

where all carbon atoms are substituted with the ^{13}C isotope (Miah et al. 2016). Comparison of $[^{14}C]$sucrose and $[^{13}C]$sucrose in successive octanol/water partitioning procedures revealed that the radiotracer showed higher lipophilicity, which declined upon serial partitioning but remained higher even after the third round (Fig. 10.8). This is compatible with the results of HPLC fractionation of brain samples obtained following IV injection of $[^{14}C]$sucrose in rats. As shown in Fig. 10.9, the majority of radioactivity in the brain was attributable to compounds other than sucrose. The contaminants are not the result of metabolism, but their chemical identity remained undetermined. Accordingly, the K_{in} values estimated for $[^{14}C]$sucrose

Table 10.1 Plasma AUC and free fraction values (Mean ± SD) of fluorescein after a short (5 min) intravenous infusion of the marker (25 mg/kg) in rats subjected to Pringle maneuver without (IR) or with (HxIR) partial hepatectomy or sham surgery (Sham), followed by 15 min or 8 h of *in vivo* reperfusion

Treatment	AUC, μg.min/ml	f_u
15-Min groups		
Sham ($n = 7$)	2700 ± 490	0.233 ± 0.056
IR ($n = 6$)	3590 ± 594	0.301 ± 0.034
HxIR ($n = 6$)	3750 ± 971[*]	0.408 ± 0.112[**]
8-h groups		
Sham ($n = 6$)	2510 ± 145	0.215 ± 0.049
IR ($n = 6$)	3470 ± 432[**]	0.245 ± 0.052
HxIR ($n = 6$)	3060 ± 446	0.356 ± 0.082[**,a]

From Miah et al. (2015), with permission
[*]$p < 0.05$, [**]$p < 0.01$: Significantly different from the corresponding Sham group (ANOVA, followed by Tukey's post-hoc analysis)
[a]$p < 0.05$: Significantly different from the corresponding IR group (ANOVA, followed by Tukey's post-hoc analysis).

were six- to sevenfold higher than corresponding values for [^{13}C]sucrose. Further evidence for the preferential uptake of lipophilic impurities of [^{14}C]sucrose into the brain tissue was provided by a recent brain microdialysis study in mice, who received IV injections of either radiolabeled sucrose or [^{13}C]sucrose (Alqahtani et al. 2018). While whole-brain tissue concentrations of ^{14}C radioactivity were 4.1-fold higher and calculated K_{in} was 3.6-fold higher compared to [^{13}C]sucrose, the concentrations of radioactivity and of [^{13}C]sucrose in brain dialysate samples were comparable. Because microdialysis only samples analytes from brain extracellular space, the majority of radioactive tracer in brain tissue was apparently in a nondialyzable compartment, likely intracellular.

After [^{13}C$_{12}$]sucrose was introduced as a superior BBB permeability marker, which is devoid of the shortcomings of radiolabeled sucrose variants, the technique has recently been expanded and refined in two aspects.

First, as outlined above and in Sect. 10.2.4 of this chapter, vascular space correction in brain uptake studies is possible by either vascular washout or inclusion of a vascular marker in the analysis. Because there are multiple variants of stable isotope labeled sucrose commercially available, using another version of sucrose as vascular marker in BBB permeability studies was an obvious opportunity. Chowdhury et al. used [^{13}C$_6$]sucrose (all carbons of the fructose moiety replaced by ^{13}C) as the vascular marker, which coelutes with [^{13}C$_{12}$]sucrose on a UPLC column and can be simultaneously quantified by MS/MS multiple reaction monitoring (Chowdhury et al. 2018). In a pharmacokinetic experiment, [^{13}C$_{12}$]sucrose, as the analyte used for permeability measurements, is administered at the start of the study, and the vascular marker [^{13}C$_6$]sucrose is injected shortly (e.g., 30 seconds) before the terminal sampling time. Comparison of a series of mice undergoing vascular washout by

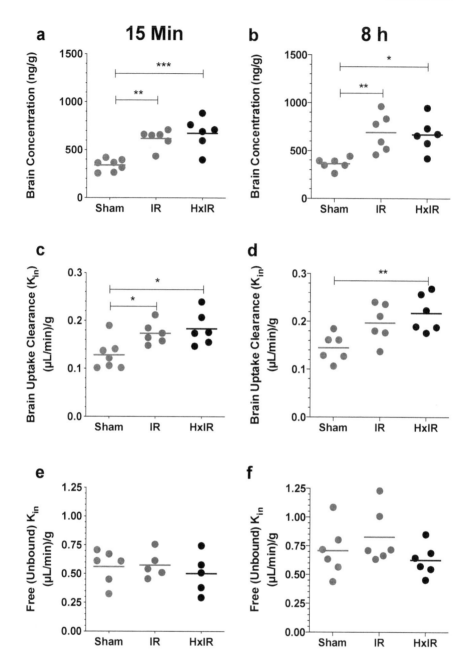

Fig. 10.7 Brain concentrations (**a** and **b**), apparent brain uptake clearance (K_{in}) based on the total AUC(**c** and **d**), and apparent brain uptake clearance (K_{in}) based on the free (unbound) AUC (**e** and **f**) of sodium fluorescein in the Sham, IR, and HxIR animals 15 min (left panels) or 8 h (right panels) after reperfusion (n = 6–7/group).The symbols and horizontal lines represent the individual and mean values, respectively. Statistical analysis is based on one-way ANOVA, followed by Tukey's post-hoc test. *p < 0.05, **p < 0.01, ***p < 0.001. From Miah et al. (2015), with permission

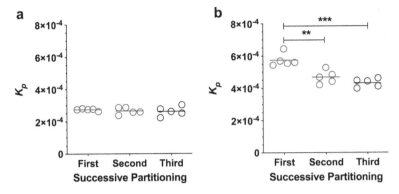

Fig. 10.8 Octanol:water partition coefficient (K_p) of [^{13}C] (**a**) and [^{14}C] (**b**) sucrose ($n = 5$/marker), which were used in the current *in vitro* and *in vivo* studies. Aqueous samples were successively partitioned for 3 times. Symbols and horizontal lines represent individual and mean values, respectively. $**p < 0.01$; $***p < 0.001$ based on one-way ANOVA, followed by Tukey's post-hoc analysis. Adapted from Miah et al. (2017), with permission

buffer perfusion via the left ventricle of the heart with a group of animals receiving the [^{13}C$_6$]sucrose vascular marker resulted in equivalent values of brain concentrations and K_{in}.

The second enhancement of the technique consisted of including mannitol as an additional marker for measurement of BBB permeability (Noorani et al. 2020). Mannitol is a small molecule (182 Da) that is about half the size of sucrose and has otherwise similar physicochemical characteristics. Furthermore, mannitol has been widely used over the last 30 years in the lactulose/mannitol (L/M) test as a common dual-sugar test to assess the intestinal barrier function (Camilleri et al. 2010). Radiolabeled versions of mannitol with ^{3}H or ^{14}C have been used for measurement of BBB integrity (Sisson and Oldendorf 1971; Amtorp 1980; Preston et al. 1984, 1995; Daniel et al. 1985; Iliff et al. 2012), but, as with radiolabeled sucrose, these require a radioactive license and special handling skills. Importantly, radiolabeled mannitol is also prone to contamination by low levels of lipophilic compounds, resulting in artifactual overestimation of the BBB permeability (Preston and Haas 1986). For implementation of the dual marker technique, the following stable isotope labeled species were selected: [^{13}C$_6$]mannitol and [^{13}C$_{12}$]sucrose as permeability markers, [^{2}H$_8$]mannitol and [^{2}H$_2$]sucrose as internal standards, and [^{13}C$_6$]sucrose as a vascular marker. Hydrophilic interaction liquid chromatography separates mannitol and sucrose, while the isotopic variants of each sugar coelute and can then be detected and quantified using suitable combinations of mass transitions and settings of the mass detector. Method development and validation followed FDA guidelines for bioanalytical methods. As the first in vivo application of the method, a comparative pharmacokinetic study was performed in two groups of anesthetized mice, which received IV injections of [^{13}C$_6$]mannitol and [^{13}C$_{12}$]sucrose at a dose of 10 mg/kg each. Repeated blood samples were drawn, and the animals were euthanized after 30 min. The first group of mice received [^{13}C$_6$]sucrose 30 seconds before

Fig. 10.9 Representative HPLC fractionation of a brain sample collected 1 h after the *in vivo* administration of [¹⁴C]sucrose (**a**), a blank brain homogenate spiked *in vitro* with [¹⁴C]sucrose (**b**), and a representative plasma sample collected after the *in vivo* administration of [¹⁴C]sucrose (**c**). For comparison, the fractionation of the dosing solution is also shown along with the brain sample (a). For *in vivo* experiments, a single intravenous dose (100 μCi) of [¹⁴C]sucrose was injected into the rats (*n* = 3), and brain (after blood removal) and plasma samples were collected at 1 h after dosing. Adapted from Miah et al. (2017), with permission

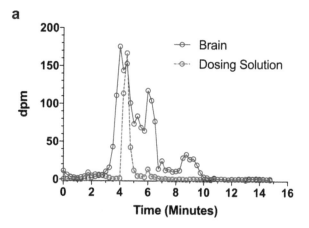

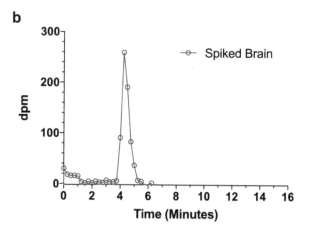

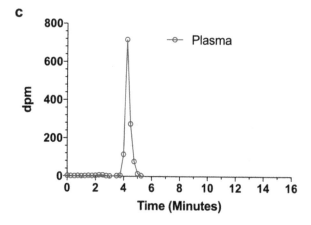

termination (vascular marker group), while the second group was subjected to vascular perfusion via the left ventricle of the heart (washout group).

The value of corrected brain concentration ($C_{br-corr}^{Analyte}$) in the vascular marker group, which received $[^{13}C_6]$sucrose, was determined as follows:

$$C_{br-corr}^{analyte} = \frac{(V_d - V_0) \times C_{pl}^{analyte}}{1 - V_0} \tag{10.8}$$

Here, V_d is the apparent volume of distribution of the BBB permeability marker, $[^{13}C_6]$mannitol and $[^{13}C_{12}]$ sucrose, V_0 is the apparent volume of distribution of the vascular marker, $[^{13}C_6]$ sucrose, and $C_{pl}^{Analyte}$ is the terminal (30 min) plasma concentration of $[^{13}C_6]$mannitol or $[^{13}C_{12}]$sucrose. V_d and V_0 values were obtained using the following two equations:

$$V_d = C_{br}^{analyte} / C_{pl}^{analyte} \tag{10.9}$$

$$V_0 = C_{br}^{vascular\ marker} / C_{pl}^{vascular\ marker} \tag{10.10}$$

where $C_{br}^{analyte}$ is the total (uncorrected) brain concentration of $[^{13}C_6]$ mannitol or $[^{13}C_{12}]$sucrose, $C_{br}^{vascular\ marker}$ is the total brain concentration of $[^{13}C_6]$sucrose at the terminal sampling time (30 min), and $C_{pl}^{vascular\ marker}$ is the terminal plasma concentration of the vascular marker at 30 min.

Brain tissue concentration values in the washout group were considered as corrected for intravascular content. Values for brain uptake clearance, K_{in}, were calculated using the following equations based on either uncorrected ($C_{br}^{analyte}$) or corrected ($C_{br-corr}^{analyte}$) brain concentrations of mannitol and sucrose:

$$K_{in} = C_{br}^{analyte} / AUC_0^T \tag{10.11}$$

$$K_{in-corr} = C_{br-corr}^{analyte} / AUC_0^T \tag{10.12}$$

where AUC_0^T denotes the area under the plasma concentration-time curve from time point 0 to the terminal sampling time (30 min) for $[^{13}C_6]$mannitol and $[^{13}C_{12}]$sucrose. AUC_0^T was estimated via the linear-logarithmic trapezoidal method.

The results of the pharmacokinetic study are shown in Figs. 10.10 and 10.11. The plasma profiles of both groups (vascular marker group and washout group) were similar for mannitol and sucrose, and the areas under the curve from 0 to 30 min were not significantly different. Moreover, the plasma profiles of mannitol and sucrose were similar, and both showed a biexponential decline (Fig. 10.10).

Comparison of the corrected brain concentrations (washout vs. vascular marker correction) showed no significant difference for both mannitol and sucrose (unpaired, two-tailed t-test) (Fig. 10.11). C_{br} in percent of injected dose per mL (%ID/mL) of mannitol was 0.071 ± 0.007 and 0.065 ± 0.009 for vascular marker and washout, respectively, whereas the C_{br} of sucrose was about half of mannitol C_{br}

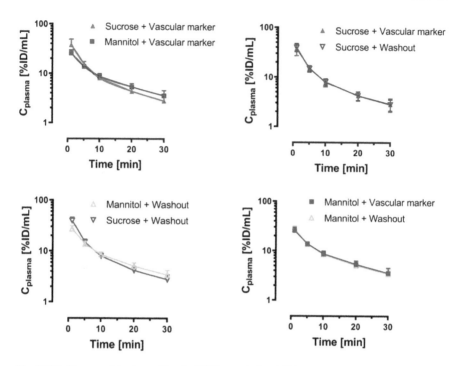

Fig. 10.10 Pharmacokinetic profiles for [^{13}C$_6$]mannitol and [^{13}C$_{12}$]sucrose in mouse plasma up to 30 min after IV bolus (mean ± SD, n = 6). From Noorani et al., with permission

values (0.035 ± 0.003 and 0.037 ± 0.005%ID/mL for vascular marker and washout group, respectively). Similarly, comparison of K$_{in}$ values for each marker showed no significant difference between these two groups (unpaired, two-tailed t-test) (Fig. 10.11). For example, the K$_{in}$ value of mannitol was 0.267 ± 0.021 µl.g^{-1}.min^{-1} and 0.245 ± 0.013 µl.g^{-1}.min^{-1}for the vascular marker and washout groups, respectively. From a practical experimental perspective, the correction by vascular marker administration could be advantageous compared to the washout method in several aspects: Technically, it is easier to perform, and brain tissue collection is attainable within seconds after the terminal blood sampling, as opposed to delays for several minutes by performing thoracotomy and perfusion (e.g., over 10 min in the present study). Furthermore, rapid sampling gains importance when, apart from measuring the BBB permeability, parts of the brain samples were needed for measurement of other analytes such as neurotransmitters or metabolites that may undergo rapid degradation.

Comparing the dual markers, the K$_{in}$ of mannitol (0.267 ± 0.021 µl.g^{-1}.min^{-1}) was more than twice that of sucrose (0.126 ± 0.025 µL.g^{-1}.min^{-1}).

The sucrose/mannitol technique is equally suitable for in vitro applications, and this was demonstrated using an in vitro BBB model with Transwells and human brain microvascular endothelial cells (BMECs) derived from the induced pluripotent stem cell line IMR90-c4. BMECs were differentiated following established

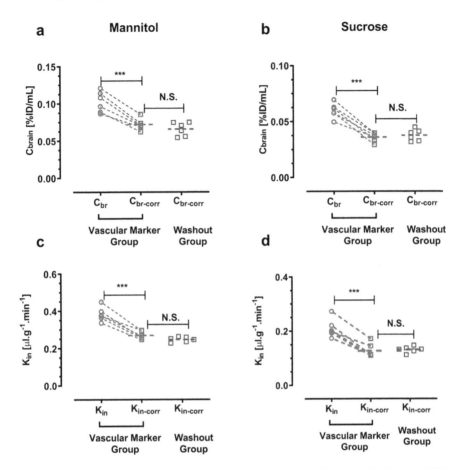

Fig. 10.11 Panels (**a**) and (**c**): Differences in brain concentration and brain uptake clearance (K_{in}) of [$^{13}C_6$]mannitol with or without correction by vascular marker. Panels (**b**) and (**d**): C_{br} and K_{in} of [$^{13}C_{12}$]sucrose with or without correction by vascular marker. ***$p < 0.001$ ($n = 6$), analyzed by Student's paired t-test (two-tailed). N.S. = not significant by Student's unpaired t-test. From Noorani et al. 2020 with permission

protocols (Lippmann, et al. 2014; Nozohouri et al. 2020). The average TEER value was 1812 ± 54 $\Omega.cm^2$, which is similar to values reported in the literature for this model (Lippmann et al. 2014; Patel and Alahmad 2016). Paracellular permeability was assessed by adding 1 mg/mL of [$^{13}C_6$] mannitol and [$^{13}C_{12}$] sucrose to the donor side (apical chamber) of the Transwell system. Then aliquots were collected repeatedly from the acceptor (basolateral) chamber up to 120 min. The clearance or permeability-surface area (PS) products for mannitol and sucrose were calculated using the following steps: First, the cleared volume up to each time point was calculated from the following equation:

$$\text{Cleared Volume} = \left(C_{\text{acceptor}} * V_{\text{acceptor}}\right) / C_{\text{donor}} \tag{10.13}$$

Here, C_{acceptor} refers to measured concentration in the acceptor compartment at a given sampling time point, and V_{acceptor} refers to the volume of the acceptor compartment. Also, C_{donor} is the concentration in the donor compartment. Then, linear regression was applied to the plotted values of cleared volume versus time for samples and blank (filter only without cells) to obtain the PS of the Transwell system. Afterward, the permeability coefficient (P) was obtained by the following equations:

$$P = PS / S \tag{10.14}$$

$$\frac{1}{P_{\text{Cells}}} = \frac{1}{P_{\text{total}}} - \frac{1}{P_{\text{blank}}} \tag{10.15}$$

where S is the surface area of the filter insert (0.33 cm^2), and then the permeability coefficient of the cell monolayer (P_{cells}) was obtained by subtracting the permeability coefficient of Transwell (P_{total}) from the permeability coefficient of the coated filter (P_{blank}). For a comparison between in vitro and in vivo data, the K_{in} values or permeability-surface area products (PS) were converted to permeability coefficients, taking 120 cm^2/g of the brain as the surface area of the BBB in vivo (Pardridge 2020). The in vitro permeability coefficients of mannitol and sucrose were $4.99 \pm 0.152 \times 10^{-7}$ and $3.12 \pm 0.176 \times 10^{-7}$ cm/s, respectively. Figure 10.12a depicts the permeability values of the two markers, with mannitol showing higher permeability compared to sucrose ($p < 0.0001$ unpaired, two-tailed t-test). The PS value of mannitol and sucrose in vivo was 0.267 ± 0.021 and 0.126 ± 0.025 μl.g^{-1}.min^{-1}, respectively, which corresponds to a permeability coefficient value of $3.71 \pm 0.296 \times 10^{-8}$ and $1.75 \pm 0.355 \times 10^{-8}$ cm/s for mannitol and sucrose,

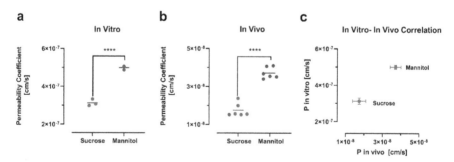

Fig. 10.12 (**a**) Permeability coefficient (P) of mannitol and sucrose in the Transwell model with TEER value of 1812 ± 54 Ω.cm^2 (n = 3). (**b**) Permeability coefficient (P) of mannitol and sucrose in the *vivo* model (n = 6). ****p < 0.0001, analyzed by Student's unpaired t-test (two-tailed). (**c**) *In vitro* and *in vivo* correlation of mannitol and sucrose based on the permeability coefficients. Adapted from Noorani et al. 2020 with permission

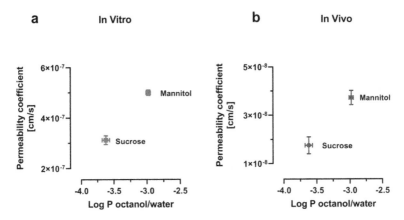

Fig. 10.13 (**a**) Correlation of *in vitro* permeability coefficient (n = 3) and Log P (n = 5). (**b**) Correlation of *in vivo* permeability coefficient (n = 6) and Log P (n = 5). From Noorani et al. 2020, with permission

respectively. Figure 10.12c shows the in vitro and in vivo correlation of the markers. Interestingly, the P values for mannitol and sucrose in vitro were only about 13-fold and 18-fold higher than the permeability coefficient in vivo. The correlations between octanol/water partition coefficient Log P and in vitro and in vivo permeability coefficients are shown in Fig. 10.13. Mannitol is about fourfold more lipophilic than sucrose (log P of -2.98 ± 0.033 vs. -3.62 ± 0.056, respectively) as analyzed by LC-MS/MS using the ^{13}C labeled substances (Noorani et al. 2020). The lower lipophilicity of sucrose likely contributes to its lower permeability coefficient compared to mannitol, if a transcellular pathway of diffusion via the plasma membrane is considered. In addition, the higher molecular weight of sucrose causes its diffusion coefficient to be about 30% lower (Peck et al. 1994), which affects both paracellular and transmembrane transport.

A sensitive method for measuring BBB permeability would be expected to detect subtle changes induced by pathological challenges, for example, inflammatory conditions elicited by pro-inflammatory cytokines. This was studied by exposure of iPSC-derived BMECs in the Transwell model to interleukin 1-beta (IL-1β). As shown in Fig. 10.14, the permeability coefficient of mannitol and sucrose significantly increased from $6.90 \pm 0.689 \times 10^{-7}$ and $4.74 \pm 0.314 \times 10^{-7}$ to $1.67 \pm 0.188 \times 10^{-6}$ and $1.23 \pm 0.163 \times 10^{-6}$, respectively, with 100 ng/mL IL-1β. Moreover, The TEER values of iPSC-derived BMECs decreased 38% after 1 day of exposure to 100 ng/mL IL-1β.

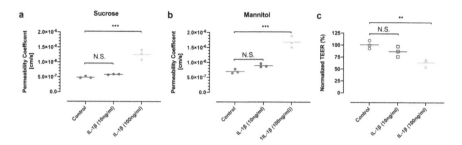

Fig. 10.14 Permeability coefficient of (**a**) mannitol, (**b**) sucrose in iPSC-BMECs following treatment with different concentrations of IL-1β. (**c**) The effect of IL-1β cytokine on TEER of iPSC-BMECs. **p < 0.01 and ***p < 0.001, 1-way ANOVA, followed by Tukey's multiple comparisons test (n = 3). Adapted from Noorani et al. 2020, with permission

10.3.4 Conclusion

In conclusion, the examples presented in the above sections illustrate that measurements of BBB permeability in vivo with poorly permeable markers require understanding of pharmacokinetic principles and application of rigorous analytical methods in order to obtain reliable data.

10.4 Going Beyond BBB Permeability to Drug Uptake and Distribution in Brain Metastases

The prior section focused on evaluations of BBB permeability with limited exposure and initial uptake of a compound to determine resistance and baseline permeability for solute passage into the CNS. This is a very important parameter that gives valuable insight into the fundamental properties of the BBB.

In this section, additional insight that can be obtained with longer dose exposure, virtually to steady-state equilibration, is highlighted. These types of experiments also give information on the degree to which drug exposure is limited and on the mechanisms involved. They highlight the importance of disease-induced changes in the barrier, selective and nonselective drug binding, as well as intracellular transport in determining the levels achieved of drug in the brain. In addition, we illustrate these considerations using preclinical models of tumors in the CNS whereby we have correlated simultaneously in time and space with high resolution, changes in BBB permeability, anticancer drug distribution, and in vivo antitumor effect in a preclinical model of brain metastases of breast cancer. The global goal is to be able to draw correlations between drug levels and induced pharmacologic effect.

10.4.1   Introduction: BBB Breakdown and Heterogeneity in Brain Tumors

The BBB is partially compromised in brain metastases, forming the blood-tumor barrier (BTB). Molecular parameters inducing and controlling the magnitude of the barrier disruption are poorly understood. Equally, the contribution of the BTB to brain metastasis permeability has been the topic of over 50 years of debate and controversy (Steeg et al. 2011). Breast cancer is one of the primary tumors that metastasize to the brain. Therefore, we determined drug delivery of the primary agents used to treat metastatic breast cancer (e.g., paclitaxel, doxorubicin, lapatinib), in different preclinical models of brain metastases of breast cancer (Lockman et al. 2010; Taskar et al. 2012; Thomas et al. 2009). This was conducted using both human tumors in immune-compromised mice and syngenic tumor models in immune-competent mice.

While BTB permeability can be greatly increased compared to the brain (2- to 20-fold), our research and that of others have found that the barrier remains sufficiently patent to limit drug access to metastases (Lockman et al. 2010; Babak et al. 2020). Consequently, the extent and impact of drug delivery to the brain and brain tumors have been difficult to interpret given a lack of knowledge regarding the level of drug necessary for activity in brain metastases (i.e., in vivo IC_{50}) and the level of drug that represents free equilibration. Our research in preclinical animal models has established that brain metastasis exposure to critical chemotherapeutics, such as paclitaxel, doxorubicin, and lapatinib, though elevated compared to uninvolved brain, is orders of magnitude less than that reaching systemic metastases (Thomas et al. 2009; Taskar et al. 2012; Lockman et al. 2010).

An additional parameter of importance in studies that look at total drug concentration, as noted previously above, is the identification of the most similar tissue or compartment without a barrier that can be used as a comparative reference. For brain metastases, this could be a systemic tumor, such as in the liver, lung, kidney, or other tissues. This issue is of importance when comparing total drug levels, or K_p, from one tissue to another, where one would like to minimize the impact of other factors such as tissue binding or active efflux or accumulation.

This discussion elegantly shows the value of high-resolution analysis of unbound drug distribution in small tissue regions. This can be performed with a variety of methods; in Sect. 10.4.2, we describe the use of autoradiography combined with other imaging methods to assess permeability and effect. This analysis allows an additional assessment of barrier restriction, i.e., the extent to which free drug in tissue equilibrates with free drug in circulation. At steady state, absence of barrier is shown by free drug concentration in the brain equivalent to that in circulation. In contrast, when the barrier is present, some drugs show extremely low free levels relative to circulation. Thus, Kp_{uu} is a very important parameter for assessing barrier function in vivo, distinct from simple permeability.

Following up, for confirmation in humans, a presurgery study of capecitabine and lapatinib is presented that has confirmed limited drug exposure in human brain metastases (Sect. 10.4.3) (Morikawa et al. 2015).

10.4.2 Case Study: Role of the Barrier in Limiting Drug Therapeutic Effect

In this section, we focus on one of the agents, vinorelbine, a widely used third-generation synthetic vinca alkaloid cytostatic agent, which binds to tubulin, thereby inducing apoptosis. Vinorelbine exhibits an improved therapeutic index compared to other anti-tubulin agents such as vincristine, vinblastine, and the taxanes (Galano et al. 2011). The methods used to perform these studies are presented in detail in Samala et al. (2016).

The design of these experiments was to administer drug to mice that had developed brain and systemic metastases and measure levels obtained. Mice were intravenously administered with vinorelbine containing ^{3}H-vinorelbine at the maximum tolerated dose (12 mg/kg). Drug was allowed to circulate for 0.5, 2, and 8 h, to allow time for drug distribution and equilibration. Ten minutes prior to the end of the exposure period, Texas red was administered intravenously to measure BBB permeability. At the end of the circulation period, the animal was anesthetized, a fresh blood sample was collected by cardiac puncture, and the residual blood was removed by perfusion (45 s). At the end of perfusion, the brain and tissues with systemic metastases were rapidly removed and flash frozen.

To assess equilibration, drug levels can be determined by a variety of methods in dissected tissue. Quantitative autoradiography (QAR) or scanning LC-MS/MS can be used to determine spatial distribution of drug in tissue slices. In contrast, when spatial distribution is not of interest, simple LC-MS/MS is the gold standard method. In our study, we used QAR, which has the advantages of extremely high sensitivity (~0.3 nCi/g) and spatial resolution of ~10 μm (based on pixel size), with signal linearity over a ~ 10^5 concentration range (Lockman et al. 2010). With this resolution, a 1 mm^2 metastasis in one section would contain on the order of 10^4 pixels and could extend over multiple sections for additional confirmation of the metastasis characteristics.

In these studies, we evaluated tracer integrity and possible contributions from metabolism of vinorelbine by two independent techniques. Vinorelbine integrity was assessed at all time points in the brain, liver, and kidney by LC-MS/MS, and radiotracer integrity was evaluated in the plasma, lung, and kidney after 2 and 8 h of circulation in vivo using HPLC coupled to a flow scintillation analyzer for inline radiotracer detection. Tracer impurities not related to drug levels were found to be negligible at all time points. The only metabolism-related change was formation of the diacetyl-vinorelbine species, which has binding and activity virtually identical to those of vinorelbine. In this case, use of radioactivity may be an advantage because both species are equally active and detected together.

Measurements were performed using adjacent coronal tissue sections to determine tissue structure (cresyl violet), drug distribution and binding, permeability (Texas red), and effect (apoptosis) (Lockman et al. 2010; Taskar et al. 2012; Samala et al. 2016). Antihuman cytokeratin staining was used as needed to confirm metastases. Each metastasis was evaluated on serial tissue sections to confirm that the metastasis was an independent unit (separated by >100 μm from other metastases). Images of the cresyl violet-stained sections were imported into Slidebook (Olympus) and MCID software (Imaging Research) to correlate with analyses of permeability by Texas red fluorescence and drug uptake by QAR, respectively. Vinorelbine exposure was imaged within and between tumors by QAR. The partition coefficient for unbound drug distribution ($K_{p,uu}$), determined with equilibrium dialysis in plasma and tissue homogenate, was applied across brain metastases to assess drug equilibration.

In addition, vinorelbine binding was assessed using an adjacent section to those used for drug distribution measurements, using the frozen brain slice binding method. In brief, a hydrophobic well was drawn around the brain sections using a Liquid Blocker Super Pap pen (Daido Sangyo Co. Ltd), which were then incubated at 37 °C for 45 min with PBS containing 1 μCi/ml of ^{3}H-vinorelbine. At the end of the incubation period, PBS was collected, and the ^{3}H-vinorelbine concentration measured in the collected PBS is referred to as the "in vitro unbound drug concentration." After rinsing the slides with cold PBS and allowing them to dry at room temperature, the slides were placed in cassettes along with tissue-calibrated standards for QAR analysis, to obtain total in vitro concentration in the brain slice (Fig. 10.15a).

This brain slice binding method was validated by comparing results from equilibrium dialysis of paclitaxel and doxorubicin to literature values and then to results from our brain slice binding technique. Once the appropriate conditions were selected, vinorelbine unbound fraction was similarly evaluated by both methods. Due to slow vinorelbine interstitial diffusion and high intracellular binding, the fresh intact brain slice technique (Fridén et al. 2009a, b, 2011) was not successful because of the long time necessary for vinorelbine to equilibrate across the 300-μm-thick tissue sections used with the method.

To obtain in vivo spatially resolved free (unbound) and bound drug concentrations, as illustrated in Fig. 10.15b, the image of measured in vivo total drug concentration (A) and the in vitro unbound drug concentration (C) were combined using MCID transformation function to obtain images of in vivo free (F) and bound (E) drug concentrations. The combination of these techniques provides regional information on drug concentration in the brain and brain metastases that had previously not been available. The steady-state tissue partition coefficient (K_p) for vinorelbine distribution was calculated from the ratio of integrated total drug concentration in tissue ($AUC_{tot, tissue}$) to that in plasma ($AUC_{tot,plasma}$). The matching partition coefficient for unbound vinorelbine distribution ($K_{p,uu}$) was calculated as the ratio of

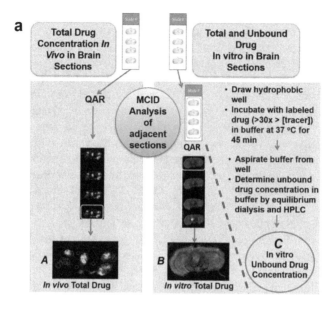

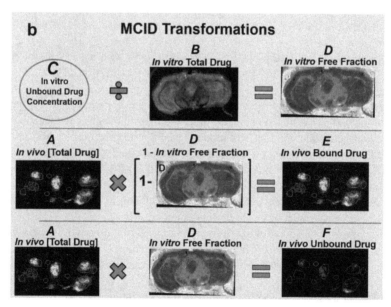

Fig. 10.15 Schematic of the brain slice binding method for the determination of *in vivo* free drug concentration. (**a**) *In vivo* total drug concentration (*A*) was quantified in one slide containing 10 μm brain slices using QAR. The slide adjacent to the QAR slide was incubated with higher concentration of radiotracer at 37 °C for 45 min and the *in vitro* unbound drug concentration (*C*) was estimated using equilibrium dialysis. After rinsing and drying, the radioactivity in the section adjacent to the last section of the previous slide was quantified using QAR to obtain the *in vitro* drug concentration (*B*). (**b**) The MCID transform function was used to calculate the *in vitro* free fraction (*D*), the *in vivo* bound drug concentration (*E*), and *in vivo* unbound drug concentration (*F*) as illustrated. Adapted from Samala et al. (2016), with permission

integrated unbound drug concentration in tissue ($AUC_{u,tissue}$) to that in plasma ($AUC_{u,plasma}$) (Hammarlund-Udenaes et al. 2008). In vivo unbound concentration was calculated as $C_u = f_u \times C_{tot}$ assuming rapid binding equilibration. $K_{p,uu}$ was calculated as $K_p \times (f_{u,brain}/f_{u,plasma})$. The barrier-free distribution coefficient $K_{p,\,tissue\,-\,no\,barrier}$ was estimated as $f_{u,plasma}/f_{u,tissue}$ assuming $K_{p,uu} = 1$.

Matching in vivo TUNEL staining, a biomarker of early apoptosis, was evaluated on brain sections of mice exposed to vinorelbine for 2 h contrasted with exposure and efficacy obtained in matching 2-h experiments in in vitro cell culture. No increase in vascular density has been observed in this model of brain metastasis of breast cancer (Lockman et al. 2010).

A striking characteristic of brain metastases is the wide range of drug uptake both within and between metastases. Median metastasis vinorelbine concentration

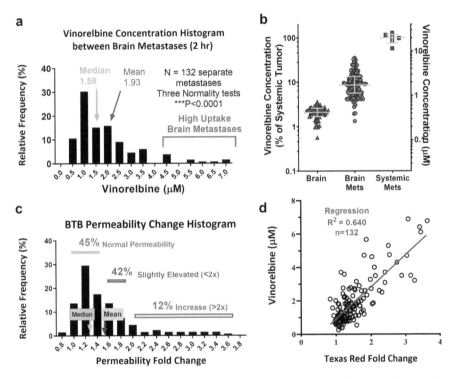

Fig. 10.16 ^{3}H-vinorelbine distribution between metastases correlates with permeability and is non-Gaussian. Data are for animals at 2 h after *i.v.* administration of ^{3}H-vinorelbine (10 mg/kg). (**a**) Histogram of average ^{3}H-vinorelbine concentrations in brain metastases, exhibiting a tail of higher uptake brain metastases. (**b**) Average ^{3}H-vinorelbine concentrations in brain, brain metastases and systemic metastases as a percent of median systemic metastasis concentration (left axis) and concentration (right axis) with median and quartiles shown by green lines. (**c**) Non-Gaussian distribution of average metastasis permeability, as measured with Texas red fluorescence, expressed as fold increase in fluorescence from brain. (**d**) Average vinorelbine concentration vs. fold increase in Texas red. Pearson correlation, shown as a solid blue line, was statistically significant (P < 0.001). Adapted from Samala et al. (2016), with permission

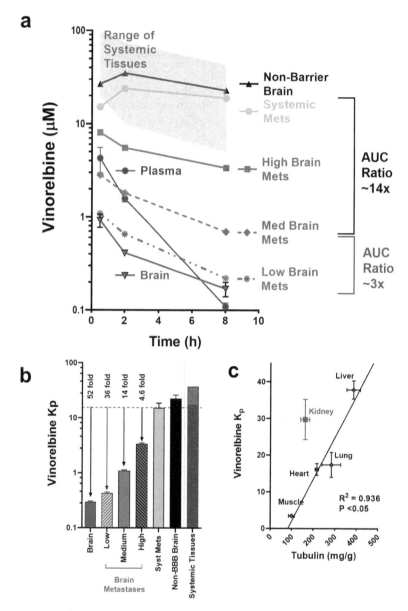

Fig. 10.17 Pharmacokinetics of ³H-vinorelbine distribution to brain, brain metastases and peripheral tissues. (**a**) Pharmacokinetics of ³H-vinorelbine in various organs over 8 h. Data represent mean ± SD for n = 3 animals per time point. Brain metastases were divided into three groups based on uptake compared to uninvolved brain: low metastases being less than brain +3SD, medium mets ≤10 × brain, and high mets ≥10 × brain. Results for peripheral tissues (e.g., liver, kidney, heart, lung, spleen) are shown as a band indicating the range of values. (**b**) Vinorelbine partition coefficient in various tissues showing greatly reduced distribution into brain metastases compared to peripheral tissues and non-barrier brain. (**c**) Projected Kp based on tubulin content in different tissues. Tubulin concentrations were obtained from Table I of Wierzba et al. (1987). Line represents least squares fit to data. Adapted from Samala et al. (2016), with permission

varied >30 fold between brain metastases at each time point analyzed. While most metastases (85%) had vinorelbine concentrations that exceeded normal brain ($P < 0.05$), a small subset (8%) had very high levels (10–30-fold), and 15% had low values that did not differ from normal brain (Fig. 10.16a). Median brain metastasis vinorelbine concentration was fourfold greater than normal brain (Fig. 10.16b). However, this level was still only 8% of that of matching systemic metastasis concentrations (Fig. 10.16b). Neither concentration nor permeability change followed a normal distribution (Figs. 10.16a and c), but drug concentration was significantly correlated with metastasis permeability, as measured by Texas red (Fig. 10.16d). The brain had the lowest concentration of all tissues measured (Fig. 10.16b). Vinorelbine distribution also varied markedly within brain metastases, as evaluated with drug distribution imaging; it was found that localized vinorelbine concentrations ranged >80-fold, from 0.2 to 40 μM, across metastases.

Vinorelbine concentrations in tissues over time are shown in Fig. 10.17a and used to calculate apparent K_p values (Fig. 10.17b). Brain and brain metastases showed dramatically lower levels compared to most all other tissues. Because of the tremendous heterogeneity of uptake, brain metastases were divided into three groups—low, medium, and high uptake—as described in Fig. 10.17. Medium brain metastases accumulated vinorelbine at about 14x lower amounts than systemic metastases (Fig. 10.17b).

Vinorelbine concentration fell quickly away from the edge of brain metastases. In fact, a detailed microscopic analysis of the tumor border showed that vinorelbine dropped to levels approaching baseline values within ~300 μm away from the tumor border (Samala et al. 2016). This finding is similar to what we found in several other experimental breast cancer brain metastasis models with drugs of high binding capacity and lipophilicity and indicates that the brain-around-tumor region may not be as important in drug delivery and activity as it can appear when analyzed by dissection.

The unbound vinorelbine concentrations in tissues and plasma and $K_{p,uu}$ values are summarized in Fig. 10.18. The calculated $K_{p,uu}$ equaled 0.025 for the brain and 0.032–0.23 for brain metastases (Fig. 10.18b). While the great majority of brain metastases had vinorelbine $K_{p,uu}$ values indicative of substantial restriction in MDA-MB-231BR brain metastasis drug exposure, ~5% of subregions within "high uptake" brain metastases predicted $K_{p,uu} \sim 0.5 - >1$, suggesting localized substantial, if not total, compromise of the BTB (Fig. 10.19). Vinorelbine unbound concentrations of >150–300 nM were found in such areas. These regions demonstrated a strong correlation between vinorelbine unbound concentration and the permeability marker Texas red (Fig. 10.19F). Texas red was validated as a permeability marker relative to AIB for in vivo imaging experiments (Lockman et al. 2010). Brain metastasis f_u varied minimally across metastases, within a tight range of <two-fold, and followed a normal distribution. Thus, the differences in C_u correlated with BTB compromise.

TUNEL staining, an in vivo biomarker of drug activity, revealed that 42% of brain metastases fell within the range of 0–10% positive staining and 51% had intermediate (10–50%) positive staining. Only 7% of brain metastases were associated

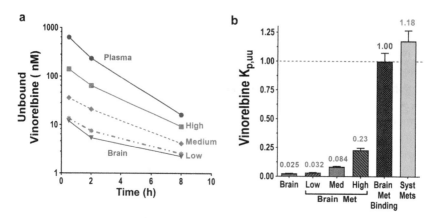

Fig. 10.18 Unbound vinorelbine time course and $K_{p,uu}$ of brain and brain metastases. Brain metastases are divided into subgroups as described in Fig. 8.15. (**a**) Unbound vinorelbine time course in plasma, brain, and brain metastases subgroups. (**b**) Vinorelbine $K_{p,uu}$ of brain, brain metastases subgroups, systemic metastases and compared to brain metastases binding obtained through frozen tissue slice binding method. Adapted from Samala et al. (2016), with permission

with >50% TUNEL staining. Together, the results provide a framework in which to assess therapeutic drug delivery and efficacy in brain metastases. Figure 10.20 shows the concentrations of free vinorelbine needed to induce significant apoptosis, based on our in vivo measurements. A fourfold difference was yielded between the in vitro and in vivo dose-response assays, with an apparent higher concentration required in vivo to attain the same level of apoptosis in vitro. The fact that the two models agreed within an order of magnitude suggests that the in vitro dose-response assay can indeed be useful for providing insight as to drug concentrations required for CNS antitumor effects.

Further, the difference may provide useful insight as to the fundamental question of whether many current anticancer drugs perform poorly against CNS tumors because of a) poor drug delivery or b) poor fundamental activity against tumors within the CNS. The higher IC_{50} value in vivo may be explained by a number of factors, including (a) tumor cell resistance to drug due to upregulation of P-gp or other efflux drug transporters at tumor cell membranes in vivo (Fig. 10.1), (b) in vivo hypoxia- or acidosis-induced drug resistance from other pathways, (c) in vivo effects of extracellular matrix, and (d) interactions between cancer cells and other cells (e.g., astrocytes) within the tumor to vary the sensitivity to vinorelbine (Fidler 2015; Blecharz et al. 2015). While in vivo expression of P-gp in brain metastasis cells in the CNS was observed, the same expression was not seen in vitro or in systemic tumors in the same animal (Samala et al. 2016). Taking this difference of expression into account, the agreement between in vitro and in vivo measurements is quite good. Our results confirm that in this model, low drug exposure is by far the primary reason for poor CNS metastasis antitumor activity. From the data derived thus far by our laboratory with the preclinical

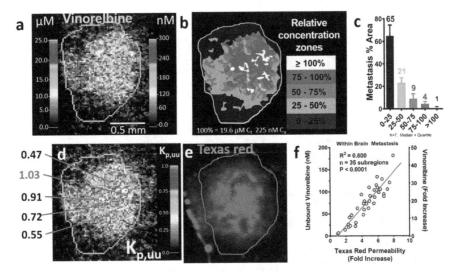

Fig. 10.19 ³H-vinorelbine distribution within a high-uptake metastasis. (**a**) Expanded QAR image of a representative high-uptake (>10 × (uninvolved brain average vinorelbine concentration)) metastasis showing localized total (left scale) and corresponding unbound (right scale) vinorelbine. (**b**) Illustration of the concentration zones observed in a relative to calculated equilibrium concentration that would be attained in brain without barrier (19.6 μM). (**c**) Areas within high uptake brain metastases (N = 7) partitioned by concentration relative to equilibrium concentration. (**d**) Calculated Kp,uu within metastasis, with select high-uptake regions shown, which could be reaching efficacious levels. (**e**) Localized Texas red fluorescence measurement in the same metastasis. (**f**) Correlation between fold increase in Texas red permeability and unbound vinorelbine concentration. Adapted from Samala et al. (2016), with permission

MDA-MB-231BR brain metastasis model, restricted vinorelbine exposure to CNS accounts for a median 14-fold compromise over that in a systemic tumor. Also, results from PK-PD modeling predict that in vivo efficacy requires 14-fold higher drug exposure for desired effect. *Therefore, based upon these numbers, BTB restriction in drug supply accounts for 80% of the compromise in vinorelbine efficacy against brain metastases, while the remaining 20% may be determined by the reduced sensitivity of the tumor in the brain microenvironment.* This is the first study that measured both local drug delivery and local drug effect to gain quantitative insight into the role of both delivery and sensitivity to therapeutic outcome of systemic chemotherapy. Results with other tumor preclinical models likely will vary, but in studies in our laboratory with over five different breast cancer cell lines, in each case the drug delivery had a major role. Interestingly, even though the changes observed in some brain metastases for BBB breakdown were at the higher range of magnitude observed for any condition in vivo (>20-fold), drug delivery invariably in the end in these tumors was inadequate locally because of the tremendous deficiency of delivery established by active efflux transport in the

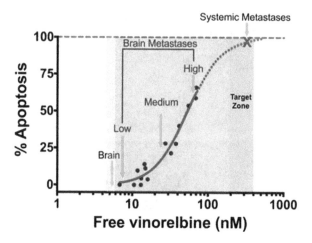

Fig. 10.20 *In vivo* dose response of cells in brain slices bearing MDA-MB-231BR metastases to vinorelbine, as measured by TUNEL staining 2 h after i.v. administration of 12 mg/kg vinorelbine. *In vivo* unbound vinorelbine concentration (nM) was calculated by the *in vivo* brain slice binding method. Extrapolation to 100% TUNEL staining was performed using Hill equation with GraphPad Prism 7.0. Data represents mean only (n = 4). Adapted from Samala et al. (2016), with permission

brain (i.e., for vinorelbine, doxorubicin, paclitaxel, lapatinib) and because of the heterogeneous nature of the barrier breakdown observed in many models. Though more drugs may have been delivered in certain parts of the tumor, these generally were limited in scope, and in the great majority of the CNS tumor delivery was far from that needed for efficacy. This raises questions regarding when disease-related barrier breakdown contributes significantly to altered drug effect and the interplay in such cases to the passive permeability barrier vs. the role played by active efflux transport. Brain perfusion data demonstrate that while the barrier permeability to passive markers, such as Texas red, urea, and AIB, may be highly compromised in brain metastases of breast cancer, the efflux barrier is maintained to a much greater extent. The increased brain metastasis uptake thus seen in such cases may represent drug that circumvents the transporter, by leaking into brain tumor via the paracellular or aqueous channel pathway. The leakage may be magnified by leakage of the fraction of the drug bound to circulating albumin. The correlation between passive permeability barrier breakdown (Texas red, AIB) and drug uptake (vinorelbine, paclitaxel, doxorubicin, and lapatinib) is striking over 20–50-fold. Ultimately, much more needs to be learned from these processes. But, the current high-resolution imaging results showing the striking correlation between barrier permeability, drug delivery, and apoptosis in vivo are the same whether one looks within tumors or between tumors (Samala et al. 2016) and indeed demonstrate that small areas within tumor exist in many cases where the barrier has completely broken down ($K_{p,uu}$). However, this is countered by the great majority of the tumor (>85%) where the delivery is far from adequate and results in minimal effect.

Lastly, the imaging results also call into possible question studies of "brain adjacent to tumor" where it has been suggested there is a partially disrupted BBB. In >99% of the studies to date, little confirmation analysis was performed to show that the samples indeed did not contain small pieces of tumor tissue, Further, by bulk dissection, it is extremely difficult to look at gradients from 0 to 0.5 mm. Such is the distance expected by diffusion from tumor to the adjacent brain. Further, to carry out such analysis requires microscopic examination of the presence of the tumor specimens over distances across three dimensions and requires extremely rapid brain extraction and freezing techniques so as to accurately trap meaningful gradients that would continue to melt away postmortem as tissue sits waiting for processing. The very sharp drug gradients observed in our studies, with accurate distance and tissue confirmation and rapid freezing, show sharply declining tissue drug levels over the first 100 μm from the edge of the tumor and continuing lower to surrounding normal brain levels by 200–300 μm, with no evidence of overall disruption or BBB compromise in tissue 0.2–0.6 mm from tumor. This suggests that the "brain adjacent to tumor" containing intermediate drug levels is far smaller in scope than previously estimated. Assuming a spherical tumor with a diameter of 2 mm, the added volume represented by the 100 μm diffusion zone with a drug concentration within a factor of 2 of the tumor was only ~33% of the original tumor itself. Such a volume is very difficult to dissect and remove accurately in tissue specimens without histologic confirmation. Thus, this suggests it would be prudent to refrain from reporting "brain adjacent to tumor" in standard studies of drug delivery to brain tumors, as the results likely contain significant contamination if proper methods were not used. The implication of this is that such tissue does exist in the brain but likely is much, much smaller than previously envisioned. Such data are important and reflect the importance of the methods used.

10.4.3 Case Study: Clinical Measurements of Brain Metastasis Drug Uptake

The ultimate goal of measurements of preclinical brain drug access in healthy and disease states is to evaluate a drug's possible efficacy in humans. However, the limitations and challenges of measuring drug levels in patients are manifold. Permeability differences in childhood brain tumors have successfully been measured with [^{11}C]L-methionine PET scanning (O'Tuama et al. 1990, 1991). We evaluated drug uptake in brain metastases of breast cancer in a prospective study involving 12 patients with medically indicated craniotomy who were administered with either capecitabine or lapatinib before surgery (Morikawa et al. 2015).

Capecitabine is an oral prodrug of 5-flurouracil (5-FU), which was designed to decrease systemic toxicity, and is converted to 5-FU via three-step enzymatic conversion. Lapatinib, an oral small-molecule tyrosine kinase inhibitor, targets the epidermal growth factor receptor and human epidermal growth factor receptor 2

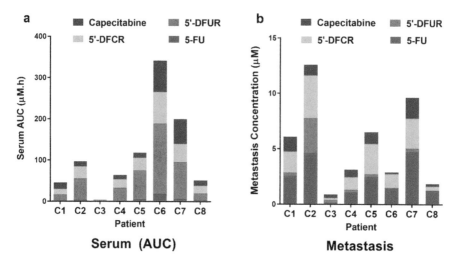

Fig. 10.21 Interpatient variability shown by serum AUC's (μM.h) and metastasis mean concentrations (μM) of capecitabine and its metabolites 5′-DFCR, 5′-DFUR, and 5-FU in (**a**) serum and (**b**) metastases in all patients C1-C8 administered capecitabine. Adapted from Morikawa et al. (2015), with permission

(HER2). In this study, patients with HER2$^+$ cancer received oral lapatinib daily 2–5 days before surgery, as clinical circumstances permitted, with the last dose administered 2–3 h before surgery. Patients with HER2$^-$ cancer received one preoperative dose of 1250 mg/m^2 capecitabine 2–3 h before surgery. Serum and samples of brain metastases were analyzed in duplicate for lapatinib or capecitabine and its prodrug metabolites 5′-deoxy-5-fluorocytidine (5′-DFCR), 5′-deoxy-5-fluorouridine (5′-DFUR), and 5-FU with LC-MS/MS. Various locations within the tumor samples were analyzed. Lapatinib is highly lipophilic; hence, the unbound fraction (f$_u$) was determined from the in vitro equilibrium distribution of lapatinib between 20 μm slices of brain metastasis tissue slices and buffered saline. The unbound concentration of lapatinib in the metastases was calculated for comparison with the extracellular lapatinib IC$_{50.}$

Values for capecitabine and its metabolites varied greatly between the eight patients studied, both in serum AUC and metastasis concentrations (Fig. 10.21). Even after removal of one patient who exhibited low systemic exposure (patient C3 in Fig. 10.21), we observed variability of concentrations of ~fivefold for 5-FU and > tenfold for 5′-DFCR and capecitabine. Heterogeneity in drug concentration within the same metastasis sample was also observed, with one patient exhibiting measured 5-FU concentration variability of ~ten-fold, while another patient showed only ~ ±20% variability. Another value of this study is the ability to compare 5-FU metastasis levels to serum levels of the active metabolites 5′-DFCR and 5′-DFUR, which may be indicative of inter-patient variability in metabolism rates, as shown in Fig. 10.22. This study, while limited

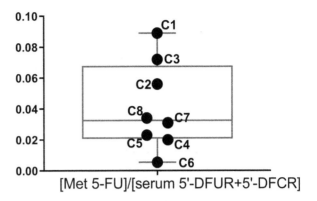

Fig. 10.22 Ratios of 5-FU metastasis levels to serum levels of the active metabolites 5′-DFCR and 5′-DFUR in each patient. Patients are identified C1–C8 as in Fig. 8.21. Adapted from Morikawa et al. (2015), with permission

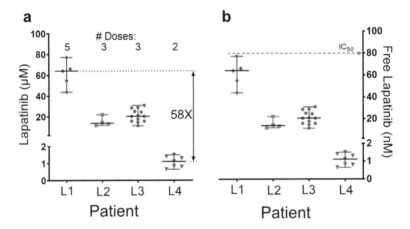

Fig. 10.23 Lapatinib concentrations (µM) (**a**) and free lapatinib concentrations (nM) (**b**) in metastases of all patients L1-L4, showing range, median, and quartile of concentration. *In vitro* lapatinib IC$_{50}$ measured in breast cancer cell line is shown as a dashed red line. Adapted from Morikawa et al. (2015), with permission

in scope, demonstrates the challenges faced in evaluating delivery of chemotherapeutics to brain metastases.

Uptake of lapatinib into brain metastases in four patients in this admittedly small study, after 2–5 consecutive daily 1250 mg oral doses, varied greatly, ranging from 1.0 to 63 µM, and was strongly correlated with the number of preoperative doses (r = 0.99, P < 01) (Fig. 10.23a). Lapatinib was expected to bind highly to tissue proteins and lipids, due to its marked lipophilicity (cLogP >5). Hence, we determined the unbound fraction (f_u) from in vitro equilibrium distribution of lapatinib between 20 µm metastasis tissue slices and buffered saline at 37 °C, in order to compare unbound drug levels in the tumor to extracellular lapatinib IC$_{50}$ (Fig. 10.23b).

Although the tissue levels were likely not at steady state, metastasis concentrations were comparable to previously reported preclinical IC_{50} of 0.1–4 µM in breast cancer cell lines (Rusnak, et al. 2001).

10.5   Future Challenges/Directions

Considerable progress has been made recently in methods to assess in vivo drug distribution and availability to the brain and in correlating drug availability with in vivo pharmacodynamics effects. Challenges continue to be the low throughput nature of the experimental setup and the complexity of factors impacting measurements. In many cases, studies are performed only in healthy subjects and do not accurately reflect circumstances in the diseased brain.

Marked continued progress is needed in dissecting out the roles of individual transporters beyond P-gp and BCRP. Recent publications highlight the importance of MRP4 in penetration of methotrexate and raltitrexed into the brain (Kanamitsu et al. 2017) and in limiting brain distribution of camptothecin analogs, together with P-gp and BCRP (Lin et al. 2013). Further, species differences in some cases can be important, where rodent models do not reflect the situation in the human brain. Cassette dosing in some cases allows the simultaneous analysis of drug distribution for a series of agents at low doses that do not alter drug transport, distribution, binding, or clearance.

10.6   Conclusions

An array of accurate and readily utilized procedures are available to investigators to assess the uptake and distribution of drugs in vivo in preclinical models and in humans. The gold standard continues to be direct in vivo assessment, which offers the advantage that all the parameters are present that have importance. Results are obtained using the true BBB, not a model, and alterations from disease are readily incorporated. More recent studies linking free drug concentration to effect are critically important as most all agents cross the BBB to some degree. By comparing the attained free drug concentration in the brain to the EC_{50} (or K_D or K_i as discussed in the introduction) for drug effect, one can more clearly assess the extent to which BBB transport is limiting and how improvement can be made by increasing brain delivery or reducing EC_{50} with more potent agents.

10.7   Points for Discussion (Questions)

- How do direct measurements of brain drug availability differ from those obtained with in vitro models?
- How do BBB PS, brain K_p, and brain $K_{p,uu}$ differ in what they tell us about brain drug availability?
- To what extent is the ratio of brain free drug concentration to EC_{50} a better index of BBB availability than BBB PS or $K_{p,uu}$?
- How do selected methods like in situ brain perfusion, microdialysis, and brain efflux index provide additional insights of brain drug distribution and transport?
- How does the presence of the BBB alter drug distribution to the brain distinct from other organs of the body?
- What impact does altered BBB function in disease have on drug distribution and availability to the CNS?

References

Alqahtani F, Chowdhury EA, Bhattacharya R, Noorani B, Mehvar R, Bickel U (2018) Brain uptake of [(13)C] and [(14)C]sucrose quantified by microdialysis and whole tissue analysis in mice. Drug Metab Dispos 46:1514–1518

Amtorp O (1980) Estimation of capillary permeability of inulin, sucrose and mannitol in rat brain cortex. Acta Physiol Scand 110:337–342

André P, Saubaméa B, Cochois-Guégan V, Marie-Claire C, Cattelotte J, Smirnova M, Schinkel AH, Scherrmann JM, Cisternino S (2013) Transport of biogenic amine neurotransmitters at the mouse blood-retina and blood-brain barriers by uptake1 and uptake2. AAPS J 32:1989–2001

Babak MV, Zalutsky MR, Balyasnikova IV (2020) Heterogeneity and vascular permeability of breast cancer brain metastases. Cancer Lett 489:174–181

Blecharz KG, Colla R, Rohde V, Vajkoczy P (2015) Control of the blood-brain barrier function in cancer cell metastasis. Biol Cell 107:342–371

Camilleri M, Nadeau A, Lamsam J, Nord SL, Ryks M, Burton D, Sweetser S, Zinsmeister AR, Singh R (2010) Understanding measurements of intestinal permeability in healthy humans with urine lactulose and mannitol excretion. Neurogastroenterol Motil 22:e15–e26

Cao Y, Ni C, Li Z, Li L, Liu Y, Wang C, Zhong Y, Cui D, Guo X (2015) Isoflurane anesthesia results in reversible ultrastructure and occludin tight junction protein expression changes in hippocampal blood-brain barrier in aged rats. Neurosci Lett 587:51–56

Cardoso FL, Brites D, Brito MA (2010) Looking at the blood-brain barrier: molecular anatomy and possible investigation approaches. Brain Res Rev 64:328–363

Chowdhury EA, Alqahtani F, Bhattacharya R, Mehvar R, Bickel U (2018) Simultaneous UPLC-MS/MS analysis of two stable isotope labeled versions of sucrose in mouse plasma and brain samples as markers of blood-brain barrier permeability and brain vascular space. J Chromatogr B Analyt Technol Biomed Life Sci 1073:19–26

Crone C (1963) Permeability of capillaries of various organs as determined using the "indicator diffusion" method. Acta Physiol Scan 58:292–305

Daniel PM, Lam DK, Pratt OE (1985) Comparison of the vascular permeability of the brain and the spinal cord to mannitol and inulin in rats. J Neurochem 45:647–649 de Lange ECM, Danhof M, de Boer AG, Breimer DD (1997) Methodological considerations of intracerebral microdialysis in pharmacokinetic studies of drug transport across the blood-brain barrier. Brain Res Brain Res Rev 25:27–49

Duncan MW, Villacreses N, Pearson PG, Wyatt L, Rapoport SI, Kopin IJ, Markey SP, Smith QR (1991) 2-Amino-3- (methylamino)-propanoic acid (BMAA) pharmacokinetics and blood-brain barrier permeability in the rat. J Pharmacol Exp Ther 258:27–35

Fidler IJ (2015) The biology of brain metastasis: challenges for therapy. Cancer J 21:284–293

Fridén M, Ducrozet F, Middleton B, Antonsson M, Bredberg U, Hammarlund-Udenaes M (2009a) Development of a high-throughput brain slice method for studying drug distribution in the central nervous system. Drug Metab Dispos 37:1226–1233

Fridén M, Winiwarter S, Jerndal G, Bengtsson O, Wan H, Bredberg U, Hammarlund-Udenaes M, Antonsson M (2009b) Structure-brain exposure relationships in rat and human using a novel set of unbound drug concentrations in brain interstitial and cerebrospinal fluids. J Med Chem 52:6233–6243

Fridén M, Bergström F, Wan H, Rehngren M, Ahlin G, Hammarlund-Udenaes M, Bredberg U (2011) Measurement of unbound drug exposure in brain: modeling of pH partitioning explains diverging results between the brain slice and brain homogenate methods. Drug Metab Dispos 39:353–362

Galano G, Caputo M, Tecce MF, Capasso A (2011) Efficacy and tolerability of vinorelbine in the cancer therapy. Curr Drug Saf 6:185–193

Hammarlund-Udenaes M, Fridén M, Syvänen S, Gupta A (2008) On the rate and extent of drug delivery to the brain. Pharm Res 25:1737–1750

Hladky SB, Barrand MA (2018) Elimination of substances from the brain parenchyma: efflux via perivascular pathways and via the blood-brain barrier. Fluids Barriers CNS 15(1):30

Iliff JJ, Wang M, Liao Y, Plogg BA, Peng W, Gundersen GA, Benveniste H, Vates GE, Deane R, Goldman SA, Nagelhus EA, Nedergaard M (2012) A paravascular pathway facilitates CSF flow through the brain parenchyma and the clearance of interstitial solutes, including amyloid β. Sci Transl Med 4:147ra111

Kakee A, Terasaki T, Sugiyama Y (1996a) Brain efflux index as a novel method of analyzing efflux transport at the blood-brain barrier. J Pharmacol Exp Ther 277:1550–1559

Kakee A, Terasaki T, Sugiyama (1996b) Selective brain to blood efflux transport of para-aminohippuric acid across the blood-brain barrier: in vivo evidence by use of the brain efflux index method. J Pharmacol Exp Ther 283:1018–1025

Kalvass JE, Olson ER, Cassidy MP, Selley DE, Pollack GM (2007) Pharmacokinetics and pharmacodynamics of seven opioids in p-glycoprotein-competent mice; assessment of unbound brain EC50,u and correlation of in vitro, preclinical, and clinical data. J Pharmacol Exp Ther 323:346–355

Kanamitsu K, Kusuhara H, Schuetz JD, Takeuchi K, Sugiyama Y (2017) Investigation of the importance of multidrug resistance-associated protein 4 (Mrp4/Abcc4) in the active efflux of anionic drugs across the blood-brain barrier. J Pharm Sci 106:2566–2575

Kaya M, Ahishali B (2011) Assessment of permeability in barrier type of endothelium in brain using tracers: evans blue, sodium fluorescein, and horseradish peroxidase. Methods Mol Biol 763:369–382

Kemper EM, van Zandbergen AE, Cleypool C, Mos HA, Booger DW, Beijnen JH, van Tellingen O (2003) Increased penetration of paclitaxel into the brain by inhibition of P-Glycoprotein. Clin Cancer Res 9:2849–2855

Lavis LD, Rutkoski TJ, Raines RT (2007) Tuning the pK(a) of fluorescein to optimize binding assays. Anal Chem 79:6775–6782

Li W, Rockey JH (1982) Fluorescein binding to normal human serum proteins demonstrated by equilibrium dialysis. Arch Ophthalmol 100:484–487

Lin F, Marchetti S, Pluim D, Iusuf D, Mazzanti R, Schellens JH, Beijnen JH, van Tellingen O (2013) Abcc4 together with abcb1 and abcg2 form a robust cooperative drug efflux system that restricts the brain entry of camptothecin analogues. Clin Cancer Res 19:2084–2095

Lippmann ES, Al-Ahmad A, Azarin SM, Palecek SP, Shusta EV (2014) A retinoic acid-enhanced, multicellular human blood-brain barrier model derived from stem cell sources. Sci Rep 4:4160

Liu X, Smith BJ, Chen C, Callegari E, Becker SL, Chen X, Cianfrogna J, Doran AC, Doran SD, Gibbs JP, Hosea N, Liu J, Nelson FR, Szewc MA, Van Deusen J (2005) Use of a physiologically based pharmacokinetic model to study the time to reach brain equilibrium: an experimental analysis of the role of blood-brain barrier permeability, plasma protein binding, and brain tissue binding. J Pharmacol Exp Ther 313:1254–1262

Liu X, Chen C, Smith BJ (2008) Progress in brain penetration evaluation in drug discovery and development. J Pharmacol Exp Ther 325:349–356

Liu X, Vilenski O, Kwan J, Apparsundaram S, Weikert R (2009) Unbound brain concentration determines receptor occupancy: a correlation of drug concentration and brain serotonin and dopamine reuptake transporter occupancy for eighteen compounds. Drug Metab Dispos 37:1548–1556

Lockman PR, Mittapalli RK, Taskar KS, Rudraraju V, Gril B, Bohn KA, Adkins CE, Roberts A, Thorsheim HR, Gaasch JA, Huang S, Palmieri D, Steeg PS, Smith QR (2010) Heterogeneous blood-tumor barrier permeability determines drug efficacy in mouse brain metastases of breast cancer. Clin Cancer Res 16:5664–5678

Mandula H, Parepally JMR, Feng R, Smith QR (2006) Role of site-specific binding to plasma albumin in drug availability to brain. J Pharmacol Exp Ther 317:667–675

Miah MK, Bickel U, Mehvar R (2016) Development and validation of a sensitive UPLC-MS/MS method for the quantitation of [(13)C]sucrose in rat plasma, blood, and brain: its application to the measurement of blood-brain barrier permeability. J Chromatogr B Analyt Technol Biomed Life Sci 1015-1016:105–110

Miah MK, Shaik IH, Bickel U, Mehvar R (2015) Effects of Pringle maneuver and partial hepatectomy on the pharmacokinetics and blood-brain barrier permeability of sodium fluorescein in rats. Brain Res 1618:249–260

Miah MK, Chowdhury EA, Bickel U, Mehvar R (2017) Evaluation of [^{14}C] and [^{13}C]sucrose as blood-brain barrier permeability markers. J Pharm Sci 106:1659–1669

Morikawa A, Peereboom DM, Thorsheim HR, Samala R, Balyan R, Murphy CG, Lockman PR, Simmons A, Weil RJ, Tabar V, Steeg PS, Smith QR, Seidman AD (2015) Capecitabine and lapatinib uptake in surgically resected brain metastases from metastatic breast cancer patients: a prospective study. Neuro-Oncology 17:289–295

Nishioku T, Yamauchi A, Takata F, Watanabe T, Furusho K, Shuto H, Dohgu S, Kataoka Y (2010) Disruption of the blood-brain barrier in collagen-induced arthritic mice. Neurosci Lett 482:208–211

Noorani B, Chowdhury EA, Alqahtani F, Ahn Y, Patel D, Al-Ahmad A, Mehvar R, Bickel U (2020) LC-MS/MS-based in vitro and in vivo investigation of blood-brain barrier integrity by simultaneous quantitation of mannitol and sucrose. Fluids Barriers CNS 17:61

Nozohouri S, Noorani B, Al-Ahmad A, Abbruscato TJ (2020) Estimating brain permeability using in vitro blood-brain barrier models. Methods Mol Biol PMID: 32789777

O'Tuama LA, Phillips PC, Strauss LC, Uno Y, Smith QR, Dannals RF, Wilson AA, Ravert HT, LaFrance ND, Wagner HN (1990) Two phase [11C]L methionine pet scanning in the diagnosis of childhood brain tumors. Pediatr Neurol 6:163–170

O'Tuama LA, Phillips PC, Smith QR, Strauss LC, Dannals RF, Wilson AA, Ravert HT, Wagner HN (1991) L methionine uptake by human cerebral cortex: maturation from infancy to old age. J Nucl Med 32:16–22

Ohno K, Pettigrew KD, Rapoport SI (1978) Lower limits of cerebrovascular permeability to non-electrolytes in the conscious rat. Am J Phys 235:H299–H307

Oppenheim HA, Lucero J, Guyot AC, Herbert LM, McDonald JD, Mabondzo A, Lund AK (2013) Exposure to vehicle emissions results in altered blood brain barrier permeability and expression of matrix metalloproteinases and tight junction proteins in mice. Part Fibre Toxicol 10:62

Pardridge WM (2020) The isolated brain microvessel: a versatile experimental model of the blood-brain barrier. Front Physiol 11:398

Parepally JMR, Mandula H, Smith QR (2006) Brain uptake of nonsteroidal anti-inflammatory drugs – ibuprofen, flurbiprofen and indomethacin. Pharm Res 23:873–881

Patel R, Alahmad AJ (2016) Growth-factor reduced Matrigel source influences stem cell derived brain microvascular endothelial cell barrier properties. Fluids Barriers CNS 13:6

Peck KD, Ghanem AH, Higuchi WI (1994) Hindered diffusion of polar molecules through and effective pore radii estimates of intact and ethanol treated human epidermal membrane. Pharm Res 11:1306–1314

Preston E, Haas N (1986) Defining the lower limits of blood-brain barrier permeability: factors affecting the magnitude and interpretation of permeability-area products. J Neurosci Res 16:709–719

Preston E, Haas N, Allen M (1984) Reduced permeation of 14C-sucrose, 3H-mannitol and 3H-inulin across blood-brain barrier in nephrectomized rats. Brain Res Bull 12:133–136

Preston JE, al-Sarraf H, Segal MB (1995) Permeability of the developing blood-brain barrier to 14C-mannitol using the rat in situ brain perfusion technique. Brain Res Dev Brain Res 87:69–76

Preston E, Foster DO, Mills PA (1998) Effects of radiochemical impurities on measurements of transfer constants for [14C]sucrose permeation of normal and injured blood-brain barrier of rats. Brain Res Bull 45:111–116

Rapoport SI, Ohno K, Pettigrew KD (1979) Drug entry into the brain. Brain Res 172:354–359

Renkin EM (1959) Transport of potassium from blood to tissue in isolated mammalian skeletal muscles. Am J Phys 197:1205–1210

Rusnak DW, Lackey K, Affleck K, Wood ER, Alligood KJ, Rhodes N, Keith BR, Murray DM, Knight WB, Mullin RJ, Gilmer TM (2001) The effects of the novel, reversible epidermal growth factor receptor/ErbB-2 tyrosine kinase inhibitor, GW2016, on the growth of human normal and tumor-derived cell lines in vitro and in vivo. Mol Cancer Ther 1:85–94

Samala R, Thorsheim HR, Goda S, Taskar K, Gril B, Steeg PS, Smith QR (2016) Vinorelbine delivery and efficacy in the MDA-MB-231BR preclinical model of brain metastases of breast cancer. Pharm Res 33:2904–2919

Sisson WB, Oldendorf WH (1971) Brain distribution spaces of mannitol-3H, inulin-14C, and dextran-14C in the rat. Am J Phys 221:214–217

Smith QR (2003) A review of blood-brain barrier transport techniques. Methods Mol Med 10:193–208

Smith QR, Rapoport SI (1986) Cerebrovascular permeability coefficients to sodium, potassium, and chloride. J Neurochem 46:1732–1742

Steeg PS, Camphausen KA, Smith QR (2011) Brain metastases as preventive and therapeutic targets. Nat Rev Cancer 11:352–363

Takasato Y, Rapoport SI, Smith QR (1984) An in situ brain perfusion technique to study cerebrovascular transport in the rat. Am J Phys 247:H484–H493

Taskar KS, Rudraraju V, Mittapalli RK, Samala R, Thorsheim HR, Lockman J, Gril B, Hua E, Palmieri D, Polli JW, Castellino S, Rubin SD, Lockman PR, Steeg PS, Smith QR (2012) Lapatinib distribution in HER2 overexpressing experimental brain metastases of breast cancer. Pharm Res 29(3):770–781

Thomas FC, Taskar K, Rudraraju V, Goda S, Thorsheim HR, Gaasch JA, Palmieri D, Steeg PS, Lockman PR, Smith QR (2009) Uptake of ANG1005 – a novel paclitaxel-peptide derivative, through the blood-brain barrier into brain and experimental brain metastases of breast cancer. Pharm Res 26:2486–2492

Tress EE, Clark RS, Foley LM, Alexander H, Hickey RW, Drabek T, Kochanek PM, Manole MD (2014) Blood brain barrier is impermeable to solutes and permeable to water after experimental pediatric cardiac arrest. Neurosci Lett 578:17–21 van Bree JB, de Boer AG, Danhof M, Ginsel LA, Breimer DD (1988) Characterization of an "in vitro" blood-brain barrier: effects of molecular size and lipophilicity on cerebrovascular endothelial transport rates of drugs. J Pharmacol Exp Ther 247:1233–1239

Wierzba K, Sugiyama Y, Okudaira K, Iga T, Hanano M (1987) Tubulin as a major determinant of tissue distribution of vincristine. J Pharm Sci 76:872–875

Zanetti-Domingues LC, Tynan CJ, Rolfe DJ, Clarke DT, Martin-Fernandez M (2013) Hydrophobic fluorescent probes introduce artifacts into single molecule tracking experiments due to non-specific binding. PLoS One 8:e74200

Chapter 11
Principles of PET and Its Role in Understanding Drug Delivery to the Brain

Stina Syvänen, Roger N. Gunn, and Lei Zhang

Abstract Positron emission tomography (PET) is a noninvasive medical imaging technique that enables the investigation of drug pharmacokinetics in vivo. The technique is especially powerful for pharmacokinetic studies of new CNS drug candidates as tissue samples from the brain are understandably difficult to obtain. The PET technique involves the administration of a radiolabeled molecule, often referred to as a PET radiotracer, whose spatiotemporal distribution can be measured using tomography. The radiolabeled molecule can be the drug under investigation, a structurally different molecule that binds to the same target as the drug candidate, or a molecule that interacts with a downstream target that is believed to be affected by the action of the drug candidate. Such radiolabeled probes allow PET to address several questions central for CNS drug development: Does the drug candidate reach the target site? Does the drug candidate interact with the desired target? Is the concentration of the drug at the target site sufficient to illicit an effect? What is the temporal nature of such an interaction? What is the relationship between the target site concentration and the administered dose and/or plasma concentrations?

Keywords Positron emission tomography (PET) · Radiotracer · Molecular imaging · Drug development · Neuroreceptors

S. Syvänen (✉)
Department of Public Health and Caring Sciences, Uppsala University, Uppsala, Sweden
e-mail: stina.syvanen@pubcare.uu.se

R. N. Gunn
Department of Brain Sciences, Imperial College London, London, UK

Invicro, London, UK

L. Zhang
Medicine Design, Pfizer Inc, Cambridge, MA, USA

© American Association of Pharmaceutical Scientists 2022
E. C. M. de Lange et al. (eds.), *Drug Delivery to the Brain*, AAPS Advances in the Pharmaceutical Sciences Series 33, https://doi.org/10.1007/978-3-030-88773-5_11

11.1 Introduction

11.1.1 Background

Positron emission tomography (PET) is a medical imaging technique that allows for the measurement of a range of biological processes involving receptors, enzymes, and transporters in addition to the biodistribution of labeled drugs. The development of PET imaging was initiated in the early 1970s with the first operational human PET scanner in 1975. Early PET work was dominated by [^{18}F]fluorodeoxyglucose ([^{18}F]FDG), a marker for glucose metabolism, which has subsequently been translated into a diagnostic tool in the clinic for the detection of tumors in oncology. [^{18}F]FDG, a glucose analog, allows the direct measurement of regional glucose consumption in the body. A high uptake of [^{18}F]FDG indicates increased metabolism in a viable tumor and is founded on the Warburg effect (Warburg 1956) which determines that cancer cells have a higher rate of glycolysis. The effect of anticancer treatments may be monitored by serial [^{18}F]FDG PET examinations, although the long-term response to treatment may not be well predicted by the short-term reduction in glycolysis for all drugs (Fernandes et al. 2017). The brain is another organ with high glucose consumption, and [^{18}F]FDG can be used to study brain physiology and function in health and disease. For example, it has been used in diagnosing Alzheimer's disease which is characterized by a decreased [^{18}F]FDG uptake particularly in temporoparietal areas of the brain.

The number of available PET radiotracers increased during the late 1980s, and PET has utilized these imaging tools to study other neurological disorders such as anxiety, epilepsy, and Parkinson's disease. In the late 1990s, many pharmaceutical companies realized the potential of PET in drug development and started to apply it particularly in neuroscience applications (Table 11.1). In the past and to some extent still today, the selection of new drug candidates for neurosciences relies mainly on in vitro techniques which are good and often preferable for studying specific drug-target interactions but which may fail to mimic the complexity of the in vivo situation. Although preclinical in vivo studies are used, the results can be confounded by species differences. The potential to actually study new drug candidates in vivo in man at an early phase in drug candidate selection was obviously appealing to the pharmaceutical industry. What hampered the use of PET for drug development in the 1990s was the availability of lab facilities sufficiently equipped for radiopharmaceutical research, the lack of radiolabeling methods, for introducing the PET radionuclide into drug candidates, and to some extent the cost. Today, the cost associated with PET is still high, but PET is available at more locations, and most small drugs or drug-like molecules can be labeled with one of the available PET radionuclides (Table 11.2) which has increased utilization of PET in the development of new CNS drugs, especially antipsychotic, anti-depressive, and anti-amyloid-β drugs (Bergström et al. 2003; Pike 2009; Matthews et al. 2012; Gunn and Rabiner 2017).

Table 11.1 Application of PET in CNS drug development

Early phase	Biodistribution studies to confirm that the drug reaches the brain or a specific target site in the brain at sufficient concentrations.
	Brain pharmacokinetics, for example, as a guide for dose selection.
	Drug-target (receptor occupancy) interactions, for example, as a guide for dose selection.
	Pharmacodynamic biomarkers for proof of concept (reasons to believe).
	Biomarkers to be used as patient inclusion criteria in early phase clinical trials.
	Translational preclinical imaging to aid candidate selection or to identify and validate biomarkers.
	In vivo measures for monitoring safety or toxicity.
Late phase	Surrogate markers of response (may be more sensitive and faster to measure than clinical outcome).
	Stratification of patients based on potential for successful treatment (personalized medicine).
	Pharmacological differentiation of competitor compounds (best in class).
	Inclusion criterion, to verify that patients in clinical trials are correctly diagnosed.
Marketed drugs	Evaluation of ongoing treatment based on biomarkers.
	Differentiation between available treatments.
	Patient stratification based on disease sub-phenotype or early treatment response.
	Improved disease classification/diagnosis.
	Earlier diagnosis.

Table 11.2 Radionuclides used in PET and their half-lives

Isotope	Half-life	Comments
^{15}O	122 seconds	Oxygen (O) is common in drug-like molecules and ^{15}O-labeling does therefore not change pharmacokinetic properties. However, the very short half-life is a drawback
^{13}N	9.97 minutes	Although not as abundant as O and C in drug-like molecules, nitrogen (N) is still fairly common in drug-like molecules and ^{13}N-labeling may also be used to avoid alteration in pharmacokinetic properties
^{11}C	20.4 minutes	Essentially all endogenous and drug-like molecules contain carbon (C) and ^{11}C-labeling, where an isotopically unmodified carbon atom is replaced by ^{11}C and is often desirable as this approach does not alter the molecule with respect to its pharmacokinetic properties
^{68}Ga	68.3 minutes	Used for labeling of peptides and antibody fragments
^{18}F	110 minutes	Can often replace hydrogen (H) without any major effects on pharmacokinetic properties.
^{89}Zr	78.4 hours	Used for labeling of macromolecules such as antibodies with slow pharmacokinetics
^{124}I	100 hours	Used for labeling of macromolecules such as antibodies with slow pharmacokinetics

11.1.2 Principles of PET

A PET radiotracer is a molecule labeled with a positron-emitting radionuclide such as ^{11}C, ^{13}N, ^{15}O, ^{18}F, or ^{68}Ga. Nearly all endogenous and drug-like molecules contain carbon (C) which makes them amenable to PET labeling with ^{11}C. By replacing a naturally occurring carbon isotope in the molecule with ¹¹C, it is possible to generate a PET tracer with a chemical structure and pharmacokinetic properties which are the same as the original compound. In addition, nitrogen (N) and oxygen (O) are also common in endogenous and drug-like molecules and may also be replaced by positron-emitting isotopes. Further, most molecules also contain hydrogen (H). Nonacidic hydrogen can often be substituted by fluorine (F) with minor changes of the molecule's pharmacological or physiochemical properties. These commonly used PET radionuclides have a relatively short half-life, 20.3 min for ¹¹C and 110 minutes for ¹⁸F (see Table 11.2), and thus the tracer synthesis time has to be short. All ¹¹C-tracers must be injected into the subject shortly after being synthesized, while ¹⁸F-tracers have the benefit of a longer half-life often resulting in a shelf life of several hours. The short half-life of these radionuclides is challenging for the synthesis and production of PET radiotracers but allows for repeated scans in single subjects on the same day.

All PET radionuclides contain an excess of protons compared to neutrons, and to increase stability, one proton is converted into a neutron and during this decay event a single positron is emitted. It is the emission of these positrons which is the basis for the PET signal. Each emitted positron will travel a few mm in the tissue until it collides with an electron causing a positron-electron annihilation that produces two photons emitted at an angle of 180 degrees. These photons are detected by a ring of detectors in the PET scanner (Fig. 11.1). The acquisition of PET data may simply be a single 3D image representing the average concentration over a particular time period (static image), or it can be a 4D image that measures the changing

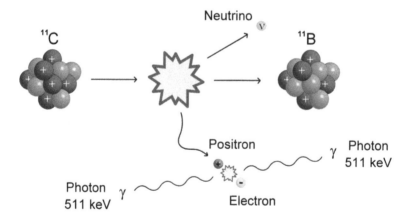

Fig. 11.1 Decay of ¹¹C. A positron-emitting nuclide, in this figure ¹¹C, is unstable due to a surplus of protons. Therefore, one proton is converted to a neutron to increase stability, and, at the same time, a positron is emitted from the nucleus. The positron travels through tissue for up to a few mm until it collides with an electron. The positron-electron annihilation produces two photons that are emitted at an angle of approximately 180 degrees. These photons are then detected by the PET scanner, and knowledge of which detector pairs registered the coincidence events and their precise timings enables the reconstruction of the spatial distribution of the emitted positrons

concentration over time (dynamic image). The raw tomographic data is reconstructed into quantitative images by applying appropriate corrections for confounding factors such as attenuation and scatter. For image quantification, a region of interest is often delineated on the PET image around the tissue or part of a tissue that is of interest for the study. The outlined region can be applied to images from different time frames to generate a dynamic time-activity curve for the particular region (Fig. 11.2). Many regions can be outlined in the set of PET images, allowing for assessment of regional differences in pharmacokinetics. The application of appropriate biomathematical models to the dynamic data allows for the estimation of quantitative biological parameters (Gunn et al. 2001; Heurling et al. 2017) such as those presented in Sect. 11.1.3.

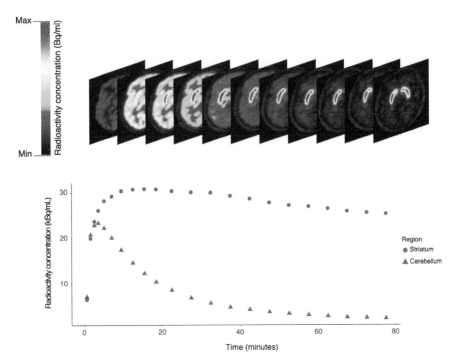

Fig. 11.2 Dynamic PET scanning. The detector recordings of counts starts at the time of injection of the tracer, and is reconstructed time-windows (frames) showing the tissue radioactivity over time. Anatomical regions of interest can be defined in the image data, and the average radioactivity concentration in that region over time is called time-activity curve (TAC), typically corrected for physical decay of the radioactivity and hence only representing the pharmacokinetic distribution phases of the PET tracer in the tissue. The graph shows the radioactivity concentration measured in striatum and in the cerebellum for the dopamine transporter ligand [^{11}C]PE2I used to visualize and quantify the integrity of dopamine neurons, for example, in Parkinson's disease and related disorders. Figure obtained from (Heurling et al. 2017) with permission from the publisher

There are three principal ways in which PET can be used to understand drug behavior in vivo:

- Using a labeled drug to understand its administration, distribution, metabolism, or elimination (ADME).
- Using a separate labeled compound which allows imaging of the target and drug action on the target.
- Using a separate labeled compound which reflects the effects of drug action on cellular or organ physiology.

In the first scenario, PET can be used to study the pharmacokinetics of the drug molecule directly, so that drug uptake in the brain, time to maximum concentrations in the brain, and brain concentrations over time can easily be obtained. However, if the drug is only slowly distributed to the brain, the information obtained in this experimental setting might have limited value since a PET investigation cannot be extended beyond three to four half-lives of the radionuclide.

In the second scenario, a radiotracer is used as a marker for specific target system. The purpose could be to simply verify that the desired target is present, as in clinical trials of amyloid-β decreasing drugs, where it is important to include only patients that are amyloid-β positive and exclude patients that might suffer from another dementia disorder with similar manifestation . Dose finding studies are, e.g., prior to larger clinical trials, are also included in this category of PET applications. In this case, it is the changes in the uptake of the radiotracer after administration of the drug that are studied in the PET investigation. For example, if a radiotracer is known to bind to a specific receptor, a PET scan before drug administration will allow for a quantitative estimate of the receptor availability in the absence of the drug. A subsequent scan following the administration of the drug then measures the change in receptor availability caused by binding of the unlabeled drug and enables the construction of a dose-occupancy relationship. By performing multiple PET scans at different time points, it is possible to measure the kinetics of the drug candidate at its target site in relation to the plasma pharmacokinetics in order to provide a more comprehensive characterization of the drug (Abanades et al. 2011). The characterization of drug-target occupancy in relation to dose (or concentration) and time in the brain provides valuable information to drug development teams that addresses both the questions of brain entry and also optimization of dose levels for larger clinical studies of efficacy.

Finally, a radiotracer might be used to monitor the effect of the drug on cellular function. For example, [^{18}F]FDG is frequently used for studying cancer tumors; in this case, a high uptake indicates extensive metabolism and a viable tumor, and decreased uptake after treatment may suggest that the treatment was successful. In clinical trials of amyloid-β decreasing drugs, [^{11}C]PIB or other amyloid binding tracers have been used increasingly to monitor intra-brain levels of amyloid-β, often at several time points during the trial (Rinne et al. 2010; Sevigny et al. 2016).

11.1.3 PET Concepts and Nomenclature

PET measures the total amount of radioactive material in the tissue of interest. At its most simplest, this can be quantified as the measured radioactivity, normalized to injected dose or normalized to injected dose per body weight, given as:

$$\% \text{injecteddose}\left(\%\text{ID}\right) = \frac{\text{Radioactivitypertissueweight}}{\text{Injectedradioactivity}} * 100 \tag{11.1}$$

or

$$\text{SUV} = \frac{\text{Radioactivitypertissueweight}}{\text{Injectedradioactivityperbodyweight}} \tag{11.2}$$

where SUV is the standardized uptake value. Both of these measurements reflect the radioactive concentration at the site of measurement in relation to the amount of radioactivity injected. However, since the amount of radioactivity in the tissue is dependent on the amount supplied to it via the plasma, further analysis is required to derive parameters that are specific for the tissue, and this involves the parallel measurement of radioactivity in the whole blood or plasma.

The most common parameter estimated from measurements of radioactivity in both tissue and plasma is the brain-to-plasma partition coefficient. The nomenclature for this parameter in PET and pharmacokinetic literature differs, although Innis et al. (Innis et al. 2007) have published a suggestion for standardization of the PET nomenclature and Summerfield et al. have presented a table clarifying the PET nomenclature in relation to standard pharmacokinetic terminology (Summerfield et al. 2008; Syvänen and Hammarlund-Udenaes 2010). In PET, the brain-to-plasma partition coefficient is often referred to as the volume of distribution (V_T), while, in pharmacokinetic studies, it is called K_p (Table 11.3). This PET nomenclature can be confusing, since the distribution volume in standard pharmacokinetics refers to the apparent volume of distribution for a drug, given in volume units. K_p (V_T) can be determined from PET data in a number of different ways: by compartmental modeling (Gunn et al. 2001), by model-independent graphical analysis (Patlak et al. 1983; Logan et al. 1990), or simply by comparing steady-state concentrations in the brain and plasma (Carson et al. 1993). In its most simple definition, it is defined as the ratio of the concentration in tissue to that in plasma at equilibrium.

The net rate of drug transfer to the brain can be measured with PET if radioactivity concentrations are measured in plasma in parallel to PET scanning. The permeability surface area product PS, which is equal to the net influx clearance CL_{in}, measured in $ml*min^{-1}*g\_brain^{-1}$, is comparable to the PET parameter K_1, measured in $ml*min^{-1}*cm^{-3}$.

In addition to V_T and K_1, other common outcome measures from PET studies are binding potential (BP) and receptor occupancy. BP is a composite parameter that includes both the density of "available" binding sites (B_{max}), e.g., receptors, and the

Table 11.3 Terms explaining relationship between PET and pharmacokinetic nomenclature in brain biodistribution studies

PET	Description	Relation to field of standard pharmacology
SUV	Standardized uptake value; total tissue concentrations normalized for injected dose per body weight	Not used
%Inj dose	Total tissue concentrations normalized for injected dose	Same
V_T	Equilibrium partition coefficient; total brain to total plasma concentration ratio at equilibrium	K_p
V_{ND}	Non-displaceable equilibrium partition coefficient; total brain to total plasma concentration ratio when no specific binding exists	K_p in a region devoid of specific binding
K_1 (mL min^{-1} cm^{-3})	Rate constant for drug transfer from arterial plasma to tissue	PS (permeability surface product) or CL_{in} (net influx clearance)
$\dfrac{f_{ND}}{f_p} \dfrac{C_{ND}}{C_p}$	Equilibrium partition coefficient; ratio of unbound brain and unbound plasma concentration, where f_{ND} and f_p are the free fractions in the brain tissue and plasma, respectively, and can be determined by equilibrium dialysis and C_{ND} and C_p are total concentrations in tissue (devoid of specific binding) and plasma and can be obtained from PET and blood sampling, respectively	Similar to $K_{p,uu}$ but also include nonspecific binding that cannot be displaced

affinity of the radiotracer for its target (K_d). BP can also be calculated from separate estimates of V_T in a target and reference region (devoid of the target site).

$$BP = \frac{B_{max}}{K_d} = \frac{V_T^{Target} - V_T^{Reference}}{V_T^{Reference}} \tag{11.3}$$

As a rule of thumb, BP should be between 0.5 and 15 for a good radiotracer candidate, as values below 0.5 indicate that the signal to noise will likely be too low, while BP above 15 indicates near irreversible kinetics and problems in accurately estimating BP. The receptor occupancy after administration of a drug candidate can be calculated based on BP before drug administration (BP$_{baseline}$) and BP after drug administration (BP$_{drug}$) according to Eq. 11.4.

$$ReceptorOccupancy\,(\%) = \frac{BP_{baseline} - BP_{drug}}{BP_{baseline}} \cdot 100 \tag{11.4}$$

11.1.4   Discovery Process of CNS PET Radiotracers

CNS PET radiotracer discovery process bears substantial similarities to a typical drug discovery effort and involves characterization of the target of interest, identification of the right radiotracer leads with required attributes, radiolabeling and subsequent in vivo PET imaging, and safety evaluation in toxicity studies prior to the trigger of investigational new drug (IND) filing and clinical studies (Zhang and Anabella 2016). Target characterization is a critical first step to determine whether PET imaging is viable for a given target. Specifically, B_{max} is a key parameter for consideration as it defines the K_d required for a PET radiotracer. In vitro BP greater than 10 is typically required (Patel and Gibson 2008). As such if a given target's B_{max} is too low (e.g., < 1 nM), it would be highly challenging to identify leads with sufficient affinity to show specific binding. From property perspective, compounds need to possess a set of specific attributes to be considered as CNS PET radiotracer leads. Structurally, a PET radiotracer lead must have a structure amenable to late-stage ^{11}C or ^{18}F radiolabeling. Pharmacologically, it needs to have sufficient affinity and selectivity toward the target of interest in order to show specific binding in vivo. Pharmacokinetically, it should be blood-brain barrier (BBB) permeable and have low nonspecific binding to brain lipids to achieve adequate signal-to-noise ratio. Furthermore, it should not form BBB permeable radioactive metabolites as PET detector is incapable to differentiate the origin of radioactivity (parent or metabolites), compromising the accuracy of specific binding quantification. Once a suitable radiotracer lead is identified, it will be radiolabeled and assessed in a preclinical PET imaging study. In vivo specificity can then be defined by brain biodistribution pattern aligned with target expression and a blocking study by co-dosing with a target-selective "cold" (nonradioactive) ligand. The latter study, if performed in a dose-responsive manner, will yield valuable target occupancy measurement. Radiotracers that demonstrate sufficient specific binding in preclinical species will then be advanced to safety assessment in preparation for exploratory IND filing. Since high specific activity of PET tracers allows for administration at sub-pharmacological doses clinically (typically in µg scale), less extensive toxicology testing is required, and candidates rarely fail due to safety reasons preclinically. At the final stage, a PET radiotracer will be validated in clinical PET imaging studies, and if successful, it will be used as a valuable translation tool to measure target engagement and facilitate the advancement of drug candidates.

Historically CNS PET ligand discovery was carried out by radiolabeling of existing literature leads and advancing directly to in vivo PET studies, often with primary focus on potency but minimal attentions to physicochemical and PK properties. This largely empirical approach suffered from low success rate, with high nonspecific binding and poor BBB permeability among most frequent causes for failure (Pike 2009). Gratifyingly, recent advances in tracer discovery strategies, particularly in the realm of rational design and cost-effective evaluation, have led to significant improvement in overall efficiency. One notable advance is around translating desired PET tracer attributes to measurable property parameters that in turn could guide rational radiotracer design and lead prioritization (Zhang et al. 2013). This was accomplished via a systematic property analysis of clinically validated PET radiotracers and failed tracers. From this analysis, a set of design parameters were identified, including a new CNS PET multiparameter optimization (MPO) tool for physicochemical properties (CNS PET MPO > 3), high passive permeability (RRCK Papp AB >5 x 10^{-6} cm/sec) and low P-gp efflux (MDR1 BA/AB <2.5) as predictive measurement for good brain permeability, and appropriate fraction unbound in the brain ($f_{u,brain}$ > 0.05) to minimize nonspecific binding. The prospective use of these parameters was demonstrated by the identification of a first-in-class phosphodiesterase 2B (PDE2B) PET radiotracer (Zhang et al. 2013). To facilitate in vivo evaluation of radiotracer leads, a novel liquid chromatography-mass spectral (LC-MS) "cold" tracer method was used as a cost-effective surrogate of in vivo PET imaging for specific binding assessment (Chernet et al. 2005). In this method, a "cold" (nonradioactive) tracer lead is dosed in vivo, and the distribution of the "cold" tracer in various brain regions is quantified by high sensitivity LC-MS instead of scintillation count. The specific binding is determined by co-dosing the cold tracer with a high dose of target-selective blocking compound or using KO animals. The main benefit of this method is that the study can be carried out inexpensively with low material requirement and fast turnaround time. The usage of this methodology was demonstrated by the discovery of a novel nociceptin/orphanin FQ peptide (NOP) PET tracer (Pike et al. 2011). Finally, incorporation of both strategies in the PET radiotracer discovery process was illustrated by the recent discovery of PDE4B (Zhang et al. 2017) and BACE1 (Zhang et al. 2018) PET radiotracers, wherein both efforts were accelerated by a focused PET-specific SAR and lead prioritization effort guided by PET-specific design parameters and subsequent in vivo specific binding assessment using the LC-MS "cold" tracer method.

11.1.5   Study Protocols

As discussed above, the pharmacokinetic parameter of interest in most PET studies is the brain-to-plasma concentration ratio K_p (V_T in PET nomenclature, Table 11.3). In the absolute majority of PET studies, K_p is usually estimated from PET experiments in which the tracer is administered as a single bolus and application of pharmacokinetic modeling (Gunn et al. 2001; Slifstein and Laruelle 2001). For example, K_p can be obtained from K_1 and k_2 for a single-tissue compartment model ($K_p = K_1/k_2$) or from K_1, k_2, k_3, and k_4 ($K_p = K_1/k_2 \cdot (1 + k_3/k_4)$) for a two-tissue compartment model (Gunn et al. 2001) or from the slope of a Logan graphical analysis ($K_p = 1/$ slope) (Logan et al. 1990). Single-tissue and two-tissue compartment models and rate constants are shown in Fig. 11.3. K_p can also be directly calculated from the steady-state concentrations of the drug in the brain and plasma. Steady-state concentrations can be obtained by appropriate infusion protocols based on elimination kinetics of the tracer. Using the steady-state approach, no assumptions have to be made regarding the pharmacokinetic model. There are advantages and disadvantages

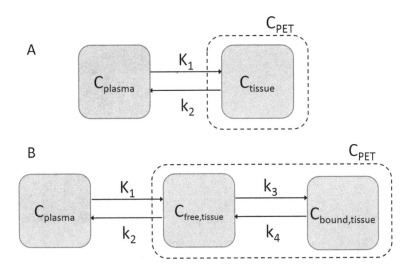

Fig. 11.3 Schematic overview of PET compartment models. (**a**) A standard single-tissue compartment model (1T2k in PET nomenclature). (**b**) A two-tissue compartment model (2T4k in PET nomenclature). K_1 is the transfer of tracer from plasma into tissue, k_2 is the rate constant describing the transport of tracer back to plasma, k_3 is the rate constant describing the association of the tracer with the specific binding sites, and k_4 is considered as the dissociation rate constant of the receptor-tracer complex. For clarity, the blood volume in the PET compartments has been omitted

with each of these designs. For example, bolus injections are technically easier than infusions as a single bolus dose of the radiotracer does not require an infusion pump and the injection can be given manually. In addition, it may not be possible to get the system into equilibrium within the duration of the scan.

The outcome of a drug intervention is often compared between individuals that are treated with the drug and individuals that act as controls (non-treated or placebo treated). If bolus protocols are used, two groups are needed or, alternatively, each individual can serve as its own control, which requires two PET scans: one before and one after intervention. When separate control and intervention groups are used, it can be difficult to know whether a possible difference between the control and the intervention group is due to interindividual/inter-occasion variability or to an effect of the intervention drug. In this setting, steady-state protocols are appealing as the dynamics of drug intervention or drug interaction can be studied in each subject. For example, chemical inhibition of efflux transporters at the BBB can be studied in real time with infusion of a radiotracer that is an efflux transporter substrate (Fig. 11.4) (Syvänen et al. 2006). Also, displacement of radiotracer from specific targets by intervention drugs can be visualized with steady-state approaches (Carson et al. 1993). The main disadvantage of using infusions is the increase in radioactive exposure to both the subject and the attending personnel. A special shield for the syringe containing the radiotracer has to be used both for protection of the subject/personnel and also to avoid any influence on the scanner. The radioactivity in the bolus component of the steady-state protocol may have to be decreased somewhat in comparison

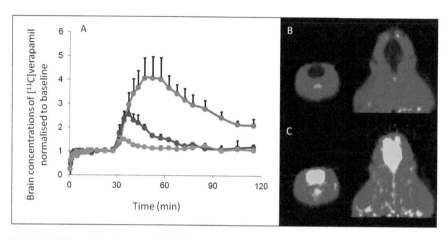

Fig. 11.4 Effect of P-glycoprotein inhibitor tariquidar on brain distribution of [¹¹C]verapamil. (**a**) [¹¹C]verapamil concentrations in the brain before (0–30 min) and after (30–120 min) P-glycoprotein inhibition with 3 (green), 10 (blue), and 25 (red) mg/kg cyclosporine. (**b**) PET images before P-glycoprotein inhibition and C. after P-glycoprotein inhibition. Intense colors indicate high concentrations. Figure obtained from (Syvänen et al. 2006 and Syvänen et al. 2008) with permission from the publisher

with the amount of radioactivity injected using a single bolus protocol since additional radioactivity is infused during the infusion component. However, compared to acquiring two separate PET scans, e.g., one before and one after intervention, the steady-state protocols normally result in less radioactivity dose. Steady-state protocols have been used in human subjects with both ¹¹C and ¹⁸F radiotracers (Pinborg et al. 2003; Lee et al. 2013).

11.2   Current Status

11.2.1   Brain Distribution Studies

PET biodistribution studies use the radiolabeled version of the drug as the radiotracer. Hence, PET can be used to study the pharmacokinetics of a drug candidate. When the radiotracer is injected in tracer dose, this setup is often referred to as microdosing (Bergström et al. 2003). The radiotracer (i.e., the labeled drug) can also be co-injected with unlabeled drug, and, since the relationship between labeled and unlabeled drug concentrations is known, the tissue pharmacokinetics can then be deduced quantitatively at clinically relevant doses. Further, such co-injections with unlabeled compound may also reveal whether any BBB efflux transporters that are active at tracer dose become saturated at therapeutic dose and allow for sufficient drug to enter the brain to elicit a pharmacological response. In addition, application of transporter inhibitors with these experimental setups may provide further confidence about transporter influence on drug distribution into the brain (Syvänen and Hammarland-Udenaes 2010; Syvänen and Eriksson 2013).

Regardless of whether the study is performed as a microdosing study or at clinical doses, a number of factors need to be considered.

First, PET measures the total radioactivity; so any metabolites carrying the radioactive label will also contribute to the signal, thus potentially confounding the results. Hence, the radiotracer data needs to be evaluated with respect to metabolism and the appearance of radioactive metabolites in the plasma and in tissue. The position of the radioactive label will determine which radioactive metabolites are produced and thus contribute to the signal. Ideally for a CNS tracer, the position of the label should be such that the label will be associated with hydrophilic metabolites only, i.e., metabolites that are unlikely to enter the lipophilic environment of the brain to the same extent as the tracer.

Second, radioactivity measured with PET will originate from radioactivity in both the brain tissue and in the vascular component of the brain. Negligence to correct for vascular activity, especially for compounds that do not enter the brain readily, may lead to overestimation of brain concentrations of the investigational compound. As initial biodistribution studies with new drug candidates are often preclinical, a third factor to consider is the choice of preclinical species.

Ultimately the drug candidates need to be effective in humans, but for different reasons, mainly toxicological, it is not always possible to directly study a new drug candidate in humans (even when using microdosing). Translation between species is complex for CNS active drugs as the passage of drugs into the brain is governed not only by passive diffusion but also by the presence of active transport mechanisms and protein binding in both plasma and brain tissue. Compared to other organs, where molecules can easily diffuse between cells, making active transport processes less important for the target site concentrations, the brain concentrations will be dependent on active influx and efflux from the brain. These processes may, in addition to systemic elimination and protein binding, differ between species, and significant differences in brain concentrations have been reported for both unlabeled drugs and radiotracers across species (Syvänen et al. 2009). In the development of new drug candidates or radiotracers, it is important to consider species differences when taking a decision on whether or not to proceed with the development of a molecule when it has been discovered that it interacts with, for example, an efflux transporter. Even when a molecule is a clear P-glycoprotein (P-gp) substrate in cell lines or rodents, it could reach relatively high brain concentrations in humans. In fact, several radiotracers, e.g., [^{11}C]2, 3, 4, 5, 6, 7-hexahydro-1{4-[1[4-(2-methoxyphenyl)-piperazinyl]]-2-phenylbutyry}-1H-azepine ([^{11}C]RWAY), 4-(2'-methoxyphenyl)-1-[2'-(N-2"-pyridinyl)-p-[^{18}F]fluorobenzamido]ethylpiperazine ([^{18}F]MPPF), [^{18}F]altanserin, and [^{11}C](S)-(2-methoxy-5-(5-trifluoromethyltetrazol-1-yl)-phenylmethylamino)-2(S)-phenylpiperidine (GR205171), have been successfully used as PET CNS tracers in humans before they were found to be P-gp substrates.

11.2.2 Drug-Target Interactions

PET studies that measure receptor occupancy of a drug candidate are usually per-formed with a PET radiotracer that is structurally different from the drug candidate but which binds to the same binding site. This is a consequence of the fact that only a small subset of drug compounds actually produces PET radiotracers with a mea-surable specific signal. The successful subset is dependent on high brain penetration and fast delivery, low nonspecific binding, and moderate to high affinity for the target of interest. For the ideal PET radiotracer, the dissociation from the target should be fairly fast in comparison with therapeutic drugs. For a ligand to be a successful thera-peutic drug, the rate of delivery across the BBB is often not as important since drugs are dosed to steady-state concentrations and often optimized for high target affinity including a slow dissociation. Development of PET probes for a novel target should proceed in parallel with the drug development program itself as the process of obtain-ing an applicable tool in humans will require at least 18 months' lead time. Candidate molecules for the new PET radiotracer can be screened in parallel as they will often originate from the same series of molecules and may benefit from concomitant medicinal chemistry support. Thus, it is important to start the development of the PET radiotracer in parallel to the development of the novel drug so that the PET imaging tools are available to be used in first time in human studies. It should also be mentioned that the importance of performing these measurements in man as early as possible is important because there may be species differences that mean that pre-clinical estimates of the in vivo IC_{50} are not applicable in humans. For example, H_3 histamine occupancy of a candidate drug has been shown to be significantly different between preclinical species and humans (Ashworth et al. 2010).

The main application of PET receptor occupancy studies is in dose finding studies which can involve exploring the relationship between temporal occupancy profiles and the plasma concentration of the drug. These studies consist, initially, of the acquisition of baseline scans in the absence of the drug to measure the baseline receptor availability and subsequently involve the acquisition of further scans at dif-ferent time points post-dosing with the drug (Fig. 11.5). Comparison of the receptor availability post-dose with the baseline values allows for the calculation of the frac-tional receptor occupancy. Combined with knowledge of the desired receptor occu-pancy, these studies provide confidence in selecting doses for larger later phase clinical studies. For example, the cost in terms of time and money (tens of millions of dollars) in performing a study at either non-pharmacological doses or at doses that produce side effects must be avoided. Receptor occupancy studies may be performed at single dose or repeat dose to characterize the relationship between the plasma concentration time course and the target occupancy. A study by Abanades et al. (Abanades et al. 2011) has demonstrated how the application of plasma-target occu-pancy models to single dose PET occupancy data can be used to predict the target occupancy data at repeat dose even if the drugs' kinetics are not direct (i.e., there is an increased target residence time of the molecule which means that the effective IC_{50} of the drug is different following single and repeat dose). This is important as

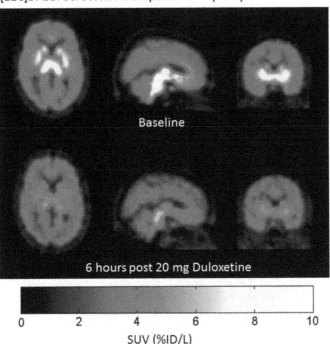

Fig. 11.5 Dose-receptor occupancy. Radioligand signal before (upper panel) and after (lower panel) administration of cold drug, duloxetine, competing for the same binding site. PET scanning after different doses of cold drug and at different time points post-dose administration enables characterization of drug-target occupancy in relation to dose (or concentration) and time

repeat dosing is the usual dosing regimen applied in patients and thus the ability to characterize this as early in the drug development process as possible is valuable.

11.2.3 Drug Effects on Cellular or Organ Physiology

Glucose metabolism in tissue is a classic example of PET as a tool to monitor cellular function, and as mentioned above, the glucose analog [^{18}F]FDG has been used to study brain glucose utilization. Certain dementia disorders are characterized by region-specific decreases in glucose turnover, and these can be visualized with PET and thus it can aid in their diagnosis. Another example is amyloid imaging with the PET radiotracer [^{11}C]PIB (N-methyl-[^{11}C]2-(4′-methylamino-phenyl)-6-hydroxybenzothiazole), a derivate of thioflavin-T, which has been the most frequently used radiotracer in the assessment of amyloid-β plaques in Alzheimer's disease. During the last decade, three ^{18}F-labelled radiotracers ([^{18}F]florbetapir,

[^{18}F]florbetaben, and [^{18}F]flutemetamol) have been FDA-approved as amyloid-imaging agents and have to some degree replaced [^{11}C]PIB in clinical trials (Morris et al. 2016). Amyloid imaging detects Alzheimer's disease pathogenesis early in the course of disease and helps distinguish Alzheimer's disease from other types of dementia, e.g., Lewy body dementia and frontotemporal dementia in the differential diagnosis (Engler et al. 2008; Rabinovici et al. 2007). However, [^{11}C]PIB imaging appears to be of limited value to measure disease progression, since the signal does not further increase as the disease progresses, i.e., there is a ceiling effect with amyloid-β levels plateauing rather early in the clinical stages of the disease process (Engler et al. 2006). Despite this, brain uptake of amyloid PET tracers has been shown to be significantly lower in patients with Alzheimer's disease treated with anti-amyloid-β antibodies compared to patients treated with placebo (Rinne et al. 2010; Sevigny et al. 2016). Recently new radiotracers binding to neurofibrillary tau tangles, i.e., another misfolded protein observed in the Alzheimer brain, have been introduced and utilized (Jack et al. 2018). It appears that the brain retention of tau tracers is better correlated to the cognitive decline than the brain retention of amyloid tracers (Schöll et al. 2016; Ossenkoppele et al. 2021). Another novel class of radiotracers, also connected to imaging of brain function, is the SV2A binding ligands that are described to enable quantification of synaptic density. The single most used tracer within this class is [^{11}C]UCB-J (Finnema et al. 2016), although ^{18}F radiolabeled ligands are appearing as well. It remains to be seen whether these ligands will contribute to our understanding of neurodegeneration and aid in the search for novel drugs.

In summary, drug development for many CNS diseases is hampered by the lack of knowledge about the disease mechanism. Imaging of cellular or organ physiology can be useful in the development of a new therapeutic when the exact target site for drug action is unknown or when no PET radiotracer is available for the target. Thus, these types of studies may provide some indirect evidence about successful drug delivery to the target tissue.

11.2.4 Challenges When Using PET for Studies of BBB Transport

The effect of a drug intended for a target inside the brain is generally related to the unbound (free) drug concentration inside the brain (Hammarlund-Udenaes et al. 2008; Jeffrey and Summerfield 2009; Watson et al. 2009). PET measures total radioactivity, including that associated with both unbound and bound drug and any metabolites carrying the positron-emitting radionuclide, and hence only K_p and $K_{p,u}$ (ratio of total concentration in the brain and unbound concentration in plasma) can be calculated. The unbound ratio of drug in brain interstitial fluid to unbound in plasma, $K_{p,uu}$, cannot be estimated from a PET investigation unless the unbound fraction in the brain ($f_{u,brain}$) is deduced by other means. This can be done by combining PET data with in vitro equilibrium dialysis assays of the free fraction of drug molecules in the brain tissue and plasma (Gunn et al. 2012; Gustafsson et al. 2017). While the free fraction in the brain may be difficult to obtain for human tissue, Summerfield et al. and Di et al. have shown that this fraction is well conserved across species (in contrast to the plasma free fraction) for a number of test compounds and thus the estimated tissue free fraction obtained in preclinical species might be used (with caution) together with clinical PET data (Summerfield et al. 2008; Di et al. 2011). The free fraction in plasma should always be estimated in the species under investigation, as species differences in plasma protein binding are common.

As a true translational technique, PET experiments can be performed according to similar protocols in vivo in animals as well as in humans. This means that preclinical studies can precede clinical studies and provide valuable parameter estimates that help refine clinical experimental designs. However, in addition to potential species differences, the need for anesthesia may limit the use of preclinical data for prediction of human response. For example, a number of studies have shown that differences in brain uptake may be a consequence of different anesthesia (Harada et al. 2004: Palner et al. 2016). It is important to make an educated choice regarding anesthesia method in preclinical studies so that the anesthetics used do not interfere with the studied target system.

Lastly, when studying a new CNS drug candidate with PET, it is always important to study the radiotracer concentrations in plasma or, if available, in a reference region, since intervention with a new drug candidate could potentially change the metabolism or plasma protein binding of the radiotracer and thus alter the fraction of radiotracer molecules available for transport into and binding within the brain. Measurement of radiotracer concentrations in the blood is also required for correction of radioactivity in the blood volume in the brain.

11.3 Future

11.3.1 Macromolecules and Biologics

For decades, the development of small molecular drugs has been the main focus for the pharmaceutical industry. Today there is an increasing interest in macromolecules and biologics such as peptides, proteins, oligonucleotides, and monoclonal antibodies. In line with this trend in the development of new drugs, there is also an increasing use of PET to explore the biodistribution of macromolecules, and given the slow kinetics of these molecules, this has led to the use of longer-lived radionuclides such as ^{89}Zr ($t_{1/2}$ = 78.4 h) and ^{124}I ($t_{1/2}$ = 100.3 h) (Hooker 2010; van Dongen and Vosjan 2010; van Dongen et al. 2012). PET studies involving biologics have mainly been applied in the field of oncology, and most studies have focused on quantitative evaluation of monoclonal antibody binding to specific targets such as B-lymphocyte antigen CD20, cMet (proto-oncogene encoding hepatocyte growth factor), and PSMA (prostate-specific membrane antigen) as a scouting procedure prior to radioimmunotherapy (van Dongen et al. 2012). For this purpose, the radiotracer should show similar biodistribution as the antibody used in the therapeutic radioimmunotherapy. As macromolecular drugs are more frequently developed also for CNS targets, PET biodistribution studies will be important for estimating brain distribution of these molecules. For example, several pharmaceutical companies are developing anti-amyloid-β antibodies that aim to reduce the cerebral amyloid-β load. However, it has been reported that only 0.1% of peripherally administered antibody reaches the brain (Bard et al. 2000; Poduslo et al. 1994), and it has even been questioned whether antibodies penetrate the brain parenchyma at all (Pardridge 2016) or if antibody concentrations measured in the brain rather reflect transport from the blood into the CSF. Hence, it is also of interest to image the levels of these antibodies inside the brain or other targets for which radiotracers based on small molecules have proven to be difficult to develop. The large size of antibodies limits the diffusion from the blood into the brain, and for this purpose the use of antibody fragments or engineered proteins like affibodies, diabodies, and nanobodies might be an option. In addition to being smaller, these proteins are also more rapidly cleared from the body and can thus be labeled with shorter-lived radionuclides. Radiolabeling with ^{68}Ga ($t_{1/2} = 1.13$ h) is attractive as generation of ^{68}Ga does not require a cyclotron but can be produced from commercially available ^{68}Ge/^{68}Ga generators. Another option, already shown to be feasible in preclinical studies, is the engineering of bispecific antibodies that show affinity for a transporter expressed at the BBB in addition to its primary intra-brain target (Yu et al. 2011). The first antibody-based PET radiotracer for an intra-brain target enabled imaging of soluble aggregates of amyloid-β (as opposed to insoluble amyloid-β plaques that are imaged with standard PET radiotracers such as [^{11}C]PIB) and utilized the transferrin receptor for receptor-mediated transcytosis across the BBB (Sehlin et al. 2016).

Taken together, PET imaging using macromolecular CNS radiotracers is still in its infancy, but interest in the area is likely to increase in parallel to focus on new CNS active macromolecular drugs and biologics.

11.3.2   Instrumentation

The development of dual modality scanners, i.e., PET combined with computerized tomography (CT) or magnetic resonance imaging (MRI), has facilitated co-registration of structural and functional data. PET scanners are today usually available with integrated high-end multi-detector-row CT scanners or MRI. Thus, PET and CT/MRI scans can be performed in immediate sequence during the same session, with the study subject not changing position between the two types of scans. The co-registered images display both functional and anatomical information so that areas of abnormality on the PET imaging can be correlated with anatomy on the CT/MRI images. The combination of PET-MRI has also enabled novel methods for correction of motion during the scans which otherwise degrade image quality but are difficult to avoid in long PET scans.

The most recent advance is the construction of human whole-body PET scanners (Cherry et al. 2018). These scanners, although very expensive, can provide whole-body images without having to change the bed position during the scan and require markedly less injected activity and/or scanning time. This in turn allows for high-throughput scanning and more efficient use of batches of radiotracers labeled with short-lived radionuclides such as the clinically preferred ^{18}F and ^{11}C.

11.3.3 PET Chemistry

During the last decade, development of new ^{11}C- and ^{18}F-radiolabeling methods has been a particularly active research area, undoubtedly driven by the significant uptake of PET imaging in clinical studies and the growing need of novel PET radiotracers. Compared to typical synthetic transformations, there are some unique challenges associated with PET radiochemistry, due to short half-lives of PET radionuclides. A viable radiolabeling method would need to be not only rapid and high yielding but also functional group tolerant as radiolabels are typically introduced at the ultimate or the penultimate step to minimize radioactivity decay.

Due to its short half-life, ^{11}C chemistry needs to be carried out in proximity to a cyclotron that produces [^{11}C]carbon dioxide, a common feedstock to all ^{11}C reagents. Synthetically, ^{11}C is most often introduced via a nucleophilic displacement reaction with an amine (NH), alcohol (OH), or SH (thiol)-bearing labeling precursor with [^{11}C]methyl iodide and, to a lesser extent, [^{11}C]methyl triflate. For [^{11}C]methyl labeling to aryl, heteroaryl, or alkenyl moieties, Pd-mediated cross coupling reactions, e.g., Suzuki and Stille coupling, with [^{11}C]methyl iodide would be needed. Recent advances also allow [^{11}C]CO$_2$ being directly used in radiolabeling reactions to introduce ^{11}C-carbonyl-labeled carboxylic acids, ureas, and carbamates with high radiochemical yield and high specific activities (amount of radioactivity per mole of molecule) (Rotstein et al. 2013). [^{11}C]Carbon monoxide ([^{11}C]CO) is interesting as a synthon (synthetic building block) primarily used for the synthesis of ^{11}C-carbonyl-labeled carboxylic acids, esters, and amides using transition metal-mediated carbonylation reactions. One issue that hampered the general use of [^{11}C]CO in PET radiotracer synthesis is its low solubility in in most organic solvents, thus confined [^{11}C]CO reaction in low volume reaction vessels. Methods have been described which will enable easy use of [^{11}C]CO for labeling (Kealey et al. 2009; Eriksson et al. 2012). [^{11}C]CN is a less commonly used synthon that has garnered more recent attention, enabled by the successful utilization of Pd- or Cu-mediated cross coupling reactions to introduce ^{11}C-labeling of aryl nitriles, a common motif in small drug-like molecules (Rotstein et al. 2016).

[^{18}F]Fluoride is the predominant reagent used for ^{18}F-labeling and can be produced from proton irradiation of ^{18}O in high specific activity and quantity. Compared to ^{11}C, ^{18}F has a relatively longer half-life (110 min vs. 20 min), which allows for multistep synthesis and reagent distribution to PET centers without on-site cyclotron. Most commonly used ^{18}F-radiolabeling strategy is to generate [^{18}F]alkyl fluoride via nucleophilic displacement of a good leaving group, such as mesylate, tosylate, triflate, or halide, with [^{18}F]fluoride. Alternatively, it can be incorporated by ring opening of an activated cyclic moiety such as epoxide or aziridine. Notable recent advances in nucleophilic ^{18}F-labeling include a new deoxyfluorination reagent [^{18}F]PyFluor for one-pot conversion of alkyl alcohol to [^{18}F]alkyl fluoride (Nielsen et al. 2015) and a robust moisture tolerate method involving TiO$_2$ catalysis (Sergeev et al. 2015). Along the same line, aryl ^{18}F-fluorination can be achieved via an aromatic nucleophilic substitution (SnAr) reaction. However, such reaction requires a suitable precursor bearing both a leaving group and an activating group (typically a strong electron-withdrawing group) to elicit adequate reactivity. Extensive research in this area has led to exciting new advances including novel activated leaving groups such as diarylsulfonium (Sander et al. 2015) as well as a novel deoxyfluorination process (Neumann et al. 2016), which have expanded the substrate scope from electron-deficient substrates to electron-rich aryl/heteroaryl precursors. More recently, ^{18}F-labeling was also demonstrated in the synthesis of [^{18}F]trifluoromethyl (-CF$_3$), [^{18}F]difluoromethyl (-CHF$_2$) groups previously not been attainable (Verhoog et al. 2016). Taken together, these new advances in ^{11}C and ^{18}F chemistry have provided much improved synthetic flexibility, which in turn will allow researchers to focus on finding radiotracer leads with right attributes instead of being limited by radiolabeling viability.

11.4   Conclusions

The development of new drugs that elicit their effect inside the brain is complicated because of the protective nature of the BBB and the technical difficulties in studying drug concentrations at the CNS target site in humans. With PET, it is possible to measure drug concentrations noninvasively in the brain, and this has meant that the method is playing an increasingly important role in drug development processes. Advances in labeling methods, novel tracers, study design, analysis of PET data, and the introduction of multimodality scanners are likely to further increase the number of PET applications in pharmaceutical research.

11.5   Points for Discussion

- How can V_T/K_p be estimated from PET measurements?
- Common radionuclides that can be used with PET.
- Discuss the advantages and disadvantages when using a radionuclide with a short half-life.
- Why are ^{11}C and ^{18}F useful for radiolabeling of endogenous and small drug-like molecules?

- When is it relevant to preform PET studies at tracer dose (microdosing)? And at pharmacological dose?
- Discuss the advantages and disadvantages of bolus only and bolus plus infusion regimens.
- Why are metabolites of PET radiotracers a potential confounding factor for the readout?
- Why is it important to correct the PET signal obtained in the brain area for radioactivity in the vascular space of the brain?
- How can PET be used in drug development?
- What information can be obtained from a receptor occupancy study?

References

Abanades S, van der Aart J, Barletta JA, Marzano C, Searle GE, Salinas CA, Ahmad JJ, Reiley RR, Pampols-Maso S, Zamuner S, Cunningham VJ, Rabiner EA, Laruelle MA, Gunn RN (2011) Prediction of repeat-dose occupancy from single-dose data: characterisation of the relationship between plasma pharmacokinetics and brain target occupancy. J Cereb Blood Flow Metab 31(3):944–952

Ashworth S, Rabiner EA, Gunn RN, Plisson C, Wilson AA, Comley RA, Lai RY, Gee AD, Laruelle M, Cunningham VJ (2010) Evaluation of 11C-GSK189254 as a novel radioligand for the H3 receptor in humans using PET. J Nucl Med 51(7):1021–1029

Bard F, Cannon C, Barbour R, Burke RL, Games D, Grajeda H, Guido T, Hu K, Huang J, Johnson-Wood K, Khan K, Kholodenko D, Lee M, Lieberburg I, Motter R, Nguyen M, Soriano F, Vasquez N, Weiss K, Welch B, Seubert P, Schenk D, Yednock T (2000) Peripherally administered antibodies against amyloid beta-peptide enter the central nervous system and reduce pathology in a mouse model of Alzheimer disease. Nat Med 6(8):916–919

Bergström M, Grahnen A, Långström B (2003) Positron emission tomography microdosing: a new concept with application in tracer and early clinical drug development. Eur J Clin Pharmacol 59(5–6):357–366

Carson RE, Channing MA, Blasberg RG, Dunn BB, Cohen RM, Rice KC, Herscovitch P (1993) Comparison of bolus and infusion methods for receptor quantitation: application to [18F]cyclofoxy and positron emission tomography. J Cereb Blood Flow Metab 13(1):24–42

Chernet E, Martin LJ, Li D, Need AB, Barth VN, Rash KS, Phebus LA (2005) Use of LC/MS to assess brain tracer distribution in preclinical, in vivo receptor occupancy studies: dopamine D2, serotonin 2A and NK-1 receptors as examples. Life Sci 78(4):340–346

Cherry SR, Jones T, Karp JS, Qi J, Moses WW, Badawi RD (2018) Total-body PET: maximizing sensitivity to create new opportunities for clinical research and patient care. J Nucl Med 59(1):3–12

Di L, Umland JP, Chang G, Huang Y, Lin Z, Scott DO, Troutman MD, Liston TE (2011) Species independence in brain tissue binding using brain homogenates. Drug Metab Dispos 39(7):1270–1277

Engler H, Forsberg A, Almkvist O, Blomquist G, Larsson E, Savitcheva I, Wall A, Ringheim A, Langström B, Nordberg A (2006) Two-year follow-up of amyloid deposition in patients with Alzheimer's disease. Brain 129(Pt 11):2856–2866

Engler H, Santillo AF, Wang SX, Lindau M, Savitcheva I, Nordberg A, Lannfelt L, Långström B, Kilander L (2008) In vivo amyloid imaging with PET in frontotemporal dementia. Eur J Nucl Med Mol Imag 35:100–106

Eriksson J, van den Hoek J, Windhorst AD (2012) Transition metal mediated synthesis using [11C]CO at low pressure – a simplified method for 11C-carbonylation. J Labelled Comp Rad

Fernandes RS, Ferreira CA, Soares DCF, Maffione AM, Townsend DM, Rubello D, de Barros ALB (2017) The role of radionuclide probes for monitoring anti-tumor drugs efficacy: a brief review. Biomed Pharmacother 95:469–476

Finnema SJ, Nabulsi NB, Eid T, Detyniecki K, Lin SF, Chen MK, Dhaher R, Matuskey D, Baum E, Holden D, Spencer DD, Mercier J, Hannestad J, Huang Y, Carson RE (2016) Sci Transl Med 8(348):348ra96

Gunn RN, Rabiner EA (2017) Imaging in central nervous system drug discovery. Semin Nucl Med 47(1):89–98

Gunn RN, Gunn SR, Cunningham VJ (2001) Positron emission tomography compartmental models. J Cereb Blood Flow Metab 21(6):635–652

Gunn RN, Summerfield SG, Salinas CA, Read KD, Guo Q, Searle GE, Parker CA, Jeffrey P, Laruelle M (2012) Combining PET biodistribution and equilibrium dialysis assays to assess the free brain concentration and BBB transport of CNS drugs. J Cereb Blood Flow Metab 25(10):1

Gustafsson S, Eriksson J, Syvanen S, Eriksson O, Hammarlund-Udenaes M, Antoni G (2017) Combined PET and microdialysis for in vivo estimation of drug blood-brain barrier transport and brain unbound concentrations. NeuroImage 155:177–186

Hammarlund-Udenaes M, Friden M, Syvänen S, Gupta A (2008) On the rate and extent of drug delivery to the brain. Pharm Res 25(8):1737–1750

Harada N, Ohba H, Fukumoto D, Kakiuchi T, Tsukada H (2004) Potential of [(18)F]beta-CFT-FE (2beta-carbomethoxy-3beta-(4-fluorophenyl)-8-(2-[(18)F]fluoroethyl)nortropane) as a dopamine transporter ligand: a PET study in the conscious monkey brain. Synapse 54(1):37–45

Heurling K, Leuzy A, Jonasson M, Frick A, Zimmer ER, Nordberg A, Lubberink M (2017) Quantitative positron emission tomography in brain research. Brain Res 1670:220–234

Hooker JM (2010) Modular strategies for PET imaging agents. Curr Opin Chem Biol 14(1):105–111

Innis RB, Cunningham VJ, Delforge J, Fujita M, Gjedde A, Gunn RN, Holden J, Houle S, Huang SC, Ichise M, Iida H, Ito H, Kimura Y, Koeppe RA, Knudsen GM, Knuuti J, Lammertsma AA, Laruelle M, Logan J, Maguire RP, Mintun MA, Morris ED, Parsey R, Price JC, Slifstein M, Sossi V, Suhara T, Votaw JR, Wong DF, Carson RE (2007) Consensus nomenclature for in vivo imaging of reversibly binding radioligands. J Cereb Blood Flow Metab 27(9):1533–1539

Jack CR Jr, Wiste HJ, Schwarz CG, Lowe VJ, Senjem ML, Vemuri P, Weigand SD, Therneau TM, Knopman DS, Gunter JL, Jones DT, Graff-Radford J, Kantarci K, Roberts RO, Mielke MM, Machulda MM, Petersen RC (2018) Longitudinal tau PET in ageing and Alzheimer's disease. Brain 141(5):1517–1528

Jeffrey P, Summerfield S (2009) Assessment of the blood-brain barrier in CNS drug discovery. Neurobiol Dis 37(1):33–37

Kealey S, Miller PW, Long NJ, Plisson C, Martarello L, Gee AD (2009) Copper(I) scorpionate complexes and their application in palladium-mediated [11C]carbonylation reactions. Chem Commun:3696–3698

Lee DE, Gallezot JD, Zheng MQ, Lim K, Ding YS, Huang Y, Carson RE, Morris ED, Cosgrove KP (2013) Test-retest reproducibility of [11C]-(+)-propyl-Hexahydro-Naphtho-Oxazin positron emission tomography using the bolus plus constant infusion paradigm. Mol Imaging 12(2):77–82

Logan J, Fowler JS, Volkow ND, Wolf AP, Dewey SL, Schlyer DJ, MacGregor RR, Hitzemann R, Bendriem B, Gatley SJ et al (1990) Graphical analysis of reversible radioligand binding from time-activity measurements applied to [N-11C-methyl]-(−)-cocaine PET studies in human subjects. J Cereb Blood Flow Metab 10(5):740–747

Matthews PM, Rabiner EA, Passchier J, Gunn RN (2012) Positron emission tomography molecular imaging for drug development. Br J Clin Pharmacol 73(2):175–186

Morris E, Chalkidou A, Hammers A, Peacock J, Summers J, Keevil S (2016) Diagnostic accuracy of 18F amyloid PET tracers for the adiagnosis of Alzheimer's disease: a systematic review and meta-analysis. Eur J Nucl Med Mol Imag 43(2):374–385

Neumann CN, Hooker JM, Ritter T (2016) Concerted nucleophilic aromatic substitution with $^{19}F^-$ and $^{18}F^-$. Nature 534(7607):369–373

Nielsen MK, Ugaz CR, Li W, Doyle AG (2015) PyFluor: a low-cost, stable and selective deoxyfluorination reagent. J Am Chem Soc 137(30):9571–9574

Ossenkoppele R et al (2021) Accuracy of tau positron emission tomography as a prognostic marker in preclinical and prodromal alzheimer disease: a head-to-head comparison against amyloid positron emission tomography and magnetic resonance imaging. JAMA Neurol 78(8):961–971. PMID: 34180956

Palner M, Beinat C, Banister S, Zanderigo F, Park JH, Shen B, Hjoernevik T, Jung JH, Lee BC, Kim SE, Fung L, Chin FT (2016) Effects of common anesthetic agents on [^{18}F]flumazenil binding to the GABA$_A$ receptor. EJNMMI Res 6(1):80

Pardridge WM (2016) Re-engineering therapeutic antibodies for Alzheimer's disease as blood-brain barrier penetrating bi-specific antibodies. Expert Opin Biol Ther 16(12):1455–1468

Patel S, Gibson R (2008) In vivo site-directed radiotracers: a mini-review. Nucl Med Biol 35(8):805–815

Patlak CS, Blasberg RG, Fenstermacher JD (1983) Graphical evaluation of blood-to-brain transfer constants from multiple-time uptake data. J Cereb Blood Flow Metab 3(1):1–7

Pike VW (2009) PET radiotracers: crossing the blood-brain barrier and surviving metabolism. Trends Pharmacol Sci 30(8):431–440

Pike VW, Rash KS, Chen Z, Pedregal C, Statnick MA, Kimura Y, Hong J, Zoghbi SS, Fujita M, Toledo MA, Diaz N, Gackenheimer SL, Tauscher JT, Barth VN, Innis RB (2011) Synthesis and evaluation of radioligands for imaging brain nociceptin/orphanin FQ peptide (NOP) receptors with positron emission tomography. J Med Chem 54(8):2687–2700

Pinborg LH, Adams KH, Svarer C, Holm S, Hasselbalch SG, Haugbol S, Madsen J, Knudsen GM (2003) Quantification of 5-HT2A receptors in the human brain using [18F]altanserin-PET and the bolus/infusion approach. J Cereb Blood Flow Metab 23(8):985–996

Poduslo JF, Curran GL, Berg CT (1994) Macromolecular permeability across the blood-nerve and blood-brain barriers. Proc Natl Acad Sci U S A 91(12):5705–5709

Rabinovici GD, Furst AJ, O'Neil JP, Racine CA, Mormino EC, Baker SL, Chetty S, Patel P, Pagliaro TA, Klunk WE, Mathis CA, Rosen HJ, Miller BL, Jagust WJ (2007) 11C-PIB PET imaging in Alzheimer's disease and frontotemporal lobar degeneration. Neurology 68(15):1205–1212

Rinne JO, Brooks DJ, Rossor MN, Fox NC, Bullock R, Klunk WE, Mathis CA, Blennow K, Barakos J, Okello AA, de Liano SRM, Liu E, Koller M, Gregg KM, Schenk D, Black R, Grundman M (2010) 11C-PiB PET assessment of change in fibrillar amyloid-beta load in patients with Alzheimer's disease treated with bapineuzumab: a phase 2, double-blind, placebo-controlled, ascending-dose study. Lancet Neurol 9(4):363–372

Rotstein, B. H., Liang SH, Holland JP, Collier TL, Hooker JM, Wilson AA, Vesdev N (2013) ^{11}CO$_2$ fixation: a renaissance in PET radiochemistry. Chem Comm 40(50): 5621–5629

Rotstein BH, Liang SH, Placzek MS, Hooker JM, Gee AD, Dolle F, Wilson AA, Vesdev N (2016) ^{11}C=O bonds made easily for positron emission tomography radiophamaceuticals. Chem Soc Rev 45(17): 4708–4726

Sander K, Gendron T, Yinnaki E, Cybulska K, Kalber TL, Lythgoe MF, Astad E (2015) Sulfonium salts as leaving groups for aromatic labeling of drug-like molecules with fluorine-18. Sci Rep 5:9941

Schöll M, Lockhart SN, Schonhaut DR, O'Neil JP, Janabi M, Ossenkoppele R, Baker SL, Vogel JW, Faria J, Schwimmer HD, Rabinovici GD, Jagust WJ (2016) PET imaging of tau deposition in the aging human brain. Neuron 89(5):971–982

Sehlin D, Fang XT, Cato L, Antoni G, Lannfelt L, Syvänen S (2016) Antibody-based PET imaging of amyloid beta in mouse models of Alzheimer's disease. Nat Commun 7:10759

Sergeev ME, Morgia F, Lazari M, Wang C, van Dam RM (2015) Titania-catalyzed radiofluorination of tosylated precursors in highly aqueous medium. J Am Chem Soc 137(17):5686–5694

Sevigny J, Chiao P, Bussiere T, Weinreb PH, Williams L, Maier M, Dunstan R, Salloway S, Chen T, Ling Y, O'Gorman J, Qian F, Arastu M, Li M, Chollate S, Brennan MS, Quintero-Monzon O, Scannevin RH, Arnold HM, Engber T, Rhodes K, Ferrero J, Hang Y, Mikulskis A, Grimm J, Hock C, Nitsch RM, Sandrock A (2016) The antibody aducanumab reduces Abeta plaques in Alzheimer's disease. Nature 537:50–56

Slifstein M, Laruelle M (2001) Models and methods for derivation of in vivo neuroreceptor parameters with PET and SPECT reversible radiotracers. Nucl Med Biol 28(5):595–608

Summerfield SG, Lucas AJ, Porter RA, Jeffrey P, Gunn RN, Read KR, Stevens AJ, Metcalf AC, Osuna MC, Kilford PJ, Passchier J, Ruffo AD (2008) Toward an improved prediction of human in vivo brain penetration. Xenobiotica 38(12):1518–1535

Syvänen S, Eriksson J (2013) Advances in PET imaging of P-glycoprotein function at the blood-brain barrier. ACS Chem Neurosci 4(2):225–237

Syvänen S, Hammarlund-Udenaes M (2010) Using PET studies of P-gp function to elucidate mechanisms underlying the disposition of drugs. Curr Top Med Chem 10(17):1799–1809

Syvänen S, Blomquist G, Sprycha M, Höglund AU, Roman M, Eriksson O, Hammarlund-Udenaes M, Långström B, Bergström M (2006) Duration and degree of cyclosporin induced P-glycoprotein inhibition in the rat blood-brain barrier can be studied with PET. NeuroImage 32(3):1134–1141

Syvänen S, Hooker A, Rahman O, Wilking H, Blomquist G, Långström B, Bergström M, Hammarlund-Udenaes M (2008) Pharmacokinetics of P-glycoprotein inhibition in the rat blood-brain barrier. J Pharm Sci 97(12):5386–5400

Syvänen S, Lindhe Ö, Palner M, Kornum BR, Rahman O, Långström B, Knudsen GM, Hammarlund-Udenaes M (2009) Species differences in blood-brain barrier transport of three positron emission tomography radioligands with emphasis on P-glycoprotein transport. Drug Metab Dispos 37(3):635–643

van Dongen GA, Vosjan MJ (2010) Immuno-positron emission tomography: shedding light on clinical antibody therapy. Cancer Biother Radiopharm 25(4):375–385

van Dongen GA, Poot AJ, Vugts DJ (2012) PET imaging with radiolabeled antibodies and tyrosine kinase inhibitors: immuno-PET and TKI-PET. Tumour Biol 33(3):607–615

Verhoog S, Pfeifer L, Khotavivattana T, Calderwood S, Collier TL, Wheelhouse K, Tredwell M, Gouverneur V (2016) Silver-mediated ^{18}F-labeling of aryl-CF$_3$ and aryl-CHF$_2$ with 18F-fluoride. Synlett 27(1):25–28

Warburg O (1956) On the origin of cancer cells. Science 123(3191):309–314

Watson J, Wright S, Lucas A, Clarke KL, Viggers J, Cheetham S, Jeffrey P, Porter R, Read KD (2009) Receptor occupancy and brain free fraction. Drug Metab Dispos 37(4):753–760

Yu YJ, Zhang Y, Kenrick M, Hoyte K, Luk W, Lu Y, Atwal J, Elliott JM, Prabhu S, Watts RJ, Dennis MS (2011) Boosting brain uptake of a therapeutic antibody by reducing its affinity for a transcutosis target. Sci Trans Med 3(84):84ra44

Zhang L, Anabella V (2016) Strategies to facilitate the discovery of novel CNS PET ligands. EJNMMI Radiopharm 1:13

Zhang L, Villalobos A, Beck EM, Bocan T, Chappie TA, Chen L, Grimwood S, Heck SD, Helal CJ, Hou X, Humphrey JM, Lu J, Skaddan MB, McCarthy TJ, Verhoest PR, Wager TT, Zasadny K (2013) Design and selection parameters to accelerate the discovery of novel central nervous system positron emission tomography (PET) ligands and their application in the development of a novel phosphodiesterase 2A PET ligand. J Med Chem 56(11):4568–4579

Zhang L, Chen L, Beck EM, Chappie TA, Coelho RV, Doran SD, Fan K-H, Helal CJ, Humphrey JM, Hughes Z, Kuszpit K, Lachapelle EA, Lazzaro JT, Lee C, Mather RJ, Patel NC, Skaddan MB, Sciabola S, Verhoest PR, Young JM, Zasadny K, Villalobos A (2017) The discovery of a novel phosphodiesterase (PDE) 4B-preferring radioligand for positron emission tomography (PET) imaging. J Med Chem 60(20):8538–8551

Zhang L, Chen L, Dutra JK, Beck EM, Nag S, Takano A, Amini N, Arakawa R, Brodney MA, Buzon LM, Doran SD, Lanyon LF, McCarthy TJ, Bales KR, Nolan CE, O'Neill BT, Schildknegt K, Halldin C, Villalobos A (2018) Identification of a novel positron emission tomography (PET) ligand for imaging β-site amyloid precursor protein cleaving enzyme 1 (BACE-1) in brain. J Med Chem 61(8):3296–3308

Chapter 12
Approaches Towards Prediction of CNS PK and PD

Elizabeth C. M. de Lange, Hsueh Yuan Chang, and Dhaval Shah

Abstract It has to be realized that a drug's pharmacokinetics and pharmacodynamics results from the combination of drug properties and biological system characteristics. This chapter will address the gross anatomy and physiology of the central nervous system (CNS) and how physiological processes play a role in CNS drug-target site distribution. This is followed by physiologically based pharmacokinetic (PBPK) model characteristics and the recently developed multi-CNS compartment PBPK models for small molecules and antibodies, with good predictive power for CNS target site distribution in human, as it explicitly distinguishes between drug and systems properties. Understanding the CNS target site concentrations (Stevens et al. 2012) further helps in the development of (PB)PK-pharmacodynamic (PD) models in healthy and disease conditions to further pave the way to predict the right drug at the *right location, right time, and right concentration* (De Lange 2013a, b).

12.1 Introdction

It has to be realized that a drug's pharmacokinetics and pharmacodynamics results from the combination of drug properties and biological system characteristics. This chapter will address the gross anatomy and physiology of the central nervous system (CNS) and how physiological processes play a role in CNS drug-target site distribution. This is followed by physiologically based pharmacokinetic (PBPK) model characteristics and the recently developed multi-CNS compartment PBPK models for small molecules and antibodies, with good predictive power for CNS target site distribution in human, as it explicitly distinguishes between drug and

E. C. M. de Lange (✉)
Research Division of Systems Pharmacology and Pharmacy, Leiden University, Leiden, The Netherlands
e-mail: ecmdelange@lacdr.leidenuniv.nl

H. Y. Chang · D. Shah
Department of Pharmaceutical Sciences, School of Pharmacy and Pharmaceutical Sciences
State University of New York at Buffalo, Buffalo, NY, USA

© American Association of Pharmaceutical Scientists 2022
E. C. M. de Lange et al. (eds.), *Drug Delivery to the Brain*, AAPS Advances in the Pharmaceutical Sciences Series 33, https://doi.org/10.1007/978-3-030-88773-5_12

systems properties. Understanding the CNS target site concentrations (Stevens et al. 2012) further helps in the development of (PB)PK-pharmacodynamic (PD) models in healthy and disease conditions to further pave the way to predict the right drug at the *right location, right time, and right concentration* (De Lange 2013a, b).

12.2 Physiology of the Brain

The brain consists of different physiological compartments. These include the brain extracellular fluid (brainECF), the brain parenchyma cells, and the different spaces of cerebrospinal fluid (CSF) being the lateral ventricles, third ventricle, fourth ventricle, cisterna magna, and the subarachnoid spaces (Segal 1993). The brain is separated from direct contact with blood by the presence of barriers. The blood-brain barrier (BBB) is situated between the blood and brainECF and is made up of endothelial cells of brain capillaries joined by tight junctions. The blood-CSF barrier (BCSFB) is mainly situated at the epithelium of the choroid plexuses (Cserr 1984; Abbott 2006; Bernacki 2008). The CSF is separated from the brain parenchyma cells by an ependymal layer without barrier function (Del Bigio 1995). Table 12.1 summarizes the values that have been reported for rat and human CNS physiological parameters for which different ratios exist between rat and human.

Apart from differences between conditions (species, gender, genotype, diet, drug treatment, etc.), in disease condition also physiological parameters can be affected. Several disease-related processes result in enhanced BBB permeability to fluid and/or solutes. These include hypertension, radiation, edema, inflammation, ischemia,

Table 12.1 Human and rat approximate values for CNS physiological parameters (Yamamoto et al. 2017b) and their human/rat ratios

Parameter	Rat value (R)	Human value (H)	Ratio (H/R)
Blood volume	20 ml	5000 ml	250
Plasma volume	10.6 ml	2900 ml	273
Brain weight	1.8 g	1400 g	777
Cerebral blood flow/ g brain	1.1 ml/min	0.4 ml/min	0.36
Brain ECF volume	290 µl	240–280 ml	1000
Brain ECF bulk flow	0.2–0.5 µl/min	0.15–0.20 ml/min	896
CSF production	2.2 µl/min	0.35–0.4 ml/min	168
CSF turnover	11 times/day	4 times/day	0.36
CSF volume	250–300 µl	140–150 ml	575
CSF volume lateral ventricle	50 µl	22.5 ml	450
CSF volume cisterna magna	17 µl	7.5 ml	441
CSF volume 3rd and 4th ventricle	50 µl	22.5 ml	250
CSF volume subarachnoid space (SAS)	180 µl	90 ml	500
BBB surface area	155 cm^2	10–20 m^2	967
Choroid plexus surface area	75 cm^2	0.021 m^2	2.8

and reperfusion (reoxygenation) (Banks and Kastin 1996; Abbott 2010). Changes in cerebral blood flow, BBB functionality, BCSFB functionality, plasma protein binding, brain tissue binding, CSF flow, and enzyme functionality may all have their effects on drug brain distribution and elimination (De Lange 2013a).

12.3 Physiological Processes Involved in CNS Drug Distribution

A number of factors play a role in the relationship between CNS drug dose and resulting CNS effects, and it is important to realize that these processes occur in parallel and do influence each other (De Lange 2013a). The main processes that determine CNS target site distribution include plasma pharmacokinetics and plasma protein binding (Schmidt et al. 2010), cerebral blood flow, passive (paracellular and transcellular) and active transport across the BBB, transport across the BCSFB, brainECF bulk flow, CSF turnover, brain tissue binding, brain metabolism, and brain degradation (Fenstermacher et al. 1974; Spector et al. 1977). The brainECF bulk flow and the CSF production and elimination contribute to elimination clearance of drugs from the brain, especially for the drugs with low permeability, and should explicitly be distinguished from actual BBB transport. Also, intracellular and extracellular exchange of drugs should be considered (Fenstermacher 1974). This may include both passive and active transport. As to our current knowledge, passive membrane transport only occurs for unionized molecules, and the pH gradient from the extracellular space (pH=7.3) and the intracellular space (pH =7.0) is of importance for weak bases and acids. In addition, transport between brainECF and brain cells may be governed by active transport processes (Loryan et al. 2014). Also, brain metabolism at the barriers (Gazzin et al. 2012; Strazielle and Ghersi-Egea 2013) but also in brainECF and/or cells might affect CNS PK.

An important feature is that the BBB is under continuous physiologic control by astrocytes, pericytes, neurons, and plasma components. All together, these factors determine the delicate homeostasis of the brain environment. This dynamic regulation of the BBB indicates that different situations may result in different BBB functionalities and changes in pathological conditions (De Lange et al 2005; Bell et al 2009; Bengtsson et al. 2009; Zlokovic et al 2010) and species differences (Syvanen et al. 2009).

All these abovementioned processes occur in parallel and are interconnected. Thus, it can be seen that CNS target site drug delivery includes a complex combination of processes. Oversimplification of these processes has significantly contributed to the very high attrition rate in the development of CNS drugs. Thus, we need to put additional effort into performing the type of investigations that provide data that we learn from in having the right CNS drug "at the right place, at the right time, and at the right concentration" (De Lange 2013a).

12.4   Small Molecules

12.4.1   *Physiologically Based Pharmacokinetic (PBPK) Model Characteristics*

The PBPK modeling approach is not based on concentration-time profiles of drugs in virtual body compartments (classical compartmental pharmacokinetic approach) that are just based on their rate of equilibrium with the plasma compartment (Dedrick and Bisschoff 1980; Davies and Morris 1993). In conjunction with physicochemical properties, these physiological and biochemical parameter values determine the pharmacokinetics of the drug. To that end, the PBPK modeling mathematically describes mass transport of the drug between true body (physiological) compartments, using quantitative parameter values of physiological volumes of tissues, tissue components, tissue blood flow, ECF bulk flow, as well as expression of transporters (Shawana et al. 2011), expression of enzymes, pH values, etc. All together this results in values for tissue permeability of a drug (PS value, which is an expression for rate of transport) and tissue distribution (Kp value, expressed as a ratio of total drug concentrations in tissue divided by that in blood or plasma at equilibrium, being an expression for extent) (Rowland et al. 2011; Hammarlund-Udenaes et al. 2008). In principle, this is the strongest approach to derive and to predict the impact of a change in a physiological value on the pharmacokinetics of a drug, but a lot of data and, therewith, time is needed for development of a PBPK model. Because the unbound drug concentrations drive membrane transport and interaction with targets and enzymes (Watson et al. 2009; Hammarlund-Udenaes et al. 2008; Hammarlund-Udenaes 2009), PBPK models should include unbound drug concentrations (Yamamoto et al. 2017a).

The aspect of time to equilibrium between plasma and brain concentrations (rate) has been specifically addressed by Liu et al. (2005). It was demonstrated that a high BBB permeability alone does not necessarily result in a rapid brain equilibration but actually requires a combination of high BBB permeability and low brain tissue binding. So, if looking for a drug with rapid brain equilibration, drug discovery should look for compounds with high BBB permeability and low nonspecific binding in brain tissue.

The extent of equilibration is often viewed between plasma and brain, as a ratio of their steady-state concentrations, clearance-in over clearance-out of the brain, or AUC values. Also, here it is the unbound concentration that should be taken as the basic input (Hammarlund-Udenaes et al. 2008). Ratio values being larger than 1 may indicate active BBB transport into the brain, while values smaller than 1 indicate active BBB transport out of the brain, metabolism, and/or other elimination processes, such as brainECF bulk flow, CSF turnover, and degradation in lysosomes.

For good CNS PBPK models, unbound and total drug concentration and time resolved data combined with (patho-)physiological data are needed, to be combined with the drug's physicochemical and biological properties.

Serial blood sampling, (parallel) microdialysis sampling, and end-of-experiment brain tissue can be used to provide (unbound) plasma pharmacokinetics (PK), unbound PK at different CNS locations, and total brain drug concentrations, respectively. The samples may even provide biomarkers of the effects (PD) and/or disease. Advanced mathematical modeling of such data may give a useful set of parameters that provide insight into the interrelationship of the processes that govern plasma PK, CNS drug distribution, target site PK, PD, and/or disease. This is called the *mastermind research approach* (De Lange 2013a). "Mastermind" is a game in which Player 1 makes a code of pins with different colors and positions that are hidden from being seen by Player 2. Player 2 has to "crack the code" within a limited number of trials. For being successful, this has to be done by serially and strategically choosing combination and positions of the pins, with feedback of Player 1 on each trial. The *mastermind research approach* entails the allegory with this game on how we should design our experiments and need advanced mathematical modeling to crack the codes of the biological systems (e.g., the interrelationships of body processes). In this way the explicit distinction between the role of drug-specific properties and the characteristics of the system is "stored" in physiologically based PK(PD) models.

12.4.2 Current Status

12.4.2.1 The Multi-CNS Compartment CNS PBPK Model for Small Molecules

Our ultimate aim is to develop models with higher predictive power of CNS target site distribution in human and connect to binding kinetics and PD models (such as the neural circuit quantitative systems pharmacology model) (Geerts et al. 2015) to predict CNS drug effects in human. To that end, information on species- and/or condition-dependent differences in the body and CNS processes is essential. The first step is having a CNS PBPK model that predicts drug concentrations in multiple CNS compartments. The available literature information on CNS drug distribution appeared far too scarce for this purpose, and we extended our historical data to generate our own dataset, making use of intracerebral microdialysis at multiple sites in the brain and serial plasma sampling to obtain unbound concentration-time profiles at these sites, in individual rats. Using advanced mathematical modeling, we could develop drug-specific semi-physiological CNS drug distribution models for three drugs with distinctive physicochemical biological properties: acetaminophen (neutral, moderately lipophilic), methotrexate (hydrophilic acid, MRP-OATP substrate), and quinidine (base, lipophilic, P-gp substrate) (Westerhout et al. 2012, 2013, 2014). The dataset was extended for atenolol (De Lange et al, 1995), methotrexate (Westerhout et al. 2014; De Lange, 1995), remoxipride (Stevens et al. 2012; Van der Brink et al., 2017), risperidone, paliperidone, phenytoin (Yamamoto et al. 2017b), and morphine (Bouw et al. 2001; Groenendaal et al. 2007), and the first versions of

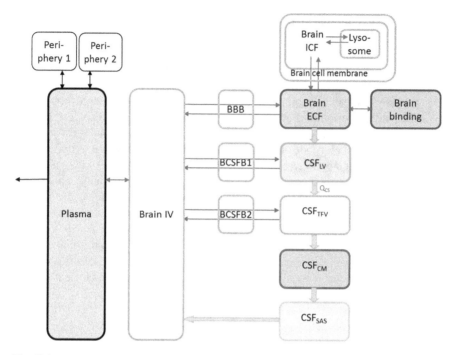

Fig. 12.1 Peripheral, plasma, and CNS compartments, with exchange routes for small molecules and flow routes, as part of the multi-CNS compartment CNS PBPK model for small molecules. IV = intravascular, ICF = interstitial fluid, ECF = extracellular fluid, CSF = cerebrospinal fluid, LV = lateral ventricle, TFV = third and fourth ventricle, CM = cisterna magna, SAS = subarachnoid space

the multi-CNS compartment CNS PBPK model for small molecules in rat and in human have been published (Yamamoto et al. 2017a, b, 2018). The CNS PBPK model structure is shown in Fig. 12.1.

The overlay of the rat and human model predictions for different drugs and observed data are shown in Fig. 12.2. An important note is that a total of 335 rats were used for the development and validation of this multi-CNS compartment rat model. Now, this model can be used to help save animals from being used for CNS drug distribution studies.

12.4.2.2 CNS Target Site Distribution and Target Binding Kinetics

Drug-target binding kinetics has received increasing interest in drug discovery, due to its influence on the time course of the drug effect (PD). Thus, apart from CNS target site distribution, the time course of target binding (interaction, affinity) is the next step to be integrated. This is distinctively different from the efficacy of the drug (see operational model of agonism; Kenakin 2004, 2011). Recently, we published a study on how drug-target binding kinetics, target concentrations, and tissue

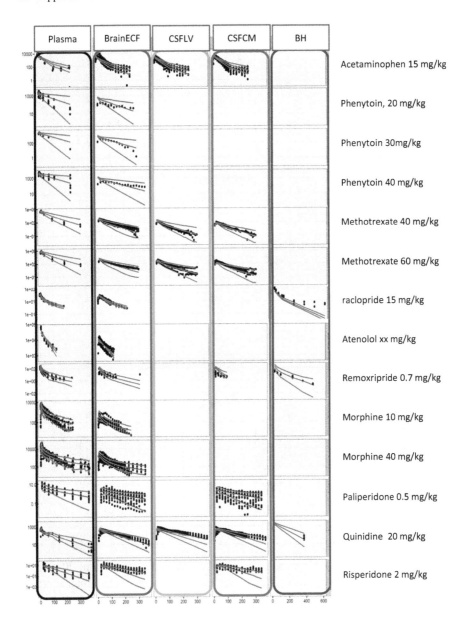

Fig. 12.2 The overlay of the rat and human CNS PBPK model predictions for different drugs and actual data for rat (above) and human (below), respectively. ECF = extracellular fluid, CSF = cerebrospinal fluid, LV = lateral ventricle, TFV = third and fourth ventricle, CM = cisterna magna, BH = brain homogenate

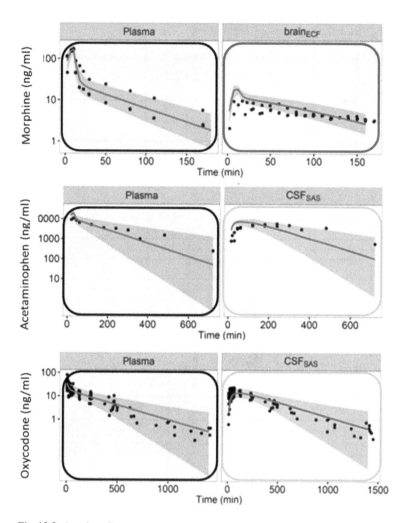

Fig. 12.2 (continued)

characteristics can be used to predict selectivity over time (Vlot et al. 2017). We investigated the determinants of selectivity by using simulations and mathematical model analysis with a minimal PBPK model, which incorporated target binding of two targets in the same or one target in different tissues (Fig. 12.3). This was investigated both for selective occupancy of the primary target compared to the secondary target (thereby investigating target selectivity) and for selective occupancy of the primary target in the targeted tissue compared to a nontargeted tissue (thereby investigating tissue selectivity).

Our simulations showed that target selectivity is determined by affinity (K_D) values, partition coefficients, and target concentrations for the two targets. Also, target selectivity within a single tissue is only increasing over time when the drug-target

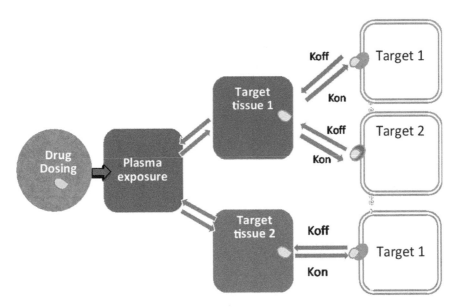

Fig. 12.3 Representation of a minimal PBPK model, which includes target binding of two targets in the same tissue and one target in different tissues, to explore influence of tissue distribution and affinity on selectivity of target binding

dissociation rate constants (K_{off} values) are different from each other and the lowest dissociation rate is rate-limiting for the decline of target occupancy (Fig. 12.4).

Then, under typical conditions, tissue selectivity is determined by target concentrations and partition coefficients. Moreover, tissue selectivity between tissues is only increasing over time if distribution out of one of the tissues is rate-limiting and different from the other tissue (Fig. 12.5).

These results indicate that the prediction of both target and tissue selectivity requires integration of drug elimination, drug distribution, and drug-target binding. However, target concentration and drug-target binding kinetics are the main determinants for the development of selectivity over time.

In conclusion, factors that govern target occupancy are plasma PK, target tissue (site) distribution, target binding kinetics, and target concentration. Actual target occupancies depend on the values of all these parameters. Mathematical models are very helpful (needed) to understand the determinants of target occupancy under specific conditions.

12.4.2.3 Translational PKPD Modeling

Many times, brain distribution is studied without measuring associated (biomarkers of the) effects. It would be of great added value if PK and associated PD would be obtained in a single experimental subject or at least single experimental context. Therefore, it is of importance to learn more about factors in target activation. Here

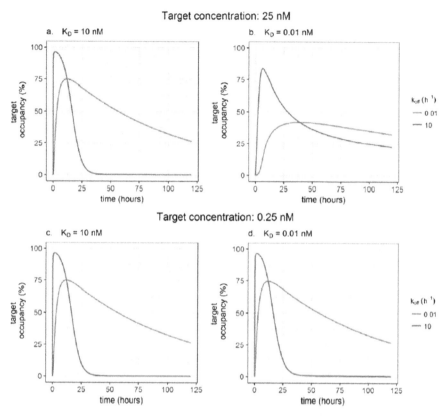

Fig. 12.4 Target selectivity
Target binding of two targets in the same tissue. Target concentration and K_D determine the extent of in vivo kinetic target selectivity. Target selectivity is characterized by a difference in target occupancy between a target with a high affinity ($K_{off}=0.01$ h^{-1}, green) and a target with a low affinity ($K_{off} = 10$ h^{-1}, orange)

we assume the target being a receptor. At equilibrium, the relationship between agonist concentration ([A]) and agonist-occupied receptor ([AR]) is described by Equation 12.1:

$$[AR] = ([RT]*[A]) / ([A]+KA) \qquad (12.1)$$

in which [RT] represents total receptor concentration and KA represents the agonist-receptor equilibrium dissociation constant.

Receptor theory as included in the operational model of agonism assigns mathematical rules to biological systems in order to quantify drug effects and define what biological systems can and cannot do, leading to the design of experiments that may further modify the model. For the relation between agonist-occupied receptors [AR] and receptor activation (Black and Leff 1983; Watson et al. 2009) derived a practical or "operational" equation. If agonist binding to the target is

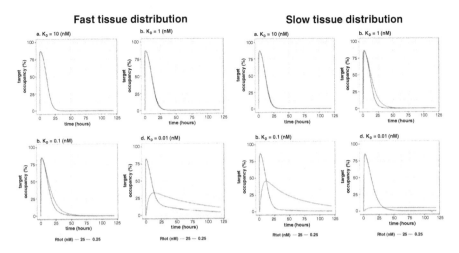

Fig. 12.5 Target occupancy
Target occupancy for fast (left) and slow (right) tissue distribution of a drug. For slow target tissue distribution, tissue selectivity reverses to off-target tissue selectivity as K_D decreases. Tissue selectivity is characterized by a difference in target occupancy for a single target in different tissues having a total receptor concentration (Rtot) which is high (25 nM, green) and low (0.25 nM, orange)

hyperbolic and the concentration-response curve has a Hill slope of 1.0, the equation linking the concentration of "agonist-occupied receptors" to the response must also be hyperbolic. This leads to the "transducer function," as the mathematical representation of the transduction of receptor occupation into a response, in Equation 12.2:

$$E = (Em*[AR])/([AR]+KE)$$ (12.2)

The parameter, Em, is the maximum response possible in the system (tissue). It is important to note that this is not necessarily equal to the maximum response that a particular agonist actually produces. The parameter KE is the concentration of [AR] that elicits half the maximal tissue response, Em. The efficacy of an agonist is determined by both KE and the total receptor density of the tissue ([RT]). Black and Leff (1983) combined those two parameters into a ratio ([RT]/KE) and called this parameter tau (τ), the "transducer constant."

It indicates that two agents in a setting with equivalent sets of receptors may not produce equal degrees of effect even if both agents are given in maximally effective doses. This is due to differences in "intrinsic activity" (or efficacy) that can be defined as the property of a drug that determines the amount of biological effect produced per unit of drug-receptor complex formed. Thus, the drug that produces the greater maximum effect has the greater intrinsic activity. It is important to note that intrinsic activity is not the same as "potency" and may be completely independent of it.

Activation of the receptor should be "transduced" to elicit the response. Combining the hyperbolic occupancy equation with the hyperbolic transducer function yields an explicit Equation (12.3) describing the effect at any concentration of agonist:

$$E = (Em * \ddot{A}n * [A]n) / ((KA + [A]n) + \ddot{A}n * [A]n) \qquad (12.3)$$

in which E = effect, Em = maximum response achievable in system, KA = agonist dissociation equilibrium constant, and n = slope index of the receptor occupancy effect function. It actually describes a three-dimensional interrelationship. Intrinsic activity—like affinity—depends on the characteristics of both the drug and the receptor, but intrinsic activity and affinity apparently can vary independently. This means that the EC50 does not equal KA but rather KA/(1+ τ). As an example, having a strong agonist that reaches a 50% response upon binding fewer than half the available receptors, its EC50 will be much less than KA.

Receptor affinity and intrinsic activity are "drug- specific" properties and can be estimated in in vitro bioassays, with the maximal response of the drug being determined, not from single dose-response curves but from using pairs of dose-response curves (usually treatment and control) for a particular tissue, here CNS, sharing some parameters.

Subsequent simultaneous analysis of the resulting different PKPD relationships must be performed to build a mechanism-based model that explicitly distinguishes between the drug-specific and the system-specific properties to allow prediction of the intrinsic activity and potency of another drug for a particular pharmacological effect or response. These different PKPD relationships may be obtained in different ways.

- Studying one agonist under control conditions and conditions in which the number of receptors available for binding is reduced (Furchgott et al. 1966; Garrido et al. 2000)
- Studying series of chemically similar drugs with varying degrees of agonism for the specific receptor and simultaneous analysis of the PKPD relationships (Cox et al. 1998; Groenendaal et al. 2008)

The operational model of agonism has been successfully applied in numerous in vitro studies and later also in mechanism-based PKPD analysis of in vivo drug effects (Kenakin 2004; Danhof et al. 2005; Danhof et al. 2007). For adenosine A1 receptor agonists, a good correlation was observed between the in vivo pKA and the in vitro pKi and also between the in vivo efficacy parameter (τ) and the in vitro GTP shift (as measure for intrinsic activity), thus enabling the prediction of in vivo concentration-effect relationships (Van der Graaf et al., 1997a, b). In addition, excellent in vitro-in vivo correlations have also been observed for benzodiazepines (Tuk et al. 1999, 2002; Visser et al. 2003) and neuroactive steroids (Visser et al. 2003).

Taken together, incorporation of receptor theory into PKPD models on in vivo concentration-effect relationships could provide information on:

- Tissue selectivity of drug effects (Van Schaick et al. 1998)
- Interspecies differences in concentration-effect relationships
- Tolerance and sensitization (Cleton et al. 2000)
- Intra- and interindividual variability

Of course, life is not that simple that in all cases the incorporation of receptor theory in mechanism-based PKPD models is successful. For the opioids alfentanil, fentanyl, and sufentanil, it was shown by simulation that the concentration-effect relationships could be explained by the operational model of agonism under the assumption of a considerable receptor reserve (Cox et al. 1998), while also, a shift in the concentration-effect relationship of alfentanil was observed following pre-treatment with the irreversible μ-opioid receptor antagonist β-funaltrexamine, which was consistent with the 40–60% reduction in the available number of specific μ-opioid binding sites as shown in an in vitro receptor bioassay (Garrido et al. 2000). However, a proper incorporation of the receptor theory in a mechanism-based PKPD model of the opioid receptor agonists could not been accomplished.

Also, for the 5-HT1A receptor agonists, a rather poor correlation was found between the in vivo pKA and the in vitro pKi, despite a good correlation between in vivo and in vitro GTP shift (Zuideveld et al. 2007). Failure of successful inclusion of the receptor theory in the PKPD models of the opioid and 5-HT1A agonists could be due to complexities at the level of blood-brain transport and intracerebral distribution which was not addressed in these studies, as estimates of hypothetical target site concentrations were made using the link model.

When solving shortcomings in knowledge on target site distribution of drugs, the principles of the operational model will provide the basis for future developments in drug development by classifying drugs and predicting their mechanism of action in pharmacology (Kenakin 2011)

When drug-specific and biological system-specific parameters are quantified in a PKPD model, it provides the opportunity to scale the system-specific parameters from animal to human to translate PKPD relationship to man. Allometric scaling of drug pharmacokinetic properties and biological system-specific parameters has been used in translational investigations, with reasonable degree of success, to predict drug effects in humans (Yassen et al. 2007; Zuideveld et al. 2007). But, PD properties are more difficult to scale compared to PK properties, since PD parameters are often not related to bodyweight (e.g., receptor occupancy). Such information may be available by in vitro bioassays.

Preclinically Derived Translational Human PKPD Model for Remoxipride

Stevens et al. (2012) successfully predicted human CNS effects of remoxipride, on the basis of preclinical experiments with information on plasma PK, BBB transport, and brainECF concentrations and using prolactin plasma concentrations as a readout of the CNS effects. For many drugs and endogenous compounds, clinical information is often readily available, like for target binding characteristics of dopaminergic compounds (Kvernmo et al. 2008) and on prolactin in animals and human (Ben Jonathan et al. 2008).

For the human situation, it was assumed that the BBB transport of remoxipride in humans is comparable to that in rat (in essence based on passive diffusion). The preclinical derived translational human PKPD model successfully predicted the system prolactin response in humans, indicating that positive feedback on prolactin synthesis and allometric scaling thereof could be a new feature in describing complex homeostatic mechanisms (Stevens et al. 2012).

Multivariate PKPD Analysis with Metabolomics for Systems-Wide Effects

The ultimate goal is to predict CNS effects in human. Often, the focus has been on a single biomarker to reflect the CNS drug effect. However, given the complexity of brain diseases, it can be seen that the search for a single biomarker to explain the disease relative to the healthy condition, and/or changes in the disease condition by (drug) treatment, will never lead to a success. Actually, we do not deal with "the" effect, but a composite of effects. The search should therefore be on "fingerprints" of multiple biomarkers, in a time-dependent manner, for investigations on the "effect spectrum." With metabolomics as an emerging scientific tool, many more compounds in brain fluids and in plasma can be measured in parallel, in a quantitative and time-dependent manner. Furthermore, the emphasis should lie on measures that can be obtained both preclinically and clinically, to enhance translational insights and, therewith, predictive power of preclinically obtained information. Knowledge of human brain target side concentrations will then be useful for further development of PBPKPD models in health and disease conditions (Westerhout et al. 2011; Stevens et al. 2012), to further pave the way to predict the right drug at the *right location, right time, and right concentration* (De Lange 2013a).

While currently biomarker discovery is still typically driven by the known pharmacological mechanisms, metabolomics fingerprinting is less biased to these pathways. Composite biomarkers enhance the evaluation of the proof of pharmacology of CNS drugs, which is crucial for successful drug development. It is particularly important to dynamically evaluate the biomarker responses in relation to the systems PK of the drug (i.e., not only plasma PK but PK in multiple body compartments), given that the interaction between PK and PD typically is nonlinear and time-dependent.

Metabolomics analysis has revealed multiple new biochemical pathways in relation to drug responses. One of the techniques being useful also in CNS biomarker discovery is intracerebral microdialysis. It has been successfully applied to study drug concentrations as well as drug response biomarkers in brainECF to evaluate CNS PK and PD. Therefore, microdialysis is the method of choice to dynamically evaluate a metabolomics fingerprint in brainECF simultaneously upon CNS drug treatment. Such dynamical evaluation would improve the quantitative insights into systems-wide responses (i.e., changes in biomarker concentrations), thereby shifting CNS drug development from an empirical toward a mechanistic discipline.

This has been shown by a multivariate (PKPD) evaluation of a metabolomics response in plasma revealing multiple dynamics underlying a systems response upon treatment with remoxipride [Van den Brink et al. 2018) (Fig. 12.6), as well as for quinpirole, where also the metabolomics response in both plasma and brainECF was evaluated (van der Brink et al. 2019). Overall, the purpose was to provide insight into the systems-wide biochemical responses of CNS drugs, combined with PKPD modeling as a new approach to discover blood-based biomarkers of central responses. For quinpirole, the multivariate PKPD evaluation was applied to describe the biomarker responses and to explore the target site of drug action, and the metabolic effects of 8-day relative to 1-day treatment were investigated by covariate analysis. Of 23 biomarkers in plasma, 19 were also reflected in brainECF. Vice versa, 5 of the 15 brainECF biomarkers were reflected in plasma. The range of estimated potencies (EC_{50}) differed substantially both in plasma and in brainECF. Quinpirole affected dopamine and glutamate signaling in the brainECF and the branched-chain amino acid metabolism and the immune system in plasma. No effect of 8-day administration was observed. To our knowledge this was for the first time that multivariate PKPD describing dynamical drug systems response in

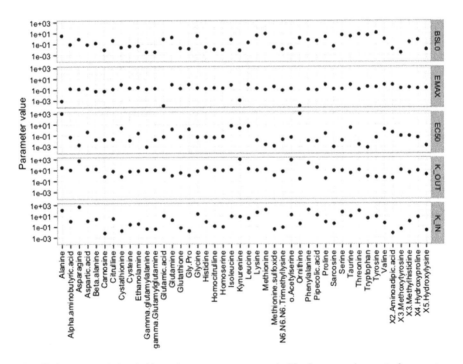

Fig. 12.6 The metabolomic biomarker response to remoxipride shows a unique set of parameter values for each metabolite: the biomarker fingerprint. Kin = production rate constant, Kout = degradation rate constant

brainECF and plasma has been shown to identify the target site of drug action. Further development of this method is envisioned to provide an important connection between drug discovery and early drug development.

12.5 Antibodies

12.5.1 Introduction

Unlike small molecules, the pharmacokinetics of antibodies in CNS has not been investigated extensively. This is mainly because of the assumptions of impermeability of antibodies across the BBB and the concept of immune privilege nature of the brain. However, in the past decades, these assumptions have been challenged by observations such as the efficacy of anti-amyloid β (Aβ) antibody in mouse model of Alzheimer's disease (AD) following peripheral administration (Bard et al. 2000; Bacskai et al. 2001; Syvänen et al. 2018). While there are increasing preclinical studies on antibody-based therapeutics for CNS disorders, our understanding of monoclonal antibody (mAb) disposition at the site-of-action in the brain, and advances in methodologies to measure the exposure of mAb in the brain, is limited (Neves et al. 2016; Pardridge 2016). These shortcomings have resulted in notable clinical failures of mAb-based therapies for the treatment of CNS disorders. To overcome this gap, it is necessary to comprehensively understand the factors that govern the exposure of mAbs in the brain and accurately translate the preclinical findings into the clinic. As such, brain microdialysis studies of mAbs have been conducted recently, and data from these studies have been used to develop translational PBPK and PKPD models of CNS targeted mAbs. These efforts are bound to refine our understanding of the factors that determine brain PK of mAbs and help optimize clinical usage and dosing regimen of CNS targeted mAbs.

(See also Chap. 3 on brain delivery of therapeutic antibodies via transcytosis.)

12.5.2 PK Prediction

12.5.2.1 Motivation

While most CNS targets are localized within the parenchyma, free antibody concentration in the brainECF is rarely measured, due to the inaccessibility of this site in the clinical setting. As a result, clinical evaluation of antibody exposure in CNS is highly dependent on the translation of the preclinical findings. A translational PBPK model for antibody has come to the picture for providing *a prior* prediction of antibody PK in different regions of the brain. It can simulate antibody distribution in the

CNS according to species-specific brain physiology, target expression in the CNS, target binding affinity, and FcRn-binding affinity. The simulation results may guide antibody engineering to acquire maximum brain delivery and minimal peripheral adverse effects. As the correlations between plasma, CSF, and brainECF can be mechanistically characterized using the modeling approach, the model may further reduce the need for cerebrospinal fluid samplings in the clinical setting.

12.5.2.2 Preclinical PK Study at the "Site-of-Action"

To build such models, it requires considerable efforts on collecting the in vivo CNS PK of antibody at different regions of the brain. The more details a model aims to account for, the more data is needed to validate the model. However, the free antibody concentration in brainECF has been hard to obtain due to technical challenges. Therefore, whole-brain homogenates and CSF samples collected via cisternal puncture are generally used as a surrogate for antibody exposure in the CNS in preclinical studies. The limitation of whole-brain measurement of antibody is that it cannot distinguish mAb distribution in brain capillary endothelium cells and at the extravascular space. To avoid any false positive readout from antibodies accumulated within brain capillary endothelial cells, brain capillary depletion method has been proposed to remove capillary EC components from the freshly collected brain tissues (Triguero et al. 1990; Pardridge et al. 1991). However, this widely applied approach cannot avoid the diffusion and redistribution of capillary-bound mAb during the sample preparation procedure and cannot guarantee there is no unremoved capillary lysates remaining in the prepared sample. It may still overestimate the antibody exposure at the brain parenchyma (Watts and Dennis 2013; Freskgård et al. 2014). The use of CSF to infer brainECF exposure of mAb has different concerns. CSF concentration of endogenous IgG seems proportional to the brainECF concentration of IgG at the steady state. Without reaching steady state, CSF concentration of mAb may not be proportional to the brainECF concentration. In fact, CSF samples collected from different ventricles or sites have different concentrations of mAbs (Chang et al. 2018). Further, the CSF concentration of delivered mAb is not a function of the BBB transport, but it is a measure of antibody transport across the choroid plexus (Pardridge 2016). The CSF concentration of antibody appears to be correlated with distributed antibody in the ependymal surface of the brain or spinal cord, but not with the deeper parenchyma or the site-of-action (Wolak et al. 2015; Martín-García et al. 2013; Yadav et al. 2017). The correlation between CSF and brainECF has not been established yet. In order to characterize antibody disposition among different regions of the brain, including the site-of-action and CSF at different sites, a large pore microdialysis approach for larger molecule (Jadhav et al. 2017) has been used to measure the brainECF PK of antibody (Chang et al. 2018).

Fig. 12.7 A three-compartment PK model for antibody disposition in the brain. Vc = volume of central compartment, Vb = volume of brain tissue, Vt = volume of peripheral tissues, CL$_{up}$ = brain uptake, CL$_{eff}$ = brain efflux), CLd = clearance of distribution, CL = plasma clearance (Abuqayyas and Balthasar 2013)

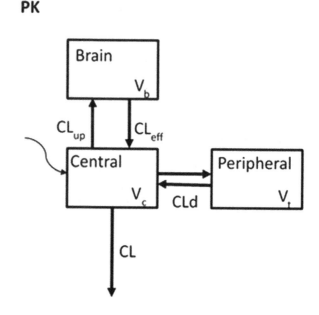

12.5.2.3 Empirical PK Model

Empirical models developed for predicting the PK of small-molecule drugs have also been applied to characterize the disposition of antibody in the CNS (Fig. 12.7). However, the translation (e.g., allometric scaling) of these models tends to provide poor accuracy due to the involvement of binding kinetic (Glassman and Balthasar 2016). The allometric scaling is based on body weight and may not accurately account for discrepancy in brain physiology among different species.

12.5.2.4 PBPK Model

Recently developed translational platform PBPK model accounts for antibody disposition in different regions of the mouse, rat, monkey, and human brains (Fig. 12.8) (Chang et al. 2019). The model accounts for known anatomy and physiology of the brain, including the presence of distinct BBB and BCSFB. At the BBB, paracellular transport of mAbs is restricted, and passive diffusion of plasma immunoglobulin Gs (IgGs) is extremely limited due to the size and polarity of these molecules. Nonspecific pinocytotic vesicles are relatively few at the BBB, compared to peripheral endothelial cells, but may still sufficiently account for the observed entry of macromolecules including mAb into the brainECF (Broadwell and Salcman 1981; Reese and Karnovsky 1967; Davis et al. 2014). The brain concentrations of mAbs have been reported to be ~0.1% of plasma concentrations in previous works (Banks et al. 2002; Garg and Balthasar 2009; Yu et al. 2011; Abuqayyas and Balthasar

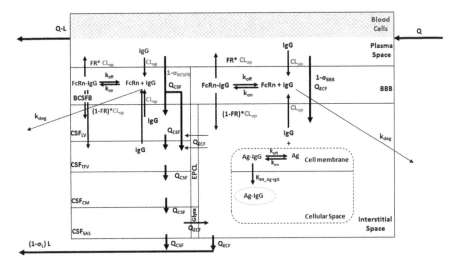

Fig. 12.8 Brain PBPK model scheme

Brain PBPK model consists of two interfaces between the blood and CNS. One is blood-brain barrier (BBB), and the other is blood-CSF barrier (BCSFB). Pinocytosis with low frequency (CL_{up}) occurs at both the BBB and BCSFB. The FcRn may mediate both influx and efflux of antibody crossing the interface. A fraction of pinocytosed antibody may go to lysosomal degradation. The leakiness of BCSFB is considered to provide nonspecific entrance of antibody into the CSF. This nonspecific entrance rate is adjusted by the formation rate of CSF (Q_{CSF}) and a reflection coefficient (σ_{BCSFB}). LV = lateral ventricle, TFV = third-forth ventricle, CM = cisterna magna, SAS = subarachnoid space with lumbar spine, Q_{ECF} = the perivascular pathway as the entrance of antibody from CSF_{SAS} to the brain interstitial space and to the lymphatic system (L) (Chang et al. 2019)

2013; Chang et al. 2018). At the choroid plexus, the secretion site of the CSF, a tiny fraction of plasma IgG may cross the BCSFB into the ventricles because of the relative leakiness of the epithelial cellular barrier (Pardridge 2016). The pinocytosed IgG in the endosome of brain capillary endothelial cells (BCECs) and choroid plexus epithelial cells may bind to the neonatal Fc receptor (FcRn) and be transported along the FcRn. The role of FcRn in brain disposition of antibodies remains controversial. Based on in vivo antibody PK studies in the brain, following intracranial injection, it has been suggested that the clearance of antibodies from the brain is a saturable process and associated with the FcRn-binding affinity (Zhang and Pardridge 2001; Caram-Salas et al. 2011; Cooper et al. 2013). This FcRn-mediated transport has also been examined in the in vitro endothelial cells (Turksen et al. 2011). It is considered one of the mechanisms for the classical "CNS immune privilege" hypothesis. While a majority of pinocytosed antibody will be recycled back to the blood via FcRn-mediated efflux (Cooper et al. 2013), a small fraction of antibody may undergo transcytosis, reaching extravascular space of the brain (Kozlowski 1992). Although this FcRn-mediated mechanism is still not validated due to the lack of dedicated experiments, brain uptake of human intravenous immunoglobulin (IVIG) in rats has been shown to be dose-dependent following in situ brain perfusion, which supports the role of FcRn in brain uptake of antibodies (St-Amour et al.

2013). Of note, a substantial amount of vesicular endogenous mouse IgG is still found to traffic to the lysosomes for degradation (Villaseñor et al. 2016). However, brain exposure of mAbs does not show significant difference between wild-type and FcRn-knockout mice following systemic administration, which brings into question the significance of FcRn in brain disposition of mAbs (Abuqayyas and Balthasar 2013).

The entrance of antibody into the brain parenchyma is not limited via BBB. The newly identified perivascular pathway has been proposed as a new distribution pathway for macromolecules from CSF at the subarachnoid space (SAS) to the brain (Iliff et al. 2012; Pizzo et al. 2018). Previous microdialysis study for mAb disposition in the brain has shown a significant entrance of antibodies into the CNS at the lateral ventricle (LV), one of the formation sites of CSF, as well as a relatively high concentration of endogenous IgG accumulating in the cisterna magna (CM) (Chang et al. 2018). Based on current understanding of brain physiology, once entering ventricles across choroid plexus epithelial cells via the blood-CSF interface, antibodies may travel along the bulk flow of CSF circulation to the CM, lumbar space, and the subarachnoid (SAS). Then, antibodies in the CSF at the SAS can flow along perivascular space (PVS) via fenestrated leptomeningeal cellular layer. Although the diffusion of antibodies is extremely limited, antibodies in PVS may further reach brain parenchyma (Pizzo et al. 2018). A mechanistic understanding of perivascular transport pathway for macromolecular entry into the brain is still under investigation.

The clearance of antibodies from CNS has not been fully investigated. The antibodies in brainECF have been considered to be cleared via FcRn-mediated efflux from the brain to blood or local degradation in lysosome. Brain CSF antibodies may be cleared via lymphatic vessels (Aspelund et al. 2015; Louveau et al. 2015) into the lymph nodes, which eventually empty into the systemic circulation.

The PBPK model parameters were estimated using a rich CNS PK dataset within one species (i.e., rat) and translated to other species simply by changing the values of physiological parameters corresponding to each species (e.g., ventricle volumes, CSF flow rate, estimated brain capillary EC endosomal volume, FcRn-binding affinity, etc.). Using this PBPK model, the antibody can be administered via various systemic or CNS routes, and the disposition of antibody in the whole brain, CSF, and brainECF can be estimated (Fig. 12.9). The model can be further expanded to account for target engagement, disease pathophysiology, and novel mechanisms, such as receptor-mediated transcytosis (RMT) and adsorption-mediated transcytosis (AMT) (Kumagai et al. 1987; Triguero et al. 1990; Girod et al. 1999; Herve 2008), to support the development of novel CNS targeting mAbs. The PBPK model also allows for characterization and prediction of target engagement at the site-of-action within the brain, which can be used to develop PBPKPD models for mAbs.

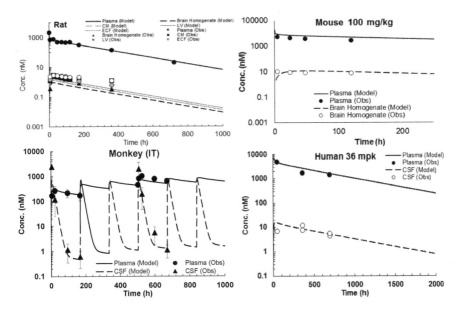

Fig. 12.9 Development and validation of brain PBPK model for antibody

The PBPK model was first fitted to the rat brain PK data collected from diverse literature studies including in vivo microdialysis study of antibody in different regions of the rat brain (only one set of raw data is demonstrated here in the figure). The fitted model was then translated to other species by only changing physiological values, such as the tissue volume, blood flow, and the CSF flow rate. The translated model for mouse, monkey, and human can predict the brain exposure of antibody in different regions of the brain following different routes of administration (Chang et al. 2019)

12.5.3 PD Prediction: Case Study

12.5.3.1 PKPD Modeling of Bispecific TfR/BACE1 Antibodies in Mice

B-secretase 1 (BACE1) is an enzyme responsible for the initial cleavage of amyloid precursor protein (APP) to a soluble extracellular fragment (sAPP β) and a cell membrane-bound fragment (C99). The C99 is further cleaved to soluble amyloid beta peptides (A β_{40} & A β_{42}) by γ-secretase. The accumulation of the Aβ aggregates is believed to be the Alzheimer's disease (AD) pathology (Panza et al. 2019). Blocking BACE1 may reduce the formation of A β peptides and reduce, in theory, the neurotoxicity. However, the anti-BACE1 antibody itself may have very limited brain exposure and could not achieve the required therapeutic concentration in CNS.

Receptor-mediated transcytosis is one of the established delivery strategies to enhance brain penetration of the antibody-based protein therapeutics. A group of endogenous receptors expressed on the luminal surface of BCECs undergo constitutive or stimulated internalization. Antibodies can be engineered to bind to these receptors to enhance their brain uptake. Once the antibodies bind to the receptor, they will stimulate internalization and enter the endosomes. A fraction of them may be recycled back to the plasma via early endosome. Another fraction of antibodies may undergo transcytosis and reach the abluminal surface. They may have a chance to dissociate from the bound receptor and readily engage with their CNS target in the brain parenchyma. The rest of the antibodies, including those which did not have enough chance to dissociate from the receptor on the abluminal side, may go to lysosomes for degradation. These receptors include transferrin, insulin, insulin-like growth factors, or lipid receptors like LRP1 (Lajoie and Shusta 2015). This delivery strategy has been tested in a clinical trial (Okuyama et al. 2018). The most studied receptor-mediated transport receptor on the BBB is transferrin receptor (TfR) (Freskgård and Urich 2017). Bispecific TfR/BACE1 antibodies have been developed to enhance their brain delivery (Yu et al. 2011; Couch et al. 2013; Bien-Ly et al. 2014). They can bind to the transferrin receptor (TfR) expressed on brain capillary ECs and be carried to the brainECF using RMT (Fig. 12.10). Once the bispecific antibodies are disassociated from the receptor at the abluminal site and reach neurons, the antibodies will bind to BACE1 with the second arm and block their activity.

Interestingly, it has been revealed that the intracellular trafficking of anti-TfR antibodies is determined by both binding affinity and the valency (Yu et al. 2011; Sade et al. 2014; Bien-Ly et al. 2014; Niewoehner et al. 2014;

Haqqani et al. 2018; Thom et al. 2018). Based on in vivo observation, strong binding affinity to TfR does not show the highest brain exposure. Rather, an anti-TfR antibody with optimal binding affinity shows a higher efficiency in transcytosis. If the binding affinity is reduced to an extreme, the antibody will not undergo RMT (Yu et al. 2011).

To estimate the optimal binding affinity to have the highest A β reduction and safety, Gadkar et al. have developed an empirical PKPD model (Fig. 12.11). This model accounts for BACE1-mediated plasma clearance, TfR-mediated plasma clearance, TfR-mediated brain transcytosis, and nonspecific antibody clearance

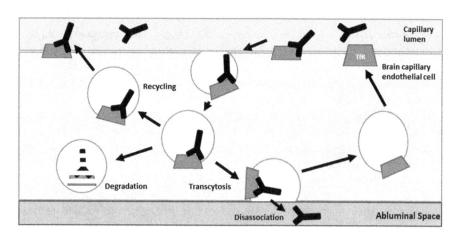

Fig. 12.10 Receptor-mediated transcytosis (RMT) of an anti-TfR antibody

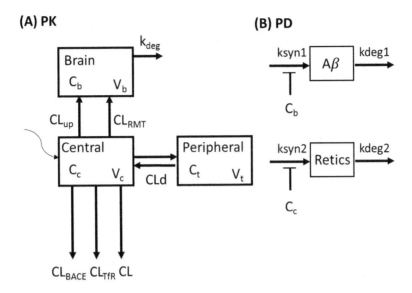

Fig. 12.11 Anti-BACE1 antibody PKPD model for mouse

(a) A compartmental PK model is used to describe plasma and whole-brain PK. (b) The reduction of amyloid beta (A β) in CSF is driven by the Cb. Vc, Vb, Vt = the volume of the central, brain, and peripheral compartments, respectively. Cb and Cc = brain and plasma concentration, respectively. CLup = nonspecific brain uptake of antibody, CLRMT = receptor-mediated transcytosis of antibody, kdeg = regional degradation of antibody in CNS. CL, CLBACE, CLTfR = nonspecific antibody plasma clearance, BACE1-mediated plasma clearance, TfR-mediated plasma clearance, respectively (Gadkar et al. 2016)

(Gadkar et al. 2016). An inhibitory drug effect on the synthesis rate of A β proteins (ksyn1) in the brain is applied to characterize the reduction of the A β proteins, and the adverse effect on reticulocytes is estimated by the inhibitory effect on synthesis rate of reticulocytes (ksyn2). The model reasonably captures both serum and whole-brain PK with different anti-TfR binding affinities. For the PD component, the model could capture the reduction of brain Aβ burden and the adverse event of acute reduction in circulating reticulocytes after the anti-TfR treatment in mice (Couch et al. 2013). The model can be used to guide the selection of optimal TfR binding affinity. The selection criteria may depend on the acceptable level of reduction in reticulocytes and/or the required reduction in A β levels.

12.5.3.2   PKPD Modeling of Anti-TfR antibody and A β Protein Reduction in Monkey

To further expand the brain PKPD model to include TfR binding kinetics, Kanodia et al. have developed a more mechanistic brain PKPD model in monkey (Fig. 12.12) (Kanodia et al. 2016).

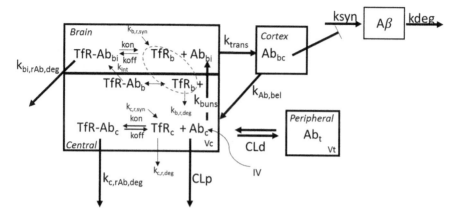

Fig. 12.12 The scheme of brain PKPD model with TfR binding kinetics
The PKPD model has considered the regional distribution of antibody in a transitional brain compartment where antibody has not yet reached their targets in the site-of-action (as the cortex compartment). The kinetic binding and RMT of antibody is accounted in the model (Kanodia et al. 2016)

This model accounts for bound/unbound fraction of antibody, internalization process in endosome, and TfR turnover. The complexity of the model is increased to address the process of RMT. One of the advantages of this model is the direct application of kon and koff rate constants, along with physiological parameters such as internalization rate, brain endosomal volume, and TfR turnover rate, which could be measured using in vitro methods in different species. This may help the translation of the PKPD model from monkey to human. The model can characterize the correlation between plasma antibody concentration and the CSF A β reduction, in the absence of brainECF or CSF antibody concentrations (Fig. 12.13). If such correlation is well validated, brain CSF sampling of antibody could be avoided.

The calibrated model can not only predict plasma PK of antibody and A β reduction in the CSF compartment by different anti-TfR variants in the monkeys but can also estimate A β reduction in humans after scaling up the model parameters (Fig. 12.14). Based on the simulation result, one may find an optimal anti-TfR affinity and design a suitable dosing regimen for drug development purpose.

12.6 Future Directions

While important steps have been made in the direction of prediction of human CNS drug effects, there are still many steps to be taken. For the multi-compartment CNS PBPK model for small and large molecules, the brain tissue (brainECF and cells) is assumed to be homogenous, while regional differences in tissue characteristics may be present (Loryan et al. 2016; Yadav et al. 2017; Pizzo et al. 2018). Furthermore,

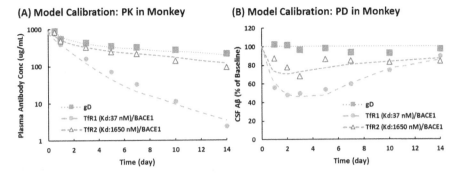

Fig. 12.13 Calibration of PKPD model developed for bispecific anti-TfR/BACE1 antibody using monkey

(**a**) Plasma PK profile of antibodies. Higher plasma clearance of strong anti-TfR binding antibody can be well characterized by the model. (**b**) Amyloid beta (A β) protein reduction in CSF. gD = control antibody, TfR1/BACE1 = bispecific anti-TfR/BACE1 antibody with strong binding TfR affinity, TfR2/BACE1 = bispecific antibody with lower TfR binding affinity (Kanodia et al. 2016)

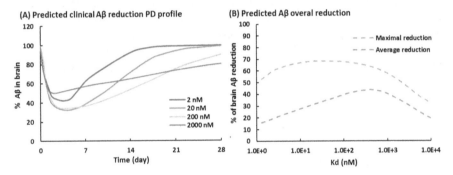

Fig. 12.14 Prediction of clinical amyloid beta (A β) reduction in CNS with new anti-TfR treatments

(**a**) PD profile following single i.v. administration of antibodies with four different anti-TfR K_D values. (**b**) A β protein overall reduction for 4 weeks (Kanodia et al. 2016)

disease-specific versions of the model need to be developed, as disease may influence physiology in a disease-specific manner (Gustafsson et al. 2019). Such efforts are currently under way.

While the drug transport mechanism in CNS has been well characterized for small molecules, our current understanding of how antibody and/or protein therapeutics are delivered into the CNS is gradually growing with considerable efforts in the field. At the same time, many delivery strategies have been developed to enhance the brain exposure of protein therapeutics, which dramatically makes protein therapeutic a viable option for CNS disorders. As an increasing attention on developing novel protein therapeutics for CNS disorders, an advanced CNS PKPD model that accounts for novel delivery mechanism and is generalized to different sizes of protein therapeutics will be required in a near future.

12.7   Challenges

The development of a PKPD model to predict the CNS drug disposition and the efficacy in humans on the basis of preclinical data requires 1) correct assumptions on the mechanism of drug absorption, distribution, metabolism, and elimination (ADME), in the whole body and in the CNS; 2) adequate datasets to calibrate the model; 3) sufficient physiological information to build a more mechanistic model; 4) the PKPD relationship; 5) the change of the states in disease conditions; and 6) identified correlation among different species.

The ADME is the foundation for predicting the drug exposure at the site-of-action and determining a proper PK model structure. Missing information about ADME processes may significantly reduce the accuracy when scaling up the model from animals to humans. For example, if an efflux transport or a potential protein binding has not been identified before the model development, the model will not be able to address those saturable processes at different doses and under different protein expression levels between the animal and the human. The estimation of model parameters requires adequate data to support. The more complex a model is to build, the more sample collection sites should be included in a study. The quantification of unbound drug concentration at the site-of-action is very crucial when our aim is to capture the in vivo PKPD relationship in the CNS since the correlation between the unbound plasma concentration and unbound brainECF concentration is not simply proportional. The binding protein and drug transporter often play a role here. However, it should be noted that the approach for measuring unbound drug concentration in the brainECF is highly limited in both preclinical and clinical studies. Microdialysis and brain capillary depletion method (Triguero et al. 1990) have been used to quantify the drug disposition in the brainECF and brain parenchyma. With increasing complexity of experiments, the chance of failures will increase. Thus, from the people perspective, performing advanced surgeries, complex experimentation, and the use of apparatus needed for monitoring techniques can only be performed by well-trained and skilled persons. CSF sampling may be available in the clinical setting to present a surrogate for the CNS drug exposure, but the collected data tends to be sparse. The validation of the model is challenging. Without a rich dataset to evaluate the drug exposure in the CNS in humans, the interpretation of failing to meet the primary endpoint also becomes hard to draw a conclusion: if the dose should be increased or if the target of drug is wrong.

The fluid movement in the CNS may drive drug regional distribution, especially for macromolecules with low diffusion coefficients. For predicting antibody disposition in different regions of the brain, associated physiological parameters should be determined (e.g., the flow rates of the brainECF, CSF, and lymphatic flow; the volumes of brain ventricles and meningeal lymphatic vessels) in different species.

Since biological systems operate at different set points in the body under different conditions, the ability to predict drug effects under a variety of circumstances is important, and more advanced experimental designs are needed to decipher and learn more on the factors that govern plasma pharmacokinetics, BBB transport, intra-brain distribution as well as their interrelationships, and consequences for CNS effects in different settings (Garrido et al. 2000; Grime 2006; Gabrielsson and Green 2009; Danhof et al. 2007, 2008; Ploeger et al. 2009; Kenakin 2011).

12.8 Conclusions

We have to accept that CNS drug delivery and CNS disease research are complex, and we need to (continue to) put efforts in performing the type of investigations that provide data that we learn from in having a CNS drug "at the right place, at the right time, and at the right concentration."

The mathematical modeling approach can help us compile all our prior knowledge of drug disposition in the CNS, and the PKPD relationship, to predict therapeutic responses under new and different doses, routes of administration, and even different properties of drug. These kinds of models can bridge the information between the preclinical and clinical setting and not only help in optimizing the dosage regimen based on preclinical studies but can also guild the design of better drug candidates that can treat CNS disorders with fewer side effects. It is envisioned that integration of PKPD modeling and simulation approaches in CNS drug development and clinical studies would result in the need for fewer individuals and less samples per individual to establish the proof-of-concept in humans.

12.9 Points for Discussion

- The body is a total system in which processes are interdependent. Studies need to be designed such that mutual dependence gets clear. How can studies be best designed to have the most valuable data collected?
- What concentrations in humans can be assessed and used best to predict CNS target site concentrations?
- Can we address sources of variability between drug responses in human populations, aiming at personalized CNS medicine?

References

Abbott NJ, Patabendige AA, Dolman DE, Yusof SR, Begley DJ (2010) Structure and function of the blood-brain barrier. Neurobiol Dis 37(1):13–25

Abbott NJ, Rönnbäck L, Hansson E (2006) Astrocyte-endothelial interactions at the blood-brain barrier. Nat Rev Neurosci 7:41–53

Abuqayyas L, Balthasar JP (2013) Investigation of the role of FcγR and FcRn in mAb distribution to the brain. Mol Pharm 10(5):1505–1513

Aspelund A, Antila S, Proulx ST, Karlsen TV, Karaman S, Detmar M, Wiig H, Alitalo K (2015) A dural lymphatic vascular system that drains brain interstitial fluid and macromolecules. J Exp Med 212:991–999

Bacskai BJ, Kajdasz ST, Christie RH, Carter C, Games D, Seubert P, Schenk D, Hyman BT (2001) Imaging of amyloid-β deposits in brains of living mice permits direct observation of clearance of plaques with immunotherapy. Nat Med 7:369–372

Banks WA, Kastin AJ (1996) Passage of peptides across the blood-brain barrier: pathophysiological perspectives. Life Sci 59(23):1923–1943

Banks WA, Terrell B, Farr SA, Robinson SM, Nonaka N, Morley JE (2002) Passage of amyloid β protein antibody across the blood–brain barrier in a mouse model of Alzheimer's disease. Peptides 23(12):2223–2226

Bard F, Cannon C, Barbour R, Burke RL, Games D, Grajeda H, Guido T, Hu K, Huang J, Johnson-Wood K et al (2000) Peripherally administered antibodies against amyloid β-peptide enter the central nervous system and reduce pathology in a mouse model of Alzheimer's disease. Nat Med 6:916–919

Bell RD, Zlokovic BV (2009) Neurovascular mechanisms and blood-brain barrier disorder in Alzheimer's disease. Acta Neuropathol 118(1):103–113

Ben Jonathan N, LaPensee CR, LaPensee EW (2008) What Can We Learn from Rodents about Prolactin in Humans? Endocr Rev 29:1–41

Bengtsson J, Ederoth P, Ley D, Hansson S, Amer-Wåhlin I, Hellström-Westas L, Marsál K, Nordström CH, Hammarlund-Udenaes M (2009) The influence of age on the distribution of morphine and morphine-3-glucuronide across the blood-brain barrier in sheep. Br J Pharmacol 157(6):1085–1096

Bernacki J, Dobrowolska A, Nierwińska K, Małecki A (2008) Physiology and pharmacological role of the blood-brain barrier. Pharmacol Rep 60:600–622

Black J, Leff P (1983) Operational model of pharmacological agonism. Proc R Soc Lond B 220:141–162

Bien-Ly N, Yu YJ, Bumbaca D, Elstrott J, Boswell CA, Zhang Y, Luk W, Lu Y, Dennis MS, Weimer RM et al (2014) Transferrin receptor (TfR) trafficking determines brain uptake of TfR antibody affinity variants. J Exp Med 211:233–244

Bouw MR, Ederoth P, Lundberg J, Ungerstedt U, Nordstrom CH, Hammarlund-Udenaes M (2001) Increased blood-brain barrier permeability of morphine in a patient with severe brain lesions as determined by microdialysis. Acta Anest Scand 45:390–392

Broadwell RD, Salcman M (1981) Expanding the definition of the blood-brain barrier to protein. Proc Natl Acad Sci U S A 72(20):7820–7824

Caram-Salas N, Boileau E, Farrington GK, Garber E, Brunette E, Abulrob A, Stanimirovic D (2011) In vitro and in vivo methods for assessing FcRn-mediated reverse transcytosis across the blood–brain barrier. Methods in Molecular Biology Permeability Barrier:383–401

Chang HY, Morrow K, Bonacquisti E, Zhang W, Shah DK (2018) Antibody pharmacokinetics in rat brain determined using microdialysis. MAbs 10(6):843–853

Chang HY, Wu S, Meno-Tetang G, Shah DK (2019) A translational platform PBPK model for antibody disposition in the brain. J Pharmacokinet Pharmacodyn 46(4):319–338

Cleton A, Odman J, Van der Graaf PH, Ghijsen W, Voskuyl R, Danhof M (2000) Mechanism-based modeling of functional adaptation upon chronic treatment with midazolam. Pharm Res 17:321–327

Cooper PR, Ciambrone GJ, Kliwinski CM, Maze E, Johnson L, Li Q, Feng Y, Hornby PJ (2013) Efflux of monoclonal antibodies from rat brain by neonatal Fc receptor. FcRn Brain Res 1534:13–21

Couch JA, Yu YJ, Zhang Y, Tarrant JM, Fuji RN, Meilandt WJ, Solanoy H, Tong RK, Hoyte K, Luk W et al (2013) Addressing safety liabilities of TfR bispecific antibodies that cross the blood-brain barrier. Sci Transl Med 5:1–12

Cox EH, Kerbusch T, van der Graaf PH, Danhof M (1998) Pharmacokinetic-pharmacodynamic modeling of the electroencephalogram effect of synthetic opioids in the rat. Correlation with binding at the mu-opioid receptor. J Pharmacol Exp Ther 284:1095–1103

Cserr HF, Bundgaard M (1984) Blood-brain interfaces in vertebrates: a comparative approach. Am J Physiol 246:R277–R288

Danhof M, Alvan G, Dahl SG, Kuhlmann J, Paintaud G (2005) Mechanism-based pharmacokinetic-pharmacodynamic modeling-a new classification of biomarkers. Pharm Res 22(9):1432–1437. https://doi.org/10.1007/s11095-005-5882-3

Danhof M, de Jongh J, de Lange ECM, Della Pasqua OE, Ploeger BA, Voskuyl RA (2007) Mechanism-based pharmacokinetic-pharmacodynamic modeling: biophase distribution, receptor theory, and dynamical systems analysis. Annu Rev Pharmacol Toxicol 47:357–400

Danhof M, de Lange EC, Della Pasqua OE, Ploeger BA, Voskuyl RA (2008) Mechanism-based pharmacokinetic-pharmacodynamic (PKPD) modeling in translational drug research. Trends Pharmacol Sci 29:186–191

Davies B, Morris T (1993) Physiological parameters in laboratory animals and humans. Pharm Res 10:1093–1095

Davis TP, Preston JE, Abbott J, Begley DJ (2014) Chapter five - transcytosis of macromolecules at the blood–brain barrier. In: Pharmacology of the blood brain barrier: targeting CNS disorders, vol 71. Elsevier, Waltham, MA, pp 147–163

de Lange EC, Bouw MR, Mandema JW, Danhof M, de Boer AG, Breimer DD (1995) Application of intracerebral microdialysis to study regional distribution kinetics of drugs in rat brain. Br J Pharmacol 116(5):2538–2544

de Lange EC, Ravenstijn PG, Groenendaal D, van Steeg TJ (2005) Toward the prediction of CNS drug-effect profiles in physiological and pathological conditions using microdialysis and mechanism-based pharmacokinetic-pharmacodynamic modeling. AAPS J 7(3):E532–43. https://doi.org/10.1208/aapsj070354

De Lange ECM, Hesselink MB, Danhof M, De Boer AG, Breimer DD (1995) The use of intracerebral microdialysis to determine changes in blood-brain barrier transport characteristics. Pharm Res 12:129–133

De Lange ECM (2013a) The mastermind approach to CNS drug therapy: translational prediction of human brain distribution, target site kinetics, and therapeutic effects. Fluid Barrier CNS 10:12

De Lange ECM (2013b) Utility of CSF in translational neuroscience. J Pharmacokinet Pharmacodyn 40(3):315–326

Dedrick RL, Bisschoff KB (1980) Species similarities in pharmacokinetics. Fed Proc 39:54–59

Del Bigio MR (1995) The ependyma: a protective barrier between brain and cerebrospinal fluid. Glia 14(1):1–13. https://doi.org/10.1002/glia.440140102

Fenstermacher JD, Patlak CS, Blasberg RG (1974) Transport of material between brain extracellular fluid, brain cells and blood. Fed Proc 33:2070–2074

Freskgård P-O, Niewoehner J, Urich E (2014) Time to open the blood–brain barrier gate for biologics? Future Neurol 9:243–245

Freskgård P-O, Urich E (2017) Antibody therapies in CNS diseases. Neuropharmacology 120:38–55

Furchgott RF (1966) The use of β-haloalkylamines in the differentiation of receptors and in the determination of dissociation constants of receptor-agonist complexes. Adv Drug Res 3:21–55

Gabrielsson J, Green AR (2009) Quantitative pharmacology or pharmacokinetic pharmacodynamic integration should be a vital component in integrative pharmacology. J Pharmacol Exp Ther 331:767–774

Gadkar K, Yadav DB, Zuchero JY, Couch JA, Kanodia J, Kenrick MK, Atwal JK, Dennis MS, Prabhu S, Watts RJ et al (2016) Mathematical PKPD and safety model of bispecific TfR/BACE1 antibodies for the optimization of antibody uptake in brain. Eur J Pharm Biopharm 101:53–61

Garg A, Balthasar JP (2009) Investigation of the influence of FcRn on the distribution of IgG to the brain. AAPS J 11:553–557

Garrido M, Gubbens-Stibbe J, Tukker E, Cox E, von Frijtag J, Künzel DM, Ijzerman A, Danhof M, Van der Graaf PH (2000) Pharmacokinetic-pharmacodynamic analysis of the EEG effect of alfentanil in rats following beta-funaltrexamine-induced mu-opioid receptor "knockdown" in vivo. Pharm Res 17:653–659

Gazzin S, Strazielle N, Tiribelli C, Ghersi-Egea JF (2012) Transport and metabolism at blood-brain interfaces and in neural cells: relevance to bilirubin-induced encephalopathy. Front Pharmacol 18;3:89. https://doi.org/10.3389/fphar.2012.00089

Geerts H, Roberts P, Spiros A (2015) Assessing the synergy between cholinomimetics and memantine as augmentation therapy in cognitive impairment in schizophrenia. A virtual human patient trial using quantitative systems pharmacology. Front Pharmacol 6:198

Glassman P, Balthasar JP (2016) Physiologically-based pharmacokinetic modeling to predict the clinical pharmacokinetics of monoclonal antibodies. J Pharmacokinet Pharmacodyn 43(4):427–446

Girod J, Fenart L, Regina A, Dehouck M, Hong G, Scherrmann J, Cecchelli R, Roux F (1999) Transport of cationized anti-tetanus Fab2 fragments across an in vitro blood-brain barrier model: involvement of the transcytosis pathway. J Neurochem 73:2002–2008

Grime K, Riley RJ (2006) The impact of in vitro binding on in vitro-in vivo extrapolations, projections of metabolic clearance and clinical drug-drug interactions. Curr Drug Metab 7(3):251–264

Groenendaal D, Freijer J, de Mik D, Bouw MR, Danhof M, De Lange EC (2007) Population pharmacokinetic modelling of non-linear brain distribution of morphine: influence of active saturable influx and P-glycoprotein mediated efflux. Br J Pharmacol 151(5):701–712

Groenendaal D, Freijer J, Rosier A, de Mik D, Nicholls G, Hersey A, Ayrton AD, Danhof M, de Lange EC (2008) Pharmacokinetic/pharmacodynamic modelling of the EEG effects of opioids: the role of complex biophase distribution kinetics. Eur J Pharm Sci 34(2-3):149–163

Gustafsson S, Sehlin D, Lampa E, Hammarlund-Udenaes M, Loryan I (2019) Heterogeneous drug tissue binding in brain regions of rats, Alzheimer's patients and controls: impact on translational drug development. Sci Rep 9(1):5308

Hammarlund-Udenaes M, Fridén M, Syvänen S, Gupta A (2008) On the rate and extent of drug delivery to the brain. Pharm Res 25:1737–1750

Hammarlund-Udenaes M (2009) Active-site concentrations of chemicals – are they a better predictor of effect than plasma/organ/tissue concentrations? Basic Clin Pharmacol Toxicol 106:215–220

Haqqani AS, Thom G, Burrell M, Delaney CE, Brunette E, Baumann E, Sodja C, Jezierski A, Webster C, Stanimirovic DB (2018) Intracellular sorting and transcytosis of the rat transferrin receptor antibody OX26 across the blood-brain barrier in vitro is dependent on its binding affinity. J Neurochem 146:735–752

Herve F, Ghinea N, Scherrmann J-M (2008) CNS delivery via adsorptive transcytosis. The AAPS J 10(3):455–472

Iliff JJ, Wang M, Liao Y, Plogg BA, Peng W, Gundersen GA, Benveniste H, Vates GE, Deane R, Goldman SA, Nagelhus EA, Nedergaard M (2012) A paravascular pathway facilitates CSF flow through the brain parenchyma and the clearance of interstitial solutes, including amyloid beta. Sci Transl Med 4:147ra111

Jadhav SB, Khaowroongrueng V, Fueth M, Otteneder MB, Richter W, Derendorf H (2017) Tissue distribution of a therapeutic monoclonal antibody determined by large pore microdialysis. J Pharm Sci 106:2853–2859

Kanodia J, Gadkar K, Bumbaca D, Zhang Y, Tong R, Luk W, Hoyte K, Lu Y, Wildsmith KR, Couch JA et al (2016) Prospective design of anti-transferrin receptor bispecific antibodies for optimal delivery into the human brain. CPT Pharmacometrics Syst Pharmacol 5:283–291

Kenakin T, Christopoulos A (2011) Analytical pharmacology: the impact of numbers on pharmacology. Trends Pharmacol Sci 32(4):189–196

Kenakin T (2004) Principles: receptor theory in pharmacology. Trends Pharmacol Sci 25(4):186–192

Kozlowski GP (1992) Localization patterns for immunoglobulins and albumins in the brain suggest diverse mechanisms for their transport across the blood-brain barrier (BBB). Prog Brain Res 91C:149–154

Kumagai AK, Eisenberg JB, Pardridge WM (1987) Absorptive-mediated endocytosis of cationized albumin and a beta-endorphin-cationized albumin chimeric peptide by isolated brain capillaries. Model system of blood-brain barrier transport. J Biol Chem 262(31):15214–15219

Kvernmo T, Houben J, Sylte I (2008) Receptor-binding and pharmacokinetic properties of dopaminergic agonists. Curr Top Med Chem 8(12):1049–1067

Lajoie JM, Shusta EV (2015) Targeting receptor-mediated transport for delivery of biologics across the blood-brain barrier. Annu Rev Pharmacol Toxicol 55:613–631

Liu X, Smith BJ, Chen C, Callegari E, Becker SL, Chen X, Cianfrogna J, Doran AC, Doran SD, Gibbs JPN et al (2005) Use of a physiologically based pharmacokinetic model to study the time to reach brain equilibrium: an experimental analysis of the role of blood–brain barrier permeability, plasma protein binding, and brain tissue binding. J Pharmacol Exp Ther 313:1254–1262

Loryan I, Melander E, Svensson M, Payan M, König F, Jansson B, Hammarlund-Udenaes M (2016) In-depth neuropharmacokinetic analysis of antipsychotics based on a novel approach to estimate unbound target-site concentration in CNS regions: link to spatial receptor occupancy. Mol Psychiatry 21(11):1527–1536

Loryan I, Sinha V, Mackie C, Van Peer A, Drinkenburg W, Vermeulen A, Morrison D, Monshouwer M, Heald D, Hammarlund-Udenaes M (2014) Mechanistic understanding of brain drug disposition to optimize the selection of potential neurotherapeutics in drug discovery. Pharm Res 31(8):2203–2219

Louveau A, Smirnov I, Keyes TJ, Eccles JD, Rouhani SJ, Peske JD, Derecki NC, Castle D, Mandell JW, Lee KS et al (2015) Structural and functional features of central nervous system lymphatic vessels. Nature 523:337–341

Martín-García E, Mannara F, Gutiérrez-Cuesta J, Sabater L, Dalmau J, Maldonado R, Graus F (2013) Intrathecal injection of P/Q type voltage-gated calcium channel antibodies from paraneoplastic cerebellar degeneration cause ataxia in mice. J Neuroimmunol 261:53–59

Neves V, Aires-Da-Silva F, Corte-Real S, Castanho MA (2016) Antibody approaches to treat brain diseases. Trends Biotechnol 34:36–48

Niewoehner J, Bohrmann B, Collin L, Urich E, Sade H, Maier P, Rueger P, Stracke JO, Lau W, Tissot AC et al (2014) Increased brain penetration and potency of a therapeutic antibody using a monovalent molecular shuttle. Neuron 81:49–60

Okuyama T, Sakai N, Yamamoto T, Yamaoka M (2018) Tomio T. Novel blood-brain barrier delivery system to treat CNS in MPS II: First clinical trial of anti-transferrin receptor antibody fused enzyme therapy. Mol Genet Metab 123(2):S109

Panza F, Lozupone M, Logroscino G, Imbimbo BP (2019) A critical appraisal of amyloid-β-targeting therapies for Alzheimer disease. Nat Rev Neurol 15:73–88

Pardridge WM (2016) CSF, blood-brain barrier, and brain drug delivery. Expert Opin Drug Deliv 13(7):963–975

Pardridge WM, Buciak JL, Friden PM (1991) Selective transport of an anti-transferrin receptor antibody through the blood-brain barrier in vivo. J Pharmacol Exp Ther 259(1):66–70

Pizzo ME, Wolak DJ, Kumar NN, Brunette E, Brunnquell CL, Hannocks M, Abbott NJ, Meyerand ME, Sorokin L, Stanimirovic DB, Thome RG (2018) Intrathecal antibody distribution in the rat brain: surface diffusion, perivascular transport and osmotic enhancement of delivery. J Physiol 596(3):445–475

Ploeger BA, van der Graaf PH, Danhof M (2009) Incorporating receptor theory in mechanism-based pharmacokinetic-pharmacodynamic (PK-PD) modeling. Drug Metab Pharmacokinet 24:3–15

Reese TS, Karnovsky MJ (1967) Fine structural localization of a bloodbrain barrier to exogenous peroxidase. J Cell Biol 34:207–217

Rowland M, Peck C, Tucker G (2011) Physiologically-based pharmacokinetics in drug development and regulatory science. Annu Rev Pharmacol Toxicol 51:45–73

Sade H, Baumgartner C, Hugenmatter A, Moessner E, Freskgård P, Niewoehner J (2014) A human blood-brain barrier transcytosis assay reveals antibody transcytosis influenced by pH-dependent receptor binding. PLoS One 9(4):e96340

Schmidt S, Gonzalez D, Derendorf H (2010) Significance of protein binding in pharmacokinetics and pharmacodynamics. J Pharm Sci 99(3):1107–1122

Segal MB (1993) Extracellular and cerebrospinal fluids. J Inherit Metab Dis 16:617–638

Shawahna R, Uchida Y, Declèves X, Ohtsuki S, Yousif S, Dauchy S, Jacob A, Chassoux F, Daumas-Duport C, Couraud PO et al (2011) Transcriptomic and quantitative proteomic analysis of transporters and drug metabolizing enzymes in freshly isolated human brain microvessels. Mol Pharm 8(4):1332–1341

Spector R, Spector AZ, Snodgrass SR (1977) Model for transport in the central nervous system. Am J Physiol 1:R73–R79

St-Amour I, Paré I, Alata W, Coulombe K, Ringuette-Goulet C, Drouin-Ouellet J, Vandal M, Soulet D, Bazin R, Calon F (2013) Brain bioavailability of human intravenous immunoglobulin and its transport through the murine blood–brain barrier. J Cereb Blood Flow Metab 33:1983–1992

Stevens J, Ploeger B, Hammarlund-Udenaes M, Osswald G, Graaf PH, Danhof M, de Lange ECM (2012) Mechanism-based PK–PD model for the prolactin biological system response following an acute dopamine inhibition challenge: quantitative extrapolation to humans. J Pharmacokinet Pharmacodyn 39(5):463–477

Strazielle N, Ghersi-Egea JF (2013) Physiology of blood-brain interfaces in relation to brain disposition of small compounds and macromolecules. Mol Pharm 10(5):1473–1491

Syvänen S, Lindhe Ö, Palner M, Kornum BR, Rahman O, Långström B, Knudsen GM, Hammarlund-Udenaes M (2009) Species differences in blood-brain barrier transport of three positron emission tomography radioligands with emphasis on P-glycoprotein transport. Drug Metab Dispos 37:635–643

Syvänen S, Hultqvist G, Gustavsson T, Gumucio A, Laudon H, Söderberg L, Ingelsson M, Lannfelt L, Sehlin D (2018) Efficient clearance of Aβ protofibrils in AβPP-transgenic mice treated with a brain-penetrating bifunctional antibody. Alzheimer Res Therap 10(49):1–10

Thom G, Burrell M, Haqqani AS, Yogi A, Lessard E, Brunette E, Delaney C, Baumann E, Callaghan D, Rodrigo N et al (2018) Enhanced delivery of Galanin conjugates to the brain through bioengineering of the anti-transferrin receptor antibody OX26. Mol Pharm 15:1420–1431

Triguero D, Buciak J, Pardridge WM (1990) Capillary depletion method for quantification of blood-brain barrier transport of circulating peptides and plasma proteins. J Neurochem 54(6):1882–1888

Tuk B, van Gool T, Danhof M (2002) Mechanism-based pharmacodynamic modeling of the interaction of midazolam, bretazenil, and zolpidem with ethanol. J Pharmacokinet Pharmacodyn 29(3):235–250

Tuk B, van Oostenbruggen MF, Herben VM, Mandema JW, Danhof M (1999) Characterization of the pharmacodynamic interaction between parent drug and active metabolite in vivo: midazolam and alpha-OH-midazolam. J Pharmacol Exp Ther 289(2):1067–1074

Turksen K, Caram-Salas N, Boileau E, Farrington GK, Garber E, Brunette E, Abulrob A, Stanimirovic D (2011) In vitro and in vivo methods for assessing FcRn-mediated reverse transcytosis across the blood–brain barrier. In: Permeability barrier: methods and protocols, vol 763. Humana Press/Springer, New York, pp 383–401

Van den Brink W, van den Berg DJ, Bonsel F, Hartman R, Wong YC, van der Graaf PH, De Lange ECM (2019) Blood-based biomarkers of quinpirole pharmacology: multivariate PK/PD and metabolomics to unravel the underlying dynamics in plasma and brain. CPT Pharmacometrics Syst Pharmacol 8(2):107–117

van den Brink WJ, Elassais-Schaap J, Gonzalez B, Harms A, van der Graaf PH, Hankemeier T, de Lange ECM (2017) Remoxipride causes multiple pharmacokinetic/pharmacodynamic response patterns in pharmacometabolomics in rats. Eur J Pharm Sci 109:431–440

Van den Brink WJ, van den Berg DJ, Bonsel FEM, Hartman R, Wong YC, van der Graaf PH, de Lange ECM (2018) Fingerprints of CNS drug effects: a plasma neuroendocrine reflection of D2 receptor activation using multi-biomarker pharmacokinetic/pharmacodynamic modelling. Br J Pharmacol 175(19):3832–3843

Van der Graaf PH, Danhof M (1997a) Analysis of drug-receptor interactions in vivo: a new approach in pharmacokinetic-pharmacodynamic modelling. Int J Clin Pharmacol Ther 35:442–446

Van der Graaf PH, Danhof M (1997b) On the reliability of affinity and efficacy estimates obtained by direct operational model fitting of agonist concentration-effect curves following irreversible receptor inactivation. J Pharmacol Toxicol Methods 38(2):81–85

Van Schaick EA, Tukker HE, Roelen HCPF, IJzerman AP, Danhof M (1998) Selectivity of action of 8-alkylamino analogues of N6-cyclopentyladenosine in vivo: haemodynamic versus anti-lipolytic responses in rats. Br J Pharmacol 124(3):607–618

Villaseñor R, Ozmen L, Messaddeq N, Grüninger F, Loetscher H, Keller A, Betsholtz C, Freskgard P-O, Collin L (2016) Trafficking of endogenous immunoglobulins by endothelial cells at the blood-brain barrier. Sci Rep 6:25658

Visser SA, Wolters FL, Gubbens-Stibbe JM, Tukker E, Van Der Graaf PH, Peletier LA, Danhof M (2003) Mechanism-based pharmacokinetic/pharmacodynamic modeling of the electroencephalogram effects of GABAA receptor modulators: in vitro-in vivo correlations. J Pharmacol Exp Ther 304(1):88–101

Vlot AHC, Witte WEA, Danhof M, van der Graaf PH, van Westen GJP, de Lange ECM (2017) Target and tissue selectivity prediction by integrated mechanistic pharmacokinetic-target binding and quantitative structure activity modelling. AAPSJ 20(1):11

Watson J, Wright S, Lucas A, Clarke KL, Viggers J, Cheetham S, Jeffrey P, Porter R, Read KD (2009) Receptor occupancy and brain free fraction. Drug Metab Dispos 37:753–760

Watts RJ, Dennis MS (2013) Bispecific antibodies for delivery into the brain. Curr Opin Chem Biol 17:393–399

Westerhout J, Danhof M, De Lange EC (2011) Preclinical prediction of human brain target site concentrations: considerations in extrapolating to the clinical setting. J Pharm Sci 100(9):3577–3593. https://doi.org/10.1002/jps.22604

Westerhout J, Ploeger B, Smeets J, Danhof M, de Lange ECM (2012) Physiologically based pharmacokinetic modeling to investigate regional brain distribution kinetics in rats. AAPS J 14(3):543–553

Westerhout J, Smeets J, Danhof M, de Lange ECM (2013) The impact of P-gp functionality on non-steady state relationships between CSF and brain extracellular fluid. J Pharmacokinet Pharmacodyn 40:327–342

Westerhout J, van den Berg DJ, Hartman R, Danhof M, de Lange ECM (2014) Prediction of methotrexate CNS distribution in different species and the influence of disease conditions. Eur J Pharm Sci 57:11–24

Wolak DJ, Pizzo ME, Thorne RG (2015) Probing the extracellular diffusion of antibodies in brain using in vivo integrative optical imaging and ex vivo fluorescence imaging. J Control Release 197:78–86

Yadav DB, Maloney JA, Wildsmith KR, Fuji RN, Meilandt WJ, Solanoy H, Lu Y, Peng K, Wilson B, Chan P et al (2017) Widespread brain distribution and activity following i.c.v. infusion of anti-β-secretase (BACE1) in nonhuman primates. Br J Pharmacol 174(22):4173–4185

Yamamoto Y, Danhof M, de Lange EC (2017a) Microdialysis: the key to physiologically based model prediction of human CNS target site concentrations. AAPS J 19(4):891–909

Yamamoto Y, Välitalo PA, Huntjens DR, Proost JH, Vermeulen A, Krauwinkel W, Beukers MW, van den Berg DJ, Hartman RH, Wong YC et al (2017b) Predicting drug concentration-time profiles in multiple CNS compartments using a comprehensive physiologically-based pharmacokinetic model. CPT Pharmacometrics Syst Pharmacol 6(11):765–777

Yamamoto Y, Välitalo PA, Wong YC, Huntjens DR, Proost JH, Vermeulen A, Krauwinkel W, Beukers MW, van den Berg DJ, Hartman RH et al (2018) Prediction of human CNS pharmacokinetics using a physiologically-based pharmacokinetic modeling approach. Eur J Pharm Sci 112:168–179

Yassen A, Olofsen E, Kan J, Dahan A, Danhof M (2007) Animal-to-human extrapolation of the pharmacokinetic and pharmacodynamic properties of buprenorphine. Clin Pharmacokinet 46:433–447

Yu YJ, Zhang Y, Kenrick M, Hoyte K, Luk W, Lu Y, Atwal J, Elliott JM, Prabhu S, Watts RJ et al (2011) Boosting brain uptake of a therapeutic antibody by reducing its affinity for a transcytosis target. Sci Transl Med 3:84ra44

Zhang Y, Pardridge WM (2001) Mediated efflux of IgG molecules from brain to blood across the blood–brain barrier. J Neuroimmunol 114:168–172

Zlokovic BV (2010) Neurodegeneration and the neurovascular unit. Nat Med 16(12):1370–1371. https://doi.org/10.1038/nm1210-1370

Zuideveld KP, van der Graaf PH, Peletier LA, Danhof M (2007) Allometric scaling of pharmacodynamic responses: application to 5-Ht1A receptor mediated responses from rat to man. Pharm Res 24:2031–2039

Part III
Industrial Approaches for Investigation of Potential CNS Drugs

Chapter 13
Drug Discovery Methods for Studying Brain Drug Delivery and Distribution

Irena Loryan and Margareta Hammarlund-Udenaes

Abstract Methods used in drug discovery laboratories for assessing the delivery of small molecules to the brain have changed significantly in recent years. There is now more focus on measuring or estimating target unbound drug concentrations in the brain and evaluating the quantitative aspects of drug transport across the blood-brain barrier (BBB). The techniques for the investigation of the rate and extent of BBB transport of new chemical entities (NCEs) are discussed in this chapter. Combinatory methodology for rapid mapping of the extent of brain drug delivery via assessment of the unbound drug brain partitioning coefficient is presented. The chapter also explains the procedures for approximation of subcellular distribution of NCEs, particularly into the lysosomes. The principles, technical issues, advantages, and potential applications of techniques for evaluation of intra-brain distribution, i.e., equilibrium dialysis-based brain homogenate and brain slice methods, are described. The assessment of the extent of BBB transport and intracellular distribution of NCEs, the identification of intra-brain distribution patterns, and their integration with pharmacodynamic measurements are valuable implements for candidate evaluation and selection in drug discovery and development.

Keywords Brain homogenate method · Brain slice method · Lysosomal trapping · $V_{u,brain}$ · Combinatory mapping approach · Translation

Abbreviations

A_{brain}	Amount of drug in brain tissue
$AUC_{tot,brain}$	Area under the total brain concentration-time curve
$AUC_{tot,plasma}$	Area under the total plasma concentration-time curve
BBB	Blood-brain barrier

I. Loryan (✉) · M. Hammarlund-Udenaes
Translational PKPD Research Group, Department of Pharmacy, Uppsala University, Uppsala, Sweden

SciLife Lab, Solna, Sweden
e-mail: Irena.loryan@farmaci.uu.se; mhu@farmaci.uu.se

© American Association of Pharmaceutical Scientists 2022
E. C. M. de Lange et al. (eds.), *Drug Delivery to the Brain*, AAPS Advances in the Pharmaceutical Sciences Series 33, https://doi.org/10.1007/978-3-030-88773-5_13

BCRP	Breast cancer resistance-associated protein
BCSFB	Blood-CSF barrier
CB	Cellular barrier
C_{buffer}	Concentration of compound in the buffer (brain slice method)
$C_{tot,blood}$	Total drug concentration in blood
CNS	Central nervous system
CSF	Cerebrospinal fluid
$C_{tot,plasma}$	Total drug concentration in plasma
$C_{tot,brain}$	Total drug concentration in brain
$C_{u,brainISF}$	Unbound drug concentration in brain interstitial fluid
$C_{u,cell}$	Unbound drug concentration in intracellular fluid
$C_{u,cyto}$	Unbound drug concentration in cytosol
$C_{u,lyso}$	Unbound drug concentration in lysosomes
$C_{u,plasma}$	Unbound drug concentration in plasma
DMPK	Drug metabolism and pharmacokinetics
ECF	Extracellular fluid (same as ISF)
ED	Equilibrium dialysis
ER	Efflux ratio
$f_{u,brain}$	Fraction of unbound drug in brain homogenate
$f_{u,brain,corrected}$	$f_{u,brain}$ corrected for pH partitioning into cells
$f_{u,hD}$	Fraction of unbound drug in diluted brain homogenate
$f_{u,plasma}$	Fraction of unbound drug in plasma
HTS	High-throughput screening
ICF	Intracellular fluid in the brain
ISF	Interstitial fluid in the brain
K_d	Equilibrium dissociation constant
$K_{p,brain}$	Ratio of total-brain-to-total plasma drug concentrations (also abbreviated as BB)
$K_{p,uu,brain}$	Ratio of brain ISF-to-plasma unbound drug concentrations
$K_{p,uu,cell}$	Ratio of brain ICF-to-ISF unbound drug concentrations
$K_{p,uu,cyto}$	Ratio of cytosolic-to-extracellular unbound drug concentrations
$K_{p,uu,lyso}$	Ratio of lysosomic-to-cytosolic unbound drug concentrations
LC-MS/MS	Liquid chromatography tandem mass spectrometry
logBB	Logarithm of $K_{p,brain}$ (BB)
MWCO	Molecular weight cut-off
NCE	New chemical entity
neuroPK	Neuropharmacokinetics
P_{app}	Unidirectional apparent permeability coefficient measured in the apical-to-basolateral direction (cm/s)
PBS	Phosphate-buffered saline
PD	Pharmacodynamics
PET	Positron emission tomography
P-gp	P-glycoprotein
PK	Pharmacokinetics
PLD	Drug-induced phospholipidosis

PS   Permeability surface area product (μL/min $\cdot$ g brain^{-1})
V_{ss} Apparent volume of distribution at steady state
$V_{u,brain}$ Volume of distribution of unbound drug in brain (mL $\cdot$ g brain^{-1})

13.1 Introduction

The existing situation in the discovery and development of drugs for CNS-related conditions is unprecedentedly desperate, in the face of enormous unmet medical need (Eaton et al. 2008; Schoepp 2011; Schwab and Buchli 2012; Butlen-Ducuing et al. 2016; Cummings et al. 2016). The probability of success with emerging breakthrough first-in-class CNS drugs is small. Further, because neurotherapeutic drugs move more slowly in the development pipeline (compared to, e.g., AIDS antivirals), they require a relatively extended time to get to the market (Kaitin and DiMasi 2011). Despite immense efforts from the drug industry and academia, it could be thought that CNS drug discovery is currently almost in a blind alley. In contrast, however, Weaver and Weaver have used molecular modeling to reach the conclusion that the pharmaceutical industry is still in its infancy when it comes to exploring the neuroactive chemical space (Weaver and Weaver 2011). In addition, multiple pharmaceutical companies are on the way of the development of various biologicals including antibodies for treatment of neurological diseases (Farrington et al. 2014; Freskgard and Urich 2017; Stanimirovic et al. 2018).

The reasons for the apparent failure of CNS drug discovery, such as lack of clinically translatable animal disease models, lack of relevant biomarkers, and inadequate exposure of the CNS to potential drugs because of the blood-brain barrier (BBB), are generally acknowledged and are challenging to resolve (Jeffrey and Summerfield 2007; Hammarlund-Udenaes et al. 2008; Neuwelt et al. 2008; Kelly 2009; Reichel 2009; Abbott et al. 2010; Brunner et al. 2012; Mehta et al. 2017).

This chapter is dedicated to the quantitative aspects of drug transport across the BBB and contemporary methods of assessing CNS exposure to NCEs in drug discovery and development programs. From drug discovery perspectives, it is important to mention that the BBB per se is not the only obstacle to drug delivery to the brain. Inadequate understanding of the principles of drug transport at the BBB and a lack of appropriate interpretation of target exposure could also be seen as hindrances to progression (Hammarlund-Udenaes et al. 2009).

As explained in the article by Elebring and colleagues, it is becoming more and more imperative to separate and define two crucial aspects of drug discovery: efficacy (i.e., doing the right things) and efficiency (i.e., doing things right) (Elebring et al. 2012). In the modern pharmaceutical industry, we often observe the problems associated with "high-throughput" thinking (high efficiency) which typically biases biopharmaceutical scientists toward simple "one-fits-all" solutions. Alternatively, a tailored specific approach could be more effective. If this approach is to be applied to brain drug delivery, it is important initially to define what is meant by brain drug delivery and subsequently to identify the relevant core neuropharmacokinetic (neuroPK) parameters and applicable methods for the assessment of CNS exposure.

Because the novel strategies available for CNS drug delivery differ widely (invasive, noninvasive), the definitions of brain drug delivery, and consequently the choice of appropriate neuroPK parameters, are also divergent (Pardridge et al. 1992; Thorne et al. 1995; Begley 1996, 2004; Huwyler et al. 1996; Pardridge 1997; Li et al. 1999; Scherrmann 2002; Reichel et al. 2003; Garberg et al. 2005; Garcia-Garcia et al. 2005; Terasaki and Ohtsuki 2005; Pardridge 2006; de Boer and Gaillard 2007; Hammarlund-Udenaes et al. 2008; Wang et al. 2009; Gaillard et al. 2012; Stevens et al. 2012; de Lange 2013a). This chapter focuses on "classical" blood-to-brain delivery of small molecules, where drug delivery from the blood to the brain through the BBB can be described by *rate* and *extent* parameters (see Chap. 7, which discusses the pharmacokinetic concepts of brain drug delivery).

The *rate* of BBB transport is commonly characterized by the **permeability surface area product** (PS, mL/min/kg body weight). Being unidirectional, the PS describes the speed at which the drug enters the brain (Fenstermacher 1992; Tanaka and Mizojiri 1999; Gaillard and de Boer 2000; Summerfield et al. 2007; Liu et al. 2008; Zhao et al. 2009). Generally, fast permeation is a key requirement for drugs when rapid CNS onset is wanted, e.g., for general anesthetics and analgesics. Although only a limited number of compounds in a few pharmacological classes are required to permeate the brain quickly, the apparent BBB permeability (P_{app}; measured in vitro) is among the parameters considered by pharmaceutical industry to be essential for evaluation of BBB penetration in drug development programs (Liu et al. 2005; Jeffrey and Summerfield 2007; Summerfield et al. 2007). Moreover, combined with an in vitro P-glycoprotein (P-gp) assay, it is used as a basis for guiding the lead optimization and candidate selection (Di et al. 2012a). To make this point more explicit, it is worth mentioning that permeability-limited drug distribution in the brain (<10% of cerebral blood flow or logPS < −2.9) is a very rare phenomenon associated with a slow equilibration time in the brain and is not a matter of concern for potential CNS drugs intended for chronic administration (Abraham 2011; Kell et al. 2011, 2013; Deo et al. 2013). It is obvious that permeability as a test for BBB penetration is overpromoted in the pharmaceutical industry. The methods used for permeability measurements are not covered in this chapter, but are thoroughly discussed in Chaps. 7 and 8.

In the drug discovery setting, the *extent* of BBB transport is traditionally evaluated in rodents using the steady-state ratio of total-brain-to-total-plasma drug concentrations ($K_{p,brain}$, BB, or logBB). Many generations of CNS drug discovery programs have been driven by optimizing $K_{p,brain}$, which has led to mass production of CNS compounds with high lipophilicity and development of the phenomenon known as the "lipidization trap": higher lipophilicity-higher $K_{p,brain}$ value-higher brain tissue binding-lower fraction of unbound drug in the brain (Deo et al. 2013). Because it is affected by nonspecific binding of the drug to plasma proteins and brain tissue, $K_{p,brain}$ masks the actual BBB net flux value (Lin et al. 1982; Lin and Lin 1990; Kalvass and Maurer 2002; Summerfield et al. 2007; Wan et al. 2007; Hammarlund-Udenaes et al. 2008; Read and Braggio 2010; Friden et al. 2011;

Longhi et al. 2011). The use of $K_{p,brain}$ for optimizing novel neurotherapeutics has thus created further confusion in the field. In this regard, the steady-state **ratio of brain interstitial fluid (ISF) to plasma unbound drug concentrations** ($K_{p,uu,brain}$) is currently considered to be the most relevant measure of BBB function (Gupta et al. 2006; Jeffrey and Summerfield 2007; Hammarlund-Udenaes et al. 2008, 2009; Liu et al. 2009b; Reichel 2009; Read and Braggio 2010; Di et al. 2012a; Doran et al. 2012; Loryan et al. 2014, 2016; Schou et al. 2015).

$K_{p,uu,brain}$, the unbound drug brain partitioning coefficient, allows the assessment of the concentration of cerebral unbound drug, which is the main pharmacokinetic determinant of CNS activity of neurotherapeutics, based on a given plasma concentration (Harashima et al. 1984; Gupta et al. 2006; Kalvass et al. 2007a; Liu et al. 2009b; Watson et al. 2009; Hammarlund-Udenaes 2010; Bundgaard et al. 2012b). Thus far, cerebral microdialysis has been the "gold" standard for the measurement of unbound cerebral concentrations in the brains of animals and humans (Elmquist and Sawchuk 1997, 2000; Hammarlund-Udenaes et al. 1997; de Lange et al. 1999; Kitamura et al. 2016; Hammarlund-Udenaes 2017). However, the practice of microdialysis for evaluation of BBB penetration in a drug discovery setup is limited mainly due to extensive adsorption to plastic tubing and probe. Nevertheless, a clinically relevant picture of the extent of brain drug delivery can be achieved using the combinatory mapping approach (CMA, Fig. 13.1) by means of evaluation of

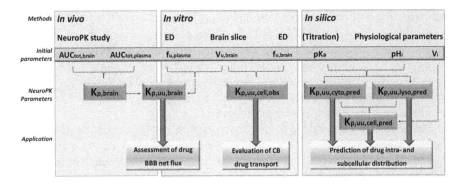

Fig. 13.1 An illustration of the combinatory mapping approach (CMA) in the form of a screening toolbox for the evaluation of unbound drug CNS exposure required for the selection of novel drug candidates. Figure obtained from Loryan et al. (2014). The platform comprising of in vivo, in vitro, and in silico toolboxes. Total drug brain and plasma exposure (e.g., by means of area under the curve of concentration-time profiles, $AUC_{tot,brain}$, and $AUC_{tot,plasma}$) determined in an in vivo neuroPK study is essential for the assessment of the brain partitioning coefficient $K_{p,brain}$. In vitro measurements of drug plasma and brain tissue binding properties using equilibrium dialysis (ED) and brain slice techniques are required for the estimation of $K_{p,uu,brain}$ and $K_{p,uu,cell}$ neuroPK parameters. Compound-specific pKa values in combination with the physiological estimates of pH (pHi) of the relevant compartments (i = plasma, interstitial fluid, cytosol, or lysosomes) are used for in silico calculation of drug subcellular distribution, i.e., $K_{p,uu,cyto,pred}$ and $K_{p,uu,lyso,pred}$. Physiological volumes (Vi) of interstitial fluid, cytosol, and lysosomes with $K_{p,uu,cyto,pred}$ and $K_{p,uu,lyso,pred}$ are used for the calculation of $K_{p,uu,cell,pred}$. Assessed neuroPK parameters in conjunction with relevant pharmacodynamics readouts are recommended to be used for evaluation and selection of novel drug candidates

pharmacokinetic (PK) parameters such as $K_{p,brain}$, the volume of distribution of unbound drug in the brain ($V_{u,brain}$), and the fraction of unbound drug in the plasma ($f_{u,plasma}$) (Chap. 7 and Sect. 13.5).

A very important element of brain drug disposition, although it is unrelated to the BBB, is the *intracerebral distribution* of the drug, which is discussed in Sects. 13.2, 13.3 and 13.4. Enhanced understanding of the distribution of the drug in the brain provides new perspectives on the pharmacodynamics of neurotherapeutics. Typically, brain tissue binding is measured as the **fraction of unbound drug in the brain** ($f_{u,brain}$) using an equilibrium dialysis (ED) technique to assess the extent of nonspecific binding to the brain tissue (Kalvass and Maurer 2002; Kalvass et al. 2007a; Wan et al. 2007; Friden et al. 2011; Longhi et al. 2011; Di et al. 2012b). The method is mainly assessing intracellular binding (Friden et al. 2007; 2011).

Alternatively, the **volume of distribution of unbound drug in the brain** ($V_{u,brain}$), estimated using the fresh brain slice method, can allow assessment of the overall uptake by the brain tissue (Kakee et al. 1997; Liu et al. 2006; Benkwitz et al. 2007; Friden et al. 2009a; Kodaira et al. 2011; Uchida et al. 2011a). In this chapter, we have chosen to express information from brain homogenate studies as $f_{u,brain}$ and information from brain slice studies as $V_{u,brain}$ to differentiate and clarify the information as much as possible. Both these parameters, $V_{u,brain}$ and $f_{u,brain}$, permit the estimation of the concentration of unbound drug in brain ISF ($C_{u,brainISF}$) using total brain concentration ($C_{tot,brain}$) measurements and give an indication of the probable extracellular target engagement. However, the intracellular concentration of unbound drug is also of great interest. In view of this, approximation of the **ratio of brain intracellular fluid (ICF) to ISF unbound drug concentrations** ($K_{p,uu,cell}$) may be beneficial for understanding the pharmacological query related to intracellular targets and may be strategically influential (Friden et al. 2007). The $K_{p,uu,cell}$ concept is innovative, as it provides the basis for an increased awareness of the impact of cellular barrier function on intracerebral drug distribution, which has hitherto been neglected in drug discovery programs.

The approaches applied for prediction, assessment, and optimization (Chap. 12) of the BBB transport of NCEs, such as in silico (Chap. 14), in vitro (Chaps. 8 and 9), and in vivo methods (Chaps. 10 and 11), depend on the development phase of the drug and the questions of interest.

13.2 The Brain Homogenate Method for $f_{u,brain}$

The concentration of unbound drug in the brain, estimated using $C_{tot,brain}$ corrected for brain tissue binding, is a surrogate for $C_{u,brainISF}$. $C_{u,brainISF}$ is currently considered to be the most relevant parameter for measuring the pharmacological response of neurotherapeutics (Bouw et al. 2001; Bostrom et al. 2006; Bundgaard et al. 2007, 2012b; Kalvass et al. 2007b; Liu et al. 2009b; Watson et al. 2009; Hammarlund-Udenaes 2010; Smith et al. 2010; Westerhout et al. 2011).

Brain tissue binding can be determined by various methods, including ED, stepwise ED, ultrafiltration, ultracentrifugation, gel filtration, and absorption by brain lipid membrane vesicles stabilized on silica beads (TRANSIL brain absorption kit) (Fichtl et al. 1991a; Kalvass and Maurer 2002; Mano et al. 2002; Vuignier et al. 2010; Longhi et al. 2011). This section focuses on the ED technique for the estimation of $f_{u,brain}$, which is presently used in drug discovery programs in a high-throughput manner.

13.2.1 Equilibrium Dialysis

In 2001, Kariv et al. presented the successful development of a 96-well equilibrium dialysis (ED) plate suitable for evaluation of plasma protein binding for large numbers of biologically active NCEs during high-throughput screening (HTS) (Kariv et al. 2001). Contemporary 96-well ED apparatus allows the researcher to examine a large number of samples, time points, or replicates in the same experiment.

Using a similar approach, Kalvass and Maurer introduced a high-throughput ED technique designed for the determination of brain tissue binding (Kalvass and Maurer 2002). The method rapidly became standard, and it is currently widely used for the estimation of $f_{u,brain}$ for a large number of chemically diverse compounds (Summerfield et al. 2006; Wan et al. 2007; Di et al. 2011; Friden et al. 2011; Longhi et al. 2011). The need for protein binding data in combination with the large number of compounds created from combinatorial chemistry has stimulated the development of a novel cassette-based pooling approach which allows simultaneous assessment of $f_{u,brain}$ or $f_{u,plasma}$ for more than five compounds per sample (Fung et al. 2003; Wan et al. 2007; Plise et al. 2010; Longhi et al. 2011).

Several research groups and pharmaceutical companies have validated the compatibility of the high-throughput ED techniques (96-, 48-well formats) with most standard laboratory supplies and robotics (Banker et al. 2003; van Liempd et al. 2011). Several devices based on a 96-well format are currently on the market (e.g., the Equilibrium Dialyzer-96 from Harvard Biosciences (Holliston, MA, USA), the Rapid Equilibrium Device from Thermo Scientific/Pierce (Rockford, IL, USA), and the Micro-Equilibrium Dialysis Device from HTdialysis LLC (Gales Ferry, CT, USA)).

13.2.1.1 Principles

The semipermeable membrane between the buffer and the homogenate compartments in the ED apparatus acts as a molecular filter permitting diffusion against the concentration gradient of molecules smaller than a definite molecular weight. The drug (1–5 μM) is added to the brain homogenate (donor side) and is sampled from both the donor and the buffer (receiver) sides. To be able to perform ED, the brain homogenate needs to be diluted with phosphate-buffered saline (PBS; Sect.

13.2.1.2.2), commonly with dilution factors of either three (Kalvass and Maurer 2002) or five (Di et al. 2012b). As a general rule, the drug-tissue protein interaction is reversible, and, in the majority of cases, equilibrium rapidly occurs between the unbound and bound molecular species. At equilibrium, the unbound fraction in diluted brain homogenate can be calculated as

$$f_{u,hD} = \frac{C_{receiver}}{C_{donor}} \tag{13.1}$$

where $f_{u,hD}$ is the measured experimental fraction of unbound compound in diluted (D) brain homogenate, $C_{receiver}$ is the concentration of the compound in the buffer, and C_{donor} is the concentration of compound in the donor chamber at equilibrium.

The interaction between the compound/drug and brain tissue is, in most cases, a rapid and reversible process governed by the law of mass action, given that binding does not alter the drug or protein (Klotz 1973). The model assumes that binding between drug and brain tissue takes place in a single step and that the drug interacts with only one binding site on the protein. The equilibrium is described as

$$[D] + [B] \leftrightarrow [DB] \tag{13.2}$$

where [D] and [B] represent the unbound drug and brain tissue protein concentrations and [DB] represents the concentration of the drug-brain tissue protein complex.

The equilibrium dissociation constant (K_d) characterizes the concentration of unbound drug that occupies half of the binding sites on the protein at equilibrium:

$$K_d = \frac{k_{off}}{k_{on}} = \frac{[D][B]}{[DB]} \tag{13.3}$$

Accordingly, the fraction of unbound drug can be described as

$$f_{u,hD} = \frac{[DB]}{[B] + [DB]} = \frac{K_d}{[B] + K_d} \tag{13.4}$$

Rearranging Eq. 13.4 gives

$$K_d = \frac{f_{u,hD}[B]}{1 - f_{u,hD}} \tag{13.5}$$

The unbound drug fraction usually increases as the brain homogenate is diluted. Therefore, $f_{u,hD}$ in the brain homogenate has to be corrected for dilution (Kurz and Fichtl 1983). There are several issues related to the dilution of the brain homogenate and subsequent adjustment methods (Fichtl et al. 1991b). The relationship between the measured unbound drug fraction and the dilution factor is typically not linear (Kurz and Fichtl 1983). The relative impact of dilution of the brain homogenate on the formation of drug-brain tissue protein complexes has been thoroughly discussed by Romer and Bickel (Romer and Bickel 1979). Assuming two different concentrations of brain tissue binding components $[B]_1$ and $[B]_2$ with unbound drug fractions $f_{u,brain}$ and $f_{u,hD}$, Eq. 13.5 can be rewritten as

$$1 = \frac{f_{u,brain} \cdot [B]_1 \cdot \left(1 - f_{u,hD}\right)}{f_{u,hD} \cdot [B]_2 \cdot \left(1 - f_{u,brain}\right)} \tag{13.6}$$

The ratio of $[B]_1/[B]_2$ is projected as the brain homogenate dilution factor D. Hence, Eq. 13.6 can be reorganized to obtain the fraction of unbound drug in the undiluted brain tissue homogenate, which is used to calculate $f_{u,brain}$:

$$f_{u,brain} = \frac{\dfrac{1}{D}}{\left(\left(\dfrac{1}{f_{u,hD}}\right) - 1\right) + \dfrac{1}{D}} \tag{13.7}$$

13.2.1.2 Technical Challenges

The implementation of a 96-well ED plate improved the robustness of the ED method and allowed the use of volumes of brain homogenate and/or plasma as small as 30 µl (e.g., HTdialysis LLC). Although ED is regarded as a "gold" standard method, it has drawbacks which need to be discussed along with the advantages of the method. The equilibration time, concentration of drugs and proteins, membrane surface area, membrane features, and molecular charges can all crucially affect the rate of dialysis.

Selection of Dialysis Membrane

Dialysis membranes consist of a spongy matrix of cross-linked polymers with different pore ratings or molecular weight cut-off (MWCO) points. The MWCO is defined by the molecular weight of solute that is 90% retained by the membrane during a 17-h period. Various membranes (e.g., cellulose ester, regenerated cellulose, and polyvinylidene difluoride) with a range of MWCOs from 3.5K to 50K are applicable for ED. The most commonly used MWCO range is 12–14K.

A potential caveat of the ED method is the risk of nonspecific adsorption of drugs or proteins onto the chamber walls and the dialysis membrane (Vuignier et al. 2010). The use of an inert reusable 96-well Teflon construction minimizes nonspecific binding of test compounds to the apparatus. However, the investigation of different types of dialysis membranes could be beneficial for the selection of the most suitable material.

Recovery (also called mass balance) is traditionally evaluated to account for non-specific binding and is used as an acceptance criterion for ED-based experiments. However, a recent investigation found that recovery had no influence on $f_{u,brain}$ or $f_{u,plasma}$ (Di et al. 2012b). These researchers recommended focusing on stability issues as a main cause of uncertainty in the binding experiments instead.

Preparation of Brain Homogenate

Because an undiluted brain tissue homogenate is paste-like in consistency and difficult to handle, it is diluted with PBS pH7.4. However, this raises several questions concerning the trustworthy conversion of the brain tissue binding values estimated from diluted homogenate into values for the original protein concentrations in the brain tissue. The dilution factor may not affect the final $f_{u,brain}$ measurement (unpublished observations), and various dilution factors have been used. For example, Kalvass and colleagues diluted with two volumes of PBS (Kalvass and Maurer 2002; Liu et al. 2005; Friden et al. 2007; Summerfield et al. 2007; Wan et al. 2007), while Di and co-authors diluted with four volumes of Dulbecco's PBS (Di et al. 2011).

Either frozen or fresh brain tissue can be used to prepare the brain homogenate. However, because of limited supplies of fresh brain homogenate, frozen brain homogenate is often used in drug discovery programs (Di et al. 2012b). To date, no systematic study has been carried out to confirm or reject the existence of differences in brain tissue binding measured using fresh and frozen brain homogenates.

Depending on the method of exsanguination, brain tissue may contain some serum albumin as a result of the residual blood left in the tissue (Glees and Voth 1988). The presence of residual blood in the brain homogenate could affect $f_{u,brain}$ measurement, predominantly for compounds with high affinity for serum albumin (Longhi et al. 2011). Friden and co-workers demonstrated that the procedure of exsanguination of the animal before sampling the brain tissue could influence the residual volume of blood in the brain (Sect. 13.6) (Friden et al. 2010). Thus, the method of sacrificing animals should be standardized with the aim of reducing the residual volume of blood in the brain tissue. As a precautionary action, intracardial perfusion with cold phosphate-buffered saline (PBS) before extraction of the brain could be useful (Longhi et al. 2011). It has been proposed that determination of the serum albumin and total protein content in a brain tissue homogenate could aid the characterization and normalization of different batches (Kodaira et al. 2011; Longhi et al. 2011).

Equilibration Process

After spiking the diluted brain homogenate with the compound(s) of interest, usually up to 150 µL, aliquots are usually loaded into the 96-well ED apparatus and dialyzed against an equal volume of PBS. Compounds with poor aqueous solubility are typically considered to be problematic and limit the use of ED. Equilibrium is generally achieved by incubating the 96-well ED apparatus in a 37°C incubator at 155 rpm for 4–6 h (Kalvass and Maurer 2002). However, if more exact information is wanted, it could be an advantage to perform an initial set of studies to determine the time required for the system to reach equilibrium, as slow drug-protein dissociation may occur.

The equilibration time needed in ED, normally 4–6 h, is considered to be one of the drawbacks of the method if the compounds studied are unstable in the plasma or brain homogenate. Moreover, the equilibration time is associated with a volume shift that takes place because of the semipermeable membrane and the presence of proteins. This volume shift can be as large as 10–30% for ED with plasma (Huang 1983). Measuring drug concentrations on both sides of the membrane is therefore required.

Bioanalysis

During the equilibration period, the buffer side becomes enriched with ions, amino acids, lipids, carbohydrates, and any other molecules smaller than the MWCO of the dialysis membrane that are not already present in the buffer. The brain homogenate composition also changes as a result of osmotic pressure. The modifications in the composition of the buffer and brain homogenate could result in a "matrix" effect during subsequent liquid chromatography tandem mass spectrometry (LC-MS/MS) analysis (e.g., ion suppression, enhancement of analyte signal) (Van Eeckhaut et al. 2009). Mixed-matrix and semi-automated mixed-matrix methods are currently being developed to decrease mass spectrometer run times and reduce the probability of experimental artifacts (Plise et al. 2010). For semi-automated mixed-matrix methods with a cassette-based approach, a single matrix is prepared following dialysis by mixing dialyzed plasma and buffer containing different test compounds from the same dialysis plate. The method should eliminate the need for standard curves, and increase the consistency of the sample matrix for LC-MS/MS analysis. This approach could easily be adopted when running the ED-based brain homogenate method and can be considered as a step toward further optimization of ED.

In conclusion, ED-based determination of $f_{u,brain}$ can be considered a proficient method. However, the biological and pharmacological meaning of the obtained values must be critically evaluated in relation to other neuroPK parameters (Sects. 13.4 and 13.5).

Recently, ED measures of the unbound fraction of drugs in plasma and brain were used as additional parameters for the interpretation of in vivo positron emission tomography (PET) results, particularly for the estimation of unbound drug concentrations in the CNS and accurate quantification of receptor binding (Gunn et al. 2012).

13.3   The Brain Slice Method for $V_{u,brain}$

With respect to assessing the intracerebral distribution of small drug molecules, the ED-based brain homogenate method has drawbacks that are primarily linked to the disruption of brain parenchymal cells (Becker and Liu 2006; Liu et al. 2006; Friden et al. 2007, 2011). In this regard, the brain slice method is an advanced, well-functioning approach to the evaluation of the overall uptake of drugs into the brain tissue via determination of the volume of distribution of unbound drug in the brain ($V_{u,brain}$; mL · g brain^{-1}). This method has the benefits of being used in a regulated in vitro environment, while at the same time, preserving much of the cellular complex integrity, including cellular barriers and circuitry, and as a result conserving the functionality of the in vivo brain. As a result, the technique delivers information that is directly relevant to issues such as nonspecific binding to tissues, lysosomal trapping (Sect. 13.4.3), and active uptake into the cells.

The brain slice method was implemented by Henry McIlwain more than six decades ago and is nowadays widely used in neurobiology, electrophysiology, and quantitative neuropharmacology (McIlwain 1951b; Collingridge 1995). The first use of this method for evaluation of intracerebral distribution of substances aimed to estimate the uptake of nutrients such as glucose and amino acids into the brain (McIlwain 1951a; Blasberg et al. 1970; Newman et al. 1988a, 1991; Smith 1991). Later, the method was proposed for in vitro investigation of the distribution of drugs in the brain (Van Peer et al. 1981; Kakee et al. 1996, 1997; Ooie et al. 1997). There have been several efforts to establish mechanistic pharmacokinetic/pharmacodynamics links using brain slice methodology, e.g., for propofol (Gredell et al. 2004), etomidate (Benkwitz et al. 2007), and volatile agents (Chesney et al. 2003).

$V_{u,brain}$ can also be measured using cerebral microdialysis and total brain concentration measurements; this is currently accepted as an in vivo reference method for evaluating intracerebral drug distribution. When the fresh brain slice method was validated against microdialysis, $V_{u,brain}$ was within a threefold range of the microdialysis results for 14 of 15 investigated compounds (Friden et al. 2007). In contrast, when $V_{u,brain}$ was recalculated using data from the brain homogenate method for the same list of compounds, the results were less accurate. In particular, the brain homogenate method overpredicted in vivo $V_{u,brain}$ for compounds limited to intracerebral ISF distribution (e.g., morphine-3- and morphine-6-glucuronide, R- and S-cetirizine) and underpredicted the distribution of gabapentin, which has predominantly active cellular uptake (Friden et al. 2007). However, these results have been challenged. Liu and colleagues demonstrated that, for eight of the nine studied compounds   (carbamazepine,   citalopram,   ganciclovir,   metoclopramide, N-desmethylclozapine,   quinidine,   risperidone,   9-hydroxyrisperidone,   and thiopental), the $C_{u,brainISF}$ estimated using the brain homogenate method was within a threefold range of that obtained using cerebral microdialysis (Liu et al. 2009a). Nonetheless, these contrasting results should still be critically evaluated, since the microdialysis probes were calibrated using only in vitro recovery. Determination of $V_{u,brain}$ values that are more relevant to the in vivo situation, using fresh brain slices instead of brain homogenate, appears to be associated with more accurate assessment of $C_{u,brainISF}$ (i.e., $C_{u,buffer}$).

Despite the obvious benefits of the fresh brain slice method, it has not yet received wide acceptance in the drug industry compared to the brain homogenate method. The arguments against acceptance include that the method requires greater labor intensity. However, a high-throughput brain slice method has now been developed to fit the drug discovery format, thus offering new possibilities for the utilization of the method (Friden et al. 2009a; Loryan et al. 2013). Once the brain slice technique is established in a laboratory, one skilled assistant can perform up to four experiments per day. Up to ten compounds in one cassette can be tested simultaneously (prior consultation with an analytical chemist is obligatory). A series of three experiments is enough to obtain consistent results for one cassette. The detailed protocol of how to perform brain slice studies can be found in the publication by Loryan et al. (Loryan et al. 2013).

13.3.1   Section Heading 13.3.1

13.3.1.1   Principles

The use of the apparent $V_{u,brain}$, obtained in vivo using cerebral microdialysis (Eq. 13.8), to assess the distribution of drugs in the brain was first suggested by Wang and Welty (they used the abbreviation $V_{e,app}$) (Wang and Welty 1996). $V_{u,brain}$ describes the relationship between the total drug concentration in the brain and the unbound drug concentration in the brain ISF, regardless of BBB function.

Assessment of $V_{u,brain}$ using the in vitro fresh brain slice method is based on the assumption that *at equilibrium*, $C_{u,brainISF}$ is equal to the drug concentration in protein-free artificial extracellular fluid buffer (aECF). Thus, $V_{u,brain}$ (mL · g brain^{-1}) is calculated as the ratio of the amount of drug in the brain slice (A_{brain}, nanomoles · g brain^{-1}) to the measured final aECF after reaching equilibrium (C_{buffer}, micromoles · L^{-1}):

$$V_{u,brain} = \frac{A_{brain}}{C_{u,brainISF}} = \frac{A_{brain}}{C_{buffer}} \qquad (13.8)$$

Because a certain volume of the aECF remains on the surface of the brain slice (V_i, mL · g slice^{-1}), even after removing the excess with filter paper, this has to be accounted for. V_i is estimated in a separate experiment using [^{14}C] inulin as described in Friden et al. (2009a). Equation 13.8 is then rearranged to obtain $V_{u,brain}$ corrected for V_i $(1-V_i)$:

$$V_{u,brain} = \frac{A_{brain} - V_i \cdot C_{buffer}}{C_{buffer} \cdot (1 - V_i)} \tag{13.9}$$

As outlined by Wang and Welty, a $V_{u,brain}$ value that is higher than 1 mL · g brain^{-1} indicates intracellular accumulation or excessive brain tissue binding because it exceeds the total volume of water in the brain which is 0.8 mL · g brain^{-1} (Wang and Welty 1996). $V_{u,brain}$ values between 1 and 0.2 mL g brain^{-1} indicate limited distribution of drug in the brain ECF and ICF (Nicholson and Sykova 1998; Sykova and Nicholson 2008). As the volume of healthy adult rat brain ISF is 0.2 mL · g brain^{-1}, a volume below 0.2 mL · g brain^{-1} is not possible. However, it should be kept in mind that this technique does not account for possible intracerebral metabolism (Chap. 6).

In the literature, $V_{u,brain}$ is sometimes expressed as $f_{u,brain,slice}$ (Kodaira et al. 2011; Uchida et al. 2011a). It is important to keep in mind that $f_{u,brain,slice}$ could be considerably different from $f_{u,brain}$, as they obtained using different matrices, i.e., brain slice and brain tissue homogenate (Sect. 13.4).

13.3.1.2   Technical Challenges

Artificial Extracellular Fluid and Formation of Cassettes

It is important to preserve the viability of brain slices during the experiment and to mirror the in vivo cellular milieu as closely as possible. There are two main approaches to achieving this, regarding the medium used. One approach is based on the use of either fresh or thawed *plasma* as a medium for the incubation, with subsequent evaluation of the brain slice-to-plasma drug concentration ratio (Becker and Liu 2006). The second and more commonly applied approach is to use a *protein-free* artificial cerebrospinal fluid (CSF) or aECF as an incubation medium (Kakee et al. 1996, 1997). The latter simplifies the interpretations of the results obtained. A large number of formulations for aECF can be found in the literature (Newman et al. 1991; Kakee et al. 1996, 1997; Gredell et al. 2004; Friden et al. 2009a; Uchida et al. 2011a). In many of these, ascorbic acid is used as a natural free radical scavenger to protect cell membranes from lipid peroxidation and swelling of the brain slices (Rice 1999). The HEPES-buffered aECF containing 129 mM NaCl, 3 mM KCl, 1.4 mM CaCl$_2$, 1.2 mM MgSO$_4$, 0.4 mM K$_2$HPO$_4$, 25 mM HEPES, 10 mM glucose, and 0.4 mM ascorbic acid is a robust and practical formulation for sustaining the physiological pH (around 7.3 at 37 °C after 5-h incubation) for the high-throughput setup (Friden et al. 2009a).

Another critical requirement is an adequate oxygen supply. Either 100% humidified oxygen or carbogen (a mixture of 95% oxygen and 5% CO$_2$) can be used.

The brain slice method allows examination of up to ten compounds per experiment, covering a wide range of physicochemical properties and pharmacological targets, mixed together in the same *cassette* (the mixture of compounds under investigation is called the cassette) (Friden et al. 2009a; Kodaira et al. 2011). Low concentrations of compounds (e.g., 0.1–0.2 µM) are preferable. The summed concentration of all the drugs in the cassette should not exceed 1 µM (Friden et al. 2009a). Application of higher concentrations of various compounds can lead to accumulation of compounds in the acidic compartments of the cells (i.e., lysosomes) or competition for specific cell membrane transporters with subsequent incorrect values for $V_{u,brain}$. For instance, it is recognized that interactions between two weak bases are regulated by the free concentrations of the compounds in the cassette and the ability of these compounds to elevate intralysosomal pH (Daniel and Wojcikowski 1999b). Potential bioanalytical issues should be addressed when assembling the cassettes for investigation, so as to avoid technical hitches.

Preparation of Brain Slices and Incubation

It is important that the fresh brain slices are of high quality if the $V_{u,brain}$ values are to be relevant to the in vivo situation. This can be accomplished by keeping strictly to the protocol for preparation and maintenance of the brain slices during the experiment (Friden et al. 2009a; Loryan et al. 2013). The key steps of the brain slice method are illustrated in Fig. 13.2.

Fig. 13.2 An illustration of the main steps in the preparation of brain slices. (**a**) Schematic representation of the cutting direction. (**b**) The brain glued to the slicing platform in a coronal position. (**c**) Brain slices transferred into the Ø80-mm flat-bottomed glass beaker. (**d**) A beaker covered by the custom-fabricated lid fitted with a Teflon-fluorinated ethylene-propylene film. (**e**) The setup for the incubation-equilibration period. (Reprinted with permission from BioMed Central (Loryan et al. 2013))

The protocol for the fresh brain slice method (also called in vitro brain slice uptake technique) has not been unified among research laboratories, which makes comparison and interpretation of the results challenging. For instance, the brain can be sliced using a brain microslicer (Ooie et al. 1997; Benkwitz et al. 2007; Kodaira et al. 2011), a McIlwain tissue chopper (Becker and Liu 2006), or a vibratome (Friden et al. 2009a; Loryan et al. 2013). Moreover, researchers have used slices from different planes of the brain, such as the hypothalamic (Kakee et al. 1997), cortical (Kodaira et al. 2011) or striatal (Friden et al. 2009a, b). The thickness of the brain slices also differs between protocols: 300 μm (Kakee et al. 1996, 1997; Friden et al. 2009a, b), 400 μm (Becker and Liu 2006), or even 1000 μm (Van Peer et al. 1981). Accordingly, the incubation time (time required to reach equilibrium) varies and could be 8 h or longer, which may be too long to sustain the viability of the slices.

The time needed to reach equilibrium is influenced by various factors such as the amount of brain tissue per unit of the buffer volume, the stirring speed, and the initial concentration of the compound (Gredell et al. 2004; Benkwitz et al. 2007; Friden et al. 2009a). The ratio of six/ten (rat/mouse) 300 μm sequential brain slices to 15/10 mL (rat/mouse) of aECF has been found to be the most optimal for various diverse compounds to reach equilibrium in about 5 h (Friden et al. 2009a; Loryan et al. 2013). Very lipophilic compounds may require a longer equilibration time in some experimental setups, and this could compromise the viability of the brain slices. In this case, mathematical modeling of the data could be a reasonable alternative (Kodaira et al. 2011).

Sufficient viability of the brain slices is a critical prerequisite. The viability can be assessed indirectly by measuring the pH of the aECF (acidification of the medium is linked to low viability of the slices). However, more advanced methods such as measuring the ATP content of the slices (Friden et al. 2007; Kodaira et al. 2011; Uchida et al. 2011a) or the activity of released lactate dehydrogenase (Dos-Anjos et al. 2008; Loryan et al. 2013) are now recommended.

Bioanalysis

The drug concentrations in brain slices and aECF samples taken at equilibrium can be analyzed after homogenization using high-throughput techniques and LC-MS/MS as discussed for brain homogenate samples in Sect. 13.2.1.2.4. To avoid the preparation of calibration curves, 10- and 100-fold dilutions of the samples are preferable (Friden et al. 2009a). Several groups normalize the protein concentrations to correct for the dilution of brain homogenate (Kodaira et al. 2011).

In summary, the fresh brain slice method is a precise and robust technique for estimating the overall uptake of drugs into the brain tissue. This method is recommended for the estimation of target-site PK and toxicokinetics in the early drug discovery process in order to guide candidate selection (Friden et al. 2014; Loryan et al. 2014, 2017). One of the attractive features of the brain slice method is that it can be developed to investigate compound-specific molecular mechanisms of the intracerebral distribution of compounds (Friden et al. 2011; Chen et al. 2014; Puris et al. 2019). For instance, fresh brain slices could be prepared from different strains of wild-type or genetically modified mice and rats to elucidate the effects of intra-cerebral transporters on the distribution of drugs within the brain (BBB transporters cannot be directly mapped with this technique). Furthermore, the brain slices could be manipulated genetically using various methods such as viral infection (Stokes et al. 2003) or biolistics (Wellmann et al. 1999). Disease models could also be used to study the diffusion and distribution of drugs or radiotracers within the brain (Newman et al. 1988b; Patlak et al. 1998). In addition, pharmacological inhibition or stimulation could be used to investigate particular distributional mechanisms, e.g., monensin or nigericin, to study the impact of lysosomal accumulation on the intracellular distribution of drugs (Friden et al. 2011; Logan et al. 2012).

13.4 Intracellular Distribution

Historically, it has been presumed that the transport of small molecules between intra- and extracellular neurocompartments is more efficient than BBB transport, which is considered to be a rate-limiting step for drug distribution to the brain (Wang and Welty 1996). Accordingly, from a PK point of view, the assessment of the intracerebral distribution of NCEs is usually less prioritized, is often inadequate because of a lack of reliable methods, and is narrowed to estimation of the unbound drug fraction in a brain homogenate, with subsequent evaluation of its half-life in the brain tissue (Liu et al. 2005). However, awareness of compound-specific intra-cerebral distributional mechanisms in early drug discovery could allow better directed evaluation and selection of drug candidates, based on the location of the potential CNS target (i.e., extra- or intracellular) and the probable side effects.

After passing the BBB, drugs are distributed in the extracellular space mainly by diffusion and convection (see Chap. 5 for a comprehensive analysis of the transport processes of drugs within the CNS). As pointed out in the state-of-the-art review by Wolak and Thorne (2013), the diffusion of molecules is governed by the features of the extracellular space (i.e., width, volume fraction, viscosity, geometry) as well as by any potential binding to the extracellular matrix or cellular membrane compo-nents (Fenstermacher and Kaye 1988). It should be highlighted that the diffusion of compounds in the extracellular neurocompartment is a potentially limiting step for macromolecules, nanoparticles, and viral vectors (Thorne et al. 2004, 2008; Thorne and Nicholson 2006). The bulk flow of the ISF should be accounted for in addition to the diffusion and hydraulic permeability (see Chap. 1 and Table 13.1). However, although it can be influential for poorly penetrating compounds, it is not a matter of concern for small highly lipophilic compounds (Cserr 1992; Davson 1995; Abbott 2004; Abbott et al. 2018). The bulk flow of the ISF has been measured as ~0.1–0.3 μL min^{-1} g^{-1} in the rat brain, but the actual value may be greater than this (Chap. 1 and Joan Abbott personal communication).

Because ISF is virtually protein free (Davson et al. 1970; Davson 1995), the drug present in the ISF can be measured as unbound and accessible for interactions at a

Table 13.1 Key components affecting drug distribution to and from the different compartments in the brain[a]

Extracellular neurocompartment	Cellular membranes	Intracellular neurocompartment
Diffusion in extracellular space Hydraulic permeability ISF bulk flow	Membrane permeation Active influx (e.g., organic cation transporters, L-type amino acid transporters) Active efflux Nonspecific binding to cell membrane components (often quantitatively insignificant) Specific binding to the target (often quantitatively insignificant)	Nonspecific binding to intracellular membrane components (often quantitatively significant) pH differences causing accumulation of weak bases in acidic compartments (e.g., lysosomes, endosomes) Specific binding to the target (e.g., tubulin, enzymes) Drug metabolism (often insignificant)[b]

[a]Differences between the types of brain parenchymal cells and brain subregions are not taken into account
[b]Drug metabolism is discussed in more detail in Chap. 6

cellular membrane level (Hammarlund-Udenaes et al. 1997; Ooie et al. 1997; Kalvass and Maurer 2002; Mano et al. 2002; Shen et al. 2004; Doran et al. 2005; Liu et al. 2005; Summerfield et al. 2006; Friden et al. 2007; Watson et al. 2009). The permeation of unbound, unionized drug through the cell membrane could be defined as the most significant distributional process of small molecules into the cell. Accumulation is a distributional process that is associated with asymmetry at the cellular barrier, is linked to the physiological pH gradient, and is driven by acidic intracellular compartments such as lysosomes, endosomes, peroxisomes, and the trans-Golgi network (Sect. 13.4.2). Asymmetry at the cellular barrier level can also occur as a consequence of active transport such as influx processes governed by organic cation transporters (e.g., 1-methyl-4-phenylpyridinium, tetraethylammonium, metformin) and L-type amino acid transporters (e.g., gabapentin), or efflux processes (Lee et al. 2001a, b; Bendayan et al. 2002; Kusuhara and Sugiyama 2002; Ohtsuki et al. 2004; Syvanen et al. 2012). The specific and nonspecific binding of compounds to extracellular constituents of the cell membrane can be ignored because of their much smaller surface areas, i.e., the external surface area of a typical human cell membrane represents less than 0.5% of the total cell membrane surface area (Freitas 1999).

After passing the cellular barrier, compounds can bind reversibly to intracellular constituents such as lipoproteins, phospholipids of the inner cellular membrane, or organelles. Nonspecific binding is often the dominant distributional component for small lipophilic compounds. In most cases, specific intracellular binding is irrelevant from a distributional perspective because of the low expression levels of the targets in relation to the extent of nonspecific binding. However, there are some exceptions; these are discussed at the end of Sect. 13.4.

Off-target or nonspecific binding of the drug to the cellular membranes is often not associated with any pharmacological response. However, progress has been made in recent decades toward an understanding of the interactions between the ligand and the target (i.e., receptor, ion channel, enzyme). Primarily, the "passive" role of the cell membrane in target-binding kinetics has been questioned (Vauquelin and Packeu 2009). Novel membrane-connected concepts that reexamine the notion of the so-called nonspecific plasma membrane partitioning are being proposed (Sargent et al. 1988; Vauquelin and Van Liefde 2005). It has been recognized that nonspecific ligand-membrane interactions could be favorable, although not in all cases, for ligand-target interactions (Sargent and Schwyzer 1986; Bean et al. 1988; Vauquelin et al. 2012). This process could be very important for peptide-target interactions (Sargent and Schwyzer 1986). Another crucial aspect of membrane partitioning is the increased in vivo residence time of hydrophobic ligands. Slow release from the cell membranes is commonly acknowledged to be strongly associated with the long-lasting effects of highly lipophilic compounds (e.g., salmeterol). In other words, the cell membrane can be perceived as a depot/reservoir for hydrophobic ligands.

13.4.1 Using $K_{p,uu,cell}$ to Estimate the Extent of Cellular Barrier Transport

Frequently, as with plasma protein binding, scientists define the binding of drugs to brain tissue as "nonspecific." However, in comparison with plasma protein binding, less is known about the drug-brain tissue interaction, mainly because of technical difficulties in obtaining data on the tissue-binding components and in the quantification of intracellular drug concentrations.

In most cases, intracerebral distribution is assessed by either the ED-based brain homogenate method, with evaluation of $f_{u,brain}$, or the fresh brain slice method, with assessment of $V_{u,brain}$. Combining the two methods provides further information on intracellular distribution. The main determinant of $f_{u,brain}$ is *brain tissue binding* which primarily consists of nonspecific binding of the drug to various *intracellular* lipids and proteins. $V_{u,brain}$ then provides complementary data on intracerebral distribution factors other than binding. The importance of $K_{p,uu,cell}$ in this respect has been discussed by Friden et al. (2007, 2009a, 2011). $K_{p,uu,cell}$ can be estimated by combining $f_{u,brain}$ (brain homogenate) with $V_{u,brain}$ (brain slice) using Eq. 13.10 (Friden et al. 2007):

$$K_{p,uu,cell} = V_{u,brain} \cdot f_{u,brain} \qquad (13.10)$$

$K_{p,uu,cell}$ describes the *steady-state* relationship of intracellular-to-extracellular unbound drug concentrations and provides the average concentration ratio for all cell types within the brain. The assumptions behind the $K_{p,uu,cell}$ concept are the following (Friden et al. 2007):

1. The ISF concentration is assumed to describe unbound drug (ISF is a practically protein-free fluid).

2. $C_{u,brainISF}$ represents the concentration of unbound drug in brain ISF from the entire brain (cranioregional and cell-type dissimilarities are not accounted for).
3. Membrane passive permeation and binding to intra- and extracellular constituents are the key distributional processes.
4. Intracellular drug molecules can be unbound or bound to intracellular components.
5. Drug binding to the outer part (surface) of the cell is negligible. However, this assumption could be incorrect for molecules with distribution entirely restricted to the ISF (e.g., large molecules) and/or those that are significantly bound to cellular membranes.

The derivation of the equations presented below is based on the definition of $V_{u,brain}$ as the ratio of the total amount of drug in the brain excluding the blood (A_{brain}) to the concentration of unbound drug in brain ISF (Eq. 13.8). According to the proposed distributional model (Friden et al. 2007), the total amount of drug in the brain can be presented as

$$A_{brain} = V_{brainISF} \cdot C_{u,brainISF} + V_{cell} \cdot V_{u,cell} \cdot C_{u,cell} \tag{13.11}$$

where $V_{brainISF}$ and V_{cell} are the physiological fractional volumes of ISF (~0.2 mL · g brain^{-1}) (Nicholson and Sykova 1998; Sykova and Nicholson 2008) and brain parenchymal cells (~0.8 mL · g brain^{-1}) and the density of brain tissue is assumed to be 1. $V_{u,cell}$ describes the volume of distribution of unbound drug in the cell (mL ICF · mL cell^{-1}) and relates the total amount of drug in the cell to the intracellular concentration of unbound drug; $C_{u,cell}$. $V_{u,cell}$ can be compared with $V_{u,brain}$, describing the whole brain drug distribution.

Another way of explaining $V_{u,cell}$ is that it describes the affinity of the drug to bind inside the cell. The more drug is bound, the higher the value of $V_{u,cell}$. It can be estimated using the ED-based brain homogenate method:

$$V_{u,cell} = 1 + \frac{D}{V_{cell}} \cdot \left(\frac{1}{f_{u,hD}} - 1 \right) \tag{13.12}$$

where $f_{u,hD}$ is the buffer-to-brain homogenate concentration ratio measured using ED and D is the dilution factor associated with homogenate preparation (Sect.13.2). Rewriting Eq. 13.8 using Eq. 13.11 and dividing both sides by $C_{u,brainISF}$ give

$$V_{u,brain} = V_{brainISF} + V_{cell} \cdot V_{u,cell} \cdot \frac{C_{u,cell}}{C_{u,brainISF}} \tag{13.13}$$

Consequently, the ratio of brain ICF to ISF unbound drug concentrations ($K_{p,uu,cell}$) can be derived as

$$K_{p,uu,cell} = \frac{C_{u,cell}}{C_{u,brainISF}} = \frac{V_{u,brain} - V_{brainISF}}{V_{cell} \cdot V_{u,cell}} \qquad (13.14)$$

When analyzing numerical values of $K_{p,uu,cell}$, it is important to remember its meaning. When cellular membrane permeation is predominantly passive, the unbound drug intra- and extracellular concentrations are the same, giving a $K_{p,uu,cell}$ equal to unity. $K_{p,uu,cell}$ values higher than unity indicate intracellular accumulation, and $K_{p,uu,cell}$ values below unity could indicate active efflux at the cellular barrier. The estimation of $K_{p,uu,cell}$ is valuable for interpreting and understanding the processes governing the distribution of drugs into the brain parenchymal cells. It should be remembered, however, that the numbers obtained are average values from all the cell types in the brain.

13.4.2 Lysosomal Trapping

Although they were discovered in the early 1970s, the role of lysosomes in drug tissue distribution kinetics can still be considered as *terra incognita* (De Duve 1971). Lysosomes are conventionally acknowledged as the cell's "garbage disposal units." They are membrane-bound organelles containing about 50 hydrolytic enzymes that function at pH 4.5. Vacuolar-type H^+-ATPase embedded in the lysosomal membrane maintains the intralysosomal acidic environment.

Lysosomotropism or *lysosomal trapping* is a phenomenon where compounds (*lysosomotropic agents*) with both a lipophilic moiety and a basic moiety are accumulated in acidic intracellular compartments mainly in lysosomes (Fig. 13.3) (De Duve 1970; Nadanaciva et al. 2011).

Lysosomal trapping is governed by the large physiological pH gradient between ICF and lysosomes. The process of lysosomal trapping is saturable, energy-dependent (necessary for the normal function of the H^+-ATPase), and requires cellular integrity (De Duve 1970; MacIntyre and Cutler 1988; Daniel and Wojcikowski 1999a). Weak bases in their unionized state permeate cellular and lysosomal membranes and accumulate in the acidic compartment of lysosomes (Fig. 13.4). Diacidic bases are trapped more easily than monoacidic bases, with a subsequent impact on their distribution (MacIntyre and Cutler 1988). Because they are protonated within the lysosomes, the bases are not able to diffuse back into the cytosol (MacIntyre and Cutler 1988; Lloyd 2000; Kaufmann and Krise 2007). The intralysosomal concentrations of trapped compounds can reach high levels, with lysosome-to-cytosol accumulation ratios as high as 100:1 (Daniel and Wojcikowski 1997). Moreover, because the weak bases interact with phospholipids within the lysosome, the apparent lysosomal volume measured indirectly could be substantially greater than the physical (i.e., actual) lysosomal volume (MacIntyre and Cutler 1988; Duvvuri and Krise 2005). The physical volume of the lysosomes can also increase with time due

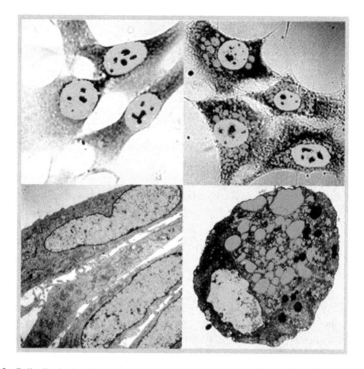

Fig. 13.3 Cells displaying the lysosomal trapping phenomenon. Picture from Boya et al. (2003). In contrast to controls (top left and bottom left panels), cells treated with the lysosomotropic drug ciprofloxacin (top right and bottom right panels) manifest multiple autophagic vacuoles (colored pink) in the cytoplasm, before undergoing apoptosis. The bottom microphotographs have been obtained by electron microscopy, while the top ones result from conventional light microscopy, after Giemsa staining. Nuclei are colored blue. (Reprinted with permission from Rockefeller University Press (picture appeared on the cover page of J Exp Med, May 19, 2003))

to vesicle-mediated trafficking and fusion of lysosomes with the cell membrane (Kaufmann and Krise 2007; Logan et al. 2012).

Consequently, despite the very small physiological volume of the lysosomes (~ 0.01 mL · g brain^{-1}), lysosomotropic compounds show extensive tissue accumulation (e.g., in the lungs, liver, and brain) which is reflected by a high apparent volume of distribution (Daniel and Wojcikowski 1999b). Moreover, lysosomal trapping can result in drug-drug interactions (Daniel et al. 1995, 1998, 2000; Daniel and Wojcikowski 1999b; Logan et al. 2012). For instance, because the process of lysosomal trapping is saturable, the lysosomal uptake of co-administered drugs could decline. All this suggests that lysosomal trapping is an important mechanism of drug distribution with potential impact on systemic PK.

Although the brain tissue is not as lysosome-rich as the lungs, liver, and kidneys, lysosomal trapping could also influence the brain PK. Many marketed and novel neurotherapeutics are cationic amphiphilic compounds; it is thus not surprising that they are lysosomotropic (Daniel 2003; Nadanaciva et al. 2011). Hence, it is

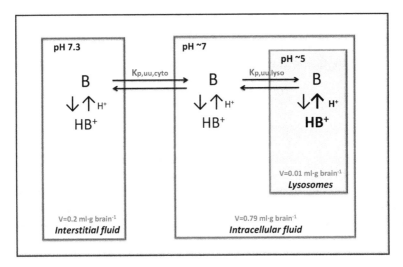

Fig. 13.4 Graphic illustration of the pH partitioning of a basic drug between extra- and intracellular compartments, i.e., interstitial fluid, intracellular fluid, and lysosomes. Accumulation of the protonated form (HB+) of the basic drug (B) in the compartments is driven by the physiological pH gradient. The cytosolic-to-interstitial fluid unbound drug concentration ratio ($K_{p,uu,cyto}$) and the lysosomic-to-cytosolic unbound drug concentration ratio ($K_{p,uu,lyso}$) can be estimated using a three-compartment pH partitioning model (Friden et al. 2011)

recommended that particular attention be paid to lysosomotropism in CNS drug development programs.

Lysosomotropism is also interesting in that there are several lysosomal acidic hydrolases that may be useful pharmacological CNS targets (de Duve 1975; Boya and Kroemer 2008; Schultz et al. 2011). For instance, acid sphingomyelinase affects ceramide levels in several psychiatric and neurological disorders such as major depression, morphine antinociceptive tolerance, and Alzheimer's disease (Schwarz et al. 2008; Ndengele et al. 2009; He et al. 2010; Kornhuber et al. 2011). The inhibition of acid sphingomyelinase results in anti-apoptotic, proliferative, and anti-inflammatory effects. Consequently, functional acid sphingomyelinase inhibitors have potential in a number of new clinical therapies (Muehlbacher et al. 2012).

13.4.2.1 Compensation for pH Partitioning

Several researchers have suggested that lysosomal accumulation is a potential explanation for dissimilarities between in vitro (homogenates) and in vivo measurements when describing the distribution of acidic, neutral, and basic drugs in tissues other than the brain (Harashima et al. 1984; Sawada et al. 1984; MacIntyre and Cutler 1988; Daniel and Wojcikowski 1997; Yokogawa et al. 2002; Maurer et al. 2005). For instance, it has been documented that predictions of the pharmacokinetic parameter apparent volume of distribution at steady state (V_{ss}) for 36 compounds, based on measurement of the unbound drug fraction in 15 different tissues, were less accurate for acidic and strongly basic substances (Berry et al. 2010). However, after making allowance for the ionic effects of tissue-to-blood pH gradients, the predictions for V_{ss} were accurate within a threefold range for 81% of the compounds studied.

Inconsistencies between $C_{u,brainISF}$ values obtained using cerebral microdialysis and those projected from A_{brain} corrected for nonspecific binding using $f_{u,brain}$ for weak bases and acids are also thought to be linked to lysosomotropism (Friden et al. 2007). Lysosomotropism in the brain tissue is also important when comparing brain slice and brain homogenate data (Friden et al. 2011; Loryan et al. 2015, 2017).

The cell partitioning coefficient frequently deviates from unity. Intracellular accumulation as a result of the pH gradient is often suggested as one of the main reasons for the mismatch between the brain homogenate and brain slice data, i.e., $f_{u,brain} \neq 1/V_{u,brain}$. The lack of agreement is mainly due to the different properties of the two methods; cell and organelle membranes are retained in the slices, and pH differences are preserved. If the intracellular unbound drug concentration is similar to the brain ISF unbound drug concentration (i.e., $K_{p,uu,cell}$ is close to unity and $V_{u,brain}$ exceeding 1 mL · g brain^{-1}), it can be assumed that intracellular nonspecific binding to membrane constituents is a major, quantitatively significant, distributional mechanism.

If only $K_{p,uu,brain}$ is of interest and brain homogenate data are used, the $f_{u,brain}$ values can be corrected to more in vivo-like values by compensating for pH partitioning according to the pKa of the drug (Friden et al. 2007, 2011; Loryan et al. 2014).

A three-compartment (ISF, cytosol, and lysosomes) pH partitioning model for $K_{p,uu,cell}$ based on the strong relationship between drug accumulation in acidic compartments due to lysosomal trapping and the pKa values of the compound has been developed (Fig. 13.4) (Friden et al. 2011; Loryan et al. 2015). The starting point is described by Eq. 13.15:

$$K_{p,uu,cell} = V_{u,brain} \cdot f_{u,brain} = \frac{A_{brain}}{C_{u,brainISF}} \cdot f_{u,brain} \qquad (13.15)$$

The total amount of drug in the brain can be described as the sum of the total amounts in the ISF, the cytosol, and the lysosomes, denoted as A_{ISF}, A_{cyto}, and A_{lyso}, respectively. Each compartment is described by its physiological volume multiplied by the concentration of unbound drug in the compartment, divided by $f_{u,brain}$:

$$A_{brain} = A_{ISF} + A_{cyto} + A_{lyso} = \frac{V_{brainISF} \cdot C_{u,brainISF} + V_{cyto} \cdot C_{u,cyto} + V_{lyso} \cdot C_{u,lyso}}{f_{u,brain}} \qquad (13.16)$$

V_{ISF}, V_{cyto}, and V_{lyso} are the physiological volumes of the ISF (0.20 mL · g brain^{-1}), cytosol (0.79 mL · g brain^{-1}), and lysosomes (0.01 mL · g brain^{-1}), respectively. $C_{u,cyto}$ and $C_{u,lyso}$ describe the unbound drug concentrations in cytoplasm and lysosomes, respectively. If Eqs. 13.15 and 13.16 are combined, $K_{p,uu,cell}$, predicted from the three-compartment pH partition model, can be defined as

$$K_{p,uu,cell} = V_{ISF} + K_{p,uu,cyto} \cdot \left(V_{cyto} + V_{lyso} \cdot K_{p,uu,lyso} \right) \quad (13.17)$$

The estimation of the cytosolic-to-interstitial fluid unbound drug concentration ratio ($K_{p,uu,cyto}$) and the lysosomic-to-cytosolic unbound drug concentration ratio ($K_{p,uu,lyso}$) can be computed by introducing the pKa values of the compounds (i.e., bases) in Eqs. 13.18 and 13.19, respectively:

$$K_{p,uu,cyto} = \frac{C_{u,cyto}}{C_{u,brainISF}} = \frac{10^{pKa-pH_{cyto}} + 1}{10^{pKa-pH_{ISF}} + 1} \quad (13.18)$$

$$K_{p,uu,lyso} = \frac{C_{u,lyso}}{C_{u,brainISF}} = \frac{10^{pKa-pH_{lyso}} + 1}{10^{pKa-pH_{cyto}} + 1} \quad (13.19)$$

where $pH_{cyto} = 7.06$, and $pH_{lyso} = 5.18$, as determined by Friden and co-workers (Friden et al. 2011).

The main application of the pH partitioning model is related to $f_{u,brain}$, measured using the brain homogenate method. Based on pH partitioning, $V_{u,brain}$ ($1/f_{u,brain,corrected}$) can be estimated from $f_{u,brain}$ using Eq. 13.15, i.e., by dividing the calculated $K_{p,uu,cell}$ by $f_{u,brain}$ (Eq. 13.17). As demonstrated after the correction for pH partitioning, the discrepancy between brain homogenate and brain slice methods was practically abolished in a dataset consisting of 56 compounds (Friden et al. 2011). However, the pH partitioning model was still incapable of identifying and/or correcting other processes governing the dissimilarities between the brain slice and homogenate methods, such as active uptake into the cells.

The three-compartment pH partitioning model can also be used for the preliminary evaluation of $K_{p,uu,cell}$ and identification of potential lysosomotropic compounds already in the lead optimization phase (Friden et al. 2011; Loryan et al. 2014, 2017). pKa values are frequently calculated in silico in the early discovery stages, and a critical approach is recommended since they may not reflect the real pKa values. pKa values measured at 25 °C can also diverge from actual in vivo values when using the pH partitioning model (Sun and Avdeef 2011). This can lead to some differences in experimental $K_{p,uu,cell}$ and computed $K_{p,uu,cell}$ values.

13.4.3 Intracerebral Distributional Patterns

Because of the physicochemical features and character of the pharmacological targets, the patterns of intracerebral distribution can differ for different drugs (Fig. 13.5). Thioridazine, salicylic acid, and gabapentin are used as model drugs and are discussed in detail in this section.

Figure 13.5a shows the intracerebral distribution of thioridazine. Thioridazine is a base, with a pKa of 8.9 and pronounced plasma and brain tissue binding mainly as

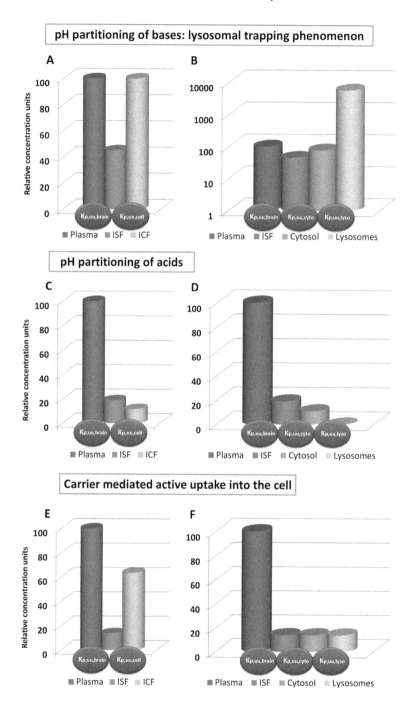

Fig. 13.5 The intracerebral unbound drug distribution patterns of prototypical drugs (**a** and **b**, thioridazine; **c** and **d**, salicylic acid; **e** and **f**, gabapentin). The distributional pattern depends on both the physicochemical properties of the compound and the functional characteristics of the

a result of its high lipophilicity (ClogP 6.0). The experimental $f_{u,plasma,rat}$ is 0.002, and the $V_{u,brain}$ is around 3000 mL · g brain^{-1}, the highest $V_{u,brain}$ observed so far (Friden et al. 2009a; Loryan et al. 2013). As a result, the determined $K_{p,brain}$ of 3.75 is significantly influenced by nonspecific binding to the brain tissue and plasma proteins. The $K_{p,uu,brain}$ is 0.45 (Friden et al. 2009b). *Lysosomal trapping* is the main reason for thioridazine accumulating in the cells. When the intracellular compartment is viewed as one unit, there is a 2.24-fold higher intracellular concentration of unbound thioridazine than in the brain ISF. Moreover, because of the presence of the physiological pH gradient, thioridazine as a base accumulates in the cytosol and then becomes trapped in the acidic intracellular compartments (Fig. 13.5b). The calculations (Eq. 13.19) indicate that, when the cytosolic compartment is separated from the lysosomal compartment, thioridazine will reach a 75-fold higher intralysosomal than cytosolic concentration. This type of distribution could be considered as a *signature pattern* for basic compounds.

Acidic compounds such as salicylic acid (see Fig. 13.5c) have a different distribution pattern in the brain. Only 19% of the unbound salicylic acid in the plasma crosses the BBB ($K_{p,uu,brain}$ = 0.19). Moreover, about 60% of the unbound salicylic acid in brain ISF equilibrates across the cellular barrier. Using the three-compartment pH partitioning model (Fig. 13.5d), it is possible to describe the unbound cytosolic and lysosomal partitioning coefficients and identify a lysosomal exclusion phenomenon ($K_{p,uu,lyso}$ = 0.015).

Active carrier-mediated transport into the cells is an alternative process which can be observed at the cellular barrier. Gabapentin provides a classic example of a compound lacking any nonspecific binding to the brain tissue ($f_{u,brain}$ = 1) while at the same time exhibiting active uptake into the cells (Fig. 13.5e). Due to the active passage of gabapentin into the cells by the L-α-amino acid transporter (Su et al. 1995), it reaches nearly five-fold higher intracellular concentrations on average. Additional examples of compounds undergoing active cellular uptake include

Fig. 13.5 (continued) compartments and membranes, defined by $K_{p,uu,brain}$ and $K_{p,uu,cell}$. The graphs were constructed from data in Friden et al. (2007, 2011). The unbound drug plasma concentration is set at 100 arbitrary units

(a) Efflux of thioridazine at the BBB ($K_{p,uu,brain}$ = 0.45) and its accumulation in the cells as described by a $K_{p,uu,cell}$ of 2.24. Because thioridazine is a weak base with a pKa of 8.9, it is subject to lysosomal trapping and accumulation in the cells. The pH partitioning of thioridazine (**b**) is described by the unbound thioridazine cytosolic ($K_{p,uu,cyto}$=1.72) and lysosomal ($K_{p,uu,lyso}$=75) partition coefficients computed using the three-compartment pH partitioning model. (**c and d**) The distribution and pH partitioning of salicylic acid. Salicylic acid is poorly transported across the BBB ($K_{p,uu,brain}$ = 0.19) and has reduced cellular penetration ($K_{p,uu,cell}$ = 0.62). Moreover, as an acid (pKa 4.3), salicylic acid has limited distribution in the brain tissue ($V_{u,brain}$ = 1 mL · g brain^{-1}). The pH partitioning model (**d**) supports the suggestion that salicylic acid is mainly distributed in the cytosol ($K_{p,uu,cyto}$ = 0.58) and is almost completely absent from acidic compartments such as lysosomes ($K_{p,uu,lyso}$ = 0.015). (**e and f**) The zwitterion gabapentin. Gabapentin transport in the BBB is restricted ($K_{p,uu,brain}$ = 0.14). However, after passing the BBB, it is excessively accumulated in the cells ($K_{p,uu,cell}$ = 4.55). The pH partitioning model is, however, incapable of identifying its uptake in the cells since the uptake is not related to lysosomal accumulation. Gabapentin is a substrate of the L-type amino acid transporter, which explains the observed active uptake into the cells. (Wang and Welty 1996; Friden et al. 2011)

1-methyl-4-phenylpyridinium (MPP, $K_{p,uu,cell}$ = 77) and tetraethylammonium (TEA, $K_{p,uu,cell}$ = 8.95) (Friden et al. 2011).

Because of the practical value of $K_{p,uu,cell}$ and its further division into $K_{p,uu,lyso}$ and $K_{p,uu,cyto}$, it is highly recommended that the unbound drug intra-to-extracellular concentration ratio be assessed in DMPK studies. Estimated neuroPK parameters are important contributors to the evaluation of the intracerebral distribution pattern of NCEs and their possible side effects.

13.5   Combinatory Mapping Approach

$K_{p,brain}$ estimated under steady-state conditions or using the area under the concentration-time curves in the brain tissue ($AUC_{tot,brain}$) and plasma ($AUC_{tot,plasma}$) after a single dose (Eq. 13.20) has historically been recognized as a driving force in CNS drug discovery screening programs (Pardridge 1989; Ghose et al. 1999, 2012) (see Chap. 7 for an explanation of the doctrines of brain PK).

$$K_{p,brain} = \frac{A_{brain}}{C_{tot,plasma}} = \frac{AUC_{tot,brain}}{AUC_{tot,plasma}} \tag{13.20}$$

The identification and selection of drug candidates with "acceptable brain penetration" has typically been based on pre-defined cut-off values for $K_{p,brain}$ (BB or often logBB); however, these vary between groups/companies. For instance, logBB = 0.3 ($K_{p,brain}$ = 2) has often been used as the cut-off point for NCE penetration of the BBB (Reichel 2006). Another approach uses an arbitrary cut-off point for $K_{p,brain}$ of greater than unity (Kalvass et al. 2007a; Padowski and Pollack 2011a). At Eli Lilly research laboratories, the cut-off point for $K_{p,brain}$, determined using a mouse brain uptake assay, was 0.3 (30%) (Raub et al. 2006). Alternatively, substances with $K_{p,brain}$ values higher than 0.04 (determined using the brain tissue with residual blood) have been considered "brain penetrants" by some, since this value exceeds the cerebral blood volume, approximated as 4% of the total brain volume (Hitchcock and Pennington 2006; Shaffer 2010). Basically, higher $K_{p,brain}$ values have frequently been considered to be favorable for CNS penetration (Young et al. 1988; Pardridge 1989; Ghose et al. 1999, 2012; Segall 2012). Despite the fact that it has been found to be inadequate for evaluation of the transport of drugs across the BBB and to be by no means foolproof, this type of "taxonomy" has been common practice in the pharmaceutical industry.

However, off-target binding of drug to plasma and brain tissues irrefutably masks the actual BBB net flux (see Chaps. 6, 13, and 14 for more detailed explanations). Currently, driven by abundant evidence supporting the "free-drug hypothesis," $K_{p,uu,brain}$ (also called $K_{p,free}$) is replacing $K_{p,brain}$.

Several scientists have tried to differentiate between the two main components of $K_{p,brain}$, i.e., nonspecific binding to tissues and free (unbound) drug (Lin et al. 1982; Kalvass and Maurer 2002; Mano et al. 2002; Maurer et al. 2005; Gupta et al. 2006;

Summerfield et al. 2006; Friden et al. 2007; Hammarlund-Udenaes et al. 2008; Liu et al. 2009a). For instance, Becker and Liu categorize the ratio of $f_{u,plasma}$ to $f_{u,brain}$ as an "intrinsic" partition coefficient between the brain and plasma ($K_{p,in}$) which could be considered a descriptor of nonspecific binding in brain and plasma (Becker and Liu 2006). It is, however, essential to bear in mind that $K_{p,in}$ and $K_{p,uu,brain}$ describe different properties of the compound, where $K_{p,in}$ describes the ratio of the binding properties without including BBB transport (if there is no observed active transport, $K_{p,brain} = K_{p,in}$), and $K_{p,uu,brain}$ specifically defines the BBB transport of unbound drug. $K_{p,in}$ cannot therefore be used to assess the $K_{p,uu,brain}$ of NCEs.

Alternative approaches to the use of microdialysis for determining $K_{p,uu,brain}$ that are based on the co-estimation of $K_{p,brain}$ and nonspecific binding to plasma and brain tissues have been established (Gupta et al. 2006; Friden et al. 2007; Hammarlund-Udenaes et al. 2008, 2009). Hence, $f_{u,plasma}$ can be used to correct $C_{tot,plasma}$ (binding to formal elements of the blood is excluded):

$$C_{u,plasma} = C_{tot,plasma} \cdot f_{u,plasma} \tag{13.21}$$

Correspondingly, $V_{u,brain}$ (mL $\cdot$ g brain^{-1}) or $f_{u,brain}$ corrected for pH partitioning ($f_{u,brain,corrected}$) is used to estimate $C_{u,brainISF}$ (μmol $\cdot$ g brain^{-1}):

$$C_{u,brainISF} = \frac{A_{brain}}{V_{u,brain}} = A_{brain} \cdot f_{u,brain,corected} \tag{13.22}$$

Accordingly, $K_{p,uu,brain}$ can be derived from Eq. 13.20 as

$$K_{p,uu,brain} = \frac{K_{p,brain}}{V_{u,brain} \cdot f_{u,plasma}} = \frac{K_{p,brain}}{\dfrac{1}{f_{u,brain,corrected}} \cdot f_{u,plasma}} \tag{13.23}$$

Because this method (Eq. 13.23) is based on several individually determined parameters obtained using various techniques, the level of uncertainty and variability in the final $K_{p,uu,brain}$ estimates is increased. Therefore, reduction of the potential uncertainty in each measurement ($K_{p,brain}$, $V_{u,brain}$, $f_{u,brain}$, $f_{u,plasma}$) will make assessment of the brain partitioning coefficient for unbound drug more secure in drug discovery. Some critical steps in determining the brain partitioning coefficient for total drug, required for the assessment of $K_{p,uu,brain}$, are described below.

Ideally, the brain partitioning coefficient would be determined using steady-state total brain and plasma concentrations after constant-rate intravenous infusion (Friden et al. 2009b; Hammarlund-Udenaes et al. 2009). However, in drug discovery and development setups, intravenous infusions can be challenging and consequently are often not an option. Alternatively, $K_{p,brain}$ can be determined as the AUC$_{tot,brain}$/AUC$_{tot,plasma}$ ratio (Eq. 13.20), using various time points (up to five animals per time point) after a single (discrete) dose. In fact, subcutaneous administration is most commonly used, because it decreases the inter-experimental variability, mainly as a result of the compounds circumventing oral absorption and first-pass metabolism. In some cases, $K_{p,brain}$ is assessed using total brain and plasma concentrations obtained at a specific point in time after drug administration. However, this approach has been heavily criticized since it is known that $K_{p,brain}$ is a time-dependent parameter (Padowski and Pollack 2011a, b). In this regard, Padowski and Pollack have suggested the use of different notations of $K_{p,brain}$ with the intention of specifying the conditions under which brain exposure has been determined, i.e., $K_{p,brain,t}$ (single time point), $K_{p,brain,DE}$ (distributional equilibrium reached), and $K_{p,brain,SS}$ (in a steady state system) (Padowski and Pollack 2011a). These researchers have also used a simulation approach to study the links between $K_{p,brain,t}$ with a sampling time prior to the point of distribution equilibrium and the experimentally obtained $K_{p,brain}$ in the presence vs absence of P-gp efflux transport. In some cases, an initial overshoot or increase in $K_{p,brain,t}$ values was followed by a decline to a value which remained constant with time. Consequently, it was concluded that the P-gp effect estimated based on a $K_{p,brain}$ value prior to reaching distribution equilibrium could be significantly inaccurate. The experimental design will thus greatly influence the conclusions made. The simulations also indicated that assessment of the P-gp effect was more precise and less variable with intravenous constant-rate infusions than with bolus administration, i.e., that $K_{p,brain SS}$ was the most appropriate choice (Gibaldi 1969; Padowski and Pollack 2011b). Although the proposed ranking of these parameters certainly introduces clarity and flags the importance of potential time-dependent differences in BBB equilibration, it has not been followed up in practice to any great extent (in this chapter $K_{p,brain}$ refers to $K_{p,brain,ss}$).

The correlation between $K_{p,brain}$ derived from a single (discrete) dose and that derived at steady-state has also been investigated in an attempt to improve throughput in neuroPK studies in the industrial setting. For instance, $K_{p,brain}$ values derived from a single dose differed maximally 2.5-fold from the steady-state values for eight of the nine commercial and two proprietary compounds tested (>2.5-fold for thiopental) (Liu et al. 2009a; Doran et al. 2012). These results give the impression that the single-dose approach, which is more time-efficient, may not compromise data quality to any great extent.

Another approach, which was introduced with the intention of reducing the use of animals and improving the efficiency of investigations into CNS exposure in drug discovery programs, uses a mixture of up to five NCEs administered together, termed a *cocktail*, *cassette*, or *Nin1* (Manitpisitkul and White 2004; Friden et al. 2009b; Liu et al. 2012). Liu and colleagues investigated the brain partitioning coefficients of 11 model compounds using discrete and cassette dosing and discovered that drug-drug interactions at the BBB level are unlikely at these low subcutaneous cassette doses (Liu et al. 2012). Nevertheless, it is advisable to administer low doses of the drugs during the experiment to prevent any interactions at the BBB as well as potential side effects. Overall, the route and duration of administration, the dose (discrete or cassette dosing), and the brain and plasma tissue sampling times should be critically evaluated prior to the experiment to avoid potential pitfalls.

Methods for correcting the residual blood in the sampled brain tissue also need to be considered. Using $V_{u,brain}$ to determine $K_{p,uu,brain}$, Friden et al. showed that the literature values for V_{blood} may be too high when used for correcting A_{brain} for the residual blood (Friden et al. 2010). This was especially observed for drugs with low $K_{p,brain}$ values. A low $K_{p,brain}$ value can be caused by either very efficient efflux at the BBB or plasma protein binding that greatly exceeds the nonspecific binding of the drug in the brain. The latter becomes a problem when using a value for V_{blood} that is too high. An improved method has been developed for this estimation (Friden et al. 2010). It should be noted that the remaining brain vascular space varies with the method used to sacrifice the animal.

The correction for residual blood can be calculated from the effective plasma space in the brain for a given drug, V_{eff}, which in turn can be calculated from the measured plasma protein binding according to

$$V_{eff} = f_{u,plasma} \cdot V_{water} + \left(1 - f_{u,plasma}\right) \cdot V_{protein} \tag{13.24}$$

V_{water} and $V_{protein}$ in rat brain capillary blood have been estimated as 7.99 µl · g brain^{-1} and 10.3 µl · g brain^{-1}, respectively (Friden et al. 2010). This equation can be used when binding to blood elements is not significant. The amount of drug in the brain tissue excluding the capillary contents, A_{brain}, can be calculated as

$$A_{brain} = \frac{C_{tot,brain} - V_{eff} \cdot C_{tot,plasma}}{1 - V_{water}} \tag{13.25}$$

$C_{tot,brain}$ is the concentration of drug in the whole brain tissue sample, and $C_{tot,plasma}$ is the drug concentration in a regular (arterial) plasma sample. The total physical volume of residual blood in the rat brain after exsanguination by severing the heart has been estimated as 12.7 µl · g brain^{-1} (Friden et al. 2010).

The complexity of the processes governing the drug concentrations in the brain requires the input of several methods, each providing a defined piece of the information required to assemble a more in-depth picture of drug disposition in the CNS on the level of the entire brain or the brain regions of interest (Loryan et al. 2014, 2016). Using the CMA, it is possible in the early drug development phases to map the concentrations of unbound drug in the main pharmacokinetic compartments relevant to drug disposition in the brain, such as plasma, ISF, ICF (and if necessary lysosomes), and CSF. The compartments and relevant concentration relationships are illustrated in Fig. 13.6, using the atypical antidepressant bupropion as a model.

The main benefit of this mapping approach is the visualization and better understanding of the target site PK. Additionally, it allows the ranking of the compounds based on the target compartment unbound drug concentration normalized by the PD parameters (EC50, IC50, Ki, etc.) as well as in the design of new PK/PD studies.

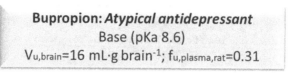

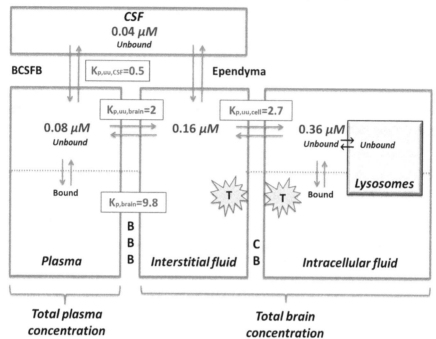

Fig. 13.6 Schematic representation of the distribution of a drug, here exemplified by the atypical antidepressant bupropion, into the different compartments (plasma, brain ISF, brain ICF, lysosomes, and CSF) involved in the disposition of drugs across the barriers (BBB, CB, and BCSFB), with the resulting concentrations obtained in each compartment. T represents the possible target sites of the drug, facing either the ISF or the ICF. The graph was constructed using steady-state total plasma, total brain, and CSF concentration determinations in rats after a 4-h constant-rate intravenous infusion of bupropion 2 (μmol/kg)/h (Friden et al. 2009b). Using this model and given the unbound drug plasma concentration, it is possible to estimate the target site concentrations. This approach can be used in drug discovery programs for establishing the link between the PK and engagement of the target. The $K_{p,uu,CSF}$ is quite different from the $K_{p,uu,brain}$ for bupropion, which means that estimations of the target site concentrations will be less valuable if based on CSF measurements

13.6 Translational Aspects of the Methods

In the drug discovery process, in vitro assays and preclinical animal studies are widely used to evaluate the potency of NCEs and to identify candidates that may have desirable clinical responses.

However, when there is no correlation between in vivo and in vitro potencies, the validity of the in vitro assay, the animal model, and the target can be questioned

(Brunner et al. 2012). Translational science is the study of the extrapolation of experimental findings to clinical solutions. It is important to improve the proficiency of clinical trial design by planning clinical doses based on nonclinical results. Animal brain PK studies are a routine tool for predicting drug behavior in humans. Thus far, it has been extremely challenging to master the translation of in vitro-to-in vivo and animal-to-human data in the drug discovery process, primarily because of the shortage of supportive data and the underlying multiple assumptions. Some translational aspects linked to methodologies described in this chapter are discussed below.

13.6.1 Translational Aspects of Brain Tissue Binding Assays

It is important to estimate the cerebral concentrations of unbound neurotherapeutic drugs in various species and related these to the potential CNS activity and target engagement of the drugs in preclinical and clinical PK studies. The $f_{u,brain}$ of drug candidates is routinely determined in several species to account for possible species dependence, as is the case with plasma protein binding, although this does not fit with experimental results demonstrating that brain tissue binding is less sensitive to interspecies dissimilarities than plasma protein binding (Summerfield et al. 2007; Wan et al. 2009; Read and Braggio 2010). In fact, when Di et al. evaluated the degree and nature of potential species differences in brain tissue binding, they found that brain tissue binding is species independent when studying healthy mammals (Di et al. 2011). This finding was very beneficial for translational medicine because it meant that a single representative species such as the rat could replace multispecies determinations of $f_{u,brain}$. However, a recent study on drug brain-regional brain tissue binding investigated in postmortem material obtained from patients with Alzheimer's disease revealed extensive intra- and interindividual variability that is more pronounced in disease conditions (Gustafsson et al. 2019). The findings highlight the need of investigation of brain tissue binding also in pathological conditions.

Laboratory studies have not found any significant dissimilarities in estimated $V_{u,brain}$ values from fresh brain slices between Sprague-Dawley rats and NMRI mice (Fig. 13.7). However, more systematic investigations are required to support the possibility of the interchangeable use of $V_{u,brain}$ measurements for translational studies.

13.6.2 Translational Aspects of Brain Exposure Assessment

In the drug industry, the translation of drug tissue distribution data between species is grounded on the assumption that the tissue-to-plasma drug partitioning coefficient for passive transport is tissue- and species-independent. However, the available information in the literature supports the existence of interspecies differences

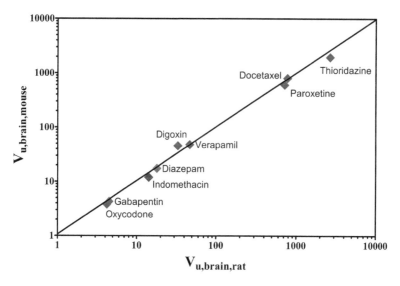

Fig. 13.7 The relationship between rat (x-axis) and mouse (y-axis) brain slices for the estimation of the volume of distribution of unbound drug in the brain of ten compounds ($V_{u,brain}$; mL · g brain^{-1}). The solid line represents the line of identity. The color of the diamonds represents the ion class of the compound (bases: thioridazine, docetaxel, paroxetine, verapamil, digoxin, oxycodone; neutral: diazepam; acid: indomethacin; zwitterion: gabapentin). Constructed from data in Loryan et al. (2013)

in the lipid composition of the tissues, which is considered to be the main factor in drug binding to tissues (Rouser et al. 1969; Simon and Rouser 1969). Elaborate investigation of tissue lipid composition with regard to drug distribution in dogs and rats has demonstrated clear differences between the animals; e.g., the proportion of neutral lipids was fivefold lower in dog brain than in rat brain (Rodgers et al. 2012). The authors suggested that the assumption of constancy in tissue-to-plasma partitioning should be used with caution when species-specific tissue distribution is of interest. Nevertheless, based on a widely accepted measure of prediction that describes the number of compounds that fall within a two- to threefold range, various groups have demonstrated the reliability of rodent-derived PK parameters for predicting BBB net flux in humans and large animals, although this has mainly been for compounds with predominantly passive transport (Friden et al. 2009b; Di et al. 2012a; Doran et al. 2012; Kielbasa and Stratford Jr. 2012; Westerhout et al. 2012). For instance, Doran and colleagues showed that preclinical rat-derived neuroPK parameters, particularly $K_{p,uu,brain}$, can be used to extrapolate $C_{u,brainISF}$ in dogs and nonhuman primates for freely permeating non-P-gp substrates (Doran et al. 2012). In contrast, the prediction of $C_{u,brainISF}$ for P-gp substrates such as risperidone and 9-hydroxyrisperidone using a similar approach was significantly flawed, with underprediction of $K_{p,uu,brain}$ in dogs and nonhuman primates from rat-derived data. Several reports describing species differences in brain exposure measurements have also been documented (Dagenais et al. 2001; Syvanen et al. 2008; Syvanen et al.

2009; Bundgaard et al. 2012a). In addition, to strengthen the translation of neuroPK parameters to patients, it is critical to investigate the CNS exposure in animal models mimicking the diseases of interest, and if possible in patients.

Issues related to the disequilibrium of drug concentrations at the BBB make it difficult to rank the importance of the PK parameters for the translation. Consequently, it is critical to assess the extent of human BBB transport and evaluate the potential impact of the degree of asymmetry on brain exposure in relation to target engagement or pharmacological activity early in drug discovery and development programs. The main reason for the observed asymmetry in BBB equilibration is the species-specific presence of efflux and influx transporters (see Chap. 4 for an overview of BBB transporters and pharmacoproteomics). There is no doubt that P-gp is one of the most important efflux transporters at the BBB (Tsuji et al. 1992, 1993; Terasaki and Hosoya 1999; Demeule et al. 2002; Mizuno et al. 2003; Lin 2004; Syvanen et al. 2008; Kodaira et al. 2011; Uchida et al. 2011a; Agarwal et al. 2012). However, the relative importance of P-gp in humans and rats was questioned after the breast cancer resistance protein (BCRP) was found to be the most abundant protein expressed in the human BBB (Uchida et al. 2011b). Nonetheless, cell lines transfected with human transporters, mostly only P-pg, are often used in lead optimization and candidate selection in the preclinical phases of drug discovery (see Chaps. 8 and 9 for a comprehensive overview of cell culture models of the BBB). Transporter knockout animals or chemically "knocked-out" animals (i.e., after the administration of P-gp or BCRP inhibitors) are used in drug discovery projects (see Chap. 10 for an exploration of in situ and in vivo animal models and Chap. 15 for the current thinking on this topic in the drug industry). Regardless of the "solid" status of in vitro and in vivo P-gp assays in drug discovery, both the rationale of the applied methods and the interpretation of the obtained results are debatable. Overall, it remains challenging to predict the BBB net flux of potential transporter substrates from rodent data. Consequently, due to the lack of translational knowledge, the recommendation not to advance efflux transporter substrates is often promoted in the drug industry (Di et al. 2012a).

The most critical issue in the assessment of brain exposure is related to using methods that can be applied for the same purpose across species including humans. In this regard, PET (see Chap. 11) has multiple advantages and is so far the most reliable technique that could be applied for translational purpose (Syvanen et al. 2009; Bauer et al. 2012; Wanek et al. 2013). The biggest challenge with PET is associated with the fact that total radioactivity is measured in both brain and blood. There are a few attempts to apply correction to total brain concentration obtained via PET using both $f_{u,brain}$ and $V_{u,brain}$ for the assessment of unbound cerebral concentrations (Gunn et al. 2012; Schou et al. 2015). Combined PET and brain microdialysis study design allows better understanding of the relationship between unbound and total concentrations and the convertibility between the methods (Gustafsson et al. 2017). In spite of multiple advantages, PET is considered too elaborate and expensive for screening purposes, and, hence, it is used only in later stages of drug development.

13.7 Current Status and Future Directions

Notwithstanding the immense progress in the understanding of drug delivery to the brain and improved screening cascades in drug discovery programs, the clinical success rate for novel neurotherapeutics is exceptionally low at present (Butlen-Ducuing et al. 2016; Cummings et al. 2016; Mehta et al. 2017; Danon et al. 2019). The approaches to the selection and optimization of compounds with sufficient delivery to the brain in drug discovery are currently stereotypical, high-throughput methods in most pharmaceutical companies. The complications associated with the measurement of active-site concentrations for potential CNS drugs have made surrogate methodologies (such as the assessment of brain ISF drug concentrations using matrices such as CSF and plasma) popular. There have been advancements in methodologies related to the assessment of $C_{u,brainISF}$, making it easier to measure the actual value rather than a surrogate. The use of the CSF as a relevant surrogate for $C_{u,brainISF}$ has been extensively investigated to support the rationale of its use in drug discovery (de Lange and Danhof 2002; de Lange et al. 2005; Liu et al. 2006, 2009a; Lin 2008; Friden et al. 2009b; Di et al. 2012a; de Lange 2013b; Loryan et al. 2014; Yamamoto et al. 2017; Ketharanathan et al. 2019; Vendel et al. 2019). Issues related to the sampling of CSF and interpretations of the data are discussed in Chaps. 4 and 15. However, despite the progress made, problems related to the veracity of the predicted values remain. It is important to remember that "you get what you measure" (Elebring et al. 2012), meaning that the definitions of the parameters and/or appropriate surrogates including critical interpretation are crucial. In this regard, understanding brain regional drug disposition in health and pathological conditions will require more focused investigations in the future, in order to improve the translational value of neuroPK parameters of NCEs.

The establishment of a PKPD relationship very early in drug development is a great advance for drug discovery (Chap. 15) (Bostrom et al. 2006; Westerhout et al. 2011; Stevens et al. 2012). Strategies to minimize neurotoxicity for non-CNS compounds are also of great interest (Wager et al. 2012). The exploration of the potential of mathematical modeling, particularly physiologically based PKPD modeling in drug discovery programs, will facilitate better understanding of the BBB transport of small molecules (Yamamoto et al. 2017; Vendel et al. 2019).

The role of the efflux and influx transporters and their potential interactions require investigation to provide further insight into active BBB transport supported by disease-specific BBB transcriptome and proteome atlases. We need to learn how to incorporate our knowledge on BBB cellular proteomics into drug transport mathematical modeling. Advancements in our understanding of pathological conditions (Part V) and their influences on the most important neuroPK parameters ($K_{p,uu,brain}$, $V_{u,brain}$, and $K_{p,uu,cell}$) will also improve the translational aspects of drug discovery.

13.8 Points for Discussion (Questions)

- What are the conceptual differences between the brain homogenate and brain slice methods?
- What allows the combination of the brain homogenate and brain slice methods to approximate the cellular unbound drug partitioning coefficient?
- Discuss the driving forces of BBB and CB drug transport.
- What is the physiological basis and pharmacokinetic impact of lysosomotropism for basic compounds?
- How can information about whether the compound is lysosomotropic influence the drug discovery and development processes?
- What are the pros and cons of using cut-off values for $K_{p,uu,brain}$ in relation to the evaluation of drug target engagement?
- Which neuroPK parameters are critical for translational medicine?
- Discuss the impact of the threefold difference cut-off point for methods in drug discovery.
- In which phase of drug discovery is it best to investigate the BBB transport and brain drug distribution of NCEs?
- What changes in neuroPK in particular diseases?

References

Abbott NJ (2004) Evidence for bulk flow of brain interstitial fluid: significance for physiology and pathology. Neurochem Int 45(4):545–552

Abbott NJ, Patabendige AA, Dolman DE, Yusof SR, Begley DJ (2010) Structure and function of the blood-brain barrier. Neurobiol Dis 37(1):13–25

Abbott NJ, Pizzo ME, Preston JE, Janigro D, Thorne RG (2018) The role of brain barriers in fluid movement in the CNS: is there a 'glymphatic' system? Acta Neuropathol 135(3):387–407

Abraham MH (2011) The permeation of neutral molecules, ions, and ionic species through membranes: brain permeation as an example. J Pharm Sci 100(5):1690–1701

Agarwal S, Uchida Y, Mittapalli RK, Sane R, Terasaki T, Elmquist WF (2012) Quantitative Proteomics of Transporter Expression in Brain Capillary Endothelial Cells Isolated from P-gp, BCRP, and P-gp/BCRP Knockout Mice. Drug Metab Dispos

Banker MJ, Clark TH, Williams JA (2003) Development and validation of a 96-well equilibrium dialysis apparatus for measuring plasma protein binding. J Pharm Sci 92(5):967–974

Bauer M, Zeitlinger M, Karch R, Matzneller P, Stanek J, Jager W, Bohmdorfer M, Wadsak W, Mitterhauser M, Bankstahl JP, Loscher W, Koepp M, Kuntner C, Muller M, Langer O (2012) Pgp-mediated interaction between (R)-[11C]verapamil and tariquidar at the human blood-brain barrier: a comparison with rat data. Clin Pharmacol Ther 91(2):227–233

Bean JW, Sargent DF, Schwyzer R (1988) Ligand/receptor interactions – the influence of the microenvironment on macroscopic properties. Electrostatic interactions with the membrane phase. J Recept Res 8(1–4):375–389

Becker S, Liu X (2006) Evaluation of the utility of brain slice methods to study brain penetration. Drug Metab Dispos 34(5):855–861

Begley DJ (1996) The blood-brain barrier: principles for targeting peptides and drugs to the central nervous system. J Pharm Pharmacol 48(2):136–146

Begley DJ (2004) Delivery of therapeutic agents to the central nervous system: the problems and the possibilities. Pharmacol Ther 104(1):29–45

Bendayan R, Lee G, Bendayan M (2002) Functional expression and localization of P-glycoprotein at the blood brain barrier. Microsc Res Tech 57(5):365–380

Benkwitz C, Liao M, Laster MJ, Sonner JM, Eger EI 2nd, Pearce RA (2007) Determination of the EC50 amnesic concentration of etomidate and its diffusion profile in brain tissue: implications for in vitro studies. Anesthesiology 106(1):114–123

Berry LM, Roberts J, Be X, Zhao Z, Lin MH (2010) Prediction of V(ss) from in vitro tissue-binding studies. Drug Metab Dispos 38(1):115–121

Blasberg R, Levi G, Lajtha A (1970) A comparison of inhibition of steady state, new transport, and exchange fluxes of amino acids in brain slices. Biochim Biophys Acta 203(3):464–483

Bostrom E, Simonsson US, Hammarlund-Udenaes M (2006) In vivo blood-brain barrier transport of oxycodone in the rat: indications for active influx and implications for pharmacokinetics/pharmacodynamics. Drug Metab Dispos 34(9):1624–1631

Bouw MR, Xie R, Tunblad K, Hammarlund-Udenaes M (2001) Blood-brain barrier transport and brain distribution of morphine-6-glucuronide in relation to the antinociceptive effect in rats--pharmacokinetic/pharmacodynamic modelling. Br J Pharmacol 134 (8):1796-1804

Boya P, Kroemer G (2008) Lysosomal membrane permeabilization in cell death. Oncogene 27(50):6434–6451

Boya P, Andreau K, Poncet D, Zamzami N, Perfettini JL, Metivier D, Ojcius DM, Jaattela M, Kroemer G (2003) Lysosomal membrane permeabilization induces cell death in a mitochondrion-dependent fashion. J Exp Med 197(10):1323–1334

Brunner D, Balci F, Ludvig EA (2012) Comparative psychology and the grand challenge of drug discovery in psychiatry and neurodegeneration. Behav Processes 89(2):187–195

Bundgaard C, Jorgensen M, Mork A (2007) An integrated microdialysis rat model for multiple pharmacokinetic/pharmacodynamic investigations of serotonergic agents. J Pharmacol Toxicol Methods 55(2):214–223

Bundgaard C, Jensen CJ, Garmer M (2012a) Species comparison of in vivo P-glycoprotein-mediated brain efflux using mdr1a-deficient rats and mice. Drug Metab Dispos 40(3):461–466

Bundgaard C, Sveigaard C, Brennum LT, Stensbol TB (2012b) Associating in vitro target binding and in vivo CNS occupancy of serotonin reuptake inhibitors in rats: the role of free drug concentrations. Xenobiotica 42(3):256–265

Butlen-Ducuing F, Petavy F, Guizzaro L, Zienowicz M, Haas M, Alteri E, Salmonson T, Corruble E (2016) Regulatory watch: Challenges in drug development for central nervous system disorders: a European Medicines Agency perspective. Nat Rev Drug Discov 15(12):813–814

Chen X, Loryan I, Payan M, Keep RF, Smith DE, Hammarlund-Udenaes M (2014) Effect of transporter inhibition on the distribution of cefadroxil in rat brain. Fluids Barriers CNS 11(1):25

Chesney MA, Perouansky M, Pearce RA (2003) Differential uptake of volatile agents into brain tissue in vitro. Measurement and application of a diffusion model to determine concentration profiles in brain slices. Anesthesiology 99(1):122–130

Collingridge GL (1995) The brain slice preparation: a tribute to the pioneer Henry McIlwain. J Neurosci Methods 59(1):5–9

Cserr HF, Patlak, C.S. (1992) Secretion and bulk flow of interstitial fluid. In: Bradbury, MWB (Ed), Physiology and Pharmacology of the Blood-Brain Barrier Springer, Berlin:245-261

Cummings J, Aisen PS, DuBois B, Frolich L, Jack CR Jr, Jones RW, Morris JC, Raskin J, Dowsett SA, Scheltens P (2016) Drug development in Alzheimer's disease: the path to 2025. Alzheimers Res Ther 8:39

Dagenais C, Zong J, Ducharme J, Pollack GM (2001) Effect of mdr1a P-glycoprotein gene disruption, gender, and substrate concentration on brain uptake of selected compounds. Pharm Res 18(7):957–963

Daniel WA (2003) Mechanisms of cellular distribution of psychotropic drugs. Significance for drug action and interactions. Prog Neuropsychopharmacol Biol Psychiatry 27(1):65–73

Daniel WA, Wojcikowski J (1997) Contribution of lysosomal trapping to the total tissue uptake of psychotropic drugs. Pharmacol Toxicol 80(2):62–68

Daniel WA, Wojcikowski J (1999a) Lysosomal trapping as an important mechanism involved in the cellular distribution of perazine and in pharmacokinetic interaction with antidepressants. Eur Neuropsychopharmacol 9(6):483–491

Daniel WA, Wojcikowski J (1999b) The role of lysosomes in the cellular distribution of thioridazine and potential drug interactions. Toxicol Appl Pharmacol 158(2):115–124

Daniel WA, Bickel MH, Honegger UE (1995) The contribution of lysosomal trapping in the uptake of desipramine and chloroquine by different tissues. Pharmacol Toxicol 77(6):402–406

Daniel WA, Syrek M, Haduch A, Wojcikowski J (1998) Pharmacokinetics of phenothiazine neuroleptics after chronic coadministration of carbamazepine. Pol J Pharmacol 50(6):431–442

Daniel WA, Syrek M, Haduch A, Wojcikowski J (2000) Different effects of amitriptyline and imipramine on the pharmacokinetics and metabolism of perazine in rats. J Pharm Pharmacol 52(12):1473–1481

Danon JJ, Reekie TA, Kassiou M (2019) Challenges and Opportunities in Central Nervous System Drug Discovery. Trends in Chemistry 1(6):612–624

Davson H, Segal, M.B. (1995) Physiology of the CSF and Blood-Brain Barriers. CRC Press, Boca Raton, USA

Davson H, Hollingsworth G, Segal MB (1970) The mechanism of drainage of the cerebrospinal fluid. Brain 93(4):665–678

de Boer AG, Gaillard PJ (2007) Drug targeting to the brain. Annu Rev Pharmacol Toxicol 47:323–355

De Duve C (1970) The role of lysosomes in cellular pathology. Triangle 9(6):200–208

De Duve C (1971) Tissue fractionation. Past and present. J Cell Biol 50(1):20d–55d

de Duve C (1975) The role of lysosomes in the pathogeny of disease. Scand J Rheumatol Suppl 12:63–66

de Lange EC (2013a) The mastermind approach to CNS drug therapy: translational prediction of human brain distribution, target site kinetics, and therapeutic effects. Fluids Barriers CNS 10(1):12

de Lange EC (2013b) Utility of CSF in translational neuroscience. J Pharmacokinet Pharmacodyn

de Lange EC, Danhof M (2002) Considerations in the use of cerebrospinal fluid pharmacokinetics to predict brain target concentrations in the clinical setting: implications of the barriers between blood and brain. Clin Pharmacokinet 41(10):691–703

de Lange EC, de Boer BA, Breimer DD (1999) Microdialysis for pharmacokinetic analysis of drug transport to the brain. Adv Drug Deliv Rev 36(2–3):211–227

de Lange EC, Ravenstijn PG, Groenendaal D, van Steeg TJ (2005) Toward the prediction of CNS drug-effect profiles in physiological and pathological conditions using microdialysis and mechanism-based pharmacokinetic-pharmacodynamic modeling. AAPS J 7(3):E532–E543

Demeule M, Regina A, Jodoin J, Laplante A, Dagenais C, Berthelet F, Moghrabi A, Beliveau R (2002) Drug transport to the brain: key roles for the efflux pump P-glycoprotein in the blood-brain barrier. Vascul Pharmacol 38(6):339–348

Deo AK, Theil FP, Nicolas JM (2013) Confounding Parameters in Preclinical Assessment of Blood-Brain Barrier Permeation: An Overview With Emphasis on Species Differences and Effect of Disease States. Mol Pharm

Di L, Umland JP, Chang G, Huang Y, Lin Z, Scott DO, Troutman MD, Liston TE (2011) Species independence in brain tissue binding using brain homogenates. Drug Metab Dispos 39(7):1270–1277

Di L, Rong H, Feng B (2012a) Demystifying Brain Penetration in Central Nervous System Drug Discovery. J Med Chem

Di L, Umland JP, Trapa PE, Maurer TS (2012b) Impact of recovery on fraction unbound using equilibrium dialysis. J Pharm Sci 101(3):1327–1335

Doran A, Obach RS, Smith BJ, Hosea NA, Becker S, Callegari E, Chen C, Chen X, Choo E, Cianfrogna J, Cox LM, Gibbs JP, Gibbs MA, Hatch H, Hop CE, Kasman IN, Laperle J, Liu

J, Liu X, Logman M, Maclin D, Nedza FM, Nelson F, Olson E, Rahematpura S, Raunig D, Rogers S, Schmidt K, Spracklin DK, Szewc M, Troutman M, Tseng E, Tu M, Van Deusen JW, Venkatakrishnan K, Walens G, Wang EQ, Wong D, Yasgar AS, Zhang C (2005) The impact of P-glycoprotein on the disposition of drugs targeted for indications of the central nervous system: evaluation using the MDR1A/1B knockout mouse model. Drug Metab Dispos 33(1):165–174

Doran AC, Osgood SM, Mancuso JY, Shaffer CL (2012) An Evaluation of Using Rat-derived Single-dose Neuropharmacokinetic Parameters to Project Accurately Large Animal Unbound Brain Drug Concentrations. Drug Metab Dispos

Dos-Anjos S, Martinez-Villayandre B, Montori S, Salas A, Perez-Garcia CC, Fernandez-Lopez A (2008) Quantitative gene expression analysis in a brain slice model: influence of temperature and incubation media. Anal Biochem 378(1):99–101

Duvvuri M, Krise JP (2005) A novel assay reveals that weakly basic model compounds concentrate in lysosomes to an extent greater than pH-partitioning theory would predict. Mol Pharm 2(6):440–448

Eaton WW, Martins SS, Nestadt G, Bienvenu OJ, Clarke D, Alexandre P (2008) The burden of mental disorders. Epidemiol Rev 30:1–14

Elebring T, Gill A, Plowright AT (2012) What is the most important approach in current drug discovery: doing the right things or doing things right? Drug Discov Today 17(21–22):1166–1169

Elmquist WF, Sawchuk RJ (1997) Application of microdialysis in pharmacokinetic studies. Pharm Res 14(3):267–288

Elmquist WF, Sawchuk RJ (2000) Use of microdialysis in drug delivery studies. Adv Drug Deliv Rev 45(2–3):123–124

Farrington GK, Caram-Salas N, Haqqani AS, Brunette E, Eldredge J, Pepinsky B, Antognetti G, Baumann E, Ding W, Garber E, Jiang S, Delaney C, Boileau E, Sisk WP, Stanimirovic DB (2014) A novel platform for engineering blood-brain barrier-crossing bispecific biologics. FASEB J 28(11):4764–4778

Fenstermacher JD (1992) The blood-brain barrier is not a "barrier" for many drugs. NIDA Res Monogr 120:108–120

Fenstermacher J, Kaye T (1988) Drug Diffusion within the Brain. Ann Ny Acad Sci 531:29–39

Fichtl B, Von Nieciecki A, Walter K (1991a) Tissue binding versus plasma binding of drugs: general principles and pharmacokinetic consequences. In: Testa B, advances in drug research, vol 20. Academic Press, pp 117–166

Fichtl B, Von Nieciecki A, Walter K (1991b) ChemInform abstract: tissue binding versus plasma binding of drugs: general principles and pharmacokinetic consequences. ChemInform 22

Freitas RA (1999) Nanomedicine, volume I: basic capabilities

Freskgard PO, Urich E (2017) Antibody therapies in CNS diseases. Neuropharmacology 120:38–55

Friden M, Gupta A, Antonsson M, Bredberg U, Hammarlund-Udenaes M (2007) In vitro methods for estimating unbound drug concentrations in the brain interstitial and intracellular fluids. Drug Metab Dispos 35(9):1711–1719

Friden M, Ducrozet F, Middleton B, Antonsson M, Bredberg U, Hammarlund-Udenaes M (2009a) Development of a high-throughput brain slice method for studying drug distribution in the central nervous system. Drug Metab Dispos 37(6):1226–1233

Friden M, Winiwarter S, Jerndal G, Bengtsson O, Wan H, Bredberg U, Hammarlund-Udenaes M, Antonsson M (2009b) Structure-brain exposure relationships in rat and human using a novel data set of unbound drug concentrations in brain interstitial and cerebrospinal fluids. J Med Chem 52(20):6233–6243

Friden M, Ljungqvist H, Middleton B, Bredberg U, Hammarlund-Udenaes M (2010) Improved measurement of drug exposure in the brain using drug-specific correction for residual blood. J Cereb Blood Flow Metab 30(1):150–161

Friden M, Bergstrom F, Wan H, Rehngren M, Ahlin G, Hammarlund-Udenaes M, Bredberg U (2011) Measurement of unbound drug exposure in brain: modeling of pH partitioning explains diverging results between the brain slice and brain homogenate methods. Drug Metab Dispos 39(3):353–362

Friden M, Wennerberg M, Antonsson M, Sandberg-Stall M, Farde L, Schou M (2014) Identification of positron emission tomography (PET) tracer candidates by prediction of the target-bound fraction in the brain. EJNMMI Res 4(1):50

Fung EN, Chen YH, Lau YY (2003) Semi-automatic high-throughput determination of plasma protein binding using a 96-well plate filtrate assembly and fast liquid chromatography-tandem mass spectrometry. J Chromatogr B Analyt Technol Biomed Life Sci 795(2):187–194

Gaillard PJ, de Boer AG (2000) Relationship between permeability status of the blood-brain barrier and in vitro permeability coefficient of a drug. Eur J Pharm Sci 12(2):95–102

Gaillard PJ, Visser CC, Appeldoorn CC, Rip J (2012) Targeted Blood-to-Brain Drug Delivery - 10 Key Development Criteria. Curr Pharm Biotechnol 13(12):2328–2339

Garberg P, Ball M, Borg N, Cecchelli R, Fenart L, Hurst RD, Lindmark T, Mabondzo A, Nilsson JE, Raub TJ, Stanimirovic D, Terasaki T, Oberg JO, Osterberg T (2005) In vitro models for the blood-brain barrier. Toxicol In Vitro 19(3):299–334

Garcia-Garcia E, Andrieux K, Gil S, Couvreur P (2005) Colloidal carriers and blood-brain barrier (BBB) translocation: a way to deliver drugs to the brain? Int J Pharm 298(2):274–292

Ghose AK, Viswanadhan VN, Wendoloski JJ (1999) A knowledge-based approach in designing combinatorial or medicinal chemistry libraries for drug discovery. 1. A qualitative and quantitative characterization of known drug databases. J Comb Chem 1(1):55–68

Ghose AK, Herbertz T, Hudkins RL, Dorsey BD, Mallamo JP (2012) Knowledge-Based, Central Nervous System (CNS) Lead Selection and Lead Optimization for CNS Drug Discovery. ACS Chem Neurosci 3(1):50–68

Gibaldi M (1969) Effect of mode of administration on drug distribution in a two-compartment open system. J Pharm Sci 58(3):327–331

Glees P, Voth D (1988) Clinical and ultrastructural observations of maturing human frontal cortex. Part I (Biopsy material of hydrocephalic infants). Neurosurg Rev 11(3–4):273–278

Gredell JA, Turnquist PA, Maciver MB, Pearce RA (2004) Determination of diffusion and partition coefficients of propofol in rat brain tissue: implications for studies of drug action in vitro. Br J Anaesth 93(6):810–817

Gunn RN, Summerfield SG, Salinas CA, Read KD, Guo Q, Searle GE, Parker CA, Jeffrey P, Laruelle M (2012) Combining PET biodistribution and equilibrium dialysis assays to assess the free brain concentration and BBB transport of CNS drugs. J Cereb Blood Flow Metab

Gupta A, Chatelain P, Massingham R, Jonsson EN, Hammarlund-Udenaes M (2006) Brain distribution of cetirizine enantiomers: comparison of three different tissue-to-plasma partition coefficients: K(p), K(p,u), and K(p,uu). Drug Metab Dispos 34 (2):318-323

Gustafsson S, Eriksson J, Syvanen S, Eriksson O, Hammarlund-Udenaes M, Antoni G (2017) Combined PET and microdialysis for in vivo estimation of drug blood-brain barrier transport and brain unbound concentrations. Neuroimage 155:177–186

Gustafsson S, Sehlin D, Lampa E, Hammarlund-Udenaes M, Loryan I (2019) Heterogeneous drug tissue binding in brain regions of rats, Alzheimer's patients and controls: impact on translational drug development. Sci Rep 9(1):5308

Hammarlund-Udenaes M (2010) Active-site concentrations of chemicals - are they a better predictor of effect than plasma/organ/tissue concentrations? Basic Clin Pharmacol Toxicol 106(3):215–220

Hammarlund-Udenaes M (2017) Microdialysis as an Important Technique in Systems Pharmacology-a Historical and Methodological Review. AAPS J 19(5):1294–1303

Hammarlund-Udenaes M, Paalzow LK, de Lange EC (1997) Drug equilibration across the blood-brain barrier--pharmacokinetic considerations based on the microdialysis method. Pharm Res 14 (2):128-134

Hammarlund-Udenaes M, Friden M, Syvanen S, Gupta A (2008) On the rate and extent of drug delivery to the brain. Pharm Res 25(8):1737–1750

Hammarlund-Udenaes M, Bredberg U, Friden M (2009) Methodologies to assess brain drug delivery in lead optimization. Curr Top Med Chem 9(2):148–162

Harashima H, Sugiyama Y, Sawada Y, Iga T, Hanano M (1984) Comparison between in-vivo and in-vitro tissue-to-plasma unbound concentration ratios (Kp,f) of quinidine in rats. J Pharm Pharmacol 36 (5):340-342

He X, Huang Y, Li B, Gong CX, Schuchman EH (2010) Deregulation of sphingolipid metabolism in Alzheimer's disease. Neurobiol Aging 31(3):398–408

Hitchcock SA, Pennington LD (2006) Structure-brain exposure relationships. J Med Chem 49(26):7559–7583

Huang JD (1983) Errors in estimating the unbound fraction of drugs due to the volume shift in equilibrium dialysis. J Pharm Sci 72(11):1368–1369

Huwyler J, Wu D, Pardridge WM (1996) Brain drug delivery of small molecules using immunoliposomes. Proc Natl Acad Sci U S A 93(24):14164–14169

Jeffrey P, Summerfield SG (2007) Challenges for blood-brain barrier (BBB) screening. Xenobiotica 37(10–11):1135–1151

Kaitin KI, DiMasi JA (2011) Pharmaceutical innovation in the 21st century: new drug approvals in the first decade, 2000-2009. Clin Pharmacol Ther 89(2):183–188

Kakee A, Terasaki T, Sugiyama Y (1996) Brain efflux index as a novel method of analyzing efflux transport at the blood-brain barrier. J Pharmacol Exp Ther 277(3):1550–1559

Kakee A, Terasaki T, Sugiyama Y (1997) Selective brain to blood efflux transport of para-aminohippuric acid across the blood-brain barrier: in vivo evidence by use of the brain efflux index method. J Pharmacol Exp Ther 283(3):1018–1025

Kalvass JC, Maurer TS (2002) Influence of nonspecific brain and plasma binding on CNS exposure: implications for rational drug discovery. Biopharm Drug Dispos 23(8):327–338

Kalvass JC, Maurer TS, Pollack GM (2007a) Use of plasma and brain unbound fractions to assess the extent of brain distribution of 34 drugs: comparison of unbound concentration ratios to in vivo p-glycoprotein efflux ratios. Drug Metab Dispos 35(4):660–666

Kalvass JC, Olson ER, Cassidy MP, Selley DE, Pollack GM (2007b) Pharmacokinetics and pharmacodynamics of seven opioids in P-glycoprotein-competent mice: assessment of unbound brain EC50,u and correlation of in vitro, preclinical, and clinical,u data. J Pharmacol Exp Ther 323(1):346–355

Kariv I, Cao H, Oldenburg KR (2001) Development of a high throughput equilibrium dialysis method. J Pharm Sci 90(5):580–587

Kaufmann AM, Krise JP (2007) Lysosomal sequestration of amine-containing drugs: analysis and therapeutic implications. J Pharm Sci 96(4):729–746

Kell DB, Dobson PD, Oliver SG (2011) Pharmaceutical drug transport: the issues and the implications that it is essentially carrier-mediated only. Drug Discov Today 16(15–16):704–714

Kell DB, Dobson PD, Bilsland E, Oliver SG (2013) The promiscuous binding of pharmaceutical drugs and their transporter-mediated uptake into cells: what we (need to) know and how we can do so. Drug Discov Today 18(5–6):218–239

Kelly J (2009) Principles of CNS Drug Development: From Test Tube to Clinic and Beyond.

Ketharanathan N, Yamamoto Y, Rohlwink UK, Wildschut ED, Mathot RAA, de Lange ECM, de Wildt SN, Argent AC, Tibboel D, Figaji AA (2019) Combining Brain Microdialysis and Translational Pharmacokinetic Modeling to Predict Drug Concentrations in Pediatric Severe Traumatic Brain Injury: The Next Step Toward Evidence-Based Pharmacotherapy? J Neurotrauma 36(1):111–117

Kielbasa W, Stratford RE Jr (2012) Exploratory translational modeling approach in drug development to predict human brain pharmacokinetics and pharmacologically relevant clinical doses. Drug Metab Dispos 40(5):877–883

Kitamura A, Okura T, Higuchi K, Deguchi Y (2016) Cocktail-Dosing Microdialysis Study to Simultaneously Assess Delivery of Multiple Organic-Cationic Drugs to the Brain. J Pharm Sci 105(2):935–940

Klotz IM (1973) Physiochemical aspects of drug-protein interactions: a general perspective. Ann N Y Acad Sci 226:18–35

Kodaira H, Kusuhara H, Fujita T, Ushiki J, Fuse E, Sugiyama Y (2011) Quantitative evaluation of the impact of active efflux by p-glycoprotein and breast cancer resistance protein at the blood-brain barrier on the predictability of the unbound concentrations of drugs in the brain using cerebrospinal fluid concentration as a surrogate. J Pharmacol Exp Ther 339(3):935–944

Kornhuber J, Muehlbacher M, Trapp S, Pechmann S, Friedl A, Reichel M, Muhle C, Terfloth L, Groemer TW, Spitzer GM, Liedl KR, Gulbins E, Tripal P (2011) Identification of novel functional inhibitors of acid sphingomyelinase. PLoS One 6(8):e23852

Kurz H, Fichtl B (1983) Binding of drugs to tissues. Drug Metab Rev 14(3):467–510

Kusuhara H, Sugiyama Y (2002) Role of transporters in the tissue-selective distribution and elimination of drugs: transporters in the liver, small intestine, brain and kidney. J Control Release 78(1–3):43–54

Lee G, Dallas S, Hong M, Bendayan R (2001a) Drug transporters in the central nervous system: brain barriers and brain parenchyma considerations. Pharmacol Rev 53(4):569–596

Lee G, Schlichter L, Bendayan M, Bendayan R (2001b) Functional expression of P-glycoprotein in rat brain microglia. J Pharmacol Exp Ther 299(1):204–212

Li JY, Sugimura K, Boado RJ, Lee HJ, Zhang C, Duebel S, Pardridge WM (1999) Genetically engineered brain drug delivery vectors: cloning, expression and in vivo application of an anti-transferrin receptor single chain antibody-streptavidin fusion gene and protein. Protein Eng 12(9):787–796

Lin JH (2004) How significant is the role of P-glycoprotein in drug absorption and brain uptake? Drugs Today (Barc) 40(1):5–22

Lin JH (2008) CSF as a surrogate for assessing CNS exposure: an industrial perspective. Curr Drug Metab 9(1):46–59

Lin TH, Lin JH (1990) Effects of protein binding and experimental disease states on brain uptake of benzodiazepines in rats. J Pharmacol Exp Ther 253(1):45–50

Lin JH, Sugiyama Y, Awazu S, Hanano M (1982) In vitro and in vivo evaluation of the tissue-to-blood partition coefficient for physiological pharmacokinetic models. J Pharmacokinet Biopharm 10(6):637–647

Liu X, Smith BJ, Chen C, Callegari E, Becker SL, Chen X, Cianfrogna J, Doran AC, Doran SD, Gibbs JP, Hosea N, Liu J, Nelson FR, Szewc MA, Van Deusen J (2005) Use of a physiologically based pharmacokinetic model to study the time to reach brain equilibrium: an experimental analysis of the role of blood-brain barrier permeability, plasma protein binding, and brain tissue binding. J Pharmacol Exp Ther 313(3):1254–1262

Liu X, Smith BJ, Chen C, Callegari E, Becker SL, Chen X, Cianfrogna J, Doran AC, Doran SD, Gibbs JP, Hosea N, Liu J, Nelson FR, Szewc MA, Van Deusen J (2006) Evaluation of cerebrospinal fluid concentration and plasma free concentration as a surrogate measurement for brain free concentration. Drug Metab Dispos 34(9):1443–1447

Liu X, Chen C, Smith BJ (2008) Progress in brain penetration evaluation in drug discovery and development. J Pharmacol Exp Ther 325(2):349–356

Liu X, Van Natta K, Yeo H, Vilenski O, Weller PE, Worboys PD, Monshouwer M (2009a) Unbound drug concentration in brain homogenate and cerebral spinal fluid at steady state as a surrogate for unbound concentration in brain interstitial fluid. Drug Metab Dispos 37(4):787–793

Liu X, Vilenski O, Kwan J, Apparsundaram S, Weikert R (2009b) Unbound brain concentration determines receptor occupancy: a correlation of drug concentration and brain serotonin and dopamine reuptake transporter occupancy for eighteen compounds in rats. Drug Metab Dispos 37(7):1548–1556

Liu X, Ding X, Deshmukh G, Liederer BM, Hop CE (2012) Use of Cassette Dosing Approach to Assess Brain Penetration in Drug Discovery. Drug Metab Dispos

Lloyd JB (2000) Lysosome membrane permeability: implications for drug delivery. Adv Drug Deliv Rev 41(2):189–200

Logan R, Funk RS, Axcell E, Krise JP (2012) Drug-drug interactions involving lysosomes: mechanisms and potential clinical implications. Expert Opin Drug Metab Toxicol 8(8):943–958

Longhi R, Corbioli S, Fontana S, Vinco F, Braggio S, Helmdach L, Schiller J, Boriss H (2011) Brain tissue binding of drugs: evaluation and validation of solid supported porcine brain membrane vesicles (TRANSIL) as a novel high-throughput method. Drug Metab Dispos 39(2):312–321

Loryan I, Friden M, Hammarlund-Udenaes M (2013) The brain slice method for studying drug distribution in the CNS. Fluids Barriers CNS 10(1):6

Loryan I, Sinha V, Mackie C, Van Peer A, Drinkenburg W, Vermeulen A, Morrison D, Monshouwer M, Heald D, Hammarlund-Udenaes M (2014) Mechanistic understanding of brain drug disposition to optimize the selection of potential neurotherapeutics in drug discovery. Pharm Res 31(8):2203–2219

Loryan I, Sinha V, Mackie C, Van Peer A, Drinkenburg WH, Vermeulen A, Heald D, Hammarlund-Udenaes M, Wassvik CM (2015) Molecular properties determining unbound intracellular and extracellular brain exposure of CNS drug candidates. Mol Pharm 12(2):520–532

Loryan I, Melander E, Svensson M, Payan M, Konig F, Jansson B, Hammarlund-Udenaes M (2016) In-depth neuropharmacokinetic analysis of antipsychotics based on a novel approach to estimate unbound target-site concentration in CNS regions: link to spatial receptor occupancy. Mol Psychiatry 21(11):1527–1536

Loryan I, Hoppe E, Hansen K, Held F, Kless A, Linz K, Marossek V, Nolte B, Ratcliffe P, Saunders D, Terlinden R, Wegert A, Welbers A, Will O, Hammarlund-Udenaes M (2017) Quantitative Assessment of Drug Delivery to Tissues and Association with Phospholipidosis: A Case Study with Two Structurally Related Diamines in Development. Mol Pharm 14(12):4362–4373

MacIntyre AC, Cutler DJ (1988) The potential role of lysosomes in tissue distribution of weak bases. Biopharm Drug Dispos 9(6):513–526

Manitpisitkul P, White RE (2004) Whatever happened to cassette-dosing pharmacokinetics? Drug Discov Today 9(15):652–658

Mano Y, Higuchi S, Kamimura H (2002) Investigation of the high partition of YM992, a novel antidepressant, in rat brain - in vitro and in vivo evidence for the high binding in brain and the high permeability at the BBB. Biopharm Drug Dispos 23(9):351–360

Maurer TS, Debartolo DB, Tess DA, Scott DO (2005) Relationship between exposure and non-specific binding of thirty-three central nervous system drugs in mice. Drug Metab Dispos 33(1):175–181

McIlwain H (1951a) Glutamic acid and glucose as substrates for mammalian brain. J Ment Sci 97(409):674–680

McIlwain H (1951b) Metabolic response in vitro to electrical stimulation of sections of mammalian brain. Biochem J 48(4):lvi

Mehta D, Jackson R, Paul G, Shi J, Sabbagh M (2017) Why do trials for Alzheimer's disease drugs keep failing? A discontinued drug perspective for 2010-2015. Expert Opinion on Investigational Drugs 26(6):735–739

Mizuno N, Niwa T, Yotsumoto Y, Sugiyama Y (2003) Impact of drug transporter studies on drug discovery and development. Pharmacol Rev 55(3):425–461

Muehlbacher M, Tripal P, Roas F, Kornhuber J (2012) Identification of Drugs Inducing Phospholipidosis by Novel in vitro Data. ChemMedChem

Nadanaciva S, Lu S, Gebhard DF, Jessen BA, Pennie WD, Will Y (2011) A high content screening assay for identifying lysosomotropic compounds. Toxicol In Vitro 25(3):715–723

Ndengele MM, Cuzzocrea S, Masini E, Vinci MC, Esposito E, Muscoli C, Petrusca DN, Mollace V, Mazzon E, Li D, Petrache I, Matuschak GM, Salvemini D (2009) Spinal ceramide modulates the development of morphine antinociceptive tolerance via peroxynitrite-mediated nitroxidative stress and neuroimmune activation. J Pharmacol Exp Ther 329(1):64–75

Neuwelt E, Abbott NJ, Abrey L, Banks WA, Blakley B, Davis T, Engelhardt B, Grammas P, Nedergaard M, Nutt J, Pardridge W, Rosenberg GA, Smith Q, Drewes LR (2008) Strategies to advance translational research into brain barriers. Lancet Neurol 7(1):84–96

Newman GC, Hospod FE, Patlak CS (1988a) Brain slice glucose utilization. J Neurochem 51(6):1783–1796

Newman GC, Hospod FE, Wu P (1988b) Thick brain slices model the ischemic penumbra. J Cereb Blood Flow Metab 8(4):586–597

Newman GC, Hospod FE, Schissel SL (1991) Ischemic brain slice glucose utilization: effects of slice thickness, acidosis, and K+. J Cereb Blood Flow Metab 11(3):398–406

Nicholson C, Sykova E (1998) Extracellular space structure revealed by diffusion analysis. Trends Neurosci 21(5):207–215

Ohtsuki S, Takizawa T, Takanaga H, Hori S, Hosoya K, Terasaki T (2004) Localization of organic anion transporting polypeptide 3 (oatp3) in mouse brain parenchymal and capillary endothelial cells. J Neurochem 90(3):743–749

Ooie T, Terasaki T, Suzuki H, Sugiyama Y (1997) Quantitative brain microdialysis study on the mechanism of quinolones distribution in the central nervous system. Drug Metab Dispos 25(7):784–789

Padowski JM, Pollack GM (2011a) The influence of distributional kinetics into a peripheral compartment on the pharmacokinetics of substrate partitioning between blood and brain tissue. J Pharmacokinet Pharmacodyn 38(6):743–767

Padowski JM, Pollack GM (2011b) Influence of time to achieve substrate distribution equilibrium between brain tissue and blood on quantitation of the blood-brain barrier P-glycoprotein effect. Brain Res 1426:1–17

Pardridge WM (1989) Strategies for drug delivery through the blood-brain barrier. Neurobiol Aging 10(5):636–637. discussion 648–650

Pardridge WM (1997) Drug delivery to the brain. J Cereb Blood Flow Metab 17(7):713–731

Pardridge WM (2006) Molecular Trojan horses for blood-brain barrier drug delivery. Curr Opin Pharmacol 6(5):494–500

Pardridge WM, Boado RJ, Black KL, Cancilla PA (1992) Blood-brain barrier and new approaches to brain drug delivery. West J Med 156(3):281–286

Patlak CS, Hospod FE, Trowbridge SD, Newman GC (1998) Diffusion of radiotracers in normal and ischemic brain slices. J Cereb Blood Flow Metab 18(7):776–802

Plise EG, Tran D, Salphati L (2010) Semi-automated protein binding methodology using equilibrium dialysis and a novel mixed-matrix cassette approach. J Pharm Sci 99(12):5070–5078

Puris E, Gynther M, de Lange ECM, Auriola S, Hammarlund-Udenaes M, Huttunen KM, Loryan I (2019) Mechanistic study on the use of the l-type amino acid transporter 1 for brain intracellular delivery of ketoprofen via prodrug: a novel approach supporting the development of prodrugs for intracellular targets. Mol Pharm 16(7):3261–3274

Raub TJ, Lutzke BS, Andrus PK, Sawada GA, Staton BA (2006) Early preclinical evaluation of brain exposure in support of hit identification and lead optimization. In: Borchardt RT, Middagh CR (eds) Optimization of drug-like properties during lead optimization, Biotechnology: pharmaceutical aspects series. Am Assoc Pharm Sci Press, Arlington

Read KD, Braggio S (2010) Assessing brain free fraction in early drug discovery. Expert Opin Drug Metab Toxicol 6(3):337–344

Reichel A (2006) The role of blood-brain barrier studies in the pharmaceutical industry. Curr Drug Metab 7(2):183–203

Reichel A (2009) Addressing central nervous system (CNS) penetration in drug discovery: basics and implications of the evolving new concept. Chem Biodivers 6(11):2030–2049

Reichel A, Begley D, Abbott N (2003) An overview of in vitro techniques for blood-brain barrier studies. In: Nag S, Methods in molecular medicine The blood-brain barrier: biology and research tools, Humana Press Inc, Totowa Volume 89

Rice ME (1999) Use of ascorbate in the preparation and maintenance of brain slices. Methods 18(2):144–149

Rodgers T, Jones HM, Rowland M (2012) Tissue lipids and drug distribution: dog versus rat. J Pharm Sci 101(12):4615–4626

Romer J, Bickel MH (1979) A method to estimate binding constants at variable protein concentrations. J Pharm Pharmacol 31(1):7–11

Rouser G, Simon G, Kritchevsky G (1969) Species variations in phospholipid class distribution of organs. I. Kidney, liver and spleen. Lipids 4(6):599–606

Sargent DF, Schwyzer R (1986) Membrane lipid phase as catalyst for peptide-receptor interactions. Proc Natl Acad Sci U S A 83(16):5774–5778

Sargent DF, Bean JW, Schwyzer R (1988) Conformation and orientation of regulatory peptides on lipid membranes. Key to the molecular mechanism of receptor selection. Biophys Chem 31(1–2):183–193

Sawada Y, Hanano M, Sugiyama Y, Harashima H, Iga T (1984) Prediction of the volumes of distribution of basic drugs in humans based on data from animals. J Pharmacokinet Biopharm 12(6):587–596

Scherrmann JM (2002) Drug delivery to brain via the blood-brain barrier. Vascul Pharmacol 38(6):349–354

Schoepp DD (2011) Where will new neuroscience therapies come from? Nat Rev Drug Discov 10(10):715–716

Schou M, Varnäs K, Lundquist S, Nakao R, Amini N, Takano A, Finnema SJ, Halldin C, Farde L (2015) Large variation in brain exposure of reference CNS drugs: a PET study in nonhuman primates. Int J Neuropsychopharmacol 18(10):pyv036

Schultz ML, Tecedor L, Chang M, Davidson BL (2011) Clarifying lysosomal storage diseases. Trends Neurosci 34(8):401–410

Schwab ME, Buchli AD (2012) Drug research: plug the real brain drain. Nature 483(7389):267–268

Schwarz E, Prabakaran S, Whitfield P, Major H, Leweke FM, Koethe D, McKenna P, Bahn S (2008) High throughput lipidomic profiling of schizophrenia and bipolar disorder brain tissue reveals alterations of free fatty acids, phosphatidylcholines, and ceramides. J Proteome Res 7(10):4266–4277

Segall MD (2012) Multi-parameter optimization: identifying high quality compounds with a balance of properties. Curr Pharm Des 18(9):1292–1310

Shaffer CL (2010) Defining Neuropharmacokinetic Parameters in CNS Drug Discovery to Determine Cross-species Pharmacologic Exposure-Response Relationships. Annu Rep Med Chem 45:55–70

Shen DD, Artru AA, Adkison KK (2004) Principles and applicability of CSF sampling for the assessment of CNS drug delivery and pharmacodynamics. Adv Drug Deliv Rev 56(12):1825–1857

Simon G, Rouser G (1969) Species variations in phospholipid class distribution of organs. II. Heart and skeletal muscle. Lipids 4(6):607–614

Smith QR (1991) The blood-brain barrier and the regulation of amino acid uptake and availability to brain. Adv Exp Med Biol 291:55–71

Smith DA, Di L, Kerns EH (2010) The effect of plasma protein binding on in vivo efficacy: misconceptions in drug discovery. Nat Rev Drug Discov 9(12):929–939

Stanimirovic DB, Sandhu JK, Costain WJ (2018) Emerging Technologies for Delivery of Biotherapeutics and Gene Therapy Across the Blood-Brain Barrier. BioDrugs 32(6):547–559

Stevens J, Ploeger BA, Hammarlund-Udenaes M, Osswald G, van der Graaf PH, Danhof M, de Lange EC (2012) Mechanism-based PK-PD model for the prolactin biological system response following an acute dopamine inhibition challenge: quantitative extrapolation to humans. J Pharmacokinet Pharmacodyn 39(5):463–477

Stokes CE, Murphy D, Paton JF, Kasparov S (2003) Dynamics of a transgene expression in acute rat brain slices transfected with adenoviral vectors. Exp Physiol 88(4):459–466

Su TZ, Lunney E, Campbell G, Oxender DL (1995) Transport of gabapentin, a gamma-amino acid drug, by system l alpha-amino acid transporters: a comparative study in astrocytes, synaptosomes, and CHO cells. J Neurochem 64(5):2125–2131

Summerfield SG, Stevens AJ, Cutler L, del Carmen OM, Hammond B, Tang SP, Hersey A, Spalding DJ, Jeffrey P (2006) Improving the in vitro prediction of in vivo central nervous system penetration: integrating permeability, P-glycoprotein efflux, and free fractions in blood and brain. J Pharmacol Exp Ther 316(3):1282–1290

Summerfield SG, Read K, Begley DJ, Obradovic T, Hidalgo IJ, Coggon S, Lewis AV, Porter RA, Jeffrey P (2007) Central nervous system drug disposition: the relationship between in situ brain permeability and brain free fraction. J Pharmacol Exp Ther 322(1):205–213

Sun N, Avdeef A (2011) Biorelevant pK(a) (37 degrees C) predicted from the 2D structure of the molecule and its pK(a) at 25 degrees C. J Pharm Biomed Anal 56(2):173–182

Sykova E, Nicholson C (2008) Diffusion in brain extracellular space. Physiol Rev 88(4):1277–1340

Syvanen S, Hooker A, Rahman O, Wilking H, Blomquist G, Langstrom B, Bergstrom M, Hammarlund-Udenaes M (2008) Pharmacokinetics of P-glycoprotein inhibition in the rat blood-brain barrier. J Pharm Sci 97(12):5386–5400

Syvanen S, Lindhe O, Palner M, Kornum BR, Rahman O, Langstrom B, Knudsen GM, Hammarlund-Udenaes M (2009) Species differences in blood-brain barrier transport of three positron emission tomography radioligands with emphasis on P-glycoprotein transport. Drug Metab Dispos 37(3):635–643

Syvanen S, Schenke M, van den Berg DJ, Voskuyl RA, de Lange EC (2012) Alteration in P-glycoprotein functionality affects intrabrain distribution of quinidine more than brain entry-a study in rats subjected to status epilepticus by kainate. AAPS J 14(1):87–96

Tanaka H, Mizojiri K (1999) Drug-protein binding and blood-brain barrier permeability. J Pharmacol Exp Ther 288(3):912–918

Terasaki T, Hosoya K (1999) The blood-brain barrier efflux transporters as a detoxifying system for the brain. Adv Drug Deliv Rev 36(2–3):195–209

Terasaki T, Ohtsuki S (2005) Brain-to-blood transporters for endogenous substrates and xenobiotics at the blood-brain barrier: an overview of biology and methodology. NeuroRx 2(1):63–72

Thorne RG, Nicholson C (2006) In vivo diffusion analysis with quantum dots and dextrans predicts the width of brain extracellular space. Proc Natl Acad Sci U S A 103(14):5567–5572

Thorne RG, Emory CR, Ala TA, Frey WH 2nd (1995) Quantitative analysis of the olfactory pathway for drug delivery to the brain. Brain Res 692(1–2):278–282

Thorne RG, Hrabetova S, Nicholson C (2004) Diffusion of epidermal growth factor in rat brain extracellular space measured by integrative optical imaging. J Neurophysiol 92(6):3471–3481

Thorne RG, Lakkaraju A, Rodriguez-Boulan E, Nicholson C (2008) In vivo diffusion of lactoferrin in brain extracellular space is regulated by interactions with heparan sulfate. Proc Natl Acad Sci U S A 105(24):8416–8421

Tsuji A, Terasaki T, Takabatake Y, Tenda Y, Tamai I, Yamashima T, Moritani S, Tsuruo T, Yamashita J (1992) P-glycoprotein as the drug efflux pump in primary cultured bovine brain capillary endothelial cells. Life Sci 51(18):1427–1437

Tsuji A, Tamai I, Sakata A, Tenda Y, Terasaki T (1993) Restricted transport of cyclosporin A across the blood-brain barrier by a multidrug transporter. P-glycoprotein. Biochem Pharmacol 46(6):1096–1099

Uchida Y, Ohtsuki S, Kamiie J, Terasaki T (2011a) Blood-brain barrier (BBB) pharmacoproteomics: reconstruction of in vivo brain distribution of 11 P-glycoprotein substrates based on the BBB transporter protein concentration, in vitro intrinsic transport activity, and unbound fraction in plasma and brain in mice. J Pharmacol Exp Ther 339(2):579–588

Uchida Y, Ohtsuki S, Katsukura Y, Ikeda C, Suzuki T, Kamiie J, Terasaki T (2011b) Quantitative targeted absolute proteomics of human blood-brain barrier transporters and receptors. J Neurochem 117(2):333–345

Van Eeckhaut A, Lanckmans K, Sarre S, Smolders I, Michotte Y (2009) Validation of bioanalytical LC-MS/MS assays: evaluation of matrix effects. J Chromatogr B Analyt Technol Biomed Life Sci 877(23):2198–2207

van Liempd S, Morrison D, Sysmans L, Nelis P, Mortishire-Smith R (2011) Development and validation of a higher-throughput equilibrium dialysis assay for plasma protein binding. J Lab Autom 16(1):56–67

Van Peer AP, Belpaire FM, Bogaert MG (1981) Binding of drugs in serum, blood cells and tissues of rabbits with experimental acute renal failure. Pharmacology 22(2):146–152

Vauquelin G, Packeu A (2009) Ligands, their receptors and … plasma membranes. Mol Cell Endocrinol:311 (1-2):1-10

Vauquelin G, Van Liefde I (2005) G protein-coupled receptors: a count of 1001 conformations. Fundam Clin Pharmacol 19(1):45–56

Vauquelin G, Bostoen S, Vanderheyden P, Seeman P (2012) Clozapine, atypical antipsychotics, and the benefits of fast-off D(2) dopamine receptor antagonism. Naunyn Schmiedebergs Arch Pharmacol 385(4):337–372

Vendel E, Rottschäfer V, de Lange ECM (2019) The need for mathematical modelling of spatial drug distribution within the brain. Fluids and barriers of the CNS 16(1):12–12

Vuignier K, Schappler J, Veuthey JL, Carrupt PA, Martel S (2010) Drug-protein binding: a critical review of analytical tools. Anal Bioanal Chem 398(1):53–66

Wager TT, Liras JL, Mente S, Trapa P (2012) Strategies to minimize CNS toxicity: in vitro high-throughput assays and computational modeling. Expert Opin Drug Metab Toxicol 8(5):531–542

Wan H, Rehngren M, Giordanetto F, Bergstrom F, Tunek A (2007) High-throughput screening of drug-brain tissue binding and in silico prediction for assessment of central nervous system drug delivery. J Med Chem 50(19):4606–4615

Wan H, Ahman M, Holmen AG (2009) Relationship between brain tissue partitioning and micro-emulsion retention factors of CNS drugs. J Med Chem 52(6):1693–1700

Wanek T, Mairinger S, Langer O (2013) Radioligands targeting P-glycoprotein and other drug efflux proteins at the blood-brain barrier. J Labelled Comp Radiopharm 56(3–4):68–77

Wang Y, Welty DF (1996) The simultaneous estimation of the influx and efflux blood-brain barrier permeabilities of gabapentin using a microdialysis-pharmacokinetic approach. Pharm Res 13(3):398–403

Wang YY, Lui PC, Li JY (2009) Receptor-mediated therapeutic transport across the blood-brain barrier. Immunotherapy 1(6):983–993

Watson J, Wright S, Lucas A, Clarke KL, Viggers J, Cheetham S, Jeffrey P, Porter R, Read KD (2009) Receptor occupancy and brain free fraction. Drug Metab Dispos 37(4):753–760

Weaver DF, Weaver CA (2011) Exploring neurotherapeutic space: how many neurological drugs exist (or could exist)? J Pharm Pharmacol 63(1):136–139

Wellmann H, Kaltschmidt B, Kaltschmidt C (1999) Optimized protocol for biolistic transfection of brain slices and dissociated cultured neurons with a hand-held gene gun. J Neurosci Methods 92(1–2):55–64

Westerhout J, Danhof M, De Lange EC (2011) Preclinical prediction of human brain target site concentrations: considerations in extrapolating to the clinical setting. J Pharm Sci 100(9):3577–3593

Westerhout J, Ploeger B, Smeets J, Danhof M, de Lange EC (2012) Physiologically based pharmacokinetic modeling to investigate regional brain distribution kinetics in rats. AAPS J 14(3):543–553

Wolak D, Thorne R (2013) Diffusion of Macromolecules in the Brain: Implications for Drug Delivery. Mol Pharm

Yamamoto Y, Valitalo PA, van den Berg DJ, Hartman R, van den Brink W, Wong YC, Huntjens DR, Proost JH, Vermeulen A, Krauwinkel W, Bakshi S, Aranzana-Climent V, Marchand S, Dahyot-Fizelier C, Couet W, Danhof M, van Hasselt JG, de Lange EC (2017) A Generic Multi-Compartment CNS Distribution Model Structure for 9 Drugs Allows Prediction of Human Brain Target Site Concentrations. Pharm Res 34(2):333–351

Yokogawa K, Ishizaki J, Ohkuma S, Miyamoto K (2002) Influence of lipophilicity and lysosomal accumulation on tissue distribution kinetics of basic drugs: a physiologically based pharmacokinetic model. Methods Find Exp Clin Pharmacol 24(2):81–93

Young RC, Mitchell RC, Brown TH, Ganellin CR, Griffiths R, Jones M, Rana KK, Saunders D, Smith IR, Sore NE et al (1988) Development of a new physicochemical model for brain penetration and its application to the design of centrally acting H2 receptor histamine antagonists. J Med Chem 31(3):656–671

Zhao R, Kalvass JC, Pollack GM (2009) Assessment of blood-brain barrier permeability using the in situ mouse brain perfusion technique. Pharm Res 26(7):1657–1664

Chapter 14
Prediction of Drug Exposure in the Brain from the Chemical Structure

Markus Fridén

Abstract The level of drug exposure in the brain is long known to relate to the physicochemical properties of the drug. The study of this relationship has attracted much attention through the years as it holds a promise that this drug property can be predicted in silico from the chemical drug structure. Various in vivo methodologies have been used to define and quantify drug exposure in the brain, the most commonly used parameter being logBB, which is the brain-to-blood ratio of the *total* drug concentrations. From datasets of logBB, it has been inferred that drug exposure in the brain is promoted by the lipophilicity, i.e. lipid solubility, of the drug and restricted by its hydrogen bonding potential. Recent work with the $K_{p,uu,brain}$ parameter, representing a pharmacologically relevant brain-to-blood ratio of *unbound* drug concentrations, has confirmed the limiting effect of hydrogen bonding on drug exposure in the brain but also indicated no dependence on lipophilicity. The challenges associated with obtaining high predictivity models for $K_{p,uu,brain}$ confirm the contemporary view of the blood-brain barrier as being not only physical and passive in nature but also involving specific carrier-mediated processes. It follows that in silico approaches need to compliment and merge with experimental methodologies to advance the field of brain exposure predictions.

Keywords Physico-chemical properties · In silico · Prediction · Hydrogen bonding potential · Molecular descriptors · Integration

14.1 Introduction

For decades it has been recognized that a drug's ability to cross the blood-brain barrier (BBB) is related to its physicochemical properties. This idea is not only supported by experimental data in animals but also by clinical notions of, e.g., the hydrophilic beta blocker atenolol having less CNS-related side effects than do the

M. Fridén (✉)
Inhalation Product Development, Pharmaceutical Technology & Development, AstraZeneca, Gothenburg, Sweden
e-mail: Markus.Friden@astrazeneca.com

© American Association of Pharmaceutical Scientists 2022   437
E. C. M. de Lange et al. (eds.), *Drug Delivery to the Brain*, AAPS Advances
in the Pharmaceutical Sciences Series 33, https://doi.org/10.1007/978-3-030-88773-5_14

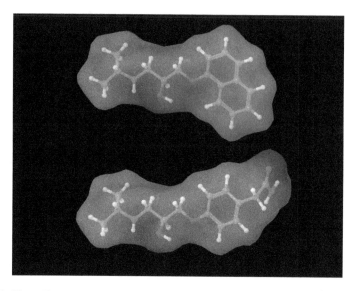

Fig. 14.1 Three-dimensional structures of propranolol (top) and atenolol (bottom) coloured according to the electrostatic potential, where red and blue areas indicate the negative and positive charges of oxygen and nitrogen atoms (MacroModel v. 8.0, MM3* force field). Note that the drugs differ only in the substitution of the aromatic ring where propranolol has a fused benzene ring, i.e. a naphthyl group, and atenolol has an amide, i.e. two additional hydrogen bond acceptors

more lipophilic propranolol (see Fig. 14.1 for molecular structures) (McAinsh and Cruickshank 1990). The relationship with lipophilicity seems to have become common knowledge even amongst clinicians, and it has nourished an idea that BBB transport is essentially predictable. This chapter critically reviews the various approaches that have been taken during the years to predict the level of drug exposure in the brain from the chemical structure of the drug. Whilst the computational approaches have reached advanced levels, the weakest spot can be considered to be the quality and relevance of the underlying experimental data that is used to derive the prediction models (Mehdipour and Hamidi 2009). Therefore, a discussion is included on various endpoints of drug exposure in the brain that are currently used as basis for in silico prediction models. The actual analysis of the relationship with chemical drug properties is focused on the parameter that is currently regarded as the most appropriate: the unbound brain-to-plasma ratio $K_{\mathrm{p,uu,brain}}$ (Sect. 14.2.2) commonly determined in the rat. The reader will also be provided with some basic knowledge of how to derive and validate prediction models (Sect. 14.1.2.5) and some tactics of how to utilize them for drug design (Sect. 14.3.2).

14.1.1 Various Measurements of Brain Exposure and Availability of Data

Needless to say the vast majority of data on drug exposure in the brain and related measurements are from rodents. The validity of these data for making predictions in humans resides with the extent to which the drug transport properties of the BBB are 'conserved' across the mammalian species. There are distinct pharmacokinetic aspects to consider for drug transport across the BBB such as the rate and extent to which it occurs. When attempting to experimentally quantify these aspects, there are associated drug properties that need to be considered such as the degree of non-specific binding in blood plasma and in the brain tissue. Whilst another chapter covers the various experimental in vivo methodologies to determine each of these aspects or properties, this section highlights those that have been commonly used for deriving the relationship with the chemical structure of the drug.

14.1.1.1 LogBB

With respect to the prediction of drug exposure in the brain, the by far most commonly used measurement has been the (steady-state) ratio of total brain-to-total plasma or blood concentrations. This total brain-to-blood or brain-to-plasma ratio is referred to as the partitioning coefficient of the brain $K_{p,brain}$ or BB and the logarithm thereof (logBB). The relative ease of its experimental determination in animals has made this parameter the choice of preference for many modellers, and the datasets are commonly as large as ~150 logBB values, whilst some workers have compiled up to 470 values (Lanevskij et al. 2011). Rather recently there has been an increased awareness that only the free, *unbound*, drug is available for transport across the BBB and binding to target proteins. This has raised serious concerns regarding any work based on logBB since this is a measure of total drug in brain (bound + unbound) relative to plasma. Notwithstanding that the logBB value contains some information about the BBB, drug compounds generally differ with respect to the balance between bound and unbound drug in brain tissue versus blood plasma, thus making logBB very misleading (van de Waterbeemd et al. 2001; Martin 2004; Hammarlund-Udenaes et al. 2008). This topic is expanded on in Sect. 14.2.3.

14.1.1.2 $K_{p,uu,brain}$

Given the acceptance of the free drug hypothesis and that the extent of drug transport is probably more important than the rate for most CNS and non-CNS drugs, it is currently argued that the most useful measurement of brain exposure is the steady-state unbound brain-to-plasma ratio $K_{p,uu,brain}$ (Gupta et al. 2006), also called the unbound partition coefficient of the brain. As described in another chapter, $K_{p,uu,brain}$ can be directly determined in vivo using microdialysis or derived from $K_{p,brain}$

(equivalent to logBB) values by determining the binding in blood plasma and the brain tissue in vitro (Friden et al. 2007). However, it is only rather recently that data has been generated for this parameter and systematically compiled to publicly available datasets (Sect. 14.2.2).

14.1.1.3 BBB Permeability Surface Area Product (PS)

It has been proposed that the rate of transport, i.e. the BBB permeability surface area product (PS) should be used to derive predictive models (Pardridge 2004); however, it is difficult to rationalize this choice of parameter since the rate and extent are not necessarily correlated. Relative to logBB it is technically more challenging to determine logPS, and as a result the datasets are smaller, as has been the interest in modelling this parameter. Liu et al. determined and modelled logPS for a set of 28 compounds (Liu et al. 2004), which is a similar number of compounds as in earlier work (Gratton et al. 1997; Abraham 2004).

14.1.1.4 Classification Approaches

In order to remedy the relatively limited availability of logBB values, larger datasets have been created by classifying marketed or investigational drugs as CNS active (CNS$^+$) or inactive (CNS$^-$) according to the presence or lack of central drug effects or side effects. The underlying assumption of this approach is that CNS$^+$ drugs 'cross' the BBB, whereas CNS$^-$ drugs do not. This is obviously correct for all CNS$^+$ drugs, but the lack of CNS effects of CNS$^-$ drugs can arguably have different backgrounds. Recently a BBB$^+$/BBB$^-$ classification scheme was proposed based on actual measurements of drug concentrations in the human brain or cerebrospinal fluid (Broccatelli et al. 2012). Experimentally determined values of rodent logBB have also been added to these datasets by using arbitrary cut-off values for classification as CNS$^+$ or CNS$^-$. A reservation to categorical modelling is that brain exposure is a continuous variable by nature, and strictly speaking, CNS$^-$ drugs do not exist since all drugs enter the brain to some extent. The size of CNS$^+$/CNS$^-$ datasets range from a few hundred to several tenths of thousands depending on the databases that are used (Abraham and Hersey 2007). More recently computational models derived from classifications of $K_{p,uu,brain}$ datasets have been employed (Varadharajan et al. 2015; Zhang et al. 2016).

14.1.2 Modelling Strategies

The significant number of different approaches to predicting drug exposure in the brain poses a challenge to scientists that enter the field of in silico modelling. To assess the value of existing approaches or to develop new methods for a particular situation, it is therefore useful to know some basics of modelling strategies. The procedure for developing predictive computational models for, e.g., brain exposure can be divided into five general steps: (1) selecting a relevant set of drug molecules; (2) generating experimental data for the drug property of interest; (3) describing the chemical structure of the molecules in terms of numerical descriptor values; (4) relating the structural description to the experimental data using a mathematical relationship and (5) validating the predictivity of the model (Matsson 2007). This section (Sect. 14.1.2) describes the general basis of developing predictive models in the context of being able to predict drug exposure in the brain for a new chemical entity. Strategies for drug design are discussed in Sect. 14.3.2.

14.1.2.1 Compound Selection

The selection of a *training set* of compounds on which to build the relationship between brain exposure and molecular structure is not an arbitrary choice, since it will define the *applicability domain* of the model. The desired applicability domain can be larger, e.g. to encompass drugs in general (global models), or small, to encompass only structures that are relevant to, e.g., a particular drug discovery program (local models). A higher level of predictivity is expected from local models than from global models though it comes at the expense of a more restricted applicability domain. Regardless of whether global or local models are considered, one should strive for a structurally diverse selection within the domain.

14.1.2.2 Molecular Descriptors

Molecular structures need to be translated into numerical representations before a mathematical relationship can be derived with the measured drug property. This is done by molecular descriptors encoding various properties of the molecule. There are several sets of descriptors which are associated with the different computational approaches or software (Winiwarter et al. 2007). Some of these are derived from the three-dimensional structure of the molecule. For the prediction of BBB transport, however, standard physicochemical descriptors have been commonly used (Table 14.1). Physicochemical descriptors provide information about the molecular size, shape, lipid solubility (lipophilicity) and information on the hydrogen bonding potential of the drug. Acid-base properties, i.e. proton dissociation constants (pKa), can also be predicted from the structure and used to classify drugs as neutrals, acids, bases or zwitterions.

Table 14.1 Commonly used molecular descriptors

Property	Abbreviation	Molecular descriptor
Lipophilicity	ClogP	Prediction of octanol/water partition coefficient for molecules in their neutral state
	ClogD (ACDLogD7.4)	Prediction of octanol/water partition coefficient at pH 7.4 using ACD/labs software
Size/shape	MW	Molecular weight
	VOL	Molecular volume as defined by a Gaussian volume
	RotBond	Number of rotatable bonds in molecule
	RingCount	Number of rings in molecule
	NPSA	Non-polar surface area in Å^2
Hydrogen bonding	HBA	Number of hydrogen bond acceptors (number of oxygen plus nitrogen atoms, N + O)
	HBD	Number of hydrogen bond donors (number of hydroxyls + amine hydrogen atoms (OH + NH)
	PSA	Polar surface area in Å^2
	TPSA	Topological surface area in Å^2
Charge/polarity	Acid	Presence of an acid function
	Base	Presence of a basic function
	Neutral	Absence of acid and basic functions
	Zwitterion	Presence of at least one acid and one basic function

14.1.2.3 Generation of Experimental Data

This step is often considered the most costly and time-demanding step of model development. There is consequently always a risk of using inadequate experimental methods or not applying sufficiently stringent criteria for the inclusion of experimental data from literature. It is well known that good-quality data are a *conditio sine qua non*, an absolutely essential condition. A prediction model can never make better predictions than the experimental data used for its generation.

14.1.2.4 Relating Experimental Data to Molecular Descriptors

There are several mathematical or statistical modelling approaches collectively referred to as machine learning that can be used in the process of identifying and describing the relationship between molecular descriptors and a measured parameter. The simplest form would be to look at the correlation between the measured parameter and individual molecular descriptors. If a strong relationship is found (linear or not), the equation describing the relationship could be used as a computational prediction model for future compounds. If a strong relationship cannot be seen with any one descriptor, it is possible that several descriptors can give a better prediction when combined in, e.g., a *multiple linear regression* (MLR) analysis. A related technique is a *partial least squares projection to latent structures* (PLS) (Wold 2001). By this method of modelling, a larger number of molecular descriptors can be reduced to a smaller number of latent super-variables or *principal components*, which are then related to experimental data. Advantages of using PLS include that descriptors that are irrelevant to the problem are handled as well as closely related (correlated) descriptors. PLS models are also easily interpreted in terms of how the molecular properties could be changed. A major drawback is that PLS is a linear method that does not handle nonlinear relationships unless variables are transformed prior to analysis. Novel and computationally more advanced machine learning algorithms include support vector machine, random forest and neural networks (Mehdipour and Hamidi 2009).

14.1.2.5 Validation of the Model

Before a computational prediction model can be taken into practice, it must be validated. Whilst the coefficient of determination (R^2) describes the correlation between observed and predicted values for the training set, it cannot be taken for granted that the predictivity is equally good for drugs not used in the training of the model. In fact, R^2 should never be used to compare prediction models or be expected to reflect the real model predictivity for new compounds. Cross-validation or *leave-many-out* is a method for validating a model (Wold 1991). By dividing the compounds in groups, a model can be generated based on all groups but one, for which the values are instead predicted. The procedure is repeated until all groups have been withheld from the model and predicted. The cross-validated coefficient of determination (Q^2) is generally the first method of validating a PLS model and is used continuously to assess the predictivity of rivalling models. Unfortunately, a high value for Q^2 is neither a guarantee for a predictive model. The only way to really validate a prediction model is to use an external *test set* of compounds which have not at all been used in the training of the model. The failure of a high Q^2 model to satisfactorily predict compounds in a test set indicates that there are unresolved issues with defining the applicability domain of the model. This highlights the importance of the compound selection procedure which, if made appropriately for the problem at hand, increases the chances of obtaining a model that is fit-for-purpose.

14.1.3 Overview of BBB Prediction Models

Whilst there are several exhaustive reviews of modelling brain exposure (Mehdipour and Hamidi 2009; Abraham and Hersey 2007; Ecker and Noe 2004; Clark 2003; Norinder and Haeberlein 2002; Hitchcock and Pennington 2006) including some more recent (Morales et al. 2017; Liu et al. 2018), this section briefly highlights some of the historical studies that have been influential. The era of computational modelling of BBB transport began in 1980 when Levin observed a strong relationship between the BBB permeability (PS) and the octanol-water partitioning coefficient (LogP) for a set of 27 compounds (Levin 1980). Interestingly, four compounds with molecular weight greater than 400 dalton were excluded from the analysis since they were considered 'extremely restricted' owing to their size. In retrospect it is realized that these were substrates of P-glycoprotein. It was, however, concluded that there exists a molecular weight cut-off for 'significant BBB passage'. A relationship between descriptors of lipophilicity and logBB was also found by Young et al. in 1988 for a set of 20 antihistamines (Young et al. 1988). Since then, the public dataset of logBB values has expanded well over a hundred compounds, and several computational approaches have been used by different groups (Abraham et al. 1994; Luco 1999; Osterberg and Norinder 2000; Abraham et al. 2002; Abraham 2004; Bendels et al. 2008). These studies taken together indicate that brain penetration as measured by logBB is negatively correlated to descriptors of hydrogen bonding, e.g. the number of hydrogen bond donors (HBD), acceptors (HBA) or polar molecular surface area (PSA). A positive correlation with logBB is seen for descriptors related to lipophilicity such as LogP. Furthermore, acids having a negative charge at physiological pH generally have lower logBB than do basic drug with a net positive charge (Clark 2003). The underlying mechanisms of these findings are identified and discussed in Sect. 14.2.3. The predictivity levels achieved using various datasets and approaches are sometimes considered to 'approach experimental error', since the predictions are on average ~ three-fold away from the measured value (Clark 2003).

With respect to classification approaches, Palm and co-workers demonstrated that orally administered drugs should not exceed a polar molecular surface area (PSA) greater than 120 Å^2 (Palm et al. 1997). Inspired by this work, Kelder and co-workers published a prediction model for logBB based on PSA together with an analysis showing that the majority of CNS^+ drugs have PSA 60 Å^2 or less (Kelder et al. 1999). The accuracy of classification of CNS^+/CNS^- datasets is generally >80% correct and slightly better for CNS^+ compounds than CNS^- compounds (Clark 2003). Table 14.2 summarizes a number classification rules of thumb that indicate the characteristics of CNS drugs.

14.2 Current Status

14.2.1 Which Parameter of Drug Exposure in the Brain Should be Used?

A challenge that is currently posed to the community of in silico modellers of drug exposure in the brain is the lack of common understanding of which parameter should be predicted. Whilst recent years' debate has highlighted again and again the pitfalls of the logBB parameter (Martin 2004; Hammarlund-Udenaes et al. 2009; Mehdipour and Hamidi 2009), the discussions seem not to have resulted in a consensus view of what is the appropriate alternative; logBB, PS, $K_{p,uu,brain}$, $f_{u,brain}$, CNS^+/CNS^- and BBB^+/BBB^- are all used in parallel. Let alone that logBB is still often

Table 14.2 Characteristics of CNS drugs

Reference	Type of dataset	Property	Favourable value for CNS drugs
van de Waterbeemd et al. (1998)	CNS+/CNS-	PSA	< 90 Å^2
		MW	< 450
		logD(7.4)	1–3
Kelder et al. (1999)	CNS+/CNS−	PSA	< 60–70 Å^2
Norinder and Haeberlein (2002)	CNS+/CNS−	N + O	≤ 5
	logBB	logP-(N + O)	> 0
Friden et al. (2009)	$K_{p,uu,brain}$	HBA	≤ 2
Chen et al. (2011)	$K_{p,uu,brain}$	Kappa2	≤ 8
Wager et al. (2010)	CNS+	ClogP	2.8[a]
		ClogD	1.7[a]
		MW	305.3[a]
		TPSA	44.8[a]
		HBD	1[a]
		pKa	8.4[a]
Broccatelli et al. (2012)	CNS+/CNS−	Volsurf+ CACO2	> −0.3
		Substrate for Pgp[b]	Not a substrate
		BDDCS[c]	Class 1
Loryan et al. (2015)	$K_{p,uu,brain}$	TPSA	< 95 Å^2
Zhang et al. (2016)	$K_{p,uu,brain}$	TSPA	< 100Å^2
		Most acidic pKa	≥ 9.5

[a]Median (optimal) values for CNS+
[b]For compounds with Volsurf+ CACO2 > −0.3
[c]For Pgp substrates with Volsurf+ CACO2 > −0.3

used; virtually every published modelling work seems to be accompanied by a new interpretation of drug transport kinetics across the blood-brain barrier. To the extent that this diversity has become problematic, one of its causes are factual misunderstandings of pharmacokinetic principles such as muddling up unbound fractions with unbound concentrations or not appreciating the (in)dependence of the two. In addition, there could be challenges in communicating the intentions of a prediction model. There can, for example, be different ways of using a logBB prediction model that are based on compounds whose BBB transport is supposedly 'governed by passive diffusion' (see, e.g., Lanevskij et al. 2011). Such a prediction model could be used either with the intention to 'optimize logBB' (not recommended) or for comparison with experimental logBB data to indirectly deduce information on drug efflux and unbound drug exposure in the brain (sound).

Recently, the focus may have shifted towards characterising CNS$^+$ datasets or modelling CNS$^+$/CNS$^-$ datasets (Sect. 14.2.4), which highlights that the optimization of CNS drugs should be done in a more holistic manner that doesn't treat drug exposure in the brain separately from other critical drug properties that need optimization. With respect to in silico modelling of experimental measurements, the fundamentals of pharmacokinetics make a clear case that the $K_{p,uu,brain}$ is an appropriate and adequate parameter. The experimental methodology for $K_{p,uu,brain}$ determination has been available since microdialysis became a quantitative methodology in the 1990s (Elmquist and Sawchuk 1997; de Lange et al. 1999; Hammarlund-Udenaes 2000). More efficient methods were described around the turn of the century (Kakee et al. 1996; Kalvass and Maurer 2002), yet there are still few reports including an explicit analysis of its relationship with the chemical properties of the drug (Friden et al. 2009; Chen et al. 2011; Loryan et al. 2015; Varadharajan et al. 2015; Zhang et al. 2016).

14.2.2 Emerging Understanding of Determinants of Unbound Drug Exposure in the Brain, $K_{p,uu,brain}$

In 2009, Fridén et al. published a dataset of $K_{p,uu,brain}$ for 41 structurally diverse compounds obtained using a combination of in vivo brain tissue sampling and in vitro brain slice method (Friden et al. 2009). With respect to the values of $K_{p,uu,brain}$, it clearly showed that active efflux dominates drug disposition in the brain, since a majority of drugs had $K_{p,uu,brain}$ values less than unity and few drugs had values greater than unity (Fig. 14.2). The range of $K_{p,uu,brain}$ was from 0.006 for methotrexate to 2.0 for bupropion, i.e. 300-fold. In contrast, $K_{p,brain}$ from the same dataset ranged from ~0.002 for sulphasalazine to 20 for amitriptyline, i.e. 10,000-fold, which is considerably larger than for $K_{p,uu,brain}$. The relationship between $K_{p,uu,brain}$ and 16 conventional molecular descriptors was analysed using PLS. The most significant molecular descriptors for the relationship with $K_{p,uu,brain}$ were those that relate to hydrogen bonding, i.e. PSA and HBA (Fig. 14.3). Most other descriptors, including those of lipophilicity, did not add to the predictivity, and a simple model was put forth that used HBA as a single descriptor (Fig. 14.4).

$$\log K_{p,uu,brain} = -0.04 - 0.14 \times HBA \qquad (14.1)$$

The model was interpreted as follows; in order to achieve a twofold increase in $K_{p,uu,brain}$, it is necessary to remove two HBAs from the molecular structure. Conversely, a twofold reduction in $K_{p,uu,brain}$ can be achieved by addition of two HBAs. The moderate predictivity of the models, explaining only 40–50% of the variability in $K_{p,uu,brain}$ between drugs was rationalized as being due to the smaller range of values compared to $K_{p,brain}$ (logBB) and the fact that any deviation of $K_{p,uu,brain}$ from unity reflects the involvement of carrier-mediated transport.

Using similar experimental methodology, Chen and co-workers expanded this dataset to 246 with AstraZeneca proprietary compounds (Chen et al. 2011). The analysis was made with 196 molecular descriptors and various machine learning

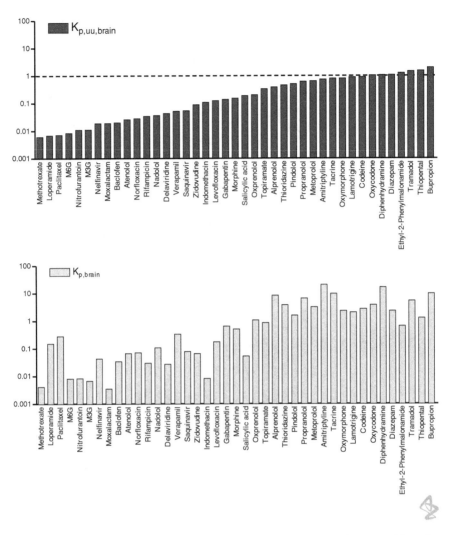

Fig. 14.2 Distribution of $K_{p,uu,brain}$ and $K_{p,brain}$ values for 41 diverse drugs. (Constructed from data in reference Friden et al. 2009)

algorithms. The best model was a consensus model that incorporated several sub-models. The predictivity on a subset (test set) of 73 compounds ($R^2 = 0.56$) suggests a substantial improvement versus the previous study (Friden et al. 2009); however, it is much worse than reported for models of logBB. The analysis of the importance of individual descriptors identified several descriptors relating to hydrogen bonding; however, the single most important descriptor was Kappa2 that describes the molecular shape in terms of its linearity. Interestingly, extensively branched compounds with Kappa2 greater than 8 had lower values for $K_{p,uu,brain}$ than did more linear compounds with Kappa2 values of 8 or below (Fig. 14.5). The importance of

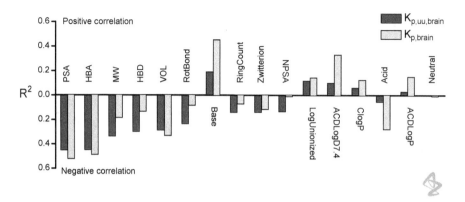

Fig. 14.3 Linear correlation coefficient, R^2, for $K_{p,uu,brain}$, $K_{p,brain}$ and each of the 16 molecular descriptors for the selected drugs in the training dataset. The upward and downward orientations of the bars represent positive and negative correlations, respectively, with $K_{p,uu,brain}$ and $K_{p,brain}$. (Constructed with permission from data in reference Friden et al. 2009)

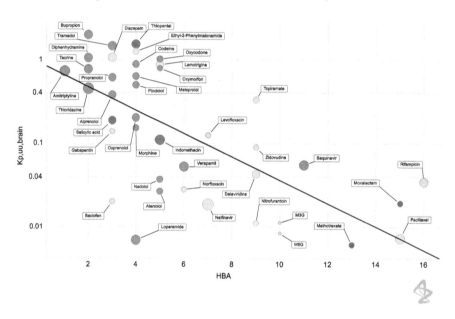

Fig. 14.4 Observed versus predicted rat $K_{p,uu,brain}$ based on HBA (Eq. 14.1). There is no obvious relationship between $K_{p,uu,brain}$ and lipophilicity (marker size by cLogP) or ion class (bases, blue; acids, red; neutral, yellow; and zwitterions, green). (Constructed form data in reference Friden et al. 2009)

polarity and hydrogen bonding was also corroborated by Loryan and colleagues with a novel dataset of 40 compounds, highlighting a strong association of $K_{p,uu,brain}$ with the topological polar surface area (TPSA) of the molecule and a descriptor called vsurf_Cw8 that represents the hydrogen bond donor and acceptor regions per

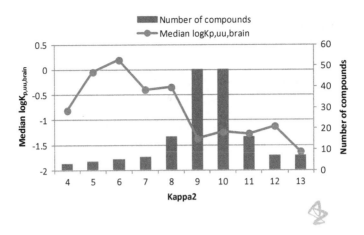

Fig. 14.5 The relationship between the median log $K_{p,uu,brain}$ and Kappa2. Note the drop in $K_{p,uu,brain}$ as Kappa2 exceeds 8 for increasingly branched molecules. (Constructed with permission from data in reference Chen et al. 2011)

surface unit of a molecule based on the 3D structure (Loryan et al. 2015). Furthermore, acidity as indicated by the highest acidic pKa has been found negatively correlated with $K_{p,uu,brain}$ (Zhang et al. 2016).

Whilst the importance of hydrogen bonding is in line with models of logBB and CNS⁺/CNS⁻ classification, it is intriguing that lipophilicity, which is normally correlated with passive transport, did not increase the value of $K_{p,uu,brain}$. This is a remarkable finding since it directly contradicts the common perception amongst clinicians that drug access to the brain is determined by the lipid solubility of the drug. It was proposed as a plausible explanation that the effect of increased passive transport by increased lipophilicity is paralleled and offset by increased efflux owing to increased drug concentrations in the membrane where the interaction takes place between drug and P-glycoprotein or other transporters (Friden et al. 2009). Hence, the dominating position of hydrogen bonding for structure-brain exposure relationships seems to arise from its additive effects on passive and active transport independently of lipophilicity; a less lipophilic drug with many HBAs has very limited passive transport and is thus sensitive to low-capacity active efflux, whilst a lipophilic drug with many HBAs is a probable transporter substrate, e.g. a Pgp substrate (Seelig and Landwojtowicz 2000).

The example of beta blockers introduced above can be analysed in this context (Fig. 14.1); it suggests that the lower $K_{p,uu,brain}$ of atenolol versus propranolol is due to its two additional hydrogen bond acceptors (and the increased potential for transporter interactions) rather than being related to reduced passive diffusion as governed by lipophilicity (ACDlogD7.4). Note the similar structures of atenolol and propranolol in Fig. 14.1.

14.2.3 The Relationship between Prediction Models for LogBB and $K_{p,uu,brain}$

The evident discrepancy between the interpretations of the classical work with logBB and the recent work with $K_{p,uu,brain}$ warrants a more detailed discussion. The key to understanding the difference is to grasp how logBB (Eq. 14.2) is a composite parameter of $K_{p,uu,brain}$ (the actual effect of the BBB), plasma protein binding ($f_{u,p}$) and nonspecific binding in the brain tissue described by the unbound volume of distribution in the brain ($V_{u,brain}$).

$$BB = K_{p,uu,brain} \times V_{u,brain} \times f_{u,p} \tag{14.2}$$

As such, any model of logBB is 'contaminated' with $V_{u,brain}$ and $f_{u,p}$; $K_{p,uu,brain}$ is in itself the pharmacologically relevant concentration gradient of unbound drug across the BBB, and its steady-state value is independent of $V_{u,brain}$ and $f_{u,p}$. A logBB PLS model was developed using the same molecular descriptors and dataset (Friden et al. 2009). The relationship between logBB and structure was dominated by hydrogen bonding similar to $K_{p,uu,brain}$ (Fig. 14.3). In contrast to $K_{p,uu,brain}$, however, logBB was also positively correlated with descriptors of lipophilicity. Furthermore, logBB was higher for basic drugs than for acidic drugs (Fig. 14.3). The PLS model that was developed contained one descriptor for hydrogen bonding (HBA), one descriptor of lipophilicity (ACDLogD7.4) and the ion class of the drug (acid or base):

$$\log BB = -0.18 - 0.097 \ HBA + 0.10 \ ACD \ \log D7.4 + 0.68 \ Base - 0.67 \ Acid \tag{14.3}$$

The predictivity of the logBB model was better than the $K_{p,uu,brain}$ model based on the comparison of Q^2 (0.693 vs. 0.426). The better Q^2 value of the logBB model should be seen in the light of the 30-fold greater range of observed values. In contrast, similar predictivity is seen based on the root of mean squared error (RMSE, 4.0-fold vs. 3.9-fold). The logBB model was mechanistically rationalized as follows: HBA accounts for the part of logBB which is, in fact, related to $K_{p,uu,brain}$; ACDLogD7.4 and the drug being basic accounts for binding to phospholipid in tissue ($V_{u,brain}$) and the drug being acidic accounts for extensive binding to albumin in plasma ($f_{u,p}$). A plot of $K_{p,brain}$ versus $K_{p,uu,brain}$ (Fig. 14.6) clearly shows that the ion class explains much of the differences between the two. The independence of $K_{p,uu,brain}$ and on basicity has been reported from additional datasets where the authors' reason that the classical view of basicity favouring CNS exposure basically stems from classical CNS drugs targeting G-protein-coupled receptors for signalling ligands containing basic nitrogen atoms (Zhang et al. 2016). The imminent risk of relying on logBB-derived prediction models is the design of drugs that are unnecessarily lipophilic or basic without improved unbound brain exposure, or if restricted brain exposure is desired, the design of albumin-bound acidic compounds that later prove to have significant CNS effects at therapeutic plasma concentrations.

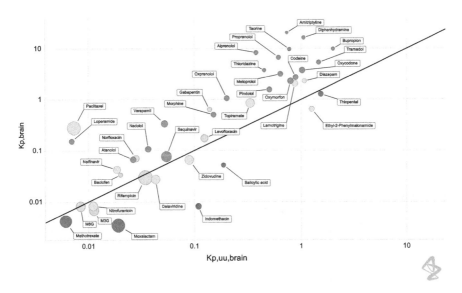

Fig. 14.6 The relationship between $K_{p,brain}$ and $K_{p,uu,brain}$ is largely dependent on the ion class of the drug; relative to their $K_{p,uu,brain}$ basic drugs (blue markers) have high $K_{p,brain}$ and acids (red markers) low $K_{p,brain}$. This is because basic drugs tend to bind to brain tissue constituents more strongly than to plasma proteins, whereas the opposite is observed for acidic drugs. For neutrals (yellow markers) and zwitterions (green markers), there is a better agreement and no clear trend. Markers are sized according to the number of hydrogen bond acceptors. The solid line represents identity. (Constructed from data in Friden et al. 2009)

14.2.4   Recent Developments in CNS⁺/CNS⁻ Classification

Possibly pushed by the debate around logBB, many workers have recently used CNS⁺/CNS⁻ classification as the parameter for in silico model development. The CNS⁺/CNS⁻ datasets have been trimmed in various ways; some have chosen not to include any investigational drugs on the basis of that the majority of investigational CNS⁺ drugs fail to reach the market (Ghose et al. 2012). Others have also carefully scrutinized the CNS⁻ class by applying a cut-off value for actual measurements in humans using positron emission tomography (PET) or sampling of cerebrospinal fluid and named the classification scheme BBB⁺/BBB⁻ (Broccatelli et al. 2012). Much work is also done to analyse the group of CNS drugs on the market as it stands. The philosophy is that there are so many other pivotal CNS drug properties in addition to brain exposure that need consideration in drug design: metabolic clearance, safety risks, etc. Wager and co-workers developed a CNS multiparameter optimization (MPO) approach to assess the alignment of a drug candidate's chemical properties to those of marketed CNS drugs (Wager et al. 2010). Using this approach, they showed that the CNS MPO score of 108 Pfizer candidates were distributed considerably less favourably, thus suggesting that the CNS MPO score could indicate the chances of a candidate becoming a registered drug. Recently, a pipeline of rules was proposed based on a combination of in silico predicted permeability, experimental classification as substrate or nonsubstrate of Pgp and the class belonging to the Biopharmaceutics Drug Disposition Classification System (BDDCS) (Broccatelli et al. 2012). The prediction model performed better than a number of purely computationally based classification models based on the same BBB^+/BBB^- dataset. Recent classification approaches are summarized in Table 14.2 together with older rules of thumb.

Carefully scrutinized CNS^+/CNS^- classification is arguably a convincing parameter of drug exposure in the brain in the sense that it is based on human in vivo information. A major drawback of classification approaches, however, is that information on brain exposure is reduced from its natural (continuous) ratio measurement to the binary CNS^+/CNS^-. By introducing cut-off values to compile datasets, one is not only removing potentially useful information but also adding in new by the arbitrary choice of cut-off. This problem is obvious when considering a group of equivalent drugs around the cut-off value.

14.3 Future Directions and Challenges

Recent work on $K_{p,uu,brain}$ datasets has shed new light on the significance of molecular properties for drug exposure in the brain; lipophilicity does not seem to be an important parameter. However, just as importantly, it has shown that the predictivity of $K_{p,uu,brain}$ models is much lower than for logBB models assessed by cross-validated Q^2 on training sets or R^2 on test sets. This is related to the smaller range of values for $K_{p,uu,brain}$ than for logBB; to date at best, only 50–60% of the variability in $K_{p,uu,brain}$ between compounds has been possible to relate to descriptors of molecular properties. There are three messages in this observation that need to be appreciated. Firstly, the smaller range of $K_{p,uu,brain}$ compared to BB or PS means that the extent to which drugs differ with respect to exposure in the brain has been somewhat exaggerated. Secondly, and related to the first, the established perception that 'brain penetration' is a predictable drug property needs reconsideration; the apparent success of in silico models of logBB or PS is due to the parameter being inclusive of additional drug properties such as nonspecific binding in the brain, which are easily predicted yet not relevant to the problem. Thirdly, the remaining hardship of predicting $K_{p,uu,brain}$ reflects the fact that any variability in $K_{p,uu,brain}$ between compounds is caused by drug-specific, multi-specific molecular interactions with the drug transporters at the BBB.

14.3.1 Improving Predictions of $K_{p,uu,brain}$ by Integration of Approaches

It is possible, however questionably probable, that the recent year's focus on artificial intelligence and increasing computer power will lead to significantly better $K_{p,uu,brain}$ prediction models based on molecular structure alone. Drug exposure in the brain is primarily determined by specific interactions with the transporters at the BBB. It therefore appears likely that the approaches used so far have already delivered close to their full potential. An interesting way forward can be molecular modelling of the BBB drug transporters. Molecular modelling of transporters can be applied either by using a transporter-based approach, which utilizes the three-dimensional crystal structure of the protein, or if this is not known as is generally the case, a substrate-based approach, e.g. comparative molecular field analysis (CoMFA) yielding a pharmacophore model of the transporter. Several review articles have featured the developments in this field (Ekins et al. 2007; Ecker et al. 2008; Winiwarter and Hilgendorf 2008; Demel et al. 2009). However successful these models may become in terms of discriminating substrates from nonsubstrates, it is just a first step of being able to make quantitative predictions of $K_{p,uu,brain}$. This is because a drug can be a substrate of several different transporters at the BBB and there is interplay with passive transport as well as additional mechanisms of drug elimination from brain including metabolism and bulk flow of the brain interstitial fluid. The integration of molecular modelling (of several transporters) with passive diffusion and physiologically based pharmacokinetic (PBPK) models would seem like a logical and appealing approach, but it is likely to remain a utopia for years to come. The success of progressing the field of $K_{p,uu,brain}$ predictions is more likely to depend on the development of new and imaginative ways of integrating different computational and experimental methodologies with machine learning algorithms. As a simple example, it would be feasible to use experimental measurements of, e.g. passive diffusion and molecular modelling 'docking scores' as variables alongside molecular descriptors in, e.g., a PLS analysis. Moreover, novel machine learning algorithms previously used to predict $K_{p,uu,brain}$ (or logBB) from computed molecular descriptors could also find an application to predict $K_{p,uu,brain}$ from a battery of in vitro transporter assays. This was recently exemplified and put forth as a general strategy to combine in silico $K_{p,uu,brain}$ modelling with in vitro measurement of the P-glycoprotein efflux ratio (Zhang et al. 2016). Various in vitro models of the BBB as well as transporter assays of nonbrain origin are covered in another chapter. Whilst for preclinical species in vivo methodologies will remain the mainstay for drug discovery, it is important to progress predictions of drug exposure in the brain using in vitro and in silico methodologies in order to conduct a proper translation to humans.

14.3.2 Drug Design Strategies

As discussed above, a default and crude strategy to 'optimize' drug exposure in the brain would be to add or remove hydrogen bond acceptors to the molecular structure; for every addition or removal of two hydrogen bond acceptors, one should expect a twofold reduction or increase, respectively, in $K_{p,uu,brain}$. Even though the accuracy of this kind of prediction is improved by using more complex modelling approaches, it comes at the expense of the model being difficult to comprehend for the modeller. This lack of transparency can be very unhelpful for the chemist who needs an idea of *how* to change the molecular structure to obtain the desired change in $K_{p,uu,brain}$. In an investigation of the trade-off between model accuracy and comprehensibility, it was shown for a set of 16 biopharmaceutical classification tasks that in general there is a limited cost of prediction accuracy when choosing a comprehensible model (Johansson et al. 2011). The lack of transparency of a model is commonly mitigated by the construction of virtual compound libraries that can be screened in the prediction model, whereby promising compound structures can be identified for synthesis.

The discussed challenges of making $K_{p,uu,brain}$ predictions based on in silico models should not shadow its utility for drug design. Frequently, more predictive models can be obtained for a set of compounds with a more limited range of properties, i.e. with a smaller applicability domain. This is typically the case when optimizing a chemical drug series for a particular target. The limited applicability domain of this situation is a double-edged sword; on the one hand, it helps the understanding of what molecular features are associated with high or low $K_{p,uu,brain}$ for the particular series, but it also does not tell the chemist how to make molecules that are different and even better than the ones already made and used in the model. An approach recently taken at AstraZeneca to extend the $K_{p,uu,brain}$ beyond the existing domain was to pay attention to structural modifications that result in compounds that deviate from the prediction model in a favourable direction (Plowright et al. 2012). The idea was that in spite of such a compound having a mediocre $K_{p,uu,brain}$, it possesses a structural element that when combined with the favourable molecular properties as described by the model, a step-change in $K_{p,uu,brain}$ can be achieved for the chemical series (Fig. 14.7). This approach exemplifies how computational methods can not only be used to derive prediction models as such but also to discover and exploit hidden patterns in the experimental data.

14.4 Conclusions

In silico prediction of a compounds' ability to efficiently cross the BBB has been an area of development for decades; computational methodologies have evolved, and experimental datasets have increased in size. Yet, the (lack of) pharmacologic meaning of the commonly used logBB measurement has been generally overlooked, and

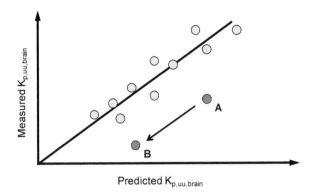

Fig. 14.7 A prediction model can be used to identify outliers with promising properties for optimization. Compound A is predicted substantially higher than the experimental value for whatever reason, but the value is not as low as desired. The structural modification of A as guided by the prediction model results in a new compound B with superior properties. (Adapted with permission from Plowright et al. 2012)

it is rather recently that the focus of modellers have either turned to classification approaches on CNS⁺/CNS⁻ datasets or to datasets of *unbound* drug exposure in the brain ($K_{p,uu,brain}$), which is considered a pharmacologically meaningful parameter. In silico modelling of $K_{p,uu,brain}$ has corroborated some findings from logBB models but disputed others; whereas the importance of hydrogen bonding stands strong, there is no evidence based on $K_{p,uu,brain}$ that compound lipophilicity or basicity has any influence on drug exposure in the brain. As should be expected from a parameter that is determined by multiple specific and unspecific molecular interactions, the success of $K_{p,uu,brain}$ predictions has been moderate even when applying state-of-the-art modelling methodologies. Hence there appears to be rather limited scope for improvement in this field with the toolbox used to date, i.e. modelling of molecular descriptors. Instead, the advancement of predictions of drug exposure in the brain from the chemical structure requires the inclusion of additional sources of information such as in vitro measurements and clever ways of integrating the data. Future work with predictions of drug exposure in the brain will facilitate the translation from the preclinical species to humans and thus raise the power of in silico modelling to the desired level for successful development of new drug treatments.

14.5   Topics for Discussion

- What is meant by the applicability domain of a model, and why is it important to define it?
- Why is it necessary to use a test set of compounds to validate a model?
- What aspects of a molecule's properties are not covered by the descriptors listed in Table 14.1?

- What are the mechanistic reasons why hydrogen bonding plays such an important role for unbound drug exposure in the brain?
- What could be the reasons for prediction models of $K_{p,uu,brain}$ not requiring descriptors of lipophilicity?
- At which stages in the drug discovery process is it more useful to have an in silico model for drug exposure in the brain?
- To what an extent can in silico models replace in vitro or in vivo experiments?
- Which measurement logPS or logBB would you expect to correlate more closely with lipophilicity and why?
- What are the strengths and weaknesses of modelling categorical CNS⁺/CNS⁻ datasets versus the continuous $K_{p,uu,brain}$ variable?

References

Abraham MH (2004) The factors that influence permeation across the blood-brain barrier. Eur J Med Chem 39:235–240

Abraham MH, Hersey A (2007) In silico models to predict brain uptake. In: Comprehensive medicinal chemistry II. Elsevier, London, p 745

Abraham MH, Chadha HS, Mitchell RC (1994) Hydrogen bonding. 33. Factors that influence the distribution of solutes between blood and brain. J Pharm Sci 83:1257–1268

Abraham MH, Ibrahim A, Zissimos AM, Zhao YH, Comer J, Reynolds DP (2002) Application of hydrogen bonding calculations in property based drug design. Drug Discov Today 7:1056–1063

Bendels S, Kansy M, Wagner B, Huwyler J (2008) In silico prediction of brain and CSF permeation of small molecules using PLS regression models. Eur J Med Chem 43:1581–1592

Broccatelli F, Larregieu CA, Cruciani G, Oprea TI, Benet LZ (2012) Improving the prediction of the brain disposition for orally administered drugs using BDDCS. Adv Drug Deliv Rev 64:95–109

Chen H, Winiwarter S, Friden M, Antonsson M, Engkvist O (2011) In silico prediction of unbound brain-to-plasma concentration ratio using machine learning algorithms. J Mol Graph Model 29:985–995

Clark DE (2003) In silico prediction of blood-brain barrier permeation. [see comment]. Drug Discov Today 8:927–933

de Lange EC, de Boer BA, Breimer DD (1999) Microdialysis for pharmacokinetic analysis of drug transport to the brain. Adv Drug Deliv Rev 36:211–227

Demel MA, Kramer O, Ettmayer P, Haaksma EE, Ecker GF (2009) Predicting ligand interactions with ABC transporters in ADME. Chem Biodivers 6:1960–1969

Ecker GF, Noe CR (2004) In silico prediction models for blood-brain barrier permeation. Curr Med Chem 11:1617–1628

Ecker GF, Stockner T, Chiba P (2008) Computational models for prediction of interactions with ABC-transporters. Drug Discov Today 13:311–317

Ekins S, Ecker GF, Chiba P, Swaan PW (2007) Future directions for drug transporter modelling. Xenobiotica 37:1152–1170

Elmquist WF, Sawchuk RJ (1997) Application of microdialysis in pharmacokinetic studies. Pharm Res 14:267–288

Friden M, Gupta A, Antonsson M, Bredberg U, Hammarlund-Udenaes M (2007) In vitro methods for estimating unbound drug concentrations in the brain interstitial and intracellular fluids. Drug Metab Dispos 35:1711–1719

Friden M, Winiwarter S, Jerndal G, Bengtsson O, Wan H, Bredberg U, Hammarlund-Udenaes M, Antonsson M (2009) Structure-brain exposure relationships in rat and human using a novel data set of unbound drug concentrations in brain interstitial and cerebrospinal fluids. J Med Chem 52:6233–6243

Ghose AK, Herbertz T, Hudkins RL, Dorsey BD, Mallamo JP (2012) Knowledge-based, central nervous system (CNS) Lead selection and Lead optimization for CNS drug discovery. ACS Chem Neurosci 3:50–68

Gratton JA, Abraham MH, Bradbury MW, Chadha HS (1997) Molecular factors influencing drug transfer across the blood-brain barrier. J Pharm Pharmacol 49:1211–1216

Gupta A, Chatelain P, Massingham R, Jonsson EN, Hammarlund-Udenaes M (2006) Brain distribution of cetirizine enantiomers: comparison of three different tissue-to-plasma partition coefficients: K(p), K(p,u), and K(p,uu). Drug Metab Dispos 34:318–323

Hammarlund-Udenaes M (2000) The use of microdialysis in CNS drug delivery studies. Pharmacokinetic perspectives and results with analgesics and antiepileptics. Adv Drug Deliv Rev 45:283–294

Hammarlund-Udenaes M, Friden M, Syvanen S, Gupta A (2008) On the rate and extent of drug delivery to the brain. Pharm Res 25:1737–1750

Hammarlund-Udenaes M, Bredberg U, Friden M (2009) Methodologies to assess brain drug delivery in lead optimization. Curr Top Med Chem 9:148–162

Hitchcock SA, Pennington LD (2006) Structure-brain exposure relationships. J Med Chem 49:7559–7583

Johansson U, Sonstrod C, Norinder U, Bostrom H (2011) Trade-off between accuracy and interpretability for predictive in silico modeling. Future Med Chem 3:647–663

Kakee A, Terasaki T, Sugiyama Y (1996) Brain efflux index as a novel method of analyzing efflux transport at the blood-brain barrier. J Pharmacol Exp Ther 277:1550–1559

Kalvass JC, Maurer TS (2002) Influence of nonspecific brain and plasma binding on CNS exposure: implications for rational drug discovery. Biopharm Drug Dispos 23:327–338

Kelder J, Grootenhuis PD, Bayada DM, Delbressine LP, Ploemen JP (1999) Polar molecular surface as a dominating determinant for oral absorption and brain penetration of drugs. Pharm Res 16:1514–1519

Lanevskij K, Dapkunas J, Juska L, Japertas P, Didziapetris R (2011) QSAR analysis of blood-brain distribution: the influence of plasma and brain tissue binding. J Pharm Sci 100:2147–2160

Levin VA (1980) Relationship of octanol/water partition coefficient and molecular weight to rat brain capillary permeability. J Med Chem 23:682–684

Liu X, Tu M, Kelly RS, Chen C, Smith BJ (2004) Development of a computational approach to predict blood-brain barrier permeability. Drug Metab Dispos 32:132–139

Liu H, Dong K, Zhang W, Summerfield SG, Terstappen GC (2018) Prediction of brain: blood unbound concentration ratios in CNS drug discovery employing in silico and in vitro model systems. Drug Discov Today 23:1357–1372

Loryan I, Sinha V, Mackie C, Van Peer A, Drinkenburg WH, Vermeulen A, Heald D, Hammarlund-Udenaes M, Wassvik CM (2015) Molecular properties determining unbound intracellular and extracellular brain exposure of CNS drug candidates. Mol Pharm 12:520–532

Luco JM (1999) Prediction of the brain-blood distribution of a large set of drugs from structurally derived descriptors using partial least-squares (PLS) modeling. J Chem Inf Comput Sci 39:396–404

Martin I (2004) Prediction of blood-brain barrier penetration: are we missing the point? [see comment] [comment]. Drug Discov Today 9:161–162

Matsson P (2007) ATP-binding cassette efflux transporters and passive membrane permeability in drug absorption and disposition. Acta Universitatis Upsaliensis, Uppsala, p 68

McAinsh J, Cruickshank JM (1990) Beta-blockers and central nervous system side effects. Pharmacol Ther 46:163–197

Mehdipour AR, Hamidi M (2009) Brain drug targeting: a computational approach for overcoming blood-brain barrier. Drug Discov Today 14:1030–1036

Morales JF, Montoto SS, Fagiolino P, Ruiz ME (2017) Current state and future perspectives in QSAR models to predict blood-brain barrier penetration in central nervous system drug R&D. Mini Rev Med Chem 17:247–257

Norinder U, Haeberlein M (2002) Computational approaches to the prediction of the blood-brain distribution. Adv Drug Deliv Rev 54:291–313

Osterberg T, Norinder U (2000) Prediction of polar surface area and drug transport processes using simple parameters and PLS statistics. J Chem Inf Comput Sci 40:1408–1411

Palm K, Stenberg P, Luthman K, Artursson P (1997) Polar molecular surface properties predict the intestinal absorption of drugs in humans. Pharm Res 14:568–571

Pardridge WM (2004) Log(BB), PS products and in silico models of drug brain penetration. [comment]. Drug Discov Today 9:392–393

Plowright AT, Nilsson K, Antonsson M, Amin K, Broddefalk J, Jensen J, Lehmann A, Jin S, St-Onge S, Tomaszewski MJ, Tremblay M, Walpole CS, Wei Z, Yang H, Ulander J (2012) Discovery of agonists of cannabinoid receptor 1 with restricted CNS penetration aimed for treatment of gastroesophageal reflux disease. J Med Chem 56(1):220–240

Seelig A, Landwojtowicz E (2000) Structure-activity relationship of P-glycoprotein substrates and modifiers. Eur J Pharm Sci 12:31–40

van de Waterbeemd H, Camenisch G, Folkers G, Chretien JR, Raevsky OA (1998) Estimation of blood-brain barrier crossing of drugs using molecular size and shape, and H-bonding descriptors. J Drug Target 6:151–165

van de Waterbeemd H, Smith DA, Jones BC (2001) Lipophilicity in PK design: methyl, ethyl, futile. J Comput Aided Mol Des 15:273–286

Varadharajan S, Winiwarter S, Carlsson L, Engkvist O, Anantha A, Kogej T, Friden M, Stalring J, Chen H (2015) Exploring in silico prediction of the unbound brain-to-plasma drug concentration ratio: model validation, renewal, and interpretation. J Pharm Sci 104:1197–1206

Wager T, Hou X, Verhoest PR, Villalobos A (2010) Moving beyond rules: the development of a central nervous system multiparameter optimization (CNS MPO) approach to enable alignment of druglike properties. ACS Chem Neurosci 1:435–449

Winiwarter S, Hilgendorf C (2008) Modeling of drug-transporter interactions using structural information. Curr Opin Drug Discov Devel 11:95–103

Winiwarter S, Ridderström M, Ungell AL, Andersson TB, Zamora I, Zamora I (2007) Use of molecular descriptors for absorption, distribution, metabolism, and excretion predictions. In: Comprehensive medicinal chemistry II. Elsevier, London, p 745

Wold S (1991) Validation of QSAR's. Quant Struct Act Relat 10:191–193

Wold S (2001) PLS-regression: a basic tool of chemometrics. Chemom Intel Lab 58:109–130

Young RC, Mitchell RC, Brown TH, Ganellin CR, Griffiths R, Jones M, Rana KK, Saunders D, Smith IR, Sore NE (1988) Development of a new physicochemical model for brain penetration and its application to the design of centrally acting H2 receptor histamine antagonists. J Med Chem 31:656–671

Zhang YY, Liu H, Summerfield SG, Luscombe CN, Sahi J (2016) Integrating in silico and in vitro approaches to predict drug accessibility to the central nervous system. Mol Pharm 13:1540–1550

Part IV
Strategies for Improved CNS Drug Delivery

Chapter 15
Intranasal Drug Delivery to the Brain

Jeffrey J. Lochhead, Niyanta N. Kumar, Geetika Nehra, Mallory J. Stenslik, Luke H. Bradley, and Robert G. Thorne

Abstract The barriers that separate the blood from brain interstitial and cerebrospinal fluids present a significant challenge to efficient and practical drug delivery into the central nervous system (CNS). New strategies to circumvent the blood-brain barrier (BBB) have long been needed to utilize polar pharmaceuticals and large biotherapeutics for CNS disease treatment because the BBB is typically impermeable to such compounds. The increasing application of biologics as therapeutics over the past several decades has brought much new interest in routes of drug delivery that may be more easily utilized for chronic dosing of large molecules, e.g., oral, subcutaneous, transdermal, pulmonary, and intranasal administration. The intranasal route in particular offers a number of advantages for chronic dosing including its noninvasiveness, efficient uptake and absorption into a highly vascular submucosa, avoidance of hepatic first-pass elimination, rapid pharmacokinetic pro-

J. J. Lochhead (✉)
Department of Pharmacology, University of Arizona College of Medicine, Tucson, AZ, USA
e-mail: lochhead@email.arizona.edu

N. N. Kumar (✉)
Pharmacokinetics, Pharmacodynamics, & Drug Metabolism, Merck & Co. Inc.,
West Point, PA, USA
e-mail: niyanta.kumar@merck.com

G. Nehra
Sanders-Brown Center on Aging, College of Medicine, University of Kentucky,
Lexington, KY, USA

M. J. Stenslik
Translational Imaging Biomarkers, MRL, Merck & Co., Inc., West Point, PA, USA

L. H. Bradley
Department of Neuroscience, University of Kentucky College of Medicine,
Lexington, KY, USA

R. G. Thorne (✉)
Denali Therapeutics, South San Francisco, CA, USA

Department of Pharmaceutics, College of Pharmacy, University of Minnesota-Twin Cities,
Minneapolis, MN, USA
e-mail: thorne@dnli.com

© American Association of Pharmaceutical Scientists 2022
E. C. M. de Lange et al. (eds.), *Drug Delivery to the Brain*, AAPS Advances
in the Pharmaceutical Sciences Series 33, https://doi.org/10.1007/978-3-030-88773-5_15

461

files, and ease of administration. Importantly, the intranasal route has also been demonstrated to potentially allow a variety of drugs direct access to the brain and/ or cerebrospinal fluid. Studies over the past few decades have shown that even large biotherapeutics may have access to the CNS along extracellular pathways associated with the olfactory and trigeminal nerves. This chapter provides an overview of the unique anatomic and physiologic attributes of the nasal mucosa and its associated cranial nerves that allow small but significant fractions of certain intranasally applied drugs to transfer across the nasal epithelia and subsequently be transported directly into the CNS. We also review some of the preclinical and clinical literature related to intranasal targeting of biologics to the CNS and comment on future directions for the further clinical translation of this route of administration.

Keywords Drug delivery · Nasal passage · Olfactory · Trigeminal · Proteins · Gene vectors · Stem cells

15.1 Introduction

The blood-brain barrier (BBB) and blood-cerebrospinal fluid barriers (BCSFB) are critical for the maintenance of central nervous system (CNS) homeostasis. Although these barriers restrict neurotoxic substances from entering the brain, they also restrict many potential therapeutics from reaching the CNS. The BBB, formed by brain endothelial cell lining microvessels, exhibits a low rate of pinocytosis and possesses tight junction (TJ) protein complexes on apposing cells that limit paracellular permeability (Reese and Karnovsky 1967). These TJ create a high transendothelial electrical resistance of 1500–2000 $\Omega{\cdot}cm^2$ compared to 3–30 $\Omega{\cdot}cm^2$ across most peripheral microvessels (Crone and Olesen 1982; Butt et al. 1990). This high resistance is associated with very low paracellular permeability, and typically, only small (<600 Da), lipophilic molecules appreciably cross the healthy BBB via transcellular passive diffusion, although some limited transport of certain peptides and peptide analogs has been reported (Banks 2009). Additionally, many potential therapeutics that would otherwise be predicted to cross the BBB based on their molecular weight (MW) and lipophilicity are restricted by the expression of drug transporters (e.g., P-glycoprotein) (Miller 2010; Ronaldson et al. 2007).

Nearly all CNS drugs in clinical use today can be categorized as small MW pharmaceuticals that either cross the BBB transcellularly (e.g., barbiturates) or utilize endogenous transporters expressed on endothelial cells (e.g., Parkinson's therapeutic levodopa). Just about all large MW substances are severely restricted from crossing the BBB under physiological conditions. Indeed, the only examples of large MW drugs approved for clinical use in treating neurological illnesses are those that act outside the CNS (e.g., type I interferons for treating multiple sclerosis), those with the chance to cross compromised endothelial barriers associated with some CNS tumors (e.g., the humanized monoclonal antibody bevacizumab for the treatment of recurrent glioblastoma); a peptide administered intrathecally to treat severe, chronic pain (the ~3 kDa cone snail toxin ziconotide); an antisense oligonucleotide administered intrathecally to treat spinal muscular atrophy; an enzyme administered intraventricularly to treat Batten disease; and a recently approved AAV9-based gene therapy. Many other large MW peptides, proteins, oligonucleotides, and gene therapy vectors have been identified as potential CNS therapeutics based on studies utilizing in vitro systems and animal models; however, new drug delivery strategies are needed to allow these potential drugs to cross or bypass the BBB and BCSFB for these studies to translate to the clinic (Neuwelt et al. 2008). It is also likely that recent advances in cerebrovascular biology, e.g., single-cell transcriptomics analyses of the brain vasculature in mice (Vanlandewijck et al. 2018) and humans (Yang et al. 2021), coupled with advances in our understanding of the complex physiology of brain fluids (Abbott et al. 2018) may yield fresh, new ideas and previously unexplored novel approaches for CNS delivery.

The central input of substances through intraparenchymal, intracerebroventricular, or intrathecal injections/infusions represent one strategy, but these routes of administration are invasive and typically not ideal for chronic administration. Increasing evidence suggests the intranasal (IN) route of administration provides a noninvasive method to bypass the BBB and directly deliver therapeutics to the CNS along extracellular pathways associated with the olfactory and trigeminal nerves (Fig. 15.1). In addition to its noninvasiveness, the IN administration route has long

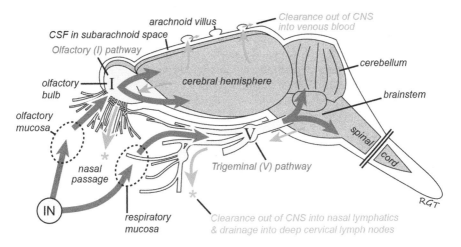

Fig. 15.1 Intranasal (IN) administration provides access to olfactory and trigeminal pathways (shown in red for the rat), potentially allowing certain peptides, proteins, and even cells to reach widespread CNS regions. Based on work utilizing radiolabeled proteins in rats and primates (Thorne et al. 2004a, b, 2008a, b), a small fraction of intranasally applied drug may be rapidly transported via components associated with the olfactory nerves (the first cranial nerve) to the olfactory bulbs and rostral brain regions or via components associated with the trigeminal nerves (the fifth cranial nerve) to the brain stem and caudal brain regions. Drug entry into the brain appears to occur rapidly following transport across the olfactory or respiratory epithelia. Other work has shown that a variety of substances may also be cleared out of the brain along possibly related pathways (shown in green) connecting CNS parenchymal tissue and cerebrospinal fluid (CSF) in the subarachnoid spaces with lymphatics in the nasal passages and, ultimately, the deep cervical lymph nodes of the neck (Bradbury and Cserr 1985; Kida et al. 1993). The principal clearance of CSF into the venous blood occurs through the arachnoid villi that extend from the subarachnoid space into the dural sinuses

been associated with a number of advantages (Lochhead and Thorne 2012), mostly based on the application of drugs with a systemic mode of action; these include typically rapid onset of effects, ease of administration by nasal drops or sprays, simple dose adjustment, avoidance of hepatic first-pass elimination, and a developing record of experience with clinically approved formulations (e.g., the nasal spray of the 3.5 kDa polypeptide hormone calcitonin has been used for many years to treat postmenopausal osteoporosis). The main disadvantages of the IN route comprise a limitation typically to potent drugs due to low nasal absorption (particularly for hydrophilic drugs, peptides, and proteins), limited solution volumes (typically, 25–200 µl in humans), active mucociliary clearance processes resulting in limited contact time with the absorptive epithelia, nasal enzymatic degradation for some drugs, interindividual variability, and low CNS delivery efficiencies (<0.05%) for most proteins measured thus far (Lochhead and Thorne 2012; Costantino et al. 2007).

The IN administration route has a long, successful history of clinical application, where it has been used to deliver a number of drugs to the systemic circulation that cannot be given orally (Lansley and Martin 2001; Costantino et al. 2007). The possibility that IN administration may also deliver potentially therapeutic amounts of large MW drugs directly from the nasal passages to the CNS was first described relatively recently (Thorne et al. 1995; Frey 2nd et al. 1997). The delivery of small molecules, macromolecules, gene vectors, and even cells from the nasal passages to the brain has now been documented in numerous animal and clinical studies (Lochhead and Thorne 2012; Dhuria et al. 2010; Baker and Genter 2003; Illum 2004). This chapter provides an overview of relevant nasal anatomy and physiology as well as the potential pathways and transport mechanisms that are involved in the distribution of therapeutics from the nasal cavity to the CNS. We also summarize some of the most relevant preclinical and clinical studies that have presented evidence of brain entry and/or efficacy following intranasal targeting of biotherapeutics to the CNS and speculate on future directions.

15.2   Nasal Anatomy and Physiology

15.2.1   General Overview

The nasal chamber is divided into two separate passages by the nasal septum, with each nasal passage principally consisting of an olfactory region (containing the olfactory epithelium) and a respiratory region (containing the respiratory epithelium) extending from the nostrils (nares) to the nasopharynx. The general organization of the rat nasal passage is shown in Fig. 15.2. The olfactory region contains olfactory sensory neurons that are responsible for the detection of airborne odorants (i.e., mediating the sense of smell). Most of the nonolfactory epithelium in the nasal passages of laboratory animals and human beings consists of a respiratory epithelium specialized for warming and humidifying inspired air as well as the removal of

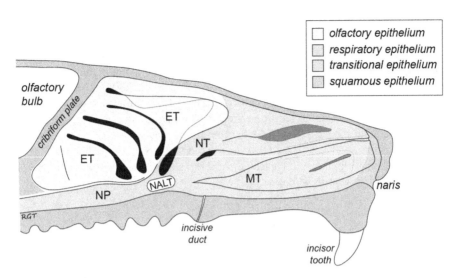

Fig. 15.2 Schematic diagram of the nasal passage showing the distribution of the surface epithelia on the lateral wall of the rat. Abbreviations: *ET* ethmoturbinates, *MT* maxilloturbinate, *NALT* nasal-associated lymphoid tissue, *NP* nasopharynx, *NT* nasoturbinate. (Figure partly based on Mery et al. 1994; Harkema et al. 2006)

allergens, microorganisms, and particulates (Harkema et al. 2006). The human nasal cavity has a large absorptive surface area of ~160 cm^2 due to three, comma-shaped bony structures called turbinates or conchae (inferior, middle, and superior) which filter, humidify, and warm inspired air (Harkema et al. 2006). The differences in nasal structure, organization, and physiology between primates and rodents may potentially be important in evaluating experimental data in support of nose-to-brain transport pathways (Lochhead and Thorne 2012), e.g., humans and monkeys are oronasal breathers, while rats are obligate nasal breathers with a turbinate architecture that is considerably more complex than that in primate species. Additionally, the olfactory region accounts for only about 10% of the total absorptive surface area in the human nasal cavity, whereas it comprises ~50% of the total nasal surface area in the rat, likely reflecting the greater importance of this sense for macrosmatic mammals such as rodents. By contrast, the absolute olfactory surface area does not differ too greatly between human beings (~12.5 cm^2), rhesus monkeys (6–9 cm^2), and rats (7 cm^2) (Lochhead and Thorne 2012). While it is not yet clear if significant differences in nose-to-brain transport occur with different species, most investigations have utilized rodents simply because it has not been practical to conduct certain types of research in monkeys and human beings; further developments in noninvasive imaging may allow better comparisons between species in the future.

In addition to the olfactory and respiratory regions, the nasal cavity also contains squamous and transitional regions, along with a small specialized area of the lymphoepithelium (Harkema et al. 2006). The squamous region extends from just inside the nares to the anterior portion of the inferior turbinates and is lined with stratified squamous epithelium containing coarse hairs in addition to sebaceous and sweat glands. The transitional region is a non-ciliated cuboidal or columnar epithelium located between the squamous and respiratory epithelia. Nasal-associated lymphoid tissue (NALT) contains the lymphoepithelium, a region on both sides of the naso-pharyngeal duct in rodents (Fig. 15.2) that appears to play a role in the induction of antigen-specific immune responses (Kiyono and Fukuyama 2004). The stimulation of protective systemic/mucosal immunity resulting from intranasal administration of specific antigens (usually requiring the co-administration of an enhancing adju-vant for adequate stimulation of NALT) provides the basis for nasal vaccine devel-opment. It is generally considered that the olfactory and respiratory epithelia are by far the most important sites for nasal absorption, so these regions will be covered individually in greater detail below.

15.2.2 Blood Supply and Lymphatic Drainage

The nasal mucosa is extremely vascular, a feature which allows efficient absorption into the systemic circulation for drugs possessing the right properties for this to occur (e.g., drugs that are sufficiently small to cross through the interendothelial clefts of nasal capillaries). Once in the systemic circulation, a substance would need to cross the BBB or BCSFB to enter the CNS. Although some nasal endothelial cells express TJ proteins such as zona occludens (ZO)-1, occludin, and claudin-5, capillaries in the nasal submucosa appear fenestrated with porous basement mem-branes, suggesting higher permeability than capillaries comprising the BBB (Cauna and Hinderer 1969; Wolburg et al. 2008). Nasal venules and arterioles are continu-ous and lack fenestrations. The vascular density and relative vascular permeability vary in different regions of the nasal mucosa and serve as important considerations when designing intranasal dosing strategies to maximize drug delivery to the CNS (Kumar et al. 2015). The caudal olfactory region has a ~ fivefold lower mean capil-lary density and lower vascular permeability to hydrophilic macromolecules than the anterior respiratory region of the nasal mucosa. Delivering drugs to the olfactory region may therefore minimize clearance into the systemic circulation and conse-quently favor more drug to access the extracellular cranial nerve-associated path-ways to the CNS (Kumar et al. 2015). Indeed, intranasal devices have been designed to target the olfactory region with the goal of enhancing direct delivery to the CNS (Hoekman and Ho 2011a, b). Clearly, the utility of targeting different regions of the nasal passage with intranasally administered drugs for the purpose of enhancing brain targeting is an area that merits further study.

The blood supplying the nasal passages is chiefly provided by (i) branches of the ophthalmic artery, (ii) the sphenopalatine artery, and (iii) branches of the facial artery (Greene 1935; Standring 2021; Schuenke et al. 2010). The anterior and pos-terior ethmoidal arteries branch from the ophthalmic artery to supply the olfactory region, anterior septum, and anterior lateral wall. The sphenopalatine artery mostly supplies the posterior septum and posterior lateral wall with smaller branches extending to further areas. Branches of the facial artery supply the anteroinferior septum and lateral wall. Species differences between rats and humans exist upstream of the ophthalmic and sphenopalatine arteries. The internal carotid artery gives rise to the ophthalmic artery in humans, while the ophthalmic artery branches from the pterygopalatine artery in rats. In humans, the sphenopalatine artery is a branch from the maxillary artery via the external carotid artery. The rat sphenopalatine artery, however, arises from the pterygopalatine artery via the internal carotid artery. It is also relevant that both olfactory and trigeminal arteries have been described in the rat and in other mammals, including human beings (Coyle 1975; Scremin 2004; Favre et al. 1995); these vessels travel at least some distance with their respective nerve bundles and likely provide complex anastomoses between nasal arteries in the nasal passages and cerebral arterial branches from the anterior and posterior brain circulations. Venous drainage in the posterior nasal passage occurs primarily through the sphenopalatine vein, while veins accompanying the ethmoidal arteries drain the anterior nasal passage. Some veins in the nasal passage connect with cerebral veins on the frontal lobe after passing through the cribriform plate.

Although there are no lymph nodes in the CNS, several studies have shown that extracellular and cerebrospinal fluids in the brain may drain either through the arachnoid villi to the venous blood or through the cribriform plate to the nasal lamina propria and then subsequently to the deep cervical lymph nodes in the neck (Fig. 15.1) (Bradbury and Cserr 1985). Intranasally administered substances that are absorbed to the nasal lamina propria but do not enter nasal capillaries (i.e., the systemic circulation) may therefore drain to the deep cervical lymph nodes. Lymphatic vessels have been found traversing the cribriform plate (Furukawa et al. 2008; Norwood et al. 2019). The potential involvement of these lymphatic vessels in the transport or clearance of intranasally applied substances to the CNS has not been established but warrants further examination. Radiolabeled protein tracers or dyes injected into the brain or CSF are cleared to the nasal lamina propria to reach the deep cervical lymph nodes at high concentrations (Bradbury and Cserr 1985; Kida et al. 1993). Sealing the cribriform plate with kaolin or acrylate glue significantly reduced the drainage of [^{125}I]-albumin following intraventricular infusion (Bradbury and Cserr 1985). Recent studies have demonstrated the presence of functional lymphatic vessels lining the dural sinuses (Louveau et al. 2015) and cranial nerves such as the olfactory nerve as it traverses the cribriform plate to innervate the olfactory mucosa (Aspelund et al. 2015). The drainage of CSF- or parenchymally administered tracers and macromolecules like antibodies has been shown to occur along olfactory perivascular/perineural spaces and/or lymphatics (Faber 1937; Kida et al. 1993; Pizzo et al. 2018; Aspelund et al. 2015). Relatively higher drainage to the olfactory region versus the respiratory region may have functional implications due to the former's lower vascularity/vascular permeability (Kumar et al. 2015), a circumstance which favors drainage to local cervical lymph nodes and potential induction of peripheral immune responses against CNS antigens (Cserr et al. 1992).

Dyes and macromolecules like antibodies can be found in the perineural sheaths of the fila olfactoria as well as the deep cervical lymph nodes following intranasal administration (Faber 1937; Yoffey and Drinker 1938; Kumar et al. 2018a). This localization of intranasally administered dyes and macromolecules is often similar to what has been reported for dyes and macromolecules injected into the subarachnoid space CSF (Kida et al. 1993; Pizzo et al. 2018), suggesting that pseudolymphatic pathways leading out of the brain may be similar to pathways leading into the brain. These studies suggest the subarachnoid space, nasal lamina propria, and deep cervical lymph nodes are in communication. Importantly, the localization of microfil following injection into the CSF compartment of cadavers has confirmed that some of these connections also appear to be present in humans (Johnston et al. 2004).

15.2.3 The Olfactory Region of the Nasal Passage

The olfactory region consists of a pseudostratified columnar epithelium (Fig. 15.3a) located on the most superior aspect of the nasal cavity where the olfactory sensory neurons (OSN) reside. The OSN are the only first-order neurons possessing cell bodies located in a distal epithelium. The tips of their dendritic processes contain several nonmotile cilia which extend into the overlying mucus layer; odorant receptors are found in the plasma membrane of the olfactory cilia, where they are positioned to respond to olfactory stimuli in the external environment. The OSN are bipolar cells possessing unmyelinated axons which extend through the epithelial

---

Fig. 15.3 (continued) projecting to the olfactory bulb. Red arrows indicate potential pathways for drug delivery across the olfactory epithelium and into the brain following intranasal administration. Intranasally applied drugs may be transported by an *intracellular* pathway from the olfactory epithelium to the olfactory bulb within olfactory sensory neurons following adsorptive, receptor-mediated, or nonspecific fluid-phase endocytosis. Other drugs may cross the olfactory epithelial barrier by *paracellular* or *transcellular* transport to reach the lamina propria, where a number of different *extracellular* pathways for distribution are possible: (1) absorption into olfactory capillaries and entry into the general circulation; (2) absorption into olfactory lymphatics draining to the deep cervical lymph nodes of the neck; and (3) extracellular diffusion or convection in compartments associated with olfactory nerve bundles and entry into the cranial compartment. Transport within the perineural space bounded by olfactory nerve fibroblasts is shown, but other possibilities exist, e.g., transport within the fila olfactoria compartment contained by ensheathing cells, transport within the perivascular spaces of blood vessels traversing the cribriform plate with olfactory nerves (not shown), or transport within lymphatics traversing the cribriform plate with olfactory nerves (not shown). Possible pathways for distribution of substances from the perineural space into the olfactory subarachnoid space cerebrospinal fluid (CSF) or into the olfactory bulb are shown. (Figure adapted with permission from Lochhead and Thorne 2012). (**b**) The lymphatic drainage of the nasal mucosa is principally to the deep cervical lymph nodes. The deep cervical lymph nodes are present in the viscera of the neck deep into the superficial muscles and just lateral to the common carotid artery. (**c**) Rodent olfactory mucosa sections stained with hematoxylin and eosin or immunostained using an antibody to olfactory marker protein, a protein specific to mature olfactory sensory neurons (not sustentacular or basal cells). Sections show the pseudostratified layers of the olfactory epithelium with the relative positions of the cell bodies of sustentacular (S) cells and olfactory sensory (receptor, R) neurons indicated. Numerous blood vessels (BV) and Bowman's glands (BG) are also visible within the lamina propria. (Images of sections kindly provided by Professor Harriet Baker, Weill Medical College of Cornell University)

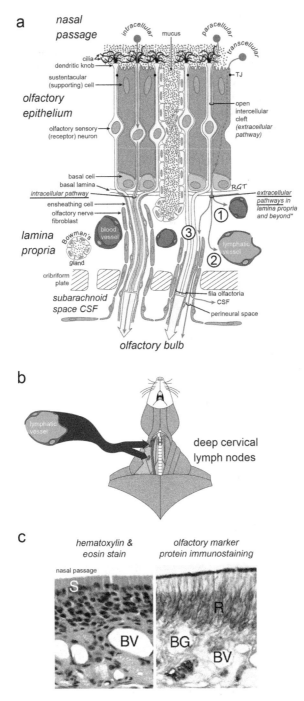

Fig. 15.3 The olfactory region: organization and histology. (**a**) The olfactory mucosa consists of the olfactory epithelium and the lamina propria. Axonal processes of olfactory sensory neurons converge into bundles (fila olfactoria), surrounded by ensheathing cells and fibroblasts, before

basal lamina and converge with axons from other OSN to form nerve bundles called fila olfactoria. Interlocking olfactory ensheathing cells (OEC) form continuous channels around the fila olfactoria from their origin to the olfactory bulb. Multicellular sheets of olfactory nerve fibroblasts enclose the OEC to form a perineural-like sheath around the fila olfactoria (Field et al. 2003). The olfactory nerve is comprised of the ensheathed fila olfactoria and travels through the cribriform plate of the ethmoid bone into the brain where its axons terminate on dendrites of mitral, periglomerular, and tufted cells in the glomeruli of the olfactory bulb. Axons of the mitral and tufted cells project to a number of areas including the anterior olfactory nucleus, olfactory tubercle, piriform cortex, amygdala, and entorhinal cortex (Carmichael et al. 1994).

In addition to OSN, several other cell types are located within the olfactory epithelium and the underlying lamina propria. Sustentacular (supporting) cells extend from the apical region of the epithelium to the basal lamina and possess long, irregular microvilli which intermingle with the cilia of the OSN (Hegg et al. 2009). In the lamina propria, the Bowman's gland forms tubular-type ducts which traverse the basal lamina to produce and secrete a serous fluid which serves as a solvent for inhaled odorants and intranasally applied drugs. Globose basal cells (GBC), located in the lamina propria, are neural progenitors which provide a source for the continuous replacement of the OSN throughout life (Caggiano et al. 1994). Horizontal basal cells are located superficial to the GBC and function as multipotent progenitors to the GBC, sustentacular cells, and cells of the Bowman's gland and ducts (Iwai et al. 2008). Microvillar cells also reside in the olfactory epithelium although their functions are not well defined (Elsaesser and Paysan 2007). Endothelial cells of blood and lymphatic vessels as well as inflammatory cells are also present in the lamina propria of the olfactory region (Fig. 15.3b and c).

15.2.4 The Respiratory Region of the Nasal Passage

The nasal respiratory region consists of a pseudostratified columnar secretory epithelium (Fig. 15.4a). Cell types of the human respiratory epithelium include goblet cells, ciliated cells, intermediate cells, and basal cells (Jafek 1983). Serous glands,

Fig. 15.4 (cintinued) of the trigeminal nerve shown together with the nasal blood supply. The cell bodies of the trigeminal nerve fibers are located in the semilunar ganglion; their axons project into the brain stem at the level of the pons and ultimately synapse with neurons in a number of areas including the principal sensory and spinal trigeminal nuclei. Of the three main trigeminal nerve divisions (V_1, the ophthalmic nerve; V_2, the maxillary nerve; and V_3, the mandibular nerve), only V_1 and V_2 send branches to the nasal epithelium. Blood supply to the nasal passages is provided by ethmoidal branches of the ophthalmic artery, sphenopalatine branches of either the external carotid artery (ECA)/maxillary artery (in humans) or the internal carotid artery (ICA)/pterygopalatine artery (in rats), and nasal branches from the ECA/facial artery. Numerous anastomoses (*) are indicated; these specialized connections between arteries may experience directional change in blood flow depending on the relative pressures within parent arteries. (Figures adapted from Lochhead and Thorne 2012 with permission)

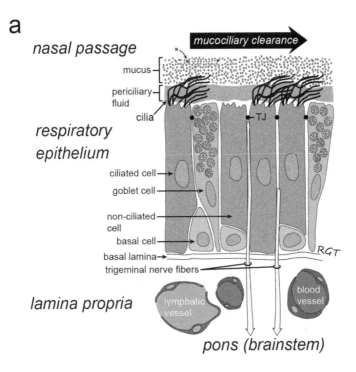

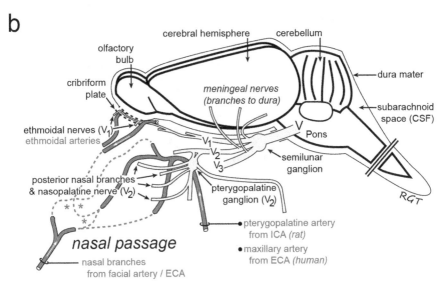

Fig. 15.4 The nasal respiratory region: general organization, trigeminal innervation, and blood supply. (**a**) The respiratory mucosa includes the respiratory epithelium and its underlying lamina propria. The trigeminal nerve, important for conveying chemosensory, nociceptive, touch, and temperature information, is found throughout the nasal epithelium; free nerve endings extend nearly to the epithelial surface, just beneath tight junctions (TJ). (**b**) Central projections

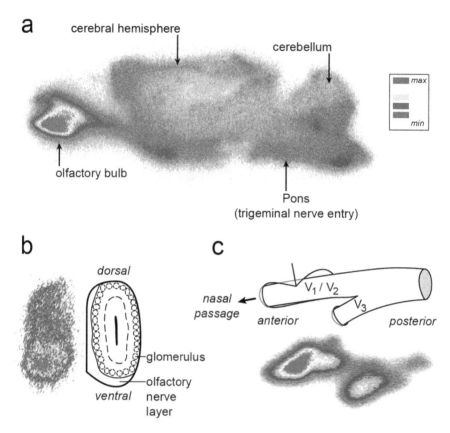

Fig. 15.5 The central distribution of [^{125}I]-labeled IGF-I following intranasal application in anesthetized adult rats is characterized by high levels within the olfactory bulbs and trigeminal nerves. (**a**) Sagittal brain section from a rat approximately 30 min following intranasal administration of a low specific activity solution of [^{125}I]-labeled IGF-I, allowing visualization of brain entry sites in the olfactory bulb (putative olfactory pathway) and pons (putative trigeminal pathway). (**b**) Coronal section through the olfactory bulb of a rat approximately 30 min following intranasal administration of a high specific activity solution of [^{125}I]-labeled IGF-I. Signal intensity is highest in the ventral portion of the bulb in closer proximity to the olfactory nerve entry sites at the cribriform plate. (**c**) Transverse section through the trigeminal nerve of a rat approximately 30 min following intranasal administration of a high specific activity solution of [^{125}I]-labeled IGF-I. Signal intensity is highest in portions of the ophthalmic (V_1) and maxillary (V_2) nerve divisions which innervate the nasal passage. (Figures adapted from Thorne et al. (2004a, b) with permission)

seromucous glands, and intraepithelial glands are also associated with the nasal respiratory epithelium. Most nasal secretions are produced by seromucous glands although goblet cells also secrete mucus. The primary role of the ciliated cells in primates is to propel mucus with their motile cilia toward the nasopharynx where it is either swallowed or expectorated. In rodents, mucus is propelled mostly in the anterior direction. Basal cells are relatively undifferentiated cells which give rise to other cell types in the nasal respiratory epithelium (Fig. 15.5).

Both the nasal respiratory and olfactory epithelia are innervated by branches of the trigeminal nerve (cranial nerve V), the largest of the 12 cranial nerves (Schuenke et al. 2010). Fibers from trigeminal ganglion cells ramify extensively within the nasal submucosa so that their free nerve endings stop at the TJ level near the epithelial surface (Finger et al. 1990). The trigeminal nerve exits the pons bilaterally and consists of a very large sensory root and a small motor root. Its motor fibers innervate the muscles of mastication, and the sensory fibers transmit information from the face, scalp, mouth, and nasal passages. The trigeminal nerve consists primarily of somatic afferent fibers which convey sensory information to nuclei located within the brain stem and spinal cord.

The trigeminal nerve is comprised of three major branches: the ophthalmic nerve (V_1), the maxillary nerve (V_2), and the mandibular nerve (V_3) (Fig. 15.4b). V_1 and V_2 are sensory nerves that also carry autonomic fibers, while V_3 contains the mixed portion of the trigeminal nerve. Importantly, ethmoidal (V_1), nasopalatine (V_2), and nasal (V_2) branches of the trigeminal nerve provide sensory innervation to the nasal passages (Tucker 1971; Bojsen-Moller 1975). A portion of trigeminal ganglion cells with sensory endings located in the nasal epithelium also send collaterals directly into the olfactory bulb in addition to the brain stem (Schaefer et al. 2002). Two other nerves, the nervus terminalus (terminal nerve; cranial nerve zero) and the vomeronasal nerve and organ (Jacobsen's organ), are also located in the nasal passages but have so far not been viewed as important for CNS delivery following intranasal administration, particularly in adult human beings where they may be vestigial or even absent.

15.3   Mechanisms and Pathways for Transport Into the CNS From the Nasal Passages

15.3.1   Transport Across the Olfactory and Respiratory Epithelial Barriers

The pathways and mechanisms governing the transport of substances from the nasal epithelium to various regions of the CNS are not fully understood. Substances which distribute throughout the CNS following intranasal administration must initially cross the nasal epithelial barrier through intracellular or extracellular (paracellular) routes. Proteins (e.g., albumin, horseradish peroxidase (HRP), wheat germ agglutinin-horseradish peroxidase (WGA-HRP)) and viruses (e.g., herpes, poliomyelitis, rhabdoviruses) endocytosed by OSN may reach the CNS (olfactory bulb) through intracellular axonal transport in the anterograde direction (Doty 2008; Kristensson and Olsson 1971; Broadwell and Balin 1985; Thorne et al. 1995; Baker and Spencer 1986; Kristensson 2011). HRP is taken up by OSN to a limited extent via pinocytosis, whereas WGA-HRP is internalized by OSN preferentially by adsorptive endocytosis (Broadwell and Balin 1985). Following intranasal administration, WGA-HRP is also endocytosed and transported intracellularly through the trigeminal nerve to the brain stem (Anton and Peppel 1991; Deatly et al. 1990). Viruses and bacteria may also be transmitted to the CNS along trigeminal nerve components within the nasal passages (Deatly et al. 1990; Jin et al. 2001). Endocytosis by peripheral trigeminal nerve processes and subsequent intracellular transport to the brain stem could potentially occur at either the olfactory or respiratory regions of the nasal epithelium.

Substances may also cross the nasal epithelial barrier through transcytosis or paracellular diffusion to access the lamina propria. Electron micrographs of nasal epithelial cells have demonstrated the existence of TJ, but the paracellular permeability of the nasal epithelia remains poorly defined (Altner and Altner-Kolnberger 1974; Kerjaschki and Horander 1976); this is partly due to the difficulty in establishing and utilizing in vitro models to predict transport for epithelia having neurons as integral components. The TJ proteins ZO-1, ZO-2, and ZO-3; occludin; and claudin-1, claudin-3, claudin-4, claudin-5, and claudin-19 are expressed at the olfactory epithelium of rats (Wolburg et al. 2008; Steinke et al. 2008). Measurements across excised rabbit nasal epithelium have yielded electrical resistance values ranging from 40 Ω•cm^2 (Hosoya et al. 1993), suggesting a relatively permeable barrier, to 261 Ω•cm^2 (Rojanasakul et al. 1992), suggesting barrier properties comparable to the intestinal epithelium. The regular turnover of cells in the nasal epithelium may lead to continual rearrangement and loosening of the TJ as basal cells replace epithelial cells throughout life (Altner and Altner-Kolnberger 1974), resulting in a relatively high permeability compared to other epithelial sites. Electron microscopic studies in the intestinal epithelium have demonstrated colloidal gold nanoparticles cross the epithelial barrier and distribute to other tissues through spaces created by single, degrading enterocytes as they are extruded from the villus in a process known as persorption (Hillyer and Albrecht 2001). The replacement of cells throughout life at the nasal epithelial barrier may create similar potential spaces which may allow paracellular transport of substances to the lamina propria. Evidence for these spaces has recently been shown following intranasal administration of prions which could be found in holes approximately 5–20 μm near the surface of the nasal epithelium (Kincaid et al. 2015). Paracellular transport of molecules across nasal epithelia can be enhanced by modulating local tight junction complexes using MMP-9 at physiological concentrations (Lochhead et al. 2015; Kumar et al. 2018a). The expression of FcRn at the nasal epithelia and differences in pH between their apical and basal sides may facilitate directional transport of IgG from the epithelial surface to the lamina propria via an FcRn-dependent mechanism (Ye et al. 2011; Heidl et al. 2015). Substances that reach the lamina propria through transcellular or paracellular routes may be absorbed into the systemic circulation, drain to the deep cervical lymph nodes, or enter the CNS by direct pathways, utilizing components of the peripheral olfactory and/or trigeminal systems.

15.3.2 Transport from the Nasal Lamina Propria to Sites of Brain Entry

IN administration of $[^{125}I]$-insulin-like growth factor I (IGF-I, MW = 7.65 kDa) and $[^{125}I]$-immunoglobulin G (IgG, MW = 150 kDa) in rats and $[^{125}I]$-interferon-β1B (IFN-β1B, MW = 18.5 kDa) in monkeys all suggest that delivery to the CNS occurs along components associated with the olfactory and trigeminal nerves, followed by widespread distribution to other sites of the CNS within 30–60 min (Thorne et al. 2004a, 2008a; Kumar et al. 2018a, b). Substances may reach the brain from the nasal mucosa intracellularly following endocytosis by OSN or neurons of the trigeminal ganglion, as discussed above. There also appear to be extracellular pathways into the brain following transcytosis or paracellular diffusion across the nasal epithelium to the lamina propria; these pathways have been proposed based on much experimental evidence obtained by a large number of different groups (reviewed in several sources, including Thorne et al. 2004b; Illum 2004; Dhuria et al. 2010; Lochhead and Thorne 2012; Kumar et al. 2018a). The extracellular pathways potentially providing nose-to-brain transport routes include diffusion or convection within perineural, perivascular, or lymphatic channels associated with olfactory and trigeminal nerve bundles extending from the lamina propria to the olfactory bulb and brain stem, respectively.

The perineural distribution around olfactory nerve bundles extending from the lamina propria to the outermost layer of the olfactory bulb has been observed following the IN administration of potassium ferrocyanide and iron ammonium citrate solutions, 3 kDa and 10 kDa dextrans as well as IgG (Faber 1937; Jansson and Bjork 2002; Lochhead et al. 2015; Kumar et al. 2018a, b). This suggests that perineural spaces may act as pathways for molecules to distribute to the CNS from the nasal cavity. OEC maintain continuous open spaces in the nerve bundles to allow regrowth of olfactory nerve fibers (Li et al. 2005). These compartments provide a potential path that substances may take to reach the brain from the perineural space of entering olfactory nerve bundles. The perineural spaces of the olfactory and trigeminal nerves appear to also allow the distribution of certain substances to the CSF of the subarachnoid space, particularly smaller peptides and proteins, although the anatomical/physiological aspects of this perineural space-to-CSF distribution remain poorly understood. Indeed, the barrier between the perineural space and the CSF may be more permeable to some substances than others. Sakane and colleagues demonstrated a size-dependent entry of intranasally administered dextrans of varying sizes (4–20 kDa) into the CSF. Certain proteins, e.g., IGF-I, have not been detected in the CSF despite experimental evidence of brain entry following intranasal administration (Thorne et al. 2004b). Although a fairly large molecule, IgG has been found to enter the CSF in trace amounts within 30 min following intranasal administration (Kumar et al. 2018a). CSF IgG concentrations were ~ 2- to 30-fold lower than in the brain parenchyma, suggesting IgG access to the parenchyma occurred via pathways that do not require access to the CSF compartment first (Kumar et al. 2018a). Entry of intranasally administered IgG into the CSF, despite its large size, may be due to the role it plays in immune surveillance and may be aided by FcRn-dependent transport mechanisms. For substances capable of accessing the CSF of the subarachnoid space following IN administration, further distribution to more distant sites of the CNS may occur along pathways of CSF flow.

The precise mechanisms underlying the rapid transport (30 min) of radiolabeled proteins from the rat nasal mucosa to widespread areas of the CNS along components of the olfactory and trigeminal nerves are at present unknown. Possibilities include intracellular (axonal) transport, extracellular diffusion, and extracellular convective (bulk) flow within perineural, perivascular, or lymphatic channels associated with olfactory and trigeminal nerve bundle. Recently, we have shown that fluorescently labeled insulin or IgG can be found within perineural and/or perivascular spaces of the trigeminal nerve within minutes after intranasal administration, suggesting bulk flow within these spaces are involved in the delivery of macromolecules to the CNS along the trigeminal route (Kumar et al. 2018a; Lochhead et al. 2019). We have previously estimated the time it would take for a molecule to reach the olfactory bulb and brain stem of rats by intracellular transport, diffusion, or convective flow (Lochhead and Thorne 2012). Intracellular (axonal) transport rates within olfactory or trigeminal nerves were estimated from experimental rates measured in fish olfactory nerves (Buchner et al. 1987). Rates of diffusion were based on experimental measurements and known correlations for protein-free diffusion coefficients (Thorne et al. 2004a). Convective flow rates were estimated from experimentally measured albumin transport within the perivascular spaces of pial arteries using an open cranial window preparation in rats (Ichimura et al. 1991).

In short, the intranasal delivery of macromolecular dextran tracers and proteins such as IGF-I and IgG, among others, resulting in transport to widespread areas of the CNS within 30 min of intranasal application, strongly indicates a convective (bulk) flow process along the olfactory and trigeminal nerve components that is likely the only plausible transport mechanism that can explain the experimental CNS distribution (Kumar et al. 2018a; Lochhead et al. 2015). This is an area clearly in need of further, careful study; more detailed discussion can be found elsewhere (Thorne et al. 2004b, 2008a, b; Lochhead and Thorne 2012; Kumar et al. 2018a; b).

15.3.3   Transport from Brain Entry Sites to Widespread Areas Within the CNS

The final distribution of substances to other CNS areas after they have reached the pial surface of the brain at the level of the olfactory bulb and brain stem has been shown to occur at least in part via bulk flow within perivascular spaces of cerebral blood vessels (Thorne and Frey 2001; Thorne et al. 2004b; Lochhead et al. 2015; Kumar et al. 2018a). It has been speculated that the normal expansion and contraction of cerebral blood vessels due to cardiac pulsatility could generate a pronounced fluid flow within the perivascular spaces. Different groups have attempted to understand the direction and characteristics for such a flow by modeling the process, but thus far the results have produced conflicting ideas as to its directionality (Bilston et al. 2003; Schley et al. 2006; Wang and Olbricht 2011). It has been shown that increasing the blood pressure and heart rate results in a larger distribution of adeno-associated virus 2 capsids or fluorescent liposomes after injection into the striatum, suggesting the involvement of arterial pulsations in the intraparenchymal distribution of these large substances via the perivascular spaces (Hadaczek et al. 2006). Several groups have also observed rapid distribution along perivascular spaces following tracer application into the CSF (Rennels et al. 1985; Iliff et al. 2012); however, it must be noted that others have seen limited perivascular distribution following injection of tracers into the subarachnoid CSF (Kida et al. 1993; Szentistvanyi et al. 1984). Pizzo et al. showed full-length IgG (150 kDa) and smaller single-domain antibodies (sdAb; ~15 kDa) distributed via diffusion at brain-CSF interfaces and throughout the brain along perivascular spaces of cerebral blood vessels of all caliber in a size-dependent manner following intrathecal infusion into the cisterna magna (Pizzo et al. 2018). Intranasal administration of fluorophore-labeled 3 kDa dextran by itself or 10 kDa dextran and IgG following nasal pre-administration of MMP-9 (a physiologic nasal permeability enhancer) has been demonstrated to result in rapid access to the brain parenchyma. Such access to the brain has been suggested to occur first via transport along perivascular compartments of cerebral blood vessels followed by diffusion out of the perivascular space and into the brain parenchyma (Kumar et al. 2018a; Lochhead et al. 2015). Notably, the extent of macromolecule access to cerebral perivascular compartments following intranasal administration appears to be size dependent (Lochhead et al. 2015; Kumar et al. 2018a). The precise role that perivascular transport plays in dictating CNS distribution in health and disease following intranasal targeting of substances to the brain certainly deserves further study.

15.4 Current Status of the Intranasal Route of Administration for CNS Targeting

IN administration has become an increasingly popular method to bypass the BBB and deliver therapeutics directly to the CNS. Numerous preclinical studies have indicated IN administration offers advantages over other routes of administration for delivery of some substances to the CNS. The published literature now includes a vast amount of animal work reporting positive effects following the intranasal administration of small molecules, peptides, proteins, oligonucleotides, gene vectors, or cell-based therapeutics using a number of different CNS disease models. Most importantly, several clinical trials involving IN administration for the treatment of CNS disorders have either been completed, are currently in progress, or are in the process of planning/recruiting. The sections below provide a brief summary of some notable preclinical and clinical work that has been conducted to date. This review is by no means exhaustive; more comprehensive summaries may be found elsewhere (Lochhead and Thorne 2012; Dhuria et al. 2009).

15.4.1 Intranasal Delivery of Small Molecules to the CNS

Intranasal delivery has long been appreciated to offer unique advantages for small molecule administration across a variety of applications: (i) when local effects are desired (e.g., as with decongestants, antibiotics, and mucolytics); (ii) when noninvasive, needle-free access to the systemic circulation is needed for rapid drug onset (e.g., in the context of illicit drug overdose); and (iii) to avoid extensive hepatic first-pass elimination (e.g., as with the application of the opioid antagonist naloxone following opioid overdose). Indeed, multiple studies have demonstrated that intranasal delivery of both small molecules (e.g., zolmitriptan, sumatriptan, butorphanol tartrate, fentanyl, nicotine, and estradiol) and low-molecular-weight peptide drugs (e.g., calcitonin, desmopressin, buserelin, oxytocin) can yield drug absorption and disposition profiles capable of producing clinically meaningful responses in a safe, patient-friendly manner. The ability to achieve significant systemic exposure for intranasally applied small molecules and peptides (as compared to proteins and other large molecules) is likely due to their relatively high paracellular permeability across the nasal epithelia and efficient absorption into the blood stream through the extensive nasal vasculature present in the underlying lamina propria (Kumar et al. 2015; Nehra et al. 2021). Interestingly, it is often questioned whether intranasal delivery can truly yield improved CSF or brain exposures for small molecule therapeutics, due partly to a lack of careful studies capable of distinguishing direct delivery to the brain/CSF versus systemic absorption followed by brain/CSF entry across the BBB and/or blood-cerebrospinal fluid barriers (Nehra et al. 2021).

Small molecules may be able to directly access the CNS through the IN route of administration. The paracellular permeability of substances across the nasal epithelium is likely inversely proportional to their size. This would favor a higher percentage of small molecules than macromolecules reaching the lamina propria following IN administration. Small molecules, however, may also be more easily absorbed into the nasal capillaries due to their smaller size. Therefore, intranasally administered small molecules may be more likely to access the nasal lamina propria than large molecules, but their size may favor absorption into the systemic circulation. Small molecules which escape absorption into the nasal vasculature may directly access the CNS through olfactory or trigeminal nerve-associated pathways. Absorption into the CSF may favor small molecules over macromolecules. Small molecules distributed in the perineural space of the olfactory or trigeminal nerve may also more easily cross the perineural barrier than large molecules. Therefore, small molecules may have greater access than large molecules to the CSF within the subarachnoid space surrounding the olfactory and trigeminal nerves. Upon entry into the CSF, small molecules may also have access to more distant sites in the

CNS; conversely, small molecules may in some cases be cleared from the CNS compartment more quickly than larger molecules.

It has been questioned whether small molecules can directly access the brain following IN administration (Merkus et al. 2003). Merkus and coworkers measured the levels of melatonin (MW = 232 Da) in the CSF after IN or intravenous (IV) administration and concluded no direct delivery to the brain occurred. However, melatonin is able to cross the BBB, making it difficult to ascertain whether its detection in the CSF represents direct delivery from the nasal mucosa or delivery across the BBB or BCSFB from the systemic circulation. Furthermore, intranasally applied macromolecules such as IGF-I and vascular endothelial growth factor have been found in the brain but not the CSF following IN delivery, suggesting drug levels in the CSF may not always correlate with brain levels (Thorne et al. 2004a, b; Yang et al. 2009). In another study, the dopamine-D2 receptor antagonist remoxipride (MW = 371 Da) was measured in the brain extracellular fluid (ECF) using a microdialysis probe placed within the striatum; the brain ECF/plasma area under the curve (AUC) ratios was found to be significantly higher in rats administered remoxipride intranasally compared to intravenous application (Stevens et al. 2011). Elegant semiphysiologically based pharmacokinetic modeling by this group suggested 75% of remoxipride entering the brain following intranasal application did so using a direct nose-to-brain transport pathway. Similar results were obtained when levels of three glycine receptor antagonists and one angiotensin antagonist with varying degrees of BBB permeability (MW = 369–611 Da) were compared following IN or IV administration (Charlton et al. 2008). CNS/plasma AUC ratios were higher following IN versus IV administration for each compound. Autoradiographs further detected the angiotensin antagonist GR138950 in the olfactory nerves, CSF, and brain within minutes following IN administration. Finally, the local anesthetic lidocaine (MW = 234 Da) has also been shown to be transported to the brain along the trigeminal nerve pathway (Johnson et al. 2010).

Several disease models have been successfully treated with intranasally administered small molecule drugs. For example, the angiotensin type II receptor antagonist losartan (MW = 423 Da), which poorly penetrates the BBB, decreased amyloid β (Aβ) plaques and inflammation without inducing hypotension in an Alzheimer's disease (AD) transgenic mouse model (Danielyan et al. 2010). The iron chelator deferoxamine (MW = 561 Da) also exhibits neuroprotection in models of Parkinson's disease (PD), AD, and ischemic stroke (Febbraro et al. 2013; Guo et al. 2013; Hanson et al. 2009).

Finally, a number of studies, including clinical trials, have suggested that perillyl alcohol (POH; MW = 152 Da), a plant-derived monocyclic terpene and chemotherapeutic agent, may hold promise for the treatment of recurrent forms of primary brain cancers, particularly low-grade glioma (NCT02704858) following intranasal administration (see Nehra et al. 2021 for review).

15.4.2 Intranasal Delivery of Peptides/Proteins to the CNS

Peptides and proteins are the most widely used drugs which have been administered intranasally to treat disorders of the CNS in both animal models and humans. Most preclinical studies utilizing the intranasal route of administration have shown behavioral or pharmacodynamic effects but not presented pharmacokinetic data indicating direct delivery of the drug to the CNS. This makes it difficult to determine if the drug entered the brain through direct pathways from the nasal cavity, crossed the BBB or accessed circumventricular areas from the systemic circulation, or exerted its effects through direct action on the BBB itself. For some peptides and proteins, there is pharmacokinetic data to support their ability to directly enter the CNS from the nasal cavity.

A pioneering study by Born and colleagues was among the first studies to obtain CNS pharmacokinetic data following IN delivery of peptides in humans. The peptides melanocortin(4–10) (MW = 980 Da), arginine-vasopressin (MW = 1.1 kDa), and insulin (5.8 kDa) were all detected in the CSF within 30 min in healthy volunteers with a lumbar puncture (Born et al. 2002). Importantly, there was no increase in plasma concentration of melanocortin(4–10), insulin, or glucose with intranasal dosing of melanocortin or insulin in this study. CSF levels of the peptides remained elevated for at least 80 min following IN administration.

Insulin is one of the most widely studied biologics with regard to its effects on the CNS following intranasal administration. A number of studies have intranasally administered insulin to treat metabolic and cognitive disorders in animal models as well as in humans. IN administration of [125I]-insulin to mice yields significantly higher CNS levels after 1 h when compared to subcutaneous administration (Francis et al. 2008). [125I]-insulin distributed widely throughout the mouse brain following IN administration, with the highest levels detected in the trigeminal nerve and the olfactory bulbs (Francis et al. 2008). Electron microscopic studies have found insulin within olfactory nerve bundles minutes following IN administration in mice (Renner et al. 2012a, b). A recently completed clinical trial showed IN insulin improved memory and preserved general cognition in patients with mild cognitive impairment or AD (Craft et al. 2012). Changes in memory and cognitive function were associated with changes in $A\beta_{42}$ levels and tau/$A\beta_{42}$ ratio in CSF (Craft et al. 2012). IN insulin has also suppressed food intake and increased brain energy levels in humans, suggesting potential as a treatment for obesity (Jauch-Chara et al. 2012). As already discussed above, IN administration of [125I]-IGF-I results in significantly higher CNS levels than comparable intravenous dosing, with widespread CNS distribution occurring via olfactory and trigeminal nerve pathways, and the activation of IGF-I signaling pathways in brain areas such as the olfactory bulb and brain stem trigeminal nuclei (Thorne et al. 2004b); IGF-I brain entry and effects following IN application may also be relevant for understanding how IN insulin exerts its central actions because the two proteins share significant structural homology. A large multicenter trial examining the effects of intranasal insulin in AD and mild cognitive impairment is now underway in the United States.

Oxytocin (MW = 1 kDa) is a neuropeptide which exhibits a wide range of effects on human behavior. Oxytocin receptors are expressed centrally in the accessory olfactory bulb, anterior olfactory nucleus, islands of Calleja, amygdala, CA1 of the hippocampus, ventral medial hypothalamus, nucleus accumbens, brain stem, and spinal cord (Stoop 2012). The BBB prevents the passage of peripheral oxytocin (Ermisch et al. 1985; Kang and Park 2000), and IN administration of oxytocin has increasingly become a popular method for assessing oxytocin's central effects. Oxytocin is currently being administered intranasally in clinical trials to treat autism spectrum disorders, schizophrenia, and alcohol withdrawal. Despite the widespread use of oxytocin in clinical settings, little is known in animals or humans about oxytocin distribution in the brain following IN administration, suggesting a need for further study in this area.

Dopamine neuron-stimulating peptide-11 (DNSP-11; MW = 1.18 kDa) is a synthetic, amidated 11-amino acid peptide derived from the pro-domain of human glial cell line-derived neurotrophic factor (GDNF) that possesses broad neuroprotective and neurorestorative properties on dopaminergic neurons both in vitro and in vivo (Bradley et al. 2010; Kelps et al. 2011; Fuqua et al. 2014; Stenslik et al. 2015, 2018). In the first of a series of studies to examine the efficacy of repeated IN administration of DNSP-11 on the dopaminergic system, Stenslik et al. reported changes in d-amphetamine-induced rotation, recovery of dopamine turnover, and tyrosine hydroxylase (TH) neuronal sparing in a severe, unilateral 6-hydroxydopamine (6-OHDA) Fisher 344 (F344) rat model of parkinsonism (Stenslik et al. 2015). In the same report, a single, IN [^{125}I]-DNSP-11 dose in naïve F344 rats resulted in rapid, widespread distribution throughout the CNS, including the nigrostriatal system, and uptake in the CSF within 30 min. Highest levels of radiolabel were observed in the olfactory bulbs at 60 min (Fig. 15.6; Stenslik et al. 2015). In a subsequent report, Stenslik et al. developed a methodology to evaluate repeated IN administration of DNSP-11 in nonhuman primates (rhesus macaques) without the need for sedation (Stenslik et al. 2018). Stenslik et al. demonstrated that DNSP-11 administered IN to awake, chair-trained 1-methyl-4-phenyl-1,2,3,6-tetrahydroryidine (MPTP)-treated rhesus macaques in a dose-escalating manner over the course of several weeks resulted in bilateral, neurochemical changes in the striatal system without observable, adverse behavioral effects or weight loss (Stenslik et al. 2018). In addition, a single, intranasal [^{125}I]-DNSP-11 dose revealed rapid, widespread distribution throughout the CNS and uptake in the CSF within 60 min, with the highest levels of radiolabel observed in the olfactory bulbs and trigeminal nerves (Fig. 15.7; Stenslik et al. 2015, 2018). These findings are consistent with other foundational nonhuman primate IN peptide/protein radiolabeled tracer studies discussed below (Thorne et al. 2008a, b). Collectively, these studies support the idea that DNSP-11 can safely and effectively deliver IN to target the dopaminergic system in both rodents and nonhuman primates.

Orexin-A (hypocretin-1, MW = 3.6 kDa) is a sleep-related peptide produced in the hypothalamus which has shown effects in monkeys and humans following IN administration. Intranasally administered orexin-A improved task performance and induced changes in the brain metabolic activity in sleep-deprived rhesus monkeys

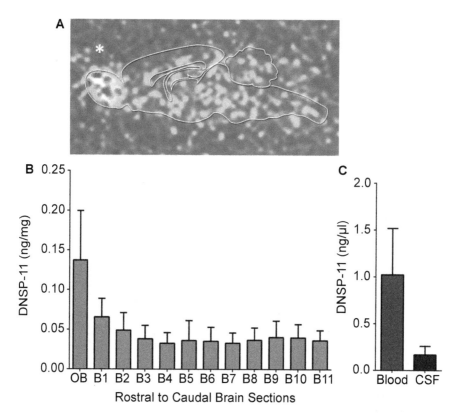

Fig. 15.6 Intranasal administration of DNSP-11 results in the delivery to the brains of Fischer 344 rats. Normal F344 rats were given a one-time intranasal dose of [^{125}I]-labeled DNSP-11 to determine the distribution in the brain. At 60 min the blood (500 µl), cerebrospinal fluid (100–120 µl), and brain tissue were collected from individual rats and processed by gamma counting (n = 3) and autoradiography (n = 1). (**a**) The distribution in a representative sagittal brain section (0.5 mm) supports a qualitative increase in radioactive signal found in the olfactory bulbs (OB) and diffuses signal throughout the brain. (**b**) Normalized DNSP-11 concentrations (ng/mg wet tissue weight) as analyzed by gamma counting were consistent with the autoradiography analysis. (**c**) Normalized DNSP-11 concentrations (ng/µl) as analyzed by gamma counting indicate the presence of radioactive signal in the blood and cerebrospinal fluid samples at the single timepoint examined. * Denotes the olfactory bulb (OB) in a representative sagittal section of the midbrain following autoradiography analysis. B1–11 denote rostral to caudal serial brain sections taken for gamma counting analysis. (Figures adapted from Stenslik et al. 2015 with permission)

(Deadwyler et al. 2007). In humans suffering from narcolepsy with cataplexy, IN administration of orexin-A attenuates olfactory dysfunction and induces and stabilizes REM sleep (Baier et al. 2008, 2011). In rats, intranasally administered orexin-A distributed to the brain within 30 min, yielding tissue-to-blood concentration ratios that were 5–8 times higher in the posterior trigeminal nerve, olfactory bulbs, hypothalamus, and cerebellum compared to rats given IV orexin-A (Dhuria et al. 2009). High levels of orexin-A were found in the cerebral blood vessel walls, and

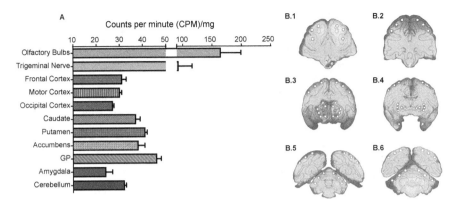

Fig. 15.7 The intranasal administration of DNSP-11 results in the delivery to the brain of a rhesus monkey (*Macaca mulatta*). To determine the distribution in the nonhuman primate brain, a single [^{125}I]-labeled DNSP-11 dose (5 mCi/10 mg DNSP-11; 2.5 mCi/5 mg/0.5 mL per naris) was administered to a rhesus macaque. (**a**) Whole olfactory bulbs and sections of trigeminal nerve were harvested, along with multiple 2-mm-diameter tissue punches from coronal sections of the frontal cortex (n = 12), motor cortex (n = 12), occipital cortex (n = 12), caudate nucleus (n = 8), putamen (n = 12), nucleus accumbens (n = 2), globus pallidus (n = 4), amygdala (n = 2), and cerebellum (n = 12) for gamma counting analysis. In the brain samples examined, normalized radioactive signal (CPM/mg) demonstrated highest signal in the olfactory bulbs and trigeminal nerves, with diffuse lower signal levels throughout the other brain regions sampled. (B1–6) The same 2-mm-thick coronal sections that were used for tissue biopsy mapping were subsequently processed for autoradiography. Qualitative visual assessment supports the highest radioactive signal in the olfactory tracts, with high levels also observed in the white matter regions. Circles represent tissue punches taken for gamma counting analysis. (Figures adapted from Stenslik et al. 2018 with permission)

low levels were found in the CSF of these rats, suggesting transport pathways may have involved distribution within the perivascular spaces.

NAP (davunetide) is an eight-amino acid neuroprotective peptide (MW = 825 Da) derived from activity-dependent neurotrophic factor. Intact levels of [^{3}H]-labeled NAP are found in the cortex and cerebellum of rats within 30 min following IN administration (Gozes et al. 2000). IN administration of NAP reduced levels of Aβ and hyperphosphorylated tau in an AD mouse model (Matsuoka et al. 2007) and decreased neurofibrillary tangles in a model of tauopathy (Shiryaev et al. 2009). IN NAP decreased hyperactivity and protected visual memory in a mouse model of schizophrenia (Powell et al. 2007). Unfortunately, IN NAP failed to show efficacy in a recent clinical trial to treat progressive supranuclear palsy. Clinical trials evaluating whether IN NAP is beneficial in the treatment of schizophrenia and tauopathies are currently in progress.

The 18.5-kDa protein interferon-β1B (IFN-β1B) is a cytokine therapeutic approved to treat the relapsing-remitting form of multiple sclerosis. Studies in rats have shown that IN application of [^{125}I]-labeled IFN-β1B results in significantly higher CNS levels than intravenous dosing (Ross et al. 2004). High IFN-β1B levels were measured in the olfactory bulbs and trigeminal nerves, with significant but lower levels in other brain regions and the spinal cord, approximately 30 min after the start of administration. A subsequent study evaluating CNS delivery following IN application of [^{125}I]-labeled IFN-β1B in cynomolgus monkeys (*Macaca fascicularis*) also demonstrated widespread distribution within the brain, with highest levels again in the olfactory bulbs and trigeminal nerves (Thorne et al. 2008a, b). Importantly, this study also showed an anatomically unique and significant central localization of [^{125}I]-labeled IFN-β1B to regions of the basal ganglia that was remarkably consistent between different animals (Fig. 15.8). This study was among the first to describe the precise distribution and concentrations achievable in the CNS of a primate species following IN administration; [^{125}I]-IFN-β1B concentrations in the olfactory bulbs, trigeminal nerves, and many other brain areas were found to be above the levels required for the antiviral, antiproliferative, and immunomodulatory actions of IFN-β1B.

Antibodies are immunoglobulin proteins which are able to bind peptides and proteins with high affinity. This property makes them attractive drug candidates to treat diseases of the CNS, but antibodies have shown limited BBB penetration when administered systemically (Banks 2004; St-Amour et al. 2013; Kumar et al. 2018a). The efficiency of IgG transport from the systemic circulation into the brain parenchyma via sites such as the circumventricular organs, across the blood-CSF barrier, or the BBB has remained largely unknown, and it is very likely that BBB transport of IgG has been overestimated due to systemically derived exogenous IgG remaining sequestered within the brain endothelial compartment with limited entry into the parenchyma itself (St-Amour et al. 2013). Anti-Aβ immunoglobulin G (IgG) has been administered intravenously in several clinical trials to treat or prevent AD (Kumar et al. 2018b). A few studies have intranasally administered antibodies or antibody fragments in mouse models of AD. Full-length IgG (MW = 150 kDa) as well as a single-chain variable fragment antibody (scFv) (MW = 26 kDa) directed against the C terminus of Aβ(1–42) reduced amyloid plaque levels following intranasal administration to APPswe/PS1dE9 transgenic mice (Cattepoel et al. 2011). The scFv was detected in the brain immunohistochemically, while the full-length antibody was not, suggesting greater delivery of the smaller fragment. Another study in 5XFAD mice showed improved spatial learning and lower levels of Aβ following IN administration of an anti-Aβ oligomer antibody (Xiao et al. 2013). Limited levels of HRP-labeled antibody have been reported in the brain following the development with diaminobenzidine (Xiao et al. 2013). Kumar et al. (2018a, b) performed a quantitative evaluation of [^{125}I]-labeled IgG delivery to the brain and CSF 30 min following intranasal delivery in rats. The highest concentrations of

---

Fig. 15.8 (continued) solution. The highest concentrations were evident in regions of the basal ganglia (putamen, caudate, and globus pallidus) with slightly lower signal in other subcortical structures (e.g., hippocampus and amygdala). (**c**) Coronal brain autoradiographs and labeled templates from two different monkeys at the same level as in (**b**) demonstrating remarkably low variability in central distribution across different subjects. The distribution for each animal is shown approximately 60 min following intranasal administration of [^{125}I]- IFN-β1b (Portions of (**a**), (**b**), and (**c**) adapted from Thorne et al. (Thorne et al. 2008a) with permission)

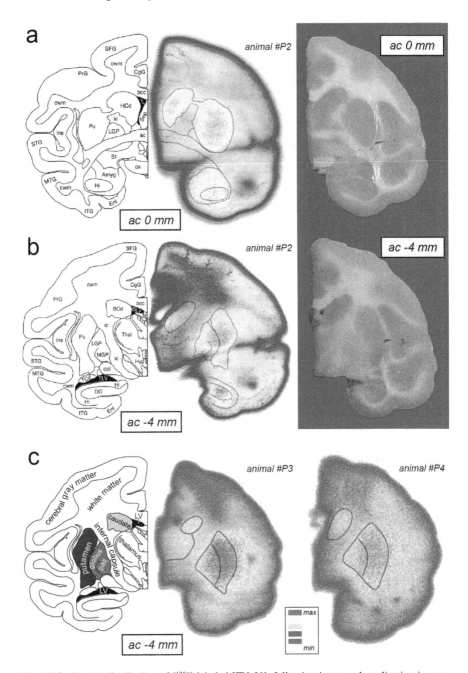

Fig. 15.8 Central distribution of [^{125}I]-labeled IFN-β1b following intranasal application in anesthetized cynomolgus monkeys (*Macaca fascicularis*). Coronal brain autoradiographs and labeled templates at (**a**) the level of the anterior commissure (ac 0 mm) or (**b**) 4 mm posterior to the anterior commissure (ac, 4 mm) with corresponding brain sections provided to illustrate the highly anatomical distribution in a single monkey receiving a very high specific activity [^{125}I]-IFN-β1b

radiolabeled IgG were observed in the olfactory bulbs, trigeminal nerves, and the walls of leptomeningeal blood vessels, supporting IgG access to perineural and perivascular pathways during nose-to-brain transport (Kumar et al. 2018a). The CNS delivery of radiolabeled IgG was significantly higher following intranasal administration versus systemic (intra-arterial) administration at doses resulting in similar endpoint blood levels (Kumar et al. 2018a). A positive dose-response was observed following intranasal radiolabeled IgG delivery with increasing doses resulting in higher CNS targeting, but, importantly, such a dose-response was not observed following intra-arterial delivery (Kumar et al. 2018a). A single intranasal radiolabeled IgG tracer dose (50 μg) in rats resulted in low-to mid-picomolar brain concentrations, while higher intranasal radiolabeled IgG doses (1 mg and 2.5 mg) resulted in low nanomolar levels in the CNS. This suggests that therapeutic levels of IgG may be achieved in the CNS following intranasal dosing, particularly at higher doses. The pre-administration of MMP-9 was used to enhance the transport of fluorophore-labeled IgG across the nasal epithelial barrier in rats and achieved a sufficiently high signal-to-noise ratio, allowing microscopic examination of IgG transport pathways to the CNS. Fluorescence microscopy images showed distribution across the nasal epithelium into the lamina propria, along fila olfactoria into the olfactory nerve layer, and along perivascular spaces surrounding perineural and cerebral blood vessels (Kumar et al. 2018a, b). Follow-up studies in C57BL6 mice and in the APPswe/PS1dE9 transgenic AD mouse model have demonstrated that the intranasal administration of [^{125}I]-IgG either alone (Fig. 15.9) or following intranasal MMP-9 pretreatment (Fig. 15.10) results in widespread brain delivery compared to systemic administration using autoradiography (*Unpublished*; Nehra et al. 2021).

15.4.3   Intranasal Delivery of Gene Vectors and Oligonucleotides to the CNS

Long-term induction or suppression of gene products in the CNS has great potential to treat many neurological disorders. Many viral vectors, plasmids, and oligonucleotides exhibit low BBB permeability, making IN administration a potentially attractive alternative route. Several studies have investigated the IN route of administration to induce or repress gene products in the CNS.

Viral vectors, such as the recombinant adenoviral vector ADRSVβgal (Draghia et al. 1995), the growth compromised herpes simplex virus type 2 mutant ΔRR (Laing and Aurelian 2008), and herpes simplex virus type 1 (Broberg et al. 2004), have all been reported to induce gene expression in widespread areas of the brain following IN administration. ΔRR encoded the anti-apoptotic gene ICP10PK and prevented kainic acid-induced seizures, neuronal loss, and inflammation in rats. IN administration was more efficient at delivering herpes simplex virus type 1 DNA in the brain than corneal or intralabial infection. A filamentous bacteriophage (1000 nm long, 6 nm wide) expressing anti-Aβ scFv has also been shown to bind amyloid

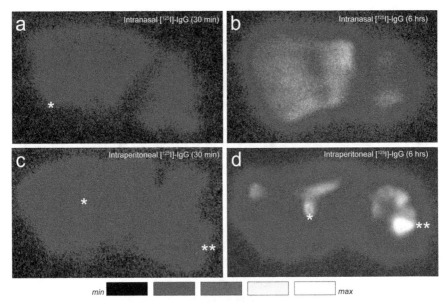

Fig. 15.9 Autoradiography images of [125I]-IgG distribution in sagittal brain sections of C57BL6 mouse after intranasal administration demonstrate brain delivery. (**a**) [^{125}I]-IgG exposure was observed at ventral cortical regions 30 min following intranasal administration of 20 μg [^{125}I]-IgG (*). (**b**) [^{125}I]-IgG distribution and intensity increased across widespread regions in the brain 6 h following intranasal [^{125}I]-IgG delivery. (**c**) Equal-dose intraperitoneal administration of 20 μg [^{125}I]-IgG resulted in [^{125}I]-IgG profiles at the ventricular region and brain stem, indicating putative access at the level of the choroid plexus (CP) in the lateral ventricle (*) and possibly also brain entry at circumventricular organs (CVOs) such as the area postrema (**) at the early time point of 30 min. (**d**) The [^{125}I]-IgG signal showed a similar distribution, albeit with higher intensity, at 6 h post intraperitoneal [125I]-IgG administration, suggesting IgG entry across the blood-CSF barrier at the level of the CP in the lateral ventricle (*) and a region near the area postrema (**) dominated the biodistribution profile

plaques in an AD mouse model following IN administration (Frenkel and Solomon 2002). While the filamentous phage was detected in the brain, a spheroid phage was not, suggesting the shape of the phage may be important for IN delivery to the brain.

Plasmid DNA has also been delivered to the brain through the IN route. A 7.2-kb pCMVβ and a 14.2-kb pN2/CMVβ (encoding the gene for β-galactosidase) have been detected in the brain within 15 min of IN administration (Han et al. 2007). β-galactosidase activity was significantly higher in brain homogenates 48 h later and the brain-to-serum AUC ratio of pCMVβ levels was ~2600-fold higher 10 min after IN administration when compared to IV administration. Higher levels of plasmid DNA were detected in brain endothelial cells than microglia, while it was unclear if neurons were transfected.

Finally, IN delivery of oligonucleotides to the CNS has also been reported. In one study, IN delivery of αB-crystallin small interfering RNA (siRNA) complexed with DharmaFECT 3 resulted in reduced expression of αB-crystallin in neurons and astrocytes in the olfactory bulb, amygdala, entorhinal cortex, and hypothalamus

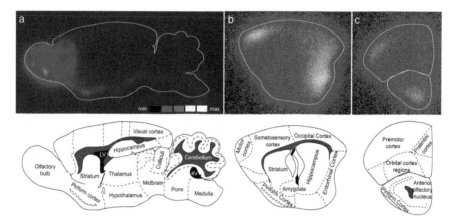

Fig. 15.10 Autoradiography images of [^{125}I]-IgG distribution in APPswe/PS1dE9 mouse brain following intranasal administration of 20 µg [^{125}I]-IgG following intranasal MMP-9 pretreatment (100 nM) to enhance absorption. (**a**) Sagittal brain section autoradiograph (approximately 1.5 mm lateral to the midline; Paxinos and Franklin 2019) demonstrating high signal associated with the olfactory bulb following MMP-9 pretreatment at 30 min post-administration. The lower schematic corresponds to the approximate position of the sagittal section. Settings optimized for high olfactory bulb signal intensity. (**b**) Sagittal brain section autoradiograph (approximately 3 mm lateral to the midline; Paxinos and Franklin 2019) demonstrating relatively high signal intensity associated with cortical regions. The lower schematic corresponds to the approximate position of the sagittal section. (**c**) Coronal brain section autoradiograph (approximately 2.5 mm anterior to the bregma; Paxinos and Franklin 2019) demonstrating relatively high signal intensity in both the rostral cortex and olfactory tract (anterior olfactory nucleus). The lower schematic corresponds to the approximate position of the coronal brain section. *LV* lateral ventral, *4 V* fourth ventricle

both 3 and 12 h following administration (Kim et al. 2009). In another study, a 21 base pair fluorescently labeled siRNA was delivered to the olfactory bulb following IN administration (Renner et al. 2012a). CNS delivery of the 22 base pair antagomir AM206 has also been reported following IN application in an AD transgenic mouse model (Lee et al. 2012). The authors concluded that AM206 increased brain-derived neurotrophic factor levels and memory function in mice by neutralizing microRNA-206.

15.4.4 Intranasal Delivery of Cell-Based Therapies to the CNS

Several recent studies have reported that the IN application of stem cells results in brain delivery and therapeutic effects in disease models. It is not known whether cells utilize the same pathways that molecular therapeutics use to reach the CNS after IN administration, but mesenchymal stem cells (MSC) have been found crossing the cribriform plate adjacent to the olfactory nerve bundles and in the olfactory nerve layer of the olfactory bulb 1 h after IN administration (Galeano et al. 2018). These observations suggest both cells and molecules may access the CNS from the nasal passages through the olfactory route. Fluorescently labeled rat mesenchymal stem cells (MSC) have been detected in the olfactory bulb, hippocampus, thalamus, cortex, and subarachnoid space of mice 1 h after IN delivery (Danielyan et al. 2009). Intranasally administered MSC have shown therapeutic potential in models of PD (Danielyan et al. 2011) and several models of stroke (Wei et al. 2012; Donega et al. 2013; van Velthoven et al. 2013). 1.5 h following IN administration, Hoechst labeled MSC could be found lining the blood vessels as well as in the parenchyma after ischemic stroke in mice (Wei et al. 2012). A study in which neural stem/progenitor cells (NSPC) were administered by the IN route found that the cells were targeted to the site of an intracerebral glioma within 6 h (Reitz et al. 2012). The enhanced green fluorescent protein expressing NSPC were located in the olfactory bulb within 6 h and the olfactory tract at 24 h. Few cells were observed in the trigeminal nerve at 24 h suggesting NSPC migrated into the brain within the first 24 h by the olfactory pathway as well as via the systemic circulation. Finally, IN administration of T cells engineered to express a chimeric antigen receptor targeting myelin oligodendrocyte glycoprotein have been reported to result in brain delivery and to suppress inflammation in a mouse model of multiple sclerosis (Fransson et al. 2012).

15.5 Future Challenges and Directions for Intranasal Drug Delivery to the Brain

15.5.1 Methods to Enhance CNS Delivery Following Intranasal Administration

A number of absorption enhancers have been used in experimental and clinical settings to enhance intranasal drug delivery to the systemic circulation. In theory, absorption enhancers may also increase the delivery of intranasally administered substances to the brain by increasing access to transport pathways from the lamina propria to the CNS. The mechanisms by which most absorption enhancers typically work are by enhancing the permeability of compounds across the nasal epithelial barrier and/or decreasing mucociliary clearance (Deli 2009; Illum 2012). Materials that have been used as intranasal absorption enhancers include surfactants, bile salts, bile salt derivatives, phospholipids, cyclodextrins, cationic polymers, proteases, and lipids (Davis and Illum 2003). Enhancing the permeability of mucosal membranes is often associated with irritation or damage (Sezaki 1995). Most absorption enhancers have not been well tolerated when administered intranasally in humans; one exception may be chitosan, which is produced by the deacetylation of chitin, a polysaccharide found in crustacean cells. Chitosan transiently opens TJ in mucosal membranes, has bioadhesive properties, and has been shown to be nonirritating with low local and systemic toxicity (Illum 2012). The development of nontoxic, physiological absorption enhancers is needed and may increase the delivery of substances to the brain following IN administration, particularly for larger substances that have difficulty crossing the nasal epithelial barriers.

15.5.2 Unresolved Questions

Despite the increasing use of IN administration as a means to bypass the BBB and deliver substances directly to the CNS, a number of basic questions remain with regard to the mechanisms governing transport from the nasal mucosa to the brain. It is unknown if there is a size limit governing what can be delivered to the brain via the intranasal route. Studies with dextrans suggest there is an inverse relationship between MW and the CSF concentration following IN administration (Sakane et al. 1995). The permeability of the nasal epithelial barrier is not well characterized and may differ between the olfactory and respiratory regions. As the trigeminal nerve innervates both the respiratory and olfactory regions, it is unclear whether preferentially targeting one of these regions favors the delivery of substances to the brain stem along trigeminal nerve-associated pathways. What are the sites and rates of bulk flow that govern the extracellular transport of substances from the lamina propria to the CNS? Is the bulk flow associated with perineural, perivascular, and perilymphatic channels or some combination of these pathways? A summary of the hypothetical pathways into the brain from the nasal passages given our current state of knowledge is depicted in Fig. 15.11.

Further questions revolve around the effect of brain diseases on the pathways and mechanisms underlying brain uptake following intranasal administration. For example, it is poorly appreciated how or, if disease states, might affect brain delivery and/or the distribution of substances in the CNS following IN administration. It will be important to establish whether the capacity for nose-to-brain transport is compromised, unaffected, or otherwise altered by specific pathology, disease, or other factors. Finally, it is not yet clear how cells reach the brain by the IN route of administration; whether cells use the same or different pathways/mechanisms to gain entry to the brain from the nasal passage that have been identified for small molecules, peptides, and proteins remains an open question. There is a clear need for further research to address these questions and advance knowledge so that clinical applications utilizing intranasal targeting of drugs to the CNS can be evaluated with the best possible opportunities for success. Recent years have witnessed new studies that have expanded our understanding of nasal physiology (e.g., Kumar et al. 2015), factors governing intranasal delivery to the brain and CSF of smaller-molecular-weight substances (e.g., Stenslik et al. 2015, 2018; Nehra et al. 2021), and factors governing the intranasal delivery to the brain and CSF of large molecules (e.g., Lochhead et al. 2015; Kumar et al. 2018a, b). Despite this recent progress, more work is clearly warranted.

15.6   Conclusions

Drug delivery to the CNS remains a challenge due to the restrictive nature of the BBB and BCSFB. A number of studies suggest the IN route of administration may allow rapid, noninvasive delivery of substances directly to the CNS along pathways associated with the olfactory or trigeminal nerves. These pathways are not yet fully characterized, presenting opportunities for further investigation. Methods to enhance the delivery of substances from the nasal cavity to the CNS are also needed due to the typically low delivery efficiencies that have thus far been measured (<0.05% for proteins). A better understanding of the mechanisms governing the transport of substances into the CNS from the nasal mucosae may lead to improvements in the efficiency of IN administration. While IN drug delivery to the brain has shown great promise in animals and humans, it is clearly an area where more research is needed to fully exploit its potential.

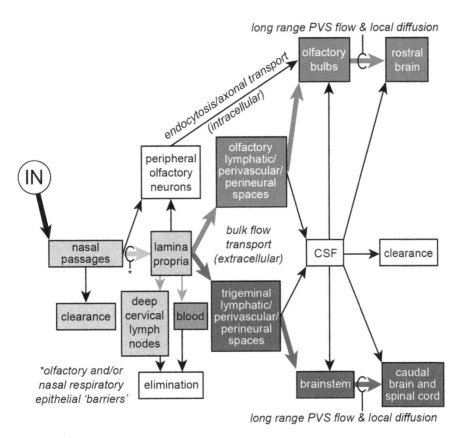

Fig. 15.11 Proposed pathways and mechanisms responsible for drug transport into the CNS following intranasal administration, based primarily upon studies utilizing radiolabeled proteins

15.7 Points for Discussion

- What advantages does the intranasal route of administration offer for chronic administration? Compare and contrast intranasal application requirements with those of other strategies designed to target biotherapeutics to the brain (e.g., are trained health professionals required for administration? Can administration be performed in an outpatient setting or is hospitalization necessary? What economic resources will likely be required to accomplish administration? Can the methods be easily applied in both developed and developing countries?).
- Why might knowledge of nasal epithelial organization and physiology be important for implementing, optimizing, and adjusting brain targeting following intranasal drug administration?
- What are the likely pathways that underlie nose-to-brain transport? What are the likely mechanisms that underlie nose-to-brain transport? What additional information is needed to more fully understand the pathways/mechanisms responsible for intranasal targeting of drugs to the CNS?
- Discuss each of the steps that may be required for an intranasally applied drug to reach CNS target sites?
- How might the transport of small molecules, macromolecules, and cell-based therapies into the brain following intranasal application be similar? How might their transport be different? What areas require further study?
- What key questions remain regarding the clinical translation of intranasal approaches targeting the CNS? How and why are rodent and monkey studies of intranasal administration different with respect to translational relevance for human studies?

Acknowledgments Portions of this work were supported by the University of Wisconsin-Madison School of Pharmacy, the Graduate School at the University of Wisconsin, the Michael J. Fox Foundation for Parkinson's Research, the Wisconsin Alzheimer's Disease Research Center (NIH P50-AG033514), and the Clinical and Translational Science Award (CTSA) program, through the NIH National Center for Advancing Translational Sciences (NCATS; grant UL1TR000427). Additional funding was provided by the NIA (T32-AG000242: M.J.S.). All content is solely the responsibility of the authors and does not necessarily represent the official views of the NIH. Robert Thorne is currently a full-time employee of Denali Therapeutics, and both Niyanta Kumar and Mallory Stenslik are currently full-time employees of Merck. Additionally, Robert Thorne acknowledges service on the scientific advisory boards for Alcyone Lifesciences and for a Lundbeck Foundation-funded Research Initiative on Brain Barriers and Drug Delivery. Jeffrey Lochhead and Robert Thorne also acknowledge being inventors on patents and/or patent applications related to intranasal drug delivery. Luke Bradley acknowledges being an inventor on patents and/or patent applications related to DNSP-11.

References

Abbott NJ, Pizzo ME, Preston JE, Janigro D, Thorne RG (2018) The role of brain barriers in fluid movement in the CNS: is there a 'glymphatic' system? Acta Neuropathol 135(3):387–407
Altner H, Altner-Kolnberger I (1974) Freeze-fracture and tracer experiments on the permeability of the zonulae occludentes in the olfactory mucosa of vertebrates. Cell Tissue Res 154(1):51–59

Anton F, Peppel P (1991) Central projections of trigeminal primary afferents innervating the nasal mucosa: a horseradish peroxidase study in the rat. Neuroscience 41(2–3):617–628

Aspelund A, Antila S, Proulx ST, Karlsen TV, Karaman S, Detmar M, Wiig H, Alitalo K (2015) A dural lymphatic vascular system that drains brain interstitial fluid and macromolecules. J Exp Med 212(7):991–999. https://doi.org/10.1084/jem.20142290

Baier PC, Weinhold SL, Huth V, Gottwald B, Ferstl R, Hinze-Selch D (2008) Olfactory dysfunction in patients with narcolepsy with cataplexy is restored by intranasal Orexin A (Hypocretin-1). Brain 131(Pt 10):2734–2741. https://doi.org/10.1093/brain/awn193

Baier PC, Hallschmid M, Seeck-Hirschner M, Weinhold SL, Burkert S, Diessner N, Goder R, Aldenhoff JB, Hinze-Selch D (2011) Effects of intranasal hypocretin-1 (orexin A) on sleep in narcolepsy with cataplexy. Sleep Med 12(10):941–946. https://doi.org/10.1016/j.sleep.2011.06.015

Baker H, Genter MB (2003) The olfactory system and the nasal mucosa as portals of entry of viruses, drugs, and other exogenous agents into the brain. In: Doty RL (ed) Handbook of olfaction and gustation. Marcel Dekker, New York, pp 549–573

Baker H, Spencer RF (1986) Transneuronal transport of peroxidase-conjugated wheat germ agglutinin (WGA-HRP) from the olfactory epithelium to the brain of the adult rat. Exp Brain Res 63(3):461–473

Banks WA (2004) Are the extracellular pathways a conduit for the delivery of therapeutics to the brain? Curr Pharm Des 10(12):1365–1370

Banks WA (2009) Characteristics of compounds that cross the blood-brain barrier. BMC Neurol 9(Suppl 1):S3. 1471-2377-9-S1-S3 [pii] https://doi.org/10.1186/1471-2377-9-S1-S3

Bilston LE, Fletcher DF, Brodbelt AR, Stoodley MA (2003) Arterial pulsation-driven cerebrospinal fluid flow in the perivascular space: a computational model. Comput Methods Biomech Biomed Engin 6(4):235–241. https://doi.org/10.1080/1025584031000160611

Bojsen-Moller F (1975) Demonstration of terminalis, olfactory, trigeminal and perivascular nerves in the rat nasal septum. J Comp Neurol 159(2):245–256. https://doi.org/10.1002/cne.901590206

Born J, Lange T, Kern W, McGregor GP, Bickel U, Fehm HL (2002) Sniffing neuropeptides: a transnasal approach to the human brain. Nat Neurosci 5(6):514–516. https://doi.org/10.1038/nn849

Bradbury MWB, Cserr HF (1985) Drainage of cerebral interstitial fluid and of cerebrospinal fluid into lymphatics. In: Johnston MG (ed) Experimental biology of the lymphatic circulation. Elsevier, Amsterdam/New York, pp 355–391

Bradley LH, Fuqua J, Richardson A, Turchan-Cholewo J, Ai Y, Kelps KA, Glass JD, He X, Zhang Z, Grondin R (2010) Dopamine neuron stimulating actions of a GDNF propeptide. PLoS One 5(3):e9752

Broadwell RD, Balin BJ (1985) Endocytic and exocytic pathways of the neuronal secretory process and trans-synaptic transfer of wheat germ agglutinin-horseradish peroxidase in vivo. J Comp Neurol 242(4):632–650. https://doi.org/10.1002/cne.902420410

Broberg EK, Peltoniemi J, Nygardas M, Vahlberg T, Roytta M, Hukkanen V (2004) Spread and replication of and immune response to gamma134.5-negative herpes simplex virus type 1 vectors in BALB/c mice. J Virol 78(23):13139–13152. https://doi.org/10.1128/JVI.78.23.13139-13152.2004

Buchner K, Seitz-Tutter D, Schönitzer K, Weiss DG (1987) A quantitative study of anterograde and retrograde axonal transport of exogenous proteins in olfactory nerve C-fibers. Neuroscience 22(2):697–707

Butt AM, Jones HC, Abbott NJ (1990) Electrical resistance across the blood-brain barrier in anaesthetized rats: a developmental study. J Physiol 429:47–62

Caggiano M, Kauer JS, Hunter DD (1994) Globose basal cells are neuronal progenitors in the olfactory epithelium: a lineage analysis using a replication-incompetent retrovirus. Neuron 13(2):339–352

Carmichael ST, Clugnet MC, Price JL (1994) Central olfactory connections in the macaque monkey. J Comp Neurol 346(3):403–434. https://doi.org/10.1002/cne.903460306

Cattepoel S, Hanenberg M, Kulic L, Nitsch RM (2011) Chronic intranasal treatment with an anti-abeta(30-42) scFv antibody ameliorates amyloid pathology in a transgenic mouse model of Alzheimer's disease. PLoS One 6(4):e18296. https://doi.org/10.1371/journal.pone.0018296

Cauna N, Hinderer KH (1969) Fine structure of blood vessels of the human nasal respiratory mucosa. Ann Otol Rhinol Laryngol 78(4):865–879

Charlton ST, Whetstone J, Fayinka ST, Read KD, Illum L, Davis SS (2008) Evaluation of direct transport pathways of glycine receptor antagonists and an angiotensin antagonist from the nasal cavity to the central nervous system in the rat model. Pharm Res 25(7):1531–1543. https://doi.org/10.1007/s11095-008-9550-2

Costantino HR, Illum L, Brandt G, Johnson PH, Quay SC (2007) Intranasal delivery: physicochemical and therapeutic aspects. Int J Pharm 337(1–2):1–24. https://doi.org/10.1016/j.ijpharm.2007.03.025

Coyle P (1975) Arterial patterns of the rat rhinencephalon and related structures. Exp Neurol 49(3):671–690

Craft S, Baker LD, Montine TJ, Minoshima S, Watson GS, Claxton A, Arbuckle M, Callaghan M, Tsai E, Plymate SR, Green PS, Leverenz J, Cross D, Gerton B (2012) Intranasal insulin therapy for Alzheimer disease and amnestic mild cognitive impairment: a pilot clinical trial. Arch Neurol 69(1):29–38. https://doi.org/10.1001/archneurol.2011.233

Crone C, Olesen SP (1982) Electrical resistance of brain microvascular endothelium. Brain Res 241 (1):49–55. 0006-8993(82)91227-6 [pii]

Cserr HF, Harling-Berg CJ, Knopf PM (1992) Drainage of brain extracellular fluid into blood and deep cervical lymph and its immunological significance. Brain Pathol 2(4):269–276

Danielyan L, Schafer R, von Ameln-Mayerhofer A, Buadze M, Geisler J, Klopfer T, Burkhardt U, Proksch B, Verleysdonk S, Ayturan M, Buniatian GH, Gleiter CH, Frey 2nd. WH (2009) Intranasal delivery of cells to the brain. Eur J Cell Biol 88 (6):315–324. S0171-9335(09)00021-1 [pii] https://doi.org/10.1016/j.ejcb.2009.02.001

Danielyan L, Klein R, Hanson LR, Buadze M, Schwab M, Gleiter CH, Frey WH (2010) Protective effects of intranasal losartan in the APP/PS1 transgenic mouse model of Alzheimer disease. Rejuvenation Res 13(2–3):195–201. https://doi.org/10.1089/rej.2009.0944

Danielyan L, Schafer R, von Ameln-Mayerhofer A, Bernhard F, Verleysdonk S, Buadze M, Lourhmati A, Klopfer T, Schaumann F, Schmid B, Koehle C, Proksch B, Weissert R, Reichardt HM, van den Brandt J, Buniatian GH, Schwab M, Gleiter CH, Frey WH 2nd. (2011) Therapeutic efficacy of intranasally delivered mesenchymal stem cells in a rat model of Parkinson disease. Rejuvenation Res 14(1):3–15. https://doi.org/10.1089/rej.2010.1130

Davis SS, Illum L (2003) Absorption enhancers for nasal drug delivery. Clin Pharmacokinet 42(13):1107–1128

Deadwyler SA, Porrino L, Siegel JM, Hampson RE (2007) Systemic and nasal delivery of orexin-A (Hypocretin-1) reduces the effects of sleep deprivation on cognitive performance in non-human primates. J Neurosci 27(52):14239-14247. 27/52/14239 [pii] https://doi.org/10.1523/JNEUROSCI.3878-07.2007

Deatly AM, Haase AT, Fewster PH, Lewis E, Ball MJ (1990) Human herpes virus infections and Alzheimer's disease. Neuropathol Appl Neurobiol 16(3):213–223

Deli MA (2009) Potential use of tight junction modulators to reversibly open membranous barriers and improve drug delivery. Biochim Biophys Acta 1788(4):892–910

Dhuria SV, Hanson LR, Frey WH 2nd. (2009) Intranasal drug targeting of hypocretin-1 (orexin-A) to the central nervous system. J Pharm Sci 98(7):2501–2515. https://doi.org/10.1002/jps.21604

Dhuria SV, Hanson LR, Frey WH 2nd. (2010) Intranasal delivery to the central nervous system: mechanisms and experimental considerations. J Pharm Sci 99(4):1654–1673. https://doi.org/10.1002/jps.21924

Donega V, van Velthoven CT, Nijboer CH, van Bel F, Kas MJ, Kavelaars A, Heijnen CJ (2013) Intranasal mesenchymal stem cell treatment for neonatal brain damage: long-term cognitive and sensorimotor improvement. PLoS One 8(1):e51253. https://doi.org/10.1371/journal.pone.0051253

Doty RL (2008) The olfactory vector hypothesis of neurodegenerative disease: is it viable? Ann Neurol 63(1):7–15. https://doi.org/10.1002/ana.21327

Draghia R, Caillaud C, Manicom R, Pavirani A, Kahn A, Poenaru L (1995) Gene delivery into the central nervous system by nasal instillation in rats. Gene Ther 2(6):418–423

Elsaesser R, Paysan J (2007) The sense of smell, its signalling pathways, and the dichotomy of cilia and microvilli in olfactory sensory cells. BMC Neurosci 8(Suppl 3):S1. https://doi.org/10.1186/1471-2202-8-S3-S1

Ermisch A, Barth T, Ruhle HJ, Skopkova J, Hrbas P, Landgraf R (1985) On the blood-brain barrier to peptides: accumulation of labelled vasopressin, DesGlyNH2-vasopressin and oxytocin by brain regions. Endocrinol Exp 19(1):29–37

Faber WM (1937) The nasal mucosa and the subarachnoid space. Am J Anat 62(1):121–148

Favre JJ, Chaffanjon Ph, Passagia JG, Chirossel JP (1995) Blood supply of the olfactory nerve. Surg Radiol Anat 17(2):133–138

Febbraro F, Andersen KJ, Sanchez-Guajardo V, Tentillier N, Romero-Ramos M (2013) Chronic intranasal deferoxamine ameliorates motor defects and pathology in the alpha-synuclein rAAV Parkinson's model. Exp Neurol. https://doi.org/10.1016/j.expneurol.2013.03.017

Field P, Li Y, Raisman G (2003) Ensheathment of the olfactory nerves in the adult rat. J Neurocytol 32(3):317–324. https://doi.org/10.1023/B:NEUR.0000010089.37032.48

Finger TE, Jeor VLS, Kinnamon JC, Silver WL (1990) Ultrastructure of substance P- and CGRP-immunoreactive nerve fibers in the nasal epithelium of rodents. J Comp Neurol 294(2):293–305. https://doi.org/10.1002/cne.902940212

Francis GJ, Martinez JA, Liu WQ, Xu K, Ayer A, Fine J, Tuor UI, Glazner G, Hanson LR, Frey WH 2nd, Toth C (2008) Intranasal insulin prevents cognitive decline, cerebral atrophy and white matter changes in murine type I diabetic encephalopathy. Brain 131(Pt 12):3311–3334. https://doi.org/10.1093/brain/awn288

Fransson M, Piras E, Burman J, Nilsson B, Essand M, Lu B, Harris RA, Magnusson PU, Brittebo E, Loskog AS (2012) CAR/FoxP3-engineered T regulatory cells target the CNS and suppress EAE upon intranasal delivery. J Neuroinflammation 9:112. https://doi.org/10.1186/1742-2094-9-112

Frenkel D, Solomon B (2002) Filamentous phage as vector-mediated antibody delivery to the brain. Proc Natl Acad Sci USA 99(8):5675–5679. https://doi.org/10.1073/pnas.072027199

Frey WH 2nd, Liu J, Chen X, Thorne RG, Fawcett JR, Ala TA, Rahman YE (1997) Delivery of 125I-NGF to the brain via the olfactory route. Drug Deliv 4:87–92

Fuqua JL, Littrell OM, Lundblad M, Turchan-Cholewo J, Abdelmoti LG, Galperin E, Bradley LH, Cass WA, Gash DM, Gerhardt GA (2014) Dynamic changes in dopamine neuron function after DNSP-11 treatment: effects in vivo and increased ERK 1/2 phosphorylation in vitro. Peptides 54:1–8

Furukawa M, Shimoda H, Kajiwara T, Kato S, Yanagisawa S (2008) Topographic study on nerve-associated lymphatic vessels in the murine craniofacial region by immunohistochemistry and electron microscopy. Biomed Res 29(6):289–296

Galeano C, Qiu Z, Mishra A, Farnsworth SL, Hemmi JJ, Moreira A, Edenhoffer P, Hornsby PJ (2018) The route by which intranasally delivered stem cells enter the central nervous system. Cell Transplant 27(3):501–514. https://doi.org/10.1177/0963689718754561

Gozes I, Giladi E, Pinhasov A, Bardea A, Brenneman DE (2000) Activity-dependent neurotrophic factor: intranasal administration of femtomolar-acting peptides improve performance in a water maze. J Pharmacol Exp Ther 293(3):1091–1098

Gray H (2008) In: Standring S (ed) Gray's anatomy, 40th edn. Elsevier, Philadelphia

Greene EC (1935) Anatomy of the rat. Braintree: Braintree Scientific

Guo C, Wang T, Zheng W, Shan ZY, Teng WP, Wang ZY (2013) Intranasal deferoxamine reverses iron-induced memory deficits and inhibits amyloidogenic APP processing in a transgenic mouse model of Alzheimer's disease. Neurobiol Aging 34(2):562–575. https://doi.org/10.1016/j.neurobiolaging.2012.05.009

Hadaczek P, Yamashita Y, Mirek H, Tamas L, Bohn MC, Noble C, Park JW, Bankiewicz K (2006) The "perivascular pump" driven by arterial pulsation is a powerful mechanism for the distribution of therapeutic molecules within the brain. Mol Ther 14(1):69–78. https://doi.org/10.1016/j. ymthe.2006.02.018

Han IK, Kim MY, Byun HM, Hwang TS, Kim JM, Hwang KW, Park TG, Jung WW, Chun T, Jeong GJ, Oh YK (2007) Enhanced brain targeting efficiency of intranasally administered plasmid DNA: an alternative route for brain gene therapy. J Mol Med 85(1):75–83. https://doi.org/10.1007/s00109-006-0114-9

Hanson LR, Roeytenberg A, Martinez PM, Coppes VG, Sweet DC, Rao RJ, Marti DL, Hoekman JD, Matthews RB, Frey WH 2nd, Panter SS (2009) Intranasal deferoxamine provides increased brain exposure and significant protection in rat ischemic stroke. J Pharmacol Exp Ther 330(3):679–686. https://doi.org/10.1124/jpet.108.149807

Harkema JR, Carey SA, Wagner JG (2006) The nose revisited: a brief review of the comparative structure, function, and toxicologic pathology of the nasal epithelium. Toxicol Pathol 34(3):252–269. https://doi.org/10.1080/01926230600713475

Hegg CC, Irwin M, Lucero MT (2009) Calcium store-mediated signaling in sustentacular cells of the mouse olfactory epithelium. Glia 57(6):634–644. https://doi.org/10.1002/glia.20792

Heidl S, Ellinger I, Niederberger V, Waltl EE, Fuchs R (2015) Localization of the human neonatal Fc receptor (FcRn) in human nasal epithelium. Protoplasma 253(6):1557–1564. https://doi.org/10.1007/s00709-015-0918-y

Hillyer JF, Albrecht RM (2001) Gastrointestinal persorption and tissue distribution of differently sized colloidal gold nanoparticles. J Pharm Sci 90(12):1927–1936

Hoekman JD, Ho RJ (2011a) Effects of localized hydrophilic mannitol and hydrophobic nelfinavir administration targeted to olfactory epithelium on brain distribution. AAPS PharmSciTech 12(2):534–543. https://doi.org/10.1208/s12249-011-9614-1

Hoekman JD, Ho RJ (2011b) Enhanced analgesic responses after preferential delivery of morphine and fentanyl to the olfactory epithelium in rats. Anesth Analg 113(3):641–651. https://doi.org/10.1213/ANE.0b013e3182239b8c

Hosoya K, Kubo H, Natsume H, Sugibayashi K, Morimoto Y, Yamashita S (1993) The structural barrier of absorptive mucosae: site difference of the permeability of fluorescein isothiocyanate-labelled dextran in rabbits. Biopharm Drug Dispos 14(8):685–695

Ichimura T, Fraser PA, Cserr HF (1991) Distribution of extracellular tracers in perivascular spaces of the rat brain. Brain Res 545(1-2):103–113

Iliff JJ, Wang M, Liao Y, Plogg BA, Peng W, Gundersen GA, Benveniste H, Vates GE, Deane R, Goldman SA, Nagelhus EA, Nedergaard M (2012) A paravascular pathway facilitates CSF flow through the brain parenchyma and the clearance of interstitial solutes, including amyloid beta. Sci Transl Med 4(147):147ra111. https://doi.org/10.1126/scitranslmed.3003748

Illum L (2004) Is nose-to-brain transport of drugs in man a reality? J Pharm Pharmacol 56(1):3–17. https://doi.org/10.1211/0022357022539

Illum L (2012) Nasal drug delivery – recent developments and future prospects. J Control Release 161(2):254–263. https://doi.org/10.1016/j.jconrel.2012.01.024

Iwai N, Zhou Z, Roop DR, Behringer RR (2008) Horizontal basal cells are multipotent progenitors in normal and injured adult olfactory epithelium. Stem Cells 26(5):1298–1306. https://doi.org/10.1634/stemcells.2007-0891

Jafek BW (1983) Ultrastructure of human nasal mucosa. Laryngoscope 93(12):1576–1599

Jansson B, Bjork E (2002) Visualization of in vivo olfactory uptake and transfer using fluorescein dextran. J Drug Target 10(5):379–386. https://doi.org/10.1080/1061186021000001823

Jauch-Chara K, Friedrich A, Rezmer M, Melchert UH, Scholand-Engler HG, Hallschmid M, Oltmanns KM (2012) Intranasal insulin suppresses food intake via enhancement of brain energy levels in humans. Diabetes 61(9):2261–2268. https://doi.org/10.2337/db12-0025

Jin Y, Dons L, Kristensson K, Rottenberg ME (2001) Neural route of cerebral Listeria monocytogenes murine infection: role of immune response mechanisms in controlling bacterial neuroinvasion. Infect Immun 69(2):1093–1100. https://doi.org/10.1128/IAI.69.2.1093-1100.2001

Johnson NJ, Hanson LR, Frey WH (2010) Trigeminal pathways deliver a low molecular weight drug from the nose to the brain and orofacial structures. Mol Pharm 7(3):884–893. https://doi.org/10.1021/mp100029t

Johnston M, Zakharov A, Papaiconomou, C, Salmasi G, Armstrong D (2004) Evidence of connections between cerebrospinal fluid and nasal lymphatic vessels in humans, non-human primates and other mammalian species. Cerebrospinal Fluid Res 1 (1):2. 1743-8454-1-2 [pii] https://doi.org/10.1186/1743-8454-1-2

Kang YS, Park JH (2000) Brain uptake and the analgesic effect of oxytocin–its usefulness as an analgesic agent. Arch Pharm Res 23(4):391–395

Kelps KA, Turchan-Cholewo J, Hascup ER, Taylor TL, Gash DM, Gerhardt GA, Bradley LH (2011) Evaluation of the physical and in vitro protective activity of three synthetic peptides derived from the pro-and mature GDNF sequence. Neuropeptides 45(3):213–218

Kerjaschki D, Horander H (1976) The development of mouse olfactory vesicles and their cell contacts: a freeze-etching study. J Ultrastruct Res 54(3):420–444

Kida S, Pantazis A, Weller RO (1993) CSF drains directly from the subarachnoid space into nasal lymphatics in the rat. Anatomy, histology and immunological significance. Neuropathol Appl Neurobiol 19(6):480–488

Kim ID, Kim SW, Lee JK (2009) Gene knockdown in the olfactory bulb, amygdala, and hypothalamus by intranasal siRNA administration. Korean J Anat 42(4):285–292

Kincaid AE, Ayers JI, Bartz JC (2015) Specificity, size, and frequency of spaces that characterize the mechanism of bulk transepithelial transport of prions in the nasal cavities of hamsters and mice. J Virol 90(18):8293–8301. https://doi.org/10.1128/JVI.01103-15

Kiyono H, Fukuyama S (2004) NALT-versus Peyer's-patch-mediated mucosal immunity. Nat Rev Immunol 4(9):699–710. https://doi.org/10.1038/nri1439

Kristensson K (2011) Microbes' roadmap to neurons. Nat Rev Neurosci 12(6):345–357. https://doi.org/10.1038/nrn3029

Kristensson K, Olsson Y (1971) Uptake of exogenous proteins in mouse olfactory cells. Acta Neuropathol 19(2):145–154

Kumar NN, Gautam M, Lochhead JJ, Wolak DJ, Ithapu V, Singh V, Thorne RG (2015) Relative vascular permeability and vascularity across different regions of the rat nasal mucosa: implications for nasal physiology and drug delivery. Sci Rep 6:31732. https://doi.org/10.1038/srep31732

Kumar NN, Lochhead JJ, Pizzo ME, Nehra G, Boroumand S, Greene G, Thorne RG (2018a) Delivery of immunoglobulin G antibodies to the rat nervous system following intranasal administration: distribution, dose-response, and mechanisms of delivery. J Control Release 286:467–484. https://doi.org/10.1016/j.jconrel.2018.08.006

Kumar NN, Pizzo ME, Nehra G, Wilken-Resman B, Boroumand S, Thorne RG (2018b) Passive immunotherapies for central nervous system disorders: current delivery challenges and new approaches. Bioconjug Chem 29(12):3937–3966. https://doi.org/10.1021/acs.bioconjchem.8b00548

Laing JM, Aurelian L (2008) DeltaRR vaccination protects from KA-induced seizures and neuronal loss through ICP10PK-mediated modulation of the neuronal-microglial axis. Genet Vaccines Ther 6:1. https://doi.org/10.1186/1479-0556-6-1

Lansley AB, Martin GP (2001) Nasal drug delivery. In: Hillery AM, Lloyd AW, Swarbrick J (eds) Drug delivery and targeting. CRC Press, Boca Raton, pp 237–268

Lee ST, Chu K, Jung KH, Kim JH, Huh JY, Yoon H, Park DK, Lim JY, Kim JM, Jeon D, Ryu H, Lee SK, Kim M, Roh JK (2012) miR-206 regulates brain-derived neurotrophic factor in Alzheimer disease model. Ann Neurol 72(2):269–277. https://doi.org/10.1002/ana.23588

Li Y, Field PM, Raisman G (2005) Olfactory ensheathing cells and olfactory nerve fibroblasts maintain continuous open channels for regrowth of olfactory nerve fibres. Glia 52(3):245–251. https://doi.org/10.1002/glia.20241

Lochhead JJ, Thorne RG (2012) Intranasal delivery of biologics to the central nervous system. Adv Drug Deliv Rev 64(7):614–628. https://doi.org/10.1016/j.addr.2011.11.002

Lochhead JJ, Wolak DJ, Pizzo ME, Thorne RG (2015) Rapid transport within cerebral perivascular spaces underlies widespread tracer distribution in the brain after intranasal administration. J Cereb Blood Flow Metab 35(3):371–381. https://doi.org/10.1038/jcbfm.2014.215

Lochhead JJ, Kellohen KL, Ronaldson PT, Davis TP (2019) Distribution of insulin in trigeminal nerve and brain after intranasal administration. Sci Rep 9(1):2621. https://doi.org/10.1038/s41598-019-39191-5

Louveau A, Smirnov I, Keyes TJ, Eccles JD, Rouhani SJ, Peske JD, Derecki NC, Castle D, Mandell JW, Lee KS, Harris TH, Kipnis J (2015) Structural and functional features of central nervous system lymphatic vessels. Nature 523(7560):337–341. https://doi.org/10.1038/nature14432

Matsuoka Y, Gray AJ, Hirata-Fukae C, Minami SS, Waterhouse EG, Mattson MP, LaFerla FM, Gozes I, Aisen PS (2007) Intranasal NAP administration reduces accumulation of amyloid peptide and tau hyperphosphorylation in a transgenic mouse model of Alzheimer's disease at early pathological stage. J Mol Neurosci 31(2):165–170

Merkus P, Guchelaar HJ, Bosch DA, Merkus FW (2003) Direct access of drugs to the human brain after intranasal drug administration? Neurology 60(10):1669–1671

Mery S, Gross EA, Joyner DR, Godo M, Morgan KT (1994) Nasal diagrams: a tool for recording the distribution of nasal lesions in rats and mice. Toxicol Pathol 22(4):353–372

Miller DS (2010) Regulation of P-glycoprotein and other ABC drug transporters at the blood-brain barrier. Trends Pharmacol Sci 31(6):246–254. https://doi.org/10.1016/j.tips.2010.03.003

Nehra G, Andrews S, Rettig J, Gould MN, Haag JD, Howard SP, Thorne RG (2021) Intranasal administration of the chemotherapeutic perillyl alcohol results in selective delivery to the cerebrospinal fluid in rats. Sci Rep 11:6351

Neuwelt E, Abbott NJ, Abrey L, Banks WA, Blakley B, Davis T, Engelhardt B, Grammas P, Nedergaard M, Nutt J, Pardridge W, Rosenberg GA, Smith Q, Drewes LR (2008) Strategies to advance translational research into brain barriers. Lancet Neurol 7(1):84–96. https://doi.org/10.1016/S1474-4422(07)70326-5

Norwood JN, Zhang Q, Card D, Craine A, Ryan TM, Drew PJ (2019) Anatomical basis and physiological role of cerebrospinal fluid transport through the murine cribriform plate. elife 8. https://doi.org/10.7554/eLife.44278

Paxinos G, Franklin KBJ (2019) Paxinos and Franklin's the mouse brain in stereotaxic coordinates: Academic press

Pizzo ME, Wolak DJ, Kumar NN, Brunette E, Brunnquell CL, Hannocks MJ, Abbott NJ, Meyerand ME, Sorokin L, Stanimirovic DB, Thorne RG (2018) Intrathecal antibody distribution in the rat brain: surface diffusion, perivascular transport, and osmotic enhancement of delivery. J Physiol. https://doi.org/10.1113/JP275105

Powell KJ, Hori SE, Leslie R, Andrieux A, Schellinck H, Thorne M, Robertson GS (2007) Cognitive impairments in the STOP null mouse model of schizophrenia. Behav Neurosci 121(5):826–835. https://doi.org/10.1037/0735-7044.121.5.826

Reese TS, Karnovsky MJ (1967) Fine structural localization of a blood-brain barrier to exogenous peroxidase. J Cell Biol 34(1):207–217

Reitz M, Demestre M, Sedlacik J, Meissner H, Fiehler J, Kim SU, Westphal M, Schmidt NO (2012) Intranasal delivery of neural stem/progenitor cells: a noninvasive passage to target intracerebral glioma. Stem Cells Transl Med 1(12):866–873. https://doi.org/10.5966/sctm.2012-0045

Rennels ML, Gregory TF, Blaumanis OR, Fujimoto K, Grady PA (1985) Evidence for a 'para-vascular' fluid circulation in the mammalian central nervous system, provided by the rapid distribution of tracer protein throughout the brain from the subarachnoid space. Brain Res 326(1):47–63

Renner DB, Frey WH 2nd, Hanson LR (2012a) Intranasal delivery of siRNA to the olfactory bulbs of mice via the olfactory nerve pathway. Neurosci Lett 513(2):193–197. https://doi.org/10.1016/j.neulet.2012.02.037

Renner DB, Svitak AL, Gallus NJ, Ericson ME, Frey WH 2nd, Hanson LR (2012b) Intranasal delivery of insulin via the olfactory nerve pathway. J Pharm Pharmacol 64(12):1709–1714. https://doi.org/10.1111/j.2042-7158.2012.01555.x

Rojanasakul Y, Wang LY, Bhat M, Glover DD, Malanga CJ, Ma JK (1992) The transport barrier of epithelia: a comparative study on membrane permeability and charge selectivity in the rabbit. Pharm Res 9(8):1029–1034

Ronaldson PT, Babakhanian K, Bendayan R (2007) Drug transport in the brain. In: You G, Morris ME (eds) Drug transporters: molecular characterization and role in drug disposition. Wiley-Interscience, Hoboken, pp 411–461

Ross TM, Martinez PM, Renner JC, Thorne RG, Hanson LR, Frey Ii WH (2004) Intranasal administration of interferon beta bypasses the blood–brain barrier to target the central nervous system and cervical lymph nodes: a non-invasive treatment strategy for multiple sclerosis. J Neuroimmunol 151(1–2):66–77

Sakane T, Akizuki M, Taki Y, Yamashita S, Sezaki H, Nadai T (1995) Direct drug transport from the rat nasal cavity to the cerebrospinal fluid: the relation to the molecular weight of drugs. J Pharm Pharmacol 47(5):379–381

Schaefer ML, Bottger B, Silver WL, Finger TE (2002) Trigeminal collaterals in the nasal epithelium and olfactory bulb: a potential route for direct modulation of olfactory information by trigeminal stimuli. J Comp Neurol 444(3):221–226. https://doi.org/10.1002/cne.10143

Schley D, Carare-Nnadi R, Please CP, Perry VH, Weller RO (2006) Mechanisms to explain the reverse perivascular transport of solutes out of the brain. J Theor Biol 238(4):962–974. https://doi.org/10.1016/j.jtbi.2005.07.005

Schuenke M, Schulte E, Schumacher U (2010) Head and neuroanatomy. In: Ross LM, Lamperti ED, Taub E (eds) Atlas of Anatomy. Thieme, Stuttgart/New York

Scremin OU (2004) Cerebral vascular system. In: Paxinos G (ed.) The rat nervous system (1167–1202). San Diego: Elsevier, Inc.

Sezaki H (1995) Mucosal penetration enhancement. J Drug Target 3(3):175–177. https://doi.org/10.3109/10611869509015941

Shiryaev N, Jouroukhin Y, Giladi E, Polyzoidou E, Grigoriadis NC, Rosenmann H, Gozes I (2009) NAP protects memory, increases soluble tau and reduces tau hyperphosphorylation in a tauopathy model. Neurobiol Dis 34(2):381–388. https://doi.org/10.1016/j.nbd.2009.02.011

Standring S (2021) Gray's anatomy e-book: the anatomical basis of clinical practice. Elsevier Health Sciences

St-Amour I, Paré I, Alata W, Coulombe K, Ringuette-Goulet C, Drouin-Ouellet J, Vandal M, Soulet D, Bazin R, Calon F (2013) Brain bioavailability of human intravenous immunoglobulin and its transport through the murine blood-brain barrier. J Cereb Blood Flow Metab 33(12):1983–1992. https://doi.org/10.1038/jcbfm.2013.160

Steinke A, Meier-Stiegen S, Drenckhahn D, Asan E (2008) Molecular composition of tight and adherens junctions in the rat olfactory epithelium and fila. Histochem Cell Biol 130(2):339–361. https://doi.org/10.1007/s00418-008-0441-8

Stenslik MJ, Potts LF, Sonne JWH, Cass WA, Turchan-Cholewo J, Pomerleau F, Huettl P, Ai Y, Gash DM, Gerhardt GA (2015) Methodology and effects of repeated intranasal delivery of DNSP-11 in a rat model of Parkinson's disease. J Neurosci Methods 251:120–129

Stenslik MJ, Evans A, Pomerleau F, Weeks R, Huettl P, Foreman E, Turchan-Cholewo J, Andersen A, Cass WA, Zhang Z (2018) Methodology and effects of repeated intranasal delivery of DNSP-11 in awake Rhesus macaques. J Neurosci Methods 303:30–40

Stevens J, Ploeger BA, van der Graaf PH, Danhof M, de Lange EC (2011) Systemic and direct nose-to-brain transport pharmacokinetic model for remoxipride after intravenous and intranasal administration. Drug Metab Dispos 39(12):2275–2282. https://doi.org/10.1124/dmd.111.040782

Stoop R (2012) Neuromodulation by oxytocin and vasopressin. Neuron 76(1):142–159. https://doi.org/10.1016/j.neuron.2012.09.025

Szentistvanyi I, Patlak CS, Ellis RA, Cserr HF (1984) Drainage of interstitial fluid from different regions of rat brain. Am J Phys 246(6 Pt 2):F835–F844

Thorne RG, Emory CR, Ala TA, Frey WH 2nd. (1995) Quantitative analysis of the olfactory pathway for drug delivery to the brain. Brain Res 692(1–2):278–282

Thorne RG, Frey WH (2001) Delivery of neurotrophic factors to the central nervous system. Clin Pharmacokinet 40(12):907–946

Thorne RG, Hrabetova S, Nicholson C (2004a) Diffusion of epidermal growth factor in rat brain extracellular space measured by integrative optical imaging. J Neurophysiol 92(6):3471–3481. https://doi.org/10.1152/jn.00352.2004

Thorne RG, Pronk GJ, Padmanabhan V, Frey WH 2nd. (2004b) Delivery of insulin-like growth factor-I to the rat brain and spinal cord along olfactory and trigeminal pathways following intranasal administration. Neuroscience 127(2):481–496. https://doi.org/10.1016/j.neuroscience.2004.05.029

Thorne RG, Hanson LR, Ross TM, Tung D, Frey WH 2nd. (2008a) Delivery of interferon-beta to the monkey nervous system following intranasal administration. Neuroscience 152(3):785–797. https://doi.org/10.1016/j.neuroscience.2008.01.013

Thorne RG, Hanson LR, Ross TM, Tung D, Frey Ii WH (2008b) Delivery of interferon-β to the monkey nervous system following intranasal administration. Neuroscience 152(3):785–797

Tucker D (1971) Nonolfactory responses from the nasal cavity: Jacobsen's organ and the trigeminal system. In: Biedler LM (ed) Handbook of sensory physiology. Springer, New York, pp 151–181

van Velthoven CT, Sheldon RA, Kavelaars A, Derugin N, Vexler ZS, Willemen HL, Maas M, Heijnen CJ, Ferriero DM (2013) Mesenchymal stem cell transplantation attenuates brain injury after neonatal stroke. Stroke. https://doi.org/10.1161/STROKEAHA.111.000326

Vanlandewijck M, He L, Mae MA, Andrae J, Ando K, Del Gaudo F, Nahar K, Lebouvier T, Lavina B, Gouveia L, Sun Y, Raschperger E, Rasanen M, Zarb Y, Mochizuki N, Keller A, Lendahl U, Betsholtz C (2018) A molecular atlas of cell types and zonation in the brain vasculature. Nature 554:475–480

Wang P, Olbricht WL (2011) Fluid mechanics in the perivascular space. J Theor Biol 274(1):52–57. https://doi.org/10.1016/j.jtbi.2011.01.014

Wei N, Yu SP, Gu X, Taylor TM, Song D, Liu XF, Wei L (2012) Delayed intranasal delivery of hypoxic-preconditioned bone marrow mesenchymal stem cells enhanced cell homing and therapeutic benefits after ischemic stroke in mice. Cell Transplant. https://doi.org/10.3727/096368912X657251

Wolburg H, Wolburg-Buchholz K, Sam H, Horvat S, Deli MA, Mack AF (2008) Epithelial and endothelial barriers in the olfactory region of the nasal cavity of the rat. Histochem Cell Biol 130(1):127–140. https://doi.org/10.1007/s00418-008-0410-2

Xiao C, Davis FJ, Chauhan BC, Viola KL, Lacor PN, Velasco PT, Klein WL, Chauhan NB (2013) Brain transit and ameliorative effects of intranasally delivered anti-amyloid-beta oligomer antibody in 5XFAD mice. J Alzheimers Dis. https://doi.org/10.3233/JAD-122419

Yang JP, Liu HJ, Cheng SM, Wang ZL, Cheng X, Yu HX, Liu XF (2009) Direct transport of VEGF from the nasal cavity to brain. Neurosci Lett 449(2):108–111. S0304-3940(08)01510-3 [pii] https://doi.org/10.1016/j.neulet.2008.10.090

Yang AC, Vest RT, Kern F, Lee DP, Maat CA, Losada PM, Chen MB, Agam M, Schaum N, Khoury N, Calcuttawala K, Palovics R, Shin A, Wang EY, Luo J, Gate D, Siegenthaler JA, McNerney MW, Keller A, Wyss-Coray T (2021) A human brain vascular atlas reveals diverse cell mediators of Alzheimer's disease risk. bioRxiv. https://doi.org/10.1101/2021.04.26.441262

Ye L, Zeng R, Bai Y, Roopenian DC, Zhu X (2011) Efficient mucosal vaccination mediated by the neonatal Fc receptor. Nat Biotechnol 29(2):158–163. https://doi.org/10.1038/nbt.1742

Yoffey JM, Drinker CK (1938) The lymphatic pathway from the nose and pharynx: the absorption of dyes. J Exp Med 68(4):629–640

Chapter 16
Blood-to-Brain Drug Delivery Using Nanocarriers

Yang Hu, Pieter J. Gaillard, Jaap Rip, and Margareta Hammarlund-Udenaes

Abstract Central nervous system (CNS) disorders represent a large, unmet medical need. CNS drug development is hampered by the restricted transport of drug candidates across the blood-brain barrier (BBB). Current strategies to enhance brain drug delivery focus either on local injections circumventing the BBB or on global delivery through the bloodstream. In this chapter, we will discuss blood-to-brain drug delivery strategies using nanocarriers, such as liposomes and nanoparticles. The focus will be on the pharmaceutical, pharmacokinetic/pharmacodynamic, and regulatory aspects of the clinical development of nanocarriers. Clinical development of nanocarrier-based brain treatments is not as straightforward as for a single active molecule. Therefore, we will highlight the issues that should be considered when translating basic research towards clinical development. Although it remains unrealistic to expect a magic bullet for CNS drug delivery, much progress has been made towards the successful development of nanocarrier-based treatments for patients with devastating brain diseases.Keywords Central nervous system · Blood-brain barrier · Brain disease · Brain delivery · Nanocarrier · Liposomes · Nanoparticles · Pharmacokinetic/pharmacodynamic · Clinical development

Y. Hu (✉)
Discovery ADME, Drug Discovery Science, Boehringer Ingelheim RCV GmbH & Co KG, Vienna, Austria
e-mail: yang_2.hu@boehringer-ingelheim.com

P. J. Gaillard
2-BBB Medicines B.V., Leiden, CH, The Netherlands

J. Rip
20Med Therapeutics B.V., Enschede, NB, The Netherlands

M. Hammarlund-Udenaes
Translational PKPD Research Group, Department of Pharmacy, Uppsala University, Uppsala, Sweden

© American Association of Pharmaceutical Scientists 2022
E. C. M. de Lange et al. (eds.), *Drug Delivery to the Brain*, AAPS Advances in the Pharmaceutical Sciences Series 33, https://doi.org/10.1007/978-3-030-88773-5_16

16.1   Introduction

Efficient drug delivery to the brain is still an enormous challenge for the treatment of central nervous system (CNS) diseases (Juillerat-Jeanneret 2008; Masserini 2013; Tosi et al. 2016). The blood-brain barrier (BBB) is the primary impediment for brain delivery, restricting the access of many systemically administered drugs to the CNS with therapeutically effective concentrations (Abbott 2013; Strazielle and Ghersi-Egea 2013). The limited success in developing CNS drugs with excellent BBB penetrability has driven the development of various strategies aiming at improving drug delivery into the brain. Currently, these innovative strategies can mainly be divided into two categories: local delivery and global (or blood-to-brain) delivery. For local delivery, neurotherapeutics are infused directly into the brain parenchyma through, e.g., convection-enhanced delivery, thereby circumventing the BBB. The local delivery route is often used to deliver nucleic acids to the brain with viral or nonviral vectors (Wang and Huang 2019; Perez et al. 2020). Although clinical trials have shown promising results, viral gene therapy is outside the scope of this chapter. Compared to local infusion, global delivery through the bloodstream is investigated more intensively since it is less invasive and better for whole-brain treatment. The commonly used blood-to-brain delivery approaches include transiently opening the BBB tight junctions, co-administration of efflux transporter inhibitor with the therapeutic drug, and the use of drug delivery vehicles like nanocarriers.

Since strategies such as intranasal delivery and (transiently) opening the BBB are discussed elsewhere in this book (Fortin 2020; Konofagou 2020; Lochhead 2020), we will not elaborate on them. Instead, we will focus on discussing strategies that make use of nanocarriers. Although functionalizing nanocarriers with BBB-targeting ligands may affect delivery outcomes, the focus of this chapter will be on pharmaceutical development, pharmacokinetics (PK), and brain uptake of nano-therapeutics and not, or to a lesser extent, on the biology of ligand-receptor interactions at the BBB.

Many nanocarrier-based approaches to facilitate drug delivery across the BBB are under development, both by academic research groups and pharmaceutical and biotechnology companies, albeit with little clinical success. To date, there are no nanotherapeutics available on the market that specifically aim at improving brain delivery. To translate basic (academic) research into safe and effective treatments for patients with devastating brain diseases, many steps in many different research areas are required (Fig. 16.1). These aspects have been summarized as the ten key development criteria for targeted blood-to-brain drug delivery (Gaillard 2010; Gaillard et al. 2012a, b). Within the currently active developments, there are only limited numbers of novel nanoformulations that have been approved for clinical research on treating CNS diseases, e.g., glutathione PEGylated (GSH-PEG) liposomal doxorubicin (DOX) (2B3-101, now 2X-111), GSH-PEG liposomal methylprednisolone (2B3-201, now ENX-201), gold nanocrystals (CNM-Au8), gold nanoparticles (NU0129), and albumin-bound rapamycin nanoparticles (ABI-009). The majority of brain-targeted nanoformulations are still in the preclinical research and development stage. In this chapter, we will thus focus on the pharmaceutical,

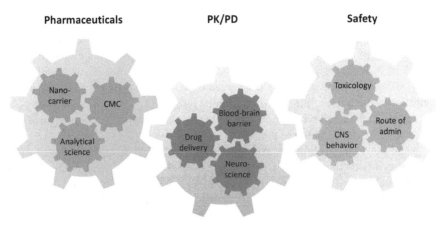

Fig. 16.1 Obtaining regulatory approval for clinical research using nanocarriers for drug delivery to the brain requires the connection of many involved research areas with clockwork precision. Turning one wheel will influence the whole development process. *CMC* Chemistry, manufacturing, and controls

pharmacokinetic/pharmacodynamic (PK/PD), and regulatory aspects of the development of nanocarriers for blood-to-brain drug delivery. Throughout this chapter, we will use the term "nanocarrier" as the European Medicine Agency (EMA) defined it: a drug formulation with the intention to form a particle in the nanoscale range. Examples of commonly used nanocarriers are liposomes, lipid nanoparticles, albumin nanoparticles, and polymeric nanoparticles.

16.2 Current Status of Therapeutic Nanocarrier Development

Currently, there are three families of therapeutic nanocarriers, i.e., liposomes, albumin nanoparticles, and lipid nanoparticles, established in clinical practice worldwide primarily for treating various cancers (Wicki et al. 2015; Anselmo and Mitragotri 2016; Moss et al. 2019). PEGylated liposomal DOX (Doxil®/Caelyx®) was the first-ever nanomedicine that was approved by the FDA in 1995 and the EMA in 1996. Thereafter, the FDA has approved several liposomal formulations including non-PEGylated liposomal daunorubicin (DaunoXome®), non-PEGylated liposomal cytarabine (DepoCyt®), non-PEGylated liposomal vincristine (Marqibo®), and most recently PEGylated liposomal irinotecan (Onivyde®). Two liposomal formulations have received EMA approval, i.e., non-PEGylated liposomal DOX (Myocet®) and non-PEGylated liposomal mifamurtide (Mepact®). Another nanoformulation used clinically is albumin-bound paclitaxel nanoparticles (Abraxane®), which was approved by the FDA in 2005 and by the EMA in 2008. Other than applications in oncology, there are also liposomal nanomedicines approved for other diseases, such as liposomal amphotericin B (Ambisome®) for fungal treatments and liposomal verteporfin (Visudyne®) for macular degeneration. Recent developments in the delivery of nucleic acids resulted in the first FDA-approved small interference RNA nanomedicine (Patisiran, a lipid nanoparticle formulation) for hereditary transthyretin-mediated familial amyloidosis (Adams et al. 2018). Although brain cancer is one of the indications for some of the approved cancer nanomedicines like Marqibo® and Onivyde® (Anselmo and Mitragotri 2016), none of the abovementioned nanomedicines are specifically designed for enhanced drug delivery across the BBB.

In addition to these clinically available nanoformulations, many therapeutic nano-carriers are undergoing clinical trials, despite limited ones designed for CNS indications (Anselmo and Mitragotri 2016). The majority of nanocarriers investigated in ongoing clinical trials for brain indications are still the marketed nanomedicines. For example, Onivyde® is currently investigated in a Phase II clinical trial to treat brain metastases in patients with breast cancer (ClinicalTrails.gov: NCT03328884). Ambisome®, in combination with fluconazole, is being tested in a Phase II trial for the treatment of cryptococcal meningitis (ClinicalTrails.gov: NCT03945448).

Apart from those already marketed nanoformulations, several novel nanocarriers are also undergoing clinical research for CNS indications. A representative one is 2B3-101, originally developed by 2-BBB Medicines BV and now as 2X-111 at Oncology venture A/S, which combines Doxil®/Caelyx® with glutathione to actively target the glutathione transporter at the BBB and thereby facilitate the brain delivery of DOX. 2B3-101 has completed a Phase I/IIa dose-escalation study in patients with glioma or brain metastases (ClinicalTrails.gov: NCT01386580). In another Phase IIa trial, the product was investigated in patients with breast cancer and leptomeningeal metastases (ClinicalTrails.gov: NCT01818713). Another promising nanoformulation is albumin-bound rapamycin (ABI-009), which is currently tested in multiple clinical trials. For example, ABI-009 is investigated in a Phase II trial in patients with recurrent high-grade glioma and newly diagnosed glioblastoma (ClinicalTrails.gov: NCT03463265). It is also evaluated in a Phase IIa study for patients with genetically confirmed Leigh or Leigh-like syndrome (ClinicalTrails.gov: NCT03747328) and in a Phase I study for patients with surgically refractory epilepsy (ClinicalTrails.gov: NCT03646240). Inorganic nanoparticles like gold nanocrystals also have their presence in clinical trials. For instance, CNM-Au8, gold nanocrystals developed by Clene Nanomedicine, are currently being assessed in five Phase II studies for treating multiple CNS disorders, including multiple sclerosis, Parkinson's diseases, and amyotrophic lateral sclerosis (ClinicalTrails.gov: NCT03993171, NCT03815916, NCT03843710, NCT04098406, and NCT03536559). NU-0129, a gold base spherical nucleic acid nanoconjugate targeting the glioblastoma oncogene BCL2L12, represents a novel class of blood-brain and blood-tumor barrier-permeable nanoformulation for suppressing gene expression in tumors of glioblastoma patients (Kumthekar et al. 2019). It is now investigated in an early Phase I clinical trial (ClinicalTrails.gov: NCT03020017).

More nanocarrier-based treatments are still in preclinical research. It would go too far to mention all of these in this chapter. Therefore, we will first discuss the general criteria that nanocarriers should follow in order to move into clinical practice. Subsequently, we will focus on the applicability of nanocarriers for drug delivery to the brain and the points to consider for regulatory approval of nanocarriers.

16.2.1 Criteria for Nanocarriers to Move into Clinical Practice

Based on FDA and EMA recommendations, product quality and safety are the most important properties to consider and guarantee to receive approval for clinical research. This is consistent with the World Medical Association (WMA) Declaration of Helsinki, in which it is stated that "All medical research involving human subjects must be preceded by careful assessment of predictable risks and burdens to the individuals and groups involved in the research in comparison with foreseeable benefits to them and to other individuals or groups affected by the condition under investigation" (WMA 2013). Alternatively, to put it plain and simple, safety comes first. For all human use drugs, and especially for neurotherapeutics, it is important to demonstrate drug safety in the CNS (FDA 2001). However, governmental regulations are hindered by a lack of toxicology data for nanocarriers (Fernandes et al. 2010; Farrell et al. 2011). In addition, Wolf and Jones recommended extra oversight for clinical research due to the uncertain but possibly significant risks of new science and technology associated with nanomedicines, including drug-loaded nanocarriers (Wolf and Jones 2011).

Clinical development of treatments using nanocarriers is not as straightforward as for a single active molecule. Although there are no published strict guidelines for treatments employing nanocarriers, both the FDA and EMA have been continuously working on guidance documents. Two Science for Policy reports by the Joint Research Centre (JRC), the European Commission's science and knowledge service, provide background information regarding the use of nanotechnology in health products (Halamoda Kenzaoui et al. 2019; Hubert et al. 2019). To date, the most advanced and relevant guidance documents are the ones on "Drug products, including biological products, that contain nanomaterials" (FDA 2017) and on "Liposome drug products" (FDA 2018). From these documents, it is clear that chemistry, manufacturing, and controls (CMC) should be investigated for each of the separate nanomaterials constituting the nanocarrier, as well as the final product.

As for every drug product, the effectiveness of a drug-loaded nanocarrier, especially the improved therapeutic index compared to unformulated drug, should be shown in (GLP) preclinical research as well as in controlled clinical trials. The benefit of nanocarriers is often the prolonged circulation of a drug in plasma (e.g., PEGylated liposomal DOX versus non-liposomal DOX), which can result in an improved therapeutic index mainly due to reduced toxicity (Ait-Oudhia et al. 2014; Petersen et al. 2016). When developing a nanocarrier-based treatment, it is of importance to investigate whether the amount of administered nanoformulated drug will lead to desired exposure of released, unbound drug at the desired site of action without inducing off-target toxicity related to drug or nanomaterials. It is noteworthy that different types of nanocarriers can lead to different drug loading efficiencies, which may ultimately influence the therapeutic effect. For example, while a drug loading of >90% can be reached for liposomes via remote loading techniques (Jiang et al. 2011), polymeric nanoparticles usually only have a loading efficiency of approximately 10% (Costantino and Boraschi 2012). Although low drug loading might be acceptable for high-potency drugs, the therapeutic effect may often not be positively influenced for drugs with lower potency, as the desired tar-get-site concentration cannot be met or the cost of goods is too high (a high dose is required).

Finally, when applying nanocarriers for CNS indications, the effect of the targeting ligands on the nanocarrier properties should be considered. When combined with certain formulations of nanocarriers, these targeting ligands can serve as homing devices to the BBB, resulting in improved brain delivery compared to nontargeted formulation and unformulated drug. Targeting ligands can be antibodies, peptides, proteins, and small molecules (Masserini 2013; Gabathuler 2014; Rip 2016). Importantly, these ligands may greatly affect the pharmaceutical properties of a nanocarrier, e.g., charge, stability, immunogenicity, and PK/PD properties, i.e., plasma protein binding, systemic PK, BBB transport, effectiveness, and safety.

16.2.2   Nanocarriers Suitable for Brain Drug Delivery

Searching for the terms "nanocarriers" and "brain" in PubMed results in just over 410 articles; however, replacing "nanocarriers" with "nanoparticles" increases the results to around 4100, and "liposomes" and "brain" results in nearly 1500 articles. Combining all this, and extending it with other specific nanocarriers, one may conclude that there is a wealth of research published on this topic. Rather than discussing all possible nanocarriers, we follow the FDA guideline on drug products, including biological products that contain nanomaterials, in which nanocarriers for human use were discussed (FDA 2017). For these nanocarriers (Fig. 16.2), we will also discuss their applicability as brain drug delivery vehicles.

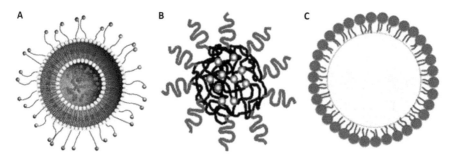

Fig. 16.2 Schematic presentation of the main nanocarriers discussed in this book chapter. (**a**) Liposomes; (**b**) polymeric nanoparticles, including albumin nanoparticles; (**c**) solid lipid nanoparticles. Liposome picture from 2-BBB Medicines BV, polymeric and solid lipid nanoparticle adapted from (Masserini 2013)

16.2.2.1 Liposomes

Liposomes have a strong presence in clinical use and research. There is not one universal liposomal formulation, as the constituents (choice of lipids) can vary. In general, by choosing the right constituents, liposomes are regarded as safe nanocarriers. Besides the more general "nanocarrier advantages" such as prolonged circulation time and the possibility to be functionalized with ligands, an additional benefit of liposomes is their ability to encapsulate both lipophilic and hydrophilic compounds ranging from small molecules to large biologics without modification of the active pharmaceutical ingredient (API) (Gaillard et al. 2012a, b).

The composition of liposomes, combined with the drug properties, can make a profound impact on multiple in vitro and in vivo properties. For example, the loading efficiency of methotrexate (MTX) was higher in PEGylated (PEG) liposomes based on egg yolk phosphatidylcholine (EYPC) than in liposomes based on hydrogenated soy phosphatidylcholine (HSPC) (Hu et al. 2017). Furthermore, it has been shown that the selection of lipids influences the plasma circulation of different liposomal formulations containing the model drug ribavirin. The choice of lipids also affects the in vivo drug release, as exemplified by our recent studies showing faster MTX release in plasma from EYPC-containing than from HSPC-containing liposomes (Hu et al. 2017, 2019a, b). When encapsulated in the same liposomal formulation, different drug payloads also showed dramatically different release profiles. For example, both in vitro and in vivo releases of diphenhydramine were much more rapid from PEG-EYPC liposomes compared to that of MTX from the same formulation (Hu et al. 2017, 2018).

Most importantly, the formulation of liposomes can determine whether and to what extent they enhance drug delivery across the BBB. When choosing an appropriate formulation, even nontargeted PEG liposomes can increase brain drug delivery. As our recent studies showed, PEG-EYPC liposomal formulations improved brain uptake of MTX with up to 15-fold increase in unbound brain-to-plasma concentration ratio ($K_{p,uu,brain}$), while PEG-HSPC liposomes had no significant impact on the uptake, compared with administering MTX itself (Hu et al. 2017, 2019a, b). The likelihood of PEG-HSPC liposomes to improve CNS drug delivery will increase when the BBB integrity is compromised, e.g., in the newly formed blood vessels in brain tumors or because of traumatic brain injury or other diseases. For instance, in patients with malignant glioblastoma multiforme, PEG-HSPC liposomal DOX (Doxil®/Caelyx®), when used alone or in combination with other chemotherapeutics, was considered to be safe and moderately effective (Fabel et al. 2001; Hau et al. 2004; Glas et al. 2007). This indicates that DOX delivered with PEG-HSPC liposomes might only have reached the center of larger brain tumors with disrupted BBB rather than all tumor regions, since the integrity of the BBB is maintained around the infiltrative growing tumor cells as well as in micrometastases (de Vries et al. 2006; Hambardzumyan and Bergers 2015; van Tellingen et al. 2015; Quail and Joyce 2017; Sarkaria et al. 2018).

Many brain-targeting strategies have been developed to increase drug delivery to the brain, which can also be used together with liposomal drug delivery. These have been extensively discussed in several reviews (Gaillard et al. 2012a, b; Lai et al.

2013; Rip 2016); therefore, only a few examples are listed to illustrate the issues that could be relevant for clinical development. Brain-targeting ligands, when combined with liposomes, can influence their systemic PK and more importantly brain uptake. For example, repeated administration of identical doses of Doxil®/Caelyx® resulted in slightly higher plasma concentrations and lower systemic clearance compared to GSH-PEG liposomal DOX (2B3-101) (both formulations containing HSPC) (Gaillard et al. 2012a, b). The DOX retention in the brain was almost three times higher after repeated 2B3-101 administration compared to Doxil®/Caelyx® at the same dose (Gaillard et al. 2012a, b). Another example is that GSH-PEG liposomal ribavirin resulted in fivefold higher unbound drug concentration in the brain with similar systemic exposure compared to PEG control liposomes (Rip et al. 2010). Another study showed that while total plasma PK profiles were comparable, the unbound brain concentration of carboxyfluorescein was increased fourfold when delivered with GSH-PEG liposomes compared to PEG controls (Rip et al. 2014). However, it is worth noting that in the abovementioned cases, HSPC was used as the choice of phospholipid in all the formulations. We have recently found that the brain-targeting effect of GSH depends highly on the liposomal formulation on which GSH is conjugated (Hu et al. 2019a, b). Compared to the PEG control formulations, GSH-PEG-HSPC liposomes increased brain delivery of MTX fourfold, while GSH coating on PEG-EYPC liposomes did not result in a further enhancement of brain uptake (Fig. 16.3). Similarly, GSH-PEG-EYPC liposomes had no additional benefit in improving brain delivery of DAMGO, an opioid peptide, compared with PEG-EYPC liposomes (Lindqvist et al. 2013). All of these findings highlight the pivotal role of formulation optimization in the development of liposomal delivery strategies for brain treatments.

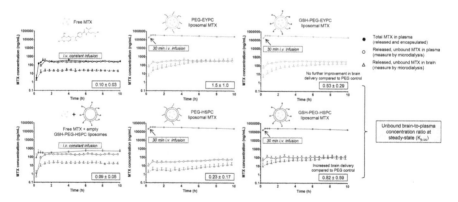

Fig. 16.3 Unbound brain-to-plasma concentration ratios at steady-state ($K_{p,uu,brain}$) and observed concentration-time profiles for unbound drug concentration in the brain interstitial fluid (open triangles) and plasma (open circles) and total drug concentration in plasma (filled circles) after intravenous administration of free MTX and free MTX + empty liposomes and different liposomal formulations (Hu et al. 2019a, b)

16.2.2.2   Albumin Nanoparticles

Albumin nanoparticles are formed by mixing albumin with a drug in an aqueous solvent and then passing the product through filters to obtain particles with a size of 100–200 nm (Kratz 2008). It was shown that enhanced uptake of albumin-bound paclitaxel nanoparticles (Abraxane®) in solid tumors is both passive through enhanced permeability and retention (EPR) and potentially active through binding to gp60 and SPARC receptors overexpressed on the tumor cell surface (Kratz 2008). As paclitaxel dissociates from the albumin very shortly after administration in the bloodstream, the main advantage of Abraxane® may be the replacement of Cremophor EL by albumin, since Cremophor EL can lead to severe side effects. Abraxane® is currently the only nanoparticle form of albumin that is approved, although the use of albumin as a drug carrier through direct conjugations with small molecules or through covalent bonds with peptides or proteins is under investigation (Kratz 2008; An and Zhang 2017). Albumin, being present in high concentrations in the circulation, is considered safe since it is biodegradable and not immunogenic (Kratz 2008; Dadparvar et al. 2011). However, safety is not guaranteed, as it is a blood-derived product. The package insert of Abraxane® states that it contains albumin derived from human blood, which has a theoretical risk of viral transmission and therefore needs to be carefully controlled.

Albumin nanoparticles are also under investigation for their capability of delivering drugs into the brain. A nontargeted albumin nanoparticle formulation (ABI-009, albumin-bound rapamycin) is being investigated in clinical research for the treatment of brain tumors, surgically refractory epilepsy, and Leigh or Leigh-like syndrome (ClinicalTrails.gov: NCT03463265, NCT03646240 and NCT03747328). From preclinical studies, nontargeted albumin nanoparticles were also shown to be effective in improving brain delivery. For instance, encapsulation in human serum albumin nanoparticles markedly increased the delivery of lapatinib into normal and metastatic mice brains compared to unformulated lapatinib, which ultimately resulted in improved antitumor efficacy in vivo (Wan et al. 2016).

There are more albumin nanoparticles functionalized with different targeting ligands tested in preclinical studies. For example, a recent study showed that cationic and mannose-modified albumin nanoparticles resulted in a marked improvement in BBB penetration and anti-glioma efficacy of DOX presumably through cationic adsorptive-mediated and also glucose transporter-mediated transcytosis (Byeon et al. 2016). Functionalization of albumin nanoparticles with a cell-penetrating peptide LMWP led to enhanced BBB penetration and improved treatment outcomes for glioma (Lin et al. 2016). Using transmission electron microscopy, Zensi et al. have shown that albumin nanoparticles conjugated with apolipoprotein A-I (Apo A-I) or apolipoprotein E (ApoE) were detected in brain capillary endothelial cells and neurons, whereas no uptake into the brain was detectable with nanoparticles without modification (Zensi et al. 2009, 2010). Conjugation of transferrin to PEGylated albumin nanoparticles containing azidothymidine (AZT, an antiviral drug) resulted in increased brain uptake of AZT compared to unformulated drug, non-PEGylated nanoparticles, and PEGylated albumin nanoparticles (Mishra et al.

2006). In addition, Ulbrich et al. have shown that modified albumin nanoparticles with transferrin and antibodies against the transferrin receptor (OX26 and R17217) were able to enhance the antinociceptive efficacy of loperamide, compared to non-targeted or albumin nanoparticles conjugated with a negative control antibody (Ulbrich et al. 2009).

16.2.2.3 Polymeric Nanoparticles

In the last few decades, polymeric nanoparticles as drug delivery vehicles have been intensively investigated in preclinical studies for the treatment of various diseases including CNS disorders. However, the clinical translation of this type of nanoparticles remains slow. To date, there is no polymer-based nanoformulation approved for clinical use. Although several polymeric nanoparticles are undergoing clinical trials, none of them is related to CNS treatments.

Most polymeric nanoparticles initially produced were based on nonbiodegradable polymers, which are unsuitable for clinical development due to their inherent chronic toxicity and immunogenic response. Since then, more nanoparticles fabricated from biocompatible, biodegradable, and bioadhesive polymers have been developed in preclinical studies for CNS treatments, as recently reviewed (Mahmoud et al. 2020; Shakeri et al. 2020). These polymers include natural ones like chitosan and gelatin and synthetic ones like polylactic acid (PLA), poly(amidoamines) (PAA), polycaprolactone (PCL), poly(butyl-cyanoacrylate) (PBCA), poly(glycolic acid) (PGA), and poly(lactic-co-glycolic acid) PLGA (Elzes et al. 2016; Mahmoud et al. 2020; Shakeri et al. 2020). Among these, the most commonly used ones are PBCA, chitosan, and PLGA. PBCA nanoparticles were the first polymer-based nanoformulation attempted for brain drug delivery (Kreuter et al. 1995). Despite the ease of production, PBCA nanoparticles feature rapid biodegradation and inefficient encapsulation of very hydrophobic and hydrophilic compounds. Chitosan is a natural product that is biodegradable and is available in different molecular weights and different degrees of deacetylation. Recent progress of using chitosan nanoparticles to treat brain diseases in preclinical studies have been reviewed (Yu et al. 2019). PLGA nanoparticles typically exhibit more sustainable release kinetics and better encapsulation than most of the other polymeric nanoparticles. PLGA is approved by the FDA and EMA and is already used in various (parenteral-topical) drug delivery systems in humans (Danhier et al. 2012). The polymers are commercially available with different molecular weights and copolymer composition.

Nanoparticles based on the above three polymers have also been investigated for CNS drug delivery. Different routes of administration have been explored including invasive direct injections in the brain parenchyma (Garbayo et al. 2009), intranasal administration (Musumeci et al. 2019), and most commonly intravenous administration (Costantino and Boraschi 2012). When focusing on intravenous administration, it is generally believed that polymeric nanoparticles cannot efficiently increase drug delivery across the BBB without the use of a BBB-targeting ligand-mediating transport. The abovementioned polymeric nanoparticles have been combined with a variety of BBB-targeting ligands for improved brain delivery, as extensively reviewed (Costantino and Boraschi 2012; Mahmoud et al. 2020; Shakeri et al. 2020). Some examples of these targeting ligands include transferrin, anti-transferrin receptor antibody, lactoferrin, Apo E, polysorbate 80, poloxamer 188, glutathione, Angiopep, and magnetic guidance Peptide (T7).

16.2.2.4 Other Nanocarriers

Several other nanocarriers have been used, aiming to deliver drugs to the brain. Although some of the nanoparticles described in this section include the use of polymers and targeting ligands related to the ones presented in Sect. 16.2.2.3, we will describe them here separately. All approaches and their applicability for brain delivery with or without the already mentioned targeting ligands have been reviewed previously. Therefore, we will limit our discussion to a brief presentation of the nanocarriers and refer the reader to the reviews for more detailed information.

The first approach is the use of dendrimers, e.g., the well-known poly(amidoamine) (PAMAM) dendrimers. Dendrimers are nanosized macromolecules featuring a hyper-branched globular structure that can be used for drug delivery (Zhu et al. 2019). Drug loading in dendrimers can be achieved by covalent conjugation, non-covalent electrostatic absorption, or by encapsulation of drug in a micellar structure formed by the dendrimers. The use of dendrimers for brain delivery has recently been reviewed (Zhu et al. 2019).

Solid lipid nanoparticles (SLNs) consist of spherical solid lipid particles in the nanometer range, which are dispersed in water or aqueous surfactant solution and have the potential to carry lipophilic or hydrophilic drug(s) or diagnostics (Bondi et al. 2012). The advances of SLN for treating brain diseases have been reviewed (Tapeinos et al. 2017).

More recently, a new type of particle was introduced, i.e., drug nanocrystals, which are particles made from 100% drug that are stabilized by surfactants. Nanocrystals have been used for oral drug administration but can potentially be used for brain drug delivery after intravenous administration (Muller and Keck 2012; Karami et al. 2019). Several more nanoformulations to be mentioned that are promising for brain delivery include nanoemulsions (Karami et al. 2019), polymeric micelles, and metal-based inorganic nanoparticles like gold nanoparticles (Khan et al. 2018).

Recent decades have witnessed remarkable progress in the development of RNA-based gene therapy. For RNA therapeutics, nanocarriers are needed to ensure the stability and efficient intracellular delivery and to prevent off-target effects. The most advanced nanodelivery systems for RNA therapeutics are lipid nanoparticles. There are several products in clinical development phase that are expected to reach the market soon (Moss et al. 2019). Although most of these RNA therapeutics are not aimed for CNS delivery, the development of nanocarriers will likely also lead to CNS delivery strategies for RNA-based neurotherapeutics to treat neurological diseases.

16.2.3 Points to Consider for Regulatory Approval

Translating basic (academic) findings into clinical practice requires regulatory approval from governmental bodies. However, it is not yet fully clear what exactly will be required for nanocarrier applications (Sanhai et al. 2008). Based on the Guidance for Industry documents from the FDA (FDA 2017, FDA 2018), we found that a nanocarrier should meet many criteria (Fig. 16.1) related to pharmaceuticals, PK/PD, and safety before receiving approval to be tested in clinical research. These criteria are discussed in the sections below along with the idea that following these criteria should result in a "definable product."

16.2.3.1 Pharmaceuticals

Besides the general CMC documentation supplied by the FDA, the FDA guidance document on liposome drug products provides the most extensive regulatory information on liposomes (FDA 2018). This document can, to a large extent, also be applied for other nanoparticles. Table 16.1 provides an overview of the criteria from this guidance document that should be taken into consideration.

A strictly controlled manufacturing process should ensure the absence of batch-to-batch differences during nanocarrier production. Reproducibility should also be demonstrated from multiple batches at different production scales. Critical manufacturing parameters (e.g., shear force, pressure, pH, temperature, etc.) should be identified and evaluated, particularly during the scale-up of the production process (FDA 2018). Obtaining a sterilized final product can be challenging as nanoparticle components can block filters or interact with the filter matrix, causing the filters to be ineffective or to cause a loss of material.

Table 16.1 Main CMC criteria for product quality and product safety of nanocarriers (based on FDA guidance document on liposome drug products (FDA 2018))

Main criteria	Details
Product quality	Characterization of physicochemical properties of the end product, such as Size, charge, and morphology Encapsulation efficiency and drug loading Phase transition temperature In vitro release of drug substance Leakage rate throughout shelf-life
Control of excipients	For each of the separate constituents, it is necessary to have: A full description and characterization Manufacturing specifications Stability data
Manufacturing process and process controls	During the manufacturing process reproducibility, purity and sterility need to be demonstrated, also during the upscaling of the production process
Control of drug product	Assays for encapsulated and free drug substance and for nanocarrier components, including degradation products
Stability	Shelf-life; physical and chemical stability

Evaluating the particle size distribution and charge are often the first characterization steps after the preparation of nanocarriers. Dynamic light scattering (DLS) technique allows relatively easy measurements of these parameters. The size of a nanocarrier will influence the capacity for drug loading (encapsulation), especially since the volume of spherical vesicles increases or decreases with the radius to the third power ($V = 4/3\ \pi r^3$). The drug loading efficiency is dependent not only on size but also on the nanocarrier constituents and specific interaction with the drug. The charge or hydrophilicity of the lipids or polymers will influence the interaction between nanocarrier constituents and the drug payload. Furthermore, the method of drug loading can greatly influence the loading efficiency. Drugs can be encapsulated into liposomes via passive loading (phospholipids dispersed in an aqueous solvent containing the drug spontaneously forming concentric bilayers separated by narrow aqueous compartments with relatively low encapsulation efficiency (Mufamadi et al. 2011). Drug loading can also be achieved through remote loading, in which drugs are loaded into preformed liposomes using a transmembrane pH or salt gradient with encapsulation efficiencies of 80–100% (Zucker et al. 2009; Tazina et al. 2011). The higher encapsulation efficiency associated with remote loading increases the drug-to-lipid ratio, which in turn increases the chance to deliver enough drug without reaching the dose limits of lipids inducing nanotoxicity.

An optimal and reproducible encapsulation efficiency needs to be demonstrated using validated analytical assays of the active substance as well as the nanocarrier components, targeting moiety, and other excipients. The analytical method needs to discriminate between the encapsulated and the released or nonencapsulated drug entities. All nanocarrier components need to be well defined. The source (synthetic or biological) and certificates of origin and analysis need to be provided together with their stability data.

Once a nanocarrier product is manufactured and characterized, physical and chemical stability and shelf-life need to be determined. Nanocarriers are susceptible to fusion, aggregation, and leakage of the drug during storage. Both the drug and the nanocarrier itself are susceptible to change, which might influence the quality, safety, and efficacy of the product. Therefore, extensive testing of all components is required. Lipids can degrade by oxidative stress or hydrolyze to form (toxic) lysolipids (Parnham and Wetzig 1993; Lutz et al. 1995). This process needs to be studied under normal storage conditions of the drug products (e.g., 2–8 °C). Also, nanocarriers should be exposed to stress conditions to characterize the physical state of the carrier and to test the drug leakage from the carrier. All of these tests should ultimately be performed in the final production scale and in the vials or containers intended for clinical use. For liposomal DOX, the FDA has suggested several conditions for this in vitro drug leakage testing, including incubation in human plasma at 37 °C, exposure to a range of pH values and temperatures, and low-frequency ultrasound disruption, all in an attempt to mimic in vivo conditions (FDA 2010).

Finally, in the case that changes in manufacturing processes are necessary, these changes should be made according to the guidance for changes to an approved new drug application (NDA) or abbreviated NDA (ANDA) to ensure that post-change products are identical to pre-change products (FDA 2004).

16.2.3.2   PK/PD

PK/PD criteria include the determination of PK (both systemic and brain disposition) and PD (exposure-response relationship) of the released (unbound) drug and preferably PK of the nanocarrier itself. It is important to determine the "fate" of the nanocarrier, the nanocarrier-associated drug, and the released drug in the body after administration in order to answer some related questions: How is the drug released in vivo? Is there a prolonged circulation because of PEGylation? How does the nanoencapsulation influence systemic PK and brain uptake of released (unbound) drug compared to unformulated drug? Are there any toxic degradation products formed? In this section, we will focus on discussing the in vitro and in vivo models that can be used to determine the transport of drugs formulated in nanocarriers across the BBB.

In vivo models are generally less suitable for detailed mechanistic studies since the number of controls needed for solid experimental proof in such studies cannot easily be performed. Compared to in vivo models, reliable in vitro models may give a better insight into the mechanism of brain uptake. However, nanocarriers, like liposomes, may exhibit a nonspecific cell uptake in vitro, complicating in vitro brain uptake assays and potentially concealing the influence of specific drug release mechanisms that operate in vivo. Brain uptake in vitro is often studied in the so-called in vitro BBB models, i.e., monoculture models of immortalized brain capillary endothelial cells and co-culture models in which endothelial cells are co-cultured with astrocytes and/or pericytes (for reviews see (Deli et al. 2005; Wilhelm et al. 2011; Helms et al. 2016)).

To determine systemic PK and biodistribution of nanocarriers in vivo, one should bear in mind that metabolic pathways and/or tissue distribution patterns can be different in different species and may be different in diseased animals compared to healthy animals. The bioanalytical methods used to determine systemic and target-site PK of nanocarriers should be validated and be capable of discriminating between the released drug and the nanocarrier-associated drug (FDA 2017).

When investigating nanocarriers for drug delivery to the brain, brain distribution is the most commonly used method in which total drug concentrations in plasma and brain homogenate are measured. Although this method is somewhat useful, there are some limitations associated with it. First, it is not possible to take multiple brain samples in one individual like blood sampling, and as a result, the time-aspects of the delivery could not be observed without a large increase in the use of animals. The second limitation is the risk of contamination of residual blood in the brain homogenates, which may lead to false-positive results especially when long-circulating nanocarriers are used. Even if the residual blood in the brain tissue is completely removed after perfusion, there may still be drugs within or attach to the endothelial cells in released and nanoencapsulated forms. Moreover, this method measures only the total concentrations without separating different drug entities including the released, unbound drug, the released drug bound to plasma protein or brain tissue, and the nanocarrier-associated drug (Hammarlund-Udenaes 2016; Rip 2016). Therefore, it is difficult to examine how much of the drug is presented in its released or encapsulated form in plasma and brain.

In vivo brain microdialysis has been proven to be one of the most suitable techniques for characterizing the influx and efflux transport functions across the BBB under physiological and pathological conditions (de Lange et al. 2000; Deguchi 2002; Chaurasia et al. 2007). A microdialysis probe has a semipermeable membrane, which allows small, water-soluble solutes to cross by passive diffusion. When studying nanocarrier-mediated brain delivery, a unique feature with microdialysis is that it only allows continuous measurement of the released, unbound drug concentrations in plasma and brain interstitial fluid (ISF) over time, which enables the separation of released, pharmacologically relevant drug from the nanocarrier-associated drug. Combined with regular blood sampling, both the rate and extent of in vivo drug release and transport at the BBB can be quantitatively assessed. As long as the delivered drug is microdialysable and the study design is proper, microdialysis is a valuable and probably the best tool for the quantitative investigation of nanodelivery to the brain, providing unique and detailed information that is not possible to obtain with other techniques (Lindqvist et al. 2012, 2013; Hammarlund-Udenaes 2016; Hu et al. 2017, 2018, 2019a, b). There are also drawbacks associated with microdialysis. For example, many lipophilic compounds are unsuitable to be studied using microdialysis due to extensive sticking to the tubing or probe material, as well as recovery issues, i.e., too low or unstable in vivo recovery. Also, the surgical implantation of the brain probe often leads to some tissue damage, which requires a certain time between surgery and experiment for the recovery of the BBB integrity to reach a reasonable level. Although microdialysis as a preclinical and clinical tool has been available for decades, there is still uncertainty about the use of microdialysis in drug research and development, both from a methodological and a regulatory point of view (Chaurasia et al. 2007). Ultimately, the acceptance of microdialysis as a regulatory tool will be dependent upon the correlation of microdialysis results with clinical responses. Thus, validation will be the key to regulatory acceptance of the methodology (Chaurasia et al. 2007).

More recently, open-flow microperfusion has been developed for continuous glucose and lactate monitoring, and subsequently for dermal drug sampling (Holmgaard et al. 2012). It is currently also used for continuous sampling in brain ISF (Birngruber et al. 2014; Birngruber and Sinner 2016; Hummer et al. 2019). Like microdialysis, cerebral open-flow microperfusion (cOFM) is based on measuring the concentration of compounds in the brain using a probe. The probe used for cOFM, however, has microscopic perforations instead of a semipermeable membrane for microdialysis, which makes it suitable for larger and lipophilic compounds, including nanocarriers. However, it is noteworthy that since cOFM sample is unfiltered and non-dialyzed, it would include both unbound drug and nanocarrier-associated drugs if intact nanocarriers cross the BBB. They need to be further separated using, i.e., ultrafiltration to measure drug concentrations in different entities (Birngruber et al. 2014).

Cerebrospinal fluid (CSF) sampling is also used to determine brain uptake, especially in large animals like primates and also humans, as direct measurement of brain ISF or total brain concentrations is not possible. However, it is important to realize that the CSF drug concentration does not always reflect unbound drug concentration in the brain ISF (de Lange and Danhof 2002; Lin 2008; Hammarlund-Udenaes 2010). Blood contamination of CSF samples is a particular problem when using long-circulating nanocarriers, as plasma can contain high concentrations of the investigated drug, creating high variability and a false-positive result of drug levels in CSF samples, even when performed at specialist contract laboratories (data not shown).

Other experimental methods to determine brain uptake of nanoformulated drugs include (invasive) cranial window and noninvasive methods such as positron emission tomography (PET), single-photon emission computed tomography (SPECT), magnetic resonance imaging (MRI), or computed X-ray tomography (CT). The cranial window technology is based on in vivo imaging of brain tissue using two-photon laser scanning microscopy and animals (often mice) with a cranial window or thinned skull (Helmchen and Denk 2005; Shih et al. 2012). Brain delivery of nanocarriers with fluorescent cargo can be imaged using this technology. However, no distinction can be made between released and encapsulated drugs other than by particle morphology. Besides, BBB integrity may be altered locally after the experimental procedure, leading to local point bleeds and extravasation of drugs from the vasculature. The noninvasive techniques require the addition of a radiolabel to be able to measure the brain uptake (Wong et al. 2012). For this, it is important to take into account which part of the nanoformulation is labeled, i.e., the encapsulated drug or (part of) the nanocarrier itself, when interpreting the outcomes.

In vitro PD models that are used for unformulated drugs, such as tests for receptor occupancy, are not suitable for nanoformulated drugs, as the drug first needs to be released from the nanocarrier. Consequently, the PD of a drug product should be determined in a relevant animal model. It would go too far to discuss all available disease models in this chapter. However, we would like to point out that models should be validated and the right controls (nanocarriers versus unformulated drug, brain-targeted versus nontargeted nanocarriers, etc.) should be used to compare the effectiveness of the investigated nanotherapeutics. Since nanocarriers influence the systemic PK and biodistribution of a drug, one should take into account that the PD parameters might then also change. For example, the time points at which the PD is measured may need to be adapted compared to investigations into the unformulated compound. Finally, from a development cost perspective, in vivo PD studies should preferably be short-term and in small animal models.

16.2.3.3 Nanotoxicity

While most publicly available information focuses on the active substance in nanocarriers, only limited information is available about possible adverse effects of the nanocarrier components themselves. Szebeni and colleagues have investigated complement activation by liposomes that could occur after intravenous administration (Szebeni et al. 2010, 2011). In most people, the symptoms remain subclinical, even though significant complement activation may occur (Szebeni et al. 2010). The addition of PEG did not decrease complement activation in pigs (Szebeni et al.

2012), although opsonization by proteins and scavenging by the reticuloendothelial system (RES) was decreased (Gabizon 2001). Complement activation-related pseudoallergy (CARPA), an acute hypersensitivity or infusion reaction, may occur after the first administration of liposomal nanocarriers and also other (polymeric) nanocarriers such as dendrimers via intravenous infusion, and its severity usually declines after repeated administration (Jiskoot et al. 2009; Duncan and Gaspar 2011; Mohamed et al. 2019). Other immunological risks include antibody formation against any of the constituents of a nanoformulation, including the targeting ligand and the active drug. These antibodies can either lead to an accelerated blood clearance (reducing the target-site exposure and efficacy) or to a burst-release through complement-mediated lysis of the nanocarrier (resulting in increased blood concentration of the active moiety and possibly toxicity) (Jiskoot et al. 2009). Immune response in itself will not be a problem as long as it is rapidly deactivated. However, a severe pathology can occur when the defense response is anomalous in extent or duration (Boraschi et al. 2012).

Nanotoxicity of polymeric nanoparticles is an important issue, yet very difficult to discuss in general, since the different polymers and methods that are used in a large number of studies undoubtedly lead to different toxicity profiles (Kean and Thanou 2010). General remarks on nanotoxicity are probably not valuable, but safety has to be determined individually for each different nanoparticle formulation. Concerning brain delivery, it is important to realize that nanotoxicity may lead to (temporary) BBB opening that will influence drug delivery to the brain and potentially even lead to neurotoxicity (Rempe et al. 2011).

For nanotherapeutics developed for brain diseases, it is also critical to show that besides general toxicities, there are no particle- or drug-induced CNS-related toxicities, such as behavioral effects. In the preclinical development of 2B3-101, EEG measurements and a modified Irwin test were therefore included to demonstrate that there was no change in neurobehavior (Gaillard et al. 2012a, b).

16.2.3.4 Therapeutic Index

The pharmacology and safety studies will together determine the therapeutic index. To evaluate the therapeutic index of a brain-directed nanoformulation, it is necessary to examine the PK profiles of the released, unbound drug in both the blood and brain. It is critical to determine whether the released, unbound drug delivered to the brain by a nanocarrier can reach the therapeutically effective level at a dose that will not lead to drug- and/or particle-induced toxicities. The therapeutic index of a nanoformulation for brain treatment will be mostly influenced by how much the $K_{p,uu,brain}$ can be improved by nanodelivery compared to unformulated drug. The $K_{p,uu,brain}$, as the ratio of target site exposure (related to CNS therapeutic effect) to off-target site exposure (linked to peripheral toxicity), is considered the most important parameter to consider and optimize when developing nanocarrier-based brain treatments (Hu et al. 2019a, b). The higher the $K_{p,uu,brain}$, the more therapeutic benefit the nanocarriers would provide (Hu et al. 2019a, b). Furthermore, side effects of the drug under investigation should be acceptable depending on the severity of the disease/condition being treated. This consideration of potential cost versus benefit will rely on the disease the drug is used for, e.g., the side effects of chemotherapy are usually more severe, yet acceptable, compared to anti-migraine treatments. To obtain regulatory approval and the possibility to continue the development of nanocarrier treatments for brain diseases, the therapeutic index is, therefore, a very important decision point.

16.3 Future Challenges/Directions

One can appreciate that if the clinical development of a single active compound already requires a stringent development plan with ample decision points, this is even more complicated for nanoformulations, consisting of multiple constituents besides the API. Previously, ten key development criteria that are important for drug delivery to the brain have been discussed, as summarized in Table 16.2 (Gaillard 2010; Gaillard et al. 2012a, b). Some important considerations have been highlighted for optimizing the nanocarrier drug development process and minimizing costs as much as possible (Gray 2014).

16.3.1 Preparation and Characterization of Nanocarriers

Currently, the FDA and EMA only have a draft guideline for drug products containing nanomaterials and a guideline for liposomal drug products (FDA 2017, 2018). The criteria in these guidelines, therefore, need to be reviewed on a case-by-case basis for the development of other nanocarriers. For instance, drug loading into liposomes can be done through remote loading approach using a salt gradient of either ammonium sulfate (used for Doxil®/Caelyx®) or calcium acetate. Although this technique is not specifically mentioned in the guideline for liposome drug products, the remaining salt concentration after completing the drug loading needs to be determined from a CMC perspective.

Table 16.2 Ten key development criteria for blood-to-brain drug delivery. (From: (Gaillard 2010; Gaillard et al. 2012a, b))

Targeting the BBB	Nanocarriers	Drug development from lab to clinic
1. Proven inherently safe receptor biology in humans	5. Favorable PK	8. Low costs and straightforward manufacturing
2. Safe and human applicable ligand	6. No modification of active ingredient	9. Activity in all animal models
3. Receptor-specific binding	7. Able to carry various classes of molecules	10. Strong IP protection
4. Applicable for acute and chronic indications		

16.3.2   Delivery and Efficacy of Brain-Targeted Nanocarriers

When it comes to evaluating the PK and therapeutic efficacy of a nanocarrier, it is important to avoid false-positive and false-negative results by selecting the most suitable read-out/models and including the appropriate controls. Due to the complex nature of nanocarriers, this will require more effort compared to single moieties.

Administration of brain-targeted nanocarriers has typically utilized either intravenous or intraparenchymal routes. For chronic disorders, it would be more patient-friendly as well as cost-effective to use other routes, such as subcutaneous, intramuscular, or oral administration. However, these routes are more complex to explore with more factors affecting the delivery outcomes, so the intravenous route is usually preferred.

16.3.3   Safety of Brain-Targeted Nanocarriers

Many of the toxicity findings associated with the use of nanocarriers per se are immunogenicity. Nanocarriers can be optimized concerning their shape, size, surface charge, and chemical composition, and these characteristics will influence whether the nanocarrier is eliminated, tolerated, or ignored by the immune system (Boraschi et al. 2012). Besides, the route of administration almost certainly influences the risk of immunological responses (Jiskoot et al. 2009). By changing the route of administration from intravenous to other routes, infusion reactions that are observed with several types of nanocarriers may also be avoided. In the clinic, infusion reactions are often mitigated by diluting the infusion solution, extending the infusion time, or applying premedication. However, it is still crucial to predict potential problems and if possible, to eliminate them (Duncan and Gaspar 2011).

16.3.4   Clinical Research

There is still a strong need to develop novel nanodelivery-based treatments for patients with devastating brain diseases. For this, clinical research of nanocarriers in human subjects is necessary. Providing preclinical proof that the investigational product is safe is one consideration, while regulatory aspects make up another. Wolf and Jones have previously reviewed whether or not it is necessary to have extra oversight, i.e., additional approval processes beyond the current institutional review boards for emerging technologies such as nanocarriers (Wolf and Jones 2011). They claim that there is heightened uncertainty regarding the risks in fast-evolving science, yielding complex, and increasingly active materials. This, together with the likelihood of research on vulnerable participants (e.g., patients with cancer) and potential risks to others beyond the research participants, could warrant the need for extra oversight, particularly for more chronic treatments.

16.4   Conclusions

Since the approval of the first nanotherapeutic (Doxil®) in 1995, much progress has been made towards the clinical development of nanoformulations. However, the clinical translation of brain-targeted nanocarriers has been lagging mainly because of the added challenges associated with delivery to the brain and other inherent difficulties in CNS drug development. Emerging gene therapy using RNA-based therapeutics need nanocarriers for efficient target-site delivery. The accelerated development of nanodelivery systems for RNA therapeutics will likely also lead to new treatments of diseases in the CNS. Currently, it is not realistic to expect the emergence of a magic bullet for CNS delivery. Nevertheless, combining safe targeting ligands with well-known and safe nanocarriers, brain-targeted nanoformulations may be capable of enhancing brain drug delivery and impacting the clinical treatment of devastating CNS diseases.

16.5   Points for Discussion

To strengthen research towards clinically applicable nanocarriers for drug delivery to the brain, we encourage a scientific discussion among researchers from industry and academia on the following points:

- Minimizing the toxicity of nanocarriers and ensuring the use of safe ligands and receptor biology, as this will improve clinical applicability.
- Taking the cost of the product into account, i.e., it is, of course, exciting to design a complicated new nanoformulation with many different components but this may come at a steep price when thinking about the translation to successful clinical development.
- Taking the route of administration into consideration, especially for chronic administration.
- How to investigate successful delivery?

References

Abbott NJ (2013) Blood-brain barrier structure and function and the challenges for CNS drug delivery. J Inherit Metab Dis 36(3):437–449

Adams D, Gonzalez-Duarte A, O'Riordan WD, Yang CC, Ueda M, Kristen AV, Tournev I, Schmidt HH, Coelho T, Berk JL, Lin KP, Vita G, Attarian S, Plante-Bordeneuve V, Mezei MM, Campistol JM, Buades J, Brannagan TH 3rd, Kim BJ, Oh J, Parman Y, Sekijima Y, Hawkins PN, Solomon SD, Polydefkis M, Dyck PJ, Gandhi PJ, Goyal S, Chen J, Strahs AL, Nochur SV, Sweetser MT, Garg PP, Vaishnaw AK, Gollob JA, Suhr OB (2018) Patisiran, an RNAi therapeutic, for hereditary transthyretin amyloidosis. N Engl J Med 379(1):11–21

Ait-Oudhia S, Mager DE, Straubinger RM (2014) Application of pharmacokinetic and pharmaco-dynamic analysis to the development of liposomal formulations for oncology. Pharmaceutics 6(1):137–174

An FF, Zhang XH (2017) Strategies for Preparing Albumin-based Nanoparticles for Multifunctional Bioimaging and Drug Delivery. Theranostics 7(15):3667–3689

Anselmo AC, Mitragotri S (2016) Nanoparticles in the clinic. Bioeng Transl Med 1(1):10–29

Birngruber T, Sinner F (2016) Cerebral open flow microperfusion (cOFM) an innovative interface to brain tissue. Drug Discov Today Technol 20:19–25

Birngruber T, Raml R, Gladdines W, Gatschelhofer C, Gander E, Ghosh A, Kroath T, Gaillard PJ, Pieber TR, Sinner F (2014) Enhanced doxorubicin delivery to the brain administered through glutathione PEGylated liposomal doxorubicin (2B3-101) as compared with generic Caelyx,((R))/Doxil((R))-a cerebral open flow microperfusion pilot study. J Pharm Sci 103(7):1945–1948

Bondi ML, Di Gesu R, Craparo EF (2012) Lipid nanoparticles for drug targeting to the brain. Methods Enzymol 508:229–251

Boraschi D, Costantino L, Italiani P (2012) Interaction of nanoparticles with immunocompetent cells: nanosafety considerations. Nanomedicine (Lond) 7(1):121–131

Byeon HJ, Thao Le Q, Lee S, Min SY, Lee ES, Shin BS, Choi HG, Youn YS (2016) Doxorubicin-loaded nanoparticles consisted of cationic- and mannose-modified-albumins for dual-targeting in brain tumors. J Control Release 225:301–313

Chaurasia CS, Muller M, Bashaw ED, Benfeldt E, Bolinder J, Bullock R, Bungay PM, DeLange EC, Derendorf H, Elmquist WF, Hammarlund-Udenaes M, Joukhadar C, Kellogg DL Jr, Lunte CE, Nordstrom CH, Rollema H, Sawchuk RJ, Cheung BW, Shah VP, Stahle L, Ungerstedt U, Welty DF, Yeo H (2007) AAPS-FDA workshop white paper: microdialysis principles, applica-tion and regulatory perspectives. Pharm Res 24(5):1014–1025

Costantino L, Boraschi D (2012) Is there a clinical future for polymeric nanoparticles as brain-targeting drug delivery agents? Drug Discov Today 17(7–8):367–378

Dadparvar M, Wagner S, Wien S, Kufleitner J, Worek F, von Briesen H, Kreuter J (2011) HI 6 human serum albumin nanoparticles–development and transport over an in vitro blood-brain barrier model. Toxicol Lett 206(1):60–66

Danhier F, Ansorena E, Silva JM, Coco R, Le Breton A, Preat V (2012) PLGA-based nanopar-ticles: an overview of biomedical applications. J Control Release 161(2):505–522

de Lange EC, Danhof M (2002) Considerations in the use of cerebrospinal fluid pharmacokinetics to predict brain target concentrations in the clinical setting: implications of the barriers between blood and brain. Clin Pharmacokinet 41(10):691–703

de Lange EC, de Boer AG, Breimer DD (2000) Methodological issues in microdialysis sampling for pharmacokinetic studies. Adv Drug Deliv Rev 45(2–3):125–148

de Vries NA, Beijnen JH, Boogerd W, van Tellingen O (2006) Blood-brain barrier and chemo-therapeutic treatment of brain tumors. Expert Rev Neurother 6(8):1199–1209

Deguchi Y (2002) Application of in vivo brain microdialysis to the study of blood-brain barrier transport of drugs. Drug Metab Pharmacokinet 17(5):395–407

Deli MA, Abraham CS, Kataoka Y, Niwa M (2005) Permeability studies on in vitro blood-brain barrier models: physiology, pathology, and pharmacology. Cell Mol Neurobiol 25(1):59–127

Duncan R, Gaspar R (2011) Nanomedicine(s) under the microscope. Mol Pharm 8(6):2101–2141

Elzes MR, Akeroyd N, Engbersen JF, Paulusse JM (2016) Disulfide-functional poly(amido amine)s with tunable degradability for gene delivery. J Control Release 244(Pt B):357–365

Fabel K, Dietrich J, Hau P, Wismeth C, Winner B, Przywara S, Steinbrecher A, Ullrich W, Bogdahn U (2001) Long-term stabilization in patients with malignant glioma after treatment with liposo-mal doxorubicin. Cancer 92(7):1936–1942

Farrell D, Ptak K, Panaro NJ, Grodzinski P (2011) Nanotechnology-based cancer therapeutics–promise and challenge–lessons learned through the NCI Alliance for Nanotechnology in Cancer. Pharm Res 28(2):273–278

FDA (2001) Guidance for industry: S7A safety pharmacology studies for human pharmaceuticals. http://www.fda.gov/downloads/Drugs/GuidanceComplianceRegulatoryInformation/Guidances/UCM074959.pdf

FDA (2004) Guidance for industry: changes to an approved NDA or ANDA. http://www.fda.gov/OHRMS/DOCKETS/98fr/1999d-0529-gdl0003.pdf

FDA (2010) Draft guidance on doxorubicin hydrochloride. http://www.fda.gov/downloads/Drugs/GuidanceComplianceRegulatoryInformation/Guidances/UCM199635.pdf

FDA (2017) Guidance for industry: drug products, including biological products, that contain nanomaterals. https://www.fda.gov/media/109910/download

FDA (2018) Guidance for industry: Liposome drug products. https://www.fda.gov/media/70837/download

Fernandes C, Soni U, Patravale V (2010) Nano-interventions for neurodegenerative disorders. Pharmacol Res 62(2):166–178

Fortin D (2020) Chapter 19: therapeutic osmotic modification of the BBB in the treatment of CNS pathologies. In: Hammarlund-Udenaes M, de Lange E, Thorne R (eds) Drug delivery to the brain – physiological concepts, methodologies and approaches. Springer, New York

Gabathuler R (2014) Chapter 16: Development of new protein vecotrs for the physiologic delivery of large therapeutic compounds to the CNS. In: Hammarlund-Udenaes M, de Lange E, Thorne R (eds) Drug delivery to the brain–physiological concepts, methodologies and approaches, 1st edn. Springer, New York

Gabizon AA (2001) Stealth liposomes and tumor targeting: one step further in the quest for the magic bullet. Clin Cancer Res 7(2):223–225

Gaillard PJ (2010) Crossing barriers from blood-to-brain and academia-to-industry. Ther Deliv 1(4):495–500

Gaillard P, Gladdines W, Appeldoorn C (2012a) Development of glutathione pegylated liposomal doxorubicin (2B3-101) for the treatment of brain cancer. In: Proceedings of the 103rd annual meeting of the american association for cancer research, 31 Mar–4 Apr 2012, Chicago, Illinois. Philadelphia (PA): AACR. Abstract nr 5687

Gaillard P, Visser C, Appeldoorn C, Rip J (2012b) Enhanced brain drug delivery: safely crossing the blood-brain barrier. Drug Discov Today Technol 9(2):e71–e174

Garbayo E, Montero-Menei CN, Ansorena E, Lanciego JL, Aymerich MS, Blanco-Prieto MJ (2009) Effective GDNF brain delivery using microspheres–a promising strategy for Parkinson's disease. J Control Release 135(2):119–126

Glas M, Koch H, Hirschmann B, Jauch T, Steinbrecher A, Herrlinger U, Bogdahn U, Hau P (2007) Pegylated liposomal doxorubicin in recurrent malignant glioma: analysis of a case series. Oncology 72(5-6):302–307

Gray D (2014) Chapter 13: pharmacoeconomical considerations of CNS drug development. In: Hammarlund-Udenaes M, de Lange E, Thorne R (eds) Drug delivery to the brain–physiological concepts, methodologies and approaches, 1st edn. Springer, New York

Halamoda Kenzaoui B, Box H, Van Elk M, Gaitan S, Geertsma R, Gainza Lafuente E, Owen A, Del Pozo A, Roesslein M, Bremer S (2019) Anticipation of regulatory needs for nanotechnology-enabled health products. Publications Office of the European Union

Hambardzumyan D, Bergers G (2015) Glioblastoma: defining tumor niches. Trends Cancer 1(4):252–265

Hammarlund-Udenaes M (2010) Active-site concentrations of chemicals – are they a better predictor of effect than plasma/organ/tissue concentrations? Basic Clin Pharmacol Toxicol 106(3):215–220

Hammarlund-Udenaes M (2016) Intracerebral microdialysis in blood-brain barrier drug research with focus on nanodelivery. Drug Discov Today Technol 20:13–18

Hau P, Fabel K, Baumgart U, Rummele P, Grauer O, Bock A, Dietmaier C, Dietmaier W, Dietrich J, Dudel C, Hubner F, Jauch T, Drechsel E, Kleiter I, Wismeth C, Zellner A, Brawanski A, Steinbrecher A, Marienhagen J, Bogdahn U (2004) Pegylated liposomal doxorubicin-efficacy in patients with recurrent high-grade glioma. Cancer 100(6):1199–1207

Helmchen F, Denk W (2005) Deep tissue two-photon microscopy. Nat Methods 2(12):932–940

Helms HC, Abbott NJ, Burek M, Cecchelli R, Couraud PO, Deli MA, Forster C, Galla HJ, Romero IA, Shusta EV, Stebbins MJ, Vandenhaute E, Weksler B, Brodin B (2016) In vitro models of the blood-brain barrier: an overview of commonly used brain endothelial cell culture models and guidelines for their use. J Cereb Blood Flow Metab 36(5):862–890

Holmgaard R, Benfeldt E, Nielsen JB, Gatschelhofer C, Sorensen JA, Hofferer C, Bodenlenz M, Pieber TR, Sinner F (2012) Comparison of open-flow microperfusion and microdialysis methodologies when sampling topically applied fentanyl and benzoic acid in human dermis ex vivo. Pharm Res 29(7):1808–1820

Hu Y, Rip J, Gaillard PJ, de Lange ECM, Hammarlund-Udenaes M (2017) The impact of liposomal formulations on the release and brain delivery of methotrexate: an in vivo microdialysis study. J Pharm Sci 106(9):2606–2613

Hu Y, Gaillard PJ, Rip J, de Lange ECM, Hammarlund-Udenaes M (2018) In vivo quantitative understanding of PEGylated liposome's influence on brain delivery of diphenhydramine. Mol Pharm 15(12):5493–5500

Hu Y, Gaillard PJ, de Lange ECM, Hammarlund-Udenaes M (2019a) Targeted brain delivery of methotrexate by glutathione PEGylated liposomes: how can the formulation make a difference? Eur J Pharm Biopharm 139:197–204

Hu Y, Hammarlund-Udenaes M, Friden M (2019b) Understanding the influence of nanocarrier-mediated brain delivery on therapeutic performance through pharmacokinetic-pharmacodynamic modeling. J Pharm Sci 108(10):3425–3433

Hubert R, Gert R, Agnieszka M, Peter G, Vikram K, Thomas L, Juan RS (2019) An overview of concepts and terms used in the European Commission's definition of nanomaterial. Publications Office of the European Union

Hummer J, Altendorfer-Kroath T, Birngruber T (2019) Cerebral open flow microperfusion to monitor drug transport across the blood-brain barrier. Curr Protoc Pharmacol 85(1):e60

Jiang W, Lionberger R, Yu LX (2011) In vitro and in vivo characterizations of PEGylated liposomal doxorubicin. Bioanalysis 3(3):333–344

Jiskoot W, van Schie RM, Carstens MG, Schellekens H (2009) Immunological risk of injectable drug delivery systems. Pharm Res 26(6):1303–1314

Juillerat-Jeanneret L (2008) The targeted delivery of cancer drugs across the blood-brain barrier: chemical modifications of drugs or drug-nanoparticles? Drug Discov Today 13(23–24):1099–1106

Karami Z, Saghatchi Zanjani MR, Hamidi M (2019) Nanoemulsions in CNS drug delivery: recent developments, impacts and challenges. Drug Discov Today 24(5):1104–1115

Kean T, Thanou M (2010) Biodegradation, biodistribution and toxicity of chitosan. Adv Drug Deliv Rev 62(1):3–11

Khan AR, Yang X, Fu M, Zhai G (2018) Recent progress of drug nanoformulations targeting to brain. J Control Release 291:37–64

Konofagou E (2020) Chapter 21: optimization of blood-brain barrier opening with focused ultrasound-the animal perspective. In: Hammarlund-Udenaes M, de Lange E, Thorne R (eds) Drug delivery to the brain – physiological concepts, methodologies and approaches. Springer, New York

Kratz F (2008) Albumin as a drug carrier: design of prodrugs, drug conjugates and nanoparticles. J Control Release 132(3):171–183

Kreuter J, Alyautdin RN, Kharkevich DA, Ivanov AA (1995) Passage of peptides through the blood-brain barrier with colloidal polymer particles (nanoparticles). Brain Res 674(1):171–174

Kumthekar P, Rademaker A, Ko C, Dixit K, Schwartz MA, Sonabend AM, Sharp L, Lukas RV, Stupp R, Horbinski C, McCortney K, Stegh AH (2019) A phase 0 first-in-human study using NU-0129: a gold base spherical nucleic acid (SNA) nanoconjugate targeting BCL2L12 in recurrent glioblastoma patients. J Clin Oncol 37(15 suppl):3012

Lai F, Fadda AM, Sinico C (2013) Liposomes for brain delivery. Expert Opin Drug Deliv 10(7):1003–1022

Lin JH (2008) CSF as a surrogate for assessing CNS exposure: an industrial perspective. Curr Drug Metab 9(1):46–59

Lin T, Zhao P, Jiang Y, Tang Y, Jin H, Pan Z, He H, Yang VC, Huang Y (2016) Blood-brain-barrier-penetrating albumin nanoparticles for biomimetic drug delivery via albumin-binding protein pathways for antiglioma therapy. ACS Nano 10(11):9999–10012

Lindqvist A, Rip J, Gaillard PJ, Bjorkman S, Hammarlund-Udenaes M (2012) Enhanced brain delivery of the opioid peptide DAMGO in glutathione pegylated liposomes: a microdialysis study. Mol Pharm 10(5):1533–1541

Lindqvist A, Rip J, van Kregten J, Gaillard PJ, Hammarlund-Udenaes M (2013) In vivo functional evaluation of increased brain delivery of the opioid peptide DAMGO by glutathione-PEGylated liposomes. Pharm Res 33(1):177–185

Lochhead J (2020) Chapter 16: intranasal drug delivery to the brain. In: Hammarlund-Udenaes M, de Lange E, Thorne R (eds) Drug delivery to the brain – physiological concepts, methodologies and approaches. Springer, New York

Lutz J, Augustin AJ, Jager LJ, Bachmann D, Brandl M (1995) Acute toxicity and depression of phagocytosis in vivo by liposomes: influence of lysophosphatidylcholine. Life Sci 56(2):99–106

Mahmoud BS, AlAmri AH, McConville C (2020) Polymeric nanoparticles for the treatment of malignant gliomas. Cancers (Basel) 12(1)

Masserini M (2013) Nanoparticles for brain drug delivery. ISRN Biochem 2013:238428

Mishra V, Mahor S, Rawat A, Gupta PN, Dubey P, Khatri K, Vyas SP (2006) Targeted brain delivery of AZT via transferrin anchored pegylated albumin nanoparticles. J Drug Target 14(1):45–53

Mohamed M, Abu Lila AS, Shimizu T, Alaaeldin E, Hussein A, Sarhan HA, Szebeni J, Ishida T (2019) PEGylated liposomes: immunological responses. Sci Technol Adv Mater 20(1):710–724

Moss KH, Popova P, Hadrup SR, Astakhova K, Taskova M (2019) Lipid nanoparticles for delivery of therapeutic RNA oligonucleotides. Mol Pharm 16(6):2265–2277

Mufamadi MS, Pillay V, Choonara YE, Du Toit LC, Modi G, Naidoo D, Ndesendo VM (2011) A review on composite liposomal technologies for specialized drug delivery. J Drug Deliv 2011:939851

Muller RH, Keck CM (2012) Twenty years of drug nanocrystals: where are we, and where do we go? Eur J Pharm Biopharm 80(1):1–3

Musumeci T, Bonaccorso A, Puglisi G (2019) Epilepsy disease and nose-to-brain delivery of polymeric nanoparticles: an overview. Pharmaceutics 11(3)

Parnham MJ, Wetzig H (1993) Toxicity screening of liposomes. Chem Phys Lipids 64(1-3):263–274

Perez BA, Shutterly A, Chan YK, Byrne BJ, Corti M (2020) Management of neuroinflammatory responses to AAV-mediated gene therapies for neurodegenerative diseases. Brain Sci 10(2)

Petersen GH, Alzghari SK, Chee W, Sankari SS, La-Beck NM (2016) Meta-analysis of clinical and preclinical studies comparing the anticancer efficacy of liposomal versus conventional non-liposomal doxorubicin. J Control Release 232:255–264

Quail DF, Joyce JA (2017) The microenvironmental landscape of brain tumors. Cancer Cell 31(3):326–341

Rempe R, Cramer S, Huwel S, Galla HJ (2011) Transport of Poly(n-butyl cyanoacrylate) nanoparticles across the blood-brain barrier in vitro and their influence on barrier integrity. Biochem Biophys Res Commun 406(1):64–69

Rip J (2016) Liposome technologies and drug delivery to the CNS. Drug Discov Today Technol 20:53–58

Rip J, Appeldoorn C, Manca F, Dorland R, van Kregten J, Gaillard P (2010) Receptor-mediated delivery of drugs across the blood–brain barrier. In: Frontiers in pharmacology. Conference abstract (2010) pharmacology and toxicology of the blood–brain barrier: state of the art, needs for future research and expected benefits for the EU

Rip J, Chen L, Hartman R, van den Heuvel A, Reijerkerk A, van Kregten J, van der Boom B, Appeldoorn C, de Boer M, Maussang D, de Lange EC, Gaillard PJ (2014) Glutathione PEGylated liposomes: pharmacokinetics and delivery of cargo across the blood-brain barrier in rats. J Drug Target 22(5):460–467

Sanhai WR, Sakamoto JH, Canady R, Ferrari M (2008) Seven challenges for nanomedicine. Nat Nanotechnol 3(5):242–244

Sarkaria JN, Hu LS, Parney IF, Pafundi DH, Brinkmann DH, Laack NN, Giannini C, Burns TC, Kizilbash SH, Laramy JK, Swanson KR, Kaufmann TJ, Brown PD, Agar NYR, Galanis E, Buckner JC, Elmquist WF (2018) Is the blood-brain barrier really disrupted in all glioblastomas? A critical assessment of existing clinical data. Neuro Oncol 20(2):184–191

Shakeri S, Ashrafizadeh M, Zarrabi A, Roghanian R, Afshar EG, Pardakhty A, Mohammadinejad R, Kumar A, Thakur VK (2020) Multifunctional polymeric nanoplatforms for brain diseases diagnosis, therapy and theranostics. Biomedicines 8(1)

Shih AY, Mateo C, Drew PJ, Tsai PS, Kleinfeld D (2012) A polished and reinforced thinned-skull window for long-term imaging of the mouse brain. J Vis Exp 61

Strazielle N, Ghersi-Egea JF (2013) Physiology of blood-brain interfaces in relation to brain disposition of small compounds and macromolecules. Mol Pharm 10(5):1473–1491

Szebeni J, Alving C, Baranyi L, Bunger R (2010) Interaction of liposomes with complement leading to adverse reactions. In: Gregoriadis G (ed) Liposome technology – volume III interactions of liposomes with the biological milieu, 3rd edn. Informa Healthcare USA, Zug

Szebeni J, Muggia F, Gabizon A, Barenholz Y (2011) Activation of complement by therapeutic liposomes and other lipid excipient-based therapeutic products: prediction and prevention. Adv Drug Deliv Rev 63(12):1020–1030

Szebeni J, Bedocs P, Rozsnyay Z, Weiszhar Z, Urbanics R, Rosivall L, Cohen R, Garbuzenko O, Bathori G, Toth M, Bunger R, Barenholz Y (2012) Liposome-induced complement activation and related cardiopulmonary distress in pigs: factors promoting reactogenicity of Doxil and AmBisome. Nanomedicine 8(2):176–184

Tapeinos C, Battaglini M, Ciofani G (2017) Advances in the design of solid lipid nanoparticles and nanostructured lipid carriers for targeting brain diseases. J Control Release 264:306–332

Tazina E, Kostin K, Oborotova N (2011) Specific features of drug encapsulation in liposomes (a review). Pharm Chem J 45(8):481–490

Tosi G, Musumeci T, Ruozi B, Carbone C, Belletti D, Pignatello R, Vandelli MA, Puglisi G (2016) The "fate" of polymeric and lipid nanoparticles for brain delivery and targeting: strategies and mechanism of blood–brain barrier crossing and trafficking into the central nervous system. J Drug Deliv Sci Technol 32:66–76

Ulbrich K, Hekmatara T, Herbert E, Kreuter J (2009) Transferrin- and transferrin-receptor-antibody-modified nanoparticles enable drug delivery across the blood-brain barrier (BBB). Eur J Pharm Biopharm 71(2):251–256

van Tellingen O, Yetkin-Arik B, de Gooijer MC, Wesseling P, Wurdinger T, de Vries HE (2015) Overcoming the blood-brain tumor barrier for effective glioblastoma treatment. Drug Resist Updat 19:1–12

Wan X, Zheng X, Pang X, Pang Z, Zhao J, Zhang Z, Jiang T, Xu W, Zhang Q, Jiang X (2016) Lapatinib-loaded human serum albumin nanoparticles for the prevention and treatment of triple-negative breast cancer metastasis to the brain. Oncotarget 7(23):34038–34051

Wang S, Huang R (2019) Non-viral nucleic acid delivery to the central nervous system and brain tumors. J Gene Med 21(7):e3091

Wicki A, Witzigmann D, Balasubramanian V, Huwyler J (2015) Nanomedicine in cancer therapy: challenges, opportunities, and clinical applications. J Control Release 200:138–157

Wilhelm I, Fazakas C, Krizbai IA (2011) In vitro models of the blood-brain barrier. Acta Neurobiol Exp (Wars) 71(1):113–128

WMA (2013) World Medical Association Declaration of Helsinki: ethical principles for medical research involving human subjects. JAMA 310(20):2191–2194

Wolf SM, Jones C (2011) Designing oversight for nanomedicine research in human subjects: systematic analysis of exceptional oversight for emerging technologies. J Nanopart Res 13(4):1449–1465

Wong HL, Wu XY, Bendayan R (2012) Nanotechnological advances for the delivery of CNS therapeutics. Adv Drug Deliv Rev 64(7):686–700

Yu S, Xu X, Feng J, Liu M, Hu K (2019) Chitosan and chitosan coating nanoparticles for the treatment of brain disease. Int J Pharm 560:282–293

Zensi A, Begley D, Pontikis C, Legros C, Mihoreanu L, Wagner S, Buchel C, von Briesen H, Kreuter J (2009) Albumin nanoparticles targeted with Apo E enter the CNS by transcytosis and are delivered to neurones. J Control Release 137(1):78–86

Zensi A, Begley D, Pontikis C, Legros C, Mihoreanu L, Buchel C, Kreuter J (2010) Human serum albumin nanoparticles modified with apolipoprotein A-I cross the blood-brain barrier and enter the rodent brain. J Drug Target 18(10):842–848

Zhu Y, Liu C, Pang Z (2019) Dendrimer-based drug delivery systems for brain targeting. Biomolecules 9(12)

Zucker D, Marcus D, Barenholz Y, Goldblum A (2009) Liposome drugs' loading efficiency: a working model based on loading conditions and drug's physicochemical properties. J Control Release 139(1):73–80

Chapter 17
Transport of Transferrin Receptor-Targeted Antibodies Through the Blood-Brain Barrier for Drug Delivery to the Brain

Torben Moos, Johann Mar Gudbergsson, and Kasper Bendix Johnsen

Abstract Entering a new era of using biologics for the treatment of diseases within the CNS, a major adjoining advance is the recent conquering of the years-long obstacle that has prevented large molecules from passing the blood-brain barrier (BBB). The BBB, consisting of brain capillary endothelial cells (BCECs) with their intermingling tight junctions, is now passable taking approaches where antibodies specifically target nutrient transporters to enter the brain. Among the nutrient transporters, the receptor for iron-carrying transferrin of the blood plasma is of particular interest, as this receptor is selectively expressed by BCECs and not by endothelial cells elsewhere in the body. Injecting antibodies targeted to the transferrin receptor of BCECs into the circulation leads to a preferential high uptake by BCECs, but the antibodies will only pass through the BBB and enter the brain when experimentally engineered to differ from the conventional binding characteristics of being monovalent with capability of high-binding affinity. Accordingly, antibodies either made bispecific or weakened in their affinity for the transferrin receptor appear able to pass the BBB, allowing such antibodies to achieve therapeutic activity within the brain either by direct antibody-mediated binding to specific targets or by delivering conjugate therapeutics. These new approaches for targeted delivery to the BBB and beyond open up new possibilities for the treatment of diseases within the CNS, and the potential for such treatments, particularly as applied to neurodegenerative diseases where the BBB is not compromised, probably never has been more optimal.

T. Moos (✉) · J. M. Gudbergsson
Neurobiology Research and Drug Delivery, Department of Health Science and Technology, Aalborg University, Aalborg, Denmark
e-mail: tmoos@hst.aau.dk; jmg@hst.aau.dk

K. B. Johnsen
Department of Health Technology, DTU Health Tech, Technical University of Denmark, Lyngby, Denmark
e-mail: kasjoh@dtu.dk

© American Association of Pharmaceutical Scientists 2022
E. C. M. de Lange et al. (eds.), *Drug Delivery to the Brain*, AAPS Advances in the Pharmaceutical Sciences Series 33, https://doi.org/10.1007/978-3-030-88773-5_17

Keywords Large molecules · Affinity · pH-sensitive · Bi-specific · Fc-fragment · Transcytosis

Abbreviations

BBB	Blood-brain barrier
BCEC	Brain capillary endothelial cell
CAM	Cell adhesion molecule
CNS	Central nervous system
CSF	Cerebrospinal fluid
DMT1	Divalent metal transporter 1 (DMT1)
IgG	Immunoglobulin
Lrp1	Low-density lipoprotein receptor-related protein 1
TfR1	Transferrin receptor 1
TfR2	Transferrin receptor 2

17.1 Introduction

The blood-brain barrier (BBB), consisting of brain capillary endothelial cells (BCECs) with their interconnecting tight junctions, protects foreign substances from entering the brain, while simultaneously maintaining the environment of the brain interior at a constant level to sustain low levels of extracellular solutes (Abbott 2013). The BCECs differ from endothelial cells of the remaining body by means of five grand characteristics, i.e., high transcellular electric resistance (TEER), low passive paracellular permeability, diminished transendothelial vesicular transport, high expression of luminal and abluminal efflux transporters, and low expression of cell adhesion molecules (Daneman and Prat 2015).

In protecting the entry of unwanted substances from the periphery, the BBB also forms a major restraint in hindering the passaging of pharmaceuticals from the blood to the brain. This is clearly unfortunate, as otherwise potent therapeutics are excluded from acting within the brain. So potent is the exclusion of pharmaceuticals by the BBB that roughly only as little as 1–2% of available CNS active drugs can enter the brain (Pardridge 2012). Such pharmaceuticals, e.g., barbiturates and benzodiazepines, will mainly be able cross the BBB by free diffusion due to their high lipophilicity and low molecular weight, hence fitting into Lipinski's rule of five, which predicts that molecules are able to cross any biological membrane by diffusion, if they have a molecular weight below 500 g/mol (Lipinsky et al. 1997). Specifically, for the membranes of BCECs, their passaging of small molecules is probably even less, hence allowing only molecules smaller than 400 g/mol to cross the BBB (Pardridge 2012). Larger and more hydrophilic pharmaceuticals undergo active transport through the BBB but also at very limited rates, which occur by mechanisms involving carrier-mediated transport following binding to BCEC cell surface components, e.g., transporters of large and neutral amino acids (Abbott et al. 2009).

Large molecules sized above 1 kDa undergo receptor-mediated endocytosis by the BCECs after binding to luminally expressed receptors with affinity for endogenous molecules such as apolipoprotein E (ApoE), insulin, and transferrin (Abbott et al. 2009; Kreuter 2014). These receptors can also be targeted by specific antibodies injected into the peripheral circulatory system (Lichota et al. 2010). Recent publications have revealed that antibodies with lower affinity or avidity targeting the targeting receptor not only get taken up by the brain endothelium but also undergo transcytosis across the brain endothelium, i.e., complete BBB transport (Yu et al. 2011; Niewoehner et al. 2014; Hultquist et al. 2017; Sonoda et al. 2018; Stocki et al. 2020; Kariolis et al. 2020; Sehlin et al. 2020; Ullman et al. 2020). Weakened in their binding to epitopes expressed by the BCECs, such antibodies are highly relevant for drug delivery to the brain by direct conjugation of therapeutic proteins, e.g., enzymes used for replacement therapies in lysosomal storage diseases (Kariolis et al. 2020; Ullman et al. 2020). They must however be injected at higher doses compared to high-affinity antibodies to enable significant uptake by BCECs and subsequent transport across the BBB (Johnsen et al. 2019a).

Antibodies developed to enable drug transport through the BBB were mainly studied in rodents and raised against transporters of nutrients like the large amino acid transporter, insulin receptor, glucose transporter 1 (solute carrier family 2, facilitated glucose transporter member 1, or SLC2A1) (Chih et al. 2016; Zuchero et al. 2016), transferrin receptor (Yu et al. 2011; Niewoehner et al. 2014; Hultquist et al. 2017; Sonoda et al. 2018; Stocki et al. 2020), or low-density lipoprotein receptor-related protein 1 (Lrp1) with affinity for ApoE (Zuchero et al. 2016). The insulin receptor was used for targeting purposes in the nonhuman primate brain (Pardridge 2012). However, uptake and subsequent transcellular transport through the BBB was mainly observed in the rodent brain for antibodies targeting the large amino acid transporter, glucose transporter 1, and transferrin receptor, and hardly for antibodies targeting the insulin receptor or Lrp1 (Zuchero et al. 2016). Other attempts for antibody transport at the BBB were made by injecting antibodies targeting cell adhesion molecules (CAMs) like intercellular cell adhesion molecule (ICAM aka CD54) and vascular cell adhesion molecule 1 (VCAM-1 aka CD106) (Manthe et al. 2020; Marcos-Contreras et al. 2020).

Taking a proteomic approach and subtracting data of BCECs by those of liver and pulmonary endothelial cells identified the transferrin receptor, CD98hc, Glut1, and basigin (CD147) as particularly promising targets for antibody-mediated approaches to enable specific transport across the BBB (Zuchero et al. 2016). The study of Zuchero et al. (2016), which is among the very few studies to compare the uptake of antibodies raised against specific epitopes, identified anti-CD98hc, and anti-transferrin receptor antibodies as having the highest uptake by the brain. Anti-transferrin receptor antibodies weakened in affinity top the uptake leading to the appearance of 0.6% of the injected dose/gram brain weight appearing in brain 24 h after injecting the antibodies in a dose of 20 mg/kg (Yu et al. 2011).

It stands out a big paradox that targeting to the transferrin receptor finally proved the most favorable suitable route for circumventing the impermeability of the BBB to enable large molecules to enter the brain. Mainly because the transferrin receptor has been regularly exploited for targeting purposes during the past three decades without clear evidences of enabling transport through the BBB (c.f. Lichota et al. 2010; Johnsen et al. 2019a) and also because current biotechnology allowing large-scale examinations of protein specificity and expression indeed have identified other novel candidates for targeting purposes and drug delivery strategies (e.g., Zuchero et al. 2016). However, identified targeting molecules have been unable to combat the quantitative uptake of antibodies targeting the transferrin receptor. Here we review the transferrin receptor for drug delivery purposes at the BBB in the healthy and pathological brain.

17.2   The Transferrin Receptor

The transferrin receptor family consists of two different members sharing 45% homology. The transferrin receptor 1 (TfR1 or CD71, encoded by the TFRC gene) is the canonical transferrin receptor with transferrin receptor 2 (TfR2) being a later identified homologous receptor with a 25 times lower affinity for transferrin as compared to TfR1, but nonetheless also capable of delivering iron-containing transferrin to cells (Trinder and Baker 2003). The significance of the transferrin receptor 2 for iron delivery in the normal brain is probably of less importance as this receptor tends to localize in inflammatory cells only in the pathological brain (Mastroberardino et al. 2009; Heidari et al. 2016). Perturbation of the transferrin receptor 2 gene causes brain pathology with iron accumulation widespread in the brain, which may relate to the handling of iron by microglia (Heidari et al. 2016). Likewise, evidence prevails showing that the elivery of iron to the mitochondria by the transferrin receptor 2 is perturbed in Parkinson's disease (Mastroberardino et al. 2009). The present review focusses on transferrin receptor 1 and uses the term transferrin receptor synonymously.

The transferrin receptor is a transmembrane glycoprotein that consists of two 90 kD subunits linked together by disulfide bonds. Each of these subunits can bind a single transferrin molecule that contains two specific high-affinity binding sites for ferric iron (Morgan 1996). The transferrin receptor is mainly localized luminally in the cellular membrane to ensure uptake of iron-containing transferrin (aka holo-transferrin) present in blood plasma. Binding of iron-containing transferrin to the transferrin receptor at the luminal membrane leads to endocytosis in clathrin-coated pits, which subsequently are uncoated and transform into endosomes. Within the endosome, the pH is around pH 5.5–6.5 maintained by an H1-ATPase that exchanges protons with sodium across the endosomal membrane. This lowering of the pH leads to uncoupling of iron from transferrin. Iron is then reduced and subsequently transported across the endosomal membrane mediated by divalent metal transporter 1 (DMT1), while the resulting iron-free transferrin (aka apo-transferrin) loses its affinity for the transferrin receptor and is released into the cellular exterior, as the endosome recycles and fuses with the cellular membrane (Morgan 1996; Chua et al. 2007).

The cellular need for iron is particularly high when cells are mitotically active (Laskey et al. 1988), and the expression of the transferrin receptor is accordingly high in dividing cells like cells of the bone marrow, gastrointestinal tract, and skin. In the bone marrow, the transferrin receptor expression is particularly high in red cell precursors. Moreover, the transferrin receptor is highly expressed in the nondividing cells of the brain. This includes the cell types forming the BBB (i.e., BCECs) and the blood-CSF barriers (i.e., choroid plexus epithelial cells) (Moos 1996). The concentration of transferrin receptor protein in BCECs varies between rodents and man ranging from 5.2 pmol/mg protein (mouse) and 7.8 pmol/mg protein (rat) to 2.3 pmol/mg protein (man) (Pardridge 2020). The high expression of transferrin receptors by BCECs, however, is not necessarily reflected in the physiological iron uptake, as BCECs contain a prominent pool of spare receptors available for recruitment, e.g., in conditions with low iron availability (van Gelder et al. 1995, 1997). Inside the brain, neurons are also highly transferrin receptor-expressing, whereas only a slight expression of transferrin receptors is seen in glial cell types, e.g., astrocytes, oligodendrocytes, and microglia (Moos 1996). The expression of transferrin receptors in BCECs, choroid plexus, and neurons continues throughout life (Moos and Morgan 2001) and probably reflects a life-long need for iron acquisition by the brain (Dallman et al. 1975; Dallman and Spirito 1977; Yang et al. 2020). The transferrin receptor has also affinity for another major iron-binding protein, ferritin, which is capable of binding around 5000 iron atoms per ferritin molecule (Morgan 1996; Chua et al. 2007). Cryo-electron microscopy analyses have provided evidence showing that ferritin also binds to the transferrin receptor but at a site different from that of transferrin (Montemiglio et al. 2019). A ferritin nanocarrier was demonstrated to pass cerebral capillaries in glioma tissue (Fan et al. 2018), but it remains to be clarified if the binding of ferritin to the transferrin receptor in vivo also leads to internalization and subsequent iron transport across the BBB (Fisher et al. 2007), and if so, how the quantitative aspect of such iron transport will compare to that mediated by holo-transferrin.

17.3   The Significance of the Transferrin Receptor for Iron Uptake and Transport at the Blood-Brain Barrier

Hemochromatosis caused by abnormally high uptake of iron in the gut often surpasses the iron-binding capacity of transferrin in blood plasma, which leads to the presence of non-transferrin bound iron (Chua et al. 2007; Heidari et al. 2016). The exclusive binding of iron to transferrin in blood plasma in physiological conditions with no signs of hemochromatosis ensures that iron only enters BCECs bound to transferrin (Moos et al. 2000). The succeeding intracellular transport has been much more clarified concerning the ligand-receptor interaction between transferrin and its receptor. Therefore, the interaction of the transferrin receptor and transferrin receptor-targeted ligands such as transferrin are described separately from the interaction between the transferrin receptor and targeted antibodies, which is covered in Sect. 17.4.

Initiating the uptake and transport into BCECs, transferrin receptor-targeted ligands and antibodies both attach to the luminally expressed transferrin receptor, which is followed by receptor internalization and formation of endosomes (Broadwell et al. 1996). In the highly polarized BCECs, the transferrin receptor is prominently expressed on the luminal side, whereas several attempts to determine the presence of the transferrin receptor on the abluminal side of the BCECs have been largely unsuccessful. Many experiments have focused on transport through BCECs of antibodies targeting the transferrin receptor, whereas the morphological detection of the transferrin receptor at the subcellular level within the BCECs in vivo remains clearly understudied and has only been addressed in just a few studies (Roberts et al. 1993; Broadwell et al. 1996; Simpson et al. 2015). Using pre-embedding immuno-electron microscopy, the study of Roberts et al. (1993) observed that the transferrin receptor is localized to vesicular structures at the luminal side without signs of the receptor being approximated to the abluminal side of the BCECs. A possible limitation of this pre-embedding technique is the lack of penetration of the primary antibody during the staining procedure, and probably the post-embedding immuno-detection of frozen sections would be more sensitive for detection of epitopes hidden in, e.g., organelles, situated deeper within the cell. A more recent immuno-electron microscopy study by Simpson and Connor did observe transferrin receptors on the abluminal surface of the brain endothelium in vivo both in wild-type Sprague-Dawley rats and in the Belgrade rat model (DMT1 mutation) of iron deficiency (Simpson et al. 2015). Nonetheless, supporting the notion of a preferential localization of the transferrin receptor predominantly confined to the luminal side of the BCECs, confocal imaging of primary rat BCECs co-cultured in vitro with rat astrocytes under polarized condition revealed a subcellular localization of transferrin receptor at the luminal side above the cellular nucleus (Burkhart et al. 2016; Johnsen et al. 2017). Furthermore, morphological attempts to detect the transport of human transferrin with high affinity for the rat transferrin receptor (Taylor and Morgan 1991) also failed with the transferrin being confined to the BCECs without passage through the BBB (Moos and Morgan 2004).

The conclusions of the morphological studies on the transport of transferrin and its receptor within the BCECs, which all stand out quite qualitatively, obtained quantitative support from studies using radioisotopes. This aspect was initially addressed by Evan Morgan and co-workers, who took advantage of the simultaneous examination of iron and transferrin (e.g., Taylor and Morgan 1991; Taylor et al. 1991; Crowe and Morgan 1992). This allowed for reliable estimates of the relative presence of both compounds within the brain, and evidently, results showed that the transport of radioactive iron through the BBB and blood-CSF barrier by far outnumbered that of radiolabeled transferrin even just a few hours after intravenous injection (Fig. 17.1). The difference could not be explained by the degradation of

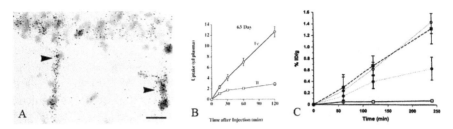

Fig. 17.1 (a) Transferrin receptor mRNA expression in neonatal brain capillary endothelial cells (arrowheads) as shown by in situ hybridization. Transferrin receptor mRNA is also detectable in the developing pia-arachnoid as shown in the upper part of the illustration. Data obtained in collaboration with Dr. P.S. Oates, University of Western Australia. Scale bar = 20 μm (Moos et al. 2007). (b) Iron (Fe) and transferrin (Tf) to the brain in 63-day-old rats at different times after the intravenous injection of radiolabeled $^{59}Fe/^{125}I$-Tf. Notice the much higher uptake of iron compared to transferrin (Moos and Morgan 2001; data derived from studies of Crowe and Morgan 1992). (c) High-affinity, bivalent anti-transferrin receptor antibody (OX26) uptake in brains of differently aged rats expressed as % ID/g weight ±SD. The uptake of OX26 is age-dependent, being far-fold highest in P15 brains and lowest in adult rats. Iron deficiency does improve uptake of OX26 in P15 rats. □, adult rat; ■, P15 normal rat; O, P15 iron-deficient rat; ●, P0 rat (Moos and Morgan 2001)

transferrin within the brain (Strahan et al. 1992). Similar observations were obtained in independent experiments of another research group (Morris et al. 1992). These data led to a general conclusion that iron-containing transferrin is typically taken up by receptor-mediated endocytosis at the luminal membrane of brain capillaries, followed by the dissociation of iron from transferrin within endosomal compartments, with subsequent iron transport across the abluminal lipid bilayer of BCECs, whereas transferrin is retro-endocytosed back to the luminal membrane (Taylor et al. 1991; Morris et al. 1992).

Counteracting the notion of receptor-mediated endocytosis of transferrin at the BBB and providing some evidence favoring receptor-mediated transcytosis of transferrin, other studies have suggested that transferrin may in fact be transferred through BCECS in vivo under some conditions (e.g., Fishman et al. 1987; Bickel et al. 1994; Skarlatos et al. 1995; Broadwell et al. 1996). As a major caveat in the interpretations of these data for understanding the mechanisms of iron and transferrin transport at the BBB, these studies unfortunately solely focused on transferrin internalization and trafficking, hence leaving out the possibility of comparing transferrin transport with iron transport.

Using an animal model of functional DMT1 deficiency that prevents iron from being transferred across the endosomal membrane, iron uptake was clearly reduced when examined in isolated brain capillaries, which indicated a functional role of DMT1 for ferrying iron through BCECs without transferrin (Moos and Morgan 2004). A later study provided evidence for the expression of ferric reductase, which is necessary for the reduction of ferric iron when uncoupled from transferrin inside the endosome (Burkhart et al. 2016). Combined with the simultaneous detection of DMT1, the presence of ferroportin necessary for directed efflux of ferrous iron across the abluminal membrane of BCECs and the existence of ferrous oxidase activity by BCECs, astrocytes, and pericytes have enabled modeling to estimate iron transport through BCECs independent of transferrin binding, where iron-free transferrin is envisioned to be recycled to the luminal membrane (Johnsen et al. 2019a). Ferroportin is a candidate for the transport of iron through this abluminal membrane, as it facilitates the export of non-transferrin-bound iron (Burdo et al. 2001; McCarthy and Kosman 2013). Ferroportin is also of interest as a putative regulator of iron transport into the brain, as it acts as a receptor for hepcidin, a 25-amino acid peptide normally secreted into the circulation by hepatocytes, but which may also possibly be expressed by astrocytes and microglia (Urrutia et al. 2013). Hepcidin binding to ferroportin ultimately leads to the internalization of the complex and ferroportin degradation, allowing hepcidin to act as a negative regulator of the total body iron influx across the gut. Another obvious possible role for hepcidin is in the regulation of iron transport to the brain via degradation of ferroportin at the abluminal membrane of brain endothelial cells (Simpson et al. 2015); however, hepcidin appears to be minimally expressed in the normal brain (Burkhart et al. 2016). Hepcidin may well play a role for limiting iron transport across the BBB in the pathological brain with concomitant neuroinflammation (Urrutia et al. 2013; Thomsen et al. 2015).

The weight of evidence from comparing iron and transferrin transport at the BBB suggests that iron is typically transferred across the abluminal membrane without the participation of transferrin, which more commonly undergoes recycling back to the luminal side (Johnsen et al. 2019a). There may nonetheless exist a type of directed transport of transferrin across the BBB simulating transcytosis: The transferrin receptor expression by BCECs is developmentally upregulated in the early postnatal brain (Moos and Morgan 2001), and the transferrin transport into the brain is accordingly higher following the injection of radioactive iron bound to transferrin in the developing brain than seen at later ages (Taylor and Morgan 1991). The transport of transferrin is also higher than that of albumin but still magnitudes lower than that of iron (Taylor and Morgan 1991; Morgan and Moos 2002). The cytoplasmic width of BCECs is approximately 200–400 nm. Many more endocytic vesicles, typically sized about 70 nm in diameter, are present in BCECs of the developing brain (Kniesel et al. 1996), so it follows that if even just a fraction of these vesicles by random fuse with the cellular membrane of the BCECs, there would a priori be more vesicles emptying their content into the brain during development than in adulthood. Endocytic vesicles are formed as part of transferrin receptor docking at the luminal side of the BCECs, but the transferrin attachment will also nonspecifically capture fluids from the extracellular space of the luminal side, which could explain why albumin is also transferred through the BBB to a higher degree in the developing brain (Morgan and Moos 2002) despite the fact that the BBB is already well formed without significant admission of paracellular transport at the onset of vasculogenesis in the embryological period (Møllgård et al. 1988; Kniesel et al. 1996).

17.4   Targeting the Transferrin Receptor for Transport of Anti-transferrin Receptor Antibodies Across the Blood-Brain Barrier

The concept of targeting specific receptors in the body purposely to obtain a precise effect without causing unwanted side effects ranges back to the hypothesis on the magic bullet brought forward by Paul Ehrlich in 1909 (Bosch and Rosich 2008). Nowadays, specific targeting relates to the passive immunization that occurs when performing immunotherapy in patients with antibodies directed against a specific protein (Kumar et al. 2018). Concerning immunotherapy to the brain, the concept has undergone significant modifications as specific antibodies targeting proteins in the brain in several studies have failed to yield therapeutic efficacy due to their incapability to pass the BBB (Pardridge 2020). This has spawned much research and development toward the production of bispecific antibodies, F(ab')-conjugated fusion antibodies (Fig. 17.2), hence taking advantage of engineered transport vehicles that may be capable of efficient transcytosis across the BBB and significantly increased brain exposure directed toward a number of different therapeutic targets (e.g., Yu et al. 2014; Niewoehner et al. 2014; Campos et al. 2020; Kariolis et al. 2020; Ullman et al. 2020).

The transferrin receptor has had a central place in a plethora of attempts to enable antibodies, enzymes, and other large molecule therapeutics to target and undergo transport through the BBB. The transferrin receptor expressed by BCECs was initially targeted with monoclonal antibodies in an original study by Jefferies and coworkers, who not only targeted the BCECs but also discovered that brain endothelial cells form the sole endothelial layer in the body that expresses transferrin receptors (Jefferies et al. 1984). Evidently, targeting to the transferrin receptor proves more efficient when using specific antibodies in comparison with transferrin, its natural ligand, because endogenous transferrin is present in blood plasma in high concentration due to continuous hepatic secretion, which outcompetes the binding of transferrin-drug conjugates (Pardridge 2012). An important aspect of this strategy is the ability of antibodies and other binders to target the apical domain of the transferrin receptor; the apical domain does not bind transferrin, so apical transferrin receptor targeting therefore excludes competition for binding by circulating endogenous transferrin (Testi et al. 2019).

The antibody-directed targeting approach to the transferrin receptor of the BCECs has been supported by several experimental studies using intravascular injection of antibodies to the transferrin receptor conjugated with various therapeutics in order to facilitate their transport through the BBB. These approaches, which aided to understand the possible differences and interactions between the binding of monoclonal anti-transferrin receptor and transferrin to the transferrin receptor, have contributed to bridge the fields of drug delivery to the brain and cerebral iron homeostasis (Ueda et al. 1993; Bradbury 1997; Moos et al. 2007; Freskgård and Urich 2017; Villasenor et al. 2019).

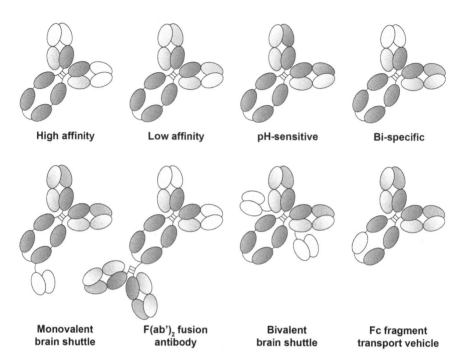

Fig. 17.2 A panel of modified anti-transferrin receptor antibodies generated to enable targeted uptake and transport through the BBB. *Upper row*: the classical high-affinity, bivalent anti-transferrin receptor antibody that does not cross the BBB (Moos and Morgan 2001) with transferrin receptor binding indicated in yellow. Next, low-affinity (orange), pH-sensitive (pink), and bispecific bivalent anti-transferrin receptor (yellow) antibodies that crosses the BBB; the latter also showing therapeutic target engagement with amyloid (green) (Yu et al. 2011, 2014). *Lower row*: antibody conjugated to a single high-affinity, monovalent transferrin receptor binding moiety (yellow) aka "brain shuttle" (Nievohner et al. 2014). F(ab')2 fusion and bivalent brain shuttles binding the transferrin receptor (yellow) and used in recent studies to target amyloid inside the brain (green) (e.g., Sehlin et al. 2016; Hultqvist et al. 2017). Right: The Fc-fragment transport vehicle where the transferrin receptor binding moiety (yellow) is moved to the Fc-near region allowing for two spare Fab parts for therapeutic binding or carrying therapeutic molecules into the brain (Kariolis et al. 2020; Ullman et al. 2020)

Early concerns centered around whether the putative high uptake of anti-transferrin receptor antibodies in the brain (Frieden et al. 1991) was simply reflecting a high concentration of antibody within BCECs rather than within the brain itself and thus calling into question whether BBB transport actually occurs (Moos et al. 2000). Such concerns inspired new investigations on the fate of monoclonal anti-transferrin receptor antibodies injected into the periphery and using an extended protocol that included developing rats with a far-fold higher expression of transferrin receptor than seen in the adult brain and the examination of isolated brain capillaries; some of these earlier studies concluded that high-affinity anti-transferrin receptor antibodies concentrate in BCECs without further passaging across the BBB (Moos and Morgan 2001). Later attempts to transfer high-affinity anti-transferrin receptor antibodies through the BBB also failed, again resulting in the antibodies being retained within the BCECs (Manich et al. 2013; Paris-Robidas et al. 2011; Alata et al. 2014). These high-affinity antibodies likely form a strong covalent interaction with the binding epitope of the transferrin receptor, preventing further detachment of the antibody from its receptor and hence making further movement of the antibody across the BBB unlikely. These observations were contrasted in ultrastructural studies made in the rat injected with high-affinity monoclonal anti-transferrin receptor antibodies, which demonstrated the presence of antibody in vesicular structures of the BCECs and even also in exocytic vesicles that appeared to establish membrane continuity with the abluminal plasmalemma suggestive of transcytosis (Broadwell et al. 1996). Unfortunately, the absence of quantitation of this observation made it very hard to adequately assess its significance, but, as suggested above, there could be a minor fraction that actually fuses with the abluminal side of the BCECs albeit in an amount too low to enable detection in most quantitative studies.

Later studies using similar or identical high-affinity antibodies conjugated to liposomes showed virtual absence of these targeted liposomes in the brain, indicating they were unable to pass the BBB (Gosk et al. 2004; Johnsen et al. 2017). It was also demonstrated that the liposomal cargo can in some cases slip out of the liposomes and enter the brain, probably because of liposomal degradation within the BCECs (Johnsen et al. 2017). A more recent in vivo multiphoton imaging study used state-of-the-art methods to show that transferrin receptor-targeted liposomes may in certain cases exit the abluminal cellular membrane at the level of postcapillary venules, as opposed to capillary endothelial cells, and subsequently enter the perivascular space compartment of the brain (Kucharz et al. 2021). The significance of the latter finding clearly warrants further exploration in order to verify if the postcapillary venules could be a probable site for entry of targeted liposomes only, or whether this also occurs for targeted antibodies without conjugation to liposomes.

The perceived failure in obtaining transport of transferrin receptor-targeted antibodies through BCECs underwent a substantial reevaluation following the introduction of low-affinity monospecific antibodies for drug delivery to the brain (Yu et al. 2011). Injected into the peripheral circulatory system, antibodies modified in their binding affinity to the transferrin receptor led to transport into the brain with accumulation in neurons, hence supporting the notion of transport through the BBB (Yu et al. 2011) (Fig. 17.2). A follow-up study utilized the synthesis of bispecific antibodies with one domain of the antibody's antigen-binding domain targeting the transferrin receptor at high affinity and another directed toward a putative therapeutic target, beta-secretase 1, for the reduction of amyloidogenic peptide formation of relevance in treating Alzheimer's disease (Yu et al. 2014). This bispecific approach (and other similar monovalent transferrin receptor-targeted antibodies) not only led to the transport across the BBB but also reduced the formation of amyloid (Yu et al. 2011; Bien-Ly et al. 2014; Niewoehner et al. 2014), a key pharmacodynamic readout. Mechanistically, the brain entry may have been facilitated by the antibodies reversibly binding to transferrin receptors within the BCECs, thereby allowing the antibodies to detach from their binding epitope within the slightly acidic pH environment of the endosomes (Bien-Ly et al. 2014; Niewoehner et al. 2014). Monovalent interactions with the transferrin receptor also likely facilitated less trafficking to BCEC lysosomes, a known liability of bivalent transferrin receptor binding thought to at least partially be a consequence of receptor cross-linking. Importantly, evidence of antibody entry into the brain for these next-generation approaches was demonstrated by anti-amyloid engagement within the brain's extracellular space (Niewoehner et al. 2014) or the neutralization of the β-secretase enzyme BACE-1, which processes amyloid precursor protein into β-amyloid peptides via direct binding at the neuronal cellular membrane (Bien-Ly et al. 2014). Most importantly, these observations made in mouse models showing that amyloid production could be halted via bispecific antibodies were later confirmed in repeat studies conducted in nonhuman primates (Yu et al. 2014), suggesting cross-species relevance and hinting at their translational potential in clinical studies. Today, clinical trials are ongoing or about to be initiated (NCT04023994, NCT04251026).

As an alternative to bispecific antibody architectures (and other bivalent, high-affinity transferrin receptor binding antibodies), a novel new approach has been described whereby the Fc portion of immunoglobulin G (IgG) has been engineered to incorporate transferrin receptor binding, hence allowing the complete sparing of IgG's Fab domains for other binding purposes (Kariolis et al. 2020; Ullman et al. 2020). This preserves full capacity for therapeutic target engagement of the resulting proteins, termed transport vehicles, when expressed in an antibody format, while maintaining the characteristics of a single binding low-avidity antibody with respect to its binding to the transferrin receptor and the capability to cross the BBB (Niewoehner et al. 2014; Kariolis et al. 2020; Ullman et al. 2020). Importantly, an enzyme fusion protein incorporating these engineered transport vehicles is now under evaluation in the clinic for the treatment of mucopolysaccharidosis type II, a rare, inherited genetic disorder with frequently severe neurologic involvement (NCT04251026). Another interesting approach to target the transferrin receptor using modification of their affinity or avidity was introduced by single-domain antibodies, replacing the CDR3-binding domain of another species with acknowledged lower affinity for the transferrin receptor, leading to substantially higher antibody transport through the BBB in vivo (Sehlin et al. 2000; Stocki et al. 2020) and in vitro (Thom et al. 2018). Finally, recent studies have revealed that nanoparticles conjugated to bi-specific antibodies can cross the BBB due, in part, to lower-avidity interactions with the transferrin receptor (Johnsen et al. 2018, 2019b).

17.5 Pharmacokinetics of Therapeutics Targeting the Transferrin Receptor at the Blood-Brain Barrier

The pharmacokinetic profiles of large molecules, including antibody-targeted pharmaceutics, often present major challenges compared to small molecules when brain targeting is needed. Advantages of targeted pharmaceutics are denoted by their long half-life in circulation combined with a high surface permeability product, which allows for single-dose regimens with days to weeks in between dosing (Pardridge

2012). Related hereto, it is noteworthy that a number of companies in this space intend to prescribe brain-targeted antibodies or fusion proteins for intravenous injection only a single time weekly. Given the various approaches are all designed to target the transferrin receptor at the BBB, it is tempting to speculate if the transferrin receptor really internalizes their targeted compound with identical efficacy over time when antibodies circulate for longer time in the plasma. The extraction of the targeted antibody and further transport through the BCECs could possibly be higher, if an alternative dose regimen, e.g., using longer intervals between dosing, was used, allowing the transferrin receptor molecules of the BBB to be completely free of interacting antibodies before the next portion of antibodies were dosed.

17.6 Does Existing Knowledge on the Expression of the Transferrin Receptor at the BBB Allow for Improved Antibody Transfer Across the BBB?

The precise mechanisms for transport of transferrin receptor-targeted antibodies through the BBB remain unclear, in part due to an incomplete understanding of transcytosis and vesicular trafficking in BCECs. The criteria for transcytosis predict that the BCECs must enclose extracellular material by invagination of the cellular membrane to form a vesicle, which, when presented to the cells' abluminal side, leads to ejection of the material carried within the vesicle. This mechanism is unlikely to take place concerning iron-containing transferrin, as described above, because this would necessarily be expected to take iron and transferrin through the BBB in equal amounts. Despite the robust existence of sorting tubular systems within BCECs that can allow trafficking through the BBB via fusion between vesicles (Siupka et al. 2017; Villasenor et al. 2017), the thin diameter of the BCEC may nonetheless allow a small yet significant fraction of coated pit-containing vesicles to nonspecifically fuse with the abluminal membrane, hence giving the impression of transcytosis. This process would however be nonspecific and probably unlikely to be a target for improved vesicle release at the abluminal side that could be improved. Rather than true transcytosis, the transport through the endothelium of transferrin receptor-targeted therapeutics may be better characterized as a two-stage process, i.e., a blood-to-endothelium transport that is specific and directed from the luminal side toward the endosomes and a nonspecific endothelium-to-brain transport of which much more knowledge is necessary to learn whether and how this transport might be pharmacologically manipulated or enhanced (Gosk et al. 2004). The latter supposedly nonspecific transport process may well limit both the speed and capacity of transport across BBB (Freskgård and Urich 2017; Villasenor et al. 2019).

Reversal to an earlier developmental state for BCECs, while potentially risky with respect to physiological outcomes and unanticipated consequences, may provide another mechanism to improve transport of transferrin receptor-targeted pharmaceutics. The expression of the transferrin receptor by BCECs is far-fold higher in the developing postnatal brain than seen in adulthood (Moos and Morgan 2001; Lichota et al. 2010). The expression is higher earlier in development for at least two reasons: (i) the vascular bed is still actively proliferating in earlier postnatal periods, and (ii) there is a developmentally higher iron uptake in differentiating neurons and particularly with the proliferation of glial cells around birth (Moos and Morgan 2004). Another interesting fact is that the neurovascular unit is less mature at this developmental stage (Al Ahmad et al. 2011). The ingrowth of pericytes along the cerebral capillaries supposedly downscales the vesicular trafficking through the BCECs (Danemann and Prat 2015). The interactions between BCECs and pericytes that cause a lower vesicular trafficking may occur via direct membranous contacts and humoral secretion (Armulik et al. 2011). Lowering pericyte contacts with BCECs or reducing their supporting secretion of growth factors like basic fibroblast growth factor (bFGF) (Chen et al. 2020) could reverse brain endothelial cells to a more immature state with potentially a higher expression of the transferrin receptor, although it is hard to envision how such a manipulation would not come without other untoward consequences.

Iron deficiency leads to higher uptake of iron following intravascular injection of iron transferrin (Taylor et al. 1991). Interestingly however, this increased uptake in iron deficiency is not reflected in a higher expression of the transferrin receptor by BCECs, which tend to follow the developmental rather than nutritional stage (Moos et al. 1998; Moos and Morgan 2001). Combining the developmental stage with iron deficiency does not lead to higher expression of transferrin receptor than can be observed in developing rodents without iron deficiency, again identifying development as the most important regulator of transferrin receptors at the BBB. It may be that transferrin receptor expression closely follows a developmental pattern because of physiological iron deficiency present during earlier development, in effect requiring expansion of the system to its maximal capacity (Moos and Morgan 2001). Opposite to the developing brain, the aged brain likely lowers the expression of transferrin receptors (Yang et al. 2020), which is unfortunate as many patients in this age group are relevant for targeted therapy. Going forward, it will be important for the field to better understand age-dependent changes in transferrin receptor expression and their possible impact on transferrin receptor-targeted approaches for increasing therapeutic brain exposure. The BCECs do not upregulate their expression of transferrin receptors in Alzheimer's disease (Bourassa et al. 2019).

Looking more into the trafficking of transferrin receptor-containing vesicles within BCECs, the identification of a sorting tubular system that may be manipulated to traffic with higher efficiency toward the abluminal side is another promising approach that could aid the understanding of whether trafficking through the BCECs can be optimized for a particular pharmacological purpose (Villasenor et al. 2017). Studies on the identification of transferrin receptor-containing vesicles in co-cultures have also enhanced the understanding of trafficking within BCECs (Haqqani et al. 2018a, b). In all, the understanding of the mechanisms responsible for trafficking transferrin receptor-containing vesicles in BCECs and the possible transfer of the endothelium into a higher expressional state could enhance uptake of targeted transferrin receptor-targeted pharmaceutics and trafficking through the BBB.

17.7   The Fate of Transferrin Receptor-Targeted Antibodies Within the Brain

The use of targeted therapeutics to treat diseases within the brain is particularly relevant in chronic neurodegenerative diseases, e.g., Alzheimer's disease, Parkinson's disease, amyotrophic lateral sclerosis, and Huntington's disease, as well as in rare, inherited orphan diseases with clinical manifestations beginning in early life. Opposed to a commonly held belief claiming that the BBB is compromised in chronic neuropathology, there is really no solid evidence to claim that the passive permeability of the BBB is higher to circulating antibodies (Bien-Ly et al. 2015).

Our understanding of the specific requirements for antibody actions (and off-target effects) within the brain is still incomplete, and fundamentals like information about the number of antibodies that can enter the brain and how they distribute within the brain are important issues that remain up for debate. In passing the BBB, transferrin receptor-targeted antibodies of typical size (~10 nm; Wolak et al. 2015) can diffuse through the mesh formed by the vascular basement membrane and further through the extracellular space. Their clearance out of the brain microenvironment is much less well understood. Nontargeted antibodies are believed to undergo some degree of transport into the ventricular space across choroid plexus epithelial cells, where complex relationships govern to what degree CSF concentrations reflect the percentage of the dose entering the brain (Routhe et al. 2020). As part of the normal circulation of the CSF along major vessels of the brain, some fraction of antibodies entering into the CSF are expected to circulate into the brain once present in the subarachnoid space along cerebral perivascular pathways (Pizzo et al. 2018). However, limited diffusion into the brain from the CSF and additional potential barriers to antibody access from the perivascular spaces under normal conditions (Abbott et al. 2018; Pizzo et al. 2018) may effectively limit antibody exposure to the brain extracellular space and targeting to neurons. Furthermore, the clearance and turnover of CSF are expected to result in a significant antibody clearance as well. With systemic administration of transferrin receptor-targeted antibodies, transport at the microvessel level is expected to more efficiently allow for contact with neurons as compared to delivery from the CSF (Pardridge 2012). Furthermore, transferrin receptor-targeted antibodies with high affinity for neurons will likely only minimally appear in CSF because their exchange between brain interstitial fluid and CSF will be opposed by neuronal binding and internalization, making it difficult to predict the concentration within the brain.

Concerning pharmacodynamic properties, a fraction of anti-transferrin receptor-targeted antibodies that pass the BBB is likely to engage with the neuronal transferrin receptor and enter neurons (Fig. 17.3) (Kariolis et al. 2020; Stocki et al. 2020). The entry of targeted antibodies into neurons is not necessarily a preferred route and may hamper the intended use of antibodies to exert their action at other sites, e.g., in the extracellular space to neutralize deposition of protein aggregates like amyloid or prevent protein seeding like tau and alpha-synuclein, which all are thought to be very important to halt disease progression in Alzheimer's disease and Parkinson's

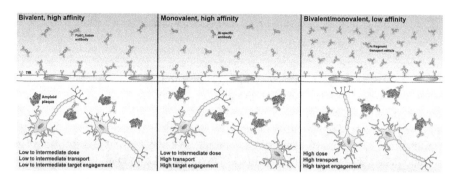

Fig. 17.3 Examples of the fate of engineered anti-transferrin receptor antibodies targeting the BBB. *Left*: Bivalent, high-affinity antibodies indeed engage at the luminal side of brain capillary endothelial cells but hardly undergo transport through the BBB. *Middle*: Monovalent, high-affinity antibodies like bispecific antibodies also engage at the luminal side of brain capillary endothelial cells and undergo substantial higher transport through the BBB

disease (Katsinelos et al. 2019; Visanji et al. 2019). Antibodies can also engage with proteins expressed in the neuronal membrane like BACE-1 (Kariolis et al. 2020), and they can deliver conjugate therapeutics, e.g., in diseases where proteins are lacking (Ullman et al. 2020). Ideally, multifunctional antibodies should be designed to ease the engagement with the intracerebral target by enabling a higher affinity relative to that of the transferrin receptor engagement. Principally, this would allow such antibodies to engage with their high-affinity neuronal target rather than neuronal transferrin receptors, which owing to their high neuronal expression in certain brain regions (e.g., the brain stem) could theoretically result in unwanted, off-target effects if transferrin receptor expression were to completely drive the biodistribution process (Hultquist et al. 2017)

17.8   Side Effects of Transferrin Receptor-Targeted Therapeutics Targeting the Blood-Brain Barrier

The putative side effects of targeted antibodies relates to their binding moieties. Concerning anti-transferrin receptor antibodies, possible unwanted effects within the CNS will relate to the binding to the neuronal transferrin receptors of neurons not relevant for disease treatment described above. While binding of the transferrin receptor expressed by BCECs could theoretically block the binding of transferrin and lower the transport of iron into the brain, nearly all transferrin receptor-targeted molecules in recent use have been designed so that their binding to the receptor occurs remote from that of the binding site for iron-containing transferrin (Yu et al. 2014; Kariolis et al. 2020).

The latter also relates to the otherwise possible binding of antibodies to transferrin receptors expressed on peripheral tissues, particularly red cell precursors of the bone marrow. The possibility that such antibodies might elicit anemia through the reduction of iron uptake in reticulocytes (e.g., via internalization and degradation of the transferrin receptor in these cells) or reticulocyte destruction (e.g., from antibody-dependent cell-mediated cytotoxicity mechanisms) has been a source of concern (Couch et al. 2013). However, independent studies that carefully monitored transferrin receptor-targeted antibody effects have shown that reticulocyte depletion may be minimized by lowering transferrin receptor affinity and by eliminating or attenuating effector function through the elimination or reduction of Fc-gamma receptor binding (Couch et al. 2013). An alternate approach changing the antibody's binding mode enabled the preservation of the binding and effector function in the brain while preventing peripheral effects caused by antibody infusion into the systemic circulation (Weber et al. 2018).

17.9 Points for Discussion

- The expression level of the transferrin receptor is many-fold higher in BCECs during development. It is fascinating to speculate that a possibly pharmacological prevention in the developmental downregulation of the transferrin receptor may enhance targeted delivery to the brain.
- Can the expression level of the transferrin receptor by the BCECs be increased, e.g., by the induction of a state of iron deficiency or by changing communication with other cell types of the neurovascular unit?
- There is a profound need for more information on the transport of anti-transferrin receptor antibodies at the BBB, particularly at the high-resolution level using techniques like two-photon microscopy and post-embedding immune-electron microscopy.
- Blocking the direction of anti-transferrin receptor antibodies toward lysosomes seemingly enhance their half-life in BCECs, but does this also result in higher transport into the brain?
- How do low-affinity anti-transferrin receptor antibodies undergo transport from BCECs and further transport into the brain, e.g., is the process saturable and will it be possible to enhance such transport?
- Might enhancement of fluid transport in the brain extracellular space allow further transport of antibodies from BCECs into the brain as part of a fluidic drag through the BBB?
- Does a first-pass effect exist when it comes to uptake of anti-transferrin receptor antibodies targeting the BCECs? In other words, might dosing with lower amounts given at more frequent intervals lead to higher uptake compared to a single bolus injection of a higher amount?
- To which extent will transport through the BBB of anti-transferrin receptor bispecific antibodies also bind to specific neurons or glia?

- The many approaches taken to change the affinity and/or avidity of targeted anti-transferrin receptor antibodies merit a head-to-head attempt to identify the superior formulation for transport across the BBB.
- It is important to improve understanding about how anti-transferrin receptor antibodies are transported through the brain extracellular space to enter the ventricles and subarachnoid space, as this phenomenon remains vaguely characterized.

Acknowledgments The most recent results obtained and described by the authors were generated by generous grants from the Lundbeck Foundation Research Initiative on Brain Barriers and Drug Delivery (Grant no. R155-2013-14113 and R392-2018-2266).

References

Abbott NJ (2013) Blood–brain barrier structure and function and the challenges for CNS drug delivery. J Inher Metab Dis 36:437–449

Abbott NJ, Patabendige AAK, Dolman DEM, Yusof SR, Begley DJ (2009) Structure and function of the blood–brain barrier. Neurobiol Dis 37:13–25

Abbott NJ, Pizzo ME, Preston JE, Janigro D, Thorne RG (2018) The role of brain barriers in fluid movement in the CNS: is there a 'glymphatic'' system? Acta Neuropathol 135:387–407

Al Ahmad A, Taboada CB, Gassmann M, Ogunshola OO (2011) Astrocytes and pericytes differentially modulate blood-brain barrier characteristics during development and hypoxic insult. J Cereb Blood Flow Metab 31:693–705

Alata W, Paris-Robidas S, Emond V, Bourasset F, Calon F (2014) Brain uptake of a fluorescent vector targeting the transferrin receptor: a novel application of in situ brain perfusion. Mol Pharm 11:243–253

Armulik A, Mäe M, Betsholtz C (2011) Pericytes and the blood-brain barrier: recent advances and implications for the delivery of CNS therapy. Ther Deliv 2:419–422

Bickel V, Kank Y-S, Yoshikawa T, Pardridge WM (1994) In vivo demonstration of subcellular localization of anti-transferrin receptor monoclonal antibody-colloidal gold conjugate in brain capillary endothelium. J Histochem Cytochem 42:1493–1497

Bosch F, Rosich L (2008) The contributions of Paul Ehrlich to pharmacology: a tribute on the occasion of the centenary of his Nobel Prize. Pharmacology 82:171–179

Bourassa P, Alata W, Tremblay C, Paris-Robidas S, Calon F (2019) Transferrin receptor-mediated uptake at the blood-brain barrier is not impaired by Alzheimer's disease neuropathology. Mol Pharm 16:583–594

Bradbury MW (1997) Transport of iron in the blood-brain-cerebrospinal fluid system. J Neurochem 69:443–454

Broadwell RD, Baker-Cairns BJ, Frieden PM, Oliver C, Villegas JC (1996) Transcytosis of protein through the mammalian cerebral epithelium and endothelium. Exp Neurol 142:47–65

Burdo JR, Menzies SL, Simpson IA, Garrick LM, Garrick MD, Dolan KG, Haile DJ, Beard JL, Connor JR (2001) Distribution of divalent metal transporter 1 and metal transport protein 1 in the normal and Belgrade rat (2001). J Neurosci Res 66:1198–1207

Burkhart A, Skjørringe T, Johnsen KB, Siupka P, Thomsen LB, Nielsen MS, Thomsen LL, Moos T (2016) Expression of iron-related proteins at the neurovascular unit supports reduction and reoxidation of iron for transport through the blood-brain barrier. Mol Neurobiol 53:7237–7253

Campos CR, Kemble AM, Niewoehner J, Freskgård PO, Urich E (2020) Brain shuttle neprilysin reduces central amyloid-β levels. PLoS One 15:e0229850

Chen P, Tang H, Zhang Q, Xu L, Zhou W, Hu X, Deng Y, Zhang L (2020) Basic Fibroblast Growth Factor (bFGF) protects the blood-brain barrier by binding of FGFR1 and activating the ERK signaling pathway after intra-abdominal hypertension and traumatic brain injury. Med Sci Monit 26:e922009

Chih B, Zuchero JJ, Chen X, Bien-Ly B, Bumbaca D, Tong RK, Gao Z, Zhang S, Hoyte K, Luk W, Huntley MA, Phu L, Tan C, Kallop D, Weimer RM, Lu Y, Kirkpatrick DS, Ernst J, Dennis MS, Watts RJ (2016) Validation of CD98hc as a novel blood brain barrier target. Soc Neurosci Abstr

Chua AC, Graham RM, Trinder D, Olynyk JK (2007) The regulation of cellular iron metabolism. Crit Rev Clin Lab Sci 13:413–459

Couch JA, Yu YJ, Zhang Y, Tarrant JM, Fuji RN, Meilandt WJ, Solanoy H, Tong RK, Hoyte K, Luk W, Lu Y, Gadkar K, Prabhu S, Ordonia BA, Nguyen Q, Lin Y, Lin Z, Balazs M, Scearce-Levie K, Ernst JA, Dennis MS, Watts RJ (2013) Addressing safety liabilities of TfR bispecific antibodies that cross the blood-brain barrier. Sci Transl Med 5:183ra57

Crowe A, Morgan EH (1992) Iron and transferrin uptake by brain and cerebrospinal fluid in the rat. Brain Res 592:8–16

Dallman PR, Spirito RA (1977) Brain iron in the rat: extremely slow turnover in normal rats may explain long-lasting effects of early iron deficiency. J Nutr 107:1075–1081

Dallman PR, Siimes MA, Manies EC (1975) Brain iron: persistent deficiency following short-term iron deprivation in the young rat. Br J Haematol 31:209–215

Daneman R, Prat A (2015) The blood-brain barrier. Cold Spring Harb Perspect Biol 7(1):a020412

Fan K, Jia X, Zhou M, Wang K, Conde J, He J, Tian J, Yan X (2018) Ferritin nanocarrier traverses the blood brain barrier and kills glioma. ACS Nano 12:4105–4115

Fisher J, Devraj K, Ingram J, Slagle-Webb B, Madhankumar AB, Liu X, Klinger M, Simpson IA, Connor JR (2007) Ferritin: a novel mechanism for delivery of iron to the brain and other organs. Am J Physiol Cell Physiol 293:C641–C649

Freskgård PO, Urich E (2017) Antibody therapies in CNS diseases. Neuropharmacology 120:38–55

Gosk S, Vermehren C, Storm G, Moos T (2004) Targeting anti-transferrin receptor antibody (OX26) and OX26-conjugated liposomes to brain capillary endothelial cells using in situ perfusion. J Cereb Blood Flow Metab 24:1193–1204

Haqqani AS, Thom G, Burrell M, Delaney CE, Brunette E, Baumann E, Sodja C, Jezierski A, Webster C, Stanimirovic DB (2018a) Intracellular sorting and transcytosis of the rat transferrin receptor antibody OX26 across the blood-brain barrier in vitro is dependent on its binding affinity. J Neurochem 146:735–752

Haqqani AS, Delaney CE, Brunette E, Baumann E, Farrington GK, Sisk W, Eldredge J, Ding W, Tremblay TL, Stanimirovic DB (2018b) Endosomal trafficking regulates receptor-mediated transcytosis of antibodies across the blood brain barrier. J Cereb Blood Flow Metab 38:727–740

Heidari M, Johnstone DM, Bassett B, Graham RM, Chua AC, House MJ, Collingwood JF, Bettencourt C, Houlden H, Ryten M, Olynyk JK, Trinder D, Milward EA (2016) Brain iron accumulation affects myelin-related molecular systems implicated in a rare neurogenetic disease family with neuropsychiatric features. Mol Psychiatry 21:1599–1607

Hultqvist G, Syvänen S, Fang XT, Lannfelt L, Sehlin D (2017) Bivalent brain shuttle increase antibody uptake by monovalent binding to the transferrin receptor. Theranostics 7:308–318

Jefferies WA, Brandon MR, Hunt SV, Williams AF, Gatter KC, Mason DY (1984) Transferrin receptor on endothelium of brain capillaries. Nature 312:162–163

Johnsen KB, Burkhart A, Melander F, Kempen PJ, Vejlebo JB, Siupka P, Nielsen MS, Andresen TL, Moos T (2017) Targeting transferrin receptors at the blood-brain barrier improves the uptake of immunoliposomes and subsequent cargo transport into the brain parenchyma. Sci Rep 7:10396

Johnsen KB, Bak M, Kempen PJ, Melander F, Burkhart A, Thomsen MS, Nielsen MS, Moos T, Andresen TL (2018) Antibody affinity and valency impact brain uptake of transferrin receptor-targeted gold nanoparticles. Theranostics 8:3416–3436

Johnsen KB, Burkhart A, Thomsen LB, Andresen TL, Moos T (2019a) Targeting the transferrin receptor for brain drug delivery. Prog Neurobiol 181:101665

Johnsen KB, Bak M, Melander F, Thomsen MS, Burkhart A, Kempen PJ, Andresen TL, Moos T (2019b) BBB Modulating the antibody density changes the uptake and transport at the blood-brain barrier of both transferrin receptor-targeted gold nanoparticles and liposomal cargo. J Ctrl Rel 295:237–249

Kariolis MS, Wells RC, Getz JA, Kwan W, Mahon CS, Tong R, Kim DJ, Srivastava A, Bedard C, Henne KR, Giese T, Assimon VA, Chen X, Zhang Y, Solanoy H, Jenkins K, Sanchez PE, Kane L, Miyamoto T, Chew KS, Pizzo ME, Liang N, Calvert MEK, DeVos SL, Baskaran S, Hall S, Sweeney ZK, Thorne RG, Watts RJ, Dennis MS, Silverman AP, Zuchero YJY (2020) Brain delivery of therapeutic proteins using an Fc fragment blood-brain barrier transport vehicle in mice and monkeys. Sci Transl Med 12(545):eaay1359

Katsinelos T, Tuck BJ, Mukadam AS, McEwan WA (2019) The role of antibodies and their receptors in protection against ordered protein assembly in neurodegeneration. Front Immunol 10:1139

Kniesel U, Risau W, Wolburg H (1996) Development of blood-brain barrier tight junctions in the rat cortex. Brain Res Dev Brain Res 96:229–240

Kreuter J (2014) Drug delivery to the central nervous system by polymeric nanoparticles: what do we know? Adv Drug Deliv Rev 71:2–14

Kucharz K, Kristensen K, Johnsen KB, Lund MA, Lønstrup M, Moos T, Andresen TL, Lauritzen MJ (2021) Post-capillary venules are the key locus for transcytosis-mediated brain delivery of therapeutic nanoparticles. Nat Commun. In press

Kumar NN, Lochhead JJ, Pizzo ME, Nehra G, Boroumand S, Greene G, Thorne RG (2018) Delivery of immunoglobulin G antibodies to the rat nervous system following intranasal administration: Distribution, dose-response, and mechanisms of delivery. J Ctrl Rel 286:467–484

Laskey J, Webb I, Schulman HM, Ponka P (1988) Evidence that transferrin supports cell proliferation by supplying iron for DNA synthesis. Exp Cell Res 176:87–95

Lichota J, Skjørringe T, Thomsen LB, Moos T (2010) Macromolecular drug transport into the brain using targeted therapy. J Neurochem 113:1–13

Lipinski CA, Lombardo F, Dominy BW, Feeney PJ (2001) Experimental and computational approaches to estimate solubility and permeability in drug discovery and development settings. Adv Drug Deliv Rev 46:3–26

Lochhead JJ, Wolak DJ, Pizzo ME, Thorne RG (2015) Rapid transport within cerebral perivascular spaces underlies widespread tracer distribution in the brain after intranasal administration. J Cereb Blood Flow Metab 35:371–381

Manich G, Cabezón I, del Valle J, Duran-Vilaregut J, Camins A, Pallàs M, Pelegrí C, Vilaplana J (2013) Study of the transcytosis of an anti-transferrin receptor antibody with a Fab' cargo across the blood-brain barrier in mice. Eur J Pharm Sci 49:556–564

Manthe RL, Loeck M, Bhowmick T, Solomon M, Muro S (2020) Intertwined mechanisms define transport of anti-ICAM nanocarriers across the endothelium and brain delivery of a therapeutic enzyme. J Ctrl Rel 324:181–193

Marcos-Contreras OA, Greineder CF, Kiseleva RY, Parhiz H, Walsh LR, Zuluaga-Ramirez V, Myerson JW, Hood ED, Villa CH, Tombacz I, Pardi N, Seliga A, Mui BL, Tam YK, Glassman PM, Shuvaev VV, Nong J, Brenner JS, Khoshnejad M, Madden T, Weissmann D, Persidsky Y, Muzykantov VR (2020) Selective targeting of nanomedicine to inflamed cerebral vasculature to enhance the blood-brain barrier. PNAS 117:3405–3414

Mastroberardino PG, Hoffman EK, Horowitz MP, Betarbet R, Taylor G, Cheng D, Na HM, Gutekunst CA, Gearing M, Trojanowski JQ, Anderson M, Chu CT, Peng J, Greenamyre JT (2009) A novel transferrin/TfR2-mediated mitochondrial iron transport system is disrupted in Parkinson's disease. Neurobiol Dis 34:417–431

McCarthy RC, Kosman DJ (2013) Ferroportin and exocytoplasmic ferroxidase activity are required for brain microvascular endothelial cell iron efflux. J Biol Chem 288:17932–17940

Møllgård K, Dziegielewska KM, Saunders NR, Zakut H, Soreq H (1988) Synthesis and localization of plasma proteins in the developing human brain. Integrity of the fetal blood-brain barrier to endogenous proteins of hepatic origin. Dev Biol 128:207–221

Montemiglio LC, Testi C, Ceci P, Falvo E, Pitea M, Savino C, Arcovito A, Peruzzi G, Baiocco P, Mancia F, Boffi A, des Georges A, Vallone B (2019) Cryo-EM structure of the human ferritin-transferrin receptor 1 complex. Nat Commun 10:1121

Moos T (1996) Immunohistochemical distribution of intraneuronal transferrin receptor immunore-activity in the adult mouse central nervous system. J Comp Neurol 374:675–692

Moos T, Morgan EH (2001) Restricted transport of anti-transferrin receptor antibody (OX26) through the blood-brain barrier in the rat. J Neurochem 79:119–129

Moos T, Morgan EH (2004) The significance of the mutated divalent metal transporter (DMT1) on iron transport into the Belgrade rat brain. J Neurochem 88:233–245

Moos T, Trinder D, Morgan EH (2000) Cellular distribution of ferric iron, ferritin, transferrin and divalent metal transporter 1 (DMT1) in substantia nigra and basal ganglia of normal and beta2-microglobulin deficient mouse brain. Cell Mol Biol (Noisy-le-grand) 46:549–561

Moos T, Rosengren Nielsen T, Skjørringe T, Morgan EH (2007) Iron trafficking inside the brain. J Neurochem 103:1730–1040

Morgan EH (1996) Iron metabolism and transport. In: Zakim D, Bayer T (eds) Hepatology. A textbook of liver disease, 3rd edn. Saunders, Philadelphia, pp 526–554

Morgan EH, Moos T (2002) Mechanism and developmental changes in iron transport across the blood-brain barrier. Dev Neurosci 24:106–113

Morris CM, Keith AB, Edwardson JA, Pullen RG (1992) Uptake and distribution of iron and trans-ferrin in the adult rat brain. J Neurochem 59:300–306

Niewoehner J, Bohrmann B, Collin L, Urich E, Sade H, Maier P, Rueger P, Stracke JO, Lau W, Tissot AC, Loetscher H, Ghosh A, Freskgård PO (2014) Increased brain penetration and potency of a therapeutic antibody using a monovalent molecular shuttle. Neuron 81:49–60

Pardridge WM (2012) Drug transport across the blood–brain barrier. J Cereb Blood Flow Metab 32:1959–1972

Pardridge WM (2020) The isolated brain microvessel: A versatile experimental model of the blood-brain barrier. Front Physiol 11:398

Paris-Robidas S, Emond V, Tremblay C, Soulet D, Calon F (2011) In vivo labeling of brain capil-lary endothelial cells after intravenous injection of monoclonal antibodies targeting the trans-ferrin receptor. Mol Pharmacol 80:32–39

Paris-Robidas S, Brouard D, Emond V, Parent M, Calon F (2016) Internalization of targeted quan-tum dots by brain capillary endothelial cells in vivo. J Cereb Blood Flow Metab 36:731–742

Pizzo ME, Wolak DJ, Kumar NN, Brunette E, Brunnquell CL, Hannocks MJ, Abbott NJ, Meyerand ME, Sorokin L, Stanimirovic DB, Thorne RG (2018) Intrathecal antibody distribution in the rat brain: surface diffusion, perivascular transport and osmotic enhancement of delivery. J Physiol 596:445–475

Roberts R, Fine RE, Sandra A (1993) Receptor-mediated endocytosis of transferrin at the blood-brain barrier. J Cell Sci 104:521–532

Routhe LJ, Thomsen MS, Moos T (2020) The significance of the choroid plexus for cerebral iron homeostasis. In: Praetorius J, Blazer-Yost B, Damkier H (eds) Role of the choroid plexus in health and disease. Physiology in health and disease. Springer, pp 125–148. https://doi.org/10.1007/978-1-0716-0536-3_5

Sehlin D, Fang XT, Cato L, Antoni G, Lannfelt L, Syvänen S (2016) Antibody-based PET imaging of amyloid beta in mouse models of Alzheimer's disease. Nat Commun 7:10759

Sehlin D, Stocki P, Gustavsson T, Hultqvist G, Walsh FS, Rutkowski JL, Syvänen S (2020) Brain delivery of biologics using a cross-species reactive transferrin receptor 1 VNAR shuttle. FASEB J 34:13272–13283

Simpson IA, Ponnuru P, Klinger ME, Myers RL, Devraj K, Coe CL, Lubach GR, Carruthers A, Connor JR (2015) A novel model for brain iron uptake: introducing the concept of regulation. J Cereb Blood Flow Metab 35:48–57

Siupka P, Hersom MN, Lykke-Hartmann K, Johnsen KB, Thomsen LB, Andresen TL, Moos T, Abbott NJ, Brodin B, Nielsen MS (2017) Bidirectional apical-basal traffic of the cation-independent mannose-6-phosphate receptor in brain endothelial cells. J Cereb Blood Flow Metab 37:2598–2613

Skarlatos S, Yoshikawa T, Pardridge WM (1995) Transport of [125I] transferrin through the rat blood-brain barrier. Brain Res 683:164–171

Sonoda H, Morimoto H, Yoden E, Koshimura Y, Kinoshita M, Golovina G, Takagi H, Yamamoto R, Minami K, Mizoguchi A, Tachibana K, Hirato T, Takahashi K (2018) A blood-brain-barrier-penetrating anti-human transferrin receptor antibody fusion protein for neuronopathic mucopolysaccharidosis II. Mol Ther 26:1366–1374

Stocki P, Szary J, Rasmussen CLM, Demydchuk M, Northall L, Logan DB, Gauhar A, Thei L, Moos T, Walsh FS, Rutkowski JL (2020) Blood-brain barrier transport using a high affinity, brain-selective VNAR antibody targeting transferrin receptor 1. FASEB J e21172

Strahan ME, Crowe A, Morgan EH (1992) Iron uptake in relation to transferrin degradation in brain and other tissues of rats. Am J Physiol 263:R924–R929

Taylor EM, Morgan EH (1991) Role of transferrin in iron uptake by the brain: a comparative study. J Comp Physiol B 161:521–524

Taylor EM, Crowe A, Morgan EH (1991) Transferrin and iron uptake by the brain: effects of altered iron status. J Neurochem 57:1584–1592

Testi C, Boffi A, Montemiglio LC (2019) Structural analysis of the transferrin receptor multifaceted ligand(s) interface. Biophys Chem 254:106242

Thom G, Burrell M, Haqqani AS, Yogi A, Lessard E, Brunette E, Delaney C, Baumann E, Callaghan D, Rodrigo N, Webster CI, Stanimirovic DB (2018) Enhanced delivery of Galanin conjugates to the brain through bioengineering of the anti-transferrin receptor antibody OX26. Mol Pharm 15:1420–1431

Thomsen MS, Andersen MV, Christoffersen PR, Jensen MD, Lichota J, Moos T (2015) Neurodegeneration with inflammation is accompanied by accumulation of iron and ferritin in microglia and neurons. Neurobiol Dis 81:108–118

Trinder D, Baker E (2003) Transferrin receptor 2: a new molecule in iron metabolism. Int J Biochem Cell Biol 35:292–296

Ueda F, Raja KB, Simpson RJ, Trowbridge IS, Bradbury MW (1993) Rate of 59Fe uptake into brain and cerebrospinal fluid and the influence thereon of antibodies against the transferrin receptor. J Neurochem 60:106–113

Ullman JC, Arguello A, Getz JA, Bhalla A, Mahon CS, Wang J, Giese T, Bedard C, Kim DJ, Blumenfeld JR, Liang N, Ravi R, Nugent AA, Davis SS, Ha C, Duque J, Tran HL, Wells RC, Lianoglou S, Daryani VM, Kwan W, Solanoy H, Nguyen H, Earr T, Dugas JC, Tuck MD, Harvey JL, Reyzer ML, Caprioli RM, Hall S, Poda S, Sanchez PE, Dennis MS, Gunasekaran K, Srivastava A, Sandmann T, Henne KR, Thorne RG, Di Paolo G, Astarita G, Diaz D, Silverman AP, Watts RJ, Sweeney ZK, Kariolis MS, Henry AG (2020) Brain delivery and activity of a lysosomal enzyme using a blood-brain barrier transport vehicle in mice. Sci Transl Med 12:eaay1163

Urrutia P, Aguirre P, Esparza A, Tapia V, Mena NP, Arredondo M, González-Billault C, Núñez MT (2013) Inflammation alters the expression of DMT1, FPN1 and hepcidin, and it causes iron accumulation in central nervous system cells. J Neurochem 126:541–549

Van Gelder W, Huijskes-Heins MI, van Dijk JP, Cleton-Soeteman MI, van Eijk HG (1995) Quantification of different transferrin receptor pools in primary cultures of porcine blood-brain barrier endothelial cells. J Neurochem 64:2708–2715

Van Gelder W, Cleton-Soeteman MI, Huijskes-Heins MI, van Run PR, van Eijk HG (1997) Transcytosis of 6.6-nm gold-labeled transferrin: an ultrastructural study in cultured porcine blood-brain barrier endothelial cells. Brain Res 746:105–116

Villaseñor R, Schilling M, Sundaresan J, Lutz Y, Collin L (2017) Sorting tubules regulate blood-brain barrier transcytosis. Cell Rep 21:3256–3270

Villaseñor R, Lampe J, Schwaninger M, Collin L (2019) Intracellular transport and regulation of transcytosis across the blood-brain barrier. Cell Mol Life Sci 76:1081–1092

Visanji NP, Lang AE, Kovacs GG (2019) Beyond the synucleinopathies: alpha synuclein as a driving force in neurodegenerative comorbidities. Transl Neurodegener 8:28

Weber F, Bohrmann B, Niewoehner J, Fischer JAA, Rueger P, Tiefenthaler G, Moelleken J, Bujotzek A, Brady K, Singer T, Ebeling M, Iglesias A, Freskgård PO (2018) Brain shuttle antibody for Alzheimer's disease with attenuated peripheral effector function due to an inverted binding mode. Cell Rep 22:149–162

Wolak DJ, Pizzo ME, Thorne RG (2015) Probing the extracellular diffusion of antibodies in brain using in vivo integrative optical imaging and ex vivo fluorescence imaging. J Ctrl Rel 197:78–86

Yang AC, Stevens MY, Chen MB, Lee DP, Stähli D, Gate D, Contrepois K, Chen W, Iram T, Zhang L, Vest RT, Chaney A, Lehallier B, Olsson N, du Bois H, Hsieh R, Cropper HC, Berdnik D, Li L, Wang EY, Traber GM, Bertozzi CR, Luo J, Snyder MP, Elias JE, Quake SR, James ML, Wyss-Coray T (2020) Physiological blood-brain transport is impaired with age by a shift in transcytosis. Nature 583:425–430

Yu YJ, Zhang Y, Kenrick M, Hoyte K, Luk W, Lu Y, Atwal J, Elliott JM, Prabhu S, Watts RJ, Dennis MS (2011) Boosting brain uptake of a therapeutic antibody by reducing its affinity for a transcytosis target. Sci Transl Med 3:84ra44

Yu YJ, Atwal JK, Zhang Y, Tong RK, Wildsmith KR, Tan C, Bien-Ly N, Hersom M, Maloney JA, Meilandt WJ, Bumbaca D, Gadkar K, Hoyte K, Luk W, Lu Y, Ernst JA, Scearce-Levie K, Couch JA, Dennis MS, Watts RJ (2014) Therapeutic bispecific antibodies cross the blood-brain barrier in nonhuman primates. Sci Transl Med 6:261ra154

Zuchero YJ, Chen X, Bien-Ly N, Bumbaca D, Tong RK, Gao X, Zhang S, Hoyte K, Luk W, Huntley MA, Phu L, Tan C, Kallop D, Weimer RM, Lu Y, Kirkpatrick DS, Ernst JA, Chih B, Dennis MS, Watts RJ (2016) Discovery of novel blood-brain barrier targets to enhance brain uptake of therapeutic antibodies. Neuron 89:70–82

Chapter 18
Drug Delivery to the CNS in the Treatment of Brain Tumors: The Sherbrooke Experience

David Fortin

Abstract Drug delivery to the central nervous system (CNS) remains a challenge in neuro-oncology. Despite decades of research in this field, no consensus has emerged as to the best approach to tackle this physiological limitation. Moreover, the relevance of doing so is still sometimes questioned in the community. In this paper, we present our experience with CNS delivery strategies that have been developed in the laboratory and have made their way to the clinic in a continuum of translational research. Using the intra-arterial (IA) route as an avenue to deliver chemotherapeutics in the treatment of brain tumors, complemented by an osmotic breach of the blood-brain barrier (BBB) in specific situations, we have developed over the years a comprehensive research effort on this specialized topic. Looking at preclinical work supporting the rationale for this approach and presenting results discussing the safety of the strategy, as well as results obtained in the treatment of malignant gliomas and primary CNS lymphomas, this paper intend to comprehensively summarize our work in this field.

Keywords CNS delivery strategies · Pre-clinical · Translational · Clinical · Intra-arterial · Osmotic opening

18.1 Introduction

Chemotherapeutic drug trials for brain tumor treatment have been conducted worldwide for many decades, with marginal improvement in patient's outcome. Indeed, the standard of care for first-line management of glioblastoma is the addition of temozolomide, an alkylating drug, to radiotherapy, which led to an improvement in survival of 2 months compared to radiotherapy alone (Stupp et al. 2005). Any

D. Fortin (✉)
Department of Neurosurgery, Neuro-oncology, Sherbrooke University,
Sherbrooke, QC, Canada
e-mail: David.fortin@usherbrooke.ca

© American Association of Pharmaceutical Scientists 2022
E. C. M. de Lange et al. (eds.), *Drug Delivery to the Brain*, AAPS Advances
in the Pharmaceutical Sciences Series 33, https://doi.org/10.1007/978-3-030-88773-5_18

attempts to further improve on this outcome have produced disappointing results. Although interesting molecular targets such as EGFR have been identified, the potential of EGFR targeting for brain tumors remains an unfulfilled promise, thanks to delivery impediment to the CNS (Westphal et al. 2017).

Interestingly, one of the only reported therapy with seemingly improved results is the addition of a local device emitting low-intensity, intermediate-frequency alternating electric fields (TTF) (Fabian et al. 2019). As this device is applied directly to the scalp of the patient, its mechanism of action doesn't require a specific delivery paradigm to reach the CNS bypassing the impediment of the blood-brain barrier (BBB) altogether. Indeed, among the factors that explain a lack of improvement in the care of brain tumor patients, one stands as a major culprit: the impaired delivery to the CNS, related to the presence of the blood-brain barrier (BBB) (Drapeau and Fortin 2015). Hence, in the presence of a brain tumor, the first barrier to treatment options is just that a barrier: the BBB.

18.2 The BBB as an Impediment to the Treatment of Brain Tumors: A Historic View

Ehrlich was the first to describe the blood-brain barrier in 1906 (Bellavance et al. 2008). Ironically, he also coined the term "chemotherapy" and originated the idea of the "magic bullet," a concept that has become a current focus in oncology research with the advent of the so-called targeted therapy. In 1921, Stern and Gautier observed that the barrier between the vascular compartment and the brain was selective to certain agents and called it "la barrière hémato-encéphalique" (Stern and Gautier 1921). Brodie et al. later detailed the significance of a molecule's lipid solubility as an indicator of its permeability through the barrier in 1960 (Brodie et al. 1960). A series of investigators next established that tight junctions between the cerebral capillary cells served as the prime anatomical basis for the barrier (Brightman et al. 1973). Muldoon et al. then investigated the possible role played by the basement membrane (basal lamina) as a secondary impediment to molecule delivery to the central nervous system by virtue of its negative charge (Muldoon et al. 1995). In addition to these factors, it has been found that the astrocytic foot processes, as well as the pericytes, also play a role in modulating BBB properties.

The "neurovascular unit" is a term used to refer to the complex functional barrier collectively composed of endothelial cells, pericytes, astrocytic endfeets, and neuronal cells (Davson and Oldendorf 1967; Cohen et al. 1996, 1997; Fenstermacher et al. 1988; Reese and Karnovsky 1967). This highly organized system confers unique properties to the CNS vasculature, accounting for the selective permeability of the BBB. Endothelial tight junctions, the lack of fenestrae, and low pinocytic/endosomal transport prevent the entry of hydrosoluble molecules into the brain across the BBB (Wolburg and Lippoldt 2002; Abbott et al. 2006; Deeken and Loscher 2007). Moreover, astrocytic endfeet apposition practically seals CNS vascular structures shut by covering nearly 99% of their external surface (Abbott et al. 2006; Deeken and Loscher 2007; Smith and Gumbleton 2006). This peculiar organization of the BBB restrains passive transport (non-receptor- or noncarrier-mediated transport) to the CNS compartment from the blood.

18.3   The BBB in Clinical Practice Nowadays

It has been a long way, and remains a challenge today still, to recognize the extent to which the BBB really impacts CNS delivery. This topic is often still debated in some publications, as some authors keep on arguing that the presence of contrast enhancement on CT scans (iodine-based) or on magnetic resonance scans (paramagnetic contrast) entails clear enough evidence that the integrity of the BBB is altered, and hence, free access to the CNS is granted implicitly (Fortin 2012; Van Den Bent 2003). In that context, these same authors claim that the BBB apparatus does not represent a significant impediment to therapeutic delivery of molecules to the CNS when in the presence of pathological lesions, implying that the breach in permeability is sufficient to allow adequate dissemination of therapeutics. In this scenario, the breach in BBB triggers a sufficient opening of the tight junctions so as to permit an adequate buildup of therapeutics within the CNS. This oversimplification in the conceptualization of the neurovascular unit only reveals the ignorance of some clinicians working in the field of neuro-oncology.

This type of all or none argument once again translates a lack of knowledge and understanding of the BBB and its alterations with relation to CNS delivery subtleties in the context of a lesion. Indeed, different pharmacokinetic compartments are defined by the presence of a brain tumor, with wide variation of the effects on BBB permeability, BTB (brain-tumor barrier), and ultimately, on delivery (Reichel 2009). This aspect is frequently neglected, underestimated and under-discussed in clinical research papers.

The presence of the ABC transporters such as P-glycoprotein (P-gp) efflux pump at the luminal surface of the brain capillaries (as well as at the surface of tumor cells) and the constant flow of cerebral spinal fluid (CSF) emanating from perivascular spaces toward the ventricular compartment further contribute to a decrease in the actual concentration and time of exposure of tumor cells to chemotherapy when treating CNS neoplasms and should be considered an integral part of the collective entity we refer to as the BBB (Fortin 2012).

Looking at recent data on the topic, there is no doubt that the BBB limits therapeutics entry to the CNS, even in the presence of a lesion and edema; this delivery impediment clearly limits a buildup in concentrations from reaching clinically efficient levels (Fortin et al. 2014; Fortin 2012). Part of the confusion in the literature arises from the fact that within a brain tumor and in close proximity to the main tumor nodule, the BBB is replaced by a brain-tumor barrier (BTB) which depicts entirely different pharmacokinetics properties, displaying a permeability that is classically intermediate between a normal and a breached BBB. This increase in permeability, which is a function of the breach in the integrity of the BBB and BTB, is highly variable and heterogeneous and appears to depend on tumor size and type (Kroll and Neuwelt 1998; Pardridge 2007; Bellavance et al. 2008). Another factor at play is the tumor itself. These entities are highly heterogeneous themselves, whether they are primary (gliomas) or secondary (metastasis), and will often contain areas of necrosis in which there is obviously no barrier whatsoever. Thus, within any tumoral lesion, drug distribution is inherently uneven, with a pernicious preferential accumulation in the necrotic central core areas (Tosoni et al. 2004). Hence, drug dissemination at the edge of the tumor is classically nonexistent, or marginal at best (Sato et al. 1998). As such, although the BBB and the BTB are often partially breached, there remains a significant limitation in delivery, impacting therapeutic levels of drugs that turn out insufficient within the breached areas to mount a clinically significant response as a result (Reichel 2009; Silbergeld and Chicoine 1997). In this context, achieving effective concentration profile in the brain remains a significant challenge in CNS drug development despite enormous efforts, as was so eloquently exposed by Reichel (Reichel 2009). Another intrinsic physiological factor further complicates the matter: the majority of malignant brain tumors are infiltrative (Silbergeld and Chicoine 1997). The implications are that tumor cells permeate the brain parenchyma at a distance from the main tumor nodule(s) and often away from even the most sensitive imaging MR scan sequences (FLAIR) (Silbergeld and Chicoine 1997; Chicoine and Silbergeld 1995). Obviously the BBB permeability is unaltered in these areas, and tumor cells are shielded by an intact BBB. No matter the level of breach in the BBB and BTB in the presence of a brain lesion, there hence remains most of the times tumor cells in residence behind an entirely normal and functional BBB.

18.4   The BBB in Primary Tumors

The BBB is heterogeneously and variably affected by the presence of a glial tumor as mentioned before. As most low-grade and up to 30% of high-grade glial tumors do not present contrast enhancement at MRI, a significant amount of these tumors may not greatly impact BBB permeability. However, for those that do, they can do so in an extremely inconsistent fashion.

As an illustration of the potency of the BBB to impede drug delivery in the treatment of glial tumors, Boyle and colleagues reported fascinating findings while studying vincristine penetration in a 9L rat glioma model (Boyle et al. 2004). The authors found negligible vincristine delivery in tumors despite a clear increase in permeability, as evidenced by a marked Evans blue staining. Evans blue binds to albumin once in the circulation, so a macroscopic blue discoloration of the brain parenchyma suggested the 66 kDa albumin could cross the blood-tumor barrier to reach the CNS (Blanchette and Fortin 2011). Despite the fact that the authors elected to administer the vincristine intra-arterially to improve CNS delivery, drug concentration was 6- to 11-fold higher in the liver when compared to tumor levels, and 15- to 37-fold higher in the liver compared to the normal brain levels in non-implanted animals. The authors thus concluded that this observation was likely explainable by the activity of the P-gp efflux pumps blocking the entry of the vincristine xenobiotic.

Another interesting study performed by Sato et al. also supports similar findings in humans (Sato et al. 1998). In a daring and elegant design, these investigators used fluorescein micro-angiograms to assess BBB integrity after having performed a neurosurgery in four patients. The authors found that although the BBB was partially permeable to the fluorescein, an important filling defect was observed at the immediate periphery of the tumor, suggesting competent barrier properties at that location. The administration of mannitol improved fluorescein delivery by increasing BBB permeability within the tumor, mostly at the venous capillary level, as well as at the periphery of the tumors.

18.5   The BBB in Metastatic (Secondary) Tumors

The metastatic process involves different interactions altogether between the implanting cancer cell and the brain microenvironment, the so-called seed and soil concept. Resident CNS cells such as astrocytes and microglial cells play an active role in the way metastatic tumor cells seed establish and flourish in the brain (Deeken and Loscher 2007). This active process requires recruitment of blood vessels via different mechanisms, whether it be angiogenesis, vasculogenesis, co-option, intussusception, or vascular mimicry. Notwithstanding the process, these vessels and their associated blood-tumor barrier (BTB) properties present certain morphological features: they have a significantly larger dilated diameter and a thicker basal membrane and present a lower microvessel density than the surrounding normal brain parenchyma (Eichler et al. 2011). More so, the tight junction structure is compromised, the perivascular space is increased, and fenestrations as well as pinocytic vacuoles can sometimes be found in these vessels (Deeken and Loscher 2007).

These brain tumor vessels also depict a different expression profile of efflux transporters, even though this difference is not very well characterized at the present time, as many discrepancies between the results of different studies have been reported (de Boer and Gaillard 2007; de Boer et al. 2003; Lee et al. 2001; Loscher and Potschka 2005; Guo et al. 2010). Some confusion arises because these proteins can also be expressed directly by tumor cells in addition to cerebral vessels (Shen et al. 2008). However, as a generalization, most studies looking at the expression of P-gp (and/or other ABC transporters) at the BTB in the context of brain metastasis have found either decreased or unchanged expression, relative to the levels in surrounding brain vessels (Deeken and Loscher 2007). To the contrary, tumor cells have often shown an increase in the expression profile of these transporters relative to normal glial cells (Shen et al. 2008).

All these observations support the caveat that even if the permeability of the BBB and BTB in malignant brain tumors is abnormal, it is nonetheless sufficiently preserved to represent an obstacle to delivery. Therapeutic drug levels are thereby often insufficient within brain regions protected from the systemic circulation by this partially and heterogeneously breached barrier. By steeply reducing the concentration of intravenously administered chemotherapeutic agent at the periphery of the tumor, a sink effect is yet another mechanism that can contribute to chemotherapy failure in CNS neoplasm treatment. Although the BBB is breached in primary and secondary brain tumors, other factors can limit the delivery of antineoplastic agents, such as an increase in interstitial pressure within the tumor tissue as well as in the brain surrounding the tumor. The presence of numerous efflux pump systems targeting xenobiotics must also not be neglected as an additional impediment to brain entry.

Distressingly, when dealing with brain tumor patients, several iatrogenic factors may reestablish BBB integrity, thereby further impeding drug delivery. Antiangiogenic therapies such as bevacizumab and the use of dexamethasone are commonly used medications exerting this effect.

18.6   Alternate Drug Delivery

Different approaches have been designed and tested to circumvent the CNS delivery impediment, bypassing the BBB and BTB, and maximize therapeutics delivery to the brain (Table 18.1). As was eloquently exposed in the first sections of this text, alternative delivery strategies have to be considered if we wish to increase the number of therapeutic options available to treat CNS tumors. For a detailed review on the subject covering different strategies, please consult the paper reviewed by Drapeau et al. (Drapeau and Fortin 2015). Of all the approaches we have tested in the laboratory, one is currently used in the clinic: the cerebral intra-arterial infusion of chemotherapy (CIAC) with and without BBB permeabilization. In these pages, we will give a thorough description of the efforts carried by our group to successfully deploy and integrate these strategies in the treatment of brain tumors. Indeed, we have implemented a continuum of translational research on this topic that will be described in detail herein.

18.7   The Cerebral Intra-arterial Infusion of Chemotherapy (CIAC) and the Blood-Brain Barrier Disruption (BBBD) Adjunct

When one realizes the extensiveness of the vascular network supplying the brain, it appears obvious that a global delivery strategy is rational and plausible via this vascular network as a delivery corridor (Kroll and Neuwelt 1998). The importance of this vascular system has already been detailed by Bradbury and colleagues; these authors claim that the entire network covers an area of 12 m^2/g of cerebral parenchyma (Bradbury 1986). To understand the extensiveness of the cerebral vascularization in a more prosaic way, let us just consider that the brain receives 20% of the total systemic circulation, whereas its weight amount to less than 3% of the total body weight (Kroll and Neuwelt 1998).

Table 18.1 Strategies to circumvent the blood-brain barrier

Strategy	Methods	Potential applications
Surgical		
Convection-enhanced delivery	Catheters placed around the resection cavity at the time of surgery	To improve local control rates or prevent relapse after gross total resection
Intra-arterial infusion	Direct intra-arterial infusion of chemotherapeutic, antibody, or nanoparticle drug	Treatment of all malignant brain tumors sub-types
Osmotic BBB disruption	Intra-arterial infusion of hyperosmotic agent before intra-arterial infusion of chemotherapeutic, antibody, or nanoparticle drug	Multiple brain metastases from a chemosensitive primary tumor; primary CNS lymphomas
Targeted ultrasound BBB disruption	Intravenous injection of preformed gas bubbles before pulsed ultrasound treatment	Single or limited number of brain metastases from a chemosensitive primary; single refractory or recurrent brain metastasis
Pharmacological		
Bradykinin analogs	Intravenous or intra-arterial b1 or b2 agonist delivery to transiently increase the permeability of the BTB	In combination with chemotherapy agents
Exploiting RMT: TfR, IR, IGF-1R, LRP-1	To achieve RMT, chemotherapy of choice linked to an antibody that targets the TfR, IR, IGF-1R, or LRP-1	Broad applicability for single or multiple brain metastases
PgP inhibitors	Inhibiting the drug efflux pump (e.g., HM30181A, cyclosporine A, valspodar, elacridar, zosuquidar)	Administration concurrently with chemotherapy for broad applications

BBB blood-brain barrier, *IGF-1R* insulin-like growth factor 1 receptor, *IR* insulin receptor, *LRP-1* low-density lipoprotein receptor-related protein 1, *PgP* P-glycoprotein, *RMT* receptor-mediated transcytosis, *TfR* transferrin receptor

The access of this cerebral vascular network in a patient is technically easy and actually repeatedly performed in the clinic on a regular basis (Fortin 2004). Via a simple puncture to access the femoral artery, a catheter is then introduced and navigated intra-arterially to reach one of the four major cerebral arteries. Once in the target vessel, a therapeutic can be administered via the catheter, that is later withdrawn at the conclusion of the procedure. The CIAC allows the construct of a regional chemotherapy distribution paradigm within the area deserved by the targeted vessel (Newton et al. 2003).

An increase in the local plasma peak concentration of the drug produces a significantly improved AUC (concentration of drug according to the time) through the first pass effect (Newton et al. 2003, 2006). This consequently translates in an increased local exposure of the target tissue to the therapeutic agent. Interestingly, as our lab as shown, it is also accompanied by a decreased systemic drug distribution, hence reducing systemic toxicities and potential side effects (Drapeau et al. 2017). Classically, the therapeutic concentration at the tumor cell target is increased by a 3.5- to 5-fold factor (Newton et al. 2006). This procedure is performed under local anesthesia in the angiographic suite and typically lasts around 45 min.

As an additional layer, the delivery can be further improved by adding to the procedure an osmotic blood-brain barrier disruption (BBBD) as an adjunct. This strategy is conceptually based on the cerebral intravascular infusion of a hypertonic solution to produce a transient increase in permeabilization of the BBB and BTB, prior to the administration of therapeutics. Hence, as is the case with CIAC, the parent vessel targeted is selected based on the tumor localization in the brain. This adjunct approach to CIAC is physiologically more demanding; it requires a general anesthesia and needs a careful preparation, but it does increase significantly delivery across the BBB and BTB (Drapeau et al. 2017; Kroll and Neuwelt 1998). During the procedure, an IA infusion of a hyperosmolar solution (usually mannitol) is accomplished in a flow rate sufficient to allow a complete filling of the parent vessel. In this process, two parameters are paramount in the ability to mediate a hyperosmolar modification of the barrier: the osmolality of the solution and the infusion time. Using a solution of 1.6 molal arabinose in pentobarbital-anesthetized rats, Rapoport found an interval duration of 30 s as the optimal infusion time to produce a BBBD (Kroll and Neuwelt 1998). The same infusion time was also applied to the use of mannitol with similar findings in the same animal model (Kroll and Neuwelt 1998). These parameters have made their way to the clinic; albeit the anesthetic agents used are now different. Patients are now treated under propofol anesthesia.

The combination of IA infusion of a molecule with osmotic BBBD has been shown to further increase the effect of first pass through the brain, increasing maximal peak concentration as well as AUC of the administered molecule (Kroll and Neuwelt 1998; Blanchette et al. 2009, 2014). As mentioned earlier, Sato et al. presented in vivo data showing that BBBD produces a marked increase in permeability specifically at the edge of the tumor. Interestingly, this area is typically associated with active tumor cells proliferation, whereas the permeability of the BBB and BTB tends to renormalize (Sato et al. 1998). That is why, theoretically, the concept of breaching the permeability of the BBB is quite compelling, as it could help evade the "sink effect" by providing higher and more uniform delivery to a whole CNS vascular territory, allowing a prolonged tumor cells exposure to higher concentrations of the administered therapeutics (Reicehl 2009; Kroll and Neuwelt 1998). This sink effect is triggered by areas of necrosis within the tumor, which tend to attract and concentrate the chemotherapy crossing the CNS, stealing the peripheral areas of the tumor where the drug would be most useful (Kroll and Neuwelt 1998). Obviously, this includes the neoplastic cells at the tumor edges that are often the most proliferative and protected by an intact BBB and/or BTB (Silbergeld and Chicoine 1997; Chicoine and Silbergeld 1995; Kroll and Neuwelt 1998; Pitz et al. 2011).

18.8   Preclinical Data

While numerous investigators have studied CIAC and BBBD over the years, we undertook a thorough preclinical characterization in the Fischer-F98 model to ascertain, objectivate, and measure the delivery advantage provided by both approaches. Initially, we characterized the F98-Fischer glioma model as a benchmark for our delivery studies. The model was found to be highly predictive and reproducible in terms of tumor growth dynamics and animal survival (Fig. 18.1).

Using a standardized implantation procedure, the tumor take has systematically been 100%, with a median survival of 26 + or − 2 days (Mathieu et al. 2007; Blanchard et al. 2006). Figure 18.1a shows the animal in the stereotactic frame for a consistent insertion of the needle in the brain of the animal using a precise and standardized coordinate system (Mathieu et al. 2007). This is paramount for reproducibility across experiments. Indeed, a free-hand implantation technique which is frequently employed in the literature is inadequate for this type of studies and should be discouraged. Likewise, we found that the use of a micro-infusion pump is essential in minimizing tissue damage and associated inflammatory reaction triggered by the implantation process (Mathieu et al. 2007). The slow (1 ul/min) and steady infusion rate and the low volume (100,00 cells in 5ul) insures a minimal cerebral tissue disruption and prevents backflow along the implantation track so commonly associated with implantation models (Mathieu et al. 2007; Blanchard et al. 2006). These precautions produce a constant pattern of tumor growth in the right hemisphere of the animal, where the tumor is already noticeable at day 3 post implantation (Fig. 18.1b) and starts to produce a terminal alteration in consciousness around day 26. Experimental treatments are performed at day 10 post implantation, when the tumor has reached a significant and measurable size (Fig. 18.1b), without altering the neurological functions of the animal (Charest et al. 2010, 2012, 2013).

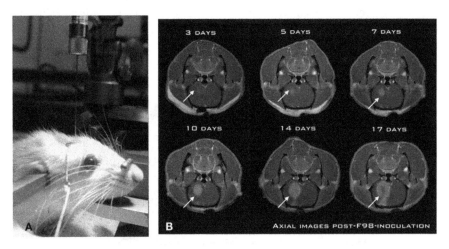

Fig. 18.1 The Fischer-F98 glioma model shows a reproducible and predictable growth pattern. (**a**) The infusion of the cell suspensions is accomplished using a slow steady perfusion with a micro-infusion pump. 10,000 cells are infused at a rate of 1 ul/min over 5 min. (**b**) Coronal views of an implanted animal at day 3,5,7,10,14 and 17 days post implantation. Notice the gradual progression of the gadolinium enhancement on the MR scans in the right hemisphere (arrows), depicting the steady tumour progression. The animal starts to develop faint subtil symptoms (lateralization) at day 14, that culminate at day 26 + or − 2 days

Using this model, and based on slight alterations of the methods described by Neuwelt and his team (Kroll and Neuwelt 1998), we developed a technique allowing the perfusion of therapeutics via the intra-arterial (IA) route in the carotid of the Fischer rats while under general anesthesia in an MR gantry. This allowed us to study the dynamics of real-time imaging during the infusion of any selected MR traceable molecule (Fig. 18.2) (Blanchette et al. 2014).

In this particular surgical montage, the right external carotid artery has been identified, incised, and cannulated using a PE50 catheter. Once in position, any solution can be perfused in a retrograde fashion via the external carotid artery back into the internal carotid artery (Fig. 18.2). When high-flow solutions are infused, such as when we perform a BBBD, a clip is secured on the common carotid artery to isolate the system from the heart and prevent downstream backflow of mannitol in the heart. After perfusion, the clip is removed, the external carotid artery is ligated, and the incision is closed.

As an initial experiment, we first characterized the baseline level of CNS entry for two paramagnetic compounds, Magnevist (743 Da) and Gadomer (17,000 Da) in tumor-bearing Fischer-F98 rats. As predicted, the smaller Magnevist displayed a greater than threefold baseline penetration in the tumor compared to Gadomer across the BTB. Penetration in the BBB around the tumor was negligible and was no different than in the contralateral hemisphere for both molecules (Blanchette et al. 2009, 2014).

Next, we studied the concentrations of different platinum drugs obtained when administered via different routes: intravenous (IV), IA, and IA+BBBD using inductively coupled plasma mass spectrometry (ICP-MS) in the Fischer-F98 rat model. We did so for five platinum compounds: cisplatin, carboplatin, oxaliplatin, Lipoplatin, and Lipoxal (Charest et al. 2013). Figure 18.3 shows the data summary of these experiments. Ten days after the F98 glioma cell implantation, the platinum

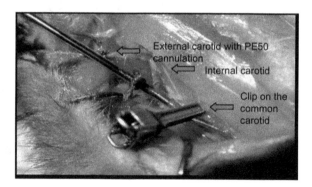

Fig. 18.2 The surgical montage for intraarterial infusion and BBBD in the Fischer rat. Once the montage is ready, the animal can than be inserted in the MR gantry for real time imaging. The intraarterial carotid perfusion is accomplished in a retrograde fashion via the external carotid artery. As can be appreciated on this image, a clip is also placed on the common carotid artery to prevent downstream back flow. As soon as the infusion is completed, the clip is removed, and the external carotid artery is sutured

drugs were administered via the selected route of administration. Equivalent doses of platinum to the one used in humans in clinic were established based on the body surface area of the animals (compounds: cisplatin, carboplatin, oxaliplatin, Lipoplatin, and Lipoxal) (Charest et al. 2013). Animals were euthanized 24 h after drug perfusion; the brains were harvested and cut in sections with a brain matrix (Charest et al. 2012). The tumor was separated in fractions and divided in a cytoplasmic and nuclear compartment using a commercial Nuclear Extract Kit (Active Motif, Carlsbad, CA) for analysis by ICP-MS (Charest et al. 2013).

When surveying specifically the concentration of platinum reaching the nucleus of the tumor cells, we observed significant differences between the different routes of administration (Fig. 18.3). Comparing IA against IV, an increase in the order of 20-fold was observed for IA carboplatin, whereas it reached 40-fold for Lipoplatin and 90-fold for Lipoxal! Interestingly these studies also depicted significant neurotoxicity when experimenting IA infusion of either oxaliplatin or cisplatin, hinting at the fact that these two drugs would never be suitable candidates for IA delivery (Charest et al. 2010, 2012, 2013). The concentrations of platinum reaching the tumor cell nuclei were even more dramatic when a BBBD was added to the IA infusion. Specifically looking at carboplatin, the IA+BBBD further increased the delivery by a 17-fold factor compared to IA alone, a 320-fold factor compared to the IV infusion (Charest et al. 2010)!

Using the same experimental design, we also assessed the delivery of temozolomide. Temozolomide is the first-line standard of care in the treatment of primary

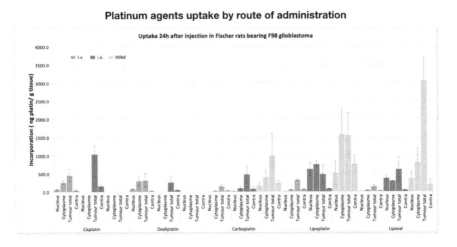

Fig. 18.3 A comparison of the 5 platinum drugs accumulations as related to the route of infusion, measured by icp-ms in the Fischer-F98 rat. As can be observed, the IA and IA +BBBD routes were not tested for Oxaliplatin and Cisplatin because of a significant toxicity. Results are reported as measurements of platinum (ng pt/g tissue) in the nucleus, cytoplasm and whole tumor. The magnitude of the increase observed for each platinum agent can be appreciated in relation to the route of delivery. There is a significant increase in platinum delivery (ng platinum/g tissue) with all molecules, except Oxaliplatin

brain tumors. As the bio-disponibility of the oral formulation is close to 100%, the IV formulation is available but rarely used in the clinic. In the present study, the IV formulation was used to emulate clinical oral administration. Hence, we tested the delivery of IA, IA+BBBD, and IV temozolomide in the Fischer-F98 glioma model. The animals were once again treated 10 days after implantation. Using liquid chromatography with tandem mass spectrometry (LC-MS\MS), we measured temozolomide in the plasma, CSF, and brain at three timepoints post-temozolomide infusion (Drapeau et al. 2017). Compared to IV, we found a fourfold increase in temozolomide peak concentrations in brain tumor tissues with IA infusion and a fivefold increase with BBBD (Fig. 18.4) (Drapeau et al. 2017).

The increase was not as dramatic using the BBBD as an adjunct with IA of temozolomide, compared to the platinum compounds. The values of c max according to the route of delivery were as follows: 10.582 (iv), 42.989 (IA), and 50.751 (IA+BBBD), respectively. In this experiment, although we could measure a significant increase in temozolomide delivery as described above, we did not observe a parallel increase in survival of the treated animals. In vitro characterization of the F-98 glioma cell line showed it later to be resistant to temozolomide (Drapeau et al. 2017). Hence, it is obvious that delivery is not the only factor at play, as will be discussed later.

These preclinical results really highlight the potency of IA and IA+BBBD as an adequate route of delivery to improve the different pharmacokinetic parameters of CNS therapeutic delivery. The preclinical research continuum to improve and maximize these procedures continues, as each therapeutic offered by this route first needs to be screened for innocuity in animal models to rule out any major toxicities. Indeed, oxaliplatin, cisplatin, and Taxol were found to be extremely toxic in

TMZ PHARMACOKINETIC AND TISSUE DISTRIBUTION PARAMETERS.

Table 1 - IV admnistration of TMZ (200 mg/m²)

	$T_{1/2}$ (h)	T_{max} (h)	C_{max}		AUC_{0-t}	
Plasma	1,08	0,25	63,581	µg/ml	53,409	h.µg/ml
CSF	0,87	0,25	7,628	µg/ml	6,658	h.µg/ml
Tumor	1,51	0,25	10,582	µg/g	9,521	h.µg/g
Ipsilateral brain	0,66	0,25	10,273	µg/g	8,530	h.µg/g
Contralateral brain	0,83	0,25	9,790	µg/g	8,547	h.µg/g

Table 2 – IA administration of TMZ (200 mg/m²)

	$T_{1/2}$ (h)	T_{max} (h)	C_{max}		AUC_{0-t}	
Plasma	n/a	0,5	40,676	µg/ml	38,759	h.µg/ml
CSF	1,89	0,25	8,436	µg/ml	7,681	h.µg/ml
Tumor	0,34	0,25	42,989	µg/g	31,934	h.µg/g
Ipsilateral brain	0,35	0,25	31,056	µg/g	23,930	h.µg/g
Contralateral brain	n/a	0,5	11,714	µg/g	10,130	h.µg/g

Fig. 18.4 Temozolomide (TMZ) pharmacokinetic parameters measured by Liquid chromatography tandem-mass spectrometry (LC-MS\MS) in Fischer-F98 rats treated 10 days after tumor implantation. Parameters are compared between the IV route (Table 1) and the IA route (Table 2)

preclinical testing, excluding de facto these drugs as eventual candidates for IA delivery. Moreover, as can be deduced from the results obtained with the temozolomide experiments, an increase in delivery is not necessarily associated with an improvement in outcome. Hence, delivery is only one of the many aspects of therapeutic success in the treatment of CNS tumors, albeit an important one. We will further discuss this issue in the next section on the clinical applications of these procedures.

18.9   Clinical Procedures

The access to the arterial system is obviously accomplished differently in humans. In humans, the arterial system is accessed via a percutaneous transfemoral puncture. Once accessed, the catheter is navigated in the arterial system using radiological imaging (fluoroscopy). As shown in Fig. 18.5a, the catheter has been placed in the left carotid artery, and a contrast infusion shows the distribution of this vessel. The human cerebral arterial system is organized in such a way that there basically are four major arteries responsible for brain irrigation (two carotids and two vertebral arteries). The vascular anatomy is variable from one individual to another, and thus the precise anatomy must be determined during the first treatment session by a formal cerebral angiography. If a lesion transgress more than one vascular distribution, or if there are multiple lesions, the treatment is delivered by splitting the chemotherapy dose in the different distributions (vessels) involved. Parameters such as catheter placement, dilution, and rate of infusion are all standardized.

The technique involves the following steps:

- *Selective catheterization* via percutaneous transfemoral puncture of the left internal carotid artery, right internal carotid artery, left vertebral artery, or right vertebral artery. The tip of the catheter is positioned at the C2–C3 vertebral level in the carotid (Fig. 18.5) or at the C6–C7 vertebral level in the vertebral artery.
- *IA infusion of the drug*. When infusing intra-arterial solutions, the concentration and the rate of infusion of the solutions are critical factors that need consideration in avoiding neurotoxicity. The phenomenon of streaming defines an inhomogeneous distribution of the administered solution because of poor mixing at the infusion site (9). The density and viscosity of fluid, lumen diameter of the infused vessels, and velocity of flow are all important determinants to control in order to avoid streaming. Caelyx, Melphalan, and Etoposide phosphate are infused at a rate of 0.12 cc/second, whereas the Carboplatin and Methotrexate are infused at a standard rate of 0.2 cc/second.
- *In the case of a BBBD*. BBBD procedures require a general anesthesia. Hence, after general anesthesia with propofol, we proceed to the same selective catheterization of the treated artery via percutaneous transfemoral puncture. We then determine the individual infusion rate of mannitol. We use iodinated contrast injection and fluoroscopy to establish the ideal infusion rate in each patient; it is

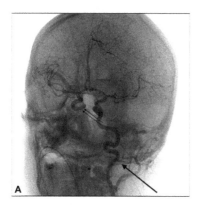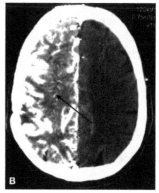

Fig. 18.5 (**a**) Catheter placement in a glioblastoma patient treated for BBBD in the left carotid artery (arrow). An iodine contrast was infused, opacifying the left carotid distribution, as well as the controlateral carotid (double arrow) via the polygon of willis (**b**) The image produced by a BBBD of the right carotid artery on a ct scan in a patient afflicted by a primary CNS lymphoma after an infusion of iv iodine contrast. As can be appreciated, the whole hemisphere is bathed by the contrast, an evidence that the BBB is breached

the rate that will fill the entire vessel distribution, without producing a significant reflux in the common carotid artery. Once established (usually between 3 and 6 cc/second for 30 s), the patient is prepared for the hemodynamic repercussions of the procedure. Indeed, the osmotic disruption is a physiologically stressful procedure hemodynamically, with occurrences of vasovagal response accompanied by bradycardia and hypotension. It can also induce focal seizures in 5% of procedures. In order to prevent the occurrence of these adverse effects, the following medications are administered just prior to the disruption: diazepam 0.2 mg/kg IV (maximum dose = 10 mg), and atropine IV, titrated to increase heart rate 10–20% from baseline (0.5 to 1 mg). We then proceed to the BBBD, after which IA infusion of chemotherapy is accomplished. Fig. 18.5b shows the concrete repercussion of BBBD on delivery in the clinical setting. In this image, an IV contrast material was infused shortly after the BBBD (within 5 min), showing a diffuse penetration of the contrast compound in the brain parenchyma (arrow) in the disrupted brain hemisphere.

18.10 CIAC or CIAC+BBBD?: A Question of Intensity of Delivery

The question of whether to use CIAC alone or with an adjunctive BBBD really is a question of intensity in the amount of delivery. There is no question that BBBD will increase delivery compared to an isolated CIAC. When studying platinum compounds, this increase has been shown to be variable for each molecule, providing a twofold increase for carboplatin (overall), up to a fivefold increase for Lipoxal compared to IA alone (Fig. 18.3). However, this increase in intensity comes at a cost: the use of BBBD requires general anesthesia and is significantly more demanding for the patient. Hence, its use can also be limited by logistics such as the availability of anesthesia and all it implies (recovery room, etc.). On the other hand, CIAC is easy to perform and virtually devoid of these burdens. The procedure is cheap, and the only limitation is the access to the angiography suite. Hence, in our institution, we have traditionally reserved the use of BBBD for patients with potentially curable disease, such as primary CNS lymphomas (PCNSL). Metastatic brain diseases, as well as glial tumors, are typically treated by CIAC. We built most of our clinical studies around a model in which the patient receives a monthly treatment session, typically up to 15 sessions. Only in patients presenting a complete response or near-complete response will we consider using BBBD to consolidate the treatment response in the last two cycles of the treatment.

18.11   Clinical Data: Safety

Neurotoxicity is a legitimate concern when deploying a strategy that increases the CNS delivery of therapeutics. Indeed, transgressing the BBB could result in an increase in neurotoxicity. The occurrence of such neurotoxicity nearly killed this treatment paradigm in the late 1980s when different trials using IA delivery of nitrosoureas reported tremendous toxicities, with stroke incidence in the order of 20% (Tonn et al. 1991). The choice of the therapeutic or infusion parameters were obviously to blame, as the advent of newer and safer molecules has rendered the treatment safer.

The first step is obviously the prior testing of each therapeutics used in the clinic in the animal model to screen for compatible drug candidates for CIAC/CIAC+ BBBD clinical use. Obviously, this does not entirely preclude the risks of toxicity. However, now looking at modern series of CIAC/CIAC+BBBD, we can confidentially claim it to be safe, when performed in expert centers.

Doolittle et al. reported on the experience of the BBBD consortium, a multi-site consortium performing CIAC with and without BBBD for malignant brain tumors (Doolittle et al. 2000). These authors concluded that with standardized protocols, CIAC was safe across multiple centers, with a low incidence of catheter-related complications. In their series of 221 patients treated between 1994 and 1997, they observed a subintimal tear rate of 5%, whereas the rate of strokes was 1.7%.

We undertook a detailed review of our own experience in terms of complications, going into further details. We analyzed our entire cohort of CIAC patients to brush the best possible picture in terms of innocuity. Between January 2000 and June 2015, a total of 3583 arteriographic procedures for CIAC/CIAC+BBBD were performed on 722 patients in the treatment of brain tumors at CHUS (Centre Hospitalier Universitaire de Sherbrooke). All these patients were afflicted by a malignant brain tumor (463 primary brain tumor, 158 metastasis, 101 lymphomas). To our knowledge, this is the largest such series reviewed in the literature (Fortin et al. 2014).

As clinical data have been cumulated prospectively in the context of clinical studies, data were extracted from centralized computerized hospitalization records for care events related to a CIAC procedure in the treatment of brain tumors (glial tumors, PCNSL, and metastatic tumors). Complications were studied and grouped under three different subheadings: vascular complications, per-procedural epileptic manifestations, and hematological toxicities. The results are detailed in Fig. 18.6.

18.12   Vascular Complications

Overall, a total of 66 vascular angiographic or MRI incidents were uncovered (1.84%). More specifically, five asymptomatic dissections were observed, nine asymptomatic carotid stenosis, and three occlusions were identified, two of which were symptomatic (Fig. 18.6).

In terms of newly described cerebral lesions, the MRI identified 5 acute hemorrhagic strokes (1 symptomatic), 38 lacunar strokes (20 symptomatic), and 6 acute ischemic lesions (4 symptomatic). One of these strokes in the posterior fossa was a catastrophic event that led to the patient's death (recurrent medulloblastoma). Overall in this series, the total number of symptomatic vascular complication rate was 27 (0.75%).

18.13   Seizure Events

The overall per-procedural seizure incidence was 2% (74 incidents) as can be appreciated in Fig. 18.6. Of these, 9 were generalized seizures, whereas 65 were partial seizures. Interestingly, a simple discontinuation of the chemotherapy infusion was sufficient to halt the seizure in all patients, but one. Most seizure fits (84%) were observed during MTX infusion for PCNSL. Only three partial seizures were observed in the treatment of glial tumors with carboplatin.

18.14   Hematological Complications

Hematological complications were classified according to the NCIC toxicity criteria. A total of 11.4% grade 3 and 7.2% grade 4 toxicities were observed.

Hence, from the analysis of this data, we feel justified to conclude that the procedure is safe, and its use is appropriate in this clinical context. This affirmation does imply that the treatments are performed in expert centers, and the therapeutics used with CIAC have been screened with an adequate methodology and are known to be devoid of neurotoxicity. While the technical details of the procedure are beyond the scope of this chapter, a few considerations need to be discussed. First, although some might argue that a supra-selective catheter placement might be of

Angiographic + vascular complications	Number of events MRI + angiographic findings)	MRI findings	Symptomatic lesions (the lesion was accompanied by clinical symptoms)	Asymptotic (lesion found at MRI or angiography without consequent symptoms)
Dissections	5	1	0	5
Stenosis	9	2	0	9
Occusions	3	2	2	1
Hemorragic lesions	5	5	1	4
Lacunar Strokes	38	38	20	18
Strokes	6	6	4	2
Total of events on 3586 procedures	66 (1.84%)	54 (1.5%)	27 (0.75%)	39 (1.08%)

Focal seizures (# of events)	Generalized seizures (# of events)	Lymphomas	Metastasis	Glial tumors	MTX	Carboplatin
65	9	23	4	12	62	12
74 seizure events (2%)		39 patients (5.4%)				

Hematologic toxicites (per NCIC toxicity criteria)	Grade 1	Garde 2	Grade 3	Grade 4	Total
Neutropenia	70 (9.7%)	67 (9.3%)	22 (3.1%)	21 (2.9%)	180 (24.9%)
Thrombocytopenia	43 (5.9%)	37 (5.1%)	35 (4.9%)	21 (2.9%)	136 (18.8%)
Anemia	115 (15.9%)	78 (10.8%)	25 (3.5%)	10 (1.4%)	228 (31.6%)
Total	228 (31.6%)	182 (25.2%)	82 (11.4%)	52 (7.2%)	

Fig. 18.6 angiographic, seizure-related and hematologic complications in the series of CIAC/CIAC+BBBD patients treated in Sherbrooke, from 2000 to 2015. A total of 3583 procedures in 722 patients were accomplished

interest to increase treatment precision, we always use a proximal position in the treated vessel (C1–C2 for the carotid, C2–C3 for the vertebral). The rationale supporting this has to do with the infiltrative nature of most brain tumors, always lending for a more widespread disease than the MR scans actually reveal. This is true for glial tumors and PCNSL but also metastasis. Because of that, we see little interest in targeting a supra-selective catheterization, especially considering that this approach would likely increase the risks of complications while limiting the actual distribution of chemotherapy to the CNS. However, this technical aspect is obviously open to discussion and counterarguments.

Secondly, each therapeutics should bear its own set of infusion parameters based on the concentration, density, and volume of the infusion solution. This is paramount to minimize the risks of neurotoxicity related to concentration and streaming.

18.15 Clinical Results

We will focus the discussion on the results obtained for GBM (grade 4 astrocytomas) and PCNSL. As a remainder, all patients with GBM were treated by CIAC, whereas 10% of these receive CIAC+BBBD as a consolidation procedure for the last two cycles. PCNSL patients were all treated with CIAC+BBBD.

18.16   Glioblastoma (GBM)

We treated 319 GBM patients by CIAC. Treatment sessions were typically performed every 4 weeks, hematologic parameters permitting. Carboplatin was the
drug of choice in this disease, either alone or combined with either melphalan or
etoposide phosphate, depending on the protocol used. All GBM patients were
treated at relapse. Seventeen percent were treated at first relapse, 68% at second
relapse, 11% at third relapse, and 4% at fourth (Fig. 18.7).

The fact that most of our patients were treated at second relapse unfortunately
negatively biases our results.

Overall, patients received a median of four cycles (1–22 cycles). The progression-
free survival was 5 months. The whole series presented an overall median survival
of 25 months, and survival from study entry was 9 months for the entire cohort
(Fig. 18.8). This is superior to most commonly reported treatment at relapse, which
usually produces median survival from treatment initiation of 4–6 months (Fortin
2016; Drapeau et al. 2017; Fabian et al. 2019).

Ten patients are still alive, with the longest survival now at 17 years. When looking exclusively at those patients treated at second relapse, median survival jump at
11 months from study entry for the entire cohort.

Looking at the best radiological responses obtained according to the RANO criteria, we found the following: 23% of patients have shown progression, 26% have
presented a stabilization of their disease, 42% have shown a partial response
(Fig. 18.9), and 6 % a complete response.

Although these results are encouraging, comparing this data with modern series
is complicated by two factors. Our data on GBM is plagued by a major weakness:
heterogeneity. Indeed, in 2016, the classification on gliomas changed, and one
major overhaul has been the inclusion of molecular markers to better stratify the
patients in prognostic classes that strongly determine their evolution (32). The IDH
marker is the stronger determinant of survival, and, nowadays, most modern series
stratify patients according to this marker, which we did not. The other source of
heterogeneity in this series pertains to the fact that a majority of patients were
exposed to multiple treatments prior to accrual (Fig. 18.7). These two factors are
obviously now considered in the design of our newer clinical studies. Recruitment
and stratification is now refined to eliminate these confounding factors.

Ideally, to avoid these heterogeneity, a randomization process should also be
utilized. We tried to launch such a study a few years back but were faced with difficulty as to what should be the randomized arm. The design of the randomized
study is now complete and submitted. It will compare carboplatin/etoposide phosphate CIAC against oral CCNU in the control arm and will constrain accrual at first
relapse as well as stratify patients against molecular status. Hopefully, this will
allow us to demonstrate the superiority of this approach, once and for all.

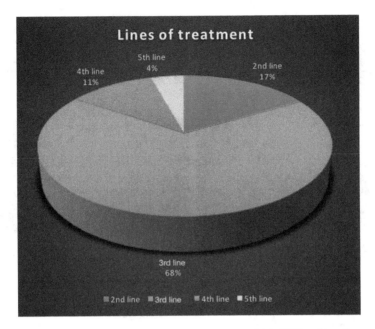

Fig. 18.7 A breakdown of the number of treatment lines to which GBM patients were exposed prior to accrual in our series. As can be appreciated, most patients were exposed to 2 lines of treatment (68%) prior to accrual

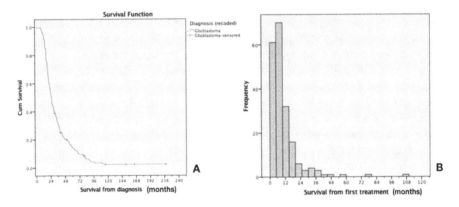

Fig. 18.8 Median survival from diagnosis (**a**) and distribution survival histogram (**b**) of GBM patients exposed to CIAC. Median survival from diagnosis was 25 months, whereas it was 8 months from study entry. As can be appraised from the distribution histogram, most patients progress and die from their disease in the first 12 months after accrual. This leaves around 25% of patients whose survival is greater than 12 months

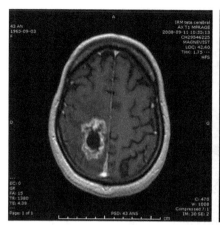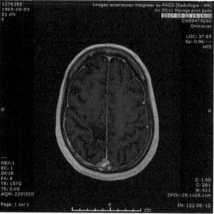

Fig. 18.9 Example of one of our best responder: A 43 year old female GBM patient treated with intraarterial Carboplatin/melphalan in 2008, who remains in complete response in 2017. This patient was treated at first relapse, without seeing other therapy than the Stupp regimen at first-line

18.17 Future Perspectives in the Treatment of Malignant Gliomas

18.17.1 Heterogeneity of Response: The Impossibility to Predict the Best Regimen for Each Patient

Carboplatin is our drug of choice, as it appears to be the most effective overall, producing responses in 70% of patients for a median PFS of 5 months at relapse. However, we do have an array of agents available for intra-arterial infusion in the clinic, and we continue to expand this list: carboplatin, methotrexate, melphalan, etoposide phosphate, and more recently Caelyx. These agents have all been used safely in CIAC for brain tumor treatment by our team over the years. Interestingly, in the case of nonresponse to carboplatin, other agents can be used in subsequent cycles of treatment for patients still presenting an adequate functional status. In these cases, we are often confronted to extremely variable results, with some long-term responses (up to 180 months) observed with other agents than carboplatin, whereas some patients show no response whatsoever to any agents. Indeed, these tumors all appear to have their own distinctive sensitivity profile to chemotherapy agents, and we believe that they should therefore all be approached as a singular disease entity requiring a personalized treatment. Molecular stratification has come a long way in the management of glial tumors (Louis et al. 2016), but its role has been so far limited to assist pathological stratification and prognosis. It is not yet used in treatment selection. We propose to combine data from in vitro drug sensitivity testing (DST) and molecular characterization using "The Cancer Genome Atlas" (TCGA) stratification, in addition to a panel of chemoresistance markers, to select the best drug candidates prior to the initiation of CIAC. Hence in accordance to this scheme in a proposed clinical study at first relapse for GBM, all patients will be reoperated prior to the beginning of CIAC. During surgery, a tumor sample will be obtained for the DST, molecular stratification, and chemoresistance panel markers, and the treatment will be tailored specifically to each patient.

18.17.2 Radio-Chemotherapy

Another area we have started to explore is the combination of IA-delivered carboplatin with radiation therapy (Choy 2003; Mamon and Tepper 2014). Indeed, radiotherapy is the most effective single-treatment modality for GBM tumors, but it controls the disease only transiently. A way to improve treatment consists of coupling radiation with a potent radiosensitizer. Carboplatin, a platinum (Pt) drug, is ideally suited for this. Our group has demonstrated that the addition of carboplatin to ionizing radiation produced significantly more DNA strand breaks (Boudaiffa et al. 2000; Zheng et al. 2008; Rezaee et al. 2013; Tippayamontria et al. 2013, 2014). In numerous cell lines, combining radiotherapy and carboplatin was found to increase cell death. In a mouse model, we observed a maximum antitumor effect with carboplatin administration at 4 h or 48 h prior to irradiation. This timing correlated to the highest levels of Pt bound to DNA (Boudaiffa et al. 2000; Zheng et al. 2008; Rezaee et al. 2013; Tippayamontria et al. 2013). Concurrent carboplatin and radiation treatment represents a common modality for treating a variety of cancers. Unfortunately, since this class of drug does not readily cross the BBB when administered via the standard IV routes, they are not used to treat GBM. We have just started accrual on a new phase II study in which we administer IA carboplatin with a re-irradiation protocol in a dose escalation scheme. We feel that this combination has the potential to improve clinical results. We have 15 patients (of a total of 35) recruited, and enrollment is ongoing.

18.17.3 Primary CNS Lymphomas

PCNSL are a rare and aggressive form of central nervous system tumors. Generally confined to the brain, eyes, and/or cerebrospinal fluid compartments, these extra nodal non-Hodgkin large B-cell lymphomas typically show no evidence of systemic diffusion (Bairey and Siegal 2018). PCNSL is an extremely aggressive disease, with a median survival time of 3 months without treatment (Bairey and Siegal 2018). It is a fairly unusual occurrence, accounting for 1% of cases of lymphoma, whereas it represents 4% of primary brain tumors (Campo et al. 2011; Darlix et al. 2017). A current trend in the treatment of this disease has been radiation therapy avoidance, as it was shown to be extremely neurotoxic to patients (Doolittle et al. 2013). Over the years, different protocols of IV high-dose methotrexate have shown encouraging results. Indeed, Da Broi et al. reported the results of 57 patients treated over 12 years with chemotherapy (Da Broi et al. 2018). Overall, they found a median OS of 35.4 months and a PFS of 15.7 months. Using CIAC+BBBD infusion protocol of high-dose methotrexate (combined to etoposide, cyclophosphamide, and/or

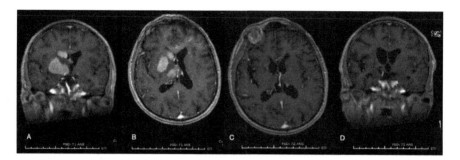

Fig. 18.10 A 71 year old man presented with rapidly progressing left-sided hemiparesis combined with a sudden decline in cognitive functioning. On T1-gadolinium enhanced MRI, multiple nodular enhancements are obvious, devoid of significant oedema. Local mass effect can be observed on the ventricular system in axial (**a**) and coronal (**b**) acquisition. Same MRI acquisition in axial (**c**) and coronal (**d**) after 8 cycles of treatment. This patient remains disease-free to this date, 26 months after the conclusion of his last cycle. Clinically, he presented a full recovery

procarbazine), Angelov et al. reported a median overall survival of 3.1 years. They also reported the neuropsychological outcome profile in 26 long-term survivors from the treatment (median follow-up of 12 years), showing the innocuity of this approach (Angelov et al. 2009). This good-quality data shows without a doubt that repeated CIAC+BBBD infusion protocol of high-dose methotrexate do not impact the neurocognitive functioning of responding patients.

Using CIAC+ BBBD carboplatin (400 mg/m^2) in addition to high-dose IA methotrexate (5 g), we treated 43 newly diagnosed PCNSL patients from 1999 to 2018. The median age of the cohort was 63, with a mean age of 60 years old. The cohort was comprised of 24 males and 19 females. Overall, remission was induced in 34 patients (79%). The Overall median survival was 46.5 months for the entire cohort. Actuarial survival was 88%, 64%, 54%, 39%, and 18% at 1, 2, 3, 5, and 10 years. The progression-free survival for the entire cohort was 43.3 months. The actuarial PFS was 83 %, 59%, 56%, 30%, and 9% at 1, 2, 3, 5, and 10 years. These are among the best results ever published in the treatment of this disease, without the use of radiation therapy. The detailed manuscript presenting these results is now published (Fig. 18.10) (Iorio-Morin et al. 2021).

18.18   Conclusion

Intra-arterial chemotherapy is a delivery vehicle allowing the increase of available therapeutics in the treatment of brain cancers. Its initial use many decades ago has been hampered by toxicity, a problem which is no more of concern. Angiographic refinements, combined to intra-arterial infusion of therapeutics carefully selected for this purpose, have rendered this approach safe and sound. The addition of an osmotic permeation of the BBB further increases the delivery of therapeutics to the CNS (Fig. 18.11). We need to acknowledge the extreme heterogeneity of GBM and eventually start tailoring treatment to each tumor for individual patient in order to improve the modest results obtained so far. Drug selection is at the core of this

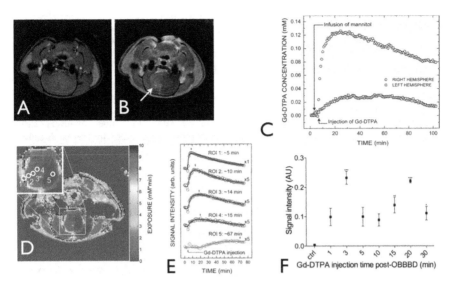

Fig. 18.11 Characterization of the osmotic blood-brain disruption in an healthy rat model. Axial MRI T_1-weighted images, (**a**) pre-contrast and (**b**) 17 min post-contrast post BBBD showing the permeabilization of the BBB in the treated hemisphere (arrow). (**c**) Comparison of the concentration of Gd-DTPA in the treated (right) and the untreated (left) hemisphere as a function of time. (**d**) Exposure (in mM × min) to Gd-DTPA for the first 17 min calculated from a series of T_1-weighted axial images. The enhancement patterns for each ROI are plotted in (**e**). Enhancement pattern for different ROIs indicated in (**d**). Each open circle represents an individual data point. The time to reach the maximum concentration (indicated by arrows pointing downwards) increases as the distance from the vascularized region increases. (**f**) Analysis of the signal intensities in a slice of the treated hemisphere 15 min after Gd-DTPA injection (mean + S.E.M.). The contrast agent was injected 1, 3, 5, 10, 15, 20 or 30 min after the BBBD procedure. In the sham BBBD experiment, saline was infused instead of mannitol; Gd-DTPA was injected 3 min later. In (**b**), (**c**) and (**d**) Gd-DTPA was administered 3 min post-BBBD

process. We also need to keep expending the pool of agents that can safely be administered via this route. As for the treatment of PCNSL, different refinements are considered to keep improving outcome in the treatment of this disease. The addition of rituximab, a CD-20 antibody, should be considered as an adjunct to the treatment protocol. In the end, the use of intra-arterial therapeutics infusion combined to osmotic blood-brain barrier permeation answers the need to adequately address an issue that is commonly underestimated: the presence of the BBB and the complex pharmacokinetic set of compartments it imposes on CNS delivery.

References

Abbott NJ, Ronnback L, Hansson E (2006) Astrocyte-endothelial interactions at the blood-brain barrier. Nat Rev Neurosci 7(1):41–53

Angelov L, Doolittle ND, Kraemer DF, Siegal T, Barnett GH, Peereboom DM et al (2009) Blood-brain barrier disruption and intra-arterial methotrexate-based therapy for newly diagnosed primary CNS lymphoma: a multi-institutional experience. J Clin Oncol 27:3503–3509. https://doi.org/10.1200/JCO.2008.19.3789

Bairey O, Siegal T (2018) The possible role of maintenance treatment for primary central nervous system lymphoma. Blood Rev 32:378–386. https://doi.org/10.1016/j.blre.2018.03.003

Bellavance MA, Blanchette M, Fortin D (2008) Recent advances in blood–brain barrier disruption as a CNS delivery strategy. AAPS J 10(1):166–177

Blanchard J, Mathieu D, Patenaude Y, Fortin D (2006) MR-pathological comparison in F98-Fischer glioma model using a human gantry. Can J Neurol Sci 33:86–91

Blanchette M, Fortin D (2011) Blood-brain barrier disruption in the treatment of brain tumors. Methods Mol Biol 686:447–463

Blanchette M, Pellerin M, Tremblay L, Lepage M, Fortin D (2009) Real-time monitoring of gadolinium Diethylenetriamine penta-acetic acid during osmotic blood-brain barrier disruption using magnetic resonance imaging in normal Wistar rats. Neurosurgery 65: 344–350. discussion 350–51. https://doi.org/10.1227/01.NEU.0000349762.17256.9E

Blanchette M, Tremblay L, Lepage M, Fortin D (2014) Impact of drug size on brain tumor and brain parenchyma delivery after a blood-brain barrier disruption. J Cereb Blood Flow Metab 34:820–826. https://doi.org/10.1038/jcbfm.2014.14

Boudaiffa B, Cloutier P, Hunting D, Huels MA, Sanche L (2000) Resonant formation of DNA strand breaks by low energy (3-20 eV) electrons. Science 287:1658–1660

Boyle FM, Eller SL, Grossman SA (2004) Penetration of intra-arterially administered vincristine in experimental brain tumor. Neuro-oncology 6(4):300–305

Bradbury MWB (1986) Appraisal of the role of endothelial cells and glia in barrier breakdown. In: Suckling AJ, Rumsby MG, Bradbury MWB (eds) The blood-brain barrier in health and disease. Ellis Horwood, Chichester, pp 128–129

Brightman MW, Hori M, Rapoport SI, Reese TS, Westergaard E (1973) Osmotic opening of tight junctions in cerebral endothelium. J Comp Neurol 152:317–325

Brodie BB, Kurz H, Schanker LS (1960) The importance of dissociation constant and lipid-solubility in influencing the passage of drugs into the cerebrospinal fluid. J Pharmacol 130:20–25

Campo E, Swerdlow SH, Harris NL, Pileri S, Stein H, Jaffe ES (2011) The 2008 WHO classification of lymphoid neoplasms and beyond: evolving concepts and practical applications. Blood 117:5019–5032

Charest G, Paquette B, Fortin D, Mathieu D, Sanche L (2010) Concomitant treatment of F98 glioma cells with new liposomal platinum compounds and ionizing radiation. J Neurooncol 97:187–193

Charest G, Sanche L, Fortin D, Mathieu D, Paquette B (2012) Glioblastoma treatment: bypassing the toxicity of platinum compounds by using liposomal formulation and increasing treatment efficiency with concomitant radiotherapy. Int J Radiat Oncol Biol Phys. https://doi.org/10.1016/j.ijrobp.2011.10.054

Charest G, Sanche L, Fortin D, Mathieu D, Paquette B (2013) Optimization of the route of platinum drugs administration to optimize the concomitant treatment with radiotherapy for glioblastoma implanted in the Fischer rat brain. J Neurooncol 115(3):365–373. https://doi.org/10.1007/s11060-013-1238-8

Chicoine MR, Silbergeld DL (1995) Assessment of brain tumor cell motility in vivo and in vitro. J Neurosurg 82:615–622. https://doi.org/10.3171/jns.1995.82.4.0615

Choy H (2003) Chemoradiation in cancer therapy. Humana Press, Totowa

Cohen Z, Bonvento G, Lacombe P, Hamel E (1996) Serotonin in the regulation of brain microcirculation. Prog Neurobiol 50(4):335–362

Cohen Z, Molinatti G, Hamel E (Aug 1997) Astroglial and vascular interactions of noradrenaline terminals in the rat cerebral cortex. J Cereb Blood Flow Metab 17(8):894–904

Da Broi M, Jahr G, Beiske K, Holte H, Meling TR (2018) Efficacy of the nordic and the MSKCC chemotherapy protocols on the overall and progression-free survival in intracranial PCNSL. Blood Cells Mol Dis 73:25–32. https://doi.org/10.1016/j.bcmd.2018.08.005

Darlix A, Zouaoui S, Rigau V, Bessaoud F, Figarella-Branger D, Mathieu-Daude H et al (2017) Epidemiology for primary brain tumors: a nationwide population-based study. J Neuro-Oncol 131:525–546

Davson H, Oldendorf WH (1967) Symposium on membrane transport. Transport in the central nervous system. Proc R Soc Med 60(4):326–329 de Boer AG, Gaillard PJ (2007) Drug targeting to the brain. Annu Rev Pharmacol Toxicol 47:323–355

de Boer AG, van der Sandt IC, Gaillard PJ (2003) The role of drug transporters at the blood-brain barrier. Annu Rev Pharmacol Toxicol 43:629–656

Deeken JF, Loscher W (2007) The blood-brain barrier and cancer: transporters, treatment, and Trojan horses. Clin Cancer Res 13(6):1663–1674

Doolittle ND, Miner ME, Hall WA et al (2000) Safety and efficacy of a multicenter study using intra-arterial chemotherapy in conjunction with osmotic opening of the blood–brain barrier for the treatment of patients with malignant brain tumors. Cancer 88(3):637–647

Doolittle ND, Dosa E, Fu R, Muldoon LL, Maron LM, Lubow MA et al (2013) Preservation of cognitive function in primary CNS lymphoma survivors a median of 12 years after enhanced chemotherapy delivery. J Clin Oncol 31:4026–4027. https://doi.org/10.1200/JCO.2013.52.7747

Drapeau A, Fortin D (2015) Chemotherapy delivery strategies to the central nervous system: neither optional nor superfluous. Curr Cancer Drug Targets 15:752–768

Drapeau A, Poirier M-B, Madugundu G-S, Wagner RJ, Fortin D (2017) Intra-arterial Temozolomide, osmotic blood-brain barrier disruption and radiotherapy in a rat F98-glioma model. Clin Cancer Drugs 4:1–11. https://doi.org/10.2174/2212697X04666170727152212

Eichler AF et al (2011) The biology of brain metastases-translation to new theapies. Nature reviews. Clinical oncology 8:344–356

Fabian D, Eibl MPGP, Alnahhas I, Sebastian N, Giglio P, Puduvalli V, Gonzalez J, Palmer JD (2019) Treatment of glioblastoma (GBM) with the addition of Tumor-Treating Fields (TTF): a review. Cancers (Basel) 11:174. https://doi.org/10.3390/cancers11020174

Fenstermacher J, Gross P, Sposito N, Acuff V, Pettersen S, Gruber K (1988) Structural and functional variations in capillary systems within the brain. Ann NY Acad Sci 529:21–30

Fortin D (2004) La barrière hémato-encéphalique : un facteur clé en neuro-oncologie. Rev Neurol 160(5):523–532

Fortin D (2012) The blood–brain barrier: its influence in the treatment of brain tumors metastases. Curr Cancer Drug Targets 12(3):247–259

Fortin D (2016) Safety of intra-arterial chemotherapy in the treatment of brain tumors. Society for neuro-oncology, presentation

Fortin D et al (2014) Intra-arterial carboplatin as a salvage strategy in the treatment of recurrent glioblastoma multiforme. J Neurooncol 119(2):397–403

Guo Z, Zhu J, Zhao L, Luo Q, Jin X (2010) Expression and clinical significance of multidrug resistance proteins in brain tumors. J Exp Clin Cancer Res 29:122

Iorio-Morin C, Gahide G, Morin C, Vanderweyen D, Roy MA, St-Pierre I, Massicotte-Tisluck K, Fortin D (2021) Management of primary central nervous system lymphoma using intra-arterial chemotherapy with osmotic blood-brain barrier disruption: retrospective analysis of the Sherbrooke cohort. Font Oncol 10:543648

Kroll RA, Neuwelt EA (1998) Outwitting the blood-brain barrier for therapeutic purposes: osmotic opening and other means. Neurosurgery 42(5):1083–1099 discussion 1099–1100

Lee G, Dallas S, Hong M, Bendayan R (2001) Drug transporters in the central nervous system: brain barriers and brain parenchyma considerations. Pharmacol Rev 53:569–596

Loscher W, Potschka H (2005) Role of drug efflux transporters in the brain for drug disposition and treatment of brain diseases. Prog Neurobiol 1:22–76

Louis DN, Perry A, Reifenberger G, Deimling A, Figarella-Branger D, Cavenee WK et al (2016) The 2016 World Health Organization classification of tumors of the central nervous system: a summary. Acta Neuropathol 131:803–820. https://doi.org/10.1007/s00401-016-1545-

Mamon HJ, Tepper JE (2014) Combination chemoradiation therapy: the whole is more than the sum of the parts. J Clin Oncol 32:367

Mathieu D, Lecomte R, Tsanaclis AM, Larouche A, Fortin D (2007) Standardization and detailed characterization of the syngeneic Fischer/F98 glioma model. Can J Neurol Sci 34:296–306

Muldoon LL, Nilaver G, Kroll RA, Pagel MA, Breakefield XO, Chiocca EA, Davidson BL, Weissleder R, Neuwelt EA (1995) Comparison of intracerebral inoculation and osmotic blood-brain barrier disruption for delivery of adenovirus, herpesvirus and iron oxide particles to normal rat brain. Am J Pathol 147:1840–1851

Newton HB, Slivka MA, Volpi C et al (2003) Intra-arterial carboplatin and intravenous etoposide for the treatment of metastatic brain tumors. J Neuro-Oncol 61:35–44

Newton HB, Figg GM, Slone HW, Bourekas E (2006) Incidence of infusion plan alterations after angiography in patients undergoing intra-arterial chemotherapy for brain tumors. J Neuro-Oncol 78:157–160

Pardridge WM (2007) Blood–brain barrier delivery. Drug Discov Today 12(1–2):54–61

Pitz MW, Desai A, Grossman SA, Blakeley JO (2011) Tissue concentration of systemically administered antineoplastic agents in human brain tumors. J Neurooncol. https://doi.org/10.1007/s11060-011-0564-y

Reese TS, Karnovsky MJ (1967) Fine structural localization of a blood-brain barrier to exogenous peroxidase. J Cell Biol 34(1):207–217

Reichel A (2009) Addressing central nervous system (CNS) penetration in drug discovery: basics and implications of the evolving new concept. Chem Biodivers 6(11):2030–2049. https://doi.org/10.1002/cbdv.200900103

Rezaee M, Hunting DJ, Sanche L (2013) New Insights into the mechanism underlying the synergistic action of ionizing radiation with platinum chemotherapeutic drugs: the role of low-energy electrons. Int J Radiat Oncol Biol Phys 87:847–853

Sato S, Kawase T, Harada S, Takayama H, Suga S (1998) Effect of hyperosmotic solutions on human brain tumor vasculature. Acta Neurochir (Wien) 140(11):1135–1141. discussion 1141–1142

Shen F, Chu S, Bence AK, Bailey B, Xue X, Erickson PA, Montrose MH, Beck WT, Erickson LC (2008) Quantitation of doxorubicin uptake, efflux, and modulation of multidrug resistance (MDR) in MDR human cancer cells. J Pharmacol Exp Ther 324(1):95–102

Silbergeld DL, Chicoine MR (1997) Isolation and characterization of human malignant glioma cells from histologically normal brain. J Neurosurg 86

Smith MW, Gumbleton M (2006) Endocytosis at the blood-brain barrier: from basic understanding to drug delivery strategies. J Drug Target. 14(4):191–214

Stern L, Gautier R (1921) Rapports entre le liquide céphalo-rachidien et la circulation sanquine. Arch Int Physiol 17:138–192

Stupp R, Mason W, van de Bent MJ et al (2005) Radiotherapy plus concomitant and adjuvant temozolomide for glioblastoma. N Engl J Med 352:987–996

Tippayamontria T, Kotb R, Paquette B, Sanche L (2013) Efficacy of Cisplatin and lipoplatin™ in combined treatment with radiation of a colorectal tumor in nude mouse. Anticancer Res 33:3005–3014

Tippayamontria T, Kotb R, Sanche L, Paquette B (2014) New therapeutic possibilities of combined treatment of radiotherapy with oxaliplatin and its liposomal formulations Lipoxal™ in rectal cancer using nude mouse xenograft. Anticancer Res 34(10):5303–5312

Tonn JC, Roosen K, Schachenmayr W (1991) Brain Necroses after Intraarterial chemotherapy and irradiation of malignant gliomas–a complication of both ACNU and BCNU? J Neuro-Oncol 11:241–242

Tosoni A, Ermani M, Brandes AA (2004) The pathogenesis and treatment of brain metastases: a comprehensive review. Crit Rev Oncol Hematol 52:199–215

Van Den Bent MJ (2003) The role of chemotherapy in brain metastases. Eur J Cancer 39(15):2114–2120

Westphal M, Maire CL, Lamszus K (2017) EGFR as a target for glioblastoma treatment: an unfulfilled promise. CNS Drugs 31:723–735. https://doi.org/10.1007/s40263-017-0456-6

Wolburg H, Lippoldt A (2002) Tight junctions of the blood-brain barrier: development, composition and regulation. Vascul Pharmacol 38(6):323–337

Zheng Y, Hunting DJ, Ayotte P, Sanche L (2008) Role of secondary low-energy electrons in the concomitant chemoradiation therapy of cancer. Phys Rev Lett 100:198101

Chapter 19
Biophysical and Clinical Perspectives on Blood-Brain Barrier Permeability Enhancement by Ultrasound and Microbubbles for Targeted Drug Delivery

Dallan McMahon and Kullervo Hynynen

Abstract The blood-brain barrier (BBB) greatly limits therapeutic treatment options for many diseases of the brain. The use of focused ultrasound (FUS) in conjunction with circulating microbubbles (MBs) provides a noninvasive means of transiently increasing BBB permeability with a high level of spatial precision. Drugs can be administered systemically to extravasate in the targeted brain regions and exert a therapeutic effect. In the hours following sonication, BBB permeability returns to baseline levels. The flexibility of this approach in facilitating the delivery of a wide range of therapeutic agents to either precise locations or large volumes, combined with efficacious results in preclinical models of disease, has motivated clinical trials to assess safety. This chapter will describe the development of FUS and MB-mediated BBB permeability enhancement as a drug delivery technique, detail several technical and biological considerations of this approach, and summarize results from the clinical trials conducted to date.

Keywords Acoustic feedback control · Blood-brain barrier · Clinical trials · Drug delivery · Focused ultrasound · Microbubbles · Ultrasound

D. McMahon (✉)
Physical Sciences Platform, Sunnybrook Research Institute, Toronto, ON, Canada
e-mail: dmcmahon@sri.utoronto.ca

K. Hynynen
Physical Sciences Platform, Sunnybrook Research Institute, Toronto, ON, Canada

Department of Medical Biophysics, University of Toronto, Toronto, ON, Canada

Institute of Biomaterials and Biomedical Engineering, University of Toronto, Toronto, ON, Canada

© American Association of Pharmaceutical Scientists 2022
E. C. M. de Lange et al. (eds.), *Drug Delivery to the Brain*, AAPS Advances in the Pharmaceutical Sciences Series 33, https://doi.org/10.1007/978-3-030-88773-5_19

19.1   Therapeutic Ultrasound

Ultrasound is one of the most widely used imaging technologies in modern medicine. As a noninvasive and nonionizing modality, diagnostic ultrasound exhibits a high safety profile and can be used to infer detailed anatomical and physiological information. This approach involves transforming electrical energy into mechanical energy via an ultrasound transducer, which acts to transmit waves of compression and rarefaction into the body (Fig. 19.1a). As these pressure waves travel through tissue, energy is scattered and absorbed. Detection of the reflected waves can be used to generate images and glean certain biological characteristics of the tissue, while the absorbed energy is a byproduct. Conversely, therapeutic ultrasound uses these same principles but focuses instead on the biological effects of depositing acoustic energy within tissue.

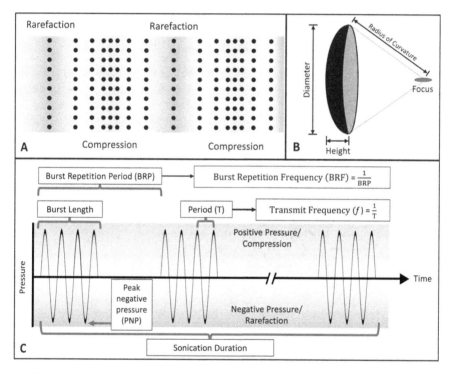

Fig. 19.1 Ultrasound nomenclature. (**a**) Ultrasound propagates through media as a pressure wave. As waves of compression (higher relative pressure) and rarefaction (lower relative pressure) travel through tissue, mechanical stresses and thermal deposition can lead to a variety of transient and long-lasting biological changes. (**b**) A spherically curved, single element-focused ultrasound transducer is depicted. The size and shape of the focal volume generated by a given transducer is influenced by its diameter, height, radius of curvature, and transmit frequency. (**c**) Ultrasound terminology is visually depicted. Much of the primary research described in this chapter utilizes burst-mode ultrasound (as opposed to continuous wave ultrasound) for which periods of ultrasound transmission alternate with periods of no transmission (off-time)

Depending on exposure parameters, ultrasound can be used to induce a range of biological responses, from enhanced vascular permeability (Vykhodtseva et al. 1995; Fyfe and Chahl 1984) to tissue necrosis (McDannold et al. 1998; Lynn et al. 1942). The amount of energy deposition required to elicit these effects is typically much greater than is achieved with diagnostic imaging, generally necessitating the use of longer bursts (i.e., milliseconds to minutes vs. microseconds) and greater peak intensities (i.e., approximately 10^2 to 10^4 W/cm^2 vs. 10^0 to 10^1 W/cm^2). Broadly speaking, the bioeffects of therapeutic ultrasound can be categorized as thermal or nonthermal, although therapies often use a combination of effects.

Of chief importance for the facilitation of brain-drug delivery is *acoustic cavitation*, which refers to the interaction of ultrasound waves with gas-filled cavities (Leighton 1994). There are two primary sources of such ultrasound-responsive cavities, the first being bubbles that are created de novo when dissolved gases coalesce under high rarefactional pressure. These bubbles are short-lived, dissolving within tens of milliseconds after an ultrasound burst (Xu et al. 2007), but can generate dramatic biological effects, including mechanical tissue ablation (e.g., histotripsy) (Roberts et al. 2006).

The second source of cavitation nuclei is exogenous; intravenously administered, encapsulated microbubbles were first developed as ultrasound contrast agents for enhanced visualization of perfused tissue (Gramiak and Shah 1968). Commercially available formulations, such as Definity™ and Optison™, consist of a protein, polymer, or phospholipid shell surrounding an air or perfluorocarbon gas core. Encapsulation enhances the stability of MBs, allowing circulation half-lives to be on the order of minutes (Grayburn 2002). The mean diameter of commercially available MBs is typically below 5 μm (Definity: 1.1–3.3 μm; Optison: 2.5–4.5 μm) but display wide size distributions (Goertz et al. 2007). While MBs were first developed for diagnostic imaging purposes, their utility in therapeutic applications is now well established (Goertz 2015).

19.2 Focused Ultrasound

To confine ultrasound-induced bioeffects to predictable volumes and to achieve high acoustic pressures deep within tissue requires the ability to focus ultrasound energy. One of the simplest approaches to achieve this is to employ a spherically curved transducer. The focal volume for a given transducer is determined by its geometry (i.e., radius of curvature and diameter) and transmit frequency. As the transducer diameter or transmit frequency increases or the radius of curvature decreases, the focal spot volume will be reduced (Hynynen et al. 1981); however, the shape of focus will remain ellipsoidal, with the long axis parallel to the direction of ultrasound propagation (Cline et al. 1994). For example, a transducer with a radius of curvature of 8 cm, diameter of 10 cm, and transmit frequency of 500 kHz (Fig. 19.1b), the lateral and axial dimensions of the focus (i.e., full-width at half-maximum pressure in water) will be approximately 3.4 mm and 16.4 mm, respectively (calculated from equations in (Cline et al. 1994)).

The use of FUS as a noninvasive alternative to neurosurgery has been explored for decades. In 1942, Lynn et al. first demonstrated the ability to transcranially ablate focal volumes in the cortex of canine and feline specimens (Lynn et al. 1942); however, subsequent work from Lynn and Putnam described extensive tissue damage extending from the inner skull surface to the focus, as well as severe skin necrosis associated with these sonications (Lynn and Putnam 1944). The challenge of producing a spatially confined ultrasound focus at depth within the brain and through the human skull was later tackled by the Fry brothers in the 1970s and 1980s. They demonstrated the ability to produce focal thermal ablations in feline brains with a human cadaver skull section placed within the path of ultrasound propagation (Fry 1977; Fry and Goss 1980). While this work showed that transcranial FUS-mediated brain tissue ablation was possible with single-element transducers driven at low frequencies (i.e., below 1 MHz), a lack of predictability in ablated volume, focal distortions, and target shifts caused by the skull restricted the advancement of this technique toward human neurosurgical applications for decades.

The development of techniques to correct for the aberrating effects of the skull was paramount to the advancement of transcranial ultrasound therapies. Using a linear imaging array (i.e., multiple transducers arranged in a line), Smith et al. (Smith et al. 1979), and later Thomas and Fink (Thomas and Fink 1996), showed that adjusting the phase and amplitude of each transducer element based on hydrophone measurements at the targeted focus could be used to improve focusing through the skull; however, it was not until this method of aberration correction was used in combination with a large-element (64 elements), high-powered 2D array that the feasibility of transcranial ablation was demonstrated (Hynynen and Jolesz 1998). The method became noninvasive with the development of a computed tomography-based correction algorithm that takes into account skull density and shape, removing the necessity of hydrophone measurements at the focus (Clement and Hynynen 2002). Since these studies, more advanced systems have been developed that employ hundreds to thousands of transducer elements arranged in spherically curved arrays. The ability to control the phase and amplitude of each individual element allows for: (1) electronic steering of the focus (Ebbini and Cain 1991) and (2) the ability to produce a spatially confined focal volume through a heterogenous skull (Hynynen and Jolesz 1998). Using this technology, current clinical hemispherical array systems have been able to achieve ablated volumes of approximately 2 mm in diameter and 4 mm in length with transcranial propagation in humans (650 kHz system, ExAblate Neuro from INSIGHTEC, Haifa, Israel) (Lipsman et al. 2013). While smaller focal volumes may be achievable with the use of higher transmit frequencies, skull heating becomes progressively prohibitive as transmit frequency is increased due to the corresponding increase in skull attenuation and aberrations.

This early work on transcranial FUS-mediated brain tissue ablation established the foundation for clinical trials that aimed to alleviate the symptoms of essential tremor through ablation of the ventral intermediate nucleus of the thalamus (Lipsman et al. 2013; Elias et al. 2013). In 2016, INSIGHTEC's high-frequency, hemispherical array system (ExAblate Neuro) received *Health Canada* approval for the treatment of essential tremor. While tissue ablation was the chief focus of much of the early brain-FUS research, careful observation revealed that the margins of these thermal lesions contained vasculature with increased permeability (Bakay et al. 1956; Ballantine Jr et al. 1960; Shealy and Crafts 1965). These observations, combined with the long-realized challenge of delivering therapeutic agents to the brain, motivated further study into how this effect could be achieved without the creation of thermal lesions. Additionally, the methodological and technological advancements that enabled transcranial tissue ablation have been essential for the progression of other therapeutic ultrasound applications in the brain.

19.3   Focused Ultrasound, Microbubbles, and the Blood-Brain Barrier

Research in the mid-1990s demonstrated that acoustic cavitation from bubbles created de novo could be used to increase BBB permeability without the formation of a thermal lesion; however, ultrasound exposure conditions which could consistently modulate vascular leakage without generating overt tissue damage could not be established and raised concerns with respect to safety and repeatability (Vykhodtseva et al. 1995). In 2001, a major advancement in the field saw the administration of encapsulated MBs into systemic circulation prior to sonication, which allowed for FUS-induced enhancement of vascular permeability (termed *FUS + MB exposure* for brevity) to be achieved at substantially lower time-averaged acoustic intensities relative to FUS alone (Hynynen et al. 2001). This reduction in intensity allowed for transcranial exposures to be employed without concern of skull heating-induced damage and largely removed cavitation from bubbles created de novo and thermal effects as mechanisms driving changes in BBB permeability. Instead, the physical forces exerted on vasculature by ultrasound-stimulated MBs are thought to be the dominant contributor to observed bioeffects (Hynynen et al. 2001). MB behavior can vary widely depending on the characteristics on the insonating wave, local environment, and MB type, resulting in a range of forces exerted on vascular walls.

When stimulated by ultrasound at low pressure, MBs oscillate volumetrically in size around their equilibrium state, a regime of activity referred to as *stable cavitation* (Leighton 1994) (Fig. 19.2). This behavior can generate microstreaming in surrounding fluid, which in turn produces shear stresses in the endothelial lining of blood vessels (Lewin and Bjørnø 1982; Wu 2002). Depending on ultrasound exposure conditions, this force may result in the activation of physiologically relevant shear stress mechanisms, including Ca^{2+} influx and subsequent nitrous oxide production, or may produce reversible membrane perforation, cell detachment, and/or lysis (van Bavel 2007). Stably cavitating MBs can also generate circumferential stress within vascular walls, which creates tension in the proteins that link endothelial cells (ECs) together (Hosseinkhah and Hynynen 2012). Additionally, acoustic

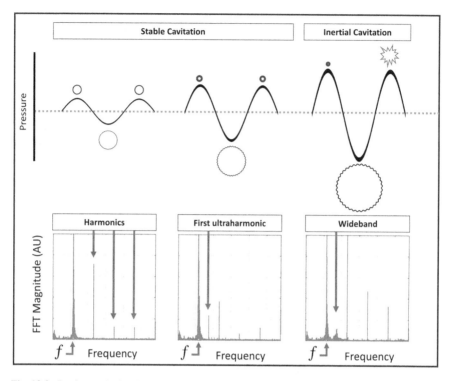

Fig. 19.2 Regimes of microbubble activity and spectral frequency content of acoustic emissions. Ultrasound-stimulated MBs can expand and contract in response to cycles of rarefaction and compression, themselves generating pressure waves referred to as acoustic emissions. Assessing the spectral frequency content of acoustic emissions from insonated MBs can give insight into their behavior. As the PNP of sonication is increased, MBs will begin to oscillate in a fashion that generates acoustic emissions at harmonics (integer multiples) of the transmit frequency (f). If the pressure amplitude is further increased above a threshold value, nonlinear volumetric oscillations will generate sub- and ultraharmonic emissions (subharmonic = $0.5f$, ultraharmonics = $1.5f$, $2.5f$, etc.). Generally speaking, MB behaviors in this regime are referred to as stable cavitation. As the PNP is further increased, MBs will begin to collapse in the compressional phase, referred to as inertial cavitation. This behavior is characterized by a sharp increase in the production of wideband emissions and is often associated with tissue damage. AU = arbitrary units, FFT = fast Fourier transform

waves exert radiation force on circulating MBs, propelling them in the direction of ultrasound propagation (Dayton et al. 1999). The force of displaced MBs on vascular walls may be sufficient to contribute to subsequent bioeffects (Dayton et al. 1999).

As acoustic pressure is further increased, MBs can collapse in the compressional phase of an ultrasound wave under the inertia of the surrounding fluid (Fig. 19.2). This behavior, referred to as *inertial cavitation*, can generate shockwaves, jet streams, free radicals, and extreme heat (Leighton 1994; Apfel and Holland 1991). The violent collapse of MBs within vasculature can result in ischemia, apoptosis, necrosis, edema, and hemorrhage (Vykhodtseva et al. 2006). Thus, efforts to reduce inertial cavitation are essential in the context of FUS + MB-mediated BBB permeability enhancement. Importantly, studies have demonstrated that BBB permeability enhancement can be achieved without acoustic emissions characteristic of inertial cavitation (Tung et al. 2010; McDannold et al. 2006).

19.3.1 Delivery of Therapeutics

Given the scarcity of drugs which permeate the BBB in therapeutically relevant concentrations and the considerable efforts required to engineer biochemical or physical solutions to this bottleneck (Pardridge 2005), the flexibility of FUS + MB exposure as a drug delivery strategy is advantageous. A large variety of therapeutic agents have been successfully delivered to targeted regions in the brains of disease models with efficacious results. This section will briefly outline the most impactful findings in this area.

Chemotherapeutics continue to be one of the most widely used classes of drugs in medical oncology; however, poor BBB transit limits their efficacy in the treatment of brain tumors (Doolittle et al. 2007). In addition, the toxicity of these drugs to healthy brain tissue, if able to cross the BBB, reinforces the need for targeted delivery. The ability to increase the concentration of doxorubicin, a commonly used chemotherapeutic agent, in brain tissue using FUS + MB exposure and systemic drug delivery was first demonstrated in healthy rats without tumors (Treat et al. 2007). Further work demonstrated efficacy in preclinical studies, reporting reduced tumor volumes and increased survival times following FUS + MB-mediated delivery of doxorubicin in a glioma rat model (Treat et al. 2012), two syngeneic glioblastoma mouse models (Kovacs et al. 2014), and others (Sun et al. 2017; Park et al. 2017; Fan et al. 2013a; Alli et al. 2018; Aryal et al. 2013; Arvanitis et al. 2018; Yang et al. 2012). Enhanced delivery of other chemotherapeutic agents, such as methotrexate (Mei et al. 2009), carmustine (Liu et al. 2010a; Fan et al. 2013b), temozolomide (Wei et al. 2013; Dong et al. 2018), and carboplatin (McDannold et al. 2019), has also been demonstrated following FUS + MB exposures.

While substantial obstacles remain (e.g., immune responses (Lotfinia et al. 2019; Büning and Mingozzi 2015) and scaling production (Clément 2019)), viral vector-based gene therapy may present a powerful avenue for controlling gene expression, providing a flexible tool in the treatment or prevention of a large number of pathologies. Thus far, the delivery of viral vectors to specific regions within the brain has primarily been achieved via intracranial injections. Proof-of-concept for FUS + MB-mediated viral vector-based gene therapy was first demonstrated in mice with the delivery of adeno-associated virus (AAV) 9-green fluorescent protein (GFP) to the striatum and hippocampus. GFP was found to be primarily expressed in neurons and astrocytes 12 days following sonication, with minimal expression in nontargeted brain areas (Thévenot et al. 2012). Others have achieved FUS + MB-mediated delivery of AAV2-GFP (Hsu et al. 2013; Stavarache et al. 2018), AAV1- and AAV2-GFP under a synapsin promoter (Wang et al. 2015), and AAV1/2-GFP under a GFAP promoter (Weber-Adrian et al. 2019). Xhima et al.

demonstrated the delivery of an AAV9 vector bearing a short hairpin RNA sequence targeting the α-synuclein gene. Authors reported at least a 50% reduction in α-synuclein protein expression in the targeted hippocampus, substantia nigra, and olfactory bulb 1 month following sonication and virus delivery, a result that may have relevance for the treatment of Parkinson's disease (Xhima et al. 2018). Nonviral gene therapy approaches have also been combined with FUS + MB exposures, leading to the enhanced delivery of small interfering RNA for huntingtin protein knockdown (Burgess et al. 2012), liposome-encapsulated plasmid DNA for the expression of trophic factors (Huang et al. 2012; Lin et al. 2015; Wang et al. 2014), and DNA-bearing nanoparticles (Mead et al. 2016).

In addition to chemotherapeutics and viral vectors, many other therapeutic agents have been shown to permeate the BBB following sonication, including neural stem cells (Burgess et al. 2011), natural killer cells (Alkins et al. 2013), anti-amyloid beta (Aβ) antibodies (Jordão et al. 2010), anti-dopamine receptor D4 antibodies (Kinoshita et al. 2006a), herceptin (Kinoshita et al. 2006b), and brain-derived neurotrophic factor (Baseri et al. 2012).

19.3.2 Nondrug Delivery Applications

Interestingly, FUS + MB exposures without therapeutic agent delivery have been shown to generate biological changes that may be beneficial in specific contexts. In a study designed to explore the impact of FUS + MB exposure in a mouse model (TgCRND8) of Alzheimer's disease (AD), Jordao et al. first described positive sonication-mediated effects on pathology, free of drug delivery. Authors reported a significant reduction in mean Aβ plaque size and total Aβ plaque surface area in the sonicated, relative to the non-sonicated, hemispheres. Additionally, they found increased microglial activation surrounding Aβ plaques, as well as greater levels of Aβ within microglia and astrocytes, suggesting that FUS + MB exposure promotes phagocytosis of Aβ (Jordão et al. 2013). Given the progressive nature of AD, research on this effect has largely focused on repeated exposures. Burgess et al. demonstrated that weekly sonications (across 3 weeks) targeted bilaterally to the hippocampus produced a significant reduction in plaque load, increased proliferation of neural progenitor cells in the dentate gyrus, and improved performance on hippocampal-dependent tasks in TgCRND8 mice (Burgess et al. 2014a). Using longitudinal in vivo two-photon microscopy, Poon et al. found that the maximal effect on plaque size occurs approximately 4–7 days following FUS + MB exposure and that plaques returned to baseline size within 3 weeks (Poon et al. 2018). This would suggest that frequent sonications without therapeutic agent delivery would be required for this treatment strategy (Poon et al. 2018). Pandit et al. demonstrated that a FUS + MB-mediated reduction in hyperphosphorylated tau in the K369I tau transgenic K3 mouse model was driven in part by the autophagy pathway in neurons, but not in glia (Pandit et al. 2019). Since the initial study by Jordao et al., others have found beneficial effects of FUS + MB exposure in APP23 (Leinenga and

Götz 2015, 2018), pR5 (Nisbet et al. 2017), and rTg4510 (Karakatsani et al. 2019) mouse models of AD .

FUS exposure has also been shown to influence neural activity, both in the presence and absence of MBs, potentially providing a more targeted, less invasive alternative to techniques like deep brain stimulation and implanted electrocortical stimulation (Bystritsky et al. 2011). Chu et al. observed the suppression somatosensory evoked potential amplitudes and blood-oxygen level-dependent responses in rat cortex for 1 week following sonication with exposure conditions that produced extensive RBC extravasation. Conversely, with parameters that produced BBB permeability enhancement without significant RBC extravasation, reductions in blood-oxygen level-dependent responses were observed at 1 h, but not 1 week, following sonication (Chu et al. 2015). Author suggests these effects may be used as an alternative to other clinical neuromodulation techniques; however, further work is required to assess the efficacy of this approach.

19.4 Biological Responses to Focused Ultrasound and Microbubble Exposures

An extensive body of literature has described the feasibility of using FUS + MB exposures to increase BBB permeability for the primary purpose of targeted therapeutic agent delivery. Also of importance for the widespread clinical implementation of this technique is a full characterization of the range of biological effects that may be expected to arise. This section will discuss the biological responses that have been observed following FUS + MB exposure, as assessed by magnetic resonance imaging (MRI) and biochemical/ histological assays.

19.4.1 Magnetic Resonance Imaging

Noninvasive imaging modalities are currently essential for precise targeting of transcranial FUS within the brain. MRI is not only effective in this regard but also allows for flexibility in how targets are located (e.g., anatomically identified brain structures (Fig. 19.3a/b), regions of abnormal blood-oxygen level-dependent responses, etc.). Additionally, MRI can also be utilized for detailed post-sonication assessment of tissue effects.

Contrast-enhanced T1-weighted (CE-T1w) imaging is commonly used to confirm BBB permeability enhancement following FUS + MB exposure (Fig. 19.3c/d); however, more quantitative MRI approaches have provided valuable insights into the duration and kinetics of this effect. Using T1-mapping and MR contrast agents of varying hydrodynamic diameters, Marty et al. observed that the time required for vascular permeability to reduce by half following sonication (i.e., half closure time) is

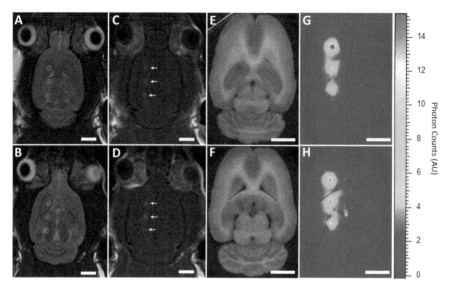

Fig. 19.3 Methods of evaluating BBB permeability enhancement. Accurate FUS targeting can be achieved through co-registration of the transducer position system with MRI spatial coordinates. (**a, b**) Sequential T2w axial slices from a rat brain are depicted, along with the targeted locations (red circles). The impact of FUS + MB exposures on BBB permeability enhancement can be assessed with a variety of methods. CE-T1w MRI is commonly used in both preclinical and clinical settings. Typically, a gadolinium-based contrast agent is administered intravenously during or shortly after sonication. (**c, d**) T1w images are then acquired to assess the magnitude and spatial distribution of vascular permeability enhancement to the contrast agent (white arrows indicate areas of contrast enhancement). Alternatively, BBB permeability enhancement can be assessed in ex vivo tissue. Evans blue dye, which under normal physiological conditions does not cross the BBB in significant quantities, can be administered intravenously during or shortly after sonication and extravasate in regions of enhanced BBB permeability. To quantitatively assess this, brain sections can first be imaged under (**e, f**) brightfield to provide an anatomical reference, then (**g, h**) fluorescently to identify regions with elevated Evans blue concentrations. Scale bars = 4 mm. AU = arbitrary units

dependent on the size of particle extravasating from systemic circulation into the brain. Half closure times were found to be approximately 1 and 5 h, for contrast agents with diameters of 7 nm and 1 nm, respectively (Marty et al. 2012). It has also been observed that, if in circulation during sonication, material as large as natural killer cells can extravasate (Alkins et al. 2013), though this is unlikely due to passive mechanisms. Similarly, dynamic contrast enhanced (DCE)-MRI, which can provide information regarding the kinetics of vascular permeability, has been used to show that the initial magnitude of BBB permeability enhancement influences the duration for which elevated permeability can be detected (Park et al. 2012). These studies emphasize the notion that the effects of FUS + MB exposure on BBB permeability are not binary (i.e., open vs. closed BBB, commonly used terms in the field) and that the extent of initial permeability enhancement is dependent on a wide range of factors relating to both sonication parameters, as well as the compounds crossing the BBB.

MRI is also commonly used to qualitatively assess tissue damage following soni-cation. Hypointensities in images collected with T2*-weighted or susceptibility-weighted sequences are indicative of microhemorrhages or hemosiderin deposits (Fazekas et al. 1999). These effects have been reported in studies that employ high peak negative pressure (PNP) (McDannold et al. 2012; Kovacs et al. 2018a) or high MB doses (Kovacs et al. 2018a; McMahon and Hynynen 2017) and have been shown to correlate with inertial cavitation (McDannold et al. 2012). Studies employing optimized acoustic feedback control strategies have demonstrated that BBB permeability enhancement can be achieved without the detection of T2* effects (McMahon and Hynynen 2017; O'Reilly et al. 2017; Jones et al. 2018); however, small regions of red blood cell (RBC) extravasation (i.e., less than 50 μm in diameter) have been noted in hematoxylin and eosin (H&E) stained sections from animals that did not display T2* effects following FUS + MB exposure (McDannold et al. 2012; Jones et al. 2018), highlighting the sensitivity limit of this detection method.

Imaging with T2w sequences, which can aid in the detection of vasogenic edema (Schwartz et al. 1992), has also provided insight into the impact of FUS + MB exposure. Abnormal hyperintensities were noted by Downs et al. in approximately 6% of targets at 30 min to 30 h following FUS + MB exposure, all of which resolved within 1 week. These effects were only seen after multiple sonications, spaced over a maximum of 20 months (Downs et al. 2015). Vasogenic edema after multiple FUS + MB exposures may be indicative of a gradual deterioration of BBB integrity caused by inappropriate exposure parameters, as others have demonstrated that vascular permeability can be repeatedly enhanced without evidence of T2 effects (O'Reilly et al. 2017). Similarly, hyperintensities have been reported following single sonications in studies that employ high PNP (Kovacs et al. 2018a) or high MB doses (Kovacs et al. 2018a; McMahon and Hynynen 2017).

19.4.2 Histological and Biochemical Assays

Much of the current understanding of how the brain responds to FUS + MB exposures has come from histological and biochemical analysis. While a complete characterization of the physical and biological processes that drive changes in BBB permeability is lacking, studies have provided detailed information on the routes of leakage and content of extravasated material, as well as changes in protein expression and cell morphology. This section will review FUS + MB studies that have explored vascular and extravascular changes using histological and biochemical assays.

19.4.2.1 Vascular Effects

Early electron microscopy studies by Sheikov et al. described an increase in the number of vesicles, vacuoles, fenestrations, and transcellular channels in ECs at 1–2 h following FUS + MB exposure (Sheikov et al. 2004). Further work demonstrated transcellular vesicular trafficking (Sheikov et al. 2006) and paracellular leakage past tight junction (TJ) complexes (Sheikov et al. 2008) of systemically administered horseradish peroxidase and lanthanum chloride (i.e., tracers that do not traverse the BBB under physiological conditions), respectively. These changes in TJ integrity were mirrored at the protein level, with a significant reduction in the immunoreactivity of occludin, claudin-5, and zonula occludens-1 in the interendothelial clefts at 1 and 2 h, but not 4–24 h, following FUS + MB exposure (Sheikov et al. 2008). While it is unclear whether this effect is due to a downregulation of TJ proteins or is the product of TJ protein trafficking away from the interendothelial cleft, it is apparent that FUS + MB exposure transiently disrupts the integrity of the link between vascular ECs. Similarly, increased density of EC vesicles may be driven by changes in protein expression, with upregulation of caveolin-1 observed 1 h following sonication (Deng et al. 2012).

These changes in transcellular and paracellular permeability appear to be nonspecific, as a variety of large molecules from systemic circulation have been observed in the brain following sonication. Significant increases in the levels of endogenous molecules, like albumin (Kovacs et al. 2017a; Alonso et al. 2011), IgG (Jordão et al. 2013; Sheikov et al. 2004; Raymond et al. 2008), and IgM (Jordão et al. 2013), as well as exogenous substances, like therapeutic agents (previously discussed) and tracer dyes (Kinoshita et al. 2006b; Sheikov et al. 2004, 2006; Yang et al. 2007) (Fig. 19.3g/h), have been observed in brain parenchyma after FUS + MB exposures. It should be emphasized that the total quantity of extravasated material following sonication appears to be related to the initial magnitude of BBB permeability enhancement (Jordão et al. 2013).

In vivo two-photon microscopy has been valuable in providing observations of the morphological changes and kinetics of vascular permeability that occur during and after FUS + MB exposure. In the very acute stages following sonication in mice (i.e., seconds after or while sonicating), Raymond et al. noted consistent vasomotor responses that typically included heterogeneous vasoconstriction (described as "lumpy" or "beaded") along the entire arterial network (mean reduction of ~60% in diameter of arteries), followed by a relaxation phase lasting several minutes, where vessels returned to baseline size. Blood flow during the constriction phase was described qualitatively as "reduced"; authors also describe instances of transiently halted blood flow (Raymond et al. 2007). This change in vascular tone may be driven by the mechanical stimulation of smooth muscle cells by oscillating MBs, as similar responses are observed following physical contact to arterial walls by guide wires or catheters during interventional radiology procedures (Takahashi et al. 1996). Vasoconstriction, during or shortly after FUS + MB exposure, has also been observed in rats; however, the frequency of this effect was found to be lower, occurring in only 25% of vessels analyzed (versus 87.5% in mice) (Cho et al. 2011).

While it is unclear if these discordant reports are the result of species differences, FUS + MB parameters, or other experimental differences, the occurrence of vaso-constriction in a substantial proportion of arteries would be expected to transiently reduce local blood flow and may initiate ischemic response mechanisms. Conversely, vasodilation has also been noted in the minutes following sonication (Burgess et al. 2014b), an effect that may be related to enhanced nitric oxide production induced by shear stress from oscillating MBs (Lu and Kassab 2011), though this has not been shown experimentally.

Quantitative and qualitative evaluation of BBB permeability enhancement using in vivo two-photon microscopy also suggests that there are at least three distinct types of leakage generated by FUS + MB exposure: (1) hemorrhagic, (2) focal dis-ruption (aka microdisruption or fast leakage), and (3) slow leakage (aka transcyto-sis) (Raymond et al. 2007; Cho et al. 2011; Burgess et al. 2014b; Nhan et al. 2013). Hemorrhagic leakage seems to be largely avoidable with the use of appropriate exposure conditions, as few regions of RBC extravasation are observed when PNP is adjusted to limit inertial cavitation (Kinoshita et al. 2006a). Focal disruptions are characterized by a rapid diffusion of dyes (i.e., K^{trans} of 0.005–0.04 min^{-1} for 10–70 kDa dextrans (Nhan et al. 2013)) into brain parenchyma at distinct points along blood vessels, evident during or shortly after sonication (Raymond et al. 2007; Cho et al. 2011; Burgess et al. 2014b; Nhan et al. 2013). This type of leakage occurs more frequently in vasculature with diameters less than 30 μm and is specu-lated to result from a widening of inter-endothelial clefts (Raymond et al. 2007; Burgess et al. 2014b; Nhan et al. 2013). The onset of slow leakage is delayed, start-ing at least 10 min after sonication, and is characterized by a gradual (i.e., K^{trans} of less than 0.005 min^{-1} for 10–70 kDa dextrans (Nhan et al. 2013)), diffuse accumula-tion of dye in the regions surrounding vessels of varying diameters (Raymond et al. 2007; Cho et al. 2011; Burgess et al. 2014b; Nhan et al. 2013). It has been widely speculated that vesicle-mediated transcytosis contributes to this type of leakage (Raymond et al. 2007; Cho et al. 2011; Burgess et al. 2014b; Nhan et al. 2013), a hypothesis supported by the electron microscopy studies previously discussed (Sheikov et al. 2004, 2006); however, there is no direct experimental evidence to link slow leakage to enhanced endocytosis. Additionally, the relative contribution of confounding factors, like laser-induced heating, cranial window-induced inflamma-tion, and prolonged exposure to anesthetics, has not been thoroughly explored in the context of FUS + MB exposures.

An alternative explanation for the different types of leakage observed following sonication may be that sonoporation of ECs contributes largely to focal disruptions and that slow leakage is chiefly the product of paracellular diffusion; the delayed presentation of slow leakage may be due to the time required for substances to dif-fuse through the inter-endothelial clefts and basement membranes surrounding cerebral vasculature and accumulate in quantities sufficient for detection. Indeed, ultrasound and MB-mediated sonoporation of cell membranes have been demon-strated extensively in vitro at MIs and burst lengths below what is typically employed to produce BBB permeability enhancement in vivo (Fan et al. 2012; Park et al. 2010, 2011). Further evidence for the occurrence of sonoporation during FUS + MB exposures comes from Sheikov et al. who observed EC fenestrations and channel formation in targeted rabbit cerebrovascular 1–2 h following sonication (Sheikov et al. 2004).

In discussing the effects of sonication on brain vasculature, it is important to consider that the BBB is more than a physical barrier. Under normal physiological conditions, efflux transporters actively contribute to the removal of foreign molecules and waste from the brain. Studies have demonstrated that FUS + MB exposure induces a reduction in the cerebrovascular immunoreactivity of P-glycoprotein (P-gp), an efflux transporter with a high affinity for many therapeutic agents, in healthy rats (Cho et al. 2016; Aryal et al. 2017; Choi et al. 2019). These observations have since been substantiated at the level of transcription with a downregulation in the expression of several BBB transporter genes, including *Abcb1a*, observed 6 h following sonication (McMahon et al. 2017). This effect appears to be related to the magnitude of BBB permeability enhancement, as reductions in the immunoreactivity of P-gp were found to return to control levels before 72 h when sonicating with a mechanical index of 0.66 but remained suppressed at this time point with more intense exposures (Aryal et al. 2017).

19.4.2.2 Extravascular Effects

Beyond direct effects on vasculature, FUS + MB exposure has also been shown to produce cellular and biochemical changes in brain parenchyma. Perhaps the best characterized effects, neurogenesis and inflammation, have been observed in several animal models and under a variety of exposure conditions.

Adult hippocampal neurogenesis can occur in the subgranular layer of the dentate gyrus and is often quantified by bromodeoxyuridine (BrdU; an exogenous molecule that is incorporated into the DNA of dividing cells) and mature neuronal marker immunodetection. Scarcelli et al. first demonstrated that unilateral sonication of the hippocampus results in a significant increase in the number of cells in the dentate gyrus that are double positive for BrdU and NeuN (i.e., mature neurons), compared to the contralateral hemisphere (Scarcelli et al. 2014). Mooney et al. showed that proliferation and survival of newborn hippocampal neurons are dependent on producing BBB permeability enhancement, as sonications at low PNP or at high PNP without MB administration (i.e., conditions with no detectable effect on BBB permeability) did not generate increases in neurogenesis (Mooney et al. 2016). Others have also shown that repeated FUS + MB exposures lead to a significant increase in dendritic branching and total dendritic length in granule neurons in the dentate gyrus (Burgess et al. 2014a).

There is a large body of literature examining changes in tissue health following FUS + MB exposures as assessed by basic histological stains, such as H&E (Alli et al. 2018; Jones et al. 2018; O'Reilly and Hynynen 2012a; Baseri et al. 2010), Prussian blue (McDannold et al. 2012), and vanadium acid fuchsin (Hynynen et al. 2003, 2005); however, the described effects on tissue health vary considerably between studies. Some note very low levels of RBC extravasations with rare occurrences of darkly stained, potentially ischemic neurons at 1–24 h following sonication (Sun et al. 2017); others have observed dilated blood vessels, astroglial scars, and metallophagocytic cells (i.e., microglia or macrophages that have phagocytosed RBCs) 13 weeks following FUS + MB exposure (Kovacs et al. 2018a). These varied observations emphasize the necessity of considering both the FUS + MB parameters employed and the amount of time that has passed between sonication and euthanasia. The former affects the magnitude of impact on tissue health, and the latter influences the opportunity for lesion formation or tissue repair.

The degree and duration of FUS + MB-induced inflammation (reviewed in McMahon et al. (McMahon et al. 2019)) have been a topic of debate (Kovacs et al. 2018a; McMahon and Hynynen 2017, 2018; Kovacs et al. 2017a; Silburt et al. 2017; Kovacs et al. 2017b, 2018b), with a wide range of effects reported. Upregulation in the transcription of proinflammatory cytokines and chemokines has been observed in microvasculature (McMahon et al. 2017) and whole tissue (McMahon and Hynynen 2017; Kovacs et al. 2017a) from targeted regions in the hours following sonication. Protein expression changes in proinflammatory markers, such as ICAM1 and MCP1, have also been noted (Kovacs et al. 2017a; McMahon et al. 2020). Of note, however, is the magnitude of these responses; as an example, upregulation in the transcription of *Icam1* in whole tissue in the focal volume relative to non-sonicated brain tissue at 6 h following FUS + MB exposure has been reported to be as high as a 6.35-fold increase (Kovacs et al. 2017a) and as low as a 1.58-fold increase (McMahon and Hynynen 2017). Differences in MB dose and PNP have been shown to influence this disparity (McMahon and Hynynen 2017).

Similarly, both morphological (Jordão et al. 2013; Leinenga and Götz 2015; McMahon et al. 2020) and biochemical indication (Jordão et al. 2013; McMahon et al. 2020) of glial cell activation have been observed in the days following sonication, with a return to basal conditions evident 4–15 days following FUS + MB exposure (Jordão et al. 2013). Conversely, with sonication parameters that produce radiological indications of hemorrhage, biochemical evidence of astrocytic and microglial activation was apparent for at least 7 weeks following sonication (Kovacs et al. 2018a). When comparing results from these studies, it is critical to consider the relationship between magnitude of initial BBB permeability enhancement and subsequent acute inflammatory processes. At 6 h following sonication, strong positive correlations between relative signal intensity changes on CE-T1w images (indicative of BBB permeability enhancement) and relative mRNA levels of proinflammatory markers, such as *Ccl5*, *Icam1*, and *Tnf*, have been observed (McMahon and Hynynen 2017). Also of note, the duration of BBB permeability enhancement, expression of proinflammatory markers, blood vessel growth (McMahon et al. 2018), and astrocyte activation can be reduced via administration of dexamethasone following FUS + MB exposure (McMahon et al. 2020).

Macrophage infiltration, as assessed by H&E staining (Kobus et al. 2016; McDannold et al. 2005, 2012; Hynynen et al. 2006), CD68 immunodetection (Leinenga and Götz 2015; Kovacs et al. 2017a, 2018a), and MRI of superparamagnetic iron oxide nanoparticle-labeled cells (Liu et al. 2010b), has been reported hours (Liu et al. 2010b) to weeks (McDannold et al. 2012; Kovacs et al. 2018a) following sonication. As with the expression of proinflammatory cytokines and chemokines, at low PNP, no or few macrophages have been detected in the brain following sonications (Kobus et al. 2016; Liu et al. 2010b). Given the goal of FUS + MB exposure in this context is to increase BBB permeability for the chief purpose of drug delivery, it is unsurprising that some level of inflammatory response has been observed. In moving toward clinical implementation, the risks associated with transient inflammation must be weighed against the benefits expected from therapeutic agent delivery.

19.5   Acoustic Feedback Control

Heterogeneity in MB dispersion and regional variance in vascularity can lead to variable effects of FUS + MB exposure on BBB permeability throughout the brain (McDannold et al. 2012; Samiotaki et al. 2017; Wu et al. 2016). Additionally, inaccuracies in estimating in situ ultrasound pressure in the focal volume (i.e., caused by skull distortions, non-normal ultrasound propagation, standing waves, etc.) can complicate the prediction of FUS + MB-associated bioeffects (Pichardo et al. 2011; Pulkkinen et al. 2014). While fixed PNPs have commonly been employed in preclinical work, the relatively narrow safety window for achieving a clinically relevant increase in BBB permeability without inducing overt tissue damage (e.g., microhemorrhages) (McDannold et al. 2012) necessitates methods of monitoring and controlling FUS + MB exposures in real time.

The volumetric oscillations and surface vibrations of ultrasound-stimulated MBs generate pressure waves that are emitted in all directions. Assessing the spectral frequency content of acoustic emissions (collected during sonication with one or many hydrophones) from insonated MBs can give insight into their in vivo behavior (Faez et al. 2011; Sijl et al. 2011) (Fig. 19.2). At PNPs that elicit stable cavitation, an increase in the magnitude of acoustic emissions at harmonics of the transmit frequency (f) can be observed (e.g., $2f$, $3f$, etc.) (Leighton 1994). If the pressure amplitude is increased above a threshold value, nonlinear volumetric oscillations can lead to the generation of sub- and ultraharmonic emissions (subharmonic = $0.5f$, first ultraharmonic = $1.5f$, second ultraharmonic = $2.5f$, etc.) (Prosperetti 2013) (Fig. 19.2). As the applied PNP is further increased, inertial cavitation will occur. This behavior is characterized by a sharp increase in the production of wideband emissions (Leighton 1994) (Fig. 19.2).

The violent collapse of MBs at high PNPs can result in ischemia, apoptosis, necrosis, edema, and hemorrhage if sustained over a sufficient number of bursts or at a high enough magnitude (Vykhodtseva et al. 2006). McDannold et al. demonstrated that as the magnitude of wideband emissions (averaged over all bursts) increases, the probability of observing RBC extravasation also increases (McDannold et al. 2006); efforts to reduce inertial cavitation are essential in the context of FUS + MB-mediated BBB permeability enhancement. Studies have also demonstrated that increased BBB permeability can be achieved without wideband emissions indicative of inertial cavitation (Tung et al. 2010; McDannold et al. 2006). While sonications that elicit stable cavitation can impact vascular permeability without overt tissue damage (Tung et al. 2010; McDannold et al. 2012), if the magnitude of stress generated by stably oscillating MBs is sufficient, blood vessel rupture can occur (Chen et al. 2011). Thus, both the prevention of wideband emissions and achieving precise in situ PNP—with well-characterized effects on tissue health—are essential for minimizing the risk of overt tissue damage induced by FUS + MB exposures.

Several strategies have been developed to control PNP in real time based on acoustic emissions, each with strengths and weaknesses. One strategy is to tune PNP in order to produce an empirically determined magnitude of harmonic emissions (Arvanitis et al. 2012). Using this concept, Sun et al. developed a closed-loop algorithm that adjusts PNP based on the magnitude of emissions at multiple harmonic frequencies and that suppresses wideband emissions. This strategy was shown to be effective in controlling the degree of BBB permeability enhancement elicited by FUS + MB exposure as well as in producing predictable delivery of doxorubicin to targeted tissue in an F98 rat tumor model (Sun et al. 2017). A potential drawback of this approach is in its reliance on an empirically determined setpoint for harmonic emissions, which may be influenced by a host of factors (e.g., animal model, MB type, MB concentration, acoustic field, transmit frequency, hydrophone sensitivity, etc.). Additionally, this control strategy relies on the magnitude of signals emitted by a population of MBs in the focal volume; thus, there is a dependence on the spatial distribution of MBs within brain vasculature. Despite these limitations, control over the degree of sonication-induced BBB permeability enhancement represents a substantial advancement in this area.

Alternatively, FUS + MB exposures can be controlled by incrementally increasing PNP until the detection of a threshold event, such as ultraharmonic (O'Reilly and Hynynen 2012b) or subharmonic emissions (Burgess et al. 2014a), then scaling to a fraction of the threshold-triggering PNP. Rather than continually modulating PNP to produce a consistent magnitude of harmonic emissions, this strategy uses the occurrence of a threshold event to establish an in situ reference pressure and then scales PNP to an empirically determined fraction. The effectiveness of this strategy in consistently inducing BBB permeability enhancement while minimizing the risk of overt tissue damage was first demonstrated by O'Reilly and Hynynen (O'Reilly and Hynynen 2012a). In this work, a linear relationship was observed between the scaling factor after a threshold event and mean intensity on CE-T1w images (i.e., indicative of vascular permeability). This general strategy has been employed in the first clinical trials to elicit changes in BBB permeability byway of transcranial ultrasound propagation (Lipsman et al. 2018; Mainprize et al. 2019; Abrahao et al. 2019). As this strategy relies on predictable MB responses to insonation at a given in situ pressure, any factors that alter this relationship will influence the resultant bioeffects induced by sonications using this approach.

Additional drawbacks include the potential need to adjust the scaling factor for MB type and hydrophone sensitivity. Technological advancements have seen spatial information incorporated into this method of acoustic feedback control through the use of three-dimensional beamforming of subharmonic emissions (Jones et al. 2018).

Regardless of the acoustic feedback control strategy employed, it is important to consider several technical details, such as the interaction between MB size distribution and the pressure at which acoustic emissions are detectable above baseline noise. For most commercially available formulations, MB size is polydispersed. Since the resonance frequency of a single MB is largely influenced by its size, as PNP is increased, a growing proportion of the MB population will generate acoustic emissions at commonly monitored bandwidths (i.e., harmonic, subharmonic, or wideband emissions). The pressure at which a sufficient number of MBs are generating signals that are detectable above baseline noise will be influenced by the sensitivity of the detector(s) implemented and will impact the biological outcomes induced by any acoustic feedback control algorithm.

Strategies for monitoring and controlling sonications in real time have been essential for improving consistency and reducing the risks associated FUS + MB exposures. This has enabled the movement of transcranial FUS + MB exposure into clinical testing. While substantial advancements have been made, efforts continue to be directed at improving the accuracy of predicting biological outcomes based on acoustic emissions (Jones et al. 2018).

19.6   Clinical Trials

As of December 2019, there are eight active clinical trials recruiting participants for ultrasound and MB exposure-mediated BBB permeability enhancement (ClinicalTrials.gov Identifiers: NCT03744026, NCT03739905, NCT03551249, NCT03714243, NCT03608553, NCT03616860, NCT03671889, and NCT03712293). Thus far, studies have demonstrated the ability to increase cerebrovascular permeability with minimal short-term side effects and no evidence of long-term side effects in human participants. Ongoing trials are focused on determining the safety of using ultrasound and MB exposures to increase BBB permeability in a variety of pathological contexts, including glioblastoma, brain metastases, AD, and Parkinson's disease. Demonstrating a high safety profile in these studies will enable future work to explore ultrasound and MB-mediated delivery of therapeutic agents, as well as the sonication of larger tissue volumes. The following sections will summarize results from clinical trials completed to date involving both transducer implantation and transcranial approaches.

19.6.1 Implanted Ultrasound Device Approach

In July of 2014, the first inhuman use of ultrasound and MB exposure for the purpose of increasing BBB permeability was demonstrated in Paris, France. This trial involved the implantation of an unfocused single element ultrasound device system (SonoCloud-1, CarThera) into the skulls of 17 participants with recurrent glioblastoma, avoiding the complications of transcranial ultrasound propagation but necessitating an invasive surgical procedure. A range of fixed PNPs were employed to induce increased BBB permeability, after which carboplatin, a chemotherapeutic agent, was administered. Sonications were repeated two to four times, monthly. Carpentier et al. reported that participants tolerated the procedure well, with no evidence of acute hemorrhage, ischemia, or edema in images acquired with susceptibility-weighted angiography, diffusion, or fluid-attenuated inversion recovery (FLAIR) sequences. CE-T1w scans showed heterogeneous BBB permeability increases in the acoustic field indicating feasibility. In the subsequent hours and days following sonication, clinical symptoms relating to the procedure were not present in any participants, including the 11 epileptic patients. Two adverse events, deemed unrelated to the procedure, occurred during the trial (Carpentier et al. 2016).

In a follow-up report, Idbaih et al. described the results from 19 participants with SonoCloud implantation and who received carboplatin following one to ten ultrasound and MB exposures. When comparing median progression-free survival and median overall survival in subjects with no or little sonication-mediated BBB permeability enhancement visible on CE-T1w images to those with clear enhancement, the later groups displayed trends toward improvement. Future studies are aimed at implanting devices that can affect larger volumes of tissue. Additionally, the authors have expressed the need to reduce the delay between sonication and the administration of chemotherapeutic agents in order to achieve a higher concentration in the targeted tissue (Idbaih et al. 2019).

19.6.2 Transcranial Approach

Phase one clinical trials conducted at *Sunnybrook Research Institute* in Toronto, Canada, were the first to utilize MRI-guided (MRIg) transcranial ultrasound exposures to increase BBB permeability in human participants. Clinical trials to date have employed the ExAblate Neuro system (INSIGHTEC, Haifa, Israel), a hemispherical phased array containing 1024 elements driven at 220 kHz (Lipsman et al. 2017). In the first trial, peritumoral tissue was targeted in five participants with malignant glioma (Mainprize et al. 2019). Sonications, in conjunction with chemotherapy (liposomal doxorubicin or temozolomide), were performed one day prior to surgical resection. The volume of tissue targeted in this study ranged from 972 to 2430 mm^3. Appropriate sonicating pressures were determined to be 50% of the input power required to detect subharmonic signals at each targeted volume, based on previous work (Huang et al. 2017). Increased signal intensity on CE-T1w images (i.e., increased BBB permeability) was observed in four of the five participants following sonication. Authors reported no new or worsening symptoms in the 24 h between FUS + MB exposure and tumor resection; neurological exams in this period were normal. Sonicated and non-sonicated regions of resected tissue were collected and analyzed in two participants, revealing increased chemotherapeutic agent concentrations in sonicated tissue in both cases (Mainprize et al. 2019).

The second phase one trial was completed in participants with AD. In this study, a presumed non-eloquent region, the superior frontal gyrus white matter of the dorsolateral prefrontal cortex, was targeted in five participants with mild to moderate AD. Two MRIgFUS procedures, separated by 1 month, were performed in each participant, with the volume of targeted tissue doubling in the second stage (approximately 350 mm^3 and 700 mm^3, respectively). This study employed the same hemispherical array and acoustic feedback control strategy as the previously discussed trial (Mainprize et al. 2019). During the study, no participant presented with clinical symptoms believed to be related to the procedure nor displayed persistent BBB permeability enhancement on CE-T1w images collected 24 h following sonication (i.e., the first follow-up time point). Hypointensities on T2*w images were observed in two participants immediately following sonication, both of which resolved within 24 h. Psychometric tests interrogating cognition and daily functioning revealed no clinically significant changes between pre- and 3 months following FUS + MB exposure (Lipsman et al. 2018). Resting state functional connectivity in the bilateral frontoparietal networks was also assessed in these participants using functional MRI (fMRI). Authors reported a transient reduction in functional connectivity in the frontoparietal network ipsilateral to sonication immediately following the procedure which recovered to baseline by the next day (Meng et al. 2019a).

The safety of MRIgFUS for increasing BBB permeability has also been assessed in participants with amyotrophic lateral sclerosis (ALS). In this trial, 350 mm^3 of tissue in the primary motor cortex was targeted from fMRI scans in four patients with ALS. Increases in BBB permeability were detected immediately following sonication in all cases and returned to baseline by 24 h (i.e., the first follow-up time point). An asymptomatic hyperintense FLAIR signal was detected in one subject 1 day following the procedure which resolved within 7 days. This signal did not correlate with new symptoms, neurological signs, or focal electroencephalography changes. No evidence of radiographic changes were noted in MR images collected with diffusion weighted imaging, gradient echo, T2w, or T2*w sequences in any participants immediately following sonication or at 1, 7, and 30 days. Similarly, no changes in neurological status were observed in the 24 h observation period following FUS + MB exposure. Three participants did, however, report moderate headaches during the MRIgFUS procedure (Abrahao et al. 2019).

Of interest, analysis of FLAIR images from subjects in the AD and ALS trials revealed indications of contrast agent clearance in the perivascular spaces (Meng et al. 2019b). In a subset of participants from these studies (3/8 AD subjects and 1/4

ALS subjects), signal intensity changes from gadobutrol (i.e., MRI contrast agent) was observed away from the sonicated regions in the perivascular spaces, subarachnoid spaces, and surrounding large draining veins. Authors suggest that this is the first evidence of local glymphatic efflux dynamics in human literature and that these findings are consistent with previous observations describing the glymphatic clearance pathway in rodents (Iliff et al. 2012).

In summary, phase one clinical trials for transcranial MRIgFUS have demonstrated that it is possible to transiently increase BBB permeability without radiologic indications of tissue damage. Study participants have reported mild to moderate headaches during the procedure (Abrahao et al. 2019) and exhibited transient reductions in resting state connectivity in the targeted location (Meng et al. 2019a), but have not presented with neurological indications believed to be related to FUS + MB exposures in the acute period following sonication (Lipsman et al. 2018; Mainprize et al. 2019; Abrahao et al. 2019). Based on limited data from the brain tumor trial, there are indications that FUS + MB-mediated increases in drug delivery are possible with this technique (Mainprize et al. 2019). Ongoing trials are aimed at sonicating larger tissue volumes and at increasing the number of exposures each participant receives. Promising results thus far encourage the movement of this technique toward combination with therapeutic agent delivery.

19.7 Conclusion

Transcranial FUS + MB exposures offer a noninvasive means of increasing BBB permeability in a targeted, transient, and controlled manner. During the hours following sonication, therapeutic agents can be systemically administered and extravate in the targeted locations. Beyond its targeted nature, the major strength of this approach lies in its flexibility, providing a means of delivering a wide range of therapeutic agents to either precise locations or large volumes. Since publication of the first experiments demonstrating BBB permeability enhancement using FUS and circulating MBs, the field has rapidly grown. Technological advances, biological characterization, efficacious preclinical results, and a great need for flexible brain-drug delivery strategies have moved this technique into clinical testing for a range of neuropathologies. Efforts continue to be directed at developing methods to improve the safety profile of FUS + MB exposures, through advancements in acoustic monitoring and control, as well as establishing more precise relationships between acoustic emissions and the biological impacts of sonications. This knowledge will not only allow detailed risk assessment and strategic treatment planning but may also encourage the design of novel therapies that utilize the FUS + MB-induced activation of specific signaling pathways.

References

Abrahao A, Meng Y, Llinas M, Huang Y, Hamani C, Mainprize T et al (2019) First-in-human trial of blood-brain barrier opening in amyotrophic lateral sclerosis using MR-guided focused ultrasound. Nat Commun 10:4373

Alkins R, Burgess A, Ganguly M, Francia G, Kerbel R, Wels WS et al (2013) Focused ultrasound delivers targeted immune cells to metastatic brain tumors. Cancer Res 73:1892–1899

Alli S, Figueiredo CA, Golbourn B, Sabha N, Wu MY, Bondoc A et al (2018) Brainstem blood brain barrier disruption using focused ultrasound: a demonstration of feasibility and enhanced doxorubicin delivery. J Control Release 281:29–41

Alonso A, Reinz E, Fatar M, Hennerici MG, Meairs S (2011) Clearance of albumin following ultrasound-induced blood–brain barrier opening is mediated by glial but not neuronal cells. Brain Res 1411:9–16

Apfel RE, Holland CK (1991) Gauging the likelihood of cavitation from short-pulse, low-duty cycle diagnostic ultrasound. Ultrasound Med Biol 17:179–185

Arvanitis CD, Livingstone MS, Vykhodtseva N, McDannold N (2012) Controlled ultrasound-induced blood-brain barrier disruption using passive acoustic emissions monitoring. PLoS One 7:e45783

Arvanitis CD, Askoxylakis V, Guo Y, Datta M, Kloepper J, Ferraro GB et al (2018) Mechanisms of enhanced drug delivery in brain metastases with focused ultrasound-induced blood-tumor barrier disruption. Proc Natl Acad Sci U S A [Internet]. Available from: https://doi.org/10.1073/pnas.1807105115

Aryal M, Vykhodtseva N, Zhang Y-Z, Park J, McDannold N (2013) Multiple treatments with liposomal doxorubicin and ultrasound-induced disruption of blood-tumor and blood-brain barriers improve outcomes in a rat glioma model. J Control Release 169:103–111

Aryal M, Fischer K, Gentile C, Gitto S, Zhang Y-Z, McDannold N (2017) Effects on P-glycoprotein expression after blood-brain barrier disruption using focused ultrasound and microbubbles. PLoS One 12:e0166061

Bakay L, Ballantine HT Jr, Hueter TF, Sosa D (1956) Ultrasonically produced changes in the blood-brain barrier. AMA Arch Neurol Psychiatry 76:457–467

Ballantine HT Jr, Bell E, Manlapaz J (1960) Progress and problems in the neurological applications of focused ultrasound. J Neurosurg 17:858–876

Baseri B, Choi JJ, Tung Y-S, Konofagou EE (2010) Multi-modality safety assessment of blood-brain barrier opening using focused ultrasound and definity microbubbles: a short-term study. Ultrasound Med Biol Elsevier 36:1445–1459

Baseri B, Choi JJ, Deffieux T, Samiotaki G, Tung Y-S, Olumolade O et al (2012) Activation of signaling pathways following localized delivery of systemically administered neurotrophic factors across the blood-brain barrier using focused ultrasound and microbubbles. Phys Med Biol 57:N65–N81

Büning H, Mingozzi F (2015) Chapter 4 - immune system obstacles to in vivo gene transfer with adeno-associated virus vectors. In: Laurence J, Franklin M (eds) Translating gene therapy to the clinic. Academic Press, Boston, pp 45–64

Burgess A, Ayala-Grosso CA, Ganguly M, Jordão JF, Aubert I, Hynynen K (2011) Targeted delivery of neural stem cells to the brain using MRI-guided focused ultrasound to disrupt the blood-brain barrier. PLoS One 6:e27877

Burgess A, Huang Y, Querbes W, Sah DW, Hynynen K (2012) Focused ultrasound for targeted delivery of siRNA and efficient knockdown of Htt expression. J Control Release 163:125–129

Burgess A, Dubey S, Yeung S, Hough O, Eterman N, Aubert I et al (2014a) Alzheimer disease in a mouse model: MR imaging-guided focused ultrasound targeted to the hippocampus opens the blood-brain barrier and improves pathologic abnormalities and behavior. Radiology 273:736–745

Burgess A, Nhan T, Moffatt C, Klibanov AL, Hynynen K (2014b) Analysis of focused ultrasound-induced blood–brain barrier permeability in a mouse model of Alzheimer's disease using two-photon microscopy. J Control Release 192:243–248

Bystritsky A, Korb AS, Douglas PK, Cohen MS, Melega WP, Mulgaonkar AP et al (2011) A review of low-intensity focused ultrasound pulsation. Brain Stimul 4:125–136

Carpentier A, Canney M, Vignot A, Reina V, Beccaria K, Horodyckid C et al (2016) Clinical trial of blood-brain barrier disruption by pulsed ultrasound. Sci Transl Med. American Association for the Advancement of Science 8:343re2–343re2

Chen H, Kreider W, Brayman AA, Bailey MR, Matula TJ (2011) Blood vessel deformations on microsecond time scales by ultrasonic cavitation. Phys Rev Lett 106:034301

Cho EE, Drazic J, Ganguly M, Stefanovic B, Hynynen K (2011) Two-photon fluorescence microscopy study of cerebrovascular dynamics in ultrasound-induced blood—brain barrier opening. J Cereb Blood Flow Metab 31:1852–1862

Cho H, Lee H-Y, Han M, Choi J-R, Ahn S, Lee T et al (2016) Localized Down-regulation of P-glycoprotein by focused ultrasound and microbubbles induced blood-brain barrier disruption in rat brain. Sci Rep 6:31201

Choi H, Lee E-H, Han M, An S-H, Park J (2019) Diminished expression of P-glycoprotein using focused ultrasound is associated with JNK-dependent signaling pathway in cerebral blood vessels. Front Neurosci 13:1350

Chu P-C, Liu H-L, Lai H-Y, Lin C-Y, Tsai H-C, Pei Y-C (2015) Neuromodulation accompanying focused ultrasound-induced blood-brain barrier opening. Sci Rep 5:15477

Clément N (2019) Large-scale clinical manufacturing of AAV vectors for systemic muscle gene therapy. In: Duan D, Mendell JR (eds) Muscle gene therapy. Springer, Cham, pp 253–273

Clement GT, Hynynen K (2002) A non-invasive method for focusing ultrasound through the human skull. Phys Med Biol 47:1219–1236

Cline HE, Hynynen K, Hardy CJ, Watkins RD, Schenck JF, Jolesz FA (1994) MR temperature mapping of focused ultrasound surgery. Magn Reson Med 31:628–636

Dayton P, Klibanov A, Brandenburger G, Ferrara K (1999) Acoustic radiation force in vivo: a mechanism to assist targeting of microbubbles. Ultrasound Med Biol 25:1195–1201

Deng J, Huang Q, Wang F, Liu Y, Wang Z, Wang Z et al (2012) The role of caveolin-1 in blood-brain barrier disruption induced by focused ultrasound combined with microbubbles. J Mol Neurosci 46:677–687

Dong Q, He L, Chen L, Deng Q (2018) Opening the blood-brain barrier and improving the efficacy of Temozolomide treatments of glioblastoma using pulsed, focused ultrasound with a micro-bubble contrast agent. Biomed Res Int 2018:6501508

Doolittle ND, Peereboom DM, Christoforidis GA, Hall WA, Palmieri D, Brock PR et al (2007) Delivery of chemotherapy and antibodies across the blood-brain barrier and the role of chemo-protection, in primary and metastatic brain tumors: report of the eleventh annual blood-brain barrier consortium meeting. J Neuro-Oncol 81:81–91

Downs ME, Buch A, Sierra C, Karakatsani ME, Teichert T, Chen S et al (2015) Long-term safety of repeated blood-brain barrier opening via focused ultrasound with microbubbles in non-human primates performing a cognitive task. PLoS One 10:e0125911

Ebbini ES, Cain CA (1991) Experimental evaluation of a prototype cylindrical section ultrasound hyperthermia phased-array applicator. IEEE Trans Ultrason Ferroelectr Freq Control 38:510–520

Elias WJ, Huss D, Voss T, Loomba J, Khaled M, Zadicario E et al (2013) A pilot study of focused ultrasound thalamotomy for essential tremor. N Engl J Med 369:640–648

Faez T, Emmer M, Docter M, Sijl J, Versluis M, de Jong N (2011) Characterizing the subharmonic response of phospholipid-coated microbubbles for carotid imaging. Ultrasound Med Biol 37:958–970

Fan Z, Liu H, Mayer M, Deng CX (2012) Spatiotemporally controlled single cell sonoporation. Proc Natl Acad Sci U S A 109:16486–16491

Fan C-H, Ting C-Y, Lin H-J, Wang C-H, Liu H-L, Yen T-C et al (2013a) SPIO-conjugated, doxorubicin-loaded microbubbles for concurrent MRI and focused-ultrasound enhanced brain-tumor drug delivery. Biomaterials 34:3706–3715

Fan C-H, Ting C-Y, Liu H-L, Huang C-Y, Hsieh H-Y, Yen T-C et al (2013b) Antiangiogenic-targeting drug-loaded microbubbles combined with focused ultrasound for glioma treatment. Biomaterials 34:2142–2155

Fazekas F, Kleinert R, Roob G, Kleinert G, Kapeller P, Schmidt R et al (1999) Histopathologic analysis of foci of signal loss on gradient-echo T2*-weighted MR images in patients with spontaneous intracerebral hemorrhage: evidence of microangiopathy-related microbleeds. AJNR Am J Neuroradiol Am Soc Neuroradiol 20:637–642

Fry FJ (1977) Transkull transmission of an intense focused ultrasonic beam. Ultrasound Med Biol 3:179–184

Fry FJ, Goss SA (1980) Further studies of the transkull transmission of an intense focused ultrasonic beam: lesion production at 500 khz. Ultrasound Med Biol 6:33–38

Fyfe MC, Chahl LA (1984) Mast cell degranulation and increased vascular permeability induced by "therapeutic" ultrasound in the rat ankle joint. Br J Exp Pathol 65:671–676

Goertz DE (2015) An overview of the influence of therapeutic ultrasound exposures on the vasculature: high intensity ultrasound and microbubble-mediated bioeffects. Int J Hyperth 31:134–144

Goertz DE, de Jong N, Van Der Steen AFW (2007) Attenuation and size distribution measurements of Definity and manipulated Definity populations. Ultrasound Med Biol 33:1376–1388

Gramiak R, Shah PM (1968) Echocardiography of the aortic root. Investig Radiol 3:356–366

Grayburn PA (2002) Current and future contrast agents. Echocardiography 19:259–265

Hosseinkhah N, Hynynen K (2012) A three-dimensional model of an ultrasound contrast agent gas bubble and its mechanical effects on microvessels. Phys Med Biol 57:785–808

Hsu P-H, Wei K-C, Huang C-Y, Wen C-J, Yen T-C, Liu C-L et al (2013) Noninvasive and targeted gene delivery into the brain using microbubble-facilitated focused ultrasound. PLoS One 8:e57682

Huang Q, Deng J, Wang F, Chen S, Liu Y, Wang Z et al (2012) Targeted gene delivery to the mouse brain by MRI-guided focused ultrasound-induced blood-brain barrier disruption. Exp Neurol 233:350–356

Huang Y, Alkins R, Schwartz ML, Hynynen K (2017) Opening the blood-brain barrier with MR imaging–guided focused ultrasound: preclinical testing on a trans–human skull porcine model. Radiology Radiol Soc North Am 282:123–130

Hynynen K, Jolesz FA (1998) Demonstration of potential noninvasive ultrasound brain therapy through an intact skull. Ultrasound Med Biol 24:275–283

Hynynen K, Watmough DJ, Mallard JR (1981) Design of ultrasonic transducers for local hyperthermia. Ultrasound Med Biol 7:397–402

Hynynen K, McDannold N, Vykhodtseva N, Jolesz FA (2001) Noninvasive MR imaging-guided focal opening of the blood-brain barrier in rabbits. Radiology 220:640–646

Hynynen K, McDannold N, Martin H, Jolesz FA, Vykhodtseva N (2003) The threshold for brain damage in rabbits induced by bursts of ultrasound in the presence of an ultrasound contrast agent (Optison). Ultrasound Med Biol 29:473–481

Hynynen K, McDannold N, Sheikov NA, Jolesz FA, Vykhodtseva N (2005) Local and reversible blood-brain barrier disruption by noninvasive focused ultrasound at frequencies suitable for trans-skull sonications. NeuroImage 24:12–20

Hynynen K, McDannold N, Vykhodtseva N, Raymond S, Weissleder R, Jolesz FA et al (2006) Focal disruption of the blood-brain barrier due to 260-kHz ultrasound bursts: a method for molecular imaging and targeted drug delivery. J Neurosurg 105:445–454

Idbaih A, Canney M, Belin L, Desseaux C, Vignot A, Bouchoux G et al (2019) Safety and feasibility of repeated and transient blood-brain barrier disruption by pulsed ultrasound in patients with recurrent glioblastoma. Clin Cancer Res:3793–3801

Iliff JJ, Wang M, Liao Y, Plogg BA, Peng W, Gundersen GA et al (2012) A paravascular pathway facilitates CSF flow through the brain parenchyma and the clearance of interstitial solutes, including amyloid β. Sci Transl Med 4:147ra111

Jones RM, Deng L, Leung K, McMahon D, O'Reilly MA, Hynynen K (2018) Three-dimensional transcranial microbubble imaging for guiding volumetric ultrasound-mediated blood-brain barrier opening. Theranostics 8:2909–2926

Jordão JF, Ayala-Grosso CA, Markham K, Huang Y, Chopra R, McLaurin J et al (2010) Antibodies targeted to the brain with image-guided focused ultrasound reduces amyloid-beta plaque load in the TgCRND8 mouse model of Alzheimer's disease. PLoS One 5:e10549

Jordão JF, Thévenot E, Markham-Coultes K, Scarcelli T, Weng Y-Q, Xhima K et al (2013) Amyloid-β plaque reduction, endogenous antibody delivery and glial activation by brain-targeted, transcranial focused ultrasound. Exp Neurol 248:16–29

Karakatsani ME, Kugelman T, Ji R, Murillo M, Wang S, Niimi Y et al (2019) Unilateral focused ultrasound-induced blood-brain barrier opening reduces phosphorylated tau from the rTg4510 mouse model. Theranostics 9:5396–5411

Kinoshita M, McDannold N, Jolesz FA, Hynynen K (2006a) Targeted delivery of antibodies through the blood–brain barrier by MRI-guided focused ultrasound. Biochem Biophys Res Commun 340:1085–1090

Kinoshita M, McDannold N, Jolesz FA, Hynynen K (2006b) Noninvasive localized delivery of Herceptin to the mouse brain by MRI-guided focused ultrasound-induced blood-brain barrier disruption. PNAS 103:11719–11723

Kobus T, Vykhodtseva N, Pilatou M, Zhang Y, McDannold N (2016) Safety validation of repeated blood-brain barrier disruption using focused ultrasound. Ultrasound Med Biol 42:481–492

Kovacs Z, Werner B, Rassi A, Sass JO, Martin-Fiori E, Bernasconi M (2014) Prolonged survival upon ultrasound-enhanced doxorubicin delivery in two syngenic glioblastoma mouse models. J Control Release 187:74–82

Kovacs ZI, Kim S, Jikaria N, Qureshi F, Milo B, Lewis BK et al (2017a) Disrupting the blood-brain barrier by focused ultrasound induces sterile inflammation. Proc Natl Acad Sci U S A 114:E75–E84

Kovacs ZI, Burks SR, Frank JA (2017b) Reply to Silburt et al.: concerning sterile inflammation following focused ultrasound and microbubbles in the brain. Proc Natl Acad Sci U S A 114:E6737–E6738

Kovacs ZI, Tu T-W, Sundby M, Qureshi F, Lewis BK, Jikaria N et al (2018a) MRI and histological evaluation of pulsed focused ultrasound and microbubbles treatment effects in the brain. Theranostics 8:4837–4855

Kovacs ZI, Burks SR, Frank JA (2018b) Focused ultrasound with microbubbles induces sterile inflammatory response proportional to the blood brain barrier opening: attention to experimental conditions. Theranostics 8:2245–2248

Leighton TG (1994) The acoustic bubble. Academic Press, London

Leinenga G, Götz J (2015) Scanning ultrasound removes amyloid-β and restores memory in an Alzheimer's disease mouse model. Sci Transl Med 7:278ra33

Leinenga G, Götz J (2018) Safety and efficacy of scanning ultrasound treatment of aged APP23 mice. Front Neurosci 12:55

Lewin PA, Bjørnø L (1982) Acoustically induced shear stresses in the vicinity of microbubbles in tissue. J Acoust Soc Am Acoustical Soc Am 71:728–734

Lin C-Y, Hsieh H-Y, Pitt WG, Huang C-Y, Tseng I-C, Yeh C-K et al (2015) Focused ultrasound-induced blood-brain barrier opening for non-viral, non-invasive, and targeted gene delivery. J Control Release 212:1–9

Lipsman N, Schwartz ML, Huang Y, Lee L, Sankar T, Chapman M et al (2013) MR-guided focused ultrasound thalamotomy for essential tremor: a proof-of-concept study. Lancet Neurol 12:462–468

Lipsman N, Ironside S, Alkins R, Bethune A, Huang Y, Perry J et al (2017) Initial experience of blood-brain barrier opening for chemotherapeutic-drug delivery to brain tumours by MR-guided focused ultrasound. Neuro-Oncology. Oxford University Press 19:vi9–vi9

Lipsman N, Meng Y, Bethune AJ, Huang Y, Lam B, Masellis M et al (2018) Blood–brain barrier opening in Alzheimer's disease using MR-guided focused ultrasound. Nat Commun 9:2336

Liu H-L, Hua M-Y, Chen P-Y, Chu P-C, Pan C-H, Yang H-W et al (2010a) Blood-brain barrier disruption with focused ultrasound enhances delivery of chemotherapeutic drugs for glioblastoma treatment. Radiology 255:415–425

Liu H-L, Wai Y-Y, Hsu P-H, Lyu L-A, Wu J-S, Shen C-R et al (2010b) In vivo assessment of macrophage CNS infiltration during disruption of the blood-brain barrier with focused ultrasound: a magnetic resonance imaging study. J Cereb Blood Flow Metab 30:177–186

Lotfinia M, Abdollahpour-Alitappeh M, Hatami B, Zali MR, Karimipoor M (2019) Adeno-associated virus as a gene therapy vector: strategies to neutralize the neutralizing antibodies. Clin Exp Med [Internet]. Available from: https://doi.org/10.1007/s10238-019-00557-8

Lu D, Kassab GS (2011) Role of shear stress and stretch in vascular mechanobiology. J R Soc Interface 8:1379–1385

Lynn JG, Putnam TJ (1944) Histology of cerebral lesions produced by focused ultrasound. Am J Pathol 20:637–649

Lynn JG, Zwemer RL, Chick AJ, Miller AE (1942) A new method for the generation and use of focused ultrasound in experimental biology. J Gen Physiol 26:179–193

Mainprize T, Lipsman N, Huang Y, Meng Y, Bethune A, Ironside S et al (2019) Blood-brain barrier opening in primary brain tumors with non-invasive MR-guided focused ultrasound: a clinical safety and feasibility study. Sci Rep 9:321

Marty B, Larrat B, Van Landeghem M, Robic C, Robert P, Port M et al (2012) Dynamic study of blood-brain barrier closure after its disruption using ultrasound: a quantitative analysis. J Cereb Blood Flow Metab 32:1948–1958

McDannold, Hynynen K, Wolf D, Wolf G, Jolesz F (1998) MRI evaluation of thermal ablation of tumors with focused ultrasound. J Magn Reson Imaging 8:91–100

McDannold N, Vykhodtseva N, Raymond S, Jolesz FA, Hynynen K (2005) MRI-guided targeted blood-brain barrier disruption with focused ultrasound: histological findings in rabbits. Ultrasound Med Biol 31:1527–1537

McDannold N, Vykhodtseva N, Hynynen K (2006) Targeted disruption of the blood-brain barrier with focused ultrasound: association with cavitation activity. Phys Med Biol 51:793–807

McDannold N, Arvanitis CD, Vykhodtseva N, Livingstone MS (2012) Temporary disruption of the blood-brain barrier by use of ultrasound and microbubbles: safety and efficacy evaluation in rhesus macaques. Cancer Res 72:3652–3663

McDannold N, Zhang Y, Supko JG, Power C, Sun T, Peng C et al (2019) Acoustic feedback enables safe and reliable carboplatin delivery across the blood-brain barrier with a clinical focused ultrasound system and improves survival in a rat glioma model. Theranostics 9:6284–6299

McMahon D, Hynynen K (2017) Acute inflammatory response following increased blood-brain barrier permeability induced by focused ultrasound is dependent on microbubble dose. Theranostics 7:3989–4000

McMahon D, Hynynen K (2018) Reply to Kovacs et al.: concerning acute inflammatory response following focused ultrasound and microbubbles in the brain. Theranostics 8:2249–2250

McMahon D, Bendayan R, Hynynen K (2017) Acute effects of focused ultrasound-induced increases in blood-brain barrier permeability on rat microvascular transcriptome. Sci Rep 7:45657

McMahon D, Mah E, Hynynen K (2018) Angiogenic response of rat hippocampal vasculature to focused ultrasound-mediated increases in blood-brain barrier permeability. Sci Rep 8:12178

McMahon D, Poon C, Hynynen K (2019) Evaluating the safety profile of focused ultrasound and microbubble-mediated treatments to increase blood-brain barrier permeability. Expert Opin Drug Deliv [Internet]. Available from: https://doi.org/10.1080/17425247.2019.1567490

McMahon D, Oakden W, Hynynen K (2020) Investigating the effects of dexamethasone on blood-brain barrier permeability and inflammatory response following focused ultrasound and microbubble exposure. Theranostics 10:1604–1618

Mead BP, Mastorakos P, Suk JS, Klibanov AL, Hanes J, Price RJ (2016) Targeted gene transfer to the brain via the delivery of brain-penetrating DNA nanoparticles with focused ultrasound. J Control Release 223:109–117

Mei J, Cheng Y, Song Y, Yang Y (2009) Experimental study on targeted methotrexate delivery to the rabbit brain via magnetic resonance imaging–guided focused ultrasound. Of ultrasound in .... Wiley Online Library 28:871–880

Meng Y, MacIntosh BJ, Shirzadi Z, Kiss A, Bethune A, Heyn C et al (2019a) Resting state functional connectivity changes after MR-guided focused ultrasound mediated blood-brain barrier opening in patients with Alzheimer's disease. NeuroImage 200:275–280

Meng Y, Abrahao A, Heyn CC, Bethune AJ, Huang Y, Pople C et al (2019b) Glymphatics visualization after focused ultrasound induced blood-brain barrier opening in humans. Ann Neurol [Internet]. Available from: https://doi.org/10.1002/ana.25604

Mooney SJ, Shah K, Yeung S, Burgess A, Aubert I, Hynynen K (2016) Focused ultrasound-induced neurogenesis requires an increase in blood-brain barrier permeability. PLoS One 11:e0159892

Nhan T, Burgess A, Cho EE, Stefanovic B, Lilge L, Hynynen K (2013) Drug delivery to the brain by focused ultrasound induced blood-brain barrier disruption: quantitative evaluation of enhanced permeability of cerebral vasculature using two-photon microscopy. J Control Release 172:274–280

Nisbet RM, Van der Jeugd A, Leinenga G, Evans HT, Janowicz PW, Götz J (2017) Combined effects of scanning ultrasound and a tau-specific single chain antibody in a tau transgenic mouse model. Brain 140:1220–1230

O'Reilly MA, Hynynen K (2012a) Blood-brain barrier: real-time feedback-controlled focused ultrasound disruption by using an acoustic emissions-based controller. Radiology 263:96–106

O'Reilly MA, Hynynen K (2012b) Blood-brain barrier: real-time feedback-controlled focused ultrasound disruption by using an acoustic emissions–based controller. Radiology 263

O'Reilly MA, Jones RM, Barrett E, Schwab A, Head E, Hynynen K (2017) Investigation of the safety of focused ultrasound-induced blood-brain barrier opening in a natural canine model of aging. Theranostics 7:3573–3584

Pandit R, Leinenga G, Götz J (2019) Repeated ultrasound treatment of tau transgenic mice clears neuronal tau by autophagy and improves behavioral functions. Theranostics 9:3754–3767

Pardridge WM (2005) The blood-brain barrier: bottleneck in brain drug development. NeuroRx 2:3–14

Park J, Fan Z, Kumon RE, El-Sayed MEH, Deng CX (2010) Modulation of intracellular Ca2+ concentration in brain microvascular endothelial cells in vitro by acoustic cavitation. Ultrasound Med Biol 36:1176–1187

Park J, Fan Z, Deng CX (2011) Effects of shear stress cultivation on cell membrane disruption and intracellular calcium concentration in sonoporation of endothelial cells. J Biomech 44:164–169

Park J, Zhang Y, Vykhodtseva N, Jolesz FA, McDannold NJ (2012) The kinetics of blood brain barrier permeability and targeted doxorubicin delivery into brain induced by focused ultrasound. J Control Release 162:134–142

Park J, Aryal M, Vykhodtseva N, Zhang Y-Z, McDannold N (2017) Evaluation of permeability, doxorubicin delivery, and drug retention in a rat brain tumor model after ultrasound-induced blood-tumor barrier disruption. J Control Release 250:77–85

Pichardo S, Sin VW, Hynynen K (2011) Multi-frequency characterization of the speed of sound and attenuation coefficient for longitudinal transmission of freshly excised human skulls. Phys Med Biol 56:219–250

Poon CT, Shah K, Lin C, Tse R, Kim KK, Mooney S et al (2018) Time course of focused ultrasound effects on β-amyloid plaque pathology in the TgCRND8 mouse model of Alzheimer's disease. Sci Rep 8:14061

Prosperetti A (2013) A general derivation of the subharmonic threshold for non-linear bubble oscillations. J Acoust Soc Am 133:3719–3726

Pulkkinen A, Werner B, Martin E, Hynynen K (2014) Numerical simulations of clinical focused ultrasound functional neurosurgery. Phys Med Biol 59:1679–1700

Raymond SB, Skoch J, Hynynen K, Bacskai BJ (2007) Multiphoton imaging of ultrasound/Optison mediated cerebrovascular effects in vivo. J Cereb Blood Flow Metab 27:393–403

Raymond SB, Treat LH, Dewey JD, McDannold NJ, Hynynen K, Bacskai BJ (2008) Ultrasound enhanced delivery of molecular imaging and therapeutic agents in Alzheimer's disease mouse models. PLoS One. journals.plos.org 3:e2175

Roberts WW, Hall TL, Ives K, Wolf JS Jr, Fowlkes JB, Cain CA (2006) Pulsed cavitational ultrasound: a noninvasive technology for controlled tissue ablation (histotripsy) in the rabbit kidney. J Urol 175:734–738

Samiotaki G, Karakatsani ME, Buch A, Papadopoulos S, Wu SY, Jambawalikar S et al (2017) Pharmacokinetic analysis and drug delivery efficiency of the focused ultrasound-induced blood-brain barrier opening in non-human primates. Magn Reson Imaging 37:273–281

Scarcelli T, Jordão JF, O'Reilly MA, Ellens N, Hynynen K, Aubert I (2014) Stimulation of hippocampal neurogenesis by transcranial focused ultrasound and microbubbles in adult mice. Brain Stimul 7:304–307

Schwartz RB, Jones KM, Kalina P, Bajakian RL, Mantello MT, Garada B et al (1992) Hypertensive encephalopathy: findings on CT, MR imaging, and SPECT imaging in 14 cases. AJR Am J Roentgenol 159:379–383

Shealy CN, Crafts D (1965) Selective alteration of the blood-brain barrier. J Neurosurg 23:484–487

Sheikov N, McDannold N, Vykhodtseva N, Jolesz F, Hynynen K (2004) Cellular mechanisms of the blood-brain barrier opening induced by ultrasound in presence of microbubbles. Ultrasound Med Biol 30:979–989

Sheikov N, McDannold N, Jolesz F, Zhang Y-Z, Tam K, Hynynen K (2006) Brain arterioles show more active vesicular transport of blood-borne tracer molecules than capillaries and venules after focused ultrasound-evoked opening of the blood-brain barrier. Ultrasound Med Biol 32:1399–1409

Sheikov N, McDannold N, Sharma S, Hynynen K (2008) Effect of focused ultrasound applied with an ultrasound contrast agent on the tight junctional integrity of the brain microvascular endothelium. Ultrasound Med Biol 34:1093–1104

Sijl J, Vos HJ, Rozendal T, de Jong N, Lohse D, Versluis M (2011) Combined optical and acoustical detection of single microbubble dynamics. J Acoust Soc Am 130:3271–3281

Silburt J, Lipsman N, Aubert I (2017) Disrupting the blood-brain barrier with focused ultrasound: perspectives on inflammation and regeneration. Proc Natl Acad Sci U S A 114:E6735–E6736

Smith SW, Phillips DJ, Von Ramm OT, Thurstone FL (1979) Some advances in acoustic imaging through the skull. Ultrasonic Tissue. books.google.com 525:209–218

Stavarache MA, Petersen N, Jurgens EM, Milstein ER, Rosenfeld ZB, Ballon DJ et al (2018) Safe and stable noninvasive focal gene delivery to the mammalian brain following focused ultrasound. J Neurosurg 1–10

Sun T, Zhang Y, Power C, Alexander PM, Sutton JT, Aryal M et al (2017) Closed-loop control of targeted ultrasound drug delivery across the blood-brain/tumor barriers in a rat glioma model. Proc Natl Acad Sci U S A 114:E10281–E10290

Takahashi M, Ikeda U, Sekiguchi H, Fujikawa H, Shimada K, Ri T (1996) Guide wire-induced coronary artery spasm during percutaneous transluminal coronary angioplasty. A case report. Angiology 47:305–309

Thévenot E, Jordão JF, O'Reilly MA, Markham K, Weng Y-Q, Foust KD et al (2012) Targeted delivery of self-complementary adeno-associated virus serotype 9 to the brain, using magnetic resonance imaging-guided focused ultrasound. Hum Gene Ther 23:1144–1155

Thomas J-L, Fink MA (1996) Ultrasonic beam focusing through tissue inhomogeneities with a time reversal mirror: application to transskull therapy. IEEE Trans Ultrason Ferroelectr Freq Control 43:1122–1129

Treat LH, McDannold N, Vykhodtseva N, Zhang Y, Tam K, Hynynen K (2007) Targeted delivery of doxorubicin to the rat brain at therapeutic levels using MRI-guided focused ultrasound. Int J Cancer 121:901–907

Treat LH, McDannold N, Zhang Y, Vykhodtseva N, Hynynen K (2012) Improved anti-tumor effect of liposomal doxorubicin after targeted blood-brain barrier disruption by MRI-guided focused ultrasound in rat glioma. Ultrasound Med Biol 38:1716–1725

Tung Y-S, Vlachos F, Choi JJ, Deffieux T (2010) In vivo transcranial cavitation threshold detection during ultrasound-induced blood–brain barrier opening in mice. Phys Med Biol 55:6141–6155

VanBavel E (2007) Effects of shear stress on endothelial cells: possible relevance for ultrasound applications. Prog Biophys Mol Biol 93:374–383

Vykhodtseva NI, Hynynen K, Damianou C (1995) Histologic effects of high intensity pulsed ultrasound exposure with subharmonic emission in rabbit brain in vivo. Ultrasound Med Biol 21:969–979

Vykhodtseva N, McDannold N, Hynynen K (2006) Induction of apoptosis in vivo in the rabbit brain with focused ultrasound and Optison®. Ultrasound Med Biol 32:1923–1929

Wang H-B, Yang L, Wu J, Sun L, Wu J, Tian H et al (2014) Reduced ischemic injury after stroke in mice by angiogenic gene delivery via ultrasound-targeted microbubble destruction. J Neuropathol Exp Neurol 73:548–558

Wang S, Olumolade OO, Sun T, Samiotaki G, Konofagou EE (2015) Noninvasive, neuron-specific gene therapy can be facilitated by focused ultrasound and recombinant adeno-associated virus. Gene Ther 22:104–110

Weber-Adrian D, Kofoed RH, Chan JWY, Silburt J, Noroozian Z, Kügler S et al (2019) Strategy to enhance transgene expression in proximity of amyloid plaques in a mouse model of Alzheimer's disease. Theranostics 9:8127–8137

Wei K-C, Chu P-C, Wang H-YJ, Huang C-Y, Chen P-Y, Tsai H-C et al (2013) Focused ultrasound-induced blood–brain barrier opening to enhance Temozolomide delivery for glioblastoma treatment: a preclinical study. PLoS One 8:e58995

Wu J (2002) Theoretical study on shear stress generated by microstreaming surrounding contrast agents attached to living cells. Ultrasound Med Biol 28:125–129

Wu S-Y, Sanchez CS, Samiotaki G, Buch A, Ferrera VP, Konofagou EE (2016) Characterizing focused-ultrasound mediated drug delivery to the heterogeneous primate brain in vivo with acoustic monitoring. Sci Rep 6:37094

Xhima K, Nabbouh F, Hynynen K, Aubert I, Tandon A (2018) Noninvasive delivery of an α-synuclein gene silencing vector with magnetic resonance-guided focused ultrasound: noninvasive knockdown of brain α-Syn. Mov Disord 14:467

Xu Z, Hall TL, Fowlkes JB, Cain CA (2007) Optical and acoustic monitoring of bubble cloud dynamics at a tissue-fluid interface in ultrasound tissue erosion. J Acoust Soc Am 121:2421–2430

Yang F-Y, Fu W-M, Yang R-S, Liou H-C, Kang K-H, Lin W-L (2007) Quantitative evaluation of focused ultrasound with a contrast agent on blood-brain barrier disruption. Ultrasound Med Biol 33:1421–1427

Yang F-Y, Wang H-E, Liu R-S, Teng M-C, Li J-J, Lu M et al (2012) Pharmacokinetic analysis of 111In-labeled liposomal doxorubicin in murine glioblastoma after blood-brain barrier disruption by focused ultrasound. PLoS one Pub Library Sci 7:e45468

Chapter 20
Optimization of Blood-Brain Barrier Opening with Focused Ultrasound: The Animal Perspective

Elisa E. Konofagou

Abstract Although great progress has been made in recent years and more than 7000 small-molecule drugs are available, few effective treatments and no cures of the central nervous system (CNS) diseases are currently available. This is mainly due to the impermeability of the blood-brain barrier (BBB) that allows only 5% of those drugs to diffuse to the brain parenchyma thereby allowing treatment of only a tiny fraction of these diseases. Safe and localized opening of the BBB has been proven to pose an equally significant challenge. Focused ultrasound (FUS), in conjunction with microbubbles, remains the sole technique that can induce localized BBB opening noninvasively. In this chapter, we demonstrate how the microbubble diameter and peak negative pressure can be optimized in order to dictate the BBB opening volume and permeability in small and large animals. We subsequently demonstrate that neuroprotection and neurorestoration in the dopaminergic neurons in the nigrostriatal pathway at the early stages of Parkinson's disease as well as amyloid and tau reduction at the early stages of Alzheimer's disease can be achieved at therapeutic levels safely.

Keywords Microbubbles · Primate · Drug delivery · Safety · Reversibility · Imaging

20.1 The Blood-Brain Barrier Physiology: Structure and Function

The brain is a unique organ. It has its own reinforced defense system because it can control several organs in the body, and therefore its compromise by any toxic molecules could be fatal. Its defense system is mainly composed by the blood-brain barrier (BBB), which, as its name denotes, indicates the barrier or "filter" that exists

E. E. Konofagou (✉)
Department of Biomedical Engineering, Columbia University, New York, NY, USA

Department of Radiology, Columbia University, New York, NY, USA
e-mail: ek2191@columbia.edu

© American Association of Pharmaceutical Scientists 2022
E. C. M. de Lange et al. (eds.), *Drug Delivery to the Brain*, AAPS Advances
in the Pharmaceutical Sciences Series 33, https://doi.org/10.1007/978-3-030-88773-5_20

between the blood circulation in the brain and the brain tissue. In other words, the brain is protected more stringently than the remaining body because of the BBB. The BBB constitutes a collection of various different cells that contribute to this structural and functional obstacle. Each one of this cells has its own function of either transporting a molecule from the blood circulation into the brain tissue, or parenchyma, or allowing it to diffuse due to its specific size, that is, exactly as a filter would do. For example, the most common feature of BBB is the tight junctions which are specialized proteins that connect the cells along inner linings of the blood vessels (also known as "endothelial cells"), thus allowing only very small molecules to traverse the BBB (<400 Da in molecular weight or <1 nm in size) (Pardridge 2015). The endothelial cells also have membranes that further filter molecules. The combination of the tight junctions and those membranes provides the filtering mechanism of the BBB. Other cells such as astrocytes and pericytes serve as mechanical absorbers and thus provide a protective mechanism of the neurons to any external effects (Abbott et al. 2006). However, even if permeating the BBB, an even more formidable obstacle exists that of the cell membrane itself so that the appropriate cascades can be triggered, for example, of neuroprotection or neurorestoration which are later discussed as part of the treatment of neurodegenerative disease.

20.2 The BBB and Neurotherapeutics

According to the US National Center for Health Statistics in (www.cdc.gov/nchs/fastats/deaths.htm), over 5.4 million Americans are currently diagnosed with Alzheimer's disease, 1 million from Parkinson's disease, 350,000 from multiple sclerosis, 20,000 from amyotrophic lateral sclerosis (ALS), and 10,000 from brain cancer. Worldwide, these diseases account for more than 25 million patients. Although great progress has been made in recent years toward understanding of neurodegenerative diseases, few effective treatments and no cures are currently available. Aging greatly increases the risk of neurodegenerative disease, and the average age of Americans is steadily increasing. Today, over 35 million Americans are over the age of 65. Within the next 30 years, this number is likely to double, putting more and more people at increased risk of neurodegenerative disease. Alzheimer's disease, which has emerged as one of the most common brain disorders, severely affects the memory center of the brain with pathology gradually spreading to most brain areas as the disease progresses; this pathology is characterized partly by deposition of protein deposits (amyloid plaques) not only in the brain tissue but also in the blood vessels themselves (Iadecola 2004).

By acting as a permeability barrier, the BBB impedes entry from blood to the brain thus rendering many potent, neurologically active substances and drugs ineffective simply because they cannot be delivered to where they are needed. As a result, traversing the BBB remains the rate-limiting factor in brain drug delivery development.

A variety of approaches have been, and are being, developed to overcome the BBB for selective therapeutic treatment of brain pathologies. Over the past decade, numerous small- and large-molecule products have been developed for treatment of neurodegenerative diseases with mixed success. When administered systemically in vivo, the BBB inhibits their delivery to the regions affected by those diseases. A review of the Comprehensive Medicinal Chemistry database indicates that only 5% of the more than 7000 small-molecule drugs treat the central nervous system (CNS) (Pardridge 2015). This does not mean that the BBB is the only reason for the dearth of effective CNS treatments but that it poses additional challenges in the subset of these drugs that could be effective after permeating through the BBB. With these, only four CNS disorders can be treated: depression, schizophrenia, epilepsy, and chronic pain (Ghose et al. 1999; Lipinski 2000). Despite the availability of pharmacological agents, potentially devastating CNS disorders including brain tumors and age-related neurodegenerative diseases such as Alzheimer's disease, Parkinson's disease, Huntington's disease, multiple sclerosis, and ALS remain undertreated mainly because of the low permeability of the BBB. A successful drug delivery system requires transient, localized, and noninvasive targeting of a specific tissue region. None of the current techniques clinically used, or currently under research, address these issues within the scope of the treatment of neurodegenerative diseases. As a result, the present situation in neurotherapeutics enjoys few successful treatments for most CNS disorders. Over the past couple of decades, several pharmaceutical companies employed the technique known as "lipidization," which is the addition of lipid groups to the polar ends of molecules to increase the permeability of the agent (Fischer et al. 1998). However, the effect was not localized as the P-glycoprotein affinity also likely increased as well as the nonspecific binding of the drugs with the side effects in nontargeted regions potentially deleterious (Fischer et al. 1998).

A second set of techniques under study are neurosurgically based drug delivery methods, which involve the invasive implantation of drugs into a region by a needle (Blasberg et al. 1975). The drug spreads through diffusion and is often localized to the targeted region because diffusion does not allow molecules to travel far from their point of release. In addition to this, invasive procedures traverse untargeted brain tissue, potentially causing unnecessary damage. Other techniques utilize solvents like mannitol mixed with drugs or adjuvants (pharmacological agents) attached to drugs to disrupt the BBB through dilation and contraction of the blood vessels (Pardridge 2015). However, this disruption is not localized within the brain, and the solvents and adjuvants used are potentially toxic. This technique may constitute a delivery method specific to the brain, but it requires special attention to each type of drug molecule and a specific transport system resulting in a time-consuming and costly process while still not being completely localized to the targeted region. As a result, none of the brain drug delivery techniques are routinely used in the clinic, and the state of the art in the treatment of brain diseases remains stagnant.

20.3   Focused Ultrasound (FUS) with Microbubbles

BBB thus remains a formidable obstacle in treating central nervous system disorders, and millions of patients are undertreated at best. Focused ultrasound provides unique combined advantage of extracorporeal application with the capability of focalization through the intact skull, offering thus an unprecedented drug delivery system. Focused ultrasound (FUS) employs curved transducers that can transmit acoustic waves which converge only at the geometric focus of the transducer (Fig. 20.1). Most of the energy delivered during sonication induces mechanical effects, thermal effects, or both. When this technology is used at high intensities, it is referred to as "high-intensity focused ultrasound" or "HIFU." Localization of brain drug delivery is extremely important because opening the BBB across the entire organ may expose critical brain regions to a drug that may have deleterious effects. In addition, most of the aforementioned brain diseases are concentrated in specific brain structures such as the hippocampus or striatum. FUS in combination with microbubbles therefore constitutes the only truly transient, localized, and non-invasive technique for opening the BBB. Due to these unique advantages over other existent techniques (Table 20.1), FUS may facilitate the delivery of already developed pharmacological agents and could significantly impact how devastating CNS diseases are treated.

Microbubbles are gas filled, protein- or lipid-shelled, formations that can be formed, activated and injected intravenously as contrast agents to enhance ultrasound imaging and especially vasculature. The combination of FUS with microbubbles allows the separation of the mechanical effect of cavitation to occur at low peak-negative pressures without incurring thermal effects and thus leading to reversible and safe opening of the blood-brain barrier (Konofagou 2012). We have shown that both the diameter and the lipid shell type can influence the BBB opening (Fig. 20.2).

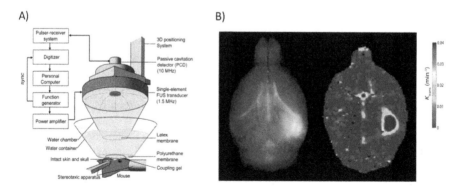

Fig. 20.1 (**a**) FUS setup for opening the BBB in mice in vivo (Tung et al. 2010); (**b**) 3D T1-weighted MRI showing the enhancement after gadolinium infusion through the BBB in the right hippocampus with corresponding Ktrans image

Table 20.1 FUS parameters and contrast delivery volumes and efficiency in NHP studies (Samiotaki et al. 2015)

PNP (kPa)	MI	Ipsilateral [Gd] (μg) or (ng/mg)	Contralateral [Gd] (μg) or (ng/mg)	Delivery efficiency (% increase)
200	0.28	6.37 ± 2.18 (10.78 ± 5.23)	0.72 ± 0.37 (1.22 ± 0.09)	784.7
300	0.42	16.34 ± 0.95 (27.66 ± 3.22)	0.90 ± 0.20 (1.52 ± 0.07)	1715.5
400	0.56	20.37 ± 2.35 (34.47 ± 5.67)	0.91 ± 0.26 (1.54 ± 0.06)	2138.5

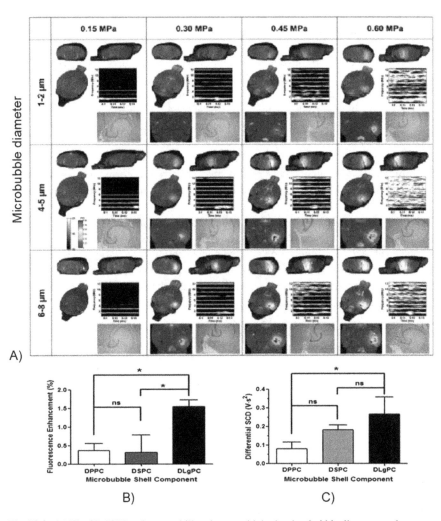

Fig. 20.2 (**a**) The 3D MRI and permeability change with both microbubble diameter and pressure magnitude. (**b**) Fluorescence enhancement. (**c**) Stable cavitation dose (SCD) increase with microbubble lipid shell component (DPP, DSPC, or DLgPC)

20.4 BBB Opening Using FUS and Microbubbles

20.4.1 Prior BBB Opening Studies Using FUS

Despite the fact that FUS is currently the only technique that can open the BBB locally, transiently, and noninvasively, several key aspects to be fully investigated in this study remain incomplete. First, initial studies involved craniotomies in rabbits and thus reported pressure amplitudes and resulting effects in the brain in the absence of the animal's skull (Vykhodtseva et al. 1995; Hynynen et al. 2001, 2005). Second, a clear correlation of BBB opening with microbubbles has been shown (Hynynen et al. 2001; McDannold et al. 2004; Choi et al. 2007, 2010b; Konofagou et al. 2008; Samiotaki et al. 2013). Our group (Tung et al. 2010, 2010b, 2011b; Tung 2012; Konofagou et al. 2012; Konofagou 2012; Wu et al. 2014, 2015) and others (McDannold et al. 2006) have indicated that BBB opening may occur without necessarily incurring inertial cavitation, that is, with stable cavitation (i.e., stable bubble oscillation) alone. Third, studies by other groups explore the brain as a whole and attempt to induce BBB opening in arbitrary, multiple locations without targeting a specific brain region. Our group focused on specific regions such as the hippocampus associated with early Alzheimer's (Choi et al. 2007, 2011; Marquet et al. 2014) evaluates the BBB properties locally. Fourth, multielement-phased arrays (with up to 1024 elements) that permit phase aberration correction (Aubry et al. 2003; Connor and Hynynen 2004) have been proposed in order to increase flexibility of the location targeted, mainly used for tumor ablation with minimal aberration. However, unlike the typical ultrasound attributes, these arrays are highly complex, inflexible, difficult to manufacture, and cumbersome in handling and positioning around a subject due to their typically bulky size and weight. More importantly, our group has shown that BBB opening occurs at low peak-rarefactional pressures comparable to diagnostic pressures and therefore phase aberration can be accounted for by using single-element transducers at lower frequencies in large animals (Deffieux and Konofagou 2010; Marquet et al. 2010, 2011, 2014). Fifth, the delivery of several agents has been shown by our group and others as follows: MRI contrast agents (Choi et al. 2007, 2007b; Vlachos et al. 2010; Samiotaki et al. 2013), Evans Blue (Kinoshita et al. 2006a), Trypan Blue (Raymond et al. 2008), Herceptin (148 kDa) (Kinoshita et al. 2006b), horseradish peroxidase (40 kDa) (Sheikov et al. 2008), doxorubicin (544 Da) (Treat et al. 2007), rabbit anti-Aβ antibodies (Raymond et al. 2008), brain-derived neurotrophic factor (BDNF) and neurturin (NTN) (Baseri et al. 2012; Samiotaki et al. 2015), and adenoviral vectors (Wang et al. 2013, 2015). However, despite the promise shown by the delivery of such a variety of compounds, several questions on their effectiveness upon delivery remain. More specifically, the bioactivity of the molecules delivered via the BBB remained largely unexplored, that is, it was not known whether the therapeutic molecules that cross through the BBB opening remain in the extracellular space or trigger downstream effects in neurons. Finally, equally unexplored by other groups and at the core of this renewal study is whether the compounds that cross the BBB through the FUS-induced

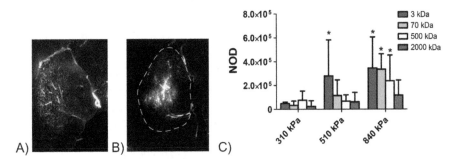

Fig. 20.3 (**a**) Protein delivery and (**b**) dextran (70 kDa) delivery in the BBB-opened hemisphere. (**c**) Normalized optical density (NOD) at distinct pressures and dextran molecular weights

opening can induce the intended therapeutic effects. Our group has shown that both neuroprotection and neurorestoration can be induced in neurons undergoing neuro-degeneration (Wang et al. 2015; Samiotaki 2016; Samiotaki et al. 2015). There have thus been several reports over the past decade or two using FUS and microbubbles to disrupt the blood-brain barrier that range from neuroprotection and neurorestora-tion in Parkinson's disease and amyloid reduction in Alzheimer's disease (Jordão et al. 2013; Leinenga et al. 2016; Karakatsani et al. 2019) to treatment of glioblas-toma in patients (Carpentier et al. 2016). The feasibility of BBB opening through intact skull and skin and successful imaging of the BBB opening in the area of the hippocampus at sub-millimeter imaging resolution has been shown in both wild-type and transgenic animals including models of glioblastoma (GBM), Alzheimer's disease, and Parkinson's disease. Due to the high spatial resolution of the FUS methodology, the beam can be focused in a specific region of the brain such as the hippocampus, a key short-term memory center and thus a drug delivery target in Alzheimer's disease or the caudate putamen, an important region for motor control and thus relevant to Parkinson's disease. Delivery of molecules of up to 20 nm in size has also been demonstrated (Chen and Konofagou 2014; Fig. 20.3). Most importantly, no neuronal or cellular damage within the range of peak rarefactional pressures of 0.3–0.45 MPa has been reported, while the barrier has been shown to close within 4–48 h under these conditions.

20.4.2   Mechanism of BBB Opening

There are two physical mechanisms for opening the BBB with FUS. The first is to use the ultrasound beam at lower pressures to induce a stable oscillation of the microbubble, also known as "stable cavitation." The second is to use higher pres-sures and increase the magnitude of oscillation of the bubble to the point that it surpasses the inertia of the fluid and collapses on itself. This is called "inertial cavi-tation." Both stable and inertial cavitation can be used to induce BBB opening. Stable cavitation has the safest profile and has been successfully monitored transcranially in real time in large animals including through the human skull (Wu et al. 2014; Karakatsani et al. 2017; Fig. 20.2c).

20.4.3 Molecular Delivery Through the Opened BBB

The delivery of both small- and large-molecule pharmacological agents using focused ultrasound (FUS) and microbubbles has been demonstrated in previous studies that include imaging contrast agents (Hynynen et al. 2001; Choi et al. 2007; Samiotaki et al. 2016), antibodies (Kinoshita et al. 2006a; Raymond et al. 2008), growth factor proteins (Baseri et al. 2012; Samiotaki et al. 2015; Fig. 20.4), stem cells (Burgess et al. 2011), and gene delivery vectors (Wang et al. 2015, 2017; Fig. 20.5). Molecular delivery studies (Choi et al. 2011; Chen and Konofagou 2014) have indicated that the size of the BBB opening increases with the ultrasonic pressure allowing larger molecules to diffuse of several orders of magnitude when the pressure is sufficiently high. The permeability of the barrier increases with both the pressure and microbubble size (Vlachos et al. 2010) indicating that the BBB opening occurs at multiple sites within the capillary tree and that the BBB opening is larger with larger microbubbles, most likely due to the larger area of contact between the bubble and the capillary wall.

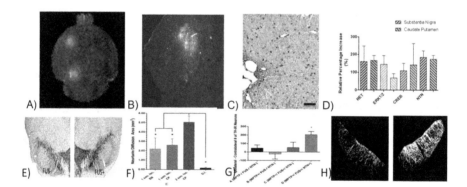

Fig. 20.4 Neurotrophic protein delivery: (**a**) contrast-enhanced MRI of a Parkinsonian mouse model. (**b**) Optical imaging showing protein uptake and (**c**) histological examination of protein uptake (in brown tint) by neurons (arrows) and (**d**) cell uptake in the striatum. [RET: cell membrane receptor, ERK1/2: cytoplasm, CREB: neuronal nucleus, NTN: overall neurturin]. (**e**) Neurorestoration occurred only in the (i) BBB-opened (ipsilateral) region (red ROI) and not in the (ii) contralateral (black ROI) region that had significant neuronal depletion due to the MPTP toxin. (**f**) Comparison of FUS with direct injection: The diffusion area is several folds smaller with the latter; (**g**) Ipsi-Contra difference in number of neurons stained. Significant increase in the number of neurons occurred only in Group D that received both FUS and the NTN neurotrophic factor (Samiotaki et al. 2015). (**h**) FUS− (left) and FUS+ (right) side showing the enhancement in the dendrites stained demonstrating neuronal survival (Samiotaki et al. 2015)

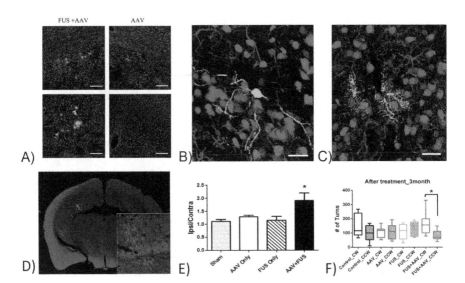

Fig. 20.5 (**a**) Gene delivery through the BBB (BBB-opened; left panel) versus contralateral (unopened; right panel) brain region. Virus transduced cells are shown in green. All scale bars indicate 150 μm. In the BBB-opened region, gene-transduced (**b**) neurons (in green) and (**c**) astrocytes (in green) (100×). (**d**) AAV delivery expressing mCherry protein in the BBB-opened (left) murine hippocampus with inset at higher magnification in the BBB-opened region (Wang et al. 2015). (**e**) Gene expression was found to be significant only in the group where both FUS and AAV were delivered. (**f**) The FUS + AAV group showed consistent opposite rotation in locomotion confirming neuroprotection only in the FUS + AAV group in behavioral studies

20.4.4   Safety and Reversibility of BBB Opening

The safe operating parameters of ultrasound exposure of brain cells have been identified (Konofagou 2012). In summary, BBB opening starts occurring at 0.3 MPa rarefactional pressure amplitude and beyond (Fig. 20.6). At pressures under 0.6 MPa, no extravasation of red blood cells (RBC) or neuronal damage was observed in the regions of the hippocampus exhibiting the most pronounced BBB opening. Beyond 0.6 MPa, RBC extravasation was detected, and beyond 0.9 MPa, neuronal damage was observed. These preliminary findings suggest that there is overlap between the feasibility and safety windows within the pressure range of 0.3–0.6 MPa, that is, the BBB can be opened throughout the entire hippocampus without endothelial or neuronal damage at those pressures (Fig. 20.6; Baseri et al. 2010). FUS-induced BBB opening was reported to close within 24 h under specific parameters in rabbits (Hynynen et al. 2001), mice (Samiotaki et al. 2016), and monkeys (Marquet et al. 2014). Behavioral studies in mice that survived over 6 months when BBB opening was performed every week for 6 months showed no evidence of behavioral or motor control damage (Olumolade et al. 2016).

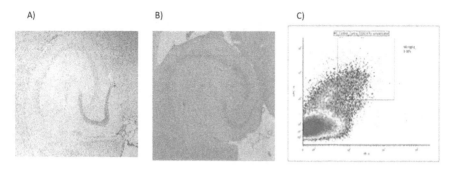

Fig. 20.6 (**a**) H&E and (**b**) TUNEL staining of the murine hippocampus. No damage by FUS is noted at the pressures used. (**c**) RNA sequencing cell distribution to assess the microglia activation as a result of AAV delivery. Microglia is identified based on their size and stain in the blue ROI as shown

20.4.5 Therapeutic Delivery Through FUS-Induced Blood-Brain Barrier Opening

20.4.5.1 An Early-Stage Parkinson's Model Used

MPTP is a neurotoxin that causes permanent symptoms of Parkinson's disease by depleting the dendrites, axons, and terminals of dopaminergic neurons in the putamen and substantia nigra of the brain (Fig. 20.2). As previously indicated, the nigrostriatal dopaminergic pathway in subacute-dose MPTP mice models clinical early-stage PD where motor symptoms are the main disease markers. Over the past few years (Samiotaki and Konofagou 2013; Karakatsani et al. 2019), we have worked extensively with this well-validated MPTP mouse model for early stage Parkinson's disease.

20.4.5.2 Protein Delivery

Neurotrophic delivery to the brain is thought to be essential in reversing neuronal degeneration processes, but so far the application of growth factors to the CNS has been hindered by the blood-brain barrier. In a recent study by our group, not only was it shown that brain-derived neurotrophic factor (BDNF) can cross the ultrasound-induced blood-brain barrier opening but also that it can trigger signaling pathways in the pyramidal neurons of mice in vivo from the membrane to the nucleus (Baseri et al. 2012). More recently, our group has shown that the neuronal morphology has been reinstated after introduction of a toxic insult (MPTP) that induces Parkinsonian symptoms in mice (Samiotaki et al. 2015; Karakatsani et al. 2019). This opens entirely new avenues in the brain drug delivery where focused ultrasound in conjunction with microbubbles can generate downstream effects at the cellular and molecular level and thus increase the drug's efficacy and potency in controlling or reversing the disease.

20.4.5.3 Adenoviral Delivery

BBB openings in MPTP mice were induced in the left caudate putamen and substantia nigra. Recombinant adeno-associated virus serotype 9 (AAV9) is what we have found to be the most efficacious in our prior studies. AAVs are of size 2–4 MDa and express green fluorescent protein (GFP) under the control of CAG promoter and NTN and were used for transduction (SignaGen Labs). As in our preliminary studies (Fig. 20.6a), mice were co-injected with microbubbles (~10^8/animal) and 100 μl AAV vectors (1.5×10^{12}GC/ml) via the tail vein and immediately followed by FUS (pulse length 6.7 ms, pulse repetition frequency 5 Hz). Tyrosine hydroxylase (TH) immunostaining revealed that AAV induced neuroprotection and neurorestoration in MPTP mice (Fig. 20.6) and a several fold increase in AAV delivery (Fig. 20.6b) (Wang et al. 2013, 2015, 2017). TH is a marker for the nigrostriatal neurons in a context where many spared neurons no longer die but remain dysfunctional as unveiled by TH downregulation. To allow gene expression to occur, 30 days after BBB opening were interleaved before animals are tested in behavioral studies. To determine neurorestoration, AAV-NTN was injected after the MPTP toxin has significantly damaged the DA neurons (~30 days).

20.4.5.4 Behavioral Assessment

Behavioral studies were performed in mice upon completion of the aforementioned experiments to assess both AAV efficacy and safety through assessment of motor control improvement in the FUS-treated mice. A behavioral facility for mice is available and has been used in the PI's laboratory for more than 5 years. The mice are placed in a custom-made open field test chamber located in a soundproof, isolated room, and free to explore the field (Fig. 20.7a). Visual tracking software (Noldus) was used to record ambulatory activity (thigmotaxis; Fig. 20.7a, b). Upon completion of the open field test, the animals were allowed to rest before being placed on the accelerating rotating rod (rotarod), where they were required to maintain balance and motor coordination for a fixed period (180 s). Behavioral results using apomorphine systemic administration indicated that only the FUS + AAV group had increased dopaminergic functionality in the ipsilateral versus the contralateral (no FUS) hemisphere (Fig. 20.5f).

20.4.5.5 Brain Preparation and Immunohistochemistry

To assess safety of the systemic viral delivery, organs (liver, heart, kidney, and muscle; Wang et al. 2015) were harvested from each animal, and AAV transductions were investigated. Previous studies by our group have shown no expression in other organs (Wang et al. 2015). H&E staining and TUNEL were performed to determine any red blood cell extravasation or dark neurons in the BBB-opened region (Baseri et al. 2010; Choi et al. 2011; Fig. 20.6).

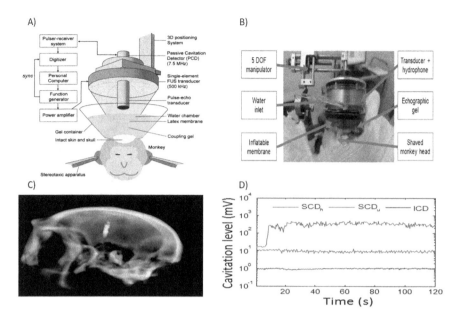

Fig. 20.7 (**a**) Illustration of current FUS NHP setup. (**b**) Picture of NHP FUS setup. (**c**) FUS focal spot (in yellow/orange) in the coronal (left) and sagittal (right) MRI planes showing targeting capability of FUS. (**d**) Real-time cavitation monitoring. The blue (harmonics) and red (ultraharmonics) lines during FUS application. The blue curve reaches a plateau indicating microbubble perfusion saturation

20.5   Large Animals

20.5.1   Rationale

The primary objective is to implement a theranostic FUS system for primates and determine the therapeutic effect of the FUS-induced gene delivery in MPTP monkeys in vivo in order to assess neurorestoration in large animals while ensuring safety through cognitive testing and thus inform future clinical (Marquet et al. 2014; Wu et al. 2014, 2016, 2018; Downs et al. 2015; Samiotaki et al. 2015) studies. It has been shown that we can open safely and noninvasively the NHP striatum and hippocampus as well as deliver gadolinium without incurring damage (Table 20.1; Samiotaki et al. 2016; Karakatsani et al. 2017).

   Table 20.1 Tracer concentration that diffused into the tissue while paired t-test revealed significant increase of gadolinium in the sonicated region ($P = 0.0002$) PNP: peak-negative pressure, MI: mechanical index. The mechanical index is maintained well within the FDA limits for ultrasound contrast imaging (MI < 0.8). Also, note that FUS delivers 3× to 7× higher concentrations (6.37 ± 2.18 to 20.37 ± 2.35 ng/mg) than those found to be efficacious in neurorestoration with multiple direct injections in NHP (2.25 ± 0.312 ng/mg; Kordower et al. 2006) and 8× to 21× higher than in the contralateral hemisphere.

20.5.2   Methods

All methods described herein have been approved by the IACUC of Columbia University.

20.5.2.1   Primate FUS System

A 0.5-MHz FUS transducer (focal size: 5.85 mm and 34 mm (depth), H-107, and Sonic Concepts, WA, USA) are attached to the Kopf stereotaxic manipulator for precise targeting of the brain structures (Figs. 20.7 and 20.8). A PC workstation (model T7600, Dell) with a customized program in MATLAB® (Mathworks, MA, USA) was developed to automatically control the sonication through a programmable function generator (model 33220A, Agilent Technologies, CA, USA) and a 50-dB amplifier (A075, ENI, NY, USA). The parameters are as follows: pressure: 450 kPa, pulse length: 10 ms, pulse repetition frequency: 2 Hz, duration: 2 min, and Definity microbubbles (1.2×10^8 bubbles/kg) with IV injection at the FDA recommended dosage of 1.2×10^8/kg. The frequency of 0.5 MHz was chosen to maximize transmission through the primate skull. The 0.5-MHz array was simulated using the NHP CT images (helical scan, resolution $0.2 \times 0.2 \times 0.6$ mm^3) to provide the acoustic properties of the skull (density and sound speed) in Hounsfield units (Deffieux and Konofagou 2010) showing feasibility of a uniform focus through the NHP skull (Fig. 20.9).

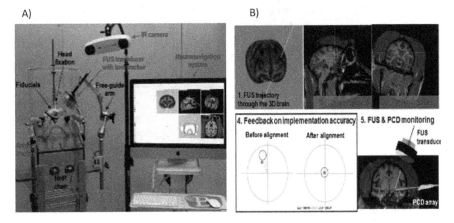

Fig. 20.8 Feasibility of BBB opening with neuronavigation: (**a**) Brainsight system integrated with FUS system and tested in an in vivo NHP. (**b**) Brainsight console with real-time feedback. The precision of targeting was assessed to be on the order of 2 mm (Wu 2017)

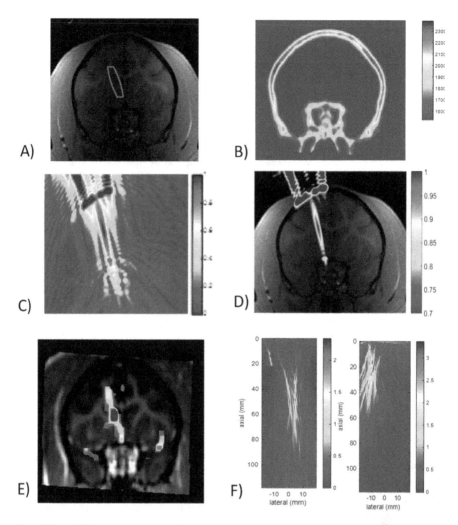

Fig. 20.9 (a) Initial targeting planned at caudate nucleus for stereotaxic FUS procedure. The focal spot is (circle) shown on the T1w image (coronal). The incidence angle to the skull and the tissue thickness (distance of the skull to the focus) were 6.6° and 16.4 mm, respectively. (b) Speed of sound map. (c) Simulated peak-negative pressure (PNP) field (normalized to the peak pressure at the focus) for the planned targeting. (d) Predicted BBB opening based on the simulated PNP field by thresholding to 0.7 of the PNP at focus for sonicating at 300 kPa. (e) Visualized BBB opening volume by overlaying the perfused Gd (comparing the post-Gd T1w images to the pre-Gd T1w images) to the post-Gd T1w image. (f) Cavitation maps through an ex vivo NHP skull (left) and in vivo NHP (right) during BBB opening

20.5.2.2   Cavitation Detection and Monitoring

Monitoring of the BBB opening through transcranial cavitation detection in monkeys was first demonstrated by our group (Tung et al. 2011; Marquet et al. 2011, 2014; Wu et al. 2014). Transmit frequencies are monitored through the center of the

FUS transducer by a flatband, spherically focused hydrophone (H-107, Sonic Concepts, WA, USA). The hydrophone is connected to a digitizer (Gage Applied Technologies, Inc. Lachine, QC, Canada) through a 20-dB amplification (5800, Olympus NDT, Waltham, MA, USA) for processing of the acoustic emissions from the microbubbles. The focal regions of the two transducers overlap within the confocal volume. The acoustic pressures at the focus of the FUS transducer with and without the skull in the beam path are calibrated before the experiments using a bullet hydrophone (ONDA, CA, USA). The monitoring technique has been implemented in real time (Fig. 20.7d). The PCD signals, frequency spectra, and spectrograms (eight-cycle Chebyshev window, 98% overlap, 4096-point fast Fourier transform) of the PCD signals are used to monitor the cavitation during BBB opening (Wu et al. 2014).

20.5.2.3 Neuronavigation

In order to tailor targeting to the specific subject, a Brainsight™ system (Rogue Research) can be used to further improve on the current targeting accuracy (0.6 mm lateral and 3 mm axial; Marquet et al. 2014; Wu et al. 2018). The neuronavigation system (Fig. 20.8) uses an anatomical routine MRI scan of the subject and fiducial facial markers as a reference in order to offer efficacy and reproducibility of patient-specific targeting outside an MRI system. Neuronavigation systems are common practice in the clinic. Our group was the first to implement such a system for FUS-induced BBB opening (Wu et al. 2016; Wu 2017, 2018). Using the targeting selected during the planning stage, the Brainsight software displays both a distance error in the X-Y axis, as well as angle of approach for positioning the transducer. This is displayed as crosshairs for easy visualization while positioning the transducer. Following alignment with the angle and X-Y axis, an offset to the center of the focal spot is selected. This offset is shown as a point in the 3D reconstruction of the MRI for real-time visualization of the focal area. The transducer is then moved in the Z axis until the point defined by the focal offset overlays with the target area and the FUS procedure can begin.

20.5.3 Imaging of BBB Opening

20.5.3.1 MRI

After BBB opening, the anesthetized animals were transported to the MR facility (2-min walk in adjacent building to the FUS setup) where T2 and T2 FLAIR (Fig. 20.10c) as well as susceptibility-weighted image (SWI) were taken to detect any potential damage caused by the sonication on a 3-T MRI scanner (Philips Healthcare, Best, NL). A high-resolution structural T1 image was recorded prior to the injection of gadodiamide (**T1 Pre**; 3D Spoiled Gradient-Echo, TR/

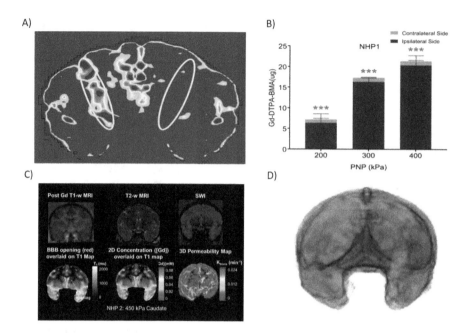

Fig. 20.10 (**a**) Gadolinium concentration map showing the BBB opening in the ipsilateral side (left) and the background signal in the contralateral side (right) as outlined by the white ellipsoids. (**b**) Increase of Gd concentration with FUS pressure; (**c**) T1w, T2w, and SWI in a NHP; bottom panel (left to right): BBB opening on a T1 map showing gray and white matter, 2D concentration [Gd] and permeability map in the same NHP. (**d**) BBB opening in the NHP hippocampus in a 3D T1w MR image

TE = 20/1.4 ms; flip angle: 30°; NEX = 2; in-plane resolution: 1×1 mm^2; slice thickness: 1 mm with no interslice gap). Thirty minutes after injection of 0.15 ml/kg gadodiamide IV, a second T1 image was acquired using identical scanning parameters (**T1 Post**). As gadodiamide does not cross the intact BBB, increased T1 signal strength (Figs. 20.10c and 11c) indicate regions with BBB opening. Three-dimensional T2-weighted (TR/TE = 3000/80 ms; flip angle: 90°; NEX = 3; spatial resolution: 400×400 mm^2; slice thickness: 2 mm with no interslice gap) and 3D SWI (TR/TE = 19/27 ms; flip angle: 15°; NEX = 1; spatial resolution: 400×400 mm^2; slice thickness: 1 mm with no gap) were applied.

20.5.3.2 DCE-MRI

DCE imaging were performed using a 3D Spoiled Gradient Echo (SPGR) T1-weighted sequence (TR/TE = 8.6 ms/4.9 ms; flip angle: 300; NSA: 4; spatial resolution: $2 \times 2 \times 2$ mm^3; scan duration: 30 min). Contrast agent (Omniscan® (574 Da)) was injected after the third dynamic acquisition similar to the mice.

20.5.4 Drug Delivery Studies

20.5.4.1 Pharmacodynamic Analysis

Our group has implemented a standard relaxometry technique for T1 mapping (Samiotaki et al. 2015) acquiring gradient-echo images with variable flip angles (VFA) with short TR. VFA-based T1 mapping has been validated in phantoms and provided T1 value estimation with high accuracy and high spatial and temporal resolution. Comparison of the T1 times before and after gadolinium injection provides the gadolinium concentration maps for pharmacodynamic analysis.

20.5.4.2 Prediction of Aberration and Targeting Correction

A structural scan using stereotactically aligned T1w MRI was acquired for each animal before the experiment, and the ellipsoidal focal spot was adjusted with its longest dimension along that of the structure of interest and the incidence angle normal to the skull in order to minimize the aberration of the ultrasound wave resulting in targeting shift and BBB opening volume (Deffieux and Konofagou 2010; Marquet et al. 2014; Wu et al. 2014; Fig. 20.9). The designed focus with the incidence angle was then used for simulation in order to predict the BBB opening as well as for the stereotactic FUS sonication after converting to the stereotactic coordinates for the 9-degree-of-freedom stereotactic frame used in our study (Marquet et al. 2014). A two-dimensional simulation of the second-order wave equation was implemented on the k-space pseudospectral method (k-Wave) (Treeby et al. 2012). Density and speed of sound of the animal skull were converted using the Hounsfield units in CT (Siemens) (Deffieux and Konofagou 2010) using 3D CT scans of the animal's skull obtained separately for each animal in vivo (GE LightSpeed Ultrafast CT) after co-registered to the stereotactically aligned aforementioned T1w images. The spatial resolution was 0.25 mm × 0.25 mm with a range of 152 mm × 152 mm, and the acoustic source and focal size were calibrated to be the same as the FUS system. Preliminary results of the simulation for planning against the actual BBB opening in the hippocampus and the in vivo BBB opening result are shown in Fig. 20.8. The overall target shift was 2.2 mm in distance and 11° in angle, which was the information used to compensate in vivo. Feasibility of targeting the hippocampus through NHP and human skulls with this single-element system without correction was assessed at the parameters identified by the simulation studies to be optimal.

20.5.4.3 Neuronavigation

In our NHP studies up to now, a generic anatomical atlas for Rhesus macaques was used for targeting. In order to tailor targeting to the specific subject, a Brainsight™ system (Rogue Research, Montreal, Quebec, Canada) was purchased to further improve on the current targeting accuracy (of 0.6 mm lateral and 3 mm axial; Marquet et al. 2014). The neuronavigation system uses an anatomical routine MRI scan of the subject and fiducial facial markers as a reference in order to offer efficacy and reproducibility of patient-specific targeting outside an MRI system. Neuronavigation systems are common practice in the clinic. This was the first time it is used for FUS-induced BBB opening. Fiducial markers are used in an array system, which provides a rigid, unambiguous set of landmarks to accurately (about 1 mm) co-register the subject to the images during intervention. The Brainsight Vet system was used that includes the Brainsight NHP software v2.3, an Apple iMac Core 2 Duo 27″ screen, mobile trolley for iMac, Northern Digital Vicra Optical Position Sensor, Passive pointer, Subject Tracker, calibration tool, replacement passive reflective tracker Spheres, Nonhuman Primate tool tracker, and adapter for ultrasound transducer. Once structural MRI of the NHP wearing the fiducials is acquired, target planning is conducted with the Brainsight software. The MRI is loaded onto the system, and the fiducial markers are identified as individual calibration points. The software generates a 3D reconstruction of the MRI for accurate selection of the target region of the brain as well as visualization of the beam path to ensure a low incidence angle with the skull. On the day of the procedure, the subject wears the dental fiducial array or goggles for calibration to the Brainsight software. The calibration tool is used to register individual fiducials in the order they were selected during the target planning of the procedure. After all fiducials are registered, the calibration error is found by pointing the tip of the calibration tool to each fiducial. The system displays a distance error in millimeter relative to the other fiducials defined in the software during the planning stage. This distance error is displayed with a color gradient indicating ranging from green (<2 mm) to red (>10 mm). Once all the fiducials have been properly registered with distance errors <2 mm, the FUS transducer with the fiducial adapter can be positioned. Using the targeting selected during the planning stage, the Brainsight software displays both a distance error in the X-Y axis, as well as angle of approach for positioning the transducer. This is displayed as crosshairs for easy visualization while positioning the transducer. Following alignment with the angle and X-Y axis, an offset to the center of the focal spot is selected in the software. This offset is shown as a point in the 3D reconstruction of the MRI for real-time visualization of the focal area. The transducer is then moved in the Z axis until the point defined by the focal offset overlays with the target area shown. After all four parameters are aligned (X, Y, Z, and angle) with the target selected during planning, the BBB opening procedure is set to start (Pouliopoulos et al. 2020).

20.5.4.4   Cavitation Mapping

Until now, we have been capable of detecting and monitoring cavitation but not mapping its actual location. More recently, we have been capable to monitor the location and extent of cavitation during sonication for BBB opening in NHP. To that extent, we were utilizing a separate ultrasound system, the Verasonics Vantage, already available in our laboratory, which was capable of acquiring the RF signals in real time. Instead of a single-element PCD detector, a 128-element linear array (L7-4, sensitivity: 2–8 MHz) confocally aligned with the FUS transducer was used to acquire cavitation emission passively. The acquired cavitation emissions saved as the channel data were reconstructed to the passive cavitation map based on the delay-and-sum algorithm and sparse-matrix calculations that our group has already published on (Hou et al. 2014). In our **preliminary studies**, cavitation maps were obtained both ex vivo and in vivo (Fig. 20.8). The cavitation map without the skull visualized the location and intensity of the cavitation during FUS (Fig. 20.9f). After placing the NHP skull, the technique was sensitive enough to detect cavitation emission and was able to visualize the location and intensity after attenuation. BBB opening at the hippocampus of a monkey in vivo was also confirmed with post-Gd T1w imaging (Fig. 20.8b) and the cavitation map (Fig. 20.8c) revealing the location and intensity of cavitation in the sonicated region.

20.6 Conclusion

Our preliminary studies have shown that diagnostic pressures and pulse lengths typically used for imaging are sufficient to open the BBB in conjunction with microbubbles. Neurotrophic delivery was also shown feasible in mice in the presence of neurodegenerative pathology using gene delivery that that triggered downstream pathways in the neuronal cell for neuroprotection and neurorestoration. This evidence strongly indicates an important opportunity to further investigate the clinical relevance of the therapeutic efficacy of the FUS-mediated brain gene delivery.

Acknowledgments The work shown here was performed by Babak Baseri, Mark Burgess, Cherry Chen, Hong Chen, James Choi, Thomas Deffieux, Vernice Jackson-Lewis, Robin Ji, Marilena Karakatsani, Fabrice Marquet, Gesthimani Samiotaki, Oluyemi Olumolade, Tobias Teichert, Yao-Sheng Tung, Shutao Wang, and Shih-Ying Wu in collaboration with the groups by Karen Duff, Vincent Ferrera, Scott Small and Serge Przersborski. The research was supported by NIH R01 EB009041, NIH R01 AG038961, NIH R21 EY018505, NSF CAREER 064471, the Focused Ultrasound Foundation, the Kinetics Foundation, and the Kavli Institute.

References

Abbott NJ, Ronnback L, Hansson E (2006) Astrocyte-endothelial interactions at the blood-brain barrier. Nat Rev Neurosci 7:41–53

Aubry JF, Tanter M, Pernot M, Thomas JL, Fink M (2003) Experimental demonstration of noninvasive transskull adaptive focusing based on prior computed tomography scans. J Acoust Soc Am 113:84–93

Baseri B, Choi JJ, Tung YS, Konofagou EE (2010) Safety assessment of blood-brain barrier opening using focused ultrasound and definity microbubbles: a short-term study. Ultrasound Med Biol 36(9):1445–1459

Baseri B, Choi JJ, Deffieux T, Samiotaki G, Tung YS, Olumolade O, Small SA, Morrison B, Konofagou EE (2012) Activation of signaling pathways following localized delivery of systemically administered neurotrophic factors across the blood-brain barrier using focused ultrasound and microbubbles. Phys Med Biol 57(7):N65-81

Blasberg RG, Patlak C, Fenstermacher JD (1975) Intrathecal chemotherapy: brain tissue profiles after ventriculocisternal perfusion. J Pharmacol Exp Ther 195:73–83

Burgess A, Ayala-Grosso CA, Ganguly M, Jordão JF, Aubert I, Hynynen K (2011) Targeted delivery of neural stem cells to the brain using MRI-guided focused ultrasound to disrupt the blood-brain barrier. PLoS One 6(11):e27877

Carpentier A, Canney M, Vignot A, Reina V, Beccaria K, Horodyckid C, Karachi C, Leclercq D, Lafon C, Chapelon JY, Capelle L, Cornu P, Sanson M, Hoang-Xuan K, Delattre JY, Idbaih A (2016) Clinical trial of blood-brain barrier disruption by pulsed ultrasound. Sci Transl Med 8(343):343re2

Chen H, Konofagou EE (2014) The size of blood-brain barrier opening induced by focused ultrasound is dictated by the acoustic pressure. J Cereb Blood Flow Metab 34(7):1197–1204

Choi JJ, Pernot M, Brown TR, Small SA, Konofagou EE (2007b) Spatio-temporal analysis of molecular delivery through the blood-brain barrier using focused ultrasound. Phys Med Biol 52:5509–5530

Choi JJ, Pernot M, Small SA, Konofagou EE (2007) Noninvasive, transcranial and localized opening of the blood-brain barrier using focused ultrasound in mice. Ultrasound Med Biol 33:95–104

Choi JJ, Selert K, Vlachos F, Wong A, Konofagou EE (2011) Noninvasive and localized neuronal delivery using short ultrasonic pulses and microbubbles. Proc Natl Acad Sci U S A 108(40):16539–16544

Connor CW, Hynynen K (2004) Patterns of thermal deposition in the skull during transcranial focused ultrasound surgery. IEEE Trans Biomed Eng 51:1693–1706

Deffieux T, Konofagou E (2010) Numerical study and experimental validation of a simple transcranial focused ultrasound system applied to blood-brain barrier opening. IEEE Trans Ultrason Ferroelectr Freq Control 212-220

Downs ME, Buch A, Sierra C, Karakatsani ME, Chen S, Konofagou EE, Ferrera VP (2015) Long-term safety of repeated blood-brain barrier opening via focused ultrasound with microbubbles in non-human primates performing a cognitive task. PLoS One 10(5):e0125911

Fischer H, Gottschlich R, Seelig A (1998) Blood-brain barrier permeation: molecular parameters governing passive diffusion. J Membr Biol 165:201–211

Ghose AK, Viswanadhan VN, Wendoloski JJ (1999) A knowledge-based approach in designing combinatorial or medicinal chemistry libraries for drug discovery. 1. A qualitative and quantitative characterization of known drug databases. J Comb Chem 1:55–68

Hou GY, Provost J, Grondin J, Wang S, Marquet F, Bunting E, Konofagou EE (2014) Sparse matrix beamforming and image reconstruction for 2-D HIFU monitoring using harmonic motion imaging for focused ultrasound (HMIFU) with in vitro validation. IEEE Trans Med Imag 33(11):2107–2117. PMCID

Hynynen K, McDannold N, Vykhodtseva N, Jolesz FA (2001) Noninvasive MR imaging-guided focal opening of the blood-brain barrier in rabbits. Radiology 220:640–646

Hynynen K, McDannold N, Sheikov NA, Jolesz FA, Vykhodtseva N (2005) Local and reversible blood-brain barrier disruption by noninvasive focused ultrasound at frequencies suitable for trans-skull sonications. NeuroImage 24:12–20

Iadecola C (2004) Neurovascular regulation in the normal brain and in Alzheimer's disease. Nat Rev Neurosci 5:347–360

Jordão JF, Thévenot E, Markham-Coultes K, Scarcelli T, Weng YQ, Xhima K, O'Reilly M, Huang Y, McLaurin J, Hynynen K, Aubert I (2013) Amyloid-β plaque reduction, endogenous antibody delivery and glial activation by brain-targeted, transcranial focused ultrasound. Exp Neurol 248:16–29

Karakatsani ME, Samiotaki G, Downs M, Ferrera V, Konofagou EE (2017) Targeting effects on the volume of the focused ultrasound induced blood-brain barrier opening in non-human primates in vivo. IEEE Tran Ultrason Ferroelectr Freq Control 64(5):798–810

Karakatsani ME, Samiotaki G, Wang S, Kugelman T, Acosta C, Jackson-Lewis V, Przedborski S, Konofagou E (2019) Focused ultrasound-facilitated drug delivery exerts neuroprotective and neurorestorative effects in a Parkinsonian mouse model. J Cont Rel 303:289–301

Kinoshita M, McDannold N, Jolesz FA, Hynynen K (2006a) Targeted delivery of antibodies through the blood-brain barrier by MRI-guided focused ultrasound. Biochem Biophys Res Commun 340:1085–1090

Kinoshita M, McDannold N, Jolesz FA, Hynynen K (2006b) Noninvasive localized delivery of Herceptin to the mouse brain by MRI-guided focused ultrasound-induced blood-brain barrier disruption. Proc Natl Acad Sci U S A 103:11719–11723

Konofagou EE (2012) Optimization of the ultrasound-induced blood-brain barrier opening. Theranostics 2(12):1223–1237

Konofagou EE, Choi J, Baseri B, Lee A (2008) Characterization and optimization of trans-blood-brain barrier diffusion in vivo. In: Emad SE (ed) 8th international symposium on therapeutic ultrasound. AIP, Minneapolis, pp 418–422

Konofagou EE, Tung YS, Choi J, Deffieux T, Baseri B, Vlachos F (2012) Ultrasound-induced blood-brain barrier opening. Curr Pharm Biotechnol 13(7):1332–1345

Kordower JH, Herzog CD, Dass B, Bakay RA, Stansell J 3rd, Gasmi M et al (2006) Delivery of neurturin by AAV2 (CERE-120)-mediated gene transfer provides structural and functional neuroprotection and neurorestoration in MPTP-treated monkeys. Ann Neurol 60:706–715

Leinenga G, Langton C, Nisbet R, Götz J (2016) Ultrasound treatment of neurological diseases – current and emerging applications. Nat Rev Neurol 12(3):161–174

Lipinski CA (2000) Drug-like properties and the causes of poor solubility and poor permeability. J Pharmacol Toxicol Methods 44:235–249

Marquet F, Tung Y-S, Konofagou EE (2010) Feasibility study of a clinical blood-brain opening ultrasound system. Nano Life 1(3 & 4):309–322. [PMCID in process]

Marquet F, Tung YS, Teichert T, Ferrera VP, Konofagou EE (2011) Noninvasive, transient and selective blood-brain barrier opening in non-human primates in vivo. PLoS One 6(7):e22598

Marquet F, Teichert T, Wu SY, Tung YS, Downs M, Wang S, Chen C, Ferrera V, Konofagou EE (2014) Real-time, transcranial monitoring of safe blood-brain barrier opening in non-human primates. PLoS One 9(2):e84310. Epub 2014/10/04. https://doi.org/10.1371/journal.pone.0084310

McDannold N, Vykhodtseva N, Hynynen K (2006) Targeted disruption of the blood-brain barrier with focused ultrasound: association with cavitation activity. Phys Med Biol 51:793–807

McDannold N, Vykhodtseva N, Jolesz FA, Hynynen K (2004) MRI investigation of the threshold for thermally induced blood-brain barrier disruption and brain tissue damage in the rabbit brain. Magn Reson Med 51:913–923

Olumolade OO, Wang S, Samiotaki G, Konofagou EE (2016) Longitudinal motor and behavioral assessment of blood-brain barrier opening with transcranial focused ultrasound. Ultrasound Med Biol S0301-5629(16):30061–30068

Pardridge WM (2015) Targeted delivery of protein and gene medicines through the blood-brain barrier. Clin Pharmacol Ther 97(4):347–361

Pouliopoulos AN, Jimenez DA, Frank A, Robertson A, Zhang L, Kline-Schoder AR, Bhaskar V, Harpale M, Caso E, Papapanou N, Anderson R, Li R, Konofagou EE (2020 May) Temporal stability of lipid-shelled microbubbles during acoustically-mediated blood-brain barrier opening. Front Phys 8:137

Raymond SB, Treat LH, Dewey JD, McDannold NJ, Hynynen K, Bacskai BJ (2008) Ultrasound enhanced delivery of molecular imaging and therapeutic agents in Alzheimer's disease mouse models. PLoS One 3(5):e2175

Samiotaki G, Konofagou EE (2013) Dependence of the reversibility of focused-ultrasound-induced blood-brain barrier opening on pressure and pulse length in vivo. IEEE Trans Ultras Ferroelect Freq Contl 60(11):2257–2265. PMCID: PMC3968797

Samiotaki G, Olumolade O, Wang S, Konofagou EE (2013) Localized delivery of the Neurturin (NTN) neurotrophic factor through focused ultrasound – mediated blood-brain barrier opening. In: IEEE international ultrasonics symposium (Prague, Czech Republic), July 21–25

Samiotaki G, Acosta C, Wang S, Konofagou EE (2015) Enhanced delivery and bioactivity of the neurturin neurotrophic factor through focused ultrasound-mediated blood–brain barrier opening in vivo. J Cereb Blood Flow Metab 35(4):611–622. PMCID: PMC4420879

Samiotaki G (2016, March) Quantitative and dynamic analysis of the focused-ultrasound induced blood-brain barrier opening in vivo for drug delivery. PhD dissertation, Columbia University

Samiotaki G, Karakatsani ME, Buch A, Papadopoulos S, Wu SY, Jambawalikar S, Konofagou EE (2016) Pharmacokinetic analysis and drug delivery efficiency of the focused ultrasound-induced blood-brain barrier opening in non-human primates. Magn Reson Imaging S0730-725X(16):30236–30233

Sheikov N, McDannold N, Sharma S, Hynynen K (2008 Jul) Effect of focused ultrasound applied with an ultrasound contrast agent on the tight junctional integrity of the brain microvascular endothelium. Ultrasound Med Biol 34(7):1093–1104

Treat LH, McDannold N, Vykhodtseva N, Zhang Y, Tam K, Hynynen K (2007) Targeted delivery of doxorubicin to the rat brain at therapeutic levels using MRI-guided focused ultrasound. Int J Cancer 121:901–907

Treeby BE, Jaros J, Rendell AP, Cox BT (2012) Modeling nonlinear ultrasound propagation in heterogeneous media with power law absorption using a k-space pseudospectral method. J Acoust Soc Am 131(6)

Tung YS, Choi JJ, Baseri B, Konofagou EE (2010) Identifying the inertial cavitation threshold and skull effects in a vessel phantom using focused ultrasound and microbubbles. Ultrasound Med Biol 36(5):840–852. PMID: 20420973 [PMCID in process]

Tung YS, Vlachos F, Choi JJ, Deffieux T, Selert K, Konofagou EE (2010b) In vivo noninvasive cavitation threshold detection during blood-brain barrier opening using FUS and Definity. Phys Med Biol 55(20):6141–6155. PMID: 20876972 [PMCID in process]

Tung YS, Marquet F, Teichert T, Ferrera V, Konofagou EE (2011) Feasibility of noninvasive cavitation-guided blood-brain barrier opening using focused ultrasound and microbubbles in nonhuman primates. Appl Phys Lett 98(16):163704. PMID: 21580802 [PMCID in process]

Tung YS, Vlachos F, Feshitan JA, Borden MA, Konofagou EE (2011b) The mechanism of interaction between focused ultrasound and microbubbles in blood-brain barrier opening in mice. J Acoust Soc Am 130(5):3059–3067. PMC3248062

Tung YS (2012) The physical mechanism of blood-brain barrier opening using focused ultrasound and microbubbles. PhD dissertation, Columbia University

Vlachos F, Tung Y, Konofagou EE (2010) Permeability assessment of the focused ultrasound-induced blood-brain barrier opening using dynamic contrast-enhanced MRI. Phys Med Biol 55:5451–5466

Vykhodtseva NI, Hynynen K, Damianou C (1995) Histologic effects of high intensity pulsed ultrasound exposure with subharmonic emission in rabbit brain in vivo. Ultrasound Med Biol 21:969–979

Wang S, Burger C, Konofagou EE (2013) Focused ultrasound induced blood-brain barrier opening in macromolecule delivery. IInternational Society of Therapeutic Ultrasound (ISTU) Meeting, Shanghai, China, Dent Abstract

Wang S, Olumolade OO, Sun T, Samiotaki G, Konofagou EE (2015) Noninvasive, neuron-specific gene therapy can be facilitated by focused ultrasound and recombinant adeno-associated virus. Gene Ther 22(1):104–110

Wang S, Buch A, Acosta C, Olumolade O, Syed H, Duff K, Konofagou EE (2017) Non-invasive, focused ultrasound-facilitated gene delivery for optogenetics. Sci Rep 7:39955

Wu SY, Tung YS, Marquet F, Downs M, Sanchez C, Chen C, Ferrera V, Konofagou EE (2014) Transcranial cavitation detection in primates during blood-brain barrier opening – a performance assessment study. IEEE Trans Ultrasonics Ferroelectr Freq Control 61(6):966–978

Wu SY, Chen CC, Tung YS, Olumolade OO, Konofagou EE (2015) Effects of the microbubble shell physicochemical properties on ultrasound-mediated drug delivery to the brain. J Cntrl Rel Off J Cntrl Rel Soc 212:30–40. PMCID: PMC4527345

Wu SY, Sanchez CS, Samiotaki G, Buch A, Ferrera VP, Konofagou EE (2016) Characterizing focused-ultrasound mediated drug delivery to the heterogeneous primate brain in vivo with acoustic monitoring. Sci Rep 6:37094. PMCID: PMC5112571

Wu S-Y (2017, February) Neuronavigation-guided transcranial ultrasound: development towards a clinical system & protocol for blood-brain barrier opening

Wu S-Y, Fix S, Arena C, Chen C, Zheng W, Olumolade O, Papadopoulou V, Novell A, Dayton P, Konofagou E (2018) Focused ultrasound-facilitated brain drug delivery using optimized nanodroplets: vaporization efficiency dictates large molecular delivery. Phys Med Biol 63(3):035002

Chapter 21
Crossing the Blood-Brain Barrier with AAVs: What's After SMA?

Yujia Alina Chan and Benjamin E. Deverman

Abstract The 2009 discovery that the AAV9 serotype can deliver genes across the blood-brain barrier (BBB) spurred the rapid development of a recently FDA-approved gene therapy for spinal muscular atrophy (SMA). The success of the SMA clinical trial alongside other promising preclinical studies are stimulating significant interest and investment into AAV-based therapies for CNS diseases. Yet, the high doses required for gene transfer into the CNS have given rise to safety and manufacturing concerns. To address this challenge, scientists are racing to develop next-generation AAVs, with proof-of-principle demonstrations that AAV capsids can be engineered to more efficiently cross the adult BBB in animal models. Nevertheless, the field awaits the development of AAVs with enhanced BBB-crossing capabilities in humans. Here, we describe the development of AAV9 for CNS gene therapy, characteristics of natural and engineered AAVs that cross the BBB, and mechanistic evidence that can inform AAV engineering.

Keywords Adeno-associated virus · AAV engineering · Capsid engineering · Gene therapy · Blood-brain barrier · BBB crossing · BBB transcytosis · SMA

21.1 AAV-mediated CNS Gene Therapy: From Discovery to Clinic in 10 years

The year 2009 marked the breakthrough discovery that the adeno-associated virus (AAV) serotype, AAV9, crosses the blood-brain barrier (BBB) (Foust et al. 2009). AAV9 was demonstrated to successfully deliver genes to motor neurons when injected intravenously into neonatal mice (Foust et al. 2009) and cats (Duque et al.

Y. A. Chan · B. E. Deverman (✉)
Stanley Center for Psychiatric Research, Broad Institute of MIT and Harvard, Cambridge, MA, USA
e-mail: bdeverma@broadinstitute.org

© American Association of Pharmaceutical Scientists 2022
E. C. M. de Lange et al. (eds.), *Drug Delivery to the Brain*, AAPS Advances in the Pharmaceutical Sciences Series 33, https://doi.org/10.1007/978-3-030-88773-5_21

2009). In the exhilarating decade since this discovery, AAV9 has been rapidly developed as a systemically delivered CNS-targeting gene therapy vector and approved by the U.S. Food and Drug Administration (FDA), as of May 2019, to treat spinal muscular atrophy type I (SMA I; the gene therapy is named Zolgensma) (Mendell et al. 2017; Center for Biologics Evaluation and Research 2019). AAV-mediated gene therapy is now recognized as a competitive alternative or complement to other state-of-the-art therapies such as antisense oligonucleotides (ASOs), which cannot cross the BBB and require repeated intrathecal injection procedures (Pattali et al. 2019).

The speed of AAV9's translation from discovery to an FDA-approved gene therapy has inspired a scientific gold rush toward applying systemic AAV-based therapies to additional indications. Indeed, due to the low pathogenicity and immunogenicity of AAVs, there have been 248 clinical trials involving AAVs registered at ClinicalTrials.gov as of March 2022. However, a major limitation for systemic AAV9-mediated CNS gene therapy is the narrow window during infancy in which it can be effectively applied (see Sect. 21.2.1), at least in the context of SMA (Foust et al. 2009; Mendell et al. 2017).

Here, we will discuss the development and current status of the AAV9-based gene therapy for SMA, ongoing evaluation of AAV9 and other vectors for additional indications, recent safety concerns that stem from the high doses of AAV required for efficient delivery following systemic administration, what is known about how AAV9 and other engineered AAVs cross the BBB, and how this mechanistic understanding may inform the development of next-generation AAVs for CNS gene therapy.

21.1.1   The History of AAV-mediated CNS Gene Delivery via Intravenous Administration

SMA is a life-threatening childhood autosomal recessive neurodegenerative disease in which the loss of motor neurons leads to progressive muscle weakness and paralysis. SMA patients also suffer systemic defects spanning the autonomic and enteric nervous system, cardiovascular system, pancreas, and other cell types. This monogenic disorder stems from the loss or dysfunction of the gene encoding survival motor neuron 1 (SMN1). Among the four SMA subtypes, SMA type 1 (SMA1) is the most severe and common genetic cause of infant death, resulting in mortality or the need for permanent ventilation support by 24 months of age in more than 90% of untreated patients (Finkel et al. 2014).

The seminal publications by Foust et al. (2010) and Valori et al. (2010) demonstrated that the postnatal vascular delivery of the *SMN* gene by AAV9 to SMA mice could rescue motor function, neuromuscular physiology, and lifespan (Foust et al. 2010; Valori et al. 2010). SMA mouse models that were treated with a self-complementary AAV9 carrying the *SMN* gene replacement (scAAV9-SMN) exhibited a substantially extended average survival of 28.5 days (low-dose cohort) or more than 250 days [high-dose cohort; $2–3.3 \times 10^{14}$ vector genomes per kilogram (vg/kg)], as compared to the 15-day average lifespan in the absence of gene therapy. Importantly, these studies determined that AAV9-mediated therapy produced positive health outcomes only when administered within a narrow window in the developmental period. Within a year, additional studies corroborated that a single intravenous injection of an optimized scAAV9-SMN could correct motor function and rescue the SMA weight loss phenotype (Dominguez et al. 2011), and showed that AAV9 could target motor neurons and glial cells in macaques (Foust et al. 2010; Bevan et al. 2011; Dehay et al. 2012). The numerous studies thereafter characterizing AAV9-mediated gene therapy targeting SMA in animal models [see summary; (Pattali et al. 2019)] galvanized the development of AAV9 for gene transfer to the CNS for pediatric disorders. In 2014, a gene therapy company, AveXis, which was scientifically co-founded by Brian Kaspar, whose group first demonstrated that AAV9 crosses the neonatal BBB, launched a clinical trial of single-dose gene-replacement therapy in SMA1 patients [ClinicalTrials. gov NCT02122952 (AveXis, Inc. 2016)].

By August 2017, AveXis reported the successful rescue of 15 SMA patients (Mendell et al. 2017), who were event-free (not requiring ventilatory support for at least 16 h a day for 14 consecutive days) at 20 months of age as compared to an 8% survival rate in a historical cohort of 34 SMA1 patients (Finkel et al. 2014). Of the 15 patients, 3 were in a low-dose cohort (6.7×10^{13} vg/kg body weight) and 12 were in a high-dose cohort (2×10^{14} vg/kg). After gene delivery, the patients in the high-dose cohort exhibited a rapid increase in CHOP INTEND score (Children's Hospital of Philadelphia Infant Test of Neuromuscular Disorders scale of motor function; 0-64, with higher scores indicating better function): a mean increase of 9.8 and 15.4 at 1 and 3 months of age, respectively, compared to a mean decline of more than 10 points between 6 and 12 months of age observed in a historical cohort (Kolb et al. 2017; Mendell et al. 2017). Remarkably, among the 12 high-dose cohort patients, 11 sat unassisted, 9 rolled over, 11 fed orally and could speak, and 2 walked independently.

Despite the astounding success of the scAAV9-SMN therapy, 4 patients exhibited elevated serum aminotransferase (SAT) levels indicative of potential liver damage. This health outcome was anticipated after its manifestation in the first patient, and was ameliorated by prednisolone treatment. Thereafter, patients 2-15 were preemptively treated with oral prednisolone at 1 mg/kg/day for 30 days starting 24 h before AAV administration. Despite the preemptive prednisolone administration, one high-dose cohort patient required additional prednisolone to return to a threshold level of liver enzymes. An additional 3 non-serious adverse events of elevated SAT levels occurring in 2 patients were judged to be treatment-related. Most importantly, the clinical trial suggested, similar to preclinical trials, that an earlier diagnosis of SMA and initiation of scAAV9-SMN therapy could improve treatment outcomes.

In 2019, Zolgensma was approved by the U.S. FDA (Office of the Commissioner 2019) and Novartis, who acquired AveXis (renamed as Novartis Gene Therapy in September 2020), has provided updates that all of the high-dose cohort patients who enrolled in long-term follow-up continue to maintain developmental milestones; the majority of the patients attained motor milestones that were not previously seen in untreated cases of SMA1 (Novartis 2020). Encouragingly, a recent report on the early outcomes of Zolgensma treatment, published in August, 2020, showed that, of 19 children with repeated outcome assessments, two exhibited stabilization and 17 displayed improved motor function (Waldrop et al. 2020). Patients aged 6 months or younger at the time of dosing tolerated the gene therapy well. However, older children more frequently exhibited serum transaminase elevations and required a higher dose of prednisolone. Notably, transient asymptomatic platelet reduction was observed in the majority of patients, but was more severe in children that had transitioned from the Biogen ASO therapy, nusinersen. Due to differences among the patients in terms of age and baseline functional testing at the time of therapy, the study was not able to discern the effect of age on therapy efficacy. Gene therapy stakeholders continue to eagerly await updates on the long-term efficacy, immune response, and potential side effects associated with the AAV9-based gene therapy. The extent to which the *SMN* gene was delivered across the BBB to the CNS and motor neurons, and how this tracks with long-term treatment outcomes remain open and important questions.

21.1.2 CNS Applications of Systemic AAV Administration Beyond SMA

The success of the Zolgensma trial has spurred pursuits to develop AAV-mediated gene therapies for other neuromuscular disorders for which the underlying genetic causes are well characterized (Aguti et al. 2018; Luxner 2019). Progress and prospects for CNS-targeted gene therapy have been reviewed recently (Hocquemiller et al. 2016; Lykken et al. 2018; Deverman et al. 2018), so we will only discuss updates from recent or ongoing clinical trials.

Duchenne muscular dystrophy (DMD), which stems from mutations in the *dystrophin* gene, is another common human genetic neuromuscular disorder that affects 1 in 5000 live births. Similar to SMA, several gene therapy approaches exist for DMD and the efficacy of AAV-mediated gene therapy for DMD has been demonstrated in dystrophin-deficient *mdx* mouse and canine models (Wang et al. 2000; Watchko et al. 2002; Gregorevic et al. 2006; Yue et al. 2015). There are now three ongoing systemic AAV gene therapy clinical trials for DMD organized by Solid Biosciences, Pfizer, and Sarepta Therapeutics. Solid Biosciences has recently resumed their IGNITE DMD clinical trial, in which patients have exhibited microdystrophin expression and potential therapeutic benefit (Solid Biosciences Inc 2019, 2020). The Pfizer DMD gene therapy has received a fast track designation from the FDA based on their phase I study that showed that the therapy was well-tolerated within the infusion period and that dystrophin expression was maintained over the first year (Pfizer 2020). The Sarepta phase I clinical trial found that their gene therapy resulted in an 81.2% increase in dystrophin expression in muscles, accompanied by marked improvements in patients' functional performance (Inacio 2019). Sarepta has been awarded a fast track designation by the FDA and is now in a phase II trial.

Systemic AAV-based gene therapies for other neuromuscular disorders including Limb-Girdle muscular dystrophy and Myotubular myopathy (XLMTM) are also underway: two Limb-Girdle muscular dystrophy trials and one XLMTM trial are registered on Clinicaltrials.gov. AAV-based gene therapy is also being considered for other systemic diseases that affect the CNS including the numerous lysosomal storage diseases (LSDs) in which the underlying mutation affects an enzyme that can be secreted and internalized by neighboring cells. The pre-clinical evaluations of gene therapies for LSDs have shown promise in a variety of rodent and large animal models (Hocquemiller et al. 2016). This potential for cross correction, where a small number of transduced cells can impact a much larger fraction of non-transduced cells, may make gene therapy for LSDs viable with existing vectors such as AAV9 at more moderate doses than those used in the DMD trials.

For many neurological disorders, the need to deliver transgenes to a large fraction of cells throughout the CNS is a significant challenge in gene therapy applications. Several monogenetic nervous system disorders, for which gene therapies are being evaluated in preclinical and clinical studies, are driven by loss-of-function (e.g., giant axonal neuropathy, *GAN*; Friedreich's ataxia, *FXN*; Rett syndrome, *MECP2*) or gain-of-function mutations (e.g., Huntington's disease, *HTT*; ALS, *SOD1*, *C9ORF*, *TARDBP*, and *FUS*; frontotemporal dementia, *MAPT* and *TARDBP*) that are thought to affect each cell independently. Given the current lack of vectors that can achieve high-efficiency gene transfer throughout the human CNS following IV administration, many CNS gene therapy programs have focused on direct intra-parenchymal injection or intrathecal (IT) administration into the CSF. Intraparenchymal delivery appears to be well suited for treating Parkinson's, in which targeted delivery of *AADC* enables neurons in the putamen to convert levodopa to dopamine (Voyager Therapeutics, Inc 2019). Intra-CSF delivery routes may also be suitable for LSDs that benefit from cross correction or for indications that are ameliorated by transduction of sensory and motor neurons.

21.1.3 High AAV Doses Can Stimulate Immune Responses That Are Detrimental to Patient Health and Therapy Efficacy

To achieve systemic gene therapy, gene therapies for SMA and muscular dystrophies have been delivered at exceptionally high doses ($1\text{-}3 \times 10^{14}$ vg/kg) that are up to 1000-fold higher than the doses used for liver-targeted gene therapy [5×10^{11} to 3×10^{13} vg/kg implemented for AAV-mediated therapies targeting hemophilia: 3×10^{13} vg/kg in the Pfizer and Sangamo Therapeutics Alta study, 2×10^{13} vg/kg in the uniQure HOPE-B trial, 6×10^{12} to 2×10^{13} vg/kg in the BioMarin hemophilia A study, 5×10^{11} to 2×10^{12} vg/kg in the Spark Therapeutics *SPK-8011* and *SPK-9001* trials (Pasi et al. 2020; uniQure Inc 2020; Sangamo 2020; Rosen et al. 2020; Spark Therapeutics, Inc 2020)]. The high doses used to target muscle and the CNS, while tolerated relatively well in the SMA trial, have stimulated dangerous immune responses and liver damage in humans and animal models.

Two of the three DMD AAV-mediated gene therapy trials have observed adverse events, which have raised safety concerns. In the DMD gene therapy trial by Solid Biosciences, 2 out of 6 patients exhibited a drop in red blood cell count, one of whom exhibited kidney injury – both were hospitalized (Offord 2019). The DMD clinical trial by Pfizer observed immune responses in all of their patients alongside adverse effects in 4 out of 6 patients; one patient developed a rapid antibody response associated with acute kidney injury, hemolysis, and reduced platelet count, and was hospitalized (Pfizer 2019). Solid Biosciences has since adjusted their trial protocol and manufacturing processes (Solid Biosciences 2020; Solid Biosciences Inc 2020), and was released from clinical hold by the FDA. In parallel to these clinical trials, there are ongoing efforts to screen 100,000 newborns per year in New York for DMD with the aim of enabling early treatment and evading immune responses (Luxner 2019).

Sadly, in the Audentes Therapeutics ASPIRO clinical trial for XLMTM, three deaths were reported among the 17 patients treated with the higher dose (3×10^{14} vg/kg) of the AAV8-mediated AT132 gene therapy (Audentes 2020). Audentes Therapeutics stated that the three patients were among the 50% of trial subjects who exhibited signs of pre-existing hepatobiliary disease. The trial is currently on hold in order to investigate the reasons for liver dysfunction in the deceased patients. Despite the serious adverse events, it is important to note that all of the high-dose systemic AAV trials have reported evidence of efficacy in terms of transgene expression or functional improvement.

Data from liver-targeted AAV-mediated gene therapy clinical trials have suggested that the administration of high AAV doses may trigger the reactivation and expansion of AAV capsid-reactive memory T cells, resulting in the loss of transgene expression and immune-mediated toxicities (Manno et al. 2006; Nathwani et al. 2011; Vandamme et al. 2017). For instance, in the hemophilia B clinical trials of rAAV2-mediated Factor IX (*FIX*) gene therapy, FIX expression reduction (down to pretreatment levels) was concomitant with a transient rise in liver transaminase levels and AAV2 capsid-specific CD8+ T cells (Manno et al. 2006). In a separate hemophilia B AAV8-based gene therapy clinical trial in which patients had been pre-screened for anti-AAV NAbs, the high-dose cohort (2×10^{12} vg/kg) exhibited a dramatic decrease in FIX expression alongside increases in serum transaminase levels and circulating capsid-specific T cells (Nathwani et al. 2011). Circulating AAV8 capsid-specific T cells were also detected in patients in the intermediate-dose cohort (6×10^{11} vg/kg), but did not appear to reduce FIX levels or increase transaminase levels. Prednisolone treatment solved the transaminitis and resulted in the maintenance of *FIX* expression (Nathwani et al. 2011). Collectively, the numerous reports from intramuscular and systemic gene transfer clinical trials indicate a correlation between vector dose and the magnitude of anti-AAV T cell responses (Manno et al. 2006; Boisgerault and Mingozzi 2015).

One practicable solution to this challenge is to exclude patients who exhibit high amounts of pre-existing anti-AAV antibodies or AAV-reactive T cells. However, this approach is unattractive because it would preclude many patients, who are generally exposed to AAVs in early childhood (Calcedo et al. 2011) and carry neutralizing antibodies (NAbs) recognizing one or more AAV serotypes (Calcedo et al. 2009). In fact, the AveXis clinical trial excluded one patient with persistently elevated anti-AAV9 antibody titers (>1:50), and preemptively treated all but the first patient with oral prednisolone prior to gene therapy (Mendell et al. 2017). Furthermore, pre-existing circulating AAV-specific T cells are difficult to systematically detect, and anti-capsid cellular responses do not necessarily translate into deleterious clinical consequences in low-dose AAV gene therapy applications (although these may reduce transgene expression) (Vandamme et al. 2017). In other words, the exclusion approach may not effectively identify and omit all patients that are likely to exhibit harmful anti-AAV responses, and, conversely, may exclude patients who may still benefit from lower doses of AAV gene therapy. In addition, AAV particles can persist in the body for years in large-animal and human tissues (Stieger et al. 2009; Mueller et al. 2013), and it is unclear whether the slow breakdown of these particles may lead to MHC class I antigen presentation and trigger capsid-directed immune responses or mimic a chronic viral infection after patients are no longer closely monitored (Pien et al. 2009; Finn et al. 2010; Gernoux et al. 2017).

The urgency of elucidating parameters other than preexisting anti-AAV immunity that predict adverse reactions is underscored by a study by Hinderer et al. that was published a year after the release of the AveXis trial results (Hinderer et al. 2018). In this study, juvenile rhesus macaques and piglets were treated with a high dose (2×10^{14} vg/kg) of AAVhu68 (a capsid that differs from AAV9 at two residues) delivering the human *SMN* gene. Shockingly, one of the three macaque subjects had to be euthanized by day 4 post-treatment, and all three piglets developed neurologic signs including hind-limb ataxia and dorsal root ganglion (DRG) neuronal cell body degeneration (these were less severe in the two older piglets) and were euthanized within 14 days of treatment. Although the severity of these reported adverse reactions stands in contrast to what has been observed in other studies and clinical trials, they serve as an important warning for the community that high-dose systemic AAV may lead to serious adverse events in a subset of subjects.

21.1.4 Navigating Immune Responses with Next-generation AAVs, Pharmacological Interventions, and Alternative Delivery Routes

To ameliorate the adverse effects triggered by high-dose AAV gene therapy, there have been extensive efforts across the gene therapy community to engineer AAVs or modulate patient immune response (Vandamme et al. 2017). In their review, Vandamme et al. propose that the most effective solution may be recombinant AAVs (rAAVs) with a higher therapeutic index; these rAAVs would ideally carry optimized therapeutic transgenes, exhibit more efficient transduction, and would not trigger immunogenicity (Vandamme et al. 2017). These advantages could stem from a CpG-depleted genome (Faust et al. 2013), contaminant-free AAV preparations, or increasing the proportion of capsids carrying the therapeutic rAAV genome (Mingozzi et al. 2013). This approach is also supported by an increasing number of studies that have successfully engineered AAV capsids with dramatically improved transduction efficiency and tropisms in animal models (see Sect. 21.3 for details on engineering BBB-crossing variants). There are also efforts to map epitopes to antibodies that recognize AAV capsids in order to engineer capsids that escape neutralization (Bartel et al. 2011; Selot et al. 2017; Giles et al. 2018).

Alternative delivery routes that limit the distribution of AAV to avoid systemic reactions are also being actively explored. Several groups have shown that intrathecally (IT)-delivered AAV9 in large animals can transduce motor neurons in the spinal cord and DRG neurons, with reduced targeting of peripheral organs when applied at lower doses; however, it is unclear whether this approach evades NAbs and reduces the required dose for efficacy (Samaranch et al. 2012; Federici et al. 2012; Gray et al. 2013; Passini et al. 2014; Meyer et al. 2015; Bailey et al. 2018; Hordeaux et al. 2019a). There is optimism that IT-delivered AAV gene therapies can address LSDs, where the gene product is often secreted from cells, and diseases such as Giant Axonal Neuropathy (GAN), which may benefit from the transduction of DRG neurons. The first clinical trial (ClinicalTrials.gov NCT02362438) harnessing IT-delivered AAV9 for gene replacement in GAN patients has been launched after promising results were obtained in patient fibroblasts and *GAN* knockout (KO) mice (Bailey et al. 2018). IT administration is also being explored in the STRONG clinical trial for SMA II although the high-dose cohort has been put on hold by the FDA due to safety concerns raised by Novartis (Gardner 2019). While promising for specific indications, IT and other CSF delivery routes are less applicable to neuromuscular disorders such as myotubular myopathy and Duchenne/Becker muscular dystrophy (DMD/BMD), which require systemic gene therapy to skeletal muscles, cardiac muscles, and motor neurons (Aguti et al. 2018) or neurodevelopmental or neurodegenerative diseases that require efficient and uniform delivery to neurons throughout the CNS.

21.2 Factors That Influence the CNS Tropism of AAV9

21.2.1 Age at Delivery

Age is one of the most important factors that influence both the transduction efficacy and the tropism of AAV serotypes in the CNS. Several AAV vectors, including AAV9, deliver genes to neurons in the CNS much more efficiently in neonates as compared to adults. This difference in cell types transduced with age has been demonstrated in rodents (Foust et al. 2009; Jackson et al. 2015), as well as in non-human primates (NHPs). In mice, only P1 and P2 administration of scAAV9-SMN resulted in efficient spinal motor neuron transduction and led to long-term survival and sustained weight gain in treated SMA mice; in contrast, mice injected at P5 or P10 had progressively more glial transduction and less motor neuron transduction (Foust et al. 2009). Similarly, in NHPs, systemic rAAV9 administration in juveniles or adults resulted in mostly glial transduction (Gray et al. 2011; Bevan et al. 2011; Samaranch et al. 2012) in contrast to neuronal transduction in neonates (Dehay et al. 2012; Mattar et al. 2013). The reduced efficiency of gene delivery to neurons and shift in tropism toward astrocytes may be influenced by the fact that the early neonatal mouse brain is not as fully populated by astrocytes (Foust and Kaspar 2009). Alternatively, the shift in biodistribution may also be influenced by changes in the extracellular space volume, which is substantially higher at birth and then rapidly declines (Lehmenkühler et al. 1993; Wolak and Thorne 2013). In addition, this difference in transduction pattern may arise from the distinct routes utilized by AAVs in neonates versus adults. In a study that tested 10 AAV serotypes in neonatal mice, the authors, Zhang et al., suggested that the AAVs may enter through the choroid plexus to result in the observed gradient of transduction near ventricles (Zhang et al. 2011). Notably, although 8 of the 10 serotypes were shown to be capable of transducing neurons, glia, choroid plexus, and endothelial cells across multiple brain regions, only a limited subset of serotypes could deliver genes to the adult CNS via the vasculature (Foust et al. 2009; Zhang et al. 2011; Yang et al. 2014). It remains to be determined how AAV serotypes transduce the neonatal CNS more effectively than the adult CNS, and why there is a preference for glial transduction in adults.

21.2.2 The Relationship Between AAV9 Receptor Binding, Persistence in the Circulation, and Its CNS Tropism

Out of the hundreds of known AAV capsids, only a handful have been reported to cross the BBB in adult animals. The characteristics that confer this unique ability remain unknown, but constitute an active area of investigation. A topic of interest and speculation has been the relationship between AAV9's delayed blood clearance and its tropism for the CNS and cardiac muscles. This extended half-life or persistence in the circulation is unique to AAV9 and absent in other tested serotypes, including the AAVrh.10 serotype that also transduces the CNS via the vasculature (Zincarelli et al. 2008; Hu et al. 2010; Kotchey et al. 2011; Tanguy et al. 2015).

Residues that contribute to this persistence phenotype have been mapped to surface-exposed regions involved in receptor binding although it remains unclear whether the two phenotypes share a causative relationship (Kotchey et al. 2011; DiMattia et al. 2012; Adachi et al. 2014). AAV9 is capable of binding to cell surface glycans with terminal β-galactose (Shen et al. 2011; Bell et al. 2011; Adachi et al. 2014), which bind to a pocket in the AAV9 capsid surrounded by three protrusions formed by residues D271, N272, Y446, N470, and W503 (Bell et al. 2012). Random mutagenesis of AAV9 resulted in the discovery of mutants, AAV9.45 and AAV9.61 with N498V or W503R mutations, that retain the parent cardiac and musculoskeletal transduction efficiencies, but have lost the ability to transduce the liver and bind to glycans (Pulicherla et al. 2011; Shen et al. 2012). In parallel, Shen et al. found that systemic administration of sialidase, which increases exposed terminally galactosylated glycans in tissues, results in a hepatotropic AAV9 biodistribution [~7-fold increase in vector genomes in the liver (Shen et al. 2012)]. They further demonstrated that AAV9 half-life, in terms of its distribution between tissue and blood and its elimination from circulation, was linked to its glycan-binding avidity; a high avidity AAV9 mutant was rapidly cleared from the blood, and a low avidity AAV9 W503R mutant had a longer half-life in the circulation compared to AAV9 (Shen et al. 2012). In support of this finding, the majority of low persistence AAV9 mutants identified by Adachi et al. in a systematic screen were found to carry mutations in this glycan-binding region and exhibited liver or globally detargeted phenotypes (Adachi et al. 2014). In a similar study of AAV2, abrogating binding to its receptor heparan sulfate proteoglycan (HSPG) resulted in liver-detargeted phenotypes (Kern et al. 2003). One of these mutants, AAV2i8, exhibits a long half-life in the circulation, alongside a robust cardiac and skeletal muscle tropism and an ability to cross the vascular endothelium (Asokan et al. 2010). How and whether glycan-binding differences, across serotypes, affect AAV half-life and tropism remains an open question.

21.2.3   Mechanistic Insights into BBB Crossing from AAV Domain Swapping

AAVrh.10 is the next best-characterized AAV that can cross the BBB in neonatal and adult mice (Zhang et al. 2011; Tanguy et al. 2015; Albright et al. 2018). To determine the structural features of AAVrh.10 that confer the ability to cross the adult mouse BBB, Albright et al. created and screened a combinatorial domain swap library using AAVrh.10 and AAV1, which shares 85% sequence homology with AAVrh.10 and transduces a fraction of brain endothelial cells but does not cross the BBB (Albright et al. 2018). This work mapped the residues of AAVrh.10 that can enhance AAV1's ability to cross the adult mouse BBB, including a region between residues 263-274 that was cloned from AAVrh.10 into AAV1 to constitute the hybrid AAV1RX [Table 21.1 (Albright et al. 2018)]. The converse swap of the same region of AAV1 inserted into AAVrh.10 was shown to reduce the BBB-crossing ability of the variant AAVRX1. Albright et al. postulated a "Goldilocks model" which links reduced interaction with sialic acid (SIA) to the enhanced BBB-crossing phenotype of their AAV1RX hybrid (Albright et al. 2019). To support this hypothesis, they showed that a W503A mutant, like AAV1RX, exhibits partially reduced dependency of cell binding on SIA and results in an improved BBB-crossing phenotype (Albright et al. 2019). Notably, the "Goldilocks model" is reminiscent of transferrin receptor antibody studies that have shown that high affinity and/or avidity interactions result in lysosomal-targeting whereas monovalent interactions result in BBB transport (Yu et al. 2011; Niewoehner et al. 2014; Bien-Ly et al. 2014; Haqqani et al. 2018).

21.3   Engineering Capsids for Enhanced Blood-Brain Barrier Crossing

Multiple studies have demonstrated that AAV vectors can be targeted to the brain vasculature [Representative capsids shown in Table 21.1; (Deverman et al. 2016; Körbelin et al. 2016; Chan et al. 2017; Hanlon et al. 2019; Ravindra Kumar et al. 2020; Nonnenmacher et al. 2021; Goertsen et al. 2022)]. In the first of these studies,

Table 21.1 AAV capsids that cross the BBB

Capsid	Transport across the BBB	Source/ modification	Characteristics: known receptors and mechanism, animal models examined	Reference
AAV9	Yes	Isolated from human	Widespread delivery after systemic administration	Foust et al. (2009) and Duque et al. (2009)
AAVrh.8	Yes	Isolated from rhesus	Widespread delivery after systemic administration	Yang et al. (2014)
AAVrh.10	Yes	Isolated from rhesus	Widespread delivery after systemic administration	Yang et al. (2014)
AAV1RX	Yes	AAVrh.10 residues 263-274 domain swap into AAV1	Improved transport across the BBB as compared with AAV1	Albright et al. (2018)
AAV-PHP.B	Yes	7-mer insertion into AAV9(K449R) after AA588 TLAVPFK	Highly efficient CNS transduction; mediated by interaction with LY6A	Deverman et al. (2016)
AAV-PHP.B2	Yes	7-mer insertion into AAV9(K449R) after AA588 SVSKPFL	Efficient CNS transduction; mediated by interaction with LY6A	Deverman et al. (2016)
AAV-PHP.B3	Yes	7-mer insertion into AAV9(K449R) after AA588 FTLTTPK	Efficient CNS transduction; mediated by interaction with LY6A	Deverman et al. (2016)
AAV-PHP.A	Yes	7-mer insertion into AAV9(K449R) after AA588 YTLSQGW	Astrocyte-selective enhancement	Deverman et al. (2016)
AAV-PHP.eB	Yes	7-mer adjacent modification AQ 587-588 to DG in PHP.B	Enhanced variant of AAV-PHP.B with improved neuronal transduction; mediated by interaction with LY6A	Chan et al. (2017)
AAV-BR1	Minimal	NRGTEWD insertion into AAV2 after AA588	Efficient and selective transduction of brain endothelial cells, minimal evidence of transcytosis	Körbelin et al. (2016)
AAV-F	Yes	FVVGQSY insertion into AAV9	Efficient transduction of C57BL/6J mice and BALB/cJ mice	Hanlon et al. (2019)

(continued)

Table 21.1 (continued)

Capsid	Transport across the BBB	Source/ modification	Characteristics: known receptors and mechanism, animal models examined	Reference
AAV-PHP. V1	Yes	7-mer insertion into AAV9(K449R) after AA588 TALKPFL	In comparison with AAV-PHP.B, more brain vasculature transduction, less astrocyte transduction; enhanced tropism compared to AAV9 not observed in BALB/cJ mice (likely interacts with LY6A); transduces human brain microvascular endothelial cell (HBMEC) culture better than AAV9 or AAV-PHP.eB	Ravindra Kumar et al. (2020)
AAV-PHP. C1	Yes	7-mer insertion into AAV9(K449R) after AA588 RYQGDSV	Broad CNS transduction, reduced neuron transduction; enhanced BBB crossing in both BALB/cJ and C57BL/6J mice	Ravindra Kumar et al. (2020)
AAV-PHP.N	Yes	7-mer adjacent modification KAQ 595-597 to SNP in PHP.B	NeuN+ neuron-specific transduction; enhanced tropism compared to AAV9 not observed in BALB/cJ mice (likely interacts with LY6A)	Ravindra Kumar et al. (2020)
AAV. CAP-B10	Yes	7-mer substitution DGAATKN (AA452-460) into AAV-PHP.eB	In mice, retained AAV-PHP.eB CNS tropism but with a bias toward neurons and increased de-targeting from liver; retains bias toward neurons and liver de-targeting in adult marmosets	Goertsen et al. (2022)
9P31	Yes	7-mer insertion into AAV9 modified with by AQ587-588DG WPTSYDA	Broad CNS transduction with neuronal bias, which is reduced in BALB/cJ mice	Nonnenmacher et al. (2021)

Deverman et al. harnessed a novel Cre-dependent *in vivo* selection method, named CREATE, to identify AAV variants, with peptide insertions in loop VIII, that cross the adult BBB in C57BL/6J mice and transduce astrocytes (Deverman et al. 2016). This resulted in the discovery of AAV variants with an enhanced CNS tropism: AAV-PHP.B, AAV-PHP.B2, and AAV-PHP.B3. What set this screen apart from prior efforts was the increased selective pressure applied by the recovery of only sequences that crossed the BBB and transduced astrocytes. In a subsequent study, additional mutagenesis of the AAV-PHP.B capsid led to the development of a further enhanced variant named AAV-PHP.eB (Chan et al. 2017). AAV-PHP.B and AAV-PHP.eB both display efficient transduction of neurons, astrocytes, oligodendrocytes, and endothelial cells after intravenous administration in adult mice. AAV-PHP.eB is capable of transducing the majority of neurons throughout the brain and spinal cord with a single intravenous injection, without requiring the use of scAAV genomes for efficient transduction. Compared to the AAV9 parent, AAV-PHP.B and AAV-PHP.eB deliver genes to the CNS at least 40- and 100-fold more efficiently, respectively (Deverman et al. 2016; Chan et al. 2017; Huang et al. 2019). In particular, AAV-PHP.eB can be systemically delivered to adult mice at a relatively low dose of ~5x10^{12} vg/kg (doses at least 10x higher are typically used for AAV9 administration) while still achieving transduction of the majority of neurons in multiple brain regions (Chan et al. 2017).

AAV-PHP.B and AAV-PHP.eB have become important tools for the neuroscience community. They have been applied across diverse experiments in mice, including multicolor sparse cell labeling (Chan et al. 2017), tracking neuronal activity (Hillier et al. 2017), manipulating brain butyrylcholinesterase levels (Gao et al. 2017), cortex-wide imaging (Allen et al. 2017), delivery of DREADDs and neuropeptides to genetically defined cell populations in the brain (Zelikowsky et al. 2018), as well as cell type-specific enhancer screening (Graybuck et al. 2021; Vormstein-Schneider et al. 2020). In addition, numerous groups are presently deploying AAV-PHP.B and AAV-PHP.eB in mouse models to interrogate genetic deficit corrections in Niemann-Pick C1 disease (Davidson et al. 2021; Gu et al. 2018; Levy et al. 2020), RETT syndrome (Luoni et al. 2020), Pompe disease (Lim et al. 2019), synucleinopathy (Morabito et al. 2017), repression of Tau (Wegmann et al. 2021), and Leigh syndrome (Reynaud-Dulaurier et al. 2020).

21.3.1 Learning from the Species and Strain Dependency of AAV-PHP.B and AAV-PHP.eB

Due to the widespread evaluation and adoption of AAV-PHP.B and AAV-PHP.eB by the research and gene therapy community, it was soon discovered that their enhanced CNS tropism does not extend to certain strains of mice [BALB/cJ mice were the first strain in which the enhanced CNS tropism was found to be absent (Hordeaux et al. 2018)] and to NHPs (Sah et al. 2018; Matsuzaki et al. 2018; Hordeaux et al. 2018). Through a genetic linkage study that interbred C57BL/6J (AAV-PHP.B permissive) and BALB/cJ (AAV-PHP.B non-permissive) mice, Hordeaux et al. were the first to report that the enhanced CNS tropism of AAV-PHP.B is a heritable trait in mice (Hordeaux et al. 2018). This observation indicated that the enhanced BBB-crossing ability of AAV-PHP.B is likely determined by a single genetic factor that varies by species and mouse strain.

A race rapidly ensued to identify and leverage this unique mechanism in the design of BBB-crossing AAVs for humans. In 2019, three groups, including ours, independently published findings that the enhanced CNS tropism of AAV-PHP.B and AAV-PHP.eB is mediated by the lymphocyte antigen 6 complex, locus A [LY6A; also known as stem cell antigen-1 (SCA-1)], a cellular protein that is highly expressed in the brain endothelial cells of certain strains of mice (Hordeaux et al. 2019b; Batista et al. 2019; Huang et al. 2019). Each of the three groups found that the overexpression of *Ly6a* in other cell types (HEK293 or CHO cells) dramatically increases binding and transduction by AAV-PHP.B and AAV-PHP.eB (Hordeaux et al. 2019b; Batista et al. 2019; Huang et al. 2019). Notably, our group showed that the interaction with LY6A allows AAV-PHP.eB to transduce endothelial cells in the absence of AAVR (Huang et al. 2019). This effectively demonstrated that AAV capsids can be engineered to utilize new modes of transduction, rendering the novel capsids less dependent on the receptors that natural AAVs rely on for transduction.

The identification of the novel LY6A-aided route of entry into the CNS raises new questions about whether proteins similar to LY6A exist on the BBB of other species, particularly in humans. LY6A is a relatively poorly characterized glycosylphosphatidylinositol (GPI)-anchored protein. However, the *Ly6* locus has previously been linked to susceptibility to mouse adenovirus (MAV1) (Spindler et al. 2010), which exhibits an endothelial cell tropism and fatal hemorrhagic encephalomyelitis in C57BL/6 but not BALB/cJ mice (Guida et al. 1995). Members of the *Ly6* family also affect susceptibility to other viruses that affect humans such as HIV1 (Loeuillet et al. 2008; Brass et al. 2008), Flaviviridae [yellow fever virus, dengue, and West Nile virus (Krishnan et al. 2008)], and Influenza A (Mar et al. 2018). These connections with viral susceptibilities motivate the investigation of the mechanism by which *Ly6* genes mediate viral infectivity, so as to design or select for AAV gene therapy vectors that leverage similar mechanisms to cross the BBB.

Although LY6A has been demonstrated to enhance transduction, the nature of its role in enhancing transcytosis remains an area of active investigation. The efficient BBB-crossing phenotype of the AAV-PHP.eB capsid suggests that LY6A may directly transport the AAV-PHP.B viruses across the BBB. However, it is unclear whether the transcytosis of AAV-PHP.eB is dependent on LY6A beyond the initial internalization into the cell, and whether it requires interactions with other AAV receptors such as AAVR for transcytosis. This question stems from the possibility that transduction and transcytosis may occur through different intracellular trafficking routes (see 21.3.4 for detailed discussion). Our observation that AAV-PHP.eB transduces brain endothelial cells but not neurons and glia in AAVR KO mice is compatible with two hypotheses. First, AAV-PHP.eB may internalize and transcytose brain endothelial cells through its interaction with LY6A, and thereby gain access to neurons and glia in the absence of AAVR. However, due to the fact that neurons and glia do not express *Ly6a*, their transduction remains dependent on AAVR. Alternatively, in the second model, LY6A only mediates entry of AAV-PHP.eB into endothelial cells, after which AAV-PHP.eB relies on an additional interaction with AAVR for transcytosis. Differentiating between these two models is challenging, but important for understanding how LY6A mediates the efficient entry of AAV-PHP.eB into the CNS. If LY6A does have a role in directing these engineered capsids for transcytosis into the CNS, it will be interesting to know whether LY6A also serves as a transporter of other ligands across the BBB. The discovery of endogenous ligands transported by LY6A could suggest functional equivalents of LY6A to be exploited for AAV engineering.

Importantly, the AAV-PHP.B strain specificity studies highlight the need to consider how the design of each AAV selection method may uniquely limit the application of the derived AAVs. Beyond species and strains, an additional challenge is that disease states can alter the BBB in a way that impacts AAV transduction and may consequently nullify the applicability of AAV variants that were selected in healthy animal models. An *in vivo* phage display screen discovered that the epitopes that bind to the brain differ between normal and disease states, and even among different disease models (Chen et al. 2009).

21.3.2 Other BBB-targeted AAVs

In 2016, Körbelin et al. described AAV-BR1, which transduces the brain vasculature with high selectivity, with minimal transduction of CNS neurons or glial cells, as well as other organs including the liver and heart (Körbelin et al. 2016). This capsid was obtained through selection of AAV2 capsid variants with peptide insertions in loop VIII in FVB mice. The authors showed that AAV-BR1 could be used to deliver reporters as well as Cre recombinase, providing a powerful new approach to selectively knockout genes of interest in adult brain endothelial cells *in vivo*. The mechanism through which AAV-BR1 selectively targets the CNS vasculature remains unreported. Given its remarkable specificity for the CNS and low efficiency of transcytosis, which differentiates it from the AAV-PHP.B capsids, it seems likely that AAV-BR1 utilizes a receptor other than LY6A for brain endothelial cell transduction. However, it may be premature to settle on this conclusion due to key differences between the parental AAVs of AAV-PHP.B and AAV-BR1, which are AAV9 and AAV2, respectively. With AAV-PHP.B, the transduction of other organs is similar to AAV9 and is likely dictated by the structural features of the parental capsid, which are often unaltered by peptide insertion into loop VIII. In contrast, loop VIII insertions in the AAV2 capsid typically disrupt capsid interaction with HSPGs, and prevent a normal route of entry for AAV2 (Müller et al. 2003; Michelfelder et al. 2009). Therefore, a hypothetical LY6A-interacting insertion in AAV2's loop VIII that disrupts HSPG binding (and possibly AAVR interactions at adjacent amino acids) would be expected to render the variant poorly efficient at transducing cells with low levels of surface-exposed LY6A (e.g., neurons, glia, liver hepatocytes, and skeletal and cardiac muscle) as is observed in the case of AAV-BR1. Moreover, the tropism of AAV-BR1 in WT mice is remarkably similar to AAV-PHP.eB in AAVR KO mice (i.e., CNS endothelial cell-specific), and AAV-BR1 was selected and tested exclusively on LY6A-positive mouse strains FVB and C57. Thus, it remains possible that AAV-BR1 uses LY6A for endocytosis into the brain endothelium. Despite the unknown mechanism by which AAV-BR1 specifically targets endothelial cells in the CNS, AAV-BR1 has been used to interrogate brain vascular biology (Tan et al. 2019), and to test gene therapies in mouse models of vascular disorders such as incontinentia pigmenti (Körbelin et al. 2016) or LSDs such as Tay-Sachs and Sandhoff disease in which it may not be necessary to transduce neuronal cells (Dogbevia et al. 2019).

In 2019, Hanlon et al. developed a new AAV9 variant called AAV-F, that is reported to exhibit transduction efficiencies that are on par with AAV-PHP.B, but with a bias toward astrocyte transduction: AAV-F targeted astrocytes (40.8%) compared to AAV-PHP.B (28.2%), and was less targeted to neurons (6.7%) compared to AAV-PHP.B (10.6%) (Hanlon et al. 2019). Although the receptor for AAV-F has not been identified, its efficient transduction of Balb/cJ mice, which have reduced LY6A levels in endothelial cells and are nonpermissive for the AAV-PHP.B capsids, supports the conclusion that the AAV-F enhanced CNS tropism does not rely on LY6A. This is encouraging as it suggests that there are multiple mechanisms that can be exploited to enhance the efficiency of CNS transduction by engineered AAV capsids (Table 21.1). Further studies are required to identify the AAV-F receptor and determine whether its CNS tropism extends to species other than the mouse.

Methods for deriving AAV capsids with enhanced *in vivo* CNS tropisms continue to be innovated. The CREATE method that derived AAV-PHP.B and its family was adapted to enable screening for 7-mer-insertion AAV variants that are enriched in a particular tissue or cell type while being depleted across other cells and organs (Ravindra Kumar et al. 2020). The multiplexed-CREATE (M-CREATE) method was used to isolate AAV9 variants that were enriched in the brain in comparison to other organs after mice were infected with an AAV9 variant library. These brain-enriched capsids were dominated by members with sequence similarity to AAV-PHP.B or AAV-PHP.B2. Of the capsids from the AAV-PHP.B family, two very similar variants, AAV-PHP.V1 and AAV-PHP.V2 exhibit enhanced transduction of the brain vasculature, but reduced transduction of neurons compared to AAV-PHP.B. AAV-PHP.B4-8 displayed similar tropism to AAV-PHP.B. Outside of the AAV-PHP.B family, several capsids, AAV-PHP.C1-AAV-PHP.C3 were found to cross the BBB more efficiently than AAV9 albeit with reduced neuron transduction compared with PHP.B. However, the AAV-PHP.C capsids differ from the AAV-PHP.B family by sequence and their ability to cross the BBB in both C57BL/6J and BALB/cJ mice. In the same work, substitution variants of AAV-PHP.B were screened for enrichment in the brain and underrepresentation in the liver. This identified a new capsid, AAV-PHP.N, that exhibits greater specificity for CNS neurons and reduced liver transduction. A following study used M-CREATE to screen capsids modified at two sites (the modification at one site was the AAV-PHP.eB loop VIII insertion) and derived the AAV.CAP-B10 capsid that provided higher levels of the transgene product in the adult CNS marmosets after IV administration without an increase in tropism measured by viral genomes (Goertsen et al. 2022) (Table 21.1). Nonnenmacher *et al.* recently identified an additional panel of BBB crossing 7-mer-insertion AAV capsid variants using a complementary screening approach, which harnessed CNS cell type-specific promoters to bypass the dependency on transgenic animals (Nonnenmacher et al. 2021). Of these new capsids, some resemble the AAV-PHP.B family and their enhanced tropism is limited to C57BL/6 mice, while others exhibit enhanced CNS tropism in both BALB/c and C57BL/6 mice. These findings demonstrate, promisingly, that there are multiple BBB-crossing mechanisms, some of which are not restricted to mouse strain (and potentially species), that can be leveraged by AAVs to enter the CNS via the vasculature.

21.3.3   Identifying and Upregulating the AAV Internalization Pathway

There is considerable interest in the scientific community to identify complementary therapies such as focused ultrasound BBB opening [FUS-BBO (Szablowski et al. 2018); see Chaps. 19 and 20 in this volume] that can enhance the efficacy of AAV gene therapy. Because intracellular transport is a limiting factor of AAV transduction (Nonnenmacher and Weber 2012), determining the cellular pathways that AAV utilizes for internalization (Fig. 21.1) can aid in the identification of physical approaches or compounds that upregulate these pathways to facilitate AAV transduction. The best-characterized AAV in terms of its dependency on different endocytosis pathways in the cell is AAV2. AAV2 transduction has been demonstrated to not rely on clathrin- or caveolae-mediated endocytosis in HeLa or 293T cells (Nonnenmacher and Weber 2011). The same study showed that, in contrast to what had been previously reported, AAV2 transduction was not influenced by the inhibition of macropinocytosis and was instead sensitive to actin (de)

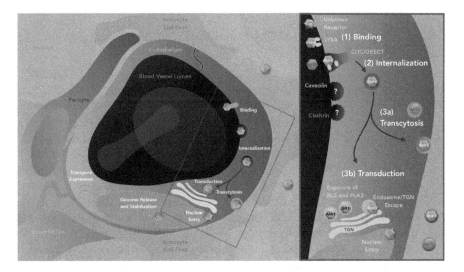

Fig. 21.1 Proposed mechanisms by which AAV9 and next-generation AAVs navigate the brain vasculature through receptor binding, internalization, transduction, or transcytosis. Systemically delivered AAVs arrive via the blood vessel lumen, (1) bind to a receptor (and co-receptor) on the endothelial cell membrane, and (2) are internalized into the cell potentially through CLIC/GEEC, clathrin-mediated, caveolar, or other modes of endocytosis. Through mechanisms that are not understood, the AAV either (3a) traffics through the endothelial cell to gain access to neighboring CNS cells, or (3b) transduces the endothelial cell by escaping the endosome and trans-Golgi network (TGN) through exposure of the AAV phospholipase A2 (PLA2) domain and nuclear localization sequence (NLS). After nuclear entry, the AAV genome is released from the capsid and stabilized, allowing for transgene expression

polymerization and cholesterol depletion. However, internalized AAV2 was resistant to cold triton X detergent extraction, suggesting that AAVs enter cells through cholesterol-rich detergent-resistant microdomains (Nonnenmacher and Weber 2011). AAV2 transduction was additionally found to depend on three main effectors of the clathrin-independent carriers/GPI-anchored-protein-enriched endosomal compartment (CLIC/GEEC) pathway: Cdc42, Arf1, and GRAF1 (Nonnenmacher and Weber 2011). AAV2 accumulates in endosomes similar to other CLIC cargos that are shuttled from the plasma membrane to perinuclear Golgi apparatus (Nonnenmacher and Weber 2011). Strikingly, inhibiting both dynamin and the CLIC/GEEC pathway resulted in a synergistic decrease in AAV2 endocytosis and transduction of HeLa and 293T cells, which suggests that AAV2 can enter cells via both endocytic pathways, albeit only endocytosis via CLIC/GEEC leads to transduction (Nonnenmacher and Weber 2011). Interestingly, the CLIC/GEEC pathway is known to be necessary for the uptake of GPI-anchored proteins (Doherty and McMahon 2009). The finding that the AAV-PHP.B vectors use LY6A, a GPI-anchored protein for endothelial transduction and entry into the mouse CNS suggests that it may be possible to harness the CLIC/GEEC pathway to enhance the efficiency of gene delivery to the CNS.

21.3.4 AAV Fates: Transduction Versus Transcytosis

It has been widely assumed that the CNS tropism of AAV9 results from its ability to cross the BBB as an intact AAV particle, likely through transcytosis (Fig. 21.1). This is supported by a recent analysis of distinct AAV fates in primary human brain microvascular endothelial cells (hBMVECs), which showed that AAV2 is more efficient at transduction whereas AAV9 is more effective at transcytosis (Merkel et al. 2017). Through live cell imaging, Merkel et al. determined that internalized AAV2 were primarily localized to the nuclei while AAV9 accumulated in smaller and tubular vesicles that could span the entire hBMVEC layer from apical to basolateral surfaces (Merkel et al. 2017). At 24 h post-incubation with the hBMVECs, AAV2 capsids remained largely inside the cells whereas the majority of AAV9 capsids had been trafficked through the cell layer (Merkel et al. 2017). In addition, Merkel et al. demonstrated that AAV9 transcytosis appeared to be mediated by active transport because it was severely reduced (89.7% reduction in the number of trafficked AAV9 particles) at 4°C compared with at 37°C (Merkel et al. 2017). Importantly, incubation with AAV9 did not compromise the integrity of the hBMVEC barrier; no loss of TEER or inflammatory response was detected (Merkel et al. 2017). Based on these findings, Merkel et al. proposed that AAV9 likely utilizes a pathway that directs viral particles toward transcytosis instead of transduction (Fig. 21.1). Therefore, AAVs that are effective at transcytosis may be less effective at the transduction of the same cells.

21.4 Concluding Remarks

In the past decade, the gene therapy community has witnessed exciting new strides toward effective AAV-mediated gene delivery to the CNS, beginning with the seminal discovery of AAV9 and, soon after, other AAV capsids isolated from humans and NHPs that can cross the BBB (Table 21.1). These BBB-crossing AAVs have now been harnessed in numerous clinical trials (e.g., NCT03306277, NCT03505099, NCT02618915, NCT02971969, NCT03461289, NCT02122952, NCT03368742, NCT03362502, NCT04240314), including the trailblazing trials for the now FDA-approved Zolgensma gene therapy for SMA. With the appearance of engineered AAVs with dramatically enhanced efficiencies and tropisms in animal models, there is mounting anticipation within the gene therapy community for the development of novel AAVs that exhibit similar improvements in humans. The innovation of AAVs with unique CNS tropisms, such as AAV-PHP.B and AAV-PHP.eB (CNS-wide transduction via LY6A-binding; more neuron-targeted as compared to AAV-F), AAV-BR1 (highly specific to brain endothelial cells), and AAV-F (CNS-wide transduction that is independent of LY6A and more astrocyte-targeted as compared to AAV-PHP.B or AAV-PHP.eB), have also provided the larger biomedical community with valuable tools for the study of the CNS and BBB.

In the coming decade, areas of research that are likely to benefit the creation of more powerful BBB-crossing AAVs include (i) the structure-function studies of AAV capsids that elucidate capsid features and receptor interactions critical for BBB-crossing, (ii) the discovery of novel mechanisms by which AAVs can achieve transcytosis across the brain endothelial layer (e.g., binding to LY6A and potentially other GPI-anchored proteins), and (iii) the discernment of cellular endocytosis pathways that can be modulated to boost AAV internalization and potentially shift the balance between AAV transcytosis versus transduction. By leveraging this improved knowledge of AAV transcytosis mechanisms, alongside next-generation capsid engineering technologies, advances in CNS enhancer or promoter discovery and engineering (to control gene dosage and cell type-specific expression), and more biologically accurate models of the human brain endothelium, there is incredible potential for the development of more potent delivery vehicles for human CNS gene therapy.

Points for Discussion

- How well will large animal models and human primary and iPSC-derived endothelial culture models (grown on transwells, in spheroids, or in microfluidic chambers; e.g., see Chap. 9 in this volume) predict the performance of AAV capsids engineered for CNS tropism in humans?
- What is the relationship between CNS tropism, AAV half-life in the blood, and receptor binding?
- What role does LY6A play at the BBB? Are there proteins with similar functions in humans that can be leveraged by next-generation AAV capsids to cross the BBB?

- What factors facilitate the CNS tropism of other engineered AAV capsids such as AAV-BR1, AAV-F, and AAV-PHP.C1-C3?
- What are some compounds or physical methods that can be used in combination therapy to enhance the efficacy of AAV-mediated gene therapy?
- Does AAV transcytosis occur through a mechanism that is distinct from transduction? Or is transcytosis related to the efficiency of endosomal escape?
- Do AAV9 and other BBB-crossing AAVs traffic across the BBB exclusively following endocytosis through the CLIC/GEEC pathway, or can they also utilize caveolae, clathrin, or other dynamin-dependent mechanisms for transcytosis?

References

Adachi K, Enoki T, Kawano Y et al (2014) Drawing a high-resolution functional map of adeno-associated virus capsid by massively parallel sequencing. Nat Commun 5:3075

Aguti S, Malerba A, Zhou H (2018) The progress of AAV-mediated gene therapy in neuromuscular disorders. Expert Opin Biol Ther 18:681–693

Albright BH, Storey CM, Murlidharan G et al (2018) Mapping the structural determinants required for AAVrh.10 transport across the blood-brain barrier. Mol Ther 26:510–523

Albright BH, Simon KE, Pillai M et al (2019) Modulation of sialic acid dependence influences the central nervous system transduction profile of adeno-associated viruses. J Virol 93. https://doi.org/10.1128/JVI.00332-19

Allen WE, Kauvar IV, Chen MZ et al (2017) Global representations of goal-directed behavior in distinct cell types of mouse neocortex. Neuron 94:891–907.e6

Asokan A, Conway JC, Phillips JL et al (2010) Reengineering a receptor footprint of adeno-associated virus enables selective and systemic gene transfer to muscle. Nat Biotechnol 28:79–82

Audentes (2020) Audentes therapeutics provides update on the ASPIRO clinical trial evaluating AT132 in patients with X-linked myotubular myopathy. In: Audentes Therapeutics. https://www.audentestx.com/press_release/audentes-therapeutics-provides-update-on-the-aspiro-clinical-trial-evaluating-at132-in-patients-with-x-linked-myotubular-myopathy/. Accessed 21 Sep 2020

AveXis, Inc. (2016) Phase I gene transfer clinical trial for spinal muscular atrophy type 1 delivering AVXS-101

Bailey RM, Armao D, Nagabhushan Kalburgi S, Gray SJ (2018) Development of intrathecal AAV9 gene therapy for giant axonal neuropathy. Mol Ther Methods Clin Dev 9:160–171

Bartel M, Schaffer D, Büning H (2011) Enhancing the clinical potential of AAV vectors by capsid engineering to evade pre-existing immunity. Front Microbiol 2:204

Batista AR, King OD, Reardon CP et al (2019) Ly6a differential expression in BBB is responsible for strain specific CNS transduction profile of AAV-PHP.B. Hum Gene Ther. https://doi.org/10.1089/hum.2019.186

Bell CL, Vandenberghe LH, Bell P et al (2011) The AAV9 receptor and its modification to improve in vivo lung gene transfer in mice. J Clin Invest 121:2427–2435

Bell CL, Gurda BL, Van Vliet K et al (2012) Identification of the galactose binding domain of the adeno-associated virus serotype 9 capsid. J Virol 86:7326–7333

Bevan AK, Duque S, Foust KD et al (2011) Systemic gene delivery in large species for targeting spinal cord, brain, and peripheral tissues for pediatric disorders. Mol Ther 19:1971–1980

Bien-Ly N, Yu YJ, Bumbaca D et al (2014) Transferrin receptor (TfR) trafficking determines brain uptake of TfR antibody affinity variants. J Exp Med 211:233–244

Boisgerault F, Mingozzi F (2015) The skeletal muscle environment and its role in immunity and tolerance to AAV vector-mediated gene transfer. Curr Gene Ther 15:381–394

Brass AL, Dykxhoorn DM, Benita Y et al (2008) Identification of host proteins required for HIV infection through a functional genomic screen. Science 319:921–926

Calcedo R, Vandenberghe LH, Gao G et al (2009) Worldwide epidemiology of neutralizing antibodies to adeno-associated viruses. J Infect Dis 199:381–390

Calcedo R, Morizono H, Wang L et al (2011) Adeno-associated virus antibody profiles in newborns, children, and adolescents. Clin Vaccine Immunol 18:1586–1588

Center for Biologics Evaluation, Research (2019) ZOLGENSMA. In: U.S. food and drug administration. https://www.fda.gov/vaccines-blood-biologics/zolgensma. Accessed 21 Nov 2019

Chan KY, Jang MJ, Yoo BB et al (2017) Engineered AAVs for efficient noninvasive gene delivery to the central and peripheral nervous systems. Nat Neurosci 20:1172–1179

Chen YH, Chang M, Davidson BL (2009) Molecular signatures of disease brain endothelia provide new sites for CNS-directed enzyme therapy. Nat Med 15:1215–1218

Davidson CD, Gibson AL, Gu T et al (2021) Improved systemic AAV gene therapy with a neurotrophic capsid in Niemann-Pick disease type C1 mice. Life Sci Alliance 4:e202101040

Dehay B, Dalkara D, Dovero S et al (2012) Systemic scAAV9 variant mediates brain transduction in newborn rhesus macaques. Sci Rep 2:253

Deverman BE, Pravdo PL, Simpson BP et al (2016) Cre-dependent selection yields AAV variants for widespread gene transfer to the adult brain. Nat Biotechnol 34:204–209

Deverman BE, Ravina BM, Bankiewicz KS et al (2018) Gene therapy for neurological disorders: progress and prospects. Nat Rev Drug Discov 17:767

DiMattia MA, Nam H-J, Van Vliet K et al (2012) Structural insight into the unique properties of adeno-associated virus serotype 9. J Virol 86:6947–6958

Dogbevia G, Grasshoff H, Othman A et al (2019) Brain endothelial specific gene therapy improves experimental Sandhoff disease. J Cereb Blood Flow Metab. 271678X19865917

Doherty GJ, McMahon HT (2009) Mechanisms of endocytosis. Annu Rev Biochem 78:857–902

Dominguez E, Marais T, Chatauret N et al (2011) Intravenous scAAV9 delivery of a codon-optimized SMN1 sequence rescues SMA mice. Hum Mol Genet 20:681–693

Duque S, Joussemet B, Riviere C et al (2009) Intravenous administration of self-complementary AAV9 enables transgene delivery to adult motor neurons. Mol Ther 17:1187–1196

Faust SM, Bell P, Cutler BJ et al (2013) CpG-depleted adeno-associated virus vectors evade immune detection. J Clin Invest 123:2994–3001

Federici T, Taub JS, Baum GR et al (2012) Robust spinal motor neuron transduction following intrathecal delivery of AAV9 in pigs. Gene Ther 19:852–859

Finkel RS, McDermott MP, Kaufmann P et al (2014) Observational study of spinal muscular atrophy type I and implications for clinical trials. Neurology 83:810–817

Finn JD, Hui D, Downey HD et al (2010) Proteasome inhibitors decrease AAV2 capsid derived peptide epitope presentation on MHC class I following transduction. Mol Ther 18:135–142

Foust KD, Kaspar BK (2009) Over the barrier and through the blood: to CNS delivery we go. Cell Cycle 8:4017–4018

Foust KD, Nurre E, Montgomery CL et al (2009) Intravascular AAV9 preferentially targets neonatal neurons and adult astrocytes. Nat Biotechnol 27:59–65

Foust KD, Wang X, McGovern VL et al (2010) Rescue of the spinal muscular atrophy phenotype in a mouse model by early postnatal delivery of SMN. Nat Biotechnol 28:271–274

Gao Y, Geng L, Chen VP, Brimijoin S (2017) Therapeutic delivery of butyrylcholinesterase by brain-wide viral gene transfer to mice. Molecules 22. https://doi.org/10.3390/molecules22071145

Gardner J (2019) Novartis gene therapy ambitions dealt another blow by FDA hold on Zolgensma. In: BioPharma Dive. https://www.biopharmadive.com/news/novartis-zolgensma-sma-strong-fda-hold-gene-therapy/566160/. Accessed 26 Nov 2019

Gernoux G, Wilson JM, Mueller C (2017) Regulatory and exhausted T cell responses to AAV capsid. Hum Gene Ther 28:338–349

Giles AR, Govindasamy L, Somanathan S, Wilson JM (2018) Mapping an adeno-associated virus 9-specific neutralizing epitope to develop next-generation gene delivery vectors. J Virol 92. https://doi.org/10.1128/JVI.01011-18

Goertsen D, Flytzanis NC, Goeden N et al (2022) AAV capsid variants with brain-wide transgene expression and decreased liver targeting after intravenous delivery in mouse and marmoset. Nat Neurosci 25:106–115

Gray SJ, Matagne V, Bachaboina L et al (2011) Preclinical differences of intravascular AAV9 delivery to neurons and glia: a comparative study of adult mice and nonhuman primates. Mol Ther 19:1058–1069

Gray SJ, Nagabhushan Kalburgi S, McCown TJ, Jude Samulski R (2013) Global CNS gene delivery and evasion of anti-AAV-neutralizing antibodies by intrathecal AAV administration in non-human primates. Gene Ther 20:450–459

Graybuck LT, Daigle TL, Sedeño-Cortés AE et al (2021) Enhancer viruses and a transgenic platform for combinatorial cell subclass-specific labeling. Neuron 109:1449–1464.e13

Gregorevic P, Allen JM, Minami E et al (2006) rAAV6-microdystrophin preserves muscle function and extends lifespan in severely dystrophic mice. Nat Med 12:787–789

Gu T, Davidson C, Gibson A et al (2018) Enhanced efficacy of gene therapy treatment for Niemann-Pick C1 disease using a novel serotype, AAV-PHP.B. Mol Ther 26:404

Guida JD, Fejer G, Pirofski LA et al (1995) Mouse adenovirus type 1 causes a fatal hemorrhagic encephalomyelitis in adult C57BL/6 but not BALB/c mice. J Virol 69:7674–7681

Hanlon KS, Meltzer JC, Buzhdygan T et al (2019) Selection of an efficient AAV vector for robust CNS transgene expression. Mol Ther Methods Clin Dev 15:320–332

Haqqani AS, Thom G, Burrell M et al (2018) Intracellular sorting and transcytosis of the rat transferrin receptor antibody OX26 across the blood-brain barrier in vitro is dependent on its binding affinity. J Neurochem 146:735–752

Hillier D, Fiscella M, Drinnenberg A et al (2017) Causal evidence for retina-dependent and -independent visual motion computations in mouse cortex. Nat Neurosci 20:960–968

Hinderer C, Katz N, Buza EL et al (2018) Severe toxicity in nonhuman primates and piglets following high-dose intravenous administration of an adeno-associated virus vector expressing human SMN. Hum Gene Ther 29:285–298

Hocquemiller M, Giersch L, Audrain M et al (2016) Adeno-associated virus-based gene therapy for CNS diseases. Hum Gene Ther 27:478–496

Hordeaux J, Wang Q, Katz N et al (2018) The neurotropic properties of AAV-PHP.B are limited to C57BL/6J mice. Mol Ther 26:664–668

Hordeaux J, Hinderer C, Buza EL et al (2019a) Safe and sustained expression of human iduronidase after intrathecal administration of adeno-associated virus serotype 9 in infant rhesus monkeys. Hum Gene Ther 30:957–966

Hordeaux J, Yuan Y, Clark PM et al (2019b) The GPI-linked protein LY6A (SCA-1) drives AAV-PHP.B transport across the blood-brain barrier. Mol Ther 27(5):912–921. https://doi.org/10.1016/j.ymthe.2019.02.013

Hu C, Busuttil RW, Lipshutz GS (2010) RH10 provides superior transgene expression in mice when compared with natural AAV serotypes for neonatal gene therapy. J Gene Med 12:766–778

Huang Q, Chan KY, Tobey IG et al (2019) Delivering genes across the blood-brain barrier: LY6A, a novel cellular receptor for AAV-PHP.B capsids. PLoS One 14(e0225206)

Inacio P (2019) DMD gene therapy showing "Very encouraging" results at 9 months in phase 1/2 study, Sarepta reports. In: Muscular Dystrophy News. https://musculardystrophynews.com/2019/04/03/dmd-gene-therapy-showing-very-encouraging-results-at-9-months-in-phase-1-2-study-sarepta-reports/. Accessed 22 Nov 2019

Jackson KL, Dayton RD, Klein RL (2015) AAV9 supports wide-scale transduction of the CNS and TDP-43 disease modeling in adult rats. Mol Ther Methods Clin Dev 2:15036

Kern A, Schmidt K, Leder C et al (2003) Identification of a heparin-binding motif on adeno-associated virus type 2 capsids. J Virol 77:11072–11081

Kolb SJ, Coffey CS, Yankey JW et al (2017) Natural history of infantile-onset spinal muscular atrophy. Ann Neurol 82:883–891

Körbelin J, Dogbevia G, Michelfelder S et al (2016) A brain microvasculature endothelial cell-specific viral vector with the potential to treat neurovascular and neurological diseases. EMBO Mol Med 8:609–625

Kotchey NM, Adachi K, Zahid M et al (2011) A potential role of distinctively delayed blood clearance of recombinant adeno-associated virus serotype 9 in robust cardiac transduction. Mol Ther 19:1079–1089

Krishnan MN, Ng A, Sukumaran B et al (2008) RNA interference screen for human genes associated with West Nile virus infection. Nature 455:242–245

Lehmenkühler A, Syková E, Svoboda J et al (1993) Extracellular space parameters in the rat neocortex and subcortical white matter during postnatal development determined by diffusion analysis. Neuroscience 55:339–351

Levy JM, Yeh W-H, Pendse N et al (2020) Cytosine and adenine base editing of the brain, liver, retina, heart and skeletal muscle of mice via adeno-associated viruses. Nat Biomed Eng 4:97–110

Lim J-A, Yi H, Gao F et al (2019) Intravenous injection of an AAV-PHP.B vector encoding human acid α-Glucosidase rescues both muscle and CNS defects in murine pompe disease. Mol Ther Methods Clin Dev 12:233–245

Loeuillet C, Deutsch S, Ciuffi A et al (2008) In vitro whole-genome analysis identifies a susceptibility locus for HIV-1. PLoS Biol 6:e32

Luoni M, Giannelli S, Indrigo M et al (2020) Whole brain delivery of an instability-prone Mecp2 transgene improves behavioral and molecular pathological defects in mouse models of Rett syndrome. elife 9. https://doi.org/10.7554/eLife.52629

Luxner L (2019) With Zolgensma's approval, scientists pursue similar gene therapies in Duchenne. In: Muscular Dystrophy News. https://musculardystrophynews.com/2019/05/29/with-zolgensmas-approval-scientists-pursue-similar-gene-therapies-in-duchenne/. Accessed 22 Nov 2019

Lykken EA, Shyng C, Edwards RJ et al (2018) Recent progress and considerations for AAV gene therapies targeting the central nervous system. J Neurodev Disord 10:16

Manno CS, Pierce GF, Arruda VR et al (2006) Successful transduction of liver in hemophilia by AAV-Factor IX and limitations imposed by the host immune response. Nat Med 12:342–347

Mar KB, Rinkenberger NR, Boys IN et al (2018) LY6E mediates an evolutionarily conserved enhancement of virus infection by targeting a late entry step. Nat Commun 9:3603

Matsuzaki Y, Konno A, Mochizuki R et al (2018) Intravenous administration of the adeno-associated virus-PHP.B capsid fails to upregulate transduction efficiency in the marmoset brain. Neurosci Lett 665:182–188

Mattar CN, Waddington SN, Biswas A et al (2013) Systemic delivery of scAAV9 in fetal macaques facilitates neuronal transduction of the central and peripheral nervous systems. Gene Ther 20:69–83

Mendell JR, Al-Zaidy S, Shell R et al (2017) Single-dose gene-replacement therapy for spinal muscular atrophy. N Engl J Med 377:1713–1722

Merkel SF, Andrews AM, Lutton EM et al (2017) Trafficking of adeno-associated virus vectors across a model of the blood-brain barrier; a comparative study of transcytosis and transduction using primary human brain endothelial cells. J Neurochem 140:216–230

Meyer K, Ferraiuolo L, Schmelzer L et al (2015) Improving single injection CSF delivery of AAV9-mediated gene therapy for SMA: a dose-response study in mice and nonhuman primates. Mol Ther 23:477–487

Michelfelder S, Kohlschütter J, Skorupa A et al (2009) Successful expansion but not complete restriction of tropism of adeno-associated virus by in vivo biopanning of random virus display peptide libraries. PLoS One 4:e5122

Mingozzi F, Anguela XM, Pavani G et al (2013) Overcoming preexisting humoral immunity to AAV using capsid decoys. Sci Transl Med 5:194ra92

Morabito G, Giannelli SG, Ordazzo G et al (2017) AAV-PHP.B-mediated global-scale expression in the mouse nervous system enables GBA1 gene therapy for wide protection from synucleinopathy. Mol Ther 25:2727–2742

Mueller C, Chulay JD, Trapnell BC et al (2013) Human Treg responses allow sustained recombinant adeno-associated virus-mediated transgene expression. J Clin Invest 123:5310–5318

Müller OJ, Kaul F, Weitzman MD et al (2003) Random peptide libraries displayed on adeno-associated virus to select for targeted gene therapy vectors. Nat Biotechnol 21:1040–1046

Nathwani AC, Tuddenham EGD, Rangarajan S et al (2011) Adenovirus-associated virus vector-mediated gene transfer in hemophilia B. N Engl J Med 365:2357–2365

Niewoehner J, Bohrmann B, Collin L et al (2014) Increased brain penetration and potency of a therapeutic antibody using a monovalent molecular shuttle. Neuron 81:49–60

Nonnenmacher M, Weber T (2011) Adeno-associated virus 2 infection requires endocytosis through the CLIC/GEEC pathway. Cell Host Microbe 10:563–576

Nonnenmacher M, Weber T (2012) Intracellular transport of recombinant adeno-associated virus vectors. Gene Ther 19:649–658

Nonnenmacher M, Wang W, Child MA et al (2021) Rapid evolution of blood-brain-barrier-penetrating AAV capsids by RNA-driven biopanning. Mol Ther Methods Clin Dev 20:366–378

Novartis (2020) Zolgensma® data including patients with more severe SMA at baseline further demonstrate therapeutic benefit, including prolonged event-free survival, increased motor function and milestone achievement. In: Novartis. https://www.biospace.com/article/releases/avexis-presents-new-data-at-epns-continuing-to-show-significant-therapeutic-benefit-of-zolgensma-in-prolonging-event-free-survival-now-up-to-5-years-of-age-in-patients-with-spinal-muscular-atrophy-sma-type-1/ Accessed 5 Oct 2020

Office of the Commissioner (2019) FDA approves innovative gene therapy to treat pediatric patients with spinal muscular atrophy, a rare disease and leading genetic cause of infant mortality. In: U.S. Food and Drug Administration. https://www.fda.gov/news-events/press-announcements/fda-approves-innovative-gene-therapy-treat-pediatric-patients-spinal-muscular-atrophy-rare-disease. Accessed 25 Nov 2019

Offord C (2019) Trial of gene therapy for Duchenne muscular dystrophy put on hold. In: The scientist magazine®. https://www.the-scientist.com/news-opinion/trial-of-gene-therapy-for-duchenne-muscular-dystrophy-put-on-hold-66711. Accessed 22 Nov 2019

Pasi KJ, Rangarajan S, Mitchell N et al (2020) Multiyear follow-up of AAV5-hFVIII-SQ gene therapy for hemophilia A. N Engl J Med 382:29–40

Passini MA, Bu J, Richards AM et al (2014) Translational fidelity of intrathecal delivery of self-complementary AAV9-survival motor neuron 1 for spinal muscular atrophy. Hum Gene Ther 25:619–630

Pattali R, Mou Y, Li X-J (2019) AAV9 Vector: a Novel modality in gene therapy for spinal muscular atrophy. Gene Ther 26:287–295

Pfizer (2019) Pfizer presents initial clinical data on phase 1b gene therapy study for Duchenne Muscular Dystrophy (DMD) | Pfizer. In: Pfizer. https://www.pfizer.com/news/press-release/press-release-detail/pfizer_presents_initial_clinical_data_on_phase_1b_gene_therapy_study_for_duchenne_muscular_dystrophy_dmd. Accessed 22 Nov 2019

Pfizer (2020) Pfizer receives FDA fast track designation for duchenne muscular dystrophy investigational gene therapy. In: Pfizer. https://www.pfizer.com/news/press-release/press-release-detail/pfizer-receives-fda-fast-track-designation-duchenne. Accessed 5 Oct 2020

Pien GC, Basner-Tschakarjan E, Hui DJ et al (2009) Capsid antigen presentation flags human hepatocytes for destruction after transduction by adeno-associated viral vectors. J Clin Invest 119:1688–1695

Pulicherla N, Shen S, Yadav S et al (2011) Engineering liver-detargeted AAV9 vectors for cardiac and musculoskeletal gene transfer. Mol Ther 19:1070–1078

Ravindra Kumar S, Miles TF, Chen X et al (2020) Multiplexed Cre-dependent selection yields systemic AAVs for targeting distinct brain cell types. Nat Methods. https://doi.org/10.1038/s41592-020-0799-7

Reynaud-Dulaurier R, Benegiamo G, Marrocco E et al (2020) Gene replacement therapy provides benefit in an adult mouse model of Leigh syndrome. Brain 143:1686–1696

Rosen S, Tiefenbacher S, Robinson M et al (2020) Activity of transgene-produced B-domain deleted factor VIII in human plasma following AAV5 gene therapy. Blood. https://doi.org/10.1182/blood.2020005683

Sah D, Mazzarelli A, Christensen E et al (2018) Safety and increased transduction efficiency in the adult nonhuman primate central nervous system with intravenous delivery of two novel adeno-associated virus capsids [abstract O661]. American Society of Gene and Cell Therapy Annual Meeting. Chicago, USA. Mol Ther

Samaranch L, Salegio EA, San Sebastian W et al (2012) Adeno-associated virus serotype 9 transduction in the central nervous system of nonhuman primates. Hum Gene Ther 23:382–389

Sangamo (2020) Pfizer and Sangamo announce updated phase 1/2 results showing sustained factor VIII activity levels and no bleeding events or factor usage in 3e13 vg/kg cohort following giroctocogene fitelparvovec (SB-525) Gene Therapy. In: Sangamo Therapeutics. https://investor.sangamo.com/news-releases/news-release-details/pfizer-and-sangamo-announce-updated-phase-12-results-showing. Accessed 5 Oct 2020

Selot R, Arumugam S, Mary B et al (2017) Optimized AAV rh.10 vectors that partially evade neutralizing antibodies during hepatic gene transfer. Front Pharmacol 8:441

Shen S, Bryant KD, Brown SM et al (2011) Terminal N-linked galactose is the primary receptor for adeno-associated virus 9. J Biol Chem 286:13532–13540

Shen S, Bryant KD, Sun J et al (2012) Glycan binding avidity determines the systemic fate of adeno-associated virus type 9. J Virol 86:10408–10417

Solid Biosciences Inc (2019) Solid Biosciences Provides Data Update from SGT-001 Development Program. In: Solid Biosciences. https://investors.solidbio.com/news-releases/news-release-details/solid-biosciences-provides-data-update-sgt-001-development. Accessed 5 Oct 2020

Solid Biosciences Inc (2020) Solid biosciences announces FDA lifts clinical hold on IGNITE DMD clinical trial. https://www.globenewswire.com/news-release/2020/10/01/2102091/0/en/Solid-Biosciences-Announces-FDA-Lifts-Clinical-Hold-on-IGNITE-DMD-Clinical-Trial.html. Accessed 4 Oct 2020

Solid Biosciences (2020) Solid biosciences announces FDA lifts clinical hold on IGNITE DMD clinical trial. In: Solid biosciences. https://www.solidbio.com/about/media/press-releases/solid-biosciences-announces-fda-lifts-clinical-hold-on-ignite-dmd-clinical-trial. Accessed 5 Oct 2020

Spark Therapeutics, Inc (2020) Spark Therapeutics Announces Updated Data on SPK-8011 from Phase 1/2 Clinical Trial in Hemophilia A at ISTH 2020 Virtual Congress. https://www.globenewswire.com/news-release/2020/07/12/2060941/0/en/Spark-Therapeutics-Announces-Updated-Data-on-SPK-8011-from-Phase-1-2-Clinical-Trial-in-Hemophilia-A-at-ISTH-2020-Virtual-Congress.html. Accessed 5 Oct 2020

Spindler KR, Welton AR, Lim ES et al (2010) The major locus for mouse adenovirus susceptibility maps to genes of the hematopoietic cell surface-expressed LY6 family. J Immunol 184:3055–3062

Stieger K, Schroeder J, Provost N et al (2009) Detection of intact rAAV particles up to 6 years after successful gene transfer in the retina of dogs and primates. Mol Ther 17:516–523

Szablowski JO, Lee-Gosselin A, Lue B et al (2018) Acoustically targeted chemogenetics for the non-invasive control of neural circuits. Nat Biomed Eng 2:475–484

Tan C, Lu N-N, Wang C-K et al (2019) Endothelium-derived Semaphorin 3G regulates hippocampal synaptic structure and plasticity via Neuropilin-2/PlexinA4. Neuron 101:920–937.e13

Tanguy Y, Biferi MG, Besse A et al (2015) Systemic AAVrh10 provides higher transgene expression than AAV9 in the brain and the spinal cord of neonatal mice. Front Mol Neurosci 8:36

uniQure Inc (2020) uniQure announces achievement of target patient dosing in HOPE-B pivotal trial of AMT-061 (Etranacogene Dezaparvovec) in Hemophilia B. https://www.globenewswire.com/news-release/2020/03/26/2006871/0/en/uniQure-Announces-Achievement-of-Target-Patient-Dosing-in-HOPE-B-Pivotal-Trial-of-AMT-061-Etranacogene-Dezaparvovec-in-Hemophilia-B.html. Accessed 5 Oct 2020

Valori CF, Ning K, Wyles M et al (2010) Systemic delivery of scAAV9 expressing SMN prolongs survival in a model of spinal muscular atrophy. Sci Transl Med 2:35ra42

Vandamme C, Adjali O, Mingozzi F (2017) Unraveling the complex story of immune responses to AAV vectors trial after trial. Hum Gene Ther 28:1061–1074

Vormstein-Schneider D, Lin JD, Pelkey KA et al (2020) Viral manipulation of functionally distinct interneurons in mice, non-human primates and humans. Nat Neurosci 23(12):1629–1636. https://doi.org/10.1038/s41593-020-0692-9

Voyager Therapeutics, Inc (2019) Voyager therapeutics provides updates for VY-AADC clinical program for parkinson's disease. In: GlobeNewswire news room. https://www.globenewswire. com/news-release/2019/01/07/1681132/0/en/Voyager-Therapeutics-Provides-Updates-for-VY-AADC-Clinical-Program-for-Parkinson-s-Disease.html. Accessed 1 Dec 2019

Waldrop MA, Karingada C, Storey MA et al (2020) Gene therapy for spinal muscular atrophy: safety and early outcomes. Pediatrics 146. https://doi.org/10.1542/peds.2020-0729

Wang B, Li J, Xiao X (2000) Adeno-associated virus vector carrying human minidystrophin genes effectively ameliorates muscular dystrophy in mdx mouse model. Proc Natl Acad Sci U S A 97:13714–13719

Watchko J, O'Day T, Wang B et al (2002) Adeno-associated virus vector-mediated minidystrophin gene therapy improves dystrophic muscle contractile function in mdx mice. Hum Gene Ther 13:1451–1460

Wegmann S, DeVos SL, Zeitler B et al (2021) Persistent repression of tau in the brain using engineered zinc finger protein transcription factors. Sci Adv 7:eabe1611

Wolak DJ, Thorne RG (2013) Diffusion of macromolecules in the brain: implications for drug delivery. Mol Pharm 10:1492–1504

Yang B, Li S, Wang H et al (2014) Global CNS transduction of adult mice by intravenously delivered rAAVrh.8 and rAAVrh.10 and nonhuman primates by rAAVrh.10. Mol Ther 22:1299–1309

Yu YJ, Zhang Y, Kenrick M et al (2011) Boosting brain uptake of a therapeutic antibody by reducing its affinity for a transcytosis target. Sci Transl Med 3:84ra44

Yue Y, Pan X, Hakim CH et al (2015) Safe and bodywide muscle transduction in young adult Duchenne muscular dystrophy dogs with adeno-associated virus. Hum Mol Genet 24:5880–5890

Zelikowsky M, Hui M, Karigo T et al (2018) The neuropeptide Tac2 controls a distributed brain state induced by chronic social isolation stress. Cell 173:1265–1279.e19

Zhang H, Yang B, Mu X et al (2011) Several rAAV vectors efficiently cross the blood-brain barrier and transduce neurons and astrocytes in the neonatal mouse central nervous system. Mol Ther 19:1440–1448

Zincarelli C, Soltys S, Rengo G, Rabinowitz JE (2008) Analysis of AAV serotypes 1-9 mediated gene expression and tropism in mice after systemic injection. Mol Ther 16:1073–1080

Part V
CNS Drug Delivery in Disease Conditions

Chapter 22
Disease Influence on BBB Transport in Neurodegeneration

Elizabeth C. M. de Lange

Abstract For the pharmacotherapy of neurodegenerative diseases, drugs must pass the blood–brain barrier (BBB). The BBB seems to play an important role in disease initiation and or progression, and many changes in BBB properties in neurodegeneration have been reported. In vivo studies including measurements of unbound drug concentrations in plasma and brain are needed for insight into BBB transport, intra-brain and target site distribution, and specific changes related to neurodegenerative conditions. However, it is surprising that only a limited number of such studies have been performed to date. This chapter summarizes the published work on these in vivo studies and provides a perspective on what is needed to advance and foster more understanding in the future . Though it is generally thought that the BBB is compromised in neurodegenerative disorders, quantitative studies indicate that this is not necessarily always the case. It is recommended to increase in vivo studies that can integrate the impact of neurodegenerative processes, to complement studies on neurodegenerative components in isolation, and to improve our understanding of target site distribution of drugs intended to treat the disease condition. As in vivo studies on human brain sampling are ethically restricted, we must rely on animal models and translational mathematical approaches to infer relevance for clinical work.

Keywords Neurodegene-rative processes · Understanding · Integration · Quantitative · Translation · Mathematical models

22.1 Introduction

Due to the aging population and perhaps also diet and lifestyle changes, we are facing a rapid increase in the prevalence of neurodegenerative diseases. Neurodegeneration can be defined as progressive loss of neuronal structure and

E. C. M. de Lange (✉)
Research Division of Systems Pharmacology and Pharmacy, Leiden University,
Leiden, The Netherlands
e-mail: ecmdelange@lacdr.leidenuniv.nl

© American Association of Pharmaceutical Scientists 2022
E. C. M. de Lange et al. (eds.), *Drug Delivery to the Brain*, AAPS Advances
in the Pharmaceutical Sciences Series 33, https://doi.org/10.1007/978-3-030-88773-5_22

function, finally culminating in neuronal cell death. Neurodegenerative diseases include Alzheimer's disease, amyotrophic lateral sclerosis, pharmacoresistant epilepsy, Huntington's disease, multiple sclerosis, Parkinson's disease, and traumatic brain injury, among others (Kalaria 2010; Cholerton et al. 2011). Most neurodegenerative diseases start in mid-life and can be characterized by motor and/or cognitive symptoms that progressively worsen with age and may reduce life expectancy.

The mechanisms of neurodegeneration are only partly understood, and thus effective treatments for neurodegenerative diseases are lacking. Despite all efforts in research to develop new CNS drugs in the last few decennia, the results of clinical trials have so far been very disappointing. This indicates that we need to learn more about neurodegenerative processes and their interrelationships in order to develop better drug treatment approaches to combat, halt, or even reverse these processes. Adequate functioning of the blood–brain barrier (BBB) is essential for efficient brain function. Structural and functional disturbances in both the neurovascular unit and CNS fluid compartments may occur with advancing age or following epileptic seizures, traumatic brain injury, or stroke (Yang et al. 2020). These disturbances may include impairment in autoregulation and neurovascular coupling, BBB leakage, a shift in transcytosis that alters the composition

of transcytosing plasma proteins, decreased cerebrospinal fluid (CSF) volume, and reduced vascular tone. Such processes make the brain vulnerable and appear to be responsible for varying degrees of neurodegeneration (Kalaria 2010; Yang et al. 2020).

As BBB dysfunction often leads to inflammatory changes such that circulating immune cells and immune mediators gain access to the brain and then contribute to the process of neurodegeneration, the BBB itself likely plays a key role in most (if not all) neurodegenerative disorders (Zlokovic 2008, 2010, 2011; Zenaro et al. 2017; Sweeney et al. 2018). Thus, for the development of drug treatment modalities for neurodegenerative conditions, we must consider the complexity of the BBB and the brain together (Palmer 2011).

Despite all research efforts in this area, effective treatments for neurodegenerative diseases are still lacking. This is probably because study designs thus far have not focused sufficiently on the interplay between the processes involved (De Lange et al. 2017). Thereby, the data obtained so far often do not provide information on the sensitivity of the obtained parameter values to the context in which they have been measured. Importantly, quantitative information on drug concentrations has been lacking, especially unbound concentrations from plasma and brain to obtain specific and quantitative information on BBB drug transport, intra-brain distribution, and target exposure. Moreover, systematic studies of disease state (or disease stage) compared to healthy conditions are almost absent.

In order to unravel the connections between neurodegeneration and transport at the BBB and intra-brain distribution, multiple quantitative measures in a single system are needed to complement studies that focus on individual processes in isolation (De Lange et al. 2017; see also Chap. 12 of this book). As the human brain is not accessible for in vivo sampling, a great deal of such studies must be performed in animal models (De Lange 2013; De Lange et al. 2017).

In this chapter, neurodegenerative processes and disorders will be discussed, followed by specific information about BBB changes and/or dysfunction in neurodegeneration. Then the current still scarce studies that have measured BBB transport, intra-brain distribution, and target site distribution (and effects) will be presented. The chapter finalizes with conclusions, points for discussion, and suggestions for future directions.

22.2 Neurodegenerative Processes and Disorders

Numerous disorders afflict the nervous system. Among those, neurodegenerative diseases are characterized by a long-lasting course of neuronal death and progressive nervous system dysfunction. The effects of neurodegenerative syndromes extend beyond cognitive function to involve key physiological processes, including eating and metabolism, autonomic nervous system function, sleep, and motor function (Ahmed et al. 2018). The strongest risk factor for brain degeneration, whether it results from vascular or neurodegenerative mechanisms or both, is age. However, several modifiable risks such as cardiovascular disease, hypertension, dyslipidemia, diabetes, and obesity enhance the rate of cognitive decline and increase, in particular, the risk of Alzheimer's disease. The ultimate accumulation of pathological CNS lesions may be modified by genetic influences, such as the apolipoprotein E ε4 allele and the environment, and multiple other potential Alzheimer disease susceptibility genes have been identified, which are ACE, CHRNB2, CST3, ESR1, GAPDHS, IDE, MTHFR, NCSTN, PRNP, PSEN1, TF, TFAM and TNF (Bertram et al. 2007), and more (Van Cauwenberghe et al. 2016). Important factors for brain protection are lifestyle measures that maintain or improve cardiovascular health, including consumption of healthy diets, moderate use of alcohol, and implementation of regular physical exercise (Mulder et al. 2001; Kalaria 2010; Sagare et al. 2012).

22.2.1 Neurodegenerative Processes

The neurodegenerative diseases have many processes in common, though these processes may be qualitatively, quantitatively, temporally, and spatially distinct. These include gene defects, intracellular calcium and oxidative stress, (toxic) protein misfolding, and accumulation that will affect different biological signaling pathways or molecular machineries to cause neuronal cell death by necrosis or apoptosis.

- *Gene defects* play a major role in the pathogenesis of neurodegenerative disorders. Knowledge gained from genetic studies has provided insight into molecular mechanisms underlying the etiology and pathogenesis of many neurodegenerative disorders (Bertram et al. 2007). Many genetic determinants for human neurodegeneration have been identified over the years, and many of these have been reflected in animal models (Qin et al. 2020; Scearce-Levie et al. 2020). In the presence of genetic defects, the course of a progressive neurodegenerative disorder can be greatly modified by environmental elements (Coppedè et al. 2006).

- *Oxidative stress* is the result of disturbances in the normal redox state of cells that can cause toxic effects through the production of peroxides and free radicals (reactive oxygen species) that damage all components of the cell, including proteins, lipids, and DNA. Furthermore, some reactive oxidative species act as cellular messengers in redox signaling. Thus, oxidative stress can cause disruptions in normal mechanisms of cellular signaling and seems to have a ubiquitous role in mechanisms that induce cell death in neurodegenerative disease states (Sayre et al. 2008; Navarro and Boveris 2010; Perez-Pinzon et al. 2012; Arnold 2012). The role of iron seems particularly prominent in oxidative stress. The so-called iron-mediated oxidative stress pathway includes a reduction in antioxidant enzymes (e.g., peroxiredoxin and cytochrome c oxidase) and an induction of ferritin (Berg and Youdim 2006; Bagwe-Parab and Kaur 2019). Potent neuroprotective compounds have often been found to reverse the effects of aging on the expression of various mitochondrial and key regulator genes involved in neurodegeneration, cell survival, synaptogenesis, oxidation, and metabolism (Weinreb et al. 2007a; Bagwe-Parab and Kaur 2019).
- *Protein misfolding and accumulation* can cause disease. Protein misfolding may happen spontaneously, or it can result when a protein follows the wrong folding pathway. The change into a toxic configuration is most likely to occur in proteins that have repetitive amino acid motifs, such as the polyglutamine expansion in the Huntingtin protein that is associated with Huntington's disease. Remarkably, the toxic configuration is often able to interact with other native copies of the same protein and catalyze their transition into a toxic state, known as an "infective conformation." The newly made toxic proteins repeat the cycle in a self-sustaining loop, amplifying the toxicity and thus leading to a catastrophic effect that eventually kills the cell or impairs its function. A prime example of proteins that catalyze their own conformational change into the toxic form is that of the prion proteins (Soto 2008; Jellinger 2012). Abnormal accumulation of proteins and organelles in neurodegenerative diseases will do further damage to the axon as part of the pathogenic process and, in particular, compromise axonal transport. It is known that disruption of axonal transport is an early and perhaps causative event in many of these diseases (Stokin and Goldstein 2006; De Vos et al. 2008).
- Finally, *cell death* occurs, as a result of necrosis and/or apoptosis. Necrosis is a form of traumatic cell death that results from acute cellular injury. In contrast, apoptosis generally confers advantages during an organism's life cycle, being instrumental in development and in homeostatic processes. Necrosis, as a passive process, does not require new protein synthesis, has only minimal energy requirements, and is not regulated by any homeostatic mechanism. Inappropriate death of cells in the nervous system is associated with multiple neurodegenerative

disorders (Price et al. 1998; Artal-Sanz and Tavernarakis 2005; Krantic et al. 2005; Bertram and Tanzi 2005; Lessing and Bonini 2009; Soto and Estrada 2008; Gorman 2008). Neuronal apoptosis, the programmed natural death of neurons, is triggered either by the activation of a death receptor upon binding of its ligand, recruitment of specific proteins at the "death domain," downstream signaling through a cascade of protein–protein interactions (extrinsic pathway), or via mitochondria and the release of pro-apoptotic factors into the cytosol with subsequent activation of executioner caspases (intrinsic pathway). While apoptosis is an important process during neurogenesis and CNS maturation, premature apoptosis and/or an aberration in apoptosis regulation is implicated in the pathogenesis of neurodegeneration. Reactive oxygen species can initiate apoptosis via the mitochondrial and death receptor pathways (Okouchi et al. 2007).

Besides the many-shared mechanisms in neurodegeneration, certain characteristic features are unique to particular diseases, such as the selective vulnerability of a neuronal population or brain structure involved in the lesion (Fu et al. 2018). The reasons for such specificity as well as the mechanisms responsible for its selective nature are largely unknown. Here the main features of different neurodegenerative diseases are shortly described together with the processes that influence BBB function for consideration of drug transport into and within the brain.

22.2.2 Alzheimer's Disease

Alzheimer's disease is the most common neurodegenerative disease. It usually starts with declarative memory loss and confusion, which is initially difficult to distinguish from normal aging. With progression of the disease, increasing behavior and personality changes are accompanied by a further decline in cognitive abilities as well as worsening problems recognizing family and friends. Alzheimer's disease ultimately leads to a severe loss of mental function. Alzheimer's disease is specifically characterized by a loss of neurons and synapses in the cerebral cortex and certain subcortical regions. This loss results in gross atrophy of the affected regions, including degeneration in the temporal lobe and parietal lobe, and parts of the frontal cortex and cingulate gyrus (Wenk 2003). There are three major hallmarks in the brain that are associated with the disease processes of Alzheimer's disease (Finder 2010).

- The *first hallmark* is the presence of amyloid (senile) plaques. Alzheimer's disease has been hypothesized to be a protein misfolding disease caused by accumulation of abnormally folded small amyloid-beta (Aβ) peptides that can vary between 39 and 43 amino acids in length. Aβ is a fragment from a larger protein called the amyloid precursor protein (APP). APP is a trans-membrane protein that penetrates through the neuron's membrane and is critical to neuron growth, survival, and post injury repair. In Alzheimer's disease, an unknown process causes APP to be divided into smaller fragments by enzymes through proteolysis.

A toxic 42 amino acid form of Aβ (Aβ1-42) gives rise to fibrils that form the core of senile plaques.

- The *second hallmark* is the presence of neurofibrillary tangles in the intracellular space of neurons, with high content of the protein "tau." Normal tau is required for healthy neurons. However, in Alzheimer's disease, hyper- phosphorylated tau aggregates as neurofibrillary tangles to cause neuronal dysfunction and eventual cell death.
- The *third hallmark* is brain atrophy and shrinkage. Neurons that lose their connection with other neurons will die; this occurs throughout the Alzheimer's disease brain, causing affected regions to atrophy and shrink.

Current treatments of Alzheimer's disease are symptomatic, that is, they affect symptoms while not slowing the progression of the disease process. These treatments include drugs such as Donepezil (Aricept), rivastigmine (Exelon), galantamine (Razadyne), and Memantine (Namenda); they mostly help patients to carry out daily tasks by maintaining thinking, memory, and/or speaking skills. Treatment modalities interfering with neurodegenerative processes of Alzheimer's disease are still elusive.

Available evidence suggests that alteration of the BBB plays an important role in Alzheimer's disease (Zlokovic 2011; Miyakawa 2010; Zenaro et al. 2017; Sweeney et al. 2018)). The BBB plays a regulatory role in the deposition of brain Aβ. Active transport of Aβ seems to occur by putative Aβ receptors that control the level of the soluble isoform of Aβ in brain. Influx of circulating Aβ is achieved via a specific receptor for advanced glycation end products (RAGE) and by gp330/megalin (LRP-2)-mediated transcytosis (Chun et al. 1999). There are also indications of transcytosis by cellular prion protein (PrP(c)) that binds Aβ(1–40) (Pflanzner et al. 2012). Aβ accumulation in the Alzheimer's affected brain is likely due to its faulty clearance from the brain (Zlokovic et al. 2000; Selkoe 2011; Tanzi et al. 2004; Holtzman and Zlokovic 2007). The BBB efflux of brain-derived Aβ into blood is accomplished by the low-density lipoprotein receptor-related protein-1 (LRP1) and P-glycoprotein (P-gp, MDR1, ABCB1) (Kuhnke et al. 2007; Bell and Zlokovic 2009; Brenn et al. 2011; Sharma et al. 2012; Sagare et al. 2012; Vogelgesang et al. 2011; Hartz et al. 2010). In plasma, a soluble form of LRP1 (sLRP1) is the major transport protein for peripheral Aβ, where it maintains a plasma "sink" activity for Aβ through binding of peripheral Aβ which in turn inhibits reentry of free plasma Aβ into the brain. LRP1 in the liver mediates systemic clearance of Aβ. In Alzheimer's disease, LRP1 expression at the BBB is reduced, and Aβ binding to circulating sLRP1 is compromised by oxidation (Sagare et al. 2012). Significantly reduced expression of P-gp, LRP1, and RAGE mRNA has been found in mice treated with Aβ(1–42), while breast cancer-resistance protein trans-porter (BCRP, ABCG2) expression was not affected; notably, expression of the four proteins was unchanged in mice treated with Aβ1-40 or reverse-sequence peptides (Brenn et al. 2011). This indicates that, in addition to the age-related decrease of P-gp expression, Aβ1-42 itself downregulates the expression of P-gp and other Aβ-transporters, which could exacerbate the intracerebral accumulation of Aβ and thereby accelerate neurodegeneration in Alzheimer's disease and cerebral β-amyloid angiopathy. Furthermore, an increased BBB permeability in Alzheimer's disease is also likely since structural damage of brain endothelial cells is quite frequently observed. Defects in LRP-1- and P-gp-mediated Aβ clearance from the brain are thought be triggered by systemic inflammation by lipopolysaccharide (LPS), leading to increased brain accumulation of Aβ (Erickson et al. 2012). This indicates that inflammation could induce and promote the disease. In addition, there are indications that ischemic events may directly contribute to enhancement of the amyloidogenic metabolism within the BBB, leading to intracellular deposition of Aβ(42), which may contribute to impaired Aβ clearance and related BBB dysfunction in Alzheimer's disease (Bulbarelli et al. 2012). Moreover, Aβ damages its own LRP1-mediated transport by oxidizing LRP1 (Owen et al. 2010).

Another contributor in risk for Alzheimer's disease is reduced insulin effectiveness. Insulin appears to play an important role in brain aging and cognitive decline that is associated with pathological brain aging (Cholerton et al. 2011).

For treatment of Alzheimer's disease, it seems that cell surface LRP1 and circulating sLRP1 represent druggable targets which can be therapeutically modified to restore the physiological mechanisms of brain Aβ homeostasis. Enhancement of P-gp functionality might also be a novel therapeutic strategy to increase Aβ clearance out of the brain (Hartz et al. 2010; Abuznait et al. 2011). In addition, lifestyle-related conditions such as insulin resistance are amenable to both pharmacologic and lifestyle interventions to reduce the deleterious impact on the aging brain (Cholerton et al. 2011). More information is needed on what processes result in impairment of the BBB functionality in Alzheimer's disease as well as in "normal aging." BBB leakage in temporal lobe cortex of human Alzheimer brain samples shows wide variation but, overall, significantly increased leakage of the BBB with progression of Alzheimer-type pathology in some studies (Viggars et al. 2011). However, other studies indicate no changes in immunoglobulin G permeability in most animal models of Alzheimer's disease (Nga Bien-Ly et al. 2015). Also, in a PET study using 11C verapamil as a P-gp functionality ligand, no evidence was found for additional BBB dysfunction of P-gp in Alzheimer's disease patients with micro-bleeds (van Assema et al. 2012). Thus, it is not entirely clear what mechanisms lead to BBB leakage in the aging brain (Viggars et al. 2011).

22.2.3 Parkinson's Disease

Parkinson's disease is a chronic, progressive neurological disorder lacking a cure. It belongs to the group of motor system disorders and results from a loss of dopamine-producing brain cells mostly in the substantia nigra, for which the cause is unknown. Parkinson's disease is the second most common neurodegenerative disease. Although Parkinson's disease is most common for ages above 60 years, many people are diagnosed at ages younger than 40 years. The core symptoms are tremor, rigidity (stiffness), bradykinesia (slowness of movement), and postural instability (balance difficulties). These symptoms become more pronounced with time. Patients may have difficulty walking, talking, or completing other simple tasks. As the disease progresses, the shaking, or tremor, which affects the majority of Parkinson's disease patients may begin to interfere with daily activities. Non-motor aspects of Parkinson's disease include depression and anxiety, cognitive impairment, sleep disturbances, sensation of inner restlessness, loss of smell (anosmia), and disturbances of autonomic function. In advanced Parkinson's disease, intellectual and behavioral deterioration, aspiration pneumonia, and bedsores (due to immobility) are common.

Current drugs available for the treatment of Parkinson's disease are L-DOPA (usually combined with a peripheral decarboxylase inhibitor), synthetic dopamine receptor agonists, centrally acting antimuscarinic drugs, amantadine, monoamine oxidase-B inhibitors, and catechol-O-methyltransferase inhibitors. These drugs unfortunately only address the symptoms of the disease, so therapeutic strategies aimed at stopping or modifying disease progression are urgently needed (Deleu et al. 2002). Usually, patients are given levodopa (L-DOPA) combined with carbidopa. L-DOPA helps in many cases of Parkinson's disease, with bradykinesia and rigidity responding best, while tremor may be only marginally reduced. Problems with balance and other symptoms may not be alleviated at all. Anticholinergics may help control tremor and rigidity. Dopamine agonists such as bromocriptine, pramipexole, and ropinirole may offer some advantages over levodopa as they likely do not require a transporter to cross the BBB (levodopa uses the large neutral amino acid transporter for this purpose) nor is enzymatic conversion necessary for their activation (levodopa must be converted to dopamine by DOPA decarboxylase once in the brain. An antiviral drug, amantadine, also appears to reduce symptoms. Animal experimentation has provided many insights into the features of Parkinson's disease. Indeed, the roles of oxidative stress, apoptosis, mitochondrial dysfunction, inflammation, and impairment of the protein degradation pathways have been highlighted by work with animal models (Grünblatt et al. 2000; Bové and Perier 2012).

The mechanism by which the brain cells in Parkinson's disease are lost may consist of an abnormal accumulation of the protein alpha-synuclein bound to ubiquitin in the damaged cells. The alpha-synuclein-ubiquitin complex cannot be directed to the proteosome. This protein accumulation forms proteinaceous cytoplasmic inclusions called Lewy bodies, which are one of the hallmarks of Parkinson's disease (De Vos et al. 2008). Impaired axonal transport of alpha-synuclein may contribute to its accumulation in the form of Lewy bodies, as reduced transport rates have been reported for both wild-type and two familial Parkinson's disease- associated mutant alpha-synucleins in cultured neurons. In addition, membrane damage by alpha-synuclein could be another Parkinson's disease mechanism (De Vos et al. 2008).

Inflammation might be a risk factor by itself and not only a factor contributing to neurodegeneration. Hernández-Romero et al. (2012) investigated the impact of mild to moderate peripheral inflammation by carrageenan on the degeneration of dopaminergic neurons by intranigral injection of lipopolysaccharide (LPS) in animals. Peripheral inflammation increased the effect of intranigral LPS on the loss of dopaminergic neurons in the substantia nigra, in addition to increasing serum levels of the inflammatory markers: TNF-α, IL-1β, IL-6, and C-reactive protein. Peripheral inflammation is also associated with damage to the BBB as well as the activation of microglia, loss of astrocytes, and the increased expression of proinflammatory cytokines, the adhesion molecule ICAM, and the enzyme iNOS.

The possible implications of BBB dysfunction for the increased loss of dopaminergic neurons has been studied using another Parkinson's disease animal model based on the intraperitoneal injection of rotenone. In this experiment, loss of dopaminergic neurons was also strengthened by carrageenan although this was achieved without obvious effects at the BBB (Hernández-Romero et al. 2012). Intracerebral injection of rotenone may provide a better model of Parkinson's disease (Ravenstijn et al. 2008), although this model does not produce concomitant changes in BBB transport for fluorescein and L-DOPA (Ravenstijn et al. 2012). The transport of bromocriptine across the BBB has been investigated in mice with MPTP-induced dopaminergic degeneration (Vautier et al. 2009); transport of the small compounds [14C]-sucrose and [3H]-inulin across the BBB was unaffected, while P-gp and BCRP functionality did not appear to change. Conversely, BCRP expression studied on brain capillaries from MPTP-treated mice was decreased (1.3-fold) and P-gp expression increased (1.4-fold). While MPTP intoxication did not seem to alter BBB permeability, bromocriptine brain distribution was increased in MPTP mice, probably by interaction with another transport mechanism. Overall, for Parkinson's disease, there is not really consensus about the changes in BBB functionality (Desai et al. 2007; Ravenstijn et al. 2008, 2012).

Although the etiology of Parkinson's disease has not yet been clarified, it is believed that aging, diet, diabetes, and adiposity all play some role (Lu and Hu 2012). Type 2 diabetes and lipid abnormalities share multiple common pathophysiological mechanisms with Parkinson's disease, as does the gradual impairment of neurovascular function with aging. Neurovascular impairment may include (focal) changes in the BBB that may result in the passage of harmful elements that would not normally be able to cross the BBB; for example, pro-inflammatory factors, reactive oxygen species, and neurotoxins may infiltrate into the brain and trigger neural injury (Reale et al. 2009).

Most recent studies suggest that both central and peripheral inflammation may be dysregulated in Parkinson's disease, not only in animal models but also Parkinson's disease patients. This strengthens and extends the idea that peripheral dysregulation in the cytokine network is associated with Parkinson's disease. Therefore, therapeutic strategies aimed at modulating systemic inflammatory reactions or energy metabolism may facilitate neuroprotection in Parkinson's disease (Reale et al. 2009; Lu and Hu 2012).

22.2.4 Pharmacoresistant Epilepsy

Epilepsy is a common and diverse set of chronic neurological disorders character-ized by recurrent seizures and/or induced brain alterations. The seizures happen when neurons, in clusters or individually, send out the wrong signals. Affected peo-ple may have strange sensations and emotions or behave strangely; with severe forms, they can exhibit violent muscle spasms and loss of consciousness. Anything that disturbs the normal pattern of neuron activity can lead to seizures, including illness, brain injury, and abnormal brain development. In many cases, however, the cause is unknown. Thus, epilepsy has many possible causes, and there are several types of seizures. Epilepsy becomes more common as people age. Onset of new cases occurs most frequently in infants and the elderly. Underlying causes of epi-lepsy may be related to brain trauma, stroke, and brain tumors.

Epilepsy is usually controlled, but not cured, with medication. For about 70 % of individuals with epilepsy, seizures can be controlled with drugs and/or surgery. Some drugs are more effective for specific types of seizures. Anti-epileptic drugs include carbamazepine for partial seizures, ethosuximide for absence seizures with-out generalized tonic-clonic seizure, and valproate for primary generalized epilep-sies as well as partial seizures. Phenytoin is used in the control of various kinds of epilepsy and of seizures associated with neurosurgery. Newer anti-epileptic drugs are often used as add-on therapies and include lamotrigine, oxcarbazepine, topira-mate, gabapentin, and levetiracetam.

Despite the availability of numerous medications for epilepsy, ~30 % of patients have seizures that remain uncontrolled. This epileptic condition is called pharmaco-resistant, drug refractory, or intractable epilepsy. In pharmacoresistant epileptic patients, status epilepticus (serious, potentially life-threatening, neurologic emer-gency characterized by prolonged seizure activity) is more common and ongoing, and uncontrolled seizure activity may result in brain damage and neurodegenera-tion, especially in young children (Bittigau et al. 2002). Seizures from (medial) temporal lobe epilepsy are most commonly pharmacoresistant (Volk et al. 2006), and the underlying mechanisms are still elusive. There are two main hypotheses for the cause of (or major contribution to) pharmacoresistance in epilepsy (Volk et al. 2006; Bethmann et al. 2008):

1. The target hypothesis—anti-epileptic drug efficacy is diminished due to reduced target sensitivity (e.g., GABAa receptor binding changes) (Volk et al. 2006).
2. The transporter hypothesis—anti-epileptic drug efficacy is diminished due to decreased brain levels resulting from localized overexpression of drug efflux transporters (mainly P-gp) in epileptogenic brain tissue.

Network alterations in response to brain damage associated with epilepsy may also result in changes in anti-epileptic drug efficiency (Bethmann et al. 2008; Ndode-Ekane et al. 2010).

Much research has implicated P-glycoprotein in epilepsy treatment inefficiency and in epileptogenesis (Marchi et al. 2004; Bankstahl et al. 2011; Löscher et al.

2011). Seizures may induce BBB transport changes (Padou et al. 1995; Sahin et al. 2003) and increased expression of P-gp at the BBB, as determined from both epileptogenic brain tissue of patients with pharmacoresistant epilepsy (Dombrowski et al. 2001) and in rodent models of temporal lobe epilepsy, including the pilocarpine model. In the latter, Bankstahl et al. (2008) found that seizure-induced glutamate release seems to be involved in the regulation of P-gp expression, which can be blocked by dizocilpine (also known as MK-801), a noncompetitive antagonist of the N-methyl-d-aspartate (NMDA) receptor. The finding that MK-801 counteracts both P-gp overexpression and neuronal damage when administered after status epilepticus may offer a clinically useful therapeutic option in patients with drug resistant status epilepticus.

In normal brain tissue, MDR1/P-gp is expressed almost exclusively by the BBB, while in epileptic cortex, it has been found that both brain endothelial cells and perivascular astrocytes express MDR1/P-gp. This change in P-gp may act as a second line of defense that may have profound implications for the pharmacokinetic properties of antiepileptic drugs and their capacity to reach neuronal targets (Marroni et al. 2003; Lee and Bendayan 2004; Bendayan et al. 2006). Using (mdr1a) P-gp knockout mice and wild-type mice, Sills et al. (2002) investigated the brain-to-serum concentration ratio for seven anti-epileptic drugs. Only topiramate yielded a higher brain-to-serum ratio in mdr1a($-/-$) mice compared to that in wild-type controls at all time points investigated. No consistent effects were observed with any of the other anti-epileptic drugs studied.

In vitro studies by Luna-Tortos et al. (2009) have indicated that topiramate is a substrate for human P-gp. Potschka et al. (2003a, b) reported that brain microdialysis concentrations of phenytoin in rats were increased by local application of the MRP transporter inhibitor probenecid; similarly, brain microdialysis concentrations of phenytoin were significantly higher in MRP2-deficient TR—rats than in normal rats. In the kindling model of epilepsy, administration of probenecid significantly increased the anticonvulsant activity of phenytoin, while in kindled MRP2-deficient rats, phenytoin exerted a markedly higher anticonvulsant activity than in normal rats. These microdialysis data indicate that MRP2 could substantially contribute to BBB function and that phenytoin appears to be a MRP2 substrate. While Hoffmann et al. (2006) did not find MRP2 expression in the brain of normal rats, clear MRP2 staining became visible in brain capillary endothelial cells and, less frequently, in perivascular astroglia and neurons after pilocarpine-induced convulsive status epilepticus (a model of temporal lobe epilepsy).

Baltes et al. (2007b) found that phenytoin and levetiracetam were transported by mouse, but not human, P-gp, and that carbamazepine was not transported by any type of P-gp. These data indicated that substrate recognition or transport efficacy by P-gp differs between human and mouse for certain anti-epileptic drugs. In vitro studies indicated that none of the common anti-epileptic drugs carbamazepine, valproate, levetiracetam, phenytoin, lamotrigine, and phenobarbital is transported by MRP1, MRP2, or MRP5, while valproate was transported by a yet unknown transporter which could be inhibited by MK571 and probenecid (Luna-Tortós et al. 2010). When specifically measuring P-gp-related BBB transport and intracerebral distribution, Syvänen et al. (2012) found in rats subjected to status epilepticus by kainate that by P-gp inhibition the intra-brain distribution of the strong and selective P-gp substrate quinidine was more affected than was BBB transport and extracellular brain concentrations. The results of this study combined with those obtained by positron emission tomography (PET) study using the same animals suggest that P-gp function in epilepsy might be altered specifically at the brain parenchymal level (Syvänen et al. 2011, 2012).

While it is established that efflux transporters are upregulated in drug-resistant epileptogenic brain tissue in humans and rodents, their role in removal of antiepileptic drugs from the brain remains controversial (Anderson and Shen 2007; Löscher et al. 2011; van Vliet et al. 2007). Nevertheless, P-gp inhibition by verapamil, administered directly into rat cerebral cortex, has been reported to modestly increase (up to twofold) the brain ECF-to-plasma concentration ratios of phenobarbital, phenytoin, lamotrigine, felbamate, carbamazepine, or oxcarbazepine (Clinckers et al. 2005a, b; Potschka et al. 2001; Potschka and Löscher 2001a, b). Furthermore, in rats with induced seizures, cyclosporine and tariquidar can reverse resistance to several antiepileptic drugs and increased their brain-to-plasma concentration ratio without changing their plasma pharmacokinetics (Brandt et al. 2006; Clinckers et al. 2005a, b; Mazarati et al. 2002).

Apart from a transport restriction and changes in multidrug efflux transporters, there might be a role for P450 metabolic enzymes in reducing brain concentrations of CNS therapeutics in drug-resistant pathologies such as refractory forms of epilepsy (Ghosh et al. 2011), and changes in cerebrovascular hemodynamic conditions can affect expression of P450 enzymes and multidrug transporter proteins.

Focal epilepsies are often associated with BBB leakage. For example, BBB leakage to albumin-bound Evans blue has been found in PTZ-induced epilepsy, with the location and pattern depending on the rat strain (Ates et al. 1999). Selective modulation of claudin expression in the brain by kindling epilepsy has also been found (Lamas et al. 2002). It has been observed during the process by which a normal brain develops epilepsy (epileptogenesis), immunoglobulin G (IgG) leakage and neuronal IgG uptake increase concomitantly with the occurrence of seizures, and IgG-positive neurons show signs of neurodegeneration, such as shrinkage and eosinophilia. This may suggest that IgG leakage is related to neuronal impairment and may be a pathogenic mechanism in epileptogenesis and chronic epilepsy (Michalak et al. 2012; Ndode-Ekane et al. 2010). Other studies point to a profound role of seizure-induced neuronal cyclooxygenase-2 (COX-2) expression in neuropathologies that accompany epileptogenesis (Serrano et al. 2011), and it is thought that epileptic seizures drive expression of the BBB efflux transporter P-gp via a glutamate/COX-2-mediated signaling pathway. Targeting this pathway may represent an innovative approach to control P-gp expression in the epileptic brain and to enhance brain delivery of antiepileptic drugs (van Vliet et al. 2010).

Many studies indicate important links to activation of the immune system with epilepsy. Zattoni et al. (2011) found that BBB disruption and neurodegeneration in the kainate-lesioned hippocampus were accompanied by sustained intercellular adhesion molecule 1 (ICAM-1) upregulation, microglial cell activation, and infiltration of cluster of differentiation 3 (+) T-cells (CD3(+) T-cells). Moreover, macrophage infiltration was selectively observed in the hippocampal dentate gyrus, where prominent granule cell dispersion was evident. Neurodegeneration was aggravated in kainate-lesioned mice lacking T and B cells (RAG1 knockout) through delayed invasion by Gr-1(+) neutrophils. The fact that these mutant mice also exhibited early onset of spontaneous recurrent seizures emphasizes the strong role immune-mediated responses can play in network excitability (Deprez et al. 2011).

ApoE isoforms exhibit diverse effects on neurodegenerative and neuroinflammatory disorders. As with other neurodegenerative diseases (e.g., Alzheimer's disease), apolipoprotein E (ApoE) genotype seems to play a significant role in epilepsy (Zhang et al. 2012). Overexpression of apoE4 has been shown to worsen KA-induced hippocampal neurodegeneration in C57BL/6 mice, possibly through an enhanced activation of microglia as compared to wild-type and apoE2 or apoE3 transgenic mice. New epilepsy treatments may utilize insulin-like growth factor-1 (IGF-I) or vitamin E, particularly where standard therapies do not show efficacy. Administration of IGF-I has been shown to decrease seizure severity, increases hippocampal neurogenesis, protects against neurodegeneration, and abolishes cognitive deficits in an animal model of temporal lobe epilepsy (Miltiadous et al. 2011). Vitamin E (α-tocopherol, α-T) is of interest as it has been proposed to alleviate glia-mediated inflammation in neurological diseases; indeed, α-T dietary supplementation was found to prevent the oxidative stress, neuroglial overactivation, and cell death that normally occurs after kainate-induced seizures (Betti et al. 2011).

22.2.5 Traumatic Brain Injury

Traumatic brain injury (TBI) occurs when head injury causes damage to the brain. The worst injuries can lead to permanent brain damage or death. Symptoms of a traumatic brain injury may not appear until days or weeks following the injury. Serious traumatic brain injuries need emergency treatment, and their long-term outcome depends on both the severity of the injury and the effectiveness of treatment. Traumatic brain injury can cause a wide range of changes affecting thinking, sensation, language, or emotions. It can also be associated with posttraumatic stress disorder. People with severe injuries usually need rehabilitation (Amenta et al. 2012; Shlosberg et al. 2010). One-third of patients, who have died of TBI, have Aβ plaques, which are pathological features of Alzheimer's disease, indicating that traumatic brain injury acts as an important risk factor for Alzheimer's disease (Sivanandam and Thakur 2012). TBI survivors also have a significantly higher risk of developing epilepsy (Christensen 2012).

The pathophysiology of traumatic brain injury consists of two main phases, a primary (mechanical) phase of damage and a secondary (delayed) phase of damage. Primary damage occurs at the moment of insult and includes contusion and laceration, diffuse axonal injury, and intracranial hemorrhage. Secondary damage includes processes that are initiated at the time of insult, but do not appear clinically for hours or even days after injury. Such processes cause brain swelling, axonal injury and hypoxia, changes in cerebral blood flow, disruption of BBB function, increased inflammatory responses, oxidative stress, neurodegeneration, and cognitive impairment (Pop and Badaut 2011; Sivanandam and Thakur 2012; Weber 2012). The calcium ion contributes greatly to the delayed cell damage and death after traumatic brain injury. A large, sustained influx of calcium into cells can initiate cell death signaling cascades, through activation of several degradative enzymes, such as proteases and endonucleases (Weber 2012). Potential influence on traumatic brain injury outcomes by polymorphisms in the BDNF gene and genes involved in dopaminergic and serotonergic system functionality have been proposed to influence six specific cognitive and social functions: working memory, executive function, decision-making, inhibition and impulsivity, aggression, and social and emotional function (Weaver et al. 2012).

While neurons have been the major focus of translational research in all types of brain injury, it has become clear that more attention is needed to treat neurovascular unit dysfunction because posttraumatic changes in the BBB are one of the major factors determining the progression of injury (Weber 2012). BBB changes observed after injury are implicated in neuronal loss, altered brain function (impaired consciousness, memory loss, and motor impairment), and alterations in the response to therapy (Chodobski et al. 2011). The disruption of tight junctions and basement membrane integrity result in increased paracellular permeability. Injury causes oxidative stress and the increased production of proinflammatory mediators. Upregulation of expression of cell adhesion molecules on the surface of the BBB promote the influx of inflammatory cells into the traumatized brain parenchyma.

There is also evidence suggesting that brain injury can change the expression and/or activity of BBB-associated monocarboxylate transporter 2 (MCT2) transporters (Prins and Giza 2006). These findings suggest that BBB breakdown and functionality changes might be useful as biomarkers in the clinic and in drug trials (Shlosberg et al. 2010; Pop and Badaut 2011).

Acute-phase treatment of traumatic brain injury has improved substantially, but prevention and management of long-term complications remain difficult (Rosenfeld et al. 2012; Shlosberg et al. 2010). Recently, lithium has been investigated for its medium-phase effect on traumatic brain injury-induced neuronal death, microglial activation, and cyclooxygenase-2 induction in mice; all of these factors were attenuated by lithium treatment, which also decreased matrix metalloproteinase-9 expression and preserved BBB integrity (Yu et al. 2012). As for behavioral outcomes, lithium treatment reduced anxiety-like behavior and improved short- and long-term motor coordination. Another recent preclinical finding is that zinc seems to play a role in resilience to traumatic brain injury, making it potentially useful in populations at risk for injury (Cope et al. 2012).

22.3 Dysfunction of the BBB in Neurodegenerative Diseases

The BBB is the regulated interface between the peripheral circulation and the CNS. The anatomical substrate of the BBB is the cerebral microvascular endothelium. Together with astrocytes, pericytes, neurons, and the extracellular matrix, it constitutes a "neurovascular unit" that is essential for the health and function of the CNS (Hawkins and Davis 2005). Dysfunction of the neurovascular unit, mostly investigated in Alzheimer's disease, is associated with both acute and chronic neurologic disorders (Sandoval and Witt 2008; Zlokovic 2008, 2010; Sweeney et al. 2018) and pathogenesis associated with BBB breakdown (Abbott et al. 2002; Zlokovic 2008, 2010, 2011; Freeman and Keller 2012; Al Ahmad et al. 2012; Zenaro et al. 2017; Sweeney et al. 2018).

22.3.1 Tight Junctions

Tight junctions regulate paracellular flux and contribute to the maintenance of homeostasis. Tight junctions are composed of transmembrane proteins such as occludin, claudin 5, claudin-8, claudin 12, and junctional adhesion molecules. Each of these transmembrane proteins is anchored into the endothelial cells by another protein complex that includes zonula occludens proteins (ZO-1, ZO-2, and ZO-3) (Aijaz et al. 2006). The components and function of tight junctions are both affected by neurodegenerative processes (Zlokovic 2011).

- *Occludin* is vulnerable to attack by matrix metalloproteinases (MMPs) (Rosenberg and Yang 2007; Yang and Rosenberg 2011), which may be activated in ischemic conditions. Accumulation of occludin in neurons, astrocytes, and microglia has also been reported in the brain tissue of Alzheimer's patients (Romanitan et al. 2007), suggesting a role for occludin in Alzheimer's disease pathogenesis. Furthermore, dephosphorylation of occludin in a multiple sclerosis mouse model precedes visible signs of disease, before changes in the BBB permeability were observed. Occludin could therefore regulate the response of the BBB to the inflammatory environment (Morgan et al. 2007).
- *Claudin-5* is degraded by MMP-2 and MMP-9 after an ischemic insult, and claudin-5 has been found in surrounding astrocytes, but not in the brain endothelium, after ischemia-related BBB disruption.
- Selective downregulation of *Claudin-8* by kindling epilepsy (Lamas et al. 2002) suggests that selective modulation of claudin expression in response to abnormal neuronal synchronization may lead to BBB breakdown and brain edema. Significant differences in the incidences of tight junction abnormalities related to a reduced ZO-1 expression have been observed between different types of lesions in multiple sclerosis and between multiple sclerosis and control white matter (Kirk et al. 2003).

22.3.2 Actin

Actin is important in the cytoskeleton for establishing and maintaining the BBB (Nico et al. 2003). Tau-induced neurotoxicity in Alzheimer's disease might be related to a direct interaction between tau and actin (Tudor et al. 2007).

22.3.3 Basal Lamina and Extracellular Matrix

The *basal lamina* that surrounds brain endothelial cells consists of laminin, fibronectin, tenascin, collagens, and proteoglycan (Paulsson 1992; Erickson and Couchman 2000; Merker 1994) and provides mechanical support for cell attachment. It also serves as a substratum for cell migration, separates surrounding tissue, and restricts the passage of macromolecules. Cell adhesion to the basal lamina involves integrins (Hynes and Lander 1992). The composition of the extracellular matrix is altered upon BBB disruption and directly affects the progression of brain diseases (Baeten and Akassoglou 2011). For example, MMPs can be activated to degrade basal lamina proteins such as fibronectin, laminin, and heparan sulfate after an ischemic insult, a process which may contribute to BBB breakdown (Cheng et al. 2006; Zlokovic 2006; Zlokovic 2011). In a recent review (Reed et al. 2019), the role of the extracellular matrix (ECM), including that of the glycocalyx at the luminal and abluminal sides of the BBB, as well as the basal lamina, has been discussed, and indicated dynamic and region and cell-type-specific regulation of the ECM during aging and neurodegeneration.

22.3.4 Perivascular Astrocytes

The specialized foot processes of perivascular astrocytes have specialized functions in inducing and regulating BBB properties. Neuronal influence may also be of importance in BBB regulation (Banerjee and Bhat 2007; Wolburg et al. 2009; Cohen-Kashi et al. 2009; Girouard et al. 2010). Astrocyte properties may be affected upon development of amyloid deposits (Yang et al. 2011; Zlokovic 2011). Abnormal astrocytic activity coupled to vascular instability has been observed in Alzheimer's disease models (Takano et al. 2007).

22.3.5 Pericytes

The impact of pericytes on BBB functionality has become more appreciated with time (Balabanov and Dore-Duffy 1998; Krueger and Bechmann 2010). In addition to providing mechanical stability, pericytes predominantly influence vessel stability by matrix deposition and by the release and activation of signals that promote brain endothelial cell differentiation and quiescence (Armulik et al. 2011a, b). Pericytes furthermore play a regulatory role in brain angiogenesis, cerebral endothelial cell tight junction formation, and BBB differentiation and contribute to microvascular structural stability. Pericytes cover 30–70% of the abluminal endothelial cell surface of brain capillaries (von Tell et al. 2006). Pericytes might have a role in the development of neuropathology in Alzheimer's disease and multiple sclerosis (Wyss-Coray et al. 2000; Allt and Lawrenson 2001; von Tell et al. 2006; Zlokovic 2011). Loss of pericytes may damage the BBB due to an associated decrease in cerebral capillary perfusion, blood flow, and blood flow responses to brain activation. This will lead to more chronic perfusion problems like hypoxia, while BBB breakdown may further lead to brain accumulation of blood proteins and several macromolecules with toxic effects on the vasculature and brain parenchyma, ultimately leading to secondary neuronal degeneration (Bell et al. 2010; Zlokovic 2011).

While it is well appreciated that APOE4 homozygosity is associated with an increased risk of sporadic Alzheimer's disease, its effects on the brain microvasculature and BBB have been less appreciated. Interestingly, APOE(4,4) is associated with thinning of the microvascular basement membrane in Alzheimer's disease (Bell et al. 2012). In APOE4 transgenic mice, a high fat diet induced deleterious effects on BBB permeability (Mulder et al. 2001). A study by Bell et al. (2012) suggested that CypA is a key target for treating APOE4-mediated neurovascular injury and the resulting neuronal dysfunction and degeneration; indeed, activating a proinflammatory CypA-nuclear factor-κB-matrix-metalloproteinase-9 pathway in pericytes is associated with increased susceptibility of the BBB to injury in APOE4 conditions.

The influence of misfolded α-synuclein has been implicated in neurodegeneration and neuroinflammation through activation of microglia and astrocytes. Dohgu et al. (2019) investigated the impact of pericytes on BBB integrity in response to monomeric α-synuclein (it did not self-assemble during experimental time) using rat brain endothelial cells (RBECs) co-cultured with rat brain pericytes (RBEC/pericyte co-culture) with luminal or abluminal exposure to α-synuclein, using sodium fluorescein as marker for BBB integrity. Added to the abluminal side, there was a significant increase in RBEC/pericyte co-culture permeability to fluorescein, while it had no marked effect when added to the luminal chamber. For RBECs alone, there was no effect on the permeability to fluorescein. It was suggested that monomeric α-synuclein-activated pericytes may contribute to BBB breakdown in Parkinson's disease (Dohgu et al. 2019). In another study, pathogenic mechanism of APOE4 in pericytes was found in an induced pluripotent stem cell-based three-dimensional model that recapitulates anatomical and physiological properties of the human BBB in vitro, by which it was revealed that dysregulation of calcineurin–nuclear factor of activated T cells (NFAT) signaling and APOE in pericyte-like mural cells induces APOE4-associated cerebral amyloid pathology (Blanchard et al. 2020).

22.3.6   Metabolic Enzymes

The BBB is rich in mitochondria and contains many metabolic enzymes that may contribute to its barrier function. These enzymes include ATPase, nicotinamide adenine dinucleotide, monoamine oxidase, acid and alkaline phosphatases, various dehydrogenases, L-DOPA decarboxylase and gamma glutamyl transpeptidase, cyto- chrome P450 hemoproteins, cytochrome P450-dependent mono-oxygenases, NADPH-cytochrome P450 reductase, epoxide hydrolase, and also conjugating enzymes such as UDP-glucuronosyltransferase and α-class glutathione S-transferase (Maxwell et al. 1987; Williams et al. 1980; Fukushima et al. 1990; Kerr et al. 1984; Tayarani et al. 1989; Volk et al. 1991; Dutheil et al. 2010; Zlokovic 2011). Apart from metabolizing compounds coming from the blood, they also help to eliminate degradation products of neurotransmitters (Baranczyk-Kuzma et al. 1989). BBB enzymes also recognize and rapidly degrade most peptides, including naturally occurring neuropeptides (Brownless and Williams 1993; Witt et al. 2001).

22.3.7   Facilitative and Active Transport Systems

Specific facilitative and active transport systems exist to transport nutrients such as hexoses; neutral, basic, and acidic amino acids; monocarboxylic acids; nucleosides; purines; amines; and vitamins, mostly toward the brain (Simpson et al. 2007; Ohtsuki and Terasaki 2007; Deeken and Loscher 2007; Spector and Johanson 2007; Spector 2009). It has been suggested that glutamate excitotoxicity is implicated in the neurodegenerative processes associated with Alzheimer's disease (Lipton 2005), amyotrophic lateral sclerosis (Van Damme et al. 2005), epilepsy (Alexander and Godwin 2006), Huntington's disease (HD) (Cowan and Raymond 2006; Fan et al. 2009), and multiple sclerosis (Vallejo-Illarramendi et al. 2006). Glutamate transporters (EAAT1, EAAT2, and EAAT3) at the BBB determine the levels of brain extracellular glutamate and are essential to prevent excitotoxicity (Lipton 2005), prompting the question of whether changes in these transporters may contribute to glutamate excess and excitotoxicity.

Facilitative glucose transport is mediated by one or more members of the closely related glucose transporter (GLUT) family. GLUT1 is the primary transporter of glucose across the BBB. Its distribution and expression in the brain is affected in different pathophysiological conditions including Alzheimer's disease, epilepsy, ischemia, and traumatic brain injury. Published work shows that GLUT1 mediates

BBB transport of some neuroactive drugs, such as glycosylated neuropeptides, low molecular weight heparin, and d-glucose derivatives (Guo et al. 2005). Protein expression of the glucose transporter GLUT1 is reduced in brain capillaries in Alzheimer's disease, without changes in GLUT1 mRNA structure (Mooradian et al. 1997) or levels of GLUT1 mRNA transcripts (Wu et al. 2005). Further, a reduction in CNS energy metabolites has been seen in several PET scanning studies of Alzheimer's patients using FDG (Samuraki et al. 2007; Mosconi et al. 2006, 2008), likely because the surface area at the BBB available for glucose transport is substantially reduced in Alzheimer's disease (Bailey et al. 2004; Wu et al. 2005).

Active efflux transporters such as the ATP-binding cassette (ABC) transporters rapidly remove ingested toxic lipophilic metabolites and many structurally unrelated, often amphipathic, cationic drugs from the brain or prevent their entry (Schinkel et al. 1994; Loscher and Potschka 2005; Hermann and Bassetti 2007; Dutheil et al. 2010). Arguably, the most important efflux transporter is P-gp. P-gp is expressed in a polarized fashion, with maximal expression at the luminal plasma membrane of brain endothelial cells; however, P-gp may also be found in the abluminal membrane of brain endothelial cells, as well as in pericytes and astrocytes (Bendayan et al. 2006). Subcellularly, P-gp is distributed along the nuclear envelope, in caveolae, cytoplasmic vesicles, the Golgi complex, and the rough endoplasmic reticulum (Bendayan et al. 2006). The possible role of the ABC transporters in the pathogenesis and treatment of neurodegenerative diseases is increasingly recognized. A positive association between the polymorphism in the MDR1 gene encoding P-gp (/ABCB1) and pharmacoresistant epilepsy has been reported in a subset of epilepsy patients (Siddiqui et al. 2003). However, the follow-up association genetics studies did not support a major role for this polymorphism, as reviewed in several publications (Tate and Sisodiya 2007; Sisodiya and Mefford 2011). Reports also suggest that P-gp-mediated elimination of Aβ from the brain may be impaired in Alzheimer's disease (Hartz et al. 2010).

22.3.8 Cerebral Blood Flow

Reduction of resting CBF or altered responses to brain activation may occur in different CNS regions in Alzheimer's disease, Parkinson's disease, and other CNS diseases (Lo et al. 2003; Iadecola 2010; Drake and Iadecola 2007; Lok et al. 2007). Even modest 20 % reductions in CBF, as seen in the aging brain, are associated with diminished cerebral protein synthesis (Hossmann 1994). Moderate regional reductions in CBF, as seen in chronic neurodegenerative disorders, lead to shifts in intracellular pH and water, and accumulation of glutamate and lactate in brain ISF (Drake and Iadecola 2007), while severe reductions in CBF (>80 %), such as that which occur in ischemic stroke, can lead to electrolyte unbalance and ischemic neuronal death.

Parodi-Rullán et al. (2019) have discussed the effects of Aβ on cerebral microvascular cell function, focusing on its impact on endothelial mitochondria from the perspective of the deposition of amyloid around cerebral vessels (cerebral amyloid angiopathy, or shortly CAA, as present in up to 90% of AD patients), with the Aβ load inducing dysfunctional hemodynamics and a potentially leaky blood-brain barrier (BBB), contributing to clearance failure and further accumulation of Aβ in the cerebrovasculature and brain parenchyma.

22.3.9   Immunological Aspects

Chronic brain inflammation is also associated with neurodegenerative processes, promoting disease pathogenesis. The BBB is indicated to have a key role in the generation and maintenance of chronic inflammation. Zenaro et al. (2017) described the interaction of the BBB with glial cells, neurons, and pericytes, together forming the NVU, and all the NVU components may undergo functional changes that contribute to neuronal injury. Based on transgenic animal studies, circulating leukocytes migrate through activated brain cerebral endothelial cells when certain adhesion molecules are expressed. These leukocytes penetrate into the brain parenchyma, interacting with other NVU components and may thereby affect their structural integrity and functionality. Adhesion molecules are of importance in this process, as blocking them results in inhibition of both Aβ deposition and tau hyperphosphorylation, while also reducing memory loss in Alzheimer's disease animal models (Zenaro et al. 2017).

Although the brain was once considered to be an immune privileged site, today it is appreciated that (a) the brain is not isolated from the immune system, (b) complex immune responses do occur within the CNS, and (c) brain microglia provide important CNS immune surveillance along with macrophages and monocytes derived from the blood and bone marrow (Prinz et al. 2011). Mononuclear phagocytes from blood are also recruited to cross the BBB and enter the CNS in multiple sclerosis and other neurodegenerative diseases like Alzheimer's disease. Chemokines in the brain can recruit immune cells from the blood or from within the brain (Britschgi and Wyss-Coray 2007) to secrete MMP-2 and MMP-9 that increase BBB permeability (Feng et al. 2011). Inhibition of this process is linked to more rapid disease progression (Dimitrijevic et al. 2007). In a recent review (Engelhardt et al. 2017), the immune privilege of the CNS was readdressed. It was reported that endothelial, epithelial, and glial brain barriers establish compartments in the CNS that differ strikingly with regard to their accessibility to immune cell subsets. Also, there is a unique system of lymphatic drainage from the CNS to the peripheral lymph nodes. It was emphasized that understanding immune privilege of the CNS requires intimate knowledge of its unique anatomy.

It can be concluded that changes in the BBB and its surrounding cells (i.e., the neurovascular unit), degeneration of brain capillaries and loss of pericytes, and reductions in resting CBF all may contribute to progression of neurodegenerative processes.

22.4 Quantitative Studies on BBB Transport and Effects of Drugs in Neurodegenerative Diseases

Many investigations have dealt with processes involved in neurodegenerative conditions, of which only a small portion have been described above. For drug treatment of such diseases, a proper CNS effect can only result from having the drug in the CNS "at the right place, at the right time, and at the right concentration." To that end, it is of importance to take into consideration the many different factors that play a role in producing CNS effects, for example, drug properties, drug concentrations in plasma, multiple BBB transport mechanisms, drug concentrations in brain and intra-brain drug distribution, target interactions, and signal transduction processes. Information on unbound drug concentrations are by far the most valuable as these provide specific information on BBB drug transport and intra-brain distribution, for which only total concentrations may be misleading, while also being the driving force in eliciting CNS drug effects. In relation to neurodegenerative diseases or neurodegenerative disease-related processes, only a few quantitative pharmacological studies on small molecules have been performed on unbound drug concentrations in brain, without/with inclusion of disease conditions and/or concomitant measures of the effects.

22.4.1 Pharmacoresistant Epilepsy

Most of the research on unbound plasma and brain pharmacokinetics in neurodegenerative diseases has been performed with regard to anti-epileptic drugs. As discussed above, pharmacoresistance in epilepsy is thought to be caused by restricted BBB transport and/or unfavorable brain distribution. It is therefore important to learn about BBB transport mechanisms of anti-epileptic drugs. Luer (1999) studied whether the fraction of gabapentin crossing the BBB is linear over a broad range of doses, using the microdialysis technique for measuring gabapentin concentrations in the brain hippocampal ECF, combined with plasma sampling. Although higher AUC brain ECF values were obtained with higher AUC plasma values, changes in AUC brain ECF were less than proportional to observed changes in AUC plasma. It seemed that BBB transport of gabapentin was saturable. Christensen et al. (2001) investigated plasma and cerebrospinal fluid (CSF) samples from epileptic adults on topiramate and lamotrigine. CSF/plasma ratios of topiramate were around 0.85, based on total concentrations of topiramate in plasma and CSF (with protein binding fractions of 84 % in plasma and 97 % in CSF). Lamotrigine concentrations were also measured, for which free concentrations in CSF were about 50 % of those in plasma. The authors' clear conclusion, based on their findings with topiramate, was that unbound plasma concentrations were most relevant for therapeutic drug monitoring. The effect of brain cerebral cortex ECF–parenchymal (intracellular) exchange has been clearly demonstrated by a study of valproate in rabbits by Scism et al. (2000). It was shown that the unfavorable brain-to-plasma gradient was the result of coupled efflux transport processes at both the parenchymal cells and the BBB. BBB transport and brain distribution of valproic acid were investigated in the absence and presence of probenecid using microdialysis and total tissue sampling during steady-state iv infusion of valproic acid. In control conditions, the intracellular brain concentration (ICC) was about 2.8 times higher than the corresponding ECF concentrations. Co-infusion of probenecid elevated the ratio of ICC over ECF concentrations to 4.2 (Table 22.1). This indicated the presence of a probenecid-sensitive efflux transporter on brain parenchymal cell membranes. The ECF to unbound plasma concentration ratio was about 0.2 and was not significantly influenced by probenecid. This study's findings therefore suggested the presence of distinctly different organic anion transporters for the efflux of valproic acid at the parenchymal cells and capillary endothelium.

Potschka et al. (2001) used in vivo microdialysis in rats to study whether the concentration of carbamazepine in brain ECF could be enhanced through P-gp inhibition by verapamil or MRP inhibition by probenecid. Local perfusion of verapamil or probenecid via the microdialysis probe increased the microdialysate concentration of carbamazepine, and the authors concluded that both P-gp (verapamil) and MRP (probenecid) participate in the regulation of brain ECF concentrations of carbamazepine. A similar study was performed for phenytoin (Potschka and Löscher

Table 22.1 Results of microdialysis studies in rabbits

Valproate	Control ($n = 5$)	+Probenecid ($n = 5$)
ECF (ug/ml)	1.72 ± 0.16	2.78 ± 0.36
ICC (ug/ml)	4.69 ± 0.27	11.6 ± 1.62
Brain: total plasma	0.069 ± 0.002	0.16 ± 0.024
Brain: free plasma	0.41 ± 0.052	0.70 ± 0.087
Brain: ECF	2.48 ± 0.23	3.66 ± 0.36
ECF: total plasma	0.029 ± 0.003	0.044 ± 0.005
ECF: free plasma	0.17 ± 0.034	0.19 ± 0.015
ICC: ECF	2.81 ± 0.28	4.24 ± 0.44
ICC: free plasma	0.46 ± 0.068	0.81 ± 0.10

Valproic acid concentrations were determined in brain extracellular fluid (ECF) of the cerebral cortex during steady-state iv infusion with valproic acid alone or with valproic acid plus probenecid. Probenecid co-infusion elevated VPA concentration in the brain tissue surrounding the tip of the microdialysis probe to a greater extent than in the ECF (230 % vs 47 %). Brain intracellular compartment (ICC) concentration was estimated. In control rabbits, the ICC concentration was 2.8 ± 0.28 times higher than the ECF concentration. Probenecid co-infusion elevated the ICC-to- ECF concentration ratio to 4.2 ± 0.44, which confirms the existence of an efflux transport system in brain parenchymal cells. The ECF-to-unbound plasma concentration ratio was well below unity (0.029), indicating an uphill efflux transport of VPA across the BBB. Co-infusion of probenecid did not have a significant effect on valproic acid efflux at the BBB as evidenced by a minimal change in the ECF-to-unbound plasma concentration ratio. This study suggests the presence of distinctly different organic anion transporters for the efflux of valproic acid at the parenchymal cells and capillary endothelium in the brain (Scism et al. 2000)

2001a, b) in which local perfusion of probenecid via the microdialysis probe significantly enhanced the microdialysate concentrations of phenytoin. The same group
later studied the influence of P-gp inhibition by verapamil on BBB transport of
phenobarbital, lamotrigine, and felbamate again using in vivo microdialysis
(Potschka et al. (2002). Verapamil was found to increase the concentration of all
three antiepileptic drugs in the cortical brain ECF. Importantly, these studies indicated that overexpression of P-gp and/or MRP in epileptic tissue might limit brain
access of many antiepileptic drugs.

For levetiracetam, a new anti-epileptic drug, the expectations were quite high as
it seemed to be an effective and well-tolerated drug in many patients with otherwise
pharmacoresistant epilepsy. Potschka et al. (2004) therefore investigated whether
the concentration of levetiracetam in the cortical brain ECF could be modulated by
inhibition of P-gp or MRPs, using the P-gp inhibitor verapamil and the MRP1/2
inhibitor probenecid. Local perfusion with verapamil or probenecid via the microdialysis probe did not increase the brain ECF concentration of levetiracetam, providing strong evidence that brain uptake of levetiracetam is not affected by P-gp or
MRP1/2. This could explain levetiracetam's antiepileptic efficacy in patients whose
seizures are poorly controlled by other anti-epileptic drugs. While the above microdialysis studies elegantly suggest the importance of transporters in determining
drug efficacy and which drugs may have potential in treating pharmacoresistant
epilepsy by virtue of their not being substrates of such transporters, some caution is
nevertheless warranted. It remains theoretically possible that other nonspecific
changes, for example, changing osmolarity, may contribute to observed results
when a drug (inhibitor) is locally perfused into the brain using in vivo microdialysis.
More work is clearly necessary in this area.

As already discussed above, by using local perfusion of the MRP inhibitor probenecid via a microdialysis probe, Potschka et al. (2003a, b) have shown an increase
in brain microdialysate levels of phenytoin in rats (reflecting, but not necessarily
equivalent to, the unbound brain concentration of phenytoin). This seems to indicate
that phenytoin is a substrate of MRP2 at the BBB. This conclusion was also supported by studies in MRP2-deficient TR-rats, in which brain microdialysate levels
of phenytoin were significantly higher than in normal background strain rats. Then,
in the kindling model of epilepsy, coadministration of probenecid significantly
increased the anticonvulsant activity of phenytoin, while in kindled MRP2-deficient
rats phenytoin exerted a markedly higher anticonvulsant activity than in normal rats.
Altogether this supports the hypothesis that MRP2 may contribute to BBB function.

The relation between brain ECF concentrations following systemic administration of oxcarbazepine and its effects on local ECF levels of dopamine and serotonin
was investigated by Clinckers et al. (2005a), including modulation of oxcarbazepine
BBB transport. The intrahippocampal perfusion of verapamil, a P-gp inhibitor, and
probenecid, a MRP inhibitor, on the BBB passage of oxcarbazepine was investigated. Simultaneously, the effects on hippocampal monoamines were studied as
pharmacodynamic markers for oxcarbazepine anticonvulsant activity in the focal
pilocarpine model for limbic seizures. Systemic oxcarbazepine administration alone
did not prevent the rats from developing seizures; however, coadministration of

verapamil or probenecid with oxcarbazepine yielded complete protection along with significant increases in hippocampal ECF levels of dopamine and serotonin. These findings indicate that oxcarbazepine is a substrate for multidrug transporters at the BBB and that coadministration of multidrug transporter inhibitors significantly potentiates oxcarbazepine anticonvulsant activity, highlighting the impact of BBB transport for the CNS effects of this antiepileptic drug.

Clinckers et al. (2005b) conducted an in vivo microdialysis study to investigate the impact of the transport kinetics of oxcarbazepine across the BBB on the observed treatment refractoriness. Also, the influence of intrahippocampal perfusion of verapamil, a P-gp inhibitor, and probenecid, a MPR inhibitor, on the BBB transport and anticonvulsant properties of oxcarbazepine was investigated, using the focal pilocarpine model for limbic seizures. Simultaneously, the effects on hippocampal monoamines were studied as pharmacodynamic markers for the anticonvulsant activity. Although systemic oxcarbazepine administration alone failed in preventing the animals from developing seizures, coadministration with verapamil or probenecid offered complete protection. Concomitantly, significant increases in extracellular hippocampal dopamine and serotonin levels were observed within our previously defined anticonvulsant monoamine range. The present data indicate that oxcarbazepine is a substrate for multidrug transporters at the blood–brain barrier. Coadministration with multidrug transporter inhibitors significantly potentiates the anticonvulsant activity of oxcarbazepine and offers opportunities for treatment of pharmacoresistant epilepsy.

Clinckers et al. (2008) examined unbound concentrations of 10,11-dihydro-10-hydroxy-carbamazepine (MHD) in plasma and in the hippocampus to study the impact of acute seizures and efflux transport mechanisms on MHD brain distribution. An integrated pharmacokinetic (PK) model describing simultaneously the PK of MHD in plasma and brain was developed. A compartmental model with combined zero- and first- order absorption, including lag time and target site distribution best described the PK of MHD. A distribution process appeared to underlie the increased brain MHD concentrations observed following seizure activity and efflux transport inhibition, as reflected by changes in the volume of distribution of the target site compartment. In contrast, no changes were observed in plasma PK.

Feng et al. (2001) have studied the BBB influx and efflux of pregabalin with microdialysis in conjunction with its anticonvulsant effects. BBB influx (CLin) and efflux (CLout) permeability for pregabalin were ~5 and 37 µL/min/g brain, respectively, following intravenous infusion. The results indicate that pregabalin can enter the brain. Interestingly, a significant delay in anticonvulsant action of pregabalin was found relative to the estimated brainECF drug concentrations. Using a PKPD link model, the counter-clockwise delay in the relationship between pregabalin brainECF concentration and the anticonvulsant effect showed that the concentration in the hypothetical effect compartment (Ce) versus effect (PD) profile exhibits a sigmoidal curve and the calculated EC50 and Keo values were 95 ng/ml and 0.0092 min^{-1}, respectively. The small value for the Keo indicates that the effect is not directly proportional to the amount of pregabalin in the brainECF compartment, possibly due to inherent delay at steps other than BBB transport.

It has also been investigated whether ABCC2 (/MRP2) is functionally involved in transport of carbamazepine, lamotrigine and felbamate across the BBB. The distribution of these drugs into the brain was determined using ABCC2-deficient TR-rats. The microdialysis results gave no evidence that ABCC2 function modulates entry of carbamazepine, lamotrigine, or felbamate into the CNS. However, ABCC2 deficiency was associated with an increased anticonvulsant response of carbamazepine in the amygdala-kindling model of epilepsy (Potschka et al. 2003a, b).

To study potential changes in brain P-gp functionality after induction of status epilepticus (SE), Syvänen et al. (2012) used a quinidine microdialysis assay in kainate-treated rats to reveal differences in brain distribution upon changes in P-gp functionality by preadministration of tariquidar, a P-gp inhibitor. In control animals, total brain quinidine concentration increased ~40-fold while quinidine ECF concentration increased ~sevenfold following tariquidar pretreatment. After kainate treatment alone, however, no difference in quinidine transport across the BBB was found compared to saline-treated (control) animals, but kainate-treated rats tended to have a lower total brain concentration but a higher brain ECF concentration of quinidine than control rats. This could be concluded using a newly developed mathematical population pharmacokinetic model that includes statistical approaches to identify sources of variability in quinidine kinetics within the whole dataset. Notably, this study did not provide evidence for the hypothesis that P-gp function at the BBB is altered at 1 week after status epilepticus induction, but rather suggests that P-gp function might be altered at the brain parenchymal level.

In many in vivo studies, P-gp expression has been determined and taken as a biomarker of P-gp functionality at the BBB in vivo. De Lange et al. (2018) performed in vivo rat studies after kainite-induced status epilepticus SE). Post-SE microdialysis experiments were performed at different days to assess BBB specific P-gp functionality using the P-gp substrate quinidine, with or without P-gp inhibition by tariquidar, while BBB specific P-gp expression was assessed ex-vivo using immunohistochemistry by harvesting the brain immediately after the microdialysis experiment. This allowed for direct comparison of P-gp expression and P-gp functionality at the BBB. It was found that changes in BBB P-gp expression are temporary and that BBB P-gp expression does not reliably indicate BBB P-gp functionality. This warrants the general use of P-gp expression as a biomarker for P-gp functionality.

22.4.2 Parkinson's Disease

To investigate potential changes in BBB transport of L-DOPA in conjunction with its intra-brain conversion in Parkinson's disease, Ravenstijn et al. (2012) used the unilateral rat rotenone model of Parkinson's disease (Ravenstijn et al. 2008). Microdialysis measurements were performed simultaneously in the control (untreated) and in the rotenone-treated cerebral hemisphere while L-DOPA was administered intravenously (10, 25 or 50 mg/kg). Serial blood samples and brain striatal microdialysates were analyzed for L-DOPA and dopamine metabolites (DOPAC and HVA). Ex-vivo brain tissue was analyzed for changes in tyrosine hydroxylase staining as a biomarker for disease model severity. An advanced mathematical model (Fig. 22.1) was developed to evaluate BBB transport of L-DOPA along with the conversion of L-DOPA into DOPAC and HVA, and the results were compared between the control and rotenone-treated diseased cerebral hemispheres. As previously found for fluorescein (Ravenstijn et al. 2008), no difference in L-DOPA BBB transport was found in the rotenone-treated diseased hemisphere as compared to the untreated hemisphere. However, basal microdialysate levels of DOPAC and HVA were substantially lower in the rotenone-treated diseased hemisphere. Upon L-DOPA administration these elimination rates were higher at the rotenone-treated hemisphere. The higher elimination rate constant as found for DOPAC and HVA would be possible if dopamine concentrations were lower in the rotenone-treated diseased hemisphere such that metabolite formation rate-dependent elimination occurs. This is also called "flip-flop kinetics," i.e. [metabolite formation rate constant × amount of metabolite remaining to be formed] is about equal to the [metabolite elimination rate constant × amount of metabolite remaining to be eliminated]). Reduced dopamine concentrations in the rotenone-treated diseased hemisphere are indeed plausible with a diminished amount of dopaminergic neurons as indicated by substantially decreased TH staining. These studies show that it is

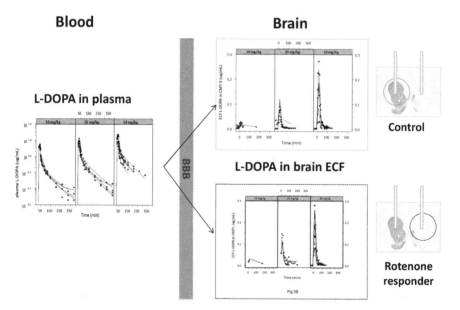

Fig. 22.1 Pharmacokinetic profiles of L-DOPA obtained after a 20-min intravenous infusion in Lewis rats in plasma (left) and in brain striatal ECF in the control cerebral hemisphere (right, upper part) and in the rotenone-treated responder cerebral hemisphere (right, lower part) for 3 doses (10, 25 and 50 mg/kg). Depicted are the observed concentrations (dots) and individual model predictions (solid lines), separated by L-DOPA dose (Ravenstijn et al. 2012)

necessary to consider both L-DOPA BBB transport and its intra-brain conversion in order to appreciate whether specific changes in BBB transport may significantly influence L-DOPA PKPD in rodent models of Parkinson's disease.

22.4.3 Alzheimers Disease

Gustafsson et al. (2018) investigated BBB integrity to large molecules in transgenic mice expressing the human Aβ precursor with the Arctic and Swedish mutations (tg-ArcSwe), and age matched wild type animals, with or without acute treatment with the murine version of the clinically investigated Aβ antibody bapineuzumab, supplemented with [^{125}I]3D6. As large molecules, 4 kDa FITC and a 150 kDa Antonia Red dextran were used to determine the dextran brain-to-blood concentration ratios as a biomarker of BBB passage. These ratios were equally low in wild type and transgenic mice (confirmed to display cerebral amyloid pathology), suggesting an intact BBB despite Aβ pathology, while also the 3D6 antibody activity in brain was not changed (Gustafsson et al. 2018). This is in line with the findings on lack of changes in Immunoglobulin G permeability in most animal models of Alzheimer's disease (Nga Bien-Ly et al. 2015) These quantitative findings are interesting and important and warrant further *quantitative* testing in other models of Alzheimer's'disease.

22.4.4 Traumatic Brain Injury

With the aim to increase knowledge of factors controlling the PK of unbound drug in the brain and related drug effect(s), Ketharanathan et al. (2019) studied plasma and brain ECF morphine concentrations in individual children with severe TBI treated with morphine. Brain ECF samples were obtained by microdialysis. Brain ECF samples were taken from "injured" and "uninjured" regions as determined by microdialysis catheter location on computed head tomography. The results were compared to predictions of the University Leiden CNS physiologically-based PK model (Yamamoto et al. 2018), *t*hat was adapted to children using pediatric physiological properties (Ketharanathan et al. 2019). The predicted brain ECF concentration-time profiles fell within a 90% prediction interval of microdialysis brain ECF drug concentrations when sampled from an uninjured area (Fig. 22.2). The "healthy" brain ECF prediction was less accurate in injured areas, which indicates the impact of CNS disease conditions on brain ECF PK in children, as also shown for adult TBI patients (Bouw et al. 2001; Yamamoto et al. 2018):

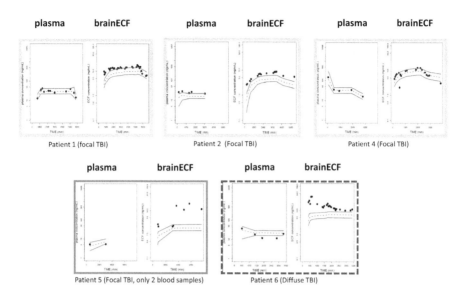

Fig. 22.2 Plasma and brain ECF morphine concentrations in individual children with severe TBI treated with morphine (black dots). Brain ECF samples were obtained by microdialysis. Brain ECF samples were taken from "injured" (orange/black) and "uninjured"(green) regions as determined by the microdialysis catheter location on computed head tomography. The results were compared to predictions of the University Leiden "healthy" physiologically based CNS PK model predictions of brain ECF under are indicated by average PK profile (dotted line) and boundaries for 90% inclusion of data (red lines) (Ketharanathan et al. 2019)

22.4.5 Role of Pericytes

As pericytes are perivascular cells that play important roles in the regulation of the BBB, it was of interest to study BBB transport of drugs in pericyte-deficient conditions. Such was done in pericyte-deficient Pdgfbret/ret mice and wild-type controls, for diazepam, digoxin, levofloxacin, oxycodone, and paliperidone in a cassette dose fashion. Total drug concentrations in brain and plasma (Kp) were determined. Equilibrium dialysis experiments were performed to estimate unbound drug fractions in brain (fu,brain) and plasma (fu,plasma), and thereby, the values of the unbound concentration ratio (Kp,uu,brain). No differences in BBB transport was found for pericyte-deficient conditions, which suggests preserved BBB features relevant for handling of these type of molecules, irrespective of pericyte presence at the brain endothelium (Mihajlica et al. 2018).

22.5 Conclusions

Most studies investigate the brain processes underlying neurodegeneration when examining neurodegenerative diseases in human beings or animal models. However, it has become clear that the BBB/neurovascular unit may play an important role in neurodegenerative diseases by a number of different mechanisms, for example, the exacerbation of neuroinflammation with increased entry of blood-borne immune cells into the diseased brain across the BBB.

For effective delivery of drugs to the target site in the brain, it is imperative to have a quantitative understanding of the influence of disease conditions on BBB transport of drugs. However, the number of studies devoted to *quantitative* measurement of BBB transport and intra-brain distribution of drugs in the context of neurodegeneration appears to be surprisingly low. The few studies that have examined BBB transport and intra-brain distribution simultaneously (with the ability to separate the two factors) have been presented. Interestingly, these studies indicate no changes in BBB transport in pharmacoresistant epilepsy and Alzheimer's disease animal conditions, while, conversely, changes were observed for TBI studies in human beings.

Overall, it can be said that conducting research on potential treatment strategies for neurodegenerative diseases is not an easy task. On the one hand, there are many processes involved in neurodegenerative diseases with associated complexity. On the other hand, it is both necessary and worthwhile to identify new ways to work around the relative inaccessibility of the human brain to commonly used research tools. This means that we must largely still rely on animal models of neurodegenerative diseases, as use of such models is, arguably, the only current method that allows for a truly integrative research approach. However, this situation may change in the future with the development of new technologies, particularly better imaging and biomarker-based inference of target engagement in human beings. Indeed, more research is urgently needed to elaborate better ways to translate findings to the clinical situation.

22.5.1 Points for Discussion

- How can studies be best designed to have the most valuable data collected?
- Why don't more studies aim to obtain quantitative and connected data?
- What biomarkers can be assessed in humans and in animals?
- What drug concentrations may be assessed in biological compartments, and, of these, which is best to predict CNS target site concentrations?
- How do animal models of disease provide valuable insights into human neurodegenerative processes and what are their limitations?
- Would the timescale of disease progression in animal models be different from the human conditions?

- Given that neurodegeneration is highly heterogeneous, how can we address sources of variability between drug response in human populations to aim at personalized CNS medicine?
- What are the advantages of a "multitarget" drug treatment (systems or network approach) compared to the more traditional "single target" drug treatment, given that neurodegenerative diseases are multifactorial?

22.5.2 Future Directions

Data on neurodegeneration so far has provided us with bits and pieces of information with valuable insights to a certain level, but much of this data is disconnected in the sense that different biological systems have been used, with disease conditions being induced by variable means. It is suggested that future studies should be designed to have a more structured and integrative nature, allowing us to learn about the interplay between processes and their sensitivity to the context in which they have been measured. The importance and challenge of performing integrative studies has already been addressed and also apply here. The following goals may facilitate future progress (De Lange et al. 2017):

- To convert qualitative data (pictures, photographs, "increase" of parameter x, "decrease" of parameter y) into quantitative data.
- To include genomics, proteomics, and metabolomics data along with traditional measures of plasma and brain fluids.
- To increase the use of neuroimaging.
- To further include the outcomes of epidemiological studies on polymorphisms.
- To search for and include biomarkers of (early) disease processes and CNS drug effects.
- To include measurements of unbound drug concentrations as it is the unbound drug concentration that drives transport processes (BBB transport, intra-brain distribution, unbound brain concentrations) and target interactions that lead to drug effects.
- To include time-dependencies for drug kinetics, drug effects, and disease stage (progression).
- To obtain information on multiple parameters in parallel in a particular context (i.e., as much as possible; to obtain "connected data" (Paweletz et al. 2010) and vary the context in a systematic manner to learn about parameter sensitivity for the context (e.g., specific inhibition of processes). For example, it has been found that even in the same strain from two breeding locations, there are differences in seizure susceptibility, pharmacological response, and basal neurochemistry (Portelli et al. 2009).
- To obtain such "connected data" in animals using both more- and less-invasive methods as well as noninvasive (imaging) techniques, the latter should also be applied in human studies (Greenhalgh et al. 2011).

- To include the use of advanced mathematical modeling to integrate all data, and by statistical approaches obtain insight into sources of variability (covariate), as this will improve interspecies extrapolation of pharmacokinetics to investigate the use of multiple drugs (a multitarget approach).

References

Abbott NJ, Khan EU, Rollinson CM, Reichel A, Janigro D, Dombrowski SM, Dobbie MS, Begley DJ (2002) Drug resistance in epilepsy: the role of the blood–brain barrier. Novartis Found Symp 243:38–47. discussion 47–53, 180–185

Abuznait AH, Cain C, Ingram D, Burk D, Kaddoumi A (2011) Up-regulation of P-glycoprotein reduces intracellular accumulation of beta amyloid: investigation of P-glycoprotein as a novel therapeutic target for Alzheimer's disease. J Pharm Pharmacol 63(8):1111–1118

Ahmed RM, Ke Y, Vucic S, Ittner LM, Seeley W, Hodges JR, Piguet O, Halliday G, Kiernan MC (2018) Physiological changes in neurodegeneration - mechanistic insights and clinical utility. Nat Rev Neurol 14:259–271

Aijaz S, Balda MS, Matter K (2006) Tight junctions: molecular architecture and function. Int Rev Cytol 248:261–298

Al Ahmad A, Gassmann M, Ogunshola OO (2012) Involvement of oxidative stress in hypoxia-induced blood–brain barrier breakdown. Microvasc Res 84(2):222–225

Alexander GM, Godwin DW (2006) Metabotropic glutamate receptors as a strategic target for the treatment of epilepsy. Epilepsy Res 71:1–22

Allt G, Lawrenson JG (2001) Pericytes: cell biology and pathology. Cells Tissues Organs 169:1–11

Amenta PS, Jallo JI, Tuma RF, Elliott MB (2012) A cannabinoid type 2 receptor agonist attenuates blood–brain barrier damage and neurodegeneration in a murine model of traumatic brain injury. J Neurosci Res 90(12):2293–2305

Anderson GD, Shen DD (2007) Where is the evidence that p-glycoprotein limits brain uptake of antiepileptic drug and contributes to drug resistance in epilepsy? Epilepsia 48(12):2372–2374

Armulik A, Genové G, Betsholtz C (2011a) Pericytes: developmental, physiological, and pathological perspectives, problems, and promises. Dev Cell 21(2):193–215

Armulik A, Mäe M, Betsholtz C (2011b) Pericytes and the blood–brain barrier: recent advances and implications for the delivery of CNS therapy. Ther Deliv 2(4):419–422

Arnold S (2012) The power of life–cytochrome c oxidase takes center stage in metabolic control, cell signalling and survival. Mitochondrion 12(1):46–56

Artal-Sanz M, Tavernarakis N (2005) Proteolytic mechanisms in necrotic cell death and neurodegeneration. FEBS Lett 579(15):3287–3296

Ates N, Esen N, Ilbay G (1999) Absence epilepsy and regional blood–brain barrier permeability: the effects of pentylenetetrazole-induced convulsions. Pharmacol Res 39(4):305–310

Baeten KM, Akassoglou K (2011) Extracellular matrix and matrix receptors in blood–brain barrier formation and stroke. Periodicals, inc. Dev Neurobiol 71:1018–1039

Bagwe-Parab S, Kaur G (2019) Molecular targets and therapeutic interventions for iron induced neurodegeneration. Brain Res Bull 156:1–9

Bailey TL, Rivara CB, Rocher AB, Hof PR (2004) The nature and effects of cortical microvascular pathology in aging and Alzheimer's disease. Neurol Res 26:573–578

Balabanov R, Dore-Duffy P (1998) Role of the CNS microvascular pericyte in the blood–brain barrier. J Neurosci Res 53(6):637–644

Baltes S, Gastens AM, Fedrowitz M, Potschka H, Kaever V, Löscher W (2007b) Differences in the transport of the antiepileptic drugs phenytoin, levetiracetam and carbamazepine by human and mouse P-glycoprotein. Neuropharmacology 52(2):333–346

Banerjee S, Bhat MA (2007) Neuron-glial interactions in blood–brain barrier formation. Annu Rev Neurosci 30:235–258

Bankstahl JP, Hoffmann K, Bethmann K, Löscher W (2008) Glutamate is critically involved in seizure-induced overexpression of P-glycoprotein in the brain. Neuropharmacology 54(6):1006–1016

Bankstahl JP, Bankstahl M, Kuntner C, Stanek J, Wanek T, Meier M, Ding XQ, Müller M, Langer O, Löscher W (2011) A novel positron emission tomography imaging protocol identifies seizure-induced regional overactivity of P-glycoprotein at the blood–brain barrier. J Neurosci 31(24):8803–8811

Baranczyk-Kuzma A, Audus KL, Borchardt RT (1989) Catecholamine-metabolizing enzymes of bovine brain microvessel endothelial cell monolayers. J Neurochem 46(6):1956–1960

Bell RD, Zlokovic BV (2009) Neurovascular mechanisms and blood–brain barrier disorder in Alzheimer's disease. Acta Neuropathol 118(1):103–113

Bell RD, Winkler EA, Sagare AP, Singh I, LaRue B, Deane R, Zlokovic BV (2010) Pericytes control key neurovascular functions and neuronal phenotype in the adult brain and during brain aging. Neuron 68(3):409–427

Bell RD, Winkler EA, Singh I, Sagare AP, Deane R, Wu Z, Holtzman DM, Betsholtz C, Armulik A, Sallstrom J, Berk BC, Zlokovic BV (2012) Apolipoprotein E controls cerebrovascular integrity via cyclophilin A. Nature 485(7399):512–516

Bendayan R, Ronaldson PT, Gingras D, Bendayan M (2006) In situ localization of P-glycoprotein (ABCB1) in human and rat brain. J Histochem Cytochem 54:1159–1167

Berg D, Youdim MB (2006) Role of iron in neurodegenerative disorders. Top Magn Reson Imaging 17(1):5–17

Bertram L, Tanzi RE (2005) The genetic epidemiology of neurodegenerative disease. J Clin Invest 115(6):1449–1457

Bertram L, McQueen MB, Mullin K, Blacker D, Tanzi RE (2007) Systematic meta- analyses of Alzheimer disease genetic association studies: the AlzGene database. Nat Genet 39(1):17–23

Bethmann K, Fritschy JM, Brandt C, Löscher W (2008) Antiepileptic drug resistant rats differ from drug responsive rats in GABA A receptor subunit expression in a model of temporal lobe epilepsy. Neurobiol Dis 31(2):169–187

Betti M, Minelli A, Ambrogini P, Ciuffoli S, Viola V, Galli F, Canonico B, Lattanzi D, Colombo E, Sestili P, Cuppini R (2011) Dietary supplementation with α-tocopherol reduces neuroinflammation and neuronal degeneration in the rat brain after kainic acid-induced status epilepticus. Free Radic Res 45(10):1136–1142

Bittigau P, Sifringer M, Genz K, Reith E, Pospischil D, Govindarajalu S, Dzietko M, Pesditschek S, Mai I, Dikranian K, Olney JW, Ikonomidou C (2002) Antiepileptic drugs and apoptotic neurodegeneration in the developing brain. Proc Natl Acad Sci U S A 99(23):15089–15094

Blanchard JW, Bula M, Davila-Velderrain J, Akay LA, Zhu L, Frank A, Victor MB, Bonner JM, Mathys H, Lin Y-T, Ko T, Bennett DA, Cam HP, Kellis M, Tsai L-H (2020) Reconstruction of the human blood–brain barrier in vitro reveals a pathogenic mechanism of APOE4 in pericytes. Nat Med 26:952–963

Bouw MR, Ederoth P, Lundberg J, Ungerstedt U, Nordstrom CH, Hammarlund-Udenaes M (2001) Increased blood-brain barrier permeability of morphine in a patient with severe brain lesions as determined by microdialysis. Acta Anest Scand 45:390–392

Bové J, Perier C (2012) Neurotoxin-based models of Parkinson's disease. Neuroscience 211:51–76

Brandt C, Bethmann K, Gastens AM, Löscher W (2006) The multidrug transporter hypothesis of drug resistance in epilepsy: proof-of-principle in a rat model of temporal lobe epilepsy. Neurobiol Dis 24(1):202–211

Brenn A, Grube M, Peters M, Fischer A, Jedlitschky G, Kroemer HK, Warzok RW, Vogelgesang S (2011) Beta-amyloid downregulates MDR1-P-glycoprotein (Abcb1) expression at the blood–brain barrier in mice. Int J Alzheimers Dis 2011:690121

Britschgi M, Wyss-Coray T (2007) Immune cells may fend off Alzheimer disease. Nat Med 13:408–409

Brownless J, Williams CH (1993) Peptidases, peptides and the mammalian blood–brain barrier. J Neurochem 60:1089–1096

Bulbarelli A, Lonati E, Brambilla A, Orlando A, Cazzaniga E, Piazza F, Ferrarese C, Masserini M, Sancini G (2012) Aβ42 production in brain capillary endothelial cells after oxygen and glucose deprivation. Mol Cell Neurosci 49(4):415–422

Cheng T, Petraglia AL, Li Z, Thiyagarajan M, Zhong Z, Wu Z, Liu D, Maggirwar SB, Deane R, Fernandez JA, LaRue B, Griffin JH, Chopp M, Zlokovic BV (2006) Activated protein C inhibits tissue plasminogen activator-induced brain hemorrhage. Nat Med 12:1278–1285

Chodobski A, Zink BJ, Szmydynger-Chodobska J (2011) Blood–brain barrier pathophysiology in traumatic brain injury. Transl Stroke Res 2(4):492–516

Cholerton B, Baker LD, Craft S (2011) Insulin resistance and pathological brain ageing. Diabet Med 28(12):1463–1475

Christensen J (2012) Traumatic brain injury: Risks of epilepsy and implications for medicolegal assessment. Epilepsia 53(Suppl 4):43–47

Christensen J, Højskov CS, Dam M, Poulsen JH (2001) Plasma concentration of topiramate correlates with cerebrospinal fluid concentration. Ther Drug Monit 23(5):529–535

Chun JT, Wang L, Pasinetti GM, Finch CE, Zlokovic BV (1999) Glycoprotein 330/megalin (LRP-2) has low prevalence as mRNA and protein in brain microvessels and choroid plexus. Exp Neurol 157(1):194–201

Clinckers R, Smolders I, Meurs A, Ebinger G, Michotte Y (2005a) Quantitative in vivo microdialysis study on the influence of multidrug transporters on the blood–brain barrier passage of oxcarbazepine: concomitant use of hippocampal monoamines as pharmacodynamic markers for the anticonvulsant activity. J Pharmacol Exp Ther 314(2):725–731

Clinckers R, Smolders I, Meurs A, Ebinger G, Michotte Y (2005b) Hippocampal dopamine and serotonin elevations as pharmacodynamic markers for the anticonvulsant efficacy of oxcarbazepine and 10,11-dihydro-10-hydroxycarbamazepine. Neurosci Lett 390(1):48–53

Clinckers R, Smolders I, Michotte Y, Ebinger G, Danhof M, Voskuyl RA, Della PO (2008) Impact of efflux transporters and of seizures on the pharmacokinetics of oxcarbazepine metabolite in the rat brain. Br J Pharmacol 155(7):1127–1138

Cohen-Kashi MK, Cooper I, Teichberg VI (2009) Closing the gap between the in-vivo and in-vitro blood–brain barrier tightness. Brain Res 1284:12–21

Cope EC, Morris DR, Levenson CW (2012) Improving treatments and outcomes: an emerging role for zinc in traumatic brain injury. Nutr Rev 70(7):410–413

Coppedè F, Mancuso M, Siciliano G, Migliore L, Murri L (2006) Genes and the environment in neurodegeneration. Biosci Rep 26(5):341–367

Cowan CM, Raymond LA (2006) Selective neuronal degeneration in Huntington's disease. Curr Top Dev Biol 75:25–71

De Lange EC (2013) The mastermind approach to CNS drug therapy: translational prediction of human brain distribution, target site kinetics, and therapeutic effects. Fluids Barriers CNS 10(1):12

De Lange ECM, van der Brink W, Yamamoto Y, de Witte W, Wong YC (2017) Novel CNS drug discovery and development approach: model-based integration to predict neuro-pharmacokinetics and pharmacodynamics. Expert Opin Drug Discovery 12(12):1207–1218

De Lange ECM, Vd Berg DJ, Bellanti F, Voskuyl RA, Syvänen S (2018) P-glycoprotein protein expression versus functionality at the blood–brain barrier using immunohistochemistry, microdialysis and mathematical modeling. Eur J Pharm Sci 23(124):61–70

De Vos KJ, Grierson AJ, Ackerley S, Miller CJJ (2008) Role of axonal transport in neurodegenerative diseases. Annu Rev Neurosci 31:151–173

Deeken JF, Loscher W (2007) The blood–brain barrier and cancer: transporters, treatment, and Trojan horses. Clin Cancer Res 13:1663–1674

Deleu D, Northway MG, Hanssens Y (2002) Clinical pharmacokinetic and pharmacodynamic properties of drugs used in the treatment of Parkinson's disease. Clin Pharmacokinet 41(4):261–309

Deprez F, Zattoni M, Mura ML, Frei K, Fritschy JM (2011) Adoptive transfer of T lymphocytes in immunodeficient mice influences epileptogenesis and neurodegeneration in a model of temporal lobe epilepsy. Neurobiol Dis 44(2):174–184

Desai BS, Monahan AJ, Carvey PM, Hendey B (2007) Blood–brain barrier pathology in Alzheimer's and Parkinson's disease: implications for drug therapy. Cell Transplant 16(3):285–299

Dimitrijevic OB, Stamatovic SM, Keep RF, Andjelkovic AV (2007) Absence of the chemokine receptor CCR2 protects against cerebral ischemia/reperfusion injury in mice. Stroke 38(4):1345–1353

Dohgu S, Takata F, Matsumoto J, Kimura I, Yamauchi A, Kataoka Y (2019) Monomeric α-synuclein induces blood-brain barrier dysfunction through activated brain pericytes releasing inflammatory mediators in vitro. Microvasc Res 124:61–66

Dombrowski SM, Desai SY, Marroni M, Cucullo L, Goodrich K, Bingaman W, Mayberg MR, Bengez L, Janigro D (2001) Overexpression of multiple drug resistance genes in endothelial cells from patients with refractory epilepsy. Epilepsia 42(12):1501–1506

Drake CT, Iadecola C (2007) The role of neuronal signaling in controlling cerebral blood flow. Brain Lang 102:141–152

Dutheil F, Jacob A, Dauchy S, Beaune P, Scherrmann JM, Declèves X, Loriot MA (2010) ABC transporters and cytochromes P450 in the human central nervous system: influence on brain pharmacokinetics and contribution to neurodegenerative disorders. Expert Opin Drug Metab Toxicol 6(10):1161–1174

Engelhardt B, Vajkoczy P, Weller RO (2017) The movers and shapers in immune privilege of the CNS. Nat Immunol 18(2):123–131

Erickson AC, Couchman JR (2000) Still more complexity in mammalian basement membranes. J Histochem Cytochem 48(10):1291–1306

Erickson MA, Hartvigson PE, Morofuji Y, Owen JB, Butterfield DA, Banks WA (2012) Lipopolysaccharide impairs amyloid beta efflux from brain: altered vascular sequestration, cerebrospinal fluid reabsorption, peripheral clearance and transporter function at the blood–brain barrier. J Neuroinflammation 9(1):150

Fan J, Cowan CM, Zhang LY, Hayden MR, Raymond LA (2009) Interaction of postsynaptic density protein-95 with NMDA receptors influences excitotoxicity in the yeast artificial chromosome mouse model of Huntington's disease. J Neurosci 29(35):10928–10938

Feng MR, Turluck D, Burleigh J, Lister R, Fan C, Middlebrook A, Taylor C, Su T (2001) Brain microdialysis and PK/PD correlation of pregabalin in rats. Eur J Drug Metab Pharmacokinet 26(1–2):123–128

Feng S, Cen J, Huang Y, Shen H, Yao L, Wang Y, Chen Z (2011) Matrix metalloproteinase-2 and −9 secreted by leukemic cells increase the permeability of blood–brain barrier by disrupt- ing tight junction proteins. PLoS One 6(8):e20599

Finder VH (2010) Alzheimer's disease: a general introduction and pathomechanism. J Alzheimers Dis 22(Suppl 3):5–19

Freeman LR, Keller JN (2012) Oxidative stress and cerebral endothelial cells: regulation of the blood–brain-barrier and antioxidant based interventions. Biochim Biophys Acta 1822(5):822–829

Fu H, Hardy J, Duff KE (2018) Selective vulnerability in neurodegenerative diseases. Nat Neurosci 21(10):1350–1358. https://doi.org/10.1038/s41593-018-0221-2

Fukushima H, Fujimoto M, Ide M (1990) Quantitative detection of blood–brain barrier-associated enzymes in cultured endothelial cells of porcine brain microvessels. In Vitro Cell Dev Biol 26(6):612–620

Ghosh C, Puvenna V, Gonzalez-Martinez J, Janigro D, Marchi N (2011) Blood–brain barrier P450 enzymes and multidrug transporters in drug resistance: a synergistic role in neurological diseases. Curr Drug Metab 12(8):742–749

Girouard H, Bonev AD, Hannah RM, Meredith A, Aldrich RW, Nelson MT (2010) Astrocytic endfoot Ca2+ and BK channels determine both arteriolar dilation and constriction. Proc Natl Acad Sci U S A 107(8):3811–3816

Gorman AM (2008) Neuronal cell death in neurodegenerative diseases: recurring themes around protein handling. J Cell Mol Med 12(6A):2263–2280

Greenhalgh AD, Ogungbenro K, Rothwell NJ, Galea JP (2011) Translational pharmacokinetics: challenges of an emerging approach to drug development in stroke. Expert Opin Drug Metab Toxicol 7(6):681–695

Grünblatt E, Mandel S, Youdim MB (2000) MPTP and 6-hydroxydopamine-induced neurode-gen- eration as models for Parkinson's disease: neuroprotective strategies. J Neurol 247(Suppl 2):II95–II102

Guo X, Geng M, Du G (2005) Glucose transporter 1, distribution in the brain and in neural dis-orders: its relationship with transport of neuroactive drugs through the blood–brain barrier. Biochem Genet 43(3–4):175–187

Gustafsson S, Gustavsson T, Roshanbin S, Hultqvist G, Hammarlund-Udenaes M, Sehlin D, Syvänen S (2018) Blood-brain Barrier Integrity in a Mouse Model of Alzheimer's Disease With or Without Acute 3D6 Immunotherapy. Neuropharmacology 143:1–9

Hartz AM, Miller DS, Bauer B (2010) Restoring blood–brain barrier P-glycoprotein reduces brain amyloid-beta in a mouse model of Alzheimer's disease. Mol Pharmacol 77(5):715–723

Hawkins BT, Davis TP (2005) The blood–brain barrier/neurovascular unit in health and disease. Pharmacol Rev 57(2):173–185

Hermann DM, Bassetti CL (2007) Implications of ATP-binding cassette transporters for brain pharmacotherapies. Trends Pharmacol Sci 28:128–134

Hernández-Romero MC, Delgado-Cortés MJ, Sarmiento M, de Pablos RM, Espinosa-Oliva AM, Argüelles S, Bández MJ, Villarán RF, Mauriño R, Santiago M, Venero JL, Herrera AJ, Cano J, Machado A (2012) Peripheral inflammation increases the deleterious effect of CNS inflamma-tion on the nigrostriatal dopaminergic system. Neurotoxicology 33(3):347–360

Hoffmann K, Gastens AM, Volk HA, Löscher W (2006) Expression of the multidrug transporter MRP2 in the blood–brain barrier after pilocarpine-induced seizures in rats. Epilepsy Res 69(1):1–14. Epub 2006 Feb 28

Holtzman DM, Zlokovic BV (2007) Role of Ab transport and clearance in the pathogenesis and treatment of Alzheimer's disease. In: Sisodia S, Tanzi RE (eds) Alzheimer's disease: advances in genetics, molecular and cellular biology. Springer, New York, pp 179–198

Hossmann KA (1994) Viability thresholds and the penumbra of focal ischemia. Ann Neurol 36:557–565

Hynes RO, Lander AD (1992) Contact and adhesive specificities in the associations, migrations, and targeting of cells and axons. Cell 68(2):303–322

Iadecola C (2010) The overlap between neurodegenerative and vascular factors in the pathogenesis of dementia. Acta Neuropathol 120(3):287–296

Jellinger KA (2012) Interaction between pathogenic proteins in neurodegenerative disorders. J Cell Mol Med 16(6):1166–1183

Kalaria RN (2010) Vascular basis for brain degeneration: faltering controls and risk factors for dementia. Nutr Rev 68(Suppl 2):S74–S87

Kerr IG, Zimm S, Collins JM, O'Neill D, Poplack DG (1984) Effect of intravenous dose and sched- ule on cerebrospinal fluid pharmacokinetics of 5-fluorouracil in the monkey. Cancer Res 44:4929–4932

Ketharanathan N, Yamamoto Y, Rohlwink U, Wildschut ED, Mathôt RAA, de Lange ECM, Wildt SN, Argent AC, Tibboel D, Figaji AA (2019) Combining Brain Microdialysis and Translational Pharmacokinetic Modeling to Predict Drug Concentrations in Pediatric Severe Traumatic Brain Injury: The Next Step Toward Evidence-Based Pharmacotherapy? J Neurotrauma 1 36(1):111–117

Kirk J, Plumb J, Mirakhur M, McQuaid S (2003) Tight junctional abnormality in multiple sclerosis white matter affects all calibres of vessel and is associated with blood–brain barrier leakage and active demyelination. J Pathol 201:319–327

Krantic S, Mechawar N, Reix S, Quirion R (2005) Molecular basis of programmed cell death involved in neurodegeneration. Trends Neurosci 28(12):670–676

Krueger M, Bechmann I (2010) CNS pericytes: concepts, misconceptions, and a way out. Glia 58(1):1–10

Kuhnke D, Jedlitschky G, Grube M, Krohn M, Jucker M, Mosyagin I, Cascorbi I, Walker LC, Kroemer HK, Warzok RW, Vogelgesang S (2007) MDR1-P-glycoprotein (ABCB1) mediates transport of Alzheimer's amyloid-beta peptides–implications for the mechanisms of abeta clearance at the blood–brain barrier. Brain Pathol 17(4):347–353

Lamas M, González-Mariscal L, Gutiérrez R (2002) Presence of claudins mRNA in the brain. Selective modulation of expression by kindling epilepsy. Brain Res Mol Brain Res 104(2):250–254

Lee G, Bendayan R (2004) Functional expression and localization of P-glycoprotein in the central nervous system: relevance to the pathogenesis and treatment of neurological disorders. Pharm Res 21(8):1313–1330

Lessing D, Bonini NM (2009) Maintaining the brain: insight into human neurodegeneration from Drosophila melanogaster mutants. Nat Rev Genet 10:359–370

Lipton SA (2005) The molecular basis of memantine action in Alzheimer's disease and other neurologic disorders: low-affinity, uncompetitive antagonism. Curr Alzheimer Res 2:155–165

Lo EH, Dalkara T, Moskowitz MA (2003) Mechanisms, challenges and opportunities in stroke. Nat Rev Neurosci 4:399–415

Lok J, Gupta P, Guo S, Kim WJ, Whalen MJ, van Leyen K, Lo EH (2007) Cell-cell signaling in the neurovascular unit. Neurochem Res 32:2032–2045

Loscher W, Potschka H (2005) Drug resistance in brain diseases and the role of drug efflux transporters. Nat Rev Neurosci 6:591–602

Löscher W, Luna-Tortós C, Römermann K, Fedrowitz M (2011) Do ATP-binding cassette transporters cause pharmacoresistance in epilepsy? Problems and approaches in determining which antiepileptic drugs are affected. Curr Pharm Des 17(26):2808–2828

Lu M, Hu G (2012) Targeting metabolic inflammation in Parkinson's disease: implications for prospective therapeutic strategies. Clin Exp Pharmacol Physiol 39(6):577–585

Luer MS (1999) Interventions to achieve tonic exposure to levodopa: delaying or preventing the onset of motor complications. Pharmacotherapy 19(11 Pt 2):169S–179S

Luna-Tortós C, Rambeck B, Jürgens UH, Löscher W (2009) The antiepileptic drug topiramate is a substrate for human P-glycoprotein but not multidrug resistance proteins. Pharm Res 26(11):2464–2470

Luna-Tortós C, Fedrowitz M, Löscher W (2010) Evaluation of transport of common antiepileptic drugs by human multidrug resistance-associated proteins (MRP1, 2 and 5) that are overexpressed in pharmacoresistant epilepsy. Neuropharmacology 58(7):1019–1032

Marchi N, Hallene KL, Kight KM, Cucullo L, Moddel G, Bingaman W, Dini G, Vezzani A, Janigro D (2004) Significance of MDR1 and multiple drug resistance in refractory human epileptic brain. BMC Med 2:37

Marroni M, Marchi N, Cucullo L, Abbott NJ, Signorelli K, Janigro D (2003) Vascular and parenchymal mechanisms in multiple drug resistance: a lesson from human epilepsy. Curr Drug Targets 4(4):297–304

Maxwell K, Berliner JA, Cancilla PA (1987) Induction of gamma glutamyltranspeptidase in cultured cerebral endothelial cells by a product released by astrocytes. Brain Res 410:309–314

Mazarati AM, Sofia RD, Wasterlain CG (2002) Anticonvulsant and antiepileptogenic effects of fluorofelbamate in experimental status epilepticus. Seizure 11(7):423–430

Merker HJ (1994) Morphology of the basement membrane. Microsc Res Tech 28(2):95–124

Michalak Z, Lebrun A, Di Miceli M, Rousset MC, Crespel A, Coubes P, Henshall DC, Lerner-Natoli M, Rigau V (2012) IgG leakage may contribute to neuronal dysfunction in drug-refractory epilepsies with blood–brain barrier disruption. J Neuropathol Exp Neurol 71(9):826–838

Mihajlica N, Betsholtz C, Hammarlund-Udenaes M (2018) Pharmacokinetics of pericyte involvement in small-molecular drug transport across the blood-brain barrier. Eur J Pharm Sci 122:77–84

Miltiadous P, Stamatakis A, Koutsoudaki PN, Tiniakos DG, Stylianopoulou F (2011) IGF-I ameliorates hippocampal neurodegeneration and protects against cognitive deficits in an animal model of temporal lobe epilepsy. Exp Neurol 231(2):223–235

Miyakawa T (2010) Vascular pathology in Alzheimer's disease. Psychogeriatrics 10(1):39–44

Mooradian AD, Chung HC, Shah GN (1997) GLUT-1 expression in the cerebra of patients with Alzheimer's disease. Neurobiol Aging 18:469–474

Morgan L, Shah B, Rivers LE, Barden L, Groom AJ, Chung R, Higazi D, Desmond H, Smith T, Staddon JM (2007) Inflammation and dephosphorylation of the tight junction protein occludin in an experimental model of multiple sclerosis. Neuroscience 147:664–673

Mosconi L, Sorbi S, de Leon MJ, Li Y, Nacmias B, Myoung PS, Tsui W, Ginestroni A, Bessi V, Fayyazz M, Caffarra P, Pupi A (2006) Hypometabolism exceeds atrophy in presymptomatic early-onset familial Alzheimer's disease. J Nucl Med 47(11):1778–1786

Mosconi L, De Santi S, Li J, Tsui WH, Li Y, Boppana M, Laska E, Rusinek H, de Leon MJ (2008) Hippocampal hypometabolism predicts cognitive decline from normal aging. Neurobiol Aging 29(5):676–692

Mulder M, Blokland A, van den Berg DJ, Schulten H, Bakker AHF, Terwel D, Honig W, de Kloet ER, Havekes LM, Steinbusch HWM, de Lange ECM (2001) Apolipoprotein E protects against neuropathology induced by a high-fat diet and maintains the integrity of the blood–brain barrier during aging. Lab Investig 81(7):953–960

Navarro A, Boveris A (2010) Brain mitochondrial dysfunction in aging, neurodegeneration, and Parkinson's disease. Front Aging Neurosci 2:34

Ndode-Ekane XE, Hayward N, Gröhn O, Pitkänen A (2010) Vascular changes in epilepsy: functional consequences and association with network plasticity in pilocarpine-induced experimental epilepsy. Neuroscience 166(1):312–332

Nga Bien-Ly C, Boswell A, Jeet S, Beach TG, Hoyte K, Luk W, Shihadeh V, Ulufatu S, Foreman O, Lu Y, DeVoss J, van der Brug M, Watts RJ (2015) Lack of widespread BBB disruption in alzheimer's disease models: focus on therapeutic antibodies. Neuron 88(2):289–297

Nico B, Frigeri A, Nicchia GP, Corsi P, Ribatti D, Quondamatteo F, Herken R, Girolamo F, Marzullo A, Svelto M, Svelto M, Roncali L (2003) Severe alterations of endothelial and glial cells in the blood–brain barrier of dystrophic mdx mice. Glia 42:235–251

Ohtsuki S, Terasaki T (2007) Contribution of carrier-mediated transport systems to the blood–brain barrier as a supporting and protecting interface for the brain; importance for CNS drug discovery and development. Pharm Res 24:1745–1758

Okouchi M, Ekshyyan O, Maracine M, Aw TY (2007) Neuronal apoptosis in neurodegeneration. Antioxid Redox Signal 9(8):1059–1096

Owen JB, Sultana R, Aluise CD, Erickson MA, Price TO, Bu G, Banks WA, Butterfield DA (2010) Oxidative modification to LDL receptor-related protein 1 in hippocampus from subjects with Alzheimer disease: implications for Aβ accumulation in AD brain. Free Radic Biol Med 49(11):1798–1803

Padou V, Boyet S, Nehlig A (1995) Changes in transport of [14C] alpha-aminoisobutyric acid across the blood–brain barrier during pentylenetetrazol-induced status epilepticus in the immature rat. Epilepsy Res 22(3):175–183

Palmer AM (2011) The role of the blood brain barrier in neurodegenerative disorders and their treatment. J Alzheimers Dis 24(4):643–656

Parodi-Rullán R, Sone JY, Fossati S (2019) Endothelial Mitochondrial Dysfunction in Cerebral Amyloid Angiopathy and Alzheimer's Disease. J Alzheimers Dis 72(4):1019–1039

Paulsson M (1992) Basement membrane proteins: structure, assembly, and cellular interactions. Crit Rev Biochem Mol Biol 27(1–2):93–127

Paweletz CP, Wiener MC, Bondarenko AY, Yates NA, Song Q, Liaw A, Lee AY, Hunt BT, Henle ES, Meng F, Sleph HF, Holahan M, Sankaranarayanan S, Simon AJ, Settlage RE, Sachs JR, Shearman M, Sachs AB, Cook JJ, Hendrickson RC (2010) Application of an end-to-end biomarker discovery platform to identify target engagement markers in cerebrospinal fluid by high resolution differential mass spectrometry. J Proteome Res 9(3):1392–1401

Perez-Pinzon MA, Stetler RA, Fiskum G (2012) Novel mitochondrial targets for neuroprotection. J Cereb Blood Flow Metab 32(7):1362–1376

Pflanzner T, Petsch B, André-Dohmen B, Müller-Schiffmann A, Tschickardt S, Weggen S, Stitz L, Korth C, Pietrzik CU (2012) Cellular prion protein participates in amyloid-β transcytosis across the blood–brain barrier. J Cereb Blood Flow Metab 32(4):628–632

Pop V, Badaut J (2011) A neurovascular perspective for long-term changes after brain trauma. Transl Stroke Res 2(4):533–545

Portelli J, Aourz N, De Bundel D, Meurs A, Smolders I, Michotte Y, Clinckers R (2009) Intrastrain differences in seizure susceptibility, pharmacological response and basal neurochemistry of Wistar rats. Epilepsy Res 87(2–3):234–246

Potschka H, Löscher W (2001a) In vivo evidence for P-glycoprotein-mediated transport of phenytoin at the blood brain barrier of rats. Epilepsia 42:1231–1240

Potschka H, Löscher W (2001b) Multidrug resistance-associated protein is involved in the regulation of extracellular levels of phenytoin in the brain. Neuroreport 12(11):2387–2389

Potschka H, Fedrowitz M, Löscher W (2001) P-glycoprotein and multidrug resistance-associated protein are involved in the regulation of extracellular levels of the major antiepileptic drug carbamazepine in the brain. Neuroreport 12:3557–3560

Potschka H, Fedrowitz M, Loscher W (2002) P-Glycoprotein-mediated efflux of phenobarbital, lamotrigine, and felbamate at the blood–brain barrier, evidence from microdialysis experiments in rats. Neurosci Lett 327(3):173–176

Potschka H, Fedrowitz M, Loscher W (2003a) Multidrug resistance protein MRP2 contributes to the blood–brain barrier function and restricts antiepileptic drug activity. J Pharmacol Exp Ther 306:124–131

Potschka H, Fedrowitz M, Loscher W (2003b) Brain access and anticonvulsant efficacy of carba- mazepine, lamotrigine, and felbamate in ABCC2/MRP2-deficient TR- rats. Epilepsia 44(12):1479–1486

Potschka H, Baltes S, Loscher W (2004) Inhibition of multidrug transporters by verapamil or probenecid does not alter blood–brain barrier penetration of levetiracetam in rats. Epilepsy Res 58(2–3):85–91

Price DL, Sisodia SS, Borchelt DR (1998) Genetic neurodegenerative diseases: the human illness and transgenic models. Science 282:1079–1083

Prins ML, Giza CC (2006) Induction of monocarboxylate transporter 2 expression and ketone transport following traumatic brain injury in juvenile and adult rats. Dev Neurosci 28(4–5):447–456

Prinz M, Priller J, Sisodia SS, Ransohoff RM (2011) Heterogeneity of CNS myeloid cells and their roles in neurodegeneration. Nat Neurosci 14(10):1227–1235

Qin T, Prins S, Groeneveld GJ, Van Westen G, de Vries HE, Wong YC, Bischoff LJM, de Lange ECM (2020) Utility of Animal Models to Understand Human Alzheimer's Disease, Using the Mastermind Research Approach to Avoid Unnecessary Further Sacrifices of Animals. Int J Mol Sci Apr 30:21(9)

Ravenstijn PG, Merlini M, Hameetman M, Murray TK, Ward MA, Lewis H, Ball G, Mottart C, de de Ville de Goyet C, Lemarchand T, van Belle K, O'Neill MJ, Danhof M, de Lange EC (2008) The exploration of rotenone as a toxin for inducing Parkinson's disease in rats, for application in BBB transport and PK-PD experiments. J Pharmacol Toxicol Methods 57(2):114–130

Ravenstijn PGM, Drenth H, Baatje MS, O'Neill MJ, Danhof M, de Lange ECM (2012) Evaluation of BBB transport and CNS drug metabolism in diseased and control brain after intravenous l-DOPA in a unilateral rat model of Parkinson's disease. Fluids Barriers CNS 9:4

Reale M, Iarlori C, Thomas A, Gambi D, Perfetti B, Di Nicola M, Onofrj M (2009) Peripheral cytokines profile in Parkinson's disease. Brain Behav Immun 23(1):55–63

Reed M, Damodarasamy M, Banks WA (2019) The extracellular matrix of the blood-brain barrier: structural and functional roles in health, aging, and Alzheimer's disease. Tissue Barriers 7(4):1651157

Romanitan MO, Popescu BO, Winblad B, Bajenaru OA, Bogdanovic N (2007) Occludin is over-expressed in Alzheimer's disease and vascular dementia. J Cell Mol Med 11(3):569–579

Rosenberg GA, Yang Y (2007) Vasogenic edema due to tight junction disruption by matrix metalloproteinases in cerebral ischemia. Neurosurg Focus 22:E4

Rosenfeld JV, Maas AI, Bragge P, Morganti-Kossmann MC, Manley GT, Gruen RL (2012) Early management of severe traumatic brain injury. Lancet 380(9847):1088–1098

Sagare AP, Deane R, Zlokovic BV (2012) Low-density lipoprotein receptor-related protein 1: A physiological Aβ homeostatic mechanism with multiple therapeutic opportunities. Pharmacol Ther 136(1):94–105

Sahin D, Ilbay G, Ates N (2003) Changes in the blood–brain barrier permeability and in the brain tissue trace element concentrations after single and repeated pentylenetetrazole-induced seizures in rats. Pharmacol Res 48(1):69–73

Samuraki M, Matsunari I, Chen WP, Yajima K, Yanase D, Fujikawa A, Takeda N, Nishimura S, Matsuda H, Yamada M (2007) Partial volume effect-corrected FDG PET and grey matter volume loss in patients with mild Alzheimer's disease. Eur J Nucl Med Mol Imaging 34:1658–1669

Sandoval KE, Witt KA (2008) Blood–brain barrier tight junction permeability and ischemic stroke. Neurobiol Dis 32(2):200–219

Sayre LM, Perry G, Smith MA (2008) Oxidative stress and neurotoxicity. Chem Res Toxicol 21(1):172–188

Scearce-Levie K, Sanchez PE, Lewcock JW (2020) Leveraging preclinical models for the development of Alzheimer disease therapeutics. Nat Rev Drug Discov 19:447–462

Schinkel A, Smit J, van Tellingen O, Beijnen J, Wagenaar E, van Deemter L et al (1994) Disruption of the mouse mdr1a P-glycoprotein gene leads to a deficiency in the blood–brain barrier and to increased sensitivity to drugs. Cell 77:491–502

Scism JL, Powers KM, Artru AA, Lewis L, Shen DD (2000) Probenecid-inhibitable efflux transport of valproic acid in the brain parenchymal cells of rabbits: a microdialysis study. Brain Res 884(1–2):77–86

Selkoe DJ (2011) Alzheimer's disease. Cold Spring Harb Perspect Biol 1:3–7

Serrano GE, Lelutiu N, Rojas A, Cochi S, Shaw R, Makinson CD, Wang D, FitzGerald GA, Dingledine R (2011) Ablation of cyclooxygenase-2 in forebrain neurons is neuroprotective and dampens brain inflammation after status epilepticus. J Neurosci 31(42):14850–14860

Sharma HS, Castellani RJ, Smith MA, Sharma A (2012) The blood–brain barrier in Alzheimer's disease: novel therapeutic targets and nanodrug delivery. Int Rev Neurobiol 102:47–90

Shlosberg D, Benifla M, Kaufer D, Friedman A (2010) Blood–brain barrier breakdown as a therapeutic target in traumatic brain injury. Nat Rev Neurol 6(7):393–403

Siddiqui A, Kerb R, Weale ME, Brinkmann U, Smith A, Goldstein DB, Wood NW, Sisodiya SM (2003) Association of multidrug resistance in epilepsy with a polmorhism in the drug- transporter gene ABCB1. N Engl J Med 348:1442–1448

Sills GJ, Kwan P, Butler E, de Lange EC, van den Berg DJ, Brodie MJ (2002) P-glycoprotein-mediated efflux of antiepileptic drugs: preliminary studies in mdr1a knockout mice. Epilepsy Behav 3(5):427–432

Simpson IA, Carruthers A, Vannucci SJ (2007) Supply and demand in cerebral energy metabolism: the role of nutrient transporters. J Cereb Blood Flow Metab 27(11):1766–1791

Sisodiya SM, Mefford HC (2011) Genetic contribution to common epilepsies. Curr Opin Neurol 24(2):140–145

Sivanandam TM, Thakur MK (2012) Traumatic brain injury: a risk factor for Alzheimer's disease. Neurosci Biobehav Rev 36(5):1376–1381

Soto C (2008) Endoplasmic reticulum stress, PrP trafficking, and neurodegeneration. Dev Cell 15(3):339–341

Soto C, Estrada LD (2008) Protein misfolding and neurodegeneration. Arch Neurol 65(2):184–189

Spector R (2009) Nutrient transport systems in brain: 40 years of progress. J Neurochem 111(2):315–320

Spector R, Johanson CE (2007) Vitamin transport and homeostasis in mammalian brain: focus on Vitamins B and E. J Neurochem 103:425–438

Stokin GB, Goldstein LSB (2006) Axonal transport and Alzheimer's disease. Annu Rev Biochem 75:607–627

Sweeney MD, Sagare AP, Zlokovic BV (2018) Blood-brain Barrier Breakdown in Alzheimer Disease and Other Neurodegenerative Disorders. Nat. Rev Neurol 14(3):133–150

Syvänen S, Luurtsema G, Molthoff CF, Windhorst AD, Huisman MC, Lammertsma AA, Voskuyl RA, de Lange EC (2011) (R)-[11C]verapamil PET studies to assess changes in P-glycoprotein expression and functionality in rat blood–brain barrier after exposure to kainate-induced status epilepticus. BMC Med Imaging 11:1

Syvänen S, Schenke M, van den Berg DJ, Voskuyl RA, de Lange EC (2012) Alteration in P-glycoprotein functionality affects intrabrain distribution of quinidine more than brain entry-a study in rats subjected to status epilepticus by kainate. AAPS J 14(1):87–96

Takano T, Han X, Deane R, Zlokovic B, Nedergaard M (2007) Two-photon imaging of astrocytic Ca2+ signaling and the microvasculature in experimental mice models of Alzheimer's disease. Ann N Y Acad Sci 1097:40–50

Tanzi RE, Moir RD, Wagner SL (2004) Clearance of Alzheimer's A beta peptide: the many roads to perdition. Neuron 43:605–608

Tate SK, Sisodiya SM (2007) Multidrug resistance in epilepsy: a pharmacogenomic update. Expert Opin Pharmacother 8:1441–1449

Tayarani I, Cloez I, Clément M, Bourre JM (1989) Antioxidant enzymes and related trace elements in aging brain capillaries and choroid plexus. J Neurochem 53:817–824

Tudor AF, Elson-Schwab I, Khurana V, Steinhilb ML, Spires TL, Hyman BT, Feany MB (2007) Abnormal bundling and accumulation of F-actin mediates tau-induced neuronal degeneration in vivo. Nat Cell Biol 9:139–148

Vallejo-Illarramendi A, Domercq M, Pérez-Cerdá F, Ravid R, Matute C (2006) Increased expression and function of glutamate transporters in multiple sclerosis. Neurobiol Dis 21(1):154–164

Van Assema DM, Goos JD, van der Flier WM, Lubberink M, Boellaard R, Windhorst AD, Scheltens P, Lammertsma AA, van Berckel BN (2012) No evidence for additional blood–brain barrier P-glycoprotein dysfunction in Alzheimer's disease patients with microbleeds. J Cereb Blood Flow Metab 32(8):1468–1471

Van Cauwenberghe C, Van Broeckhoven C, Sleegers K (2016) The genetic landscape of Alzheimer disease: clinical implications and perspectives. Genet Med 18:421–430

Van Damme P, Dewil M, Robberecht W, Van Den Bosch L (2005) Excitotoxicity and amyotrophic lateral sclerosis. Neurodegener Dis 2:147–159

Van Vliet EA, van Schaik R, Edelbroek PM, Voskuyl RA, Redeker S, Aronica E, Wadman WJ, Gorter JA (2007) Region-specific overexpression of P-glycoprotein at the blood–brain barrier affects brain uptake of phenytoin in epileptic rats. J Pharmacol Exp Ther 322(1):141–147

Van Vliet EA, Zibell G, Pekcec A, Schlichtiger J, Edelbroek PM, Holtman L, Aronica E, Gorter JA (2010) Potschka. COX-2 inhibition controls P-glycoprotein expression and promotes brain delivery of phenytoin in chronic epileptic rats. Neuropharmacology 58(2):404–412

Vautier S, Milane A, Fernandez C, Chacun H, Lacomblez L, Farinotti R (2009) Role of two efflux proteins, ABCB1 and ABCG2 in blood–brain barrier transport of bromocriptine in a murine model of MPTP-induced dopaminergic degeneration. J Pharm Pharm Sci 12(2):199–208

Viggars AP, Wharton SB, Simpson JE, Matthews FE, Brayne C, Savva GM, Garwood C, Drew D, Shaw PJ, Ince PG (2011) Alterations in the blood brain barrier in ageing cerebral cortex in relationship to Alzheimer-type pathology: a study in the MRC-CFAS population neuropathology cohort. Neurosci Lett 505(1):25–30

Vogelgesang S, Jedlitschky G, Brenn A, Walker LC (2011) The role of the ATP-binding cassette transporter P-glycoprotein in the transport of β-amyloid across the blood–brain barrier. Curr Pharm Des 17(26):2778–2786

Volk B, Hettmansperger U, Papp TH, Amelizad Z, Oesch F, Knoth R (1991) Mapping of phenytoin- inducible cytochrome P450 immunoreactivity in the mouse central nervous system. Neuroscience 42:215–235

Volk HA, Arabadzisz D, Fritschy JM, Brandt C, Bethmann K, Löscher W (2006) Antiepileptic drug-resistant rats differ from drug-responsive rats in hippocampal neurodegeneration and

GABA(A) receptor ligand binding in a model of temporal lobe epilepsy. Neurobiol Dis 21(3):633–646

Von Tell D, Armulik A, Betsholtz C (2006) Pericytes and vascular stability. Exp Cell Res 312:623–629

Weaver SM, Chau A, Portelli JN, Grafman J (2012) Genetic polymorphisms influence recovery from traumatic brain injury. Neuroscientist 18(6):631–644

Weber JT (2012) Altered calcium signaling following traumatic brain injury. Front Pharmacol 3:60

Weinreb O, Amit T, Bar-Am O, Youdim MB (2007a) Induction of neurotrophic factors GDNF and BDNF associated with the mechanism of neurorescue action of rasagiline and ladostigil: new insights and implications for therapy. Ann N Y Acad Sci 1122:155–168

Wenk GL (2003) Neuropathologic changes in Alzheimer's disease. J Clin Psychiatry 64(Suppl 9):7–10

Williams SK, Gillis JF, Matthews MA, Wagnert RC, Bitensky MW (1980) Isolation and char-acteriza- tion of brain endothelial cells: morphology and enzyme activity. J Neurochem 35(2):374–381

Witt KA, Gillespie TJ, Huber JD, Egleton RD, Davis TP (2001) Peptide drug modifications to enhance bioavailability and blood–brain barrier permeability. Peptides 22:2329–2343

Wolburg H, Noell S, Mack A, Wolburg-Buchholz K, Fallier-Becker P (2009) Brain endothelial cells and the glio-vascular complex. Cell Tissue Res 335(1):75–96

Wu Z, Guo H, Chow N, Sallstrom J, Bell RD, Deane R, Brooks AI, Kanagala S, Rubio A, Sagare A, Liu D, Li F, Armstrong D, Gasiewicz T, Zidovetzki R, Song X, Hofman F, Zlokovic BV (2005) Role of the MEOX2 homeobox gene in neurovascular dysfunction in Alzheimer dis-ease. Nat Med 11(9):959–965

Wyss-Coray T, Lin C, Sanan DA, Mucke L, Masliah E (2000) Chronic overproduction of trans-forming growth factor-beta1 by astrocytes promotes Alzheimer's disease-like microvascular degeneration in transgenic mice. Am J Pathol 156(1):139–150

Yamamoto Y, Välitalo PA, Wong YC, Huntjens DR, Proost JH, Vermeulen A, Krauwinkel W, Beukers MW, van den Berg DJ, Hartman RH, Wong YC, Danhof M, Kokkif H, Kokkif M, van Hasselt JGC, de Lange ECM (2018) Prediction of human CNS pharmacokinetics using a physiologically-based pharmacokinetic modeling approach. Eur J Pharm Sci 15(112):168–179

Yang Y, Rosenberg GA (2011) MMP-mediated disruption of claudin-5 in the blood–brain barrier of rat brain after cerebral ischemia. Methods Mol Biol 762:333–345

Yang J, Lunde LK, Nuntagij P, Oguchi T, Camassa LM, Nilsson LN, Lannfelt L, Xu Y, Amiry-Moghaddam M, Ottersen OP, Torp R (2011) Loss of astrocyte polarization in the tg-ArcSwe mouse model of Alzheimer's disease. J Alzheimers Dis 27(4):711–722

Yang AC, Stevens MY, Chen MB, Lee DP, Stähli D, Gate D, Contrepois K, Chen W, Iram T, Zhang L, Vest RT, Chaney A, Lehallier B, Olsson N, du Bois H, Hsieh R, Cropper HC, Berdnik D, Li L, Wang EY, Traber GM, Bertozzi CR, Luo J, Snyder MP, Elias JE, Quake SR, James ML, Wyss-Coray T (2020) Physiological blood-brain transport is impaired with age by a shift in transcytosis. Nature

Yu F, Wang Z, Tchantchou F, Chiu CT, Zhang Y, Chuang DM (2012) Lithium ameliorates neuro-degeneration, suppresses neuroinflammation, and improves behavioral performance in a mouse model of traumatic brain injury. J Neurotrauma 29(2):362–374

Zattoni M, Mura ML, Deprez F, Schwendener RA, Engelhardt B, Frei K, Fritschy JM (2011) Brain infiltration of leukocytes contributes to the pathophysiology of temporal lobe epilepsy. J Neurosci 31(11):4037–4050

Zenaro E, Piacentino G, Constantin G (2017) The blood-brain barrier in Alzheimer's disease. Neurobiol Dis 107:41–56

Zhang XM, Mao XJ, Zhang HL, Zheng XY, Pham T, Adem A, Winblad B, Mix E, Zhu J (2012) Overexpression of apolipoprotein E4 increases kainic-acid-induced hippocampal neurodegen-eration. Exp Neurol 233(1):323–332

Zlokovic BV (2006) Remodeling after stroke. Nat Med 12:390–391

Zlokovic BV (2008) The blood–brain barrier in health and chronic neurodegenerative disorders. Neuron 57(2):178–201

Zlokovic BV (2010) Neurodegeneration and the neurovascular unit. Nat Med 16:1370–1371

Zlokovic BV (2011) Neurovascular pathways to neurodegeneration in Alzheimer's disease and other disorders. Nat Rev Neurosci 12(12):723–738

Zlokovic BV, Yamada S, Holtzman D, Ghiso J, Frangione B (2000) Clearance of amyloid beta-peptide from brain: transport or metabolism? Nat Med 6(7):718–719

Chapter 23
The Blood-Brain Barrier in Stroke and Trauma and How to Enhance Drug Delivery

Richard F. Keep, Jianming Xiang, Ningna Zhou, and Anuska V. Andjelkovic

Abstract Ischemic or hemorrhagic stroke and traumatic brain injury (TBI) cause marked changes in blood-brain barrier (BBB) function. Such changes increase barrier permeability and induce vasogenic edema and leukocyte extravasation into the brain. In addition, BBB dysfunction affects the entry of therapeutics into the brain. This chapter describes changes in BBB function after brain injury, how stroke and TBI affect drug delivery, the BBB as a therapeutic target, and enhancing drug delivery in stroke and TBI.

Keywords Endo/transcytosis · Traumatic brain injury · Metabolic barrier · Circumventing the BBB · Liposomes · Exosomes · Transport

Abbreviations

AAVs	adeno-associated viruses
ABC transporters	ATP-binding cassette transporters
BBB	blood-brain barrier

R. F. Keep (✉)
Department of Neurosurgery, University of Michigan, Ann Arbor, MI, USA

Department of Molecular and Integrative Physiology, University of Michigan, Ann Arbor, MI, USA
e-mail: rkeep@umich.edu

J. Xiang
Department of Neurosurgery, University of Michigan, Ann Arbor, MI, USA

N. Zhou
Department of Pharmacology, Yunnan University of Traditional Chinese Medicine, Kunming, China

A. V. Andjelkovic
Department of Neurosurgery, University of Michigan, Ann Arbor, MI, USA

Department of Pathology, University of Michigan, Ann Arbor, MI, USA

© American Association of Pharmaceutical Scientists 2022
E. C. M. de Lange et al. (eds.), *Drug Delivery to the Brain*, AAPS Advances in the Pharmaceutical Sciences Series 33, https://doi.org/10.1007/978-3-030-88773-5_23

BDNF	brain-derived neurotrophic factor
bFGF	basic fibroblast growth factor
CBF	cerebral blood flow
CSF	cerebrospinal fluid
Gd-DPTA	gadolinium-diethylenetriaminepentacetate
ICH	intracerebral hemorrhage
Nrf2	nuclear factor erythroid 2-related factor 2
NVU	neurovascular unit
Oat3	organic anion transporter-3
PS product	permeability surface area product
SAH	subarachnoid hemorrhage
shRNA	short hairpin RNA
SVCT2	Na-dependent vitamin C transporter 2
TBI	traumatic brain injury
TJ	tight junction
tPA	tissue plasminogen activator

23.1 Introduction

Stroke and traumatic brain injury (TBI) are enormous personal, societal, and economic burdens. For example, ~5.5 million people die of stroke globally each year (GBD2016 2017), and 27 million people suffer a TBI (Injury and Spinal Cord Injury 2019). Stroke and TBI result in profound changes in blood-brain barrier (BBB) function (Jiang et al. 2018; Keep et al. 2014; Logsdon et al. 2015), and there is growing evidence that such changes (resulting, e.g., in edema formation and leukocyte infiltration) contribute to brain parenchymal injury and are a therapeutic target (Jiang et al. 2018; Shi et al. 2016). In addition, stroke- and TBI-induced BBB dysfunction impacts drug delivery of potential therapeutics to brain parenchyma.

In cerebral ischemia, reductions in blood flow to a brain area cause neural dysfunction. Reductions in flow are caused by thrombosis within a cerebral vessel, the lodging of emboli generated by a distant site, or temporary heart failure. The first two events cause focal cerebral ischemia, the latter global cerebral ischemia. Focal ischemic events occur with and without restoration of blood flow (transient and permanent ischemia). Cerebral ischemia accounts for most strokes, but ~15% of strokes in the United States and 20–30% in Asia are hemorrhagic (Adeoye and Broderick 2010; Roger et al. 2012). The initial symptoms in hemorrhagic stroke are similar to cerebral ischemia, but the underlying cause, blood vessel rupture, is different as are the mechanisms involved in brain injury (Keep et al. 2012). The components of brain injury after trauma are heterogeneous with physical damage to neural components, cerebral ischemia and cerebral hemorrhage, and they vary with closed and penetrating brain injury (Maas et al. 2008).

The aim of this chapter is to describe the changes that occur in BBB function during stroke and TBI, the effects of stroke and TBI on delivery of current and potential therapeutics, the BBB as a therapeutic target, and methods of enhancing drug delivery in stroke and TBI. While prior chapters have described the effects of the normal BBB on drug delivery and methods that are being used to enhance such delivery, this chapter focuses specifically on brain injury.

23.2 The Blood-Brain Barrier During Stroke and Trauma

23.2.1 Blood-Brain Barrier and Neurovascular Unit (NVU) Changes

Cerebral ischemia (permanent or transient), cerebral hemorrhage, and TBI all have major impacts on BBB function (Table 23.1). This is evinced by marked increases in permeability (Fig. 23.1), vasogenic edema formation, and leukocyte infiltration into the brain (Jiang et al. 2018; Keep et al. 2014; Shlosberg et al. 2010). These changes gradually resolve within 2–4 weeks (Menzies et al. 1993), although there is evidence of prolonged low level BBB leakiness after stroke which may be a therapeutic target (Sladojevic et al. 2019; Topakian et al. 2010). The impact of stroke on BBB function is worsened by many comorbidities including age, hyperglycemia, and hypertension (Jiang et al. 2018). Stroke and TBI can also impact blood-CSF barrier function (Johanson et al. 2000; Karimy et al. 2017; Szmydynger-Chodobska et al. 2012; Xiang et al. 2017), but that is beyond the scope of the present chapter.

Table 23.1 Changes in the blood-brain barrier after stroke and traumatic brain injury

Mechanism affected	Change	Consequence
Cerebral blood flow	Reduction[a]	Reduced uptake of highly permeable (flow dependent) compounds Potential reduction in endothelial ATP
Tight junction integrity	Disruption	Enhanced BBB permeability to small and large molecules
Endocytosis/transcytosis	Increase	Enhanced BBB permeability, particularly to large molecules
Leukocyte adhesion and transmigration	Increase	Inflammation and BBB disruption
Progenitor cell transmigration	Increase	Possible angiogenesis, reduced injury
NVU exosome production	Increase	Cellular signaling, neuroprotection, angiogenesis
ABC transporters	Increase	Reduced uptake of substrates into the brain
Other transporters	Mixed	Altered uptake of substrates into the brain

[a]In ischemic stroke, the reductions in blood flow are marked. In traumatic brain injury and subarachnoid hemorrhage, the reductions vary in magnitude between patients. In intracerebral hemorrhage, the reductions are limited
ABC transporters ATP-binding cassette transporters, *BBB* blood-brain barrier, *NVU* neurovascular unit

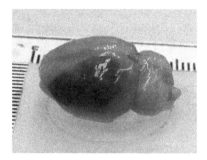

Fig. 23.1 Distribution of Evan's blue into brain after 2 h of focal middle cerebral artery occlusion with 2 h of reperfusion (transient focal ischemia) in an acutely hyperglycemic rat. Hyperglycemia exacerbates ischemia-induced BBB damage. Evan's blue binds to albumin within the bloodstream, and the blue color in the ipsilateral hemisphere reflects increased BBB permeability to protein in the middle cerebral artery territory

Stroke and TBI may have direct effects on cerebral endothelial cells, the primary site of the BBB. However, the endothelium is also regulated by the surrounding cells (e.g., astrocytes, pericytes, perivascular macrophages, and neurons) and extracellular components (e.g., the endothelial basement membrane) that form the NVU (Iadecola 2017; Sweeney et al. 2016; Yao 2019), and brain injury causes major changes within the NVU that can impact endothelial function. That includes release of cytokines, chemokines, and matrix metalloproteinases, all of which impact barrier function (Jiang et al. 2018). It should be noted that communication between the endothelium and the rest of the NVU is bidirectional and the endothelium secretes factors that impact other cells with the NVU. For example, after stroke, brain endothelial cell secrete exosomes containing a wide array of proteins and RNAs that are involved in brain repair (Zhang and Chopp 2016).

As well as participating in brain injury, events at the cerebrovasculature during stroke and TBI will also impact drug delivery. Those events include alterations in cerebral blood flow (CBF; drug delivery to brain) and alterations in barrier functions that can impact the movement of drugs between blood and brain. This section describes those underlying changes, while Sect. 23.3 describes how they impact the delivery of different types of therapeutics.

23.2.2 Blood Supply

In cerebral ischemia, the extent and duration of reductions in blood flow depend on the underlying cause. In a heart attack, CBF falls to zero and will result in death unless the heart resumes beating. The reduced blood flow, with the concomitant reductions in oxygen and glucose supply, is the underlying cause of ischemic brain injury. In focal events, the degree of ischemia depends upon the blood vessel blocked and the degree of collateral blood flow. Prolonged reductions in flow to less than ~20 ml/100 g/min result in permanent brain damage (Jones et al. 1981).

The extent of blood flow reductions in other forms of stroke and TBI varies. In intracerebral hemorrhage (ICH), most evidence indicates that there are not pronounced reductions in flow (except probably for very large hemorrhages) (Keep et al. 2012). In subarachnoid hemorrhage (SAH), early and delayed cerebral ischemia is a large part of the injury (Etminan et al. 2011; Schubert and Thome 2008). In TBI, the extent of ischemia varies between different injuries, and it can be widespread or perilesional (Maas et al. 2008).

23.2.3 Endothelial Tight Junctions

Tight junctions (TJs) link cerebral endothelial cells limiting paracellular permeability. They are comprised of transmembrane proteins (claudin 5, occludin, and junctional adhesion molecules) that occlude the paracellular space and cytoplasmic plaque proteins (e.g., zonula occludens (ZO)-1) that stabilize and regulate the TJs (Stamatovic et al. 2016). Many studies have shown alterations in BBB TJ structure after cerebral ischemia (Dimitrijevic et al. 2006; Jiao et al. 2011; Rosenberg and Yang 2007), cerebral hemorrhage (Keep et al. 2018), and TBI (Higashida et al. 2011; Walker et al. 2010) (Figs. 23.2 and 23.3). Changes in TJ structure can involve the loss of TJ proteins, protein modifications such as phosphorylation that affect

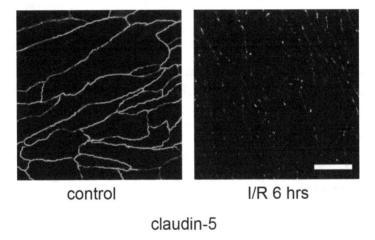

control   I/R 6 hrs

claudin-5

Fig. 23.2 Distribution of a tight junction protein, claudin-5, in mouse brain microvascular endothelial cells (mBMEC) in culture as determined by immunofluorescence. mBMEC were co-cultured with astrocytes and then either exposed to normal culture conditions (control) or exposed to 5 h of oxygen glucose deprivation (OGD; a model of in vitro ischemia) and then returned to normal oxygen and glucose for 6 h ("ischemia/reperfusion," I/R). (**a**) Under control conditions, claudin-5 is located at the cell membrane between adjacent endothelial cells. (**b**) After OGD with reperfusion, claudin-5 is lost from the cell membrane (fragmented staining) and this is associated with increased permeability of the endothelial cell monolayers (Dimitrijevic et al. 2006). Scale bar = 50 μm

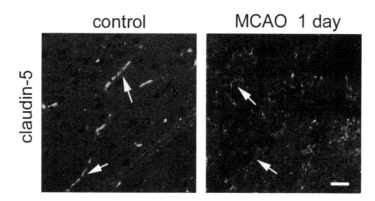

Fig. 23.3 Claudin-5 immunohistochemistry in the brain in control mice and in animals that have undergone 30 min of middle cerebral artery occlusion (MCAO) with 1 day of reperfusion. Vessels are indicated by arrows. There is a decrease in vascular claudin-5 staining after ischemia with reperfusion. Scale bar = 50 μm

protein:protein interactions, and/or the relocation of proteins from their normal site at the plasma membrane (Dimitrijevic et al. 2006; Kago et al. 2006; Stamatovic et al. 2009; Yang et al. 2007). The role of each type of change varies dependent upon the type and duration of injury.

Such TJ changes cause an increase in paracellular permeability that allows entry of large compounds (e.g., plasma albumin (Menzies et al. 1993)) into the brain. It also enhances the permeability of low molecular weight polar compounds that normally have a low brain uptake across the BBB (e.g., sucrose (Preston and Webster 2002)). Thus, these changes in TJ structure can alter the uptake of a wide range of therapeutics, effects that are dependent on the size of the molecule and the location and evolution of the injury (Nagaraja et al. 2007; Preston and Webster 2002).

It should be noted that in experimental autoimmune encephalomyelitis, an animal model of multiple sclerosis astrocytes start to express the TJ proteins claudin 1, claudin 4, and junctional adhesion molecule A in the inflammatory lesions forming a second barrier that limits protein extravasation and leukocyte infiltration into brain (Horng et al. 2017). Stroke and TBI induce neuroinflammation and whether a second barrier is induced in those conditions and how it may impact drug delivery merits investigation.

23.2.4   Endocytosis/Transcytosis

Under normal conditions, the number of vesicles in the cerebral endothelium is low compared to other endothelia (Abbott et al. 2010). At the BBB some vesicular trafficking is involved in the movement of compounds across the endothelium, that is, transcytosis. For example, transcytosis plays an important role in the transport of compounds such as transferrin, LDL-receptor-related protein 1 (LRP1), and insulin between blood and brain (Abbott et al. 2010). Cerebral ischemia and other forms of brain injury markedly increase endothelial vesicle number and transcytosis (Haley and Lawrence 2017; Knowland et al. 2014). Enhanced transcytosis may increase the brain uptake of a wide range of therapeutics.

23.2.5 Cell Trafficking

Ischemic and hemorrhagic stroke and TBI all cause an influx of leukocytes into the brain via a coordinated action of adhesion molecules, cytokines, and chemokines, with the cerebral endothelium that plays a central role (del Zoppo 2010; Iadecola and Anrather 2011; Rhodes 2011; Wang 2010). In preclinical models, inhibiting inflammation and leukocyte trafficking into brain have commonly been shown to reduce brain and cerebrovascular injury after stroke and TBI (del Zoppo 2010; Rhodes 2011). However, some leukocyte populations (e.g., Treg and B cell lymphocytes) or subpopulations (e.g., "M2" macrophages) may be beneficial (Rayasam et al. 2018) complicating this therapeutic approach.

After stroke or TBI, there is also a migration of endogenous progenitor cells within the brain (e.g., from the subventricular zone) (Kernie and Parent 2010) and from the bloodstream to the site of injury (Borlongan et al. 2011). The latter cells can integrate into the cerebrovasculature (e.g., participating in angiogenesis) and migrate into brain parenchyma (Borlongan et al. 2011). Such brain or bloodstream endogenous progenitors do not significantly integrate into the brain long term, and indeed, there may be aberrant integration (Kernie and Parent 2010). Many forms of exogenous stem cell are being examined as potential stroke therapies, including in the clinic (Stonesifer et al. 2017).

23.2.6 Transport

The effects of stroke and TBI on BBB transport have received relatively little attention. There is evidence of upregulation of the ATP-binding cassette (ABC) transporter, p-glycoprotein (ABCB-1), after transient cerebral ischemia (Cen et al. 2013; DeMars et al. 2017; Patak and Hermann 2011). An upregulation of breast cancer-related protein (Abcg2) has also been reported (Dazert et al. 2006). Such changes may limit the access of potential therapeutics into the brain. There is also evidence for increased Na-dependent vitamin C transporter 2 (SVCT2) mRNA and activity after cerebral ischemia, and it has been suggested that the increase in activity might be used to enhance drug delivery to the ischemic brain (Gess et al. 2011). It should be noted that reduced oxygen and glucose supply depletes brain ATP levels during cerebral ischemia, and this may inhibit transporters that depend directly or indirectly (secondary active transport) on ATP.

23.2.7   Metabolic Barrier

The BBB and the blood-CSF barrier possess a wide variety of enzymes that metabolize neuroactive compounds, for example, glutathione S-transferases, glutathione peroxidases, and epoxide hydrolase (el-Bacha and Minn 1999; Ghersi-Egea et al. 2006). These contribute to barrier function by degrading the compounds before they can enter the brain or by converting compounds so they become substrates for brain efflux transporters. The effect of stroke and TBI on such enzymes, which might affect drug delivery, has received little attention. There is, though, evidence for the importance of a transcription factor named nuclear factor erythroid 2-related factor 2 (nrf2) after cerebrovascular injury (Alfieri et al. 2011; Zhao et al. 2007). Nrf2 is a master regulator of antioxidant defense mechanisms, including a wide variety of xenobiotic metabolizing enzymes including glutathione S-transferase, epoxide hydrolase, and NADPH: quinone reductase (Kohle and Bock 2007; Thimmulappa et al. 2002).

23.3   Enhancing Brain Delivery of Potential Therapeutics in Stroke and Trauma

In the United States, the only current therapies for ischemic stroke are tissue plasminogen activator or mechanical thrombectomy to restore blood flow to the impacted area of the brain. In TBI, there are even less therapeutic options, although the recent CRASH-3 trial suggests that tranexamic acid reduces TBI-induced mortality by reducing cerebral hemorrhage (CRASH-3 Trial Investigators 2019). It should be noted that for both ischemic stroke and TBI, the therapies target the vasculature rather than the brain parenchyma. As testified by the rest of this book, delivering therapeutics to brain poses many challenges. This is further amplified in the setting of stroke and TBI. While there might be enhanced drug delivery due to TJ disruption and increased transcytosis at the brain endothelium, those effects may be offset by other factors (e.g., altered transport). Effects on the BBB after injury are also spatially and temporally inhomogeneous. For example, in focal cerebral ischemia, the degree of BBB "disruption" varies from the core of the infarct (with unsalvageable tissue) to the penumbra (with potentially salvageable tissue) (Astrup et al. 1981). In addition, the degree of BBB disruption initially increases with time, while a drug may need to be given very early for maximal effect. This spatial and temporal inhomogeneity in drug delivery impedes determining efficacious dosing and potential side effects (Fig. 23.4).

The impact of stroke and TBI on the delivery of therapies will depend upon the nature of the therapeutic. Whether there is a need to devise methods to enhance delivery and how to achieve that will depend on a multitude of factors. These are discussed below with examples particularly focusing on approaches that are used clinically or are in clinical trial.

Inhomogeneity in BBB disruption

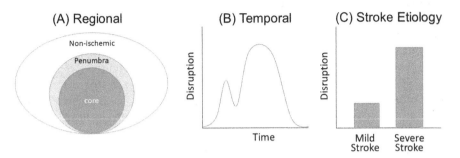

Fig. 23.4 Inhomogeneity in drug delivery after stroke. The tissue drug concentration of that will be achieved after stroke due to BBB disruption may be very heterogeneous varying by (**a**) location of the tissue (i.e., ischemic core vs ischemic penumbra vs non-ischemic tissue), (**b**) time after stroke, and (**c**) stroke etiology. This greatly complicates drug dosing, particularly if there are toxicity issues (in the brain or systemic)

23.3.1 Highly Lipophilic Compounds with No or Limited Efflux Transport

Some compounds are so lipophilic they have BBB permeability surface area (PS) products (measured in ml/g/min) that approach or exceed normal CBF (~0.6 ml/g/min). The brain uptake of such compounds is "flow limited" as their plasma concentration will markedly decrease as they pass through the cerebrovasculature due to brain uptake (Robinson 1990). Two such compounds are caffeine (PS product ~0.7 ml/g/min (Tanaka and Mizojiri 1999)) and ethanol (PS product ~1.8 ml/g/min (Raichle et al. 1976)). Caffeine in combination with ethanol (caffeinol) has undergone clinical trials for stroke in combination with thrombolysis and hypothermia (Martin-Schild et al. 2009; Piriyawat et al. 2003) and has been shown to be safe. The uptake of each drug into brain will be decreased in ischemic stroke due to reduced CBF with the degree of reduction being inhomogeneous as the severity of ischemia will differ between patients and between different brain areas in the same patient.

For some cases, this reduction in brain uptake may be compensated for by increasing drug dose, but that will depend on the safety profile of the drug (systemic and brain). In addition, drug delivery may be improved by combination with tPA- or thrombectomy-induced reperfusion therapy to increase CBF.

23.3.2 Highly Lipophilic Compounds with Efflux Transport

Many lipophilic drugs are substrates for ABC efflux transporters at the BBB, such as p-glycoprotein and breast cancer-related protein. Such efflux transporters can greatly limit drug delivery to the brain. Glucocorticoids, particularly dexamethasone, are used to treat brain edema in tumor patients. They probably act via reducing capillary permeability, although glucocorticoids also have profound anti-inflammatory effects. These steroids have been extensively examined in patients with cerebral ischemia, cerebral hemorrhage, and TBI with no evidence of benefit. However, there is evidence of benefit of high-dose methylprednisolone (another glucocorticoid) in spinal cord injury, although that is controversial (Gomes et al. 2005). While steroids are lipophilic, studies have shown they can undergo efflux transport by p-glycoprotein (Yates et al. 2003). Thus, dexamethasone shows increased brain penetration in the Mdr-1 (p-glycoprotein) knockout mouse (Meijer et al. 1998; Schinkel et al. 1995; Uchida et al. 2011). In addition, dexamethasone actually induces p-glycoprotein expression at the BBB (Bauer et al. 2004). That, along with ischemia induced p-glycoprotein upregulation (Cen et al. 2013; DeMars et al. 2017; Patak and Hermann 2011), may limit dexamethasone uptake into the brain after injury. Similarly, methylprednisolone has only a small but significant permeability at the uninjured BBB (Zlokovic et al. 1993) that may reflect that it too is a p-glycoprotein substrate (Yates et al. 2003). This may contribute to the high dose of methylprednisolone required for treating spinal cord injury.

The calcium channel antagonist, nimodipine, is approved for use in SAH in the United States (Bederson et al. 2009). Whether it acts as a vascular (e.g., preventing large vessel or microvessel-related ischemia) or a neuronal protectant is not certain (Bederson et al. 2009). Nimodipine has a relatively high BBB permeability (although it has significant protein binding) (Zlokovic et al. 1993), but it is a p-glycoprotein substrate (Hollt et al. 1992; Liu et al. 2003) that may limit brain uptake and, potentially, effect.

Spudich et al. (2006) found that inhibiting p-glycoprotein with tariquidar increased the efficacy of FK506 and rifampicin, both of which are p-glycoprotein substrates, in protecting against focal cerebral ischemia in mice. Importantly, not only did tariquidar cause an increase in the brain uptake of both drugs, that increase was greater in the ischemic compared to the non-ischemic hemisphere suggesting that p-glycoprotein was upregulated in the ischemic tissue.

While ABC transporter inhibition may be one approach to increasing the delivery of potential neuroprotectants to the brain, this has been tried in other disease states without, as yet, clinical success. Another approach may be to modulate the pathways that regulate ABC transporter expression and activity at the BBB. Those pathways are being delineated (Miller 2015), but the effects of manipulating them in the setting of stroke and TBI remain to be studied.

23.3.3 Hydrophilic Compounds

Many drugs are hydrophilic, and the permeability of hydrophilic compounds is generally enhanced after stroke and TBI due to TJ disruption and/or increased transcytosis. That includes small molecular weight compounds such as sucrose and gadolinium-diethylenetriaminepentacetate (Gd-DPTA) to large molecular weight proteins (Ewing et al. 2003; Menzies et al. 1993; Fig. 23.1). At least for stroke, the impact of changes in TJ structure on brain uptake depends on the molecular weight of the compound. Thus, the absolute increases in uptake are larger for small molecular weight compounds, but the percent increases are greater for large compounds suggesting that the increases are predominantly paracellular as the transcytosis route is not expected to show size selectivity (Preston and Foster 1997; Preston and Webster 2002).

Whether the enhanced BBB permeability provides sufficient drug uptake into the brain for efficacy depends on the pharmacokinetics of the drug and the brain concentrations required for efficacy. If it is insufficient, there are multiple potential approaches to try and enhance brain drug uptake. These include improving the pharmacokinetic profile of the drug by chemical modification (e.g., increasing plasma half-life by reducing excretion and/or metabolism) or increasing dosing of the parent drug if toxicology allows. Another potential approach is to further increase the BBB permeability of the drug. This might be achieved by several mechanisms (Table 23.2) as outlined below.

23.3.3.1   Increasing Paracellular Diffusion

Enhancing tight junction disruption. This might be achieved by hyperosmotic stress (Doolittle et al. 2014), downregulating TJ proteins by short hairpin RNA (shRNA, (Menard et al. 2017)), claudin peptides causing TJ protein internalization and

Table 23.2 Methods of enhancing brain uptake of drugs across the blood-brain barrier in stroke and traumatic brain injury[a]

Type of drug	Route	Approach
Highly lipophilic (no efflux transport)	Transcellular	Increase (restore) cerebral blood flow
Highly lipophilic (with efflux transport)	Transcellular	Efflux transporter inhibitors Downregulating efflux transporter expression Downregulating efflux transporter activity
Hydrophilic	Paracellular	Hyperosmotic-induced TJ disruption Focused ultrasound-induced TJ disruption Down regulating TJ proteins (e.g., shRNA) Inducing TJ protein degradation (e.g., claudin peptides) Modifying signaling involved in TJ regulation
Hydrophilic	Transcellular	Chemical modification—Increase lipophilicity Conjugation so target transcytosis pathways Targeting influx transporters (rare) Inhibiting efflux transporters Packaging in liposomes/exosomes/nanoparticles Adeno-associated virus expression Transducing migrating stem cells

[a]All drugs might also be administered directly into brain parenchyma or CSF. There is also the potential for intranasal administration
shRNA short hairpin RNA, *TJ* Tight junction

degradation (Dithmer et al. 2017), modifying signaling involved in TJ regulation (Stamatovic et al. 2008) and applying focused ultrasound (Leinenga et al. 2016). However, there is a concern for all these approaches that they themselves may exacerbate stroke- and TBI-induced brain damage by increasing vasogenic edema or increasing the brain entry of potentially neurotoxic compounds. Even in normal brain, there is recent evidence that claudin-5 downregulation by shRNA can result in depression-like behavior in mice (Menard et al. 2017). It is possible that short-term barrier or focused disruption may limit such potential side effects. Thus, for example, focused ultrasound with microbubbles increased the delivery of erythropoietin to the brain and improved outcome in a rat cerebral ischemia model (Wu et al. 2014).

23.3.3.2 Increasing Transcellular Movement

One potential approach to increase brain uptake of a drug to treat stroke or TBI is chemical modification to increase lipophilicity and, thus, BBB permeability. For example, the lipophilic simvastatin lactone showed protection, while the hydrophilic simvastatin acid showed no protection in a guinea pig cerebral ischemia model (Beretta et al. 2011).

Another approach is conjugate drugs to molecules that are transported across the BBB (molecular Trojan horses (Pardridge 2006)). Examples of this approach are the conjugation of brain-derived neurotrophic factor (BDNF) and basic fibroblast growth factor (bFGF) to an antibody recognizing the transferrin receptor which undergoes endocytosis at the BBB. While BDNF and bFGF alone were not protective in a rat model of cerebral ischemia, conjugating these proteins to an antibody targeting the transferrin receptor resulted in reduced injury (Song et al. 2002; Zhang and Pardridge 2001). Similarly, nanoparticles loaded with a free radical scavenger, edaravone, have been linked to Angiopep-2, a peptide that targets receptor-mediated transcytosis. That increases brain uptake and protects against ischemic brain injury in rats (Bao et al. 2018).

The cerebral endothelium possesses a wide array of transporters involved in transporting substrates from blood to brain, and these might be for drug transport. For example, levodopa (L-DOPA) utilizes the system-L amino acid transport at the BBB (Kageyama et al. 2000). SVCT2 transport is upregulated at the cerebral endothelium after cerebral ischemia (Gess et al. 2011). SVCT2 transports ascorbate, and recent evidence indicates that ascorbate administration is neuroprotective in cerebral ischemia (Morris-Blanco et al. 2019). It should be noted that the size of molecules that may be transported via such mechanisms is limited, due to steric hindrance, as compared to transcytosis mechanisms.

Efflux (brain to blood) transport at the BBB is not limited to the ABC transporters. For example, the organic anion transporter-3 (Oat3) is also involved in clearing substrates from brain at the cerebral endothelium. One Oat3 substrate is the Na/K/Cl transporter inhibitor, bumetanide, and blocking Oat3 has been shown to increase brain bumetanide concentrations (Romermann et al. 2017). There is recent evidence of the importance of Na/K/Cl transport in the development of hydrocephalus after intraventricular hemorrhage (Karimy et al. 2017) and a combination of bumetanide with an Oat3 inhibitor might be a therapeutic approach.

There are a number of methods being employed to "package" potential protective agents. These include liposomes, for a variety of compounds including proteins, neurotropic adeno-associated viruses (AAVs) for genetic material, and progenitor cells transduced to express protective factors. The packaging not only may allow entry of the protective agent into the brain but also protect the agents from metabolism and excretion and may limit side effects.

Some surface modifications to liposomes, such as expression of transferrin and lactoferrin, allow them to better target and cross the BBB (Agrawal et al. 2017). They have been extensively used to deliver potential therapeutics in preclinical stroke models, including primates (reviewed in Bruch et al. 2019; Fernandes et al. 2018). There have also been significant advances in developing AAVs that cross the BBB and target specific neural populations (Juttner et al. 2019). Such AAVs may express not only protein-coding complementary DNAs but also noncoding RNAs (e.g., microRNAs) and short-hairpin RNAs to silence genes (Zacchigna et al. 2014). AAVs have been used in preclinical stroke models (Gan et al. 2013; Kurinami et al. 2014). One potential drawback as a therapy relates to the time before the AAV, and its cargo will affect protein expression. It may be that AAVs may have a therapeutic role in long-term recovery after stroke. Stem cells are being examined as therapeutic in stroke, and they can be transduced to express different protective factors (see Sect. 23.3.4 below).

There is a rapidly growing interest in the role of exosomes (cell-derived microvesicles) in stroke and TBI. These are produced by multiple cell types within the NVU having a role in intercellular signaling (Zagrean et al. 2018). Exosomes contain a wide range of molecules including protein, lipids, mRNAs, and microRNAs (Chopp and Zhang 2015). Intravascular delivery of exosomes can reduce brain injury after stroke and TBI in preclinical models and assist in brain recovery (Chen and Chopp 2018; Zhang et al. 2019). Less is known about the potential of exosomes as drug delivery systems; for example, preloading exosomes with growth factors prior to delivery. A greater insight into the mechanisms by which exosomes are recognized and taken up by different cell types within the brain, including the endothelium, would be of great assistance (e.g., in creating artificial exosomes as vectors).

23.3.3.3 Circumventing the Blood-Brain Barrier

Another possibility to enhance drug delivery is to circumvent the BBB by direct parenchymal or CSF administration, or by using intranasal delivery (de Lange et al. 1995, 1997; Stevens et al. 2009). Direct intraparenchymal injections of therapeutics circumvent the BBB, and it avoids potential systemic clearance, metabolism, and toxicity, and the injections can be administered to localized brain regions. However, in general, direct brain administration raises a concern because of the need for surgery. In addition, single injections or slow release capsules that rely on diffusion may only result in limited penetration into nearby parenchyma (1–2 mm (Vogelbaum and Iannotti 2012)). To increase dispersion of agents, groups have used convection-enhanced delivery with a prolonged infusion of the agent, and bulk fluid flow serves to carry the drug away from the site of administration (Bobo et al. 1994; Vogelbaum and Iannotti 2012). The potential use of convection-enhanced delivery has been examined in cerebral ischemia with Gd-DPTA (Haar et al. 2010). They found that the Gd-DPTA distributed into a larger volume of the brain at lower concentrations following ischemia compared to normal conditions. These changes probably reflect the effects of cellular edema (swelling) which results in a shrinkage of the extracellular space (Hrabetova et al. 2003; Sykova 1997).

Intraventricular catheters are commonly placed in patients for CSF drainage. This makes this a more attractive route than direct parenchymal injection. The ependyma cells that line the cerebral ventricles are not linked by TJs permitting penetration of agents from CSF to brain parenchyma. However, it should be noted that CSF (particularly lumbar CSF) and brain interstitial drug concentrations may differ considerably (Yamamoto et al. 2017). With intraventricular administration, the extent of penetration into brain parenchyma may be limited without continuous drug administration (Smith et al. 2011), and this is likely to be exacerbated in the large human brain. Even with continuous infusion, there may be marked differences in drug concentration between the ependymal surface and brain areas distant from the ventricles (Milhorat et al. 1971). It should be noted that in some cases the CSF system, rather than the brain parenchyma, may be the therapeutic target. For example, intraventricular administration of tissue plasminogen activator (tPA) has been used to accelerate clearance of intraventricular blood after cerebral hemorrhage (Hanley et al. 2017).

Intranasal administration avoids the BBB by allowing entry into the brain across the olfactory epithelium along the olfactory and trigeminal neural pathways (Hanson and Frey 2008). In animal models of cerebral ischemia, intranasal administration of neuroprotectants such as insulin, insulin-like growth factor and Fas-blocking peptide have resulted in reduced ischemic brain damage (Lioutas et al. 2015; Ullah et al. 2018). A potential caveat to the use of intranasal delivery in human stroke/TBI trials relate to whether drug delivery will be homogeneous within the area affected by the brain injury. Thus, some potential therapeutics have a fairly narrow therapeutic range and may even be detrimental at high concentrations. There is, therefore, the potential for some areas of the brain to have sub-therapeutic or harmful drug delivery. This may be particularly important in TBI where areas of brain may be affected that are distant from the initial injury site. Another potential caveat is whether the injury will significantly affect the drug delivery to the injured tissue (e.g., by affecting the extracellular space/CSF flow).

23.3.4   Cell-Based Therapies

Multiple types of stem cells have been examined preclinically for efficacy in treating stroke and TBI (e.g., neural, mesenchymal, hematopoietic, and endothelial progenitor cells) with each having different pros and cons as potential treatments as outlined in Stonesifer et al. (2017). Stem cells are thought to improve outcome via reducing neuroinflammation and secondary cell death (Stonesifer et al. 2017) and are currently undergoing clinical trial (Tuazon et al. 2019).

There are questions over the best delivery route for stem cells to treat brain injury. It is interesting that different clinical trials in stroke have used different routes: Intracerebral, intravenous, intra-arterial, and intranasal delivery have all been employed (Tuazon et al. 2019). As with large molecular weight drugs, the intracerebral route avoids the need to cross the BBB but requires surgery. The ability of stem cells to migrate toward sites of injury may limit the number of sites that need to be injected. The intravenous and intra-arterial routes have relative ease of administration and a more uniform brain delivery, but the cells still need to cross the BBB. The advent of thrombectomy using intra-arterial devices as a treatment for stroke (Powers et al. 2018) raises the possibility that such devices might also deliver stem cells (or other drugs) close to the site of injury. Intranasal delivered stem cells can gain access to the brain (Lochhead and Thorne 2012), and intranasal mesenchymal stem cells administration improved recovery in a rat subarachnoid hemorrhage model (Nijboer et al. 2018). This is the least invasive approach although there are concerns over uniformity of delivery.

One concern over using stem cells as a treatment is their relatively short life span in the injured brain with its hostile microenvironment. There have been efforts to increase that preconditioning the cells prior to administration which may increase their tolerance to ischemic environments (Cai et al. 2014). It is also possible to use gene transduction to induce overexpression of potentially protective proteins (van Velthoven et al. 2009) either to protect the administered stem cell or the surrounding brain.

23.4   The Blood-Brain Barrier as Therapeutic Targeting

While the prior section focused on drug delivery to brain parenchyma in stroke and TBI, it should be noted that some targets may be intravascular (e.g., thrombolysis with tPA) or the cerebral endothelium itself. The importance of the latter is demonstrated in recent stroke studies where specifically targeting to prevent brain BBB dysfunction reduced parenchymal injury and behavioral deficits (Shi et al. 2016, 2017). Because the endothelium is directly exposed to the bloodstream, it is more accessible target than the brain parenchyma. It should be noted, however, that the "blood-brain barrier properties" of the cerebral endothelium may still limit efficacy if the site of action is other than at the luminal membrane. Thus, for example, p-glycoprotein or drug metabolizing enzymes may still reduce the concentration of substrates within the brain endothelial cell cytoplasm. Thus, several of the approaches described in the prior section may be required to enhance brain endothelial uptake.

The Glyburide Advantage in Malignant Edema and Stroke (GAMES-RP; NCT01794182) is a recent clinical trial targeting the cerebral endothelium, at least in part. Glyburide is an inhibitor of the sulfonylurea receptor 1 (Sur1)-transient receptor potential melastatin 4 (Trpm4) channel, a nonselective cation channel, that is upregulated at the cerebral endothelium after stroke and involved in brain edema formation (King et al. 2018). While that trial did not meet its primary endpoint (reducing neurological deficits), it did significantly reduce midline shift, a measure of edema formation (Sheth et al. 2016).

Endothelial protection is also the subject of much research into developing adjunct therapies for tPA (Andjelkovic et al. 2019). That agent is currently the only Food and Drug Administration (FDA)-approved drug for ischemic stroke. It acts to restore CBF in ischemic stroke patients by lysing the intravascular clot, but it can cause symptomatic ICH (Anonymous 1995), and that has limited its use. A number of adjunct therapies to limit ICH are currently being tested clinically including an activated protein C analog (3K3A-APC) ((Lyden et al. 2019); NCT02222714), an inhibitor of platelet-derived growth factor signaling, imatinib ((Wahlgren et al. 2017); NCT03639922), and a free radical scavenger, edaravone (Yamaguchi et al. 2017).

Although the beneficial actions of tPA are intravascular, there is substantial, although contentious, evidence that extravascular tPA increases brain damage (Zhu et al. 2019). This is an example of a therapeutic compound where brain penetration needs to be limited to avoid unwanted side effects. To increase its plasma half-life, tPA has been conjugated to nanoparticles (Deng et al. 2018). This might be one approach to potentially prevent passage across the BBB and limit unwanted extravascular effects.

23.5   Future Directions and Challenges

As with multiple neurological diseases, there is soul searching over the failure of so many clinical trials in stroke and TBI, particularly with regard to neuroprotectants. Is this due to inappropriate targets and agents (failure in animal modeling and understanding of underlying mechanisms of brain injury), inadequate agent delivery, lack of quantitative and time-course data, or problems with clinical trial design? Unlike animal models that are designed to be very reproducible, stroke and TBI are very heterogeneous greatly increasing the complexity of translating preclinical data. This patient-to-patient variability may alter therapeutic targets and delivery. Brain imaging has had an important impact on the use of reperfusion therapy to treat stroke (Ma et al. 2019), and it may also provide insight into which drugs should be given to brain injury patients and when. Similarly, blood biomarkers give insight into brain injury that might impact therapeutic choices. For example, blood concentrations of glial fibrillary acidic protein, neurofilament light chain, ubiquitin c-terminal hydrolase L1, and tau (proteins released from the injured brain) can currently predict which patients will have a cranial abnormality on a CT scan after TBI (Korley et al. 2018). As those biomarkers move from brain to blood, they may also give insight into the degree of BBB disruption after injury. Both imaging and blood biomarkers may be used to decide on inclusion/exclusion in particular clinical trials which may decrease variability in patient response.

There have been major advances in enhancing the delivery of therapeutics across the BBB (Table 23.2). However, there are challenges in using these approaches in the setting of stroke and TBI, including the potential impact of TJ modulation on brain injury and how decreases in the size of the extracellular space due to cytotoxic edema (Marmarou 2007) impact drug movement after crossing the BBB. A potentially even bigger challenge is how the right drug concentrations can be achieved in the right brain areas without being sub-optimal or toxic in other areas. That is very difficult when the degree of brain injury-induced BBB disruption varies with etiology, time, and location. Having agents that respond to particular brain injury signals (e.g., reactive oxygen species generation (Ballance et al. 2019)) may be one approach as might focused delivery (e.g., local delivery of ultrasound).

For cerebral ischemia, the predominant form of stroke, there are two reperfusion therapies currently available that improve outcomes, tPA and thrombectomy. For tPA, there is a 4.5 h time window for treatment, and there is evidence that this window may be even longer for thrombectomy in certain patients identified with perfusion imaging (Fisher and Albers 2013; Ma et al. 2019). As these therapies are becoming standard of care, it has changed the landscape for potential new therapeutics with a new focus on therapies that can be given in conjunction with reperfusion therapy.

Traditionally, much preclinical research has focused on acute neuronal injury after stroke and TBI. This has been gradually changing. There has been an increased focus on the events involved in long-term recovery after brain injury and how these may be modulated by therapeutics and/or rehabilitation (Chabriat et al. 2020; Hermann and Chopp 2014). Similarly, there has been a switch from specific neuronal targets to other or multiple cell types, including the NVU and endothelium. More work is needed on how to specifically target those cell types and how to use combination of therapies targeting multiple cell types. Packaging different therapeutics in liposomes/exosomes may be one approach, so might agents designed to hit multiple targets (e.g., bifunctional drugs).

23.6   Conclusion

Stroke and TBI cause BBB dysfunction that may result in marked increases in the uptake of compounds from the bloodstream into the brain. Understanding those changes and how they impact specific therapeutics may provide opportunities to enhance efficacy. However, broad statements about the BBB being disrupted after stroke and TBI and, therefore, drugs should have access to the brain and are a vast oversimplification. Indeed, the impact of stroke and TBI on drug delivery likely varies from patient to patient, changes between different brain areas, and evolves temporally within a single patient. The impact of such changes in delivery on past (failed) clinical trials for stroke and TBI and future trials needs to be addressed.

23.7 Points for Discussion

- What were the underlying causes of the failure (to date) of all non-reperfusion related clinical trials in stroke and TBI?
- To what extent was BBB delivery an underlying factor in the failure of those trials?
- Is cerebral endothelial dysfunction in human stroke and TBI a therapeutic target?
- Should combination therapies, with each targeting a specific cell type, be tested for treatment of stroke and TBI?
- Can brain imaging and blood biomarkers be developed that will help guide decisions on drug delivery for individual patients (precision medicine)?
- How safe is inducing BBB disruption in the setting of stroke and TBI?
- What are the relative benefits of intravascular, intraparenchymal, and intranasal drug delivery in the setting of stroke and TBI?

References

Abbott NJ et al (2010) Structure and function of the blood-brain barrier. Neurobiol Dis 37:13–25
Adeoye O, Broderick JP (2010) Advances in the management of intracerebral hemorrhage. Nat Rev Neurol 6:593–601
Agrawal M et al (2017) Recent advancements in liposomes targeting strategies to cross blood-brain barrier (BBB) for the treatment of Alzheimer's disease. J Control Release 260:61–77
Alfieri A et al (2011) Targeting the Nrf2-Keap1 antioxidant defence pathway for neurovascular protection in stroke. J Physiol 589:4125–4136
Andjelkovic AV et al (2019) Endothelial targets in stroke: translating animal models to human. Arterioscler Thromb Vasc Biol 39:2240–2247
Anonymous (1995) Tissue plasminogen activator for acute ischemic stroke. The National Institute of Neurological Disorders and Stroke rt-PA stroke study group. N Engl J Med 333:1581–1587
Astrup J, Siesjo BK, Symon L (1981) Thresholds in cerebral ischemia – the ischemic penumbra. Stroke 12:723–725
Ballance WC et al (2019) Reactive oxygen species-responsive drug delivery systems for the treatment of neurodegenerative diseases. Biomaterials 217:119292
Bao Q et al (2018) Simultaneous blood-brain barrier crossing and protection for stroke treatment based on edaravone-loaded ceria nanoparticles. ACS Nano 12:6794–6805
Bauer B et al (2004) Pregnane X receptor up-regulation of P-glycoprotein expression and transport function at the blood-brain barrier. Mol Pharmacol 66:413–419
Bederson JB et al (2009) Guidelines for the management of aneurysmal subarachnoid hemorrhage: a statement for healthcare professionals from a special writing group of the Stroke Council, American Heart Association. Stroke 40:994–1025

Beretta S et al (2011) Acute lipophilicity-dependent effect of intravascular simvastatin in the early phase of focal cerebral ischemia. Neuropharmacology 60:878–885

Bobo RH et al (1994) Convection-enhanced delivery of macromolecules in the brain. Proc Natl Acad Sci U S A 91:2076–2080

Borlongan CV et al (2011) The great migration of bone marrow-derived stem cells toward the ischemic brain: therapeutic implications for stroke and other neurological disorders. Prog Neurobiol 95:213–228

Bruch GE et al (2019) Liposomes for drug delivery in stroke. Brain Res Bull 152:246–256

Cai H, Zhang Z, Yang GY (2014) Preconditioned stem cells: a promising strategy for cell-based ischemic stroke therapy. Curr Drug Targets 15:771–779

Cen J et al (2013) Alteration in P-glycoprotein at the blood-brain barrier in the early period of MCAO in rats. J Pharm Pharmacol 65:665–672

Chabriat H et al (2020) Safety and efficacy of GABAA alpha5 antagonist S44819 in patients with ischaemic stroke: a multicentre, double-blind, randomised, placebo-controlled trial. Lancet Neurol 19:226–233

Chen J, Chopp M (2018) Exosome therapy for stroke. Stroke 49:1083–1090

Chopp M, Zhang ZG (2015) Emerging potential of exosomes and noncoding microRNAs for the treatment of neurological injury/diseases. Expert Opin Emerg Drugs 20:523–526

CRASH-3 Trial Investigators (2019) Effects of tranexamic acid on death, disability, vascular occlusive events and other morbidities in patients with acute traumatic brain injury (CRASH-3): a randomised, placebo-controlled trial. Lancet 394:1713–1723

Dazert P et al (2006) Differential regulation of transport proteins in the periinfarct region following reversible middle cerebral artery occlusion in rats. Neuroscience 142:1071–1079

de Lange EC et al (1995) Application of intracerebral microdialysis to study regional distribution kinetics of drugs in rat brain. Br J Pharmacol 116:2538–2544

de Lange EC et al (1997) Methodological considerations of intracerebral microdialysis in pharmacokinetic studies on drug transport across the blood-brain barrier. Brain Res Brain Res Rev 25:27–49

del Zoppo GJ (2010) Acute anti-inflammatory approaches to ischemic stroke. Ann N Y Acad Sci 1207:143–148

DeMars KM et al (2017) Spatiotemporal changes in P-glycoprotein levels in brain and peripheral tissues following ischemic stroke in rats. J Exp Neurosci 11:1179069517701741

Deng J et al (2018) Recombinant tissue plasminogen activator-conjugated nanoparticles effectively targets thrombolysis in a rat model of middle cerebral artery occlusion. Curr Med Sci 38:427–435

Dimitrijevic OB et al (2006) Effects of the chemokine CCL2 on blood-brain barrier permeability during ischemia-reperfusion injury. J Cereb Blood Flow Metab 26:797–810

Dithmer S et al (2017) Claudin peptidomimetics modulate tissue barriers for enhanced drug delivery. Ann N Y Acad Sci 1397:169–184

Doolittle ND et al (2014) Delivery of chemotherapeutics across the blood-brain barrier: challenges and advances. Adv Pharmacol 71:203–243

el-Bacha RS, Minn A (1999) Drug metabolizing enzymes in cerebrovascular endothelial cells afford a metabolic protection to the brain. Cell Mol Biol 45:15–23

Etminan N et al (2011) Effect of pharmaceutical treatment on vasospasm, delayed cerebral ischemia, and clinical outcome in patients with aneurysmal subarachnoid hemorrhage: a systematic review and meta-analysis. J Cereb Blood Flow Metab 31:1443–1451

Ewing JR et al (2003) Patlak plots of Gd-DTPA MRI data yield blood-brain transfer constants concordant with those of 14C-sucrose in areas of blood-brain opening. Magn Reson Med 50:283–292

Fernandes LF et al (2018) Recent advances in the therapeutic and diagnostic use of liposomes and carbon nanomaterials in ischemic stroke. Front Neurosci 12:453

Fisher M, Albers GW (2013) Advanced imaging to extend the therapeutic time window of acute ischemic stroke. Ann Neurol 73:4–9

Gan Y et al (2013) Gene delivery with viral vectors for cerebrovascular diseases. Front Biosci (Elite Ed) 5:188–203

GBD2016 (2017) Global, regional, and national age-sex specific mortality for 264 causes of death, 1980-2016: a systematic analysis for the Global Burden of Disease Study 2016. Lancet 390:1151–1210

Gess B et al (2011) Sodium-dependent vitamin C transporter 2 (SVCT2) expression and activity in brain capillary endothelial cells after transient ischemia in mice. PLoS ONE [Electronic Resource] 6:e17139

Ghersi-Egea J-F et al (2006) Brain protection at the blood-cerebrospinal fluid interface involves a glutathione-dependent metabolic barrier mechanism. J Cereb Blood Flow Metab 26:1165–1175

Gomes JA et al (2005) Glucocorticoid therapy in neurologic critical care. Crit Care Med 33:1214–1224

Haar PJ et al (2010) Quantification of convection-enhanced delivery to the ischemic brain. Physiol Meas 31:1075–1089

Haley MJ, Lawrence CB (2017) The blood-brain barrier after stroke: structural studies and the role of transcytotic vesicles. J Cereb Blood Flow Metab 37:456–470

Hanley DF et al (2017) Thrombolytic removal of intraventricular haemorrhage in treatment of severe stroke: results of the randomised, multicentre, multiregion, placebo-controlled CLEAR III trial. Lancet 389:603–611

Hanson LR, Frey WH 2nd (2008) Intranasal delivery bypasses the blood-brain barrier to target therapeutic agents to the central nervous system and treat neurodegenerative disease. BMC Neurosci 9(Suppl 3):S5

Hermann DM, Chopp M (2014) Promoting neurological recovery in the post-acute stroke phase: benefits and challenges. Eur Neurol 72:317–325

Higashida T et al (2011) The role of hypoxia-inducible factor-1alpha, aquaporin-4, and matrix metalloproteinase-9 in blood-brain barrier disruption and brain edema after traumatic brain injury. J Neurosurg 114:92–101

Hollt V et al (1992) Stereoisomers of calcium antagonists which differ markedly in their potencies as calcium blockers are equally effective in modulating drug transport by P-glycoprotein. Biochem Pharmacol 43:2601–2608

Horng S et al (2017) Astrocytic tight junctions control inflammatory CNS lesion pathogenesis. J Clin Invest 127:3136–3151

Hrabetova S, Hrabe J, Nicholson C (2003) Dead-space microdomains hinder extracellular diffusion in rat neocortex during ischemia. J Neurosci 23:8351–8359

Iadecola C (2017) The neurovascular unit coming of age: a journey through neurovascular coupling in health and disease. Neuron 96:17–42

Iadecola C, Anrather J (2011) The immunology of stroke: from mechanisms to translation. Nat Med 17:796–808

Injury GBDTB, Spinal Cord Injury C (2019) Global, regional, and national burden of traumatic brain injury and spinal cord injury, 1990–2016: a systematic analysis for the Global Burden of Disease Study 2016. Lancet Neurol 18:56–87

Jiang X et al (2018) Blood-brain barrier dysfunction and recovery after ischemic stroke. Prog Neurobiol 163-164:144–171

Jiao H et al (2011) Specific role of tight junction proteins claudin-5, occludin, and ZO-1 of the blood-brain barrier in a focal cerebral ischemic insult. J Mol Neurosci 44:130–139

Johanson CE et al (2000) Choroid plexus recovery after transient forebrain ischemia: role of growth factors and other repair mechanisms. Cell Mol Neurobiol 20:197–216

Jones TH et al (1981) Thresholds of focal cerebral ischemia in awake monkeys. J Neurosurg 54:773–782

Juttner J et al (2019) Targeting neuronal and glial cell types with synthetic promoter AAVs in mice, non-human primates and humans. Nat Neurosci 22:1345–1356

Kageyama T et al (2000) The 4F2hc/LAT1 complex transports L-DOPA across the blood-brain barrier. Brain Res 879:115–121

Kago T et al (2006) Cerebral ischemia enhances tyrosine phosphorylation of occludin in brain capillaries. Biochem Biophys Res Commun 339:1197–1203

Karimy JK et al (2017) Inflammation-dependent cerebrospinal fluid hypersecretion by the choroid plexus epithelium in posthemorrhagic hydrocephalus. Nat Med 23:997–1003

Keep RF, Hua Y, Xi G (2012) Intracerebral haemorrhage: mechanisms of injury and therapeutic targets. Lancet Neurol 11:720–731

Keep RF et al (2014) Vascular disruption and blood-brain barrier dysfunction in intracerebral hemorrhage. Fluids Barriers CNS 11:18

Keep RF et al (2018) Brain endothelial cell junctions after cerebral hemorrhage: changes, mechanisms and therapeutic targets. J Cereb Blood Flow Metab 38:1255–1275

Kernie SG, Parent JM (2010) Forebrain neurogenesis after focal ischemic and traumatic brain injury. Neurobiol Dis 37:267–274

King ZA et al (2018) Profile of intravenous glyburide for the prevention of cerebral edema following large hemispheric infarction: evidence to date. Drug Des Devel Ther 12:2539–2552

Knowland D et al (2014) Stepwise recruitment of transcellular and paracellular pathways underlies blood-brain barrier breakdown in stroke. Neuron 82:603–617

Kohle C, Bock KW (2007) Coordinate regulation of phase I and II xenobiotic metabolisms by the Ah receptor and Nrf2. Biochem Pharmacol 73:1853–1862

Korley FK et al (2018) Performance evaluation of a multiplex assay for simultaneous detection of four clinically relevant traumatic brain injury biomarkers. J Neurotrauma 36(1):182–187

Kurinami H et al (2014) Prohibitin viral gene transfer protects hippocampal CA1 neurons from ischemia and ameliorates postischemic hippocampal dysfunction. Stroke 45:1131–1138

Leinenga G et al (2016) Ultrasound treatment of neurological diseases – current and emerging applications. Nat Rev Neurol 12:161–174

Lioutas VA et al (2015) Intranasal insulin and insulin-like growth factor 1 as neuroprotectants in acute ischemic stroke. Transl Stroke Res 6:264–275

Liu XD, Zhang L, Xie L (2003) Effect of P-glycoprotein inhibitors erythromycin and cyclosporin A on brain pharmacokinetics of nimodipine in rats. Eur J Drug Metab Pharmacokinet 28:309–313

Lochhead JJ, Thorne RG (2012) Intranasal delivery of biologics to the central nervous system. Adv Drug Deliv Rev 64:614–628

Logsdon AF et al (2015) Role of microvascular disruption in brain damage from traumatic brain Injury. Compr Physiol 5:1147–1160

Lyden P et al (2019) Final results of the RHAPSODY trial: a multi-center, phase 2 trial using a continual reassessment method to determine the safety and tolerability of 3K3A-APC, a recombinant variant of human activated protein C, in combination with tissue plasminogen activator, mechanical thrombectomy or both in moderate to severe acute ischemic stroke. Ann Neurol 85:125–136

Ma H et al (2019) Thrombolysis guided by perfusion imaging up to 9 hours after onset of stroke. N Engl J Med 380:1795–1803

Maas AI, Stocchetti N, Bullock R (2008) Moderate and severe traumatic brain injury in adults. Lancet Neurol 7:728–741

Marmarou A (2007) A review of progress in understanding the pathophysiology and treatment of brain edema. Neurosurg Focus 22:E1

Martin-Schild S et al (2009) Combined neuroprotective modalities coupled with thrombolysis in acute ischemic stroke: a pilot study of caffeinol and mild hypothermia. J Stroke Cerebrovasc Dis 18:86–96

Meijer OC et al (1998) Penetration of dexamethasone into brain glucocorticoid targets is enhanced in mdr1A P-glycoprotein knockout mice. Endocrinology 139:1789–1793

Menard C et al (2017) Social stress induces neurovascular pathology promoting depression. Nat Neurosci 20:1752–1760

Menzies SA, Betz AL, Hoff JT (1993) Contributions of ions and albumin to the formation and resolution of ischemic brain edema. J Neurosurg 78:257–266

Milhorat TH et al (1971) Cerebrospinal fluid production by the choroid plexus and brain. Science 173:330–332

Miller DS (2015) Regulation of ABC transporters blood-brain barrier: the good, the bad, and the ugly. Adv Cancer Res 125:43–70

Morris-Blanco KC et al (2019) Induction of DNA hydroxymethylation protects the brain after stroke. Stroke. https://doi.org/10.1161/STROKEAHA.119.025665

Nagaraja TN et al (2007) Relative distribution of plasma flow markers and red blood cells across BBB openings in acute cerebral ischemia. Neurol Res 29:78–80

Nijboer CH et al (2018) Intranasal stem cell treatment as a novel therapy for subarachnoid hemorrhage. Stem Cells Dev 27:313–325

Pardridge WM (2006) Molecular Trojan horses for blood-brain barrier drug delivery. Curr Opin Pharmacol 6:494–500

Patak P, Hermann DM (2011) ATP-binding cassette transporters at the blood-brain barrier in ischaemic stroke. Curr Pharm Des 17:2787–2792

Piriyawat P et al (2003) Pilot dose-escalation study of caffeine plus ethanol (caffeinol) in acute ischemic stroke. Stroke 34:1242–1245

Powers WJ et al (2018) 2018 guidelines for the early management of patients with acute ischemic stroke: a guideline for healthcare professionals from the American Heart Association/American Stroke Association. Stroke 49:e46–e110

Preston E, Foster DO (1997) Evidence for pore-like opening of the blood-brain barrier following forebrain ischemia in rats. Brain Res 761:4–10

Preston E, Webster J (2002) Differential passage of [14C]sucrose and [3H]inulin across rat blood-brain barrier after cerebral ischemia. Acta Neuropathol 103:237–242

Raichle ME et al (1976) Blood-brain barrier permeability of 11C-labeled alcohols and 15O-labeled water. Am J Phys 230:543–552

Rayasam A et al (2018) Immune responses in stroke: how the immune system contributes to damage and healing after stroke and how this knowledge could be translated to better cures? Immunology 154:363–376

Rhodes J (2011) Peripheral immune cells in the pathology of traumatic brain injury? Curr Opin Crit Care 17:122–130

Robinson PJ (1990) Measurement of blood-brain barrier permeability. Clin Exp Pharmacol Physiol 17:829–840

Roger VL et al (2012) Heart disease and stroke statistics – 2012 update: a report from the American Heart Association. Circulation 125:e2–e220

Romermann K et al (2017) Multiple blood-brain barrier transport mechanisms limit bumetanide accumulation, and therapeutic potential, in the mammalian brain. Neuropharmacology 117:182–194

Rosenberg GA, Yang Y (2007) Vasogenic edema due to tight junction disruption by matrix metalloproteinases in cerebral ischemia. Neurosurg Focus 22:E4

Schinkel AH et al (1995) Absence of the mdr1a P-glycoprotein in mice affects tissue distribution and pharmacokinetics of dexamethasone, digoxin, and cyclosporin A. J Clin Investig 96:1698–1705

Schubert GA, Thome C (2008) Cerebral blood flow changes in acute subarachnoid hemorrhage. Front Biosci 13:1594–1603

Sheth KN et al (2016) Safety and efficacy of intravenous glyburide on brain swelling after large hemispheric infarction (GAMES-RP): a randomised, double-blind, placebo-controlled phase 2 trial. Lancet Neurol 15:1160–1169

Shi Y et al (2016) Rapid endothelial cytoskeletal reorganization enables early blood-brain barrier disruption and long-term ischaemic reperfusion brain injury. Nat Commun 7:10523

Shi Y et al (2017) Endothelium-targeted overexpression of heat shock protein 27 ameliorates blood-brain barrier disruption after ischemic brain injury. Proc Natl Acad Sci U S A 114:E1243–E1252

Shlosberg D et al (2010) Blood-brain barrier breakdown as a therapeutic target in traumatic brain injury. Nat Rev Neurosci 6:393–403

Sladojevic N et al (2019) Claudin-1-dependent destabilization of the blood-brain barrier in chronic stroke. J Neurosci 39:743–757

Smith DE et al (2011) Distribution of glycylsarcosine and cefadroxil among cerebrospinal fluid, choroid plexus, and brain parenchyma after intracerebroventricular injection is markedly different between wild-type and Pept2 null mice. J Cereb Blood Flow Metab 31:250–261

Song B-W et al (2002) Enhanced neuroprotective effects of basic fibroblast growth factor in regional brain ischemia after conjugation to a blood-brain barrier delivery vector. J Pharmacol Exp Ther 301:605–610

Spudich A et al (2006) Inhibition of multidrug resistance transporter-1 facilitates neuroprotective therapies after focal cerebral ischemia. Nat Neurosci 9:487–488

Stamatovic SM, Keep RF, Andjelkovic AV (2008) Brain endothelial cell-cell junctions: how to "open" the blood brain barrier. Curr Neuropharmacol 6:179–192

Stamatovic SM et al (2009) Caveolae-mediated internalization of occludin and claudin-5 during CCL2-induced tight junction remodeling in brain endothelial cells. J Biol Chem 284:19053–19066

Stamatovic SM et al (2016) Junctional proteins of the blood-brain barrier: new insights into function and dysfunction. Tissue Barriers 4:e1154641

Stevens J et al (2009) A new minimal-stress freely-moving rat model for preclinical studies on intranasal administration of CNS drugs. Pharm Res 26:1911–1917

Stonesifer C et al (2017) Stem cell therapy for abrogating stroke-induced neuroinflammation and relevant secondary cell death mechanisms. Prog Neurobiol 158:94–131

Sweeney MD, Ayyadurai S, Zlokovic BV (2016) Pericytes of the neurovascular unit: key functions and signaling pathways. Nat Neurosci 19:771–783

Sykova E (1997) Extracellular space volume and geometry of the rat brain after ischemia and central injury. Adv Neurol 73:121–135

Szmydynger-Chodobska J et al (2012) Posttraumatic invasion of monocytes across the blood-cerebrospinal fluid barrier. J Cereb Blood Flow Metab 32:93–104

Tanaka H, Mizojiri K (1999) Drug-protein binding and blood-brain barrier permeability. J Pharmacol Exp Ther 288:912–918

Thimmulappa RK et al (2002) Identification of Nrf2-regulated genes induced by the chemopreventive agent sulforaphane by oligonucleotide microarray. Cancer Res 62:5196–5203

Topakian R et al (2010) Blood-brain barrier permeability is increased in normal-appearing white matter in patients with lacunar stroke and leucoaraiosis. J Neurol Neurosurg Psychiatry 81:192–197

Tuazon JP, Castelli V, Borlongan CV (2019) Drug-like delivery methods of stem cells as biologics for stroke. Expert Opin Drug Deliv 16(8):823–833

Uchida Y et al (2011) Blood-brain barrier (BBB) pharmacoproteomics: reconstruction of in vivo brain distribution of 11 P-glycoprotein substrates based on the BBB transporter protein concentration, in vitro intrinsic transport activity, and unbound fraction in plasma and brain in mice. J Pharmacol Exp Ther 339:579–588

Ullah I et al (2018) Intranasal delivery of a Fas-blocking peptide attenuates Fas-mediated apoptosis in brain ischemia. Sci Rep 8:15041

van Velthoven CTJ et al (2009) Regeneration of the ischemic brain by engineered stem cells: fuelling endogenous repair processes. Brain Res Rev 61:1–13

Vogelbaum MA, Iannotti CA (2012) Convection-enhanced delivery of therapeutic agents into the brain. Handb Clin Neurol 104:355–362

Wahlgren N et al (2017) Randomized assessment of imatinib in patients with acute ischaemic stroke treated with intravenous thrombolysis. J Intern Med 281:273–283

Walker PA et al (2010) Intravenous multipotent adult progenitor cell therapy for traumatic brain injury: preserving the blood brain barrier via an interaction with splenocytes. Exp Neurol 225:341–352

Wang J (2010) Preclinical and clinical research on inflammation after intracerebral hemorrhage. Prog Neurobiol 92:463–477

Wu SK et al (2014) Targeted delivery of erythropoietin by transcranial focused ultrasound for neuroprotection against ischemia/reperfusion-induced neuronal injury: a long-term and short-term study. PLoS One 9:e90107

Xiang J et al (2017) The choroid plexus as a site of damage in hemorrhagic and ischemic stroke and its role in responding to injury. Fluids Barriers CNS 14:8

Yamaguchi T et al (2017) Edaravone with and without. 6 Mg/Kg Alteplase within 4.5 hours after ischemic stroke: a prospective cohort study (PROTECT4.5). J Stroke Cerebrovasc Dis 26:756–765

Yamamoto Y, Danhof M, de Lange ECM (2017) Microdialysis: the key to physiologically based model prediction of human CNS target site concentrations. AAPS J 19:891–909

Yang Y et al (2007) Matrix metalloproteinase-mediated disruption of tight junction proteins in cerebral vessels is reversed by synthetic matrix metalloproteinase inhibitor in focal ischemia in rat. J Cereb Blood Flow Metab 27:697–709

Yao Y (2019) Basement membrane and stroke. J Cereb Blood Flow Metab 39:3–19

Yates CR et al (2003) Structural determinants of P-glycoprotein-mediated transport of glucocorticoids. Pharm Res 20:1794–1803

Zacchigna S, Zentilin L, Giacca M (2014) Adeno-associated virus vectors as therapeutic and investigational tools in the cardiovascular system. Circ Res 114:1827–1846

Zagrean AM et al (2018) Multicellular crosstalk between exosomes and the neurovascular unit after cerebral ischemia. Ther Implications Front Neurosci 12:811

Zhang ZG, Chopp M (2016) Exosomes in stroke pathogenesis and therapy. J Clin Invest 126:1190–1197

Zhang Y, Pardridge WM (2001) Neuroprotection in transient focal brain ischemia after delayed intravenous administration of brain-derived neurotrophic factor conjugated to a blood-brain barrier drug targeting system. Stroke 32:1378–1384

Zhang ZG, Buller B, Chopp M (2019) Exosomes – beyond stem cells for restorative therapy in stroke and neurological injury. Nat Rev Neurol 15:193–203

Zhao J et al (2007) Enhancing expression of Nrf2-driven genes protects the blood brain barrier after brain injury. J Neurosci 27:10240–10248

Zhu J et al (2019) The role of endogenous tissue-type plasminogen activator in neuronal survival after ischemic stroke: friend or foe? Cell Mol Life Sci 76:1489–1506

Zlokovic BV et al (1993) Differential brain penetration of cerebroprotective drugs. Adv Exp Med Biol 331:117–120

Chapter 24
Drug Delivery to Primary and Metastatic Brain Tumors: Challenges and Opportunities

Surabhi Talele, Afroz S. Mohammad, Julia A. Schulz, Bjoern Bauer, Anika M. S. Hartz, Jann N. Sarkaria, and William F. Elmquist

Abstract The effective treatment of brain tumors is a considerable challenge in part due to the presence of the blood-brain barrier (BBB) that limits drug delivery. Multiple hurdles pose challenges in identifying drugs that may be effective in treating brain tumors, including limited central nervous system (CNS) distribution of therapeutics, heterogeneous disruption of the blood-brain barrier in the regions of the tumor that lead to heterogenous drug distribution within the tumor, and genetic heterogeneity of tumor drivers. This chapter discusses the current standard of care and its limitations, as well as complex challenges in the treatment of primary and metastatic brain tumors. We review a variety of prospective delivery solutions of therapeutics to the brain and CNS for the treatment of brain tumors that will in the future lead to opening new doors for more effective treatments.

Keywords Blood-tumor barrier · Glioblastoma · Active transporters · Receptor mediated transport · Antibody drug conjugates · Immunotherapy

S. Talele · A. S. Mohammad · W. F. Elmquist (✉)
Department of Pharmaceutics, Brain Barriers Research Center, University of Minnesota, Minneapolis, Minnesota, USA
e-mail: talel005@umn.edu; elmqu011@umn.edu

J. A. Schulz · B. Bauer
Department of Pharmaceutical Sciences, College of Pharmacy, University of Kentucky, Lexington, Kentucky, USA
e-mail: julia.schulz@uky.edu; bjoern.bauer@uky.edu

A. M. S. Hartz
Sanders-Brown Center on Aging, University of Kentucky, Lexington, Kentucky, USA

Department of Pharmacology and Nutritional Sciences, College of Medicine, University of Kentucky, Lexington, Kentucky, USA
e-mail: anika.hartz@uky.edu

J. N. Sarkaria
Department of Radiation Oncology, Mayo Clinic, Rochester, Minnesota, USA
e-mail: Sarkaria.Jann@mayo.edu

© American Association of Pharmaceutical Scientists 2022
E. C. M. de Lange et al. (eds.), *Drug Delivery to the Brain*, AAPS Advances in the Pharmaceutical Sciences Series 33, https://doi.org/10.1007/978-3-030-88773-5_24

24.1 Tumors of the CNS: The Disease

More than 23,500 new cases of primary brain and CNS tumors are expected in the United States in 2020, which will account for approximately 1.3% of overall cancer cases and represent the 10th leading cause of death for men and women (Siegel et al. 2019). In addition, an estimated 10%–20% of all cancer patients will develop brain metastases (Lin and DeAngelis 2015). In the United States, an estimated 98,000–170,000 cases of brain metastases occur each year (Amsbaugh and Kim 2019). Although brain and CNS tumors are a rare occurrence in adults, they are a significant cause of mortality and are the most common solid tumors in infants and children (McNeill 2016). Brain tumors are broadly classified into two types based on their site of origin—primary brain tumors and secondary/metastatic brain tumors. Primary brain tumors are those that originate within the brain or the surrounding areas of the CNS like the meninges or spinal cord. Conversely, secondary or metastatic tumors are those that originate elsewhere in the body and later spread to the brain. Diagnosis of brain malignancies and their treatment are often very complex and are associated with serious cognitive and functional impairment of patients and psychological stress to the patients as well as their families. These tumors have a grim prognosis with a median survival ranging between 4 and 15 months after diagnosis (Parrish et al. 2015; Pan-Weisz et al. 2019). Many experimental therapies that have shown promise in preclinical studies ultimately fail clinical trials for CNS tumors, and therefore the incidences of primary and metastatic brain tumors continue to rise. Most of the drugs in the pipeline do not cross the formidable hurdle of the blood-brain barrier (BBB) to be effectively delivered to the tumor site. Therefore, it is imperative to develop therapies that take into consideration the presence of an intact BBB in the invasive regions surrounding the tumor which continue to grow even after surgical resection (Sarkaria et al. 2018). Advancements should also be made to develop novel drug delivery systems exploiting various aspects of BBB anatomy and physiology in and around the tumor. Moreover, it will be necessary to better understand the complex cellular signaling pathways that lead to tumor proliferation and invasiveness. Finally, novel technologies may be utilized to modify the BBB to deliver therapeutics across CNS barriers to the tumor site. In this chapter, we will discuss the challenges to effective treatment for both primary and metastatic brain tumors.

24.1.1 Primary Brain Tumors

Primary brain tumors can be classified as malignant or nonmalignant based on the presence of proliferative and invasive cancer cells within the tumor. The five-year survival rate in adults following the diagnosis of a malignant brain or other CNS tumor is 35.8%; in contrast, the five-year survival rate following diagnosis of a nonmalignant brain or other CNS tumor is 91.5%, based on the Central Brain Tumor

Registry of the United States (CBTRUS) report compiling cases from 2012 to 2016 (Ostrom et al. 2019). The most common CNS tumors in children are pilocytic astrocytoma, embryonal tumors, and malignant gliomas, whereas meningiomas, pituitary tumors, and malignant gliomas are the most common brain tumor types in adults (McNeill 2016).

Gliomas are primary brain tumors that are thought to originate from neuroglial stem cells or progenitor cells. On the basis of their histological appearance, they have been traditionally classified as astrocytic, oligodendroglial, or ependymal tumors and assigned WHO grades I-IV, indicating different degrees of malignancy based on genomic, transcriptomic, and epigenetic profiling (Weller et al. 2015). These tumors vary widely in histology from benign and potentially surgically curable grade I tumors (pilocytic astrocytoma) to locally aggressive infiltrative grade IV tumors with a high risk of recurrence (glioblastoma). Survival varies by histology, with pilocytic astrocytoma having a 10-year survival of greater than 90%, whereas only about 5% of patients with glioblastoma survive up to 5 years (McNeill 2016).

Glioblastoma (GBM) is the most common primary malignant brain tumor with ~14,000 new cases per year in the United States with a 2-year survival rate of 16.9% (Ostrom et al. 2014). Currently, the United States has approximately 50,000 GBM patients. In other developed countries worldwide, approximately 3.5 GBM cases per 100,000 people are newly diagnosed each year (Porter et al. 2010). According to the World Health Organization classification system, GBMs are grade IV neoplasms (where grade I refers to the least severe and grade IV to the most severe), reflecting their highly malignant behavior (Perkins and Liu 2016). GBMs are highly infiltrative and therefore not a surgically curable disease. Tumor cells invade the surrounding brain regions and have a diffused nature making complete surgical resection impossible (Cloughesy et al. 2014; Sarkaria et al. 2018).

24.1.2 Metastatic Brain Tumors

Cancer metastasis from primary tumors to the brain is a significant concern in cancer patient management (Sperduto et al. 2012). Brain metastases are difficult to detect and diagnose, especially in early stages of the disease and have an extremely grim prognosis (Bruzzone et al. 2012). Lung cancer, breast cancer, melanoma, renal cell carcinoma, and colorectal cancer are among the tumor types associated with high brain-metastatic prevalence (Achrol et al. 2019). While lung cancer has been reported to have the largest incidence rate of brain metastases, melanoma has the highest likelihood of metastasizing to the brain (Nayak et al. 2012). Rising incidences of brain metastases can be attributed to the improvement of advanced imaging techniques for early detection as well as effective systemic treatment of the peripheral disease that extends patient survival (Fokas et al. 2013). Another reason for limited success in the therapies for brain metastases is the restricted entry of systemically active therapeutic agents into the brain because of the BBB. The BBB creates a pharmacological sanctuary that allows the growth and development of the tumor cells within the brain (Kim et al. 2018). The mechanisms by which brain metastases occur have not been well described; however, the prevalence of these metastases for a variable duration before being detected poses a treatment challenge. In addition, after the initial detection of brain metastases, there is a high likelihood of undetected "micrometastases" that will be protected by a relatively intact BBB at those locations within the brain (Oberoi et al. 2016). Therefore, to advance treatments for brain metastases, consideration of the condition of the BBB in these regions is essential, especially in the non-contrast enhancing regions of the micrometastases where the BBB can impede the delivery of anticancer agents to the tumor cells.

24.2 Standard of Care for Primary Brain Tumors

The current standard of care for primary brain tumors reflects the need to develop more effective treatments that have improved delivery to the tumor target sites. Clinical signs and symptoms of primary brain tumors progress from early symptoms like headaches and seizures due to increased intracranial pressure to more focal symptoms like dizziness and change in personality traits as the tumor grows in size and infiltrates to different areas of the brain (Perkins and Liu 2016). The diagnosis of these tumors is done with the help of gadolinium enhanced magnetic resonance imaging (MRI) or computed tomography (CT). Advanced imaging techniques combined with MRI significantly help in the diagnosis of tumor subtype. Treatment decisions are individualized by an experienced multidisciplinary team consisting of medical oncology, radiation oncology, and neurosurgery. Treatment decisions are based on tumor type and location, malignancy potential, and the patient's age and physical condition. Treatment options include a combination of surgery, chemotherapy, and radiation therapies (Alifieris and Trafalis 2015).

The current standardized treatment for GBM involves a multidisciplinary approach with maximal safe surgical resection possible, followed by concurrent radiation with temozolomide (TMZ), an oral DNA alkylating agent, followed by adjuvant chemotherapy with TMZ (McClelland et al. 2018). Following surgical resection, the chemoradiation schedule begins 4 weeks after the patient's recovery from the surgery. Radiation using three-dimensional conformal beam or intensity-modulated radiotherapy (RT) is now the standard of care, where typical total dose delivered is 60 Gray (Gy), in 1.8–2 Gy fractions administered 5 days per week for 6 weeks (J.G. et al., 2011). A clear survival advantage has been demonstrated with postoperative RT doses to 60 Gy, but dose escalation beyond this has resulted in increased toxicity without additional survival benefits (Barani and Larson 2015). Concurrent with RT, TMZ is typically given at a dose of 75 mg/m^2 daily for 6 weeks, followed by a rest period of about one month after RT is completed. When restarted, TMZ is dosed at 150 mg/m^2 daily for 5 days for the first month (usually days 1–5 of 28). If tolerated, the dose is escalated up to 200 mg/m^2 for five consecutive days per month for the rest of the treatment period (Davis 2016). The importance of the methylation of the O^6-methylguanine DNA methyltransferase (MGMT) gene in standard GBM therapy has been demonstrated by Stupp et al. in 2008. MGMT codes for an enzyme involved with DNA repair. Patients who have methylated (not activated) MGMT exhibit compromised DNA repair. When the MGMT enzyme is activated, it can interfere with the effects of treatment. RT and alkylating chemotherapy exert their therapeutic effects by causing DNA damage and cytotoxicity and triggering apoptosis. Therefore, the expression of methylated MGMT is beneficial for patients undergoing TMZ chemotherapy and RT (Stupp et al. 2008). As one can see from above, the standard of care comprised of radiation and TMZ represents a limited choice of therapy even in light of our improved knowledge of the biology of GBM. It is important to note that radiation is a highly brain penetrant therapy and TMZ, a small molecule alkylating agent, also has comparatively good brain penetration (Portnow et al. 2009).

In addition, a humanized monoclonal antibody, bevacizumab, targeting vascular endothelial growth factor (VEGF) has been approved for the treatment of recurrent GBM, but it has not shown any improvement in the overall survival of patients (Chowdhary and Chamberlain 2013). In 2015, the FDA approved another local treatment option called Optune for newly diagnosed and recurrent GBM with concomitant TMZ. Optune is a device delivering electrical fields to the brain. It emits low intensity electricity (100–300 kHz frequency) delivered through a series of transducer arrays placed regionally around the tumor region. These electrical fields have been shown to selectively disrupt cell division in the case of brain tumors. Patients with a 90% or greater compliance rate of using Optune had a median overall survival of 24.9 months (28.7 months from diagnosis) and a 5-year survival rate of 29.3% (Toms et al. 2019). Again, similar to the above standard of care, this treatment option clearly has excellent BBB penetration.

24.3 Standard of Care for Metastatic Brain Tumors

Approximately 80% of brain metastases are localized in the cerebral hemispheres (Delattre et al. 1988). Initial symptoms range from seizures and headaches to cognitive dysfunction and neurological deficits; however in some early stages, asymptomatic brain metastases are also commonly found using imaging techniques (Kim et al. 2018). Clinical treatment in most cases is mostly palliative and rarely ever curative. The prognosis and treatment modalities are affected by a variety of factors, including size, number, and location of metastases; age and performance status of the patient; type of the tumor; and active extracranial disease presence (Arvanitis et al. 2018). Given the prevalence and grim prognosis of metastatic tumors in the brain, there is a great unmet need in improving specific treatments that will require adequate penetration across the BBB.

Treatment of brain metastases closely mirrors the treatment of primary brain tumors (Fig. 24.1). Stereotactic radiosurgery or gamma knife radiation can be used

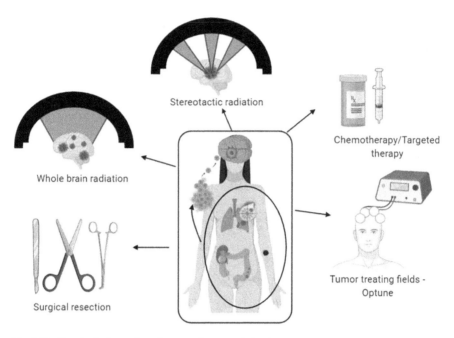

Fig. 24.1 Standard of care for primary and metastatic brain tumors

as the first option for the maximum safe surgical resection where there are few (typically <4) metastases present (Oberoi et al. 2016; Stupp 2019). In many cases, due to the size, number, or location of the tumor, surgery is not possible, and hence patients are treated with whole brain radiation therapy (WBRT). TMZ is the first-line chemotherapeutic used for GBM; however, no such proven chemotherapeutic options have been specifically effective in brain metastases (Oberoi et al. 2016). CNS metastases often express similar characteristics and sensitivities to their primary tumors and hence are treated based on their subtype and primary source of origin (Rick et al. 2019). The use of molecularly targeted agents has been on the rise in the case of non-small cell lung carcinoma (NSLC) metastases over the last 15 years in cases with evidence of drug sensitivity for specific tumor mutations (Lim et al. 2019). The use of epidermal growth factor receptor tyrosine kinase (EGFR-TKI) inhibitors gefitinib, erlotinib, and more recently the effective use of a third-generation EGFR-TKI inhibitor, osimertinib, on EGFR-mutated NSLC brain metastases are examples of such therapy (Dempke et al. 2015; Reungwetwattana et al. 2018; Soria et al. 2018; Ramalingam et al. 2020). In anaplastic lymphoma kinase (ALK) fusion protein-positive NSLC metastases, which are rare, two inhibitors—crizotinib and alectinib—have demonstrated treatment benefit (Shaw et al. 2017; Tran and Klempner 2017; Gadgeel et al. 2018). Importantly, an ALK inhibitor, lorlatinib, a molecule designed for improved BBB penetration through decreased efflux liability, showed substantial intracranial activity in a phase II study in patients with pretreated ALK-positive NSCLC (with or without baseline CNS metastases), whose disease had progressed on crizotinib or other second-generation ALK TKIs (Bauer et al. 2020). CNS metastases from breast cancer have been very difficult to effectively treat. There have been no FDA-approved systemic therapies until April 2020 with the approval of tucatinib, an oral small-molecule tyrosine kinase inhibitor of HER2 in combination with trastuzumab and capecitabine (Murthy et al. 2018, 2020). In the case of melanoma metastases, FDA approvals in recent years have included BRAF inhibitors vemurafenib and dabrafenib; a MEK inhibitor, trametinib; and an anti-CTLA4 antibody, ipilimumab (Parrish et al. 2015). However, for patients with brain tumor metastasis, the standard of care remains radiation and surgery due to limited brain distribution of these agents.

24.4 Challenges in the Treatment of Brain Tumors

Despite the aggressive multimodal approach of surgery, chemotherapy, and radiation for the treatment of brain tumors, the expected survival for patients with GBM is approximately 15 months, and for patients with brain metastases, it is approximately 4–6 months (Bi and Beroukhim 2014; Liu, Tong and Wang, 2019). As described earlier, extensive and complete surgical resection of brain tumors is difficult because they are frequently invasive and are often in areas of the brain that control speech, motor function, and the senses. TMZ, used as the first-line chemotherapeutic for the treatment of GBM, is only beneficial for a subset of patients (~50%) having the MGMT promoter methylation, and this limits its effectiveness in a broad patient population (Lee 2016). In the case of radiation, the side effects range from short-term conditions like inflammation and edema to long-term effects like radiation necrosis, blindness, and cognitive dysfunction (Laack and Brown 2004).

The identification and development of drug delivery strategies that can be used with the current standard of care of radiation and chemotherapy is a significant challenge in oncology, with multiple hurdles to be overcome. These hurdles are depicted in Fig 24.2. First, an important reason drug molecules often have limited brain penetration is due to the presence of efflux transporters at the blood-brain barrier (BBB) and the blood-tumor barrier (Omidi and Barar 2012). Second, the complex tumor microenvironment communicates with other cells in the brain environment in a manner that leads to the promotion of tumor progression and resistance to treatments (Trédan et al. 2007; Perus and Walsh 2019). Third, spatial heterogeneity of drug distribution is a critical consideration in the context of brain tumors, many of which exhibit both a partially intact BBB as well as heterogenous BBB disruption in different regions of the tumor and area surrounding the tumor (Sarkaria et al. 2018). Fourth, the highly heterogenous genetic makeup of GBM from patient to patient as well as within the tumor of a single patient presents significant additional challenges. This highlights the need to understand these complexities to be able to successfully identify agents that can selectively and significantly benefit a subset of the GBM population (Bastien et al. 2015). Finally, given the limited understanding of how molecularly targeted agents assist radiation and chemotherapy,

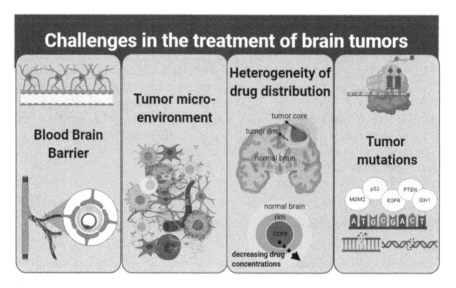

Fig. 24.2 Challenges in the treatment of brain tumors

understanding heterogenous distribution is critical to avoid the development of resistance. Moreover, determining the delivery of these agents to normal tissues leading to toxic side effects needs to be examined in conjunction with measuring specific pharmacodynamic effects that can demonstrate efficacy in tumor cells and toxicity in normal tissues.

24.5 Transporter Expression in Brain Tumors

The brain depends on nutrients for its growth and development and also needs to be protected from circulating xenobiotics and toxins. This selective entry into the brain is modulated by the presence of membrane-embedded receptors that act as transport systems (Cardoso et al. 2010). While active influx transporters and facilitated carriers are necessary for the transport of essential nutrients and growth factors, a second type of transporters, critical for brain delivery of therapeutic agents, is the efflux transport systems that are mainly comprised of the ABC (ATP binding cassette) super family that uses ATP hydrolysis to provide energy to efflux molecules from the brain back to the blood. The most relevant ABC transporters expressed in the brain endothelial cells are P-glycoprotein (P-gp), breast cancer resistance protein (BCRP), and the multidrug resistance-associated proteins (MRP) (Löscher and Potschka 2005). The expression of these transporters is depicted in Fig. 24.3.

P-glycoprotein (P-gp, ABCB1) P-gp expression was first detected in the BBB by Cordon-Cardo et al. (Cordon-Cardo et al. 1989) using immunohistochemistry. Thereafter several groups have demonstrated increased P-gp protein and ABCB1

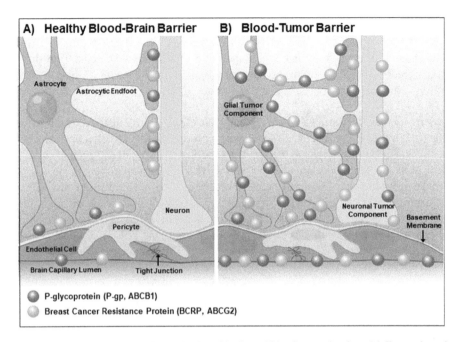

A) **Healthy Blood-Brain Barrier**   B) **Blood-Tumor Barrier**

Astrocyte
Astrocytic Endfoot
Glial Tumor Component
Neuronal Tumor Component
Basement Membrane
Neuron
Pericyte
Endothelial Cell
Brain Capillary Lumen   Tight Junction

P-glycoprotein (P-gp, ABCB1)
Breast Cancer Resistance Protein (BCRP, ABCG2)

Fig. 24.3 Transporter expression at the blood-brain and blood-tumor barriers. (**a**) Expression of P-glycoprotein and breast cancer resistance protein at the healthy blood-brain barrier. (**b**) Expression of P-glycoprotein and breast cancer resistance protein at the blood-tumor barrier and in brain tumor cells of different origin

mRNA expression levels using western blotting or quantitative PCR in whole tumor lysate from a wide range of primary and secondary human brain tumors (Demeule et al. 2001; Spiegl-Kreinecker et al. 2002; Ginguene et al. 2010; Uchida et al. 2011). Some immunohistochemistry studies have demonstrated that increase in P-gp protein expression levels was due to P-gp expression in tumor-associated brain capillary endothelial cells and not due to P-gp expression in tumor cells (Tanaka et al. 1994; Korshunov et al. 1999; Tews et al. 2000; Ginguene et al. 2010; Veringa et al. 2013). Toth et al. showed a particularly heterogenous P-gp expression pattern in patient GBM samples and demonstrated that P-gp expression was significantly decreased in capillary endothelial cells surrounding necrotic areas of the tumor core and in areas with high angiogenesis such as the tumor rim (Toth et al. 1996; Demeule et al. 2001; Bhagavathi and Wilson 2008). While P-gp protein expression is increased in brain tumor cells when compared to their healthy counterparts, the overall transporter expression has been reported to be relatively low in the tumor cell (Marroni et al. 2003). Therefore, unlike the blood-brain barrier where P-gp expression levels are high and correlate with low survival, expression in brain tumor cells did not appear to correlate with tumor grade, survival, or chemoresistance (Abe et al. 1998; Tews et al. 2000; Valera et al. 2007; Wu et al. 2019; Yan et al. 2019). However, a contrasting report does show that P-gp expressed by endothelial cells may be a negligible component of the human GBM multidrug resistance (MDR). In this report the authors indicate that the tumor perivascular astrocytes may dedifferentiate and resume a progenitor-like P-gp activity and contribute to the MDR profile of GBM vessels as well as perivascular P-gp expressing glioma stemlike cells. This study lends credence to P-gp efflux activity contributing to therapeutic failure in both vascular and parenchymal cells (de Trizio et al. 2020).

Breast Cancer Resistance Protein (BCRP, ABCG2) ABCG2 mRNA expression at the BBB was first detected in 2002 in primary porcine endothelial cells (Eisenblaetter and Galla 2002). Cooray et al. (2002) were the first to show BCRP protein expression at the human blood-brain barrier, where BCRP is located in the luminal membrane of endothelial cells and actively contributes to outwardly directed efflux transport (Cooray et al. 2002; Zhang et al. 2003; Aronica et al. 2005). In brain cancer tissue resected from patients, BCRP expression is mainly restricted to the brain tumor barrier (BTB) (Aronica et al. 2005; Bhagavathi and Wilson 2008; Ginguene et al. 2010; Sakata et al. 2011; Shawahna et al. 2011; Bhatia et al. 2012; Veringa et al. 2013).

In contrast to capillary endothelial cells of the BTB, most brain tumor cells in patient samples do not express BCRP (Sakata et al. 2011; Veringa et al. 2013). However, in those cases where BCRP is expressed in brain tumor cells, these cells often display stem cell characteristics and BCRP expression correlates with poor prognosis (Bleau et al. 2009; Emery et al. 2017). Given these studies, BCRP may be more critical in brain tumor cells compared to P-gp. However, anticancer drug efflux from tumor cells appears to be secondary to efflux at the BBB and BTB as a mechanism of drug resistance in brain tumors (Emery et al. 2017).

24.6   Transporter Regulation

A newly emerging strategy to overcome BBB P-gp/BCRP is targeting transporter regulation. Targeting the signaling pathways that regulate P-gp/BCRP and result in decreased transporter expression and activity at the BBB can potentially be exploited to improve brain delivery of anticancer drugs, which have been described in Fig. 24.4.

24.6.1   Transcriptional Regulation

24.6.1.1   Transporter Regulation Through p53

The tumor suppressor p53 (wildtype) binds to the p53 response element in the promotor region of its target genes, which stops the cell cycle and thus cell division. p53 binds to the ABCB1 promotor suppressing its activation (Johnson et al. 2001).

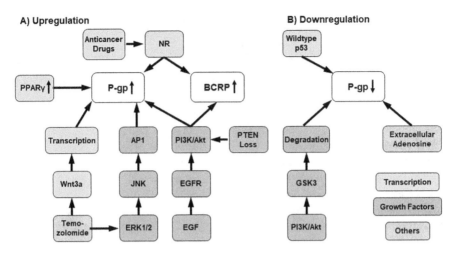

Fig. 24.4 Transporter regulation in primary brain tumors. (**a**) Mechanisms that increase the expression of P-glycoprotein and breast cancer resistance protein in brain tumors. (**b**) Mechanisms that decrease the expression of P-glycoprotein in brain tumors

Mutant p53, however, acts as an activator of the ABCB1 promotor, stimulating transcription and resulting in increased P-gp expression and activity levels (Sampath et al. 2001).

Marroni et al. showed that wild-type p53 inhibits ABCB1 and ABCG2 transcription resulting in decreased P-gp and BCRP expression levels in healthy human astrocytes (Marroni et al. 2003). In contrast, inactivation or loss of p53 increased P-gp/BCRP expression levels in several human glioma cell lines (El-Osta et al. 2002; Sarkadi et al. 2006). Kondo et al. showed that expression levels of murine double minute 2 mRNA (Mdm2), a negative regulator of p53, are increased in human U87 cells in vitro. Mdm2 overexpression inhibited p53, resulting in increased P-gp expression. On the other hand, transfecting U87 cells with antisense Mdm2 microRNA reduced P-gp expression. Thus, mutant p53 increases P-gp and BCRP expression and activity, thereby contributing to chemoresistance (Kondo et al. 1996). Understanding mutant p53 functions will lead to the development of novel approaches to restore p53 activity or promote mutant p53 degradation for future GBM therapies.

24.6.1.2 Transporter Regulation by Nuclear Receptors

Nuclear receptors are ligand-activated transcription factors that target genes including ABCB1 and ABCG2 (Nakanishi and Ross 2012, Sugawara et al. 2010, Hellmann-Regen et al. 2012, Mani et al. 2013). Nuclear receptor activation has been shown to increase P-gp/BCRP expression and activity, which reduces anticancer drug bioavailability and lowers anticancer drug levels in the brain, resulting in decreased drug efficacy (Sarkadi et al. 2006; Nakanishi and Ross 2012).

The nuclear receptor pregnane X receptor (PXR, NR1I2) is activated by a number of xenobiotics. This includes the anticancer drugs cisplatin, carboplatin, tamoxifen, and etoposide, as well as small-molecule tyrosine kinase inhibitors (e.g., lapatinib, sorafenib, and dasatinib) that have been demonstrated to activate PXR, thereby inducing P-gp expression in several human brain, colon, and liver cancer cell lines in vitro (Mani et al. 2005, Harmsen et al. 2013, Yasuda et al. 2019). Han et al. (2015) have shown a similar mechanism for peroxisome proliferator-activated receptor γ (PPARγ) in cisplatin-resistant human U87 glioblastoma cells, where PPARγ activation increased P-gp expression and activity levels, which contributed to anticancer drug resistance in vitro.

These studies may indicate that anticancer drugs can increase P-gp and BCRP mRNA and protein expression levels through nuclear receptor activation. While this phenomenon has been demonstrated in various glioma, glioblastoma, and neuroblastoma cancers, there are currently no in vivo data showing that this restricts anticancer drug uptake into the brain and brain tumor tissue.

24.6.2   Growth Factors

Growth factors stimulate proliferation and tumor growth and regulate the expression and activity of P-gp/BCRP both at the BBB and BTB (Takada et al. 2005; Zhou et al. 2006; Bleau et al. 2009; Nakanishi and Ross 2012; Munoz et al. 2014). One growth factor that is a major regulator of P-gp and BCRP is endothelial growth factor (EGF) acting through endothelial growth factor receptor (EGFR) (Chen et al. 2006; Nakanishi et al. 2006). In 57% of glioblastoma, EGFR is either mutated, amplified, or both, leading to constitutive activation of downstream signaling (Brennan et al. 2013; Eskilsson et al. 2018). Nakanishi et al. demonstrated that stimulation of EGF signaling increased the number of BCRP-positive glioma cells in vitro, making it a likely cause for drug resistance in glioblastoma cells (Nakanishi et al. 2006). Additionally, the EGFR inhibitor gefitinib decreased BCRP expression and activity levels in vitro, opening an avenue for overcoming BCRP-mediated drug resistance as well as the treatment of glioblastoma.

24.6.3   PI3K/Akt Signaling

In many cancers, overactivity of growth factor signaling overstimulates downstream targets including the phosphoinositide-3 kinase (PI3K, PIK3 genes)/protein kinase B (Akt, AKT1/2) pathway (Cancer Genome Atlas Research, 2008) (Brennan et al. 2013). Additionally, 90% of GBM patients have at least one alteration in the PI3K/Akt pathway, including loss of the tumor suppressor and negative regulator phosphatase and tensin homolog (PTEN) (Brennan et al. 2013). Bleau et al. demonstrated that the PI3K/Akt pathway is overactive in a subpopulation of primary human glioma cells with stem cell characteristics leading to increased BCRP protein levels (Bleau et al. 2009). Several groups have published corroborating evidence demonstrating that this regulatory pathway is present in brain tumors as well as at the healthy BBB (Takada et al. 2005; Bleau et al. 2009; Hartz et al. 2010b; Nakanishi and Ross 2012; Huang et al. 2013, 2014). We have shown that inhibiting PI3K/Akt in isolated brain capillaries decreased P-gp and BCRP protein expression and transport activity levels, potentially opening a window in time for anticancer drug delivery into the brain (Hartz et al. 2010a, b). Thus, inhibition of PI3K/Akt is a potential promising strategy to overcome P-gp/BCRP-mediated efflux at the BBB and BTB.

24.6.4   Adenosine Signaling

Several groups have demonstrated that the FDA-approved adenosine receptor A2B agonist regadenoson increases P-gp ubiquitination, thereby inducing P-gp proteasomal degradation (Kim and Bynoe 2015, 2016; Yan et al. 2019). Jackson et al. developed a therapeutic strategy using regadenoson to decrease P-gp protein expression and activity at the BBB and BTB (Jackson et al. 2016). The authors showed, in rats, that regadenoson co-administration significantly increased temozolomide brain levels compared to control animals that only received temozolomide. When regadenoson was administered to patients with angina or previous heart attacks (no brain tumors) that underwent cardiac stress testing, brain levels of the P-gp substrate [99]mTc-sestamibi were increased (Jackson et al. 2017). Despite these promising results, a phase I clinical trial in patients with recurrent GBM testing TMZ with and without regadenoson was unsuccessful (Jackson et al. 2018).

24.6.5   Temozolomide

Riganti et al. (2013) found that Wnt3a and P-gp protein expression levels are higher in glioblastoma stem cells compared to healthy astrocytes. They also found that activating Wnt signaling increased P-gp expression levels in glioblastoma cells. However, following temozolomide treatment of primary glioblastoma cells in vitro, Wnt signaling was decreased, resulting in decreased P-gp expression levels. From these data the authors concluded that temozolomide reversed drug resistance by decreasing P-gp protein expression through the Wnt pathway (Riganti et al. 2013). In contrast, Munoz et al. showed that temozolomide increased P-gp expression and activity in U87 and T98G glioblastoma cells in a biphasic manner. In the early treatment phase, temozolomide induced P-gp trafficking to the cell membrane and, therefore, increased P-gp efflux function in glioblastoma cells in vitro. During later stages of treatment, temozolomide activated ERK1/2-JNK-AP1 signaling, which increased ABCB1 mRNA and P-gp protein expression levels (Munoz et al. 2014).

To date, the effect of temozolomide on drug resistance remains controversial and needs further evaluation.

24.7 Strategies to Improve Treatment of Brain Tumors

As outlined in the previous sections, the delivery of adequate concentrations of anticancer-targeted therapies to tumor cells residing in the brain has proven to be a significant challenge. Various approaches to overcome the delivery barrier have been studied, and some are described in the following. These approaches are depicted in Fig. 24.5.

24.7.1 Designing Molecules with Increased Brain Penetration and Reduced Efflux Liability

Designing drug molecules that can permeate the BBB and attain effective concentrations in the brain should be a priority for CNS drug discovery programs. This can be achieved by incorporating key physicochemical properties that aid in BBB

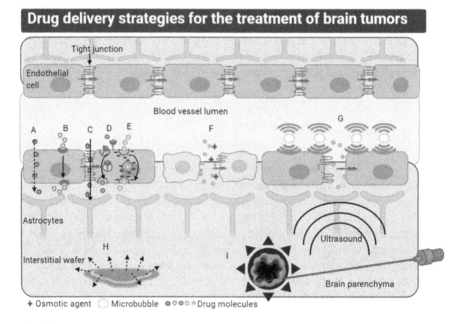

Fig. 24.5 Drug delivery strategies for the treatment of brain tumors. (**a**) Transcellular diffusion, (**b**) utilizing influx transporters, (**c**) paracellular transport, (**d**) receptor-mediated endocytosis, (**e**) adsorptive endocytosis, (**f**) osmotic BBB disruption, (**g**) focused ultrasound using microbubbles, (**h**) local delivery of cytotoxic agents, (**i**) convection enhanced delivery

penetration and rational structural modifications. Critical physicochemical properties have been identified and computational models developed to optimize these properties for successful brain delivery (Rankovic 2015, 2017; Heffron 2016). Wager et al. (Wager et al. 2010) have developed a multiparameter optimization (MPO) approach to screen molecules for optimal neuro-pharmacokinetic and safety profiles. The key physicochemical properties were: (1) lipophilicity, with a partition coefficient (ClogP) $\leq$ 3 being desirable; (2) a distribution coefficient (ClogD) $\leq$ 2; (3) molecular weight (MW) $\leq$ 360 Daltons; (4) topological polar surface area (tPSA) between 40 and 90$\mathring{A}^2$; (5) number of hydrogen bond donors (HBD) $\leq$ 0.5; and (6) most basic functional group with a pKa $\leq$ 8. The six properties were equally weighted with a score between 0 and 1, resulting in a final CNS MPO score ranging from 0 to 6, thereby allowing multiple combinations of the parameters to achieve a particular MPO score. This algorithm was applied to 119 marketed CNS drugs, and 74% of those drugs showed high (>4.5) CNS MPO scores. The compounds with a high MPO score also displayed desirable ADME properties like high permeability, low P-gp efflux liability, and higher stability as might be expected for drugs that have been approved for CNS indications (Wager et al. 2010). Additional refinements have been made to this original algorithm, using the same six parameters for optimization, to improve the structural design enhancement and quality of compounds nominated for clinical development of CNS therapeutics (Wager et al. 2016).

The use of an algorithm to optimize key physicochemical properties, in conjunction with rational structural modifications to reduce efflux liability, led to the identification of brain penetrant PI3K inhibitors, GNE-317 and GDC-0084. These molecules showed significantly greater tumor growth inhibition in GBM mouse models as compared to BBB impenetrant PI3K inhibitors (Sutherlin et al. 2010; Salphati et al. 2012; Heffron et al. 2016). Importantly, the PI3K inhibitor GDC-0084 showed promising initial results in a phase I study that was conducted in patients with recurrent high-grade glioma (Wen et al. 2020). Another example is AZD3759, a potent brain penetrant EGFR inhibitor, which was developed using gefitinib as the initial lead. Techniques like repositioning of fluoro moiety and reduction of rotatable side chain were employed for overcoming P-gp and BCRP efflux to improve brain penetration, while maintaining the quinazoline scaffold necessary for activity (Zeng et al. 2015). In a study using cassette dosing to examine the brain penetration of eight EGFR TKIs, AZD 3759 showed the greatest brain penetration (Kim et al., 2019a, b). AZD3759 is now in a phase I clinical trial to assess its safety, tolerability, and primary efficacy in patients with advanced NSLC (NCT02228369). Considering the high propensity of developing brain metastases from NSLC, the development of AZD3759 can be a significant step in the treatment of these patients. One more example of structural modification to improve BBB permeability to evade efflux is of crizotinib, an ALK inhibitor, leading to the development of BBB penetrant lorlatinib (PF-06463922) (Basit et al. 2017). This was achieved by cyclization of crizotinib to form the macrocyclic lorlatinib leading to a reduction in the effective HBDs through the formation of intramolecular hydrogen bonds. This strategy and a reduction in rotatable bond count decreased its interaction with efflux transporters and improved CNS distribution (Basit et al. 2017). In an ongoing phase II study

(NCT01970865), lorlatinib showed substantial intracranial activity in patients with recurrent ALK-positive NSCLC, with or without baseline CNS metastases, whose disease progressed on crizotinib or other second-generation ALK TKIs (Bauer et al. 2020). And finally, AZD1390, a selective and potent ATM inhibitor, synthesized to be brain penetrant using strategies informed by AZD0156 another potent ATM inhibitor that is a substrate of efflux transporters. This compound is now in early clinical trials for use as a radiosensitizer in CNS malignancies (Durant et al. 2018). Taken together, these examples clearly demonstrate that computational models and structure-guided drug design early in CNS drug discovery programs can support the development of brain penetrant drugs for brain tumors, with structural modifications to reduce the affinity for efflux transporters, a key component.

24.7.2   Inhibition of Efflux Transporters at the BBB

Inhibition of transporters particularly P-gp and BCRP as a strategy to overcome transporter-mediated drug delivery limitations has been investigated (Huisman et al. 2003; Baumert and Hilgeroth 2012). The first-generation inhibitors were comprised of marketed drugs known to inhibit efflux transporters, which includes verapamil, cyclosporine-A, and quinidine (Shen et al. 2008; Bui et al. 2016). However, these inhibitors have low potency and selectivity and require high doses. An analog of cyclosporine-A, valspodar (PSC-833), was developed as a second-generation inhibitor with more potent inhibition of P-gp, but it also interfered with cytochrome P450 function. As a consequence, third-generation inhibitors, including tariquidar, elacridar, and zosuquidar, were developed (Gampa et al. 2020). Although co-administration of tariquidar improved the brain exposure of targeted agents and corresponding efficacy in preclinical studies without any toxicity concerns, two phase III clinical trials in NSCLC patients were terminated due to toxicity when used in combination with paclitaxel/carboplatin or vinorelbine (Fox and Bates 2007). Similarly, toxicity concerns were reported in clinical studies investigating the use of zosuquidar and elacridar (Sandler et al. 2004; Kuppens et al. 2007). The clinical efficacy of pharmacological inhibition of efflux transporters to increase brain distribution clearly requires a potent efflux influx transport inhibitor that does not increase the toxicity of the CNS active agents. If very potent inhibitors are used to improve CNS delivery of toxic compounds, a careful assessment of CNS toxicity due to increase in brain delivery will be required.

24.7.3   Utilizing Influx Transporters at the BBB

An alternative to overcoming efflux transporters is designing drugs to take advantage of innate influx transporter systems already expressed at the BBB. Targeting a transport system at the BBB for drug development and improved delivery can be used in the treatment of primary brain tumors. Glucose transporters (GLUT) are known to facilitate transport of glucose from blood to the brain. It was observed that when a mannose derivative was incorporated onto a liposome, the delivery system exhibited better penetration across the BBB via the glucose transporter (GLUT1) into the mouse brain (Wei et al. 2014). Choline transporters are another group of transport systems responsible for binding with positively charged quaternary ammonium groups or simple cations. A 60-nm size particles coated with quaternary ammonium ligands have shown enhanced penetrability across an in vitro BBB model (bovine BCEC) (Gil et al. 2009). Histidine/peptide (peptide/histidine transporter), large neutral amino acid transporter (LAT1), and vitamin transporters [sodium-dependent multivitamin transporter (SMVT) and sodium-dependent vitamin C transporter (SVCT)] are some of the influx transporters that have gained attention (Castro et al. 2001; Bhardwaj et al. 2006; Uchida et al. 2015; Puris et al. 2020) These transporters are being studied extensively for targeted drug delivery to the brain.

24.7.4 Targeting Receptor-Mediated Transport Systems at the BBB

Receptor-mediated transcytosis is one of the promising strategies for targeted delivery across the BBB with high specificity, selectivity, and affinity (Xu et al. 2013). However, there might be a possibility of competition between endogenous substrates and drug ligands for the same receptor leading to reduced targeting efficiency. Receptors expressed on the brain capillary endothelium include transferrin receptor (TfR) (Pardridge et al. 1987), low-density lipoprotein receptor (Ueno et al. 2010), insulin receptor (IR), and nicotinic acetylcholine receptors (Pardridge et al. 1985; Vu et al. 2014). Targeting with endogenous ligands as well as ligands based on phage display or structure-guided design can be exploited for receptor-mediated transcytosis.

An example of this is GRN1005, an angiopep-2 peptide conjugated to paclitaxel, which gets across the BBB via transcytosis using the lipoprotein receptor-related protein 1 (LRP1) (Kurzrock et al. 2012; Drappatz et al. 2013). Another example is 2B3-101, which is a pegylated liposome conjugated with glutathione and actively transported across the BBB. This formulation showed enhancement in the uptake and delivery when compared to the conventional doxorubicin liposomal formulation (Gaillard et al. 2014). T7, targeting TfR1, has been investigated to deliver antisense oligonucleotides to gliomas (Kuang et al. 2013; Zong et al. 2014).

Monoclonal antibodies (mAbs) are another class of molecules that are being currently investigated to inhibit tumor growth driver pathways. Bevacizumab, targeting VEGF, as mentioned earlier received accelerated FDA approval for newly diagnosed and recurrent GBM. Cetuximab, another mAb targeted to EGFR failed to show survival benefit in a phase II trial (Neyns et al. 2009). Antibodies, being large

(~150 kDa) molecules, do not generally cross the BBB and hence despite showing effectiveness in case of peripheral tumors need enhanced delivery mechanisms to cross the BBB and be effective in case of brain tumors (Zhang and Pardridge 2001; St-Amour et al. 2013). With recent advances in antibody engineering and use of antibody fragments, the structure of these large molecules is being exploited to modify and utilize different domains to promote receptor-mediated transcytosis. TfRs as well as IRs have been shown to be widely used targets for therapeutic antibodies as well as nanocarriers linked to antibodies for brain delivery (Boado et al. 2010; Kim et al. 2019a, b). In a recently published study, a nanocarrier loaded with p53 gene therapy, decorated with anti-TfR1 single-chain variable fragments, SGT-53, showed success in GBM preclinical models and has moved into clinical trials (Kim et al. 2019a, b). In another study from AbbVie, dual-variable-domain IgG molecules with dual affinity (TfR for receptor mediated transcytosis and HER2 for HER2+ brain tumors) have been developed for precision targeting (Karaoglu Hanzatian et al. 2018).

24.7.5   Antibody Drug Conjugates (ADCs)

ADCs are composed of an antibody acting as a targeting agent linked to cytotoxic compounds to enable their delivery into the cells. Ado-trastuzumab emtansine (T-DM1) is an ADC which is trastuzumab (mAb targeting HER2) linked to the maytansinoid DM-1 (microtubule inhibitor) using a stable linker (Lambert and Chari 2014). A series of studies have shown prolonged progression free survival as well as treatment effect in case of breast cancer brain metastases using T-DM1 (Bartsch et al. 2015; Keith et al. 2016; Okines et al. 2018; Ricciardi et al. 2018). Depatuxizumab mafodotin (ABT-414) is composed of an antibody targeted to cells with EGFR amplifications and releases monomethyl auristatin F (microtubule toxin). ABT-414 was studied in phase II trials for recurrent GBM in combination with TMZ; however, the phase III trial was halted as no overall survival benefit was observed (Van Den Bent et al. 2020). Other EGFR targeting ADCs, ABBV-221 and ABBV-321, are being evaluated in phase I trials for GBM (NCT02365662, NCT03234712).

24.7.6   Immunotherapy

Immunotherapy involves harnessing the body's own immune system to identify, target, and kill tumor cells. This approach is particularly effective in tumors with high tumor mutational burden but has not been effective in brain tumors despite their highly heterogenous nature (Liu et al. 2020). A variety of immunotherapies are being explored for brain tumors using multiple strategies—checkpoint inhibition, utilizing chimeric t-cell receptors, dendritic cell, and peptide vaccines as well as using viral vectors for gene therapy. However, none of these have been approved for treatment. The reader is directed to a comprehensive review of these strategies as well as their challenges in the following reviews (Lyon et al. 2017; Liu et al. 2020).

24.7.7 Development of Radiosensitization Strategies with Current Standard of Care

DNA damage response signaling pathways play a critical role in DNA repair and cell survival following radiation therapy, and the inhibition of these pathways could augment the cytotoxicity associated with radiation providing a sensitizing effect. DNA damage occurs continually through various mechanisms. Environmental factors such as ultraviolet (UV) radiation, x-rays, and smoking, as well as endogenous factors, including replication errors, reactive oxygen and nitrogen species, and hydrolysis of bases are some examples through which DNA damage may occur (Hoeijmakers 2009). High proliferation rates inherent to tumor cells may also lead to an amplification of errors and DNA damage. Evolution has led to the development of complex cellular mechanisms that detect and repair such defects, and these have been collectively termed the DNA damage response (DDR) (Harper and Elledge 2007). Several pathways have been identified within the DDR, each distinct in their mechanism of repairing DNA. Core DDR pathways include nonhomologous end joining (NHEJ), homologous recombination, base excision repair, nucleotide excision repair, mismatch repair, and interstrand cross-link repair (Lord and Ashworth 2012). These pathways are activated by a cascade of events initiated by DNA damage sensor proteins that engage signaling networks and regulate cell cycle progression allowing for DNA repair to occur (O'Connor 2015). An active DDR machinery is essential for the healthy physiology of the cell, ensuring its survival, and is an important mechanism of resistance to cytotoxic approaches. Accordingly, the inhibition of the DDR in tumor cells provides an excellent therapeutic opportunity (Sun et al. 2018).

The response to DNA damage will be different depending on the cell cycle status providing a varied range of cell cycle pathways for targeting for the sensitizing effect. For example, cells in G1 will not have sister chromatid DNA available as an undamaged template and therefore will be dependent upon NHEJ pathways for the repair of DSBs. In addition, there are important differences in the primary roles of checkpoints at different stages of the cell cycle and in the DDR factors that are involved. For example, the G1/S checkpoint allows the repair of DNA damage prior to the start of DNA replication in order to remove obstacles to DNA synthesis, and key DDR factors regulating this checkpoint include ATM, CHK2, and p53. The intra-S phase checkpoint proteins ATR, CHK1, DNA-PK, and WEE1 can delay replication origin firing to provide time to deal with any unrepaired DNA damage that has occurred, thus preventing under-replicated DNA regions being taken beyond S-phase. The activities of the G2/M checkpoint proteins including CHK1, MYT1, and WEE1 lead to an increase in phosphorylated CDK1, thereby keeping it in its inactive state and delaying mitotic entry. The G2/M checkpoint really represents the last major opportunity for preventing DNA damage being taken into mitosis where unrepaired DSBs and under-replicated DNA may result in mitotic catastrophe and cell death (Castedo et al. 2004). Recent analyses suggest that there are at least 450 proteins integral to DDR (Pearl et al. 2015), and the choice of optimal drug targets within DDR will be based on what type of DNA damage repair is to be inhibited and where in the cell cycle that damage is likely to occur. Major drug development efforts are being directed to take the DDR inhibitors into the clinic as radiation and chemotherapy sensitizers.

24.7.8   Modification of Tight Junctions at the BBB

A selective disruption of the BBB followed by administration of anticancer agents provides for a promising approach to enhance drug delivery to the brain in the treatment of brain tumors. Various techniques have been employed to cause transient BBB disruption, as briefly discussed below.

24.7.8.1   Osmotic Disruption of the BBB

The administration of hypertonic solutions causes disruption of the BBB due to shrinkage of endothelial cells, leading to the alteration of tight junctions between them, thereby allowing paracellular movement of drugs. This method was first proposed by Rapoport et al. in 1972 and later was translated to the clinic with the first phase I clinical trial in 1979 (Rapoport et al. 1972; Levin et al. 1979). The hypertonic solution of 1.4 M mannitol infusion is FDA approved for administration to patients for transient BBB disruption (Neuwelt 1980). Other agents investigated include saline, arabinose, urea, lactamide, and a variety of radiographic contrast agents (Kroll et al. 1998). In a clinical study in the 1980s by Neuwelt et al., improved survival and long-term remission were observed in patients with primary CNS lymphoma following osmotic BBB disruption plus methotrexate (Neuwelt 1980). Agents in addition to methotrexate that have been used in the clinic with osmotic BBB disruption include etoposide, cyclophosphamide, carboplatin, and melphalan. The transient BBB disruption followed by administration of anticancer agents has been employed as a strategy to overcome brain drug delivery limitations (Rapoport 2000; Kemper et al. 2004). However, this approach is invasive, and complex to perform and is associated with adverse effects (Bellavance et al. 2008).

24.7.8.2 Focused Ultrasound

Focused ultrasound (FUS) is based on a concentration of acoustic energy onto a focal area that results in BBB disruption. Microbubble (MB)-enhanced FUS involves the oscillation of MBs in the presence of FUS to cause BBB disruption. These microbubbles are FDA approved for use as contrast agents in ultrasound imaging and in the context of drug delivery and are used to lower the energy threshold for BBB disruption (Timbie et al. 2015). This approach is local, transient, and reversible and has demonstrated improvements in delivery and efficacy of anticancer agents in glioma models (Liu et al. 2014; Deng et al. 2019). The delivery of small molecules like TMZ, doxorubicin, and 1,3-bis (2-chloroethyl)-1-nitrosourea (BCNU) to large molecules like bevacizumab and trastuzumab as well as cell therapy, viral therapy, and nanoparticle delivery has been facilitated by FUS with microbubbles in glioma and brain metastases (Meng et al. 2018; Bunevicius et al. 2020). Also, significant downregulation of localized P-gp expression with no apparent damage to brain endothelial cells was observed, suggesting the potential use of MB-FUS for targeted brain delivery of drugs that are liable to efflux by P-gp (Cho et al. 2016). However, the long-term effect of FUS on the brain microvasculature has not been investigated. A thorough investigation of safety due to repeated FUS treatments as well as safe and appropriate ultrasound settings has to be conducted for drug delivery applications. The reader is guided to two comprehensive reviews for the use of FUS in brain tumors and the ongoing clinical trials (Meng et al. 2018; Bunevicius et al. 2020).

24.7.8.3 Photodynamic Therapy Approaches

Photodynamic therapy (PDT) involves the administration of a photosensitizing agent that localizes in the tumor followed by photoactivation that can result in a direct inhibitory effect on tumor cells and also a localized disruption of BBB that can aid in the delivery of other anticancer agents to the brain tumor (Akimoto 2016). An early report of PDT was by Perria et al. that utilized a hematoporphyrin derivative injected i.v. as a sensitizing drug with a helium-neon laser to trigger the photodynamic process (Perria et al. 1980). First-generation photosensitizers include hematoporphyrin and its derivatives. Chlorins (talaporfin sodium and temoporfin) and 5-aminolevulinic acid (5-ALA) are examples of second-generation photosensitizers that were developed to be more potent. 5-ALA is the most commonly used photosensitizer due to its high oral bioavailability, favorable safety profile, and preferential accumulation in malignant gliomas (Mahmoudi et al. 2019). The recent FDA approval of 5-aminolevulinic acid (5-ALA) for fluorescence-guided resection (FGR) of tumors has generated immense interest in leveraging this agent as a means to administer photodynamic therapy (PDT). The joint clinical application of fluorescence-guided surgery (FGS) and PDT confers the ability to both visualize tumor cells and selectively destroy them. Clinical studies of PDT using porfimer sodium, talaporfin sodium, 5-ALA, boronated porphyrin, and temoporfin in GBM have been reported (Cramer and Chen 2020). Third-generation photosensitizers were developed for enhanced tumor cell selectivity achieved through the conjugation of modifiers including nanoparticles and antibodies (Allison and Sibata 2010). Lack of clear efficacy in overall survival, technical limitations in light delivery, and photosensitizer design as well as unclear safety profiles of varied photosensitizers have hindered the impact that PDT can have in brain tumor treatment. Exploration of novel photosensitizer agents and safe photosensitization strategies in brain tumors is warranted for incorporation of PDT into current standard of care (Cramer and Chen 2020).

24.7.9   Local Delivery Methods

Local drug administration directly into the CNS has been employed as a strategy to precisely deliver drug to the target site in the brain. These local delivery methods include biodegradable wafers placed in the tumor cavity post resection, convection-enhanced delivery (CED), and intrathecal delivery into CSF cavities (Blakeley 2008; Calias et al. 2014).

24.7.9.1   Biodegradable Wafers

Polymer-based biodegradable wafers have been available for patients with brain tumors as one of the earliest treatment options with Gliadel (BCNU/carmustine) approved by the FDA in 1996 for recurrent high-grade gliomas. These wafers are placed in the tumor cavity post resection for sustained drug release over a few days and have also been considered for improving drug delivery to brain tumors post resection. This approach provides local control of disease but is limited by the modest distribution of BCNU away from the resection cavity. Gliadel has also been used in patients with brain metastases where patients with single brain metastases underwent surgical resection followed by Gliadel implantation and whole brain radiation treatment (Ewend et al. 2007). In an effort by Domb et al., co-loading of BCNU and TMZ within poly(lactic acid-glycolic acid) (PLGA) wafers in rat glioma models led to a 25% enhancement in survival (Shapira-Furman et al. 2019). Lee et al. developed a novel material and device technology consisting of a flexible, sticky, and biodegradable wireless device loaded with doxorubicin for controlled intracranial delivery using mild-thermal actuation. In mouse and canine models of GBM, this device showed tumor volume suppression and improved survival indicating its potential to be translated to humans utilizing a variety of other potent anticancer agents for intracranial delivery (Lee et al. 2019). A major challenge for this technology is to ensure biocompatibility and biodegradation in a reasonable time period, as incompletely biodegraded material can lead to inflammatory responses in patients. The success of these therapies is limited due to their inability to reach to the invasive and dense tumor cells due to poor diffusion characteristics (Wolinsky et al. 2012).

24.7.9.2 Convection-Enhanced Delivery

Convection-enhanced delivery (CED) is a bulk-flow (hydrostatic pressure differential)-driven invasive technique that affords the continuous delivery of small and large molecular weight compounds into the brain parenchymal tissue through infusion catheters implanted during surgery (Debinski and Tatter 2009). It was first proposed by Bobo et al. in 1994 for the delivery of macromolecules to the brain (Bobo et al. 1994). Two phase III trials were initiated in participants with GBM. One trial utilizing Tf-CRM107 was aborted, with data available from a phase II trial (Weaver and Laske 2003). The other phase III trial, the PRECISE trial, compared the infusion of citredekin besudotox (PE38QQR) with recombinant human interleukin-13 delivered by CED. The study did not reveal statistically significant improvement in survival for patients with recurrent GBM (Kunwar et al. 2010). CED is being widely studied in preclinical and clinical studies for GBM as well as diffuse intrinsic pontine glioma (DIPG) (Vogelbaum and Aghi 2015; Zhou et al. 2016). Limitations of CED include limited area of distribution, requirement of surgery, and increased risk of neurotoxicity due to elevated intracranial pressure (Blakeley 2008). The brain tissue near the catheter may receive effective drug delivery, but the concentrations can decrease steeply as the distance from the catheter tip increases due to competing forces of convective flow through brain parenchyma and drug diffusion into capillaries. One problem with any local drug delivery technique is that molecules with a high permeability or active efflux liability efficiently clear from the brain tissue into blood capillaries following local brain delivery. This phenomenon, the "sink effect," can influence the volume of brain tissue captured for drug distribution. The brain, a highly perfused organ, has a dense capillary network; therefore, the probability of drug diffusion into the capillary bed can be high, depending on the physicochemical characteristics of the compound (Lonser et al. 2015). Thus, the selection of a suitable drug candidate that has minimal liability for the sink effect and optimization of delivery parameters (such as infusion parameters for CED) to capture the required brain tissue volume (e.g., brain tumor) are critical to achieve beneficial responses with local delivery methods.

24.7.9.3 Intrathecal Delivery

Intrathecal (IT) administration typically refers to the infusion of drug into the subarachnoid space in the lumbar region. Intrathecal chemotherapy is administered directly into the lumbar thecal sac via lumbar puncture or infused into the lateral ventricle through a subcutaneous reservoir and a ventricular catheter (Ommaya reservoir), thus allowing the drug to distribute into the target sites via diffusion. A phase I trial using implanted ventricular catheter has been reported where chlorotoxin was coupled with the radioisotope [131]I (^{131}I-TM-601) infusing radioactive therapy into the tumor resection cavity via an Ommaya reservoir in patients with recurrent malignant glioma (Mamelak et al. 2006). In cases of breast cancer brain metastases, studies employing intrathecal administration of rituximab, trastuzumab alone, and with other cytotoxic agents like methotrexate and cytarabine have been reported (Perissinotti and Reeves 2010; Oliveira et al. 2011; Niwińska et al. 2015; Mack et al. 2016). IT administration suffers from a variety of drawbacks, like limited drug delivery to tumors despite high CSF concentrations, slow rate of drug diffusion, and rapid CSF turnover compared to rate of diffusion leading to rapid clearance of drugs. In addition, the idea that high CSF concentrations correspond to high drug levels in the brain and tumor have led to an impediment in the advancement of IT therapies (Pardridge 2016).

24.8 How Much Is Enough? Drug Pharmacokinetic-Pharmacodynamic (PK-PD) Relationships in Brain Tumors

The intricate architecture of the CNS as well as the complex tumor microenvironment necessitates careful application of pharmacokinetic principles in the determination of drug distribution to brain tumors and, hence, understanding the PK→PD→Efficacy relationship. A schematic depicting the relationship between PK→PD→Efficacy of novel drug molecules to be used for clinical translation is depicted in Fig. 24.6.

24.8.1 Drug in Plasma Versus Drug in the Brain Versus Drug in Tumor

The drug concentrations in blood or plasma are routinely measured as surrogates for concentrations at the site of action due to ease of sampling. While drug concentration in the systemic circulation may somewhat reflect the concentration at the site of action when the target is in a peripheral, more accessible tissue, their use as a surrogate for brain drug concentrations can be misleading and even more so for tumor drug concentrations. This is particularly important in the context of the brain when compared to other organs due to the presence of the BBB, which can severely restrict drug distribution to the target site in the brain (Hawkins et al. 2010). The misconception that drug delivery to brain tumors is not impeded by the BBB due to disruption of the tumor vasculature has been furthered by studies that fail to consider the invasive nature of brain tumors. These studies use the "tumor core" concentrations to indicate effective drug delivery to the tumor (Blakeley et al. 2009; Grossman et al. 2013; Sarkaria et al. 2018). The BBB in the tumor core is often leaky, and therefore delivery to the tumor core alone is insufficient to improve patient outcomes, since the invasive cells remain untreated (Sarkaria et al. 2018). These differences in BBB integrity at the tumor core, tumor rim (area of tumor infiltration adjacent to the core), and in the normal brain have been depicted in

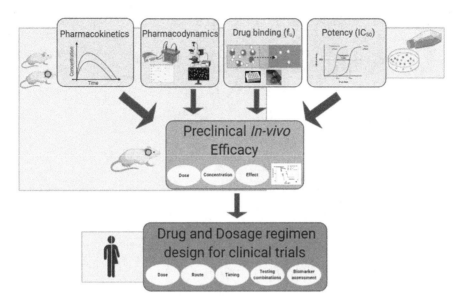

Fig. 24.6 Understanding the relationship between PK-PD-Efficacy of novel drug molecules to be used for clinical translation

Fig. 24.7. Concentrations in the tumor core can be inadequate to predict a useful concentration–response relationship, and an "adequate" concentration achieved around the invasive cells is critical for improved response. Accepting the importance of drug delivery across an intact BBB into the brain is the first critical step to develop novel therapies for brain tumors (Agarwal et al. 2012; Sarkaria et al. 2018).

The current standard of care for the treatment of brain tumors involves radiation (a highly BBB penetrant treatment) that can have serious long-term side effects that range from cognitive decline to other serious effects like blindness, local tumor recurrence, and radiation-associated tumor (Amelio and Amichetti 2012). Therefore, understanding the importance of spatial differences in BBB permeability on drug levels, particularly those drugs that are radiation sensitizers, in various regions of the CNS, and the periphery, is a critical factor in the assessment of novel therapies.

24.8.2   Utilizing Appropriate Preclinical Models to Determine Effective Drug Concentration

Slow progress in the approval of novel therapeutics for the treatment of brain tumors can be attributed to two major factors: (1) inadequate, that is, non-predictive, in vitro systems and (2) the use of preclinical models that fail to address critical aspects of the tumor in the patient (Aldape et al. 2019). A useful in vivo system should include the heterogeneity of BBB permeability and genetic makeup of the tumors. For

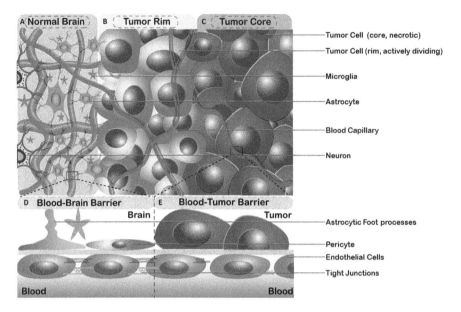

Fig. 24.7 Differences in blood capillaries and BBB in normal brain and brain tumor. (**a**) Figure representing normal brain, (**b**) tumor rim, (**c**) tumor core, (**d**) blood-brain barrier in normal brain, and (**e**) blood-tumor barrier in brain tumor

instance, many patient-derived xenograft (PDXs) orthotopic models of brain tumors can recapitulate genetic drivers and invasive growth leading to vast differences in BBB permeability to drug treatment and as such will enable a more predictive assessment of the benefit of new treatments. Development of PDXs and genetically engineered mouse models (GEMMs) are often suitable for this use. Defining the preclinical spectrum of response for novel agents/therapies across a representative panel of genetically diverse brain tumors with the necessary complexity in the tumor microenvironment including the condition of the BBB can provide important information and guide optimal clinical drug development.

24.8.3   Impact of Drug Binding in Brain Tumor Treatment

Crucial to the efficacy of any antitumor agent is adequate exposure of target cells to effective concentrations of active drug. However, reducing toxicity to normal cells often depends on limiting exposure to active drug. Many drugs are highly bound in both plasma and in brain tissue. The free drug hypothesis states that the driving force concentration for distribution into tissues is the free concentration in the blood (Dubey et al. 1989; Hammarlund-Udenaes et al. 2008). Therefore, in addition to determining the total brain-to-plasma ratio (Kp, a tissue partition coefficient) of drug molecules, their binding in plasma, brain, and tumor tissues must be evaluated in order to determine tumor exposure to free drug. New therapies will only be effective if they are able to penetrate the BBB and elicit their effects in the tumor cells. Therefore, in addition to a pharmacokinetic assessment of total drug (bound plus unbound) delivery to the brain and the tumor regions, free brain partitioning of the drugs, defined as Kp_{uu}, must also be determined (Fridén et al. 2009, 2011; Loryan et al. 2013). The reader is directed at the review article by Hammarlund-Udenaes et al. for an extensive description of factors to be considered for the rate and extent of delivery to the brain (Hammarlund-Udenaes et al. 2008). Knowledge of drug pharmacokinetics combined with drug binding in plasma and brain, as well as tumor tissues, can help explain the concentration effect relationship with respect to binding as a determinant of an effective drug concentration.

24.8.4 Dosage Regimen Design for Achieving Target Drug Concentration and Desired Pharmacodynamic (PD) Effect

A comprehensive understanding of drug distribution into tumor and normal tissues, and associated pharmacodynamic effects, is critical for defining which drugs to move forward into phase I dose-seeking studies. It is therefore important for novel treatment options to define key parameters, such as the determinants of free- and bound-drug exposure in plasma, normal brain, and brain tumor as described above, and to relate these metrics to a dose range associated with an effective endpoint, that is, tumor growth reduction, progression free survival, and overall survival. Although new combination drug regimens have generated excitement in the field and initial positive responses, they ultimately fail to demonstrate efficacy due to drug resistance mechanisms and limited brain delivery (Gottesman 2002; Trédan et al. 2007; Van Den Bent et al. 2009; Chamberlain et al. 2014; Sarkaria et al. 2018). These failures can be attributed to not only pharmacokinetic and drug delivery aspects but also pharmacodynamic or cellular mechanism responses within the tumor cells that can compensate for targeted antitumor mechanisms (Wang et al. 2008). These PD-based treatment failures can also be attributed to inadequate drug levels within the brain tumors that in turn lead to poor efficacy. Brain metastases as well as GBM have been shown to have extensive intra- and intertumoral heterogeneity in terms of genetic composition and protein expression. This genetic heterogeneity as well as the heterogeneity in drug distribution contribute to wide ranging responses to drug therapy (Perus and Walsh 2019). Therefore, as mentioned above, the determination of free drug levels within the tumor as well as normal brain is essential to understand if exposures associated with efficacious pharmacodynamic responses can be achieved in and around the tumor where invasive cancer cells are present (Laramy et al. 2017). All these considerations lead to the establishment of a therapeutic window, a key consideration when developing novel treatments for brain tumors. Time-dependent responses of pharmacodynamic biomarkers in response to a dosage regimen are critical and can drive the design of monotherapy and combination treatments. Therefore, PK-PD modeling efforts for novel agents using the predictive preclinical models are being explored to guide dosage regimen selection in humans (Sharma et al. 2012, 2013; Li et al. 2017). Optimizing dose as well as timing is necessary in predicting pharmacodynamic effects, and decisions to move forward with efficacy testing in phase II/III trials then can be made by a combined understanding of drug potency, mechanism of action, PD effects, and drug distribution to invasive tumor.

24.9   Conclusions

Despite improvements in the management of cancers over the last decade, treatments for brain tumors have not seen significant advances. A diagnosis of either primary or metastatic brain tumor is associated with a grim prognosis and none of the currently available therapies have long-term efficacy. The diffuse and infiltrative nature of these tumors, their location within the brain, and the highly heterogenetic makeup with a variety of mutations make it even more difficult to design effective therapeutics. In this chapter, we describe different drug delivery approaches for the treatment of brain tumor. These drug delivery approaches include both systemic and local delivery options. Key considerations in the PK→PD→Efficacy relationships have been included to inform the development of effective treatments for brain tumors. It is clear that consideration of drug delivery to the brain tumor needs to be incorporated at all levels of research and development in an effort to discover effective treatments.

24.10   Points for Discussion

- What factors may be limiting successful clinical translation of therapies demonstrating positive preclinical results in animal models of primary and metastatic brain cancer?
- Contrast similarities and differences in the standard of care for primary and metastatic brain cancer?
- List at least four hurdles to developing new drug delivery strategies for chemotherapeutics targeted to brain cancers.
- Describe at least two ways in which P-gp/BCRP expression is upregulated in brain cancer and at least two ways in which P-gp/BCRP expression may be pharmacologically downregulated.
- List several reasons why small molecule inhibitors of P-gp/BCRP have not yet been successfully applied for the clinical enhancement of chemotherapeutic brain exposure.

- Describe at least three methods where brain endothelial cell tight junction integrity can be altered to deliver circulating drugs to brain tumors and describe how they are thought to work. Also discuss limitations, safety issues, and drawbacks associated with each method.
- Describe at least three methods for local delivery of drugs to brain tumors and describe how they are thought to work. Also discuss limitations, safety issues, and drawbacks associated with each method.
- Discuss the following concepts and their impact on our understanding of PK→PD→efficacy relationships with respect to the effective treatment of brain tumors:
  - differences in BBB integrity at the tumor core, tumor rim, and in the normal brain
  - major factors associated with preclinical models that have limited development and approval of novel therapeutics for the treatment of brain tumors
  - drug binding in plasma and brain and the difference between Kp and Kp_{uu}
  - common reasons behind PD-based treatment failures

Acknowledgments This work was supported by the National Institutes of Health [Grants RO1 NS107548, U54 CA210180, U01 CA227954]. The authors would also like to thank the editors of the book for their comments and valuable inputs.

References

Abe T et al (1998) Expression of multidrug resistance protein gene in patients with glioma after chemotherapy. J Neurooncol. https://doi.org/10.1023/A:1005954406809

Achrol AS et al (2019) Brain metastases. Nat Rev Dis Primers. https://doi.org/10.1038/s41572-018-0055-y

Agarwal S et al (2012) Breast cancer resistance protein and P-glycoprotein in brain cancer: two gatekeepers team up. Curr Pharm Des. https://doi.org/10.2174/138161211797440186

Akimoto J (2016) Photodynamic therapy for malignant brain tumors. Neurol Med Chir. https://doi.org/10.2176/nmc.ra.2015-0296

Aldape K et al (2019) Challenges to curing primary brain tumours. Nat Rev Clin Oncol. https://doi.org/10.1038/s41571-019-0177-5

Alifieris C, Trafalis DT (2015) Glioblastoma multiforme: pathogenesis and treatment. Pharmacol Therapeutic. https://doi.org/10.1016/j.pharmthera.2015.05.005

Allison RR, Sibata CH (2010) Oncologic photodynamic therapy photosensitizers: a clinical review. Photodiagnosis Photodyn Ther. https://doi.org/10.1016/j.pdpdt.2010.02.001

Amelio D, Amichetti MA (2012) Radiation therapy for the treatment of recurrent glioblastoma: an overview. Cancer. https://doi.org/10.3390/cancers4010257

Amsbaugh, M. J. and Kim, C. S. (2019) Cancer, brain metastasis, StatPearls.

Aronica E et al (2005) Localization of breast cancer resistance protein (BCRP) in microvessel endothelium of human control and epileptic brain. Epilepsia 46(6):849–857

Arvanitis CD et al (2018) Mechanisms of enhanced drug delivery in brain metastases with focused ultrasound-induced blood–tumor barrier disruption. Proc Natl Acad Sci U S A. https://doi.org/10.1073/pnas.1807105115

Barani IJ, Larson DA (2015) Radiation therapy of glioblastoma. Cancer Treat Res. https://doi.
    org/10.1007/978-3-319-12048-5_4
Bartsch R et al (2015) Activity of T-DM1 in Her2-positive breast cancer brain metastases. Clin Exp
    Metastasis. https://doi.org/10.1007/s10585-015-9740-3
Basit S et al (2017) European journal of medicinal chemistry first macrocyclic 3rd-generation
    ALK inhibitor for treatment of ALK / ROS1 cancer: clinical and designing strategy update
    of lorlatinib. Eur J Med Chem 134:348–356. https://doi.org/10.1016/j.ejmech.2017.04.032.
    Elsevier Masson SAS
Bastien JIL, McNeill KA, Fine HA (2015) Molecular characterizations of glioblastoma, targeted
    therapy, and clinical results to date. Cancer. https://doi.org/10.1002/cncr.28968
Bauer TM et al (2020) Brain penetration of lorlatinib: cumulative incidences of CNS and non-
    CNS progression with lorlatinib in patients with previously treated ALK-positive non-small-
    cell lung cancer. Target Oncol. https://doi.org/10.1007/s11523-020-00702-4
Baumert C, Hilgeroth A (2012) Recent advances in the development of P-gp inhibitors. Anticancer
    Agents Med Chem. https://doi.org/10.2174/1871520610909040415
Bellavance MA, Blanchette M, Fortin D (2008) Recent advances in blood-brain barrier disruption
    as a CNS delivery strategy. AAPS J. https://doi.org/10.1208/s12248-008-9018-7
Bhagavathi S, Wilson JD (2008) Primary central nervous system lymphoma. Arch Pathol Lab
    Med. https://doi.org/10.1043/1543-2165-132.11.1830
Bhardwaj RK et al (2006) The functional evaluation of human peptide/histidine transporter 1
    (hPHT1) in transiently transfected COS-7 cells. Eur J Pharm Sci. https://doi.org/10.1016/j.
    ejps.2005.09.014
Bhatia P et al (2012) Breast cancer resistance protein (BCRP/ABCG2) localises to the nucleus in
    glioblastoma multiforme cells. Xenobiotica. 2012/03/10 42(8):748–755. https://doi.org/10.310
    9/00498254.2012.662724
Bi WL, Beroukhim R (2014) Beating the odds: extreme long-term survival with glioblastoma.
    Neuro Oncol. https://doi.org/10.1093/neuonc/nou166
Blakeley J (2008) Drug delivery to brain tumors. Curr Neurol Neurosci Rep. https://doi.
    org/10.1007/s11910-008-0036-8
Blakeley JO et al (2009) Effect of blood brain barrier permeability in recurrent high grade gliomas
    on the intratumoral pharmacokinetics of methotrexate: a microdialysis study. J Neurooncol.
    https://doi.org/10.1007/s11060-008-9678-2
Bleau AM et al (2009) PTEN/PI3K/Akt pathway regulates the side population phenotype and
    ABCG2 activity in glioma tumor stem-like cells. Cell Stem Cell 4(3):226–235. https://doi.
    org/10.1016/j.stem.2009.01.007
Boado RJ et al (2010) IgG-single chain Fv fusion protein therapeutic for Alzheimer's disease:
    expression in CHO cells and pharmacokinetics and brain delivery in the Rhesus monkey.
    Biotechnol Bioeng. https://doi.org/10.1002/bit.22576
Brennan CW et al (2013) The somatic genomic landscape of glioblastoma. Cell. 2013/10/15
    155(2):462–477. https://doi.org/10.1016/j.cell.2013.09.034
Bruzzone MG et al (2012) CT and MRI of brain tumors. Q J Nucl Med Mol Imaging
Bui K et al (2016) The effect of quinidine, a strong P-glycoprotein inhibitor, on the pharmaco-
    kinetics and central nervous system distribution of naloxegol. J Clin Pharmacol. https://doi.
    org/10.1002/jcph.613
Bunevicius A, McDannold NJ, Golby AJ (2020) Focused ultrasound strategies for brain tumor
    therapy. Oper Neurosurg. https://doi.org/10.1093/ons/opz374
Calias P et al (2014) Intrathecal delivery of protein therapeutics to the brain: a critical reassess-
    ment. Pharmacol Therapeutic. https://doi.org/10.1016/j.pharmthera.2014.05.009
Cardoso FL, Brites D, Brito MA (2010) Looking at the blood-brain barrier: molecular anatomy
    and possible investigation approaches. Brain Res Rev. Elsevier B.V 64(2):328–363. https://doi.
    org/10.1016/j.brainresrev.2010.05.003
Castedo M et al (2004) Cell death by mitotic catastrophe: a molecular definition. Oncogene.
    https://doi.org/10.1038/sj.onc.1207528

Castro M et al (2001) High-affinity sodium-vitamin C co-transporters (SVCT) expression in embryonic mouse neurons. J Neurochem. https://doi.org/10.1046/j.1471-4159.2001.00461.x

Chamberlain MC et al (2014) A phase 2 trial of verubulin for recurrent glioblastoma: a prospective study by the brain tumor investigational consortium (BTIC). J Neurooncol. https://doi.org/10.1007/s11060-014-1437-y

Chen JS et al (2006) EGFR regulates the side population in head and neck squamous cell carcinoma. Laryngoscope. 2006/03/17 116(3):401–406. https://doi.org/10.1097/01.mlg.0000195075.14093.fb

Cho HS et al (2016) Localized down-regulation of P-glycoprotein by focused ultrasound and microbubbles induced blood-brain barrier disruption in rat brain. Sci Rep. Nature Publishing Group 6(May):1–10. https://doi.org/10.1038/srep31201

Chowdhary S, Chamberlain M (2013) Bevacizumab for the treatment of glioblastoma. Expert Rev Neurother. https://doi.org/10.1586/14737175.2013.827414

Cloughesy TF, Cavenee WK, Mischel PS (2014) Glioblastoma: from molecular pathology to targeted treatment. https://doi.org/10.1146/annurev-pathol-011110-130324

Cooray HC et al (2002) Localisation of breast cancer resistance protein in microvessel endothelium of human brain. Neuroreport. https://doi.org/10.1097/00001756-200211150-00014

Cordon-Cardo C et al (1989) Multidrug-resistance gene (P-glycoprotein) is expressed by endothelial cells at blood-brain barrier sites. Proc Natl Acad Sci 86:695–698. Available at: https://www.ncbi.nlm.nih.gov/pmc/articles/PMC286540/pdf/pnas00242-0297.pdf

Cramer SW, Chen CC (2020) Photodynamic therapy for the treatment of glioblastoma. Front Surg. https://doi.org/10.3389/fsurg.2019.00081

Davis ME (2016) HHS public access. 20(5):1–14. https://doi.org/10.1188/16.CJON.S1.2-8. Glioblastoma

de Trizio I et al (2020) Expression of P-gp in glioblastoma: what we can learn from brain development. Curr Pharm Des. https://doi.org/10.2174/1381612826666200318130625

Debinski W, Tatter SB (2009) Convection-enhanced delivery for the treatment of brain tumors. Expert Rev Neurother. https://doi.org/10.1586/ern.09.99

Delattre JY et al (1988) Distribution of brain metastases. Arch Neurol. https://doi.org/10.1001/archneur.1988.00520310047016

Demeule M et al (2001) Expression of multidrug-resistance p-glycoprotein (MDR1) in human brain tumors. Int J Cancer 93:62–66

Dempke WCM et al (2015) Brain metastases in NSCLC-are TKIs changing the treatment strategy? Anticancer Res

Deng Z, Sheng Z, Yan F (2019) Ultrasound-induced blood-brain-barrier opening enhances anti-cancer efficacy in the treatment of glioblastoma: current status and future prospects. J Oncol. https://doi.org/10.1155/2019/2345203

Drappatz J et al (2013) Phase I study of GRN1005 in recurrent malignant glioma. Clin Cancer Res 19(6):1567–1576. Available at: http://clincancerres.aacrjournals.org/content/19/6/1567.abstract

Dubey RK et al (1989) Plasma binding and transport of diazepam across the blood-brain barrier. No evidence for in vivo enhanced dissociation. J Clin Investig. https://doi.org/10.1172/JCI114279

Durant ST et al (2018) The brain-penetrant clinical ATM inhibitor AZD1390 radiosensitizes and improves survival of preclinical brain tumor models. Sci Adv. https://doi.org/10.1126/sciadv.aat1719

Eisenblaetter T, Galla H-J (2002) A new multidrug resistance protein at the blood–brain barrier. Biochem Biophys Res Commun 293:1273–1278

El-Osta A et al (2002) Precipitous release of methyl-CpG binding protein 2 and histone deacetylase 1 from the methylated human multidrug resistance gene (MDR1) on activation. Mol Cell Biol 22(6):1844–1857. https://doi.org/10.1128/mcb.22.6.1844-1857.2002

Emery IF et al (2017) Expression and function of ABCG2 and XIAP in glioblastomas. J Neurooncol. 2017/04/23 133(1):47–57. https://doi.org/10.1007/s11060-017-2422-z

Eskilsson E et al (2018) EGFR heterogeneity and implications for therapeutic intervention in glioblastoma. Neuro Oncol. 2017/10/19 20(6):743–752. https://doi.org/10.1093/neuonc/nox191

Ewend MG et al (2007) Treatment of single brain metastasis with resection, intracavity carmustine polymer wafers, and radiation therapy is safe and provides excellent local control. Clin Cancer Res. https://doi.org/10.1158/1078-0432.CCR-06-2095

Fokas E, Steinbach JP, Rödel C (2013) Biology of brain metastases and novel targeted therapies: time to translate the research. Biochim Biophys Acta Rev Cancer. https://doi.org/10.1016/j.bbcan.2012.10.005

Fox E, Bates SE (2007) Tariquidar (XR9576): a P-glycoprotein drug efflux pump inhibitor. Expert Rev Anticancer Ther. https://doi.org/10.1586/14737140.7.4.447

Fridén M et al (2009) Development of a high-throughput brain slice method for studying drug distribution in the central nervous system. Drug Metab Dispos. https://doi.org/10.1124/dmd.108.026377

Fridén M et al (2011) Measurement of unbound drug exposure in brain: modeling of pH partitioning explains diverging results between the brain slice and brain homogenate methods. Drug Metab Dispos. https://doi.org/10.1124/dmd.110.035998

Gadgeel S et al (2018) Alectinib versus crizotinib in treatment-naive anaplastic lymphoma kinase-positive (ALKþ) non-small-cell lung cancer: CNS efficacy results from the ALEX study. Ann Oncol. https://doi.org/10.1093/annonc/mdy405

Gaillard PJ et al (2014) Pharmacokinetics, brain delivery, and efficacy in brain tumor-bearing mice of glutathione pegylated liposomal doxorubicin (2B3-101). PLoS One. https://doi.org/10.1371/journal.pone.0082331

Gampa G et al (2020) Influence of transporters in treating cancers in the CNS. In: Drug efflux pumps in cancer resistance pathways: from molecular recognition and characterization to possible inhibition strategies in chemotherapy. https://doi.org/10.1016/b978-0-12-816434-1.00009-7

Gil ES et al (2009) Quaternary ammonium β-cyclodextrin nanoparticles for enhancing doxorubicin permeability across the in vitro blood-brain barrier. Biomacromolecules. https://doi.org/10.1021/bm801026k

Ginguene C et al (2010) P-glycoprotein (ABCB1) and breast cancer resistance protein (ABCG2) localize in the microvessels forming the blood-tumor barrier in ependymomas. Brain Pathol. 2010/04/22 20(5):926–935. https://doi.org/10.1111/j.1750-3639.2010.00389.x

Gottesman MM (2002) Mechanisms of cancer drug resistance. Annu Rev Med. https://doi.org/10.1146/annurev.med.53.082901.103929

Grossman R et al (2013) Microdialysis measurement of intratumoral temozolomide concentration after cediranib, a pan-VEGF receptor tyrosine kinase inhibitor, in a U87 glioma model. Cancer Chemother Pharmacol. https://doi.org/10.1007/s00280-013-2172-3

Hammarlund-Udenaes M et al (2008) On the rate and extent of drug delivery to the brain. Pharm Res. https://doi.org/10.1007/s11095-007-9502-2

Han SR et al (2015) Effect and mechanism of peroxisome proliferator-activated receptor-gamma on the drug resistance of the U-87 MG/CDDP human malignant glioma cell line. Molecular Mol Med Rep 12(2):2239–2246. https://doi.org/10.3892/mmr.2015.3625

Harmsen S et al (2013) PXR-mediated P-glycoprotein induction by small molecule tyrosine kinase inhibitors. Eur J Pharm Sci 48(4–5):644–649. https://doi.org/10.1016/j.ejps.2012.12.019

Harper JW, Elledge SJ (2007) The DNA damage response: ten years after. Mol Cell. https://doi.org/10.1016/j.molcel.2007.11.015

Hartz AMS, Mahringer A et al (2010a) 17-beta-Estradiol: a powerful modulator of blood-brain barrier BCRP activity. J Cereb Blood Flow Metab 30(10):1742–1755. https://doi.org/10.1038/jcbfm.2010.36

Hartz AMS, Madole EK et al (2010b) Estrogen receptor beta signaling through phosphatase and tensin homolog/phosphoinositide 3-kinase/Akt/glycogen synthase kinase 3 down-regulates blood-brain barrier breast cancer resistance protein. J Pharmacol Exp Ther 334(2):467–476. https://doi.org/10.1124/jpet.110.168930

Hawkins BT et al (2010) The blood-brain barrier / neurovascular unit in health and disease. Neurobiol Dis. Elsevier B.V 64(1):13–25. https://doi.org/10.1016/j.brainresrev.2010.05.003

Heffron TP (2016) Small molecule kinase inhibitors for the treatment of brain cancer. J Med Chem:10030–10066. https://doi.org/10.1021/acs.jmedchem.6b00618

Heffron TP et al (2016) Discovery of clinical development candidate GDC-0084, a brain penetrant inhibitor of PI3K and mTOR. ACS Med Chem Lett. https://doi.org/10.1021/acsmedchemlett.6b00005

Hellmann-Regen J et al (2012) Retinoic acid as target for local pharmacokinetic interaction with modafinil in neural cells. Eur Arch Psychiatry Clin Neurosci 262(8):697–704. https://doi.org/10.1007/s00406-012-0309-8

Hoeijmakers JHJ (2009) DNA damage, aging, and cancer. N Engl J Med. https://doi.org/10.1056/NEJMra0804615

Huang FF et al (2013) Inactivation of PTEN increases ABCG2 expression and the side population through the PI3K/Akt pathway in adult acute leukemia. Cancer Lett. 2013/04/23 336(1):96–105. https://doi.org/10.1016/j.canlet.2013.04.006

Huang J et al (2014) Regenerating gene family member 4 promotes growth and migration of gastric cancer through protein kinase B pathway. Int J Clin Exp Med 7(9):3037–3044

Huisman MT et al (2003) Assessing safety and efficacy of directed P-glycoprotein inhibition to improve the pharmacokinetic properties of saquinavir coadministered with ritonavir. J Pharmacol Experiment Therapeutic. https://doi.org/10.1124/jpet.102.044388

Hunt Bobo R et al (1994) Convection-enhanced delivery of macromolecules in the brain. Proc Natl Acad Sci U S A. https://doi.org/10.1073/pnas.91.6.2076

J.G., P.-L et al (2011) Temozolomide in elderly patients with newly diagnosed glioblastoma and poor performance status: An ANOCEF phase II trial. J Clin Oncol. https://doi.org/10.1200/JCO.2011.34.8086LK. http://onesearch.unifi.it/openurl/39UFI/39UFI_Services?&sid=EMBASE&issn=0732183X&id=doi:10.1200%2FJCO.2011.34.8086&atitle=Temozolomide+in+elderly+patients+with+newly+diagnosed+glioblastoma+and+poor+performance+status%3A+An+ANOCEF+phase+II+trial&stitle=J.+Clin.+Oncol.&title=Journal+of+Clinical+Oncology&volume=29&issue=22&spage=3050&epage=3055&aulast=P%C3%A9rez-Larraya&aufirst=Jaime+G%C3%A1llego&auinit=J.G.&aufull=P%C3%A9rez-Larraya+J.G.&coden=JCOND&isbn=&pages=3050-

Jackson S et al (2016) The effect of regadenoson-induced transient disruption of the blood-brain barrier on temozolomide delivery to normal rat brain. J Neurooncol. 2015/12/03 126(3):433–439. https://doi.org/10.1007/s11060-015-1998-4

Jackson S et al (2017) The effect of regadenoson on the integrity of the human blood–brain barrier, a pilot study. J Neurooncol. https://doi.org/10.1007/s11060-017-2404-1

Jackson S et al (2018) The effect of an adenosine A2A agonist on intra-tumoral concentrations of temozolomide in patients with recurrent glioblastoma. Fluid Barrier CNS. 2018/01/16 15(1):2. https://doi.org/10.1186/s12987-017-0088-8

Johnson RA, Ince TA, Scotto KW (2001) Transcriptional repression by p53 through direct binding to a novel DNA element. J Biol Chem. 2001/05/15 276(29):27716–27720. https://doi.org/10.1074/jbc.C100121200

Karaoglu Hanzatian D et al (2018) Brain uptake of multivalent and multi-specific DVD-Ig proteins after systemic administration. MAbs. https://doi.org/10.1080/19420862.2018.1465159

Keith KC et al (2016) Activity of trastuzumab-emtansine (TDM1) in HER2-positive breast cancer brain metastases: A case series. Cancer Treat Commun. https://doi.org/10.1016/j.ctrc.2016.03.005

Kemper EM et al (2004) Modulation of the blood-brain barrier in oncology: therapeutic opportunities for the treatment of brain tumours? Cancer Treat Rev. https://doi.org/10.1016/j.ctrv.2004.04.001

Kim DG, Bynoe MS (2015) A2A Adenosine Receptor Regulates the Human Blood-Brain Barrier Permeability. Mol Neurobiol. 2014/09/30 52(1):664–678. https://doi.org/10.1007/s12035-014-8879-2

Kim DG, Bynoe MS (2016) A2A adenosine receptor modulates drug efflux transporter P-glycoprotein at the blood-brain barrier. J Clin Invest. 2016/04/05 126(5):1717–1733. https://doi.org/10.1172/JCI76207

Kim M et al (2018) Barriers to effective drug treatment for brain metastases: a multifactorial problem in the delivery of precision medicine. Pharm Res. https://doi.org/10.1007/s11095-018-2455-9

Kim M et al (2019a) Brain distribution of a panel of epidermal growth factor receptor inhibitors using cassette dosing in wild-type and ABCB1/ABCG2-deficient mice. Drug Metab Dispos. https://doi.org/10.1124/dmd.118.084210

Kim SS et al (2019b) A tumor-targeting nanomedicine carrying the p53 gene crosses the blood–brain barrier and enhances anti-PD-1 immunotherapy in mouse models of glioblastoma. Int J Cancer. https://doi.org/10.1002/ijc.32531

Kondo S et al (1996) mdm2 gene mediates the expression of mdrl gene and P-glycoprotein in a human glioblastoma cell line. Br J Cancer 74:1263–1268

Korshunov A et al (1999) Prognostic value of immunoexpression of the chemoresistance-related proteins in ependymomas: an analysis of 76 cases. J Neurooncol 45:219–227

Kroll RA et al (1998) Improving drug delivery to intracerebral tumor and surrounding brain in a rodent model: a comparison of osmotic versus bradykinin modification of the blood-brain and/or blood-tumor barriers. Neurosurgery. United States 43(4):879

Kuang Y et al (2013) T7 peptide-functionalized nanoparticles utilizing RNA interference for glioma dual targeting. Int J Pharm. https://doi.org/10.1016/j.ijpharm.2013.07.019

Kunwar S et al (2010) Phase III randomized trial of CED of IL13-PE38QQR vs Gliadel wafers for recurrent glioblastoma. Neuro Oncol. https://doi.org/10.1093/neuonc/nop054

Kuppens IELM et al (2007) A phase I, randomized, open-label, parallel-cohort, dose-finding study of elacridar (GF120918) and oral topotecan in cancer patients. Clin Cancer Res. https://doi.org/10.1158/1078-0432.CCR-06-2414

Kurzrock R et al (2012) Safety, pharmacokinetics, and activity of GRN1005, a novel conjugate of angiopep-2, a peptide facilitating brain penetration, and paclitaxel, in patients with advanced solid tumors. Mol Cancer Ther. https://doi.org/10.1158/1535-7163.MCT-11-0566

Laack NN, Brown PD (2004) Cognitive sequelae of brain radiation in adults. Semin Oncol. https://doi.org/10.1053/j.seminoncol.2004.07.013

Lambert JM, Chari RVJ (2014) Ado-trastuzumab emtansine (T-DM1): an antibody-drug conjugate (ADC) for HER2-positive breast cancer. J Med Chem. https://doi.org/10.1021/jm500766w

Laramy JK et al (2017) Heterogeneous binding and central nervous system distribution of the multitargeted kinase inhibitor ponatinib restrict orthotopic efficacy in a patient-derived xenograft model of glioblastoma. J Pharmacol Exp Ther. https://doi.org/10.1124/jpet.117.243477

Lee SY (2016) Temozolomide resistance in glioblastoma multiforme. Genes and Diseases. https://doi.org/10.1016/j.gendis.2016.04.007

Lee J et al (2019) Flexible, sticky, and biodegradable wireless device for drug delivery to brain tumors. Nat Commun. https://doi.org/10.1038/s41467-019-13198-y

Levin AB, Duff TA, Javid MJ (1979) Treatment of increased intracranial pressure: a comparison of different hyperosmotic agents and the use of thiopental. Neurosurgery. https://doi.org/10.1227/00006123-197911000-00005

Li J et al (2017) Quantitative and mechanistic understanding of AZD1775 penetration across human blood-brain barrier in glioblastoma patients using an IVIVE-PBPK modeling approach. Clin Cancer Res. https://doi.org/10.1158/1078-0432.CCR-17-0983

Lim M et al (2019) Innovative therapeutic strategies for effective treatment of brain metastases. Int J Mol Sci. https://doi.org/10.3390/ijms20061280

Lin X, DeAngelis LM (2015) Treatment of brain metastases. J Clin Oncol. https://doi.org/10.1200/JCO.2015.60.9503

Liu HL et al (2014) Combining microbubbles and ultrasound for drug delivery to brain tumors: current progress and overview. Theranostics. https://doi.org/10.7150/thno.8074

Liu Q, Tong X, Wang J (2019) Management of brain metastases: history and the present. Chin Neurosurg J. https://doi.org/10.1186/s41016-018-0149-0

Liu EK et al (2020) Novel therapies for glioblastoma. Curr Neurol Neurosci Rep. https://doi.org/10.1007/s11910-020-01042-6

Lonser RR et al (2015) Convection-enhanced delivery to the central nervous system. J Neurosurg. https://doi.org/10.3171/2014.10.JNS14229

Lord CJ, Ashworth A (2012) The DNA damage response and cancer therapy. Nature. https://doi.org/10.1038/nature10760

Loryan I, Fridén M, Hammarlund-Udenaes M (2013) The brain slice method for studying drug distribution in the CNS. Fluid Barrier CNS. https://doi.org/10.1186/2045-8118-10-6

Löscher W, Potschka H (2005) Blood-brain barrier active efflux transporters: ATP-binding cassette gene family. NeuroRx: J Am Soc Experiment NeuroTherapeutic 2(1):86–98. https://doi.org/10.1602/neurorx.2.1.86

Lyon JG et al (2017) Engineering challenges for brain tumor immunotherapy. Adv Drug Deliv Rev. https://doi.org/10.1016/j.addr.2017.06.006

Mack F et al (2016) Therapy of leptomeningeal metastasis in solid tumors. Cancer Treat Rev. https://doi.org/10.1016/j.ctrv.2015.12.004

Mahmoudi K et al (2019) 5-aminolevulinic acid photodynamic therapy for the treatment of high-grade gliomas. J Neurooncol. https://doi.org/10.1007/s11060-019-03103-4

Mamelak AN et al (2006) Phase I single-dose study of intracavitary-administered iodine-131-TM-601 in adults with recurrent high-grade glioma. J Clin Oncol. https://doi.org/10.1200/JCO.2005.05.4569

Mani S et al (2005) Activation of the steroid and xenobiotic receptor (human pregnane X receptor) by nontaxane microtubule-stabilizing agents. Clin Cancer Res 11(17):6359–6369. https://doi.org/10.1158/1078-0432.CCR-05-0252

Mani S, Dou W, Redinbo MR (2013) PXR antagonists and implication in drug metabolism. Drug Metab Rev 45(1):60–72. https://doi.org/10.3109/03602532.2012.746363

Marroni M et al (2003) Relationship between expression of multiple drug resistance proteins and p53 tumor suppressor gene proteins in human brain astrocytes. Neuroscience. https://doi.org/10.1016/S0306-4522(03)00515-3

McClelland S et al (2018) Application of tumor treating fields for newly diagnosed glioblastoma: understanding of nationwide practice patterns. J Neurooncol. https://doi.org/10.1007/s11060-018-2945-y

McNeill KA (2016) Epidemiology of brain tumors. Neurol Clin. https://doi.org/10.1016/j.ncl.2016.06.014

Meng Y et al (2018) Low-intensity MR-guided focused ultrasound mediated disruption of the blood-brain barrier for intracranial metastatic diseases. Front Oncol. https://doi.org/10.3389/fonc.2018.00338

Munoz JL et al (2014) Temozolomide induces the production of epidermal growth factor to regulate MDR1 expression in glioblastoma cells. Mol Cancer Ther 13(10):2399–2411. https://doi.org/10.1158/1535-7163.Mct-14-0011

Murthy R et al (2018) Tucatinib with capecitabine and trastuzumab in advanced HER2-positive metastatic breast cancer with and without brain metastases: a non-randomised, open-label, phase 1b study. Lancet Oncol. https://doi.org/10.1016/S1470-2045(18)30256-0

Murthy RK et al (2020) Tucatinib, trastuzumab, and capecitabine for HER2-positive metastatic breast cancer. N Engl J Med. https://doi.org/10.1056/NEJMoa1914609

Nakanishi T, Ross DD (2012) Breast cancer resistance protein (BCRP/ABCG2): its role in multi-drug resistance and regulation of its gene expression. Chin J Cancer 31(2):73–99. https://doi.org/10.5732/cjc.011.10320

Nakanishi T, Shiozawa K, Hamburger AW (2006) Bcrp expression is functionally upregulated by epidermal growth factor receptor (EGFR, ERBB1) mediated signaling in human ovarian cancer cell lines, but not in human breast cancer cell lines. AACR Meeti Abstract 146

Nayak L, Lee EQ, Wen PY (2012) Epidemiology of brain metastases. Curr Oncol Rep. https://doi.org/10.1007/s11912-011-0203-y

Neuwelt EA (1980) Reversible osmotic blood-brain barrier disruption in humans: implications for the chemotherapy of malignant brain tumors. Neurosurgery:204. https://doi.org/10.1097/00006123-198008000-00018

Neyns B et al (2009) Stratified phase II trial of cetuximab in patients with recurrent high-grade glioma. Ann Oncol. https://doi.org/10.1093/annonc/mdp032

Niwińska A, Rudnicka H, Murawska M (2015) Breast cancer leptomeningeal metastasis: The results of combined treatment and the comparison of methotrexate and liposomal cytarabine as intra-cerebrospinal fluid chemotherapy. Clin Breast Cancer. https://doi.org/10.1016/j.clbc.2014.07.004

O'Connor MJ (2015) Targeting the DNA damage response in cancer. Mol Cell. https://doi.org/10.1016/j.molcel.2015.10.040

Oberoi RK et al (2016) Strategies to improve delivery of anticancer drugs across the blood-brain barrier to treat glioblastoma. Neuro-Oncology:27–36. https://doi.org/10.1093/neuonc/nov164

Okines A et al (2018) Development and responses of brain metastases during treatment with trastuzumab emtansine (T-DM1) for HER2 positive advanced breast cancer: A single institution experience. Breast J. https://doi.org/10.1111/tbj.12906

Oliveira M et al (2011) Complete response in HER2+ leptomeningeal carcinomatosis from breast cancer with intrathecal trastuzumab. Breast Cancer Res Treat. https://doi.org/10.1007/s10549-011-1417-2

Omidi Y, Barar J (2012) Impacts of blood-brain barrier in drug delivery and targeting of brain tumors. Bioimpacts. https://doi.org/10.5681/bi.2012.002

Ostrom QT et al (2014) The epidemiology of glioma in adults: a state of the science review. Neuro Oncol. https://doi.org/10.1093/neuonc/nou087

Ostrom QT et al (2019) CBTRUS statistical report: primary brain and other central nervous system tumors diagnosed in the United States in 2012–2016. Neuro Oncol. https://doi.org/10.1093/neuonc/noz150

Pan-Weisz TM et al (2019) Patient-reported health-related quality of life outcomes in supportive-care interventions for adults with brain tumors: a systematic review. Psychooncology. https://doi.org/10.1002/pon.4906

Pardridge WM (2016) CSF, blood-brain barrier, and brain drug delivery. Expert Opin Drug Deliv. https://doi.org/10.1517/17425247.2016.1171315

Pardridge WM, Eisenberg J, Yang J (1985) Human blood—brain barrier insulin receptor. J Neurochem. https://doi.org/10.1111/j.1471-4159.1985.tb07167.x

Pardridge WM, Eisenberg J, Yang J (1987) Human blood-brain barrier transferrin receptor. Metabolism. https://doi.org/10.1016/0026-0495(87)90099-0

Parrish KE, Sarkaria JN, Elmquist WF (2015) Improving drug delivery to primary and metastatic brain tumors: strategies to overcome the blood-brain barrier. Clinic Pharmacol Therapeutic:336–346. https://doi.org/10.1002/cpt.71

Pearl LH et al (2015) Therapeutic opportunities within the DNA damage response. Nat Rev Cancer. https://doi.org/10.1038/nrc3891

Perissinotti AJ, Reeves DJ (2010) Role of intrathecal rituximab and trastuzumab in the management of leptomeningeal carcinomatosis. Ann Pharmacother. https://doi.org/10.1345/aph.1P197

Perkins A, Liu G (2016) Primary brain tumors in adults: Diagnosis and treatment. Am Fam Physician

Perria C, Capuzzo T, Cavagnaro G (1980) First attempts at the photodynamic treatment of human gliomas. J Neurosurg Sci

Perus LJM, Walsh LA (2019) Microenvironmental heterogeneity in brain malignancies. Front Immunol. https://doi.org/10.3389/fimmu.2019.02294

Porter KR et al (2010) Prevalence estimates for primary brain tumors in the United States by age, gender, behavior, and histology. Neuro Oncol. https://doi.org/10.1093/neuonc/nop066

Portnow J et al (2009) The neuropharmacokinetics of temozolomide in patients with resectable brain tumors: potential implications for the current approach to chemoradiation. Clin Cancer Res. https://doi.org/10.1158/1078-0432.CCR-09-1349

Puris E et al (2020) L-Type amino acid transporter 1 as a target for drug delivery. Pharm Res. https://doi.org/10.1007/s11095-020-02826-8

Ramalingam SS et al (2020) Overall survival with osimertinib in untreated, EGFR-mutated advanced NSCLC. N Engl J Med. https://doi.org/10.1056/NEJMoa1913662

Rankovic Z (2015) CNS drug design: balancing physicochemical properties for optimal brain exposure. J Med Chem:2584–2608. https://doi.org/10.1021/jm501535r

Rankovic Z (2017) CNS physicochemical property space shaped by a diverse set of molecules with experimentally determined exposure in the mouse brain. J Med Chem:5943–5954. https://doi.org/10.1021/acs.jmedchem.6b01469

Rapoport SI (2000) Osmotic opening of the blood-brain barrier: principles, mechanism, and therapeutic applications. Cell Mol Neurobiol. https://doi.org/10.1023/A:1007049806660

Rapoport SI, Hori M, Klatzo I (1972) Testing of a hypothesis for osmotic opening of the blood-brain barrier. Am J Physiol. https://doi.org/10.1152/ajplegacy.1972.223.2.323

Reungwetwattana T et al (2018) CNS response to osimertinib versus standard epidermal growth factor receptor tyrosine kinase inhibitors in patients with untreated EGFR-mutated advanced non-small-cell lung cancer', Journal of Clinical Oncology. https://doi.org/10.1200/JCO.2018.78.3118

Ricciardi GRR et al (2018) Efficacy of T-DM1 for leptomeningeal and brain metastases in a HER2 positive metastatic breast cancer patient: New directions for systemic therapy - a case report and literature review. BMC Cancer. https://doi.org/10.1186/s12885-018-3994-5

Rick JW et al (2019) Systemic therapy for brain metastases. Crit Rev Oncol Hematol. https://doi.org/10.1016/j.critrevonc.2019.07.012

Riganti C et al (2013) Temozolomide downregulates P-glycoprotein expression in glioblastoma stem cells by interfering with the Wnt3a/glycogen synthase-3 kinase/beta-catenin pathway. Neuro Oncol. 2013/07/31 15(11):1502–1517. https://doi.org/10.1093/neuonc/not104

Sakata S et al (2011) ATP-binding cassette transporters in primary central nervous system lymphoma: Decreased expression of MDR1 P-glycoprotein and breast cancer resistance protein in tumor capillary endothelial cells. Oncol Rep 25(2):333–339. https://doi.org/10.3892/or.2010.1102

Salphati L et al (2012) Targeting the PI3K pathway in the brain - efficacy of a PI3K inhibitor optimized to cross the blood-brain barrier. Clin Cancer Res. https://doi.org/10.1158/1078-0432.CCR-12-0720

Sampath J et al (2001) Mutant p53 cooperates with ETS and selectively up-regulates human MDR1 Not MRP1. J Biol Chem. https://doi.org/10.1074/jbc.M103429200

Sandler A et al (2004) A phase I trial of a potent P-glycoprotein inhibitor, zosuquidar trihydrochloride (LY335979), administered intravenously in combination with doxorubicin in patients with advanced malignancy. Clin Cancer Res. https://doi.org/10.1158/1078-0432.CCR-03-0644

Sarkadi B et al (2006) Human multidrug resistance ABCB and ABCG transporters: participation in a chemoimmunity defense system. Physiol Rev. https://doi.org/10.1152/physrev.00037.2005

Sarkaria JN et al (2018) Is the blood-brain barrier really disrupted in all glioblastomas? A critical assessment of existing clinical data. Neuro Oncol. https://doi.org/10.1093/neuonc/nox175

Shapira-Furman T et al (2019) Biodegradable wafers releasing Temozolomide and Carmustine for the treatment of brain cancer. J Control Release. https://doi.org/10.1016/j.jconrel.2018.12.048

Sharma J, Lv H, Gallo JM (2012) Analytical approach to characterize the intratumoral pharmacokinetics and pharmacodynamics of gefitinib in a glioblastoma model. J Pharm Sci. https://doi.org/10.1002/jps.23283

Sharma J, Lv H, Gallo JM (2013) Intratumoral modeling of gefitinib pharmacokinetics and pharmacodynamics in an orthotopic mouse model of glioblastoma. Cancer Res. https://doi.org/10.1158/0008-5472.CAN-13-0690

Shaw AT et al (2017) Alectinib versus crizotinib in treatment-naive advanced ALK -positive non-small cell lung cancer (NSCLC): primary results of the global phase III ALEX study. J Clin Oncol. https://doi.org/10.1200/jco.2017.35.18_suppl.lba9008

Shawahna R et al (2011) Transcriptomic and quantitative proteomic analysis of transporters and drug metabolizing enzymes in freshly isolated human brain microvessels. Mol Pharm. https://doi.org/10.1021/mp200129p

Shen F et al (2008) Quantitation of doxorubicin uptake, efflux, and modulation of multidrug resistance (MDR) in MDR human cancer cells. J Pharmacol Experiment Therapeutic. https://doi.org/10.1124/jpet.107.127704

Siegel RL, Miller KD, Jemal A (2019) Cancer statistics, 2019. CA Cancer J Clin. https://doi.org/10.3322/caac.21551

Soria JC et al (2018) Osimertinib in untreated EGFR-Mutated advanced non-small-cell lung cancer. N Engl J Med. https://doi.org/10.1056/NEJMoa1713137

Sperduto PW et al (2012) Summary report on the graded prognostic assessment: An accurate and facile diagnosis-specific tool to estimate survival for patients with brain metastases. J Clin Oncol. https://doi.org/10.1200/JCO.2011.38.0527

Spiegl-Kreinecker S et al (2002) Expression and functional activity of the ABC-transporter proteins P-glycoprotein and multidrug-resistance protein 1 in human brain tumor cells and astrocytes. J Neurooncol 57:27–36

St-Amour I et al (2013) Brain bioavailability of human intravenous immunoglobulin and its transport through the murine blood-brain barrier. J Cereb Blood Flow Metab. https://doi.org/10.1038/jcbfm.2013.160

Stupp R (2019) Changing paradigms in the treatment of brain metastases. J Oncol Pract. https://doi.org/10.1200/JOP.19.00602

Stupp R et al (2008) Radiotherapy plus concomitant and adjuvant temozolomide for glioblastoma – a critical review. Clinic Med Oncol. https://doi.org/10.1056/NEJMoa043330

Sugawara M et al (2010) Expressions of cytochrome P450, UDP-glucuronosyltranferase, and transporter genes in monolayer carcinoma cells change in subcutaneous tumors grown as xenografts in immunodeficient nude mice. Drug Metab Dispos 38(3):526–533. https://doi.org/10.1124/dmd.109.030668

Sun M, Chen Y, Hu B (2018) The inhibition of DNA damage response contributes to cancer radiotherapy. 5(1):149–156. https://doi.org/10.17554/j.issn.2313-3406.2018.05.47

Sutherlin DP et al (2010) Discovery of (thienopyrimidin-2-yl)aminopyrimidines as potent, selective, and orally available Pan-PI3-kinase and dual Pan-PI3-kinase/mTOR inhibitors for the treatment of cancer. J Med Chem 53(3):1086–1097. https://doi.org/10.1021/jm901284w

Takada T et al (2005) Regulation of the cell surface expression of human BCRP/ABCG2 by the phosphorylation state of Akt in polarized cells. Drug Metab Dispos. 2005/04/22 33(7):905–909. https://doi.org/10.1124/dmd.104.003228

Tanaka Y et al (1994) Ultrastructural localization of P-glycoprotein on capillary endothelial cells in human gliomas. Virchows Arch 425:133–138

Tews DS et al (2000) Drug resistance-associated factors in primary and secondary glioblastomas and their precursor tumors. J Neurooncol 50:227–237

Timbie KF, Mead BP, Price RJ (2015) Drug and gene delivery across the blood-brain barrier with focused ultrasound. J Control Release. https://doi.org/10.1016/j.jconrel.2015.08.059

Toms SA et al (2019) Increased compliance with tumor treating fields therapy is prognostic for improved survival in the treatment of glioblastoma: a subgroup analysis of the EF-14 phase III trial. J Neurooncol. https://doi.org/10.1007/s11060-018-03057-z

Toth K et al (1996) MDR1 P-Glycoprotein Is Expressed by Endothelial Cells of Newly Formed Capillaries in Human Gliomas but Is Not Expressed in the Neovasculature of Other Primary Tumors. Am J Pathol 149:853–858

Tran PN, Klempner SJ (2017) ALK on my mind: Alectinib takes an early lead in managing intracranial disease in non-small cell lung cancer with ALK rearrangements. Ann Trans Med. https://doi.org/10.21037/atm.2017.03.47

Trédan O et al (2007) Drug resistance and the solid tumor microenvironment. J Natl Cancer Inst. https://doi.org/10.1093/jnci/djm135

Uchida Y et al (2011) Blood-Brain Barrier (BBB) pharmacoproteomics: Reconstruction of in vivo brain distribution of 11 P-glycoprotein substrates based on the BBB transporter protein concentration, in vitro intrinsic transport activity, and unbound fraction in plasma and brain. J Pharmacol Experiment Therapeutic. https://doi.org/10.1124/jpet.111.184200

Uchida Y et al (2015) Major involvement of Na + -dependent multivitamin transporter (SLC5A6/SMVT) in uptake of biotin and pantothenic acid by human brain capillary endothelial cells. J Neurochem. https://doi.org/10.1111/jnc.13092

Ueno M et al (2010) The expression of LDL receptor in vessels with blood-brain barrier impairment in a stroke-prone hypertensive model. Histochem Cell Biol. https://doi.org/10.1007/s00418-010-0705-y

Valera ET et al (2007) Quantitative PCR analysis of the expression profile of genes related to multiple drug resistance in tumors of the central nervous system. J Neurooncol. 2007/04/13 85(1):1–10. https://doi.org/10.1007/s11060-007-9382-7

Van Den Bent MJ et al (2009) Randomized phase II trial of erlotinib versus temozolomide or carmustine in recurrent glioblastoma: EORTC brain tumor group study 26034. J Clin Oncol. https://doi.org/10.1200/JCO.2008.17.5984

Van Den Bent M et al (2020) INTELLANCE 2/EORTC 1410 randomized phase II study of Depatux-M alone and with temozolomide vs temozolomide or lomustine in recurrent EGFR amplified glioblastoma. Neuro Oncol. https://doi.org/10.1093/neuonc/noz222

Veringa SJ et al (2013) In vitro drug response and efflux transporters associated with drug resistance in pediatric high grade glioma and diffuse intrinsic pontine glioma. PLoS One. 2013/05/03 8(4):e61512. https://doi.org/10.1371/journal.pone.0061512

Vogelbaum MA, Aghi MK (2015) Convection-enhanced delivery for the treatment of glioblastoma. Neuro Oncol. https://doi.org/10.1093/neuonc/nou354

Vu CU et al (2014) Nicotinic acetylcholine receptors in glucose homeostasis: the acute hyperglycemic and chronic insulin-sensitive effects of nicotine suggest dual opposing roles of the receptors in male mice. Endocrinology. https://doi.org/10.1210/en.2014-1320

Wager TT et al (2010) Moving beyond rules: the development of a central nervous system multiparameter optimization (CNS MPO) approach to enable alignment of druglike properties. ACS Chem Nerosci 1(6):435–449. https://doi.org/10.1021/cn100008c

Wager TT et al (2016) Central nervous system multiparameter optimization desirability: application in drug discovery. ACS Chem Nerosci. https://doi.org/10.1021/acschemneuro.6b00029

Wang S et al (2008) Preclinical pharmacokinetic/pharmacodynamic models of gefitinib and the design of equivalent dosing regimens in EGFR wild-type and mutant tumor models. Mol Cancer Ther. https://doi.org/10.1158/1535-7163.MCT-07-2070

Weaver M, Laske DW (2003) Transferrin receptor ligand-targeted toxin conjugate (Tf-CRM107) therapy of malignant gliomas. J Neurooncol. https://doi.org/10.1023/A:1026246500788

Wei X et al (2014) Brain tumor-targeted drug delivery strategies. Acta Pharm Sin B. https://doi.org/10.1016/j.apsb.2014.03.001

Weller M et al (2015) Glioma. Nat Rev Dis Primers. https://doi.org/10.1038/nrdp.2015.17

Wen PY et al (2020) First-in-human phase i study to evaluate the brain-penetrant PI3K/mTOR inhibitor GDC-0084 in patients with progressive or recurrent high-grade glioma. Clin Cancer Res. https://doi.org/10.1158/1078-0432.CCR-19-2808

Wolinsky JB, Colson YL, Grinstaff MW (2012) Local drug delivery strategies for cancer treatment: gels, nanoparticles, polymeric films, rods, and wafers. J Control Release. https://doi.org/10.1016/j.jconrel.2011.11.031

Wu Y et al (2019) MicroRNA-302c enhances the chemosensitivity of human glioma cells to temozolomide by suppressing P-gp expression. Biosci Rep. 2019/08/15 39(9). https://doi.org/10.1042/BSR20190421

Xu S et al (2013) Targeting receptor-mediated endocytotic pathways with nanoparticles: rationale and advances. Adv Drug Deliv Rev. https://doi.org/10.1016/j.addr.2012.09.041

Yan A et al (2019) CD73 promotes glioblastoma pathogenesis and enhances its chemoresistance via A(2B) adenosine receptor signaling. J Neurosci 39(22):4387–4402. https://doi.org/10.1523/jneurosci.1118-18.2019

Yasuda M et al (2019) The Involvement of Pregnane X Receptor-regulated Pathways in the Antitumor Activity of Cisplatin. Anticancer Res 39(7):3601–3608. https://doi.org/10.21873/anticanres.13507

Zeng Q et al (2015) Discovery and evaluation of clinical candidate AZD3759, a potent, oral active, central nervous system-penetrant, epidermal growth factor receptor tyrosine kinase inhibitor. J Med Chem 58(20):8200–8215. https://doi.org/10.1021/acs.jmedchem.5b01073

Zhang Y, Pardridge WM (2001) Mediated efflux of IgG molecules from brain to blood across the blood-brain barrier. J Neuroimmunol. https://doi.org/10.1016/S0165-5728(01)00242-9

Zhang W et al (2003) Expression and functional characterization of ABCG2 in brain endothelial cells and vessels. FASEB J. https://doi.org/10.1096/fj.02-1131fje

Zhou J et al (2006) Reversal of P-glycoprotein-mediated multidrug resistance in cancer cells by the c-Jun NH2-terminal kinase. Cancer Res 66(1):445–452. https://doi.org/10.1158/0008-5472.CAN-05-1779

Zhou Z, Singh R, Souweidane M (2016) Convection-enhanced delivery for diffuse intrinsic pontine glioma treatment. Curr Neuropharmacol. https://doi.org/10.2174/1570159x14666160614093615

Zong T et al (2014) Enhanced glioma targeting and penetration by dual-targeting liposome co-modified with T7 and TAT. J Pharm Sci. https://doi.org/10.1002/jps.24186

Appendix: Central Nervous System Anatomy and Physiology: Structure-Function Relationships, Blood Supply, Ventricles, and Brain Fluids

Robert G. Thorne

Abstract Sophisticated consideration of the many different approaches for drug delivery to the brain and spinal cord requires at least a working knowledge of central nervous system (CNS) anatomy and the basics of neurophysiology. The brain is different than other organs of the body in that it may not accurately be considered as a single compartment; its complex, heterogeneous structure is responsible for a multitude of functions with many different potential target sites for drug therapy. The cerebrovasculature is critically important from a drug delivery perspective because drugs will in many cases first reach the brain from the bloodstream. Additionally, stroke, Alzheimer's disease and other conditions that affect the brain's blood vessels are a major cause of morbidity and mortality. The cerebrospinal fluid (CSF) of the brain is contained within the ventricles and the subarachnoid spaces and often assumes importance for drug delivery or as a sampling compartment to measure drug levels (pharmacokinetics) or biomarkers (pharmacodynamics). Mechanisms governing CSF and brain interstitial fluid (ISF) exchange and drainage are increasingly recognized to play a critical role in the central biodistribution of drugs and CNS biomarkers. Brain ISF is highly regulated to provide a stable environment for optimal neuronal function, efficient signaling and the avoidance of neurotoxicity, yet it is also necessary for neuronal and glial waste products to be promptly and continuously removed over the entire lifespan. However, CSF and brain ISF are in somewhat limited contact and may rarely be assumed equivalent, particularly in the context of drug delivery or with respect to the measurement of biomarkers and their interpretation. This chapter reviews the basic organization, function, blood supply, and fluids of the brain and spinal cord that relate to considerations of drug delivery and the determination of drug or biomarker levels in the central compartment.

R. G. Thorne (✉)
Biology Discovery, Denali Therapeutics, South San Francisco, CA, USA

Department of Pharmaceutics, University of Minnesota (Twin Cities), Minneapolis, MN, USA
e-mail: thorne@dnli.com

© American Association of Pharmaceutical Scientists 2022   763
E. C. M. de Lange et al. (eds.), *Drug Delivery to the Brain*, AAPS Advances in the Pharmaceutical Sciences Series 33, https://doi.org/10.1007/978-3-030-88773-5

Keywords Anatomy · Function · Physiology · Frontal lobe · Parietal lobe ·
Occipital lobe · Temporal lobe · Cranial nerves · Blood supply · Cerebral arteries ·
Anterior circulation · Posterior circulation · Circle of willis · Interstitial fluid ·
Cerebrospinal fluid · Extracellular space

*[We] ought to know that from the brain, and from the brain only, arise our pleasures, joys,
laughter and jests as well as our sorrows, pains, griefs and tears. Through it, in particular,
we think, see, hear, and distinguish the ugly from the beautiful, the bad from the good, the
pleasant from the unpleasant.*
                                        – Attributed to Hippocrates, circa Fifth Century, B.C.

*Among the various parts of an animated Body, which are subject to anatomical disposition,
none is presumed to be easier or better known than the Brain; yet in the mean time, there is
none less or more imperfectly understood.*
                                                                    – Thomas Willis, 1681

*A great deal remains to be learned about the brain and spinal cord, a task that will take
centuries, not years, to complete.*
                                                        – Santiago Ramón y Cajal, 1909

*Progress depends on our brain. The most important part of our brain, that which is neocor-
tical, must be used to help others and not just to make discoveries.*
                                                            – Rita Levi-Montalcini, 2008

A.1   Introduction to Neuroanatomy

Human brains weigh about 400 g at birth, nearly tripling in size during the first three
years of life due primarily to the growth of neuronal processes and glia. The vast
majority of human central nervous system (CNS) tissue is accounted for by the
brain, which ranges from 1050 to 1800 g in normal young adults; by contrast, the
spinal cord weighs only about 35 g. On average, the adult human male brain weighs
1350 g, and the adult human female brain weighs 1250 g. This slight difference may
be explained by the observation that brain weight positively correlates with body
size both within and across most species, for example, an elephant weighing many
thousand kilograms has a brain that weighs approximately 5 kg (although, interestingly, human brains tend to be smaller than expected relative to body size when
compared to dolphins, rodents, and certain fish and primate species). Human cognition likely is shaped by our capacity for higher-level processing. Indeed, our brains
typically contain a much higher proportion of cerebral cortex than that found in
lower mammals, for example, the cerebral cortex accounts for 77% of the human
brain's volume compared with only 31% in the rat.

Microscopically, the CNS is principally composed of two types of cells: neurons
and glia. Generally, neurons process information and signal to other neurons at synapses. Glia assist in the regulation of neuronal information by modulating synaptic
activity as well as provide electrical insulation (myelin) to neuronal processes
(axons). The cell bodies (somas) of neurons vary greatly in size, with diameters

ranging from 5 to 10 μm (e.g., cerebellar granule cells) up to as large as ~100 μm for Betz cells in the primary motor cortex. Cell bodies are typically much larger than neuronal processes (axons and dendrites, collectively referred to as neurites), which range as small as 0.2 μm. Glia consist of macroglia (astrocytes and oligodendrocytes) and microglia (the resident immune cells of the brain). Glial sizes vary greatly, particularly across species. For example, human cortical protoplasmic astrocytes typically possess somas about ~10 μm in diameter and processes extending out 50–100 μm, both being several-fold larger than those found in the rodent. Glial cells outnumber neurons in most brain areas, with the exception of the cerebellum, for example, there are likely well over ten times more glia than neurons in the thalamus and white matter.

Much progress has been made in more accurately characterizing neurons and non-neuronal cells in brains across species, in part due to methodological advances in whole hemisphere nonstereological counting, tissue clearing, immunolabeling, imaging (e.g., lightsheet microscopy), and software-based quantification/segmentation. Adult human brains contain on average 86.1 billion neurons, along with 84.6 billion other cells (i.e., 50.5% of total cells in the human brain are neurons). The brains of cynomolgus monkeys (*Macaca fascicularis*), arguably the most commonly utilized nonhuman primate species for preclinical drug delivery studies, exhibit neuron:non-neuron proportions quite similar to humans (3.44 billion neurons and 3.15 billion other cells, with neurons making up 52.2% of total brain cells). However, rodent brains have higher relative proportions of neurons compared to humans (Rat, *Rattus norvegicus*: 189 million neurons, 122 million other cells, 60.7% neurons; Mouse, *Mus musculus*: 67.9 million neurons, 33.9 million other cells, 65.3% neurons), one of many features to be kept in mind when extrapolating findings from rodent studies to clinical work. It should also be noted that mouse whole brains contain 2.3 million microglia (2.3% of total cells), leaving approximately 31.6 million astrocytes, oligodendrocytes, and brain endothelial cells to go along with 67.9 million neurons. Focusing even more specifically on cell density measurements for the cerebral cortex, the mouse neocortex has been reported to contain approximately 93,000 neurons/mm^3, 20,000 astrocytes/mm^3, and 8500 microglia/mm^3, along with perhaps as many as 70,000 endothelial cells/mm^3. The above numbers represent the normal adult healthy condition; disease states can markedly change the picture, for example, neurodegenerative conditions accompanied by neuroinflammation often result in a dramatic increase in the number of microglia. Other recent advances have yielded much more granular information about the expression profiles of different cell types across brains of different species (e.g., http://www.brainrnaseq.org/; http://mouse.brain-map.org/; http://mousebrain.org/; http://betsholtzlab.org/VascularSingleCells/database.html; https://twc-stanford.shinyapps.io/human_bbb/).

Macroscopically, the CNS is divided into the brain and the spinal cord; major parts of the brain (the forebrain, midbrain, and hindbrain) can be divided based on its embryological development (Fig. A.1). Further CNS divisions may be described through the use of several common terms for direction/orientation (Table A.1) and planes of section (Table A.2), for example, it is often quite helpful to be able to use terms such as *rostral* (nearer to the front end of the neural axis, i.e., the front of the brain) and *caudal* (nearer to the tail end of the neural axis, i.e., the end of the spinal cord) when referring to specific CNS areas. For animals that move through the world horizontally and thus maintain a horizontal or linearly oriented CNS, for example, fish, reptiles, and rodents, the *superior/inferior* and *anterior/posterior* terms are always equivalent to *dorsal/ventral* and *rostral/caudal*, respectively.

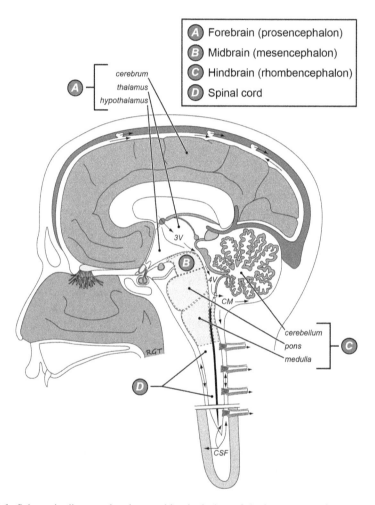

Fig. A.1 Schematic diagram showing a midsagittal view of the human central nervous system with the location of the brain and spinal cord in the cranial compartment. The cephalic flexure at the midbrain-diencephalic junction (diencephalon = thalamus + hypothalamus) is indicated by a dashed red line. Cerebrospinal fluid (CSF) flow from one of the lateral ventricles (not shown) through the third (3 V) and fourth (4 V) ventricles of the ventricular system and further circulation from the cisterna magna (CM) in the subarachnoid space is depicted with arrows

Table A.1 Directional terms used to refer to parts of the CNS

Direction /orientation	Latin	Meaning
Superior	Superus = "above"	Situated above
Inferior	Inferus = "below"	Situated below
Anterior	Ante = "before"	Situated in front
Posterior	Post = "after"	Situated behind
Dorsal	Dorsum = "back"	Toward the back
Ventral	Venter = "belly"	Toward the belly
Rostral	Rostrum = "beak"	Toward the snout
Caudal	Cauda = "tail"	Toward the tail

Table A.2 Planes used to refer to parts of the CNS

Planes	Latin	Meaning
Coronal	Corona = "crown"	Section in the plane of a tiara-like crown
Sagittal	Sagitta = "arrow"	Section in the plane of an arrow shot by an archer
Midsagittal		Sagittal section passing through the mid-line, dividing the brain into two halves
Parasagittal		Sagittal section parallel to the midsagittal plane
Horizontal		Section in the plane parallel to the horizon or floor (also called transverse or axial in humans, i.e., perpendicular to the long axis of the body)

Human beings have an upright posture so it follows that the CNS contains a prominent bend (the cephalic flexure occurring at the level of the midbrain changes the rotation of the CNS by 80–90°); this bend results in different equivalencies whether we are above or below the midbrain. Above the human midbrain, *anterior = rostral, posterior = caudal, superior = dorsal,* and *inferior = ventral.* Below the human midbrain, *anterior = ventral, posterior = dorsal, superior = rostral,* and *inferior = caudal.* Examining a schematic view of the human CNS sectioned along the midsagittal plane (Fig. A.1), we can see the location of the cephalic flexure at the junction between the midbrain and the diencephalon.

The largest portion of the nervous system in human beings is the forebrain, represented by the telencephalon (*Greek,* "end brain") and the diencephalon (thalamus, hypothalamus, and associated structures). The telencephalon contains the cerebral cortex (tissue appearing gray in gross sections due to a relative abundance of cell bodies), white matter (made up mainly of myelinated axons, imparting a white appearance in gross sections), and subcortical structures such as the hippocampal formation, amygdala, and basal ganglia. The forebrain is connected to the hindbrain by the midbrain, and the hindbrain is in turn connected to the spinal cord. One may think of the forebrain sitting on top of the midbrain, pons and medulla as broccoli or cauliflower would sit upon a stalk or stem; indeed, the midbrain, pons, and medulla together are commonly referred to as the brain stem. As with many other parts of the body, the brain exhibits a high degree of bilateral symmetry. Dividing the brain longitudinally along the midsagittal plane yields two similar appearing cerebral hemispheres that share a common pattern of surface landmarks between them. While this feature conveniently allows us to consider both hemispheres by learning the landmarks of only one, it must be kept in mind that important differences exist in the localization of function between the left and right sides, for example, the majority of language processing is accomplished in the left hemisphere of most individuals.

Observing the human brain from the lateral surface (Fig. A.2) allows us to visualize the four major brain lobes, areas of cerebral cortex with specific functions that are separated from one another by identifiable surface landmarks. The surface of the human brain contains numerous folds with ridges that are termed gyri (singular, gyrus). These folds are often absent or much less elaborate in lower mammals because their cerebral cortex is less developed than in higher mammals. Separating the gyri are furrows or grooves termed sulci (singular, sulcus); particularly deep sulci are often termed fissures. The largest sulcus is the lateral sulcus (Sylvian fissure) which runs horizontally and separates the frontal and parietal lobes from the temporal lobe. The central sulcus (Rolandic fissure) runs vertically and separates the frontal from the parietal lobe. An imaginary line extending between the parieto-occipital sulcus (best seen on the brain's medial surface) and the preoccipital notch, an indentation in the brain created by the petrous part of the temporal bone, separates the occipital lobe from the temporal and parietal lobes; a second imaginary line extended from the middle of this first line to the lateral sulcus further separates the temporal and parietal lobes from each other. In addition to the four major lobes seen on the lateral surface, another region of tissue called the insular cortex is buried within the depths of the lateral sulcus, concealed from view by portions (termed opercula; *Latin*, "lid") of the frontal, parietal, and temporal lobes. Finally,

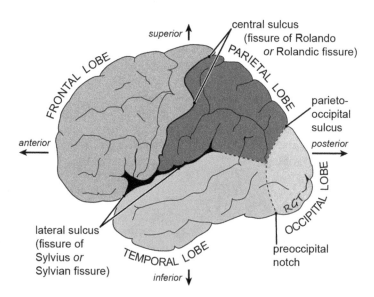

Fig. A.2 Diagram of the major lobes and sulci of the brain as seen from the lateral surface

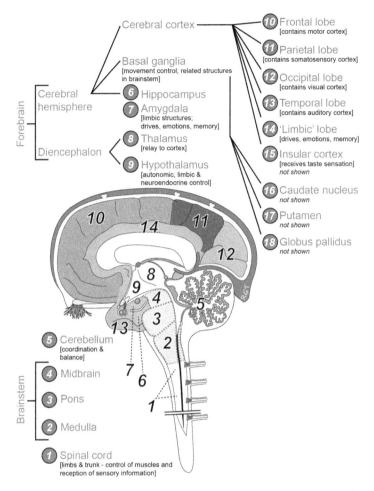

Fig. A.3 Functional subdivisions of the central nervous system. (Adapted and redrawn from Nolte 1999)

references are often made to yet another lobe, the "limbic" lobe, which is separated from the frontal and parietal lobes by the cingulate sulcus; it is best appreciated when examining the cerebral hemisphere on its medial surface (Fig. A.3).

A.2 Central Nervous System Functions

A vast array of functions may be identified for the many different brain and spinal cord areas that constitute the CNS. A simplified overview of some of the more important functional subdivisions is provided in Fig. A.3. Proceeding from caudal to rostral along the neural axis, we first encounter the spinal cord, a tubular structure which contains numerous nuclei (clusters or groups of cell bodies) corresponding to the spinal gray matter along with a large number of tracts or fasciculi (bundles of axons) corresponding to the spinal white matter. The spinal cord is concerned with the limbs and trunk of the body, primarily in the motor control of voluntary muscles and the reception of sensory information, although it also participates in the regulation of visceral functions. The human spinal cord has a total of 31 segments from which the motor (ventral) and sensory (dorsal) roots of spinal nerves arise (in order, from caudal to rostral): coccygeal (1), sacral (5), lumbar (5), thoracic (12), and cervical (8). The spinal gray matter is noticeably larger in two places that correspond to the lower and upper limbs, the lumbosacral (L2-S2) and cervical (C5-T1) enlargements, respectively.

Rostral to the spinal cord is the brain stem, a highly complex structure which regulates many basic physiological functions important for survival including arousal, blood pressure, and respiration. The brain stem is also associated with most of the 12 cranial nerves (only the olfactory and optic nerves are excluded), which provide cranial sensory information and allow for the control of head muscles, for example, the extraocular muscles that move the eyes. The attachment sites for the cranial nerves are best seen in a ventral view of the brain (Fig. A.4); their varied

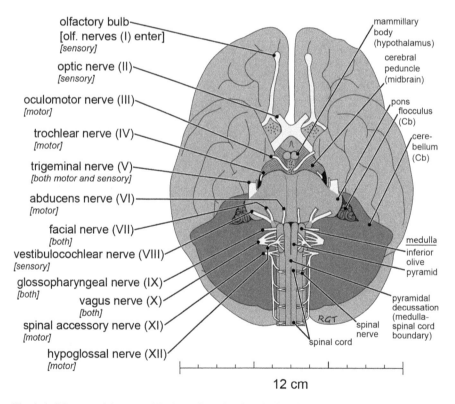

Fig. A.4 Diagram of the ventral brain surface showing the location of the twelve cranial nerves. The motor and/or sensory modality of each nerve is indicated

Table A.3 Cranial nerve functions and their evaluation

Cranial nerve	Function	Test
I (olfactory nerve)	Olfaction	Perception of an odorous substance
II (optic nerve)	Vision	Perception of a vision chart
III (oculomotor nerve)	Controls most extraocular eye muscles	Visual tracking of a moving object (e.g., following a finger)
IV (trochlear nerve)	Controls superior oblique muscle (eye)	Visual tracking toward an object (e.g., looking down at the nose)
V (trigeminal nerve)	Sensation of face, sinuses, and teeth Controls muscles of mastication	Touch, pain perception on face Ability to clench teeth
VI (abducens nerve)	Controls lateral rectus muscle (eye)	Visual tracking toward an object (e.g., looking to the side)
VII (facial nerve)	Controls muscles of facial expression Taste (ant. Tongue)	Ability to smile and raise eyebrows Perception of sugar or salt
VIII (vestibulo-cochlear nerve)	Hearing Balance	Perception of a tuning fork; Evaluation for vertigo
IX (glossopharyngeal nerve)	Pharynx sensation Swallowing Taste (post. Tongue)	Elicit gag reflex; Perception of sugar or salt
X (vagus nerve)	Controls muscles of larynx and pharynx Visceral motor control and sensation	Check for hoarseness, sound production and swallowing
XI (spinal accessory nerve)	Controls trapezius and sternocleidomastoid muscles	Shoulder raise and turning the head against resistance
XII (hypoglossal nerve)	Controls muscles of tongue	Tongue movements

functions and clinical tests commonly used to examine them are listed in Table A.3. The cerebellum, a structure embryologically related to the pons of the hindbrain, modulates motor information passing between the forebrain, brain stem and spinal cord; its function may partly be inferred by the outcome of cerebellar lesions, which result in disorders of coordination and balance (ataxia).

Rostral to the brain stem is the diencephalon, which includes the hypothalamus and thalamus as well as the epithalamus. The hypothalamus serves as a control center for the autonomic system (a part of the peripheral nervous system that provides nonconscious control over the body's organs) in addition to the neuroendocrine and limbic systems; hypothalamic nuclei are involved in a wide variety of functions that include the regulation of thirst, body temperature, hunger, satiety, and circadian rhythms. The thalamus serves as a relay center for nearly all sensory information reaching the cerebral cortex (only olfactory projections bypass it); many nonsensory pathways, for example, from the cerebellum, also reach the cerebral cortex after being processed by thalamic nuclei. The cerebrum sits above the diencephalon and consists of subcortical structures (the basal ganglia, hippocampal formation, and amygdala), the cerebral white matter and the cerebral cortex. The basal ganglia consist of a group of nuclei including the globus pallidus, caudate, and putamen. The basal ganglia modulates motor information; dysfunction in components of its circuitry (often considered to also include the substantia nigra of the midbrain and the subthalamic nucleus) leads to movement disorders such as Parkinson's and Huntington's diseases. The hippocampus and amygdala are major limbic areas located in the medial temporal lobe (although, technically, often considered part of the limbic lobe); the hippocampus is involved in learning and memory, while the amygdala is thought to be important for emotional content and social behaviors.

The various sensory and motor functions of the human cerebral cortex are best appreciated after first dividing the four lobes into component gyri (Fig. A.5a) and then considering a classification scheme such as the map developed by Korbinian Brodmann in his 1909 monograph (Fig. A.5b). Brodmann's map of the human cortex remains in wide use because his cytoarchitectonic divisions, based on differences in neuronal size, shape, and density observed in histological sections stained for cell bodies, correlate remarkably well with our current knowledge of structure-function relationships based on clinical observations, electrophysiological evidence, and neuroimaging. Brodmann divided the human cortex into 43 areas numbered between 1 and 52 (numbers 12–16 and 48–51 were not used in his map for the human brain). Numerous pathways connect the cerebral cortex with different levels of the neural axis below it; indeed, the nature of the information contained within the specific pathways arriving at and/or leaving a particular cortical area informs what is considered to be that area's function. Many pathways are longitudinally organized along the entire neural axis. For example, the dorsal column-medial lemniscal system provides information about fine touch, vibration, and position sense from the periphery, beginning at the level of the spinal cord and extending up through the brain stem and thalamus to the cerebral cortex (for this reason, it is called an ascending sensory pathway), while the corticospinal tract conveys impulses mediating voluntary movement from the cerebral cortex down to spinal cord motor neurons (for this reason, it is called a descending motor pathway). Most sensory and motor pathways cross (decussate) at some point as they travel up or down the neural axis; this crossing results in a given side of the brain controlling the muscles and receiving sensory information from the opposite side of the body.

The frontal lobe contains the primary and supplementary motor cortices, the frontal eye fields, Broca's area, and the prefrontal cortex. The primary motor cortex (Brodmann area 4), located in the precentral gyrus (the gyrus running just anterior and parallel to the central sulcus), is involved in the voluntary execution of movement for the opposite side of the body (i.e., the left primary motor cortex controls the body's right side). The supplementary motor cortex (area 6), located in the anterior part of the precentral gyrus and a portion of the adjacent superior and middle frontal gyri, is involved in the planning and initiation of movement for the opposite side of the body. The frontal eye fields (area 8) in the superior and middle frontal gyri initiate saccadic eye movements, for example, voluntary gaze toward the opposite side. Broca's area, consisting of the pars opercularis and pars triangularis of the inferior frontal gyrus (areas 44 and 45, respectively) of one hemisphere (typically the left), is important for the production of speech. The prefrontal cortex comprises

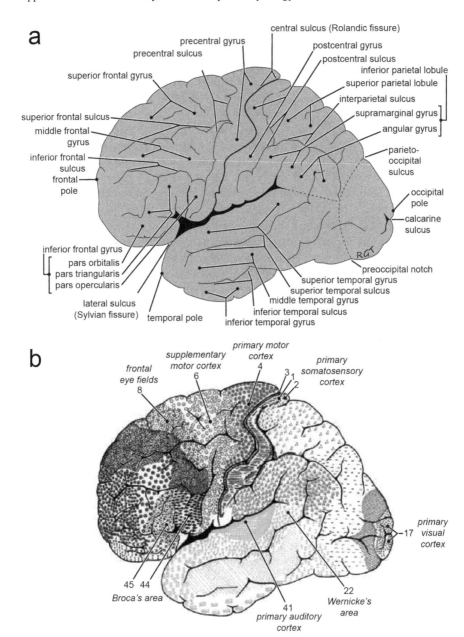

Fig. A.5 Schematic of the lateral brain surface—*detailed*. (**a**) Labeled view of the four visible lobes showing the location of the major gyri and sulci. (**b**) Brodmann's cytoarchitectonic map of the human cortex showing the location of major cortical units and their corresponding area number according to Brodmann's classification system (adapted from Zilles and Amunts 2010). The schematic in (**a**) has been drawn to closely match the view in (**b**)

most of the remainder of the frontal lobe and is generally considered to be involved with personality, thought, cognition, and planning behavior.

The parietal lobe contains the primary somatosensory cortex and other cortical areas important for the perception and integration of the senses. The primary somatosensory cortex (areas 1, 2, and 3), located in the parietal lobe's postcentral gyrus (the gyrus running just posterior and parallel to the central sulcus), is involved in the perception of touch, pain, and position for the opposite side of the body. The superior parietal lobule of the parietal lobe is important for the formation of our self-image; lesions to this area can result in complicated neurological signs including neglect of the body on the opposite side. The inferior parietal lobule of the parietal lobe is important for integrating diverse sensory information, for example, content that is heard, read, and/or visualized.

The temporal lobe contains the auditory cortex and cortical tissue for recognizing speech and for the perception of visual forms, colors, emotions, and smells. The primary auditory cortex (area 41), located within the superior temporal gyrus and a transverse gyrus extending into the lateral sulcus (not well seen from the lateral view), and the secondary auditory cortex surrounding it are important for sound perception and localization. Wernicke's area (area 22), also located in the superior temporal gyrus (posterior aspect) of one hemisphere (typically the left), mediates the recognition of spoken language. Much of the middle and inferior temporal gyri are concerned with the perception of visual form and color. Cortical areas within the anterior-most portion of the temporal lobe (the temporal pole) and the parahippocampal gyrus (observed on the brain's ventral surface; see Fig. A.6) are important for the processing of emotions and smell.

The occipital lobe contains the primary, secondary, and tertiary visual cortices subserving visual perception. The primary visual cortex (area 17), located in the banks of the calcarine sulcus (best appreciated in a medial view of the brain; see Fig. A.7), performs the initial processing of visual information. The secondary and tertiary visual cortices occupy most of the remainder of the occipital lobe and perform "higher" visual processing that allows us to perceive depth, motion, and color and to recognize faces.

Examination of a coronal section through portions of the parietal lobe, temporal lobe, and brain stem (Fig. A.8) indicates the locations of interior structures relative to surface features and cerebrospinal fluid-containing ventricular compartments (more on this below). Finally, a short word on comparative brain anatomy is warranted for some of the major species aside from humans that have been most frequently studied in drug development work. It bears emphasizing that all the regions and nerves discussed above for the human (as well as their associated functions) are present and available for study in the brains of lower species such as the rat and the mouse, provided the investigator knows where to look and can reorient as needed. A key difference between human and rodent brains is brain size (Fig. A.9), for example, the rat brain is approximately four-fold larger than the mouse brain and the human brain is approximately 800-fold and 3400-fold larger than the rat and mouse brain, respectively. Relative brain size may underlie species differences in

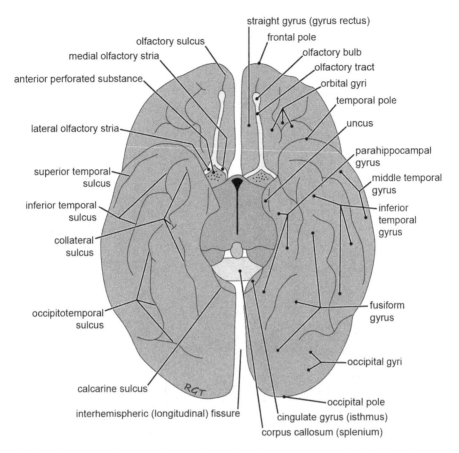

straight gyrus (gyrus rectus)
olfactory sulcus
frontal pole
medial olfactory stria
olfactory bulb
anterior perforated substance
olfactory tract
orbital gyri
temporal pole
lateral olfactory stria
uncus
parahippocampal gyrus
superior temporal sulcus
middle temporal gyrus
inferior temporal sulcus
inferior temporal gyrus
collateral sulcus
occipitotemporal sulcus
fusiform gyrus
occipital gyri
calcarine sulcus
RGT
occipital pole
interhemispheric (longitudinal) fissure
cingulate gyrus (isthmus)
corpus callosum (splenium)

Fig. A.6 Schematic of the ventral brain surface with brainstem removed—*detailed*. All major gyri and sulci are labeled

efficacy for drug delivery strategies requiring widespread distribution into the brain from the cerebrospinal fluid, where one might assume from basic transport considerations that the larger the brain, the greater the challenge to deeper brain penetration. Another obvious species difference between human and rodent brains is gyrification (cortical folding), as can be appreciated by comparing the highly folded human cortex with the relatively unfolded cortex of the rat (and mouse) in Fig. A.9. Gyrification is a feature that allows a greater surface area of brain tissue (and larger number of neurons embedded within it) to fit within a smaller cranial volume; indeed, both humans and rodents possess a highly folded cerebellum in part to accommodate an extraordinarily large number of granule cells, which account for approximately 80% of all brain neurons in these species. Lastly, differences are also apparent in the relative size of structures such as the olfactory bulbs, which may have evolved to take on greater relative importance for survival in rodents than in humans, although this remains an area of active debate.

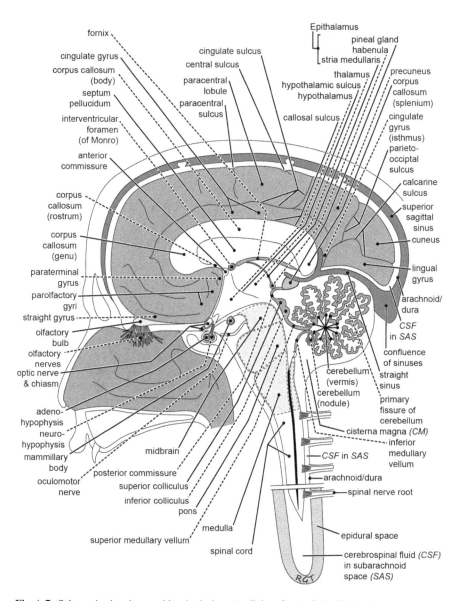

Fig. A.7 Schematic showing a midsagittal view (medial surface) of the CNS—*detailed*. All gyri, sulci, and other component structures are labeled

A.3 Cerebrovasculature

The importance of the CNS vascular supply cannot be overstated. Neurons demand tremendous metabolic resources in order to function properly, for example, a continuous supply of ATP is needed to maintain the ionic gradients essential for the

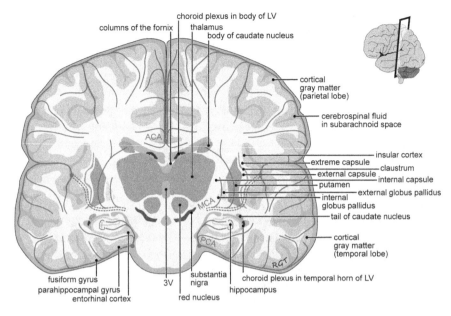

Fig. A.8 Schematic showing a coronal section through the parietal and temporal lobes—*detailed*. Labels emphasize subcortical structures and their relationships with cerebrospinal fluid-containing compartments (green). Approximate arterial blood supply regions from branches of the anterior (ACA), middle (MCA), and posterior (PCA) cerebral arteries are indicated. Inset brain shows approximate location of the section. 3 V, third ventricle; LV, lateral ventricle

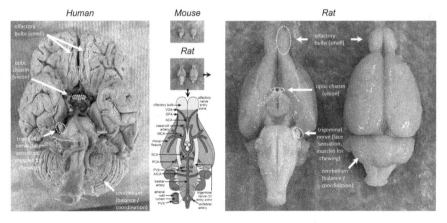

Fig. A.9 Comparative neuroanatomy of the human (ventral surface), mouse (ventral surface, left; dorsal surface, right), and rat brain (ventral surface, left; dorsal surface, right). Leftmost images of gross specimens shown approximately to relative scale. Schematic of ventral rodent brain indicates major arterial supply. Human and high magnification image of rat brains (rightmost image) illustrate the position and function of some easily identified major structures

membrane potentials underlying neurotransmission. Neurons have almost no ATP in reserve and must continuously be provided with glucose and oxygen so that aerobic metabolism can be utilized to produce the energy they require; even short periods of hypotension (low blood pressure) or ischemia (loss of blood supply) can lead to fainting or loss of consciousness; interruption of the cerebral blood supply to an area for just a few minutes can result in permanent damage. As with other body organs, CNS blood vessels may functionally be divided into distributing vessels (arteries), resistance vessels (arterioles), exchange vessels (capillaries and smaller post-capillary venules), and capacitance vessels (larger venules and veins). Brain capillaries are the smallest of these vessels, typically not larger than ~4–8 μm in diameter (approximately the same size or just a bit smaller than red blood cells). Total blood vessel length has been directly measured for the mouse brain using tissue clearing, immunolabeling and lightsheet microscopy in combination with quantitative vascular segmentation methods, yielding a whole brain vascular length of ~290 m. The largest contribution to this total in the mouse is from the cortex (~ 84 m or about 30% of the total), with the next highest contribution coming from the brain stem (~ 67 m). Measurements of total blood vessel length in the brains of humans (~ 650 km) and other species can be roughly calculated by taking into account brain weight, the specific density of the brain (1.036 g/cm^3), and the brain's average vascular density (i.e., vessel length per total tissue volume). A summary of values across species is provided in Table A.4.

In addition to facilitating the delivery of glucose, oxygen, and other endogenous blood substances to the CNS, the cerebrovasculature also provides among the most efficient routes for widespread drug access, provided the drug can pass the various barriers separating cerebral blood from the brain interstitial and cerebrospinal fluids. The most important of these barriers is the blood-brain barrier, represented by tight junctions between brain capillary endothelial cells forming the CNS microvasculature; this unique arrangement prevents nearly all but the smallest, lipophilic molecules from crossing the normal, healthy blood-brain barrier unless a specific transporter is present to facilitate their passage (e.g., as with glucose). Generally, gray matter is more highly perfused than white matter, although the normal perfusion rate of each is still much higher than that of muscle, skin, or fat. It has been estimated that gray matter energy requirements exceed that of the relatively energy efficient myelinated axon-rich white matter by at least three- to four-fold. Not surprisingly, cerebral capillary abundance can be as high as several thousand mm/mm^3 (total capillary length per tissue volume) in certain discrete areas of the adult gray matter, for example, the rat paraventricular and supraoptic nuclei of the hypothalamus, while white matter areas exhibit much lower values (e.g., about 100–300 mm/ mm^3 in the rat). This capillary abundance (also referred to as vascularity) varies dramatically over the life span, with much lower values on average at birth than in adults; data from rat and human brain have generally indicated a slight decline (~20% or less) in capillary density in older aged subjects, but results have varied across studies. Despite the brain's high capillary density, total cerebral blood volume under normal conditions across many species, including human beings and rodents, is only on the order of about 2–5% of the total tissue volume, as measured

Table A.4 Comparative parameters for CNS anatomy and physiology

Species	Body weight (g)	Brain weight (g)	Brain vascular length	Total neurons	Cortical neurons
Mouse (*Mus musculus*)	40 g	0.4 g	290 m[b] (340 m)[a]	68 M	14 M
Rat (*Rattus rattus*)	315 g	1.7 g	1400 m[a]	189 M	31 M
Cynomolgus Monkey (*Macaca fascicularis*)	5.7 kg (3.8 kg)[c]	46 g (74 g)[d]_	22 km[a] (36 km)[a]	3.4 B	801 M
Rhesus Monkey (*Macaca mulatta*)	3.9 kg	87 g	42 km[a]	6.4 B	1.7 B
Human (*Homo sapiens*)	70 kg	1350 g	650 km[a]	86.1 B	16.3 B

[a]Approximate values determined and/or inferred from published brain vascular density values (880 m/cm³ for mouse (and rat) from Tsai et al. 2009; 500 m/cm³ for human from Lauwers et al. 2008; 500 m/cm³ for other primates from Weber et al. 2008), unless indicated otherwise (e.g., [b]Kirst et al. 2020—direct measurement using immunolabeling-enabled three-dimensional imaging of solvent-cleared organs (iDISCO) combined with vascular segmentation methods on whole mouse brain). See text for further details. Other values for body weight, brain weight, and neuron numbers are as reported in Herculano-Houzel (2016) and in Gabi et al. (2010) for cynomolgus monkey. A range is indicated where variation in literature values have been reported, for example, for cynomolgus monkey body and brain weights; other values listed are from Mandikian et al. (2018)[c] and Pardo et al. (2012)[d]

using a variety of techniques (e.g., magnetic resonance imaging, positron emission tomography, and in situ brain perfusion). Cerebral blood volume is known to exhibit some regional variability, for example, measurements in rats have yielded higher values (up to ~5%) in areas such as the olfactory bulbs with lowest values in the white matter (~ 1%).

The human brain and meninges are supplied with blood derived from the common carotid and vertebral arteries (Fig. A.10). The paired internal carotid arteries arise from the common carotids and feed the *anterior circulation*, supplying most, but not all, of the forebrain. The paired vertebral arteries arise from the subclavian arteries and feed the *posterior circulation* to supply all of the hindbrain, nearly all of the midbrain and parts of the diencephalon, spinal cord, and occipital and temporal lobes. The anterior and posterior circulations meet at the circulus arteriosus, or circle of Willis (Fig. A.11); this important vascular feature bears the name of Sir Thomas Willis, among the first to accurately describe the cerebral arterial circle in 1664. The circle of Willis may be thought of as a nine-sided polygon (consisting of two each of the anterior, middle, posterior and posterior communicating arteries along with a single anterior communicating artery); it forms a complete anastomotic ring in about 50% of human beings, joining the anterior and posterior circulations. Arterial anastomoses, natural connections between two arteries, are important

Fig. A.10 Blood supply to the brain showing the origin of the anterior and posterior circulations giving rise to the major cerebral vessels: the anterior (ACA), middle (MCA), and posterior (PCA) cerebral arteries. Points of anastomosis between the internal carotid artery (ICA) and external carotid artery (ECA) are also indicated. CCA, common carotid artery

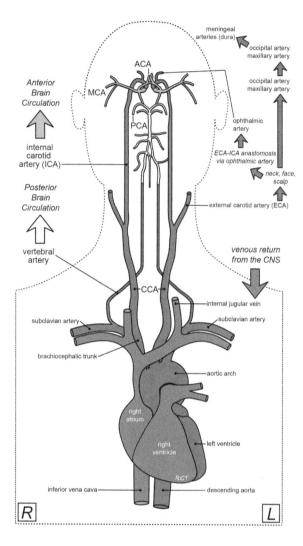

functionally because they can provide potential collateral circulation (a sort of "fail-safe system"), enlarging to compensate for occlusion or reduced supply in one of the segments. Under such conditions, a single artery may potentially supply blood to the normal territory of another in addition to its own territory following an obstruction.

All cortical areas of the cerebral hemispheres are supplied with blood via penetrating branches from one of the three main cerebral arteries (anterior, middle, and posterior); these vessels branch numerously, frequently penetrating into sulci as their leptomeningeal segments travel within the subarachnoid space just off the brain's surface (Fig. A.12). The cortical territories supplied by the cerebral arteries are shown for the lateral, medial, and ventral brain surfaces in Figs. A.13a, A.14a and b, respectively (the most common territories are shown; however, considerable

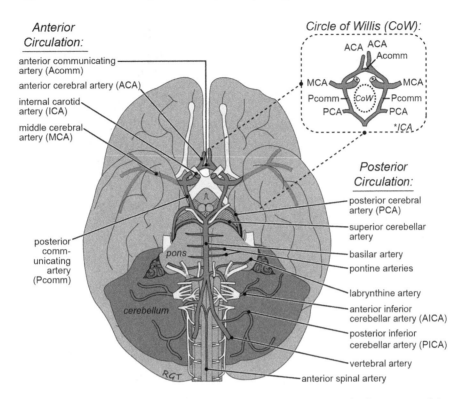

Fig. A.11 Diagram of the ventral brain surface showing the arterial supply. Components of the anterior circulation, posterior circulation, and the circle of Willis are emphasized

variability in their distribution is known to exist). In addition to the anastomotic ring at the circle of Willis, leptomeningeal anastomoses also exist between terminal branches of the cerebral arteries in areas called watershed or borderzone regions at the territorial boundaries (shown in detail in Fig. A.13a; similar watershed/border-zone regions are to be expected on the brain's medial and ventral brain surfaces but are not depicted in Fig. A.14a and b). The watershed regions are particularly vulnerable to ischemia and infarction when cerebral perfusion drops (e.g., when systemic blood pressure is dramatically reduced). Anastomoses between terminal branches of the cerebral arteries are also thought to play a role in providing collateral flow during ischemia (e.g., MCA occlusion), where they may help to save part of the penumbral tissue (potentially salvageable areas at the periphery of the core infarct). Penetrating vessels from the cerebral arteries also supply deep cerebral structures beneath the cortex (not shown); the most important of these are the lenticulostriate arteries (MCA branches that penetrate the anterior perforated substance (see Fig. A.6 and A.11), providing blood to portions of the basal ganglia and the internal capsule. The anterior choroidal artery, arising off of the internal carotid artery, and the posterior choroidal artery, arising off of the posterior cerebral artery, also supply a variety of deep cerebral structures in addition to the choroid plexus of the

Fig. A.12 Midsagittal and lateral views of a human brain with arteries attached and visible. (**a**) The medial brain surface is supplied by branches of the anterior (ACA) and posterior (PCA) cerebral arteries. (**b**) The lateral brain surface is mostly supplied by branches of the middle cerebral artery (MCA). (Adapted with permission from the Neuroanatomy Interactive Syllabus (Sundsten and Mulligan 1998))

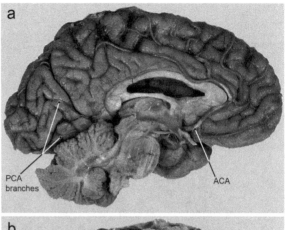

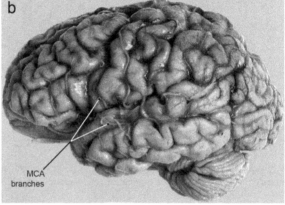

ventricular system. A summary of the arterial supply to different CNS areas is provided in Table A.5.

Cerebral veins empty into the venous sinuses, large venous channels surrounded by the dura mater, which ultimately empty into the internal jugular veins (Fig. A.10). Cerebral veins are divided into superficial groups, which lie on the brain's surface and drain into the superior sagittal sinus (Fig. A.13b), and deep groups, which drain internal structures and empty into the straight sinus (Fig. A.14a and b). Cerebral veins lack valves, contain numerous anastomoses, and do not usually run parallel to the arterial distribution.

A.4 Ventricular System and Brain Fluids

The CNS is immersed in cerebrospinal fluid (CSF), which helps to suspend the brain and avoid its distortion due to a buoyancy force that balances the downward force due to gravity. The CNS and CSF are together encased within the meninges

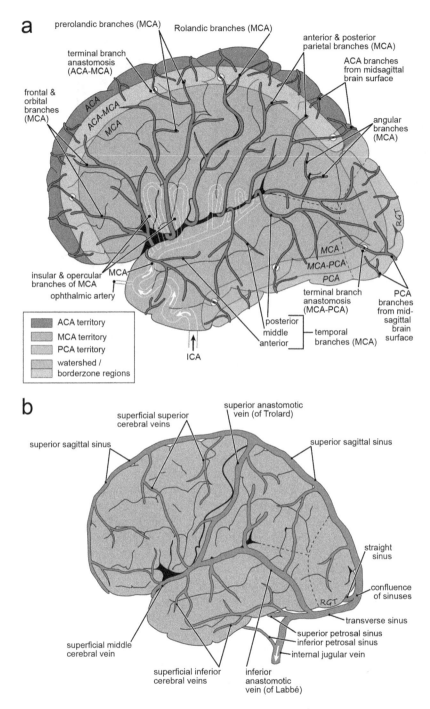

Fig. A.13 Diagram of the arterial supply (**a**) and venous drainage (**b**) on the brain's lateral surface – *detailed*. ACA, anterior cerebral artery; ICA, internal carotid artery; MCA, middle cerebral artery; PCA, posterior cerebral artery

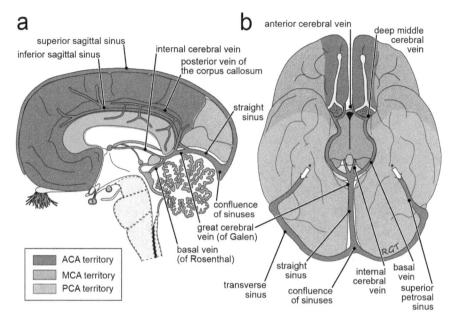

Fig. A.14 Diagram of the arterial supply and venous drainage on the brain's medial (**a**) and inferior (**b**) surfaces—*detailed*. *Venous blood draining from the cavernous sinus (not shown); *ACA* anterior cerebral artery, *MCA* middle cerebral artery, *PCA* posterior cerebral artery

(the dura mater, arachnoid, and the pia mater), which provide additional stability; the dura mater is anchored to the skull, while the arachnoid, which forms the leptomeninges with the pia, is adherent to the dura mater. Arteries and veins run within the subarachnoid space surrounded by CSF.

The CSF of mammals occupies several cavities or chambers within the brain (the ventricular system) as well as a larger volume filling the subarachnoid space that surrounds the brain and spinal cord. The human brain contains four ventricles (Fig. A.15a–c): two large, c-shaped lateral ventricles; a single third ventricle between the thalamus and hypothalamus of each hemisphere; and a single tent-shaped fourth ventricle located between the cerebellum, pons, and medulla. CSF is actively secreted by the choroid plexuses of the lateral, third, and fourth ventricles (Fig. A.15c) such that there is a brisk flow of CSF within the system. CSF flows from the lateral ventricles to the third ventricle via two interventricular foramina, then from the third ventricle to the fourth ventricle via the cerebral aqueduct, and, finally, exits into several cisterns and the subarachnoid space via three apertures, one located medially and two located laterally in the fourth ventricle (Fig. A.15d). CSF is ultimately reabsorbed back into the blood supply through arachnoid projections into the venous sinuses (Fig. A.13b and A.14) and also along cranial and spinal nerve roots to extracranial lymphatics. Additional CSF outflow may also occur along the perivascular sheaths of major blood vessels. In adult human beings, roughly 15% of the total CSF volume is present within the ventricular system, with

Table A.5 Arterial supply of the CNS

CNS area	Major arteries
Spinal cord	Anterior and posterior spinal arteries, radicular arteries
Medulla	Vertebral and posterior inferior cerebellar arteries (PICA)
Pons	Basilar and anterior inferior cerebellar arteries (AICA)
Cerebellum	*Superior surface,* superior cerebellar artery; *inferior surface,* AICA and PICA
Midbrain	Basilar, posterior cerebral, and superior cerebellar arteries, posterior and anterior choroidal arteries
Diencephalon	
Thalamus	Posterior cerebral (PCA), posterior communicating, and posterior choroidal arteries
Hypothalamus	Anterior cerebral (ACA), posterior communicating, and posterior cerebral arteries
Basal ganglia	
Globus pallidus	Anterior choroidal and middle cerebral (MCA) arteries
Putamen	(lenticulostriate arteries)
Caudate nucleus	ACA and MCA (lenticulostriate arteries)
	ACA and MCA (lenticulostriate arteries), anterior choroidal artery
Amygdala	Anterior choroidal artery
Hippocampus	PCA and anterior choroidal artery
Choroid plexus	Anterior and posterior choroidal arteries
Internal capsule Corpus	ACA, MCA, and anterior choroidal artery
callosum	ACA and PCA
Cerebral cortex	
Frontal lobe	ACA and MCA
Parietal lobe	ACA and MCA
Occipital lobe	MCA and PCA
Temporal lobe	MCA and PCA, choroidal arteries
Insular cortex	MCA

the remainder located within the fluid-filled cisterns and subarachnoid spaces outside of the brain and spinal cord; in rats, less than 5% of the total CSF volume is contained within the ventricles. Some differences in physiological parameters for the CSF and ventricular systems of adult humans and rats are listed in Table A.6. While the relative amounts of CSF obviously differ dramatically across species due to differences in brain size, with total CSF volumes ranging from ~150 ml (human) to ~290 μl (rat) to ~40 μl (mouse) to 2 nl (larval zebrafish), the general organization of the ventricular system appears to be remarkably preserved (Fig. A.15e).

It is necessary to differentiate the CSF from brain interstitial fluid. The interstitial fluid of the CNS is contained within narrow extracellular spaces (ECS), approximately 40–60 nm in width on average, that exist between neurons and glia. Interstitial fluid is in contact with the CSF at the ventricular surfaces as well as the pial surfaces facing the subarachnoid space. The ECS occupies about 20% of the total tissue volume in most brain areas of normal, adult animals. The ECS is critical to the distribution of neurotransmitters, nutrients, and all drugs within the CNS. Diffusion is

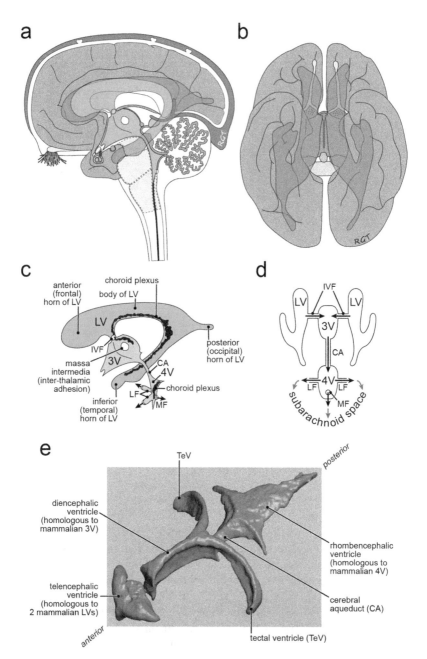

Fig. A.15 Anatomy of the ventricular system. Schematic 3-D representations of the ventricles and their drainage pathways in human beings as viewed from either the midsagittal (**a**) or inferior (**b**) brain surfaces; the approximate position of the ventricular system within the brain is depicted. Schematics of the isolated human ventricular system (**c**) and flow pathways (**d**). (**e**) 3-D rendering of larval zebrafish ventricular anatomy, based on in vivo confocal microscopy (image kindly provided by Drs. Maxwell Turner, Jeremy Richardson and Alan Kay). Homology to mammalian ventricular structures is indicated. 3 V, third ventricle; 4 V, fourth ventricle; CA, cerebral aqueduct (of Sylvius); IVF, interventricular foramen (of Monro); LF, lateral foramina (of Luschka); MF, medial foramen (of Magendie); LV, lateral ventricle

Table A.6 Approximate physiological parameters for the cerebrospinal fluids (CSF) and ventricular systems of adult humans and rats

Parameter	Human	Rat
Ventricular CSF volume	25 ml	10–12 μl
Subarachnoid space CSF volume	115 ml	190 μl
Total CSF volume	140–150 ml	200–300 μl
CSF secretion rate	350–370 μl/min	2–5 μl/min
Rate, % per minute (turnover time)	0.3–0.4 (6–7 hr)	0.7–0.75 (1–2.5 hr)

an essential mechanism for the extracellular transport of most substances through the brain interstitial fluid. Diffusion is extremely fast and efficient over short distances like the synaptic cleft (~ 15 nm) and quite effective even for distances spanning a few cell bodies (~ 10–100 μm), but it can be quite limiting over the larger distances often necessary for effective drug distribution from the ventricular or pial brain surfaces or from a syringe placed directly within the brain parenchyma. Neurons are rarely further than ~10–20 μm from their closest neighboring brain capillaries likely because the efficient diffusion of O_2, nutrients (e.g., glucose), and other molecules into the brain across the blood-brain barrier has necessitated such organization. While the composition of the CSF and brain interstitial fluid are generally thought to be quite similar, this may be strictly true only near the interface at the ventricular and pial surfaces, for at least two reasons: (i) diffusion is thought to greatly limit exchange at distances greater than a few millimeter from these surfaces and (ii) certain components of the interstitial fluid (e.g., the extracellular matrix) are bound to cell surfaces and therefore not freely available for exchange. There is some evidence that convective transport (also referred to as bulk flow) of brain extracellular and cerebrospinal fluids can occur along certain preferential pathways within the CNS, for example, within the perivascular spaces and possibly also along axon tracts, but interstitial fluid transport within the neuropil ECS of gray matter is likely restricted to diffusion.

Finally, it is important to appreciate the bony cranial compartment that the CNS tissue, cerebrospinal fluid and cerebrovasculature all occupy is rigid and unaccommodating of volume expansions except early in life. Given that the entire cerebrovascular system occupies about 2–5% of the total tissue volume and the CNS extracellular space occupies about 20% tissue volume, a normal adult human being with a brain and spinal cord weighing 1300 g will have approximately 150 ml cerebrospinal fluid, 260 ml interstitial fluid within the extracellular space, and 30–70 ml of cerebral blood. It is therefore easy to appreciate the rising intracranial pressure that often results from a significant expansion of the cerebrospinal fluid compartment (e.g., hydrocephalus), brain tissue compartment (e.g., a growing primary or metastatic brain tumor), or cerebral blood compartment (e.g., intracerebral or subarachnoid hemorrhage).

A.5 Conclusions

The brain is the most complex organ of the body. It directs our communication with the external world, what we do to our surroundings through our behaviors and what we perceive of our surroundings through our senses. It also monitors and controls our internal world, maintaining the delicate, exquisite balance among our internal organs that is necessary for sustained life. Any of these functions may be affected by disease or injury. It is in this context that it becomes necessary to consider how and where to deliver drugs to restore or improve the human condition. This chapter has attempted to summarize and briefly introduce the first considerations one must make in contemplating drug delivery to the brain, namely, to account for its diverse structure, function, and physiology.

Acknowledgments This work was supported by Denali Therapeutics, the University of Wisconsin-Madison School of Pharmacy, the Graduate School at the University of Wisconsin, the Michael J. Fox Foundation for Parkinson's Research, and the Clinical and Translational Science Award (CTSA) program, through the NIH National Center for Advancing Translational Sciences (NCATS; grant UL1TR000427). All content is solely the responsibility of the author and does not necessarily represent the official views of the NIH. The author wishes to acknowledge and thank Isha Thorne for support and advice on a portion of the chapter's content and to numerous laboratory members and students over the years for their feedback on the clarity and appropriateness of the different illustrations and sections as teaching aids in various undergraduate, graduate, and professional courses at the University of Wisconsin, University of Minnesota, and Stanford University. Finally, the author gratefully acknowledges helpful support, insights, and conversation from Professor Aparna Lakkaraju (UCSF, San Francisco, California) on the content of this chapter over the past several decades.

References

Abbott NJ (2004) Evidence for bulk flow of brain interstitial fluid: significance for physiology and pathology. Neurochem Int 45:545–552
Abbott NJ, Pizzo ME, Preston JE, Janigro D, Thorne RG (2018) The role of brain barriers in fluid movement in the CNS: is there a 'glymphatic' system? Acta Neuropathol 135:387–407
Blumenfeld H (2002) Neuroanatomy through clinical cases. Sinauer, Sunderland
Brozici M, van der Zwan A, Hillen B (2003) Anatomy and functionality of leptomeningeal anastomoses: a review. Stroke 34:2750–2762
Cai R, Pan C, Ghasemigharagoz A, Todorov MI, Forstera B, Zhao S, Bhatia HS, Parra-Damas A, Mrowka L, Theodorou D, Rempfler M, Xavier ALR, Kress BT, Benakis C, Steinke H, Liebscher S, Bechmann I, Liesz A, Menze B, Kerschensteiner M, Nedergaard M, Ertuk A (2019) Panoptic imaging of transparent mice reveals whole-body neuronal projections and skull meninges connections. Nat Neurosci 22:317–327
Cajal SRy (1909, 1911) Histologie due système nerveux de l'homme et des vertébrés. Maloine, Paris; English translation by Swanson N, Swanson L (1995) Oxford University Press, New York
Coelho-Santos V, Shih AY (2020) Postnatal development of cerebrovascular structure and the neurogliovascular unit. WIREs Develop Biol 9:e363
Davson H, Segal MB (1996) Physiology of the CSF and blood-brain barriers. CRC Press, Boca Raton

DeArmond SJ, Fusco MM, Dewey MM (1989) Structure of the human brain: a photographic atlas. Oxford University Press, New York

Gabi M, Collins CE, Wong P, Torres LB, Kass JH, Herculano-Houzel S (2010) Cellular scaling rules for the brains of an extended number of primate species. Brain Behav Evol 76:32–44

Gillilan LA (1974) Potential collateral circulation to the human cerebral cortex. Neurology 24:941–948

Haines DE (2012) Neuroanatomy: an atlas of structures, sections, and systems. Wolters Kluwer / Lippincott Williams & Wilkens, Philadelphia

Hammarlund-Udenaes M, Friden M, Syvanen S, Gupta A (2008) On the rate and extent of drug delivery to the brain. Pharm Res 25:1737–1750

Herculano-Houzel S (2015) Mammalian brains are made of these: a dataset of the numbers and densities of neuronal and nonneuronal cells in the brain of glires, primates, Scandentia, Eulipotyphlans, Afrotherians, and artiodactyls, and their relationship with body mass. Brain Behav Evol 86:145–163

Herculano-Houzel S (2016) The human advantage: a new understanding of how our brain became remarkable. The MIT Press, London

Jones HR (ed) (2005) Netter's neurology. Icon Learning Systems, Teterboro

Kandel ER, Schwartz JH, Jessel TM, Siegelbaum SA, Hudspeth AJ (2013) Principles of neural science, 5th edn. McGraw-Hill, New York

Keller D, Ero C, Markram H (2018) Cell densities in the mouse brain: a systematic review. Front Neuroanat 12:1–21

Kirst C, Skriabine S, Vietes-Prado A, Topilko T, Bertin P, Gerschenfeld G, Verny F, Topilko P, Michalski N, Tessier-Lavigne M, Renier N (2020) Mapping the fine-scale organization and plasticity of the brain vasculature. Cell 180:780–795

Lauwers F, Cassot F, Lauwers-Cances V, Puwanarajah P, Duvernoy H (2008) Morphometry of the human cerebral cortex microcirculation: general characteristics and space-related profiles. Neuroimage 39:936–948

Mandikian D, Figueroa I, Oldendorp A, Rafidi H, Ulufatu S, Schweiger MG, Couch JA, Dybdal N, Joseph SB, Prabhu S, Ferl GZ, Boswell CA (2018) Tissue physiology of Cynomolgus monkeys: cross-species comparison and implications for translational pharmacology. The AAPS J 20:107

Martin JH (1996) Neuroanatomy: text and atlas. Appleton & Lange, Stamford

Nieuwenhuys R, ten Donkelaar HJ, Nicholson C (1998) The central nervous system of vertebrates. Springer, Berlin

Nieuwenhuys R, Voogd J, van Huijzen C (2008) The human central nervous system. Springer, Berlin

Nolte J (2009) The human brain: an introduction to its functional anatomy, 6th edn. Mosby Elsevier, Philadelphia

Pardo ID, Garman RH, Weber K, Bobrowski WF, Hardisty JF, Morton D (2012) Technical guide for nervous system sampling of the cynomolgus monkey for general toxicity studies. Toxicol Pathol 40:624–636

Rapaport SI, Ohno K, Pettigrew KD (1979) Drug entry into the brain. Brain Res 172:354–359

Schuenke M, Schulte E, Schumacher U (2010) Atlas of anatomy: head and neuroanatomy. Thieme, Stuttgart

Somjen GG (2004) Ions in the brain: normal function, seizures and stroke. Oxford University Press, New York

Strazielle N, Ghersi-Egea JF (2013) Physiology of blood-brain interfaces in relation to brain disposition of small compounds and macromolecules. Mol Pharm 10:1473–1491

Sundsten JW, Mulligan KA (1998) Neuroanatomy interactive syllabus. http://www9.biostr.washington.edu/da.html. Accessed 1 Aug 2012

Swanson LW (1995) Mapping the human brain: past, present, and future. TINS 18:471–474

Swanson LW (2012) Brain architecture: understanding the basic plan. Oxford University Press, Oxford

Sykova E, Nicholson C (2008) Diffusion in brain extracellular space. Physiol Rev 88:1277–1340

Thorne RG, Nicholson C (2006) *In vivo* diffusion analysis with quantum dots and dextrans predicts the width of brain extracellular space. Proc Natl Acad Sci U S A 103:5567–5572

Tsai PS, Kaufhold JP, Blinder P, Friedman B, Drew PJ, Karten HJ, Lyden PD, Kleinfeld D (2009) Correlations of neuronal and microvascular densities in murine cortex revealed by direct counting and colocalization of nuclei and vessels. J Neurosci 29:14553–14570

Turner MH, Ullmann JFP, Kay AR (2012) A method for detecting molecular transport within the cerebral ventricles of live zebrafish (*Danio rerio*) larvae. J Physiol 590(10):2233–2240

van der Zwan A, Hillen B (1991) Review of the variability of the territories of the major cerebral arteries. Stroke 22:1078–1084

van der Zwan A, Hillen B, Tulleken CAF, Dujovny M, Dragovic L (1992) Variability of the territories of the major cerebral arteries. J Neurosurg 77:927–940

Watson C, Kirkcaldie M, Paxinos G (2010) The brain: an introduction to functional neuroanatomy. Elsevier, Amsterdam

Weber B, Keller AL, Reichold J, Logothetis NK (2008) The microvascular system of the striate and extrastriate visual cortex of the macaque. Cereb Cortex 18:2318–2330

Willis T (1664) Cerebri anatome: cui accessit nervorum descripto et usus; English translation by Samuel Pordage (1681) T. Dring, C. Harper, J. Leigh and S. Martin, London

Wolak DJ, Thorne RG (2013) Diffusion of macromolecules in the brain: implications for drug delivery. Mol Pharm 10:1492–1504

Yang AC, Vest RT, Kern F, Lee DP, Agam M, Maat CA, Losada PM, Chen MB, Schaum N, Khoury N, Toland A, Calcuttawala K, Shin H, Palovics R, Shin A, Wang EY, Luo J, Gate D, Schulz-Schaeffer WJ, Chu P, Siegenthaler JA, McNerney MW, Keller A, Wyss-Coray T (2022) A human brain vascular atlas reveals diverse mediators of Alzheimer's risk. Nature, https://doi.org/10.1038/s41586-021-04369-3

Zilles K, Amunts K (2010) Centenary of Brodmann's map – conception and fate. Nat Rev Neurosci 11:139–145

Index

© American Association of Pharmaceutical Scientists 2022
E. C. M. de Lange et al. (eds.), *Drug Delivery to the Brain*, AAPS Advances
in the Pharmaceutical Sciences Series 33,
https://doi.org/10.1007/978-3-030-88773-5